HORMONES

AUTHORS

Jean-François Bach
INSERM U.25, Université Paris VI, Paris, France

Etienne-Emile Baulieu
INSERM U.33, Université Paris XI, Le Kremlin-Bicêtre, France

Lutz Birnbaumer
Baylor College of Medicine, Houston, U.S.A

Michael S. Brown
Health Science Center, University of Texas, Dallas, U.S.A.

Pierre Chambon
INSERM U.184, Université Louis Pasteur, Strasbourg, France

Pierre Corvol
INSERM U.36, Collège de France, Paris, France

Bernard Desbuquois
INSERM U.30, Hôpital Necker, Paris, France

Robert G. Edwards
University of Cambridge, Cambridge, England

Pierre Freychet
INSERM U.145, Université de Nice, Nice, France

Jacques Glowinski
INSERM U.114, Collège de France, Paris, France

Joseph L. Goldstein
Health Science Center, University of Texas, Dallas, U.S.A.

Denis Gospodarowicz
Cancer Research Institute, University of California, San Francisco, U. S. A.

Elisabeth Granström
Karolinska Institutet, Stockholm, Sweden

Roger Guillemin
The Whittier Institute for Diabetes and Endocrinology, San Diego, U.S.A.

Jacques Hanoune
INSERM U.99, Hôpital Henri Mondor, Créteil, France

Colin M. Howles
University of Cambridge, Cambridge, England

Serge Jard
INSERM U.264, Montpellier, France

Alfred Jost
Collège de France, Paris, France

Paul A. Kelly
McGill University, Royal Victoria Hospital, Montreal, Canada

Byron Kemper
University of Illinois, Urbana, U.S.A.

Dorothy Krieger (deceased)
Mount Sinai School of Medicine, New York, U.S.A.

Fernand Labrie
Centre Hospitalier de l'Université de Laval, Quebec, Canada

Anthony Liotta
National Institute of Health, Bethesda, U.S.A.

Serge Lissitzky (deceased)
INSERM U.38, Université de Marseille, Marseille, France

Michael C. MacNamee
University of Cambridge, Cambridge, England

Yves Malthiery
INSERM U.38, Université de Marseille, Marseille, France

Anthony R. Means
Baylor College of Medicine, Houston, U.S.A

Joël Ménard
Hôpital Broussais, Paris, France

Edwin Milgrom
INSERM U.135, Université Paris XI, Le Kremlin-Bicêtre, France

Ira Pastan
National Institute of Health, National Cancer Institute, Bethesda, U.S.A.

Serge Racadot
Université Paris VII, Hôpital La Pitié Salpétrière, Paris, France

Pierre Royer
INSERM U. 30, Université Paris VI, Hôpital Necker, Paris, France

Bengt Samuelsson
Karolinska Institutet, Stockholm, Sweden

Sheldon J. Segal
Rockefeller Foundation, New York, U.S.A.

Phillip A. Sharp
Massachusetts Institute of Technology, Cambridge, U.S.A.

Pär Westlund
Karolinska Institutet, Stockholm, Sweden

Marc. C. Willingham
National Institute of Health, Bethesda, U.S.A.

Rosalyn S. Yalow
Berson Research Laboratory, New York, U.S.A.

Samuel S. C. Yen
University of California, La Jolla, U.S.A.

HORMONES

From molecules to disease

Edited by

Etienne-Emile Baulieu & Paul A. Kelly

HERMANN PUBLISHERS IN ARTS AND SCIENCE

CHAPMAN AND HALL NEW YORK AND LONDON

Published by Hermann, publishers in arts and science
293 rue Lecourbe, 75015 Paris, France
and Chapman and Hall, an imprint of
Routledge, Chapman & Hall, Inc.
29 West 35 Street, New York, NY 10001.

Published simultaneously in Great Britain by
Chapman and Hall, 11 New Fetter Lane, London EC4P 4EE

ISBN 2 7056 6030 5 (Hermann)
ISBN 0-412-02791-7 (Chapman and Hall)

Printed in France.

Library of Congress Cataloging-in-Publication Data

Hormones: from molecules to disease / edited by Etienne-Emile Baulieu
and Paul A. Kelly.
708 p. 21,5 × 28 cm.
Includes bibliographical references.
ISBN 0-412-02791-7
1. Endocrinology. 2. Hormones. I. Baulieu, Etienne-Emile.
II. Kelly, Paul A.
[DNLM: 1. Hormones. WK 102 H81192]
QP187.H595 1990
612.4'05--dc20
DNLM/DLC
for Library of Congress 90-1358
CIP

British Library Cataloguing-in-Publication Data available on request

10 9 8 7 6 5 4 3 2 1

Contents

Preface

Endocrinology is a field in which enormous advances have been made in the last decade; the rate of discovery of new hormones, hormone-like molecules, receptors, and mechanisms of action is continually advancing. The development of techniques in immunology and molecular biology has led to the possibility of describing in detail the gene structure of many of the compounds involved in hormonal systems. Remarkable homology has been shown between oncogene products and various components of the endocrine network, leading to the assertion that deregulation of hormonal function is involved in the generation and/or development of cancer. We now know that the central nervous system is both a target and a production site of many hormonal products, and that hormones, neurotransmitters, growth factors and immunopeptides all act through similar mechanisms. The only second messenger known ten years ago was cAMP; today calcium, derivatives of membrane phospholipids, and protein kinases are also known to be mediators of hormone action.

The very concept of hormonal systems has been expanded to include not only endocrine secretions but also para- and autohormones and their mechanisms of action; an understanding of their functions will be central to the immediate future of medicine.

The discovery of hormonal molecules and endocrine interactions and the subsequent understanding of hormone-related pathophysiology has led to the development of new strategies in medical treatment such as fertility control and the management of diabetes. The treatment of cancer, cardiovascular afflictions and mental disorders is undergoing modification due to advances in endocrinology, and the scope of preventive medicine has broadened.

Hormones respond to changes in brain activity, and to physiological, environmental and social influences, so endocrinology can be truly called a *humanistic* science. But since unexpected levels of complexity of hormonal systems are continually being unearthed, it seems apparent that – in spite of our optimism – any notion of "state-of-the-art" is ephemeral.

Endocrinology has entered a new era. Students, researchers, and physicians need a succinct source of classical knowledge coupled with recent information, findings, and concepts, – and a suggestion of implications for the future.

The aim of this book is to present this remarkably diverse and rapidly changing field precisely and completely, through contributions from the world's leading specialists and with the participation of experts in areas that have undergone considerable development in recent years. Major hormones and hormonal functions are described in detail in fourteen chapters, and related topics are covered in an innovative series of short, single-subject essays inserted between the chapters.

Because of the complexity and interdependence of hormonal systems, the presentation of endocrinological phenomena defies a linear approach: the same molecule may be produced at different sites in an organism and, as a function of its distribution, evoke different responses. In addition, the great majority of hormonal effects cannot be evaluated without considering multi-hormonal controls. So some of the chapters deal with chemically related hormones, while others describe hormones contributing to a common physiological function. In all cases, however, hormones are first studied in terms of their biological and cellular properties before being viewed in relation to pathophysiology and modern medical endocrinology.

Our authors were kind enough to allow us to edit their contributions; we hope we did not distort them. References provide the reader with access to detailed historical and classical, as well as recent, information, and throughout the book the many cross-references provide an extensive network designed to facilitate the integration of all aspects of a particular hormone, hormonal system, or disease.

E. -E. B., P.A.K.

CHAPTER I

Hormones: A Complex Communication Network

Etienne-Emile Baulieu

Contents, Chapter I

Hormones: A Complex Communication Network

1. Hormones Are Signal Molecules

The term *hormones* is derived from the Greek verb ωϱμειν, which means to excite, and encompasses, as a direct reflection of the generality of its origin, a vast number of biologically active compounds. The term was first used by *William Bayliss* and *Ernest Starling* in 1902[1] to describe the action of *secretin*, produced by the duodenum, which stimulates the secretion of pancreatic juice. Based more on physiological effects than on chemical structures, subsequent usage of the term has led to the definition of hormones as *signal* molecules, products of *glandular* cells, which are *secreted* into the internal milieu (*endocrine*, ενδον, meaning 'within'; and κϱινειν, 'to secrete'), most frequently into the blood[2] (Figs. I.1 and I.2). Acting in *target cells* (Fig. I.3), these chemical messengers coordinate activities of different parts of the body. Target cells, in turn, respond according to their degree of differentiation, age, functional and nutritional status, and integrate many hormonal and nervous regulatory stimuli. The *receptor* (Fig. I.4) is *the* chemical structure required for target cells to receive and recognize a messenger. Thus, by definition, it is the *coupling machine* for the transduction of external signals and the initiation of the first cellular response.

1. Bayliss, W. M. and Starling, E. H. (1902) The mechanism of pancreatic secretion. *J. Physiol.* 28: 325–353.
2. White, A. and Levine, R. (1982) History of hormones. In: *Biological Regulation and Development. Hormone Action.* (Goldberger, R. F. and Yamamoto, K. R., eds), Plenum Press, New York, 3A: 1–22.

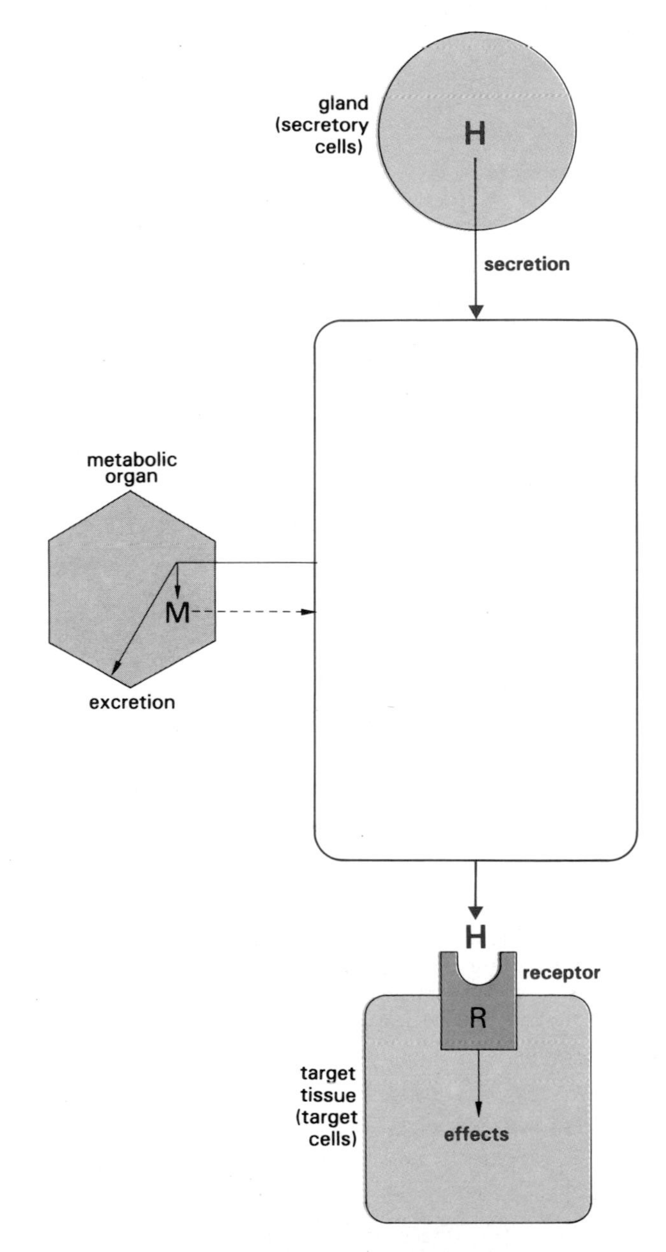

Figure I.1 The simplest representation of the endocrine system. A hormone, H, secreted by a gland, enters the circulatory system and acts on a target cell via a receptor mechanism. Circulating hormone is also metabolized and/or excreted in various places (e.g., liver, kidney). Metabolism may also occur in target cells, and conversely, metabolic organs may respond to the hormone. Note that the hormone concentration, which determines the magnitude of the hormonal response, depends mostly on secretion and metabolism (H, hormone; M, metabolite; R, receptor).

A

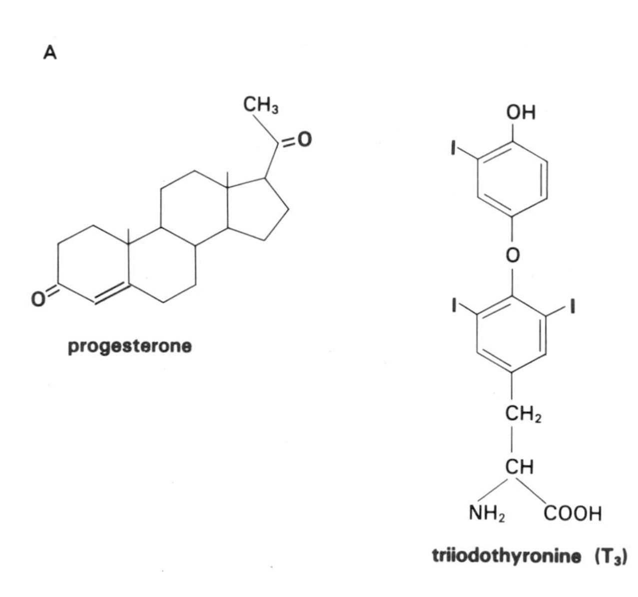

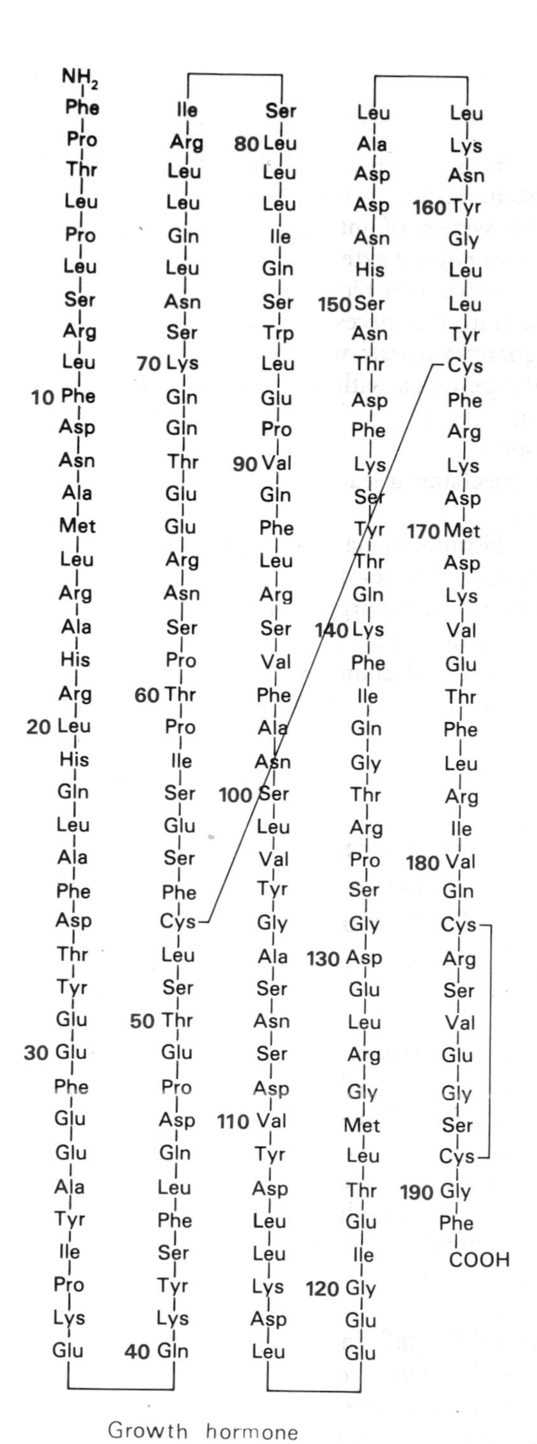

Figure I.2 Primary structure of some hormones. Only characteristic examples of different classes have been represented. Protein hormones frequently contain carbohydrate moieties (p. 22). In **B** (below), molecular models of progesterone and TRH illustrate how difficult it is to discern similarities and differences between hormones when using conventional representations as in part **A** of the figure.

B

Each of the several thousand different kinds of *cells* in the human body (a total of $\sim 10^{14}$ cells) has its *own* system of internal regulation, on which *hormones impose* external control. All cells of the body receive from the blood and extracellular fluid not only the nutrients necessary for their activity, but also *regulatory signals,* which act primarily by *direct physical interaction* with the receptor. This initial interaction between the *receptor* and hormone *triggers* a response, i.e. the hormonal effect. Thus, a non-chemical mechanism leads *to chemical changes within the cell.*

Hormones maintain the constancy of the "*milieu intérieur*"[3], necessary to insure the survival and independence of the organism confronted with the changing outside world. They are involved in *all* fundamental aspects of animal life: reproduction, growth and development, metabolism, physical mobility, and mental functions.

1.1 Non-hormonal Molecules and Non-endocrine Hormones Which Convey Information

At the frontier of the classical and the very new

All molecules have the potential to convey biological information. For this, they must interact with a cellular receptor which recognizes the hormonal message because of its three-dimensional, complementary arrangement, thus ensuring its transduction within the cell. There are *no chemical prerequisites* to the definition of such a molecule, but the concept of *structural correspondence* is essential; the interaction between the molecule and the receptor determines the biological meaning of the message, and consequently its potential significance. However, in intact living organisms, the *anatomical* distribution of hormone-secreting cells also determines the availability of hormonal molecules to their targets.

By definition, molecules of the ecosystem are not hormones, despite the fact that they may interact with specific systems. Thus, *pheromones*, involved in communication between animals, are excluded from this category, even though they are efficient at very low concentrations, acting probably through high affinity receptors.

3. Bernard, C. (1865) *Introduction à la médecine expérimentale.* Baillière, Paris.

There are also large groups of *signal compounds, not* classified as *hormones*, that are used mostly by the cells which produce them, such as cyclic nucleotides and prostaglandins. Nevertheless, cAMP is occasionally secreted from cells into the blood, and prostaglandins sometimes act as actual hormones. Indeed, the physiological significance of these and many other compounds has changed with biological evolution (p. 150). *Parahormones* and *autohormones*, which do not reach their target cells via the general circulation as do typical endocrine hormones, are discussed on p. 10 and p. 49. Molecules which pass directly from one cell to the other via specialized junctions (gap junctions), thus establishing intercellular communication, have been excluded from the definition of hormones.

1.2 Hormone Production, Distribution, and Action

"Quantitative Specificity" of the Receptor

Hormones are active at *low concentration*, much lower than the concentration of the enzyme substrates involved in general intermediary metabolism. It follows, therefore, that glandular hormone *production* is usually *small.* One may compare the hundreds of grams of food ingested per day with the milligrams or micrograms of hormonal secretions. However, even among hormones, such as steroids, a difference of 100-fold or more is recorded for the daily secretion of two essential hormones, cortisol and aldosterone (Table IX.6).

A typical endocrine system (Fig. I.1) includes a *glandular purveyor*, a system of circulatory *distribution*, and a machinery necessary for the *degradation* and *elimination* of hormone. Physiologically, the endocrine system establishes an *adequate hormone concentration* at the level of the receptors on *target cells.* Many secreted molecules (the majority, indeed) will not reach target cells during their trajectory in the circulation, but will be degraded or eliminated, mostly by the liver, adipose tissue, and kidneys. The *hormone-receptor interaction* itself, and thus the hormonal response, does not involve a *chemical reaction* modifying the structure of the hormone. However, in many cases, a hormone is metabolized at the target site, and this can play a role in its availability and in the kinetics of its action (p. 47 and p. 48). Hormone secretion is frequently independent of hormone activity in target tissues; it is regulated by blood hormone concentra-

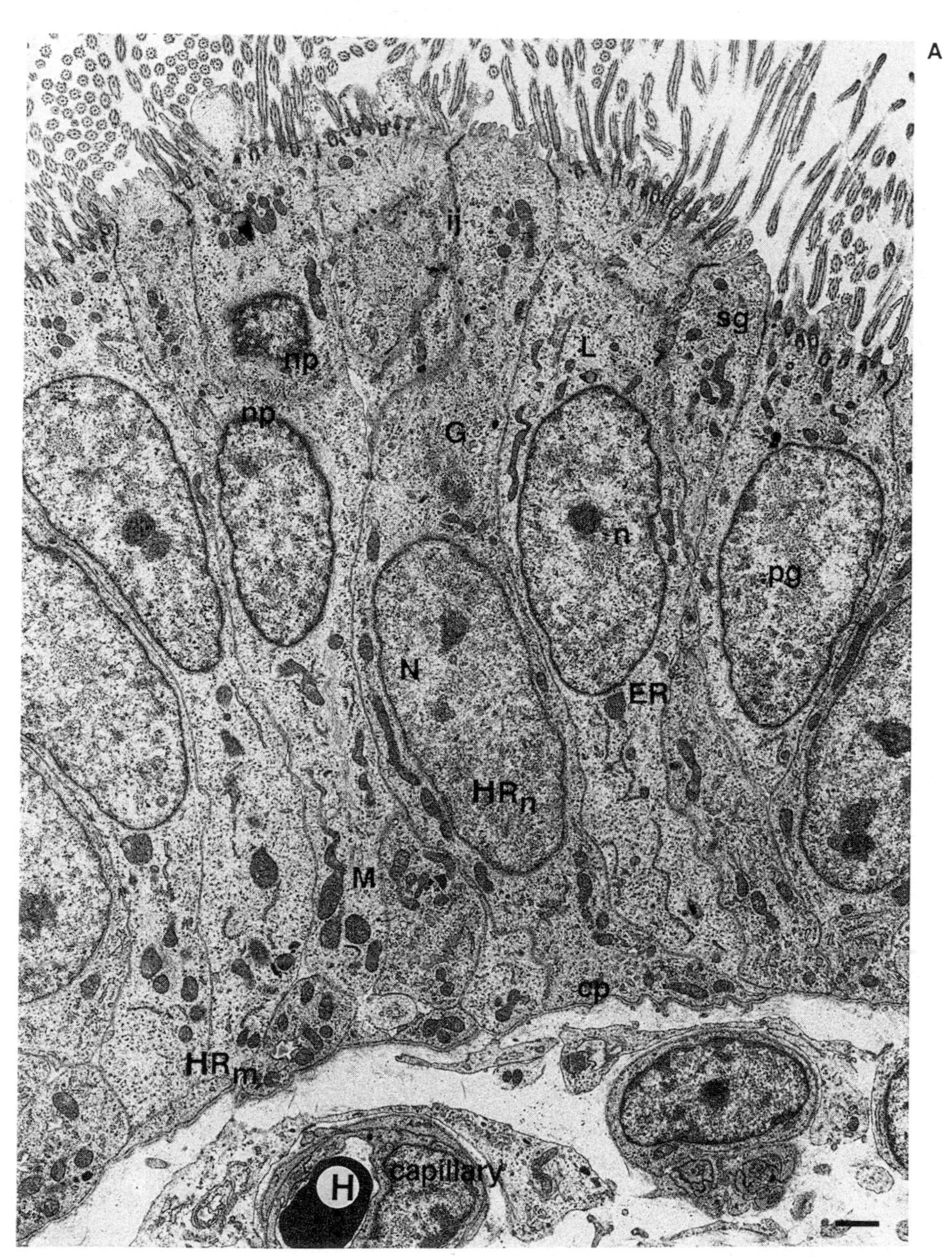

Figure I.3 Hormones and target cells. **A.** Hormones act on target cell membrane (m) or nuclear (n) receptors. To be effective, the hormone (H, in a capillary) must bind to a membrane receptor (R_m) if it is a peptide hormone, or to a nuclear receptor (R_n) if it is a steroid or thyroid hormone. On the electron micrograph of chick oviduct cells (bar=1 μm) is indicated the location of the main cellular organelles, detailed in the montage on the opposite page (**B**) and elsewhere in the book; cp, coated pits (Fig. 1, p. 225); ER, endoplasmic reticulum (**1**); G, Golgi apparatus (Fig. I.19); ij, intercellular junction; L, lysosome (Fig. 1, p. 225; Fig. 2, p. 226); M, mitochondria (with cristae, **2**; with tubules, Fig. I.12A); N, nucleus; n, nucleolus (**3**); np, nuclear pores (sectioned perpendicularly in relation to the surface, **4**; and cut tangentially, **B5**); pg, perichromatin granules (**6**); sg, secretory granules (Fig. I.12B). Ribonucleic acids are synthesized in the nucleus (**1**; bar=1 μm). Ribosomal RNA (rRNA) is transcribed from ribosomal DNA in the nucleolus (**3**; bar=1 μm), at the pars fibrosa (f) level, and accumulates within the pars granulosa (g). Messenger RNA (mRNA) is synthesized at the boundary between euchromatin and heterochromatin, where perichromatin granules (**6**; bar=0.1 μm) are observed. These RNAs are released through nuclear pores (**4** and **5**; bars=0.1 μm) into the cytoplasm, where they form polyribosomes (**7**; bar=0.1 μm), studded on the surface of the endoplasmic reticulum (**1**), which are involved in protein synthesis. Newly formed polypeptide chains can be transported by the Golgi apparatus and transformed into glycoproteins. The cavity of the endoplasmic reticulum communicates with the perinuclear cisternae and constitutes the external leaflet of the nuclear envelope. Mitochondria (**2**; bar=1 μm) are energy-producers. (Courtesy of Dr. C. Le Goascogne, Hôpital de Bicêtre).

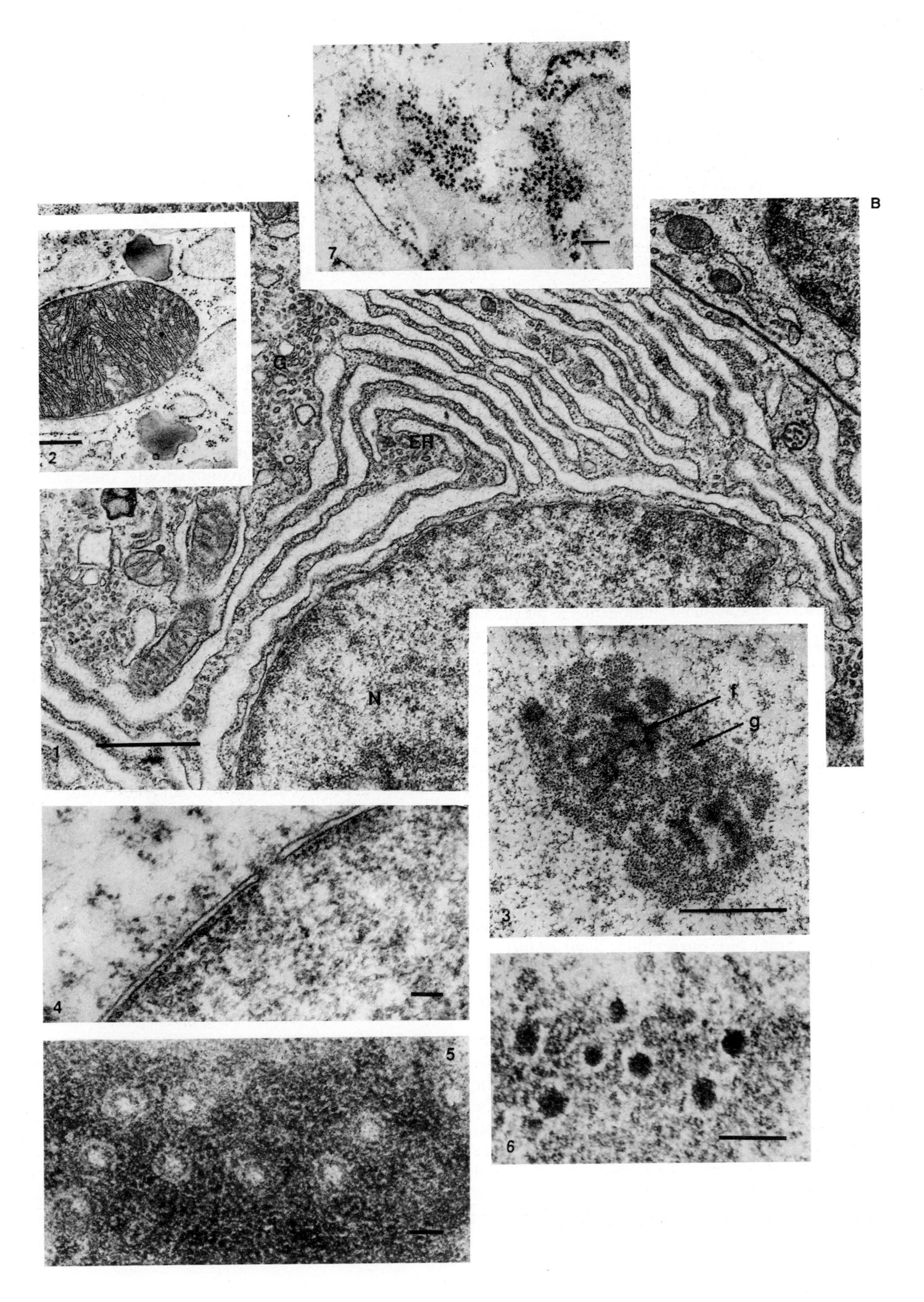
B
G
ER
N
f
g
1
2
3
4
5
6
7

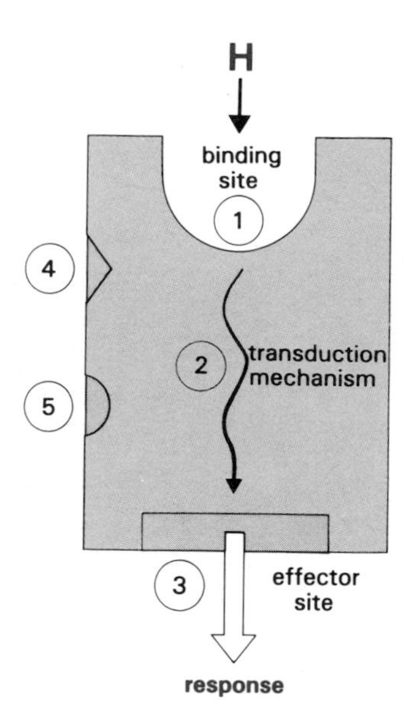

Figure I.4 The hormone receptor: a generalized representation. Three fundamental elements are necessarily involved in receptor mechanisms: **1.** A binding site which selectively recognizes the hormone, where decisive interaction with the target cell takes place. **2.** A connecting (transducing) system for the transfer of the hormonal information to the effector site. **3.** An effector site triggering the response(s) of the cellular components. Accessory sites (labelled **4** and **5**) are domains of the receptor molecule which may help position the receptor within a cellular structure, or which bind allosteric, physiological or pharmacological ligands, etc.

tion, which is dependent on degradative metabolism and on feedback at the level of the CNS (Fig. I.6). Alternatively, the metabolic change produced by the hormone may be directly operational on its own secretion (Fig. I.7 and p. 43).

In target tissues, the concentration of hormones is low, frequently in the nano- or picomolar range. Regardless of whether the *receptor* (Fig. I.4) is on the plasma membrane or is intracellular, *high affinity* and *binding specificity* are parameters which are *fundamental* to selective hormone-receptor interaction in the extra- or intracellular milieu. Thus, surrounded by a multitude of molecules, including other hormones, the receptor plays a decisive role in molecule *recognition*. Additionally, in many cases the receptor *amplifies* the hormonal signal by inducing the production of a much larger number of molecules than hormone-receptor complexes activating the event. Hormone-receptor complexes *trigger* one or several *initial* biochemical reactions which are followed by a *cascade* of events that no longer require further hormonal participation. Indeed, the molecule of hormone initiating the response is *removed rapidly*, or is degraded. At this point, we leave endocrinology proper and enter the field of cell biology, even though cell function itself may affect the availability and activity of a hormone and/or its receptor.

May all cells be target cells for all hormones? Do all cells have receptors, many or a few, for all hormones? In principle, a cell *cannot respond* to a hormone *if* the appropriate *receptor is not* present. In most cases, receptors are found at a concentration of $10^{4\pm1}$ per normal target cell, a number generally sufficient to mediate a normal hormonal response, but small enough to make receptor isolation and purification difficult. Modern means of detection have identified low levels of receptors in cells which are not physiological targets for the hormone. Thus when hormone concentration is abnormally high, either pharmacologically or pathologically, these cells may respond. In addition, certain target cells normally exposed to high concentrations of hormone are able to respond despite low physiological receptor levels. This is the case for hepatic and pituitary cells irrigated by portal blood containing, respectively, insulin or hypothalamic factors in higher concentration than in the peripheral circulation. The same is true for cells of the adrenal medulla, which are exposed to adrenal cortical venous blood rich in cortisol, or in cases where a hormone reaches target cells by non-vascular diffusion, as in para- and autocrine systems (p. 49).

Both *hormone concentration and receptor number* per cell quantitatively determine the formation of active hormone-receptor complexes and, consequently, the magnitude of the hormonal response (Fig. I.5).

$$[H] + [R] \rightleftharpoons [HR] \rightarrow \text{effect}$$

Figure I.5 The formation of the hormone-receptor complex, HR, is *reversible*. The concentration of HR is fundamental in determinating the magnitude of the hormonal effect. Hormone concentration at the target level depends on the blood concentration, and after interaction with the receptor, the hormone may return to the circulation or be metabolized in situ. The receptor is synthesized by the cell, and is either recycled or degraded as such or in the form of hormone-receptor complexes.

This *two-fold quantitative concept*, where both hormone and receptor play complementary roles, is more relevant than the qualitative notion of an all-or-none cellular receptivity. There are *qualitative nuances*,

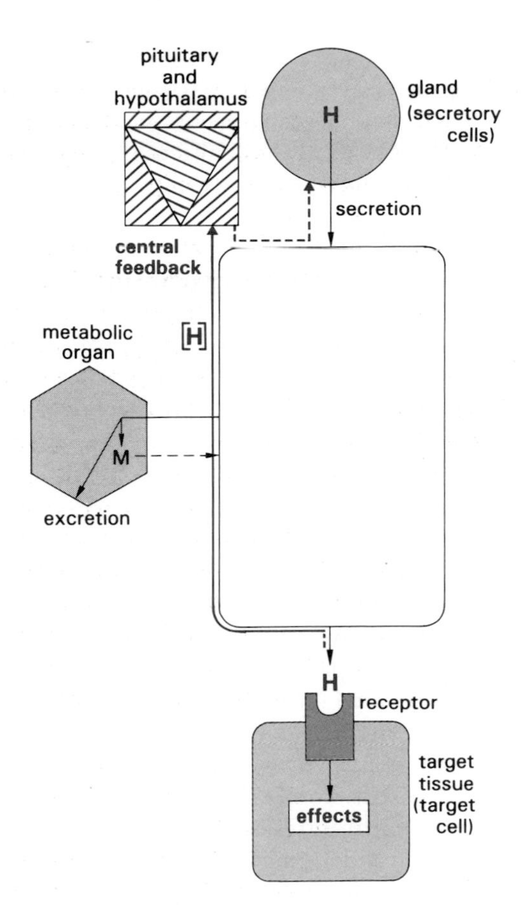

Figure I.6 *Central* feedback system regulating hormone secretion (e.g., the plasma concentration of cortisol negatively regulates hypothalamic CRF and pituitary ACTH secretion). The *concentration* of H in the blood regulates glandular secretion by a feedback mechanism operating most frequently in the hypothalamic–hypophyseal system (triangular and square symbols). Metabolism, which is essentially independent of the hormonal effect in the peripheral target cell, determines the circulating level of hormone.

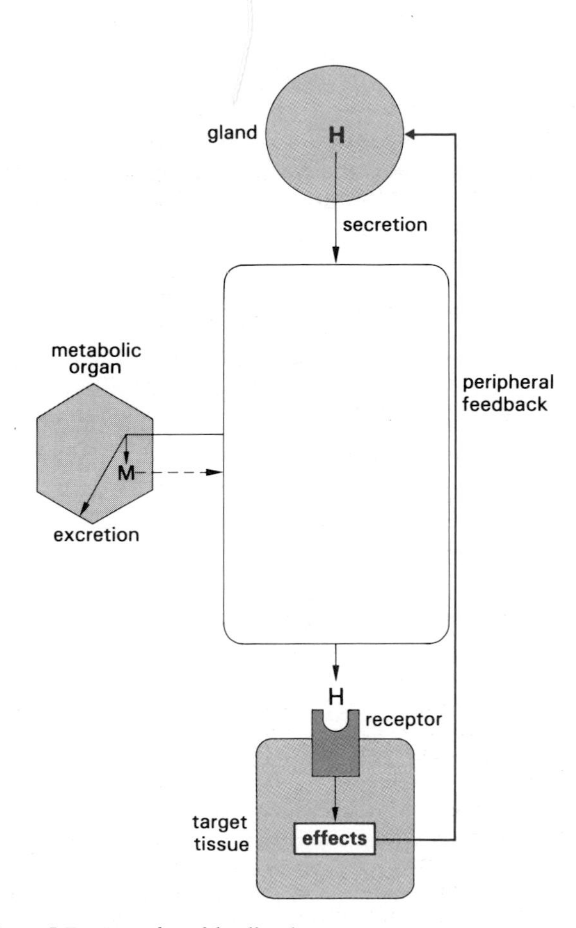

Figure I.7 *Peripheral* feedback system regulating hormone secretion (e.g., insulin secretion is regulated by the changing concentration of blood glucose). A *product* of hormone action in target cells regulates secretion by a direct mechanism operating directly in the glandular cells themselves.

however, as for example the difference between hormones that act primarily in well-defined target organs (e.g., ACTH in adrenal cortex, LH in ovary, aldosterone in kidney) and those, such as insulin, growth factors, glucocorticosteroids or thyroid hormones, which have receptors in almost all cells. The latter often play a *permissive* role (p. 41) for many fundamental cellular phenomena (metabolism, growth), allowing regulation by other agents (often hormonal in nature also).

For the physiologist or the practitioner measuring receptors in tissue extracts, the fact that a receptor concentration appears low may mean that all cells are receptor-poor, and/or that receptors are present in only a few cells (cellular heterogeneity), and/or that the receptor is abnormal (down-regulated, mutated). These different possibilities should be established both theoretically and therapeutically (e.g., in regard to some cancers, p. 147). Newer histological approaches using specific antibodies and nucleic acid probes may provide quantitative answers to these difficult questions.

1.3 Hormonal Communication: Several Patterns

In the classical endocrine concept, a gland produces a hormone which is distributed to target organs via the general circulation (Fig. I.1). This concept has been modified by the extraordinary development of analytical methods over the past 25 years, by the measurement of known hormones at every possible site, and by the discovery of hormone molecules which are not products of classical glands.

Parahormones and Autohormones

In addition to the system of general distribution, hormones are distributed regionally, either via *special vascular networks* (e.g., the hypothalamic-hypophyseal portal vessels) or simply by *local diffusion* in the intercellular fluid of the same *organ* (e.g., testosterone from testicular interstitial Leydig cells entering into seminiferous tubules; insulin regulating the secretion of glucagon in the pancreas, etc.). The transfer of steroids to the ovum from surrounding follicular cells is even more direct. When a cell secretes a compound which acts on neighboring cells, this phenomenon is defined as a *paracrine* effect (παρα: to the side). If the cell utilizes the compound it produces itself (and naturally also provides to identical neighboring cells), this is known as an *autocrine* effect (Fig. I.8A). The increasing number of recognized paracrine and autocrine systems emphasizes the physiological importance of *limited topological distribution*, which, in addition to receptor presence, is another means of ensuring cell specificity. A *parahormone* may be produced by classical glandular cells (e.g., testosterone and insulin, as seen above) and/or by other cells. Indeed, *polycentric* production and release (generally not into the circulation) and distribution of hormones have now been observed in a number of different cases; somatostatin, originally described as hypothalamic in origin, is also a pancreatic hormone with a local effect; insulin and steroids are synthesized in the brain, independently of the blood supply; cholecalciferol (or vitamin D_3), found in food or synthesized in epithelial cells, is activated by metabolism in the kidney, producing calcitriol, which acts locally as a parahormone and reaches the intestine as a true hormone. Some growth factors have a distribution similar to that of hormones (e.g., IGFs after synthesis in the liver, p. 198), but most are produced by cells, not assembled to form a gland, which may even be mobile (e.g., cells of the immune system, Fig. I.8B and p. 252). Many of these factors have typical paracrine or autocrine activities. To study them is difficult, because in contrast to the classical endocrinological approach, it is not possible to suppress the producing gland in order to establish hormone function. Thus, their actions are often presumed uniquely from in vitro experimental evidence. Care must be taken in the application of the numerous fascinating results obtained from cells in culture to the pathophysiology of the entire organism.

1.4 Hormones vs. Neurotransmitters

The *endocrine and nervous* systems are so intimately involved in the overall regulation of the body that there is probably no tissue which does not receive regulatory input from both of them. In addition, the nervous system itself is affected by hormones, and conversely, endocrine glands are partly dependent on nervous control.

The transfer of signals in the nervous system is achieved by chemical mediators, or *neurotransmitters*, which act on receptors comparable to hormone receptors (p. 85 and p. 319). The function of the *nervous* system is based on an organized cellular *network*, with axons directing information (Fig. I.9) and inducing an almost immediate, temporary effect via neurotransmitters. In contrast, in the *endocrine* system there is general *diffusion* of hormones in the extracellular space, and hormone specificity is conferred by receptors at the level of the target cells; response to stimulation is observed after a delay, and in comparison with most responses to neurotransmitters, effects persist longer. Although neurotransmitters are classically secreted into the closed space of the synapse, it is now recognized that they can diffuse outside and act as para- and autohormones (p. 49).

1.5 Homogeneous Heterogeneity

Table I.1 lists the *principal hormones* studied in this book; the *chemical heterogeneity* between them is remarkable (Fig. I.2), although there are some well-defined families. Varying periods elapsed between the discovery of a hormone and the determination of its structure. (Table I.2).

Not only are there many different hormones, each provoking a *particular response in each cell type*, but their effects are not always the same in corresponding organs in different species. Hormones are *modulators of the potential of each specific cell*, which is determined essentially by differentiation, but also *circumstantially*, by age, metabolism, activity, etc.

In conclusion, a single hormone regulates a number of different cell types, while, conversely, each cell

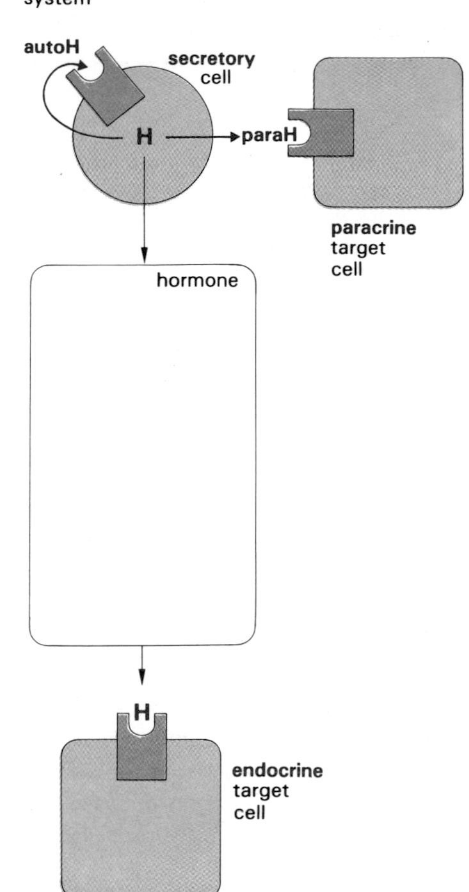

Figure I.8 New hormonal systems. **A.** *Parahormones* and *autohormones*. In addition to the classical hormones shown in Figure I.1, two additional mechanisms (paracrine and autocrine) are now recognized. A parahormone (para H) diffuses from the secretory cell to a neighboring (different) target cell. An autohormone (autoH) stimulates its own and sister secretory cells. **B.** *Neuroendocrine immunology.* Interleukin-1 (IL-1, p. 253), a monokine (lymphokine) (Fig. 4, p. 254), stimulates ACTH production at the hypothalamic and probably also hypophyseal level. The effect is more pronounced in females. In turn, cortisol inhibits IL-1 synthesis and activity. IL-2, interferon (IFN) and tumor necrosis factor (TNF) (not shown) are not active, indicating the degree of the specificity of IL-1 activity. Hypothalamic astrocytes and microglial cells produce IL-1, which may also have regulatory activity on other pituitary hormones. The mechanism of action of IL-1 includes the synthesis of prostaglandins and/or leukotrienes (p. 590). (Modified from Lumpkin, M.D. (1987) The regulation of ACTH secretion by IL-1. *Science* 238: 452–454).

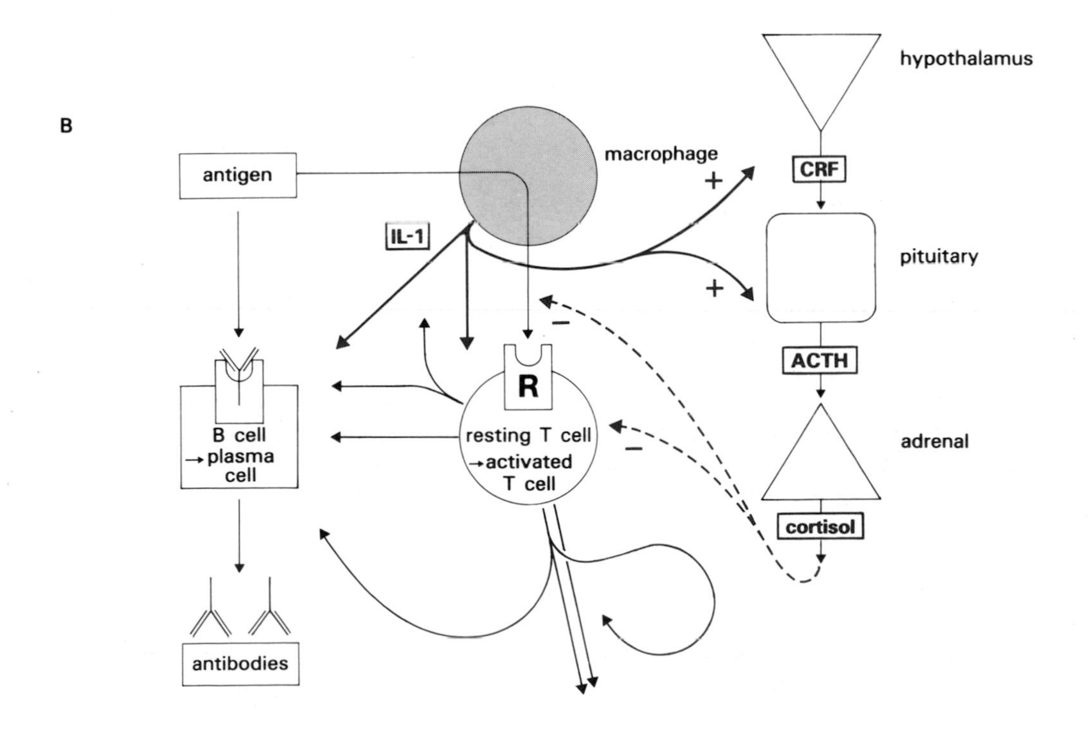

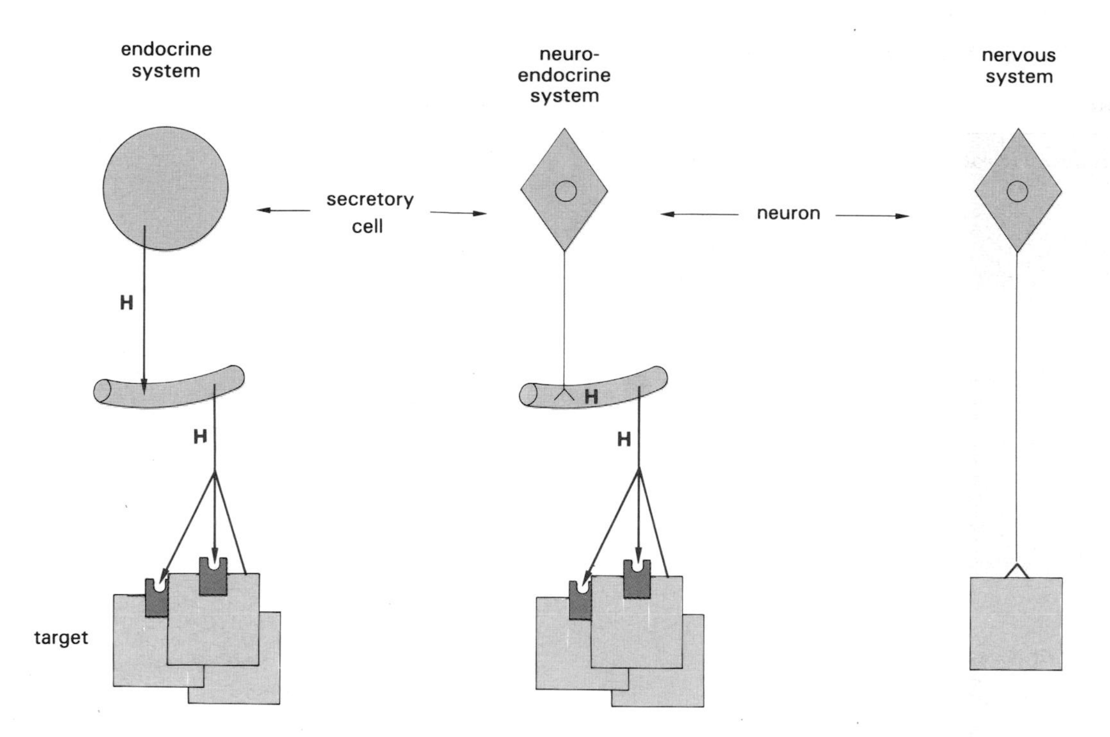

Figure I.9 The endocrine, neuroendocrine and nervous systems. Hormone secreted into the blood has access to several target cells. In contrast, a neuron innervates a single cell, and the neurotransmitter it releases is restricted to the synaptic cleft. Intermediate between the two, a neurohormone is secreted by a neuron into the blood, and is then distributed to several target cells.

type can be under the control of several hormones. This disparity accounts for the ability of the endocrine system to integrate the complex functions of the organism.

1.6 Endocrinology Is Multidisciplinary and Multifaceted

Endocrinologists study the synthesis, the regulation, and the physiological and pathological effects of hormones. It is not surprising that, depending on the circumstances, they use chemistry, and speak the language of mathematicians, physicists or computer scientists. The development of *contemporary* endocrinology can be attributed largely to the use of *radioactive elements* (and thus to the discovery of artificial radioactivity by Fréderic and Irène Joliot-Curie)[4]. Other important developments include the major technological advances in analytical and synthetic chemistry, chromatography, competition analysis, mass spectrometry, sequencing of proteins and crystallography, and, of course, the synthesis of hormones and thousands of agonistic and antagonistic analogs. Three major areas of biology, molecular and cellular biology, neurobiology and immunology, have already had a major impact on the evolution of endocrinology today. It is fair to say, however, that the converse is also true. Endocrinology, particularly the study of receptors, has contributed greatly to these fields, and continues to do so.

The evolution of endocrinological research is now largely intertwined with developments in *molecular biology*[5,6]. The first proteins synthesized by bioengineering were hormones: human insulin and GH have been commercialized, and interferons and interleukins (p. 254) are now produced in sufficient quantities by recombinant DNA technology that their therapeutic value can be assessed. Cloning techniques have already allowed background analysis of complex molecules, such as receptors for insulin, LDL, EGF and other growth factors, and steroids and thyroid hormones, an almost impossible task using classical methods of purification and direct protein analysis.

4. Joliot-Curie, I. and Joliot-Curie, F. (1934) Un nouveau type de radioactivité. *C.R. Acad. Sci. Paris* 198: 254–256.
5. Albert, B., Bray, D., Lewis, J., Raff, M., Roberts, K. and Watson, J. D., eds (1983) *Molecular Biology of the Cell.* Garland, New York.
6. Darnell, J., Lodish, H. and Baltimore, D., eds (1986) *Molecular Cell Biology.* Scientific American Books, New York.

Table I.1 Principle hormones of vertebrates.

Class	Hormones	Main (classical) source
Peptides and proteins	Melatonin	Epiphysis
	TRH	
	GnRH	
	Somatostatin	
	CRF	
	GRF	
	GH	
	ACTH	
	PRL	Hypothalamus and pituitary
	MSH	
	β-endorphin	
	TSH	
	FSH	
	LH	
	CG	(placenta)
	Oxytocin	
	Vasopressin	
	IGFs	Liver
	Insulin	
	Glucagon	
	Gastrin	Digestive glands
	Secretin	
	CCK	
	Inhibin	
	Müllerian inhibiting substance	Sex glands
	Relaxin	
	ANF	Heart
	Angiotensin	Plasma (liver and kidney)
	Erythropoietin	Kidney
	PTH	Parathyroid
	Calcitonin	C cells (thyroid)
Steroids	Calcitriol	Kidney
	Aldosterone	
	Cortisol	
	Corticosterone	Adrenal cortex
	DHEA-S	
	Androstenedione	
	Testosterone	
	Estradiol	Sex glands
	Estriol	
	Progesterone	
Amino acids	T_4	Thyroid
	T_3	
	Norepinephrine	Adrenal medulla

Hormones acting via membrane or DNA-binding receptors are listed in the pink and gray shaded areas, respectively. (No receptor has been described for DHEA-S and androstenedione).

Site-directed mutagenesis and transfection techniques (introduction of functional genes into cells) have remarkable potential. The recent observations showing homologies between some oncogene products and growth factors and receptors are striking (p. 122). In addition, molecular biology has been responsible for the discovery of complex genes encoding several hormones expressed differently in different cells, and work in this field has even suggested the existence of yet unidentified hormones (p. 29). Diagnostically, nucleic acid probes are becoming available for the measurement or cytological study of hormone receptor synthesis and hormonal responses. Finally, the spectacular results of the transfer of growth hormone coding sequences into mammalian ova (Fig. I.10) have illustrated the tremendous innovative possibilities of endocrine molecular biology.

Table I.2 Discovery of hormones.

	Hormonal function	Structure
TRH	1962	1969
GnRH	1960	1971
GH	1921	1969
ACTH	1922	1956
FSH and LH	1926	1974
PRL	1928	1969
Somatostatin	1968	1973
GRF	1964	1982
CRF	1955	1981
Inhibin	1932	1985
ANF	1981	1981
VIP	1970	1983
Oxytocin	1901	1954
Insulin	1889	1953
Gastrin and secretin	1902	1964
Calcitonin	1961	1968
Aldosterone	1934	1954
Cortisol	1935	1940
Testosterone	1889	1935
DHEA-S	1960	1935
Estradiol	1925	1931
T_4	1895	1926
T_3	1951	1953
Norepinephrine	1895	1901

Between the discovery of the hormonal principle, or function, and the establishment of the structure of one or several corresponding hormones, a variable period elapses, depending on the evolution of technical advances and concepts. For example, at the end of the nineteenth century, the principle of a secretory product with the properties of insulin was established, but isolation of the hormone by Banting and Best came only in 1922. Insulin was studied by both physiologists and clinicians who did not know its structure, until it was established by Sanger in the early 1950s. In contrast, for calcitonin, hypothalamic peptides and growth factors discovered later, the interval was much shorter, due to the great progress which was made in chemical analysis and in the synthesis of peptides. Now, the cloning of cDNAs of corresponding mRNAs is used, and it allows the establishment of the structure of hormones and precursors (and their receptors) before their actual isolation in pure form. There are (and will be) even more biochemically discovered peptides in search of a hormonal function (p. 29 and p. 264). For steroids and small molecules, knowledge progressed rapidly, with a few exceptions (aldosterone, because of its low concentration, and DHEA-S, because it was incorrectly designated as a waste product, in view of its hydrophilicity, a quality unlike other steroid hormones).

Table I.3 Peptides identified in the brain by chemical or immunological methods.

TRH
GnRH
Somatostatin-14
Somatostatin-28
CRF
GRF
MSH-inhibiting factor
ACTH
PRL
TSH
GH
α-MSH
Oxytocin
Vasopressin
β-endorphin
Dynorphins
Met-enkephalin and Leu-enkephalin
Neurotensin
Substance P
Bradykinin
Bombesin
Angiotensin II
Insulin
Glucagon
Gastrin
Motilin
Secretin
GIP
VIP
Cholecystokinin
CGRP
Carnosin

The *neurosciences*[7] and endocrinology are intimately linked for physiological reasons. *Interactions* between the nervous and endocrine systems were originally thought to be restricted to hypothalamic-

7. Kuffler, S. W. and Nicholls, J. G., eds (1976) *From Neuron to Brain*. Sinauer Associates, Sunderland, Mass.

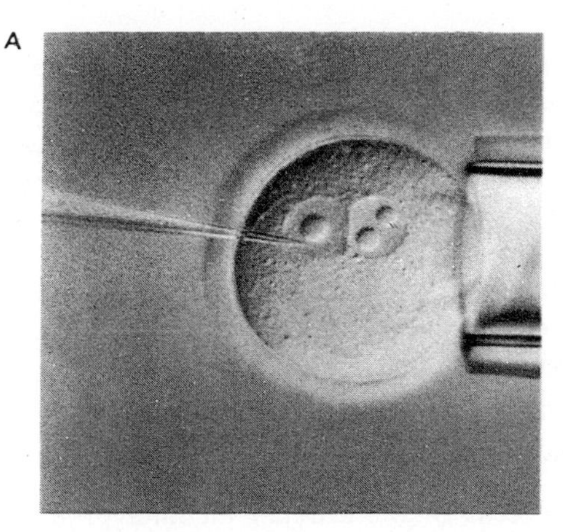

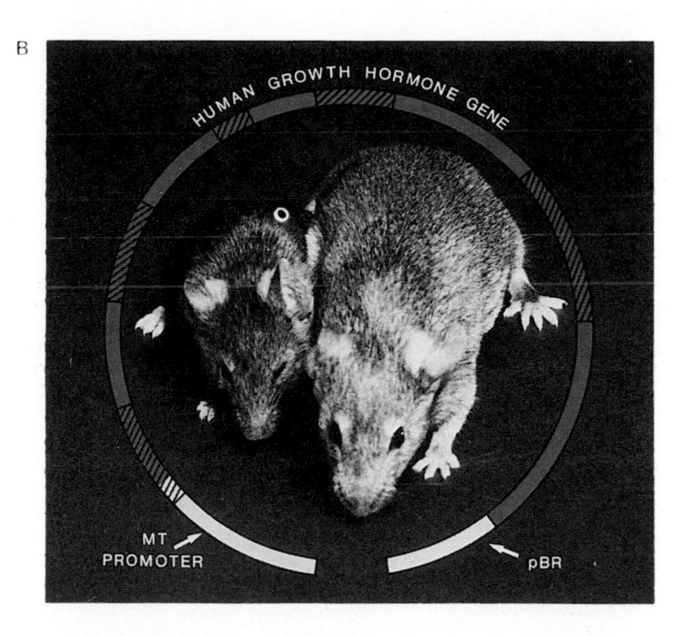

Figure I.10 Transgenic endocrinology. **A.** Transfer of genes into a fertilized mouse ovum. The 15 h fertilized egg (~ 0.1 mm diameter) is held, by suction, with a glass pipette. The male pronucleus is injected with ~ 2 pl of a solution containing ~ 2,000 copies of a foreign gene. (Courtesy of Drs. D.A. Denton, R. Richards and Mr. I. Lyons, Howard Florey Institute, Melbourne).
B. Giant transgenic mice. Two female sibling mice, approximately 24 weeks old, are shown. The male pronucleus of the mouse on the right was microinjected with a new gene composed of the mouse metallothionein promoter/regulator fused to the human growth hormone structural gene with its five exons (Fig. III.4). In general, mice that express the gene grow two to three times faster than controls, reaching a size up to twice normal. Following integration of the new gene into the mouse chromosomes, the offspring also grow larger than controls. (Courtesy of Drs. R. L. Brinster and R. E. Hammer, University of Pennsylvania, Philadelphia, and Dr. R. D. Palmiter, Washington University, Seattle; Palmiter, R. D., Norstedt, G., Gelinas, R. E., Hammer, R. E. and Brinster, R. L. (1983) Metallothionein–human growth hormone fusion genes stimulate growth of mice. *Science* 222: 809–814).

hypophyseal connections, the function of the posterior pituitary and the adrenal medulla, and the feedback control and activities of steroids and thyroid hormones in the nervous system. Numerous peptides, discovered in both the hypothalamus and gastrointestinal system, and pituitary hormones, identified in different regions of the brain (Table I.3), have effects on behavior and mental activity (Table I.4). Actually, nerve cells are capable of synthesizing hormonal peptides and steroids (p. 83 and Fig. I.13C).

Table I.4 Activity of hypothalamic, hypophysiotropic regulatory peptides.

Peptide	Effects in specific cells of the anterior pituitary	Effects in brain; control of the autonomous nervous system
TRH	Stimulates TSH secretion	Increases motor activity, heart rate, blood pressure, and sympathetic nervous system activity; produces hyperthermia; antagonizes a variety of hypnotic and sedative drug actions; inhibits growth hormone secretion
GnRH	Stimulates LH and FSH secretion	Stimulates libido and mating behavior
Somatostatin	Inhibits GH, TSH and PRL secretion	Somatostatin-28 and related analogs: inhibit adrenal epinephrine and pituitary gland ACTH secretion; increase pituitary gland TSH secretion;produce hyperthermia
CRF	Stimulates ACTH secretion	Increases sympathetic and decreases parasympathetic nervous system activity; increases heart rate and blood pressure; behavioral activation

(Brown, M.R. and Fisher, M.A. (1984) Brain peptides as intracellular messengers *JAMA* 251:1310–1315).

Immunology[8] is also essential to endocrinology. It was of extraordinary importance in the development of competition analysis (radio-, enzymo- and viroimmunological assays) for the measurement of hormones (p. 127 and p. 486). The immunosuppression of hormones has permitted a better study of their action. Immunohistochemistry of hormones, biosynthetic enzymes, receptors and endocrine responses are indispensable to endocrinology. In addition, immunological hormones (p. 250) and neuroendocrine effects on the immune system (Fig. I.8B) are currently under study.

The *medical importance* of endocrinology no longer has to be demonstrated. The use of *insulin* in diabetes is one of the highlights of contemporary medicine. *Cortisone*[9,10] and the hormonal *control of fertility*[11] have had a considerable influence on health care and even society as a whole.

2. Production and Secretion of Hormones

2.1 Hormone Synthesis

Hormone biosynthesis and secretion occurs *primarily* in specialized cells of *endocrine glands. Small* hormones consist of either modified amino acids, such as catecholamines and thyroid hormones, or steroids, derived from cholesterol. *Peptide* and *protein* hormones are basically composed of multiple amino acid units, ranging from three for TRH to 199 for human prolactin, and up to 560 for Müllerian inhibiting substance (MIS), the largest known hormone (Table I.5). The synthesis of hormonal peptides and proteins (Fig. I.11) generally begins with a relatively large precursor (preprohormone) which is then processed by cleavage and/or chemical modification to form the active, mature molecule(s).

Small Hormones

The *synthesis* of small hormones, involving *several enzymes*, is at least as complex as that of peptide and protein hormones.

Amino Acid Derivatives

The cells of the adrenal medulla produce *epinephrine* from tyrosine. Methylating enzyme, the last in the biosynthetic sequence, is increased by glucocorticosteroids which come from neighboring adrenal cortical cells[12]. The synthesis of the *hormonal iodothyronine* derivatives thyroxine (T_4) and triiodothyronine (T_3) requires the organization of the thyroid cells into follicles, where the matrix protein thyroglobulin is stored. The uptake of iodide is an important limiting factor in thyroid function.

Steroids

Physiologically, the bulk of steroid hormones is from blood cholesterol[13], synthesized in liver or ingested, and taken up by appropriate gland cells (p.

Table I.5 Sizes of some human peptide and protein hormones.

	Number of amino acids
TRH	3
Oxytocin	9
Vasopressin	9
GnRH	10
Somatostatin	14 (28)
Gastrin	17
Glucagon	29
ACTH	39
CRF	41
GRF	44
Insulin	51
PTH	84
α-LH	96
β-LH	119
GH	191
Prolactin	199
Inhibin A	250
MIS	560

8. Roitt, I. M., Brostoff, J. and Male, D. L., eds (1985) *Immunology*. Gower Medical Publ., London.
9. Reichstein, T. and Shoppee, C. W. (1943) The hormones of the adrenal cortex. *Vitams. Horm.* 1: 345–413.
10. Hench, P. S., Kendall, E. C., Slocumb, C. H. and Polley, H. F. (1949) The effect of a hormone of the adrenal cortex (17-hydroxy-11-dehydrocorticosterone: compound E) and of pituitary adrenocorticotropic hormone on rheumatoid arthritis (Preliminary report). *Proc. Staff Meetings Mayo Clinic*, 24: 181–197; *Ann. Rheum. Dis.* 8: 97–104.
11. Pincus, G., ed (1965) *The Control of Fertility*. Academic Press, New York.
12. Wurtman, R. J. and Axelrod, J. (1966) Adrenalin synthesis: control by the pituitary gland and adrenal glucocorticoids. *Science* 150: 1454–1465.
13. Bloch, K. (1965) The biological synthesis of cholesterol. *Science* 150: 19–28.

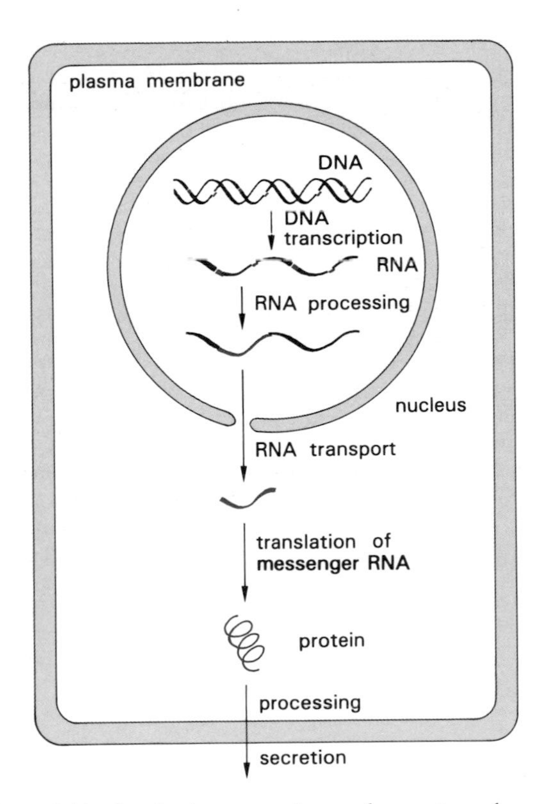

Figure I.11 Synthesis, processing and secretion of peptide and protein hormones. The sequence of events includes both the steps common to the synthesis of all proteins and those of processing and secretion which are more specific to hormones.

391). The role of lipoproteins and LDL-R is described on p. 533. Cells of the adrenals and gonads, however, may occasionally overcome a deficiency in uptake by synthesizing steroids from acetate. The structure of steroidogenic cells is characteristic (Fig. I.12A). Certain steps in the biosynthesis of steroid hormones are particularly sensitive to the influence of anterior *pituitary tropic* hormones; i.e. the *production of pregnenolone*, which corresponds structurally to cholesterol from which six carbons in the lateral chain have been removed (side-chain cleavage) (p. 392) (Figs I.13A and B). Limited amounts of steroid hormones are formed outside of the classical glands (Fig. I.13C). In some cases, secreted steroids can be considered *prohormones*[14] (with reference to precursors of peptide hormones; Fig. I.23) (Table I.6). Synthesized within the gland, and frequently not active by themselves, they are transported to other cells, where the synthesis of the active hormone is completed (Table I.6).

14. Baird, D., Horton, R., Longcope, C. and Tait, J. F. (1968) Steroid prehormones. *Perspect. Biol. Med.* 2: 384–421.

A

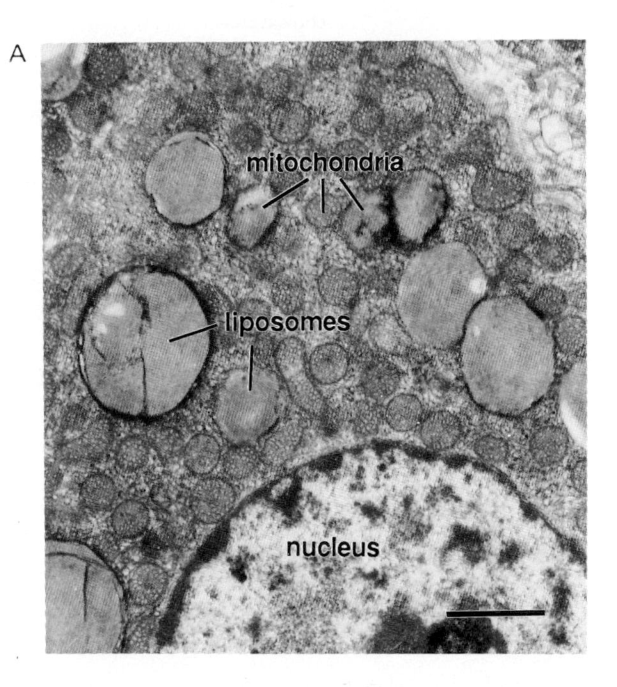

B

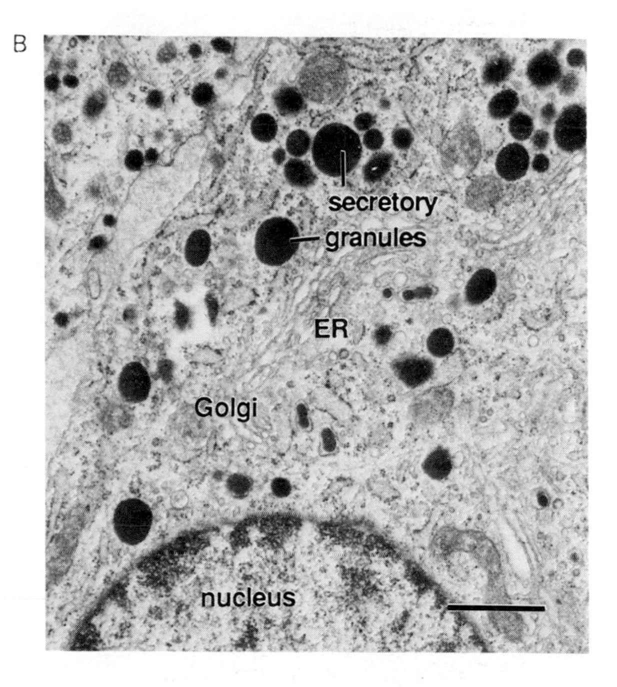

Figure I.12 Comparative structures of secretory cells secreting steroids (**A**) and proteins (**B**). **A.** Adrenocortical cell (cat) of the zona fasciculata. (Courtesy of Dr. J. Racadot, Hôpital Pitié-Salpêtrière, Paris). **B.** Lactotropic cell (rat). (Courtesy of Dr. E. Vila-Porcile, Hôpital Pitié-Salpêtrière, Paris). Bars = 1 μM. In cells that produce both steroids and proteins (e.g., Leydig cells), liposomes and rough endoplasmic reticulum, characteristics of each type of synthesis, can be found.

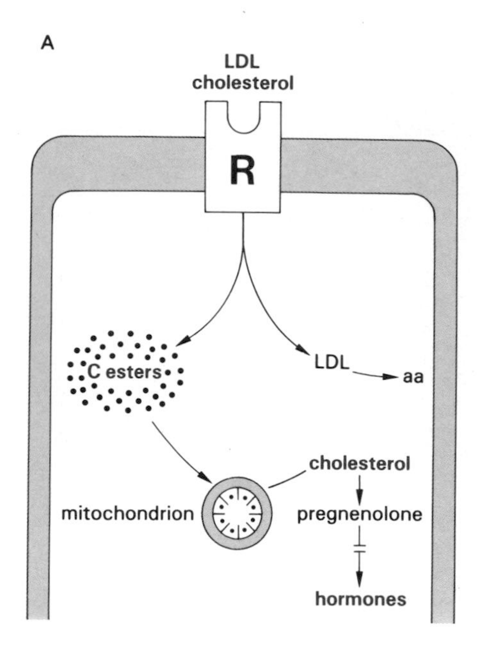

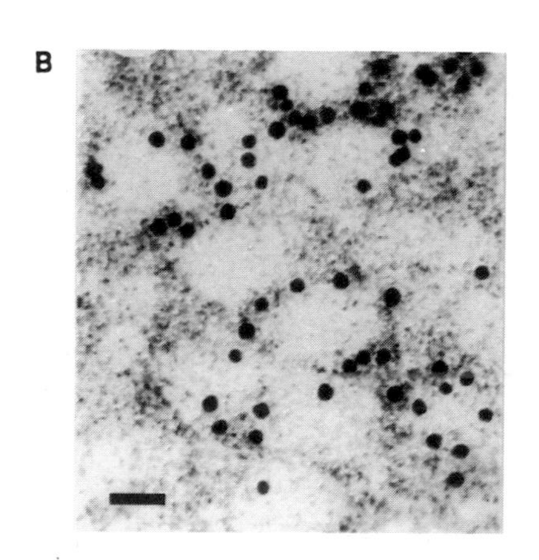

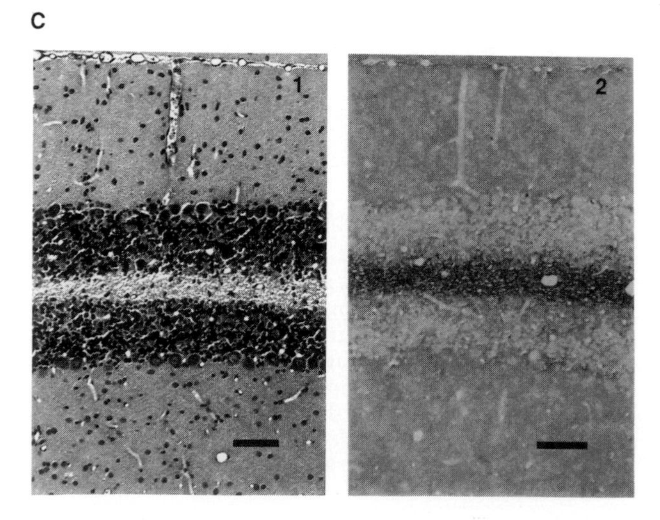

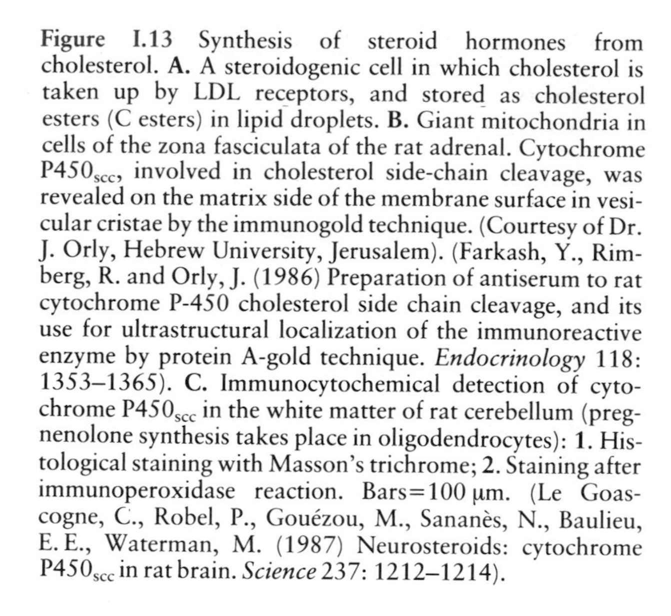
Figure I.13 Synthesis of steroid hormones from cholesterol. **A.** A steroidogenic cell in which cholesterol is taken up by LDL receptors, and stored as cholesterol esters (C esters) in lipid droplets. **B.** Giant mitochondria in cells of the zona fasciculata of the rat adrenal. Cytochrome $P450_{scc}$, involved in cholesterol side-chain cleavage, was revealed on the matrix side of the membrane surface in vesicular cristae by the immunogold technique. (Courtesy of Dr. J. Orly, Hebrew University, Jerusalem). (Farkash, Y., Rimberg, R. and Orly, J. (1986) Preparation of antiserum to rat cytochrome P-450 cholesterol side chain cleavage, and its use for ultrastructural localization of the immunoreactive enzyme by protein A-gold technique. *Endocrinology* 118: 1353–1365). **C.** Immunocytochemical detection of cytochrome $P450_{scc}$ in the white matter of rat cerebellum (pregnenolone synthesis takes place in oligodendrocytes): **1.** Histological staining with Masson's trichrome; **2.** Staining after immunoperoxidase reaction. Bars=100 μm. (Le Goascogne, C., Robel, P., Gouézou, M., Sananès, N., Baulieu, E. E., Waterman, M. (1987) Neurosteroids: cytochrome $P450_{scc}$ in rat brain. *Science* 237: 1212–1214).

Since the synthesis of small hormones involves several very specialized chemical reactions, whether it occurs in one or more cell type, it depends, despite the apparent simplicity of the structure, on the correct and coordinated expression of a number of genes encoding the enzymes involved. In contrast to peptide hormones, this makes the possibility of producing small hormones by genetic engineering methodology very remote.

Peptide Hormones

Currently, approximately 100 peptide hormones are known, and for many of them, sequencing and, more often, cDNA cloning technology have recently established their primary structure, that of their precursors, and elucidated the specific steps involved in their biosynthesis. Immunological and nucleic acid probes have also permitted the detection of hormone synthesis by localization of the peptide or its mRNA (Fig. I.14).

Peptide hormones are synthesized in specific secretory cells (Fig. I.12B). They are encoded by genes (Fig. I.15) which account not only for the segments constituting the active peptide molecule (mature), but also for others which may play a role in the *process of maturation and secretion* of the hormone, and which may permit some *biological diversity* through RNA or peptide processing (Table I.7; p. 29 and p. 27). The mechanisms involved in the synthesis and secretion of

Table I.6 Prohormones are precursors of hormones.

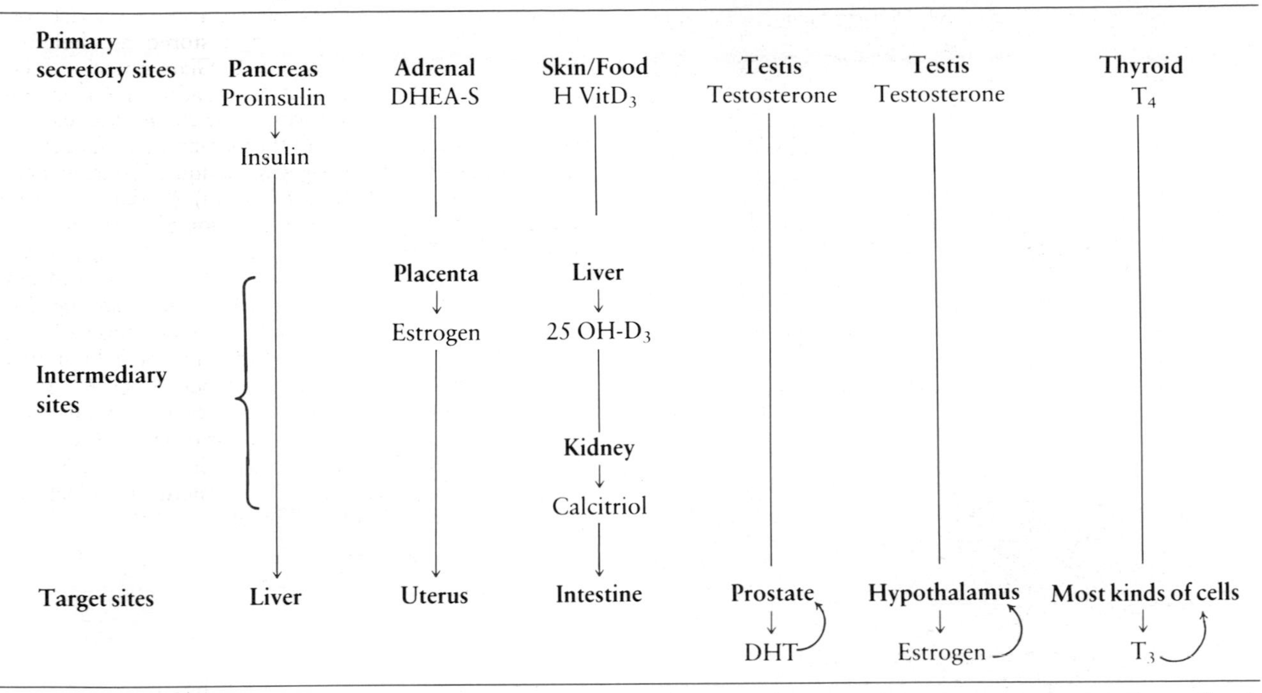

For peptide hormones, preprohormones (p. 20) are essentially inactive, and undergo processing prior to hormone secretion. For small molecules such as steroids and thyroid hormones, the prohormone is either inactive, transformed into an active product in the target cell or in one or several intermediary cells (e.g., DHEA-S, vitamin D), or it is active by itself and further activated (e.g., testosterone to DHT and T_4 to T_3).

Table I.7 Synthesis of peptide hormones.

Steps from DNA to hormone		Mechanisms involved in the generation of diversity
DNA (gene)		
↓	*Transcription*	Gene family Selective transcription
RNA (primary transcript)		
↓	*RNA processing*	Differential processing
mRNA		
↓	*Translation*	
Pre(pro)hormone		
↓	*Proteolysis* (signal peptide cleavage)	Differential proteolysis and/or chemical modification (e.g., glycosylation, phosphorylation)
(Pro)hormone		
↓	*Post-translational processing*	
Hormone		

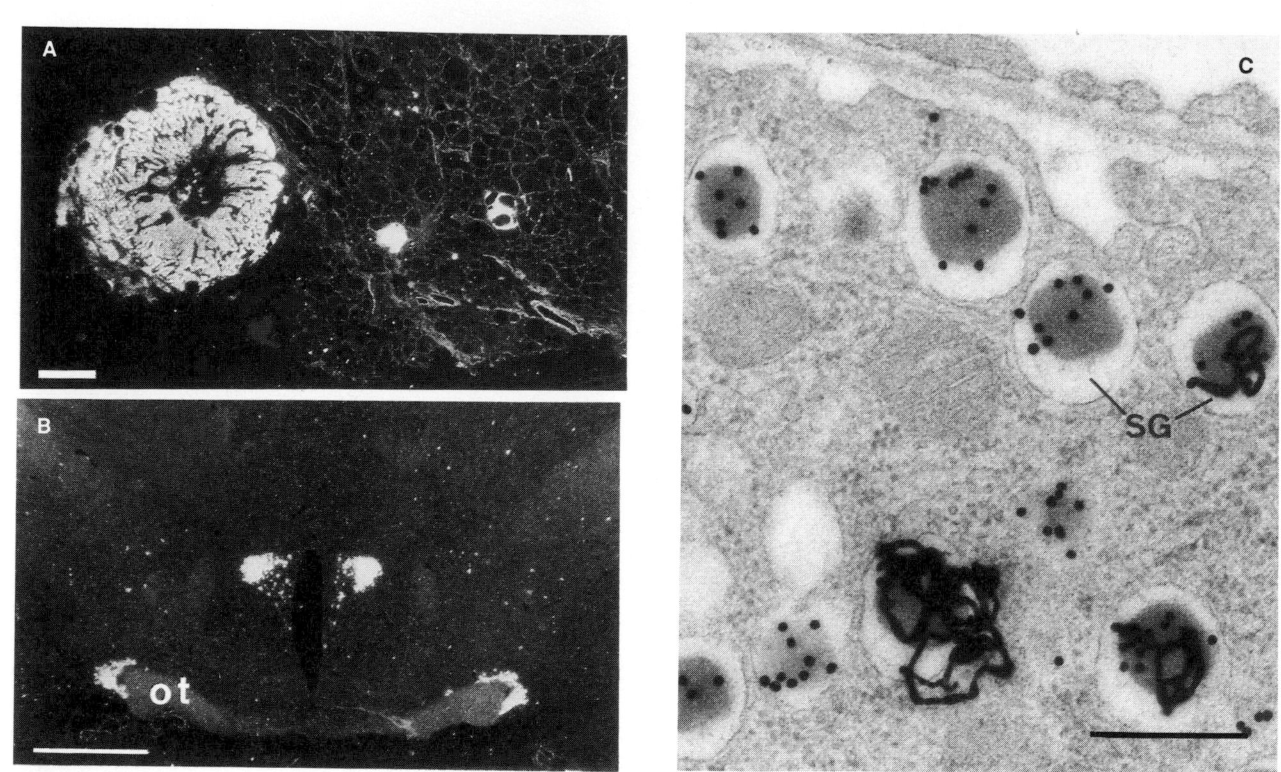

Figure I.14 Hormone production. **A.** Hormone synthesis in a gland. Detection of calcitonin mRNA in human medullary thyroid carcinoma by in situ hybridization using a double stranded ^{32}P-labelled human calcitonin cDNA. (Provided by Dr. A. Julienne, Hôpital St. Antoine, Paris). Dark-field illumination. Calcitonin mRNA is present in all cells of the tumor as well as in small islands in the adjacent thyroid. Dark areas in the tumor are amyloid substance. Bar=1 mm. (Courtesy of Drs. B. Bloch and T. Popovici, Faculté de Médecine, Besançon.) **B.** Hormone synthesis in the CNS. Detection of vasopressin mRNA in rat hypothalamus by in situ hybridization using a synthetic 45 mer oligonucleotide probe (prepared by Dr. P. Bohlen, University of Zurich), labelled with ^{35}S at its 3′ end. Dark-field illumination. Neurons containing vasopressin mRNA are present in the supraoptic and paraventricular nuclei; ot, optic tract. Bar=1 mm. (Courtesy of Drs. B. Bloch and A. F. Guitteny, Faculté de Médecine, Besançon). (Bloch, B., Popovici, D., Le Guellec, E., Normand, S., Chouham, A. F. and Bohlen, P. (1986) In situ hybridization histochemistry for the analysis of gene expression in the endocrine and central nervous system tissues: a 3-year experience. *J. Neurosci. Res.* 16: 183–200). **C.** Simultaneous detection of a protein by incorporation of a radioactive amino acid and by the protein A-gold method. A thin section of a pancreatic B cell from a [^{3}H]leucine pulse-chase experiment is first processed to reveal autoradiographic grains, and is then incubated, without photographic emulsion, with antibodies to insulin, which are revealed by the protein A-gold method. SG, secretory granule; bar=0.5 μM. (Orci, L. (1984) Patterns of cellular and subcellular organization in the endocrine pancreas. *J. Endocrinol.* 102: 3–11).

peptide hormones are in fact common to all secreted proteins.

The studies conducted by Palade and Blobel[15] revealed the principal steps that follow the translation of mRNA on ribosomes. Messenger RNA, at its 5′ end (following the initiation codon), contains a series of codons whose translation product is termed the *signal sequence* (Fig. I.15). A protein beginning by the signal (or leader) sequence is a preprotein, and for hormones it is a *prehormone*. The signal peptide differs according to the prehormone, but it is generally ~30 aa in length and contains a number of hydrophobic amino acids which probably have a fundamental role in the secretory process‡. Protein synthesis begins on an initially free, unattached ribosome, whose subunits are derived from the common cytoplasmic pool. As the signal sequence of the growing chain is synthesized, it interacts with a signal recognition particle (SRP), recognizing a specific docking protein, the SRP receptor (Fig. I.16), found on the membrane of the endoplasmic reticulum. The growing peptide chain, now on a membrane-bound polysome, passes through the membrane (the detailed mechanism by which this occurs is unknown, but it might

‡Some secreted peptides have no signal sequence (e.g., FGFs).

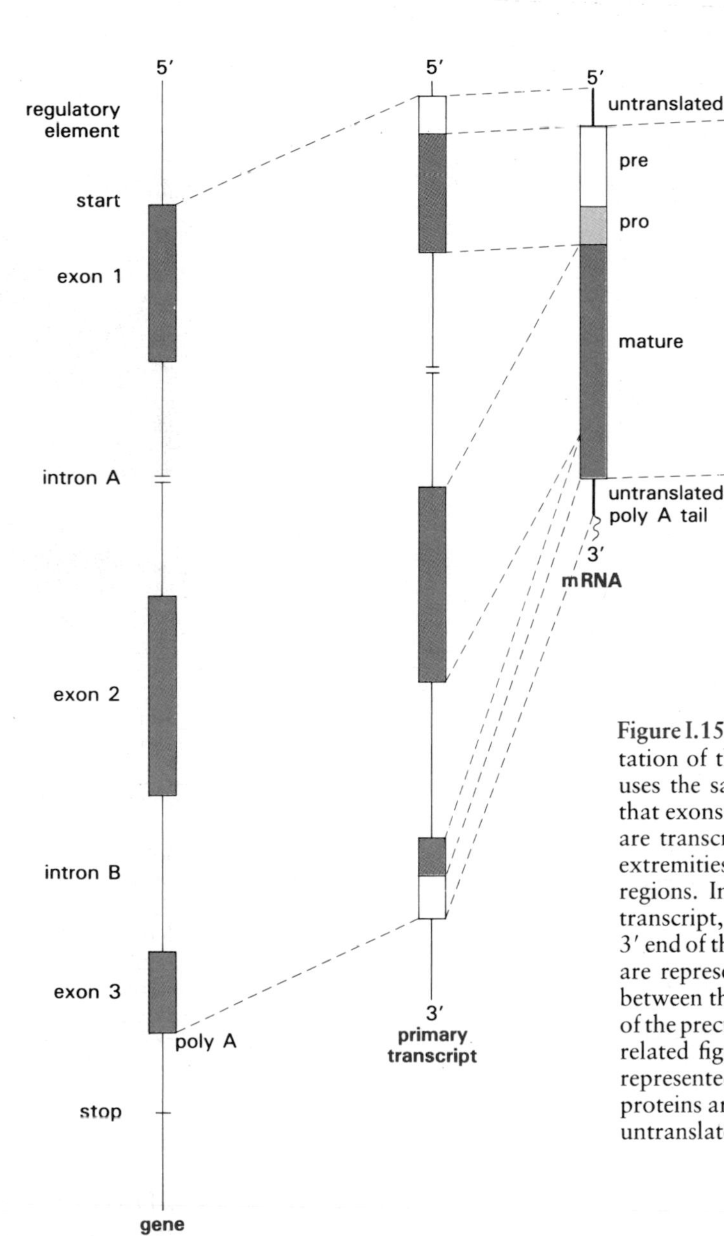

Figure I.15 From gene to protein. This schematic representation of the main sequences involved in protein synthesis uses the same symbols found throughout the book. Note that exons and introns, but not the 5′ regulatory sequences, are transcribed into RNA. The two white portions at the extremities of the primary RNA transcript are untranslated regions. Intron sequences are removed from the primary transcript, resulting in mRNA. A poly A tail is added at the 3′ end of the mRNA. The untranslated regions of the mRNA are represented by thickened black lines. Correspondence between the mRNA and the pre-, pro- and mature domains of the precursor protein (preproprotein) is indicated. In most related figures in this book, the primary transcript is not represented. Exons are in red, and the pre, pro and mature proteins are shown in white, pink and red, respectively. The untranslated mRNA has generally been omitted.

involve a "translocator protein", an oligomeric structure surrounding an aqueous channel[16]) and enters the cisternal space of the endoplasmic reticulum. The signal peptide is then cleaved by a protease located on the cisternal face of the membrane of the endoplasmic reticulum. Thus, the *half-life* of prehormones is *short*. The remaining amino acid chain becomes the definitive hormone. It will be transferred, frequently with concomitant *glycosylation* (Table I.8) and other *chemical modifications* (e.g., phosphorylation), through a series of *membrane-bound compartments* (namely, transitional elements of the endoplasmic reticulum, Golgi apparatus, condensing vacuoles, and secretory granules (Figs. I.17–I.19)), finally to be discharged from the cell after *fusion* of the granule *mem-*

15. Walter, L., Gilmore, R. and Blobel, G. (1984) Protein translocation across the endoplasmic reticulum. *Cell* 38: 5–8.
16. Singer, S.J., Maher, P.A. and Yaffe, M.P. (1987) On the translocation of proteins across membranes. *Proc. Natl. Acad. Sci., USA* 84: 1015–1019.

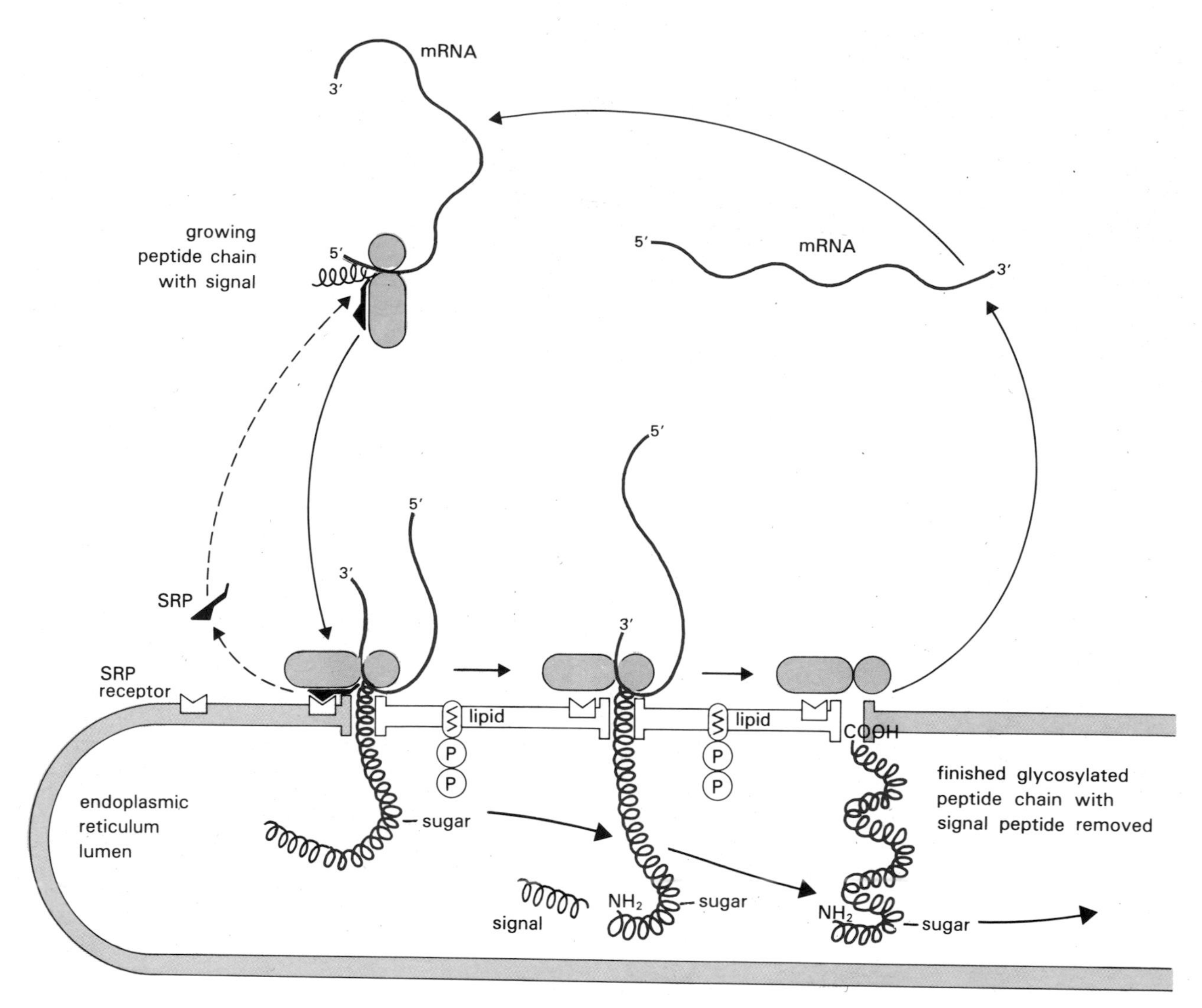

Figure I.16 Secretion of proteins through the ER, and glycosylation. The signal recognition particle (SRP) protein (shown in black) attaches to the signal peptide. It binds to an SRP receptor on the membrane of the ER, directing the peptide to traverse the lipid bilayer, probably through an oligomeric translocator protein. After entry of the protein with its signal peptide into the ER lumen, specific peptidases remove this peptide, and the glycosylation process proceeds while the protein chain grows in length. This chain may then undergo subsequent processing. The protein is released from the polyribosome and progresses through a series of membrane-bound compartments, during which time it may be further modified (e.g., proprotein to protein, phosphorylation)[15].

Table I.8 Glycosylation of hormones and receptors.

1. *N*-linked oligosaccharides: asparagine-linked; most abundant; most synthesized in the ER.
2. *O*-linked oligosaccharides: serine, threonine or hydroxylysine (collagen-linked); less abundant; synthesized in the Golgi apparatus and in the ER.

Mechanism:
The donor is a dolichol diphospho-oligosaccharide. Dolichol consists of a long, hydrophobic, 15–95 carbon atom chain firmly embedded in the membrane (Fig. I.11).
Glycosylation proceeds in the ER, and eventually in the Golgi apparatus, according to an orderly process; e.g., mannose added in the ER, sialic acid in the Golgi apparatus.

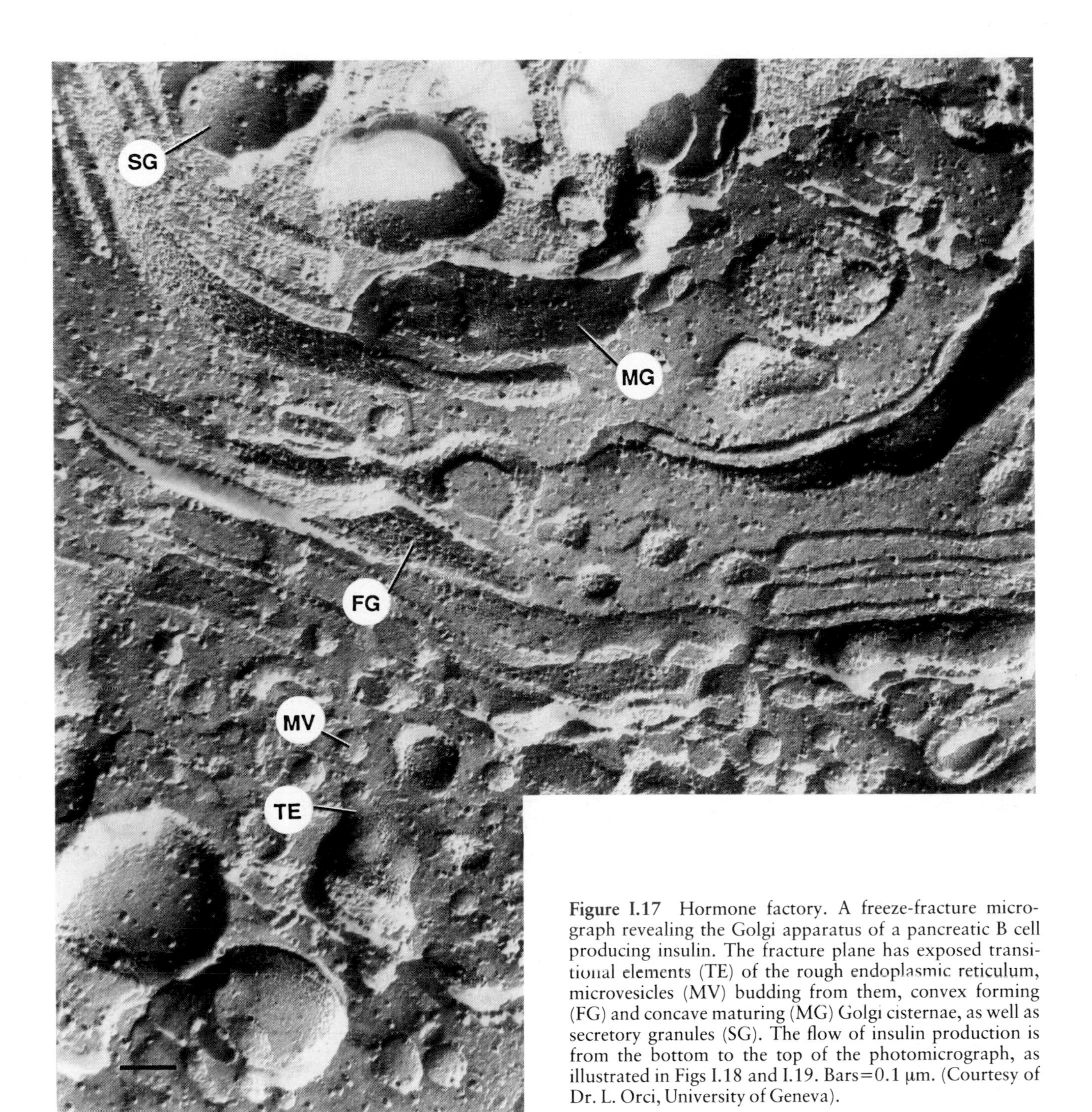

Figure I.17 Hormone factory. A freeze-fracture micrograph revealing the Golgi apparatus of a pancreatic B cell producing insulin. The fracture plane has exposed transitional elements (TE) of the rough endoplasmic reticulum, microvesicles (MV) budding from them, convex forming (FG) and concave maturing (MG) Golgi cisternae, as well as secretory granules (SG). The flow of insulin production is from the bottom to the top of the photomicrograph, as illustrated in Figs I.18 and I.19. Bars=0.1 µm. (Courtesy of Dr. L. Orci, University of Geneva).

brane with the plasma membrane (Fig. I.20). In fact, in addition to *unidirectional* events, from synthesis to secretion, there may be other distinct pathways. Protein-secreting endocrine cells are polarized, exocytosis occuring through a specialized region of the plasma membrane (an example of polarized secretion is the release of neurohormone from nerve terminals). This does not mean that secretion does not occur elsewhere on the cell surface (particularly in cells of the auto- and paracrine systems). Peptide hormone secretion can be regulated and/or constitutive (Fig. I.21)[17]. *Regulated* secretion involves the formation of secretory granules/

17. Kelly, R. B. (1985) Pathways of protein in eukaryotes. *Science* 230: 25–32.

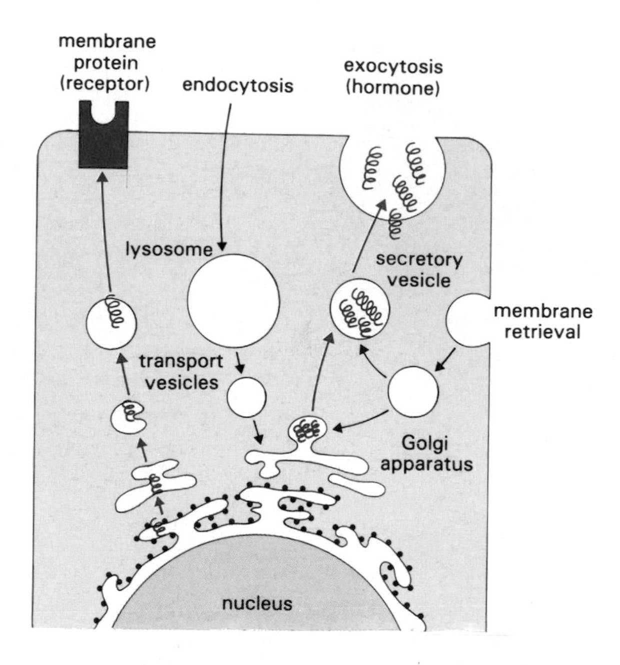

Figure I.18 Hormone secretion and insertion of the receptor into the plasma membrane. Hormones and receptors, synthesized by polyribosomes on rough ER, progress to and through the Golgi, and further, via membrane-bound compartments (vesicles). Hormones are frequently packaged in secretory vesicles before being secreted (exocytosis). Receptors are directed toward the plasma membrane. In addition, the process of endocytosis, terminating in lysosomes, and the membrane retrieval process, reutilizing membrane fragments after secretion, are represented; both are frequently involved in complex intracellular trafficking.

vesicles in which the concentration of the hormonal protein is high ($\sim 10^2$ times higher than in Golgi and in constitutive secretory vesicles that are discharged rapidly). The fact that a secreted protein molecule may be delivered either to the constitutive or to the regulated system by an intracellular transport protein or a membrane receptor implies the existence of a *sorting* mechanism, but it is not known whether the main role in sorting is played by a property inherent in the structure of the secreted protein (a sorting domain). Many other intriguing problems remain concerning the intracellular trafficking of proteins. *Constitutive* secretion depends almost exclusively on the amount of protein synthesized, while regulated secretion (and thus fusion of the granule membrane with the plasma membrane) must be initiated by a regulatory signal involving Ca^{2+} and, possibly, protein phosphorylation/dephosphorylation.

Chemically, the sequence of events in the production of peptide hormones is often complex, and the protein released from the preprotein is usually not a definitive hormone. At this stage, the protein is called a *prohormone*[18,19]. In the case of proinsulin (Fig. I.22A), synthesized in the B cells of the islets of Langerhans (p. 492), the prohormonal peptide

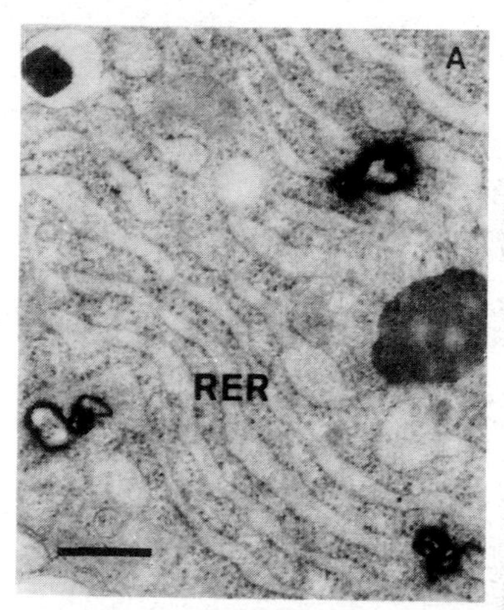

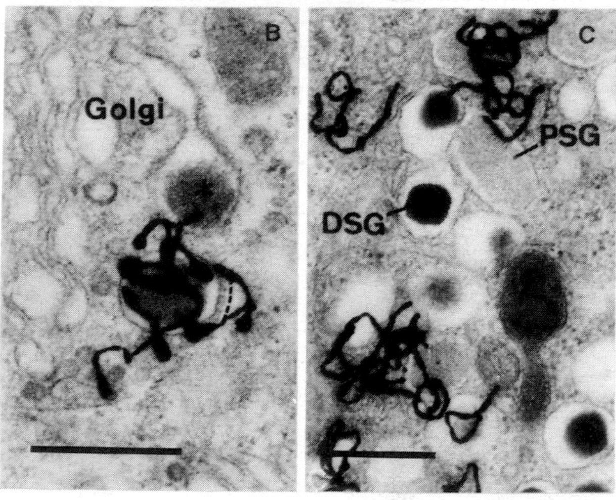

Figure I.19 Intracellular compartments involved in the synthesis and processing of insulin. Autoradiographic reaction over the rough endoplasmic reticulum (RER) (**A**), the Golgi apparatus (**B**), and pale (PSG) and dense (DSG) secretory granules (**C**). Bars=0.5 μm. (Courtesy of Dr. L. Orci, University of Geneva).

18. Docherty, K. and Steiner, D. F. (1982) Post-translational proteolysis in polypeptide hormone biosynthesis. *Annu. Rev. Physiol.* 44: 625–638.
19. Chrétien, M. and Seidah, N. G. (1984) Precursor polyproteins in endocrine and neuroendocrine systems. *Int. J. Peptide Prot. Res.* 23: 255–341.

includes, from N-terminal toward the C-terminal, the B chain sequence followed by two basic amino acids, the C peptide (31 aa) followed by two basic amino acids, and the A chain. The secondary and tertiary structures of proinsulin permit protease activity, taking place primarily in the secretory granules, to cause the release of insulin with the A and B subunits cor-

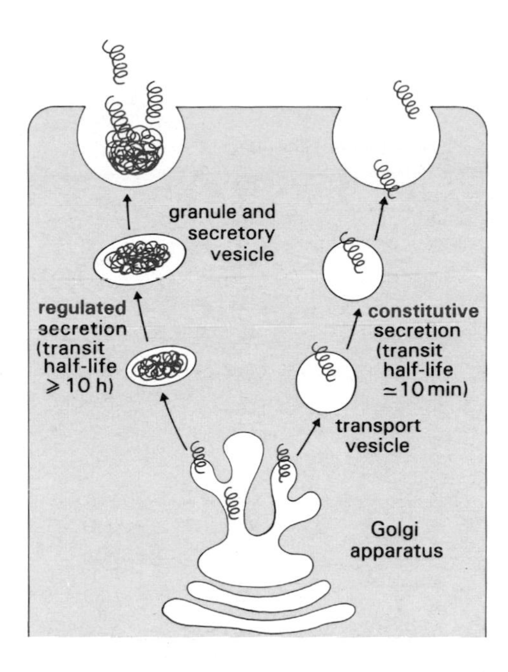

Figure I.21 Constitutive and regulated hormone secretion. In constitutive secretion, there is no storage pool, and the transit to the cell surface is short. In regulated secretion, the regulatory product accumulates in electron-dense secretory granules. Secretion involves membrane fusion, which is regulated by intracellular messenger(s) such as Ca^{2+}.

A

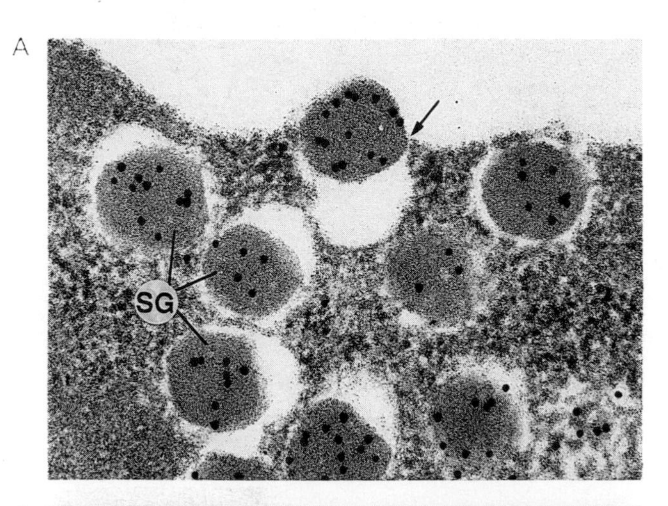

B

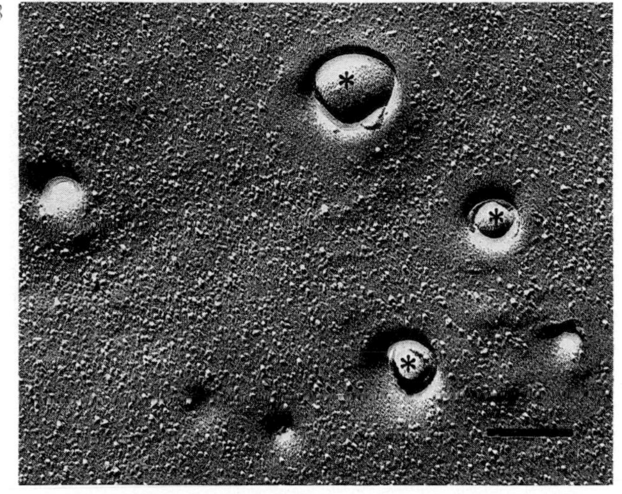

Figure I.20 Hormone secretion. Panel **A**: The peripheral cytoplasm of a B cell, where exocytosis of a secretory granule (SG) is taking place. Insulin was detected by immunostaining with an anti-insulin antiserum revealed by the protein A-gold method. A granule core is in the process of being extruded (arrow). Panel **B**: Freeze-fracture micrograph of the plasma membrane showing depressed, particle-free areas corresponding to fusion sites between plasma and secretory granule membranes. The asterisks indicate three, smooth, bulging masses, interpreted as granule cores exposed to the extracellular space. Bar=0.2 μM. (Orci, L. (1984) Patterns of cellular and subcellular organization in the endocrine pancreas. *J. Endocrinol.* 102: 3–11).

rectly aligned by two S-S bonds, the final structure which has hormonal activity (Fig. I.22B). Contrary to insulin, proinsulin has almost no affinity for the insulin receptor, and thus has no activity. Its synthesis permits a more efficient production of insulin than if the A and B chains were synthesized separately. Indeed, when the two chains are mixed directly, the efficiency of correct matching is very low (of the order of 1%). Normal or pathological variation in this mechanism may be accounted for by the presence, in the blood, of proinsulin, which may crossreact immunologically with insulin, a possibility that must be taken into consideration when interpreting radioimmunoassays (p. 487).

Almost all peptide hormones studied to date are produced initially in the form of prohormonal precursors. The example of insulin is unusual, as the C peptide fragment destined to be removed is located in the middle of the proinsulin sequence (Fig. I.22). As indicated in Figure I.23, there are different types of prohormones. The inactive *fragments* to be *eliminated* can be found before, after, or at both extremities of the bioactive sequence. In addition, a prohormone can, in

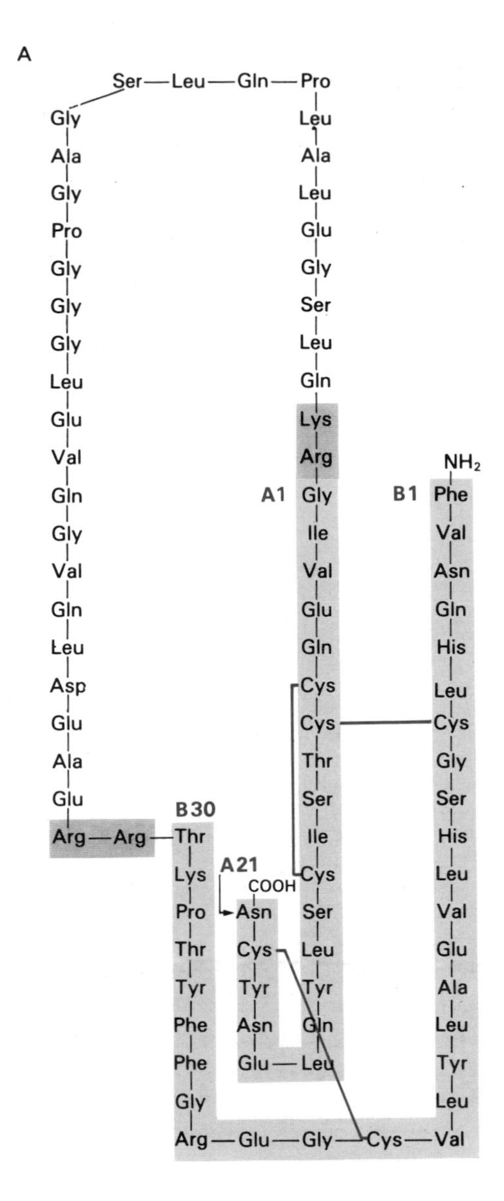

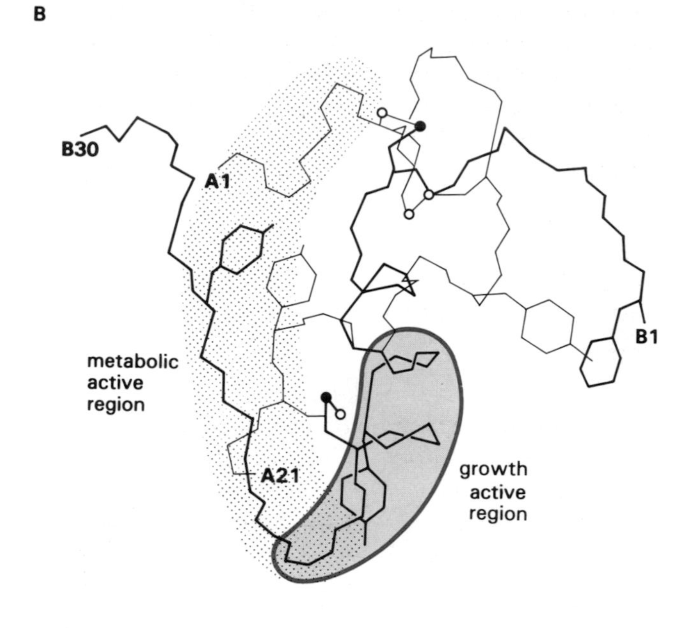

Figure I.22 The structure of proinsulin and insulin. **A.** The insulin molecule, consisting of A and B chains, is shown in red. The connecting C peptide of proinsulin is shown in black. The two pairs of successive basic amino acids, sites of proteolytic processing, are shown in gray. **B.** The three-dimensional molecular structure of insulin, obtained by X-ray crystallography (see also Fig. XI.3). Two regions are indicated. The shaded area is putatively involved in metabolic activity of the hormone, and the encircled colored area shows the hypothetical growth-active region.

fact, be a *polyprotein* (Fig. I.23)[20] containing the sequences of several identical or different hormones which will be released during proteolytic processing. The enzymes involved in the generation of peptide hormones, although incompletely identified as yet, are of major importance (p. 159). Many of the enzymes found in secretory vesicles are trypsin-like, operating at cleavage sites characterized by *pairs of successive basic amino acids* (Fig. I.22). Enzymes of either substrate *specificity* or a broad spectrum of enzymatic activity have been described, and a specific enzyme could be the target of a specific inhibitory drug (p. 603). Moreover, a given prohormone may be processed *differently* in the different *cells* which produce it. It appears, therefore, that the relatively simple synthesis described above for peptide hormones (as compared to that of small hormones) may, in fact, be more complex.

There are probably *several reasons* for the existence of prohormones. Their primary and secon-

20. Douglass, J., Civelli, O. and Herbert, E. (1984) Polyprotein gene expression: generation of diversity of neuroendocrine peptides. *Annu. Rev. Biochem.* 53: 665–715.

dary/tertiary structures may play a decisive role in the *processing* of specific peptide hormones. Their structures may also facilitate *transport* through the endoplasmic reticulum, thus protecting the future hormone from degradation before it is secreted in its bioactive form.

Diversification of Hormone Synthesis by Prohormone Processing and Post-translational Recombination

The possible *diversification* of hormone synthesis by multiple fragmentation of a prohormone, or more exactly a propolyhormone, is exemplified by the case of *proopiomelanocortin (POMC)*. This peptide contains the elements of two main hormones, ACTH and β-endorphin, and as shown in Figure I.24, those of other potential hormones. The process of cleavage differs depending on the cell type involved. A hormone such as ACTH can even be cleaved to produce another active peptide (α-MSH) (Fig. IV.3). Other notable examples include the biosynthesis of enkephalins (5 aa) (Table 1, p. 186) and TRH (3 aa) (p. 263). These very short peptides, also, are synthesized within one or several proprotein(s). In this respect, they differ from the non-hormonal tripeptide glutathion, formed by a purely enzymatic, non-ribosomal mechanism. The prohormonal precursor of short peptides may be very large, and the short bioactive sequence can be repeated within it several times. Proproteins may be multiple,

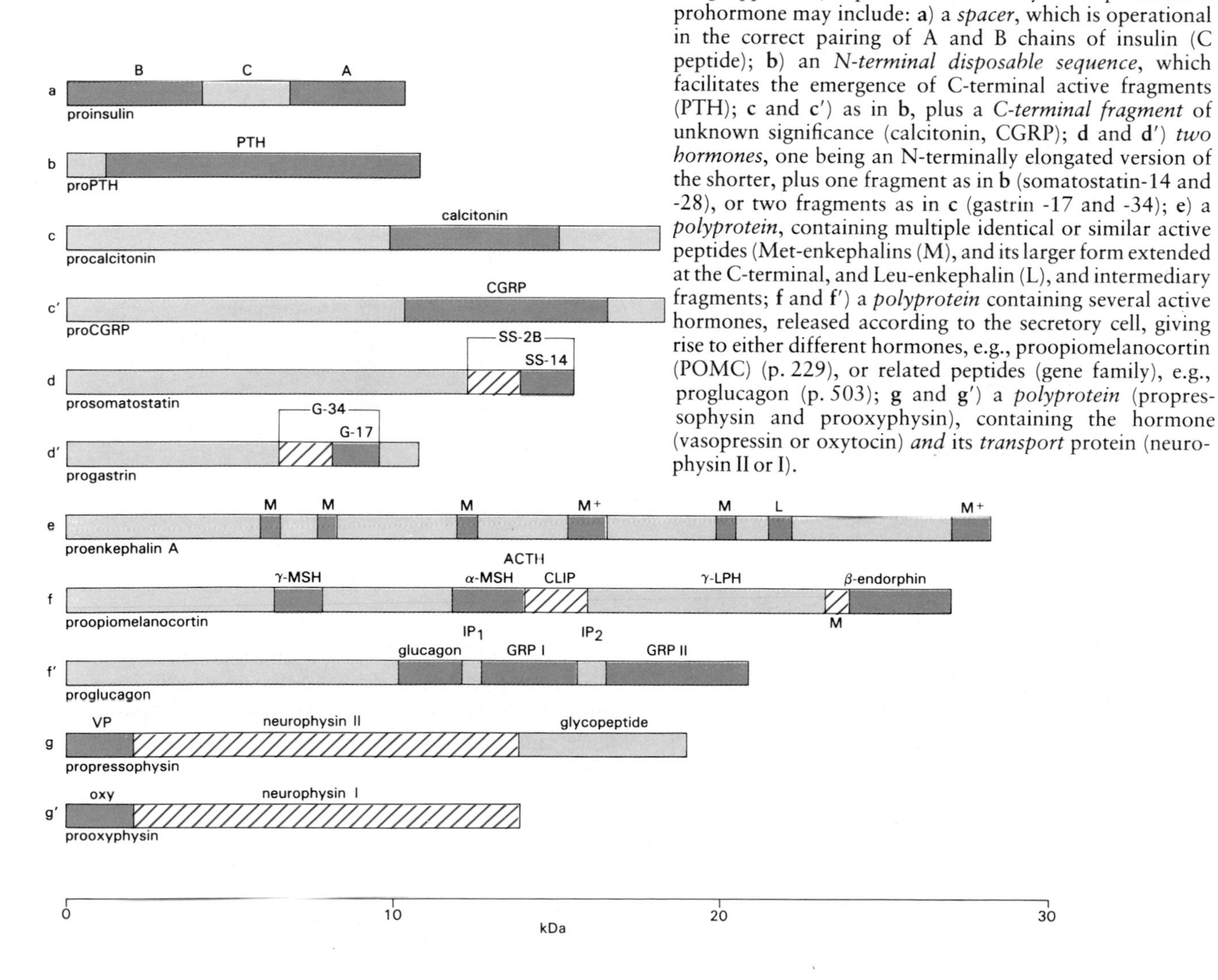

Figure I.23 *Prohormones*: precursors of peptide and protein hormones. These peptides are processed, mostly in the Golgi apparatus, to produce hormonally active products. A prohormone may include: **a**) a *spacer*, which is operational in the correct pairing of A and B chains of insulin (C peptide); **b**) an *N-terminal disposable sequence*, which facilitates the emergence of C-terminal active fragments (PTH); **c** and **c'**) as in **b**, plus a *C-terminal fragment* of unknown significance (calcitonin, CGRP); **d** and **d'**) *two hormones*, one being an N-terminally elongated version of the shorter, plus one fragment as in **b** (somatostatin-14 and -28), or two fragments as in **c** (gastrin -17 and -34); **e**) a *polyprotein*, containing multiple identical or similar active peptides (Met-enkephalins (M), and its larger form extended at the C-terminal, and Leu-enkephalin (L), and intermediary fragments; **f** and **f'**) a *polyprotein* containing several active hormones, released according to the secretory cell, giving rise to either different hormones, e.g., proopiomelanocortin (POMC) (p. 229), or related peptides (gene family), e.g., proglucagon (p. 503); **g** and **g'**) a *polyprotein* (propressophysin and prooxyphysin), containing the hormone (vasopressin or oxytocin) *and* its *transport* protein (neurophysin II or I).

such as proenkephalins (Fig. I.23 and p. 186), and two progastrin-releasing peptides are coded by mRNAs differing by 19 nucleotides[21].

The synthesis of propolyprotein may possibly facilitate the *coordination* of distinct physiological functions by producing several hormonal products, in the same or in different cells, according to an appropriate temporal pattern. This is the case for the secretion of ACTH and β-endorphin in response to stress (p. 34). The simultaneous expression of GnRH and GAP genes (p. 208) may account for an inverse relationship between the secretion of gonadotropins and prolactin (PRL): PRL levels are low at the time of ovulation, and gonadotropins are low during lactation. Another remarkable example is found in Aplysia, a marine snail, in which neuropeptides formed by the fragmentation of the polyprotein (with 10 potential cleavage sites) containing ELH (*egg-laying hormone*) are responsible for several coordinated aspects of behavior. In fact, there is a family of at least nine closely related genes involved in the control of egg laying[22].

Post-translational recombination to form hetero-oligomers of different hormonal properties with a common subunit may be another mechanism by which physiological functions are coordinated (Fig. I.25). In the anterior pituitary, the following glycoproteins: luteinizing hormone (LH), follicle stimulating hormone (FSH) and thyroid stimulating hormone (TSH) are dimers with identical α subunits (p. 257) (Fig. I.25)[23,24]. Neither the α nor any of the β subunits has biological activity alone. Glycosylation of both types of subunits is required for their combination and protection from intracellular degradation, but not for secretion. Only one α subunit gene is expressed in different cells of the pituitary and placenta, probably in coordination with the corresponding β genes. The β subunits, one for each pituitary

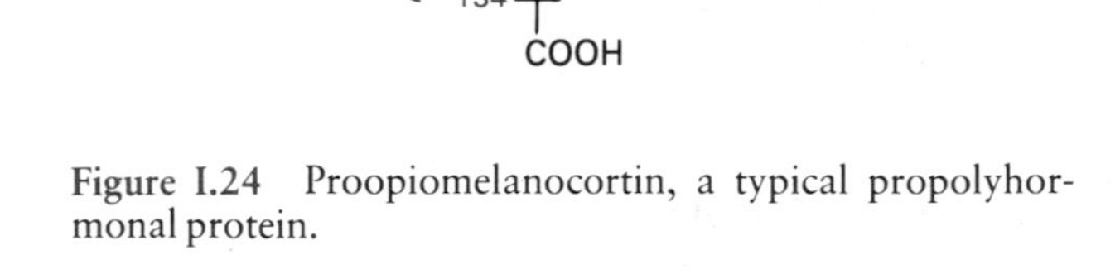

Figure I.24 Proopiomelanocortin, a typical propolyhormonal protein.

hormone and another for placental chorionic gonadotropin (CG), have limited homology with the α subunit. There is from 25 to 40% homology between them, and each has 12 cysteines located at corresponding places in their molecules. The constant α subunit has an important role in the conformation of dimeric complexes, while the β subunit confers biological specificity. LH and FSH are probably synthesized in the same gonadotropic cells, and TSH is produced in thyrotropic cells. Human CG is metabolically more stable than hLH.

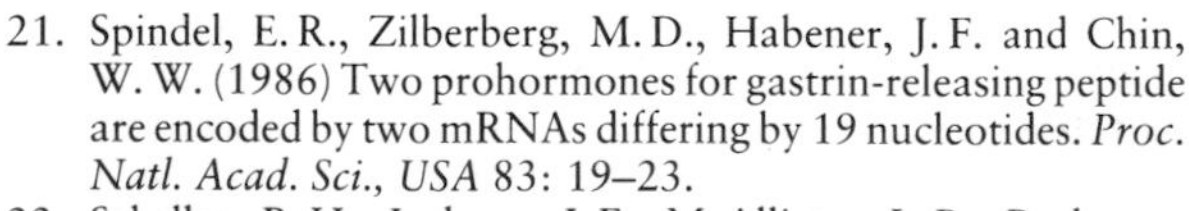

21. Spindel, E.R., Zilberberg, M.D., Habener, J.F. and Chin, W.W. (1986) Two prohormones for gastrin-releasing peptide are encoded by two mRNAs differing by 19 nucleotides. *Proc. Natl. Acad. Sci., USA* 83: 19–23.
22. Scheller, R.H., Jackson, J.F., McAllister, L.B., Rothman, B.S., Mayeri, E. and Axel, R. (1983) A single gene encodes multiple neuropeptides mediating a stereotyped behavior. *Cell* 32: 7–22.
23. Pierce, J.G., Lia, T.H., Howard, S.M., Shome, B. and Cornell, J.S. (1971) Studies on the structure of thyrotropin: its relationship to luteinizing hormone. *Rec. Prog. Horm. Res.* 27: 165–212.
24. Boothby, M., Ruddon, R.W., Anderson, C., McWilliams, D. and Boime, I. (1981) A single gonadotropin α-subunit gene in normal tissue and tumor-derived cell lines. *J. Biol. Chem.* 256: 5121–5127.

The mechanism responsible for diversity in the inhibin family is remarkable. Inhibins A and B are heterodimers with a M_r of~32,000 which have an identical α subunit of 134 aa. Each has a distinct but related β subunit, β_A (116 aa) and β_B (115 aa), linked to α by interchain, disulfide bonds (Fig. I.25). Contrary to inhibin which decreases FSH secretion, a heterodimer composed of the two β subunits has FSH-releasing activity (and no effect on LH). TGF-β, the homodimer of an inhibin β_A-like peptide (p. 219), and the β_A-β_B heterodimer also show FRP (FSH-releasing protein) or activin activity. It is remarkable that various combinations of parent gene products produce opposite biological activities[25,26]!

Diversification of Hormone Synthesis by RNA Processing

In all of the preceding cases, there is a single gene and a single primary RNA transcript, and diversification of peptide hormones is obtained at a post-translational step. *Alternative RNA processing* accounts for *diversity in the utilization of coding sequences.* The gene segment containing the coding sequence for calcitonin also contains a sequence which encodes a CGRP protein (calcitonin gene related product)[27] (Fig. XIV.10). Calcitonin, and thus its mRNA, is formed in the C cells of the thyroid (p. 651), whereas CGRP mRNA is synthesized in the nervous system, especially in sensory neurons. This is an example of a developmentally regulated, tissue-specific RNA processing mechanism. It also is representative of the discovery of *mRNA sequences which originally did not correspond to a known function* (it was found later that CGRP can turn on the gene of the α subunit of the nACh-R). Other examples are GAP, found in sequencing the 3′extremity of GnRH cDNA (Figs. III.22 and V.9), and 20K hGH (p. 195).

25. Vale, W., Rivier, J., Vaughan, J., McClintock, R., Corrigan, A., Woo, W., Karr, D. and Spiess, J. (1986) Purification and characterization of an FSH releasing protein from porcine ovarian follicular fluid. *Nature* 321: 776–779.
26. Ling, N., Ying, S. Y., Ueno, N., Shimasaki, S., Esch, F., Hotta, M. and Guillemin, R. (1986) Pituitary FSH is released by a heterodimer of the β-subunits from the two forms of inhibin. *Nature* 321: 779–782.
27. Rosenfeld, M. G., Mermod, J. J., Amara, S. G., Swanson, L. W., Sawchenko, P. E., Rivier, J., Vale, W. and Evans, R. M. (1983) Production of a novel neuropeptide encoded by the calcitonin gene via tissue-specific RNA processing. *Nature* 304: 129–135.

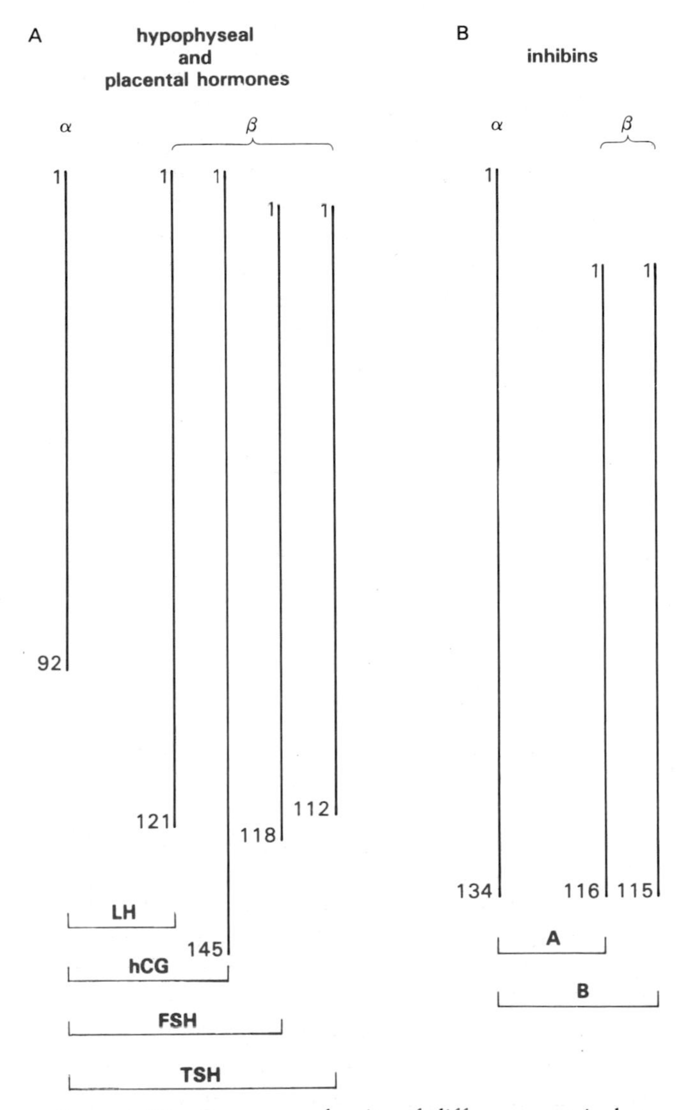

Figure I.25 Common subunits of different protein hormones: pituitary and placental hormones acting on the gonads or the thyroid (**A**) and gonadal peptides acting on the pituitary (**B**). The β sequences have been aligned for areas of maximum homology (Fig. V.1). **A.** Glycoprotein hormones have a common α subunit, while biological specificity is determined by the specific β subunit. **B.** The two inhibins also share a common α subunit. However, unlike the gonadotropin chains, inhibin α and β chains are linked by disulfide bridges.

Diversification of Hormone Synthesis at the Gene Level

Finally, polymorphism occurs at the level of *genes*, and also results in diversification. The processes of gene duplication, recombination and/or mutation are involved in the generation of *families* of related genes (congeners) encoding in most instances similar

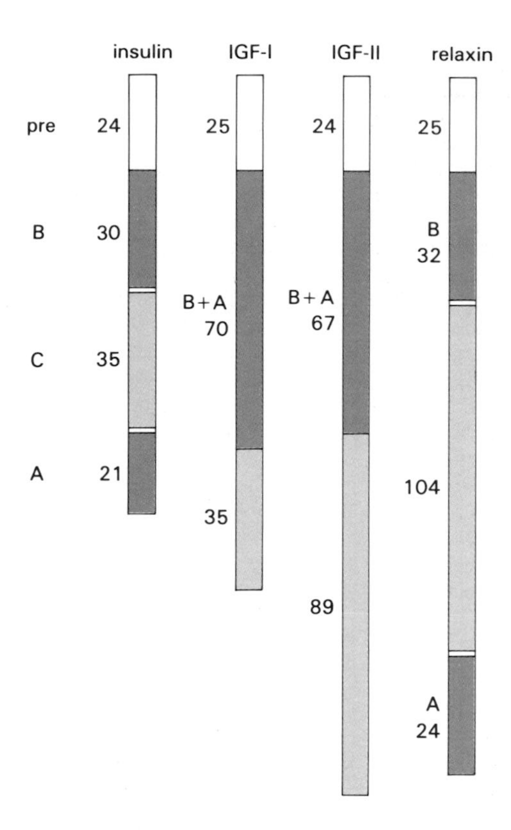

Figure I.26 Preprohormone of the insulin family. Homologous regions permit the definition of B and A domains in the four hormones, which correspond to the B and A chains of insulin and relaxin. In IGF-I and II, which are single chained mature peptides, the equivalent of the C domain, reduced to a minimum, remains a part of the final protein. The processing sites are indicated by black lines, double when two or more basic amino acids are involved. (Bell, G. I., Merryweather, J. P., Sanchez-Pescador, R., Stempien, M. M., Priestley, C., Scott, J. and Rall, L. B. (1984) Sequence of a cDNA encoding human preproinsulin-like growth factor II. *Nature* 310: 775–777).

but non-identical peptide products. The insulin family is a classical example (Fig. I.26); it includes relaxin (Fig. I.27) and IGFs (Fig. III.8). Growth hormone (GH), placental lactogen (PL) and prolactin (PRL) also form a family (p. 192). In the human, there are several hGH-related genes, but only two proteins are actually produced (hGH, and hPL, or placental lactogen, which is also called chorionic somatomammotropin or hCS). The GH variants 22 and 20K come from one gene, while a second hGH gene is activated in the placenta during the second half of pregnancy. The significance of these observations is not clear. Another partially understood example is found in the case of β-LH and β-hCG. In the human, there is only one gene for β-LH, and there are seven β-hCG genes, of which only three are expressed in placental cells. Incidentally, there is no CG gene in the rat, and CG seems to be a recent product of evolution (p. 155). The significance of the two non-allelic genes for insulin in the rat is not understood either, nor is the allelic variation in the 5′region of the single human insulin gene[28]. It is now accepted that the expression of a given gene (its transcription) depends, at least in part, on specific cellular proteins that interact with enhancers (p. 378 and p. 381) and probably other DNA segments of the promoter region of the gene. *Polymorphism of genes* for a given hormone may be associated with *polymorphism of the regulatory elements* that govern transcription in specific types of cells.

In summary, hormonal diversity can, as indicated in Table I.7, be created on three important biosynthetic levels: DNA, RNA, and protein.

Multiple Production

Classically, a hormone was thought to be synthesized by only *a single cell type* in a particular gland. It is now necessary to alter this classical concept, but the fundamental pathophysiological basis of this notion should not be forgotten. An animal or a patient has diabetes mellitus because of pancreatic insufficiency, even though some insulin is produced at other sites. In general, the *extraglandular* tissues which synthesize hormones produce them in relatively modest quantity, and frequently have an incomplete biosynthetic apparatus. In addition, they are usually not under the same type of *regulation* as normal glandular cells. However, it is clear today that classical glandular cells do not have a monopoly on the synthesis of most hormonal molecules. As a case in point, lymphocytes, in response to different stimuli (e.g., viruses, toxins), can produce POMC-derived peptides (that are feedback regulated by glucocorticosteroids!) or TRH, CG, LH, somatostatin, etc.

For *steroid hormones*, androstenedione, progesterone and 17-hydroxyprogesterone are produced in both the gonads and the adrenal cortex, and the synthesis of active androgens and estrogens occurs in the gonads of both sexes; estrogens are also formed from adrenal or gonadal precursors, in the nervous system, adipocytes, hepatocytes, breast cells, hair follicles, etc. Recently, it was demonstrated that oligodendrocytes, glial cells in the CNS, can actually produce preg-

28. Edlund, T., Walker, M. D., Barr, P. J. and Rutter, W. J. (1985) Cell-specific expression of the rat insulin gene: evidence for role of two distinct 5′ flanking elements. *Science* 230: 912–916.

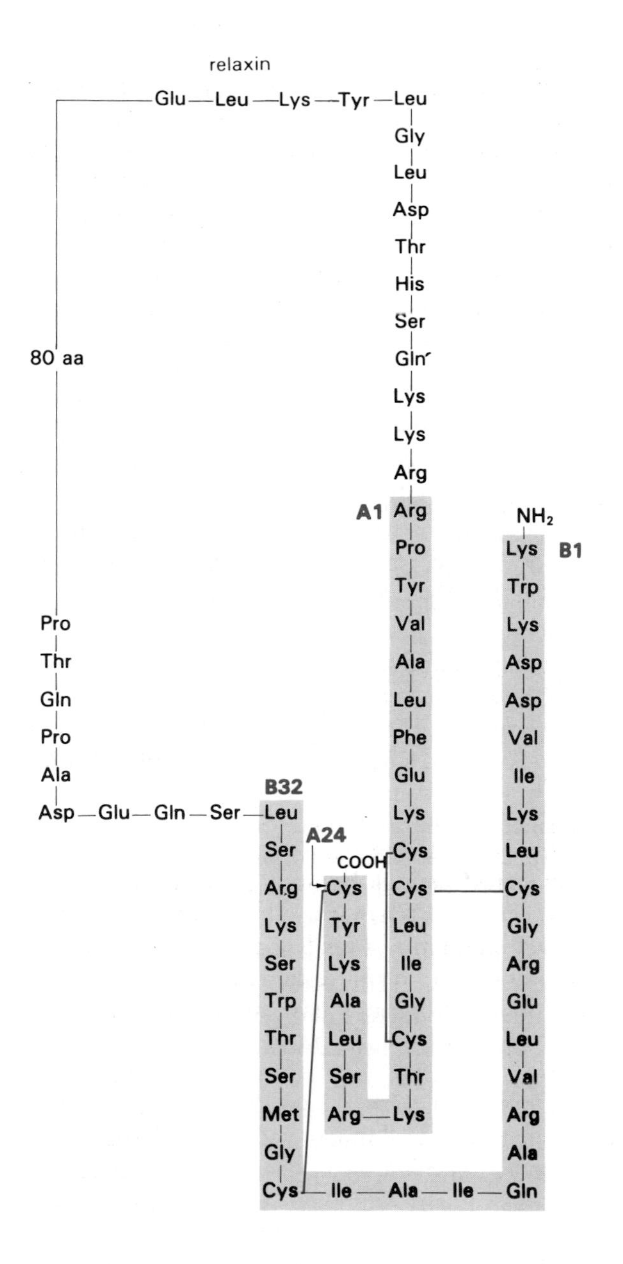

◁ Figure I.27 Relaxin is a peptide hormone which has high structural homology with insulin and IGFs. A weaker homology is also seen with β-NGF. Relaxin is produced primarily by the corpus luteum of the ovary, but also by thecal cells of the follicle and by the cytotrophoblast of the placenta. Its main function is to remodel connective tissue, principally by increasing the activity of proteolytic enzymes (collagenase and plasminogen activator), with the specific effect of softening the collagenous pelvic ligament and dilatating the uterine cervix. Relaxin may also cause the distension of fetal membranes. (Bryant-Greenwood, G. D. (1982) Relaxin as a new hormone. *Endocrine Rev.* 3: 62–90; Hudson, P., Haley, J., John, M., Cronk, M., Crawford, R., Haralambidis, J., Tregear, G., Shine, J. and Niall, H. (1983) Structure of a genomic clone encoding biologically active human relaxin. *Nature* 301: 628–631; Ullrich, A., Gray, A., Berman, C. and Dull, T. J. (1983) Human β-nerve growth factor gene sequence highly homologous to that of mouse. *Nature* 303: 821–825).

nenolone from cholesterol. There is no hormone, biosynthetic enzyme, steroid transport protein or receptor which is specific to one sex or the other; the *difference* is *not in their presence or absence*, but, is rather *quantitative* or temporal.

For *peptide hormones*, there is no apparent reason why a particular hormone cannot be made everywhere, given that all cells have a full complement of genes. Insulin has been detected in liver, human lymphocytes in culture, fibroblasts, and in the brain, where it is probably synthesized. GnRH (actually the prohormone containing GnRH and GAP) is synthesized not only in the hypothalamus, but elsewhere in the brain, and in gonads, placenta, breast and endometrium, among others. A number of identical peptides are found in cells of the nervous and digestive systems (p. 542). Low but detectable amounts of peptides similar or identical to ACTH, glucagon, parathyroid hormone (PTH), thyrotropin releasing hormone (TRH), GnRH, epidermal growth factor (EGF) and others have also been reported to be formed in the placenta (p. 476). Thus, the possibility arises that all cells make all hormones, albeit in very small quantities in most cases, since gene repression may not be an all-or-none phenomenon. Actual specific gene expression depends, probably among several other factors, on two groups of important proteins: those which regulate DNA transcription, and those, mostly proteolytic enzymes, which are engaged in the processing of proproteins. Both the quantitative importance and the possible role for the widespread production of hormonal peptides are essentially unknown, and each case should be examined individually.

Some *tumors* (primary tumors, e.g., of the lung, stomach, etc., not metastases of endocrine gland cancers) are able to produce peptide hormones identical or quite similar to those produced by the pituitary (e.g., ACTH), and especially the placenta (in particular hCG). Other proteins, frequently of an *embryonic* type, as for example α-fetoprotein, are often produced by tumors. Naturally, these hormones produce their expected effects in the body, leading eventually to clinical symptoms (e.g., the excessive production of ACTH in Cushing's syndrome). Two important characteristics should be noted. Tumoral hormones may differ structurally from their normal counterparts, mostly because the processing mechanism is incomplete in non-glandular cells. Secondly, regulatory controls do not operate in tumoral cells as in normal glands (e.g., there is a lack of negative feedback control of ACTH by glucocorticosteroids).

Storage and Release

There is *little hormonal reserve* in the body. Most often, hormone-producing cells contain less than the normal need for one day, even though secretory granules appear to serve a storage function (as for example in pancreatic and pituitary cells, etc). Thyroglobulin in the thyroid is also a form of hormone storage. The skin is a reservoir of vitamin D_3 and 7-dehydrocholesterol, precursors of the active hydroxylated derivatives.

The mechanisms involved in the secretion of steroid and thyroid hormones remain largely unknown. Apparently, the most important factor is the stimulation of biosynthesis, as if the subsequent release of the hormone were automatic (by diffusion?). There are, however, a few well documented cases in which the release of hormone can be separated from its synthesis, as for example the thyroid gland, where TSH causes an increase in T_4 secretion before increasing its synthesis.

2.2 Regulation of Hormone Secretion

All of the cells of the body must *act in concert* within a *constantly changing environment* in order to produce an integrated *response*. The hormonal network must possess the required *selectivity* and capacity in order to generate both *positive and negative* control signals. Most cells are both *receivers* (targets) and *producers* (glands) of signals, a concept substantiated recently by observations of paracrine and autocrine activities. When hormones control each other in an amplifying cascade of events (Fig. I.28), a very small quantity of neurohormone, for example, can induce the release of a larger amount of anterior pituitary hormone, which in turn stimulates the synthesis of even more peripheral hormone, finally causing the initiation of the required metabolic or anatomical response.

Despite the complexity and integration of hormone systems, they can still be roughly classified in two categories based on their connection with the nervous system (Fig. I.29). A first group includes hormones which are associated with the regulation of the *central nervous system* (CNS), and the second, those hormones whose production is for the most part *independent* of the CNS which are regulated more directly by their own effects.

Hormones Controlled by the Nervous System

The *hypothalamus-anterior pituitary system* has been studied in great detail[29–31] (Fig. I.28). Four *gland-stimulating hormones* are produced by the *anterior pituitary*: TSH, FSH, LH, and ACTH. In the hypothalamus, specific neurons secrete *releasing factors* (RF) whose best known function is to stimulate the production and secretion of pituitary hormones by corresponding secretory cells. Thus, TRH stimulates the formation of TSH and its secretion, GnRH that of LH and FSH, and CRF increases ACTH production. It is *not*, however, as *simple* as it seems at first examination, since TRH also stimulates the secretion of prolactin (p. 208). GnRH has a dual but unequal effect on LH and FSH, even though it appears that both hormones are secreted by the same type of gonadotropic cell. CRF is not the only element which controls the production of ACTH in corticotropic cells; it may also stimulate the production of diverse components of the POMC precursor (Figs. I.24 and IV.1, and p. 236).

TSH, FSH, LH and ACTH stimulate hormone production in their corresponding peripheral glandular cells, and eventually growth of the thyroid, testes or ovaries, and adrenal cortex. In response, the (peripheral) hormones released by these glands regulate the production of the corresponding pituitary hormones. This *feedback control* occurs directly at the level of the pituitary cells and/or at the level of specific neurons in the CNS, particularly in the hypothalamus, where peripheral hormones can modulate the secretion of releasing factors.

Other hormones, such as GH and PRL, are also synthesized and secreted by the anterior pituitary. Classically, it was thought that they do not stimulate

29. Harris, G. W. (1971) Humours and hormones: Sir Henry Dale lecture. *J. Endocrinol.* 53: 2–23.
30. Guillemin, R. (1978) Peptides in the brain: the new endocrinology of the neuron. *Science* 202: 390–402.
31. Schally, A. V. (1978) Aspects of hypothalamic regulation of the pituitary gland. *Science* 202: 18–28.

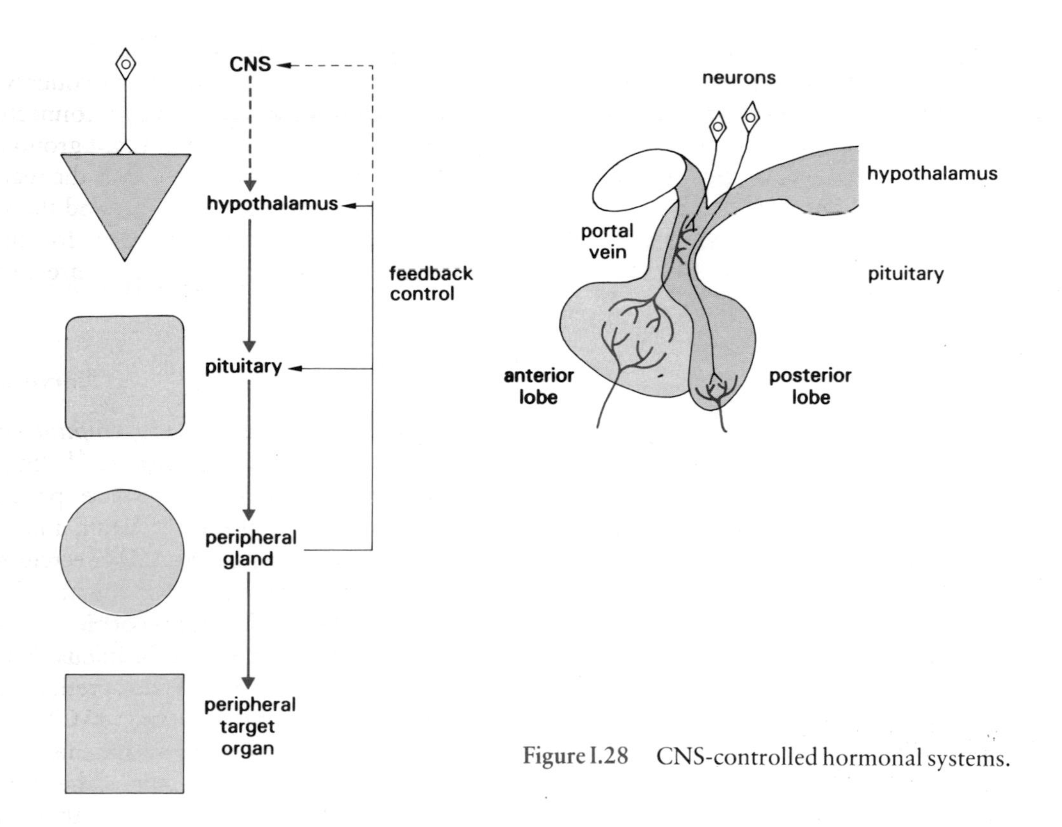

Figure I.28 CNS-controlled hormonal systems.

the secretion of peripheral hormones. However, this concept is not as clear-cut as it used to be. It is now established that GH acts, at least in part, by increasing the production of hormonal peptides in the liver (somatomedins or IGFs, p. 198)[32]. A negative feedback effect of IGF-I on GH production has been demonstrated (Fig. III.21). Hypothalamic factors are also important in the positive and negative regulation of these hormones. The stimulatory factors are GRF for GH, and probably VIP for PRL. Somatostatin inhibits GH secretion, and dopamine (and perhaps GAP) is a prolactin inhibiting factor (PIF). In any event, these systems are very complex, as is illustrated by the negative effect of somatostatin on TSH secretion (p. 269) and the modulation of GH and PRL output by steroids (p. 205).

Multiregulation of Sex Hormones

The regulation of sex hormones is highly *complex*. LH and FSH secretion (and their actions[33]) are under changing feedback controls exerted by steroids and peptides.

The *same* hormonal elements may, *depending on the species*, function *differently* (Figs. I.30 and I.31). The site where estrogen feeds back positively on the release of LH-FSH, causing subsequent ovulation, differs in the rat and monkey[34,35]. In the rat, the main site is the preoptic anterior hypothalamic area, and bypassing the arcuate nucleus, the principal GnRH pathway brings the peptide to the median eminence, to the portal blood, and to the anterior pituitary. In the monkey, in contrast, the positive feedback of estradiol occurs at the level of the pituitary, thus eliminating the need for a midcycle GnRH surge, while GnRH appears to originate principally from the oscillator in the arcuate nucleus (Fig. II.4). The primate system is more independent of the CNS in the induction

32. Daughaday, W. H., Hall, K., Raben, M. S., Salmon, W. D., Van den Brande, J. L. and Van Wyke, J. J. (1972) Somatomedin: proposed for sulphation factor. *Nature* 235: 107.
33. Hsueh, A. J. W., Dahl, K., Vaughan, J., Tucker, E., Rivier, J., Bardin, C. W. and Vale, W. (1987) Heterodimers and homodimers of inhibin subunits have different paracrine action in the modulation of luteinizing hormone-stimulated androgen biosynthesis. *Proc. Natl. Acad. Sci., USA* 84: 5082–5086.
34. Pohl, C. R. and Knobil, E. (1982) The role of the central nervous system in the control of ovarian function in higher primates. *Annu. Rev. Physiol.* 44: 583–593.
35. Ferin, M. (1983) Neuroendocrine control of ovarian function in the primate. *J. Reprod. Fertil.* 69: 369–381.

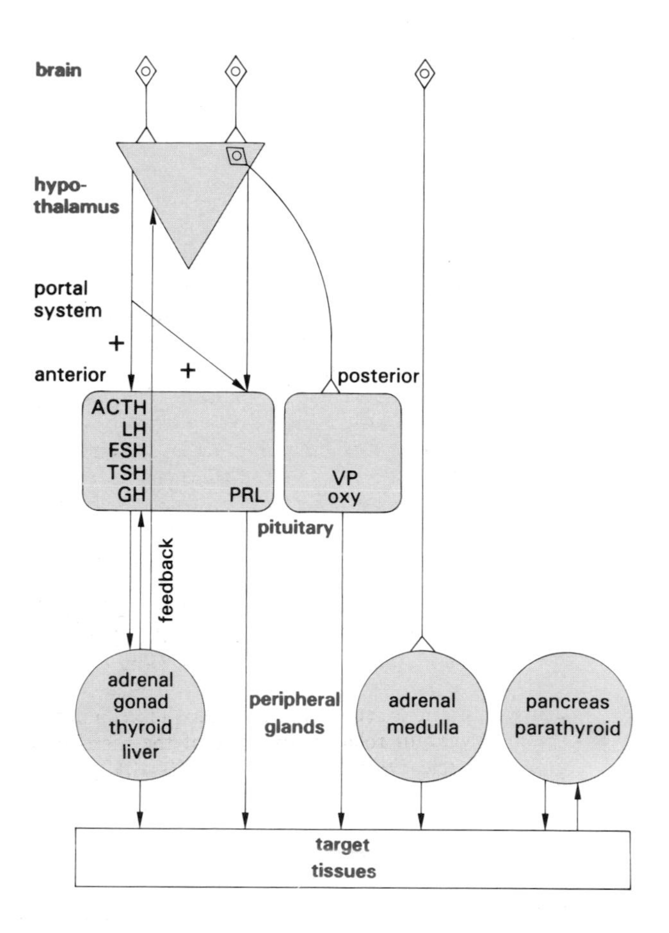

Figure I.29 Main hormonal systems in the human. The hypothalamus regulates the secretion of pituitary hormones. Most of them stimulate peripheral glands, which in turn secrete hormones acting on target cells. Note the central feedback mechanism controlling all anterior pituitary hormones except PRL. Hypothalamic neurons control posterior pituitary secretion, and other neurons directly innervate the adrenal medulla. Both the pancreas and the parathyroid glands are controlled by peripheral feedback.

of ovulation than that of the rat, and is thus less responsive to external influence. However, there is physiological and pathological evidence that higher centers can modify arcuate oscillator activity and the GnRH-gonadotropin system (p. 455). Progesterone facilitates LH secretion at the pituitary level, whereas at higher doses it has an inhibitory effect in the hypothalamus, where β-endorphin, made from POMC in the arcuate nucleus, interferes negatively with pulsatile activity (Fig. X.1) The feedback control of GnRH synthesis by sex steroids is probably indirect, since GnRH-secreting neurons do not contain receptors for progesterone (Fig. I.33) and estrogens. The characterization of the neurotransmitters involved in this feedback control is still incomplete.

Multiregulation of Stress Hormones

An intricate network of hormones and hormone-like activities is implicated in *stress*, defined as an *alarm* reaction followed by preparation for *defense*[36]. As indicated in Figure I.33, which shows the corticotropic cell of the anterior pituitary, the primary response is an increase in the secretion of ACTH, followed by an increase in cortisol. It is clear that the major positive control of ACTH secretion is effectuated by hypothalamic CRF. In addition, there are many other secretagogues working by themselves or as potentiators of CRF. According to the type of stress, central (e.g., via CRF or vasopressin) or peripheral (e.g., via epinephrine acting by both α_1 and β_2 receptors), several mechanisms are mobilized. In addition, somatostatin regulates ACTH secretion negatively. However, the major negative feedback control is of peripheral origin, and is exerted by glucocorticosteroids[37] (Fig. I.34) operating simultaneously at the level of CNS and the pituitary. In addition, and in contrast,adrenal cortisol increases the adrenal production of epinephrine, which is itself a secretagogue for ACTH at the pituitary level. POMC is produced in the intermediate lobe of rat pituitary, and ACTH is processed secondarily to produce α-MSH and CLIP (Fig. IV.1). There is no glucocorticosteroid receptor in the cells of the intermediate lobe, where POMC synthesis is not negatively regulated by steroids. Dopamine decreases the synthesis of POMC in the intermediate lobe, but has no effect on the anterior lobe.

Vasopressin and oxytocin are secreted by the *posterior pituitary*, or neurohypophysis, and anatomical and functional evidence demonstrates the direct involvement of the nervous system in their synthesis (p. 284). Catecholamines produced by the *adrenal medulla* are equivalent to the secretory product of the cells of the sympathetic nervous system (norepinephrine).

36. Axelrod, J. and Reisine, T. D. (1984) Stress hormones: their interaction and regulation. *Science* 224: 452–459.
37. Udelsman, R., Ramp, J., Gallucci, W. T., Gordon, A., Lipford, E., Norton, J. A., Loriaux, D. L. and Chrousos, G. P. (1986) Adaptation during surgical stress. A reevaluation of the role of glucocorticoids. *J. Clin. Invest.* 77: 1377–1381.

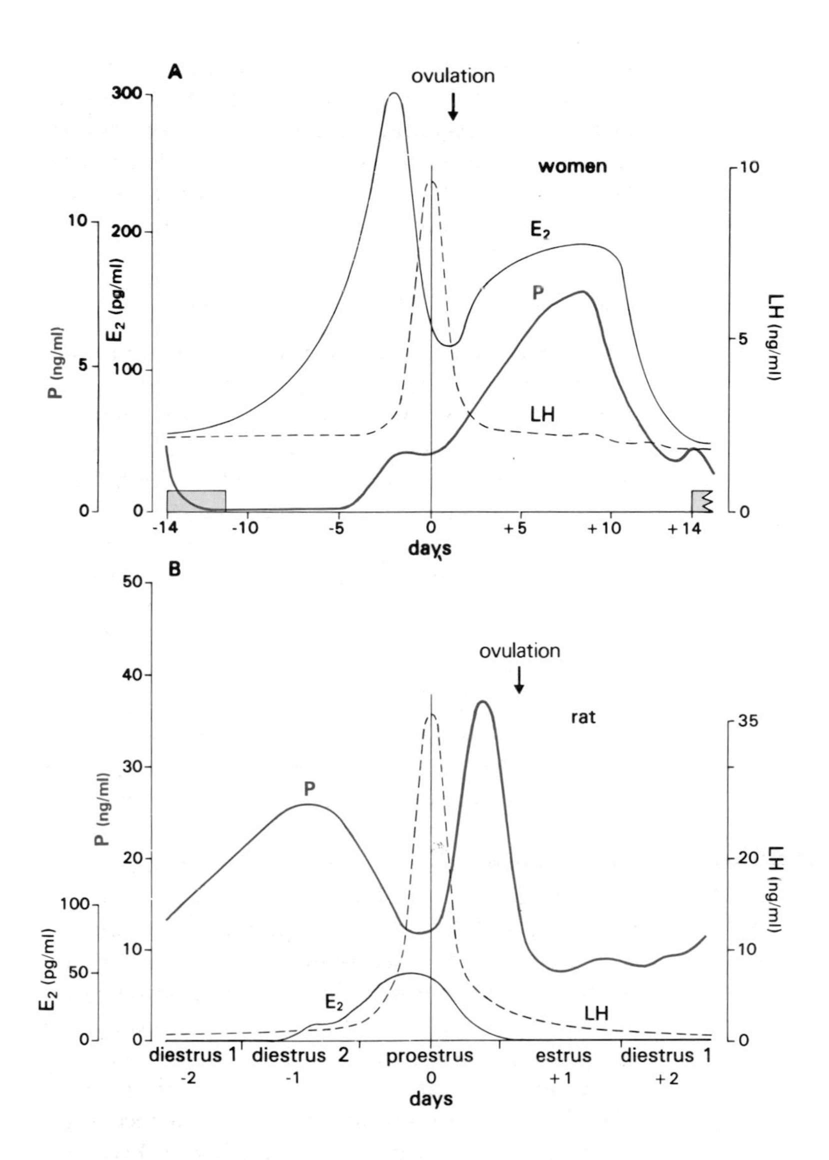

Figure I.30 Hormonal changes in the menstrual cycle of women (**A**) and in the estrous cycle of the rat (**B**). The two graphs have been centered on the LH peak and arbitrarily normalized temporally (28 d vs. 4 d). Results of LH in women are usually expressed in mU/ml (e.g., Fig. I.145). From available data obtained with the purest hLH preparations, the mid-cycle plasma LH concentration was shown to be ~10 ng/ml. (Courtesy of Dr. K. J. Catt, NIH, Bethesda, Maryland). A bioimmunoratio for plasma LH would give approximately 25 ng/ml. (Veldhuis, J. D., Beitins, I. Z., Johnson, M. L., Serabian, M. A. and Dufau, M. L. (1984) Biologically active luteinizing hormone is secreted in episodic pulsations that vary in relation to stage of the menstrual cycle. *Endocrinology* 58: 1050–1057).

The physiological control of all of these hormones by *higher centers of the nervous system* is not well understood. It has been demonstrated, by both pathological observations and physiological or pharmacological experiments, however, that all hypothalamic-hypophyseal hormones, and those produced in turn by the peripheral glands, are more or less permanently *associated with cerebral functions* of somatic integration, communication with the outside world, and elaboration of thought. This neuroendocrine network, aside from its involvement in numerous psychoneurohormonal interactions, is also responsive to *other stimuli*, as illustrated in the following examples. The regulation of aldosterone is not only under neuroendocrine control, but is also driven directly by water and salt metabolism, and consequently by the organs involved therein, such as the kidney, skin, lung, digestive tract, etc. The secretion of ACTH (and thus of glucocorticosteroids), GH and digestive hormones is dependent on the type and rhythm of food intake.

2.3 Hormones of the Peripheral System

(Figs. I.6 and I.29)

The secretion of some hormones is not controlled directly by the CNS, particularly those which are involved in major metabolic regulation. An example of *direct* control *without nervous intervention* is provided by B cells of the islets of Langerhans, in which the production of *insulin* is regulated by *glycemia*. Thus, there is a feedback system regulating the magnitude of hormone secretion. The same is true for

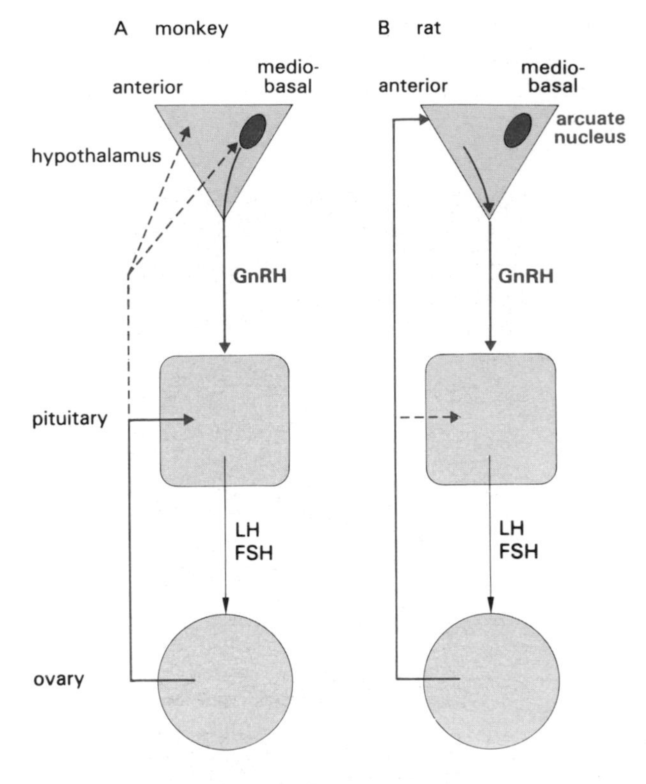

Figure I.31 Postulated differences in the site of positive feedback of estradiol on LH release in the monkey (**A**) and in the rat (**B**)[34,35]. The same hormones, estradiol, GnRH, LH and FSH, are involved in rodents and primates, but the site where estradiol feeds back positively on the system is different. In primates, pulsatile GnRH secretion governs the output of LH, and ovulation. Progression of the menstrual cycle depends primarily on the time for ovarian structures to develop and synthesize hormones (p. 451). In the rat, estradiol feeds back positively on the production of GnRH, resulting in ovulation.

the production of glucagon in A cells of the pancreas. The overall mechanism regulating glycemia, although basically peripheral, is in fact very complex, since in addition to insulin and glucagon, the two most important hormones with opposing effects, other hormones also intervene, such as cortisol (glucocorticosteroid), epinephrine and GH, all of which belong to the group of hormones under CNS control. Thyroid hormones (which regulate appetite and energy consumption), VIP (increasing insulin) and somatostatin (reducing the production of insulin and glucagon), among others, also influence glycemia indirectly. Intrapancreatic paracrine mechanisms may also be involved (p. 492).

The level of *calcemia* directly controls the secretion of *PTH* by the parathyroid gland and the secretion of *calcitonin* by the thyroid. Blood levels of calcium and phosphorus also regulate the renal hydroxylation required for the formation of calcitriol. Calcitriol also modulates the production of PTH (p. 658).

The production of *aldosterone* by the glomerulosa cells of the adrenal cortex is influenced by peripheral factors. The most important are the Na^+/K^+ ratio in plasma (itself regulated by aldosterone) and the level of angiotensin II (which depends essentially on blood pressure level as affected by renal production of renin (p. 607)). This regulatory system is made even more complex by the superimposed stimulatory effect of ACTH, perhaps that of dopamine, which down-regulates 18-hydroxylase (p. 393), and by glucocorticosteroids, which modify water and electrolyte balance, thus modulating aldosterone secretion *indirectly*. Also, a glomerulotropic factor of nervous origin has not been excluded from factors which regulate aldosterone.

Gastrointestinal (GI) hormones also are controlled by peripheral mechanisms, particularly by the intake of food. Several hormones contribute to the regulation of a single digestive function, and each hor-

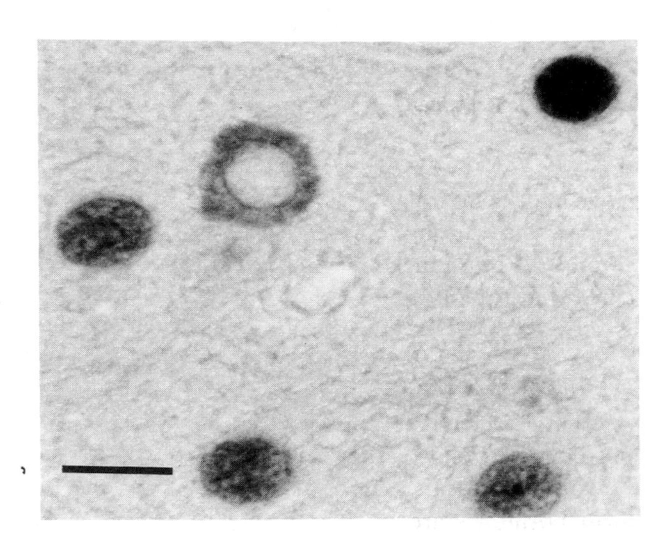

Figure I.32 GnRH-secreting neurons do not contain progesterone receptors. The preoptic area of hen hypothalamus shows GnRH immunoreactivity in the cytoplasm and progesterone immunoreactivity in the nuclei of different cells. Note the arrangement of progesterone receptor-containing cells surrounding the GnRH neurons. Bar=10 μm. (Sterling, R.J., Gasc, J.M., Sharp, P.J., Tuohimaa, P. and Baulieu, E.E. (1984) Absence of nuclear progesterone receptor in LH releasing hormone neurones in laying hens. *J. Endocrinol.* 102: R5–R7).

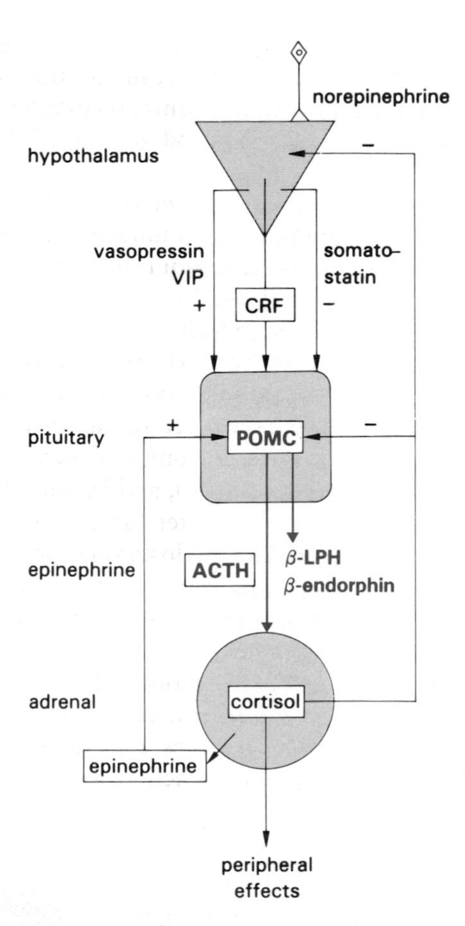

Figure I.33 Multiregulation of stress hormones.

mone may have very different effects. The significance of these intricate multiple controls is still not clear. Several GI hormones share partial sequence homology. Many are found within the CNS, and they may affect digestive secretion via the autonomic nervous system (p. 550).

2.4 Rhythms

Most hormones are secreted according to a rhythm which is apparently *programmed during development.* The pacemaker of rhythmic secretion and the major factors of its regulation are often unknown.

Circannual (or *seasonal*) rhythms have not been well studied in humans, although the influence of the seasons on thyroxine secretion and that of light on hormonal parameters of reproduction[38] (p. 304) are well known in some animals. Rhythms extending over several days or *weeks*, such as the estrous or menstrual cycles, definitely have a hypothalamic component, as has been demonstrated in the rat by neonatal androgenization experiments (p. 405).

38. Benoit, J. and Assenmacher, I. (1959) The control of visible radiations of the gonadotropic activity of the duck hypophysis. *Rec. Prog. Horm. Res.* 15: 143–164.

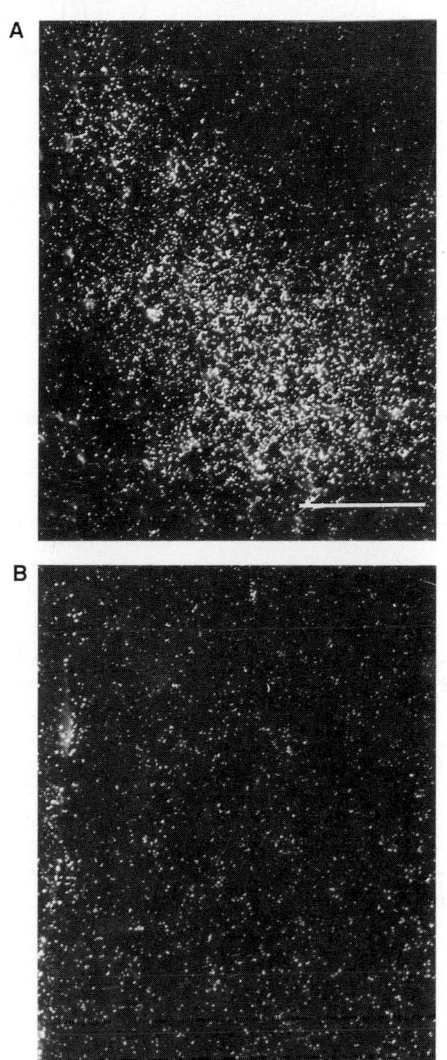

Figure I.34 Regulatory effect of glucocorticosteroid (DEX, or dexamethasone) on CRF production in the hypothalamus: an in situ mRNA hybridization study. These autoradiographs show CRF neurons in the paraventricular nucleus as revealed by hybridization of their mRNA with a radioactive cDNA corresponding to CRF. Experiments were conducted in adrenalectomized rats. **A.** Control, in which cholesterol has been implanted above the nucleus: the abundance of CRF mRNA is expected in an adrenalectomized animal. **B.** The inhibitory effect provoked by implantation of dexamethasone is evident. Bar=100 μm. (Metey, E., Young, W. S. Siegel, R. E. and Kovács, K. (1987) Neuropeptides and neurotransmitters involved in regulation of corticotropin-releasing factor-containing neurons in the rat. *Prog. Brain. Res.* 72: 119–127).

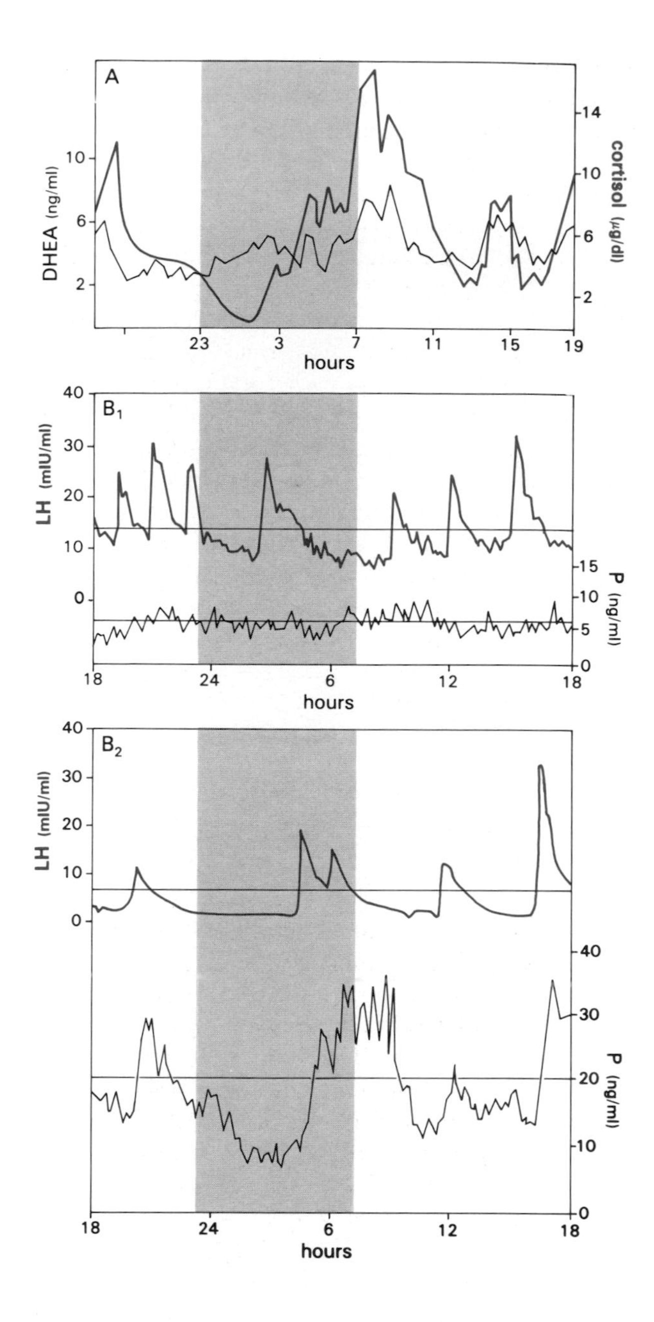

Figure I.35 Circadian and short-term oscillations of circulating levels of pituitary and peripheral hormones. **A.** Plasma cortisol and DHEA in a normal man. **B.** LH and progesterone in a normal woman two (**B_1**) and eight (**B_2**) days, respectively, after the LH midcycle surge. The horizontal lines indicate the mean concentrations of LH and progesterone the day they were studied. The mean estradiol concentrations were 99 and 150 pg/ml, respectively. (Filicori, M., Butler, J.P. and Crowley, W.F. Jr. (1984) Neuroendocrine regulation of the corpus luteum in the human. Evidence for pulsatile progesterone secretion. *J. Clin. Invest.* 73: 1638–1647).

Circadian rhythms, such as those of POMC peptides (ACTH and β-endorphin) and adrenocortical hormones (Figs. I.35A and IV.9), are probably determined by several factors, most of which originate in the CNS. Studies of circadian rhythmicity have been conducted in blind patients and in people suffering from jet-lag. Also, rhythms have been observed for GH and PRL, the secretion of which varies with different stages in the sleep cycle (Fig. III.20), and for LH, which begins to be secreted rythmically at puberty (p. 457; Figs. I.35B and X.14).

It has been shown by taking blood samples at short intervals that there are *short-term oscillations* in the secretion of both pituitary and peripheral hormones, the periodicity of which is frequently of the order of one hour (Fig. I.36), but which can range from between a few minutes to two to three hours. These pulses are interspersed within the circadian variations described above. Thus, if the half-life of the hormone is short, variations in its pulsatile secretion make measurement of plasma hormone levels of uncertain significance unless samples are obtained repeatedly. When hormone secretion is pulsatile in a system in which several hormones cooperate, oscillations recorded in the blood are reflective of both the half-life and the inertia to change of the secretion of each hormone involved. Pulsatile secretion of LH and FSH is seen in all vertebrates. Each pulse is initiated by a pulse generator, or oscillator, located in the arcuate region of the mediobasal hypothalamus, and leads to pulsed release of GnRH, from terminals of GnRH neurons, into portal blood. The effect of the pulse generator is slowed by progesterone, testosterone, opiates, α-adrenergic and dopaminergic agents.

These short oscillations can have great functional importance. The appearance of rapid sleep-entrained fluctuations of LH clearly indicates transition from childhood to puberty. In both men and women, GnRH administered to maintain a permanently elevated concentration in the plasma leads to down-regulation (desensitization) of LH secretion and, consequently, to a decrease in the production of testicular or ovarian hormones, with the inhibition of ovulation in women‡. This type of medical castration (p. 448) obtained with GnRH and its agonists has been proposed for use in contraception and in the treatment of some hormone-dependent diseases (endometriosis, prostate cancer) (p. 447 and p. 149), as well as in pubertal syndromes involving the hypersecretion of

‡ Contrary to GnRH, CRF does not desensitize the human pituitary.

hormones. In contrast, if GnRH is administered in a pulsatile fashion, usually with a pump, it is stimulatory, and is capable of reestablishing normal ovulation in anovulatory patients (p. 459). It seems that the intervals between peak agonist levels permit the reestablishment of the cellular response to GnRH.

The *nervous system* is *not* the *only* system which determines the rhythmicity of hormone secretion, even though it is always implicated and is sometimes fundamental. For example, the overall regulation of hormonal output during the menstrual cycle is the result of a complex interplay between several cerebral, hypophyseal and ovarian hormones interacting positively or negatively with each other. This rhythm depends also on the actual life-span of the cells of the corpus luteum, which is usually of fixed duration. Indeed, the feedback control exerted by ovarian steroids and peptides (inhibin/activin) emphasizes the fact that peripheral hormones, together with the CNS, play an important role in the overall cyclicity of the system (Fig. I.32). Other parameters may intervene. Starvation causes an acute decrease in the level of several anterior pituitary hormones. In contrast, if young female rats are restricted to half the normal daily food intake, their reproductive life is prolonged. Even more surprising, subjected to the same reduction in food intake, 15-month-old females with irregular cycles return temporarily to normal cyclicity[39].

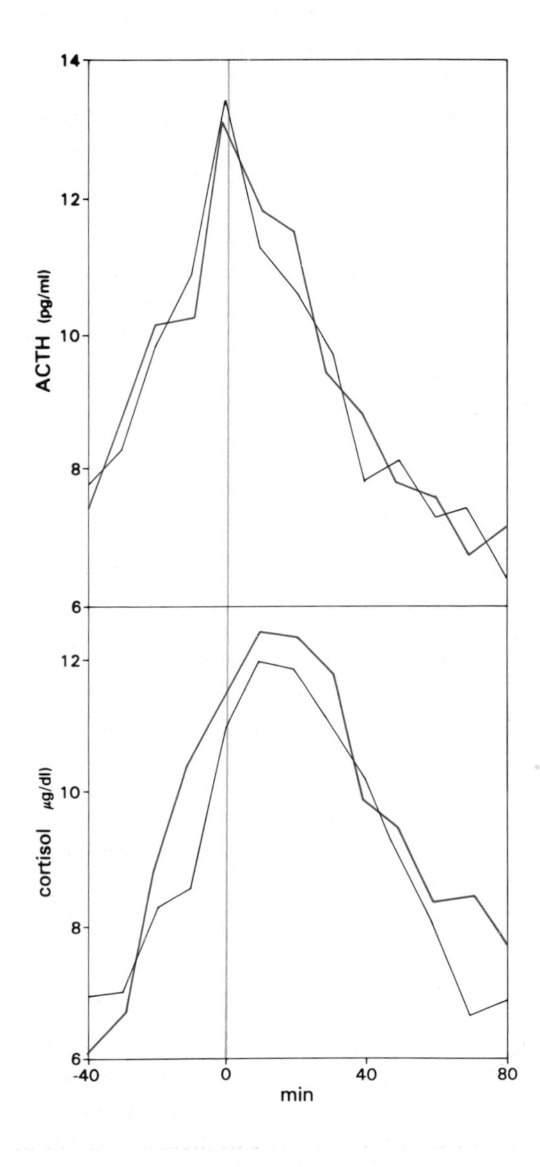

Figure I.36 Spontaneous oscillations (black line) of ACTH and cortisol secretion in a normal man and the response to human CRF (red line). Episodes (10 min) have been centered, at time zero, according to the highest ACTH values. (Schuermeyer, T. H. (1985) Pharmacologic and pharmacokinetic properties of corticotropin releasing factor in humans. Clinical applications of corticotropin releasing factor. *Ann. Intern. Med.* 102: 344–358).

2.5 Feedback Interactions and Integration of the Endocrine System

The secretion of all hormones is *integrated* in a *complex network* which is itself connected to the nervous system (and in consequence to stimuli coming from the outside world) and to a number of parameters of the general *metabolism*. There are two types of vertical integration. In one, the *CNS* has a fundamental and obligatory role. Such is the case for the CNS-pituitary-peripheral gland system, which acts almost automatically according to patterns established during development. Sometimes imprinting takes place relatively late, at the beginning of life, and is therefore highly susceptible to pathological alteration. *Growth*, reactions to *aggression*, and *reproductive functions* are the major activities controlled by CNS-dependent hormonal systems. In these cases, the hormone(s) produced by a peripheral gland act(s) to close the regulatory loop, thus establishing basal function regulated by a feedback mechanism. The *peripheral effects of a hormone do not* appear to influence its secretion (Fig. I.5), which remains under CNS influence.

39. Meites, J. (1987) Importance of the neuroendocrine system in aging processes. *Adv. Biochem. Psycho-pharmacol.* 43: 283–292.

In the second type of vertical integration, the *function* regulated ultimately by hormones of the *peripheral* or metabolic systems (blood glucose, calcemia, osmolality, or extracellular fluid volume) *feeds back itself* on the production of hormones.

The diagrams in Figures I.28 and I.29 show examples of neuroglandular and peripheral cascades with their descending and ascending interactions which are either stimulatory or inhibitory. The point to be made by illustrating "*vertical*" hormonal integration is that the *interpretation of the level of a given hormone* must take into consideration *the other hormones or metabolic compounds* which control it, and which it controls in turn; the *isolation of one hormonal measurement from the whole* would be misleading. In fact, in the proper evaluation of a vertical system, stimulatory or inhibitory *dynamic tests* must be used in order to amplify or even reveal an abnormality.

Not all endocrine products are involved in *feedback* mechanisms. The production of DHEA-S by the adrenal cortex is stimulated by ACTH, but DHEA-S does not control ACTH secretion. There is no evidence of feedback in the production of placental hormones, as if their temporary function in the body were to be accomplished independently. Finally, the secretion of certain hormones can be largely autonomous, even if all necessary elements for feedback control are in place. For example, the production of estrogen in men is physiologically insufficient to inhibit LH secretion. In women, ovarian androgens, also stimulated by LH, do not normally exert feedback control, but do so occasionally when they are pathologically elevated. The quantitative aspects of hormone production are essential to the evaluation of an endocrine system, since, qualitatively, fundamental molecular and cellular mechanisms are identical in most normal and abnormal situations.

The vertical sequences, from hormonal secretion to a response, are affected additionally by "*horizontal*" connections, thus forming an *intricate network*. For example, two hormones, H_1 and H_2, can produce the same effect simultaneously or with a different time course in a given cell. If, in addition, each of these two hormones has other, different functions (effects 1 and 2), it is clear that the network takes on a new dimension of complexity (Fig. I.37). Actually, H_1 and H_2 may have different or opposing effects on the same cellular function, and this creates considerable flexibility in regulatory processes. Examples are found in the cases of glucose and glycogen metabolism (p. 506).

The differential distribution of a hormone can be of importance (Fig. I.37B): H_1, delivered by the general circulation, may control the formation of another hormone, H_2, and H_1 can also modulate H_2's effect in a paracrine fashion! This is the case of ovarian inhibin, which decreases FSH formed in the pituitary and diminishes FSH-induced aromatization in granulosa cells (p. 453).

A member of a physiologically defined group of hormones (e.g., estradiol as a sex hormone) can modify the *secretion* of a hormone which is part of another physiologically important family (e.g., GH). A hormone can also alter the *metabolism* of other hormone(s) (thyroid hormones qualitatively and quantitatively influence the degradation of steroids) or their transport (estrogen increases CBG and SBP) (p. 43). Finally, a hormone can modify the responsiveness of tissues to another hormone by altering receptor concentration: for example, estrogen has a priming effect on progesterone action by increasing the pro-

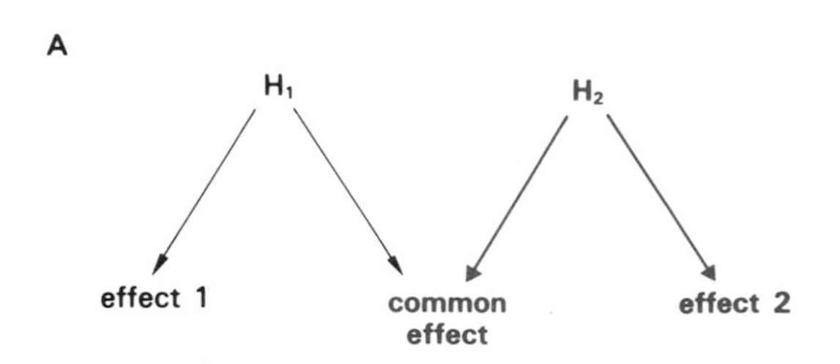

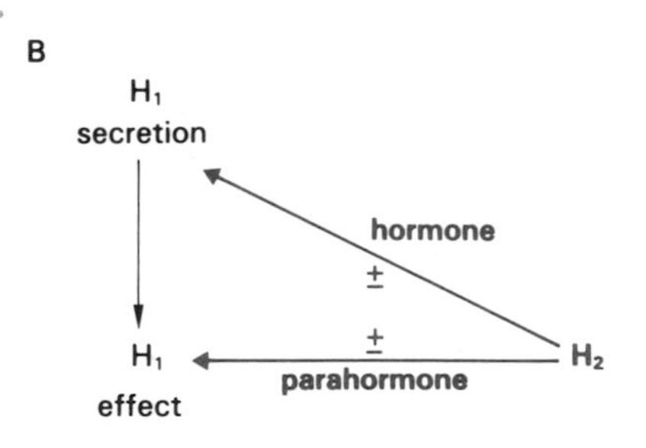

Figure I.37 Effects of two different hormones. **A.** Two hormones may both act on the same target organ, one being either synergistic or antagonistic to the other. These two hormones, in addition, provoke specific responses, denoted 1 and 2, thereby creating an integrated network of regulated activities (e.g., H_1=insulin, H_2=cortisol, common effect =glycemia, and effects 1 and 2 are the numerous other actions of these hormones). **B.** A hormone can control the secretion of another hormone and, parahormonally, its effect (e.g., H_1=testosterone; H_2=FSH; H_2 effect=synthesis of ABP in the testes (p. 44)).

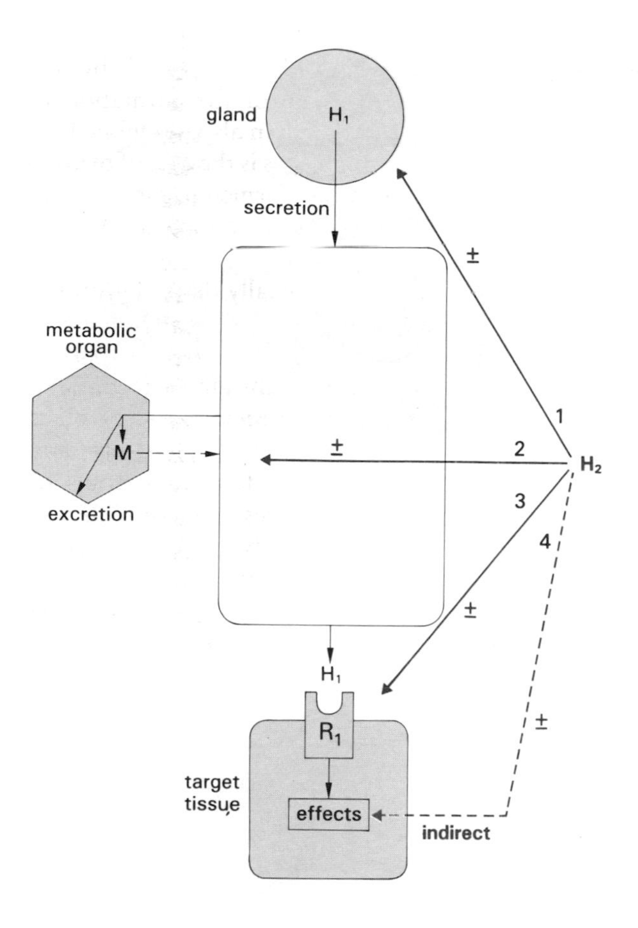

Figure I.38 Mechanisms for a hormone, H_2, to affect the secretion (**1**), metabolism (**2**) and/or action (at the receptor (**3**) or postreceptor (**4**) level) of another hormone, H_1.

gesterone receptor (Fig. I.38). More generally, *permissive* hormone action results in a change in the sensitivity of cells to the action of other hormones. This change may be operative at the level of receptors, transducing systems (e.g., G proteins, p. 338), or effector systems, which may be modified quantitatively or functionally. Glucocorticosteroids and thyroid hormones display many such permissive activities (p. 418); note that their receptors are found in practically all tissues.

Finally, the consideration of the *temporal* dimension of each component of these complex, integrated networks has to be considered. The length of time it takes to modify the synthesis and secretion of a hormone and/or its metabolic half-life makes it apparent that rational physiological or therapeutic predictions are quite difficult to make.

3. Distribution and Metabolism of Hormones

3.1 Transport in Blood

A hormone secreted into the venous capillaries of the gland (Fig. I.1) enters the general circulation rapidly (Fig. I.39), and following its transit through the lungs and heart, is distributed by arteries to the *entire organism*. Thus, a hormone reaches almost simultaneously *target cells*, upon which it acts, cells which are involved in its *metabolism*, degradation and eventually excretion, and cells of *other* tissues in which neither response (lack of receptor) nor degradation (lack of enzyme) occurs.

The study of endocrine physiology has demonstrated that the effect of the hormone *stops* when the supply of hormone ceases. Consequently, although the mechanisms implicated are poorly understood, it is clear that *renewal* of hormones in target tissues is *fundamental* to the function of the endocrine system. In itself, the interaction of a hormone with its receptor is reversible, and does *not* involve *chemical modification* of the hormone. Some in vivo observations have even demonstrated that hormone action can occur without any transformation of the hormone molecule (Fig. I.67), implying that it is returned intact to the blood after activity has taken place. In many cases, however, the hormone is *metabolized* and inactivated at the *target cell level*, following its interaction with the receptor. This process ensures the renewal of hormone, and may function to maintain a *continuous* and adequate hormone concentration. Metabolism of steroids occurs by enzymatic oxidoreduction or hydroxylation, thyroid hormones are deiodinated, and peptide hormones are internalized and degraded in lysosomes. *However*, for the most part, hormone catabolism takes place at a site *remote* from the site of hormone action. It occurs primarily in the liver, intestine and kidney, but also in a number of other organs and tissues where the hormone has no direct physiological effect (e.g., in adipose tissue for some steroid hormones).

Thus, in general terms, hormones are distributed throughout the body, are subjected to inactivation at various sites, and only a relatively *small fraction* of the secreted hormone actually gains *access to target cells*, where it carries out its specific, designed function.

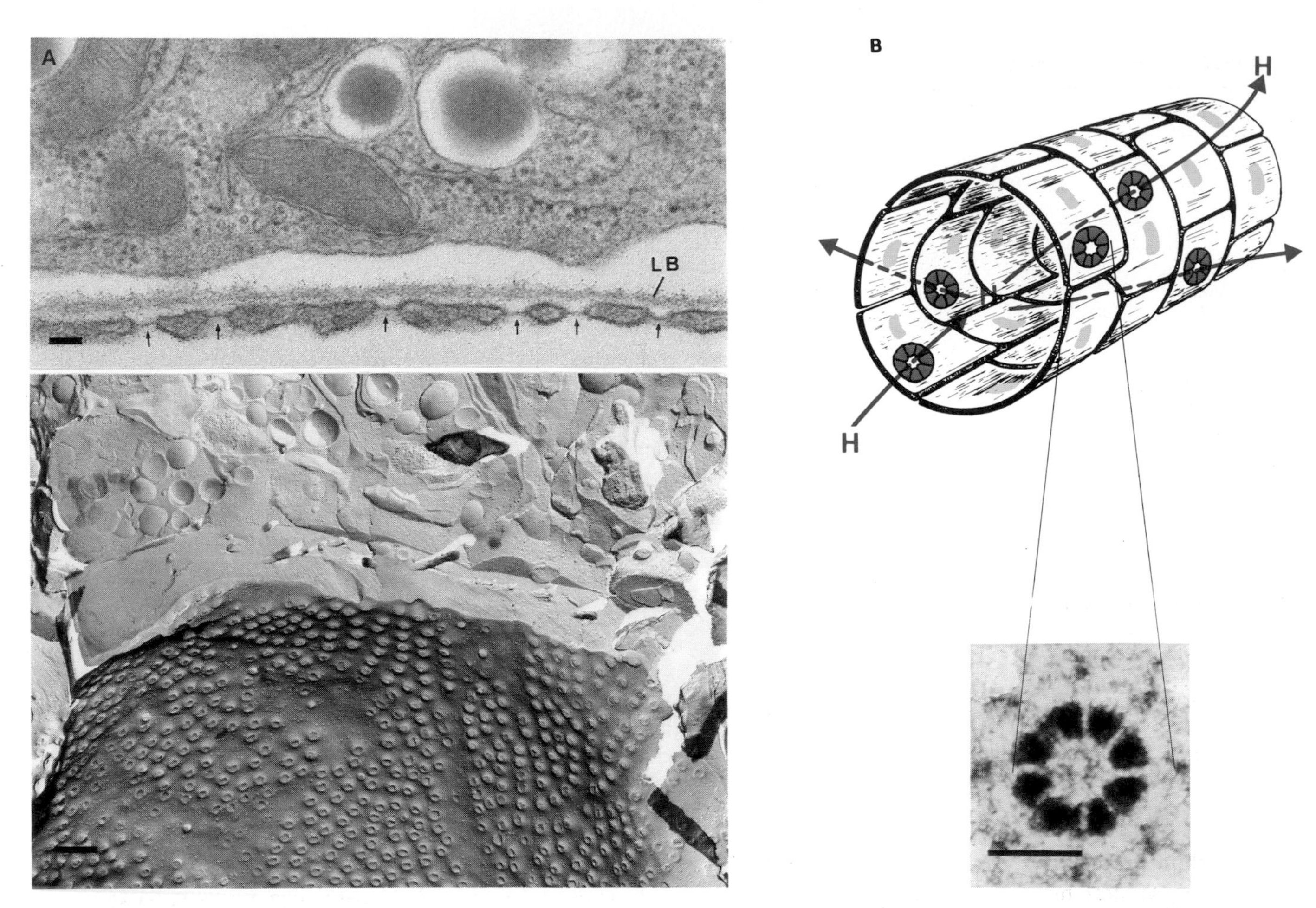

Figure I.39 How do hormones traverse the capillary wall? In addition to diffusion of lipid-soluble materials (e.g., steroids), hormones pass through openings of the capillary endothelium. Several different structures have been described. **A.** In the endocrine pancreas, on thin section perpendicular to the endothelium (upper panel), pores appear as regular openings (arrows) through the narrow endothelial cytoplasm. The pores are bridged by a thin diaphragm. A continuous lamina basalis (LB) is situated on the adluminal face of the endothelium. Freeze-fracture exposes an *en face* view of the endothelium (bottom panel), revealing the circular shape of the pores and their extensive distribution on the endothelial cell. In both images, part of the peripheral cytoplasm of an insulin-secreting cell is apparent. Bars=0.1 μm. **B.** A diagram representing the possible process of exit or entry of hormones. A fenestrated diaphragm in the endothelium of an adrenal cortex capillary is shown in the inset. Note the eight, dark, wedge-shaped communicatory channels. Bar=0.05 μm. (Bearer, E. L. and Orci, L. (1985) Endothelial fenestral diaphragms: a quick-freeze, deep-etch study. *J. Cell Biol.* 100: 418–428).

Interactions with Proteins

Primarily, hormones in the blood circulation are *dissolved* in *plasma*, although a very small fraction is adsorbed onto erythrocytes. They interact with plasma proteins, including albumin, present in the highest concentration, which has several binding sites for almost all molecules of low molecular weight. Most other proteins have lower affinity for hormones than albumin. A limited number of plasma proteins (PP), usually designated as *transport proteins*, specifically bind hormones, with a higher affinity than does albumin (K_D<0.1 μM) (Table I.9), and have only one binding site per molecule. Some of these proteins have an almost ubiquitous distribution among animals, suggesting that they represented a selective advantage during evolution. In most cases, however, their basic physiological role remains poorly understood.

Contrary to common belief, *steroid* and *thyroid hormones* are present in sufficiently low concentration (p. 397 and p. 365) that they need not be bound to a

Table I.9 Specific plasma protein binding hormones.

Binding proteins	Abbreviation	Hormones bound	Comments
Corticosteroid binding globulin	CBG	Glucocorticosteroids, progesterone	Also known as transcortin
Sex steroid binding plasma protein	SBP	Testosterone, estradiol	In humans, also known as SHBG (sex hormone binding globulin)
Progesterone binding globulin	PBP	Progesterone	Only in pregnant guinea-pig
α-fetoprotein	α-FP	Estrone	Murine, not in human
Thyroxine binding protein	TBG	T_3 and T_4	All species
Prealbumin		T_4	
Vitamin D binding protein	DBP	25OH-cholecalciferol	Gc2 (group-specific component)
α_1-glycoprotein	Orosomucoid	Progesterone (weakly)	High affinity for RU 486 in human
Growth hormone binding protein	GH-BP	GH	Extracellular part of the receptor
IGF binding proteins	IGF-BPs	IGF-I and II	Several MW proteins
Neurophysins	neurophysins I, II	Vasopressin, oxytocin	Made in hypothalamus
CRF binding protein	CRF-BP	CRF	MW ~38,000

hydrophilic protein in order to be dissolved in plasma. However, several high affinity PPs have been described[40,41]. Each binds a group of chemically related hormones: for example, corticosteroid binding globulin (CBG, or transcortin) binds not only glucocorticosteroids but also progesterone, and sex steroid binding plasma protein (SBP, or sex hormone binding globulin (SHBG)) has high affinity for testosterone and estradiol. These two proteins, produced in the liver, have no affinity for active synthetic analogs such as dexamethasone (a glucocorticosteroid) or diethylstilbestrol (a non-steroidal estrogen). Thus, their binding characteristics are not directly related to hormonal specificity, and their function is *not mandatory* for biological activity.

High affinity binding to PP is responsible for the sequestration of part of the hormone in the vascular compartment. In fact, access of hormones to cells depends not only on the half-life of hormone dissociation from plasma protein(s), but also on the capillary transit time and diffusion through cell membranes. Binding to high affinity PP *correlates negatively* with the metabolic clearance rate (*MCR*) of steroid hormones (p. 45). According to the law of mass action applied to the reaction:

$$[PP]+[H]\rightleftharpoons[H\ PP]$$

[H], the concentration of free hormone directly diffusible into target tissues will decrease as [PP] increases. This reduces negative feedback control and leads to a secondary increase in hormone secretion[42]. For example, during pregnancy, the increased synthesis of steroid and thyroid hormone-binding PPs is in part responsible for the increased production and plasma concentration of corresponding hormones. New World monkeys are relatively resistant to glucocorticosteroid action, which is probably attributable to weak receptor affinity; the low affinity of their CBG for cortisol results in a compensatory increase of free hormone in the blood.

Unlike CBG and TBG (thyroxine binding globulin), SBP is not present in all species. Other PPs

40. Forest, M. G. and Pugeat, M., eds (1986) *Binding Proteins of Steroid Hormones*. INSERM-John Libbey, London.
41. Robbins, J., Cheng, S.-Y., Gerschengorn, N. C., Glinoer, D., Cahnmann, H. J. and Edelhoch, H. (1978) Thyroxin transport proteins of plasma. Molecular properties and biosynthesis. *Rec. Prog. Horm. Res.* 34: 477–519.
42. Yates, E. F. and Urquhart, J. (1962) Control of plasma concentrations of adrenocortical hormones. *Physiol. Rev.* 42: 359–443.

appear only during certain periods in life, as is the case for progesterone binding protein (PBP) during gestation in the guinea-pig, and for estrogen binding α-fetoprotein (α-FP) in the rat fetus (α-FP does not bind steroids in humans). Thus, it is *difficult* to make qualitative and quantitative *generalizations*.

For peptide hormones, there is a protein (or a group of proteins) which binds IGF-I and possibly IGF-II, and a hGH-binding protein (which is probably the extracellular portion of the growth hormone receptor (p. 197)). Neurophysins are transport proteins which bind vasopressin and oxytocin, and are cosynthesized with their corresponding hormone (p. 285; Figs. VI.4 and VI.5).

Binding to high affinity plasma protein (Fig. I.40) may provide a hormonal reservoir capable of *buffering changes* in hormone concentration. When hormone production is low, H-PP complexes dissociate, providing free H. Conversely, when H is high, PP may limit access to target cells. In contrast, binding to PP may also amplify the increase of free hormone available to a tissue when basal secretion is increased (Fig. I.41). The possible consequence of hormone binding to albumin is still controversial; many physiologists suggest that, in vivo, both free hormone and/or *albumin*-bound hormone are delivered similarly to

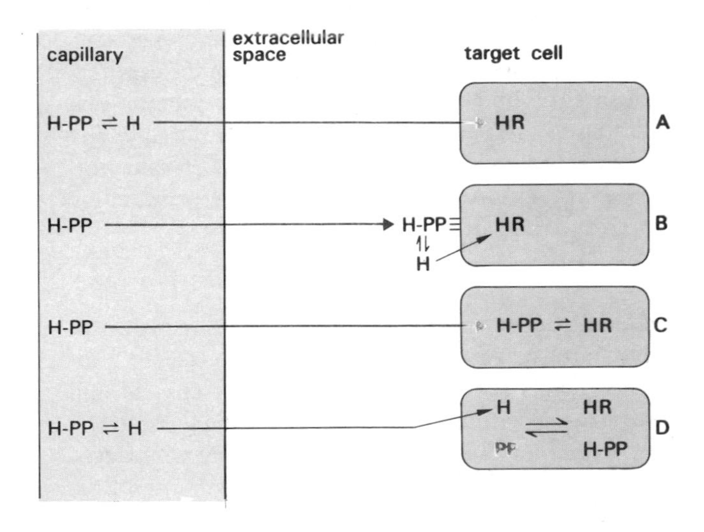

Figure I.40 Plasma proteins and the delivery of hormones to target cells. Four possibilities are represented. **A.** Classically, only free hormone has access to the target cell, where it combines with the receptor (HR). **B.** PP may specifically bind to the plasma membrane and selectively direct the hormone to the target cell. **C.** The H-PP complex enters the target cell, where released hormone will be made available to the receptor. **D.** PP or a congener is synthesized in the target cell and interferes with receptor binding.

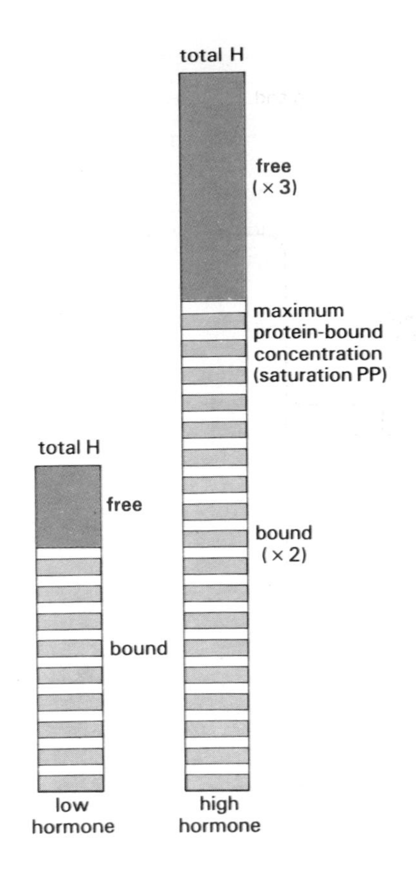

Figure I.41 Amplification of free hormone increases with PP binding. When total H concentration is low, the proportion of available free H is less than when total H is high, tending to saturate PP binding sites.

target cells, since the rate of hormone dissociation from albumin is high.

A role for PP in hormone secretion has also been postulated. It has been suggested that PP enhances the transfer of hormone from the cells of a gland to the plasma. That PP interacts with target cell membrane and is instrumental in the delivery of hormone at this level, or that there is selective entry of H-PP into target cells, has not yet been demonstrated experimentally (Fig. I.40). However, the complex, testosterone-ABP (an androgen binding protein produced in Sertoli cells which is encoded by the same gene as human SBP[43]), may be carried into epididymal cells as such

43. Petra, P. H., Titani, K., Walsh, K. A., Joseph, D. R., Hall, S. H. and French, F. S. (1986) Comparison of the amino acid sequences of the sex steroid-binding protein of human plasma (SBP) with that of the androgen-binding protein (ABP) of rat testis. In: *Binding Proteins of Steroid Hormones* (Forest, M. G. and Pugeat, M., eds), INSERM-John Libbey, London, pp. 137–142.

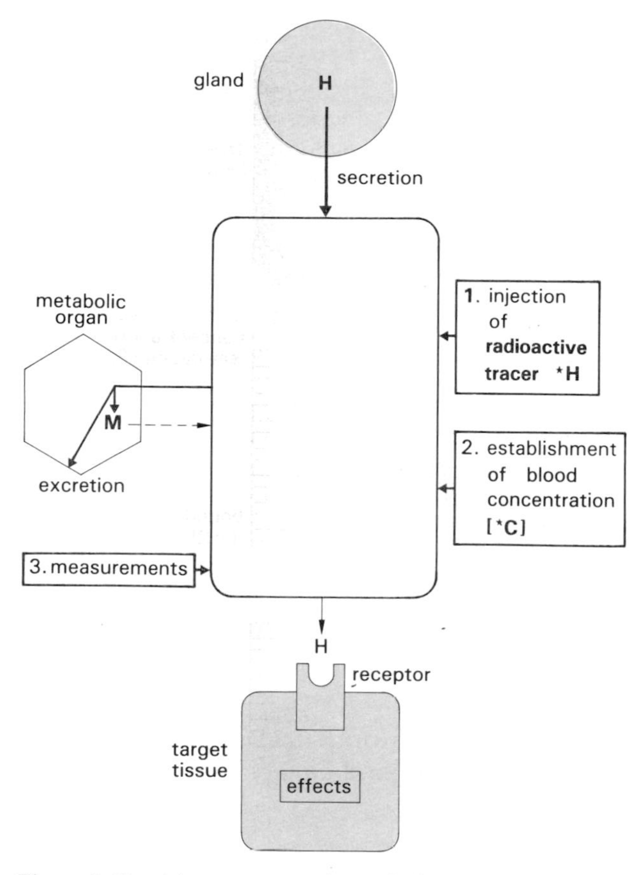

Figure I.42 Measurement of metabolic clearance rate. A radioactive hormone tracer, (★H), is injected as a single bolus or infused for a period of time (**1**), and is distributed in the circulation (**2**). Blood samples are taken (**3**) to measure the decreasing or the steady concentration [★C] of the tracer. The same organs and the same MCR are involved in the metabolism of H and ★H. Thus, MCR is calculated from the experimentally known rate of ★H perfusion and from [★C]. When MCR is known, the production rate of H can be calculated on the basis of the concentration [C] of endogenous hormone.

(incidentally, SBP/ABP is not found in the seminal fluid, which contains a small amount of testosterone, <1/10 the level in plasma). Indeed, the presence of transport proteins (or at least related antigens) has been observed *within* several types of target cells[44] where it is thought their synthesis may occur. Intracellular PP could facilitate entry of the hormone from the blood, and may play a role in hormone metabolism or in the distribution of hormone within target cells. Within target cells, such a possibility has been suggested for cellular retinoic acid and retinal binding proteins (CRABP and CRBP), which have no structural homology with the RAR (Fig. I.71). To date, there is no evidence of structural homology between PPs and corresponding steroid and thyroid hormone receptors.

44. Milgrom, E. and Baulieu, E. E. (1970) Progesterone in the uterus and the plasma. II. The role of selective binding to uterus protein. *Biochem. Biophys. Res. Commun.* 40: 723–730.
45. Tait, J. F. and Burstein, S. (1964) In vivo studies of steroid dynamics in man. In: *The Hormones* (Pincus, G., Thimann, K. V. and Astwood, E. B., eds), Academic Press, New York, vol. 5, pp. 441–557.

3.2 Metabolic Clearance and Half-life

The most important event facing a hormone is irreversible chemical transformation *into* an *inactive* product (catabolite). This occurs more frequently than direct physical elimination of a hormone by an excretory organ such as the kidney. The majority of hormone secreted into the circulation is inactivated before reaching its target cell.

The processes of degradation and excretion are known as *metabolic clearance*, the *rate* of which, MCR, is expressed as the volume of plasma cleared per unit of time[45]. The MCR determines the *plasma concentration* [C] with respect to the *secretion* rate (SR); that is, the amount of hormone which has access to the general circulation (Fig. I.42). The equation:

$$\begin{array}{cccc} & \text{SR} = & \text{MCR} \times & \text{C} \\ (\text{units:} & \text{mass/time} & \text{volume/time} & \text{mass/volume}) \end{array}$$

describes the reciprocal relationship between the three parameters.

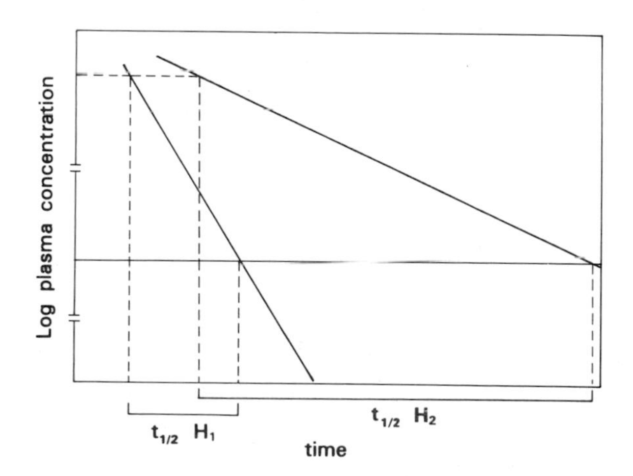

Figure I.43 MCR and half-life of two hormones. MCR H_1>MCR H_2. Half-life H_1<half-life H_2.

The *half-life* of a hormone is defined as the time required to reduce the circulating hormone concentration by one-half (in the absence of new secretion):

	half-life	=	ln2V	/	MCR
(units:	time		volume		volume/time)

'V' representing the apparent volume of distribution. The half-life of a hormone is shorter as the MCR increases and, reciprocally, is longer when metabolism is reduced (Fig. I.43). Increased binding to a plasma protein of high affinity limits the access of metabolic enzymes and thus prolongs the half-life of a hormone. A long half-life may also be related to an enzymatic defect or to a reduction of the availability of substrate (e.g., DHEA accumulating in the form of DHEA-S (Fig. I.44)).

From the *measurements* of MCR and half-life, SR can easily be calculated[45]. A method based on isotopic dilution of a radioactive tracer and the measurement of specific activity of a urinary metabolite may also be useful in the quantification of steroid hormone production[46].

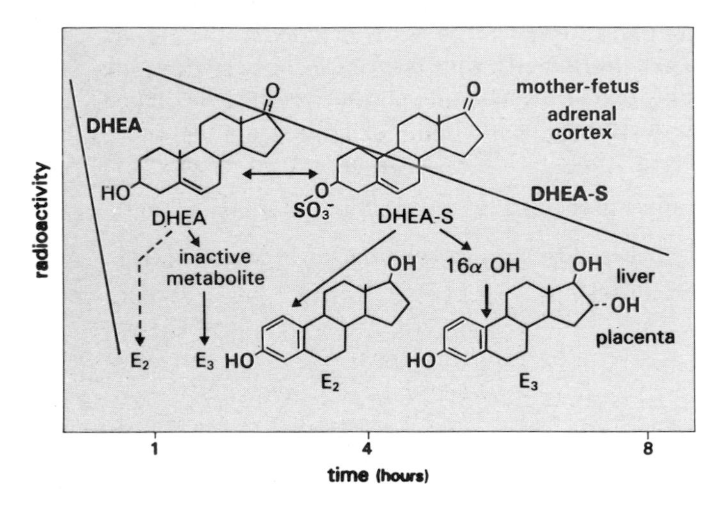

Figure I.44 Metabolism of dehydroepiandrosterone sulfate (DHEA-S). This adrenal steroid is metabolized much more slowly than DHEA, as indicated by comparison of the decreasing slopes of radioactivity (sites of modifications are indicated in red on metabolites). This accounts for its very high concentration in plasma and for the privileged synthesis of estrogens in the placenta. (Baulieu, E. E., Corpechot, C., Dray, F., Emiliozzi, R., Lebeau, M. C., Mauvais-Jarvis, P., and Robel, P. (1965) An adrenal secreted "androgen". Dehydroisoandrosterone sulfate. Its metabolism and a tentative generalization on the metabolism of other steroid conjugates in man. *Rec. Prog. Horm. Res.* 21: 411–500).

Major differences are observed in the metabolism of different hormones. For steroids, some are completely removed from the blood in a single passage through the liver (e.g., aldosterone) and therefore cannot be taken orally without losing their activity. This is even more apparent in the case of progesterone, whose MCR is greater than the rate of hepatic blood flow (Table IX.6), suggesting that a significant amount of progesterone is metabolized outside of the liver. In the clinical use of hormones, this problem has been overcome by the synthesis of analogs to natural hormones, which are made by modifying the chemical groups that are sensitive to digestive and hepatic enzymes. This is the case for synthetic progestins that ultimately led to the development of oral contraceptives[11]. For peptides, the problem is more difficult. For shorter peptides, there are synthetic analogs which resist enzymatic degradation. For larger molecules, it will probably be necessary to search for orally active compounds which retain properties of the active core of the original protein.

Administration of Hormonal Drugs: Quantitative/Temporal Effect

When a hormone or analog is administered, a concentration is established which may be equal to or above that required to saturate the corresponding receptors of target cells. However, excess hormone causes a prolonged effect directly *related* to the administered *dose*, since the duration of the exposure of the targer cell to the stimulatory effect of the hormone is longer. Not only the duration, but also the *magnitude* of the response will be increased, a phenomenon which is related to the kinetics of the response (e.g., accumulation of mRNA and protein). However, the global effect may be more complex, given that prolonged administration may lead to desensitization (p. 87).

3.3 Endocrinology of Metabolites

For steroid and thyroid hormones and catecholamines, the details of their degradative metabolism are well known. Formation of more soluble hydroxylated and/or reduced compounds (often acidic or conjugated to acids which themselves have no affinity for plasma binding proteins) favors their

46. Van de Wiele, R. L., McDonald, P. C., Gurpide, E. and Lieberman, S. (1965) Studies on the secretion and interconversion of the androgens. *Rec. Prog. Horm. Res.* 19: 275–305.

excretion by the *kidney* or the *liver-biliary system*: this is also known as *detoxification*. However, the reason for the biosynthesis of so *many* different metabolites remains unknown, despite the fact that their production is genetically determined, that it frequently differs systematically in the two sexes[47], and that it undergoes precise ontogenesis under hormonal influences[48]. Degradative metabolism may increase when synthetic steroids are administered over a long period, as a result of an augmentation of the synthesis of enzymes, which leads to secondary resistance to the treatment.

Some *metabolites* destined for excretion are potentially *active*: this is the case for 5β-steroids which induce the synthesis of △-aminolevulinate synthase involved in the formation of hemoglobin in the liver of the chick embryo. These compounds are also thermogenic in humans, due to progesterone metabolites, as is exemplified by the increase in basal body temperature (BBT) during the second part of the menstrual cycle and during pregnancy. Androsterone and some of its derivatives reduce plasma cholesterol, and have consequently been proposed for therapeutic use.

Finally, for thyroid and steroid hormones, the excretion of the unaltered, active form is minimal. A small portion is degraded to CO_2, and *inactive* metabolites, although only slightly different from the original hormone, are eliminated. A few metabolites can revert to active compounds, e.g., estrone → estradiol, cortisone → cortisol (p. 398). Since the liver is a major site of metabolism for many hormones, the question arose whether an enterohepatic cycle could contribute to the degradation and/or reactivation of hormones. In humans, this potential pathway appears to be of minor importance; however, it plays a more important role in some species, e.g., in the rat. For the most part, *peptide* and protein hormones are *irreversibly* reduced to *amino acids*. A small percentage of the intact hormone may pass though the kidney, to be eliminated in the urine. Measurement of hormones in the urine may be useful clinically, particularly in detecting urinary gonadotropin (hCG) in the diagnosis of pregnancy.

Administration of Hormonal Drugs: Effect of Route

The route of administration of a hormone is important in determining the response to a given hormone or analog. The differences between *oral* and *systemic* administration are well known. They are due, in large part, to the breakdown of hormone by enzymes in the gastrointestinal tract and metabolism in the liver. Some hormonal compounds can be administered *transdermally*. This route allows them to bypass the liver, and thus undesirable systemic side-effects may be avoided. In addition, the balance of metabolites is modified, and this may be of practical importance.

3.4 Metabolism in Target Cells (Fig. I.45)

There are a few cases of a lack of hormonal metabolism in target cells, e.g., estradiol in immature rat uterus (Figs. I.45A and I.67). Although generally less important quantitatively than catabolism occurring in the liver, kidney, etc., metabolism in target cells also consists primarily of an inactivation process (Fig. I.45B, C). This is the case for peptide and protein hormones, frequently destroyed following receptor-mediated endocytosis. It is not known whether a portion of the intact hormone is recycled, as is the case for transferrin (Fig. 1, p. 225), or if a fragment might become an active intracellular second messenger. Smaller hormones such as steroid and thyroid hormones are also biochemically inactivated, e.g., by reduction or hydroxylation for steroids, and deamination for thyroid hormones. Inactivation of hormones occurring *after* interaction with the receptor produces metabolites with low affinity for intracellular binding sites and favors the elimination of used hormone from the cell.

Between 1930 and 1950, hormones were thought to act as substrates for privileged enzymatic reactions, the products of which were presumed to modify cell metabolism. Transhydrogenation of some hydroxylated steroids was believed to allow the formation of a critical increase in NADPH[49] (Fig. I.45D). This theory of a *metabolic mechanism* of hormone function was shown unlikely by the use of synthetic derivatives which are active in spite of the fact that they cannot be metabolized.

Three types of *activating metabolism* in target cells have been described. 1. A non-active derivative can become active: e.g., in the case of androstenedione

47. Baulieu, E. E., Robel, P. and Mauvais-Jarvis, P. (1963) Différences du métabolisme des androgènes chez l'homme et chez la femme. *C.R. Acad. Sci., Paris* 256: 1016–1018.
48. Gustafsson, J. A., Mode, A., Norstedt, G. and Skett, P. (1983) Sex steroid induced changes in hepatic enzymes. *Annu. Rev. Physiol.* 45: 51–60.
49. Talalay, P. and Williams-Ashman, H. G. (1960) Participation of steroid hormones in the enzymatic transfer of hydrogen. *Rec. Prog. Horm. Res.* 16: 1–47.

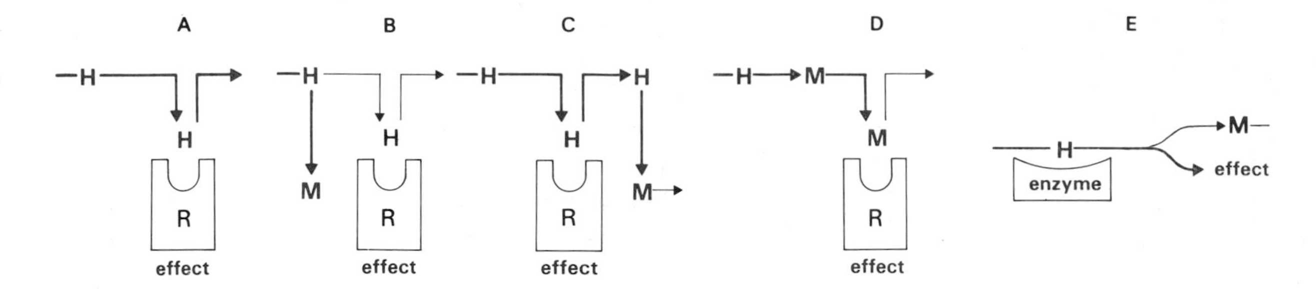

Metabolism at the target level, and hormonal activity. Five possibilities are depicted. **A.** The hormone (H) interacts directly with the receptor and does not undergo any metabolism. **B.** Metabolism may inactivate part of the hormone, decreasing its availability at the receptor level. **C.** After hormone-receptor interaction and the triggering of a response, the hormone is degraded and/or excreted out of the cell. **D.** The hormone is modified and activated, as if H were actually a prohormone. **E.** H is a substrate for an enzyme involved in a metabolic mechanism of hormone action.

which is transformed into testosterone (p. 399). 2. An active compound can become even more active (Fig. I.45E), e.g., testosterone transformed into dihydrotestosterone (DHT)[50,51], or T_4 transformed into T_3. In the above cases, the derivatives formed have higher affinity for the receptor than the original compound. 3. The hormonal compound is metabolically transformed into a compound with different activity, interacting with a different receptor: e.g., testosterone aromatized to estradiol[52] (Fig. I.46).

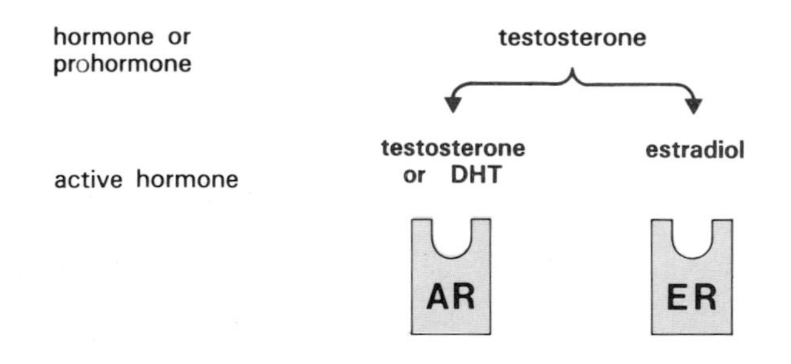

Figure I.46 Testosterone (T) action in the hypothalamus: two pathways. T is able to act directly, through the androgen receptor (AR), or indirectly, via the estrogen receptor (ER). In males, castration and the administration of antiestrogens prevent masculine differentiation, while estrogen will mimic T action (in females there is no perinatal secretion of estrogen by the ovaries). In tfm (congenital lack of AR) the hypothalamus differentiates into the male type (since T is transformed into estrogen which interacts with ER). DHT, which cannot be transformed into estrodiol, is active only via the AR.

Several of these steps occur successively, as for example when androstenedione is transformed into estradiol. The biological and physiological consequences of such metabolic transformation can be very important. As a case in point, the genetic deficit of 5α-reductase, necessary for the transformation of testosterone to DHT, is responsible for insufficient masculinization of the sex organs during fetal development (p. 442). The transformation of testosterone to estrogens occurring in the hypothalamus is essential for sexual development (in the rat). Finally, the role of the metabolism of androgens to estrogens may be important in the context of mammary tumor growth.

3.5 Alternative Aspects: Multiorgan Synthesis and Local Hormones

As compared with the classical scheme (Fig. I.1), there are *topological alternatives* to hormone synthesis, action and metabolism, even if the basic biochemical reactions remain the same.

A sequence of biosynthetic steps in different tissues

The concept of prohormones has already been introduced on p. 19 (Table I. 6). It was initially described for DHEA-S, secreted by the adrenal cortex in humans (surprisingly, since sulfo-conjugated steroids were formerly believed to be uniquely terminal forms which immediately preceded excretion). DHEA-S is a privileged, efficient substrate for the synthesis of active hor-

50. Bruchovsky, N. and Wilson, J. D. (1968) The conversion of testosterone to 5α-androstan-17β-ol-3-one by rat prostate in vivo and in vitro. *J. Biol. Chem.* 243: 2012–2021.
51. Baulieu, E. E., Lasnitzki, I. and Robel, P. (1968) Metabolism of testosterone and action of metabolites on prostate glands grown in organ culture. *Nature* 219: 1155–1156.
52. Naftolin, F., Ryan, K. J., Davies, I. J., Reddy, V. V., Flores, F., Petro, Z., Kuhn, M., White, R. J., Takaoka, Y. and Wolin, L. (1975) The transformation of estrogens by central neuroendocrine tissues. *Rec. Prog. Horm. Res.* 31: 295–319.

mones. This is attributable to the protection, provided by the sulfate moiety, from degradative enzymes (Fig. I.44). Adrenal DHEA-S and its 16α-hydroxylated derivatives are thus transformed in the placenta to estrogens, which are then secreted. Small amounts of androgens may also be transformed into estrogens in liver and adipose cells, and may then be released into the circulation. Whereas these prohormonal androgens travel a long way, androgens formed in the cells of the theca in ovarian follicles pass readily to granulosa cells, to be aromatized into estrogens (p. 394 and Fig. X.9). During pregnancy, progesterone is partially transformed in the liver to deoxycorticosterone which promotes sodium retention. Vitamin D_3 and 25OH-D_3 are also prohormones[53]. They are successively hydroxylated in the liver and kidney, respectively, and lead to the synthesis of the active derivative calcitriol (Fig. XIV.15). Finally, a hormone such as testosterone, which is active by itself in some target cells (muscles), may be a prohormone that is transformed within other target cells into an active derivative (e.g., DHT and estradiol in the prostate and brain, respectively) Table I.6; Fig. I.46). According to this definition, T_4 is also a prohormone which, when deiodinated in situ, produces the most active thyroid hormone, T_3, at the level of the target cell (p. 359).

Regional Distribution: Specific Vascular Circuits and Para- and Autohormonal Systems

Specific Vascular Arrangements

The major *portal* venous system delivers insulin and glucagon, produced by the pancreas, directly to their target cells in the liver. Interestingly, 50 and 20% of insulin and glucagon, respectively, are inactivated in the liver, but, of the two, only insulin has extrahepatic activities. The *hypothalamus-anterior pituitary* portal system carries releasing factors to their specific hypophyseal target cells. The small amount of these peptides found in the general circulation may have no functional impact on other target cells, and extrapituitary effects of CRF, GnRH or somatostatin are likely to be due to non-hypothalamic, ectopic synthesis. There is also a small portal system between the posterior and anterior pituitary lobes, but its physiological importance is not known. Blood from *adrenal capillaries* carries glucocorticosteroids directly from the adrenal cortex to cells of the adrenal medulla, where these steroids induce the enzyme which methylates norepinephrine. In addition to blood vessels, *lymphatic* connections permit the passage of hormones between organs (for example, between the uterus and the vagina).

53. DeLuca, H. F. (1971) The role of vitamin D and its relationship to parathyroid hormone and calcitonin. *Rec. Prog. Horm. Res.* 27: 479–516.

Diffusion of Hormones in the Extravascular, Extracellular Space

Instead of entering the blood circulation, hormones may diffuse through connective tissue to nearby cells.

In the *testes* (Fig. I.47), testosterone diffuses from interstitial cells into the seminiferous tubules, causing Sertoli and germ cells to be exposed to higher concentrations of hormone than those found in target cells to which hormone is transported via the general circulation. Consequently, injected testosterone generally does not increase spermatogenesis. An in situ implant of testosterone, or an injection of gonadotropin which stimulates the Leydig cells, can

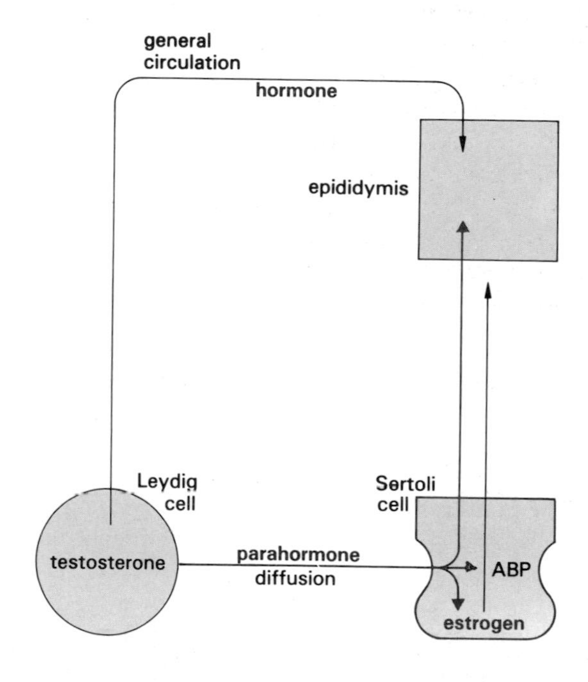

Figure I.47 Distribution of testicular testosterone by peripheral and local pathways. The epididymis, like other secondary sex organs, consists of target tissues of testosterone present in the general circulation. In addition, testosterone diffuses directly from Leydig cells into the seminiferous tubules, on which it acts as a parahormone, and where part of it binds to androgen binding protein (ABP), ultimately reaching the epididymis through the lumen. Testosterone is also partially converted to estrogen, and is involved in the control of the biosynthesis of ABP in Sertoli cells.

provide a sufficiently high, local concentration of the hormone that spermatogenesis is stimulated. The paracrine interaction between Leydig cells and Sertoli cells is not unidirectional, and peptide growth factors of Sertoli cell origin (in particular of the inhibin/activin series, Fig X.10) may play an important role in the regulation of Leydig cell function (a blood-testis barrier exists, preventing, in particular, the passage of sperm antigens into the general circulation). In amphibian *ovaries* at the time of ovulation, progesterone diffusion from surrounding follicular cells to the oocyte triggers the reinitiation of meiosis; however, such an action has not been demonstrated in mammals. In the *islets of Langerhans*, somatostatin produced in D cells inhibits insulin and glucagon production in B and A cells. Paracrine interactions between *pituitary* cells (e.g., gonadotropic and lactotropic cells) are also being discovered. The lack of development of *mammary glands* in male mice is androgen-dependent. In fact, the target cells of testosterone are mesodermal cells which in turn produce a yet poorly defined paracrine factor that prevents the growth of epithelial cells. In the *brain*, estrogens formed from androgens probably have paracrine or autocrine activity. The recently demonstrated production of pregnenolone by oligodendrocytes[54] suggests the existence of another paracrine system in the CNS. *Growth factors* also frequently act as para- or autocrine hormones.

The mechanism of *autocrine* activity does not imply that an autohormone acts only on the cell which secretes it (Fig. I.48). Sister cells producing the same autohormone have the same receptors, and thus the whole group of identical cells responds identically.

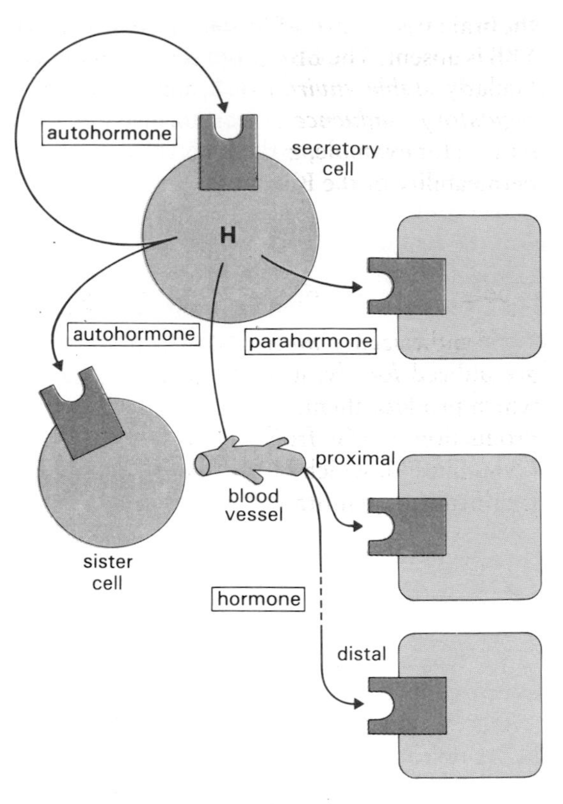

Figure I.48 Hormones, parahormones, and autohormones. This figure develops the concept presented in Figure I.8A. From the secretory cell, hormones may be delivered, by the circulatory system, to either distant target cells or, sometimes, to nearby cells, as is the case for hypothalamic factors acting on the anterior pituitary. Parahormones do not enter the blood system, but reach neighboring cells by diffusion through the extracellular space or, sometimes, via a specific canal (e.g., Fig. I.47). Autohormones, as indicated by the name, act on their own secretory cells, and probably also reach neighboring sister cells (contrary to parahormones, which by definition affect only the function of different cell types).

Blood-Brain Barrier (BBB)

The brain is in command of many physiological functions, and conversely, it receives nutrients and regulatory signals from the body. However, it is isolated from the rest of the organism by a type of second-order cell membrane, described functionally as the blood-brain barrier. Probably very few molecules synthesized in the brain enter the general circulation, even those which are potentially active hormones.

Unlike capillaries in most organs, microvessels in the brain of vertebrates are characterized by highly resistant, epithelial-like tight junctions which cement the endothelial cells together. This barrier acts principally to *prevent water-soluble and/or charged substances* from entering the brain interstitium from the blood[55]. Steroid hormones traverse the BBB by lipid mediation, and probably diffuse freely through endothelial cells. Other hormones, such as T_4 and essential nutrients for brain metabolism (glucose, amino acids, purine, and others), penetrate the BBB by carrier mediation. This also seems to be the case for some hormonal peptides, which, however, may also penetrate

54. Hu, Z. Y., Bourreau, E., Jung-Testas, I., Robel, P. and Baulieu, E. E. (1987) Neurosteroids: oligodendrocyte mitochondria convert cholesterol to pregnenolone. *Proc. Natl. Acad. Sci., USA* 84: 8215–8219.
55. Partridge, W. (1983) Neuropeptides and the blood–brain barrier. *Annu. Rev. Physiol.* 45: 73–82.
56. Bergstrom, S. (1967) Prostaglandins: members of a new hormonal system. *Science* 157: 382–391.

the brain freely, through a few discrete areas where the BBB is absent. The BBB provides the brain with a particularly *stable environment*, but is itself subject to *regulatory influences*, particularly of hormonal nature; for example, glucocorticosteroids increase the permeability of the BBB to proteins.

Prostaglandins: Often Parahormones

Synthesized in *most* cells, prostaglandins (PGs)[56] are utilized *locally*, most frequently within the cells which produce them, or in their near proximity. The production of PGs from *polyunsaturated fatty acids* ("vitamin F"), which are not synthesized in the body, is difficult to quantitate. The measurement of prostaglandins in the blood is not very meaningful, although the assay of urinary metabolites may provide an index of global production. Prostaglandins and related compounds, their biosynthesis and primary functions, many of which are connected to hormone production and action, are discussed on p. 590. PGs appear to play a major role in *membrane* regulatory mechanisms involved in the mediation of hormone activities. Aspirin and other non-steroidal anti-inflammatory drugs inhibit cyclooxygenase function, while glucocorticosteroids increase the synthesis of lipocortin, a protein blocking the synthesis of $PL\text{-}A_2$, the phospholipase which releases arachidonic acid and triggers the cascade of events of PG synthesis.

Reproductive functions offer particularly striking examples of the *operational versatility* of PGs and

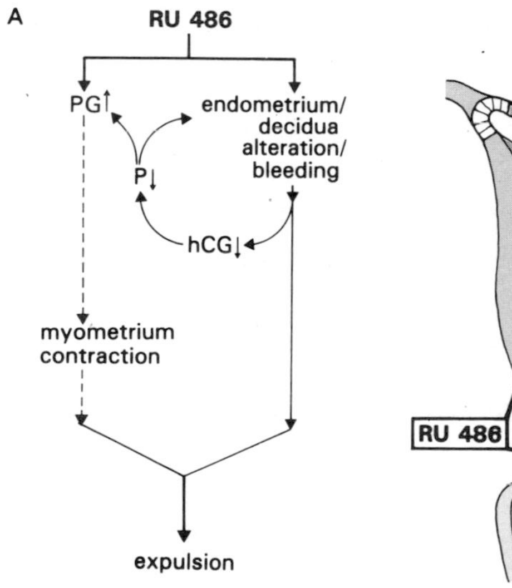

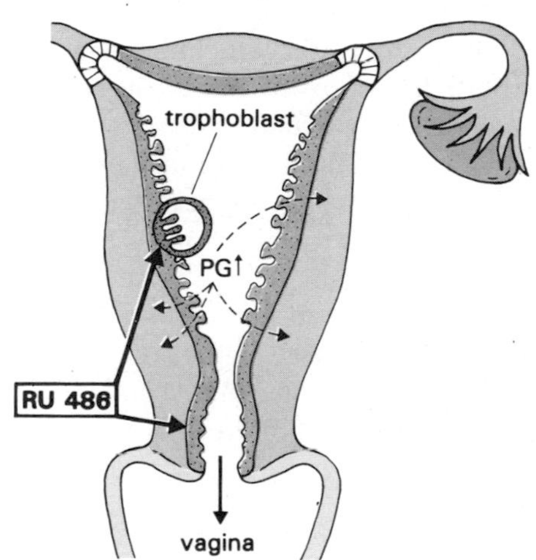

Figure I.49 The mechanism of antiprogesterone (RU 486) action during the interruption of pregnancy. **A.** RU 486 acts at the endometrium/decidua level. PG release is involved in the increase of the myocontractility of the uterus (see part **B**) and the opening of the cervix; luteolysis is a secondary event. (Baulieu, E. E. (1985) RU 486: An antiprogestin steroid with contragestive activity in women. In: *The Antiprogestin RU 486 and Human Fertility Control* (Baulieu, E. E. and Segal, S. J., eds), Plenum Press, New York, pp. 249–258). **B.** An injection of a small dose of PG has almost no effect on the contractions (black line) of an eight week gravid uterus, while it induces numerous and strong contractions when the antiprogesterone RU 486 is present (red line). (Bygdeman, M. and Swahn, M. -L. (1985) Progesterone receptor blockage. Effect on uterine contractility and early pregnancy. *Contraception* 32: 45–51).

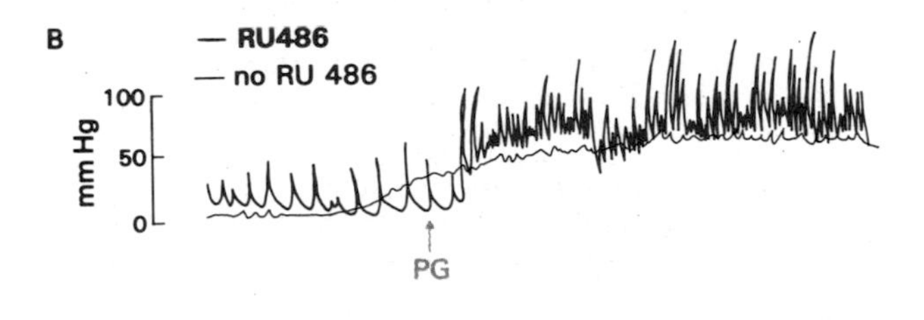

its derivatives. A type of leukotriene (p. 593) may be involved in LH secretion[57]. At the time of parturition, prostaglandins (of the PGF and PGE series) play an important role in uterine contractility. They are produced as a consequence of hormonal changes. The propensity of prostaglandins to increase myocontractility in the uterus has also been used to induce abortion[58]. They are especially efficient if the calming effect of progesterone is suppressed by an antiprogesterone (RU 486) (Fig. I.49). Prostaglandins also favor the softening and dilatation of the cervix. In addition to these local hormonal effects observed in most mammals, PGs (especially $PGF_{2\alpha}$) produced in the uterus reach the ovary of the sheep and cow directly, via a special circulatory pathway[59], and have luteolytic activity. However, PGs are not luteolytic in humans, and thus cannot be used to induce menstruation. Current studies of PGs are showing that they are important regulators of an increasing number of endocrine-related reproductive processes.

As indicated on p. 591, the arachidonic acid cascade produces both thromboxane and prostacyclin (PGI_2). Since some types of vascular disease may be associated with a reduction of the formation of prostacyclin, the administration of which might prove to be useful in medical treatment. Prostacyclin is a potent vasodilator generated in the wall of blood vessels, particularly by endothelial cells, and is the most potent endogenous inhibitor of platelet aggregation discovered thus far. It acts by increasing cAMP, and functions in a manner opposite to that of thromboxane A_2 made by platelets.

57. Hulting, A. L., Lindgren, J. A., Hökfelt, T., Eneroth, P., Werner, S., Patrono, C. and Samuelsson, B. (1985) Leukotriene C_4 as a mediator of luteinizing hormone release from rat anterior pituitary cells. *Proc. Natl. Acad. Sci., USA* 82: 3834–3838.
58. Bygdeman, M., Christensen, N., Grenn, K., Zheng, S. and Lundstrom, V. (1983) Termination of early pregnancy–future development. *Acta Obstet. Gynecol. Scand.* 113: 125–129.
59. McCracken, J. A., Baird, D. and Goding, J. R. (1971) Factors affecting the secretion of steroids from the transplanted ovary in the sheep. *Rec. Prog. Horm. Res.* 27: 537–582.
60. Hechter, O. and Halkerson, I. D. K. (1964) On the action of mammalian hormones. In: *The Hormones* (Pincus, G., Thimann, K. V. and Atswood, E. B., eds), Academic Press, New York, vol. 5, pp. 697–825.

4. Receptors

4.1 Generalities

Long ago, pharmacologists and endocrinologists postulated the existence of receptors capable of mediating the activities of agonistic and antagonistic molecules. Biochemical studies performed by endocrinologists, pharmacologists, immunologists and neurobiologists in the last 20 years have complemented the original *phenomenological* observation, and have greatly advanced the understanding of how receptors are involved in hormone action.

A hormone first interacts with a specific structure of *target cells*, called the *receptor*, which then triggers a response[60] (Figs. I.50 and I.51). A receptor consists of at least two basic constituents, a receiving or recognition site to which the *hormone binds* (the *hormone binding site*), and an *effector* (or executive) *site*, where the first response which modifies cellular function occurs. Between the two sites, there is a mechanism of *transduction*, or *coupling*. Upon hormone binding, the effector site is allosterically modified in such a way that a *response* is initiated.

The effector site may interact with a cellular constituent (e.g., with DNA or a G protein (p. 92 and p. 99)). Alternatively, it can be a catalytic site, and, in this case, the receptor is an allosteric enzyme whose binding site is a regulatory site‡. The receptor may also contain or be in contact with a transport/transfer system (*channels* for ions or selected small molecules). In addition, the receptor may bind to cytoskeletal elements, membrane lipids, or chromatin proteins. The same receptor may interact with similar but distinct regulatory elements, as for example a given steroid receptor with several promoters of regulated genes, establishing within a given cell a specific functional response pattern. Finally, receptor molecules may form oligomers (e.g., by S-S bonding), and may be more or less glycosylated or phosphorylated, modifications which affect their activity.

In principle, having triggered a response, the unaltered hormone molecule dissociates from the receptor, since the nature of binding is *reversible*, *noncovalent*, and thus purely physical. The fate of both the hormone and receptor varies, but frequently the hormone molecule is irreversibly altered by some

‡In cases where the receptor interacts with a G protein which in turn interacts with an enzyme, the receptor-G-protein ensemble is the regulatory portion of a complex allosteric enzyme system regulated by the hormone (p. 334).

enzymatic system of the target cell. The receptor may be endocytosed (membrane receptors), and is either degraded or recycled. The fact that the hormone and/or receptor is "consumed" may be understood on the basis of a fundamental principle of endocrine systems: physiologically, *new hormonal molecules* and/or new or reactivated *receptor molecules* are *necessary* for hormone action to continue. The receptor alone is *insufficient* to promote a complete response. Subsequent to the initial interaction, a complex series of steps must follow in an orderly and integrated manner. By definition, the receptor is *the* locus capable of receiving a signal and turning it into action. Thus, the causal relationship between *binding* and a subsequent hormonal *response* should be demonstrated *experimentally*. However, for many hormones, the first intracellular steps of the response remain unknown.

The study of the mechanism of action of receptors is, indeed, central to contemporary endocrinology.

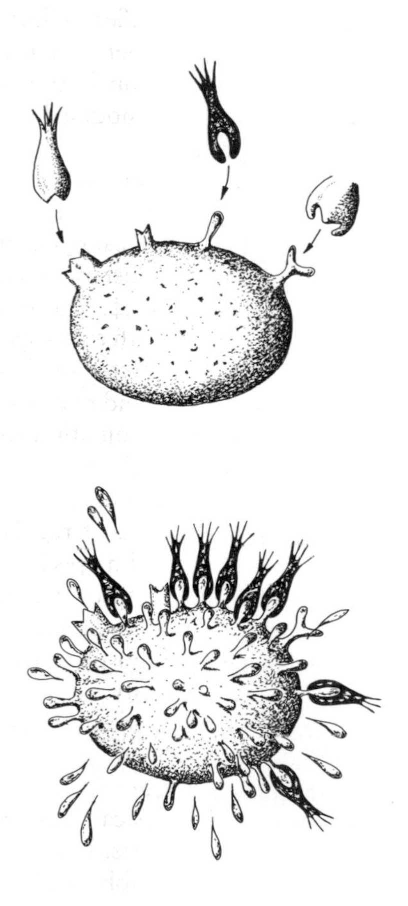

Figure I.50 The concept of cellular receptors, based on drawings by Paul Erlich, ca. 1900.

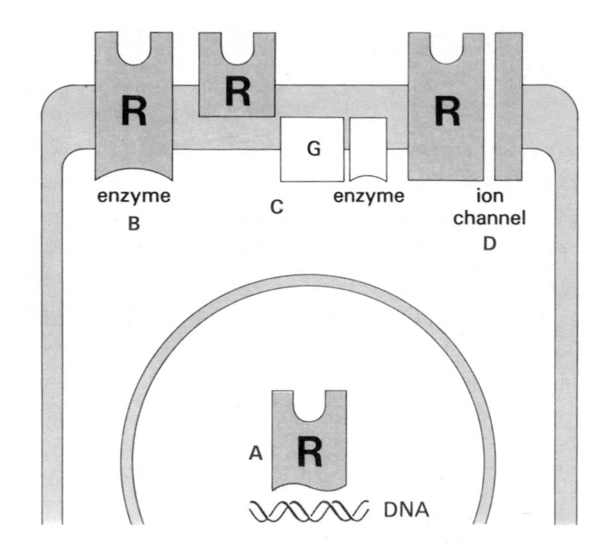

Figure I.51 Four classes of hormone receptors. Intracellular receptors interact with genes (**A**). Membrane receptors are enzymes themselves (**B**) or are coupled to an enzyme via a G protein (**C**), or are connected to or comprise a channel system (**D**).

Throughout this book, membrane receptors have been represented schematically, with the hormone binding site facing the extracellular environment and a segment inserted in the membrane. The transmembrane passage into the cytoplasm, as described for several receptors, is often not represented, except when there is a need to show the cytoplasmic domain. The subunit composition of G proteins is not usually represented, and the membrane-spanning adenylate cyclase is represented only by its catalytic region on the cytoplasmic side of the plasma membrane.

Until now, active receptors have been found *in only two strategic fractions* of the cellular machinery, plasma *membranes and chromatin*. In its approach to potential target cells, a molecule of hormone first encounters the plasma membrane, and, upon contact, receptors facing the extracellular environment mediate transmembrane signalling. Some small (lipophilic) hormones pass directly through membranes and proceed to the nucleus where the receptor-hormone complex turns on specific gene expression. There is no experimental evidence, to date, demonstrating the existence of two sets of bona fide receptors – one in the plasma membrane and a second in the nucleus – for the same hormone in the same cell. This could constitute a mechanism to ensure, for example, that membrane permeability to nutrients, gene transcription and protein synthesis function in a concerted manner. However, this does not mean that a hormone may not interact with several receptors in different cells or even in the same cell (e.g., p. 82 and Fig. I.125B). Finally, receptors, like most cellular constituents, are physically mobile (e.g., the process of recycling), and are present in varying concentrations (p. 85).

Research Strategies: "Forward" and "Backward"

Two experimental approaches which can be designated as "forward" and "backward" have been used to describe and study receptors.

The *forward* approach consists of the study of the *hormone* en route to its *biologically specific* interaction with the target cell. Since hormonal concentrations are in the nM range, a 1000-fold increase in sensitivity was needed for such studies as compared to biochemical methods used between 1930 and 1950 for enzymes and substrates whose K_M values and concentrations are generally >1 μM. *Radioactive labelling*, essentially with tritium or iodine atoms, permitted *tracers* with high specific radioactivity retaining all hormonal properties of the "cold" (nonradioactive), parent compound to be obtained‡. The synthesis of *affinity labelling* compounds which covalently bind receptors has also been very useful in receptor characterization. *Purification* methods, particularly those utilizing affinity chromatography (Fig. I.52), have progressed enormously and have led to partial determination of receptor sequences. These sequences have enabled the synthesis of oligodeoxynucleotides, used as probes in *cloning technology*. Polyclonal and monoclonal *antibodies* were also obtained with purified receptors, and have already been used for molecular mapping and localization within target cells. Antibodies have also been important in the selection of cDNA clones from expression libraries. To date, the complete amino acid sequence of hormone receptors has only been established on the basis of the corresponding cDNA.

The *backward* study of hormone action consists of working from a given stage in the hormonal effect to the point where the immediate consequence of hormone-receptor interaction is identified. For steroid and thyroid hormones, which act at the level of gene expression, continuous advances are being made in understanding their mechanism of action. Currently, direct experiments are being performed with regulatory segments of DNA that interact with receptors. For membrane receptors, regulation of the membrane enzyme adenylate cyclase, which produces the second messenger, cAMP, as well as changes in the turnover of membrane inositol phospholipids, calcium and other ions, is widely studied. Paradoxically, at present, these initial mechanisms are better defined (at least in most cases) than subsequent effects on cell metabolism. Several response mechanisms may, in principle, be activated simultaneously by the same hormone-receptor complexes, despite the fact that only a few cases can be cited (p. 242, ACTH; p. 267, GnRH).

‡A specific radioactivity >10 Ci/mmol is necessary for the measurement of both hormones and receptors. For tritium, the incorporation of one molecule per molecule of hormone or receptor results in a specific activity of ~50 Ci/mmol, while for ^{14}C, the specific activity is 1000-fold less for one atom per molecule. With one atom of ^{125}I or ^{131}I, values >1000 Ci/mmol are easily obtained.

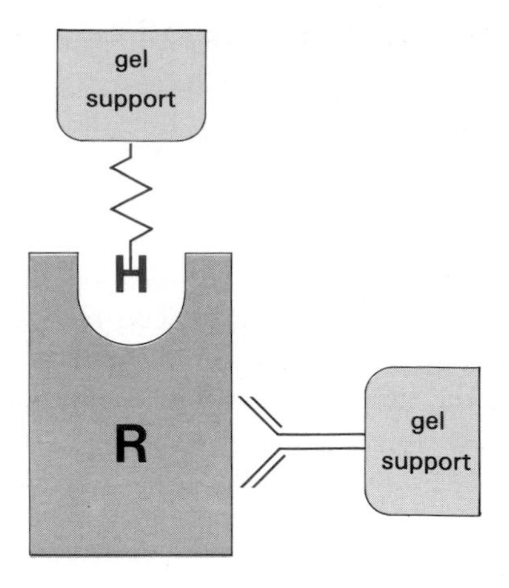

Figure I.52 Principle of two of the main techniques for the purification of hormone receptors. Both involve affinity chromatography, one through specific hormone binding and the other through a specific antibody.

4.2 Hormone Receptors Are Specific Binding Proteins

All hormone receptors which have been described thus far are *proteins*. The structural diversity of proteins determines both the high affinity and stereospecificity of hormone binding and the mechanism by which hormone is coupled to the effector site. Most membrane receptors are glycosylated, and the possible involvement of a carbohydrate moiety in the binding process is still controversial. In the case of *cholera toxin*, a hormone-like molecule, the receptor consists of a specific GM_1 ganglioside which binds α subunits of the toxin, while the released β subunit enters the cytoplasm and acts by ADP-ribosylation of a G_s subunit (p. 336).

Binding Site Specificity: Four Properties

All criteria of the specificity of hormone binding by receptors have a double, *physical and biological, connotation.*

Hormone Specificity

Based on a strict stereochemical accommodation of ligand to the binding site, hormone specificity meets an essential physiological requisite and underscores the *raison d'être* of hormone-specific receptors. It means, in principle, that a receptor can bind a given hormone, but not others. All *agonists* and *antagonists* of this hormone also bind to the receptor. It is interesting to look at the apparent twists in the double-basic rule of "one receptor for one hormone, and one hormone per receptor" (Fig. I.53). Most *neurohormonal ligands* (Table I.10) have *several* distinct receptors, which were assessed by binding studies and pharmacological properties. Indeed, these receptors may belong to the same molecular families (e.g., adrenergic receptors) or to very different classes (e.g., nicotinic and muscarinic acetylcholine receptors). *Hormones* generally have only one "primary" receptor, even though they may interact with the receptor(s) of other hormones, which in principle reduces their functional specificity. For example, the androgen receptor also binds certain estrogens and progestins (Table I.11); the glucocorticosteroid receptor binds aldosterone, a mineralocorticosteroid; the insulin receptor binds IGFs (Table I.12); the vasopressin receptor, oxytocin; and the prolactin receptor, growth hormone. Indeed, absolute specificity does not exist, which is easy to understand on the grounds of structural homology between hormones and between receptors. Thus, the *molecular code* for *receptor recognition is degenerate,* and provides for a number of interactions of a given ligand with several receptors. In most cases, *however,* hormonal specificity is insured *physiologically* by the *quantitative* component of these interactions. For example, in the binding of estrogens, aldosterone and IGF-II to the androgen, glucocorticosteroid and insulin receptors, respectively, physiological concentrations as well as the affinities for the "secondary" receptors are sufficiently low that hormonal specificity is normally respected. Pharmacologically or pathologically, of course, the case may be different.

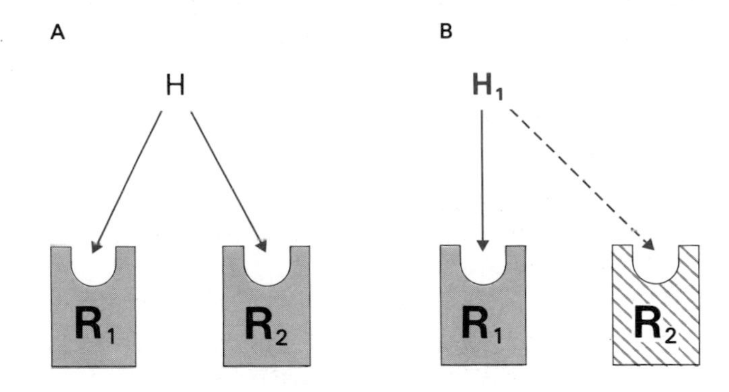

Multiple receptors. **A.** One hormone, two receptors. A hormone, H, has two receptors, R_1 and R_2. The hormone has two physiological functions through these two receptors, and has high affinity for each. **B.** Crossover binding: a hormone, H_1, whose receptor is R_1, can also interact (with relatively low affinity) with receptor(s) R_2, etc., of (an)other hormone(s).

Table I.10 One neurohormone, several distinct receptors.

Hormone	Receptors
Epinephrine	α_1-, α_2-, β_1-, β_2-adrenergic
Dopamine	D_1-, D_2-dopaminergic
Enkephalin	μ, δ, κ
Acetylcholine	Nicotinic (striated muscles) Muscarinic (smooth muscles)
Histamine	H_1, H_2

Distinct receptors for the same hormone either belong to the same structural family (adrenergic receptors) *or* are unrelated (n- and m-ACh-R, Fig. I.85 and I.81). In addition, as seen in Tables I.11 and I.12, a given hormone receptor often may bind other hormones, with lower affinity.

High Affinity

Receptor affinity corresponds to the concentration of hormone in the milieu, which is the *requisite* for high functional specificity of receptors within the organism. The dissociation constant, K_D, of hormone-receptor complexes at equilibrium is the reciprocal of the affinity constant, K_A (p. 57), and is usually very low, of the order of 1 nM to 10 pM. Generally, the very high affinity of receptors is due to a slow dissociation rate. When the magnitude of the cellular response is compared to the binding affinity of a receptor for different ligands within the same series of agonists or of antagonists, a clear correlation can be demonstrated. There is a *lack of correlation* between the affinity of the receptor and the *agonistic or antagonistic* properties of the ligand (p. 117).

Table I.11 Affinity of steroid receptors for hormones and analogs.

Receptor	Hormone					
	E_2	DES/Tam	T/DHT	P	Dex	Aldo
ER	+++	+++/+	±	0	0	0
AR	+	0	+++/+	+	0	0
PR	0	0	+	+++	±	0
GR	0	0	±	++	++++	++
MR	0	0	0	++	±	+++

+++ indicates the affinity of the corresponding physiological hormone; ++++, +++, ++, +, ±, and 0 are decreasing affinities. Differential binding specificity of synthetic ligands for receptors may be used to develop drugs more selective than endogenous hormones.

Non-specific binding sites of relatively high K_D (weak affinity) are not saturated at physiological hormone concentrations, and function as a partition system with a coefficient unaffected by ligand concentration (Fig. I.57). Whether non-specific binding of hormones decreases their availability to receptors is controversial, but this probably depends on kinetic parameters such as the rate of dissociation and the turnover of the ligand in target cells.

Limited Number of Sites

In all known cases, there is one specific binding site *per molecule* of receptor, and the number of these molecules *per diploid cell* is of the order of $10^{4\pm1}$.

The concept of a *limited number of sites* should not be confused with that of *saturability*. The latter depends on binding affinity and ligand concentration. High affinity leads to saturation of sites at relatively low physiological concentration of hormone, in contrast to what is observed in relation to non-specific sites.

Table I.12 Affinity of insulin and insulin-like growth factor receptors for insulin, IGF-I, and IGF-II.

Receptor	Hormone		
	I	IGF-I	IGF-II
I-R	+++	+	±
IGF-I-R	+	+++	+
IGF-II-R	0	0	+++

+++ indicates the affinity of the corresponding physiological hormone; ++, +, ± and 0 are decreasing affinities.

Tissue Specificity

The fourth principle of receptor function is related to the number of binding sites available to the hormone. It states simply that receptors are *present* in *all* target organs which *respond* to a hormone, and that, in their absence, tissues are insensitive (p. 419 and below). The presence of a receptor as an isolated molecule is insufficient, per se, to provoke the response of a cell. It is remarkable, therefore, that, in receptor-containing cells, there is always an effector system, as if differentiation were coordinated to provide both appropriate signal reception and signal transduction. Pathologically, however, this dual system may be disrupted.

Receptors *are sometimes difficult to detect* in tissues. For example, it was long known that skeletal muscle is sensitive to androgens, but a corresponding receptor could not be identified. The possibility remained that the anabolic effect of androgens was due entirely to a metabolic or nervous mechanism which indirectly affected muscles. It was only by improving homogenization methods and by blocking the inactivating metabolism of the radioactive ligand that the androgen receptor could be identified[61]. Initially, the β-adrenergic receptor was improperly identified because high afinity binding proved to be non-stereospecific. In fact, it was probably due to an altered form of a membrane enzyme (catechol O-methyltransferase[62]) and to the use of more selective ligands, including those of the antagonist series, that

61. Michel, G. and Baulieu, E. E. (1974) Récepteur cytosoluble des androgènes dans un muscle strié squelettique. *C. R. Acad. Sci., Paris* 279: 421–424.
62. Cuatrecasas, P., Tell, G. P. E., Sica, V., Parikh, I. and Kwen-Jen Chang. (1974) Noradrenaline binding and the search for catecholamine receptors. *Nature* 247: 92–97.

the β-adrenergic receptor was finally identified physically (Fig. I.81).

The *absence* of receptors or the presence of *defective* receptors in cells which do *not respond* to hormones has been assessed in pathological cases and experimentally. The genetic disorder known as testicular feminizing syndrome (tfm) is a remarkable case in point (p. 419). It has been observed in animals of several species in which androgens do not act for lack of a functional receptor. The precise mechanism of the disorder, whether there is complete or near absence of receptor, or whether there is only an abnormality in the receptor, causing inadequate or lack of binding, is not yet known. Variants of lymphoma cells insensitive to glucocorticosteroids have been produced by selective pressure in vitro. Some apparently contain a mutated receptor which does not bind hormone, and others have abnormal receptors with normal or subnormal hormone binding properties. Other cell types bind hormone, but are unresponsive because of a postreceptor defect in the intracellular machinery.

Cells with $<1/100$ the normal number of receptor binding sites fail to respond. The quantitative relationship between the number of sites and the intensity of the response is discussed below (p. 64). With modern techniques, receptors can be found almost everywhere, even in tissues which were not thought to be target tissues for a particular hormone. Thus, a discrete hormone function might be identifiable if the concentration of ligand were sufficiently high. Alternatively, a low receptor level measured in tissue extracts may be reflective of cellular *heterogeneity*, e.g., their presence in only a few target cells among a majority of unresponsive cells devoid of receptors.

The four criteria characterizing receptors, i.e., hormone specificity and affinity, receptor number and tissue specificity, point out the *differences* between *specific* receptor binding and *non-specific* binding. The latter has low affinity and no hormonal or tissue specificity, and as a result accurate measurement of the number of non-saturable sites is made difficult. Specific plasma steroid binding proteins, despite their high affinity, do not demonstrate the necessary binding biospecificity which would suggest direct involvement in hormone action (p. 43). Conversely, if *applied rigorously*, these four criteria are sufficient to substantiate the existence of a potentially functional receptor in cells previously not thought to be target cells[61]. Conversely, when a hormone receptor is discovered unexpectedly (e.g., the insulin receptor in lymphocytes), it is important to ensure that this receptor alone accounts for the effects being studied. Specific receptor binding studies permit a type of bioassay for natural compounds whose structures are unknown, or the screening of synthetic derivatives with undefined biological activity (p. 127). Competitive binding with the natural hormone or with a well characterized parent ligand can indicate whether an unknown substance is potentially active as either an agonist or an antagonist (p. 117). This investigational strategy permitted the discovery of endogenous enkephalins, by studying the binding of exogenous morphine analogs to opiate receptors (p. 188).

Reversibility

The *reversibility* of hormone-receptor interaction could be a *fifth* characteristic of hormone receptor binding in that it implies that the triggering of regulated cell function is short-lived. While reversibility is an important physiological feature of hormone-receptor interaction, irreversible binding is useful, experimentally, to characterize receptors. It involves *affinity labelling* (cross-linking) that covalenty attaches the ligand to the receptor.

4.3 Binding of Hormones to Receptors

Physical Definition

The binding of a hormone (H) to its receptor (R) is *non-covalent* and *reversible*. The interaction between H and R follows the *law of mass of action* and depends, once H and R are qualitatively defined, on the concentration of the two reactants:

$$[H] + [R] \underset{k_d}{\overset{k_a}{\rightleftharpoons}} [HR] \qquad (1)$$

$$K_A = \frac{k_a}{k_d} = \frac{[HR]}{[H][R]} \qquad (2)$$

$$K_D = \frac{k_d}{k_a} = \frac{[H][R]}{[HR]} \qquad (3)$$

[H] is the concentration of free hormone (in M; that is, in mol/l);
[R] is the concentration of free receptor sites;
[HR] is the concentration of hormone-receptor complexes;
K_A is the association constant (in M^{-1}) at equilibrium;
K_D is the dissociation constant (in M) at equilibrium;
k_a is the association rate constant (in $M^{-1}\ sec^{-1}$);
k_d is the dissociation rate constant (in sec^{-1}).

If R_T represents the total number of receptor sites, it is clear that:

$$[R_T] = [R] + [HR]$$

If H_T represents the total amount of hormone, at all times then:

$$[H_T] = [H] + [HR]$$
$$\text{or} = [F] + [B]$$

with F and B representing the free and bound forms of the hormone, respectively. R_T is frequently designated by N (number of sites), and it is possible, from (3), to calculate [B], the concentration of hormone bound at equilibrium (Fig. I.54):

$$[B] = \frac{[N][F]}{K_D + [F]} \qquad (4)$$

Thus, a binding system at equilibrium is defined when [N] and K_D are known. Receptor concentration ([N]/n) can only be established if the number of binding sites, n, per receptor molecule is available (n=1 in all cases studied thus far).

Hormone-receptor interaction (equation 1), which is identical to the first part of an enzymatic reaction (i.e. substrate plus enzyme give enzyme-substrate complexes which are formed prior to the formation of a product), can be described by the Michaëlis-Menton equation. The graphical representation is an asymptotic hyperbola, with [B] maximum equal to [N] (Figs. I.54, I.55, and I.56A). The concentration required for half saturation is defined as K_D, equivalent to K_M of enzymatic reactions. When K_D is measured at *equilibrium*, it can be indexed eq (K_{Deq}); the same K_D value calculated *kinetically* by the ratio of the rate constants is K_{Dkin} (p. 63).

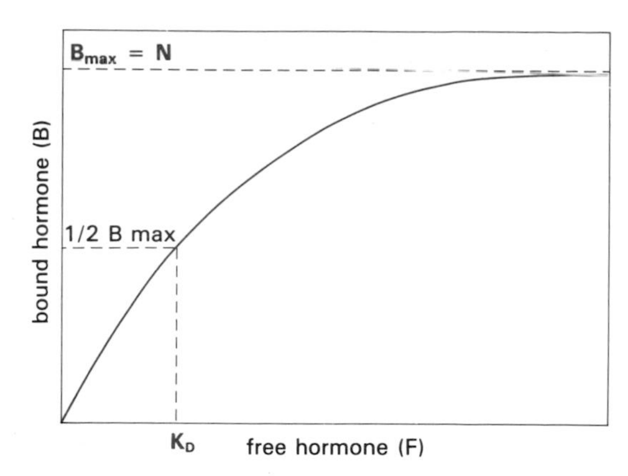

Figure I.54 Representation of the concentration of receptor-bound hormone as a function of the concentration of free hormone.

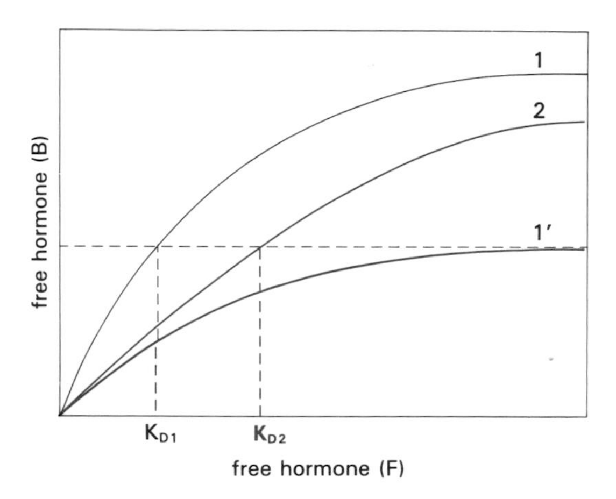

Figure I.55 Hormone binding: three curves. Curve 1 is the same as in Figure I.54, and represents the binding of a hormone, H_1, to a receptor, R_1, with an affinity corresponding to K_{D1}. Curve 1′ represents binding of the same ligand, H_1, to the same receptor, R_1, therefore with the same K_{D1}; however, the number of sites of the receptor is half that represented by curve 1. Curve 2 represents either the binding of H_1 to another receptor, R_2, with the same concentration of sites as R_1, but with a lower affinity ($K_{D2} > K_{D1}$), or the binding of another ligand, H_2, to the first receptor, R_1, with lower affinity (also, $K_{D2} > K_{D1}$).

Representation of Interactions at Equilibrium

Specific Binding (B_S)‡

To those who have already studied enzymology, the binding reactions indicated above are familiar. However, the concentration of receptor binding sites in most biological structures or extracts is not negligible with respect to that of the hormonal ligand, whereas in enzymatic reactions, the substrate (ligand) is generally much more abundant than the enzyme. Therefore, in the first approximation, the amount of bound ligand is negligible, making the total and free concentrations of substrate practically equal.

In order to linearize the results of binding experiments, it is assumed that the *law of mass action* applies (equations 1 to 3), that there is only *one type* of binding site, and that there is *no interaction* between sites.

‡The following discussion applies to specific binding sites of the receptors and plasma binding proteins mentioned above. Here, "*specific*" has a quantitative connotation, indicating an affinity sufficient for saturation to be obtained physiologically, or at least at a pharmacological or pathological concentration of the ligand (p. 56).

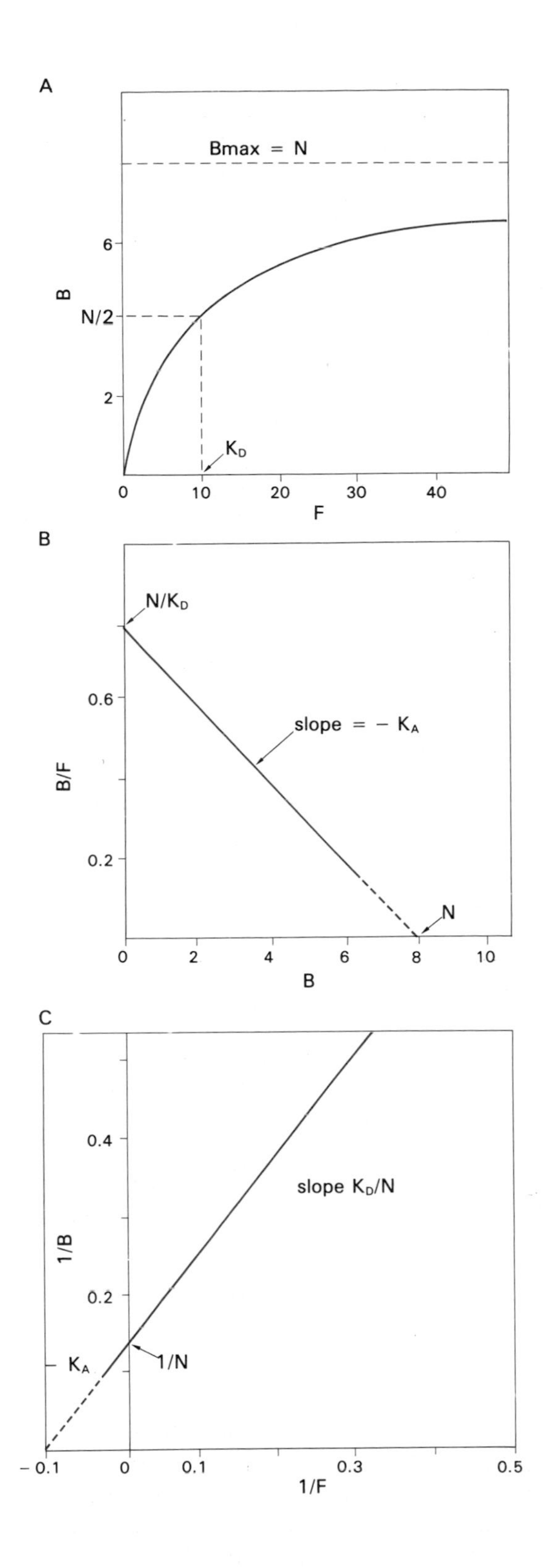

Figure I.56 Hormone binding: three different representations. **A.** Direct plot of the Michaëlis–Menten equation. **B.** Scatchard plot. (Scatchard, G. (1949) An attraction of proteins for small molecules and ions. *Ann. N.Y. Acad. Sci.* 51: 660–672). **C.** Double reciprocal plot of Lineweaver–Burk.

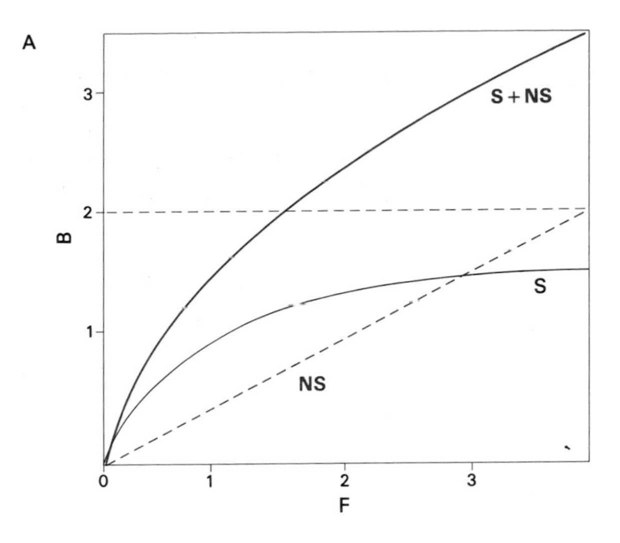

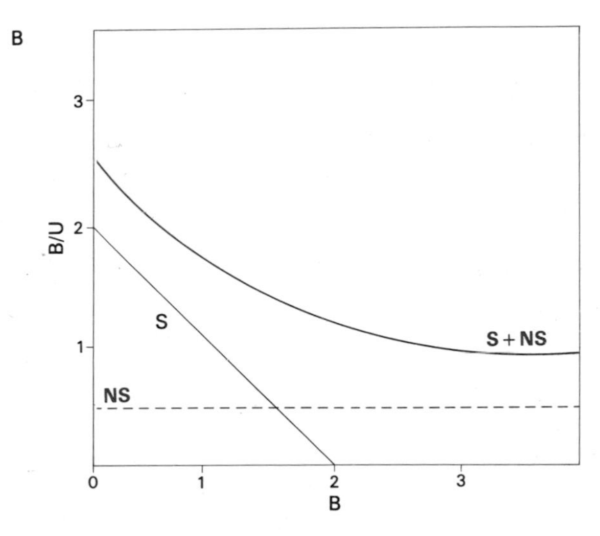

Figure I.57 Influence of non-specific binding of hormone, B_{NS}, on the representation of specific binding, B_S, to the receptor. **A.** Direct representation. **B.** Scatchard plot.

Scatchard analysis, which is the most frequently used due to its simplicity, is based, as is the method of Eadie for enzymatic reactions, on an equation derived from the preceding considerations:

$$\frac{[HR]}{[H]} = K_A([R_T] - [HR])$$

$$\frac{[B]}{[F]} = K_A([N] - [B])$$

The negative slope$=-K_A$ of the representative straight line is calculated from [B]/[F] versus [B] (Fig. I.56B). Determining the values of [B] and [F] at different hormone concentrations, it can be seen that, for elevated values, [F] increases and [B]/[F] approaches zero. [B], therefore, is found at the intersection of the straight line with the abscissa. The Lineweaver-Burk plot (Fig. I.56C) can also be used for linearization, but it is frequently less precise, because values measured at high ligand concentrations produce points on the graph which are not clearly separated.

Linear methods are very useful, since values can be obtained rapidly (and graphical representation of values is satisfactory), even though they often fail to meet rigorous statistical standards. Indeed, statistical criteria should be applied to the linearization of results. Computer assisted analysis is now used routinely in the assessment of binding models. Consequently, it is no longer always necessary to linearize data. Double logarithmic methods, such as the proportional graph[63], can be useful, especially when there are several simultaneous forms of specific binding for the same ligand in the extract. Logarithmic coordinates permit the study of a large range of hormonal concentrations, and particular aspects of binding data can be emphasized by expressing binding in terms of proportion. With linearization methods, as that of Scatchard, it is not possible to transform data into corresponding straight segments when several binding sites with different affinities for the same ligand are present in the preparation. Any binding isotherm defined with a single ligand is insensitive to site multiplicity unless affinities differ greatly ($\geqslant$1 order of magnitude). Several graphical manipulations have been proposed, but most are unconvincing. *No* method can *prove* definitively that the binding being studied follows a specific model. At best, a model facilitates the process of elimination of other hypotheses[64]. The specificity of a binding site should be assessed by competitive studies (p. 127), using several ligands, in order to determine a "binding selectivity profile"[65]

Non-specific Binding (B_{NS})

In most biological extracts, hormones bind not only to receptors or other specific protein sites, but also to many other macromolecules whose *sites* are called *non-specific*, when the steric involvement is weak and when their affinity is relatively modest (e.g., K_D>1 μM). [N] is frequently (but not necessarily) very large, and is poorly defined due to the diverse and imprecise binding regions overlapping each other on the surface of large molecules. The hyperbola which results from the direct representation of the law of mass action is replaced experimentally by a straight line, up to very high concentrations of free ligand, as predicted by the high K_D. These concentrations have no biological significance, and are sometimes not even attainable because of the solubility problem. K_D and [N] cannot be calculated separately, and total binding ($K_D \times$ [N]) is calculated from straight lines (Fig. I.57). When S and NS sites are present simultaneously, the hormone is distributed according to respective affinities and concentrations, and the representation of total binding, $[B_T]=[B_S]+[B_{NS}]$, does not form a simple curve (direct plot) or a straight line (Scatchard), but $[B_S]$ can be calculated by subtracting the corresponding value of $[B_{NS}]$ from $[B_T]$.

4.4 Techniques for Measuring Binding

Radioactively labelled hormones of high specific activity are used whenever possible. In order to measure the affinity of a non-radioactive ligand, the *inhibition* of binding of a radioactive ligand is measured in a *competition* test. The *inhibition constant* (K_I) at equilibrium has a very similar significance to that of K_D (equation 3, p.57):

$$K_I = \frac{[I][R]}{[IR]}$$

[I] is the concentration of free inhibitory compound (in M; that is, mol/l).
[IR] is the concentration of inhibitory compound-receptor complexes.

The radioactive ligand should have the *same affinity* as the non-labelled hormone, and this should be controlled carefully. For the measurement of binding parameters, it is necessary, of course, to wait long enough for *equilibrium* to be reached. Numerous separation techniques are available (Fig. I.58).

To fulfill the conditions required for measuring *binding at equilibrium*, the classical method is *equilibrium dialysis*, provided that the size of the hormone and the reliability of the semipermeable membrane allow it. At equilibrium, the concentration of free ligand, [F], is measured within one compartment, and in the other, [B+F] (usually within the dialysis sac). By simple subtraction, the value of [B] can be calculated. The partition method is useful in the study of

63. Baulieu, E.E. and Raynaud, J.P. (1970) A "proportion graph" method for measuring binding systems. *Eur. J. Biochem.* 13: 293–304.
64. Munson, P.J. and Rodbard, D. (1980) Ligand: a versatile computerized approach for characterization of ligand-binding systems. *Anal. Biochem.* 107: 220–239.
65. Goldstein, A. (1987) Binding selectivity profiles for ligands of multiple receptor types: focus on opioid receptors. *Trends Pharmacol. Sci.* 8: 456–459.

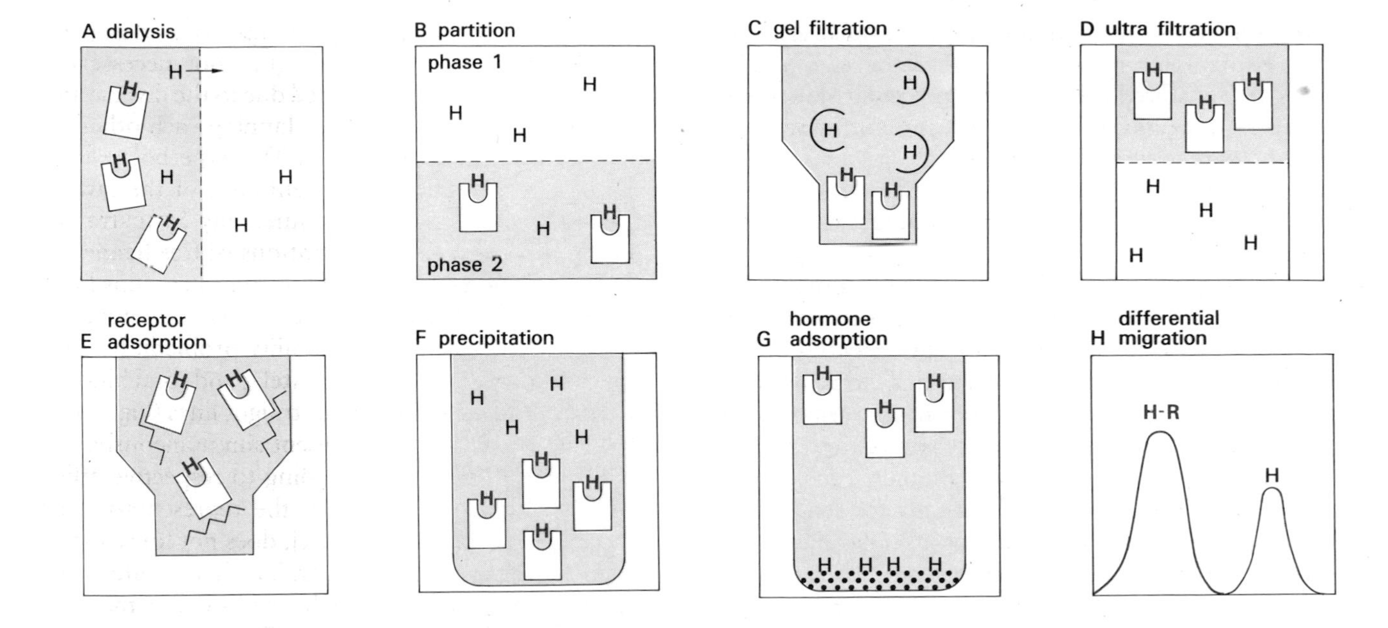

Figure I.58 Techniques for separating hormone–receptor complexes from free hormone. In E, the adsorbent may be an ion exchanger; in G, activated charcoal is often used; in H, methods include electrophoresis, isoelectric focusing, or ultracentrifugation.

equilibrium binding between soluble macromolecules (Fig. I.59).

Other techniques involve the *physical separation* of the hormone-receptor complexes, implying that the *system* is *no longer at equilibrium*. This transition, although short, should be assessed with respect to the dissociation rate of the complexes (see the principle of differential dissociation, p. 62). The free ligand can be removed by an adsorbent such as activated charcoal (dextran-coated), kaolin, talc, or by an antibody attached to or forming a solid phase. The hormone-protein complexes remain in the soluble phase, and are easily separated from the adsorbent. In contrast, the hormone-receptor complexes can be adsorbed onto a filter or a specific column (e.g., immunoadsorption), and free ligand which passes through can be measured as well as the retained complexes. Hormone-receptor complexes can also be precipitated by forming a complex with another macromolecule (such as a basic protein, like protamine, for steroid hormone receptors), or by increasing ionic concentration. The free and bound fractions can also be separated by partition (Fig. I.59), by ultracentrifugation in a den-

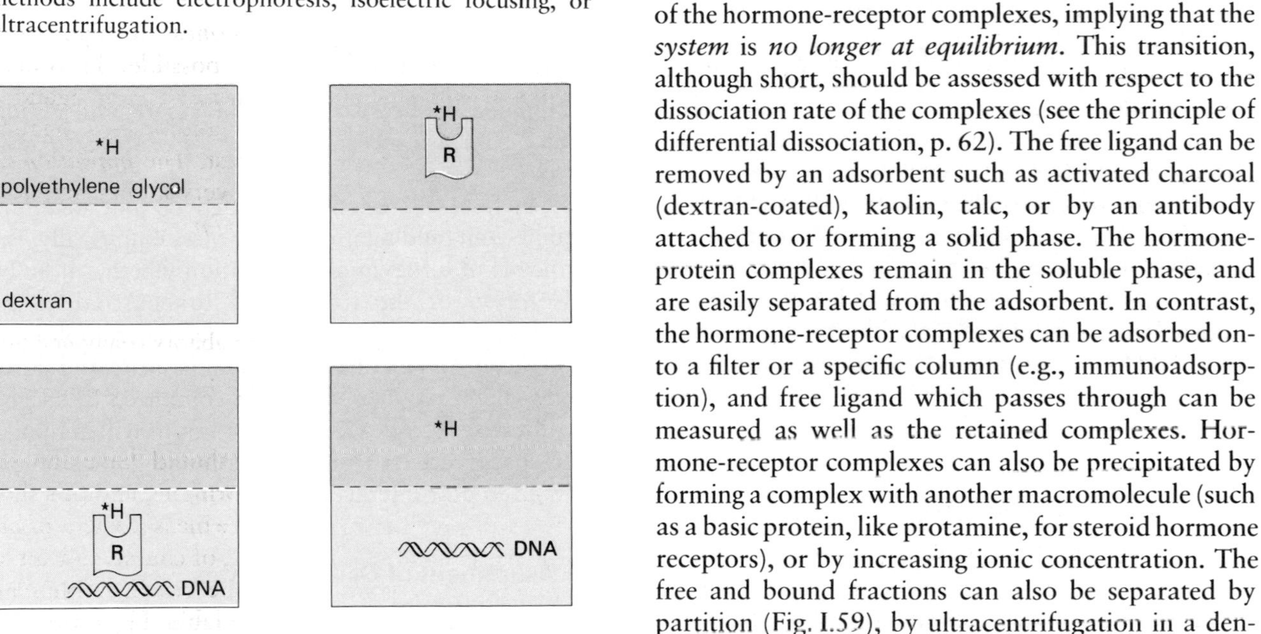

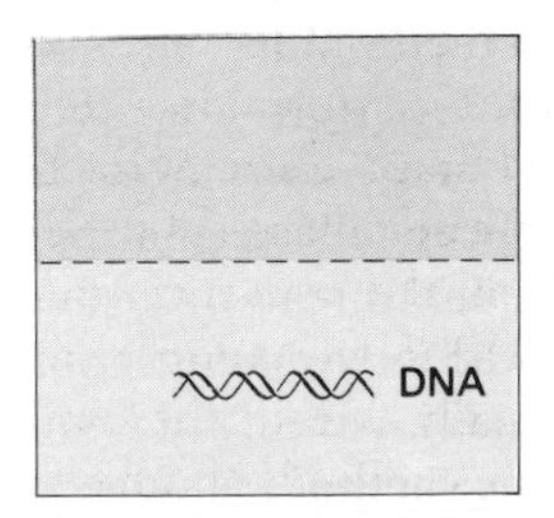

Figure I.59 Phase system for studying binding equilibrium between two soluble macromolecules. Labelled hormone (*H), receptor (R) and DNA, separately and in various combinations, are partitioned in a two-phase system of polyethylene glycol and dextran. DNA is found almost entirely in the dextran phase. *H alone, *HR, and *H in presence of DNA are found in the polyethene glycol phase. As DNA, *HR complexes bound to DNA are found in the other phase, allowing calculation of the receptor–DNA interaction. (Alberga, A., Ferrez, M. and Baulieu, E. E. (1976) Estradiol–receptor–DNA interaction: liquid polymer phase partition. *FEBS Lett.* 61: 223–226).

sity gradient, or by differential electrophoretic or chromatographic migration. In the latter case, proteins are separated according to molecular size or to charge. The recent use of high pressure increases the speed and precision of fractionation.

For receptors *not in soluble* form (e.g., in membrane or chromatin fractions), filtration, centrifugation or adsorbtion methods can be used to separate free hormone from hormone-receptor complexes. However, calculations may suffer from the heterogeneity of the system. When there are *several* specific binding systems, they can be identified either by calculation[64] or, alternatively, by the physical separation of the complexes (Fig. I.58). Since ligands, receptors, and tissues *all differ* one from another, *each* given technique should be *checked* not only for sensitivity and reproducibility, but also for its relevance, i.e. does it actually measure receptor sites? "Data" are too easy to accumulate when using radioactive tracers. *All* receptor sites and *only* receptor sites should be measured (the stability of the binding protein and the determination of non-specific binding, etc., must be verified).

The Problem of Non-specific Binding

Non-specific complexes are present in all biological extracts, and B_{NS} may be larger than B_S. Two principles can be used to calculate B_S.

Isotope dilution of the radioactive ligand (utilized between $\sim K_D$ and saturating concentration) is obtained by adding an excess of non-radioactive hormone (e.g., 100 times or more), which reduces specific radioactivity. Labelling of high affinity sites is practically abolished as compared to the result of the test which is carried out in parallel with tracer alone. In contrast, sites of lower affinity (100 times and less) are still labelled in the presence of an increased concentration of unlabelled hormone. The value of $[^\star B_S]$ is obtained by subtracting the amount of bound radioactivity remaining after isotope dilution $[^\star B_{NS}]$ from the value obtained in the presence of tracer alone.

Differential dissociation of the hormone-receptor complexes permits the determination of interactions on the basis of the slowest dissociation of tight binding (e.g., from receptors or plasma proteins). In principle, all hormone-protein complexes dissociate according to a first order reaction, linear on a semilogarithmic scale[66] (Fig. I.60). The slope of the

66. Milgrom, E. and Baulieu, E. E. (1969) A method for studying binding proteins, based upon differential dissociation of small ligand. *Biochim. Biophys. Acta* 194: 602–605.

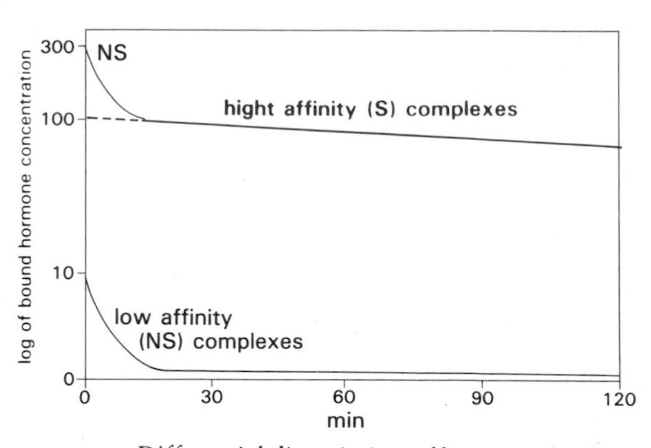

Figure I.60 Differential dissociation of hormone bound to high affinity (specific) and low affinity (non-specific) binding proteins. The hormone dissociates at a slow and rapid rate, respectively. For example, radioactive cortisol is bound to high affinity CBG (p. 396) and low affinity albumin. The free (dissociated) hormone is removed continuously by adsorption to charcoal. The radioactivity in the remaining complexes is measured as a function of time: it decreases according to a first order reaction (semilog scale)[66].

straight line becomes steeper as the dissociation rate increases. At a convenient temperature, the experiment can be followed over a period of time sufficient for the weaker complexes (non-specific) to be virtually eliminated, while specific complexes remain almost intact, and are measured by extrapolation to time zero.

In fact, all techniques which do not maintain equilibrium binding use, more or less empirically, the principle of differential dissociation whether it be by adsorption of the free ligand to charcoal or by chromatographic separation of the macromolecular complexes. Depending on the temperature and duration of the reaction, non-specific complexes dissociate to different degrees. Optimal conditions will minimize non-specific effects while maintaining significant hormone-receptor interactions.

Measurement of Occupied Sites

An additional problem is encountered when receptor sites are to be measured under varying pathophysiological conditions. Receptors are often partially or totally occupied by *endogenous* hormone (or hormone administered therapeutically or pharmacologically). By reducing the temperature to 0°C, sites may remain occupied by non-radioactive endogenous hormone, and the addition of radioactive ligand will only measure available sites. Thus, methods have been developed to *liberate occupied sites* so that they may be measured subsequently with radioactive hormone. The general concept is to isolate hormone-receptor complexes as they exist, to eliminate free or easily dissociated (B_{NS}) endogenous hormone, and then to use a sufficiently high concentration of radioactive hormone[67,68]. Endogenous hormone may be either successively removed, and the receptor sites then occupied by the labelled ligand, or it may be directly exchanged for the radioactive hormone by incubation of the endogenous hormone-receptor complexes. Exchange techniques are not easily applicable to membrane receptors, and dissociation of the complexes is usually accomplished by exposure to chaotropic agents or changes in pH. Temperature and duration of incubation are critical and should be rigorously controlled (particularly for receptor stability) (Fig. I.61).

The measurement of the number of binding sites, *alone*, does *not* provide definitive information about the biological activity of receptors, since another part of the molecule involved in the response may be altered (p. 87). Conversely, there are some *cryptic* or temporarily inactive receptors which do not bind hormone, and which are not identifiable by such studies.

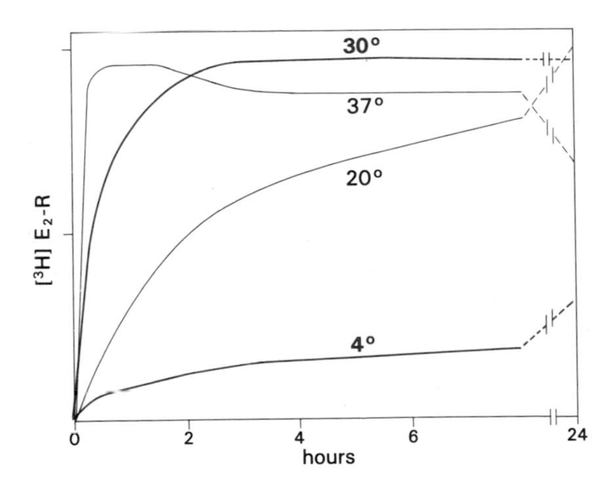

Figure I.61 In vitro labelling of nuclear receptor sites occupied by endogenous hormone: the exchange method. Radioactive hormone replaces non-radioactive endogenous hormone more rapidly as temperature increases. In this case, nuclear estrogen receptors from hen oviduct are incubated with $[^3H]E_2$. (Sutherland, R. L. and Baulieu, E. E. (1976) Quantitative estimates of cytoplasmic and nuclear oestrogen receptors in chick oviduct. Effect of oestrogen on receptor concentration and subcellular distribution. *Eur. J. Biochem.* 70: 531–541).

Kinetics Studies

Direct methods of measuring the association and dissociation rate constants k_a and k_d (p. 57) are based on the hypothesis of reversible bimolecular reaction. This model can be tested by calculating K_{Akin} and K_{Dkin}, which can be compared to K_A and K_D measured at equilibrium (p. 57).

To *determine* k_a, the concentration of sites (N) should be known. This is not always easy, however (see above). There are *two methods that are* used to *measure* the dissociation rate constant k_d. The reaction mixture can simply be diluted, which nullifies the concentration of free ligand and minimizes its reassociation, or, alternatively, non-radioactive hormone can be added at a relatively high concentration, which causes an isotope dilution of the dissociating radioactive ligand and precludes its reassociation with the receptor. These two methods do not give exactly the same results in certain cases where interactions are complex, as for example in negative cooperativity. Dissociation of the ligand sometimes appears more complex than predicted by the first order reaction (the reason remains obscure), and thus k_d may be difficult to define. The stability of hormone-receptor complexes should always be verified under the particular experimental conditions used.

Within experimental error, K_{Deq} *and* K_{Dkin} should, by definition, be identical. However, this is not always the case, especially when affinity is very high and dissociation very slow, and also when the concentrations of the ligand and binding protein are low. In these cases, results obtained at equilibrium can underestimate the affinity of the interaction. This may be due to a failure to reach equilibrium, in which case the value of K_{Deq} is artificially high; alternatively, the simple interaction model may not correctly define the system being studied.

4.5 Are Receptors Allosteric Proteins?

Cooperative Binding of Hormones by Receptors

Studies of hormone binding by receptors (and plasma transport proteins) have indicated that it fol-

67. Anderson, J., Clark, J. H. and Peck, E. J. (1972) Oestrogen and nuclear binding sites. Determination of specific sites by (^{3}H) oestradiol exchange. *Biochem. J.* 126: 561–567.
68. Milgrom, E., Perrot, M., Atger, M. and Baulieu, E. E. (1972) Progesterone in uterus and plasma. V. An assay of the progesterone cytosol receptor of the guinea pig uterus. *Endocrinology* 90: 1064–1070.

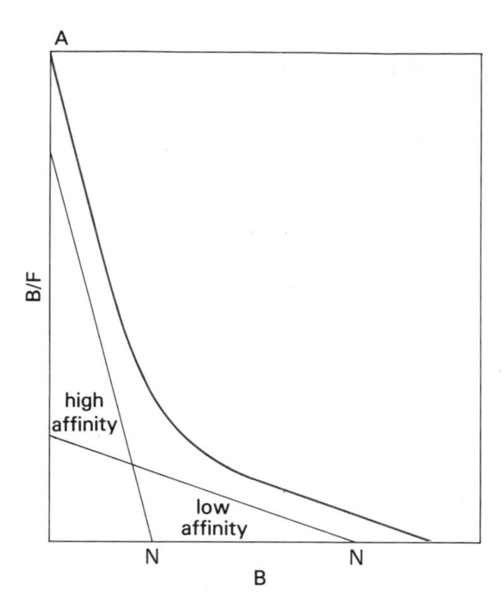

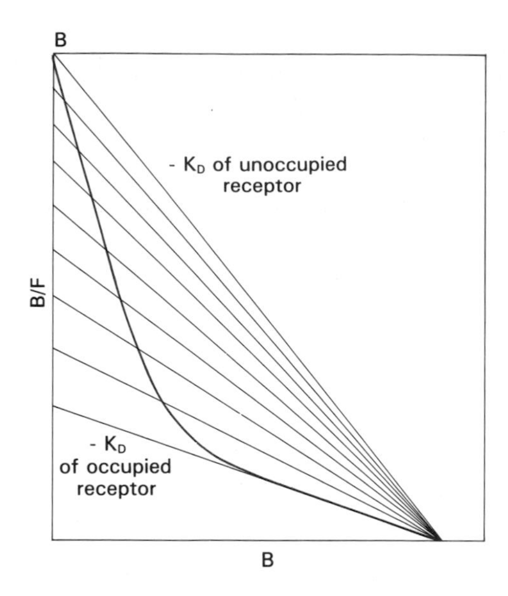

Figure I.62 Two independent binding sites, or negative cooperativity? In both **A** and **B**, the red line represents a curvilinear Scatchard plot of insulin binding to the insulin receptor. **A.** The curve is interpreted as the sum of two independent binding sites of higher and lower affinity. **B.** Alternatively, the curve represents a single population of binding sites with an affinity that is inversely related to receptor occupancy, i.e. negative cooperativity. The black lines have a slope equal to the negative average affinity at different levels of occupancy, ranging from unoccupied to fully occupied receptor sites. C. Experimental approach to the demonstration of negative cooperativity: the dissociation of insulin from binding sites is faster in the presence of an excess of non-radioactive insulin than in the case of simple addition of medium. (De Meyts, P., Roth, J., Neville, D. M. Jr., Gavin, J. R. and Lesniak, M. A. (1973) Insulin interactions with the receptors: experimental evidence for negative cooperativity. *Biochem. Biophys. Res. Commun.* 55: 154–161).

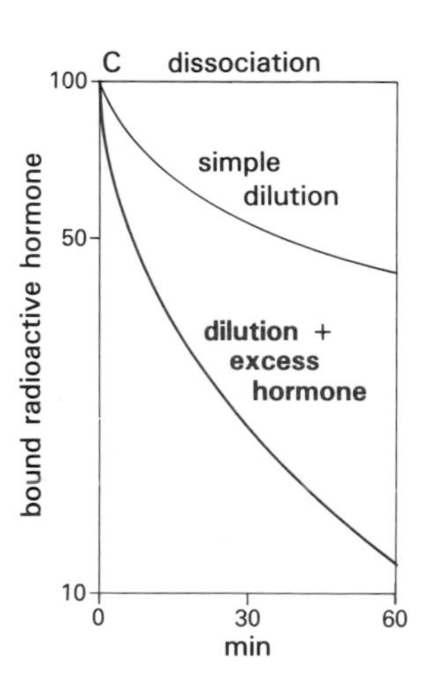

lows the law of mass action, which implies *identical* and *independent* binding sites. The linearity of Scatchard plots concurs with this concept (p. 59). Positive binding cooperativity has been claimed when, at low concentrations of hormone, the Scatchard curve increases or remains horizontal instead of readily descending. However, most reports are not convincing, except in the case of the dimeric form of the estradiol receptor[69]. *Negative cooperativity* has been described for insulin receptor binding (Fig. I.62), using concave Scatchard plots, with decreased affinity at high concentration of hormone[70]. Whether the dimeric structure of the insulin receptor (I-R) (p. 78) is implicated in this binding characteristic is not yet known. Negative cooperativity may appear as a *protective mechanism* in case of hormone excess. This decreased binding affinity by negative cooperativity should not be confused with a decrease in membrane receptors (down regulation), observable only in intact

69. Notides, A. C. and Sasson, S. (1983) The positive cooperativity of the estrogen receptor and its relationship to receptor activation. In: *Steroid Hormones Receptors. Structures and Function* (Eriksson, H. and Gustafsson, J. A., eds), Elsevier, Amsterdam, pp. 103–118.
70. De Meyts, P. (1976) Cooperative properties of hormone receptors in cell membrane. *J. Supramolecular Structure* 4: 241–258.

cells, which is due to endocytosis and/or degradation of the receptor following the formation of hormone-receptor complexes (p. 224).

Cooperative Activity of Hormone-Receptor Complexes

Receptors are *allosteric proteins* in the sense that their function is regulated by a *ligand* which intervenes at the level of a binding site *distinct from the effector site*. Hormones are *allosteric regulators* of receptor function. Since several allosteric proteins described initially display a *cooperative* mechanism of action[71,72], this question was also posed for receptors (this differs from the *cooperative* binding of the ligand examined in the previous section). There are very few data available, however. No cooperativity has been described in the modulation of adenylate cyclase activity by hormones. However, it has been suggested that receptor aggregation at partial binding site occupation could result in the maximum effect obtainable with all sites being stimulated separately[73]. Studies with intracellular receptors are difficult, since no cell-free system exists permitting the detailed study of the initial response to steroid or thyroid hormones. However, steroid antagonists (Fig. I.132) appear to stabilize an inactive form of the receptor, suggesting[74] that the receptor may exist in two conformations, active and inactive. A molecular model for the steroid receptor which does not imply cooperativity is discussed on p. 118. This model may explain the double activity of many derivatives which are both partially agonistic and antagonistic (Fig. I.129). In any case, it may be overly simplistic to propose single agonistic or antagonistic forms for the receptor (Fig. 1.133). It is conceivable that there may be subsets of hormone and antihormone-receptor complexes, which would result in subtle functional differences that are not yet predictable physicochemically.

71. Monod, J., Wyman, J. and Changeux, J.P. (1965) On the nature of allosteric transitions. *J. Mol. Biol.* 12: 88–118.
72. Koshland, D. E. Jr., Nemethy, G. and Filmer, D. (1966) Comparison of experimental binding data and theoretical models in proteins containing subunits. *Biochemistry* 5: 365–385.
73. Levitzki, A. (1974) Negative co-operativity in clustered receptors as a possible basis for membrane action. *J. Theor. Biol.* 44: 367–373.
74. Samuels, H.H. and Tomkins, G.M. (1970) Relation of steroid structure to enzyme induction in hepatoma tissue culture cells. *J. Mol. Biol.* 52: 57–74.

4.6 Spare Receptors

Pharmacologists have discussed the respective roles of the rate of receptor occupancy and the number of sites occupied as determinants of drug activities. It has been shown that maximal pharmacological responses may be obtained when only a fraction of all sites are

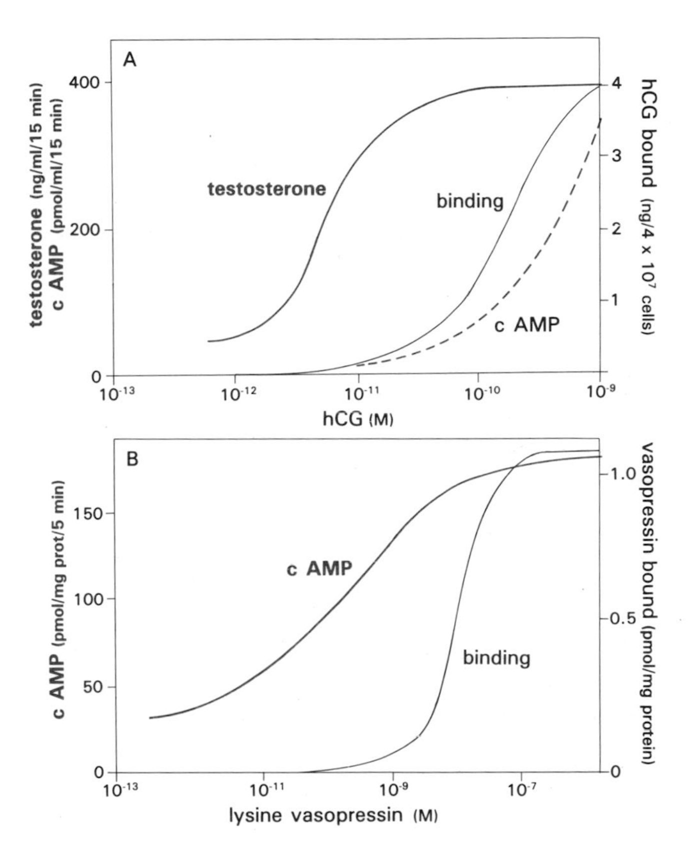

Figure I.63 Receptor occupancy, stimulation of adenylate cyclase, biological response, and the concept of spare receptors. **A.** The release of testosterone from rat Leydig cells occurs at a much lower concentration of hCG than either the increase in cAMP or binding to the gonadotropin receptor. (Catt, K. J. and Dufau, M. L. (1973) Interactions of LH and hCG with testicular gonadotropin receptors. In: *Receptors for Reproductive Hormones. Advances in Experimental Medicine and Biology* (O'Malley, B. W. and Means, A. R., eds), Academic Press, N.Y., vol. 36, pp. 379–418). **B.** Cyclic AMP is increased by vasopressin, in a membrane fraction obtained from pig kidney cells, at a concentration lower than that required for hormone binding. Note that, in both examples, the near maximal effect is obtained at low receptor occupancy. (Jard, S., Butlan, D. and Roy, C. (1977) The vasopressin-sensitive adenylate cyclase from mammalian kidney: mechanisms of activation and regulation of hormonal responsiveness. In: *Hormones and Cell Regulation* (Dumont, J. and Nunez, J., eds), Elsevier, Amsterdam, vol. 1, pp. 15–30).

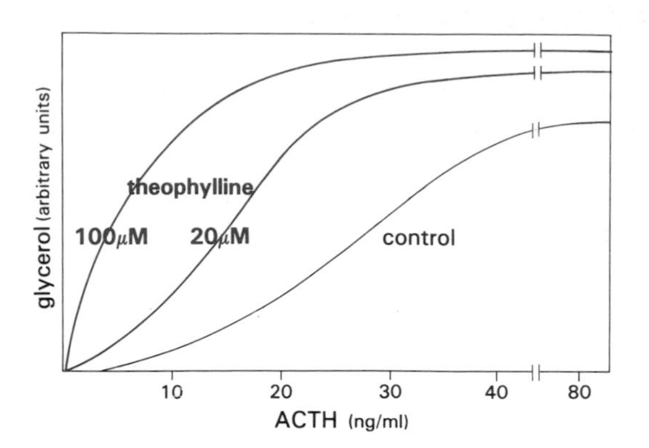

Figure I.64 Potentiation of a hormonal response and the apparent increase in spare receptors. In the presence of increasing concentrations of theophylline, a phosphodiesterase inhibitor, the lipolytic response to ACTH in rat adipocytes is enhanced. Since less hormone is required, it appears that fewer sites need be occupied to obtain the same effect. (Cuatrecasas, P., Hollenberg, M. D., Kwen-Jen Chang and Bennett, V. (1975) Hormone receptor complexes and their modulation of membrane function. *Rec. Prog. Horm. Res.* 31: 37–94).

occupied. Similar results have been observed with hormones. Sites unoccupied when a maximum effect is observed have been called *spare receptors*. Two examples are shown in Figure I.63. In Figure I.64, the increase in ACTH activity, and thus the apparent affinity of ACTH for its receptor, occurs in the presence of theophylline, which increases cAMP, the second messenger in the response. Fewer sites need be occupied at an identical ACTH concentration in order for the same biological response to be observed. Therefore, while the pharmacological effect is clear, a molecular interpretation involving higher affinity would be incorrect.

In all cases, the magnitude of the biological response is a function of the concentration of hormone-receptor complexes formed [HR], and thus of [H] and [R] (Fig. I.5). This means that it is related to the *total* number of receptor sites, and that *all* sites are involved in the response. In fact, the notion of spare receptors has *not* yet found real pathophysiological meaning.

Cryptic receptors have been described only in the case of membrane receptors. Discrepancies are probably due to the kinetics, the topology of biosynthesis, and to the maturation and recycling of receptors (Fig. 1 p. 225 and Fig. 3, p. 535).

Autoradiography of Hormone Receptors

Based on high affinity and selectivity of radioligand binding, autoradiographical studies have been performed both in vivo and in vitro. Intracellular receptors were first visualized in in vivo experiments, and were observed in the nuclei of target cells[75,76]

Figure I.65 Semiquantitative autoradiographic localization of the neurotensin receptor. A frozen section of rat brain was incubated with [^{125}I]neurotensin. The regions rich in receptor binding sites are the darkest. (Courtesy of Dr. W. Rostène, Hôpital St. Antoine, Paris).

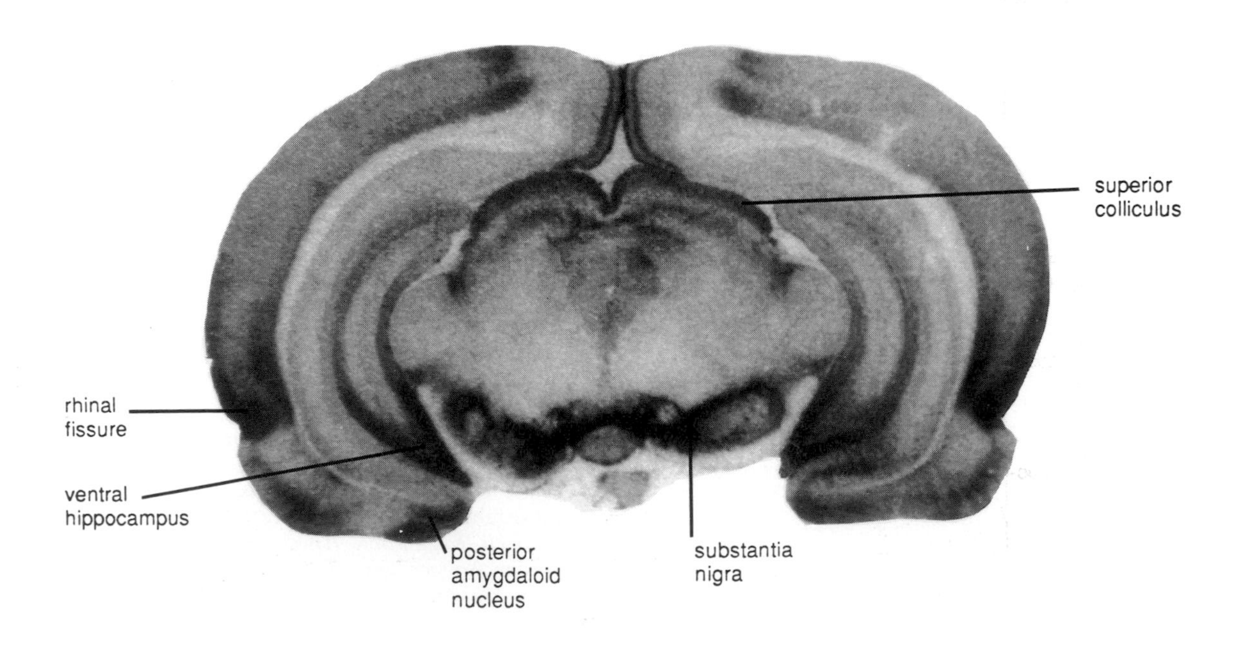

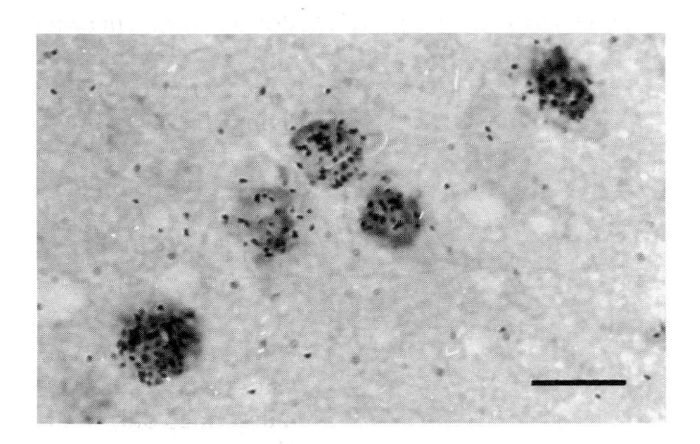

Figure I.66 Steroid hormone receptor identified by autoradiography and immunocytochemistry. [^{3}H]progestin was injected into a chick, and the pituitary was processed for autoradiography, with simultaneous detection of the progesterone receptor with an antiprogesterone receptor antibody, using the immunoperoxidase technique. (Stumpf, W. E., Gasc, J. M. and Baulieu, E. E. (1983) Progesterone receptors in pituitary and brain: combined autoradiography–immunohistochemistry with tritium-labelled ligand and receptor antibodies. *Mikroskopie* 40: 359–363).

(never in nucleoli) (Fig. I.66). Membrane receptors have also been demonstrated by autoradiography, using in vitro exposure of tissue slices to radioactive ligands (Figs. I.65 and I.79).

75. Edelman, I. S., Bogoroch, R. and Porter, G. A. (1963) On the mechanism of action of aldosterone on sodium transport: the role of protein synthesis. *Proc. Natl. Acad. Sci., USA* 50: 1169–1177.
76. Stumpf, W. E. and Sar, M. (1975) Autoradiographic techniques for localizing steroid hormones. In: *Methods in Enzymology. Hormone Action* (O'Malley, B. W. and Hardman, J. G., eds), Academic Press, New York, pp. 135–156.

4.7 DNA-Binding Receptors

Receptors for steroid and thyroid hormones interact directly with DNA. This is in contrast with receptors for other hormones which are located in the plasma membranc, where most if not all of the hormonal response is initiated. (Table I.1).

Receptors for steroid hormones were the first to be discovered. Tritiated estradiol of high specific activity was used at the end of the 1950s by Jensen and collaborators, who observed that, after injection, the labelled hormone was *concentrated* and retained specifically in the appropriate target organs[39] (Fig. I.67). The interaction of estradiol with the target tissues was *reversible*, the hormone decreasing in concentration slowly but completely in a few hours. These initial biochemical studies were confirmed by autoradiography, which demonstrated that radioactive hormone was found almost exclusively in the nuclei of target cells, a result later shown to be the same for all steroid hormones (Fig. I.66).

The subcellular distribution of steroid receptors has been a controversial matter. In the absence of hormone, most of the receptor can be extracted from

Figure I.67 The first evidence for a hormone receptor. Castrated or immature rats were injected with a physiological dose of [^{3}H]estradiol. Incorporation of radioactivity into various organs was measured as a function of time. Only the uterus and vagina, the main targets of the hormone, showed prolonged retention. It was demonstrated that estradiol is not metabolized in target organs. (Jensen, E. V. and Jacobson, H. I. (1962) Basic guides to the mechansim of estrogen action. *Rec. Prog. Horm. Res.* 18: 387–414).

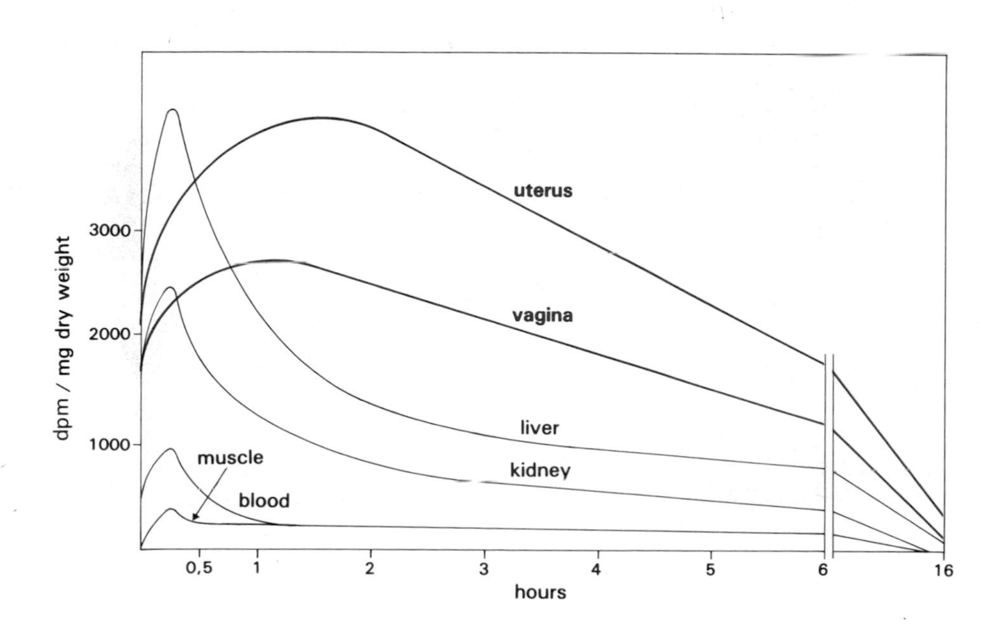

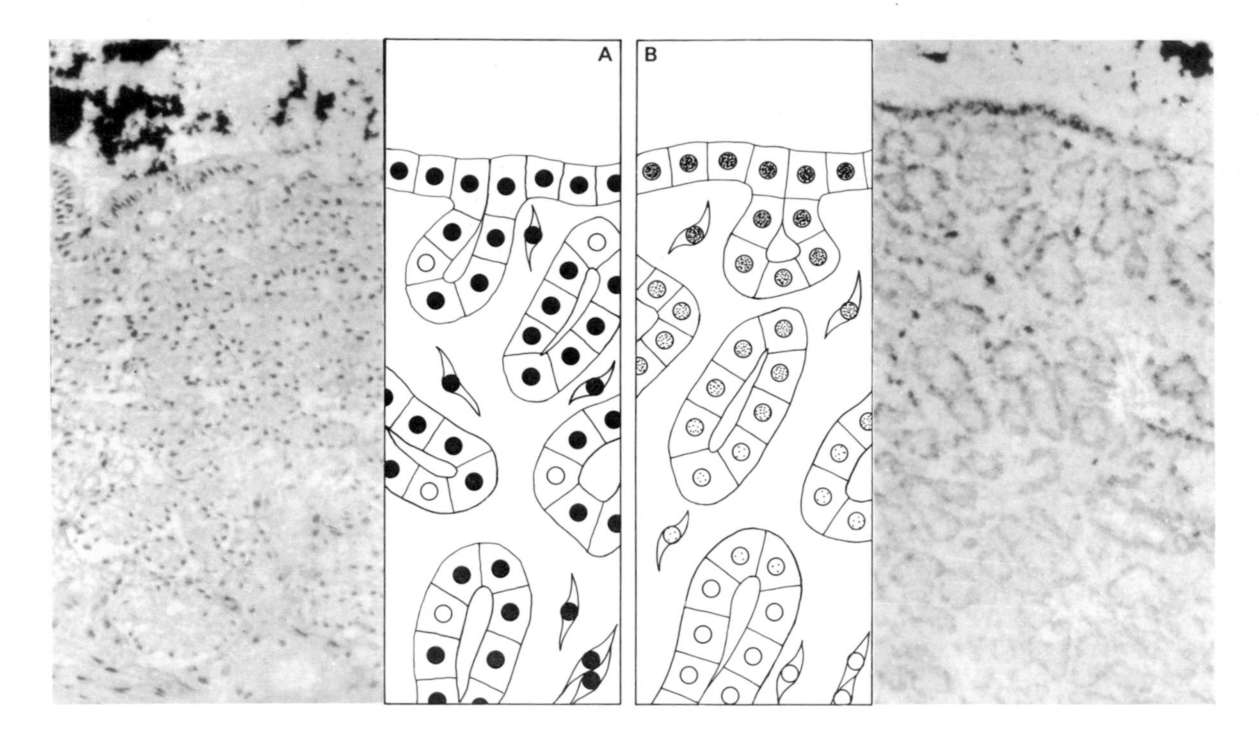

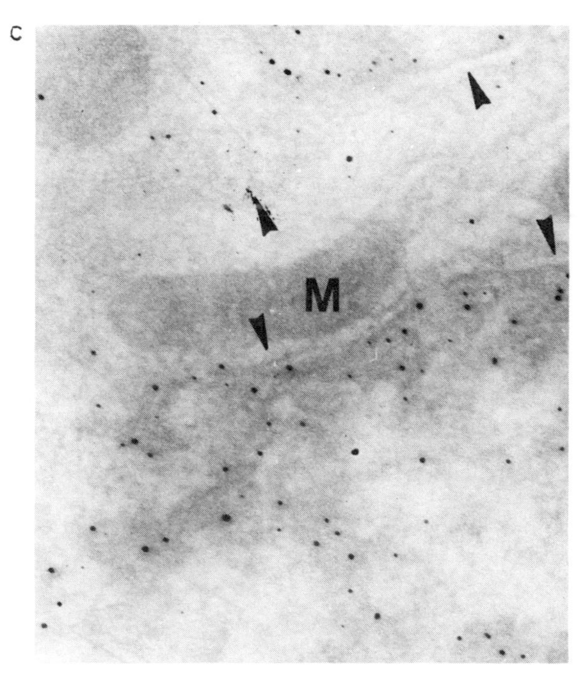

Figure I.68 The progesterone receptor is nuclear even in absence of hormone. Chick oviduct explants were incubated for a short period of time in a medium containing [^{3}H]progestin. Diffusion of the hormone from exterior to interior is indicated by the decreasing radioactive labelling in glandular epithelial cell nuclei (**B**, autoradiography). However, as was revealed with anti-PR antibodies (peroxidase technique), the receptor is already present in nuclei of the glandular cells which have not yet been reached by the progestin (**A**). The estrogen receptor is also found in nuclei of hormone-deprived target cells, while the glucocorticosteroid receptor is partly cytoplasmic. (Gasc, J. M., Ennis, B. W., Baulieu, E. E., Stumpf, W. E. (1983) Récepteur de la progestérone dans l'oviducte de poulet. Double révélation par immunohistochimie avec des anticorps antirécepteur et par autoradiographie à l'aide d'un progestagène tritié. *C.R. Acad. Sci. Paris* 297: 975–977). **C**. An electron micrograph showing the ultrastructural localization of the PR in an epithelial cell of an untreated chick oviduct. The labelling is most intense on condensed chromatin. M, nuclear membrane. (Courtesy of Dr. J. Isola, University of Tampere, Finland). (Isola, J. (1987) The effect of progesterone on localization of progesterone receptor in the nuclei of chick oviduct cells. *Cell Tiss. Res.* 249: 317–323).

target cells with a low concentration salt buffer, suggesting that the receptor is cytoplasmic. Indeed, if the tissue has been exposed previously to steroid hormone, the corresponding receptor is no longer obtained in the soluble fraction of tissue extract (*cytosol*), but is found tightly bound in nuclei. This resulted in the concept of "receptor translocation", signifying that, after hormone binding, receptors move to the nucleus, from their original position in the cytoplasm. It is now recognized that the receptors for estradiol and progesterone, which can be detected immunohistochemically, are always located in the *nucleus* of responsive cells, whether the hormone is present or not (Fig. I.68). It is now accepted, therefore, that, in the absence of hormone, all or part of the steroid receptors are initially weakly bound in the nucleus, from which they are easily extractable ("cytosol receptor"), and that their affinity for nuclear constituents increases once they have bound the hormone ("nuclear receptor"). Thus, translocation from the cytoplasm to the nucleus is clearly not a limiting step in hormone action. In any case, the majority of steroid-receptor complexes accumulate in the nucleus, where they interact with DNA regardless of the previous distribution of receptor before exposure to steroid. Thyroid hormone and calcitriol receptors are nuclear, and no cytoplasmic form has been demonstrated.

In all species and target organs, the cytosol estrogen (ER), progesterone (PR), androgen (AR), glucocorticosteroid (GR) and mineralocorticosteroid (MR) receptors have a molecular weight of approximately 300,000 and a sedimentation coefficient of 8 to 10 S when they are prepared in the absence of corresponding ligand. This large form is usually referred as the "*8 S*" receptor (Fig. I.69). It is an hetero-oligomer which includes one or two receptor molecules and two molecules of non-steroid binding *heat-shock* (or *stress*) protein, MW=90,000 (*hsp 90*)[77,78]. This particular hsp is evolutionarily highly conserved, is phosphorylated, and is found mostly in the cytoplasm. It is a constituent of most cells, and the production of it increases with heat or other forms of insult. A negatively charged sequence of the molecule may bind to the positively charged DNA binding site of receptors (Fig I.98), maintaining them in an inactive state. When hormone binds to the receptor, it provokes the release of hsp 90 (Fig. I.70). Consequently, the receptor molecule is *transformed* from the 8 S form to a smaller form. The latter has a sedimentation coefficient of ~4 S and affinity for non-specific DNA and hormone response element (HRE) DNA (p. 92), justifying the term "acidophilic activation"[79]. Indeed, hormone-receptor complexes are thus not only transformed, but *activated.* A hormone-dependent increase in receptor affinity for the hormone itself is also observed.

77. Joab, I., Radanyi, C., Renoir, J. M., Buchou, T., Catelli, M. G., Binart, N., Mester, J. and Baulieu, E. E. (1984) Immunological evidence for a common non hormone-binding component in "non-transformed" chick oviduct receptors of four steroid hormones. *Nature* 308: 850–853.
78. Catelli, M. G., Binart, N., Jung-Testas, I., Renoir, J. M., Baulieu, E. E., Feramisco, J. R. and Welch, W. J. (1985) The common 90-kd protein component of non-transformed "8S" steroid receptors is a heat-shock protein. *EMBO J.* 4: 3131–3135.
79. Milgrom, F., Atger, M. and Baulieu, E. E. (1973) Acidophilic activation of steroid hormone receptors. *Biochemistry* 12: 5198–5205.

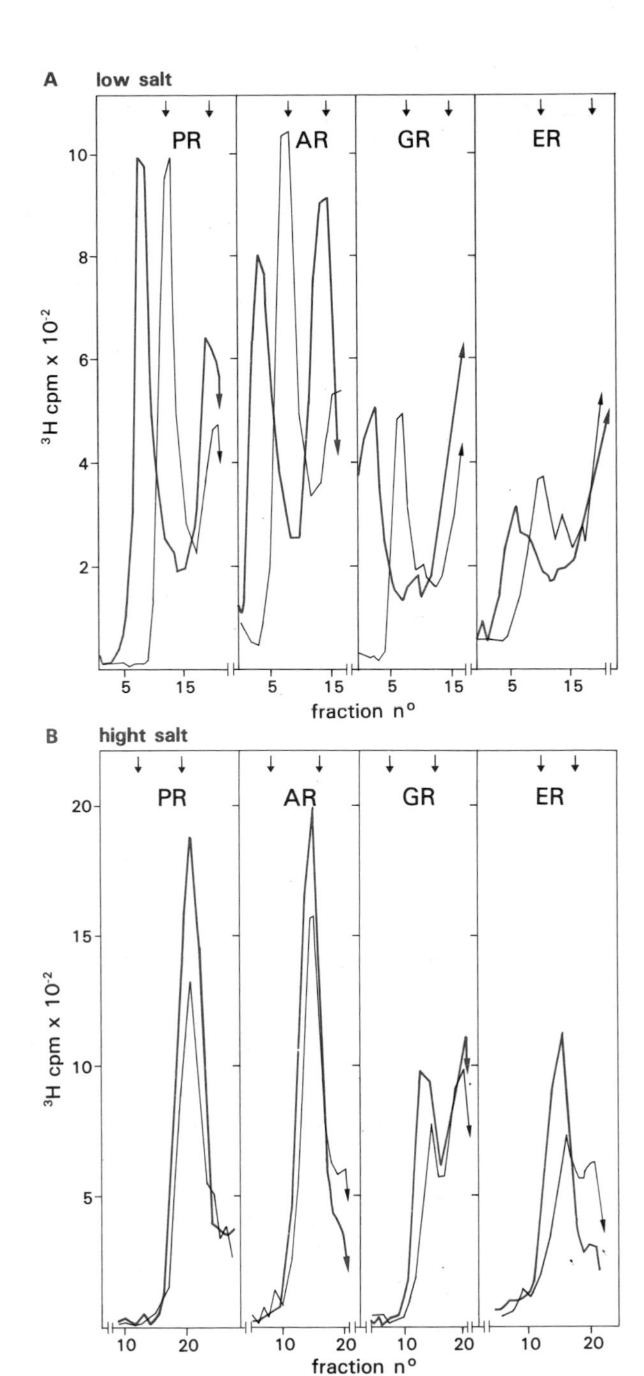

Figure I.69 The native (non-activated, non-transformed) 8 S form of all steroid receptors contains hsp 90. **A.** Chick oviduct receptors for progesterone, androgen, glucocorticosteroid and estrogen all sediment at approximately 8 S in low salt gradient ultracentrifugation experiments (black line). They all react with a monoclonal antibody directed against hsp 90, thus becoming heavier (red line). **B.** The receptor molecules themselves (~ 4 S), obtained after high salt treatment, do not interact with the monoclonal antibody.

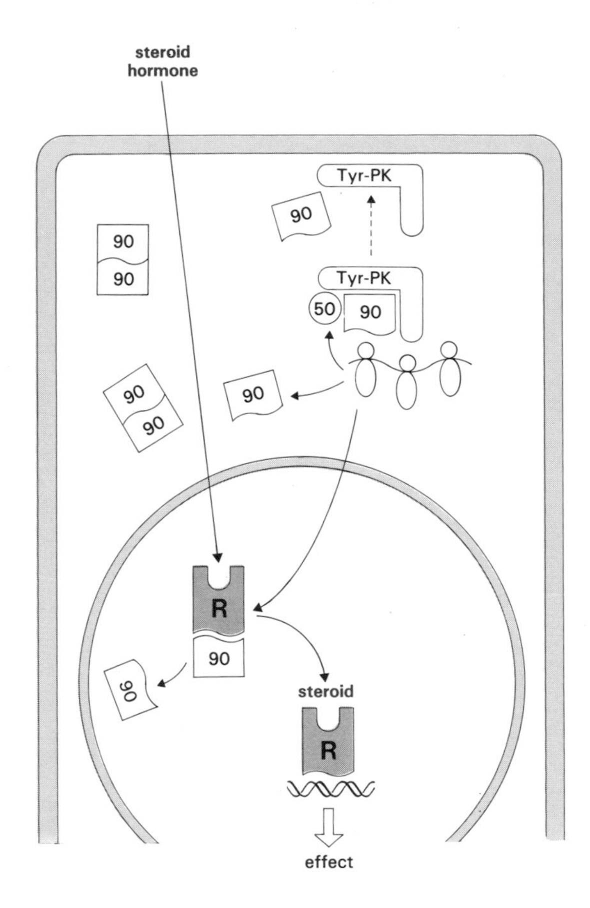

Figure I.70 Steroid hormone receptor (R) and hsp 90. In this highly schematized representation of a target cell, the entering hormone (H) diffuses to the nucleus, where it binds to the receptor. The latter is transformed, hsp 90 is released, and the HR complexes bind to specific response segments (HREs) of the DNA (p. 94). Note that R and hsp 90, both synthesized on polyribosomes in the cytoplasm, move to the nucleus, the majority of hsp 90 remaining cytoplasmic. Hsp 90 (and another 50 kDa protein) can interact with the Tyr–PK product of *src* or similar oncogenes before its attachment and activation at the membrane level. The stoichiometry of R-hsp 90 is arbitrary. Free hsp 90 is found mostly in the form of dimers. Other interactions of receptor and chromatin elements are not represented. (Baulieu, E. E. (1987) Steroid hormone antagonists at the receptor level. A role for the heat-shock protein MW 90,000 (hsp 90) *J. Cell. Biochem.* 35; 161–174).

Complete sequencing of cloned cDNAs in several species and in the human has revealed similar primary structures for all steroid hormone receptors and the receptors of calcitriol, the active metabolite of vitamin D (DR), thyroid hormones (TR), and retinoic acid (RAR) (Fig. I.71). Actually, TR is the product of the proto-oncogene c-*erb*-A (p. 124), confirming the high homology found between GR and v-*erb*-A[80]. Thus,

Figure I.71 Primary structure of DNA-binding hormone ▷ receptors. A "consensus receptor" has been represented. Homology between receptors for the same hormone in different species, and between receptors for different hormones in the same species, are indicated. 'DNA' stands for the putative DNA binding domain (DBD), and 'hormone', for the ligand binding domain (LBD) which contains a conserved 20 aa segment. (A, B, C, D, E and F refer to the nomenclature proposed by Chambon for the ER. General review: Evans, R.M. (1988) *Science* 240: 889–895; Human receptors for glucocorticosteroids (hGR): Hollenberg, S. M. et al. (1985) *Nature* 318: 635–641; for estrogens (hER): Green, S., et al. (1986) *Nature* 320: 134–139 and Greene, G. L., et al. (1986) *Science* 231: 1150–1154; for progesterone (hPR): Misrahi, M., et al. (1987) *Biochem. Biophys. Res. Commun.* 143: 740–748; for mineralocorticosteroid (hMR): Arriza, J. L., et al. (1987) *Science* 237: 268–275; for thyroid hormones (TR): Weinberger, C., et al. (1986) *Nature* 324: 641–646 and Sap, J., et al. (1986) *Nature* 324: 635–640; for the calcitriol receptor (DR): Baker, A. R., et al. (1988) *Proc. Natl. Acad. Sci. USA* 85:3294–3298. For androgen (hAR), reviewed by Tilley, W., et al. (1989) *Proc. Natl. Acad. Sci., USA* 86:327–331, different sizes have been found by Chang, C., et al. (1988) *Science* 240:324–326; Faber, P.W. et al. (1989) *Cell* 61:257–262; Lubahn, D.B., et al. (1988) *Mol. Endrocrinol.* 2:1265–1275; Tilley, W., et al. and by Govindan, M.V. (personal communication). The human retinoic acid receptor (RAR) (not shown) is 432aa long and organized similarly. (Petkovitch, M., et al. (1987) *Nature* 330:444–450; and Giguere, V., et al. (1987) *Nature* 330:624–629). HAP is a gene encoding the putative receptor of an unknown ligand; it has much homology with RAR, and is encoded in DNA immediately downstream from the insertion site of the hepatitis virus in human hepatoma. Not expressed in normal liver (contrary to many other tissues), it may be related to hepatocellular carcinogenesis. (de Thé, H., et al. (1987) *Nature* 330: 667–670).

the family of steroid and thyroid receptors has the potential to include enhancer-binding oncogenes[81].

Sequence data and transfection experiments with receptor cDNAs (Fig. I.72) have delineated several functional regions. All receptors are characterized by a Cys and Lys-Arg rich sequence defining a putative *DNA-binding domain (DBD)* of ~70 aa, approximately 300 aa away from the C-terminal extremity of the molecule. The overall pattern of amino acid sequences is highly conserved (Fig. I.73), allowing the formation of two *fingers*, each stabilized by a Zn^{2+} ion interacting tetrahydrally with four cysteines. Such "Zn fingers" may permit intervening amino acids to

80. Weinberger, C., Hollenberg, S. M., Rosenfeld, M. G. and Evans, R. M. (1985) Domain structure of human glucocorticoid receptor and its relationship to the v-*erb*-A oncogene product. *Nature* 318: 670–672.
81. Green, S. and Chambon, P. (1986) A superfamily of potentially oncogenic hormone receptors. *Nature* 324: 615–617.

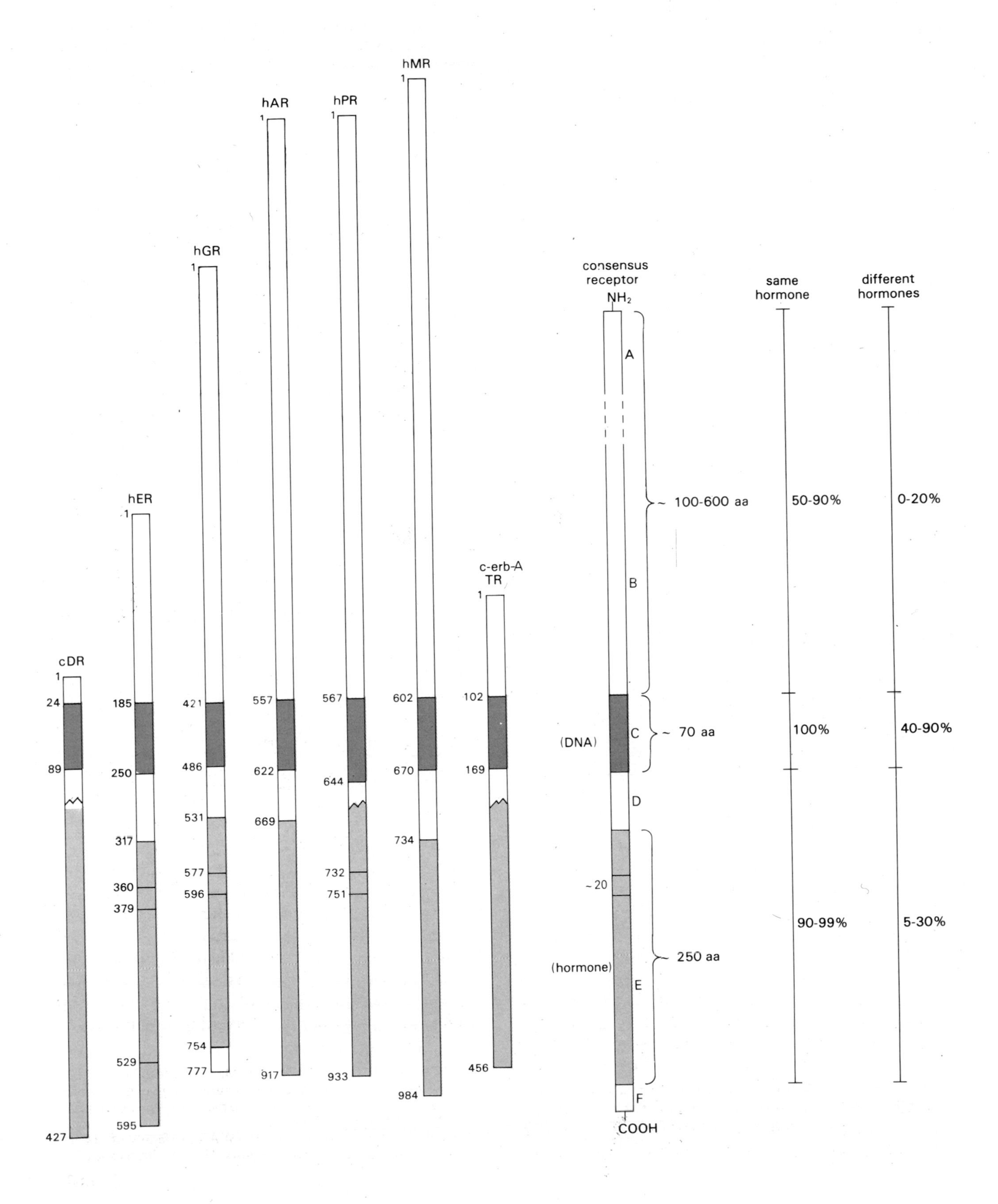

cDR
hER
hGR
hAR
hPR
hMR
c-erb-A
TR
1
24
89
427
185
250
317
360
379
529
595
421
486
531
577
596
754
777
557
622
669
917
567
644
732
751
933
602
670
734
984
102
169
456
consensus
receptor
NH2
A
B
C
D
E
F
COOH
(DNA)
(hormone)
~20
~ 100-600 aa
~ 70 aa
250 aa
same
hormone
different
hormones
50-90%
100%
90-99%
0-20%
40-90%
5-30%

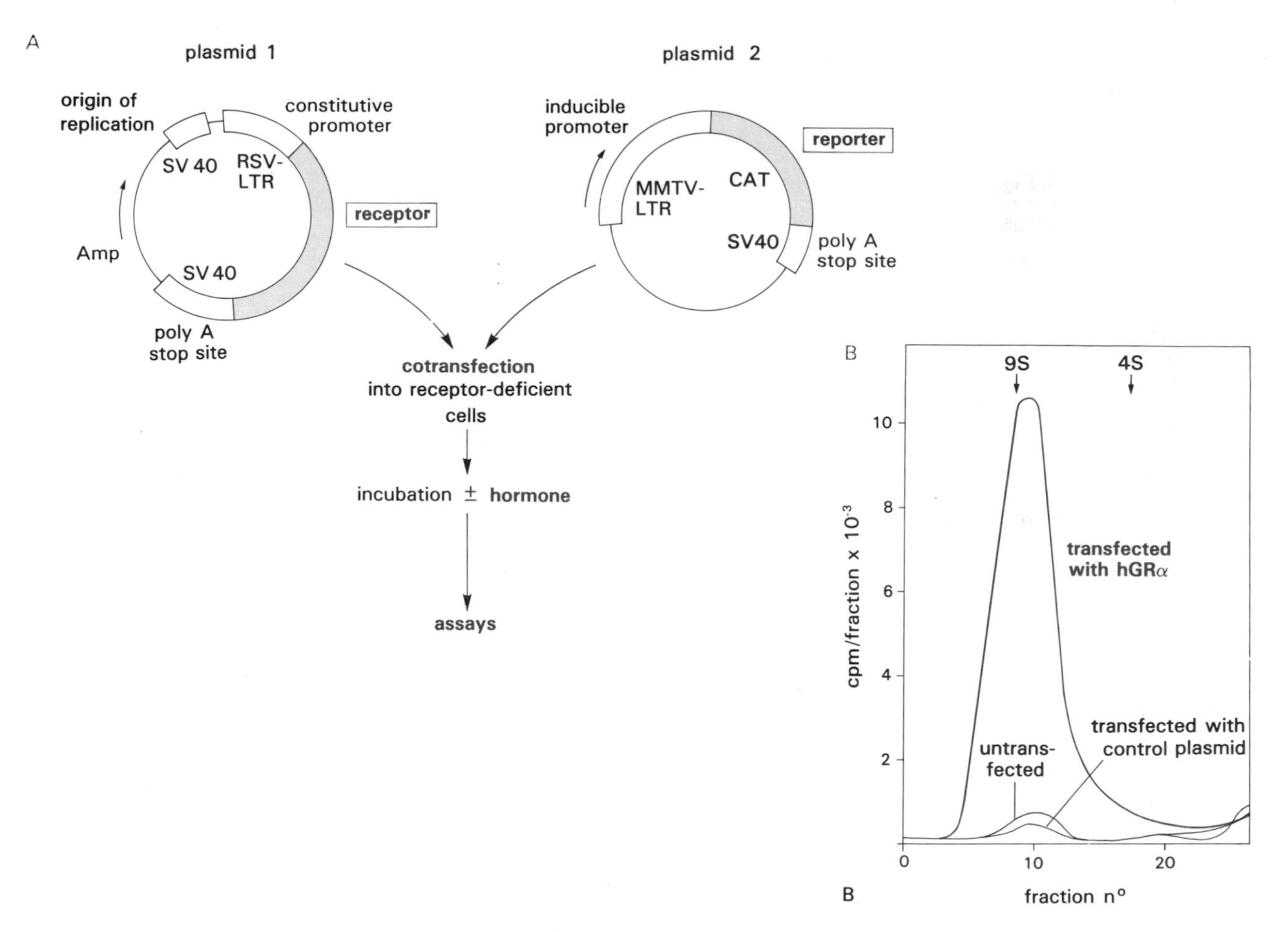

Figure I.72 Transfection experiments used in the study of receptor-regulated gene interaction. **A.** In this functional assay, an *expression vector* containing *receptor* cDNA (plasmid 1) is cotransfected, into a receptor-deficient cell (e.g., COS-1 or CV-1, which are monkey kidney cells), with a *reporter gene* preceded by an inducible promoter (plasmid 2). The cells are cultured in the absence or in the presence of *hormone*, and the receptor can then be studied (hormone binding, size, etc.) (see **B**), as well as the induction of a response, e.g., using chloramphenicol acetyltransferase (CAT) activity. RSV, Rous sarcoma virus; MMTV, mouse mammary tumor virus; Amp, gene for ampicillin resistance. (Giguere, V., Hollenberg, S. M., Rosenfeld, M. G. and Evans, R. M. (1986) Functional domains of the human glucocorticosteroid receptor. *Cell* 46: 645–652). **B.** Wild-type hGRα obtained after transfection of a plasmid containing cDNA (plasmid 1) into COS cells: formation of the 8–9 S complex. Measurement of endogenous GR in untransfected cells and in cells transfected with the plasmid devoid of receptor cDNA served as controls. (Pratt, W. B., Jolly, D. J., Pratt, D. V., Schweizer-Groyer, G., Catelli, M. G., Evans, R. M. and Baulieu, E. E. (1988) A region in the steroid binding domain determines formation of the non-DNA binding, 9 S glucocorticoid receptor complex. *J. Biol. Chem.* 263: 267–273).

make specific contacts with DNA[82]; a finger may interact with five base pairs. Similar Zn fingers have been found in the Xenopus laevis 5 S transcription factor (TFIIIA)[83] and in several other DNA-binding proteins. The intracellular receptors thus belong to a *family of transcriptional regulatory factors*, probably diversified from a superfamily of related *DNA-binding proteins*, that are characterized by their metal finger-type structures. Other subfamilies can be distinguished based upon the number and structure of their fingers (2 Cys-2 His, in contrast to 4 Cys in receptors). Interestingly, with the ER, GR and PR cloned thus far, this Cys-rich region is 100% conserved in receptors of the same hormone in different species. Paired comparisons between the receptors for different ligands (and v-*erb* A) reveal less homology (only 50% in most cases). These results, as well as those of hybrid experiments (Fig. I.90 and p. 87), suggest that the highly conserved Cys-rich sequence provides both the scaffolding and the amino acid specificity for HRE-receptor interactions (Fig. I.73). In situ mutagenesis experiments (Fig. I.74) have indi-

82. Scheidereit, C., Westphal, H. M., Carlson, C., Bosshard, H. and Beato, M. (1986) Molecular model of the interaction between the glucocorticoid receptor and the regulatory elements of inducible genes. *DNA* 5: 383–391.
83. Miller, J., McLachlan, A. D. and Klug, A. (1985) Repetitive zinc-binding domains in the protein transcription factor IIIA from Xenopus oocytes. *EMBO J.* 4: 1609–1614.

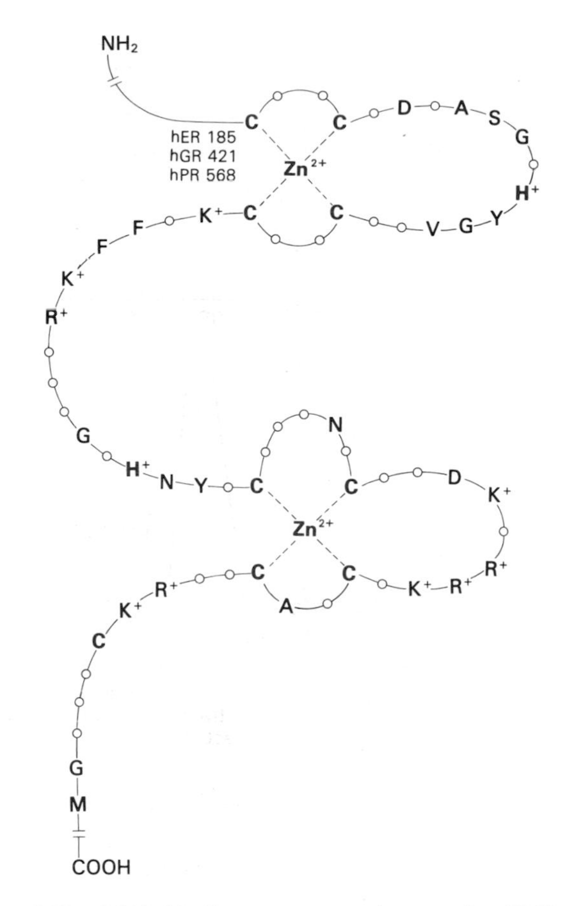

Figure I.73 DNA binding receptors: the putative DNA binding domain. The single letter code has been used to indicate amino acids. Amino acid positive charges are indicated. The invariant nine cysteines (C) and two histidines (H) are represented in red. Note the other conserved amino acids, particularly the positively charged amino acids in the second finger (arginine R, lysine K) (see references in Fig. I.71; Sabbah, M., Redeuilh, G., Secco, C. and Baulieu, E. E. (1987) The binding activity of estrogen receptor to DNA and heat shock protein (Mr 90,000) is dependent on receptor bound metal. *J. Biol. Chem.* 262: 8631–8635). A structural model is discussed in: Berg, J. M. (1988) Proposed structure for the zinc-binding domains from transcription factor IIIA and related proteins. *Proc. Natl. Acad. Sci., USA* 85: 99–102).

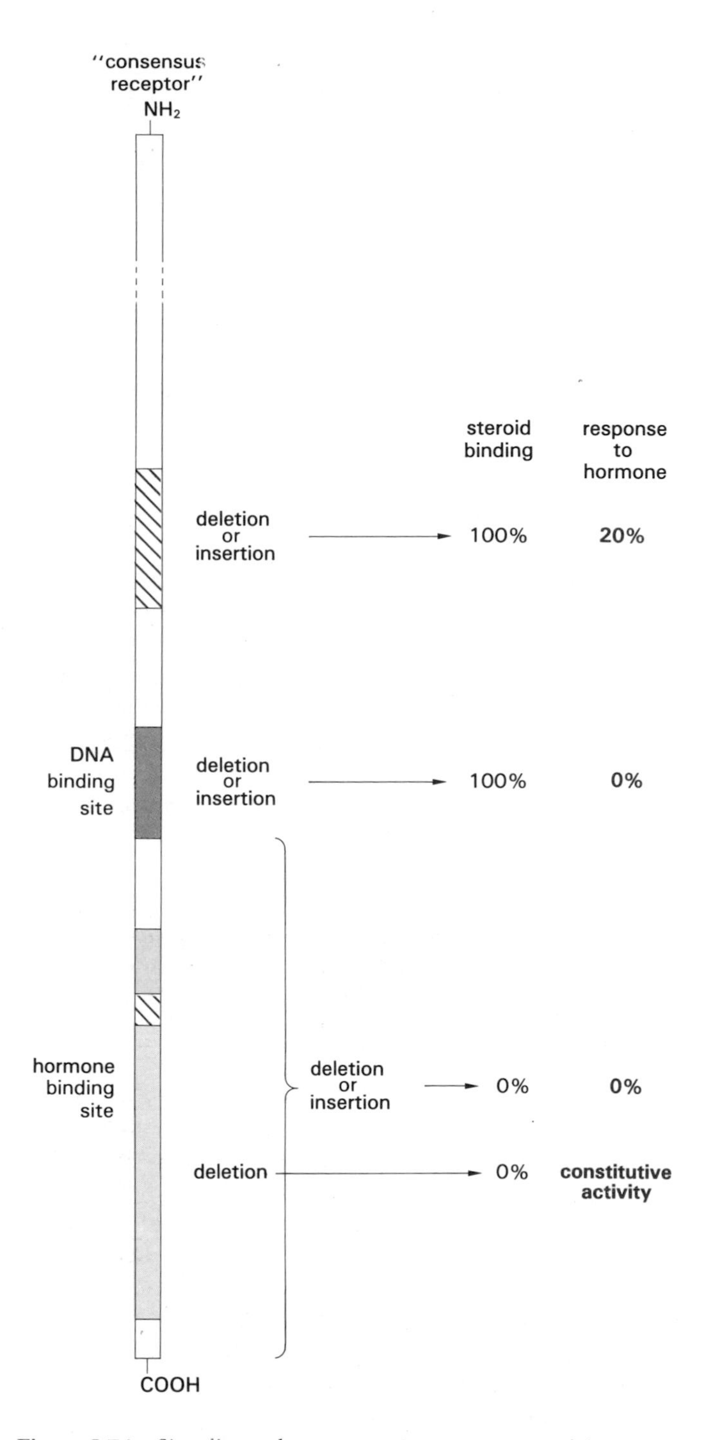

Figure I.74 Site-directed mutagenesis experiments delineate functional regions of steroid receptors. The mutated cDNAs were transfected and expressed in receptor-deficient cells (Fig. I.72). Steroid binding and response are expressed as a percent of that of the wild-type receptor. (Hollenberg, S. M., Giguere, V., Segui, P. and Evans, R. M. (1987) Colocalization of DNA-binding and transcriptional activation functions in the human glucocorticoid receptor. *Cell* 49: 39–46; Kumar, V., Green, S., Staub, A. and Chambon, P. (1986) Localisation of the oestradiol-binding and putative DNA-binding domains of the human oestrogen receptor. *EMBO J.* 5: 2231–2236).

cated the crucial nature of this DNA-binding sequence for the mediation of hormone action, although the putative binding site does not define, per se, a domain actually involved in the activation of transcription.

In the C-terminal portion of the receptor molecule, another characteristic sequence of ~250 aa is involved in *ligand binding* (LBD). This domain begins ~50 aa after the DBD, and is rich in hydrophobic amino acids. Its predicted tertiary structure suggests a combination of α helices and β strands that are able to form a *hydrophobic pocket* which can bind steroid, thyroid and retinoic ligands. For the same hormone, differences between species are limited (homology ranging between 70 and 95.5%), which concords with known binding specificities (Table I.11). In contrast, homology is only 10 to 20% when different ligand receptors are compared.

All other regions of the receptor molecules are much less well conserved. The *N-terminal region* is primarily responsible for variations in size. The sequences responsible for other specific functions, particularly through interaction with other regulatory chromatin proteins, such as transcription factors, are not yet defined. The primary sequence of the GR N-terminal region is strongly conserved among species, and the first ~ 400 aa are coded by a single exon which is far upstream (>30 kb) from the multiple exon/introns specifying the C-terminal region.

Receptor phosphorylation has been demonstrated, but its significance remains unknown; it may be involved in hormone and DNA-binding, receptor activation, or transcription efficiency. In any case, neither receptors nor hsp 90 have intrinsic protein kinase activity.

There are more homologies between PR, GR and MR, in both DNA and steroid-binding domains, than between any of these receptors and ER, confirming binding data with both steroid hormones and DNA[84]. The AR DNA-binding domain has a high degree of homology with the corresponding region of the PR. So far, the ER is the only one which is found as a dimer in the non-transformed 8 S hetero-oligomer, as after transformation ("5 S" form, $M_r \sim 130,000$), but how this dimer interacts with DNA is not yet known. There are two described hGRs: α of 777 aa which binds glucocorticosteroid, and β of 742 aa, which does not, differs at the C-terminal extremity, and is of undetermined function. There are several *c-erb*-A genes that encode distinct thyroid hormone receptors which are expressed differently in different tissues (TRβ in liver and TRα in brain). The two RARs have more homology with TR than with other hormone receptors, and together with the TR receptor, they may constitute a subfamily having diverged from steroid hormone receptors early in evolution.

4.8 Membrane Receptors

(Figs. I.75 and I.76)

Membrane receptors were first demonstrated using the backward approach (p. 54) for the study of the mechanism of action of epinephrine and glucagon. Sutherland observed adenylate cyclase activation in a plasma membrane fraction, and thus concluded that hormone receptors were present. Later, with purified membrane preparations and radioactive peptides, receptors were characterized biochemically. In order to ensure that the measured interaction is of physiological importance, the preparation of ligands with *high specific radioactivity* (labelled with tritium or iodine) which retain *full biological activity* is fundamental. Purification has been difficult, not only because receptors are limited in number and are susceptible to proteolysis, but also because of the necessity of using detergents which interfere with many biochemical techniques.

Three distinct regions can be clearly identified in membrane receptors, i.e. extracellular, membrane-spanning and cytoplasmic (intracellular) regions. The N-terminal portion of the molecules forms the *extracellular region,* which always has potential *glycosylation sites*. Most are of the Asn-X-Ser/Thr type for N-

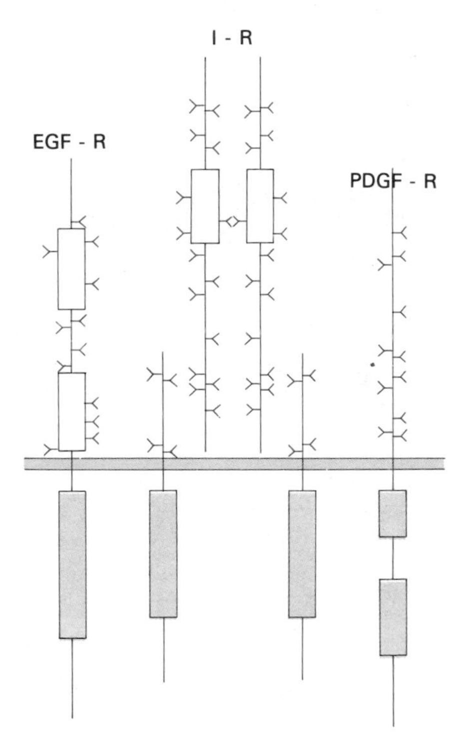

Figure I.75 Four classes of membrane hormone/neurotransmitter receptors: single membrane-spanning receptors without Tyr–PK activity (**A**) and with Tyr–PK activity (**B**) (Fig. I.78), and mutiple membrane-spanning receptors, i.e. single-chain, seven-helix coupled to G protein (**C**), and ligand-gated channels (**D**) (Fig. I.85).

84. Von der Ahe, D., Janich, S., Scheidereit, C., Renkawitz, R., Schütz, G. and Beato, G. (1985) Glucocorticoid and progesterone receptors bind to the same sites in two hormonally regulated promoters. *Nature* 313: 706–709.

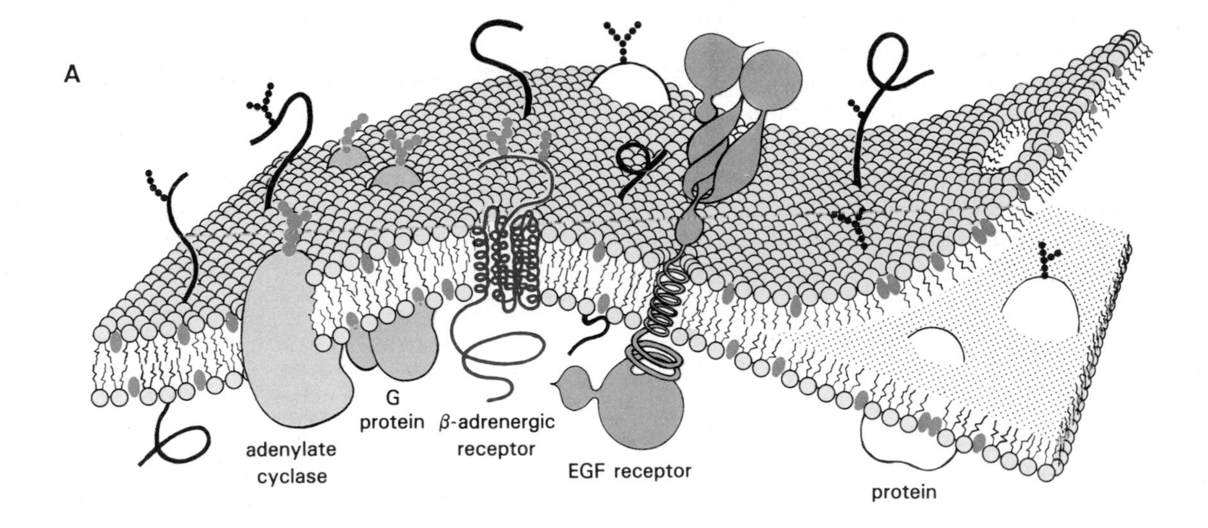

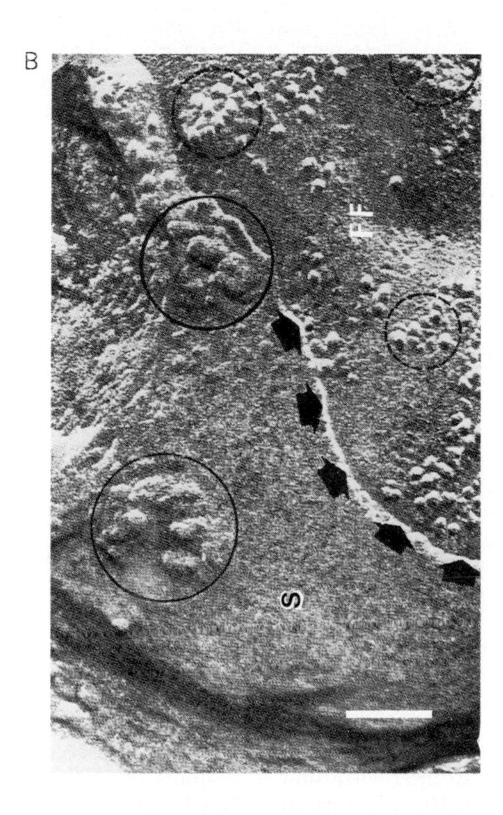

Figure I.76 Membrane receptors. **A.** Composite diagram of the fluid mosaic plasma membrane. (Singer, S. J. and Nicolson, G. L. (1972) The fluid mosaic model of the structure of cell membranes. *Science* 175: 720–731). The lipid bilayer is composed of phospholipids, with a number of cholesterol molecules (note the glycolipids in the external layer). Many intrinsic proteins traverse one or both layers, and most are glycosylated outside of the cell. Hormone receptors are represented in red. The EGF-R is typical of a receptor with Tyr–PK activity in its intracytoplasmic segment, and the ligand binding site is well out of the membrane plane. The β-adrenergic receptor has several transmembrane segments, and the ligand probably binds within this transmembrane complex. In addition, this receptor interacts with a G protein, itself connected to the enzyme adenylate cyclase. In the right-hand portion of the figure, the membrane has been opened, as occurs with freeze-fracture. **B.** *Electron micrograph*. The limit between the external surface (FF, freeze-fracture surface) and the fracture surface (S, external surface) is indicated by arrows. The black circles surround ferritin molecules, revealing the hormone receptors, while the dotted circle indicates particle aggregates on the fracture surface. Bar=0.1 μm. (Orci, L., Rufener, C., Malaisse-Lagae, F., Blonbel, B., Amherdt, M., Bataille, D., Freychet, P. and Perrelet, A. (1975) A morphological approach to surface receptors in islet and liver cells. *Isr. J. Med. Sci.* 11: 639–655).

linked oligosaccharides (Fig. I.77). It is not known whether carbohydrates are involved in cellular antigenicity and/or whether they play a role in hormone binding. Many amino acids are *hydrophilic*, and in some cases (see the single membrane-spanning class of receptors, p. 78) this region is rich in *cysteines*, sometimes grouped in clusters, which may form, via S-S bonds, *rigid pockets* involved in hormone binding. Cysteines, either in clusters or isolated, may also be implicated in *linking* different parts of the same or different receptor chains.

Transmembrane segments consist of one or several sequences of ~20 to 25 aa (mostly *hydrophobic and uncharged*) which form a *helical* structure. In some cases, the charged amino acids bound immediately before or after the transmembrane segment(s) may facilitate the anchorage of receptors to the plasma membrane. Within the membrane, cysteines may be involved in holding several helices together. In this case, as in the case where the receptor consists of several membrane-spanning subunits (p. 79), a channel may be formed, permitting the transfer of ions between extra- and intracellular compartments. In multiple membrane-spanning receptors, the hormone molecule may actually be bound within the hydrophobic intramembrane section of the receptor.

Intracytoplasmic segments of the receptor are primarily involved in *effector* functions. Two classes of relatively hydrophilic intracytoplasmic receptor components have been identified, one having intrinsic *Tyr-PK* (tyrosine protein kinase) *activity*, and the other being coupled to a *G protein*. Specific differences between tyrosine kinase regions and between the sites which interact with G proteins probably account for differences in the activities of various receptors in each class. The intracytoplasmic component of the receptor also contains regulatory element(s), such as Tyr or Ser/Thr *phosphorylation* sites.

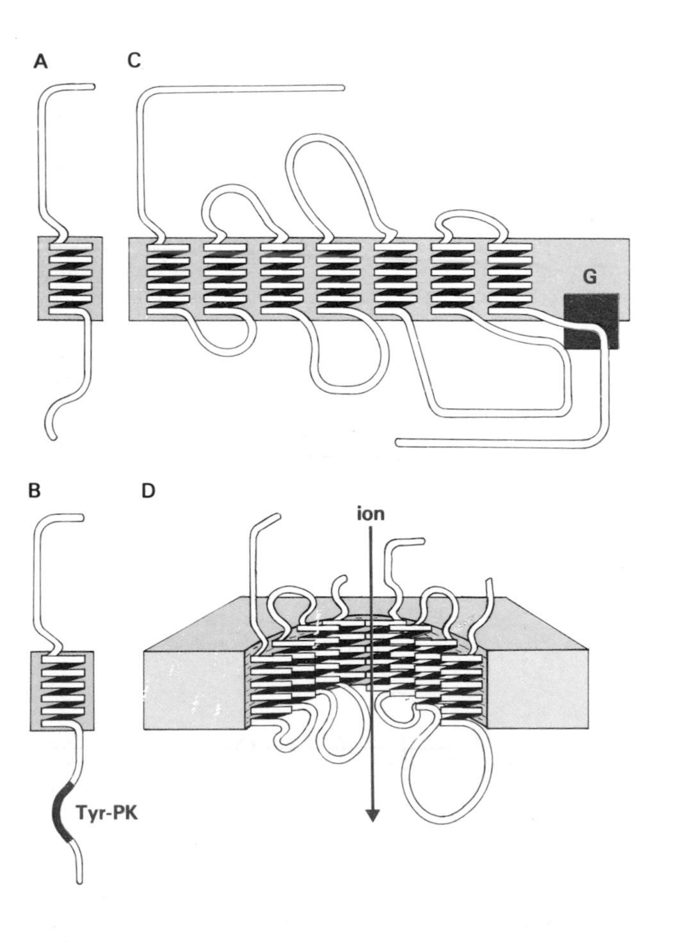

Figure I.77 Glycosylation sites in the extracellular regions of three membrane receptors (see also Fig. I.78).

Figure I.78 Single membrane-spanning receptors. The ▷ extracellular portion of these receptors contains the hormone binding site. Tyrosine protein kinase domains are shown in red, with insertions in the case of the PDGF-R and CSF-I-R. Dots represent cysteines, and cross-hatched boxes, the cysteine-rich domains. All N-terminal extremities are extracellular, with the exception of the transferrin receptor. All proteins are drawn to the same scale. The I-R and IGF-I-R consist of two S-S-linked chains. (EGF-R: Downward, J., et al. (1984) *Nature* 307: 521–527; IR: Czech, M.P. (1985) *Annu. Rev. Physiol.* 47: 357–381, Ebina, Y., et al. (1985) *Cell* 40: 747–758 and Ullrich, A., et al. (1985) *Nature* 313: 756–761; PDGF-R: Yarden, Y., et al. (1986) *Nature* 323: 226–232). There is another putative Tyr-PK receptor, encoded by the *eph* gene, which also has a cysteine-rich region (not shown). (Hirai, H., et al. (1987) *Science* 238: 1717–1720). The GH-R and PRL-R are represented (see Fig. III.6), as well as the nerve growth factor receptor (NGF-R), IGF-II receptor (IGF-II-R) and, for the sake of comparison, the transferrin receptor (T-R) (p. 83), the LDL-R (p. 533) and the EGF-precursor (pro-EGF or EGF-p). NGF-R (399 aa), the receptor of the first growth factor to be identified, includes four 40 aa repeats, with six Cys at conserved positions. It is a homodimer with neither Tyr-PK activity nor interaction with a G protein. (Johnson, D., et al. (1986) *Cell* 47: 545–554). The IGF-II-R is comprised of 2264 aa, of which only 164 are intracellular. It has 15 repeated sequences, 80% homology with a mannose 6-phosphate receptors (M6P-R), which transfer lysosomal enzymes from the Golgi to lysosomes. The type II fibronectin-like region participates in IGF-II binding (shown in black)). (Morgan, D.O., et al. (1987) *Nature* 329: 301–307). The T-R (760 aa), a transmembrane protein of 760 aa, is abundant in fast-growing cells. Constitutively phosphorylated and preclustered, the major function of the T-R is to specifically bind and internalize iron-loaded transferrin. Two moles of transferrin are bound per S-S-linked dimer, and thus each receptor molecule can internalize four iron atoms (1×10^6 atoms/min/cell with 10^5 receptors/cell). (McClelland, A., et al. (1984) *Cell* 39: 267–274). EGF-p (1217 aa) includes a 21 aa transmembrane segment, the sequence of EGF itself (53 aa) being extracellular. (Rall, L.B., et al. (1985) *Nature* 313: 228–231). It has eight extracellular cysteine-rich sequences and a segment (~ 150 aa) very homologous to part of the LDL-R. It would be striking if EGF-p were processed differently in different tissues, to produce either a hormone or a receptor (e.g., there is no processing in distal renal tubule cells in the mouse). The TGF-α precursor also has a putative transmembrane sequence in a similar position relative to the mature growth factor sequence.

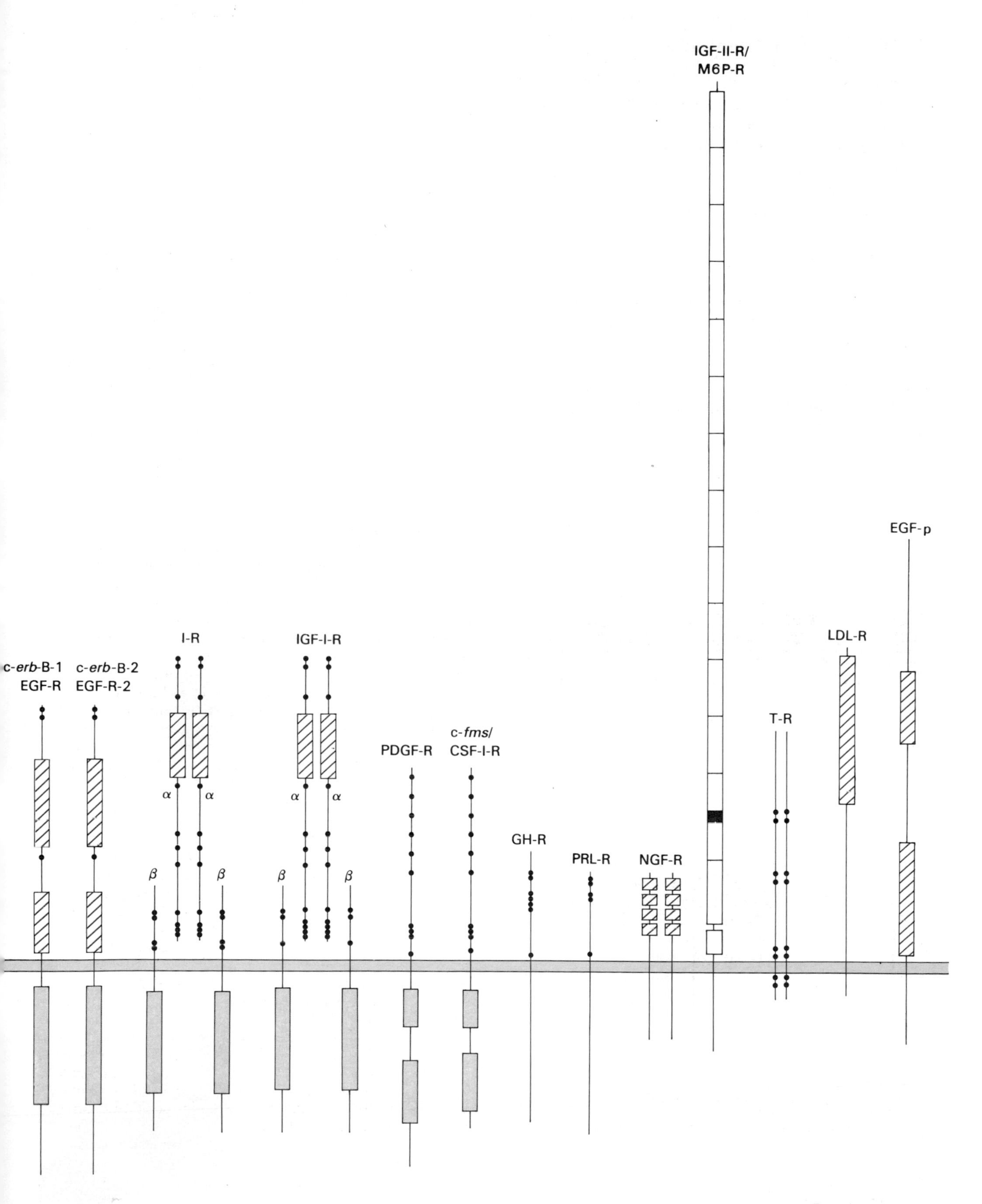

IGF-II-R/
M6P-R
EGF-p
LDL-R
I-R
IGF-I-R
c-erb-B-1
EGF-R
c-erb-B-2
EGF-R-2
T-R
c-fms/
CSF-I-R
PDGF-R
α
α
α
α
GH-R
PRL-R
NGF-R
β
β
β
β

Presently, two broad categories of membrane receptors have been defined: those which span the membrane *once*, and those which cross the membrane *several* times.

Single Membrane-Spanning Receptors
(Fig. I.78)

The receptors for epidermal growth factor (*EGF-R*), insulin (*I-R*) and platelet-derived growth factor (*PDGF-R*) are typical of the group of single membrane-spanning receptors which have Tyr-PK activity. The receptors for nerve growth factor (*NGF-R*), insulin-like growth factor II (*IGF-R-2*), growth hormone (*GH-R*), and prolactin (*PRL-R*) (Fig. III.6) have no Tyr-PK activity.

Single Membrane-Spanning Receptors with Tyr-PK Activity

The EGF-R was the first of the membrane-spanning receptors with Tyr-PK activity to be described (Fig. I.79). The very similar *EGF-II-R* does not bind EGF, and corresponds to an unknown growth factor; it is also the proto-oncogene of *neu* (Fig. I.137). The I-R is composed of two double-chained units linked by S-S bonds. The α and β chains of each unit are derived from a unique proprotein (Fig. XI.17), and only the β chain spans the membrane and has Tyr-PK activity. The insulin-like growth factor I receptor (*IGF-I-R*) is structurally very similar to the I-R. The Tyr-PK domains of single-chained PDGF-R and the related

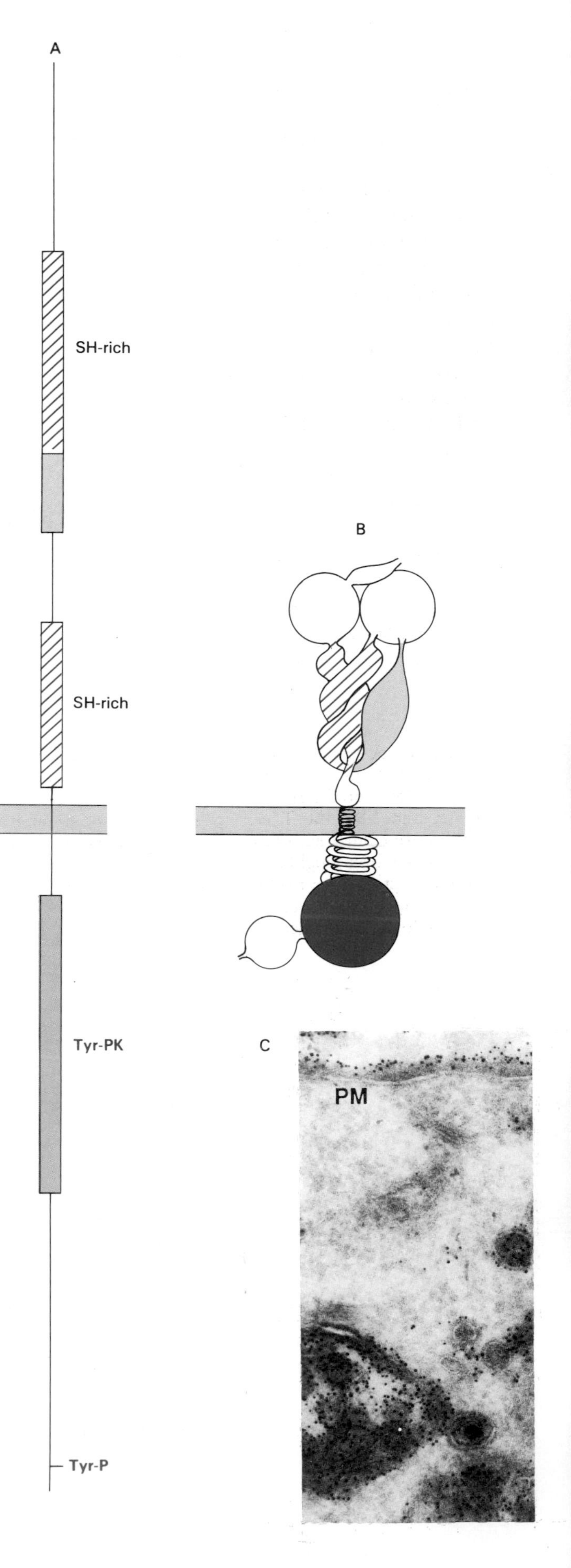

Figure I.79 EGF-receptor. **A.** The EGF-R is represented ▷ linearly (as in Fig. I.78). **B.** Conformational model provided by Dr. J. Garnier, Université Paris-Sud. (Fisleigh, R.V., Robson, B., Garnier, J. and Finn, P.W. (1987) Studies on rationales for an expert system approach to the interpretation of protein sequence data. *FEBS Lett.* 214: 219–225). **C.** Electron micrograph showing the localization of the EGF-R in a cryosection of human A431 cells by the immunogold technique. Note that part of the receptor is internalized, appearing in intracellular vesicles. PM, plasma membrane (Courtesy of Dr. J. Boonstra, State University, Utrecht; Boonstra, J., van Maurik, P., Libert, H.K., de Laat, S.W., Leunissen, J.L.M. and Verkley, A.J. (1986) Visualization of epidermal growth factor receptor in cryosections of cultured A431 cells by immuno-gold labeling. *Eur. J. Cell Biol.* 36: 209–216).

colony stimulating factor I receptor (*CSF-I-R*) are split into two parts.

The Tyr-PK regions of all of these receptors are comprised of ~250 aa and show a high degree of homology with Tyr-PKs of the *src* oncogene product family. There are ~50 aa between the membrane and the ATP-binding site (p. 106), including Thr-654 of the EGF-R, which can be phosphorylated by PK-C (p. 107). Following the C-terminal region of the Tyr-PK site, there is at least one Tyr which is preferentially autophosphorylated, playing an inportant autocatalytic role in the activation of the receptor (p. 108). The insertion of ~100 and 70 aa in the Tyr-PK site of the PDGF-R and CSF-I-R, respectively, is critical for the specificity of receptor enzymatic activity.

The extracellular regions of these receptors have the common feature of being rich in cysteines. The cysteine-rich regions, found also in other classes of membrane receptors (Fig. I.80), are probably involved in the binding of hormones which themselves have intramolecular S-S bonds (three intramolecular disulfide bonds for EGF (p. 219); insulin, Fig. I.22; and PDGF). In the EGF-R, the two cysteine-rich regions cooperate to form a single binding site. For the I-R, it is still not clear whether the heterotetramer receptor binds one or two molecules of insulin. Although the I-R and IGF-I-R are very similar, the binding properties of insulin and IGF-I are different (Table I.12). There are no cysteine clusters in the PDGF-R and CSF-I-R, but isolated cysteines in these receptors are in the same positions as in the EGF-R and I-R.

Interestingly, the PDGF-R and GH-R are covalently linked to ubiquitin, a 8.5 kDa hsp implicated in protein degradation.

Mutiple Membrane-Spanning Receptors

Mutiple membrane-spanning receptors belong to a large group of intrinsic membrane proteins that control cell exitability and which span the plasma membrane several times. This class of proteins also includes ion pumps (such as Na^+-K^+ ATPase (Fig. I.84), H^+-K^+ ATPase, and Ca^{2+}-ATPase. Single-chain, seven-helix receptors for adrenergic substances (*β-A-R*), acetylcholine (muscarinic acetylcholine receptor,

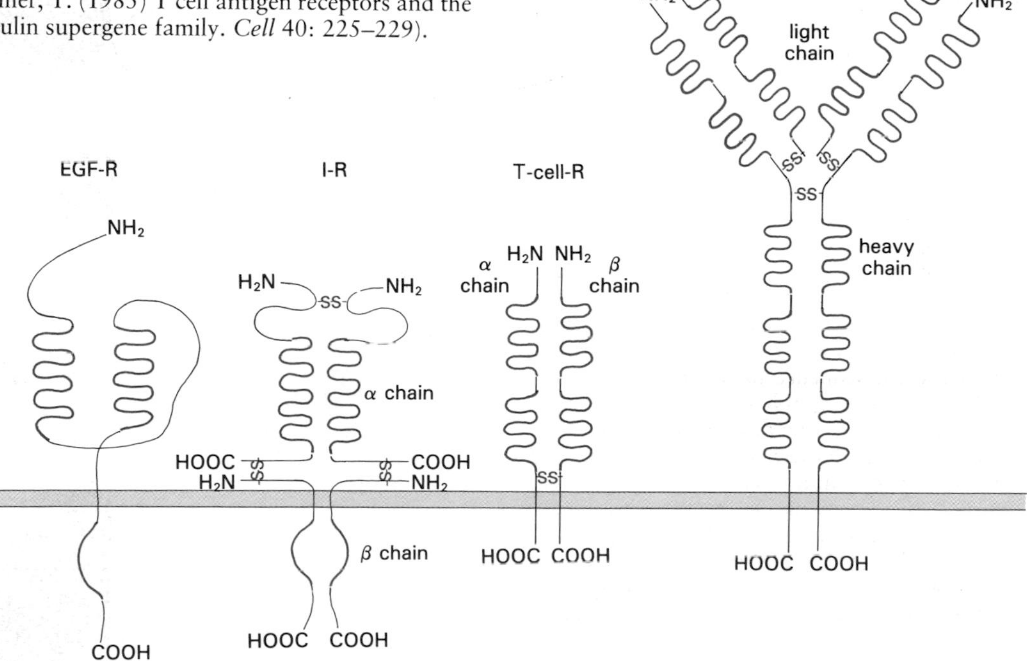

Figure I.80 A structural comparison between membrane receptors and immunoglobulin. In this very schematic drawing, the overall structures of two hormone receptors (EGF-R and I-R), the T cell receptor and an immunoglobulin are compared. Only S-S bonds between different chains have been highlighted. Repeated loops (in red) represent cysteine-rich regions. (Hood, L., Kronenberg, M. and Hunkapiller, T. (1985) T cell antigen receptors and the immunoglobulin supergene family. *Cell* 40: 225–229).

A

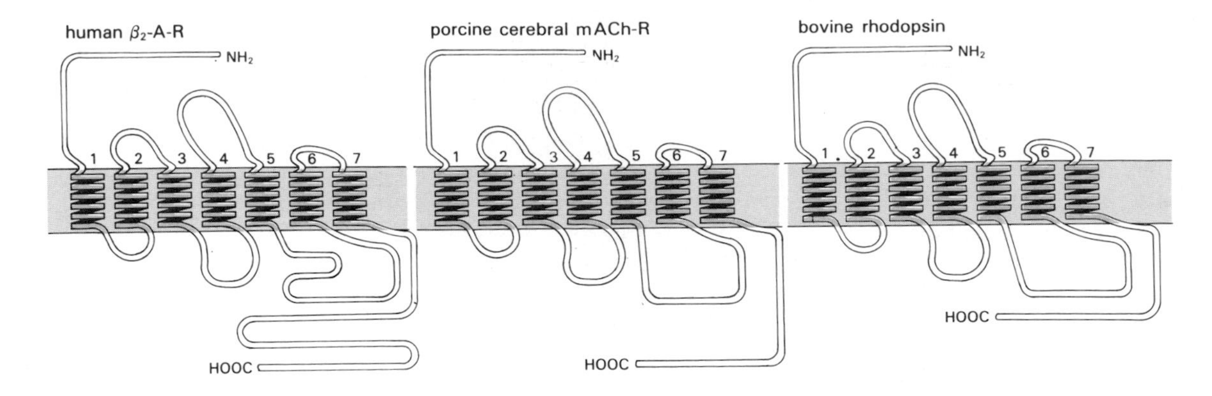

B

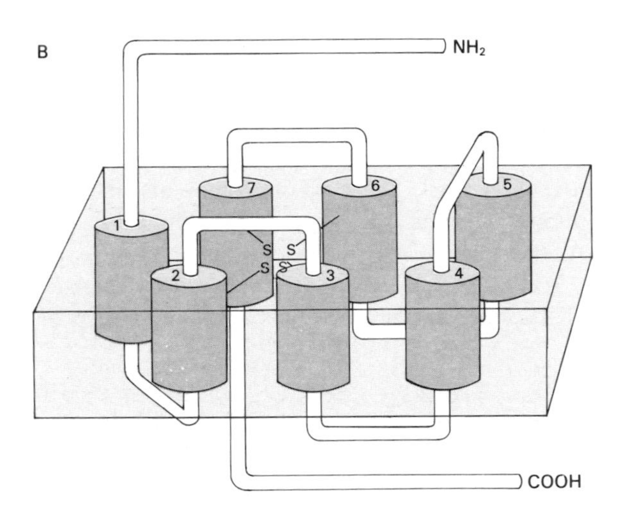

C

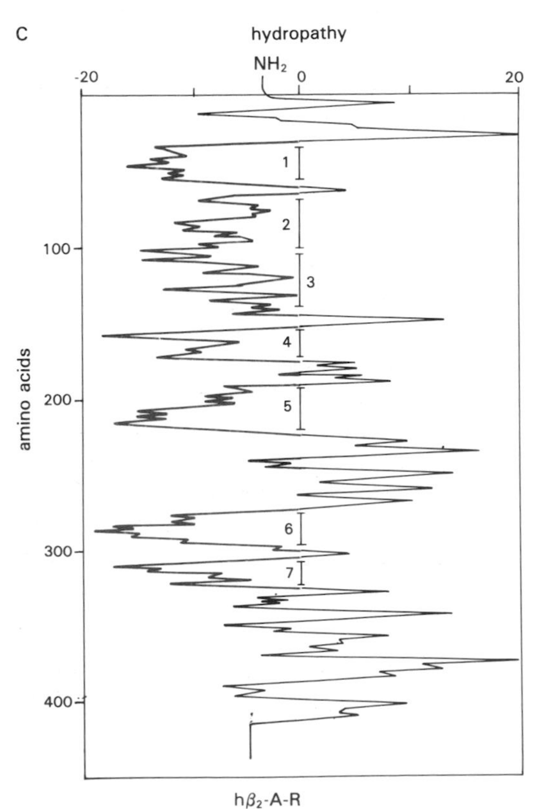

Figure I.81 Single chain, seven-helix receptors coupled to G protein. **A.** Amino acid sequences deduced from cDNAs suggest a very similar model for the three classes of receptors (β-adrenergic receptor (hβ-A-R): Kobilka, B.K., et al. (1987) *Proc. Natl. Acad. Sci., USA* 84: 46–50; muscarinic acetylcholine receptor (mACh-R): Kubo, T., et al. (1986) *Nature* 323: 411–416; rhodopsin: Ovchinnikov, Y.A. (1982) *FEBS Lett.* 148: 179–191; Nathans, J. and Hogness, D.S. (1984) Isolation and nucleotide sequence of the gene encoding human rhodopsin. *Proc. Natl. Acad. Sci., USA* 81: 4851–4855). **B.** A three-dimensional representation of a receptor indicates, in part arbitrarily, the possible interactions between several transmembrane segments, via S-S bonds. **C.** The hydrophobicity profile is shown for the hβ-A-R; the hydrophobic sequences, which correspond to the transmembrane portions, are shown in red. (Hopp, T.P. and Woods, K.R. (1981) Prediction of protein antigenic determinants from amino acid sequences. *Proc. Natl. Acad. Sci., USA* 78: 3824–3828). **D.** A three-dimensional representation of rhodopsin (opsin plus retinal). Retinal is deeply inserted within the transmembrane complex.

D

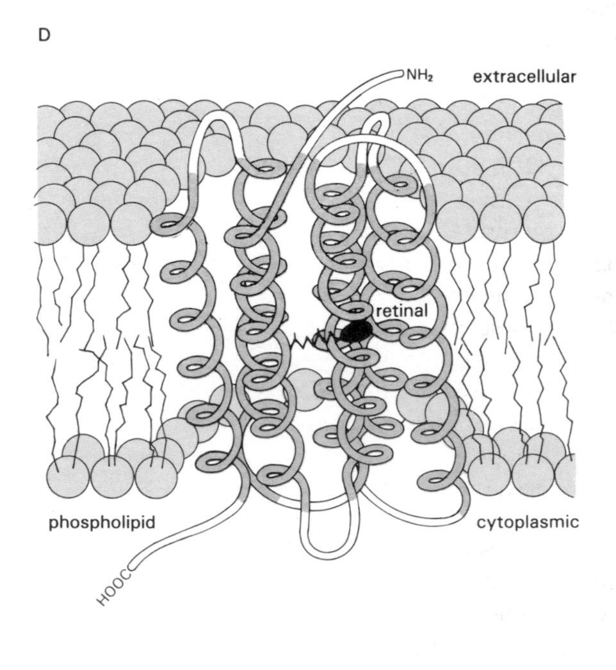

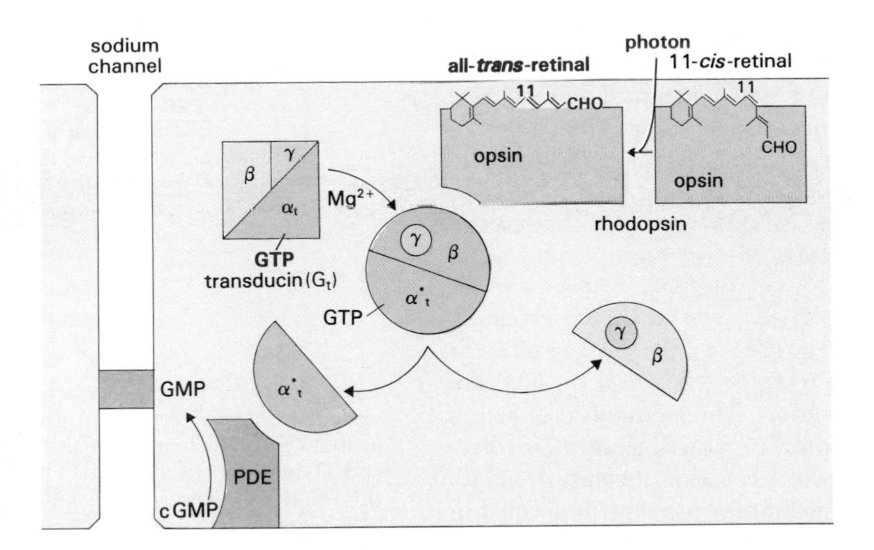

Figure I.82 A molecular mechanism for the transfer of sensory information: the photoreceptor. In the retina, there are one billion rods, photoreceptor cells which convert light into atomic motion and then into a nerve impulse. A single photon can excite a rod cell. In the outer segment of a rod, there is a stack of approximately 1000 disc sacs whose membrane is densely packed with photoreceptor molecules. The photosensitive molecule in rods is rhodopsin, consisting of opsin ($M_r \sim 38{,}000$) and 11-*cis*-retinal, a very effective chromophore. Light isomerizes retinal to all-*trans*-retinal within a few psec, resulting in a conformational change in rhodopsin, which in turn closes a Na^+ channel. A single photon can block the flow of $>10^6$ Na^+ ions. The mechanism includes a G_t protein (transducin)-dependent activation of a specific cGMP phosphodiesterase (PDE), and thus there is a consequent decrease in cGMP (Stryer, L. (1986) Cyclic GMP cascade of vision. *Ann. Rev. of Neurosci.* 9:87–119).

mACh-R (Fig. I.81)), serotonin and substance K, and the visual, "light receptor" (*rhodopsin*) are coupled to G protein. Other receptors, for acetylcholine (nicotinic acetylcholine receptor, *nACh-R* (Fig. I.85)), GABA (γ-aminobutyric acid receptor A, $GABA_A$-R) and glycine (*Gly-R*) (Fig. I.85), are *ligand-gated channels.*

Seven-Helix Receptors Coupled to G Protein
(Fig. I.81)

When it was discovered that many hormones and neurotransmitters act through a mechanism involving the different α subunits of G proteins (Table I.16), it was postulated that both structural similarities and differences would be found in their receptors. However, nobody would have predicted so much structural homology between such a diverse group of proteins as the β-A-R (p. 323), mACh-R and rhodopsin, each of which interacts with its respective G protein. Elucidation of the structure of *bacteriorhodopsin*[85], a distantly related bacterial protein, was crucial for the understanding of the structure of this class of receptor.

Hydrophobic sequences of 20 to 25 aa constitute each of seven membrane-spanning helices. The relatively short extracellular N-terminal sequence (intradisc for rhodopsin) has glycosylation sites (usually two). Short loops separate the intramembrane segments: the longest, between transmembrane helices 5 and 6, is cytoplasmic, as is the relatively long C-terminal portion rich in Ser/Thr.

Hydrophobic domains of these receptors have a high degree of homology. Several cysteines participate in interhelical connections. 11-*cis*-retinal is covalently attached to a Lys in the conserved middle portion of the transmembrane helix 7 of opsins. Preliminary evidence indicates that adrenergic and cholinergic agonists *bind to hydrophobic* regions of their receptors. Opsins may originally have been receptors for a retinal-like molecule which later became covalently attached during evolution, while retinal itself was the receptor for photons (Fig. I.82). Rhodopsin, after stimulation by light, and the β-A-R, following the binding of an adrenergic agonist, are activated and become substrates for phosphorylation,

85. Khorana, H. G. (1988) Bacteriorhodopsin, a membrane protein that uses light to translocate protons. *J.Biol. Chem.* 263: 7439–7442.

a mechanism for desensitization, which decreases interaction with the G protein (p. 108).

The β-A-R, mACh-R and rhodopsin have similar size (~410 to 480 aa), >30% identical amino acids, and no signal sequences. Each is typical of a group of related proteins that were first described by physiologists and pharmacologists. There are several different adrenergic receptors (p. 316) and two muscarinic acetylcholine receptors, M_1 (in brain) and M_2 (in heart), connected to G_i and G_p, respectively. A dopamine receptor (D_2-R), two serotonin receptors (5HT-R) (references p. 185) and the $GABA_B$ receptor and an angiotensin receptor (= c-*mas*) (p. 605) are also seven-helix receptors. The mechanism of action of the olfactory receptor[86,87] also suggests the involvement of a G protein which causes an increase in cAMP. Have a number of membrane proteins implicated in nerve transmission evolved from the same precursor?

All of these receptors might have derived from a unique ancesteral gene. Coding regions of their genes (except for that of rhodopsin) lack introns. These receptors are also homologous with mating factor (a and α) receptors of Saccharomyces cervisiae and, as confirmed with the discovery of the structure of the angiotensin receptor, with the human *mas* oncogene product (Table I.21).

As an aside, it is remarkable to note that the two neurotransmitters ACh and GABA each interacts with two receptors of two different types: G-protein-coupled and ion-gated channels, respectively.

Receptor Occupancy by Viruses

Unforseen "activity" of membrane receptors

Receptors for certain viruses have been identified in cells as being receptors for various other specific ligands. The rabies virus may utilize the ACh-R (acetylcholine receptor), and the reovirus type 3 receptor is a member of the β-adrenergic receptor family[88]. Occupation of the receptor by EGF, or with a synthetic decapeptide EGF-antagonist of the EGF-R, inhibits vaccinia virus infection. This is because there is a 19 aa sequence homology between EGF, TGF-α, and an early protein encoded by the vaccinia virus (Fig. I.83).

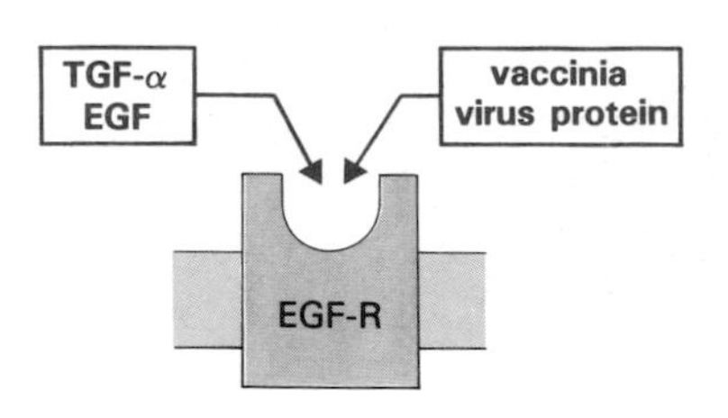

Figure I.83 Surprising interaction between the vaccinia virus and the EGF-receptor. A protein of the vaccinia virus includes a region that has high homology with both EGF and TGF-α, the latter of which binds to the EGF-receptor (EGF and TGF-α are also related to a cell surface receptor-like protein) (Fig. I.78). There are several examples of viruses which may utilize a cellular receptor specific for a physiological ligand. (Eppstein, D.A., Marsh, Y.V., Schreiber, A.B., Newman, S.R., Todaro, G.J. and Nestor, J.J. (1985) Epidermal growth factor receptor occupancy inhibits vaccinia virus infection. *Nature* 318: 663–665).

Membrane Receptors for Steroid Hormones

A membrane receptor for steroid hormones has been characterized *only* in Xenopus laevis *oocytes*, in which meiosis is reinitiated by progesterone (p. 98). Progesterone decreases adenylate cyclase and protein kinase activity in isolated membrane systems. Affinity labelling has demonstrated a protein whose binding specificity and affinity fit the biological responses to the hormone. This receptor protein is different from that of the intracellular progesterone receptor. A similar type of receptor has been detected in oocytes of other amphibians and fish.

A membrane receptor for steroid and thyroid hormones has *not yet* been clearly demonstrated in any other cell type. However, the activity of some membrane-bound enzymes suggests that such small hormone membrane receptors do exist; pharmacologically, PK-C (p. 107) can be activated by phorbol esters; Na^+-K^+ ATPase (the sodium pump) is

86. Pace, U. and Lancet, D. (1986) Olfactory GTP-binding protein: signal-transducing polypeptide of vertebrate chemosensory neurons. *Proc. Natl. Acad. Sci., USA* 83: 4947–4951.
87. Weinstock, R.S., Wright, H.N., Spiegel, A.M., Levine, M.A. and Moses, A.M. (1986) Olfactory dysfunction in humans with deficient guanine nucleotide-binding protein. *Nature* 322: 635–636.
88. Co, M.S., Gaulton, G.N., Field, D.N. and Greene, M.I. (1985) Isolation and biochemical characterization of the mammalian reovirus type 3 cell-surface receptor. *Proc. Natl. Acad. Sci., USA* 42: 1494–1498.

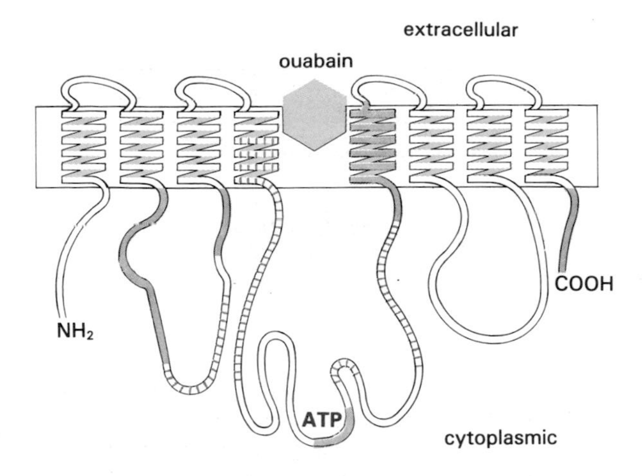

Figure I.84 A pharmacological steroid membrane receptor: models for the Na^+-K^+ ATPase catalytic subunit and Ca^{2+}-ATPase. The sites of interaction with steroidal ouabain and ATP are indicated. Homology with sarcoplasmic reticulum Ca^{2+}-ATPase is shown in red. Striped areas indicate additional homology with K^+-ATPase.

regulated by ouabain, a steroidal drug (Fig. I.84); and myocardial membrane Ca^{2+}-ATPase activity is stimulated directly by thyroid hormone[89]. Recently, pregnenolone sulfate produced in the brain (p. 17) has been shown to bind specifically, and to inhibit, the $GABA_A$ receptor (Fig. I.85B).

Nuclear Receptors for Polypeptide Hormones and Growth Factors?

A number of hormones (e.g., insulin, angiotensin, GnRH, NGF, EGF, PDGF) have been identified in target cell nuclei, and receptors have been detected, by binding studies, in the nuclear membrane and/or chromatin. Are the receptors initially nuclear, or are they delivered to the nucleus along with their ligands following the process of receptor-mediated endocytosis (p. 224)? In both cases, are hormone-receptor complexes active in the nucleus, for example in bringing their specific protein kinase activity to the nucleus?

Other Membrane Receptors of Endocrine-Related Systems

The low-density lipoprotein receptor (*LDL-R*) (p. 53) resembles growth factor receptors insofar as they all have a cysteine-rich extracellular region. It has one transmembrane segment and a very short cytoplasmic portion with no Tyr-PK activity. The transferrin receptor (*T-R*), also, has a very short cytoplasmic segment and no Tyr-PK activity (Fig. I.78). However, its orientation is unique, in reference to receptors described thus far, in that the *C-terminal region is extracellular*. Each of the two identical subunits binds two iron atoms. Both the LDL-R and T-R are models for receptor-mediated endocytosis (p. 224 and Fig. 3, p. 535) and for other *cellular nutrient receptors*.

Other receptors appear to function primarily as *ion channels*, directly gated by the ligand, which allow fast conductance changes (in the microsecond range). This is the case for the *nicotinic acetylcholine receptor* (nACh-R) (Fig. I.85A). This receptor, whose structure was the first to be clearly determined[90], is completely unrelated to the mACh-R, a fact which illustrates how different transducers could be implicated in different activities of the same ligand. Other neurotransmitter receptors, e.g., the *$GABA_A$-receptor* and another anion channel, the glycine receptor (Gly-R) (Fig. I.85B), also have a multiunit structure forming an ion channel. These receptors, beside binding their regulatory ligand, have allosteric sites for pharmacological molecules of medical importance (e.g., for the $GABA_A$-R, benzodiazepin, chlorpromazine, barbiturate, picrotoxin)[91]. Even though no detailed three-dimensional structure is known yet, it is clear that the mutually exclusive ion selectivity of channels for *anions and cations* does not exclude the sharing of common features of transmembrane sequences implicated in channel function. These *channel receptors* belong to a *superfamily* distinct from all other groups of receptors described to date. Again, as for other receptors (p. 87), there are (at least) two discrete and separable regions in these proteins, one serving as an integral ion channel and the other for ligand binding.

89. Mylotte, K. M., Cody, V., Davis, P. J., Davis, F. B., Blas, S. D. and Schoenl, M. (1985) Milrinone and thyroid hormone stimulate myocardial membrane Ca^{2+}-ATPase activity and share structural homologies. *Proc. Natl. Acad. Sci., USA* 82: 7974–7978.
90. Sakmann, B., Methfessel, C., Mishina, M., Takahashi, T., Takai, T., Kurasaki, M., Fukuda, K. and Numa, S. (1985) Role of acetylcholine receptor subunits in gating of the channel. *Nature* 318: 538–543.
91. Changeux, J. P., Giraudat, J. and Dennis, M. (1987) The nicotinic acetylcholine receptor: molecular architecture of a ligand-regulated ion channel. *Trends Pharmacol. Sci.* 8: 459–465.

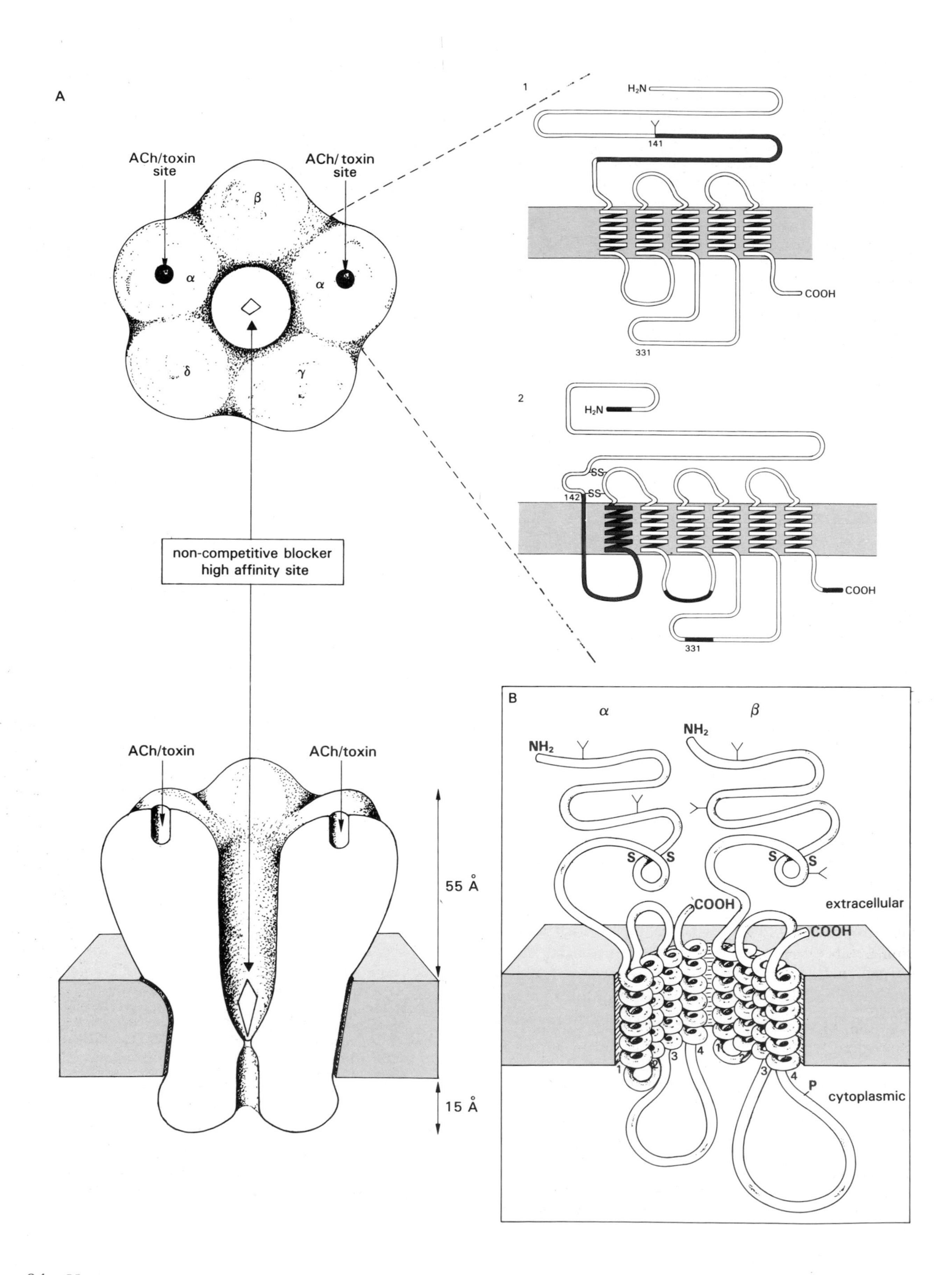
A
ACh/toxin site
ACh/toxin site
β
α
α
δ
γ
non-competitive blocker high affinity site
ACh/toxin
ACh/toxin
55 Å
15 Å
1
H₂N
141
COOH
331
2
H₂N
SS
142
SS
COOH
331
B
α
β
NH₂
NH₂
S
S
S
S
COOH
COOH
extracellular
1
2
3
4
1
2
3
4
P
cytoplasmic

◁ **Figure I.85** Mutiple membrane-spanning, ligand-gated channels. **A.** Nicotinic acetylcholine receptor (nACh-R). The most thoroughly studied of neurotransmitter receptors opens a ligand-gated, cation-selective *channel* in the plasma membrane, through which principally Na^+ and K^+ can diffuse. This is initiated by the *cooperative binding* of the neuromediator *to* the two *α subunits*, the smallest of the *5* $\alpha_2,\beta,\gamma,\delta$*-assembled units* which form a ring around the channel. The complete primary structure of the *four-subunit* precursors has been elucidated by the cloning and sequencing of cDNAs, using the receptor from the electric organ of the ray *Torpedo californica*[91]. These results now appear to be valid for similar receptors of neuromuscular junctions in mammals, including humans. Comparison of the amino acid sequences of the four subunits reveals marked homology between them. Their structure suggests that the polypeptides are oriented in a pseudosymmetrical fashion across the membrane. Two diagrammatical representations of models for the transmembrane structure of the α subunit of the acetylcholine receptor are shown: **1.** a model based on the sequence of amino acids derived from the cDNA sequence, and **2.** a model based on further studies with antibodies against selected peptides (solid regions of the chain); two additional transmembrane segments have thus been proposed. (Ratnam, M., Le Nguyen, D., Rivier, J., Sargent, P. and Lindstrom, J. (1986) Transmembrane topography of nicotinic acetylcholine receptor: immunochemical tests contradict theoretical predictions based on hydrophobicity profiles. *Biochemistry* 25: 2633–2643). It has been shown that the α subunit of the ACh-R contains the ACh binding site (in the region around aa 160), and, also, that it is involved in the binding of α-bungarotoxin (snake neurotoxin, which is an inhibitor). The δ subunit may be involved in the channel-closing step. All subunits are required to elicit a normal electrophysiological response to acetylcholine. The five subunits are contained within a cylindrical shell approximately 110 Å×80 Å in diameter, and the overall shape indicates a much larger synaptic side than cytoplasmic side. The channel is also 25–30 Å wider in diameter at the extracellular end than at the cytoplasmic end. Myasthenia gravis, an autoimmune disease, is related to antigenic determinants of the ACh-R. **B.** Inhibitory neurotransmitter receptors. The $GABA_A$ receptor ($GABA_A$-R) is a chloride channel which is opened by an $\alpha_2\beta_2$-hetero-oligomer which has binding sites for GABA (on the β subunit) and benzodiazepine (on the α subunit). These subunits, α and β, share considerable structural identity. In addition, there is similarity between domains of the $GABA_A$-R and nACh-R. In particular, there is an extracellular β-structural loop formed by disulfide bonding of two conserved Cys placed similarly in $GABA_A$-R and nACh-R. (Schofield, P. R., Darlison, M. G., Fujita, N., Burt, D. R., Stephenson, F. A., Rodriguez, H., Rhee, L. M., Ramachandran, J., Reale, V., Glencorse, T. A., Seeburg, P. H. and Barnard, E. A. (1987) Sequence and functional expression of the $GABA_A$ receptor shows a ligand-gated receptor super-family. *Nature* 328: 221-227). Similar homologies have been found for the strychnine-binding subunit of the glycine receptor. (Grenningloh, G., Rienitz, A., Schmitt, B., Methfessel, C., Zensen, M., Beyreuther, K., Gundelfinger, E. D. and Betz, H. (1987) The strychnine-binding subunit of the glycine receptor shows homology with nicotinic acetylcholine receptors. *Nature* 328: 215–220).

4.9 Regulation of Hormone Receptors

Receptors have a central role in hormone action. Consequently, any qualitative or quantitative alteration of the receptor will have an important impact on cellular responses to hormones.

Qualitative changes, particularly in either the *hormone binding site* or the *effector site*, may affect receptor function. Affinity for the ligand may be decreased by a mechanism of negative cooperativity, as has been described for the I-R (Fig. I.62). The affinity of a ligand for its receptor can be modified by changes in pH (e.g., the I-R in case of acidosis, p. 524), extracellular calcium concentration (ACTH-R, p. 242), and by the intracellular concentration of GTP, for the many receptors coupled to G proteins. Phosphorylation of membrane receptors can also modify their function (p. 108 and p. 324), and capping of DNA binding sites by hsp 90 obliterates steroid receptor activity (p. 70).

The *concentration* of receptors is also subject to modification. The number of physiological *intracellular* receptor molecules in target cells (a few thousand) is in great excess of the number of hormone-regulated genes (a few hundred). The distance between *membrane* receptors distributed randomly at the surface of the cell would be of the order of several hundred nm. However, membrane proteins can *move* several μm in seconds. Following binding of their ligand, surface receptors *cluster* within specialized regions of the plasma membrane, termed *coated pits*. Coated pits containing ligand-receptor complexes pinch off from the plasmalemma and become coated vesicles, resulting in simultaneous internalization of ligand and receptor (p. 224 and p. 535), and separation from the effector, e.g., G protein. While this process is physiologically important for the entry of nutrients into cells (cholesterol brought by LDL, iron by transferrin, vitamin B12 by transcobalamine, and vitellogenin by itself), the significance of hormone and receptor endocytosis is largely unknown. It does serve, however, as a mechanism regulating receptor turnover, thus receptor concentration on the cell surface, and consequently the magnitude of the cellular response.

Whether receptors are intracellular or in the plasma membrane, their number can be either increased or decreased. In each case, the process may be homologous, promoted by the hormone itself, or heterologous, due to (an)other regulatory factor(s). Evidence for both up- and down-regulation has often

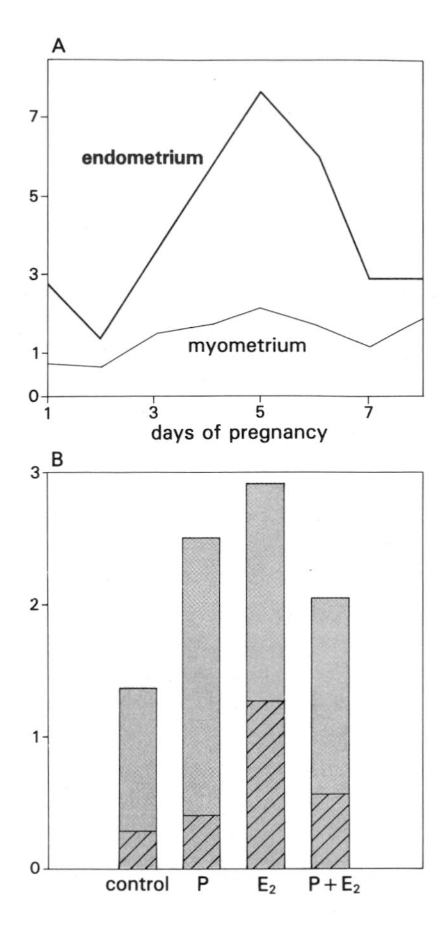

Figure I.86 Regulation of the estrogen receptor in the rat uterus. **A.** Early in gestation, the concentration of estradiol receptor varies differently in the endometrium and myometrium. **B.** The respective regulatory role of estradiol and/or progesterone has been studied following the administration of these hormones to castrated animals. Cross-hatched red bars, endometrium; gray bars, myometrium. (Mester, J., Martel, D., Psychoyos, A. and Baulieu, E.E. (1974) Hormonal control of estrogen receptor in uterus and receptivity for ovoimplantation in the rat. *Nature* 250: 776–778).

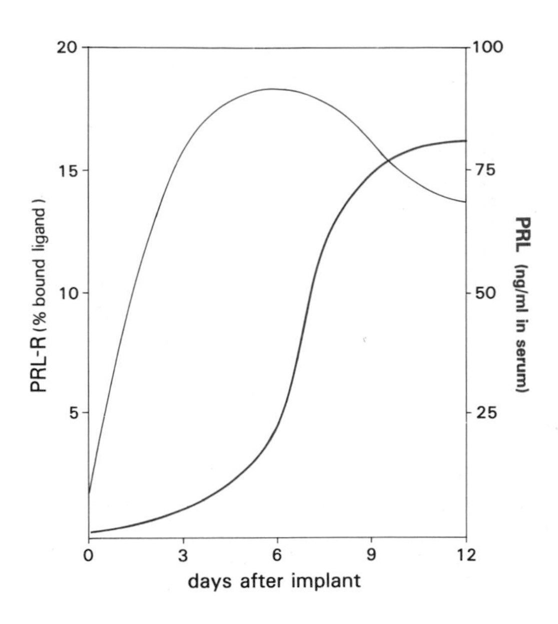

Figure I.87 Up-regulation of the prolactin receptor by prolactin. Hypophysectomized rats were implanted with pituitaries under the kidney capsule. The increase of the hormone in serum preceded the increase of the receptor in liver. (Posner, B.I., Kelly, P.A. and Friesen, H.G. (1975) Prolactin receptors in rat liver: possible induction by prolactin. *Science* 188: 57–59).

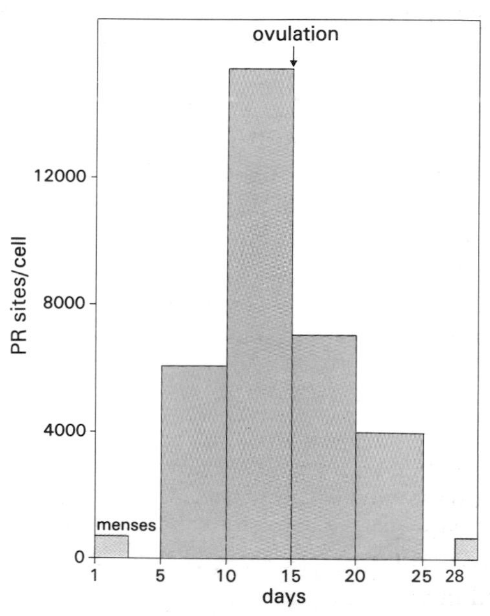

Figure I.88 Progesterone receptor in the human endometrium: five-day pooled samples at four different periods in the normal menstrual cycle. In decidua of pregnancies of ~ eight weeks of amenorrhea, obtained after voluntary interruption, the increase in PR is probably due to increased estrogen secretion. (Bayard, F., Damilano, S., Robel, P. and Baulieu, E. E. (1978) Cytoplasmic and nuclear estradiol and progesterone receptors in human endometrium. *J. Clin. Endocrinol. Metab.* 46: 635–648).

been difficult to demonstrate, since these processes frequently occur simultaneously and/or successively. In addition, the study of homologous regulation provoked by the hormone itself requires either removal of the hormone or the use of an exchange technique (p. 62) in order to measure receptor number.

Up-regulation. Homologous increase of receptors is observed with steroid hormones (e.g., ER increased by E_2 (Fig. I.86) and AR increased by testosterone) as well as protein hormones (PRL-R increased by PRL (Fig. I.87) and EGF-R increased by

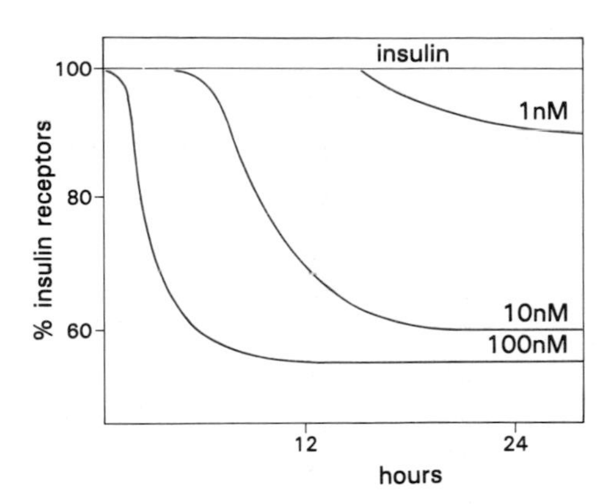

Figure I.89 Down-regulation of the insulin receptor by insulin. Cultured lymphocytes were exposed to various concentrations of hormone, and total insulin receptor levels were determined. (Gavin, J. R., Roth, J., Neville, D. M., De Meyts, P. and Buelle, D. N. (1974) Insulin-dependent regulation of insulin receptor concentration: a direct demonstration in cell culture. *Proc. Natl. Acad. Sci., USA* 71: 84–88).

EGF). *Heterologous* up-regulation is a common regulatory event. The examples of the increase of PR in the uterus, stimulated by estradiol (Fig. I.88 and p. 420) and LH-R in granulosa cells, by FSH (Fig. X.9), are classical, and both play an important physiological role in the reproductive cycle. Receptor synthesis is always involved when there is up-regulation of the receptor.

Down-regulation. Although probably true for all steroid receptors, the clearest example of *homologous* down-regulation is that of the PR by progestins[92] (Fig. IX.29), which are probably involved in the decrease of PR during the luteal phase. There is also evidence for tissue-specific down-regulation of glucocorticosteroid receptor synthesis by glucocorticosteroids. Down-regulation of *membrane receptors* was demonstrated initially with the I-R (Fig. I.89), and then for practically *all* receptors; the basic mechanism is by *endocytosis* (p. 224). A role for receptor *phosphorylation* in the down-regulation of receptors has been reported (p. 108). However, the fate of the receptor varies; it is either recycled (the case particularly for receptors involved in the transport of nutritive compounds) or degraded (Fig. I, p. 225). An interesting example of the importance of the kinetics of receptor regulation is offered by GnRH, which, as described on p. 447, down-regulates its receptor in gonadotropic cells when administered continuously. The physiological maintenance of GnRH receptor concentration is dependent on the pulsatile secretion of GnRH, because this discontinuous secretion probably allows appropriate periods of receptor synthesis. Examples of *heterologous* down-regulation are described in several other chapters (p. 108, p. 324 and p. 420). Since an agonist can down-regulate its receptor, could an antagonist stimulate an increase in receptor synthesis? After treatment with anti-β-adrenergic propanolol, an increase in the β-A-R is associated with a syndrome of deprivation. The decrease in receptor function, whether due to qualitative (e.g., by phosphorylation) or quantitative changes (e.g., by down-regulation), is responsible for the majority of reduced cellular responses, known as *tachyphylaxis, desensitization, refractoriness,* or *tolerance.*

Receptors are *constitutively* present in target cells. When the corresponding *hormone* is *absent*, however, receptor numbers are much *lower* than when the hormone is present. This is the case for steroid receptors in sex organs of castrated animals, as well as for prolactin and growth hormone receptors in the liver of hypophysectomized rats. The comparative ontogeny of a hormone and its receptor is very difficult to study, mainly because of the very low concentrations of these components in embryonic or fetal cells. Thus, it remains difficult to ascertain *which is the first* to appear during development, *Hormone and receptor* (p. 152–160).

4.10 Hormone Receptors As Composite Proteins

Highly specific and made of independent units

That receptors are *mosaic proteins* has also been illustrated for the LDL-R (Fig. 2, p. 534). The structural analysis of both DNA-binding and membrane receptors suggests that they are composed of several functional domains[93].

Two sets of experiments have indicated both the *independence* and the *specificity* of the two main *functional regions* of receptors: the hormone-binding and effector sites. Experimental chimeras have been constructed consisting of the cytoplasmic Tyr-PK of the EGF-R and the extracellular hormone binding region of the I-R (Fig. I.90A). Insulin is then able to stimulate EGF-R Tyr-PK activity. Another hybrid was

92. Milgrom, E., Luu Thi, M., Atger, M. and Baulieu, E. E. (1973) Mechanisms regulating the concentration and the conformation of progesterone receptor(s) in the uterus. *J. Biol. Chem.* 248: 6366–6374.
93. Kearin, D. M. and Shapiro, D. J. (1988) Persistent estrogen induction of hepatic *Xenopus laevis* serum retinal binding protein mRNA. *J. Biol. Chem.* 263: 3261–3265.

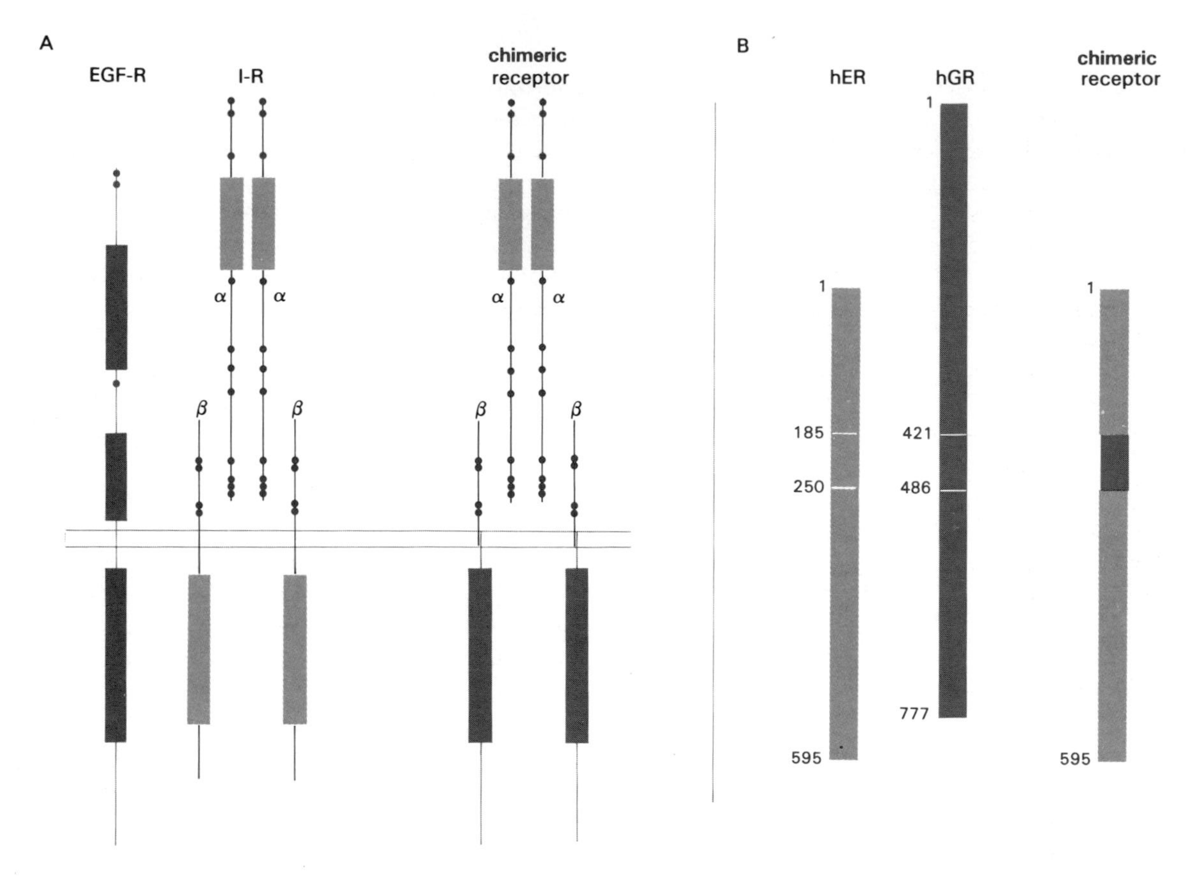

Figure I.90 Chimeric receptors. **A.** This representation of the EGF-R and the I-R suggests a similar arrangement for the extracellular portions, involved in hormone binding. Note that the two cysteine-rich domains of the EGF-R correspond to the cysteine-rich domains of the two subunits of the I-R. The *EGF-R/I-R hybrid*, stimulated by insulin, demonstrates Tyr-PK activity of the EGF-R type. (Ullrich, A. U., Riedel, H., Yarden, Y., Coussens, L., Gray, A., Dull, T., Schlessinger, J., Waterfield, M. D. and Parker, P. J. (1986) Protein kinases in cellular signal transduction: tyrosine kinase growth factor receptors and protein kinase C. *Cold Spring Harbor Symp. Quant. Biol.* 51: 713–724). **B.** The *ER/GR hybrid*, stimulated by estradiol, demonstrates glucocorticosteroid-like activity (and no estrogenic response). (Green, S. and Chambon, P. (1987) Oestradiol induction of a glucocorticoid-responsive gene by a chimaeric receptor. *Nature* 325: 75–78). Thus, in both cases, the hormone binding region and the effector domain behave specifically, and are able to exert their characteristic activity within a molecule different from the original receptor.

formed, with the DNA-binding cassette (DBD) of the GR and the hormone binding region (LBD) of the ER (Fig. I.90B), which retained the same binding specificity and affinity as the intact receptor. MMTV induction, a classical response to glucocorticosteroids (p. 415), was caused by estrogen, while an estrogen-inducible gene was no longer stimulated. Therefore, both hormone binding and effector activity display independent properties. When the hormone binding and effector sites are present in a receptor, hormone is mandatory for the induction of activity. In the *absence of hormone*, the binding region negatively influences the effector site. The constitutive activity of v-*erb*-B (homologous to hormone-site deprived EGF-R) and of experimentally truncated receptors is consistent with these observations. Phylogenetically, enzymes (e.g., Tyr-PK) and DNA-binding proteins (e.g., transcription enhancers) may have acquired a hormone-dependent regulatory machinery secondarily. The structure of steroid receptor genes known presently suggests that their N-terminal segments may be encoded by a single large exon, in contrast to the multiple exons-introns which specify the C-terminal half. The significance of small differences between receptors for the same steroid in different species, as well as genetic polymorphism which might be expressed differentially in diverse tissues of the same individual, are currently under study.

5. Mechanisms of Hormone Action

Rather than by affecting the function of soluble components of the blood plasma or extracellular fluid, hormones are molecules which act on or in *cells*. Cells are complex, highly structured machines which function, as a result of the efficiency of enzymes, at relatively constant temperature and pH. Enzymes transform substrates into products, sometimes in cooperation with coenzymes which can be considered cosubstrates. Both substrates and cosubstrates are chemically modified by enzymes. Hormones, operating via their reversible interaction with receptors, are *in a class apart from enzymes, substrates, and cosubstrates*. This does not mean, however, that they are not enzymatically altered at their site of action (p. 47).

Basically, hormones are *chemical signals* regulating *already established cellular mechanisms. In general*, hormone action is relatively *brief*, depending on the presence of ligand (exceptionally, it may persist for months[93]). However, in *some cases*, a hormone may provoke a change which persists even though the hormone is no longer present, thereby establishing *a new commitment* or state of *differentiation* (e.g., sex differentiation in the central nervous system[94]).

By definition, the study of hormone action includes the analysis of the *immediate postreceptor consequences* of hormone-receptor interaction. Since there are so many different hormones, and since there is such diversity in cell function, it is not surprising that *no single* mechanism has been found to account for the activities of all hormones. Moreover, it is remarkable that only plasma membrane and nuclear receptors have been shown to be involved in hormone action. Basic mechanisms of cell communication and gene expression occur essentially at the *cell surface* and at the level of *genes* which in turn command the remaining processes involved in cell physiology. Strikingly, *transduction* systems, found almost universally in living organisms, are *limited* in number. Consequently, each of these systems must be utilized by several different hormones, while any given hormone is capable of performing different functions in different cells. Ultimately, the *diversification* of hormonal responses is determined by *cell differentiation*.

Generalizations in biology should be made cautiously. There are *fashions in science*, related to technologies available at a given point in time. Indeed,

94. Arnold, A. P. and Gorski, R. A. (1984) Gonadal steroid induction of structural sex differences in the central nervous system. *Annu. Rev. Neurosci.* 7: 413–442.
95. Knox, W. E., Auerbach, V. H. and Lin, E. C. C. (1956) Enzymatic and metabolic adaptation in animals. *Physiol. Rev.* 36: 164–254.
96. Mueller, G. G., Herranen, A. M. and Jervell, K. F. (1958) Studies on the mechanism of action of estrogens. *Rec. Prog. Horm. Res.* 14: 95–139.

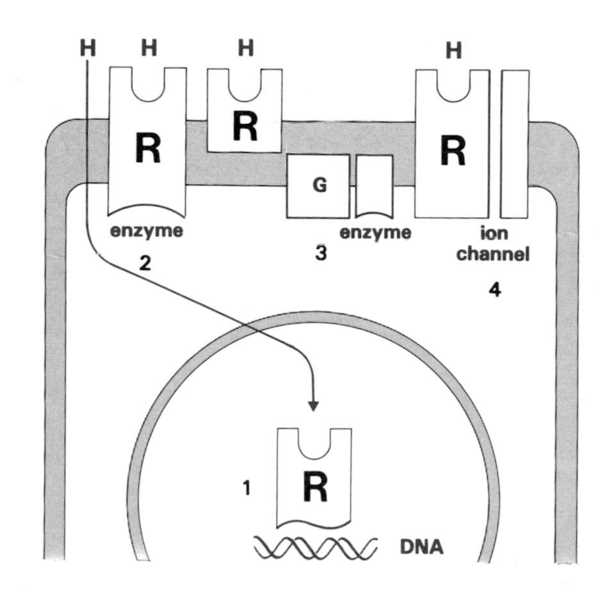

Figure I.91 Hormone receptors: four mechanisms of action. Intracellular receptors interact with genes (**1**). Membrane receptors are enzymes themselves (**2**), or are coupled to an enzyme via a G protein (**3**), or are connected to, or form, a channel system (**4**).

the last 50 years have witnessed the development of initially simplistic concepts. First, hormone solubility in membranes and direct DNA-hormone interactions were studied. When enzymology became a dominant discipline, it was suggested that hormones could be substrates or cosubstrates. Today, as shown in Figure I.91, it is believed that genes (to which receptors bind) and membrane-associated enzymes or channels (directly connected to receptors or operating via a coupling protein) are the primary levels of hormone action. Certain receptors are themselves enzymes or channels, of which hormones are allosteric regulators. These mechanisms ensure cellular responses to hormones, of which protein synthesis, stimulation of enzymatic activity, and change in membrane permeability are primary components.

5.1 Hormonal Regulation of Gene Transcription

The regulation of gene transcription is the essential mechanism of action of *steroid* and *thyroid* hormones.

In the early 1960s, several observations drew attention to the implication of *gene transcription* in steroid hormone action. First, it was established that steroid effects were dependent on *protein synthesis*, itself preceded by an increase in RNA synthesis[95, 96] (Figs. I.92, I.93 and I.94). Concurrently, it was shown

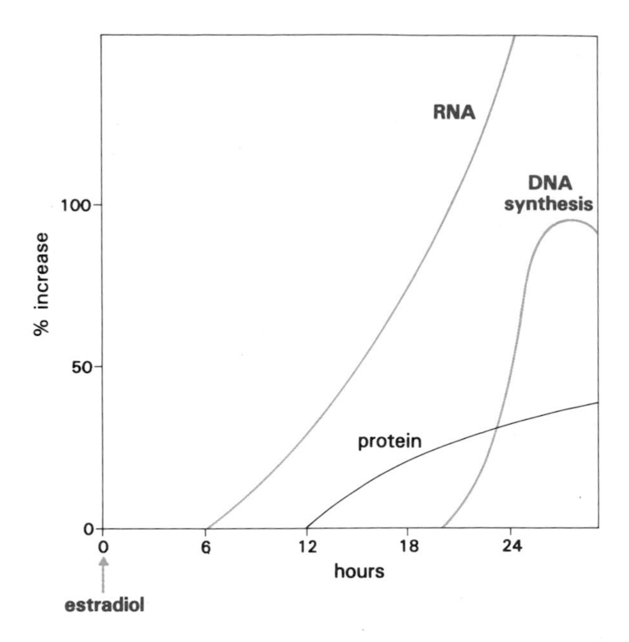

Figure I.92 Hormonal regulation of gene transcription: uterine growth stimulated by estrogen. The sequential increase in RNA, protein and DNA is represented, after a single administration of estradiol to immature or castrated rats. (Knox, W. E., Auerback, V. H. and Lin, E. C. C. (1956) Enzymatic and metabolic adaption in animals. *Physiol. Rev.* 36: 164–254; and Kaye, A. M., Sheratzky, D. and Lindner, H. R. (1972) Kinetics of DNA synthesis in immature rat uterus: age dependence and estradiol stimulation. *Biochim. Biophys. Acta.* 261: 475–486).

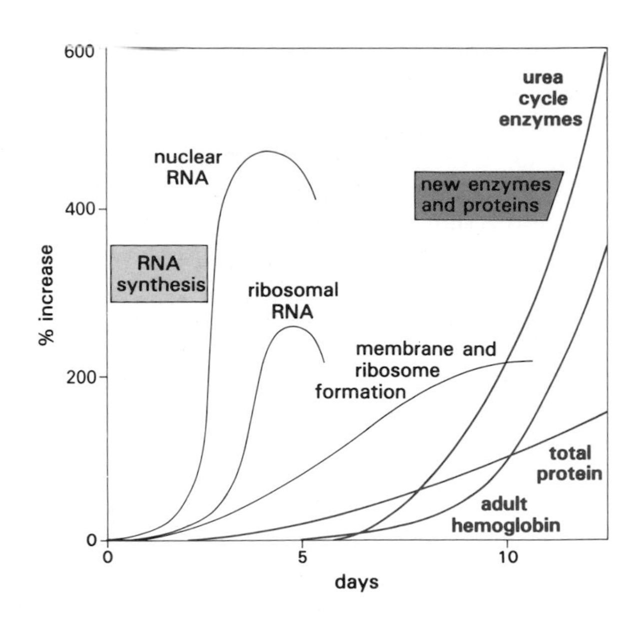

Figure I.93 Developmental response to a hormone: metamorphosis of tadpoles, induced by triiodothyronine. Note the temporal relationship between RNA synthesis and protein synthesis, and the induction of new proteins (i.e. adult hemoglobin). (Tata, J. R. (1986) The search for the mechanism of hormone action. *Persp. Biol. Med.* 29: 184–204).

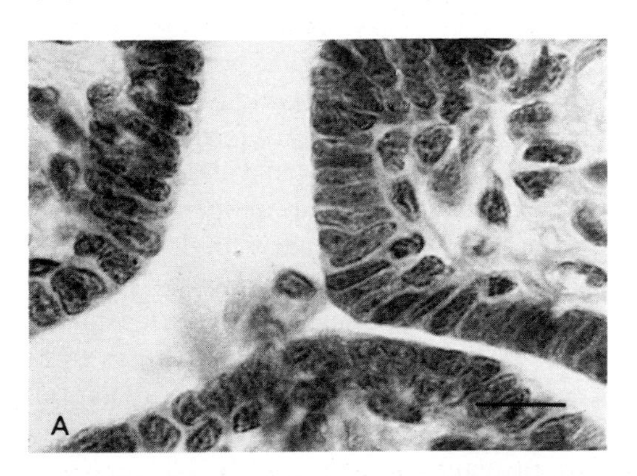

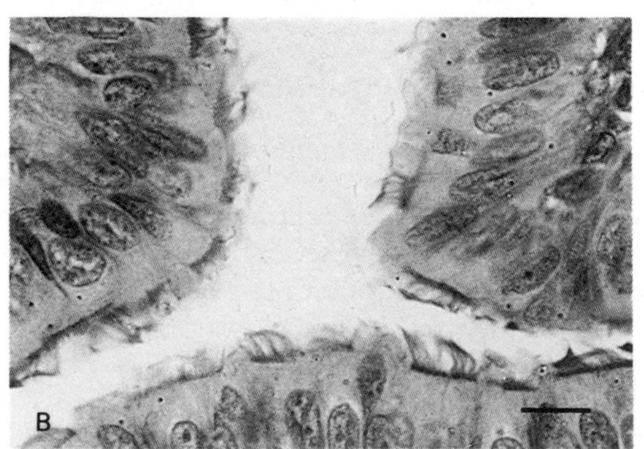

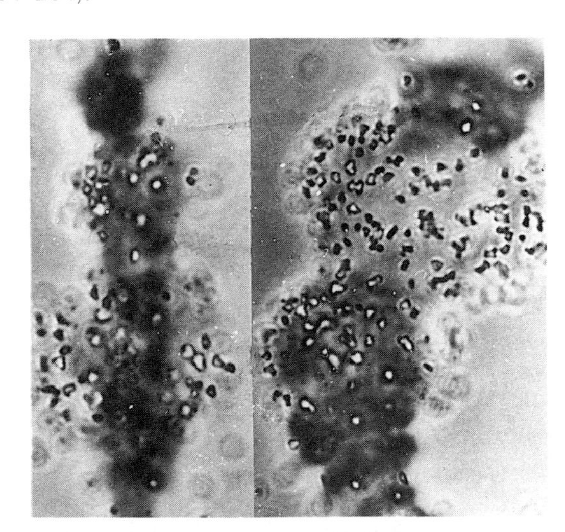

Figure I.95 The effect of ecdysone at the chromosomal level. A "puff" on a chromosome from an insect salivary gland cell is the site of active RNA synthesis (identified by autoradiography after [^{3}H]uridine incorporation). (Lezzi, M. and Gilbert, L. I. (1969) Control of gene activities in the polytene chromosomes of Chironomus Tentans by ecdysone and juvenile hormone. *Proc. Natl. Acad. Sci., USA* 64: 498–503.

◁ **Figure I.94** Steroid as a growth-promoting hormone. Effect of estrogen on immature cat endometrium. **A**, before, and **B**, after treatment. Bar=10 μm. (Courtesy of Dr. J. Racadot, Hôpital Pitié-Salpêtrière, Paris).

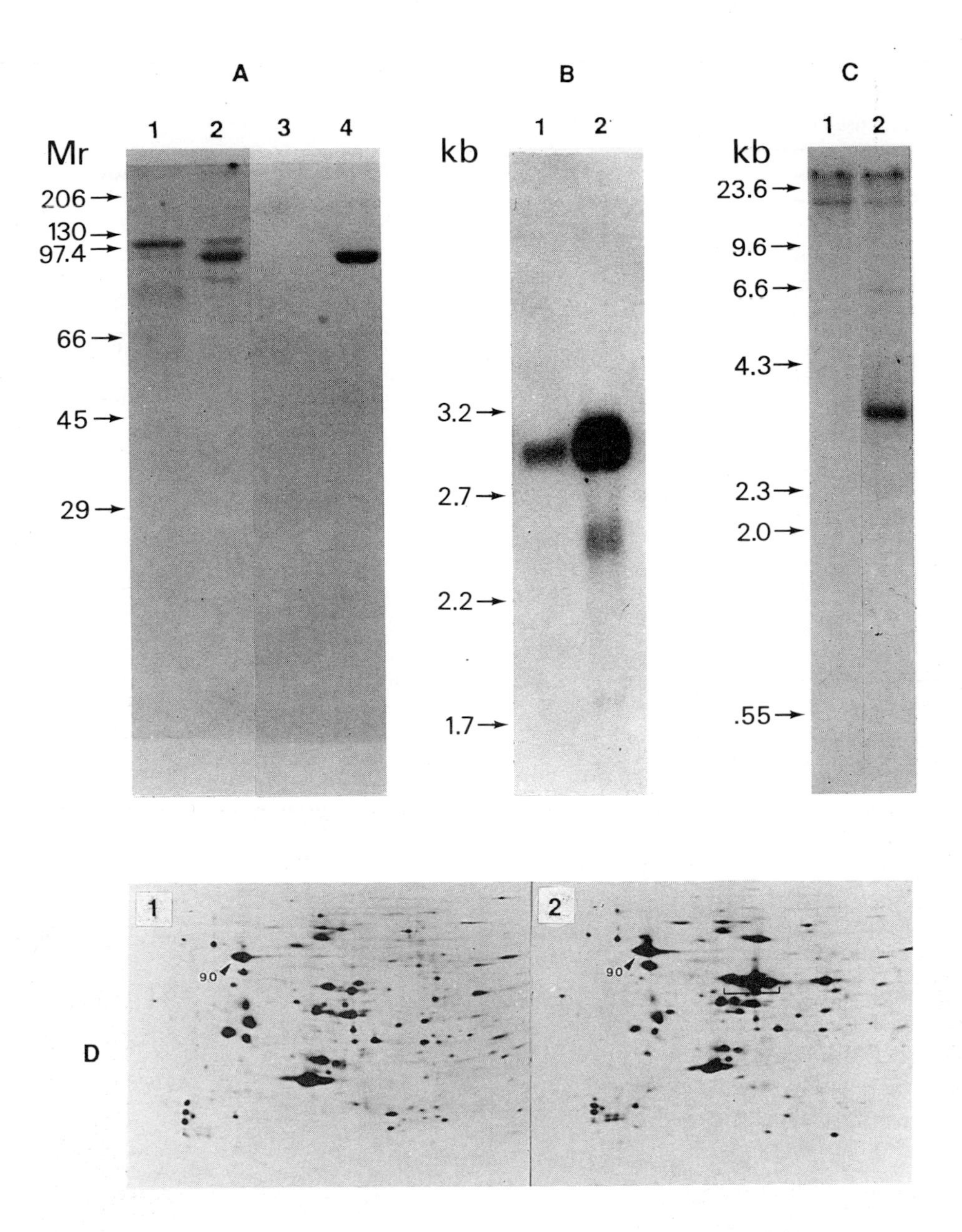

Figure I.96 Techniques used in molecular endocrinology. **A.** *Western blot* (antibody-protein interaction). Structure of 8 S progesterone receptor (PR). Purified fraction of chick oviduct 8 S-PR (MW ~ 300,000) and 4 S-PR (MW ~ 110,000) were electrophoresed on a SDS-gel. The proteins were transferred to a nitrocellulose filter and identified with specific antibodies. A polyclonal antibody recognized the 4 S-PR (B subunit, MW ~ 110,000) (1) and three components of the 8 S-PR (B–110 and 79 kDa subunits, hsp 90) (2). A monoclonal antibody detected only hsp 90 in the 8 S-PR (4), and no protein in the 4 S-PR (3). (Renoir, J. M., Buchou, T., Mester, J., Radanyi, C. and Baulieu, E. E. (1984) Oligomeric structure of molybdate-stabilized, non-transformed 8 S progesterone receptor from chicken oviduct cytosol. *Biochemistry* 23: 6016–6023). **B.** *Northern blot* (DNA–RNA hybridization). Heat shock increases hsp 90 mRNA. Poly A^+ RNA from chicken fibroblasts was obtained after incubation for 3 h at 37°C (1) or 44°C (2). RNA was electrophoresed on agarose, transferred to a nitrocellulose filter, and was identified by hybridization with a cDNA probe. (Catelli, M. G., Binart, N., Feramisco, J. R. and Helfman, D. (1985) Cloning of the chick hsp 90 cDNA in expression vector. *Nucl. Acids Res.* 13: 6035–6047). **C.** *Southern blot* (DNA–DNA hybridization). Female cells do not contain Y-DNA. DNA from normal female (1) and male (2) was electrophoresed on polyacrylamide gel, transferred to a nitrocellulose filter, and was identified by hybridization with a cloned genomic DNA probe. (pY3.4, courtesy of Dr. D. Jolly, Hôpital de Bicêtre). **D.** *2D-electrophoresis* (of proteins). Heat-shock increases hsp. An immunopurified preparation of [^{3}H]Leu-labelled proteins of Hela cells, maintained 3 h at 37°C (1) or 42°C (2), was analyzed by isoelectric focusing (pH 5–7 gradient) in the first dimension (acidic end to the left) and electrophoresed in SDS-polyacrylamide gel in the second dimension. Fluorograms are shown. Hsp 90 is indicated by arrows. Note also the increase of hsp 72 on panel 2, as indicated by the horizontal bracket[78].

Table I.13 Techniques used in molecular endocrinology.

Agents used	Methods	Components of the endocrine system
Ligand	RIA Binding Affinity purification Autoradiography	Hormones and receptors
Proteins	Rate of synthesis 2D electrophoresis‡ Enzymatic studies	Hormones, receptors, and other proteins
Antibodies	RIA Affinity purification Electrophoresis (Western blot‡) Immunohistochemistry	Hormones, receptors, and other proteins
cDNAs cRNAs	Electrophoresis (Southern blot‡, Northern blot‡) Hybridization in situ	Genes and mRNAs of hormones, of receptors and of other proteins

‡See examples in Fig. I.96.

that, in insects, ecdysone (p. 160) induces chromosome puffs, which designate loci of specific RNA synthesis (Fig. I.95). It was also demonstrated by autoradiography that radioactive steroid hormones accumulate in the nucleus. In the last 15 years, technical developments in mammalian cell culture‡, protein and mRNA synthesis, DNA sequencing and gene transfer, chromatin structure and DNA-receptor interactions (Table I.13 and Fig. I.96) have led to important advances. Both the general molecular mechanism and target cell specificity of hormone action must be considered successively.

General Mechanism[97,98]

Steroid and thyroid hormones modify protein synthesis primarily by affecting *mRNA synthesis* at the level of the *initiation* of *transcription* (Fig. I.97), although an increased rate of RNA elongation has also been reported. This implies that important events occur in the promoter region of hormone-regulated genes (Fig. I.15). In fact, hormone-receptor complexes bind to DNA segments which are now known as *hormone response elements* (*HREs*). HREs are specific *enhancers* (p. 378 and p. 381), i.e. DNA segments *cis*-acting positively, in both orientations, on transcription, which are regulated by hormone-receptor complexes acting as *trans*-acting factors. They demonstrate *two* main characteristics. First, they *bind* the appropriate *hormone-receptor* complexes

‡Serum-free culture of mammalian cells requires, in addition to nutrients, the presence of at least one hormone that acts at the level of the plasma membrane in addition to one that acts in the nucleus[197].

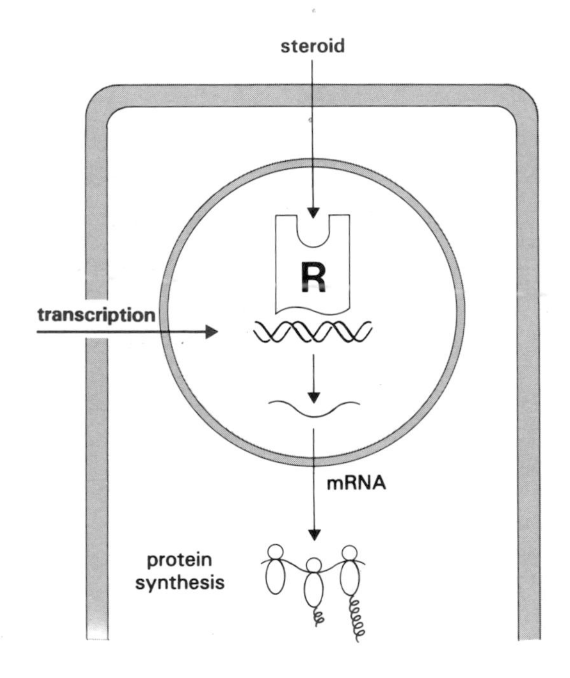

Figure I.97 Mechanism of action of steroid hormones. Regulation is achieved mainly at the level of transcription.

97. Yamamoto, K. R. (1985) Steroid receptor regulated transcription of specific genes and gene networks. *Annu. Rev. Genet.* 19: 209–252.
98. Ringold, G. M. (1985) Steroid hormone regulation of gene expression. *Annu. Rev. Pharmacol. Toxicol.* 25: 529–566.

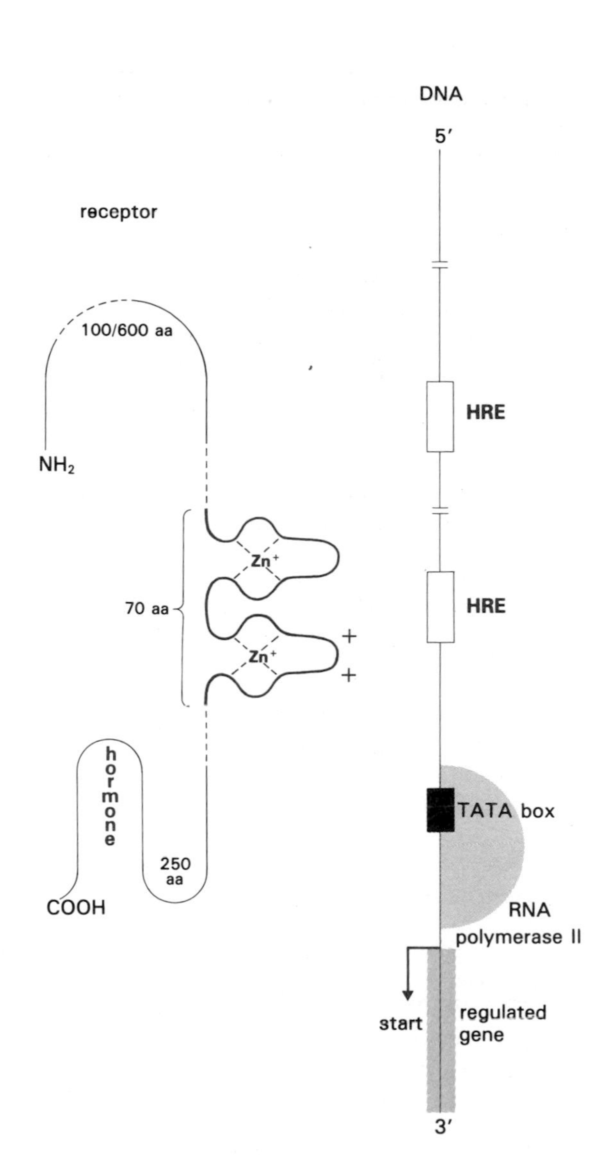

Figure I.98 Interaction between the steroid receptor and DNA. A consensus receptor (left) has been drawn to show the three domains, N-terminal, DNA-binding, and C-terminal steroid-binding. The proposed structure for the DNA-binding region (two fingers held by Zn^{2+} atoms) (Fig. I.73) interacts with HRE.

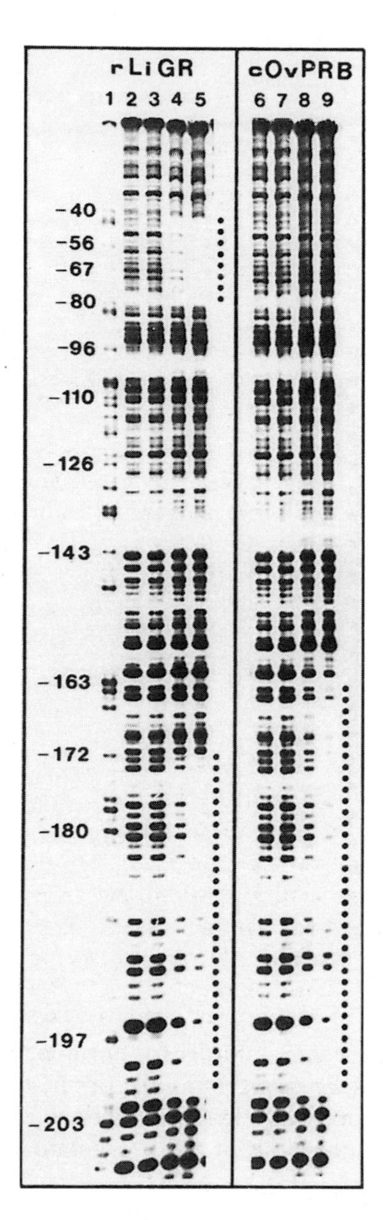

Figure I.99 Specific interactions of steroid-receptor complexes with the promoter region of a steroid-regulated gene. The promoter region of the chicken lysosyme gene was exposed to purified preparations of glucocorticosteroid and progesterone receptors bound to their respective hormones. Digestion by DNase I, followed by electrophoresis of the DNA fragments, revealed protected sequences (denoted by the dots). Lanes 4 and 5, and 8 and 9, respectively, show results after incubation with different concentrations of the enzyme. Lanes 1, 2, 3, 6 and 7 are controls. Note the similar, but not identical, binding of both receptors in the 160–200 bp region upstream of the transcription initiation point, while there is another region of binding in the 50–80 bp upstream region for the glucocorticosteroid receptor. (Von der Ahe, D., Renoir, J. M., Buchou, T., Baulieu, E. E. and Beato, M. (1986) Receptors for glucocorticosteroid and progesterone recognize distinct features of a DNA regulatory element. *Proc. Natl. Acad. Sci., USA* 83: 2817–2821).

(Fig. I.98), with high affinity, although the affinity of random, non-specific DNA for these complexes is far from negligible. The nucleotides involved in binding have been identified by techniques based on the protection of the DNA segment, by hormone-receptor complexes, from enzymatic degradation or chemical modification (Figs. I.99, IX.24, and IX.26). Secondly,

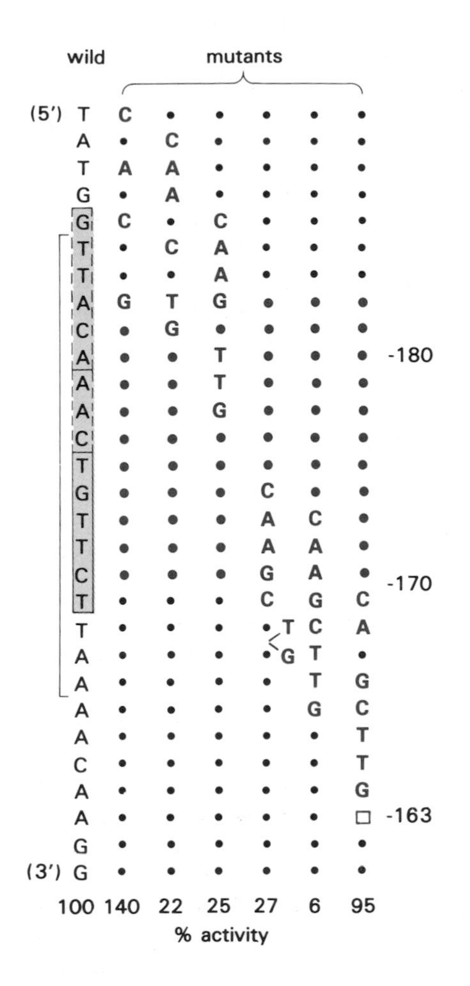

Figure I.100 Identification of HRE by mutation/transfection experiments. Deletions of segments of varying length were made in the promoter region, within the long terminal repeat (LTR), of the mouse mammary tumor virus. Plasmids were constructed, linking the hormone control region to the coding portion of the *Herpes simplex* virus thymidine kinase (TK) gene, and were introduced into LTK$^-$ cells. Mutated bp are in red and the deleted bp is indicated by a red square. Transcription was quantified in the presence or absence of glucocorticosteroid; numbers indicate the percent of activity of stable transfectants as compared to wild-type. The bracket indicates the segment protected from DNase I. A 15mer consensus sequence of glucocorticosteroid HREs is highlighted, consisting of a highly conserved hexanucleotide (boxed in solid line), an intermediate, variable, three base segment, and another partially symmetrical hexanucleotide (boxed in broken line), with the consensus hexanucleotide indicated as a solid line. Other mutation-sensitive regions defining different control elements are also indicated. Besides the receptor binding sites (distal HRE, bp 180–172, proximal HRE, bp121–117), there is a site at around −70 bp for nuclear factor I, and a TATA box at around −30 bp. (Scheidereit, C., Geisse, S., Westphal, H. M. and Beato, M. (1983) The glucocorticoid receptor binds to defined nucleotide sequence near the promoter of mouse mammary tumour virus. *Nature* 304: 749–754; Buetti, E. and Kühnel, B. (1986) Distinct sequence elements involved in the glucocorticoid regulation of the mouse mammary tumor virus promoter identified by linker scanning mutagenesis. *J. Mol. Biol.* 190: 379–389; Klock, G., Strähle, U. and Schütz, G. (1987) Oestrogen and glucocorticosteroid responsive elements are closely related but distinct. *Nature* 329: 734–736; and Beato, M. (1987) Induction of transcription by steroid hormones. *Biochem. Biophys. Acta* 910: 95–102).

HREs are *necessary* in order for hormonally-induced *changes* of *gene transcription* to occur, as has been delineated precisely by transfection experiments involving mutagenesis of the 5′ regulatory region of the genes (Fig. I.100).

The location of HREs varies greatly, between 2.6 kb upstream (uteroglobin gene) and 0.1 kb downstream (growth hormone gene) of the transcription initiation site. In most of the few genes studied so far, HREs are present in more than one copy, although the contribution of each individual element to the hormonal response is highly variable. For example, a 15-mer consensus sequence which shows an imperfect inverted symmetry centered around base pair number 8 (Fig. I.100) has been derived from the nucleotide sequence of 22 DNA binding sites for glucocorticosteroid receptor. A hexanucleotide is very well preserved, and may determine the primordial step in receptor binding to DNA, while there could be a weaker interaction of the second molecule of a dimeric binding form of the receptor (as found for ER) with six other nucleotides.

HREs for different steroids constitute a family of related DNA sequences which exhibit partial symmetry. A substitution of one or two base pairs at homologous sites in the palindrome can convert specificity from one steroid receptor to another. However, the variable numbers and positions of HREs do not help clarify the general molecular mechanism by which hormone-receptor complexes modify gene expression. Receptor-HRE interactions have been studied by using mutations in both receptors and HREs. The putative DNA-binding site of the receptor is necessary for hormone action, and its specificity is assessed by the results of hybridization experiments exchanging the cassette of one receptor for that of another (p. 87 and Fig. I.90B). The steroid-binding mutant of GR "nt^{i}" (nuclear transfer increase)[99] binds to DNA, but does not mediate a response, suggesting that the missing N-terminal region is implicated in the hor-

99. Dellweg, H. G., Hotz, A., Mugele, K. and Gehring, U. (1982) Active domains in wild-type and mutant glucocorticoid. *EMBO J.* 1: 285–289.

mone response. Conversely, truncation of the GR, suppressing the steroid binding site, results in a constitutively active DNA-binding receptor (not regulated by hormone). So far, no receptor region distinct from the DNA-binding region of the receptor appears neccesary for positive regulation of transcription, as has been demonstrated in procaryotic and yeast transcriptional systems.

In the absence of hormone, a steroid receptor is inactive, as it is when it binds an antihormone (p. 116). This may be due to the preclusion of binding of the receptor to DNA, brought about by its interaction with hsp 90 (p. 69). Experimentally, the receptor can bind to HRE, even in the absence of hormone or in the presence of an antihormone, provided that hsp 90 is removed. Thus, two conclusions may be drawn. First, the presence of *hormone*, in contrast to either its absence or the presence of antihormone, *permits* the efficient interaction between HRE and the receptor. Secondly, specific binding of the receptor to HRE is *necessary* but probably not *sufficient* to trigger a response to the hormone. A hormone agonist may stabilize a specific and appropriate conformation of the receptor, either of its DNA-binding site and/or of another (hypothetical) site, permitting interaction with nuclear transcription factor(s)[100–103]. When they bind to DNA, receptors may affect the structure of supercoiled DNA and/or chromatin (Fig. I.101), propagating a change that ultimately reaches the site of transcription initiation. Alternatively, the receptor may bind and slide towards the transcription machinery. The receptor may form a complex comprised of two or more receptor units, possibly binding to different HREs and/or nuclear proteins which themselves interact with other regulatory DNA segments. It is unlikely that there is direct interaction between hormone receptors and RNA polymerase II.

The formation of large multiprotein complexes[104] necessary for the initiation of transcription depends on the *structure of chromatin*, in order that steric hindrance, which would preclude DNA availability, be avoided. Susceptibility of chromatin DNA to *deoxyribonuclease I* (DNase I) *digestion* has been observed in and near HREs in contact with hormone-receptor complexes, and it disappears upon the withdrawal of hormone. A straightforward hypothesis suggests that, upon binding of receptor, a modification of the relevant nucleosome would occur, thereby making DNA available for the binding of nuclear transcription factor(s) (Fig. I.101B); however, this does not easily explain *cis*-acting, long-distance effects. Soon after exposure to steroid, changes in chromatin are observable by electron microscopy (Fig. I.102).

Besides HREs, which are enhancers, that is, positive regulators, other negative regulatory DNA segments, "silencers" or "blockers"[105], may play a role in the hormonal control of gene expression. They prevent the transcription of certain genes, as for example that of the chick ovalbumin gene, which is not expressed in the absence of hormone. Hormone-receptor complexes would remove their inhibitory *cis*-effect and increase gene transcription.

Mechanism to Ensure Cellular Specificity of a Response

In general, steroid or thyroid hormones regulate the transcription of *approximately 10* to *one thousand genes* in a given target cell. These include a number of important housekeeping genes, which are involved in the common features of protein synthesis and general metabolism, *as well as* genes coding for proteins involved in the differentiation and specific function of target cells. Contrary to bacteria, animal cells do not have operons, i.e. groups of sequential genes on the DNA which are under the control of a single operator gene. The coding sequences of protein subunits within the same molecule (e.g., hemoglobin) are even situated on different chromosomes. There must be a type of *sorting* mechanism that permits the *appropriate* interaction of HREs with some of the several thousand nuclear receptor molecules. In a differentiated cell, *specific genes* to be regulated have to be made available for receptor binding, and thus must be made "competent". This is probably achieved by regulatory developmental factors which may determine a parti-

100. Kadonaga, J. T., Jones, K. A. and Tjian, R. (1986) Promoter-specific activation of RNA polymerase II transcription by Sp. *TIBS* 11: 20–23.
101. Keegan, L., Gill, G. and Ptashne, M. (1986) Separation of DNA binding from the transcription-activating function of a eukaryotic regulatory protein. *Science* 231: 699–704.
102. Maniatis, T., Goodbourn, S. and Fischer, J. A. (1987) Regulation of inducible and tissue-specific gene expression. *Science* 236: 1237–1245.
103. Cordingley, M. G., Riegel, A. T. and Hager, G. L. (1987) Steroid-dependent interaction of transcription factors with the inducible promoter of mouse mammary tumor virus in vivo. *Cell* 48: 267–270.
104. Brown, D. D. (1984) The role of stable complexes that repress and activate eucaryotic genes. *Cell* 37: 359–365.
105. Chambon, P., Dierich, A., Gaub, M. P., Jakowlev, S., Jongstra, J., Krust, A., LePennec, J. P., Oudet, P. and Reudelhuber, T. (1984) Promoter elements of genes coding for proteins and modulation of transcription by estrogens and progesterone. *Rec. Prog. Horm. Res.* 40: 1–39.

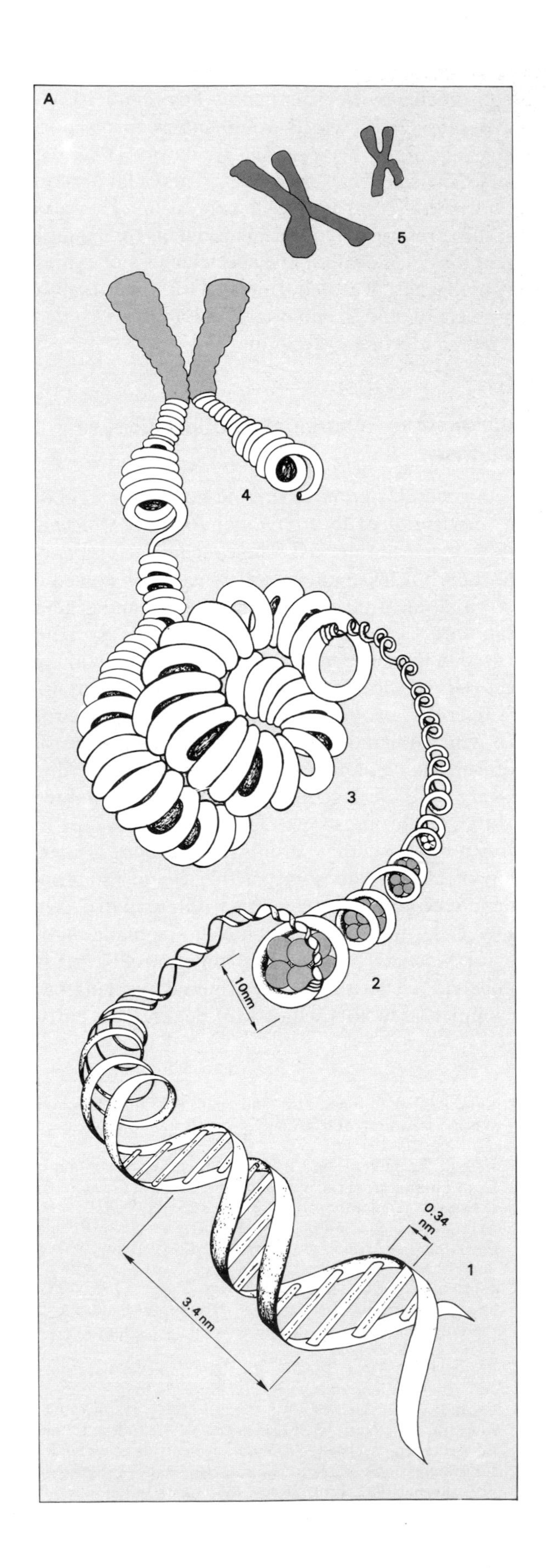

Figure I.101 Steroid action in chromatin. **A.** Chromatin: from DNA to metaphase chromosome. The double-stranded DNA (**1**) forms nucleosomes around histones (**2**), and nucleosomes are compacted in solenoid-like fibers (**3**). Chromatin fibers are condensed (**4**), forming prophasic chromosomes and then chromatids in the metaphasic chromosomes (**5**) observed during mitosis. **B.** Steroid receptor and initiation of gene transcription in chromatin. To date, studies have only been conducted at the nucleosome level. The receptor interacts with DNA of the nucleosome (**B_1**) which is then modified (**B_2**), and thus (at least) one nuclear transcription factor (TF) could bind DNA and initiate transcription. (Cordingley, M.G., Riegel, A.T. and Hager, G.L. (1987) Steroid-dependent interaction of transcription factors with the inducible promoter of mouse mammary tumour virus in vitro. *Cell* 48: 267–270; and Richard-Foy, H. and Hager, G.L. (1987) Sequence-specific positioning of nucleosomes over the steroid inducible MMTV promoter. *EMBO J.* 6: 2321–2328).

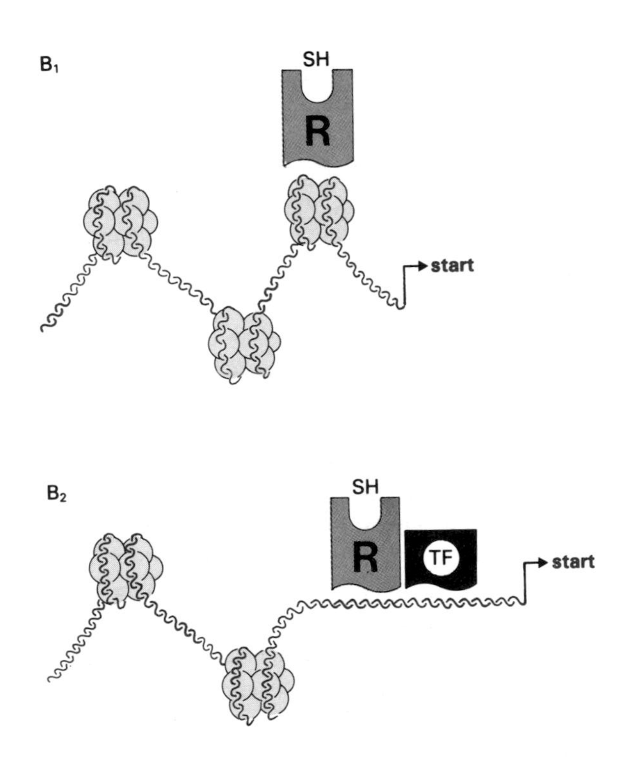

cular arrangement of chromatin. Sites which are hypersensitive to DNase I have been detected, indicating the commitment of nearby genes to be responsive prior to actual stimulation by the hormone. It has been proposed that available regulated genes are situated in *loops* of DNA specifically attached to a scaffold structure[106], or that they are present in the *nuclear matrix*

106. Gasser, S.M. and Laemmli, U.K. (1987) A glimpse at chromosomal order. *TIG* 3: 16–22.

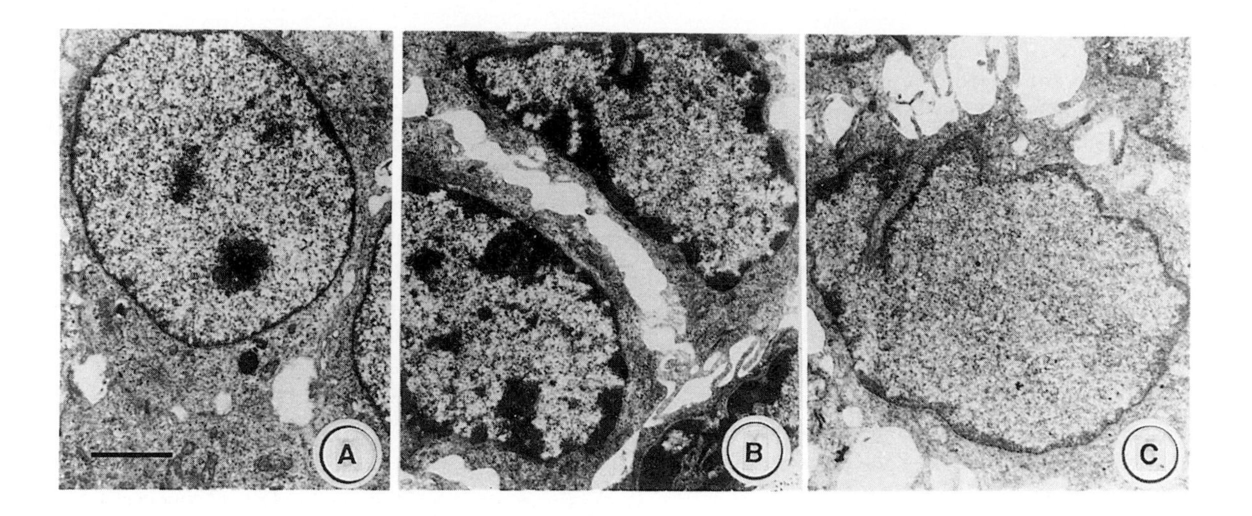

Figure I.102 Target cell nucleus: the effect of estradiol on chromatin. DMBA-induced mammary carcinoma in rat in estrus (**A**): seven days after castration (**B**), and in castrated rat 90 min after estradiol injection (**C**). Note the dispersion of clumps of condensed chromatin (**B**) due to endogenous (**A**) or administered (**C**) estrogen. The nucleus becomes more spherical in shape, and pronounced nuclear swelling has occurred. Bar=2 μm. (Vic, P., Garcia, M., André, J., Humeau, C. and Rochefort, H. (1978). Effet précoce de l'oestradiol sur l'ultrastructure de la chromatine dans l'endomètre et les tumeurs mammaires hormonodépendantes. *C.R. Acad. Sci., Paris* 287: 141–144).

where replication and transcription may begin. The state of *DNA methylation* could be involved in gene expression (active genes are hypomethylated), either directly or via changes in chromatin structure. The biological significance of acetylation, phosphorylation, poly-ADP-ribosylation and redistribution of *chromatin proteins* during differentiation and hormone action remains poorly understood. As indicated above, cell specificity is partly dependent on *specific transcription factors*, and transfection experiments have indicated that regulated genes transferred into different recipient cells respond differently to hormone.

A number of genes (probably many) are regulated by *several hormones*, at the level of transcription. Each gene of egg-white proteins is stimulated by several steroids in chick oviduct cells (Fig. I.103), and cortisol and thyroid hormones stimulate GH production in pituitary cells in culture. The same hormone, estradiol, stimulates the synthesis of ovalbumin and conalbumin in the chick oviduct but not in another organ, the liver, in which one of the proteins (ovalbumin) is not expressed and the other (conalbumin) is constitutively synthesized. The same HRE

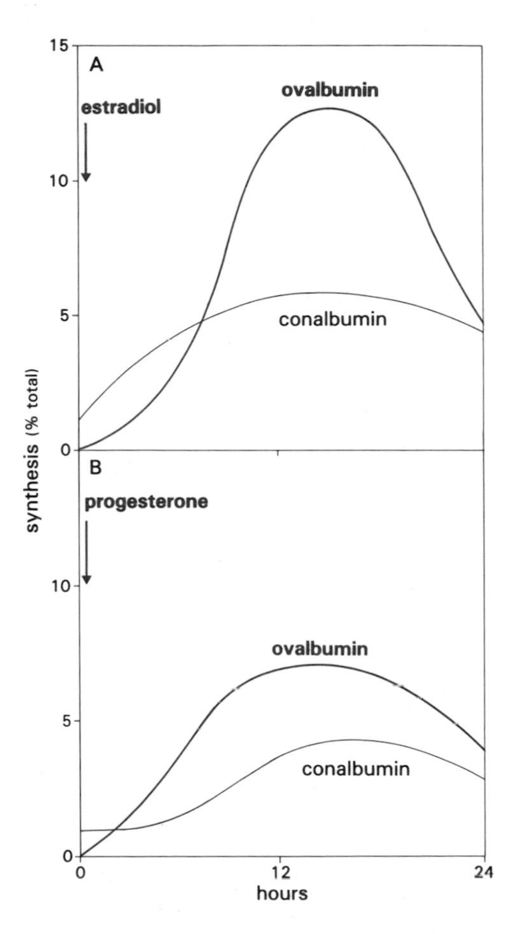

Figure I.103 Multiple hormonal regulation of protein synthesis. Immature chickens were treated with estrogen to induce growth and differentiation of the oviduct. After withdrawal, they received either estrogen (**A**) or progesterone (**B**), and the relative rates of ovalbumin and conalbumin synthesis (two egg-white proteins) were measured. (Sutherland, R. L., Geynet, C., Binart, N., Catelli, M. G., Schmelck, P. H., Mester, J., Lebeau, M. C. and Baulieu, E. E. (1980) Steroid receptors and effects of oestradiol and progesterone on chick oviduct proteins. *Eur. J. Biochem.* 107: 155–164).

may bind to different steroid receptors and be activated by their corresponding hormones. Precise, limited homologies and differences between HREs as well as between DNA-binding sites of receptors are probably responsible for the mutihormonal regulation of certain genes.

Polymorphism of the nucleotide sequences of *HREs* for the same hormone may also be responsible for differences in affinity for the corresponding receptor and for the *selective sensitivity* of different genes to the *same* concentration of hormone-receptor complexes. A spectrum of hormonal, genetic and tissular specificities combine to make each physiological regulatory system unique.

Steroids can regulate cell functions *negatively*. For example, glucocorticosteroids decrease the synthesis of prolactin in the pituitary and α fetoprotein in developing liver. In the case of POMC, the transcription of which is also decreased (p. 240), a repressor mechanism may be involved, since HRE and transcription factor binding sites may be superimposed. Differently, the *cytolytic* effect of glucocorticosteroids on lymphocytes may be related to the activation of a "lytic gene" not yet biochemically defined[107] (thus, an overall negative effect would be triggered by the enhancement of gene transcription).

A "*memory effect*" in the transcription response (Fig. I.104) has been observed in cases of secondary hormonal stimulation following a period of withdrawal. The more rapid and more efficient transcription of certain genes may be related to changes in chromatin structure, DNA methylation, or to an increase in receptor number[108].

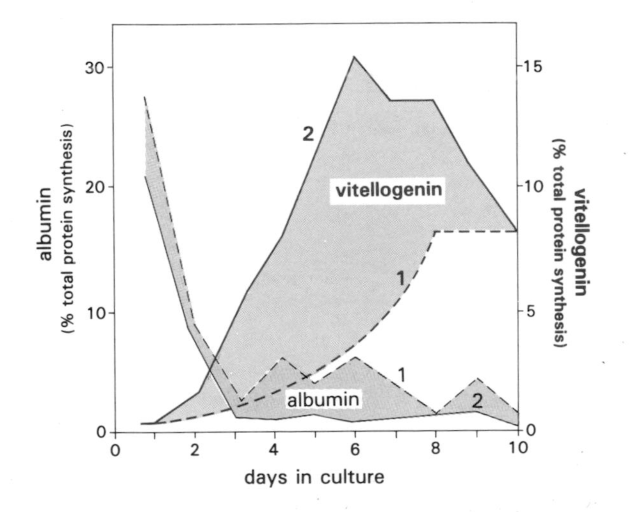

Figure I.104 Memory effect of hormone action. Male *Xenopus laevis* toads were injected with estradiol. Liver explants from untreated and 35-day estrogen-withdrawn animals were cultured in the presence of estradiol (primary (**1**) and secondary (**2**) stimulation, respectively). Vitellogenin and albumin synthesis was measured. Note the opposing effect of the steroid on the synthesis of the two proteins. Dotted line: primary response; solid line: secondary response. (Tata, J. R. and Smith, D. F. (1979) Vitellogenesis: a versatile model for hormonal regulation of gene expression. *Rec. Prog. Horm. Res.* 35: 47–95).

5.2 Post-transcriptional Mechanism of Steroid and Thyroid Hormone Action

Frequently, the increase in the synthesis of protein induced by steroid hormones is greater than can be explained by the observed increase in transcription. This increase in protein synthesis may be the result of the stabilization of mRNA, and/or may be related to another mechanism increasing the efficiency of translation. However, there are few data to substantiate *direct* effects of hormone-receptor complexes on any of the post-transcriptional steps involved in protein synthesis. More probably, in most cases the hormonal effect takes place at the level of *transcription*, for the synthesis of factors and enzymes determining translational and post-translational events[109, 110].

Xenopus Laevis Oocytes

The Xenopus laevis oocyte provides the only well documented case of protein synthesis regulated *exclusively at the post-transcriptional* level by steroid hormones. Oocytes are arrested in the prophase of the first meiotic division. Meiosis resumes when there is a gonadotropin surge with corresponding secretion of steroids, particularly progesterone, by follicular cells surrounding oocytes. After meiosis, the egg is ready for fertilization. The *reinitiation of meiosis* can be obtained in vitro by incubating oocytes with pro-

107. Gasson, J. C. and Bourgeois, S. (1983) A new determinant of glucocorticoid sensitivity in lymphoid cell lines. *J. Cell Biol.* 96: 409–415.
108. Jost, J. P., Moncharmont, B., Jiricny, J., Saluz, H. and Hertner, T. (1986) In vitro secondary activation (memory effect) of avian vitellogenin II gene in isolated liver nuclei. *Proc. Natl. Acad. Sci., USA* 83: 43–47.
109. Baulieu, E. E., Alberga, A., Raynaud-Jammet, C. and Wira, C. R. (1972) New look at the very early steps of oestrogen action in uterus. *Nature* 236: 236–239.
110. Paek, I. and Axel, R. (1987) Glucocorticoids enhance stability of human growth hormone mRNA. *Mol. Cell. Endocrinol.* 7: 1490–1507.

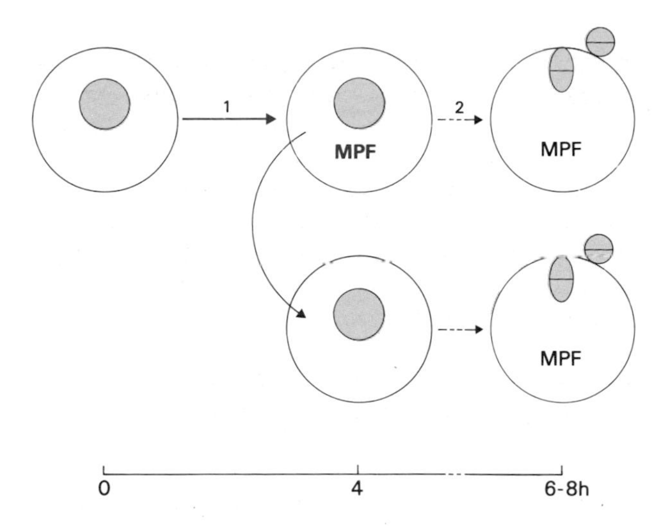

Figure I.105 Non-genomic effect of a steroid hormone: progesterone action at the membrane level in Xenopus laevis oocytes. Meiosis (or maturation) promoting factor (MPF) is formed after 4–6 h of exposure to the hormone. It can act after transfer into an oocyte not previously exposed to hormone. Phase 1 is dependent on protein synthesis, but not phase 2. MPF is also formed[111] in the absence of the nucleus. (Smith, L. D. and Wasserman, W. J. (1979) Non-nuclear steroid regulation in amphibian oocytes. Eucaryotic gene regulation. *ICN–UCLA Symposium*. Academic Press, New York, vol. 14, pp. 229–230; Masui, Y. and Clarke, H. J. (1979) Oocyte maturation. *Int. Rev. Cytol.* 57: 185–223). Cloning has indicated that MPF is identical to the fission yeast cdc2 protein. (Arion, D., et al. (1988) *Cell* 55: 371–378).

gesterone, even when it is bound to a macromolecule which precludes the entry of the hormone into the cells, but *not* if the hormone is injected directly into the oocytes. Even in the absence of the nucleus (germinal vesicle), the oocyte responds to the presence of progesterone by synthesizing and phosphorylating proteins. Protein synthesis is required for the formation of MPF (meiosis promoting factor), a necessary intermediary product which can, following its transfer into oocytes, reinitiate meiosis in the absence of hormone (Fig. I.105). Interestingly, MPF is neither species-specific nor meiosis-specific; it can activate mitosis in somatic cells of vertebrates and even yeast cells. The effect of progesterone is associated with a *decrease in cAMP* concentration which results from the *inhibition of adenylate cyclase activity* (which is not abolished by pertussis toxin (p. 336)). It is also associated with *decreased phosphorylation* of a specific membrane protein. *Insulin* and *IGF* can also induce meiosis, but the mechanism differs from that of steroids; the inhibition of their receptor tyrosine kinase activity suppresses their effect, but not that of progesterone. In fact, the oocyte system provides a clear case of potentialization between steroid and peptide hormones[111]. *Similar mechanisms* have been found in other amphibian and fish oocytes, and also in starfish oocytes, which are stimulated by methyladenine, a neurohormone. However, there is no evidence for steroid action in mammalian oocytes.

The changes in ion distribution and oxidative phosphorylation occurring after exposure of target cells to *thyroid hormones* was long thought to involve direct mechanisms at the level of the plasma membrane or mitochondria. However, most of the effects of thyroid hormones are probably secondary to transcriptional changes in the synthesis of relevant enzymes.

111. Baulieu, E. E. and Schorderet-Slatkine, S. (1983) Steroid and peptide control mechanisms in membrane of Xenopus laevis oocytes resuming meiotic division. In: *Molecular Biology of Egg Maturation. Ciba Found. Symp. 98* (Porter, R. and Whelan, J., eds), Pitman Press, Bath, pp. 137–158.

5.3 Mechanisms of Receptor-Mediated Transmembrane Signalling Implicating G Proteins

Hormone action initiated at the level of the plasma membrane generates *secondary signals* (Table I.14), most of which have been discovered recently (Table I.15). They involve a *G protein-dependent mechanism* (Table I.16 and p. 337). G proteins appear to interact with the cytoskeleton. The transmembrane signalling

Table I.14 Specific steps in hormone action at the membrane level.

Compartment	Step	Specificity
Exterior	Hormone	+++
	Receptor	+++
Cell membrane	Transduction mechanism	+
	Secondary signal	+
Interior	Differentiated response	+++

+++ and + indicate high and relatively low specificity, respectively.
Transduction mechanisms and secondary signals are limited in number as compared to hormones, receptors and responses. A higher degree of specificity is acquired by their diverse combined and temporal patterns.

Table I.15 Progress in the study of hormone action acting at the membrane level.

	Sutherland, 1958‡	Today
Hormone	Glucagon/epinephrine	~50 hormones
Transducing mechanism	Adenylate cyclase	G proteins, phosphoinositide system, arachidonic acid systems, channel/pump/antiport systems, PKs
Secondary signals	cAMP	cAMP, cGMP, DG, IP_3, Ca^{2+}, PGs, leukotrienes, glycophospholipid derivatives

‡Rall, T. W. and Sutherland, E. W. (1958) Formation of cyclic adenine ribonucleotide by tissue particles. *J. Biol. Chem.*, 232: 1065–1076.

Table I.16 G proteins involved in signal transduction.

Function
G_s Adenylate cyclase stimulator
G_i Adenylate cyclase inhibitor‡
G_t PDE-cGMPstimulator (transducing)‡
G_p Phospholipase-C stimulator‡
G_k K^+-channel regulator‡
G_c Ca^{2+}-channel regulator

‡Only these G proteins are inhibited by pertussis toxin (IAP) (p. 336).

system is not involved in the clustering, internalization and degradation of hormone-receptor complexes (p. 224).

Cyclic AMP

While looking for a mechanism which would explain how enzymes are affected by hormones, Sutherland made a fundamental discovery. He found that the addition of either glucagon (active in the liver) or epinephrine (active in muscles) to purified phosphorylase (p. 506) had no effect, in contrast to hormones which were added to intact cells or administered in vivo. Moreover, both hormones were active when added to total homogenates, but not if a preliminary low speed centrifugation removed the membrane-containing fraction. This experiment suggested that the site of hormone action was in the particulate fraction and, therefore, that the hormone did not act directly on phosphorylase. In fact, hormones were shown to activate a membrane enzyme, adenylate cyclase, forming a mediator, cAMP, which was

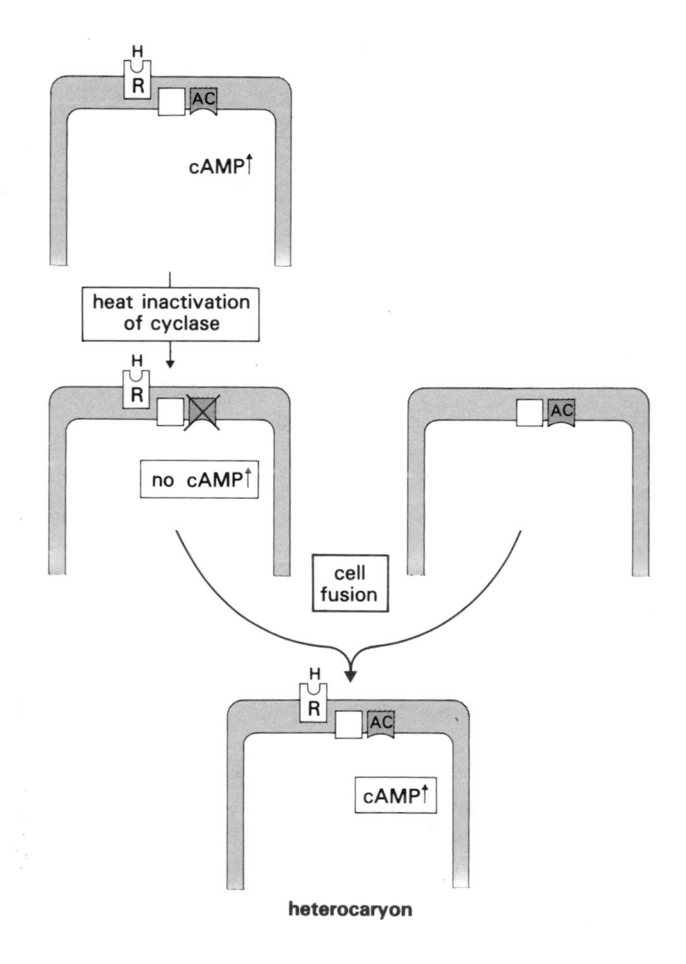

Figure I.106 Membrane receptors and adenylate cyclase are separate entities. Cells lacking functional adenylate cyclase and cells devoid of receptor are not responsive to hormone. After fusion, heterocaryons respond to hormone. (Schramm, M., Orly, J., Eimerl, S. and Korner, M. (1977) Coupling of hormone receptors to adenylate cyclase of different cells by cell fusion. *Nature* 268: 310–313).

Table I.17 Hormones acting via a cAMP system.

cAMP increase	cAMP decrease
(Cholera toxin, p. 336)	Somatostatin
ACTH	Opioids
α-MSH	(Nor)epinephrine (α_2-adrenergic)
TSH	Dopamine (D_2)
FSH	Adenosine (R_i, A_1)
LH	Bradykinin
Vasopressin (V_2)	Progesterone‡
Glucagon	
Gastrin	
Secretin	
PTH	
Calcitonin	
(Nor)epinephrine (β_1- and β_2-adrenergic)	
PGE_1	
Dopamine (D_1)	
Histamine (H_2)	
Adenosine (R_a, A_2)	
VIP	
Serotonin	
GnRH	

The subtype of receptors is indicated in parentheses.
‡No steroid or thyroid hormone is included in the table, with the exception of progesterone which binds to the membrane receptor of the amphibian oocyte (p. 82).

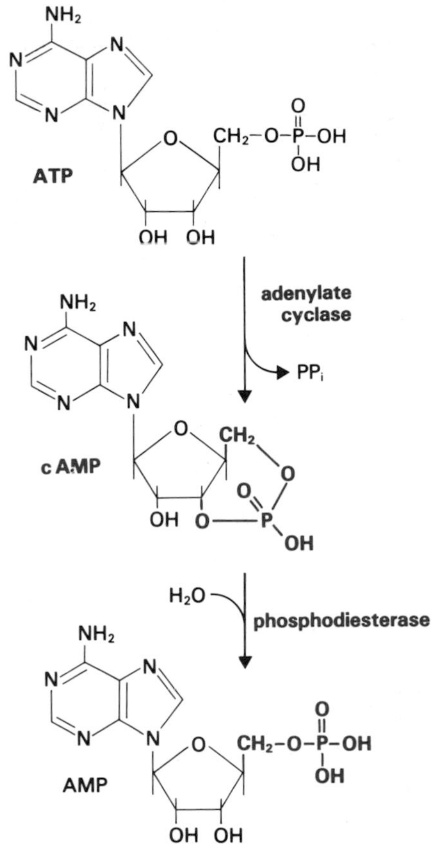

ATP
adenylate cyclase
PP_i
cAMP
H_2O
phosphodiesterase
AMP

Figure I.107 Synthesis and degradation of cAMP.

called a *second messenger*. A number of hormones (Table I.17) have also been shown to act via the generation of cAMP, while others decrease adenylate cyclase activity. Receptors and adenylate cyclase were shown to be separate entities (Fig. I.106).

Adenylate cyclase is an intrinsic constituent of almost all eucaryotic cells. It spans the lipid bilayer of plasma membranes, and is glycosylated at an extracellular site. Its substrate, ATP-Mg^{2+}, is stoichiometrically converted to cAMP (Fig. I.107) and pyrophosphate (PP_i), both of which are released in the cytoplasm. In fact, adenylate cyclase does *not interact directly* with hormone receptors in the membrane (p. 79). An *intermediary G protein*[112] couples the receptor to the enzyme, and two classes of proteins, called G_s (*stimulatory*) and G_i (*inhibitory*), induce a respective increase or decrease in adenylate cyclase activity.

Several criteria define *cAMP as second messenger* in the mechanism of action of many hormones. The hormone should modulate the formation of cAMP in its target cells, including in in vitro experiments, and should do so according to the kinetics of hormone-induced activity. Inhibitors of phosphodiesterase, an enzyme which destroys cAMP in all tissues, should act synergistically with the hormone. Finally, the addition of cAMP, or one of its lipophilic acyl derivatives which enters cells more easily (e.g., dibutyryl-cAMP), should mimic the activity of the hormone.

The *quasi-unique mode of action* of cAMP involves the *regulation of a specific protein kinase* called *cAMP-dependent protein kinase (PK-A)* (Fig. I.108) (p. 106)[113,114]. Many protein substrates of PK-A are themselves enzymes, and the cAMP system provides a powerful means of amplification when G_s is involved (Fig. I.109). The demonstration of the

112. Rodbell, M. (1980) The role of hormone receptors and GTP-regulatory proteins in membrane transduction. *Nature* 284: 17–22.
113. Krebs, E. G. (1979) Phosphorylation–dephosphorylation of enzymes. *Annu. Rev. Biochem.* 48: 923–959.
114. Greengard, P. (1978) Phosphorylated proteins as physiological effectors. *Science* 199: 146–152.

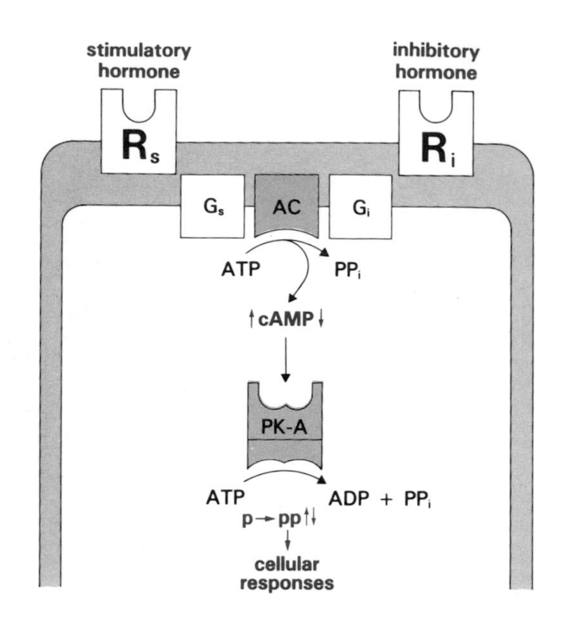

Figure I.108 Cyclic AMP, the classical second messenger. Many hormones act via receptor-G-protein-adenylate cyclase systems, which ultimately leads to an increase or a decrease in cAMP in the cytoplasm. The effector is a protein kinase (PK-A) which phosphorylates Ser/Thr residues of the proteins (p→pp) specifically involved in the cellular response.

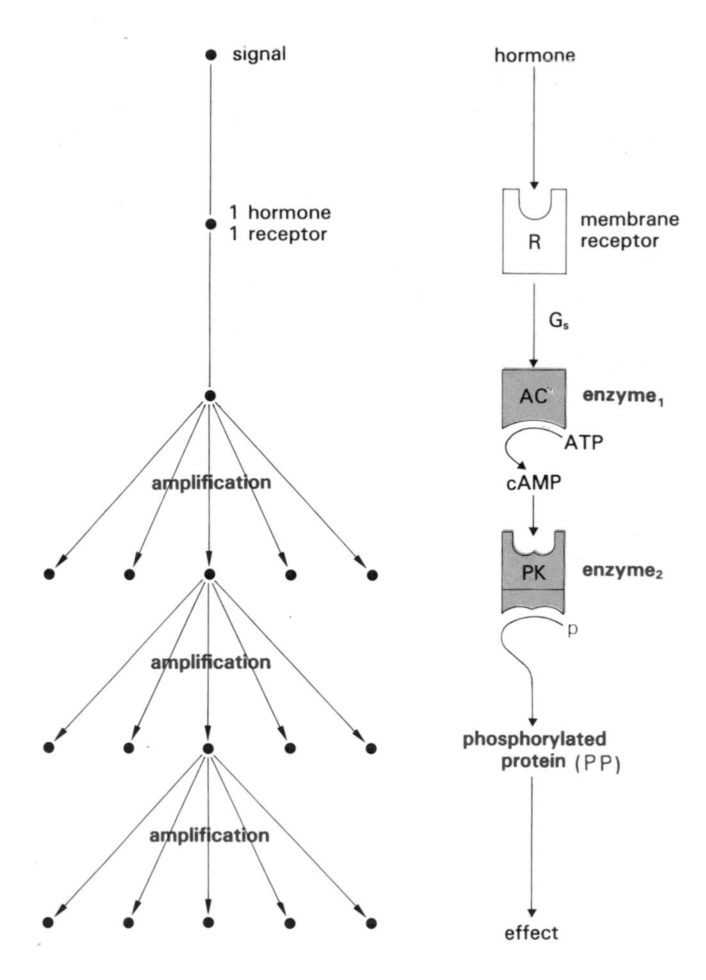

Figure I.109 The cAMP-second messenger system provides hormone amplification. The protein (p) is an enzyme activated by phosphorylation (pp).

second messenger activity of cAMP has indicated, for the first time, a regulatory mechanism implicating the chemical, post-translational modification of an enzymatic function. However, in *many* cases the physiological *substrates* of PK-A *remain unknown*. There is only one other enzyme with which cAMP interacts: S-adenosyl 1-homocysteine hydrolase, implicated in methionine-dependent methylation.

The cellular concentration of cAMP is not only dependent on adenylate cyclase, but also on specific *phosphodiesterases* (*PDE*), very active and almost ubiquitous enzymes which generally are not hormonally regulated. Indeed, the very concept of a second messenger *implies* that it is *rapidly inactivated* and that it ensures the *versatility* of the regulatory mechanisms triggered by the hormone. In a few cases, undegraded cAMP is eliminated from target cells. Clinically, the measurement of plasma cAMP following injection of a stimulatory hormone, such as glucagon or PTH (p. 646), is used as an indication of hormone function.

Cyclic GMP

The formation of another cyclic nucleotide, cGMP, is catalyzed by a different enzyme, *guanylate cyclase*. In a number of cases, its activation is correlated with a decrease in adenylate cyclase activity. The only effect of physiological importance reported thus far is its stimulation by ANF (p. 619).

The Phosphoinositide System[115,116]

The regulatory activities of many hormones are exerted by changing the *metabolism* of a class of phospholipids, the phosphoinositides (Fig. I.110 and Table I.18), which are present in *limited amounts* in plasma membranes (a few percent of membrane lipids). The metabolism of *phosphatidylinositol* (*PI*) gives rise to

115. Michell, R.H. (1975) Inositol phospholipids and cell surface receptor function. *Biochim. Biophys. Acta* 415: 81–147.
116. Berridge, M.J. and Irvine, R.F. (1984) Inositol trisphosphate, a novel second messenger in cellular signal transduction. *Nature* 312: 315–321.

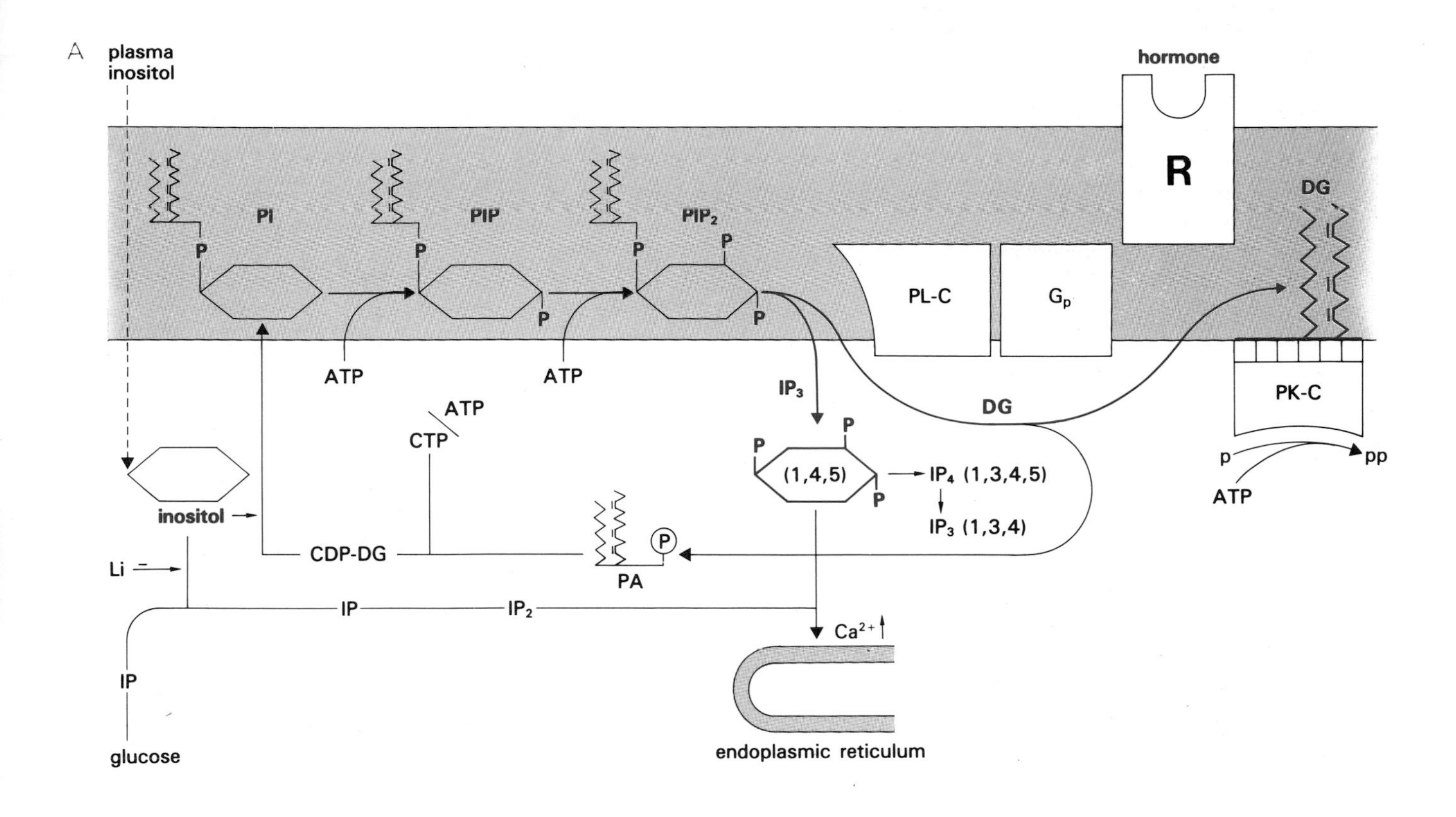

Figure I.110 The phosphoinositide system. **A.** Metabolism of phosphoinositides in the membrane: formation of two second messengers, diacylglycerol (DG) and inositol triphosphate (IP_3). All abbreviations as in the text (frequently, phosphatidylinositol is also abbreviated as PtdIns). **B.** Three main components of the phosphoinositide system. Hydroxyls or phosphate esters in 1, 2, 3 and 5, and in 4 and 6 are all *cis* within each group, and the two sets are *trans* to each other. (Stereochemistry and numbering of *myo*-inositol: Parthasarathy, R. and Eisenberg, F. Jr. (1986) The inositol phospholipids: a stereochemical view of biological activity. *Biochem. J.* 235: 313–322. Note that the tumor promoter TPA (12 O-tetradecanylphorbol 13-acetate) contains a DG-like structure (in red).

Table I.18 Hormones acting via the phosphoinositide system.

TRH
GnRH
Vasopressin (V_1)
Cholecystokinin
Angiotensin II
Norepinephrine (α_1-adrenergic)
Epinephrine
Histamine (H_1)
Serotonin (2)
Acetylcholine (mACh)
Dopamine (D_2)
VIP
NGF
EGF
PDGF
Bombesin
Platelet activating factor
Substance P
$PGF_{2\alpha}$
Thromboxanes
(Glucose in pancreatic B cells)

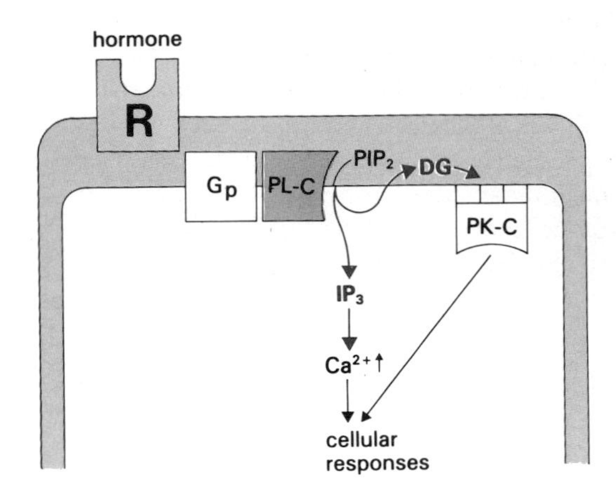

Figure I.111 DG and IP_3: mechanism of action. Some hormones act via a receptor-G-protein-phospholipase-C (PL-C) system, acting upon PIP_2 to produce IP_3 and DG. Subsequent effectors triggering specific cellular responses include increases of free Ca^{2+} concentration in the cytoplasm and of PK-C activity.

two products, *(1,2)-diacylglycerol* (*DG*) and *inositol (1,4,5)-trisphosphate* (*IP_3*), which are now recognized as two interacting second messengers for many hormones (Fig. I.111). This dual signalling system is remarkable in its ability to control *immediate* cellular events (e.g., ionic changes involving Na^+-H^+ antiporter for growth factors, and K^+ channels for neurotransmitters) *and long-term* responses (e.g., genomic effects implicated in growth and possibly memory).

The key compound formed in the membrane is *phosphatidylinositol (4,5)-bisphosphate* (*PIP_2*), which, quantitatively, is always a minor component of total phosphoinositides. It is formed from PI by sequential phosphorylation. A *specific phospholipase C (PL-C)* ("phosphoinositase") hydrolyzes PIP_2 to produce DG and IP_3‡. PL-C is the *hormone-regulated* enzyme of the phosphoinositide system. Multiple forms of PL-C of similar function have been detected. Membrane lipids regulate PL-C activity. *PL-C activation by receptors* leads to an acceleration of the hydrolysis of PIP_2 and thus to an increase in the production of the two second messengers. As for adenylate cyclase, there is a *G protein*, called *G_p* (Table I.16), which mediates the response of the receptor to PL-C.

DG is an *activator of protein kinase C* (PK-C)[117] (p. 107). It is converted secondarily, by a membrane kinase, to phosphatidic acid (PA), which in turn reacts with CTP to produce CDP-DG.

IP_3 diffuses into the cytoplasm and acts as a second messenger, causing the *release of Ca^{2+}* from intracellular stores, primarily from the *endoplasmic reticulum*. IP_3 binding sites are particularly abundant in selected regions of the brain, not always stoichiometrical with PK-C. As would be expected of a potent second messenger, IP_3 is rapidly metabolized by conversion to inositol (1,4)-bisphosphate (IP_2), which is inactive on Ca^{2+} release. Further dephosphorylation to inositol 1-phosphate and inositol can be blocked by lithium. Inositol may also come from glucose 6-phosphate, via inositol 1-phosphate. Therefore, *Li^{2+} ions* cause a decrease in the formation of inositol and thus of phosphoinositide synthesis. This effect may dimin-

‡The hydrolysis of an unrelated inositol glycophospholipid (releasing inositol phosphoglycan, IPG) seems to be involved in the mechanism of insulin action (p. 515).

117. Nishizuka, Y. (1986) Studies and perspectives of protein kinase C. *Science* 233: 305–313.
118. Rasmussen, H., Apfeldorf, W., Barrett, P., Takuwa, N., Zawalich, W., Kreutter, D., Park, S. and Takuwa, Y., eds (1986) Inositol lipids: integration of cellular signalling systems. In: *Phosphoinositides and Receptor Mechanisms*. Alan Liss, New York, pp. 109–147.

ish the activity of certain neurotransmitters, and may be beneficial in the treatment of depression. Apart from IP_3, other polyphosphorylated inositols, such as inositol (1,3,4,5)-tetrakisphosphate, its dephosphorylated product inositol (1,3,4)-trisphosphate, penta- and hexaphosphates, and inositol (1,2-cyc,4,5)-trisphosphate (a cyclic nucleotide), have been identified, but their physiological importance has not been demonstrated.

5.4 Hormonal Control of Cellular Metabolism

Calcium and Hormone Action[118]

Calcium is an intracellular regulatory signal. The control of its concentration is particularly complex, because of the several rich intracellular stores of calcium, and because of the bidirectional exchange with the exterior, through multiple channel systems.

The *cellular concentration* of free Ca^{2+} is low, maintained tightly between *~0.1* and *0.2 μM* (Fig. I.112). The concentration of extracellular Ca^{2+} is >1 mM. Ca^{2+} *enters* cells through selective Ca^{2+} channels in the plasma membrane which differ in their functional dependency on membrane potential, and through the Na^+ channel or a Na^+-Ca^+ exchanger. Calcium is transported out of the cell by two *exit* mechanisms which function to maintain relatively low intracellular concentrations of the ion: a Ca^{2+}-H^+ ATPase (a Ca^{2+} pump with high affinity and low capacity) and a Na^+-driven Ca^{2+} antiporter (a Na^+-Ca^{2+} exchanger) driven by the sodium gradient which is itself dependent on Na^+-K^+ ATPase (the sodium pump). Both systems are more active as Ca^{2+} concentration increases, and the activity of the Ca^{2+} pump is increased by interaction with calmodulin and through phosphorylation by various protein kinases.

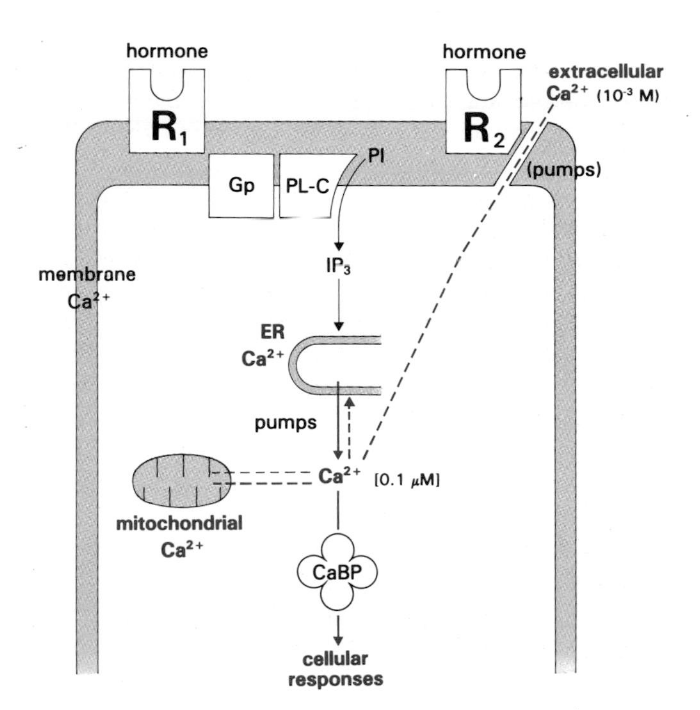

Figure I.113 Free cytoplasmic Ca^{2+} ions: are they second or third messengers? Some hormones act via IP_3, which in turn causes the release of Ca^{2+} from endoplasmic reticulum stores. Other hormones interact with gated Ca^{2+}channels in the membrane. The maintenance of free cytoplasmic calcium concentration depends mostly on the activity of mitochondrial and plasma membrane pumps. In fact, most cytoplasmic calcium is strongly bound to calcium binding proteins (CaBPs), particularly calmodulin, which are themselves mediators of calcium effects.

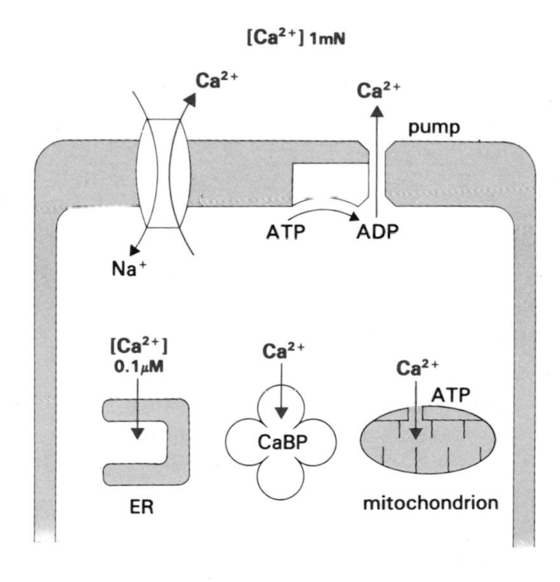

Figure I.112 Principle mechanisms of the maintenance of low intracellular Ca^{2+} concentration.

Cytoplasmic organelles also contain Ca^{2+}, mostly in the non-ionic form (phosphates), at a concentration between 1 and 10 μM. The *endoplasmic reticulum* (sarcoplasmic reticulum in muscle cells) is quantitatively the most important site of Ca^{2+} storage, and its uptake of calcium depends on ATP. Mitochondria and the plasma membrane accumulate calcium via a specific carrier.

Generally, hormones *cause a rapid* increase in cytosol Ca^{2+} concentration (within seconds). This is transitory, and is *followed* by an increase in the activity of the calcium *pump* and a *sustained, elevated rate of transfer* across the plasma membrane.

As a direct result of hormone-receptor interaction, the opening of calcium channels in the plasma membrane permits the flow of Ca^{2+} into the cell, where it affects cellular function. Thus, it acts as a *second messenger* (Fig. I.113). Alternatively, it may be designated as a *third messenger* when it is released from intracellular stores by IP_3. Regardless of whether Ca^{2+} functions as a second or third messenger, an increase in Ca^{2+} may affect *cellular function in two ways. Initially*, either Ca^{2+} itself or Ca^{2+}-calmodulin complexes (Fig. I.114) are implicated, and effects are generally brief. *Secondarily*, a sustained response can occur, which often involves the activation of PK-C. The detailed mechanism of calcium action may involve oscillating concentrations of this cation; the frequency of oscillations would determine the intensity of the response.

Physiologically, Ca^{2+}-mediated responses are central in muscular contraction and in neurosecretion, as well as in the secretory activity of exocrine and endocrine glands. It is also involved in many enzymatic activities and in cell division. Thus, the diversity of Ca^{2+}-regulated activities contrasts with the limited number of molecular mechanisms involved.

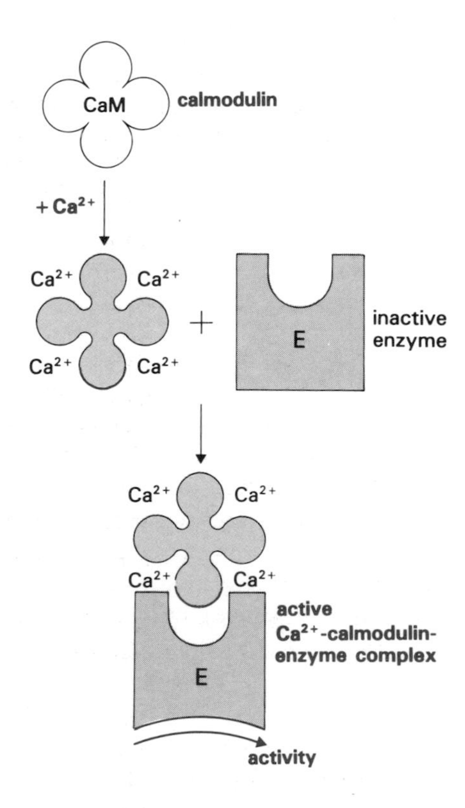

Figure I.114 Calmodulin mediates the activation of enzymes by calcium. Ca^{2+}-calmodulin complexes also modulate the activity of many other proteins.

Protein Phosphorylation[119]

The first demonstration that phosphorylation could activate an enzyme involved in cellular metabolism was that of the transition from inactive phosphorylase *b* to phosphorylase *a*, which is involved in the degradation of glycogen (p. 506). Two series of important studies followed. First, *several* distinct *protein kinases* (*PKs*) were discovered (Table I.19), all of which use ATP-Mg^{2+} as a source of phosphate and energy to phosphorylate proteins. It is now thought that there may be as many as 1000 protein kinases![120] Their primary structures include several highly conserved sequences, such as Gly-X-Gly-Y-X-Gly, followed by a Lys, ~20 aa residues downstream; they constitute part of the ATP-binding site. At a point 80 to 180 aa residues from this Lys toward the C-terminal extremity, there is a region of ~60 aa which contains conserved Arg-Asp-Leu, Asp-Phe-Gly and Ala-Pro-Glu sequences. In tyrosine protein kinases, between the two last triplets there is a major autophosphorylation site and a Tyr characteristically surrounded by acidic amino acids. Several *protein phosphatases* were also described. Secondly, a number of protein kinases were found to be *regulated by hormones*. The best known are PK-A, PK-C, a group of Ca^{2+}-dependent kinases, and kinases which phosphorylate receptors, all of which esterify Ser/Thr residues. Some *membrane receptors* (p. 78) and *oncogenes* (Table I.21) are Tyr-PKs. Removal of the ATP binding site of I-R suppresses its activity.

Table I.19 Protein kinases involved in signal transduction.

Class	Regulatory agents
PK-A	cAMP
PK-G	cGMP
PK-C	Ca^{2+} and DG (phorbol ester)
PK-Ca	Ca^{2+}
PK-CaM	Ca^{2+}-CaM
Ser/Thr-PKs (many, most of unknown substrates)	(various hormones, by unknown mechanisms)
Tyr-PKs:	
receptors	Specific ligand
oncogene products	Constitutively active

All protein kinases, with the exception of Tyr-PKs, phosphorylate Ser/Thr residues. Each PK cited actually represents a *class* of enzymes.

PK-A

PK-A is a heterotetramer (Fig. I.115) composed in its inactive form of *two identical heterodimers*, each of which is comprised of a *regulatory* subunit (also designated as the cAMP receptor) and a *catalytic* subunit. There are two types of regulatory subunits: type I (M_r~48,000) and type II (M_r~55,000). Each regulatory subunit binds two molecules of cAMP, the two sites having slightly different binding specificity but a similar affinity (K_D~10 nM). The catalytic subunit (M_r~38,000) shows sequence homology with many other protein kinases, including Tyr-PKs. The *binding of cAMP* is cooperative, and provokes the *dissociation* of the oligomeric structure, with release of the active catalytic subunits, while the regulatory subunits are autophosphorylated on serine. Most of the protein substrates of PK-As have not yet been identified, with the classical exception of enzymes involved in glycogenolysis and glycogenogenesis (p. 506). Although many enzyme, membrane, ribosome and chromatin proteins can be phosphorylated by PK-A, it is not yet clear whether all of these reactions have physiological significance. PK-A, like cAMP itself, may be a non-specific piece of the cellular machinery, conveying information to any of the protein substrates present in the target cell. The cellular concentration of cAMP is ~1 μM, a concentration much higher than the K_D of cAMP binding to PK-A. However, this enzyme is very sensitive to small changes in cAMP concentration caused by hormones, which is probably due to a *cooperative* effect of ligand binding. The reason for the variable proportion of the two isoforms I and II is unknown. PK-A is essentially *cytoplasmic*; whether a small amount is present in the nucleus remains conjectural.

A *cGMP-dependent protein kinase*, *PK-G*, has been described. Formed of a single polypeptide chain, its physiological importance has not been established.

119. Rosen, O.M. and Krebs, E.G., eds (1981) *Protein Phosphorylation*. Cold Spring Harbor Lab., N.Y., vol. 8A.
120. Hunter, J. (1987) A thousand and one protein kinases. *Cell* 50: 823–829.
121. Parker, P.J., Coussens, L., Totty, N., Rhee, L., Young, S., Chen, E., Stabel, S., Waterfield, M.D. and Ullrich, A. (1986) The complete primary structure of protein kinase C – the major phorbol ester receptor. *Science* 233: 853–866.

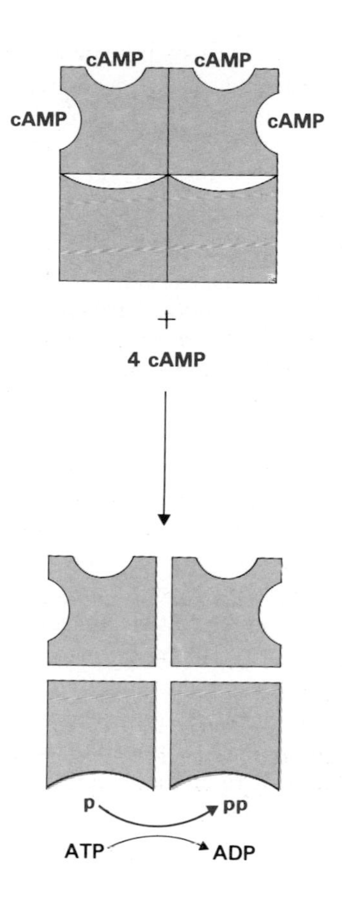

Figure I.115 Protein kinase A (PK-A) is a cAMP-dependent protein kinase. Its proposed structure indicates two cAMP binding sites per binding subunit. The catalytic sites are obliterated by cAMP binding subunits when cAMP is absent.

PK-C

PK-C[121] (Fig. I.116) depends upon Ca^{2+} and phospholipid for its activity. The phospholipid is usually *phosphatidylserine*, a constituent of the plasma membrane. The requirement of PK-C for Ca^{2+} is *absolute*, and *DG* (containing at least one unsaturated fatty acid) is required for activity *at low physiological* concentrations of *calcium* in the cytosol, since it increases the affinity of the enzyme for the divalent ion. This is the basis for defining DG as a second messenger responsible for the activation of PK-C (p. 104). A *family* of protein kinase C-related genes has been identified in different species, including humans[121]. Alpha, β and γ sequences are highly homologous, but are encoded by different chromosomes and are expressed tissue-specifically. Their structure includes a kinase domain, potential calcium binding site(s), interspersed variable regions, and a cysteine-rich N-

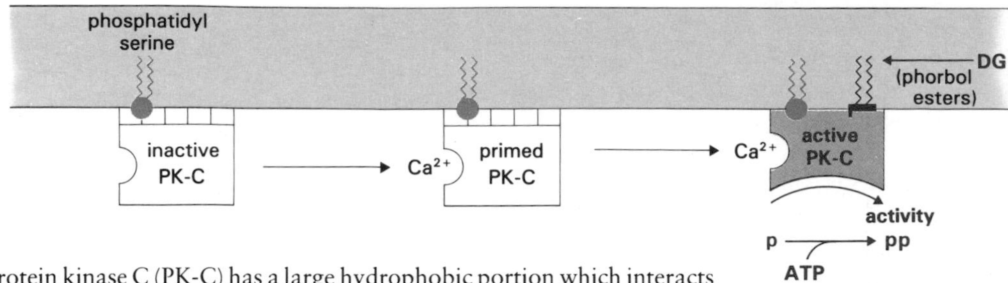

Figure I.116 Protein kinase C (PK-C) has a large hydrophobic portion which interacts with phosphatidylserine molecules of the inner layer of the plasma membrane. Its activity is dependent on Ca^{2+} and is stimulated by diacylglycerol (DG). Pharmacologically, phorbol esters can mimic DG. The specificity of the protein substrate (p) is unknown. The phosphorylated protein (pp) is esterified on specific Ser and/or Thr residues.

terminal sequence. Their molecular weight is ~80,000. Because of its *requirement for lipid*, PK-C cannot function in the soluble phase of the cytoplasm, and thus substrates must migrate to the vicinity of the membrane. Myosin light chains, glycogen synthase, and several receptors have been shown to be substrates in vivo. PK-C phosphorylation of the I-R, EGF-R and β-A-R reduces the binding affinity of their respective hormones. When bound to the membrane and thus activated, PK-C continues to be dependent on Ca^{2+}, frequently provided by an influx from the exterior. *Tumor-promoting* agents, such as phorbol esters, are structurally similarly to DG (Fig. I.110B), and can mimic its effect on enzyme activation. Thus, PK-C functions pharmacologically as the principal *receptor for phorbol ester*, and is responsible, at least in part, for the ability of tumor promoters to induce hyperplasia.

Other Calcium-Dependent Protein Kinases

Besides PK-C, *many* protein kinases are activated by calcium, and some of them have precise substrate specificity. In most cases, a *calcium-calmodulin* complex is involved (PK-CaM) (p. 630). In fact, calmodulin is a subunit of phosphorylase kinase (Fig. I.117). Activation of protein kinase is the principal means by which Ca^{2+} regulates cellular function.

Phosphorylation of Receptors

Receptor phosphorylation is involved in the regulation of *transmembrane signalling* and in *cross-talk of receptors*[122] (Fig. I.118).

PK-A and *PK-C* can phosphorylate all receptors acting via a *G protein*, a process generally involved in agonist-induced *desensitization*. There is also a *receptor protein kinase*, different from all other known kinases, which phosphorylates several receptors coupled to adenylate cyclase[122]. Specificity is ensured by the fact that *only receptors which are occupied* by an agonist are substrates.

Phosphorylation[123] of the *I-R* is complex. Insulin binding stimulates Ser/Thr phosphorylation, near the C-terminal extension, by PK-A/PK-C or other kinases, which results in down-regulation and decreased Tyr-PK activity. However, as was demonstrated initially in vitro, the I-R (β subunit) can be phosphorylated by its own Tyr-PK in two domains: one in the C-terminal and the other on the mid-portion of the β subunit (Fig. I. 78), and an increase in Tyr-PK activity on exogenous substrates is observed. Proleolytic truncation of the α subunit, like insulin binding, provokes autophosphorylation activation. Stimulation of *EGF-R* (Fig. I.119)[124] causes autophosphorylation on Tyr-1173 (major site), Tyr-1148 and Tyr-1068, at the C-terminal extremity of the protein, but the effect of EGF-R on Tyr-PK activity is uncertain[125]. The phosphorylation of Thr-654 is catalyzed by PK-C, which leads to the internalization of the receptor and the inhibition of Tyr-PK.

122. Sibley, D.R., Benovic, J.L., Caron, M.G. and Lefkowitz, R.J. (1987) Regulation of transmembrane signalling by receptor phosphorylation. *Cell* 48: 913–922.
123. Kahn, C.R. (1985) The molecular mechanism of insulin action. *Annu. Rev. Med.* 36: 429–451.
124. Carpenter, G. and Cohen, S. (1979) Epidermal growth factor. *Annu. Rev. Biochem.* 48: 193–216.
125. Bertics, P.J. and Gill, G.N. (1985) Self-phosphorylation enhances the protein–tyrosine kinase activity of the epidermal growth factor receptor. *J. Biol. Chem.* 260: 14642–14647.

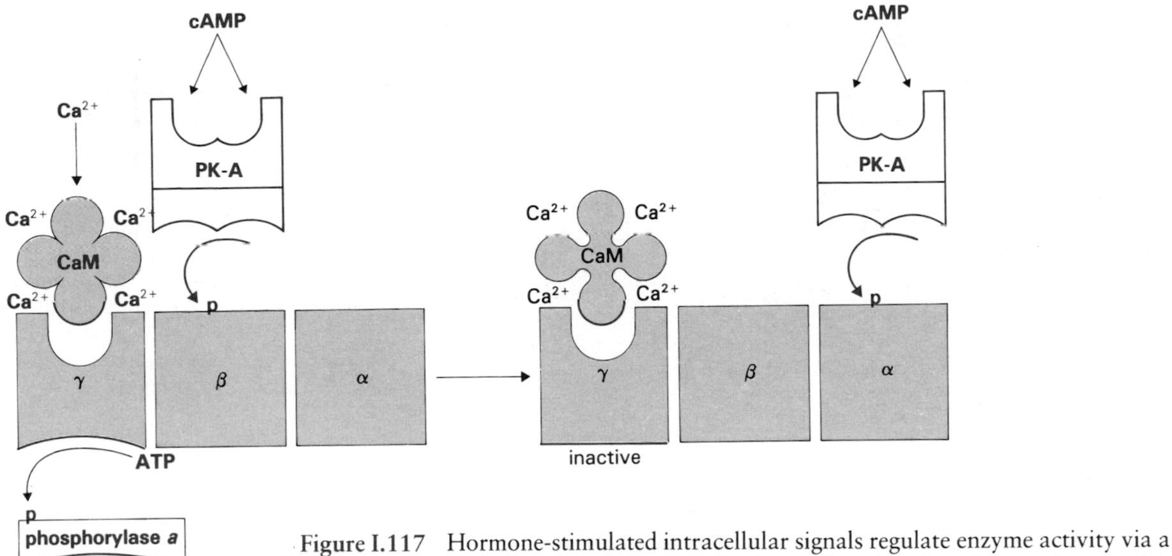

Figure I.117 Hormone-stimulated intracellular signals regulate enzyme activity via a phosphorylation/dephosphorylation mechanism. Phosphorylase kinase is a multi-subunit enzyme that activates phosphorylase by phosphorylation. The catalytic γ subunit is activated when the regulatory δ subunit (calmodulin) binds Ca^{2+}, and the regulatory β subunit is phosphorylated by PK-A (activated by cAMP). Secondarily, PK-A phosphorylates the regulatory α subunit, thus favoring the dephosphorylation of the β subunit (leading to its inactivation) by a phosphoprotein phosphatase.

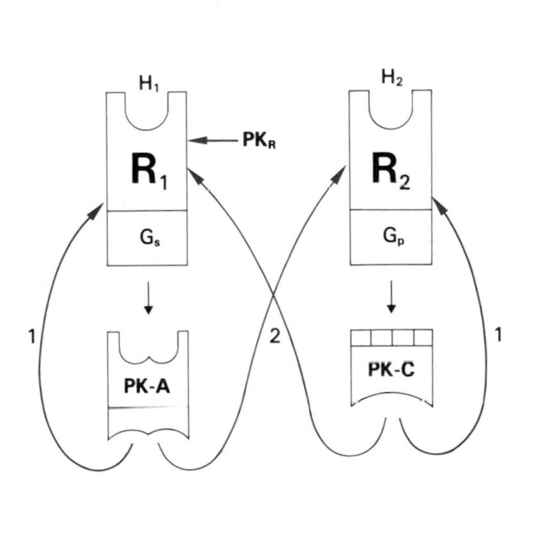

Figure I.118 Phosphorylation of receptors: a complex network. Both PK-A and PK-C stimulated by the H_1-R_1-G_s and the H_2-R_2-G_p systems can phosphorylate their receptors homologously (**1**) and heterologously (**2**, receptor "cross-talk"). In addition, a receptor protein kinase, PK_R, is able to phosphorylate agonist-coupled receptors coupled to adenylate cyclase. Functional consequences of (1) and (2) are usually negative.

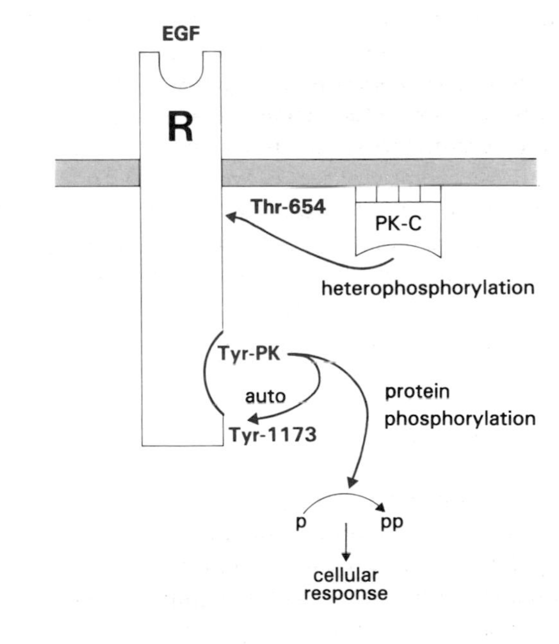

Figure I.119 Auto- and heterophosphorylation of the EGF receptor.

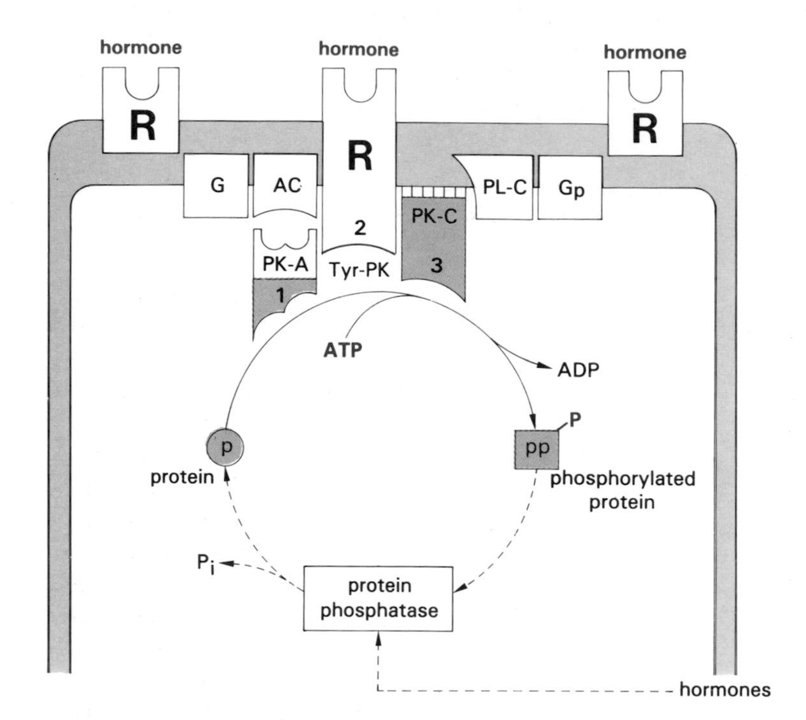

Figure I.120 Hormone-regulated phosphorylation of proteins (p→pp). Three important types of protein kinases are represented. The cycle of protein phosphorylation and dephosphorylation includes a protein phosphatase which may itself be under hormonal control.

Thus, several phosphorylation pathways operate by providing a series of positive/negative, homologous/heterologous regulating circuits. Often, receptor phosphorylation is associated with the *internalization* of the *receptor* in sequestered compartments where dephosphorylation occurs. Generally, receptor phosphorylation *regulates receptor function negatively*: decreased interaction with G protein, lower agonist binding affinity and decreased enzymatic activity (for EGF-R and I-R) are associated with Ser and/or Thr phosphorylation. Conversely, *Tyr-autophosphorylation* of receptors is a mechanism of *positive* regulation.

Protein Phosphatases

Cycles of protein phosphorylation and dephosphorylation are involved in many hormone activities (Figs. I.117 and I.120). The role of protein phosphatases is therefore very important[126]. Many are activated by Ca^{2+}*-calmodulin* complexes or *cAMP* (p. 632).

The complex and reciprocal activation of protein kinases and protein phosphatases seems to be fundamental to the precise control of metabolic activities. Despite remarkable advances made recently, however, specific substrates of kinases and phosphatases are not identifiable in most cases.

Membrane Lipids and Hormone Action

(Fig. I.121)

Arachidonic acid, or another polyunsaturated fatty acid, is often found in the β position of hormonally regulated phospholipids (including phos-

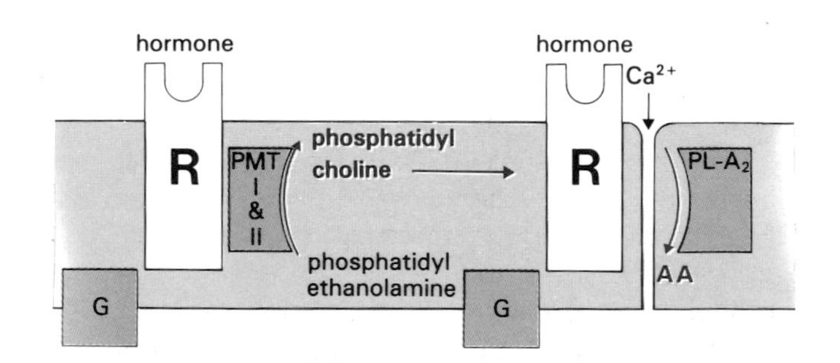

Figure I.121 A schematic and hypothetical representation of phospholipid methylation in hormonal activities at the membrane level. PMTI and PMTII are phospholipid methyltransferases I and II. Some *S*-adenosyl L-methionine analogs (e.g., SAH: 5′-adenosylhomocysteine and SIBA: 5′-deoxy-5′*S*-isobutylthioadenosine) are potent enzymatic inhibitors. In addition to the phospholipid methylation process, the release of arachidonic acid (AA), by $PL-A_2$, from phospholipid is also indicated, but its potential metabolism (Fig. I.110A) is not represented.

phatidylinositol and phosphatidic acid). When released by *phospholipase* A_2 (PL-A_2), arachidonic acid may be subsequently metabolized to produce potentially active *prostaglandins* and other related compounds (p. 590).

Changes in the structure of membrane lipids and in their fluidity have been proposed as additional mechanisms affecting the function of receptors in the plasma membrane. The enzymatic *methylation of phospholipids* may be a response to several hormone and hormone-like agents (β-adrenergic agonists, bradykinin, and chemotactic peptides)[127]. There are *two methyltransferase* enzymes that are distributed asymmetrically in the membrane, but they account for a relatively minor fraction of phosphatidylcholine synthesis. Phospholipid methyltransferase I, found near the cytoplasmic side of the membrane, methylates phosphatidylethanolamine, while, subsequently, phospholipid methyltransferase II produces phosphatidylcholine on the outer part of the membrane. Methylated phospholipids are then translocated (flip-flop) toward the external surface. The resulting *increase in fluidity* may facilitate the interaction between the receptor and the appropriate components of the membrane (for example, a G protein or a Ca^{2+}-channel), which in turn stimulates PL-A_2. The physiological importance of phospholipid methylation has not yet been assessed quantitatively, but in any case other mechanisms of hormone action are not excluded.

Hormone Action and Protein Synthesis

While clearly *central* to the mechanism of hormone action functioning via DNA-binding receptors (p. 89), transcriptional events *also* occur in response to hormones acting at the level of the plasma membrane. There is no evidence that these hormones have access to a nuclear receptor. Increased protein synthesis is probably *indirectly* provoked by hormonally regulated changes in cAMP-dependent *phosphorylation* of specific factors involved in either transcription and/or translation (Figs. I.105, I.122, and I.123).

The synthesis of specific proteins is often controlled by several hormones of the same or different classes (Fig. I.104), e.g., casein is positively regulated by prolactin, an effect that cortisol potentiates by a post-transcriptional mechanism, whereas progesterone is antagonistic.

Early and Late Activities: The Pleiotypic Response to Hormones and the Control of Mitosis

The term *pleiotypic response*[128] has been used to encompass the numerous and coordinated effects of insulin – one of the first discovered and still most enigmatic hormones in terms of its mechanism of action. Positive and negative pleiotypic responses are observed in a variety of cells exposed to an environment which affects their growth, as in bacteria subjected to stringent conditions (deprivation of amino acids). *Early responses* (within minutes) include: an increase in the transport of amino acids, sugars and ions across the plasma membrane, almost always an elevation of cytosol Ca^{2+} concentration, phosphorylation of receptors, and activation or inhibition of a number of enzymes, including adenylate cyclase and protein kinases. These effects are followed by a *progressive increase in RNA and protein synthesis*, which determines phenotypic changes. These changes may lead to modifications of the morphology of the cell, and may possibly be followed by *DNA replication* and *cell division*. In most studies, it appears that any stimulus must be present continuously for at least *6 to 10 h* in order to maintain a *permissive alkaline pH* following the activation of the *Na^+/H^+ antiporter*, (Fig. I.124) and thus commit cells to divide. The latter eliminates protons from the cells and increases the entry of Na^+. Its activation by growth factors involves PK-C. The influx of Na^+ stimulates the Na^+-K^+ ATPase pump which secondarily increases intracellular K^+ and restores the electrochemical gradient for Na^+. Thus, the antiporter system maintains *pH* and *K^+ above critical threshold*. Whether a specific intracellular signal is required for the stimulation of mitosis remains to be elucidated.

Most hormones display pleiotypic effects, although these effects may not constitute a principal physiological response. ACTH rapidly increases glucocorticosteroid output (main effect), and secondarily stimulates the growth of cells of the adrenal cortex; estrogen first accelerates water, amino acid and sugar uptake in uterine cells, and then promotes cell growth and proliferation, which is its most important role; ecdysone modifies the ionic balance in the nuclei of

126. Cohen, P. and Ingebritsen, T.S. (1983) Protein phosphatases: properties and role in cellular regulation. *Science* 221: 331–338.
127. Hirata, F. and Axelrod, J. (1980) Phospholipid methylation and biological signal transmission. *Science* 209: 1082–1090.
128. Hershko, A., Mamont, P., Shields, R. and Tomkins, G.M. (1971) Pleiotypic response. *Nature* 232: 206–211.

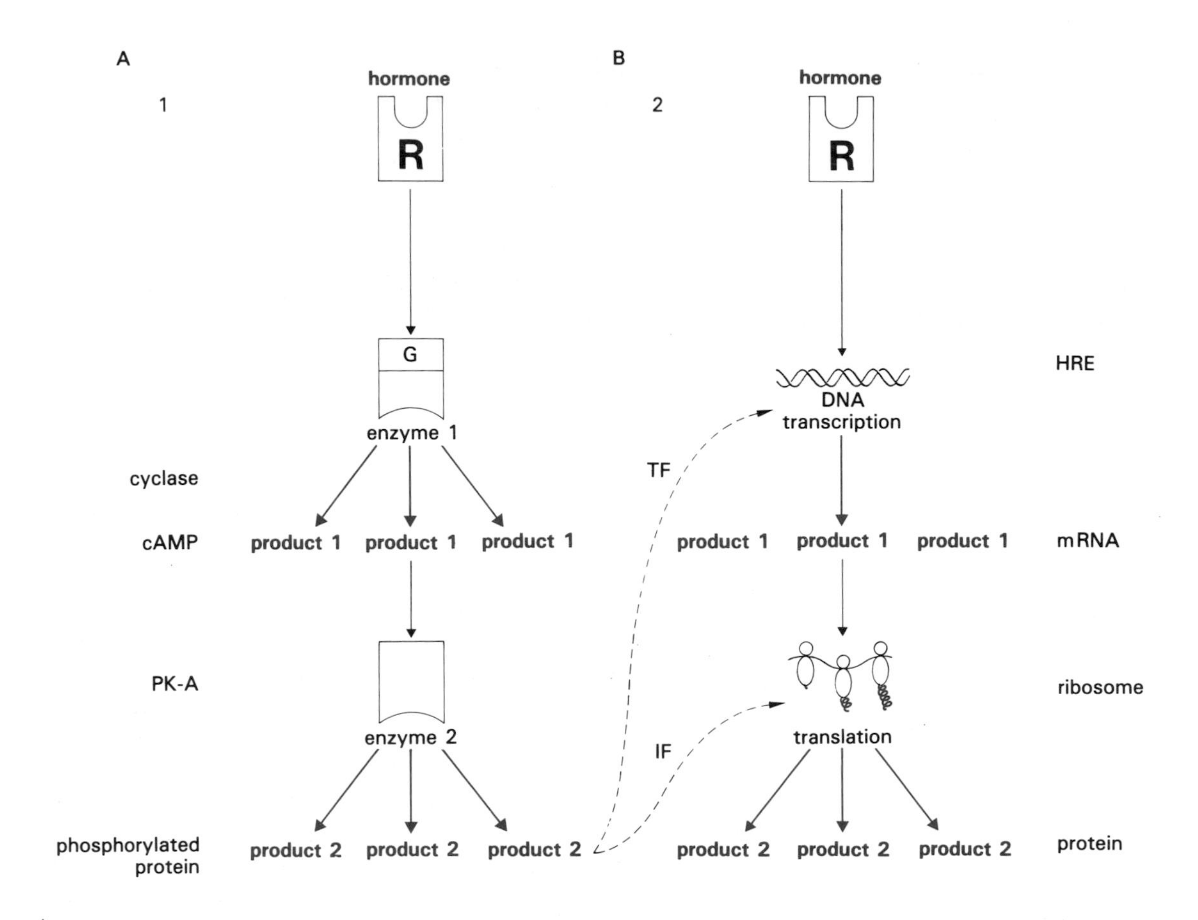

△
Figure I.122 Transduction and amplification of a hormonal signal. Two main pathways of hormone action at the cellular level are represented: transmembrane signalling (**A**) and gene transcription (**B**). In both cases, one hormone molecule results in many product molecules. The dotted line indicates that a hormone-stimulated activity, initiated at the membrane level, may operate ultimately at the level of transcription (e.g., via phosphorylation of a nuclear transcription factor (TF)) or translation (e.g., via phosphorylation of an initiation factor, IF). This mechanism would account for peptide hormone-regulated protein synthesis and growth, and at the same time would represent a means for very efficient amplification.

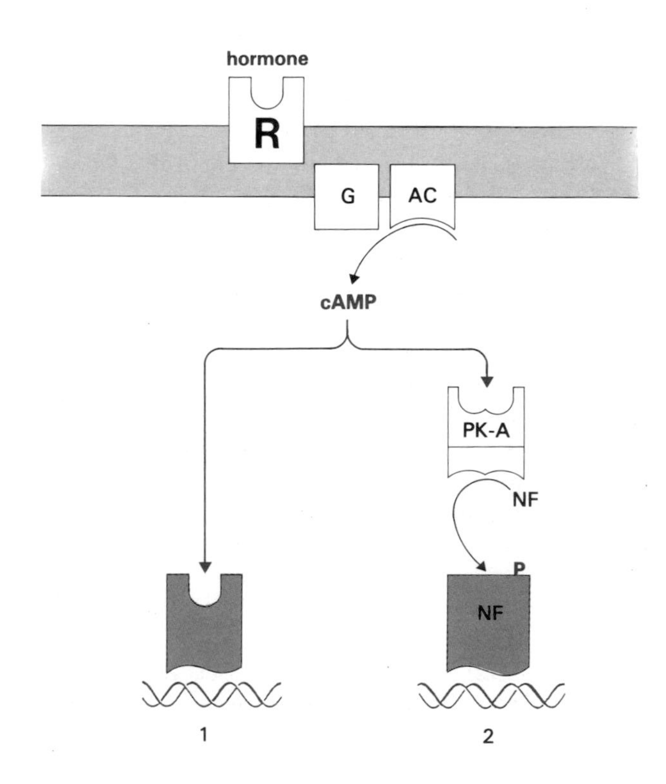

▷
Figure I.123 Cyclic AMP regulates transcription: two possible mechanisms. For both, there is a *cis*-acting DNA segment (cAMP-dependent response element (CRE)) with which either a cAMP binding protein (**1**)[203] or, more probably, a phosphorylated (P) nuclear transcription factor (NF) (**2**) may interact. (Montminy, M. R. and Bilezikjian, L. M. (1987) Binding of a nuclear protein to the cAMP response element of the somatostatin gene. *Nature* 328: 175–178).

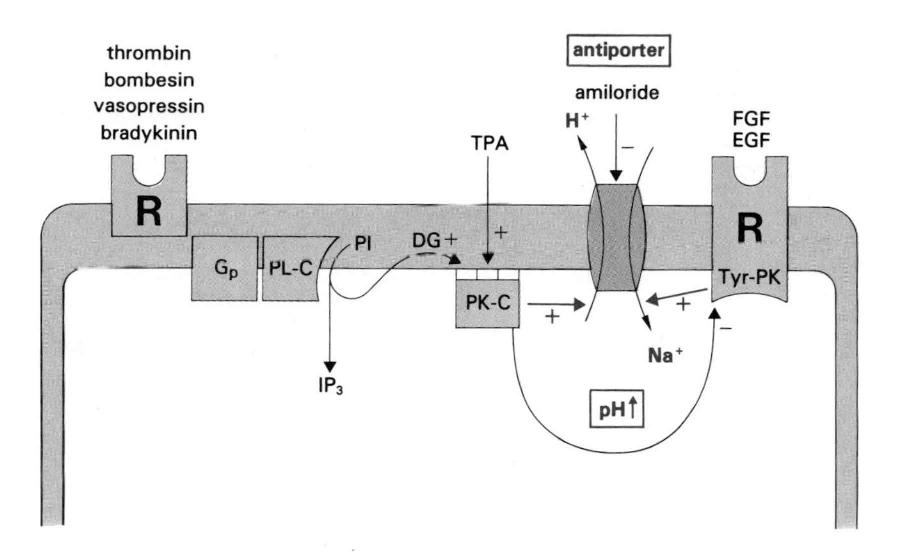

Figure I.124 Na^+/H^+ antiporter system. The activation of a protein kinase, PK-C or Tyr-PK, as the result of stimulation by a growth-promoting agent, can in turn activate the antiporter system and lead to an increase in intracellular pH. Amiloride is a pharmacological antagonist. In addition to its stimulation by growth factor receptors, the Na^+/H^+ antiporter system can be activated by the α_2-adrenergic receptor. The human antiporter has been cloned. (Sardet, C., Franchi, A. and Pouyssegur, J. (1989) Molecular cloning, primary structure and expression of the human growth factor activatable Na^+/H^+ antiporter. *Cell* 56:271–280).

insect salivary glands before increasing gene transcription[129]; insulin very rapidly increases glucose transport into cells, changes secondarily the synthesis of enzymes in several metabolic pathways, and may even be growth promoting. Modifications of enzyme activity and transport systems are rapid and transitory. Changes in protein synthesis and cell division take place after a lag period and, at least in principle, last longer. In fact, the network of rapid modifications affecting calcium, nucleotides and protein phosphorylation, often induced by PK-C, may ensure sustained activities. Conversely, transcription may occur almost immediately, as in the synthesis of MMTV RNA, or in the transcription of "early genes" implicated in growth (Fig. I.125A), e.g., vimentin (protein of intermediary filaments), c-*myc*, c-*fos*, and proliferin (which has homology with preprolactin). Thus, the type of mechanism involved in hormone action cannot be described by a simple *temporal* scheme. A common set of fundamental regulatory changes governing cellular activities operate according to patterns which are specific to each hormone and each cell system.

Frequently, *multiple pathways* are involved in the control of DNA synthesis. Each hormone generates intracellular signals differently (Fig. I.125B), and, *in concert*, these signals produce a complete growth response[130]. Steroids and growth factors stimulate cell division by different mechanisms (p. 99) (e.g., growth factor stimulation of the ribosomal protein S6, in contrast to the effect of steroid hormones). Some hormones and phorbol esters primarily stimulate the phosphoinositide system and PK-C, while others selectively modify adenylate cyclase activity. Complementarity between these two classes of "incomplete" stimuli has been demonstrated experimentally; synergism between a hormone of either class and insulin or EGF has been observed, and this is probably related to the Tyr-PK activity of insulin/EGF receptors. Indeed, *tyrosine phosphorylation* of the receptor and other cellular proteins is an almost constant *feature of growth factor* action[119], although the mechanism remains obscure. Hormones such as PDGF, FGF or bombesin are sometimes able to fully stimulate all of the components necessary for the growth response.

Apart from acting as stimulatory factors, some peptides *inhibit the growth* of many cells. TGF-β (p. 219) and tumor necrosis factor (TNF) are examples of inhibitory peptides. However, like growth factors in general, they may alternatively play positive and negative roles in the control of growth, depending on the

129. Ashburner, M. and Cherbas, P. (1976) The control of puffing by ions – the Kroeger hypothesis: a critical review. *Mol. Cell. Endocrinol.* 5: 89–107.
130. Rozengurt, E. (1986) Early signals in the mitogenic response. *Science* 234: 161–166.

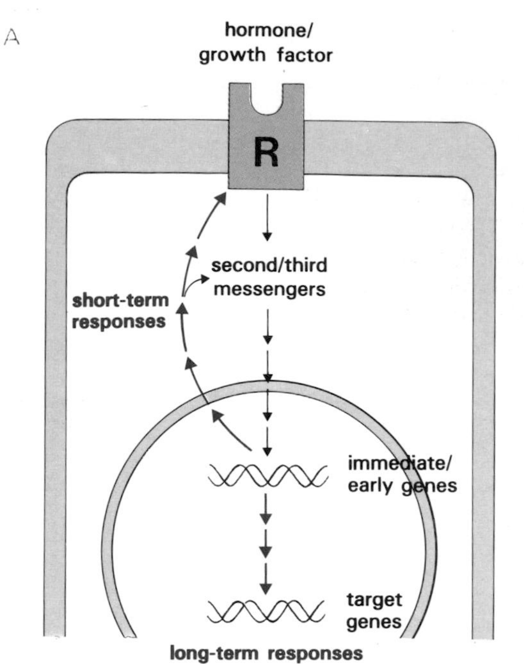

Figure I.125 Multistep and multihormal controls in cell metabolism and division, operating via multiple intracellular signals. **A.** Second/third messengers elicit short-term responses. In addition, they may contribute to the induction of transcription of immediate-early genes. The latter may intervene in the signal transduction cascade (e.g., modifying the response to repeated stimulation). They may also promote the transcription of other target genes involved in long-term, specific responses. (Curran, T. N., Morgan, J. I. (1987) Memories of *fos*. BioEssays 7:255–258). **B.** Three hormones (H_1,H_2,H_3) and their receptors are represented, each of which has a particular mode of action. A single hormone-receptor complex may activate more than one cellular mechanism. Most of the intracellular signals represented are involved in events leading to cell division regardless of whether they are initiated by one or several synergistic hormone-receptor complexes. A similar representation could apply to the regulation of peptide secretion.

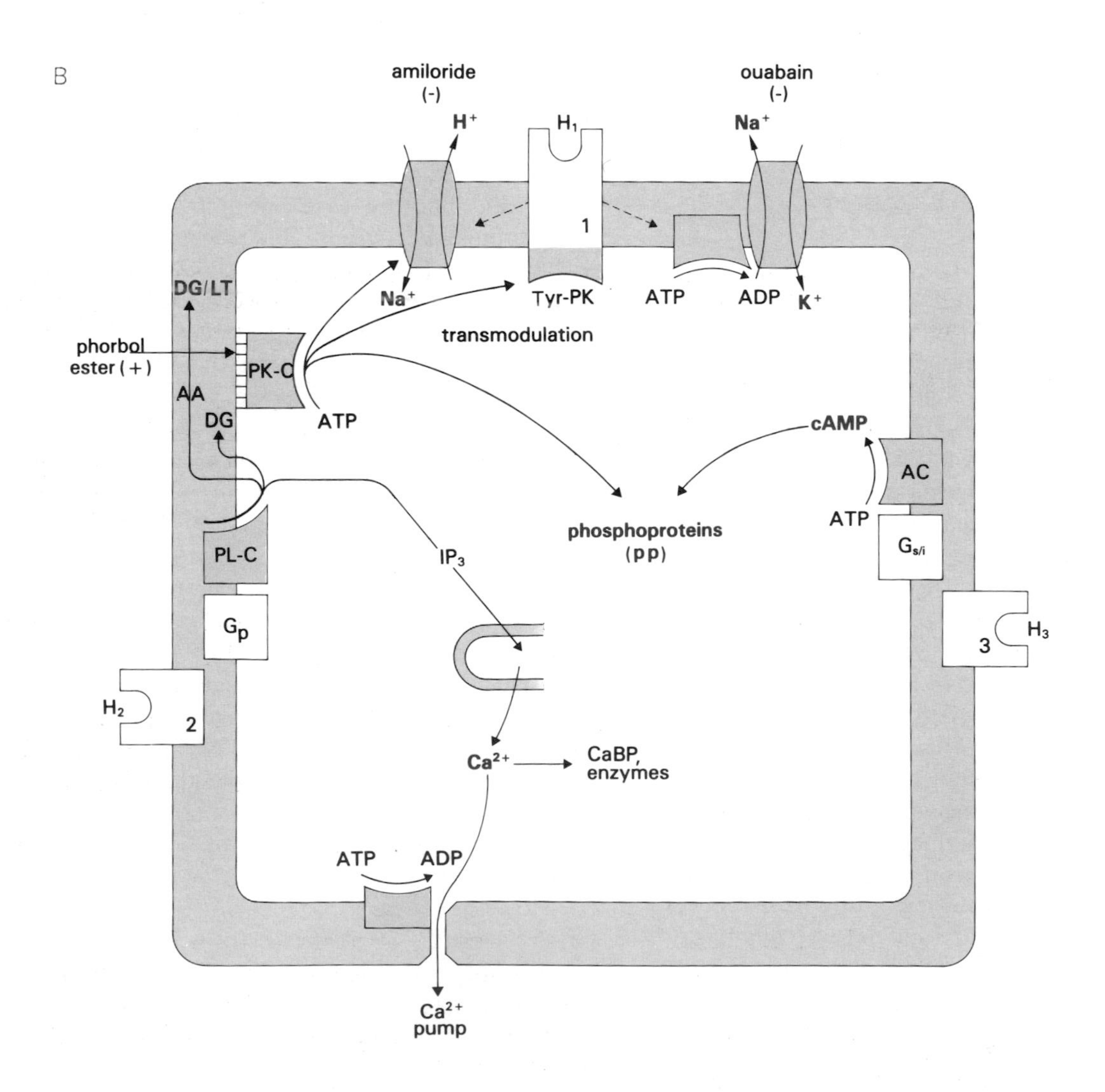

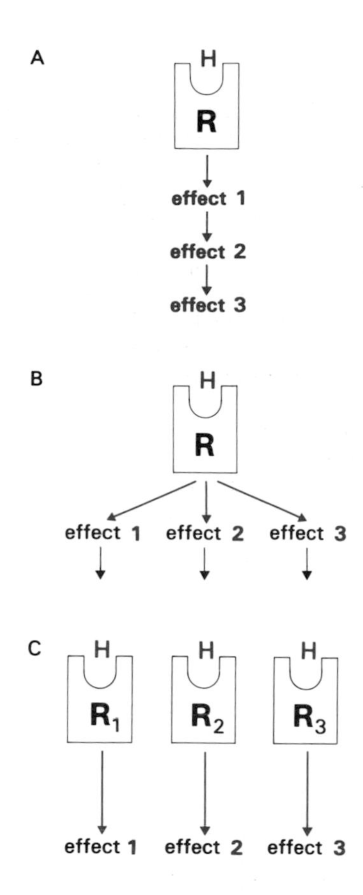

Figure I.126 Several effects of a single hormone on a target cell. **A.** The simplest case: the first effect produced by the formation of a HR complex leads to a cascade of successive effects. **B.** The hormone-receptor complex directly induces multiple effects. **C.** Each of several distinct receptors for the same hormone triggers a single effect.

biological system considered (e.g., TGF-β stimulates fibroblasts).

Thus, hormones control the two opposed processes of cell division and differentiation. It is interesting to note that, to date, *no* direct effect of hormones on *gene amplification*[131] has been recorded.

Intricacy and Complexity

A single hormone may have several effects by the intermediary of one or several receptors (Figs. I.126 and VII.11). However, transduction systems are relatively limited in number, such as a G protein for many membrane receptors, or a specific DNA sequence (HRE) for several intracellular receptors.

131. Schimke, R. T., ed (1982) Summary. In: *Gene Amplification*. Cold Spring Harbor Lab., New York, pp. 317–333.

Table I.20 Enzymes coupled to hormone receptors.

Enzymes
Adenylate cyclase (AC)
Guanylate cyclase (GC)
Phosphodiesterase (PDE)
GPTases
Protein kinases (PK)
Phospholipase-A_2 (PL-A_2)
Phospholipase-C (PL-C)
Lipid methylases

The only PDE coupled to a receptor is a cGMP-specific enzyme (Fig. I.82). G proteins are GTPases (p. 334). Some receptors are themselves PKs.

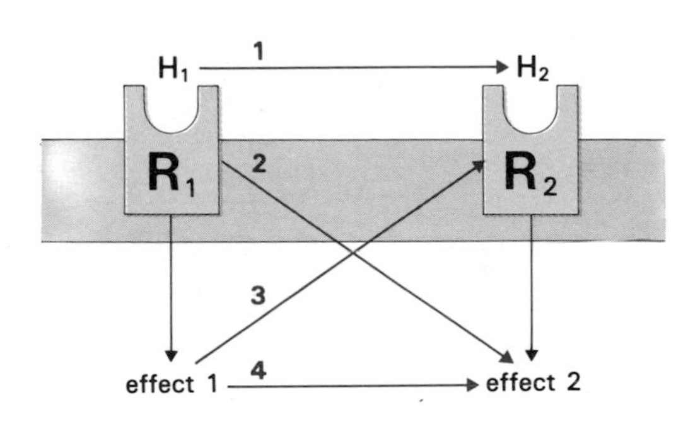

Figure I.127 Two hormones: four ways for one to modify the activity of the other. **1.** H_1 may compete for H_2 binding to R_2. **2.** The H_1R_1 complex may interact with effector 2 (e.g., synergistically). **3.** The effect 1 of H_1 may modify R_2 function (e.g., phosphorylation). **4.** Effect 1 may modulate effect 2. The mechanism whereby one hormone modifies the binding of another hormone to its receptor may take place in the membrane (e.g., ACh via mACh-R increases VIP binding), or it could follow a more complex pathway implicating the intracellular machinery (e.g., IGF-II, which increases insulin binding).

The intracellular signals generated by the hormone at the level of the membrane allosterically regulate enzymes, in a broad sense (Table I.20). In addition, each hormone, besides its primary mechanism of action, directly or indirectly modifies components of other regulatory circuits, including the function of receptors for other hormones (Fig. I.127). Each cell, not to mention each function of each cell, is regulated *multihormonally* (Fig. I.128). Thus, the control of cellular activities is achieved by an extremely complex and finely tuned cooperation of physiological inputs.

The *factor of time* is also involved in hormone activities, although it is not yet fully appreciated. For example, the kinetics of the complex sequential

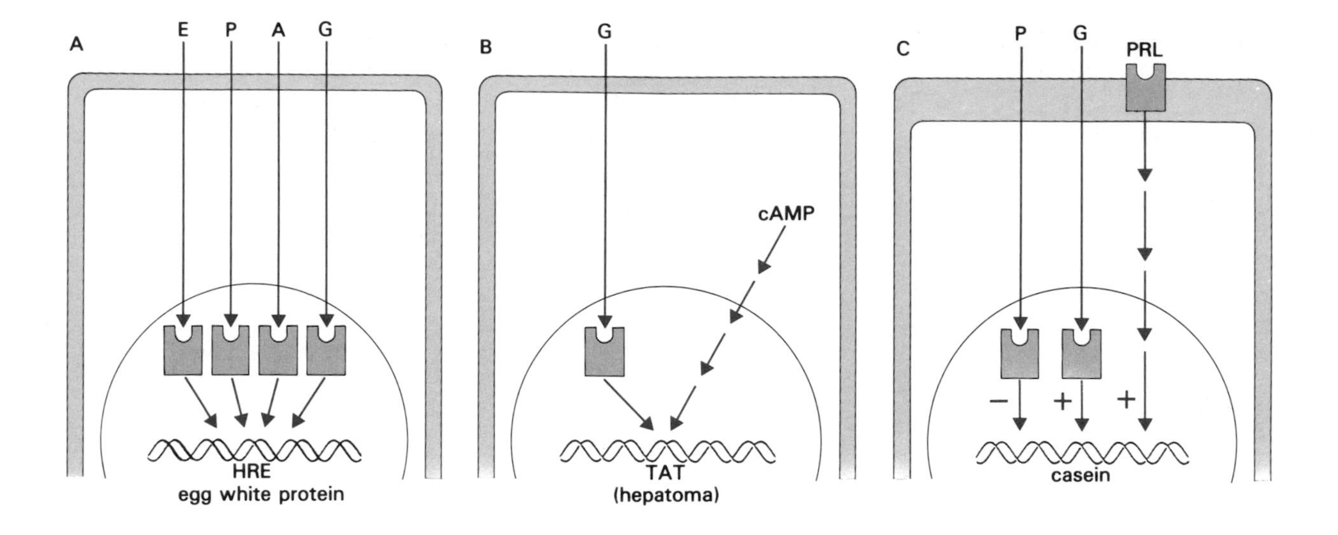

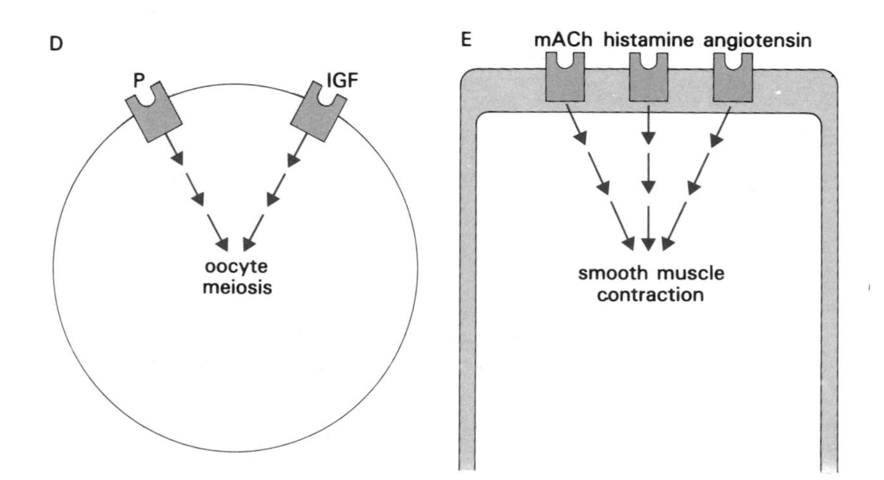

Figure I.128 Several hormones regulate the same function. These five examples demonstrate the generality of multihormonal control. **A.** Several steroids (E, estradiol; P, progesterone; A, androgen; G, glucocorticosteroid) regulate the synthesis of chick oviduct proteins (p. 415). **B.** Steroid (G) and cAMP (via different mechanisms) regulate the synthesis of liver tyrosine aminotransferase (TAT) (p. 417). **C.** Two steroids (P and G) and prolactin (PRL) regulate the synthesis of casein in mammary cells (p. 200). **D.** Steroid (P) and growth factor (IGF) control meiosis in amphibian oocytes (p. 98). **E.** Acetylcholine (ACh), histamine and angiotensin can jointly regulate smooth muscle contraction (p. 605).

changes in the movement of Ca^{2+}, the cyclic activation and deactivation of G protein, and changes in chromatin structure, among others, are fundamental determinants of intracellular responses to hormones. At the level of the entire organism, it is clear that the same cell is exposed to different hormonal and metabolic environments at different periods. *The target cells* of a given hormone in a given organ may *differ* at different periods in *development*. The prostate gland is a good example of this phenomenon. It develops from epithelial buds projecting from the sinus epithelium into the surrounding mesenchyme. The genotype of the epithelium, whether male or female, is irrelevant; what is important is a normal androgen-responsive mesenchyme[132]. Only secondarily to the presence of hormone and the response of mesenchymal cells does epithelial differentiation become androgen-dependent. Androgen-sensitive cells are also found in the mesenchyme of the fetal mouse mammary gland; they are able to suppress mammary development if stimulated in the embryonic stage[133]. It appears, therefore, that, in addition to the initiation of a cascade of events within a given target cell, the overall action of a hormone may include multiple steps in other cell types.

5.5 Antihormones

Antihormones (*hormone antagonists*) *counteract* hormone action at the *receptor level*[134]. However, there are other ways to decrease hormone activity, e.g., by *suppressing secretion* either surgically or pharmacologically. Specific drugs can alter steroid synthesis, such as aminoglutethimide which suppresses cholesterol side-chain cleavage in steroidogenic cells, metapyrone which inhibits 11β-hydroxylase activity (p. 427), and several competitive inhibitors of the aromatizing enzymes which transform androgens into estrogens (p. 394). Antithyroid drugs are not hormone antagonists but in fact are suppressors of thyroid hormone formation (p. 357). It is also possible to *decrease the access* of hormones to their target cells, as for example via antibodies in "vaccination against pregnancy" (p. 279).

Antagonism of hormone action at a *postreceptor* level has been tested experimentally using antibodies directed against the Tyr-PK of growth factor receptors. Heat shock protein (hsp 90) impedes the binding of steroid receptors to DNA (Fig. I.70). However, it is probable that intracellular, postreceptor mechanisms are not specific enough to be used therapeutically.

The best target for antihormone action is at the *receptor* level. A *decrease* in receptor concentration and function reduces hormone activity (p. 8 and Table XI.5). *No non-competitive* inhibitor for hormone receptors, such as the numerous drugs and toxins which act on neurotransmitter receptors, has been reported thus far. Antihormones bind to the *hormone binding site* of receptors, and consequently compete with agonist ligands. Thus, if their binding does not trigger a response, the net result is a full antihormonal effect.

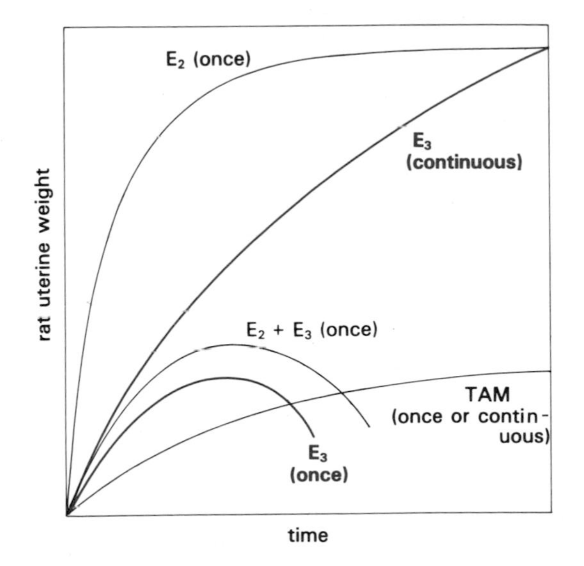

Figure I.129 Impeded estrogen is a pseudoantihormone. A single injection of estriol (E_3), regardless of the dose, never produces the maximum estrogenic response observed with a single injection of E_2. However, multiple or continuous administration of E_3 results in a maximal effect. This is not the behavior of a "real" antiestrogen, such as tamoxifen (Tam), which is also a weak agonist.

There are some *physiological* antihormones, such as progesterone, which has an antimineralocorticosteroid effect, or DHT, which is antiestrogenic at the ER level[135]. However, most antihormones are *synthetic analogs* of steroid hormones (Fig. IX.34). Their antagonistic property is now known *not* to be directly related to the affinity of the compound. On the grounds of pharmacological experiments, it was believed that an antihormone must be of lower affinity than the natural ligand. The activity of estradiol is reduced when coadministered with estriol, which has weak affinity for the ER (Fig. I.129). However, estriol is a real estrogen, since it exhibits full estrogenic activity when administered repeatedly or continuously or when it is secreted in pregnancy. Administered once, it is "impeded"[136] – that is, it is unable to reach maximum estrogenic activity, regardless of the dose, since it dissociates rapidly from the receptor and is excreted. On the contrary, tamoxifen, a synthetic non-steroidal compound with weak affinity for the ER also, is a *bona fide* antiestrogen. Although it is a partial agonist, it never develops full estrogenic activity, and its remarkable efficiency is related to the slowness with which it is metabolized and its long half-life. The metabolite 4-hydroxytamoxifen has high affinity for the ER. It is a good antiestrogen in vitro, but, in vivo, is not as useful as tamoxifen because of the rapidity with which it is metabolized. These examples demonstrate that appropriate *pharmacological selection* of an antagonistic drug should take into consideration both receptor affinity and overall metabolism. The highly active compound RU 486 (Fig. I.130) has high affinity for progesterone and glucocorticosteroid receptors, and a long half-life. This antisteroid, as well as other antihormones used in vivo, must be administered at much higher doses than would be expected from calcula-

132. Takeda, H., Mizuno, T. and Lasnitzki, I. (1985) Autoradiographic studies of androgen-binding sites in the rat urogenital sinus and postnatal prostate. *J. Endocrinol.* 104: 87–92.
133. Kratochwill, K. and Schwartz, P. (1976) Tissue interaction in androgen response of embryonic mammary rudiment of mouse: identification of target tissue for testosterone. *Proc. Natl. Acad. Sci., USA* 73: 4041–4044.
134. Furr, B. J. A. and Wakeling, A. E., eds (1987) *Pharmacological and Clinical Uses of Inhibitor of Hormone Action and Secretion.* Baillière–Tindall, Eastbourne.
135. Casey, R. W. and Wilson, J. D. (1984) Antiestrogenic action of dihydrotestosterone in mouse breast. *J. Clin. Invest.* 74: 2272–2278.
136. Huggins, G. and Jensen, E. V. (1955) The depression of estrone-induced uterine growth by phenolic estrogens with oxygenated functions at positions 6 or 16. The impeded estrogens. *J. Exper. Med.* 102: 335–346.

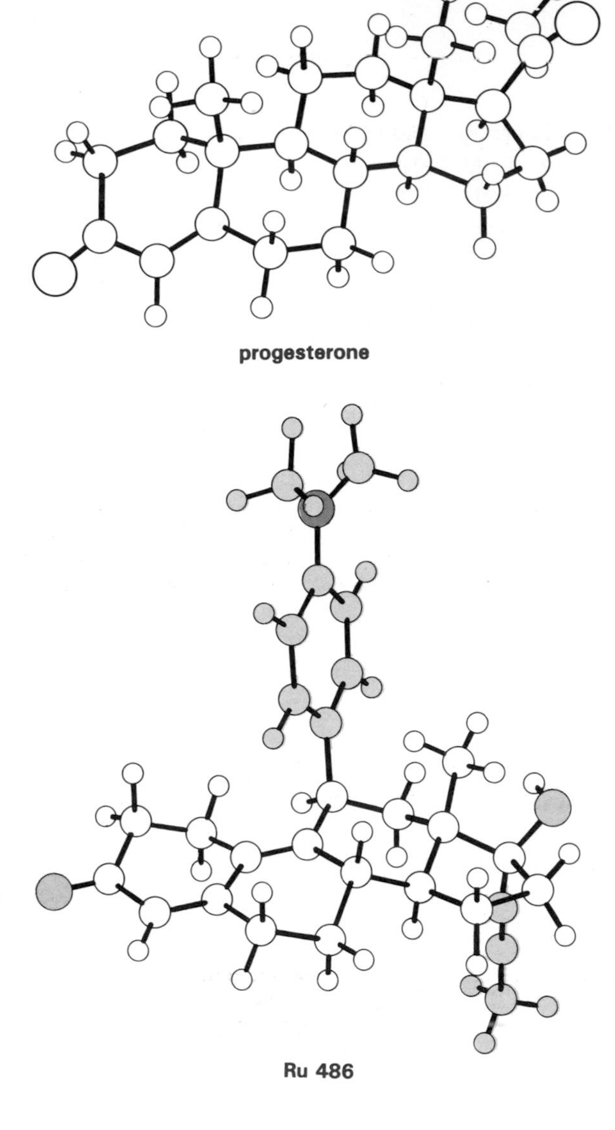

Figure I.130 Structure of progesterone and RU 486.

tions based on affinity, for reasons that in general remain poorly understood. In addition, due to the complexity of the in vivo situation, the actual activity of the antihormone should be considered within the totality of the hormone system. For example, there is no feedback regulation of progesterone secretion when RU 486 is used as a "contragestive" agent[137]. However, as an antiglucocorticosteroid, it increases the secretion of ACTH and cortisol by suppressing negative feedback controls at the level of the pituitary and hypothalamus (Fig. I.131). This compensatory effect of the compound's antiglucocorticosteroid

137. Baulieu, E. E., Ulmann, A. and Philibert, D. (1987) Contragestion by antiprogestin RU 486: a novel approach to human fertility control. In: *Fertility Regulation Today and Tomorrow. Serono Symposia* (Diczfalusy, E. and Bygdeman, M., eds), Raven Press, New York, vol. 36, pp. 55–73.

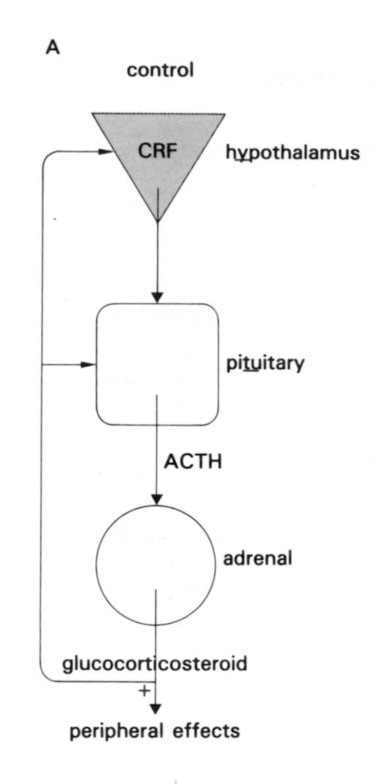

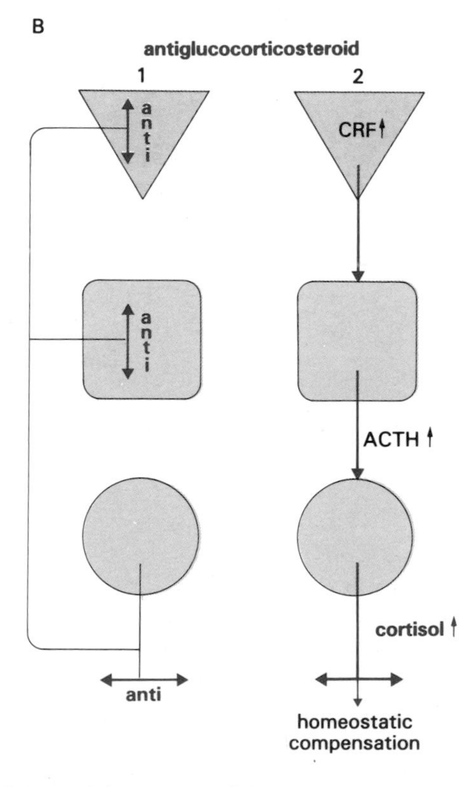

Figure I.131 Suppression of the negative feedback control of the hypothalamic-hypophyseal-adrenal system by an antiglucocorticosteroid (RU 486). **A.** Normally, cortisol regulates CRF and ACTH production negatively. It also exerts peripheral effects on various metabolic parameters. **B.** Antiglucocorticosteroid activity (1) decreases the feedback mechanism, and CRF and ACTH are increased, leading to elevated cortisol secretion (2).

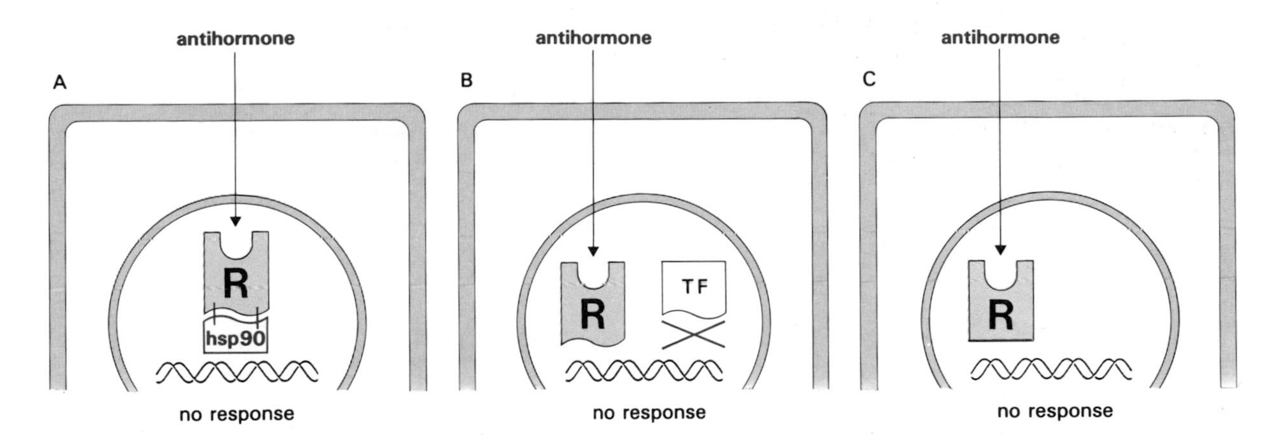

Figure I.132 Steroid hormone receptors and antihormones: three possible mechanisms of action. **A.** Antihormone (AH) stabilizes the receptor hsp 90 oligomer (Fig. I.70), thus precluding the binding of DNA by the receptor. **B.** The hormonal response requires, in addition to the binding of R to DNA, a hormone-dependent interaction with a nuclear transcription factor (TF). The latter activity is precluded if AH is present. **C.** Allosterically, AH renders the DNA binding site of the receptor inactive.

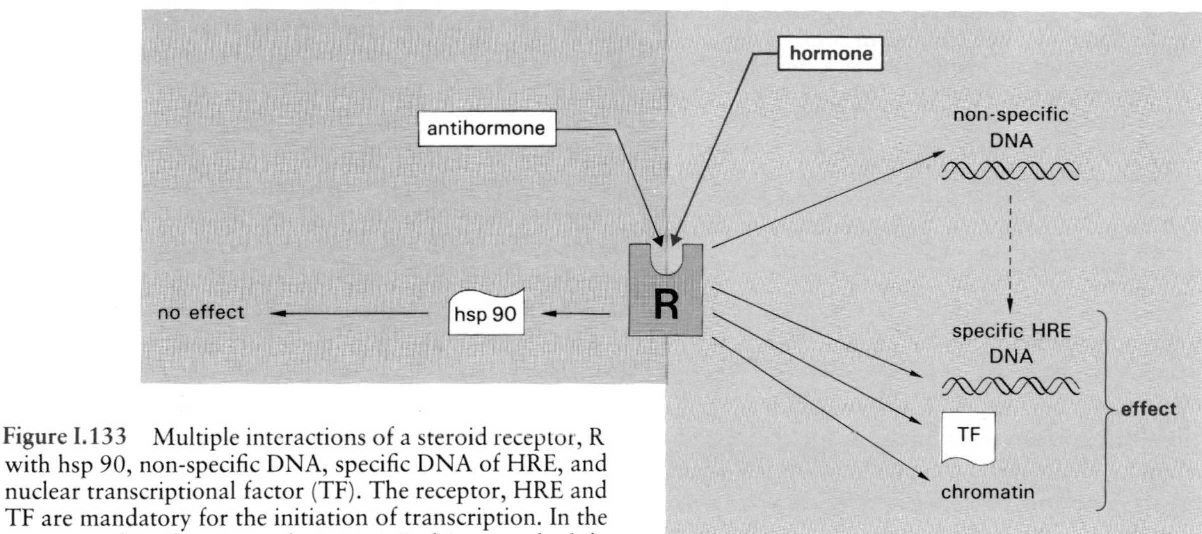

Figure I.133 Multiple interactions of a steroid receptor, R with hsp 90, non-specific DNA, specific DNA of HRE, and nuclear transcriptional factor (TF). The receptor, HRE and TF are mandatory for the initiation of transcription. In the presence of antihormone, the system is driven to the left, while the binding of hormone drives it to the right.

activity permits the use of the drug in fertility control.

The *molecular mechanism* of antihormone action probably differs as a function of the particular compound. RU 486 stabilizes the interaction of the receptor with hsp 90, which has the potential to preclude receptor interaction with HRE[138] (see Fig. I.132 for other possible mechanisms). Most anti-steroid hormones often display partial agonistic effects in addition to their antihormonal action. There may be an equilibrium between the distinct active and non-active forms of the receptor; they are stabilized by agonist and antagonist ligands, respectively, and their ratio may also depend on the intracellular environment (Fig. I.133).

The case of *tamoxifen* illustrates some of the difficulties encountered in understanding the mechanism of antihormone action (Fig. I.134). It is a pure anti-estrogen in the chick, a powerful estrogen in mice, and is both estrogenic and antiestrogenic in the rat and in humans. When tamoxifen is administered to the chick in combination with either progesterone or gluco-corticosteroid, it is estrogenic in the oviduct but not in the liver! In addition to binding to the ER, tamoxifen also interacts with other poorly defined proteins pos-

138. Groyer, A., Schweizer-Groyer, G., Cadepond, F., Mariller, M. and Baulieu, E. E. (1987) Antiglucocorticosteroid effects suggests why steroid hormone is required for receptors to bind DNA in vivo but not in vitro. *Nature* 328: 624–626.

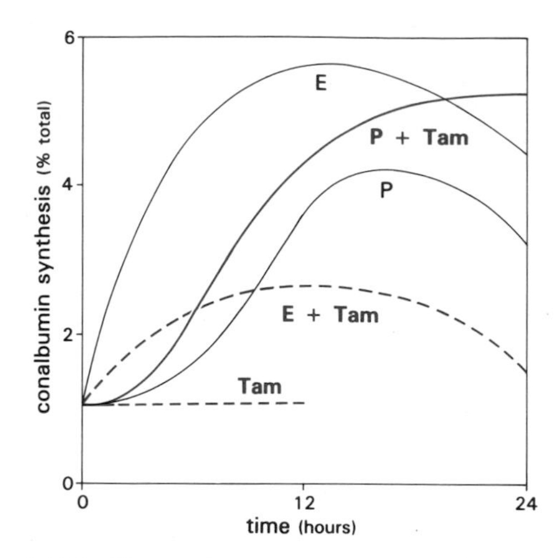

Figure I.134 Tamoxifen: an antiestrogen in the chick. The relative rate of conalbumin synthesis (see also Fig. I.104) was measured after the administration of estrogen (E), progesterone (P), and tamoxifen alone or in combination with estrogen or progesterone. Note the lack of estrogenic activity of tamoxifen alone, its antiestrogenic effect when administered with estradiol, and its synergistic effect with progesterone, suggesting newly-acquired estrogenic activity. (Catelli, M. G., Binart, N., Elkik, F. and Baulieu, E. E. (1980) Effect of tamoxifen on oestradiol and progesterone-induced synthesis of ovalbumin and conalbumin in chick oviduct. *Eur. J. Biochem.* 107: 165–172).

sibly involved in antiproliferative action[139] (possibly PK-C itself). Despite a lack of full knowledge concerning antisteroids, they are clinically useful. It is probably unrealistic to search for antagonistic therapeutic agents that covalently bind receptors and which might therefore have prolonged activity, since, in vivo, such molecules would probably react chemically with many other proteins.

Antihormones at the Membrane Receptor Level

Antagonists of ligand which act at the level of the membrane are available, such as antagonists of epinephrine (Tables VII.3 and VII.4), dopamine and histamine. Medically, they are very useful. The *antipeptide antagonists* which have been tested clinically are antiGnRH (p. 266), and naloxone, which is an antiopioid used to treat drug addiction. Like morphine itself (Table 2, p. 188), its efficacy at the receptor level suggests that non-peptide analogs of natural hormonal peptides may become available and be of therapeutic interest.

Experimentally and pathologically (p. 529), *antireceptor antibodies* can preclude binding of protein hormones to their receptors, and thus can be *antagonists*. However, in other cases, an agonistic response, at least partial, is obtained with antibodies (p. 529), probably via clustering and/or change of conformation of the receptor.

6. Oncogenes: The Hormone Connection

The concept of oncogenes has dominated basic cancer research for the past ten years. It is now possible to isolate, sequence, experimentally modify, and express sequences of DNA involved in human and animal cancers which occur either spontaneously or following exposure to viruses or chemical or physical carcinogens. Thousands of publications continue to reveal astonishing information in this domain. Despite reports of an increasing variety and complexity of cancers occuring in different species, organs, and tissues, some important results can now be summarized.

Two major types of experiments demonstrated the existence of cancer genes, known as oncogenes (*onc*). First, it was shown that certain gene segments of retroviruses are responsible for causing malignant diseases, and they were called v-*onc*[140]. The RNA which forms the genome of retroviruses is transformed into DNA in the host cell by reverse transcriptase. Oncogene products are pathogenic proteins made from this DNA. More than 30 v-*onc* have been characterized (Table I.21). Segments of tumorigenic DNA viruses have the same properties as v-*onc*. Secondly, it was shown that the *transfer of DNA* from cancer cells into receiving cells (particularly from human tumors which do not appear to be of viral origin) leads to oncogenesis. Some of the oncogenes described during these transfection experiments are very similar, or are identical, to v-*onc* (an observation of great theoretical importance in that it suggested common mechanisms in the generation of cancers of different origins). An additional 20 oncogenes, different from v-*onc*, were identified by these experiments

139. Sutherland, R. L., Watts, C. K. W. and Ruenitz, P. C. (1986) Definition of two distinct mechanisms of action of antiestrogens on human breast cancer cell proliferation using hydroxytriphenylethylenes with high affinity for the estrogen receptor. *Biochem. Biophys. Res. Commun.* 140: 523–529.
140. Bishop, J. M. (1985) Viral Oncogenes. *Cell* 42: 23–30.
141. Bishop, J. M. and Varmus, H. E., eds (1986) *Cancer Survey* 5.
142. Varmus, J. M. (1984) The molecular genetics of cellular oncogenes. *Annu. Rev. Genet.* 18: 553–612.

Table I.21

Proto-oncogene/ oncogene	Subcellular locations of gene product	Function of gene product
abl	Plasma and cytoplasmic membranes	Tyr-PK
erb-A	Nucleus	Thyroid hormone receptor
erb-B	Plasma and cytoplasmic membranes	EGF-R like
ets	Nucleus (fused with product of v-*myb*)	?
fes/fps	Plasma and cytoplasmic membranes	Tyr-PK
fgr	Plasma and cytoplasmic membranes	Tyr-PK
fms	Plasma and cytoplasmic membranes	CSF-I-R like
fos	Nucleus	?
kit	Cytoplasm	Tyr-PK
mas	Plasma and cytoplasmic membranes	Angiotensin receptor
mil/raf	Cytoplasm	Ser/Thr-PK
mos	Cytoplasm	Ser/Thr-PK
myb	Nucleus	?
myc	Nucleus	?
neu	Plasma and cytoplasmic membranes	EGF-R-2
ras	Plasma membrane	Regulator of adenylate cyclase
rel	Cytoplasm	?
ros	Plasma and cytoplasmic membranes	Tyr-PK
sis	Cytoplasm/secreted	Analog of PDGF-B
ski	Nucleus	?
src	Plasma and cytoplasmic membranes	Tyr-PK
yes	Plasma and cytoplasmic membranes	Tyr-PK
E1A	Nucleus	Regulates transcription
E1B	Nucleus and cytoplasm	?
Ps-sT	Cytoplasm	?
Py-mT	Plasma and intracellular membranes	Stimulates $pp60^{c\text{-}src}$
Py-lT	Nucleus	DNA synthesis
SV40-lT	Nucleus and membrane	Regulates transcription

Normal and pathogenic alleles of each gene have been identified, the former (proto-oncogene) in metazoan genomes, the latter (oncogenes) in retroviruses. Py, Polyoma; SV, SV 40; T, T antigen; s, small; m, middle; l, large; ?, uncertain or unknown. Approximately 20 other *onc* are currently undergoing characterization.

and were called transforming oncogenes (t-*onc*). Each oncogene shares homology with *proto-oncogene sequences* of normal cellular DNA, called cellular oncogenes (c-*onc*)[141,142], which are constitutive genes of non-cancerous cells. The transformation of c-*onc* into v- or t-*onc* occurs by two major pathways which are not mutually exclusive, i.e. either by a change in their structure (mutation) or their regulation. *Mutations* may be limited to the alteration of a single amino acid, or may involve deletion, addition or recombination with another gene. *Dysfunction* can arise from a new connection between the c-*onc* and a regulatory element of the virus or the genome of reinsertion, which leads to increased expression of the gene. When transferred to a cell infected with a virus, v-*onc* interferes with the primary function of the corresponding c-*onc*. Occasionally, the genome of a virus may not include an oncogene itself, but may provide a strong signal modifying the function of the c-*onc* in the cell. The same qualitative and quantitative changes occur when c-*onc* is modified within the cell by chemical or physical (e.g., radiation) processes. Other possible mechanisms include *amplification*, i.e. the excessive replication of the involved gene, and *chromosomal translocation* of c-*onc*, which is placed under the control of another gene (this begins to provide an explanation for the chromosomal abnormalities observed for some time in association with several cancers).

6.1 Research in Endocrinology and Oncology: Crossroads

We are far from knowing the structure and function of the majority of c-*onc*, but those which have been identified are almost always related to molecules involved in the fundamental mechanisms controlling cellular functions[140–142]. They include analogs of hormones, membrane and DNA-binding receptors, G proteins, and protein kinases.

6.2 The Triple Connection between Oncogenes and Hormones

Examples of the various classes of oncogene products numbered in red in Figure I.135 correspond to descriptions in the following paragraphs.

Some Oncogenes Have Homologies with Hormones (or Growth Factors)

Some oncogenes are hormone-like (1 in Fig. 1.135). The oncogene of the simian sarcoma virus, called v-*sis* (Fig. 1.136), is very similar to PDGF-B, one of the two structurally related subunits, A and B, of *platelet-derived growth factor* (p. 220) (the homodimer BB also has properties of a growth factor). Some cells secrete v-*sis* in excess, which stimulates the PDGF-R and causes *continuous* pathogenic *autocrine activity*. That *sis* is secreted by the cell is suggested by the effect of anti-PDGF antibodies which block the growth of infected cells, even though v-*sis* may also have intracellular activity. An oncogene of the feline sarcoma virus (*fsv*) also functions as a growth factor.

Only a few of these oncogenic "hormones" are known. A mysterious partial homology (100 aa) exists between *gastrin* and middle T antigen of the polyoma virus.

Some Oncogenes Have Homologies with Hormone Receptors

Membrane receptors (2a in Fig. I.135). One of the two oncogenes of the avian erythroblastosis virus is known as v-*erb*-B. Its product is a *truncated version of the EGF-R* (p. 78) (Fig. I.137), which has a Tyr-PK sequence, but which lacks the extracellular portion of the receptor. There is also deletion of some C-terminal amino acids, sites for autophosphorylation in the cytoplasmic segment, which, in the normal EGF-R, may act as competitive inhibitors (alternative

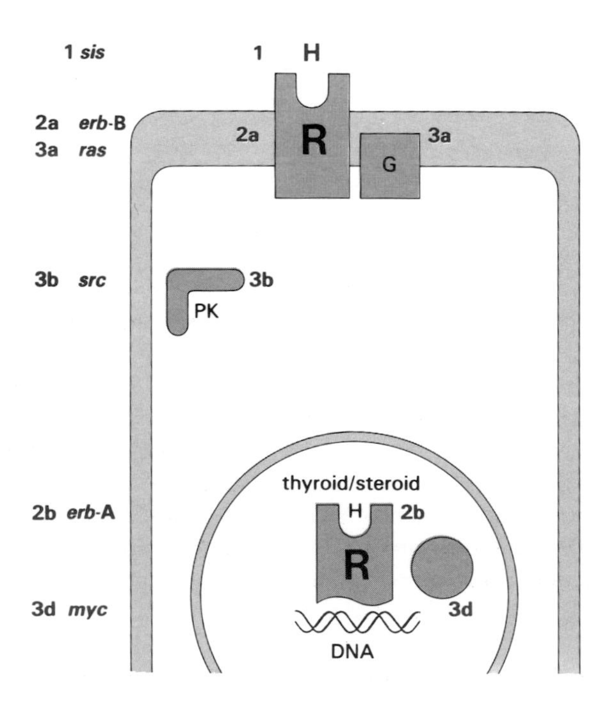

Figure I.135 Schematic representation of the location and function of proto-oncogenes. H, hormone; R, receptor; G, G protein; PK, protein kinase. Numbers in red refer to corresponding sections in the text.

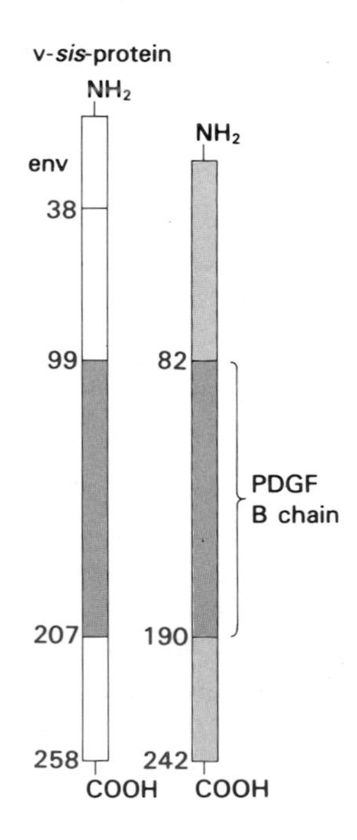

Figure I.136 Protein structure of v-*sis* and PDGF-B. The viral envelope is indicated by 'env'. Pink zones are parts of the biosynthetic precursor of PDGF-B.

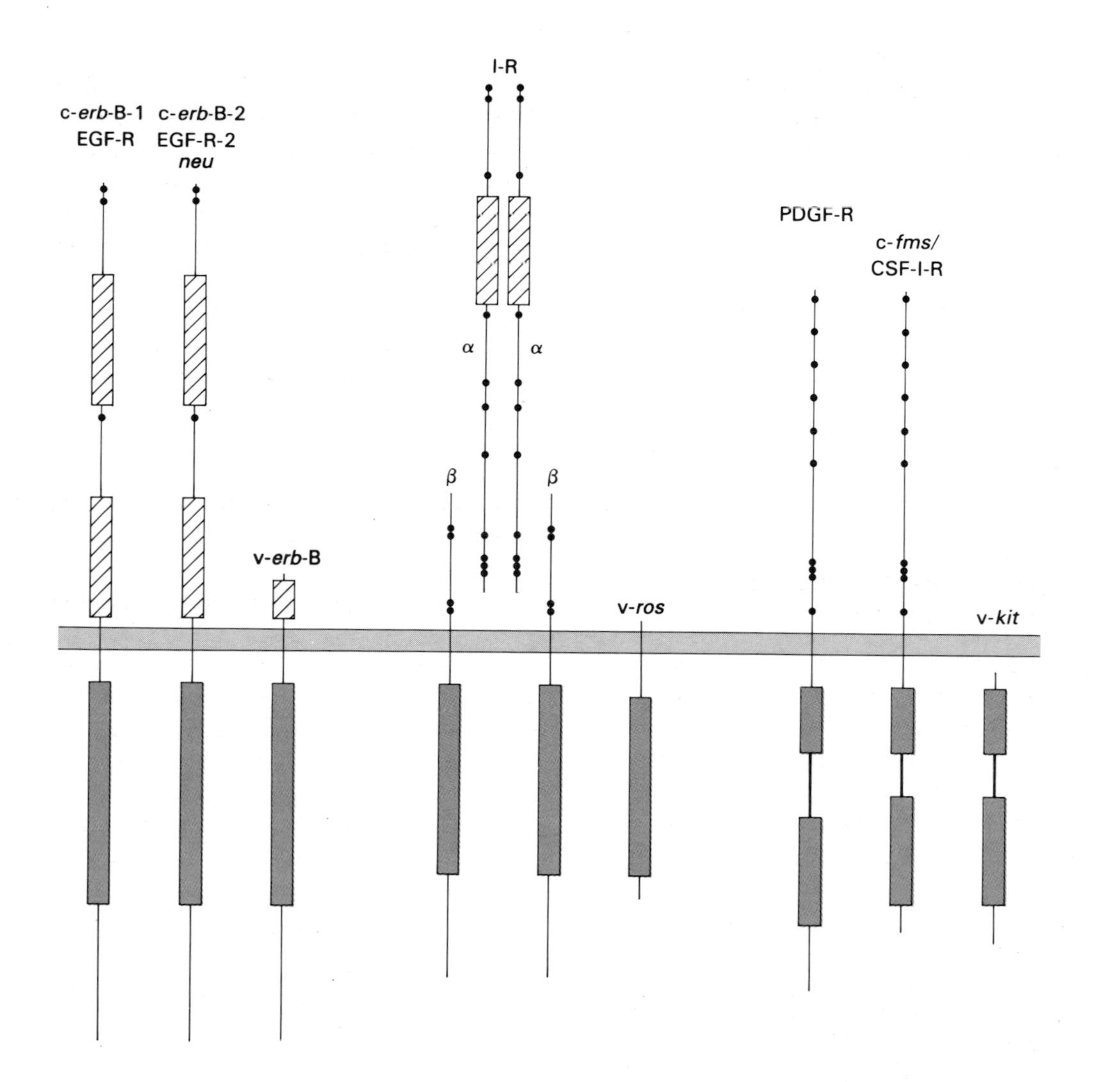

Figure I.137 Membrane receptors and oncogene products (see also Fig. I.78). The *neu* or *erb*-B-2 oncogene product is very similar to that of *erb*-B-1. The v-*erb*-B product is a truncated version of c-*erb*-B-1. Homology is high between the Tyr-PK domains of the I-R and v-*ros*. The c-*fms* product is the receptor for colony stimulating factor 1. It has high homology with the PDGF-R and v-*kit*. Red boxes indicate the Tyr-PK domains, dots represent cysteines, and cross-hatched boxes, the cysteine-rich domains.

substrates) of other peptides[143], and may also modulate hormone binding. The oncogene product of *erb*-B cannot bind EGF, and its Tyr-PK activity is constitutively and permanently active. However, autophosphorylation is not necessary for transforming activity. A fraction of *erb*-B protein remains in the Golgi, where it may have some function.

143. Gill, G. N., Bertics, P. J. and Santon, J. B. (1987) Epidermal growth factor and its receptor. *Mol. Cell. Endocrinol.* 51: 169–186.

Two other oncogenes have products with a structure similar to that of v-*erb*-B. The oncogene of the feline sarcoma virus, *fms*, encodes a receptor-like product of which the ligand would be CSF-I and GM-CSF (colony stimulating factor or macrophage-specific hemopoietic growth factor). The rat *neu* oncogene product‡, isolated initially from neuro- or glioblastomas, has a human counterpart, called *erb*-B-2 or EGF-R-2, which is similar to the EGF-R, but which does not bind EGF. There is also limited homology between the β chain of the I-R, the IGF-I-R, and v-*ros* (from avian sarcoma virus).

Uncontrolled and permanent activity of Tyr-PK of these membrane receptor-related oncogene products may lead to excessive phosphorylation of pathological proteins, and this could be *responsible* for cellular transformation.

‡Change of Val to Glu in the transmembrane domain converts c-*neu* into the transforming protein t-*neu*.

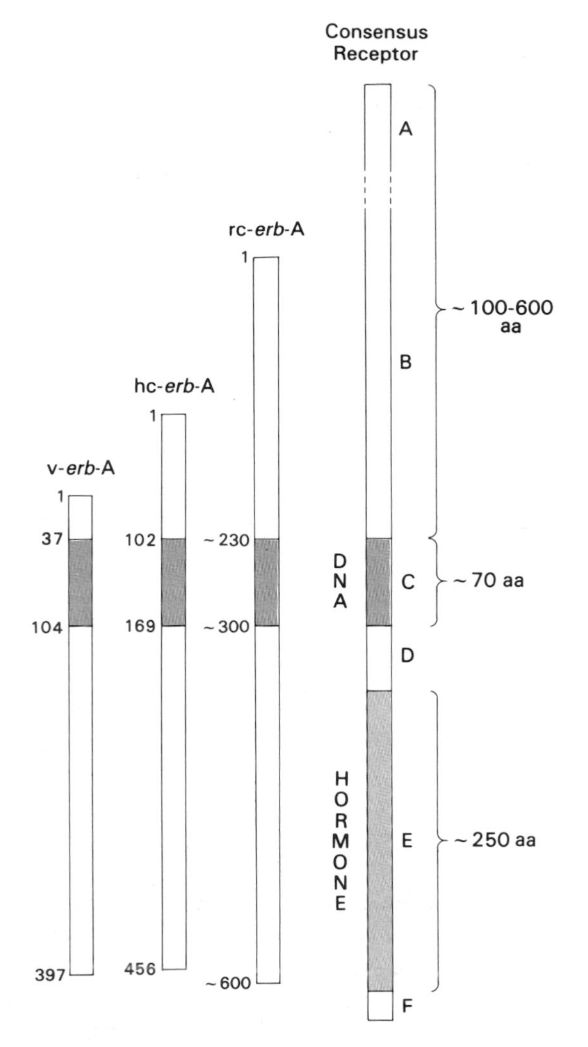

Figure I.138 Nuclear receptors and *erb*-A. The composite concensus receptor is representative of all steroid hormone receptors (see Fig. I.71).

Nuclear receptors (2b in Fig. I.135). The second *onc* product from the avian erythroblastoma virus, v-*erb*-A, favors the oncogenic activity of v-*erb*-B. It has strong sequence homology with steroid receptors (p. 70). The corresponding c-*onc* is a thyroid hormone receptor, and there is also strong homology with steroid, calcitriol, and retinoic acid receptors. In paired comparison, homology between *DNA binding sites* is ⩾50%, and between C-terminal hormone-binding regions is ~20% (Fig. I.138). Defective in hormone binding, v-*erb*-A pathological activity may be due to competition with c-*erb*-A, i.e. the thyroid hormone receptor, involved normally in erythroblast differentiation. It is remarkable that both v-*erb*-A and v-*erb*-B cooperate in the pathological activity of the erythroblastosis virus at the levels of the nucleus and the plasma membrane, respectively.

Homologies with Postreceptor Components of Hormonal Systems

Transduction proteins (3a in Fig. I.135). The membrane proteins specified by oncogenes of the *ras* class (which come from murine viruses, e.g., Ha-*ras*, or from human cancer DNA, e.g., N-*ras*) have homology with the α subunits of membrane G proteins which act as transducers in several hormonal systems and actually bind GTP (p. 99 and p. 334). These oncogene products have a M_r of ~21,000 (approximately half that of most normal α subunits), and are generally referred to as p21 transforming proteins. Ten to 15% of all human tumors express *ras*, which is under complete transcriptional regulation[144]. In yeast, *ras* products can replace the α subunit of adenylate cyclase G proteins. The p21 transforming proteins cannot hydrolyze GTP as do normal G proteins. The situation may resemble that created by a non-hydrolyzable analog of GTP or by cholera toxin, stimulating the prolonged formation of cAMP (p. 336). However, experiments substituting α-G_s or α-G_i by *ras* proteins have not been successful in higher organisms. In cancer cells expressing *ras*, the production of cAMP is actually decreased. The malignant transformation of cells infected by membrane receptor-like oncogene products may require the presence of *ras*, in contrast to the case of cytoplasmic oncogenic Tyr-PKs. Finally, *ras*, lipocortin (p. 51), GDP-binding microbial elongation factor and EF-Tu share some structural identity [145].

Membrane-bound Tyr-PKs (3b in Fig. I.135). The products of v-*src* (Rous sarcoma virus), v-*abl* (Abelson murine leukemia virus) and v-*fes*/*fps* (feline sarcoma virus) are cytoplasmic proteins with Tyr-PK activity (the catalytic domain of each represents ~30 kDa and has structural homology with the receptors described above). The phosphorylated protein (pp) of the 60 kDa product of *src* ($pp60^{v\text{-}src}$) was the first transforming retroviral protein isolated, and the first to which a biochemical activity was attributed. Tyr-PKs *bind to the plasma membrane* as well as to the *endoplasmic reticulum*. The mechanism of attachment involves the myristillation of the protein at its N-terminal extremity, a reaction specified by the first 10 to 15 aa. The enzymatic activity of Tyr-

144. Ishii, S., Kadonaga, J. T., Tjian, R., Brady, J. N., Merlino, G. T. and Pastan, I. (1986) Binding of the Sp1 transcription factor by the human Harvey *ras*1 proto-oncogene promoter. *Science* 232: 1410–1413.
145. Jurnak, F. (1985) Structure of the GDP domain of EF-Tu and location of the amino acids homologous to *ras* oncogene proteins. *Science* 230: 32–36.

PKs, exacerbated by specific mutations, is responsible for transforming activity. The *crucial substrates* of Tyr-PK involved in the transformation are *not known*.

Following their synthesis, the products of the *src* gene and of approximately 10 other retroviral genes interact temporarily with hsp 90 (p. 69) and another protein ($M_r \sim 50{,}000$), thus forming a complex in which enzymatic activity is inhibited until the dissociation of the complex releases Tyr-PK at the level of the plasma membrane.

Other protein kinases. With the possible exception of v-*mil/raf* and v-*mos*, the oncogene products which have been studied thus far do not have Ser-Thr-PK activity (as do PK-A and PK-C). An oncogene protein kinase has not been shown to be involved in phosphoinositide metabolism.

Nuclear proteins of oncogene products (4 in Fig. I.135). Oncogenes v-*fos* (mouse osteosarcoma virus), v-*myb* (avian myeloblastoma virus), v-*myc* (avian myelocytoma virus) and v-*ski* (avian virus), the large T antigen of papovavirus (SV 40, and polyoma), and the products of E1A (adenovirus) are nuclear. The first four bind DNA, and have been observed frequently in the nuclear matrix. During the cell cycle, the increase of *myc* prior to DNA replication is regulated by growth factors specific to the cell type. The *myc*, *myb* and E1A products share structural homology and act on DNA synthesis and transcription via an undefined mechanism. Oncogene products lead to the production of other active proteins which are possibly analogous to growth factors; p53 is a protein belonging to this category.

6.3 Complexity

In order to produce an experimental *cancer phenotype*, *at least two* categories of oncogenes must be present to produce both the *immortalization* and *transformation* which distinguish normal cells from cancerous cells. This transformation is the cause of contact-independent growth (Table I.22)[146,147], and, when injected into animals, transformed cells have the potential to cause tumors. However, the notion that the presence of two oncogenes is necessary for tumor development may be an oversimplification. It is probable that, in many cases, a large number of oncogenes is required. Such a complexity could account for differences between cancers, particularly with respect to their immunological properties and metastatic potential (p. 147). The involvement of several oncogenes could be the reason why there is a delay in the appearance of cancers after exposure to the initial carcinogen, and would also correspond to the general notion of multihormonal control in the regulation of most cellular functions.

Table I.22 Cooperation of oncogenes.

Class I: **"Transforming"** (acting at the membrane and/or cytoplasmic level)	**Class II:** **"Immortalizing"** (acting at the nuclear level)
ras	*myc*
E1B (adenovirus)	E1A (adenovirus)
Py-mT (polyoma virus)	Py-lT (polyoma virus)
	SV 40-lT

Do the two functional classes of oncogenes really cooperate – one to ensure unlimited growth of transformed cells (immortalization) and the other to provide other aspects of the cancer phenotype? Cooperation is possible, but the concept that certain properties are related exclusively to one type of oncogene suffers from numerous experimental contradictions[148]. Quantitative studies may be necessary to elucidate this question still under debate. For example, complete transformation of normal cells, including immortalization, may be produced by a large amount of a single oncogene, and in experiments involving two types of oncogenes, it is not clear whether carcinogenesis is attributable to qualitative complementarity or simply to quantitative summation. Many studies have been performed with NIH 3T3 cells, which are precancerous cells already immortalized by an oncogene, and therefore the results of these experiments have been difficult to interpret. The cellular specificity, and thus the dependency of oncogenes on differentiation, must also be considered, since, contrary to *myc* and *ras*, many oncogenes are expressed only selectively in a few cell types. Conversely, deregulation of c-*myc* expression in transgenic mice leads to multiple neoplasia[149].

146. Razzoulzadegan, M., Cowie, A., Carr, A.,Glaichenhaus, N., Kamen, R. and Cuzin, F. (1982) The roles of individual polyoma virus early proteins in oncogenic transformation. *Nature* 300: 713–718.
147. Land, H., Parada, L. F. and Weinberg, R. A. (1983) Cellular oncogenes and multistep carcinogenesis. *Science* 222: 771–778.
148. Duesberg, P. H. (1987) Retroviruses as carcinogens and pathogens: expectations and reality. *Cancer Res.* 47: 1199–1220.
149. Leder, A., Pattengale, P. K., Kuo, A., Stewart, T. A. and Leder, P. (1986) Consequences of widespread deregulation of the c-*myc* gene in transgenic mice: multiple neoplasms and normal development. *Cell* 45: 485–495.

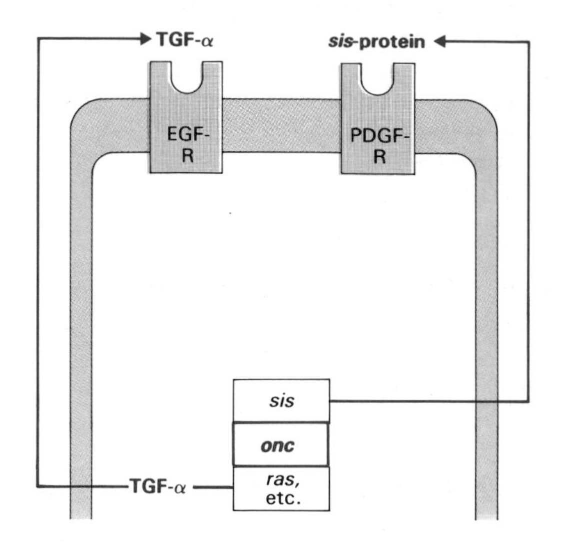

Figure I.139 Autocrine mechanism of cell growth stimulation. (Sporn, M.B. and Todaro, G.J. (1980) Autocrine secretion and malignant transformation of cells. *New Engl. J. Med.* 303: 878–880).

In conclusion, many (most, maybe all) oncogene products are clearly cognates of normal components of hormonal systems, especially those regulating cell growth, which probably constitute a relatively well defined network made up of few constituents. Tumor cells can produce growth factors that stimulate cell replication and which make growth independent of external control. The production of TGF-α (p. 219) is increased by a number of oncogenes (*ras*, *fes*, *mos*, *abl*). The *sis*-protein, homologous to PDGF-B, interacts with PDGF-R, demonstrating that an oncogene product can be directly responsible for autocrine activity (Fig. I.139). Actually, oncogene products are not the only factors capable of producing cell transformation. Overproduction of EGF obtained in transfection experiments[150] also results in cell transformation by autocrine stimulation. The interactions between various oncogene systems are innumerable. The expression of c-*fos* is stimulated by PDGF, but this increase is not necessary for the subsequent increase in c-*myc*, and c-*myc* is not indispensable for a growth response to PDGF (although in the stimulation of mitosis it can substitute for growth factor). In fact, there are numerous points of convergence between various growth factors and several oncogenes: anti-*ras* product antibodies block the response of the cell to many growth factors, and conversely, c-*ras* can be activated by different hormones acting via different receptors, such as those of insulin and EGF.

The potential complexity of the matter is considerable, given that there are still undefined and poorly defined growth factors, such as those controlling the synthesis of extracellular matrix, which may play a role in the invasive and metastatic character of tumors. The clinical involvement of oncogenes in cancer is discussed on p. 148.

Indeed, in order for a cancer to develop, a number of events must occur at specific times. In 1940, Huggins showed that NMU (nitrosomethylurea) was ineffective in generating tumors in mice which were castrated at the time of exposure to this carcinogen, indicating the complementary but necessary role of hormones.

A few years ago, the association of the words oncogenes, hormones and receptors did not have any concrete significance. In contrast, today, the integrated study of oncogenes, hormones and receptors represents one of the most productive areas of cancer research.

7. Assessment of Hormonal Function in Pathophysiological Studies‡

The development in the early 1960s of *competitive binding assays*, also known as *saturation analysis*[151], permitted the measurement of the levels of most hormones, their metabolites and synthetic analogs, in plasma or serum, causing a true revolution in endocrinology. *Radioimmunoassays (RIAs)*, using appropriate radioactively labelled hormones and specific antibodies, constitute the most important technique in pathophysiological studies[152]. The sensitivity and specificity of RIAs is still essentially unmatched, provided that they are appropriately designed for the specific requirements of the particular biological system under study (p. 486). The intrinsic specificity of radioimmunoassay is well suited to the numerous stimulation

150. Stern, D. F., Hare, D. L., Cecchini, M. A. and Weinberg, R. A. (1987) Construction of a novel oncogene based on synthetic sequences encoding epidermal growth factor. *Science* 235: 321–324.
151. Ekins, R. P. (1960) The estimation of thyroxine in human plasma by an electrophoretic technique. *Clin. Chim. Acta* 5: 453–459.
152. Yalow, R. S. and Berson, S. A. (1960) Immunoassay of endogenous plasma insulin in man. *J. Clin. Invest.* 39: 1157–1175.

‡I would like to thank Dr. Jean Prédine (Faculté de Médecine, Université Paris XI) and Dr. Fernand Dray (Institut Pasteur, Paris) for their help in preparing this section.

and suppression tests designed to assess the basic feedback mechanisms of various endocrine systems. Such specificity is also necessary for the identification of new or abnormal secretion patterns. However, it is usually not possible to measure hormones or hormone receptors directly in human tissues.

Endocrinologists thus possess potent analytical techniques for the diagnosis and medical treatment of hormone-related diseases. Not only are they able to determine the source of a disease and to situate it in terms of global endocrine function, but they are often in a position to provide information necessary to avoid inappropriate and possibly deleterious hormonal "treatment".

7.1 Before Competitive Binding Assays

Initially, hormones were measured by means of *bioassays*, since this was the only possible approach prior to chemical characterization. Some of these assays are still used to determine the *biological activity* of new molecules and radioactive ligands, and to establish their relative potency with respect to corresponding endogenous compounds. Biological assays are frequently difficult to perform, because a large sample size or volume is usually required; values are often expressed as arbitrary *international units* (IU). Until recently, biodetection of gonadotropins in urine was useful in the diagnosis of pregnancy as well as chorio-epithelioma[153]. International units are no longer used for reasons of practicality, with the exception of LH and FSH which continue to be reported in clinical studies in terms of defined standards (e.g., Fig. I.30).

Radioreceptor assays have been proposed as new *biological indices*, since all molecules that specifically bind a receptor are taken into account in measurements. In addition to technical problems due to defined receptor instability, the principle of such assays is questionable, given that receptors bind antagonists as well as agonists. These assays are still used, however, to detect unknown compounds for which antibodies are not available (e.g., metabolites of synthetic androgens used for athletic doping, or DES or other anabolic agents in meat, or new members of a peptide hormone family, etc.).

Steroid metabolites have been measured chemically[154] for some time, and urinary pregnandiol is still a good indicator of progesterone secretion during the luteal phase or during pregnancy. Blood catecholamines can be measured by fluorescence or, more recently, by high pressure (performance) liquid chromatography (HPLC) coupled to a chemical detector. Double isotope dilution is a technique used for the accurate measurement of compounds in low concentration, such as plasma aldosterone which was first measured by this technique [155]. Most but not all of the aforementioned techniques have been discontinued with the development of RIAs.

7.2 Immunoassays: Saturation Analysis (Competitive Binding Assays) and Immunometric Analysis (Fig. I.140)

While Ekins' measurement of thyroxine with TBG in plasma[151] foresaw the future of saturation analysis, the initial observations of Berson and Yalow, using anti-insulin antibodies[152] (p. 486), offered endocrinologists a precise, sensitive, and easy method of measuring hormones in blood, with their development of radioimmunoassays (RIAs). Thus, both physics and immunology were associated for the most important technique used in endocrinology today.

Competitive Binding Assay

The general principal of competitive binding is based on the reversible interaction of molecules. Binding sites (B) for the hormone (H) are provided by a protein (usually an antibody), establishing a concentration $[B_T]$. They are exposed to a low concentration of labelled tracer $[\star H]$ and to the non-labelled hormone (H) to be measured. The concentration of $\star H$ bound to the protein, $[\star B]$, is a function of the dilution of $\star H$ by H (isotopic dilution if radioactivity is used) and the law of mass action. The reactive elements are H, $\star H$, B_T, H_B et $\star H_B$, which are related as follows:

$$[H]+[\star H]+[B] \rightleftharpoons [HB]+[\star HB]$$

with:

$$[B]+[HB]+[\star HB] = [B_T]$$

In practice, the process of separating bound and free ligand is similar to that described previously on p. 60.

153. Aschheim, S. and Zondek, B. (1928) Die Schwangerschaftsdiagnose aus dem Harn durch Nachweis des Hypophysenvorderlappenhormons. *Klin. Wochschr.* 7: 1453–1457.
154. Jayle, M. F. ed (1961, 1962, 1965) *Analyse des Steroides Hormonaux*. Masson, Paris.
155. Kliman, B. and Peterson, R. E. (1960) Double isotope derivative assay of aldosterone in biological extracts. *J. Biol. Chem.* 235: 1639–1643.

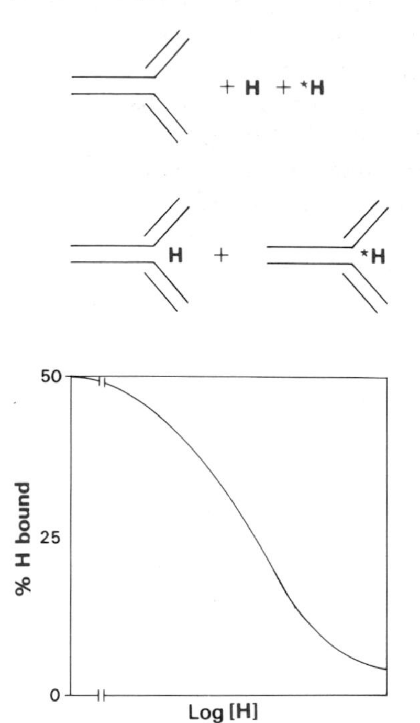

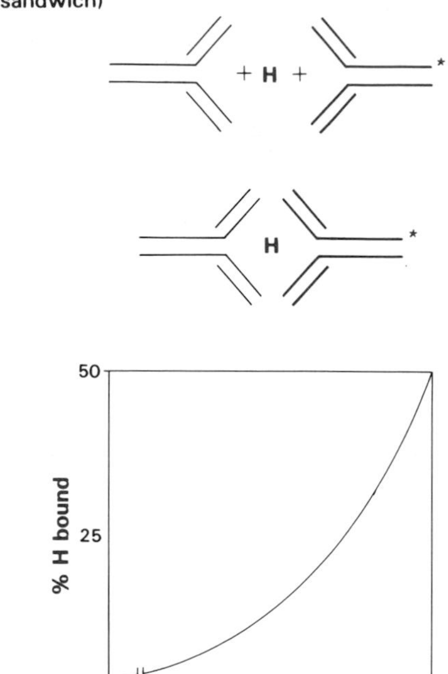

Figure I.140 Principle and typical standard curves of the two main techniques used in immunoassays of hormones. ★=label (radioactive atom, fluorescent moiety, enzyme, etc.). (Beastall, G. H. (1985) Non-isotopic immunoassay methods. *Laboratory Practice*, May issue).

Results are deduced from a standard curve (Fig. I.141A). The curve representing the concentration of [★H] bound, [★B], as a function of the concentration of [H], is not linear[65]. When the system is defined by the total concentration of tracer [★H] and hormone [H], radioactivity measured in the absence of H gives $[\star B_o]$ and the corresponding free radioactivity $[\star F_o]$, while [★B] and [★F] represent the concentrations of tracer, bound or free, in the presence of unlabelled hormone [H]. To linearize at least a part of the standard curve, the following equations can be used:

$$[\star F]/[\star B]=f[H]$$

$$[\star H]/[\star B]=f[H]$$

$$[\star B]/[\star B_o]=f[H]$$

or the logit-log transformation, with $\text{logit } Y=\log(Y/(1-Y))=a+b \log[X]$, where $Y=([\star B]-[\star NS])/([\star B_o]-[\star NS])$. (★NS) represents non-specifically bound radioactivity.

Microcomputer programs perform iteration of the standard curve, such as approximation by the spline function or that of a complex hyperbola. Such programs permit the optimization of assay conditions, including the calculation of sensitivity, the quantity of measurable hormone which is significantly different from zero, at a fixed level of probability, and the relative error of the assay, as a function of the amount of hormone assayed (Fig. I.141B).

Development of Immunoassays

Polyclonal antibodies used classically in radioimmunoassays are produced by the hyperimmunization of a rabbit, goat or sheep, using naturally antigenic substances such as protein hormones, or molecules such as steroids[156], or small peptides made antigenic by chemical coupling to a protein.

Today, there is an increased use of *monoclonal antibodies* that are produced by the fusion of lymphocytes from an immunized animal with immortal tumoral cells producing immunoglobulins[157]. This

156. Lieberman, S., Erlanger, B. F., Beiser, S. M. and Agate, F. J. Jr. (1959) Steroid–protein conjugates: their chemical, immunochemical and endocrinological properties. *Rec. Prog. Horm. Res.* 15: 165–200.
157. Milstein, C. (1986) From antibody structure to immunological diversification of immune response. *Science* 231: 1261–1268.

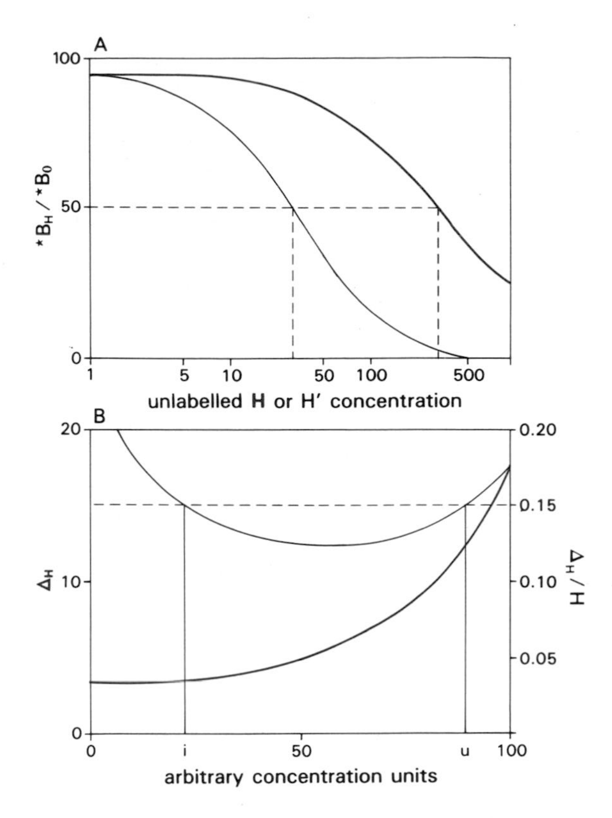

Figure I.141 Measurements by radiocompetition. **A.** Displacement curves of labelled hormone (binding at the concentration established for the experiment=[$\star B_o$]), as a function of unlabelled ligand concentration [H]. With the second, unlabelled hormone, (H′), the ratio of the concentration of H′ to that of H, both of which decrease [$\star B_o$] by 50%, gives an estimate of antibody specificity (percent of cross-binding). **B.** Total (ΔH) or relative (ΔH/H) error in measurement as a function of the concentration of product to be measured. For a relative error (e.g., 15%), there is a lower limit, i, indicating the sensitivity of the assay, and an upper limit, u, indicating the range of the amount of hormone measurable. The 50% decrease in ligand binding does not rigorously give the K_D of ligand–antibody interaction.

technique requires the identification of clones producing antibody with high affinity and specificity. Once this selection process has been made, however, such antibodies can be produced in practically unlimited quantities by injecting these hybrids into animals and provoking ascites. A large quantity of monoclonal antibodies can be produced, and their qualitative constancy is one of the most important advantages of this technique. In addition, a precise epitope can be chosen in order to discern peptides that share a common region, such as hLH and hCG or proinsulin and insulin. Some problems remain, however, as indicated on p. 488.

Tritium is usually used to label most low molecular weight compounds, and gives an adequate indication of specific activity. *Iodination* increases specific activity, permitting a reduction in the quantity of labelled ligand and greater sensitivity. However, steric hindrance caused by the size of the iodine atom is such that antigen-antibody binding and the separation of bound and free forms of the antigen may be affected. The coupling of an antigen to an *enzyme* circumvents the inconvenience of working with radioactive isotopes, but the problem of another molecule affecting the recognition between the antigen and the antibody is still present. Small molecules can be labelled with a *fluorescent* compound. The probability that polarized fluorescence emits in a particular direction differs depending on whether the labelled molecule is free or is bound to an immunoglobulin. Analysis of fluorescence in both directions permits the evaluation of the proportion of bound and free forms, and does not necessitate physical separation which would interfere with binding equilibrium. Such non-radioactive methods do not require a radioactive counter, which is relatively expensive. Automation of such non-radioactive techniques will make them very attractive for large-scale use, especially in developing countries[158]. The comparative evaluation of these different methods should take into account the class of hormone to be measured and the practical conditions in which they are to be used.

Immunometric Methods

The separation of bound and free forms of hormone is a crucial phase in competitive binding analysis since it interrupts binding equilibrium. Techniques with liquid phases use the adsorption of small molecules onto activated charcoal, or the precipitation of a first antibody by a second antibody directed against the first. Such approaches, however, are being replaced by *solid phase* techniques (sandwich), known as immunometric analysis. A first antibody, frequently monoclonal, is bound to a solid phase, and is incubated in the presence of an antigen (Fig. I.140B). Once equilibrium has been reached, and subsequent to a washing step, the second antibody directed against a different epitope interacts with the bound antigen. One or the other antibody can be labelled, depending on the technique (Table I.23), using an enzyme which

158. Kohen, E., Pazzagli, M., Serio, M., De Boevers, J. and Vandekerckhoves, D. (1985) Chemiluminescence and bioluminescence immunoassay. In: *Alternative Immunoassays* (Collins, P., ed), John Wiley, New York, pp. 103–121.

Table I.23 Competitive binding analyses.

Two Principles

A. Saturation analysis (competition). The *binding* antibody is *limited in amount*, in relation to the hormone molecules (H) to be measured. Thus, at equilibrium, free (F) and bound (B) hormone can be conveniently measured.

B. Immunometric analysis (sandwich). A *first* antibody, *in excess*, linked to or arranged as a solid phase support, binds H. A second labelled antibody, recognizing a different epitope of H, allows the bound H to be measured. In this methodology, two distinct epitopes of H must be available.

Glossary of Several Techniques

RIA: radioimmunoassay. A liquid or solid phase system involving an antibody and a radioactive ligand.

EIA: enzymoimmunoassay. A liquid or solid phase system using an antigen or an antibody labelled with an enzyme.

FIA: fluorescent immunoassay. As EIA, but the label is a fluorescent molecule.

FPIA: fluorescence polarization immunoassay. A liquid phase system in which the antigen is labelled by fluorescein. The emission of polarized fluorescence is dependent on whether the antigen is bound or free, and thus measurement does not require separation of the two forms.

ELISA: enzyme-linked immunosorbent assay. A solid phase system in which the antigen or the first antibody is immobilized, and a second enzyme-labelled antibody is used.

IRMA: immunoradiometric assay. Similar to ELISA, but the second antibody is labelled with a radioactive atom.

DELFIA: dissociated enhanced lanthanide fluoroimmunoassay. The second antibody is labelled with a europium chelate which fluoresces after dissociation of the complex from the first antibody. This method is very sensitive, and permits the measurement of ligands over a large range of concentrations.

releases a colored product (ELISA): radioactive iodine (IRMA) or a fluorescent molecule (FIA).

7.3 Other Aspects of Endocrine Measurements: In Tissues

Hormones distributed regionally (i.e. via the hypothalamic-hypophyseal-portal system) have limited access to the general circulation, and therefore their measurement in plasma is probably of limited significance. Problems involved in endocrine measurements can be compounded by multiple sources of hormone secretion (e.g., GnRH of gonadal origin). Parahormones and autohormones are difficult, if not impossible, to measure in the blood. Tissue samples are frequently unobtainable in humans. It is still not feasible to visualize hormones in situ by atomic activation, although there is promise that this might become possible in the future. Hormone measurements in *accessible fluids*, as in *cerebrospinal fluid* or *semen*, have not yet produced much meaningful information. In contrast, the measurement of steroids in *saliva* is often useful[159], but salivary steroid concentrations appear to be reflective only of plasma levels (transsudate). In specific cases, when the clinician is seeking a total integrated evaluation of a given endocrine product, measurements of *urinary levels* are frequently more helpful than plasma concentrations (e.g., some urinary steroids; EGF, also called urogastrone (p. 219); cAMP in parathyroid diseases (p. 646)).

It would be useful to be able to quantitate *receptor levels in humans* simply by measuring the emission of the appropriately labelled, bound hormone. Unfortunately, this technology is not yet available, although technical improvements have resulted in many more receptor measurements in tissue samples, as for example for steroid[160] (p. 412), insulin, prolactin and growth factor receptors. Up until now, primarily mammary, uterine and leukemic cells or tissues have been studied. Some assays can be carried out using needle biopsies, as a result of assay *miniaturization*. Immunohistochemical techniques can be applied to the detection of receptors (Fig. I.68), but they are difficult to quantitate. In any case, both the determination and the interpretation of receptor concentration are *problematic*, not only because of cellular heterogeneity, but also because both hormone binding and/or immunological detection do not necessarily imply that the receptor is functional. Frequently, the *clearest results* are those which are *negative*, thus exluding the suspected hormonal component.

Finally, there is growing interest in the use of nucleic acid probes which *hybridize mRNAs* of receptors, biosynthetic enzymes, and regulated proteins. The approach can be *biochemical* (by Northern blot, Fig. I.96) or *histological* (*in situ hybridization*)

159. Riad-Fahmy, D., Read, G. F., Walker, F. and Griffiths, K. (1982) Steroids in saliva for assessing endocrine function. *Endocrine Rev.* 3: 367–395.

160. Symposium on Estrogen Receptor Determination with Monoclonal Antibodies (1986). *Cancer Res.* 46: 4231s–4313s.

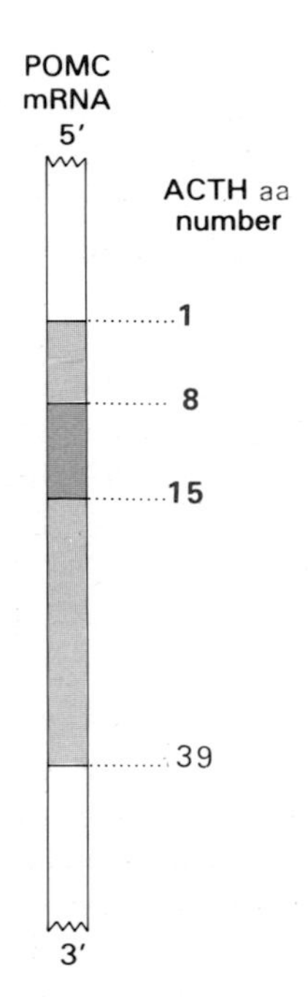

Figure I.142 Oligonucleotide probe: a 24 bp synthetic cDNA corresponding to amino acids 8–15 of ACTH. (Lewis, M. E., Sherman, T. G., Burke, S., Akil, H., Davis, L. G., Arentzen, R. and Watson, S. J. (1986) Detection of proopiomelanocortin mRNA by in situ hybridization with an oligonucleotide probe. *Proc. Natl. Acad. Sci., USA* 83: 5419–5423).

(Fig. I.32). As more sequence information becomes available, it is not only possible to use *cDNAs* encoding a particular protein of interest, but also *oligonucleotides*, which are easy to produce, and which can be specifically tailored to identify the product in question. A number of regulated steps involved in the production of such mRNAs will be identified, as has already been achieved for POMC (Fig. I.142). These recent technical developments, together with direct measurements of hormones and receptors, should facilitate the understanding of *specific cellular responses* to hormonal controls.

7.4 Investigating Integrated Endocrine Functions: Paired Measurements and Dynamic Tests

The most specific and sensitive of hormone assays cannot provide more than a single value at a time. In many cases, this is enough to confirm a diagnosis. However, most hormones are not secreted at a constant rate, and their concentration in the blood varies, often episodically, throughout the day (p. 38). *Serial determinations* frequently improve assay interpretation. The best means of establishing the actual concentration of the hormone responsible for an observed effect at the level of the target organ is provided by measurements of *plasma* hormone concentration. Plasma hormone concentration is dependent on metabolism and other factors (p. 45), and thus interpretation of hormone measurements is often complex with respect to secretion. Urinary assays, when they are possible, integrate these variations and provide information concerning the global pattern of hormone secretion.

Paired determinations (Table I.24) can contribute greatly to the interpretation of assays. In addition to measuring the hormone itself, one of its effects or another hormone that regulates its production (Fig. I.143) may also be measured. For example, both TSH and T_4 or PTH and Ca^{2+} can be measured.

Table I.24 Paired measurements: central and peripheral hormones.

Conditions	Hormones	
	Central (pituitary)	Peripheral
Primary peripheral gland deficiency	High	Low
Primary pituitary excess	High	High
Resistance to feedback control by peripheral hormone	High	High
Normal	Normal	Normal
Primary pituitary failure	Low	Low
Primary gland excess	Low	High

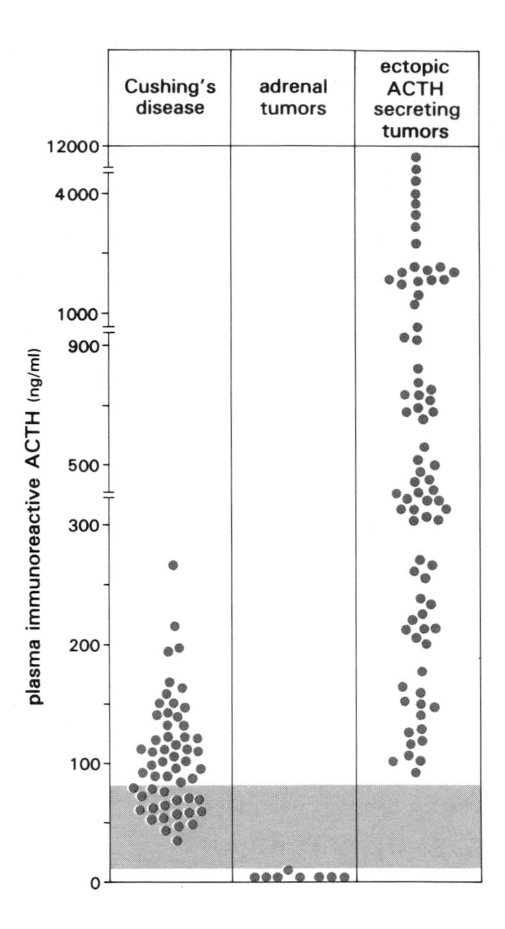

Figure I.143 Plasma ACTH levels in patients with elevated plasma cortisol (Cushing's syndrome). Gray area represents the normal range. (Symington, T. S. and Carter, R. L., eds) (1976) *Scientific Foundations of Oncology.* Heinemann Medical Books, London).

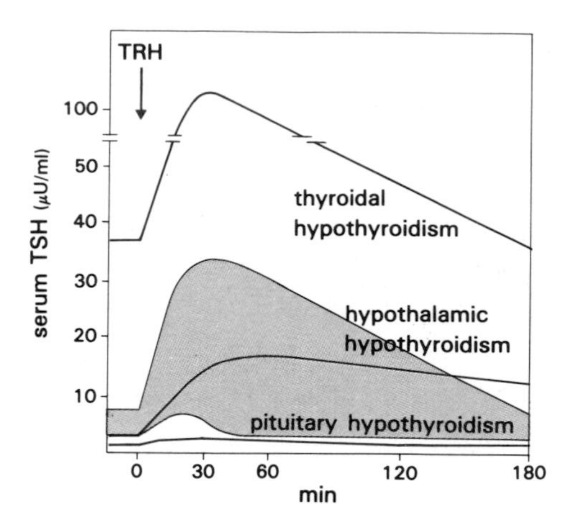

Figure I.144 TSH response to TRH. Gray area represents the normal range. (Spaulding, S. W. and Utiger, R. D. (1981) The thyroid: physiology, hyperthyroidism, hypothyroidism, and the painful thyroid. In: *Endocrinology and Metabolism* (Felig, P., Baxter, J. D., Broadus, A. E. and Frohman, L. A., eds), McGraw-Hill Books, New York, pp. 335–351).

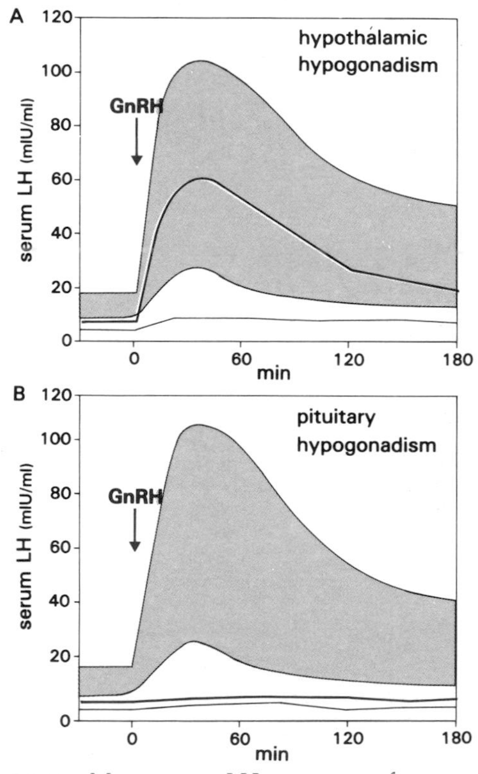

Figure I.145 Mean serum LH responses of ten men with hypogonadotropic hypogonadism to a 250 μg bolus of GnRH before (dark line) and after (red line) daily administration of GnRH for one week. Gray area represents the normal range. **A.** Five cases of hypothalamic disease. **B.** Five cases of pituitary disease. (Snyder, P. J., Rudenstein, R. S., Gardner, D. F. and Rothman, J. G. (1979) Repetitive infusion of gonadotropin-releasing hormone distinguishes hypothalamic from pituitary hypogonadism. *J. Clin. Endocrinol. Metab.* 48: 864–868).

The use of paired determinations is based on the concept that each hormone is part of an integrated system.

Dynamic tests are frequently used to facilitate the interpretation of a basal hormone level, or to identify a problem not revealed by a hormone level at the limit of normal. All types of *stimulation* and *suppression tests* affecting the various elements of the endocrine system are included in this category. Sometimes it may not even be necessary to measure the hormone itself. For example, in the water deprivation test, the resulting decrease in blood volume activates a sensor in the hypothalamus, causing an increase in the secretion of vasopressin which in turn acts on renal cells to limit loss of water. However, the measurement of urinary volume does *not* identify the *level* of a potential abnormality. Thus, it is important to select very precisely a sector of the endocrine network, and to test each of its

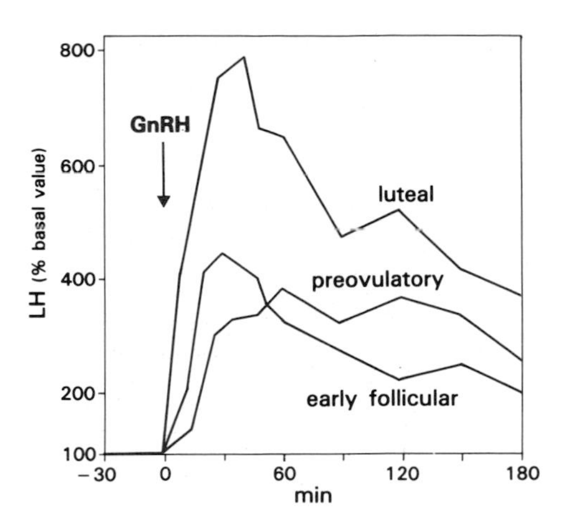

Figure I.146 LH response to GnRH administration in normal women at varying phases in the menstrual cycle. (Wollesen, F., Swerdloff, R. S. and Odell, W. D. (1976) LH and FSH responses to luteinizing releasing hormone in normal fertile women. *Metabolism* 25: 1275–1285).

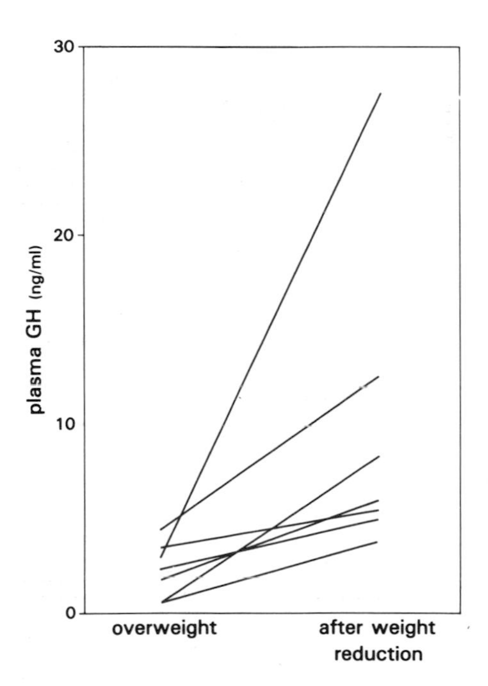

Figure I.147 Effect of weight reduction on GH response to arginine. Each line connects the peak plasma GH concentration for an individual patient before and after weight reduction. (El-Khodary, A. Z., Ball, M. F., Stein, B. and Canary, J. J. (1971) Effect of weight loss on the growth hormone response to arginine infusion in obesity. *J. Clin. Endocrinol. Metab.* 32: 42–51).

sequential components, using stimulation and suppression tests. For a hormone with multiple sites of production (e.g., androgens from the ovary and the adrenal), the complexity and the contribution of each gland can be analyzed. Figures I.144 through I.147 illustrate various approaches to the localization of the site of an endocrine disorder (e.g., at the level of secretion or regulation, Fig. I.144), to the restoration of a functionally diminished response (Fig. I.145), and to the establishment of how physiological status affects a hormonal response (Fig. I.146). In addition, for both non-stimulatory and dynamic tests, it is important to consider the overall clinical and therapeutic context, since it is likely to have a bearing on test results (e.g., obesity can result in reversible endocrine abnormalities (Fig. I.147); depression can cause abnormal secretion of GH, TSH and ACTH; the intake of drugs affects the nervous system, e.g., l-DOPA, chloropromazine, or even aspirin, which can reduce the response of TSH to TRH). Table I.25 provides a partial list of dynamic tests, and Figure I.148 shows numerous ways to intervene in the highly refined CRF-ACTH-cortisol system.

7.5 Mass Spectrometry

Today a reference method, tomorrow a major technology?

Although mass spectrometry (MS) is the best analytical means for determining the molecular structure of a hormone, it is not used routinely. Since the appearance of the first publications, immense progress has been made[161,162] in separation methods (usually gas chromatography, GC) which precede analysis, as well as in MS itself. The technique of MS involves the generation of ions which are then separated, on the basis of their mass-to-charge ratio (m/z), and measured. Many instruments with different levels of sophistication are used, depending on specific needs. Data bank and computer-assisted interpretation of spectra are basic to the use of mass spectrometry in steroid and peptide research. In the 1970s, progress was made in the refinement of gas chromatography–mass spectrometry (GC/MS) and analytical methods, following the introduction of new capillary columns

161. Shackleton, C. H. L. (1985) Mass spectrometry: application to steroid and peptide research. *Endocrine Rev.* 6: 441–486.
162. Sjövall, J. and Axelson, M. (1982) Newer approaches to the isolation, identification and quantification of steroids in biological materials. *Vitams. Horm.* 39: 31–144.

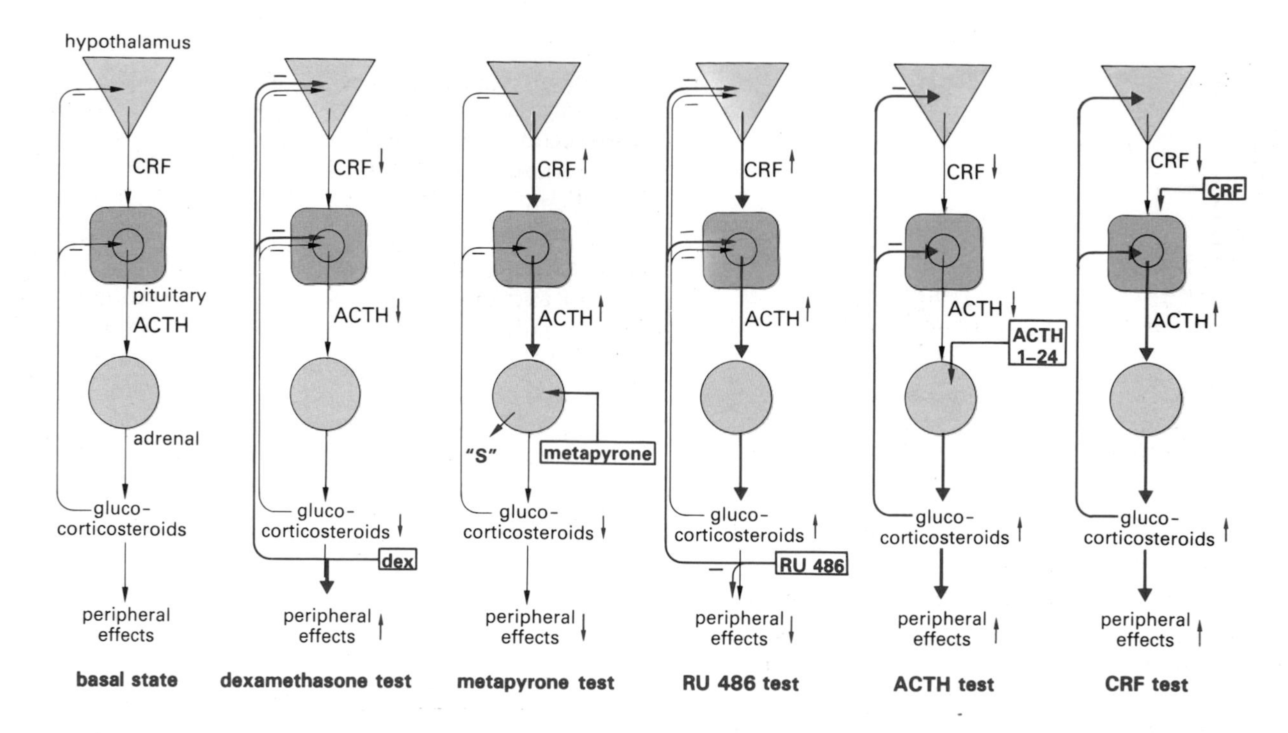

Figure I.148 Stimulation and suppression tests for the hypothalamic-hypophyseal-adrenal system. *Basal state*: under the effect of ACTH, the adrenal cortex secretes cortisol, which is responsible for the negative feedback of CRF and ACTH. *Cortrosyn or synacthen test*: 1–24 aa ACTH stimulates adrenal production, and consequently endogenous ACTH is reduced. *Dexamethasone test*: this synthetic glucocorticosteroid has a negative feedback effect on CRF and ACTH, and cortisol secretion is reduced. *Metapyrone test*: this compound, an inhibitor of 11β-hydroxylase, reduces the production of cortisol, removing the negative feedback effect on ACTH, and in turn increases adrenal androgens and compound S. *CRF test*: increases ACTH production and, consequently, cortisol production. *RU 486 test*: RU 486 blocks the effects of cortisol at the receptor level, and inhibits negative feedback on CRF and ACTH, leading to an increase in plasma cortisol.

Table I.25 Dynamic endocrine tests.

A. Stimulation

Hormones measured	Stimulatory agents
LH/FSH	GnRH, chlomiphene
Testosterone	LH/hCG
ACTH/β-endorphin	CRF, metapyrone, RU 486
Cortisol/11-deoxycortisol	CRF, ACTH, metapyrone, RU 486
Aldosterone/renin	Upright posture
PRL	TRH, chlorpromazine, metoclopramide
GH	Insulin/hypoglycemia, GRF, exercise, l-arginine, l-DOPA
TSH	TRH
Thyroid uptake of ⋆ iodine	TSH
Vasopressin/urine concentration	Water deprivation
(Penta)gastrin	Calcitonin
Ca^{2+}	Calcitonin
cAMP/phosphate excretion	PTH
Insulin	Glucose

B. Suppression

Hormones measured	Suppressive agents
GH	Glucose
Prolactin	Bromocryptine, l-DOPA
ACTH/cortisol	Dexamethasone
Aldosterone/renin	Salt
Thyroid uptake of ⋆iodine	Thyroxine
Insulin/glucose	Fasting
Norepinephrine	Clonidine

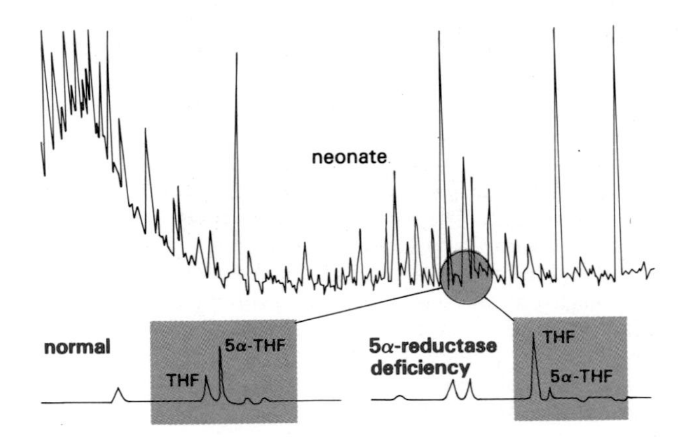

◁ **Figure I.149** Detection of minor metabolites in a complex mixture: the use of selected ion monitoring. The upper graph represents the conventional GC chromatogram of urinary steroids separated as methyloxime–trimethylsilyl ethers. In contrast to adults, neonates excrete almost no tetrahydrocortisol (THF) and 5α-tetrahydrocortisol (5α-THF), and peaks representing these steroids cannot be distinguished using a non-selective technique, as for example a flame-ionization detector. These compounds would be buried in the area circled. Adjusting a mass spectrometer to detect only ions specific for the tetrahydrocortisols (e.g., in this case, an ion with a mass of 652) enables a tracing which effectively ignores all compounds except the tetrahydrocortisols. In normal infants, 5α-THF always predominates, but the peak given by this compound in patients affected by 5α-reductase deficiency is very small. (Courtesy of Dr. C. Shackleton, Children's Hospital, Oakland, California).

and derivatization techniques. Novel ionization techniques and HPLC interfaced with MS are recent advances. The introduction of fast atom bombardment (FAD) as a new ionization method permits routine analysis of polar, charged, high mass molecules. HPLC offers the important possibility of separating hormones whose mass, polarity and/or ionic nature make isolation difficult. MS offers a means of serially analyzing media containing related hormones with slightly different structures (Figs. I.149 and I.150). It is also the best way to identify very small amounts (pmol level) of new compounds, or compounds found for the first time at a previously unsuspected site (Fig. I.151).

Presently unmatched for definitive physicochemical characterization of small molecules, MS will also be developed for the analysis of complex molecules, including peptides. It will be particularly useful in cases where the N-terminal amino acid is blocked, or when there are peptide or amino acid impurities. The quantitation of peptides is still difficult with MS. Although today instruments are still very expensive, MS may become a technique used routinely in the future for serial clinical measurements of many hormones.

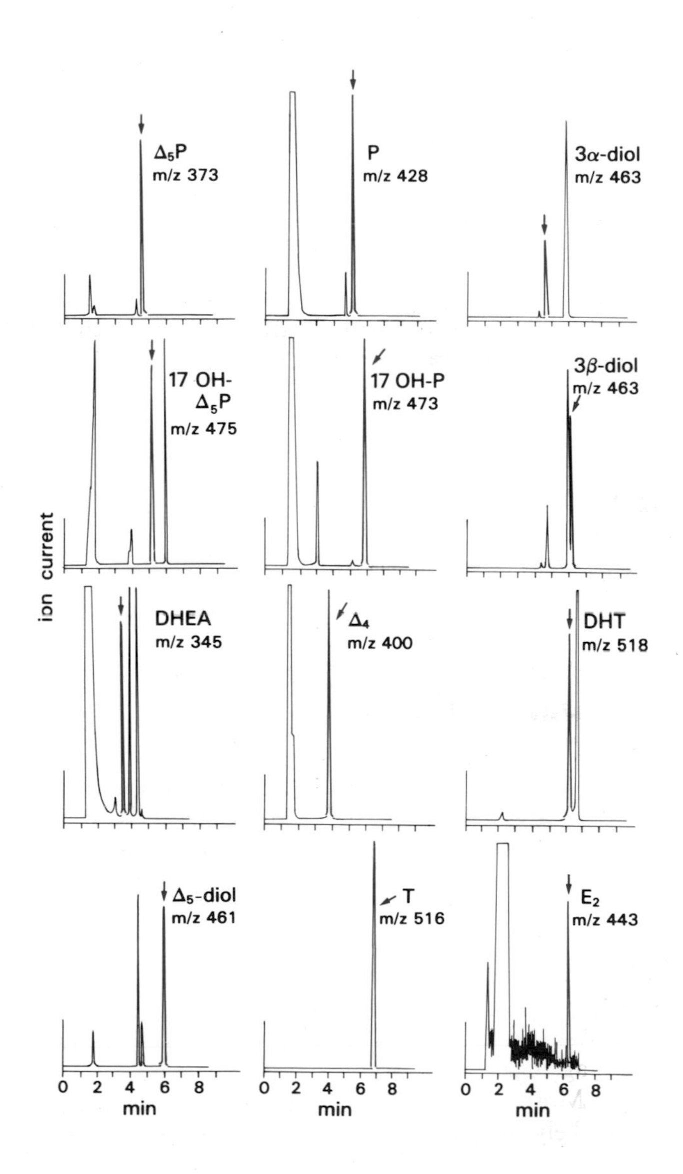

▷ **Figure I.150** Selected ion chromatograms obtained in a gas chromatographic–mass spectrometric analysis of unconjugated steroids of rat testis. Steroids were extracted and purified by gel and high-performance liquid chromatography (HPLC) and derivatized to enol-*tert*-butyldimethylsilyl ethers which were then purified by HPLC. Derivatives of Δ_5P, P, and their hydroxy analogs (17 OH), 3α-diol and 3β-diol, DHEA, Δ_4, Δ_5-diol, T, DHT and E_2 are indicated by arrows in the chromatograms of the respective *m/z* values monitored. Analogs labelled with ^{14}C were used as internal standards, and are seen in some chromatograms (^{14}C-Δ_5-diol in *m/z* 463, ^{14}C-T in *m/z* 518). (Courtesy of Dr. J. Sjövall, Karolinska Institute, Stockholm).

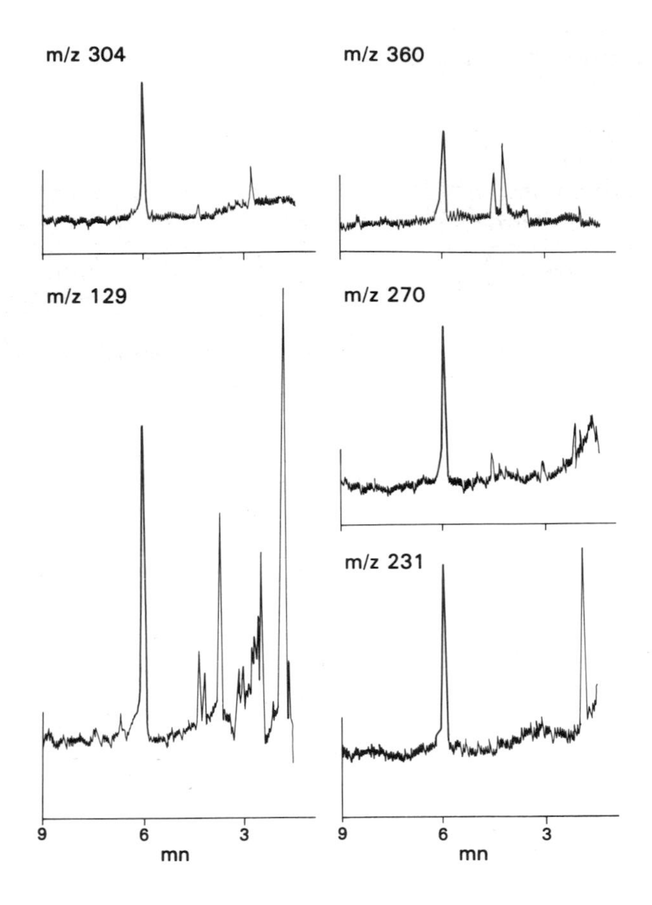

Figure I.151 Identification of dehydroepiandrosterone in rat brain by gas chromatography–mass spectrometry: single ion current chromatograms obtained by the analysis of trimethylsilyl (TMS) ether of purified steroid. Major ions given off by dehydroepiandrosterone TMS ether eluted at 6.1 min were selected. The signal of *m/z* 129 was attenuated four times. The peaks at 6.1 min represent approximately 200 pg of steroids. The relative intensities of the peaks were the same, within experimental error, for the extracted steroid and the reference dehydroepiandrosterone, and calculation gave a value which concorded with that obtained by radioimmunoassay. (Corpéchot, C., Robel, P., Axelson, M., Sjövall, J. and Baulieu, E. E. (1981) Characterization and measurement of dehydroepiandrosterone sulfate in the rat brain. *Proc. Natl. Acad. Sci., USA* 78: 4704–4707).

8. Aspects of Endocrine Pathophysiology

Endocrine pathophysiology is extremely *diverse*, since hormones are involved in *all processes* of the body, and since defects can occur at many levels (Table I.26). Endocrine diseases may be "simple", caused by a single parameter bearing on hormone availability or receptor function. In contrast, it is much more difficult to understand the part played by hormones when intricate processes are involved, as for example in cancer, mental illness, abnormal growth, hypertension, and ageing.

More and more frequently, subtle abnormalities are likely to be discovered at the molecular level. For instance, limited mutation may lead to non-functional hormone, yet the mutated and inactive hormone appears to be present in normal concentration, as measured by most immunoassays. At the level of target cells, abnormal metabolism may selectively inactivate a hormone and create hormone-resistance undetectable by blood measurements or receptor studies. A single amino acid mutation at the receptor level may leave the hormone binding function intact, but alter part or all of the receptor activity, e.g., Tyr-PK activity (p. 78), interaction with G protein (p. 99) or HRE (p. 92). Hormone receptors belong to families of proteins which include oncogenic products (p. 122), and, between the normal c-*onc* protein and its constitutively active *onc* counterpart, differences in amino acid sequences may be very limited.

However, despite growing evidence that the mechanisms involved in pathological processes are complex and interrelated, biochemical knowledge cannot replace physiological observations. The quasi-

Table I.26 Different levels of endocrine disorders.

Abnormal hormones
Synthesis (gene mutation, altered processing)
Secretion
Abnormal transport or metabolism
Degradation (increased/decreased)
Binding to transport protein or antibodies
Abnormal receptor
Binding
Transduction
Effector site
Concentration
Postreceptor defect

ubiquitous presence of insulin, the cerebral synthesis of numerous peptides and even steroids, and the synthesis of neurotransmitters in numerous peripheral organs should not lead to an underestimation of the functional importance of endocrine glands; it is clearly adrenal insufficiency that causes Addison's disease, and dysfunction of the pancreas results in diabetes. Nevertheless, physicians should not neglect the newly described, multiple sites of hormone production, since what would normally be considered insignificant hormone synthesis could be important in understanding a pathological process.

In summary, a precise evaluation of the role of hormones in pathological disorders is often very complex, and consequently the establishment of an effective medical treatment can be difficult. Three principles should always be considered. First, *many* hormones can act together on the same cells, with potentiating and/or inhibiting effects as a function of hormone concentration and kinetics. Secondly, there is frequently a *permissive* effect, of one or several hormones, that is independent of hormone concentration above given basal level. Finally, there are *auto*- and *paracrine* secretions which do not result in changes in circulating hormone concentration. Thus, in relation to these new concepts, new methods are needed in order to estimate and correct abnormalities.

8.1 Alterations of Hormone Secretion and Availability

Hormone *deficiencies* result in clinical and biological symptoms directly related to the hormone involved. Berthold (1849) first demonstrated the compensatory effect of testis transplantation in castrated animals. In 1855, Addison described a disease related to the destruction of the adrenal cortex which is easily treated now by steroid replacement therapy (p. 428). Myxedema, of unknown etiology in most cases (Fig. VIII.26A), and rickets, due to a deficiency in vitamin D (p. 660), can be prevented or completely cured by appropriate treatment. Diabetes mellitus, whatever its cause (p. 522), can be managed successfully with insulin. The physiological suppression of estrogen associated with menopause (p. 143) can also be corrected by hormones. *Functional* insufficiency of peripheral glands may be due to pituitary failure. This failure may itself be functional (secondary to altered hypothalamic regulation), or it may be organic in nature, as in Sheehan's syndrome (postpartum necrosis of the pituitary gland). Abnormality may also be the result of a modified secretory pattern (e.g., the negative consequence of continuous exposure of gonadotropic cells to GnRH, p. 447). The consequences of reduced hormone levels are not always easy to identify, e.g., estrogen deficiency during the luteal phase or in heavy smokers (osteoporosis is favored in smokers, probably due to increased metabolism of estrogens, and a lower incidence of endometrial cancer has been reported!). Hormonal deficiencies of immunologic or genetic origin are discussed in section 8.3 (p. 141).

Hypersecretion of hormones can be due to malignant or benign *tumors*, which can frequently be successfully removed surgically since they are usually detected early. Almost all glands, but strangely not the testes, can be the site of tumors secreting excessive amounts of hormone. *Multiple endocrine neoplasia* (MEN) depend on a single autosomal gene mutation transmitted in a dominant mode‡.

Hormones may be produced *ectopically* by tumors consisting of cells which normally produce little or no hormone. Such is the case for lung cancers producing ACTH or an ACTH-like substance, leading to secondary hypersecretion of the adrenals; there is no negative feedback regulation exerted by the excess of cortisol. Frequently, in both tumors of endocrine glands and ectopic hormone-producing cancers, there is an excessive production of immature forms of hormones (such as prohormones, Fig. I.23).

Hypersecretion can also be due to *non-tumoral* glandular hyperactivity. The symptoms of Cushing's disease (Fig. I.152A) are due to hypercortisolism, but the primary cause is at the pituitary and hypothalamic levels (p. 247). Other mechanisms may be involved, such as hormonally active antibodies, e.g., long-acting thyroid stimulating factors (LATS), an anti-TSH receptor antibody responsible for Graves' disease (p. 371). Functional disturbances can become auton-

‡Three major syndromes have been identified[163]. *MEN-1* (Wermer's syndrome) includes parathyroid hyperplasia, pancreatic islet cell adenoma or carcinoma (frequently gastrin-secreting), and adenoma or hyperplasia of the anterior pituitary. *MEN-2* (Sipple's syndrome) includes medullary carcinoma of the thyroid, pheochromocytoma, and parathyroid hyperplasia. *MEN-3* includes medullary carcinoma of the thyroid, pheochromocytoma, and multiple mucosal neuroma.

Neoplasia is used as a general term, and encompasses a range of lesions, from hyperplasia to adenoma and carcinoma. It is important to screen members of a family who are at high risk of developing endocrine tumors.

163. Leshin, M. (1985) Multiple endocrine neoplasia. In: *Textbook of Endocrinology*, 7th edition (Wilson, J. D. and Foster, D. W., eds), W.B. Saunders Co., Philadelphia, pp. 1274–1289.

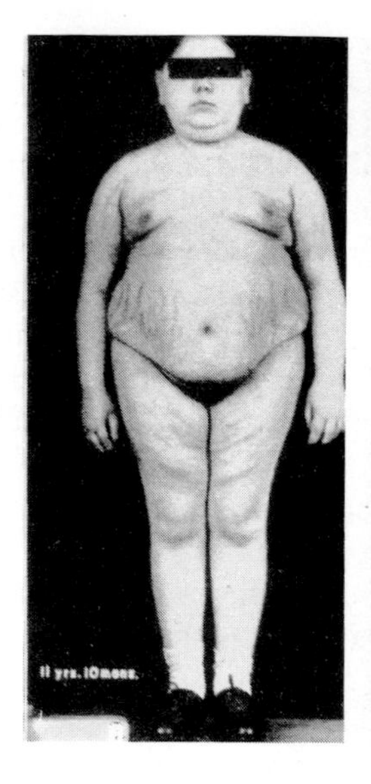
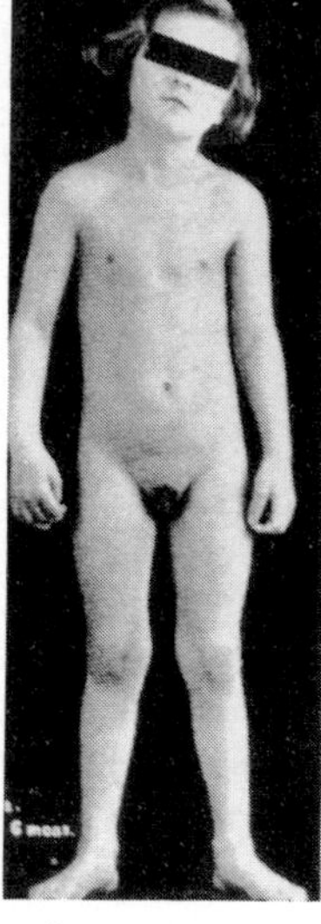

Figure I.152 Adrenocortical hyperactivity. **A.** Cushing's disease. **B.** Adrenal hyperplasia. (Albright, F. (1942–1943) Cushing's syndrome. *Harvey Lect.* 38: 123–127).

omous, e.g., when parathyroid glands have been stimulated over a prolonged period of time by abnormal calcemia (p. 644). A functional alteration may easily be reversible, as in the hypersecretion of corticosteroid associated with obesity, which regresses after weight loss.

The complexity of the clinical picture varies. A *single* enzymatic defect, as in congenital adrenal hyperplasia (p. 433), may have *several consequences*: the hypersecretion of androgens, and often of mineralocorticosteroids, is the result of a deficiency in glucocorticosteroids and a secondary excess of ACTH (Figs. I.152B and IX.37). Autoimmune diseases (p. 141) are also associated with hyper- and hypoactivity of various glands.

Abnormal Metabolism of Hormones

Changes in liver metabolism, such as those which are recorded during pregnancy or in liver disease, may not have pathological repercussions, since feedback mechanisms normally readjust the production rate and circulating level of a hormone (Fig. I.5). However, the administration of hormone may be dangerous in patients with hepatic dysfunction.

Endocrine pathophysiology may also be associated with abnormal metabolism of hormones in *target cells*. A *defect* in the transformation of testosterone to DHT[164] due to a *deficiency* in *5α-reductase* results in a feminine external phenotype in affected males. These patients lack the prostate gland, but have normally virilized Wolffian structures that terminate in the vagina. They have testes, and produce normal amounts of testosterone. At puberty, the external genitalia are partially virilized, and have a normal distribution of pubic hair. This disease is linked to the homozygotic state of an autosomal recessive gene, manifested only in males. Conversely, the role of an *increase* in the conversion of testosterone to DHT in prostate cells has been implicated in the development of prostate tumors. The formation of estrogens from adrenal androgen precursors is suspected to play a role in the evolution of breast cancer (p. 148).

A severe (genetic) deficiency of *transport proteins* is rare, and no clinical syndrome has been clearly described in relation to such a disorder, perhaps because there is a mechanism of adaptation to deficient plasma binding.

8.2 Receptor Abnormalities

Hormone Resistance

Clinically, the concept of hormone resistance, defined as a *deficiency in the response* of the target tissue to a hormone, was first proposed by Fuller Albright in 1942[165], when he described pseudohypoparathyroidism (p. 649). This disorder involves a genetically determined alteration of the G_s protein (p. 334) mediating PTH effects, which accounts for the associated disturbances in glucagon and TSH function. Indeed, a number of abnormalities may lead to hormone resistance. An apparent resistance may even be related to the hormone itself, if the rythmicity of its production is inadequate (e.g., GnRH; Table I.27), or if it is not transformed into the appropriate metabolite (e.g., testosterone to DHT). However, receptors, either qualitatively or quantitatively, are often implicated in hormone resistance, a contention substantiated by experimental evidence[166]. Two classical examples illustrate receptor-dependent diseases.

In *testicular feminization syndrome (tfm)*, there is absence of a functional androgen receptor. These patients are genetically male, with a normal testosterone level and a female phenotype (Fig. I.153).

164. Imperato-McGinley, J. and Gautier, T. (1986) Inherited 5α-reductase deficiency in man. *TIG* 2: 130–133.
165. Albright, F., Burnett, C. H., Smith, P. H., et al. (1942) Pseudohypoparathyroidism – an example of Seabright's bantam syndrome. *Endocrinology* 30: 922–932.
166. Sibley, C. H. and Tomkins, G. M. (1974) Mechanisms of steroid resistance. *Cell* 2: 221–227.

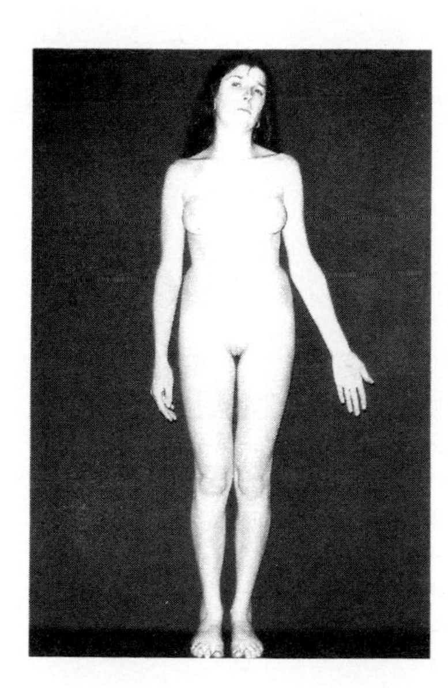

Figure I.153 Testicular feminizing syndrome (tfm). A 30-year-old "man". (Courtesy of Dr. G. Schaison, Hôpital de Bicêtre).

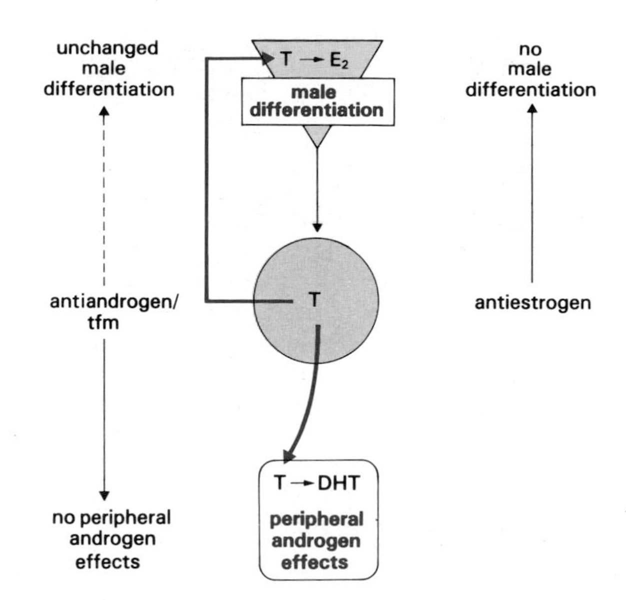

Figure I.154 Testosterone and androgen effects in the hypothalamus and in peripheral sex organs. Antiandrogen at the receptor level, or mutation of the receptor as in tfm, suppresses peripheral androgen activity. Antiestrogen abolishes the effect of testosterone in the hypothalamus.

Table I.27 The versatile medical use of GnRH.

Pulsatile administration (↑ LH)
Ovulation in hypothalamic amenorrhea
Spermatogenesis and ovulation at puberty in hypogonadotrophic hypogonadism
Continuous administration (↓ LH)
Prostatic carcinoma
Polycystic ovaries
Endometriosis
Central precocious puberty

Testes remain in the abdomen (cryptorchidism), and Wolffian ducts are atrophic (p. 441); the defect is linked to a mutation on the X chromosome. Androgen mRNA has not been detectable in kidney and liver from androgen-insensitive tfm mice. In rats with tfm, testosterone can be transformed into estradiol; the estrogen receptor is normal, and thus hypothalamic differentiation is normal for a male (Fig. I.154). A number of hereditary defects of the androgen receptor cause androgen resistance during embryogenesis and in later life. They produce varied phenotypes in males, ranging from infertile men to ambiguous genitalia,

Table I.28 Androgen receptor defects in androgen resistance.

Phenotypic spectrum:	**Male** ———			→ **Female**
Diagnostic:	Infertile male	Reinfenstein‡ syndrome	Incomplete testicular feminization	Complete testicular feminization
Androgen receptor defect				
No binding	0	0	3	14
Qualitatively abnormal	6	12	6	9
Decreased	4	5	2	0
Abnormality unidentified	1	4	4	1

‡Phenotypic male with incomplete development of the external genitalia.

The number of families is indicated in each group. In 61 families, there is a defect in the AR. In ten, the lack of receptor abnormality and 5α-reductase deficiency suggests postreceptor resistance. However, with the detailed structure of the AR now known, many unexplained cases may be shown to be receptor defects. (Wilson, J. D., Griffin, J. E. (1985). Mutations that impair androgen action. *TIG* 1: 335–339).

and to phenotypic women (Table I.28). A more complete classification will soon become available as a result of the cloning of the androgen receptor (Fig. I.71).

Familial hypercholesterolemia (FH) is a dominantly inherited genetic disease. Elevated blood cholesterol is responsible for heart attacks, often occurring early in life. It is not an endocrine disease, but the remarkable abnormalities discovered at the receptor level (p. 536) and in the feedback regulation of cholesterol homeostasis[167] make it a unique model for further studies with hormonal systems. Studies of homozygotic patients have revealed at least 10 different mutations of the LDL receptor[167]. Fifty percent of these patients do not synthesize the receptor at all. Among the others, mutations may account for an altered cytoplasmic receptor tail (Fig. I.155). Deficiency in the LDL-R causes cholesterol-containing LDL to accumulate in the blood.

Receptor abnormalities are also involved in a number of other pathophysiological events (Fig. I.156).

Normal receptors but pathogenic spillover binding. The fact that receptors are not strictly hormone-specific may be the basis for some endocrine syndromes. For example, *fetuses* developing in *diabetic* mothers are exposed to high concentrations of glucose, and as a result their secretion of insulin is increased. Fetal IGF receptors are stimulated, resulting in the *excessive weight* of *newborns*. In *breast cancer*, 3β-androstenediol, a metabolite of adrenal DHEA (p. 393), binds to the estradiol receptor, and may be involved in tumor evolution.

Normal receptors stimulated or blocked by antibodies. Anti-TSH-R antibodies (LATS) have TSH-like activity (p. 371). Conversely, anti-nACh-R antibodies block the receptor, and are involved in *myasthenia gravis*. Anti-I-R antibodies are responsible for some cases of extreme *insulin resistance*. Circulating androgen receptor antibodies, found in patients with *prostate* diseases, do not lead to clinical manifestations, perhaps due to the fact that the steroid receptor is intracellular.

Changes in receptor concentration. Whether observed changes in receptor concentration in diverse pathological conditions are primary or secondary is often difficult to establish (Table I.29). For example, in type 2 *diabetes* (non-insulin-dependent, p. 522) and in *obese* patients, the number of I-Rs is decreased.

167. Brown, M. S. and Goldstein, J. L. (1986) A receptor-mediated pathway for cholesterol homeostasis. *Science* 232: 34–47.

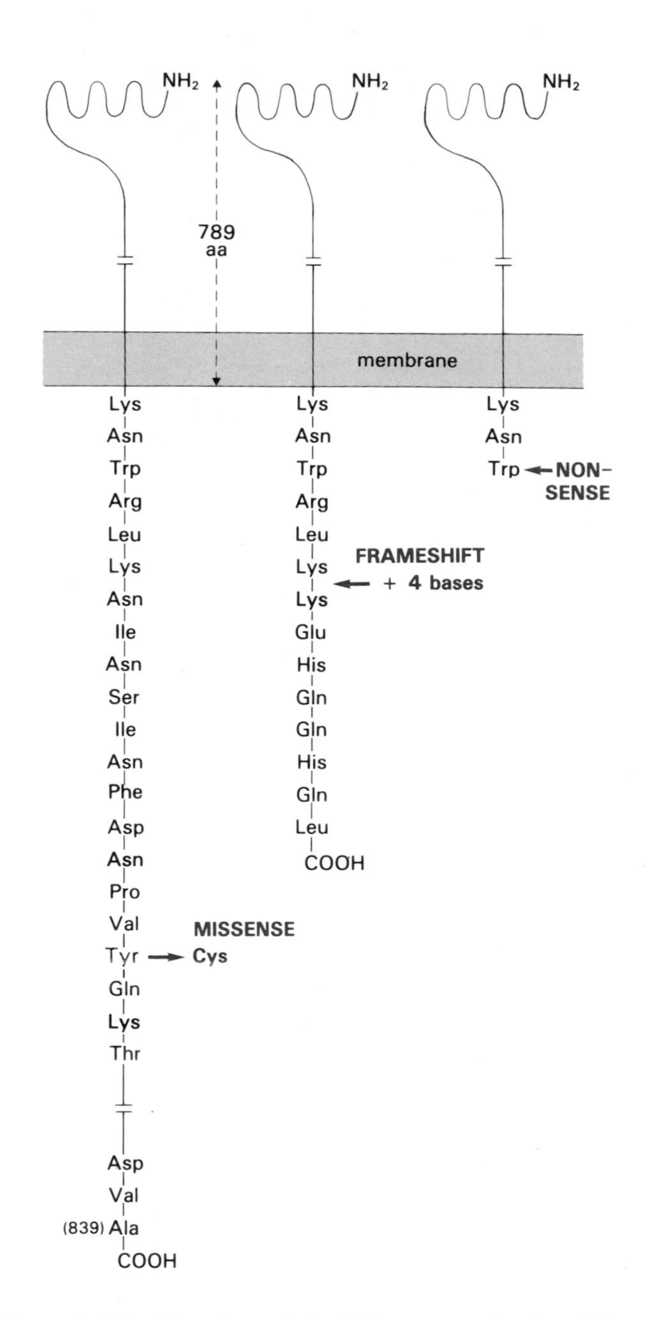

Figure I.155 Mutations of the LDL receptor in class IV familial hypercholesterolemia (FH).

There is a secondary increase in insulin secretion, followed by a decreased sensitivity to the hormone, and hyperglycemia, which itself leads to a further increase of insulin secretion and a decrease in I-R, etc. In contrast, in type 1 diabetes (insulin-dependent, p. 521), I-R concentration is increased, and the abnormality is at the postreceptor level. In *anorexia nervosa*, there is also an increase in I-R, with a resulting increase in sensitivity to insulin.

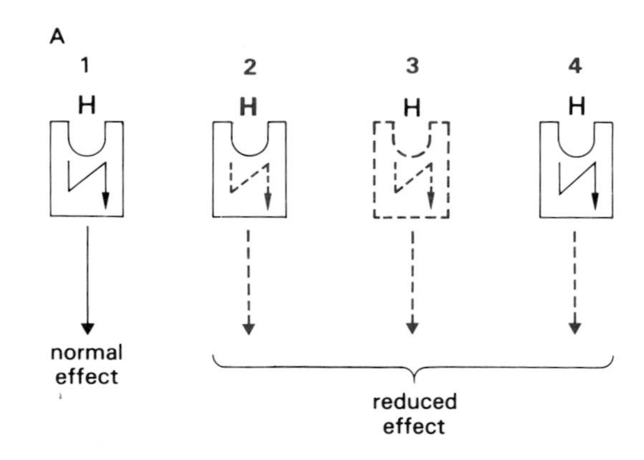

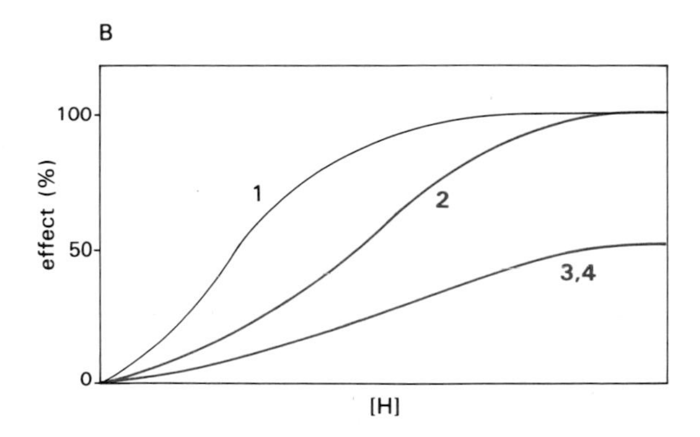

Figure I.156 Receptor and reduced hormonal effect: four situations. In **A**, the components of the hormonal system are drawn in black, and those of the deficient system, in dotted red. In **B**, the corresponding concentration-dependent response curves are represented. 1. Normal situation. 2. Hormone is either low in concentration or is of low affinity. Consequently, receptor transduction and response are decreased. 3. The hormone is normal, but the receptor is either decreased in concentration or abnormal. Consequently, receptor transduction and response are decreased. 4. Postreceptor effect: hormone and receptor are normal, but the response is decreased. (Freychet, P. (1984) Résistances à l'insuline. Aspects physio pathologiques et biochimiques. *Ann. Endocrinol.* 45: 107–114).

Change in receptor function due to abnormal glycosylation. In diabetes, this may decrease LDL-R function and be involved in atherosclerosis, since altered LDLs are taken up abnormally by macrophages.

8.3 Immunogenetic Endocrinopathies[168]

Autoimmune diseases. Immunological endocrine diseases are poorly understood. The thyroid, adrenal and pancreas are frequently affected. LATS, the *autoimmune* component of *Graves'* disease, has been mentioned. Several types of exophthalmos are associated with high titers of thyroglobulin antibodies. *Hashimoto's thyroiditis* and a number of cases of adrenal insufficiencies (*Addison's* syndrome) are autoimmune diseases.

Insulin-dependent diabetes (p. 521) is frequent in individuals under 20 years of age (1/300 in the USA); it has a clear genetic component and a strong association with the major *histocompatibility* complex, in particular with DR3 and/or DR4 antigens on chromosome 6. The definition of insulin-dependent diabetes as an autoimmunological disorder, although largely circumstantial, is suggested by characteristic pancreatic pathology, cell-mediated immunity, and antibodies directed against islet cells. Attempts are currently being made to treat diabetes at the onset of the disease with an active (but not always innocuous) immunosuppressor (cyclosporin)[169].

168. Eisenbarth, G. S. (1985) The immunoendocrinopathy syndromes. In: *Textbook of Endocrinology*, 7th edition (Wilson, J. D. and Foster, D. W., eds) W. B. Saunders, Philadelphia, pp. 1290–1300.

169. Stiller, C. R., Dupre, J., Gent, M., Jenner, M. R., Keown, P. A., Laupacis, A., Martell, R., Rodger, N. W., Graffenried, B. V. and Wolfe, B. M. J. (1984) Effects of cyclosporine immunosuppression in insulin-dependent diabetes mellitus of recent onset. *Science* 223: 1362–1367.

Table I.29 Receptor and hormone resistance: the case of insulin.

	Insulin receptor		Postreceptor/insulin activity‡
	Concentration	Affinity	
Obesity	↓	→	→ or ↓
Diabetes 1	↓	→	↓
2	↑	→	↓
Ketoacidosis	↑	↓	↓
Acromegaly	↓	↑	↓
Hypercortisolism	→	↓	↓

‡Postreceptor actually means "postinsulin binding". Defects include both abnormalities at the receptor level and in the responsive cellular machinery. They may be induced in some cases by other hormones. (↑ = increase; ↓ = decrease; → = effect maintained). (Freychet, P. (1984) Résistances à l'insuline. Aspects physio-pathologiques et biologiques. *Ann. Endocrinol.* 45: 107–114).

Two types of *autoimmune polyglandular syndromes* have been described. The most common, *type II*, is associated with two or more of the following disorders: hyperthyroidism, primary hypothyroidism, insulin-dependent diabetes, adrenal insufficiency, myasthenia gravis, and celiac disease. It is frequently associated with pernicious anemia (with antibodies to gastrin) and vitiligo. This disease has been linked to inheritance, especially to HLA-B8 and DR3 on chromosome 6. In *type I*, which is unrelated to the HLA, hypoparathyroidism and adrenal insufficiency are often associated with mucocutaneous candidiasis. Chronic hepatitis is a serious problem in many cases. Glandular polyadenomatosis, immunogenetic in nature, has also been described in relation to type I hypoparathyroidism (p. 649).

Genetic diseases. Diabetic states may be caused by a single amino acid modification of insulin, which results in a diminution of its potency, or by a defect in the processing of proinsulin. In *congenital adrenal hyperplasia*, the most common gene defect bears on 21-hydroxylase (p. 433). The classical disease is linked to HLA-Bw,47 on chromosome 6, and the cryptic and late-onset forms are associated with HLA-B14 and DR1[170] (Fig. I.157). However, deficiency in 11β-hydroxylase is not linked to the HLA system. Nucleic probes permit prenatal diagnosis and thus more efficient therapy. The genetic endocrine abnormalities which result in pseudohermaphroditism have nothing in common with the spectacular and enigmatic cases of *true hermaphroditism* (Fig. I.158).

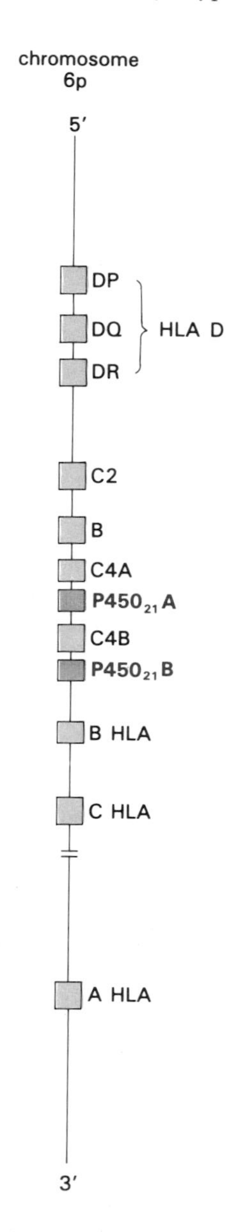

Figure I.157 Congenital adrenal hyperplasia: genes of HLA, complement factor (C) and cytochrome $P450_{21OH}$ on the small arm of chromosome 6. There are two series of C_4 and $P450_{21OH}$ genes, one next to the other (A and B). Both B genes are deleted in 21-hydroxylation deficit associated with haplotype B47, and both A genes are deleted in normal individuals with haplotype B8. In forms revealed late in life, the cytochrome $P450_{21OH}$ A gene is duplicated, and the haplotype is B14 in 75% of patients. (Boué, A., Mornet, E., Couillin, P. (1987) Génétique moléculaire du déficit en 21-hydroxylase. *Ann. Endocr.* 48: 24–30).

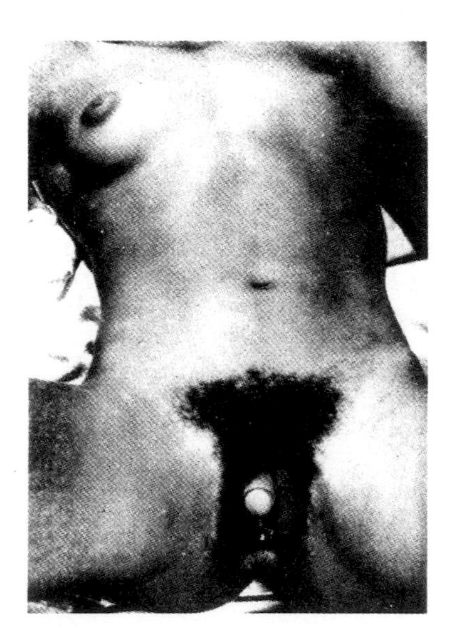

Figure I.158 True hermaphroditism: a 14-year-old subject with one testicle and one ovary. (Drs. W. Goodwin, P. Scardino and W. Scott, Brady Urological Institute, In: Wilkins, L. (1950) *The diagnosis and treatment of endocrine disorders in childhood and adolescence.* Charles C. Thomas, Springfield, Ill.).

The defects of *5α-reductase* and those of the *androgen receptor* mentioned in sections 8.1 and 8.2 are also genetic in nature.

Hereditary features participate in the natural history of many hormonal diseases. For example, among approximately 20% of *obese women* in whom fat is distributed largely in the upper body, 60% show signs of preclinical diabetes; that is, eight times the frequency in normal women. Their adipocytes are large, relatively poor in I-R, and their condition tends to respond well to weight loss. In contrast, in association with the more common, lower body type of obesity, diabetes is rare, the number of adipocytes is increased, and weight loss is difficult[171]. The basic mechanism of these hereditary differences remains unknown.

8.4 Pathophysiological Ontogenesis

Hormones play a major role in *growth* and *development*. During the *embryonic and fetal periods*, hormones are extensively involved in the differentiation and growth of many organs. Indeed, some endocrine systems are unique to the fetus, e.g., the para-aortic chromaffin ganglia, the intermediate pituitary, the ectopic production of pituitary-like hormones and neuropeptides, the synthesis of large amounts of DHEA-S by the adrenal zona reticularis (p. 393), and the placental production of progesterone [172]. Remarkably, for a period of time during fetal development, the circulating concentration of a number of hormones may be of the same order of magnitude as that of an adult (Fig. I.159). Fetal tissue becomes responsive to hormone often at a *precise period* in development, frequently coinciding with a wave of hormone production. In addition, and including in humans, abnormal exposure to hormone during fetal life may create definitive, irreversible conditions. For example, androgens may cause pseudohermaphroditism, and if the rat model applies to the human, "androgenization" of the brain may irreversibly disturb the function of the female hypothalamus

170. Kohn, B., Levine, L. S., Pollack, M. S., Pang, S., Lorenzen, F., Levy, D., Lerner, A. J., Rondanini, G. F., Dupont, B. and New, M. I. (1982) Late-onset steroid 21-hydroxylase deficiency: a variant of classical congenital adrenal hyperplasia. *J. Clin. Endocrinol. Metab.* 55: 817–827.
171. Vague, J. (1956) The degree of masculine differentiation of obesities. *Amer. J. Clin. Nutrition* 4: 20–34.
172. Fisher, D. A. (1986) The unique endocrine milieu of the fetus. *J. Clin. Invest.* 78: 603–611.
173. Ryan, K. J. (1978) Diethylstilbestrol: twenty-five years later. *N. Engl. J. Med.* 298: 794–795.

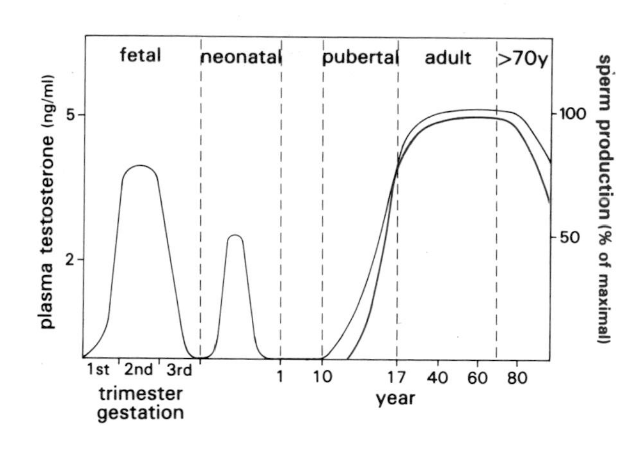

Figure I.159 Male sexual function: mean plasma testosterone level and sperm production at different periods in life. (Griffin, J. E., Wilson, J. D. (1980) The testis. In: *Metabolic control and disease.* (Bondy, P. K. and Rosenberg, L. E., eds) W. B. Saunders, Philadelphia, pp. 1535–1578).

(Fig. I.46). In the past, *diethylstilbestrol* (DES) was unfortunately given very early in pregnancy to prevent spontaneous miscarriages. This or any other estrogen causes abnormal differentiation of the fetal vagina and uterine cervix, and is responsible for adenomatosis, observed later in life, which is associated with an increased risk of cancer. Some discrete developmental abnormalities are also found in boys[173].

Puberty is a landmark of development. *Adrenarche* starts in both sexes, at the age of seven, with an increase in DHEA-S secretion. Puberty itself begins later, with increased and nocturnal pulsatile GnRH secretion (p. 457). Menses start at the age of 13 in industrialized countries, which in present day is earlier than in developing countries, due possibly to nutritional differences. The beginning of mammary development precedes menstruation by two years. Ovulation, and thus progesterone secretion, occurs one year after the first menstrual period. Girls, therefore, are fertile relatively late, when they are physically well developed, while in 11-year-old boys, spermatozoa can already be detected. In both sexes, puberty precedes the age of intellectual maturity.

Menopause provokes an abrupt loss of sex hormones. The estrogen component of substitutive therapy should be carefully adjusted, and progestin should be prescribed in order to insure a cyclical development of the endometrium and regular menses, and decrease the risk of endometrial cancer. The physical and psychological advantages of hormonal treatment of menopause are now recognized. One of the most impressive benefits is the prevention, in age-

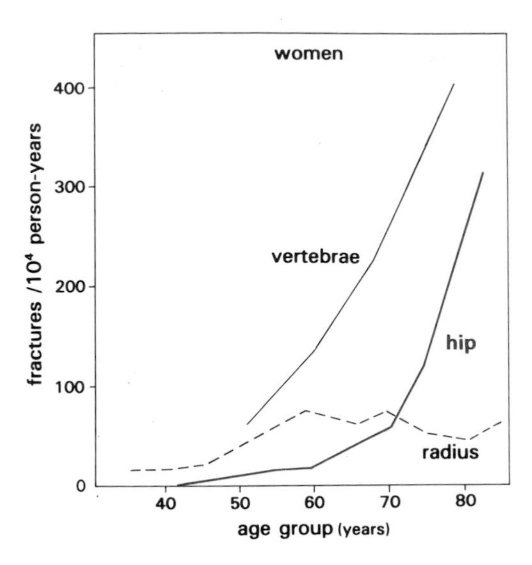

Figure I.160 Frequency of three common osteoporotic fractures, as a function of age, in women. (Riggs, B. L. and Melton III, L. J. (1986) Involutional osteoporosis. *N. Engl. J. Med.* 314: 1676–1686).

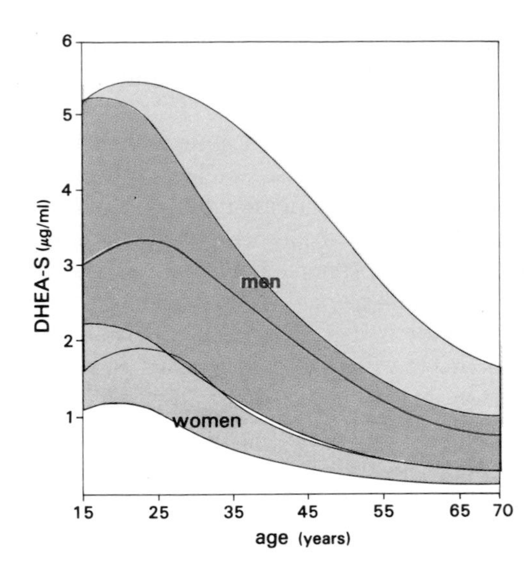

Figure I.161 DHEA-S in the serum of normal men and women. Colored areas delineate normal ranges for men and women. (Orentreich, N., Brind, J. L., Rizer, R. L. and Vogelman, J. H. (1984) Age changes and sex differences in serum dehydroepiandrosterone sulfate concentrations throughout adulthood. *J. Clin. Endocrinol. Metab.* 59: 551–555).

ing women, of osteoporosis, which is associated with increased bone fractures (Fig. I.160), of vertebrae, of the radius, and of the hip (700,000 cases per year in the USA correspond to medical costs of 2 to 4 billion dollars!). Severe coronary problems may also be avoided. However, controversy still divides practitioners who may be prejudiced against the concept of "feminine forever". In *men*, a *decrease* in testosterone is generally progressive, and is often modest until the age of 70 (Fig. I.159). Should a supplement of testosterone be given in order to maintain anabolism and sexual potency (which is probably often dependent on a vascular factor)? In *both sexes*, plasma *DHEA-S* decreases gradually with age (Fig. I.161), contrary to the concentration of glucocorticosteroids, which remains stable (mostly by decreased metabolism). Again, should replacement therapy be envisaged, and, if so, for how long should it be administered? How much consideration should be given to the many unknowns of such hormonal intervention and to the potential formation of active androgens and/or estrogens involved in hormone-responsive tumors? In elderly people, the receptors and cellular mechanisms which are necessary for a proper response to hormones in target cells may no longer be adequate. Therefore, substitutive treatment of age-related hormonal deficiencies should be undertaken cautiously.

Life expectancy was about 45 years at the turn of the century. In industrialized countries, it is now approximately 80 years for women, and about 72 years for men (Table I.30). Very little is known about ageing. Do particular cells die first, as occurs early in development for a number of neurons? Is there a general signal for the body, or does each type of cell have its own schedule? It is not known whether diet and physical and mental activities, with their hormonal consequences, have an effect on the temporal pattern of ageing and its quality. Intuitively more than scientifically, much has been attributed to the effect of changes in brain function on pituitary hormone production. Indeed, pituitary pathophysiology is dependent on age and sex (Fig. I.162)[174]. Many of the circadian rhythms of hormone secretion are almost completely suppressed with age. GH secretion decreases, and when GH is challenged by GRF, the response is shown to be decreased (Fig. I.163). Alterations of glycoprotein hormone activity may be related to abnormal glycosylation. Presently, a 60-year-old person can expect to live more than an additional 15 years (Table I.30). How will this extension of life, which may even increase in the next

174. Meites, J. (1987) Importance of the neuroendocrine system in aging process. In: *Hypothalamic Dysfunction in Neuropsychiatric Disorders* (Nerozzi, D., Goodwin, F. K. and Costa, E., eds), Raven Press, New York, vol. 43, pp. 283–292.

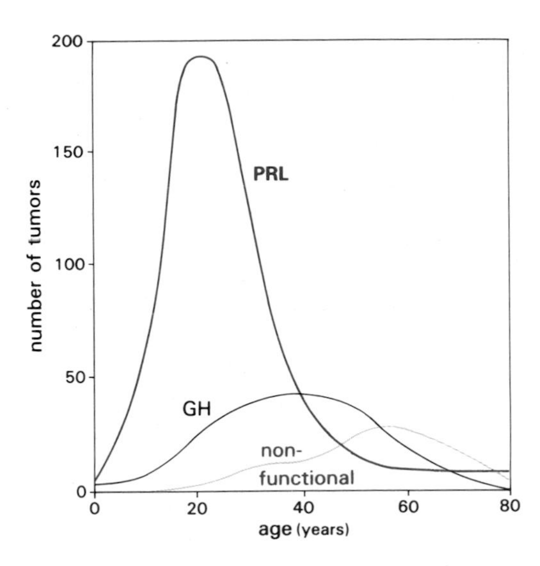

Figure I.162 Pituitary adenomas: frequency of lactotropic, somatotropic and non-functional tumors, as a function of age. (Racadot, J. (1985) Histophysiologie et histopathologie de l'hypophyse chez le sujet agé. *Actualité Gérontologie*, 12–17, Sandoz Editions, Basel).

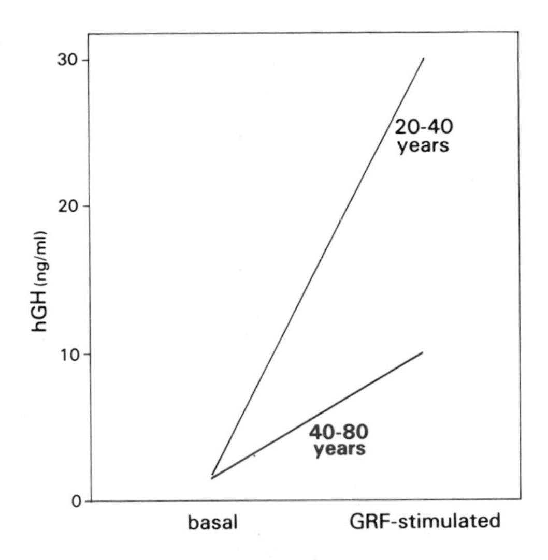

Figure I.163 Growth hormone stimulation by GRF decreases with age. (Shibasaki, T., Shizume, K., Nakahara, M., Masuda, A., Jibiki, K., Demura, H., Wakabayashi, T. and Ling, N. (1984) Age-related changes in plasma growth hormone response to growth hormone-releasing factor in man. *J. Clin. Endocrinol. Metab.* 58: 212–214).

Table I.30 Life expectancy (expressed as a function of age) in industrialized countries.

Age	Mean life expectancy in years	
	Men	Women
0	72	80
60	17	22
70	11	14
80	6	8

decades, be handled both individually and sociologically?

Hormones, in combination with the genetic background of each individual, constitute only one of many factors which influence the quality and length of life. Indeed, the difference in longevity *between men and women* has not yet been explained by hormonal mechanisms. Until the nineteenth century, women died earlier than men, mostly due to accidents related to reproductive function. Currently, smoking partially explains the shorter life expectancy of men, but with the change of sex ratio in smoking habits, the situation may be modified in the next 20 years. Will hormonal treatment help human beings reach the "theoretical" age limit of 115 years?

8.5 Pathophysiological Neuroendocrinology

The brain is not only in command of the endocrine system, but is also the processing center where information coming from both the external world and the body *merges*. The brain contains cells which *make* several compounds that are also synthesized and secreted as hormones by peripheral glands. In fact, a definitive demonstration of the structural identity of peptides synthesized in glands and neurons *(neuropeptides)* has been provided in only a very few cases. Considering the large number of gene families and multistep processing mechanisms, it is entirely possible that cognates of the same hormone are expressed at different sites. This has been suggested for GnRH which regulates gonadotropic cells and GnRH which directly affects sexual behavior.

Neurosteroids are produced in neurons and glial cells. Oligodendrocytes (glial cells) convert cholesterol to pregnenolone[55]. DHEA, estrogens and other steroid derivatives are formed in brain cells which possess the appropriate enzymes for biosynthesis.

The potential involvement of hormones in nervous processes is infinite. The effects of *ACTH-related* peptides and *vasopressin* on learning, memory and brain ageing are currently under study, although their physiological significance is controversial and

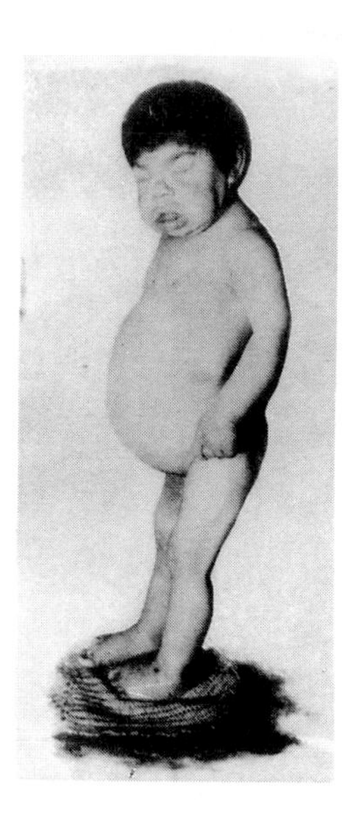

Figure I.164 Congenital juvenile myxedema with cretinism (a 22-year-old young man, 0.92 m tall). (Jeandelize, P., ed (1903) *Insuffisance Thyroidienne et Parathyroidienne*. Baillière, Paris, p. 484).

their mechanism of action uncertain, since most peptides do not cross the blood-brain barrier[56]. *Thyroid* hormones are of prime importance in brain development[175] (Fig. I.164), whereas *steroids* have more specific roles in brain differentiation. Testosterone slows neuronal formation in the left hemisphere in the fetus, and is possibly responsible for left-handedness[176].

There is a complex balance between the *CRF-ACTH-β-endorphin-glucocorticosteroid* axis and the *GnRH-LH-sex-steroid* cascade, one often predominating over the other, for example when the second is repressed by the first during stress. In *states of depression*, the central defect involves an increase in CRF and decreased feedback by glucocorticosteroids[177]; there is also evidence of excess CRF in the cerebrospinal fluid and decreased responsiveness of the pituitary to CRF. In addition, and possibly of pathogenic significance, a disturbance of circadian rhythms, with an early nadir of ACTH and cortisol (Fig. I.165), and an increase in GH, with suppression of its nocturnal peak, is observed in unipolar depression[178]. CRF plays a role in the fundamental activating system responsive to environmental challenges[179] not only in stimulating the production of ACTH and β-endorphin, but also via its direct action in the brain (there are CRF receptors in the cortex and in the limbic system) and the sympathetic nervous system. *Anorexia nervosa*[180], a syndrome generally afflicting female adolescents, is also associated with an increase in CRF, which may be responsible for an augmentation of endogenous opioids, and may also be related to the suppression of GnRH and gonadotropin

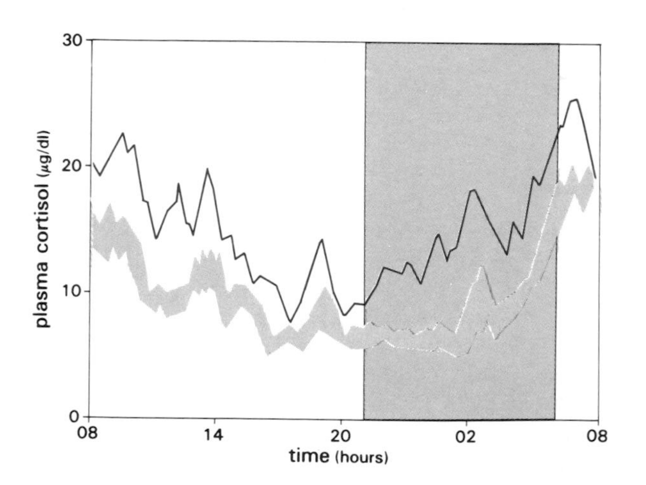

Figure I.165 Depression: disorder of circadian rhythm in endogenous depression. Plasma cortisol was collected at 15 min intervals in eight men with unipolar depression (red line). The pink area illustrates the normal range for age-matched normal men. Note the overall hypercortisolism and the earlier nadir of cortisol secretion than in normal subjects. (Linkowski, P., Mendlewicz, J., Leclercq, R., Brasseur, M., Hubain, P., Goldstein, J., Copinschi, G. and VanCauter, E. (1985) The 24-hour profile of adrenocorticotropin and cortisol in major depressive illness. *J. Clin. Endocrinol. Metab.* 61: 429–438).

175. Nunez, J. (1984) Effects of thyroid hormones during brain differentiation. *Mol. Cell. Endocrinol.* 37: 125–132.
176. Geschwind, N. and Behan, P. (1982) Left-handedness: association with immune disease, migraine, and developmental learning disorder. *Proc. Natl. Acad. Sci., USA* 79: 5097–5100.
177. Gold, P. W., Loriaux, D. L., Roy, A., Kling, M. A., Calabrese, J. R., Kellner, C. H., Nieman, L. K., Post, R. M., Pickar, D., Gallucci, W., Avgerinos, P., Paul, S., Oldfield, E. H., Cutler, G. B. and Chrousos, G. P. (1986) Responses to corticotropin-releasing hormone in the hypercortisolism of depression and Cushing's disease. *N. Engl. J. Med.* 314: 1329–1335.
178. Mendlewicz, J., Linkowski, P., Kerkhofs, M., Desmedt, D., Goldstein, J., Copinschi, G. and VanCauter, E. (1985) Diurnal hypersecretion of growth hormone in depression. *J. Clin. Endocrinol. Metab.* 60: 505–512.
179. Sutton, R. E., Koob, G. F., LeMoal, M., Rivier, J. and Vale, W. (1982) Corticotropin releasing factor produces behavioural activation in rats. *Nature* 297: 331–333.
180. Gold, M. D., Gwirtsman, H., Avgerinos, P. D., Nieman, L. K., Gallucci, W. T., Kaye, W., Jimerson, D., Ebert, M., Rittmaster, R., Loriaux, D. L. and Chrousos, G. P. (1986) Abnormal hypothalamic–pituitary–adrenal function in anorexia nervosa. *N. Engl. J. Med.* 314: 1335–1342.

secretion. Whether symptoms are due primarily to starvation or to faulty central signals is unclear.

In the brain of *Alzheimer's* patients, profound cholinergic deficiency is found mostly in the forebrain, whereas a major decrease in CRF is observed in the occipital cortex and in the caudate nucleus. Cerebral somatostatin is also low, whereas VIP and CCK are increased, suggesting that specific biochemical alterations are associated with the neuritic plaques and neurofibrillary tangles which characterize the disease histologically[181]. Alzheimer's disease differs from the *ordinary ageing* process of the brain in that it is associated with cell loss in the hippocampus, high levels of glucocorticosteroids, and reduced responsiveness to the dexamethasone suppression test.

Endocrine abnormalities may be either the cause or the consequence of brain disease. A better understanding of the deregulated hormonal circuits, inclusive of temporal aspects[182], may lead to more efficient therapeutic intervention.

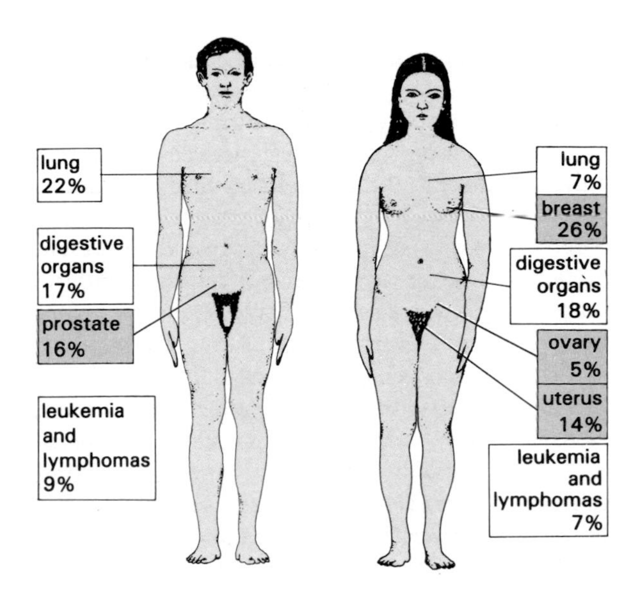

Figure I.166 Incidence of cancers in men and women (frequency in sex steroid-related target organs is highlighted in red).

8.6 Hormones and Cancers: Some Clinical Connections[183]

Cancer cells are *transformed* cells which are able to grow in culture without contact inhibition. They are unlimited in terms of the number of times they can divide *(immortalization)*, and have the potential to generate tumors when injected into animals. One of the other main characteristics of cancer cells is that they have a *decreased need for exogenous growth factors.* Are endogenous hormones or their synthetic analogs *directly carcinogenic*? Estrogens, such as diethylstilbestrol (DES), have been reported to interact with DNA, and are thought to be mutagenic, but this has never been demonstrated in vivo. The proviral DNA of the mouse mammary tumor virus, MMTV, which induces mammary tumors in mice, is preferentially integrated in a similar region. Persistent exposure to an excessive amount of a normal growth factor, such as EGF, can induce transformation, but an obligatory role of an associated oncogene has not been excluded. Abnormal exposure to estrogen during embryogenesis may alter the normal process of differentiation and favor cancer, as in the example of "children of DES" (p. 143); however, the delayed onset and the very small number of malignancies related to exposure to DES during intrauterine development indicate that a second event has probably taken place. In any case, hormones can play a *permissive* (p. 418) and/or *promoter*[184] role in cancer development. Excessive or abnormal hormone secretion is rarely involved. In fact, the expression and the effect of oncogene products depends greatly on the functional state of the cellular machinery as regulated by hormones. Are hormones determinant in the *progression* and *spreading* of the disease, including the development of marked cellular heterogeneity?

Breast cancer affects 10% of women in Western countries, and constitutes approximately 25% of all cancers afflicting women (Fig. I.166). An *estrogen window*, exposure to estrogen *unopposed* by progesterone for a period of time, is frequently found in the history of the patient with breast cancer. It may be related to a prolonged preovulatory period at puberty, to a late first pregnancy, and/or to obesity, which is associated with excessive estrogen synthesis from adrenal precursors in adipocytes. In addition, estrogens may be formed, in tumors, from adrenal androgens. Hormone receptor measurements have been

181. Fine, A. (1986) Peptides and Alzheimer's disease. *Nature* 319: 537–538.
182. Halberg, F., Reinberg, A. and Lagoguey, J. M. (1983) Human circannual rhythms in a broad spectral structure. *Int. J. Chronobiol.* 8: 225–268.
183. Huggins, C. (1967) Endocrine-induced regression of cancers. *Science* 156: 1050–1054.
184. Berenblum, I. (1954) A speculative review. The probable nature of promoting action and its significance in understanding the mechanism of carcinogenesis. *Cancer Res.* 14: 471–479.

used to determine the relevance of endocrine therapy, which is counterindicated in cases where tumors lack receptors. Results obtained during the last two decades show a high correlation between the absence of the ER and PR and lack of response to endocrine therapy (p. 413). In fact, *prognosis* is better when receptors are present, possibly because their presence is reflective of a greater degree of cell differentiation and thus of slower growth (Table I.31); the disease-free interval and overall survival time are increased. Since the antiestrogen currently used in chemotherapy (tamoxifen) is essentially free of side-effects, and since it may have additional beneficial properties (p. 119)[133], it can be given to all patients. A modest but definite improvement in overall survival has been observed in patients treated with tamoxifen. The progression of breast cancer is probably dependent on several hormones. Receptors are often present for androgen and glucocorticosteroids, GnRH (which can be directly inhibitory of growth), insulin, IGF-I, EGF (stimulating growth), TGF-β (probably inhibiting progression), etc. Clinically, it may become important to measure several hormone receptors and products of oncogene expression, e.g., EGF-R and EGF-R-2 (Fig. I.137). Mammary cancer cells secrete specific peptides, some of which probably act by auto- and parahormonal mechanisms. It is possible that they are engaged in the progression of the tumor and in the development of metastases (Fig. I.167). The transition

Table I.31 Breast cancer: multivariate survival analyses.

	Disease-free survival	Overall survival
Positive nodes	<0.001	<0.0001
PR status	0.008	0.002
ER status	0.014	0.0003
Size primary tumor	0.022	0.13
Chemotherapy	0.21	0.004
Endocrine therapy	0.14	0.08
Age	0.44	0.56
Menopause status	0.98	0.68

Note the preponderant pronostic values of positive nodes, the pronostic help of PR values in addition to those of ER, and the clear effect of chemotherapy and the borderline result of endocrine adjuvant treatment on overall survival. (McGuire, W. L. and Clark, G. M. (1984) Primary breast cancer prognostic factors. In: *Endocrinology* (Labrie, F., Proulx, L., eds), Elsevier, Amsterdam, pp. 1061–1063). Additional indices may provide new prediction of the evolution of the disease. Amplification of *neu* expression has negative prognostic value. (Slamon, D. J., Clark, G. M., Wong, S. G., Levin, W. J., Ullrich, A. and McGuire, W. L. (1987) Human breast cancer: correlation of relapse and survival with amplification of the HER-2/*neu* oncogene. *Science* 235: 117-181).

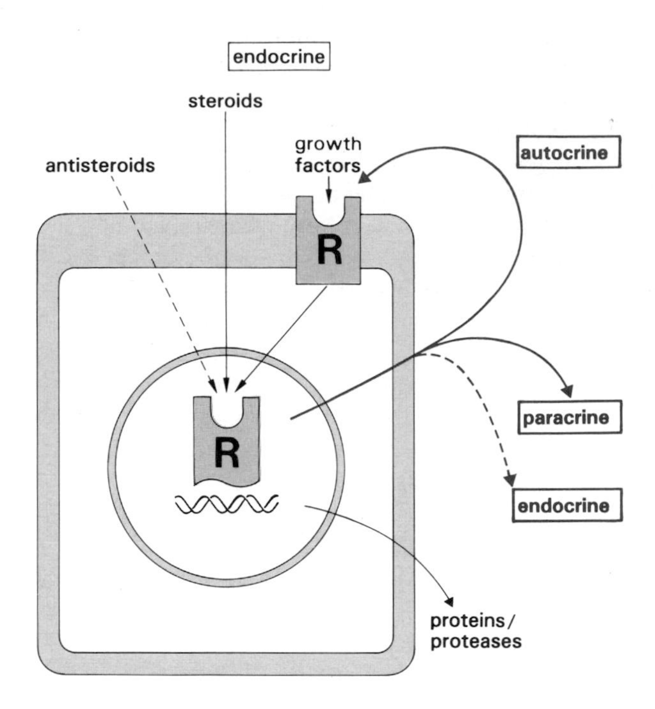

Figure I.167 Cancer cell: regulatory aspects and multihormonal activity. As is true for normal cells, exogenous hormones influence cancer cells at both the gene and membrane levels. In addition, and possibly in response to these hormones, the cancer cell may synthesize growth factors which can be active via autocrine, paracrine or even endocrine mechanisms. Growth factors such as TGF-β may be negative regulators of cell growth. Specific non-hormonal proteins, e.g., proteases and hydrolases, are also produced, and may play a role in the progression and the metastatic spread of the disease. (Rochefort, H., Capony, F., Cavalié-Barthez, G., Chambon, M., Garcia, M., Massot, O., Morisset, M., Touitou, I., Vignon, F. and Westley, B. (1986) Estrogen-regulated proteins and autocrine control of cell growth in breast cancer. In: *Breast Cancer: Origins, Detection and Treatment* (Rich, M. A., ed), Martinus Nijhoff, Boston, pp. 57–68).

from hormone sensitivity to hormone resistance in the course of the disease, a major clinical problem, may be related to various mechanisms. First, resistance to hormone may simply be a normal property of the evolution of the cancer phenotype, as has been indicated by experiments with cloned cells that always reproduce the same pattern of unresponsiveness seemingly unrelated to a receptor defect[185]. Second, a receptor for a steroid or growth factor may mutate and acquire

185. Darbre, P. and King, R. J. B. (1987) Progression to steroid insensitivity can occur irrespective of the presence of functional steroid receptors *Cell* 51: 521–528.

abnormal properties, sometimes including constitutive activity (for example *erb*-B, p. 122). Third, another oncogene (e.g., *ras*[186]) can be expressed secondarily, and may take over normal hormonal regulatory mechanisms (in most cases, however, both responsiveness to hormone and oncogene expression are observed simultaneously). Fourth, the suppression of hormone activity (e.g., by antihormone) may permit the selective growth of insensitive cells in a heterogenous tumor. This last possibility has led to the proposal that a reduction, but not an entire suppression, of hormone may maintain an equilibrium between hormone-responsive and non-responsive cells, particularly in elderly patients[187]. In any case, the type of treatment should be different for an older patient who might be expected to survive for 20 years than for a young individual.

A tremendous amount of empirical information has been collected from millions of women taking *oral contraceptive* pills. It appears now that the frequency of *breast cancer* in these women has *not* increased significantly. In addition, *endometrial and ovarian cancers* have decreased (Fig. I.168). Oral contraceptives are responsible for a small increase in liver adenomas, but no effect on prolactinomas[188] has been reported.

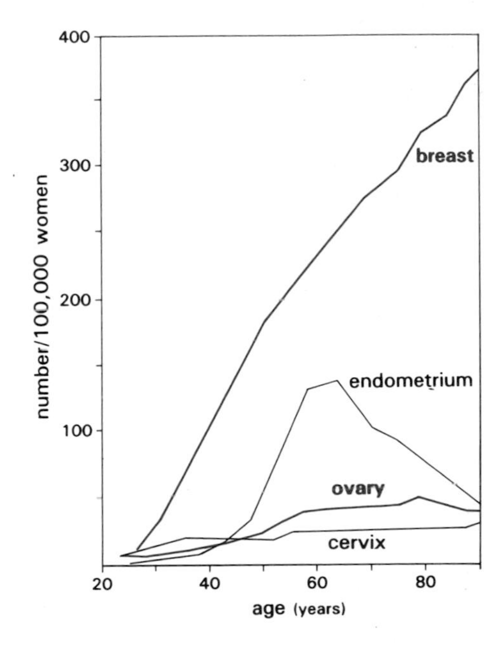

Figure I.168 Incidence of gynecological cancers, as a function of age. Breast cancer becomes more frequent with age, in striking contrast to cervical cancer.

In contrast to mammary cancers, derived from a tissue relatively insensitive to estrogens, cancers of the *endometrium* and *prostate*‡, like corresponding normal tissues, are, at least for a period of time, *totally dependent* on estrogen and androgen, respectively. Therefore, total hormone suppression may arrest the disease completely (Fig. I. 169 and Table I.32). Continuous administration of GnRH results in medical castration (p. 38); this treatment can be associated with an antiandrogen, in prostate cancer[189], to suppress the effect of testosterone formed from adrenal androgens, or with an aromatase inhibitor, in endometrial cancer, to reduce estrogen formation from adrenal precursors.

In summary, hormone therapy has had only limited success in the treatment of cancer. Although hormone receptors have been identified in a number of malignancies, hormone therapy has rarely been clearly beneficial. The fascinating discovery of oncogenes has not yet changed therapeutic strategy in the treatment of cancer. The importance of the distinction between immortalizing and transforming oncogenes is tempered by work showing that each cell reacts uniquely. The incidence of cancer and the effects of chemical and/or viral carcinogens differ according to species and to physiological conditions. In addition to initial cellular transformation, the parameters which contribute to *progression* toward malignancy remain *medically* crucial[190]. They are highly dependent on the extracellular environment, including hormonal status.

‡Curiously, the seminal vesicles and the bulbourethral gland, which are also androgen dependent, remain nearly exempt from neoplastic change. Note also that a most common disease, benign prostatic hyperplasia (88% of men by the ninth decade), is not responsive to antiandrogen treatment.

186. Kasid, A., Lippman, M. E., Papageorge, D. R., Lowy, D. R. and Gelmann, E. P. (1985) Transfection of v-ras^H DNA into MCF-7 human breast cancer cells bypasses dependence on estrogen for tumorigenicity. *Science* 228: 725–728.
187. Noble, R. L. (1977) Hormonal control of growth and progression in tumors of Nb rats following prolonged sex hormone administration. *Cancer Res.* 37: 82–94.
188. Vessey, M. P. (1984) Exogenous hormones in the aetiology of cancer in women. *J. Roy. Soc. Med.* 77: 542–549.
189. Labrie, F., Dupont, A., Belanger, A., Emond, J. and Monfette, G. (1984) Simultaneous administration of pure antiandrogens, a combination necessary for the use of luteinizing hormone-releasing hormone agonists in the treatment of prostate cancer. *Proc. Natl. Acad. Sci., USA* 81: 3861–3863.
190. Dulbecco, R. (1986) A turning point in cancer research: sequencing the human genome. *Science* 231: 1055–1056.

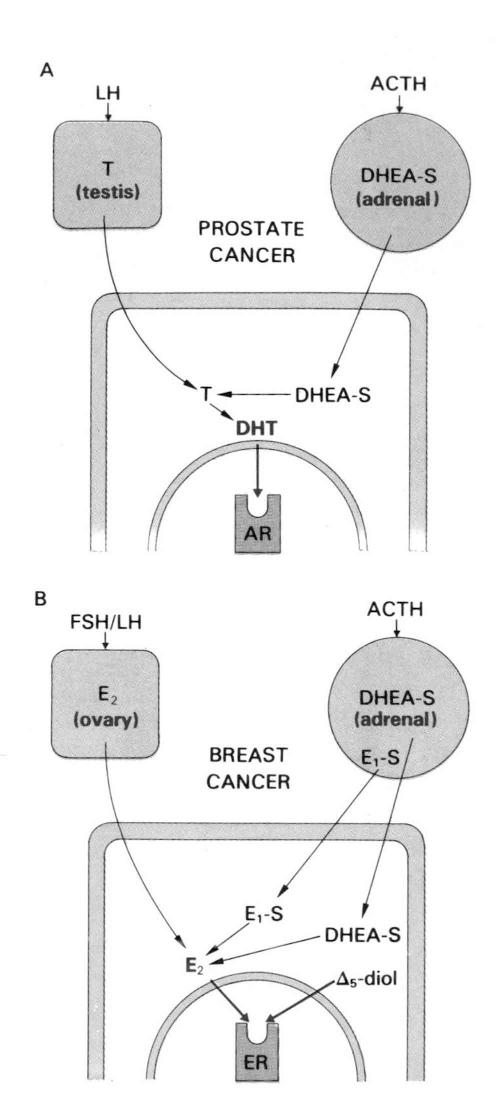

Figure I.169 Cancers and hormone production. **A.** In prostate cancer, active androgens are derived from both the testis and the adrenal. **B.** In breast cancer, active estrogens are derived from both the ovary and the adrenal.

8.7 Hormones in Medicine: New Problems?

Beyond the wonder of contemporary endocrinology, which is providing novel insights into the mechanisms involved in physical and mental processes, and even beyond medical achievements typified by insulin, cortisone and oral contraceptives, new and provocative issues are being made apparent. Matters related to reproduction (Fig. I.49 and p. 483) are brought up almost daily by the media. Medical problems posed by increased life expectancy and its management by hormones is of prime importance to both the individual and society. Should "small" children be "treated" with GRF or GH? Should the "sex" of a pseudohermaphrodite be altered to coincide with the individual's genotype?

Scientific discovery should not be limited, provided that research is not conducted directly in humans. However, *all* findings should be open to general evaluation, especially to physicians who are responsible for their application in treatment.

9. Comparative and Evolutionary Aspects of Endocrinology

9.1 Generalities

The *evolution* of interactions between living beings and the ecosystem is of importance to endocrinologists. The predominant role of the *nervous system*, interface between the exterior world and the endo-

Table I.32 Hormonal treatment of prostate cancer.

Treatments	Effects: Effect on (pro)hormone production from: Testis	Adrenal	Antiandrogen effect at R level	Global effect on androgen activity
1. Estrogen	↓	→	0	↓
2. Castration	↓	→	0	↓
3. GnRH	↓	→	0	↓
4. Antiandrogen	↑	→	+	→
5. 2+4 or 3+4	↓	→	+	↓ ↓

↑ increase; ↓ decrease; 0, effect maintained, O, no effect. (Courtesy of Dr. F. Labrie, Endocrinologie Moléculaire, Université Laval, Québec).

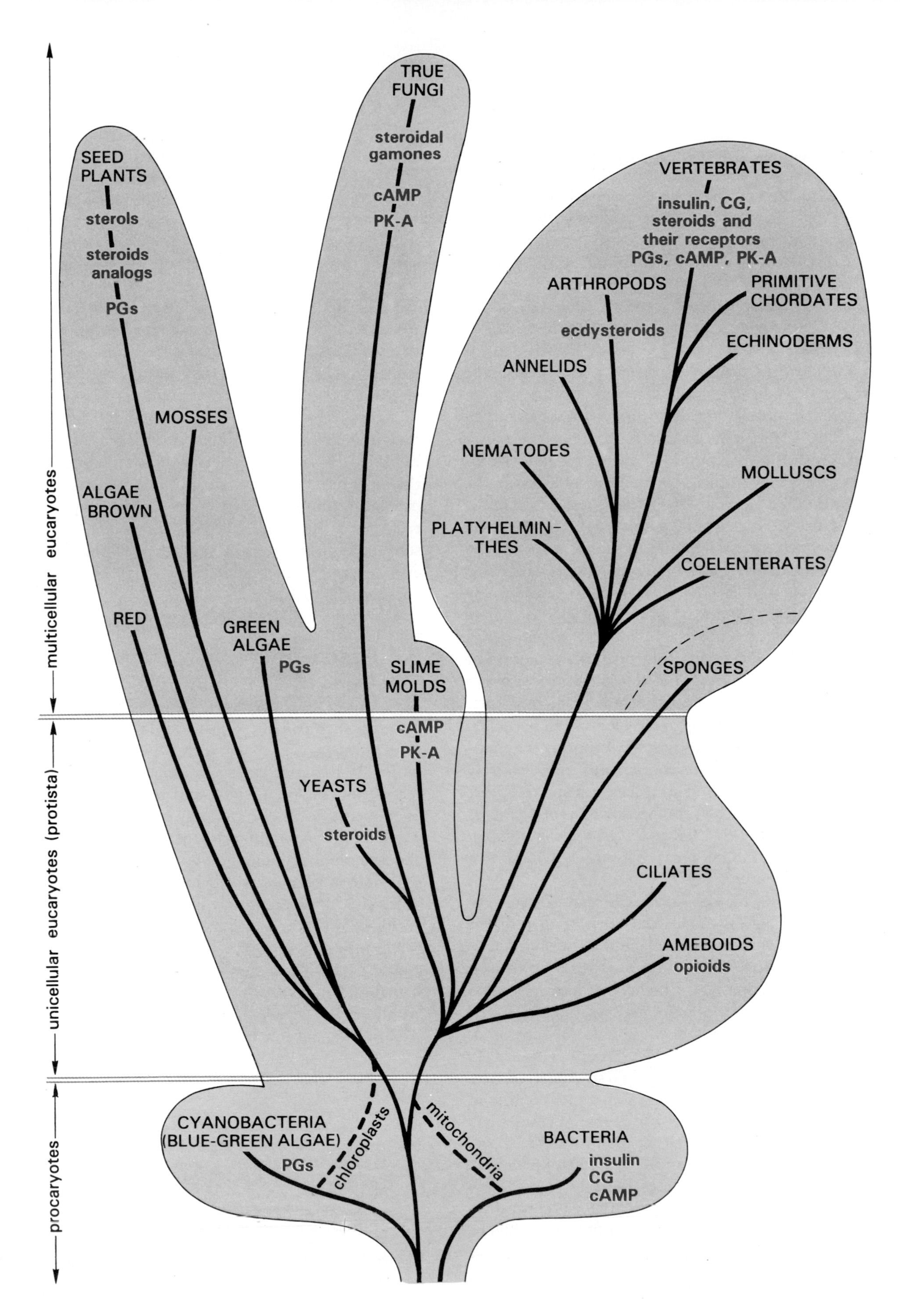

Figure I.170 Emergence of regulatory "hormonal" molecules: a phylogenetic chart of living organisms. Only a few examples have been illustrated (note that neurons are observed in species (pink area) having emerged well after those which already synthesized neurohormonal molecules).

crine system, has been confirmed in all vertebrates and invertebrates (with the exception of sponges, the simplest of metazoa). The main landmark in evolution is now placed between eucaryotes and procaryotes, even though their phylogenetic relationship is still controversial. Indeed, what is known about hormones has not helped clarify the issue (Fig. I.170).

The development of modern analytical techniques should have simplified the general understanding of the evolutionary processes which led to endocrine regulation as we know it, but in fact the more is known, the more hormonal systems seem complex. Endocrinology is no longer limited simply to the study of glandular production and function, since the same substances, subject to diverse regulatory processes, may be synthesized at different sites in the organism. The fact that *no obvious, general scheme* of endocrine function can be formulated is striking. By way of illustrating this point, the *same molecule* can be chemically *conserved* between evolutionarily distant species and *retain either similar functionality* (e.g., estrogens are always growth factors) or, alternatively, have a completely *different* physiological role (e.g., thyroxine and prolactin). The evolution of the endocrine system *cannot* be accounted for *simply* by changes in the structure of hormones and receptors. The *topological* evolution of the synthesis of both hormones and receptors can completely modify the endocrine system and place its components under entirely different regulatory control. Hormones and receptors have been described by Peter Medawar as *instruments* of an overall program: "*Endocrine evolution is not an evolution of hormones, but an evolution of the uses to which they are put*"[191].

Among the chemical structures retained throughout evolution, *small* molecules, such as steroids or cAMP, imply the conservation of the enzymes necessary for their biosynthesis. In addition, analogs of certain *peptide hormones* of higher animals have also been found in *unicellular organisms* (bacilli, fungi, protozoa), e.g., insulin, chorionic gonadotropin, somatostatin, glucagon, calcitonin, TSH, CCK, ACTH, and β-endorphin). First detected by immunological techniques, and sometimes confirmed physicochemically, the presence of these peptides has not yet been confirmed with genetic probes[192]. If corresponding genes are found in the DNA of primitive organisms, several hypotheses must be considered. The most interesting one suggests that these peptides play a role similar to that of hormones in higher animals, in the sociobiology and function of single-cell organisms. For example, opioid peptides have been shown to inhibit pinocytosis in the amoeba, an effect which can be reversed by naloxone[193].

Another possibility is that some nucleic acid and/or peptide sequences have been conserved for several hundred million years simply because they are more stable chemically, regardless of the physiological context. For instance, peptides related to insulin and its receptor, and to TGF-β, appear to play an important role in early insect development. Yet another alternative is that, secondarily, a segment of the genome of higher organisms was inserted into that of lower species (somewhat like viral oncogenes, p. 120).

Discovery rarely means full understanding. The function identified initially for a hormone may not be reflective of its principal role. For example, hypothalamic releasing factors could originally have been neurotransmitters, and may have acquired a regulatory role in pituitary hormone secretion secondarily. The notion that "everything is made everywhere"[194] does not necessarily have a functional counterpart (p. 10).

9.2 Peptide Hormones: Evolutionary Aspects

In contrast to paleontologists, biologists do not have the equivalent of fossils with which to study hormones. With the exception of some relatively well conserved human specimens found recently, predictions of endocrine evolution can only be made based on the sequences of contemporary genes. In other words, endocrinology is limited to the study of those mechanisms and substances which have been evolutionarily successful. The recent and impressive description of *families* of hormones and receptors suggests that there is extraordinary mobility among genes, and that this mobility produced new functions, sequences and structural configurations. The existence of families of hormones, as for example that of insulin and related growth factors, and that of TGF-β (ref. 3, p. 223),

191. Medawar, P. (1953) Some immunological and endocrinological problems raised by the evolution of viviparity in vertebrates. *Symp. Soc. Exp. Biol. Med.* 7: 320–338.
192. Le Roith, D., Shiloach, J., Roth, J. and Lesniak, M. A. (1980) Evolutionary origins of vertebrate hormones: substances similar to mammalian insulins are native to unicellular eukaryotes. *Proc. Natl. Acad. Sci., USA* 77: 6184–6188.
193. Csaba, G. (1980) Phylogeny and ontogeny of hormone receptors: the selection theory of receptor formation and hormonal imprinting. *Biol. Rev.* 55: 47–63.
194. Niall, H. D. (1984) The evolution of peptide hormones. *Annu. Rev. Physiol.* 44: 615–624.

growth hormone, pituitary glycoproteins and vasopressin, can probably be best explained by a process involving gene duplication. Since the original hormone still retains its initial function, the gene to be *duplicated* is free to mutate, and this mutation may result in new hormonal activities. While at the same time the basic structure of the hormone is maintained, the new products can utilize the already established mechanisms involved in the processing, distribution, and action of hormones. Thus, gene duplication is a relatively conservative process. Non-functional pseudogenes are also formed. It is not clear why some hormones which are evolutionarily relatively well-conserved accomplish similar functions in different species whereas others diverge greatly (calcitonin, relaxin). It is not clear, either, why GH has retained the same growth-promoting capacity among species, whereas prolactin, structurally related to GH, displays diversified activities (Fig. I.171).

Some examples, many questions

Vasopressin and its analogs (p. 289 and Fig. I.172). The 20-atom structure of vasopressin, closed by the formation of a cystine between residues 1 and 6, is well conserved. Arg-vasopressin is present in all mammals except hogs and other members of the suborder Suina, in which Lys-vasopressin is found. In lower vertebrates, vasotocin is the corresponding hormone (with the oxytocin ring and the C-terminal extremity of Arg-vasopressin).

Calcitonin. This hormone, although it has evolved greatly throughout various species, is identical in humans, amphibians, and even protochordates. There is no explanation for either the structural divergence or the similarity of calcitonin in different species. Is the "final" gene found in higher organisms actually present but not expressed in lower forms? And if this is the case, why?

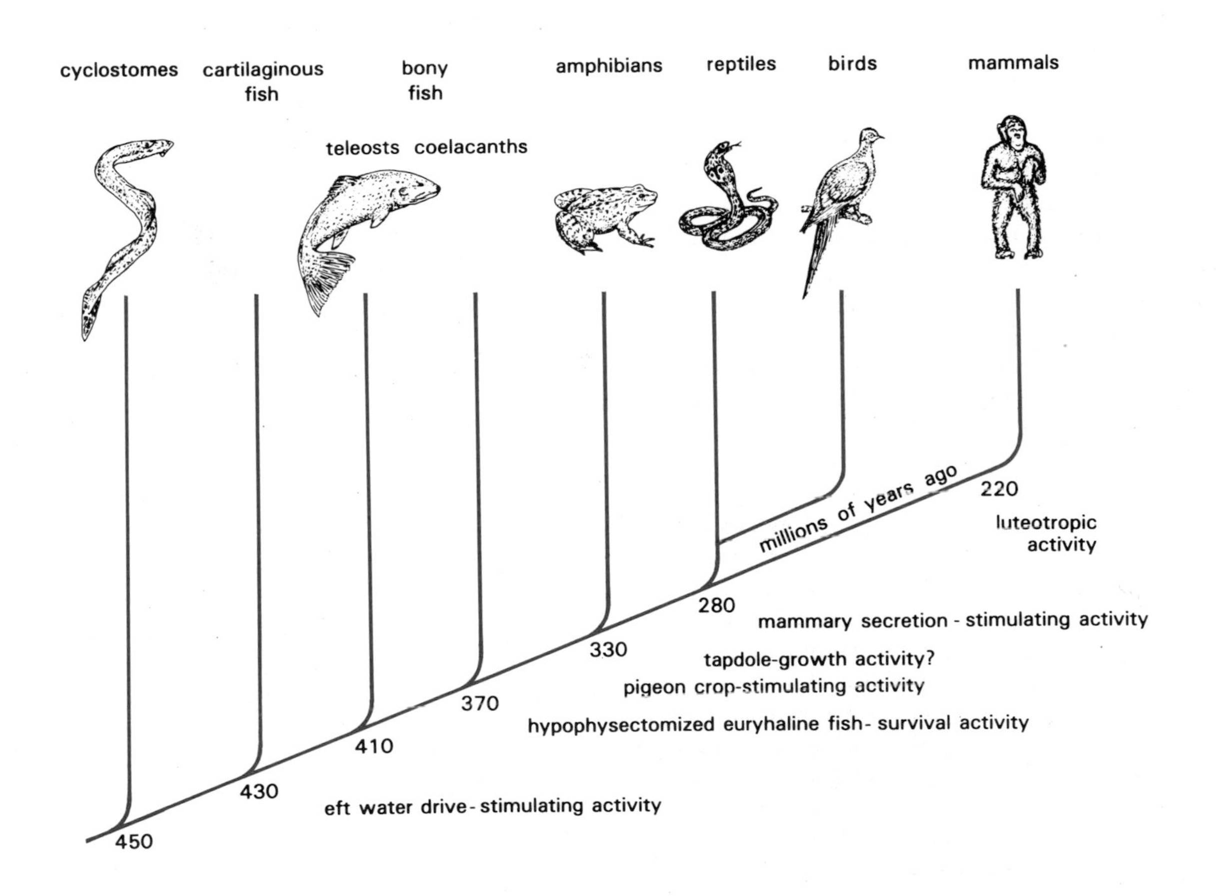

Figure I.171 Evolution of prolactin. The evolving biological properties of the molecules are indicated. (Bern, H. A. (1967) Hormone and endocrine glands of fishes. *Science* 158: 455–462; Nicoll, C. S. (1982) Prolactin and growth hormone specialists on one hand and mutual mimics on the other. *Perspect. Biol. Med.* 25: 369–381).

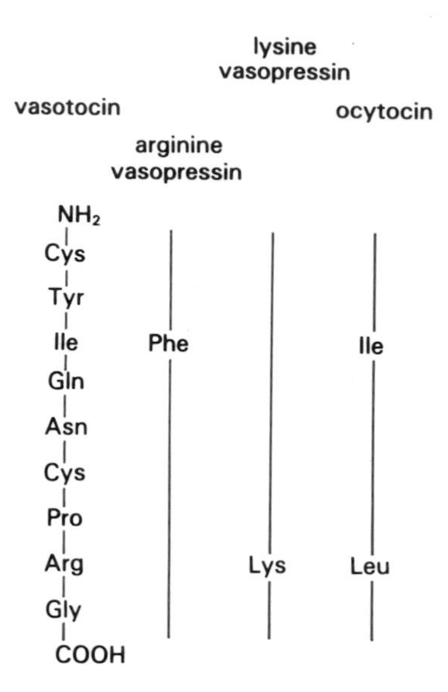

Figure I.172 The amino acid sequence of four major posterior pituitary hormones in vertebrates.

Insulin. In comparison with many other proteins, insulin is evolutionarily highly conserved, with an amino acid substitution rate of about 1×10^{-9} loci/year[195]. However, the connecting C peptide (p. 496) has undergone many modifications. Remarkably, insulin receptors are functionally more highly conserved than insulin itself (all receptors from different species bind insulin and its analogs identically). It is interesting to note that the number of insulin receptors per cell is reciprocally correlated to the biopotency of homologous insulin. In humans, the biological meaning of the polymorphism of the insulin gene has not been fully explained (p. 496).

Prolactin. Present in all vertebrates (Fig. I.171), prolactin and most of its functions (over 100) appear to be more or less related to reproduction. For example, in mammals, it acts on the mammary gland; in certain other species, it is luteotropic, and can even influence parental behavior. The classical bioassay of prolactin activity is to test the stimulation of the pigeon crop sac; however, prolactin of poikilotherms is not active in this assay. In contrast, prolactin from all species always has an effect on osmoregulation, which can be observed in salt water fish placed in fresh water (survival test), or in the return of some amphibians (tritons) to water. Tending to limit water loss, prolactin acts on osmoregulatory structures of ectodermal origin. Evolutionarily, it has favored successful reproduction (egg-laying) and terrestrial development in young aquatic animals. Indeed, throughout evolution both the structure of prolactin and the presence of receptors in diverse target organs have undergone modification.

Glycoprotein hormones. Pituitary thyrotropic and gonadotropic hormones, as well as chorionic gonadotropin, are derived entirely from an ancestral gene which was duplicated very early on in evolution. On the basis of homologies between α and β chains (p. 28 and Fig. V.1), the following evolutionary pattern may be proposed‡ (Fig. I.173). In cyclostomes, there is a single gonadotropic hormone, the α and β subunit genes having already diverged. Thereafter, the α subunit remained independent, although it under-

195. Blundell, T. L. and Wood, S. P. (1975) Is the evolution of insulin Darwinian or due to selectively neutral mutation? *Nature* 257: 197–203.

‡I thank Dr. P. Y. Fontaine (Museum d'Histoire Naturelle) for his helpful discussion on this subject.

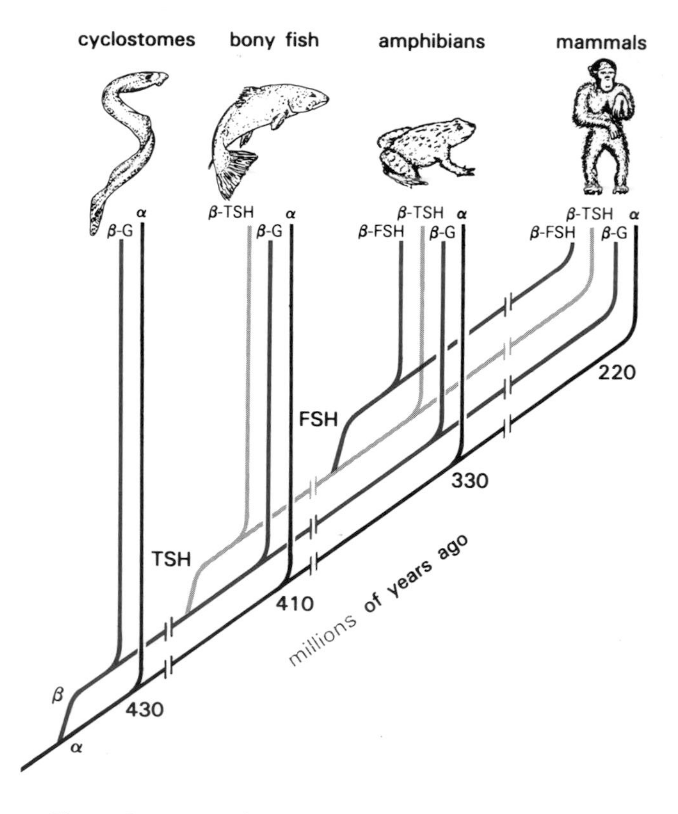

Figure I.173 Evolution of glycoprotein hormones. This simplified representation shows that the same α subunit protein is found throughout evolution, although its sequence underwent modification.

went many mutations during evolution. In bony fish, the β subunit gave rise to β-TSH and β-gonadotropin, and a separate cellular evolution took place for thyrotropic and gonadotropic cells. Later, coincidental with egg-laying as a separate event, when animals moved to land, β-TSH gave rise to β-FSH, responsible for the development of follicles, while LH remained the gonadotropin for egg-laying. However, two distinct gonadotropins have been identified in certain fish.

Differences between human β-LH and β-CG are very limited. The seven β-CG genes might have evolved from a unique ancestral β-LH gene by a series of selected changes, with very little neutral drift. The 24 aa C-terminal extension of the β-CG subunit appears to have arisen from the deletion of a single base, and has some of the 3′ untranslated region of the ancestral β-LH gene incorporated in the coding region[196].

9.3 Small Hormone Molecules: Evolutionary Aspects

The *conservation* throughout evolution of the steroid and thyronine structures and of cAMP is quite remarkable. This is perhaps even more impressive than the preservation of protein hormones in that it required the maintenance of a series of biosynthetic enzymes, each of which is at least as structurally complex as a peptide hormone (Fig. I.174). It is interesting that the successful culture of cells in a serum-free medium almost always requires the presence of a "small hormone"[197]. T_3 *and/or* T_4 are found as early as in sponges, and in vertebrates they control numerous metabolic functions and differentiation (Table I.33). *Steroid hormones*[198] are more diversified. Glucocorticosteroids resemble thyroid hormones in the ubiquitous distribution of their target cells and in the diversity of their metabolic functions. In mammals, mineralocorticosteroids constitute a separate class, with a receptor structure closely related to that of glucocorticosteroids (Fig. I.71); in primitive fish, however, a single molecule is active (cortisol), acting somewhat as a glucocorticosteroid, but principally as a mineralocorticosteroid. It is remarkable that, depending on the species, the sodium transport system has evolved by changing the location of receptors; thus it functions in organs as different as amphibian skin, reptilian bladder, avian intestine, and mammalian kidney.

196. Talmadge, K., Vamvakopoulos, N. C. and Fiddes, J. C. (1984) Evolution of the genes for the β subunits of human chorionic gonadotropin and luteinizing hormone. *Nature* 307: 37–40.
197. Sato, G. H., Pardee, A. B. and Sirbasku, D. A., eds (1982) *Growth of Cells in Hormonally Defined Media. Cold Spring Harbor Conference on Cell Proliferation.* Cold Spring Harbor Lab., New York, vol. 9.
198. Feldman, D., Tökes, L. G., Stathis, P. A., Miller, S. C., Kurz, W. and Harvey, D. (1984) Identification of 17β-estradiol as the estrogenic substance in Saccharomyces cerevisiae. *Proc. Natl. Acad. Sci., USA* 81: 4722–4726.

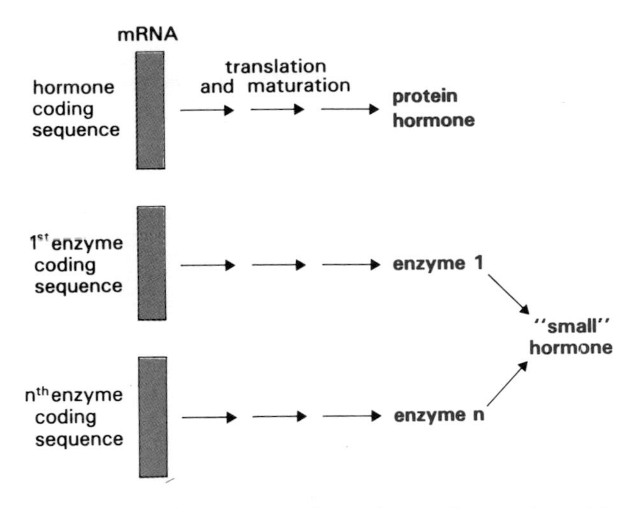

Figure I.174 Steps required for the synthesis of peptide hormones and small hormones.

Table I.33 Multiplicity of physiological actions of thyroid hormones.

Growth-promoting and developmental actions	Metabolic effects
Rate of growth of mammalian and avian tissues	Regulation of basal metabolic rate
Maturation of central nervous system and bones	Regulation of water and ion transport
All processes of amphibian metamorphosis	Calcium and phosphorus metabolism
Regulation of synthesis of mitochondrial respiratory enzymes and structural elements	Regulation of cholesterol and fat metabolism
	Nitrogen (urea, creatine) metabolism

(Tata, J. R. (1986) The search for the mechanism of hormone action. *Perspect. Biol. Med.*, 29: 184–204).

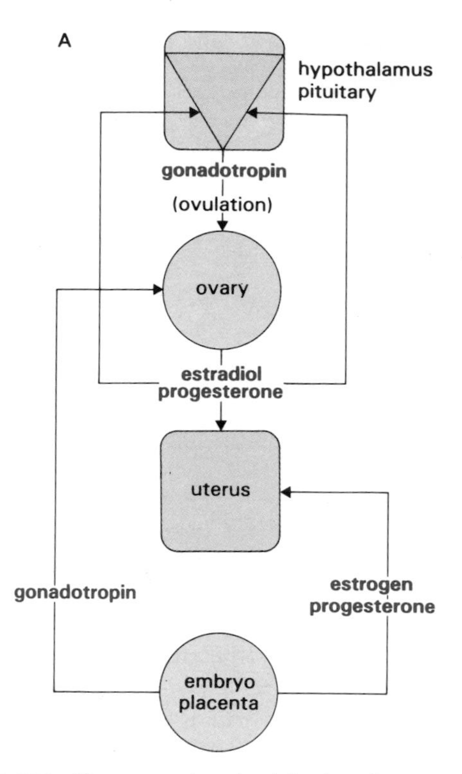

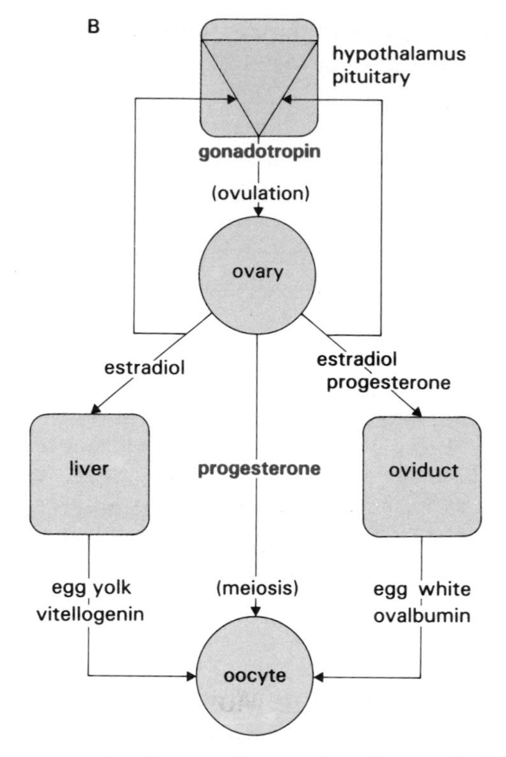

Figure I.175 Hormones involved in female mammalian (**A**) and avian/amphibian (**B**) reproductive processes.

The functional evolution of *sex steroids* is also of interest. Androgens have a similar role in birds and mammals, despite the inverse sex chromosome systems. In *oviparous* amphibians and in birds, estradiol and progesterone together control the synthesis of the nutritive material of the egg: in the oviduct, that of ovalbunim (for the egg white) and in the liver, that of vitellogenin (for the egg yolk) (Fig. I.175). In *viviparous* mammals, these two hormones are associated with the implantation of the embryo in the uterus. Thus, the female reproductive system exemplifies the remarkable evolution of target organs as opposed to hormones and receptors.

The receptor for *antheridiol* and its analogs (gamons, derived from cholesterol, which control gametogenesis in mushrooms), observed in the aquatic fungus *Achlya ambisexualis*, is structurally very similar to that of steroid hormones (Fig. I.176)[199]. *Ecdysteroids,* observed in numerous lower animals, arthropods, molluscs, worms, starfish, livestock and in some primitive plant families (ferns, conifers), are thought to have evolved from an early polyhydroxylated cholesterol derivative[200]. The important role of ecdysone in insects is discussed in section 9.6 (p. 160). The presence of ecdysteroids permits the detection of parasitic infestation of animals, as for example that of rodents by *Schistosoma mansoni*. The use of antiecdysone antihormones or antibodies has been proposed for use in the treatment of infected animals. Cholesterol is almost universally present in aerobic organisms (molecular oxygen is necessary for its synthesis), and cholesterol derivatives have functioned as signals throughout evolution. Their receptors also appear to be highly conserved. It is unclear, however, why ecdysteroids disappeared relatively early in evolution. In fact, they are not biosynthesized in vertebrates, in which no pharmacological effect of these molecules has been detected thus far.

Figure I.176 Antheridiol.

199. Riehl, R. M. and Toft, D. O. (1984) Analysis of the steroid receptor of Achlya ambisexualis. *J. Biol. Chem.* 259: 15324–15330.
200. Karlson, P. (1983) Why are so many hormones steroids? *Hoppe-Seyler's Z. Physiol. Chem.* 364: 1067–1087.

9.4 Cyclic AMP: A Molecule with (at least) Three Different Modes of Action throughout Evolution

In *eucaryotic organisms*, variation in protein phosphorylation constitutes an important mechanism in the modulation of cellular activity in response to hormones (p. 106). Variations in cAMP concentration and PK-A activity (p. 100) are frequently involved. Cyclic AMP-dependent protein phosphorylation appeared early in eucaryotic organisms.

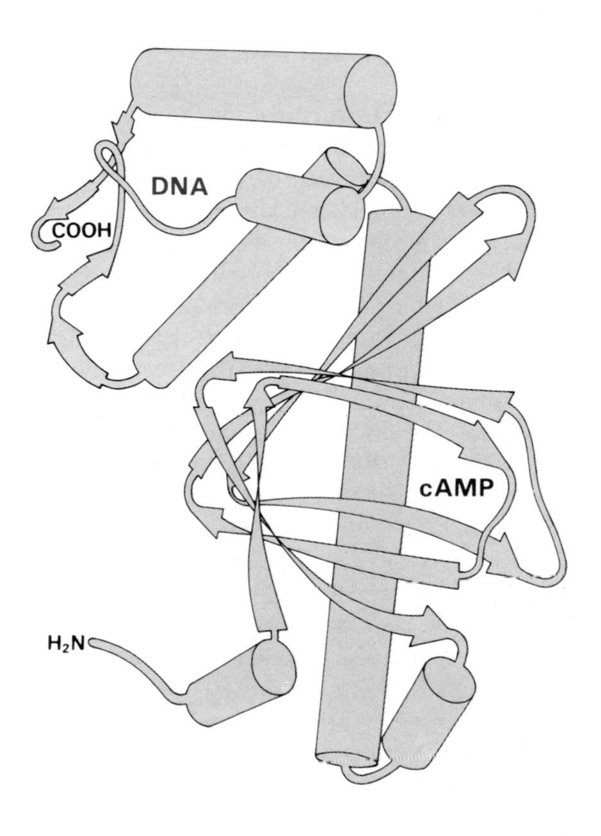

Figure I.177 CRP, a binding protein with two domains, one for cAMP and one for DNA. This schematic drawing was made based on 2.9 Å resolution of the crystalline structure of the E. coli protein (α-helices are depicted as cylinders, and β-pleated sheets as flat arrows). The N-terminal region (gray) contains the cAMP binding site, and the C-terminal region (red), the DNA-binding site. The two domains are easily separated by limited proteolysis (as in steroid receptors). (McKay, D. B. and Steitz, T. A. (1981) Structure of catabolite gene activator protein at 2.9 Å resolution suggests binding to left-handed B-DNA. *Nature* 290: 744–749). The actual CRP form which interacts with DNA is dimeric, corresponding to the two-fold symmetry in the DNA sequence of the CRP binding region of the *lac* operon.

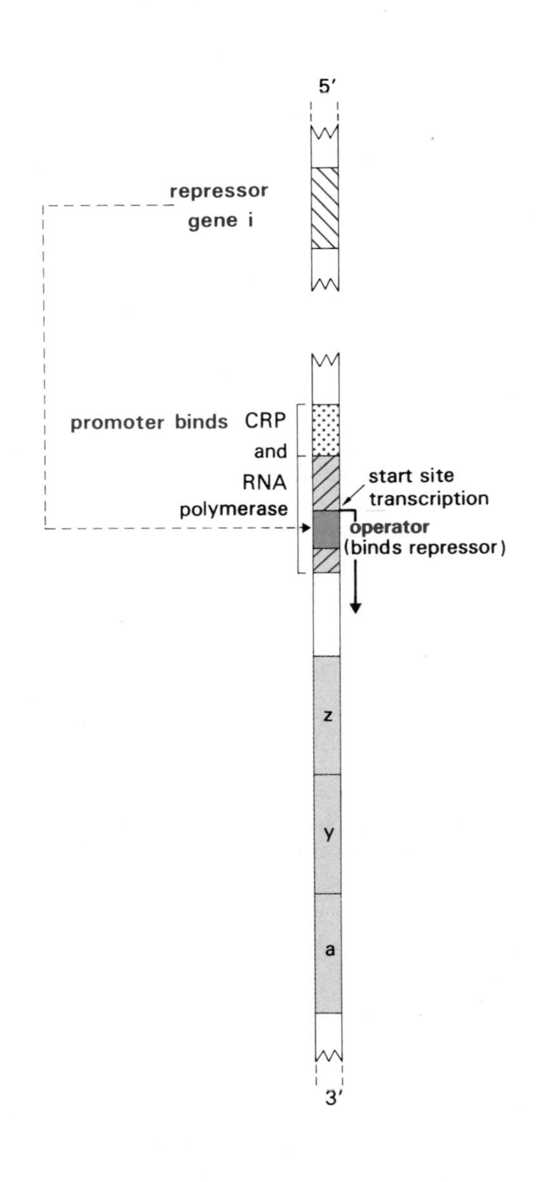

Figure I.178 Schematic representation of the E. coli *lac* operon and its regulation by CRP. The *lac* repressor, encoded by regulatory gene i (i for inducible), binds to operator DNA. Note the overlap of operator and promoter sequences. RNA polymerase binds to the promoter. The operator sequences are transcribed. Genes z, y and a are structural genes of β-galactosidase, permease, and transacetylase. The binding site for CRP is contiguous with the RNA-polymerase binding site. In the absence of glucose, cAMP is high, and if the *lac* repressor has been removed (in the presence of lactose), maximal transcription of the *lac* operon occurs. Therefore, the repressor and the cAMP-complex exert, respectively, positive and negative control on *lac* operon transcription (an operon is a cluster of genes controlled by one promoter site).

In *bacteria*, the only well established action of cAMP is to specifically induce the *expression* of some *genes*[201]. Cyclic AMP binds to a receptor protein (called catabolite repressor protein (*CRP*) or cAMP binding protein (*CAP*) (Fig. I.177). The cAMP-CRP complex increases gene transcription via its binding to a specific region of the promoter, near the RNA polymerase site (Fig. I.178). While bacteria lack cAMP-dependent protein phosphorylation, which seems to have been a later evolutionary development, the transcription of a number of genes in higher eucaryotes, such as those of tyrosine aminotransferase, prolactin, etc., has been shown to be stimulated by intracellular cAMP[202]. Regulation may be related to the phosphorylation of protein(s) which in turn would specifically increase transcription through an unknown cascade of events (Fig. I.122). Alternatively, it has been proposed recently that the R subunit of PK-A (type II) could be a DNA-binding protein which interacts with a consensus nucleotide sequence found in regulatory segments of several eucaryotic cAMP-regulated genes and in the 5′ region of several genes in bacteria[203]. Transcription, RNA processing and protein synthesis require several minutes, in contrast to the more rapid, cytoplasmic reactions, which are completed in a few seconds.

Cyclic AMP is also a signal to correct for food exhaustion in unicellular organisms intermediary between bacteria and higher eucaryotes, such as slime mold (e.g., *Dictostelium discoideum*). These single-cell animals secrete cAMP into the external medium in a pulsatile manner, and the nucleotide which binds to a membrane receptor is a quasi-*pheromone*, provoking the formation of a pseudopod which pulls the cell toward the source of nucleotide[204].

Whether in bacteria, liver and muscle, or in *Discostelium discoidium*, cAMP is a *hunger signal*, indicating an absence of glucose, which leads to the synthesis of enzymes that can exploit other energy sources. Cyclic AMP, however, is involved in the mechanism of many other hormonal responses, through protein phosphorylation by PK-A. It may also contribute to yet unknown pathways in some lower eucaryotes in which a cAMP binding protein (known to be neither CRP nor PK-A) is found.

Why has cAMP been conserved throughout evolution? Perhaps because it is easily made from universally available ATP, and also, although it is sufficiently complex to be bound with high affinity by proteins, because it is a very stable molecule. As was proposed by Gordon Tomkins, cAMP may be a prototype for a generalized "metabolic code"[205], suggesting that the chemical "language" of cellular metabolism is highly conserved.

201. Pastan, I. and Adhia, S. (1976) Cyclic adenosine 3′-5′monophosphate in Escherichia coli. *Bacteriol. Rev.* 40: 527–551.
202. Gruol, D. J., Campbell, N. F. and Bourgeois, S. (1986) Cyclic AMP-dependent protein kinase promotes glucocorticoid receptor function. *J. Biol. Chem.* 261: 4909–4914.
203. Nagamine, Y. and Reich, E. (1985) Gene expression and cAMP. *Proc. Natl. Acad. Sci., USA* 82: 4606–4610.
204. Gerisch, G. (1982) Chemotaxis in Dictyostellum. *Annu. Rev. Physiol.* 44: 535–552.
205. Tomkins, G. M. (1975) The metabolic code. *Science* 189: 760–763.

9.5 Evolution of Receptors

The hormone receptor structures which are known are mosaics, as has been shown also for the LDL-R (p. 534). Chimeras, reported on p. 87 and in Figure I.90, demonstrate biochemically that peptide and steroid hormone receptors consist of independent sections that can function autonomously but which nevertheless are able to regulate each other's function when linked together. The further recruitment of exons during evolution resulted in the development of functional domains. The Drosophila EGF-R homolog contains both the hormone binding and Tyr-PK domains, indicating that these activities have resided in the same molecule for over 800 million years.

Receptors belong to a few supergene families of related proteins which include oncogenes, e.g., the EGF-R family and *erb*-B, and the steroid receptor family and *erb*-A. The possibility of a common evolutionary origin of receptors involved in growth (EGF-R) and nutrient delivery (LDL-R) is not excluded. Common features of their primary structures and functions suggest that channel/receptors (nACh-R, $GABA_A$-R, and Gly-R, Fig. 1.85) have evolved from a common ancestral receptor, probably a homo-oligomer. Since small hormones such as steroid and thyroid hormones and catecholamines have maintained the same structure throughout vertebrate evolution, it is likely that the hormone binding sites of their respective receptors evolved very little.

Which evolved first, hormones or receptors?

The presence in unicellular organisms of compounds with chemical structures related to or even identical to hormones but which have no known regulatory role suggests that, in higher animals, the evolution of *receptors followed* that of hormones. It has been proposed that an insulin-like molecule with structural similarity to proteases of microorganisms may have favored the formation of a receptor which would undergo very limited change in evolution, while insulin itself diverged greatly among species, despite the fact that its three-dimensional active region remained the same. This implies that hormones may have existed before they were functional.

Nevertheless, the presence in lower organisms of a binding protein specific to a hormone which did not yet exist provides evidence that *receptors evolved first*, since, in this case, selection of the ligand appears to have occurred secondarily. An interesting example substantiating this view is an estrogen binding protein found in *Paracoccidionides brasiliensis*[206]. The presence of this protein explains why the development of the parasite is dependent on estrogens of the host. It has even been suggested that, in animals, some nutritive products, such as non-steroidal plant estrogens, might have favored the selection of binding proteins, which in turn may have become receptors utilized by hormonal steroids derived from cholesterol, and secondarily, by other ligands (e.g., retinoic acid, thyroid hormones).

A remarkable similarity has been observed recently between the molecules of hormones and those of receptors. In Figure I.179, the complementary RNA of the mRNA of a hormone is represented. It has been shown that its encoded amino acid sequence has high affinity for the hormone. Antibodies to this complementary peptide are themselves able to bind to the corresponding hormone receptor. Whether *hormone-receptor complementarity* is involved in evolution remains to be established.

In order to ensure the proper structure of hormones and receptors, the biochemical steps involved in their synthesis must have undergone systematic evolution, although paradoxical results have been reported, e.g., for the maturation of peptide hormones. If the POMC gene is transfected into mammalian cells that normally do not synthesize POMC-derived hormones, the correct peptide is produced, but in contrast to specialized pituitary cells, the hormone remains uncleaved (p. 231). On the other hand, if the gene of the *kex2* enzyme[207], a protease which cleaves the yeast mating factor proprotein, is

206. Loose, D.S., Stover, E.B., Restrepo, A., Stevens, D.A. and Feldman, D. (1983) Estradiol binds to a receptor-like cytosol binding protein and initiates a biological response in Paracoccidioides brasiliensis. *Proc. Natl. Acad. Sci., USA* 80: 7659–7663.
207. Thomas, G. and Herbert, E.; cited in *Science* 235: 285–286.

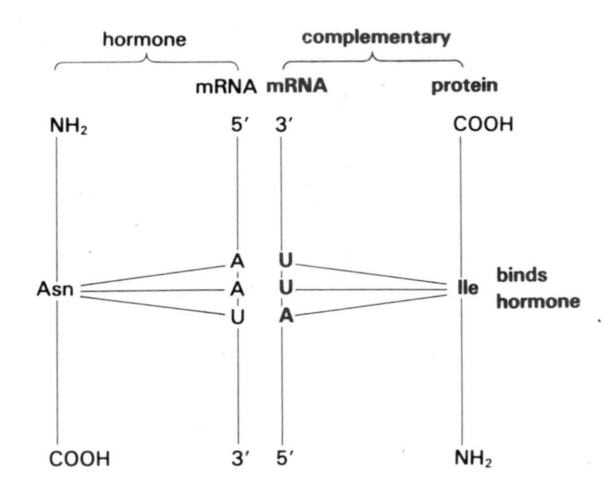

Figure I.179 Complementarity between mRNAs for hormones and their respective receptors. For large hormones, the complementary relationship may be restricted to only the specific region involved in hormone–receptor interaction.

cotransfected, the appropriate POMC-derived peptides are produced. Yet, the *kex2* enzyme has never "seen" the prohormone POMC, and thus is completely different structurally from the known maturation enzymes isolated in the specialized cells.

It is striking that central hormones belonging to a large family (e.g., glycoproteins) (Fig. I.180) act ultimately through receptor proteins forming a subset of another, completely unrelated superfamily (steroid and thyroid receptors) *via* a distinct but parallel cascade of endocrine pathways (Fig. I.180). In fact,

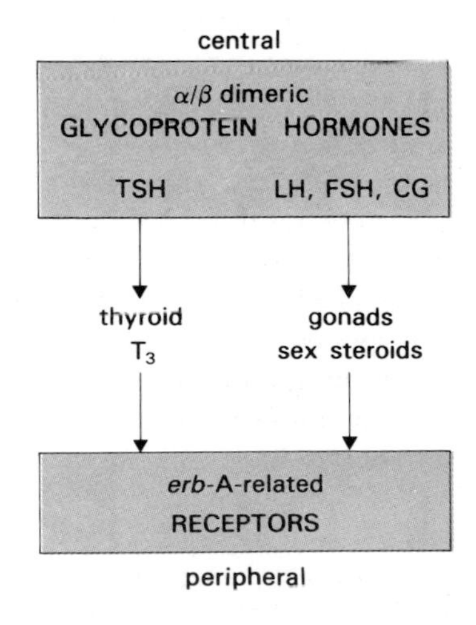

Figure I.180 Families of central glycoprotein hormones and peripheral hormone receptors.

evolution appears to have involved more diversification of hormonally regulated physiological functions than reciprocal molecular evolution of either hormone and/or receptors at the molecular level. It is still necessary to postulate the evolution of *other genes*, perhaps belonging to the same class as homeotic[208] or "gap" genes, which may have regulated the organization of the body. They would have been responsible for the *integration* of functional hormones and receptors. The genes for related hormones and receptors have never been found at the same chromosomal location, suggesting that some *trans*-coordinating mechanisms were involved in evolution.

These examples illustrate that, at both the biochemical and physiological levels, an understanding of the trend of evolution still eludes endocrinologists.

9.6 Neuroendocrine Systems in Insects‡

Linnaen classification, based on Aristotle's concept, grouped separately animals with warm blood, cold blood, and hemolymph. At the end of the eighteenth century, Lamarck was the first to make the distinction between vertebrate and invertebrate animals. The neuroendocrine systems of these two classes have much in common, with parallelism between the intercerebralis cardia allatum system of insects and the hypothalamic-hypophyseal complex of vertebrates (Table I.34 and Fig. I.181). In both cases, neural control in these animals involves neurotransmitters and neurohormones, hormones of several classes (peptide and others), multistep neuroendocrine events, expression of the same structural gene in different organs, and, sometimes, diversity in the processing of the same primary product in different cells (p. 27).

Insect endocrinology is the most clearly understood among invertebrates, and important advances have also been made recently in molluscs. The adaptation of insects to different biospheres, and their tremendous rate of reproduction (with more than 1 million known species), illustrates the success of this zoological group. By the mid-paleozoic period, most insects existed as we know them now, whereas the first terrestrial vertebrates had just begun development,

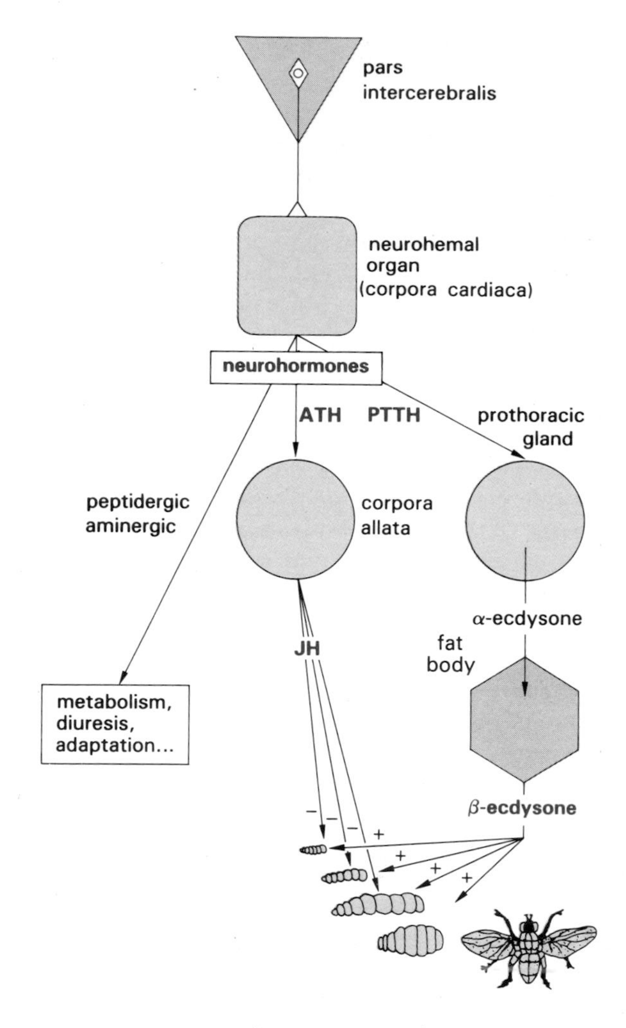

Figure I.181 Neuroendocrine system of insects.

long before mammals appeared. Similarities between the neuroendocrine organization of mammals and insects are more amazing than the differences.

Instead of the endoskeleton found in vertebrates, insects have a thick cuticular envelope (carapace). Such an exoskeleton must be shed during growth, once it has become too small, and a new, larger one is synthesized (Fig. I.182) by the hypodermis. This process is known as *molting*, or *ecdysis*. Molting cycles are under endocrine control of the steroid hormone *ecdysone*[209]‡‡ (Fig. I.183). Postembryonic development of insects consists of a series of moltings at varying intervals, and, in most insects, the reproductive phase is separated from the growth phase.

‡I thank Dr. J. Hoffman (CNRS, Strasbourg) for his contribution to this section.

‡‡A non-steroidal ecdysone agonist (RH 5849: 1,2-dibenzoyl-1-*tert*-butylhydrazine) has been described. (Wing, K. D. (1988) *Science* 241: 467–469).

208. Gehring, W. J. (1987) Homeo boxes in the study of development. *Science* 236: 1245–1252.

209. Hoffmann, J. A., ed (1980) *Progress in Ecdysone Research*. Elsevier, Amsterdam.

Table I.34 Neuroendocrine systems.

Mammals	Insects		Crustacea	Molluscs
I. Endocrine neurons (hypothalamus) *hypothalamic factors*	I. Neurohemal organ (corpora allata/ cardiaca) *prothoracicotropic hormone*	I. Endocrine neurons *allatotropic factor*	I. Neurohemal gland (X organ) *molt-inhibiting hormone*	I. Atrial gland *peptides A/B*
II. Anterior pituitary *tropic hormones*	II. Prothoracic gland *ecdysteroids*	II. Corpora allata *JH*	II. Molt gland (Y organ) *ecdysteroids*	II. Endocrine neurons (bag cells) *egg-laying hormone (ELH)*
III. Peripheral glands *peripheral hormones*				

(Acher, R. (1986) Common patterns of neuroendocrine integration in vertebrates and invertebrates. *Gen. Comp. Endocrinol.*, 61: 452–458).

Figure I.182 Exuviation.

In insects which undergo *metamorphosis*, it is possible to observe several sorts of molting: larva-larva, corresponding in general to a simple phenomenon of growth, larva-nymph and, thereafter, metamorphosis, bringing the nymph to an adult state. The adult is generally winged, and essentially serves a reproductive function. *Each* of the three types of molting is triggered by ecdysone. Larva-larva moltings without any differentiation to adult features occur as long as *juvenile hormone (JH)*, which inhibits metamorphosis, is present in large quantities (Fig. I.184). Following a reduction in the level of JH, metamorphosis will proceed.

Neurosecretory cells of the brain regulate the activity of the glands that produce ecdysone and juvenile hormone (Fig. I.181). Neurosecretory cells

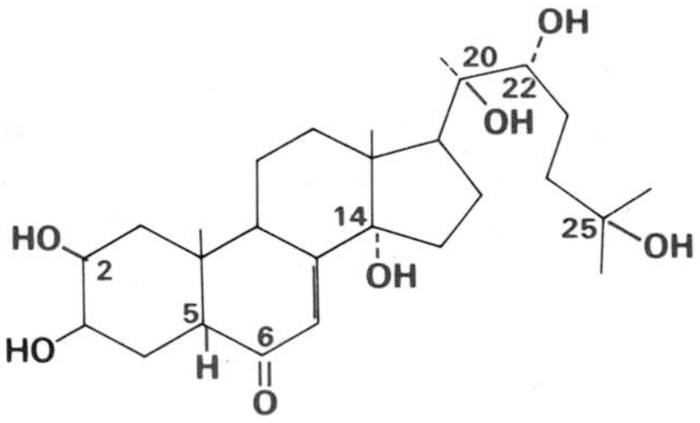

Figure I.183 β-ecdysone.

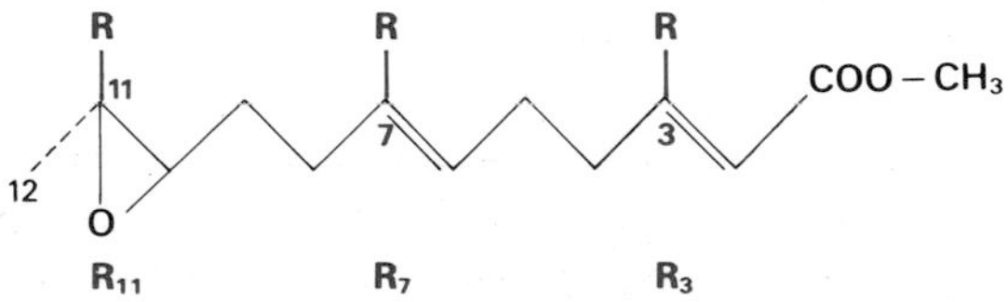

	R_{11}	R_7	R_3
JH_0	$-CH_2-CH_3$	$-CH_2-CH_3$	$-CH_2-CH_3$
JH_I	$-CH_2-CH_3$	$-CH_2-CH_3$	$-CH_3$
JH_{II}	$-CH_2-CH_3$	$-CH_3$	$-CH_3$
JH_{III}	$-CH_3$	$-CH_3$	$-CH_3$

Figure I.184 Juvenile hormones.

are limited in number, and are located precisely between the two lobes of the protocerebron which form the *pars intercerebralis*. This region corresponds to the hypothalamus in vertebrates. *Neurohormones* stimulating the glands are carried out along the axons of neurosecretory cells to the paired *neurohemal organs*. Located near the cardiac vessels (they are also known as corpora cardiaca), the latter are reminiscent of the posterior pituitary gland of vertebrates in that they secrete hormones directly into the blood. Peripherally, *prothoracic glands* produce ecdysone. These two small bodies, situated in the prothorax, transform cholesterol or a similar sterol present in food (insects do not synthesize the steroid nucleus) into hormone. Ecdysone (or α-ecdysone) is probably not the most active molecule to exert an effect on the hypodermis. It is transformed, for the most part, into *20-hydroxyecdysone* (β-ecdysone) by monooxygenases, including a cytochrome P450 hydroxylase of the fat body (the functional equivalent of the liver in vertebrates) and Malpighian tubes (kidneys). The concentration of 20-hydroxyecdysone can attain very high levels (up to 20 μmol/l) in the body fluid of certain insects. Nuclear receptors have been detected in the hypodermis, and 20-hydroxyecdysone acts primarily at level of transcription.

JH is synthesized in the *corpora allata*, a pair of organs located next to the corpora cardiaca in the heads of insects. JH is a derivative of farnesoic acid. In most orders of insects, there is only a single JH (JH_{III} or C_{16}-JH), while Lepidoptera synthesizes up to five types of JH, differing by the number and position of methyl or ethyl substitutions on the common carbon skeleton (Fig. I.184). The mechanism of the *repressive action* of JH on *differentiation* of adult characteristics is not known. Nuclear receptors have, however, been detected in target tissues. JH has *not* been identified in *other invertebrates*. On the contrary, *ecdysone* is found in *almost all invertebrates*, such as annelides, nematodes, cestodes, trematodes, molluscs, and all arthropods[209].

The *neurohormones* that regulate the activity of the prothoracic glands (prothoracicotropic hormone, PTTH) and the corpora allata (ATH, allatotropic hormone) are *peptides* (Fig. I.185).

The *corpora cardiaca* secrete *other peptides* into the blood which regulate blood pressure, lipid metabolism (AKH, at octapeptide, adipokinetic hormone), diuresis, changes in color, etc. In the *corpora cardiaca* and in the *nervous system*, aminergic *neurotransmitters* such as dopamine, norepinephrine and serotonin are produced and secreted. As in vertebrates, the distinction between neurotransmitters and neurohormones is difficult to make. In the last few years, immunocytochemical studies conducted in many invertebrates have demonstrated the presence of molecules similar to most neuropeptides found in vertebrates (e.g., gastrin, CCK, insulin, glucagon, arginine-vasopressin, endorphin, enkephalin). In certain insects, insulin and vasopressin-like molecules perform functions similar to those observed in vertebrates.

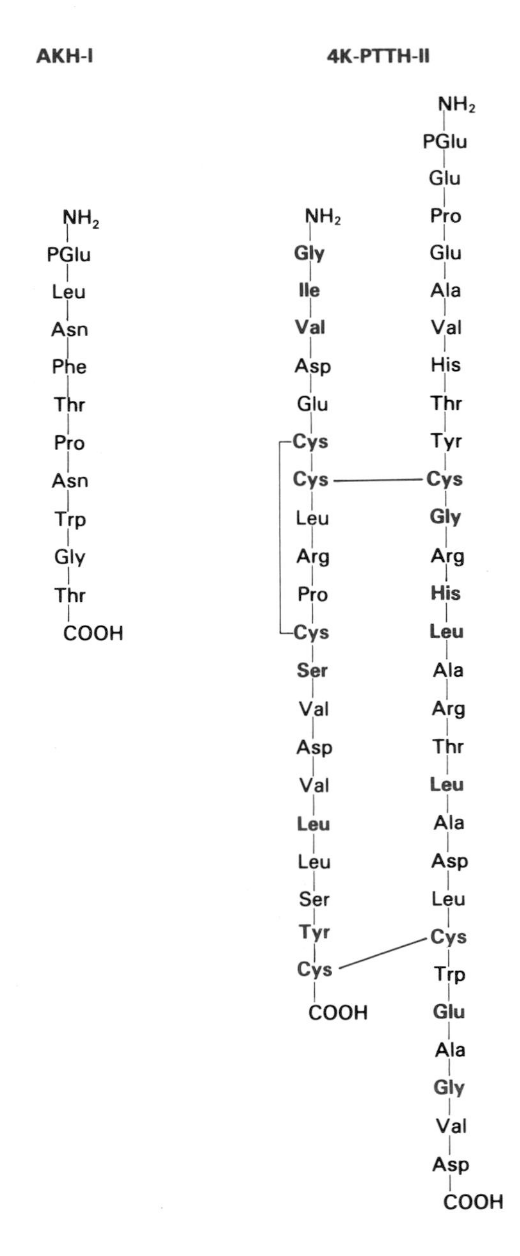

Figure I.185 Two peptide hormones in insects: AKHI is an adipokinetic hormone; 4K-PTTH-II is a prothoracicotropic hormone which has some homology with human insulin (red aa) (Fig. XI.2).

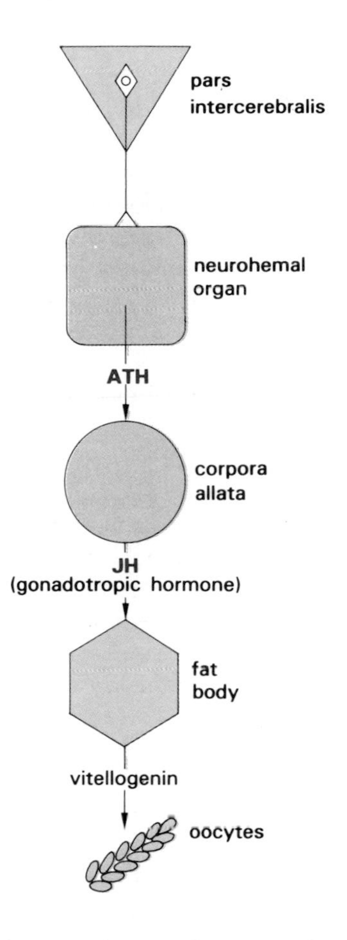

Figure I.186 Neurohormonal system of reproduction in insects.

In *reproduction* (Fig. I.186), JH functions as a *gonadotropic* hormone, in contrast to its effect on maintenance of the juvenile state during larval development. It disappears later on, allowing metamorphosis to proceed, and finally returns to exert an important effect on genital function. It is then involved in the development of the accessory sexual glands in males, and stimulates the synthesis of vitellogenin in the fat body of females. Vitellogenin enters the blood and is taken up by oocytes, a process similar to that induced by estrogens in fish, amphibians, reptiles, and birds.

It has been shown recently that insects synthesize steroids in follicular cells during the reproductive phase; these hormones are ecdysteroids, the same molecules which control molting. They are found mostly in conjugated forms (phosphoric esters, AMP esters), accumulating in the oocyte, where they are hydrolyzed in the egg during embryogenesis. The release of ecdysone and 20-hydroxyecdysone controls the cyclicity of embryonic cuticle formation, a characteristic of embryogenesis in numerous insects. This case is quite unique in that the mother transfers hormonal molecules to the gamete, which are subsequently utilized by the embryo during development.

9.7 Plant Hormones

Many biological events in higher plants are regulated by extracellular products, or plant hormones. They involve cellular growth, enlargement and differentiation, and the induction of protein synthesis. Plant hormones are divided into *five principal classes*: auxins, gibberellins, cytokinins, abscisic acid, and the gas, ethylene (Fig. I.187). *Several* hormones are often necessary to control a given, specific process. Depending on the target cell, the active concentrations of these

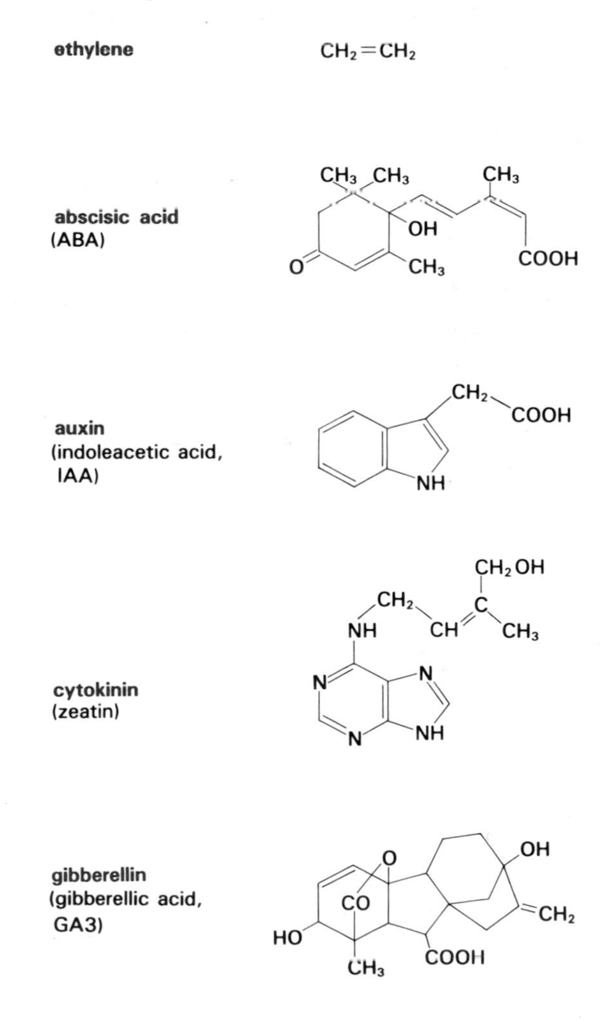

Figure I.187 Five groups of plant hormones.

hormones may differ by four to five orders of magnitude. They circulate in plants along well established pathways. Their rapid effects are probably initiated at the membrane level. Neither a membrane nor an intracellular receptor has been identified for a plant hormone[210].

Auxins, which affect the growth of higher plants, have been thoroughly studied. Their most important effect is to cause the elongation of plant cells by loosening the cell wall. The mechanism of this phenomenon involves a decrease in pH which affects the activity of membrane enzymes, thus contributing to the loosening of the wall; internal turgor caused by water uptake can expand the cell. The detailed mechanism of proton pump activation by auxin is unknown. Auxin can also massively activate the synthesis of RNA and protein in several plants, but it is not known whether it has a primary action at the DNA level. *Gibberellins* may act in a fashion similar to steroid hormones in animals. The stimulation of protein synthesis seems to be one of its primary effects.

The concentration of hormones in plants varies depending on the particular part examined and on its stage in development. Changes have also been described in the sensitivity of a plant to hormone, but the circumstances of these changes are unknown. As has already been mentioned, hormone-like molecules are synthesized in some plants (e.g., coumarol derivatives). They are of no recognized physiological importance for the plant, but probably affect the commercial quality of herbivorous animals.

9.8 Tomorrow

As is generally the case for living matter[211,212], a relatively small number of molecules and chemical mechanisms appear to be involved in the establishment of endocrine systems. Physiological function is determined by at least two levels of gene expression. One is *molecular*, establishing the primary sequence and thus the structure of the components of living matter, while the other is responsible for the *anatomical* distribution of an organism.

The life of an organism cannot be separated from the influence of its environment. In higher animals, the brain is the main intermediary between exterior and interior signals and functions, and because of its integrative role, endocrine activities can be profoundly affected by the external, social environment.

Are hormones and receptors still in the process of evolving, despite the fact that they already display a high level of complexity? To answer this question, studies of the classical hormonal system must be supplemented by an evaluation of recently described functions of the para- and autocrine systems, which reveals a largely decentralized hormonal system of intercellular communication. Is there coordinated evolution of hormones and parahormones? Will experiments with transgenic animals expressing hormone and receptor genes allow a better understanding of the specific organization of the multitude of hormonally regulated functions?

210. Kende, H. and Gardner, G. (1976) Hormone binding in plants. *Annu. Rev. Plant Physiol.* 27: 267–290.
211. Jacob, F. (1977) Evolution and tinkering. *Science* 196: 1161–1166.
212. Ohno, S. (1982) Evolution is condemned to reply upon variations of the same theme: the one ancestral sequence for genes and spacers. *Perspect. Biol. Med.* 25: 559–572.

10. Summary

1. Hormones are *endocrine signals* for target cells. Classically, these *chemical messengers* are produced by *glands*, are distributed mainly by the general blood circulation, and act on *cells* possessing *receptors*; that is, molecular structures where the hormonal response is initiated. This picture is now complemented by the discovery of hormone synthesis in non-glandular cells and by the notion of limited topological distribution of some hormones, i.e. *parahormonal* delivery to neighboring cells differing from the cells having secreted the hormone, and *autohormonal* delivery of hormone to the secretory cells themselves. The number of hormones and related factors (>100), the polymorphism of their distribution, and their many interactions at both the levels of secretion and action make the *hormonal network extremely complex*, as would be expected of a major integrative system intervening in *all functions* of the body.

Endocrinology, or the study of hormones, is dependent on *physicochemical techniques* in order to separate, label, visualize, characterize and measure biomolecules in very small amounts. Contemporary progress in *immunology*, in the *neurosciences* and in *molecular biology* provides endocrinologists new tools and concepts. Conversely, endocrinology offers molecular and physiological explanations for a number of biological and pathological phenomena.

2. *Chemical heterogeneity* among bona fide hormones, in addition to growth factors, opioids, prostaglandins, neuromediators, etc., is remarkable. There are amino acid derivatives (thyroid hormones and catecholamines), peptides of all sizes (from the smallest, hypothalamic thyrotropin releasing hormone (TRH), with three amino acids, to holoproteins and glycoproteins), as well as steroids. Peptide hormone families are created by gene families originating from the duplication of ancestral genes. The insulin family includes insulin-like growth factors (IGFs) and relaxin; the glycoproteins have a common α subunit, and different β subunits are found in thyroid stimulating hormone (TSH), luteinizing hormone (LH), follicle stimulating hormone (FSH), and chorionic gonadotropin (CG); members of the transforming growth factor β (TGF-β) family have inhibin and activin effects; growth hormone (GH) and prolactin (PRL) belong to the same group of proteins. Most hormonal peptides are synthesized initially as *prohormones*, which are then processed enzymatically before they are secreted as active hormones. Some prohormones are indeed *polyproteins* which give rise to several identical or different hormones after processing. They are often related functionally, but sometimes differentially, according to the secretory cell type (e.g., proopiomelanocortin (POMC) is a potential source of adrenocorticotropin (ACTH), melanocyte stimulating hormone (MSH), and enkephalin). Another way to create diversity is through the variable *mode of peptide chain combination* (e.g., in inhibin and activin, which are dimers with opposite actions on FSH production). Certain steroids (e.g., dehydroepiandrosterone sulfate (DHEA-S)) and thyroid products (thyroxine (T_4)) are *prohormones* that are transformed into the active compounds in an intermediary organ or even in target cells, which again, may differ (e.g., DHEA-S, is transformed into androgens or estrogens). The placenta, as well as cancers of non-glandular organs, can produce hormones very similar or identical to those synthesized in classical glandular cells.

The secretion of many hormones is controlled by the *central nervous system*. The secretion of TSH, LH, FSH, ACTH, GH, and PRL depends on releasing factors/hormones produced in the hypothalamus: TRH, GnRH (gonadotropin releasing hormone for FSH and LH), corticotropin releasing factor (CRF), growth hormone releasing factor (GRF) and vasointestinal peptide (VIP), respectively. Thyroid and steroid hormones, inhibin and IGF-I, all of which are made in peripheral glands, participate in closed negative control loops that regulate pituitary hormone secretion at the level of the hypothalamus and/or pituitary. In addition, GH and PRL are negatively controlled from the hypothalamus by somatostatin and dopamine, respectively. Vasopressin and oxytocin, made by the posterior pituitary, and epinephrine, produced in the adrenal medulla, are neurohormones. The central nervous system is responsible for the *rhythmicity* of the secretion of most of these hormones, including circadian variations and short-term oscillations. The pulsatile delivery of hormone to target cells may be fundamental to hormone action, as exemplified by the agonistic effect of pulsatile GnRH secretion; permanent delivery of this peptide desensitizes the gonadotropic cell. Thus, this same peptide may be used for the stimulation of ovulation and for the induction of reversible, medical castration.

Peripheral hormonal systems include insulin and glucagon, parathyroid hormone (PTH), calcitonin, the 1,25-dihydroxyderivative of vitamin D_3 (calcitriol), and aldosterone. The production of these hormones is controlled largely by direct feedback via the metabolic consequences of their actions, e.g., changes in the concentration of glucose, Ca^{2+}, or K^+. Whether dealing with a system controlled by the brain or not, hormonal secretions are *always* engaged in a *feedback* circuit. In addition to such "vertical" and cybernetical arrangements, there are "horizontal" connections, including those of parahormonal nature, e.g., cortisol from the adrenal cortex, which stimulates the methylation of norepinephrine; or somatostatin from D cells, which decreases the production of insulin in the B cells of the islets of Langerhans.

3. In the plasma, thyroid and steroid hormones, and some hormonal peptides, are largely bound by specific *"transport" proteins*, whose physiological significance is uncertain. These proteins, or congeners, are also involved in the distribution of hormones within some target cells. Generally, hormones are degraded rapidly, a biochemical event which makes replacement secretion mandatory, and which allows regulatory changes to be made in the concentration of active hormone. Quantitatively, the catabolism of hormone is expressed as *metabolic clearance rate* (MCR), an increase in which leads to shorter half-life. The *half-lives* of a hormone and its analog(s) vary from a few minutes to several hours. Consequently, concentrations established at the level of target cells is dependent on both the secretion and the metabolism of the hormone, a level maintained usually between 10 pM and 10 nM. No precise quantitative data are available for parahormones and autohormones which circulate in a local vascular system (e.g., portal vessels from the hypothalamus to the anterior pituitary) or which diffuse into the intercellular compartment (e.g., steroids and peptides in the same or different compartments, of testes, ovaries, or the placenta). Quantitative data are also scarce for opioid peptides and prostanoids, active derivatives of polyunsaturated fatty acids, which in most instances do not act far from their sites of production. Globally, not only the molecular structure of hormones, but also their *availability* to target cells, and thus the *anatomical arrangement* determining their distribution, are crucial parameters of their functionality.

4. Once having reached its target cell, a hormone interacts selectively with a specific molecular structure, the *receptor* (R), which triggers a *cellular response* at its *effector site*. The non-covalent binding of hormone to a receptor is of high affinity and strict stereospecificity. There is one binding site per receptor molecule, and the number of molecules of a given receptor is of the order of $10^{4\pm1}$ per target cell. Receptors are *necessary* for the hormonal response; in their absence, cells are not target cells. However, the mere presence of receptors on or in the target cells, and even their subsequent activation by molecules of hormone, is insufficient to evoke a response. An intact intracellular machinery is necessary to carry out a sequence of events which will create the overall response of the cell. Responses to hormones are *quantitatively* related to both H and R concentrations.

In most cases, H binds to R according to the law of mass action, permitting the convenient linearization of data. Binding measurements should always be interpreted cautiously. The presence of lower affinity, non-specific binding sites lacking biospecificity can create technical difficulties. When measuring receptors occupied by endogenous or pharmacological ligand, it is necessary to use an exchange technique in order to measure occupied sites. Binding affinity does not distinguish between agonistic and antagonistic ligands, and as a general rule binding does not fully describe the overall function of the receptor. In many cases, not all cellular receptors need be occupied in order to obtain a maximum response, but it is the totality of binding sites which determines the magnitude of the response, in contrast to the concept of spare receptors. Hormones may also be *"permissive"*, and an above-threshold level may make cells responsive to other agents (including other hormones).

The primary *structure* of several hormone receptors is now available from the sequence of corresponding cDNAs. *Steroid and thyroid* hormone receptors are intracellular, and, upon binding of the appropriate ligand, they interact with specific DNA *hormone response elements* (HRE) in target cell nuclei. The native "8 S" form of all steroid hormone receptors is hetero-oligomeric, and includes the heat shock protein MW=90,000 (hsp 90) which prevents receptor-DNA binding by the effector domain of the receptor when ligand is absent. Upon ligand binding, the transformation and activation of the receptor involves the release of hsp 90. Receptors for all steroids (glucocorticosteroids, mineralocorticosteroids, sex steroids, calcitriol, ecdysone), thyroid hormones, and the receptor for retinoic acid are members of a superfamily related to the oncogene *erb*-A. Their C-terminal extremities contain a hydrophobic *hormone-binding* region (~250 aa) which is 10 to 30% homologous in paired comparison. This region is preceded at its N-terminal extremity by a short hinge and an ~70 aa putative *DNA-binding* domain which includes two "Zn fingers". This last region of the receptor is mandatory for the initiation of transcription, is specific for each hormone class, and is approximately 50% homologous in paired comparison. The N-terminal region is most variable in length and sequence.

Generally, *membrane* receptors are *glycosylated* at their extracellular region, and the *transmembrane* segment(s) consist(s) of hydrophobic, uncharged helical structure(s). Receptors for epidermal growth factor (EGF), insulin (I) and platelet-derived growth factor (PDGF) are typical of *single membrane-spanning* receptors which have cytoplasmic tyrosine protein kinase (*Tyr-PK*) activity. The extracellular portion is relatively hydrophilic and rich in Cys residues, and is probably involved in *ligand binding* and receptor–receptor interaction. The I-R is composed of two S-S bound, double-chained units. Growth hormone (GH) and prolactin (PRL) have single membrane-spanning receptors lacking Tyr-PK activity. Two groups of *mutiple membrane-spanning* receptors have been identified: single-chain and seven-helix receptors, which are coupled to GTP binding proteins (*G proteins*). Adrenergic receptors (A-R) bind to G_s and G_i, and the muscarinic acetylcholine receptor (mACh-R) binds G_p. Considerable homology is found between these receptors and the visual receptor rhodopsin (coupled to G_t). For all, the regulatory ligand binds to a hydrophobic intramembrane region. Mutiple-chain receptors such as those of γ-aminobutyric acid ($GABA_A$-R) and nicotinic acetylcholine (nACh-R) are *ligand-gated channels*. The molecular organization of parent receptors, as for example those of low-density lipoprotein (LDL-R) and transferrin (T-R), resembles that of some hormone receptors. It is clear that most known receptor molecules are *mosaic proteins* composed of independent functional regions. In several receptors (e.g. I-R, EGF and intracellular steroid/thyroid receptors), the hormone binding site is a repressor of the effector site, and binding of the ligand relieves this inhibition. Truncated receptors deprived of the hormone binding site demonstrate constitutive activity.

Both the concentration and some of the properties of receptors may vary. Their *number* per cell changes ontogenetically and physiologically. This variation is dependent on the binding of the hormone itself (homologous regulation) or of other regulatory agents (heterologous regulation). In many cases, *down-regulation* of membrane receptors involves their migration into coated-pits and their internalization by endocytosis. Binding *affinity* and/or *enzymatic* (Tyr-PK) activity of receptors may be affected by receptor phosphorylation.

5. Hormones essentially regulate already established cellular mechanisms. They operate at the cell surface and in the gene machinery where basic events involved in cell communication and gene expression occur. A *relatively limited* number of immediately postreceptor *intracellular signals* are generated which relay the specific formation of hormone–receptor complexes and trigger a variety of characteristic cellular responses.

The hormonal *regulation of gene transcription* by steroid/thyroid hormones involves specific enhancers, HREs, which are found mostly 5′-upstream of the transcription initiation site of regulated genes. The binding of *trans*-regulating hormone-receptor complexes to these *cis*-regulatory elements is associated with a change in chromatin structure, with the unfolding of nucleosomes, the binding of nuclear transcription factors, and with an increase in mRNA synthesis. With the exception of progesterone in Xenopus laevis oocytes, for which there is a membrane receptor, no direct post-transcriptional effect of steroid/thyroid hormones has been demonstrated.

Several *membrane receptors* interact with a G protein connected to *adenylate cyclase*, a membrane enzyme acting to form *cAMP* from ATP. Cyclic AMP was the first *"second messenger"* to be described. It is the regulator of cAMP-dependent protein kinase (PK-A), a heterotetramer comprised of two cAMP binding regulatory subunits and two catalytic subunits which, upon cyclic nucleotide binding, are separated and activated. Depending on the particular hormone-receptor system, adenylate cyclase can be stimulated via a G_s protein or inhibited by a G_i protein.

In the membrane, an important set of receptors activate, via a G_P protein, *phospholipase-C* (PL-C) which hydrolyzes phosphatidylinositol bisphosphate (PIP_2) to produce inositol (1,4,5)-trisphosphate *(IP_3)* and diglyceride *(DG)*. IP_3 is a second messenger, releasing Ca^{2+} from stores in the endoplasmic reticulum. DG is a specific activator of protein kinase-C (PK-C), a family of membrane-anchored Ser/Thr PKs which are Ca^{2+} and lipid-dependent. The phosphoinositide cycle, which reconstitutes PIP_2, involves a lithium-sensitive step. Hormones may also activate a methylase, modifying membrane phospholipids, and a phospholipase-A_2 (PL-A_2), releasing arachidonic acid which is transformed thereafter into prostaglandins and other related compounds. *Ca^{2+}*, acting as a "third messenger" following IP_3 increase, is also a second messenger for hormones whose receptors open Ca^{2+}-channels to extracellular calcium. Cytosol Ca^{2+}, mainly in the form of calcium–calmodulin complexes, is involved in the regulation of many enzymatic and non-enzymatic proteins, including protein kinases. Indeed, *protein phosphorylation* and dephosphorylation are involved in most hormonal responses. The auto- and heterologous phosphorylation of receptors can modify their function. Tyr-PK activity is involved in insulin and growth factor activities. Phosphorylation of chromatin proteins may be involved in the transcriptional response to membrane receptors. Many hormones, particularly those involved in the regulation of cell growth, have *pleiotypic effects*, with early and late responses in which several of the above-mentioned intracellular signals are implicated.

Antihormones essentially compete for endogenous hormone binding of the receptor. Steroid hormone antagonists are already used successfully in medicine. *RU 486*, a powerful contragestive agent and an antiglucocorticosteroid in vitro, stabilizes the hsp 90 containing, non-DNA binding, hetero-oligomeric forms of PR and GR. *Tamoxifen*, widely used in the treatment of breast cancer, has a beneficial effect surpassing that which would be expected from a simple antiestrogen. Only a few peptide hormone antagonists are available presently.

6. *Oncogenes (onc)*, which at present dominate basic cancer research, and their physiological counterparts, protooncogenes (cellular oncogenes, c-*onc*), correspond remarkably to most important components of hormonal systems. Many are homologous with hormones/growth factors (e.g., *sis*), others with membrane (*erb*-B, *neu*) and intracellular (*erb*-A) receptors, and yet others are related to G proteins (as is *ras*) or protein kinases (as is *src*); the synthesis of DNA-binding *onc* (*myc*, *fos*) is under hormone/growth factor regulation. In general, any given cellular function is regulated by several hormonal systems. Similarly, the generation of a cancer often requires the cooperation of at least two *oncogenes* for the transformation and immortalization of cells. Cellular oncogenes become pathological mostly by mutation and/or overproduction, regardless of the origin of the cancer. Hormones are *promoters* of the progression of cancers.

7. Progress in endocrine pathophysiology is highly dependent on *techniques* used to measure hormones, receptors, and the most significant parameters of the hormonal response. The use of *competitive binding* assays, and principally radioimmunoassays, has been revolutionary. It is now possible to measure hormones, analogs, and metabolites, with sufficient sensitivity and specificity, in most body fluids and in some tissues. Recent advances include the use of solid phase techniques, monoclonal antibodies, and non-radioactive methods (e.g., immunoenzymatic assay). Quantitation of receptors is obtained mostly by binding assays, using radioactive ligands, but the availability of appropriate tissue samples and the lack of sensitivity of assays may remain problems. Both *antibodies* (and especially monoclonal antibodies) against hormone receptors and hormone-regulated proteins, and *cDNA probes*, have greatly assisted recent studies in endocrine pathophysiology. In particular, they have permitted the histological detection and quantitative assessment of hormone receptors and endocrine responses. Direct examination of secretory and target cells is of particular interest, considering that the production of parahormones and autohormones cannot be evaluated properly by blood measurements, and also because tissues are heterogeneous (especially those which are pathological), and therefore better analyzed at the cellular level than in total homogenates. Hormone measurements are also limited by a time factor, since a single assay cannot account for variations in hormone levels occurring over a period of time. In addition, the entire hormonal network is so complex that, frequently, determinations of *several* hormones and/or related parameters (e.g., ACTH and cortisol, PTH and Ca^{2+}) are necessary in order to assess the significance of a borderline value. Indeed, *dynamic tests* are often useful, since, in stimulating or inhibiting endocrine function, they may reveal abnormalities. In the future, computer-assisted *mass spectrometry* may become an alternative to present methods, in providing strict physicochemical identification of measured compounds.

8. Classical hormonal *pathophysiology* encompasses most well characterized syndromes where *hyposecretion or hypersecretion* of one or several hormones is due to an organic or functional alteration of a gland. A number of endocrine diseases are *genetic* in origin. They are characterized by secretory abnormalities (frequently in the thyroid and adrenal glands), by abnormal metabolism (e.g., pseudohermaphroditism, caused by a deficiency in 5α-reductase), by a *receptor defect* (as in the case of testicular feminization (tfm) and other abnormalities of male phenotype where the androgen receptor is deficient), or by postreceptor defects (e.g., in pseudohypoparathyroidism, which is associated with a modification of a G protein). In contrast to abnormal hormone production, defects in hormone metabolism and in receptor or postreceptor components are causes of *"hormone resistance"*. Genetic disorders are probably involved in such syndromes as diabetes and obesity. The causes of several *autoimmune* diseases have been related to the activity of hormone-like antibodies (long-acting thyroid stimulator (LATS) in Graves' disease) and to a blockade of a hormone receptor (nACh-R in myasthenia gravis). Since insulin-dependent

diabetes mellitus (type 1) may be autoimmune in origin, early treatment by immunodepression has been suggested. New findings in *neuroendocrinology*, including the recent elucidation of the structure of neuromediator receptors and the synthesis of neurosteroids, open the way for a better understanding of such neurological and psychiatric disorders as Alzheimer's disease and depression. The hormonal parameters of *development*, particularly those which are related to puberty, menopause and ageing, are progressively being unveiled. Before assuming that there is a causal relationship between hormonal abnormalities and a given pathological feature which would be suggestive of a specific treatment, it is important to take into account the concept of *hormone permissiveness*. The evolution of some cancers (e.g., of the breast, endometrium, and prostate) is clearly influenced by hormones, which are promoters. Studies of hormone receptors and oncogene expression may result in improved disease management. Non-glandular, "ectopic" cancers produce hormones (e.g., lung tumors, which may secrete ACTH or ACTH-like peptides). Therapy and even prophylaxis, using hormones, is being proposed for "natural" processes, such as ageing, short stature, etc. Hormonal gene therapy has already been used effectively in some cases of infertility in animals. Major medical decisions are ahead of us which undoubtedly will raise important ethical issues.

9. *Evolutionary and comparative endocrinology studies* suggest that, while steroid and thyroid hormones are the same in all vertebrates, peptide hormones have evolved to a greater or lesser degree. Molecular structures which are homologous to hormones present in higher animals are found very early in phylogenesis, even in unicellular organisms where their biological significance is unknown. Receptors have also evolved. Among hormones, as among receptors, it is clear that whole families of similar molecules were derived from a single ancestral gene. Moreover, apart from *molecular evolution*, the *functional and anatomical evolution* of hormonal systems is quite extraordinary. Different functions can be accomplished by the same hormone in different species. In insects, hormones are relatively limited in number, and thus each is expected to play a diversified role. However, there is remarkable unity among hormonal systems in the animal kingdom. They control practically all biological functions in all animal species, notably those of reproduction, nutrition, and physical and behavioral activities. The coordinated evolution of hormones and receptors (in terms of structure and distribution) is striking. However, it must be remembered that "endocrine evolution is not an evolution of hormones, but an evolution of the uses to which they are put".

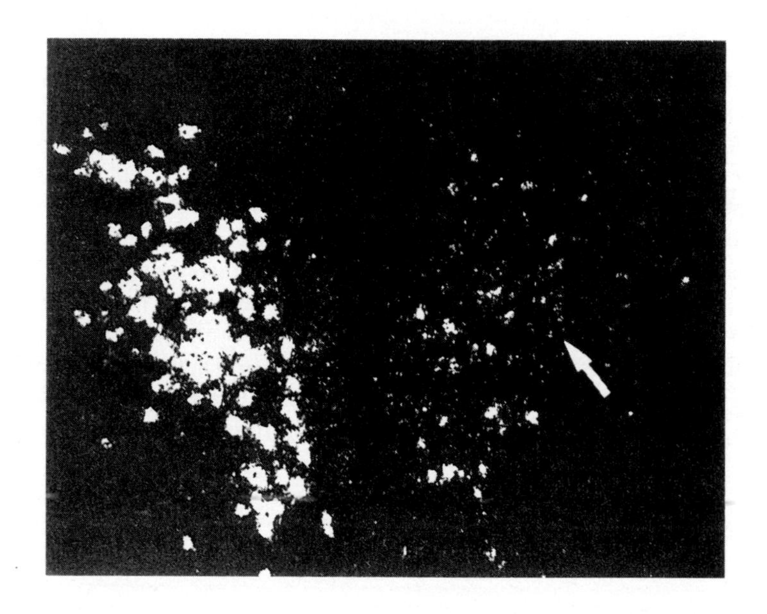

Hormonal feedback control visualized at the molecular level. In situ hybridization (darkfield illumination) of a specific probe to pro-TRH mRNA in the paraventricular nucleus of the rat. On the contralateral left side, the signals are greatly reduced due to a local implant of thyroid hormone (Dyess, E.M., Segerson, T.P., Liposits, Z., Paull, W.K., Kaplan, M.M., Wu, P., Jackson, I.M.D., and Lechan, R.M. (1988)). Triiodothyronine exerts direct cell-specific regulation of thyrotropin-releasing hormone gene expression in the hypothalamic paraventricular nucleus. *Endrocrinology* 123: 2291–2298.

Roger Guillemin

Commentary on Neuroendocrinology

Neuroendocrinology began, as a discipline of that name, in the 1930s, essentially due to the interest of a relatively small group of morphologists intrigued by their observations that some large neurons in the hypothalamus of all vertebrates showed a constant or cyclic appearance of granules. These were reminiscent of the protein secretory granules seen in well known secretory cells, such as those of the anterior pituitry gland, or of the endocrine or exocrine pancreas.

Neurohormones

Neurosecretion referred to what eventually was to be recognized as the secretory process for vasopressin and oxytocin. *Vasopressin* and *oxytocin*, two structurally homologous nonapeptides (described chemically in 1952), are each synthesized as part of a large precursor protein molecule, which also comprises the carrier proteins known as neurophysins. The neurophysins are packaged with the hormone, in secretory granules, in large neurons of the supraoptic and paraventricular nuclei of the hypothalamus. They are then carried by axoplasmic flow towards nerve endings as far as one cm away in the posterior lobe of the pituitary. There, in *nerve-to-capillary vessel contacts*, the neurosecretory granules are released into blood vessels (described in the 1950s) when triggered by an electrical nerve impulse originating, most probably, from the same neurons that synthesize the neurohormones.

Thus, there are otherwise classical neurons which are able to synthesize specific peptides that are released directly into the bloodstream, reaching receptors far away, in the distal tubule of the kidney, in myoepithelial cells of the breast tissue, or in the smooth muscle of the uterus. This anatomical and physiological arrangement has been the fundamental unit defining neuro-endocrinology. Actually, morphologists had also proposed, on the basis of their observations in *invertebrates* in the 1930s, that the process of neurosecretion was possibly *far more general* throughout the nervous system than the limited hypothalamo-hypo-physeal connection of the vertebrate brain.

The Discovery of Releasing Factors

In the mid-1940s, physiologists became interested in explaining the mechanisms controlling the secretions from the anterior lobe of the pituitary, attempting to understand their cyclic changes (ovarian or menstrual cycle, breeding seasons), secretory response to acute stress (Selye's alarm reaction), cold or heat (for the secretion of thyrotropin), etc. Knowledge of neurosecretion from the two magnocellular nuclei of the hypothalamus was, of course, the basis for the early working hypothesis proposed. After a good deal of confusion, it was concluded, in the 1970s, that neurons in the ventral hypothalamus (different from the magnocellular variety and hence called parvocellular) do synthesize and release *hypophysiotropic peptides* (unrelated to vasopressin or oxy-tocin) into a privileged *local portal system* of capillaries extending from the floor of the third ventricle (the median eminence) and terminating in the anterior pituitary. Since the early 1970s, several peptidic factors of hypothalamic origin have been isolated which have either positive or negative regulatory effects on the secretion of anterior pituitary hormones. Each is released into the portal capillaries, probably upon some electrical impulse originating from the neurosecretory neuron, itself triggered by multiple afferent fibers.

The releasing factors act, following binding to high affinity *plasma membrane receptors, on pituitary cells* (described from 1972), resulting in changes in both the synthesis and release of hypophyseal hormones.

The existence of *negative feedback* mechanisms is one of the fundamental tenets of neuroendocrinology. Peripheral hormones (p. 32) exhibit feedback at the level of both pituitary cells and hypothalamic and/or higher brain neurons. Many rather complex examples are found throughout this book.

The Hypothalamus Has Lost Its Privileged Position

Neuropeptides are *now* also detected in the *gastrointestinal tract* and *digestive glands*, and their distribution in the CNS is far more *widespread* than their original hypothalamic extraction had suggested (from 1973). Conversely, peptides originally recognized in the gastrointestinal tract are found distributed widely throughout the CNS. As remarkable as it may seem, we have essentially no knowledge of the function of these peptides in the central nervous system, but they obviously have a different role from that in the gastrointestinal tract. Characterization of opioid peptides (starting in 1975) and their subsequent localization in practically all systems of neurons and fibers involved in pain have demonstrated that they must serve some physiological functions. However, opioid peptides that are apparently unrelated to pain pathways have also been located in neuronal systems.

Multiple Neural Signals

Since 1980, many neurons have been shown to *contain*, and we must presume, to release and utilize functionally, *more than one biosynthetically unrelated peptide* (e.g., angiotensin, enkephalin, CRF). Most peptide-containing neurons *also* have the enzymatic equipment necessary to make classical *neurotransmitters*, such as acetylcholine, norepinephrine, serotonin, and dopamine, among others. The functional relationship between these neurotransmitters and peptides in the function of the neurons is currently unknown. It is possible that *peptides* act as *true neurotransmitters*, with biological *half-lives far longer*, and hence with a *distribution* space far *greater*, than those of the classic neurotransmitters, or they may act as *modulators* of the release of the classical neurotransmitters, or, yet, they may contribute to the fine control of the metabolism of a few neurons when released in their vicinity.

Diversity and Interrelationship

For the most part, the function in the CNS of these hundred or so peptides is not known. What is clear is that two remarkable and novel concepts are emerging. First, *a given molecule* may have totally different and possibly *unrelated biological activities, depending on its anatomical location* (for example, the different activities of somatostatin in the anterior pituitary, hippocampus, and pancreas, p. 574). Second, complementary to the previous notion, these *apparently unrelated* activities at different sites may be actually *correlated in some systematic way*, the parsimony being in the singularity of the single effector. For example, GnRH affects *reproductive functions* via the production of sex steroids and the direct modulation of *sexual behavior* (p. 267); CRF influences several components of responses to *stress* (p. 34 and p. 245), and angiotensin is involved in many components of the *sodium-water-blood pressure system* (p. 607).

Today and Tomorrow

So, what is neuroendocrinology today, and in which direction is it going? Neuroendocrinology has *almost* become a *redundant term* in itself. Indeed, we see the neuron as having all the capabilities of the endocrine cell to biosynthesize and release peptides. Local (paracrine) secretion, long-distance (hormonal) secretion and synaptic (neuromediator) secretion all appear to be *variations on the common theme of a means of conveying encoded information in the form of released molecules*. The neuron has the additional feature that it also rapidly transmits electrical impulses to localized targets. Current research will certainly unveil the connections between neurotransmitters and neurohormones, the various aspects of the processing of RNAs and polyproteins made by different groups of neurons, and contemporary molecular biology is already discovering many peptides of potential neurohormonal character. *Neuropharmacology*, specifically through the use of competitive antagonists, should help in the study of biochemical processes, including some that are involved in behavior.

In biomedicine, phenomenology is expanding, and reductionism is explaining more and more. The ultimate beauty of neuroendocrinology is that it may eventually begin to explain even the most complex brain functions.

CHAPTER II

The Hypothalamus-Pituitary System: An Anatomical Overview

Jean Racadot

Contents, Chapter II

The Hypothalamus-Pituitary System: An Anatomical Overview

1. Communication between the Hypothalamus and the Anterior Pituitary

Lack of Classical Innervation

The pituitary gland is in close contact with the base of the hypothalamus (Fig. II.1). The posterior lobe, although part of the pituitary, is made up of a ventral extension of the hypothalamus (Fig. II.2). As physiological and clinical studies have demonstrated an important influence of the nervous system on pituitary function, it was long thought that there was direct innervation of the pituitary gland originating in the diencephalon.

Innervation of visceral organs generally does not involve a typical synapse, but rather a less close and less structured arrangement. In all cases, an effect is triggered by the release of a neurotransmitter from an axon, which either stimulates or inhibits the target organ. The distance a neurotransmitter molecule travels is greater than the distance between the nerve terminal and target cell, but remains <10 μm.

Such innervation has never been found at the level of the anterior lobe. The only innervation is of cervical sympathetic origin for control of afferent vessels.

A Vascular Portal System

A portal system exists between the base of the hypothalamus and the *anterior* pituitary (Fig. II.3). It collects neurosecretory products and distributes them to the various cells which secrete anterior pituitary hormones.

Endocrine Secretion of Some Hypothalamic Neurons

The hypothalamus contains *two categories* of neurons. The *first* includes neurons, similar to those of other autonomic centers, which perform analogous functions: reception, integration, conduction and transmission of messages, by mechanisms which characterize the nervous system. The *other* group is different: although the neurons have the same general properties with respect to reception and conduction of messages, they also function as *endocrine* cells. Rather than small neurotransmitters, they synthesize peptides, and their axons terminate in a capillary network, allowing their secretory products to be released directly into the blood.

This is the *only* mechanism by which hypothalamic messages are transferred to the anterior pituitary. It includes numerous axons, coming from various cell groups in the hypothalamus, which converge in the median eminence. Blood from the capillaries then enters *portal veins, along the stalk*, within the pars tuberalis, and is finally distributed in the capillary network of the anterior lobe. Thus, *neurohormones* are transferred very rapidly by this vascular system to pituitary cells, and either stimulate or inhibit pituitary secretion.

Such an arrangement, as compared to that of classical autonomic innervation, involves a greater dilution of the messenger molecules. Not only are the distances traversed in the vascular network much longer (several mm), but their area of distribution is broader. The surface of the capillary bed of the anterior pituitary is as large as ~ 100 cm^2.

The system controlling the anterior pituitary consists of several groups of neurons. Their position in the

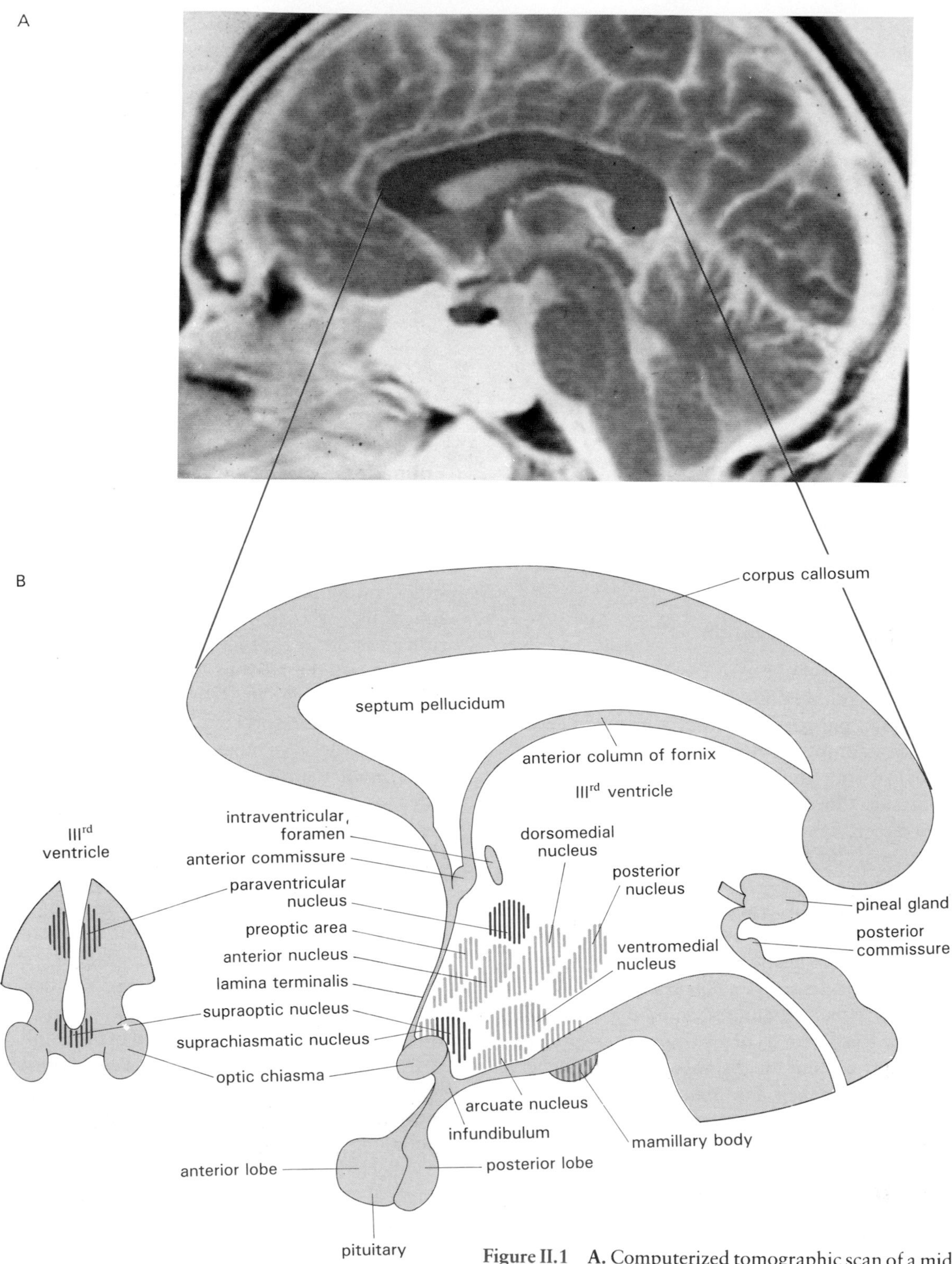

Figure II.1 **A.** Computerized tomographic scan of a midsagittal section of the human brain. **B.** Schematic representation of the human midbrain. To the left is a frontal section, demonstrating the topological relationship between the third ventricle and the supraoptic and paraventricular nuclei.

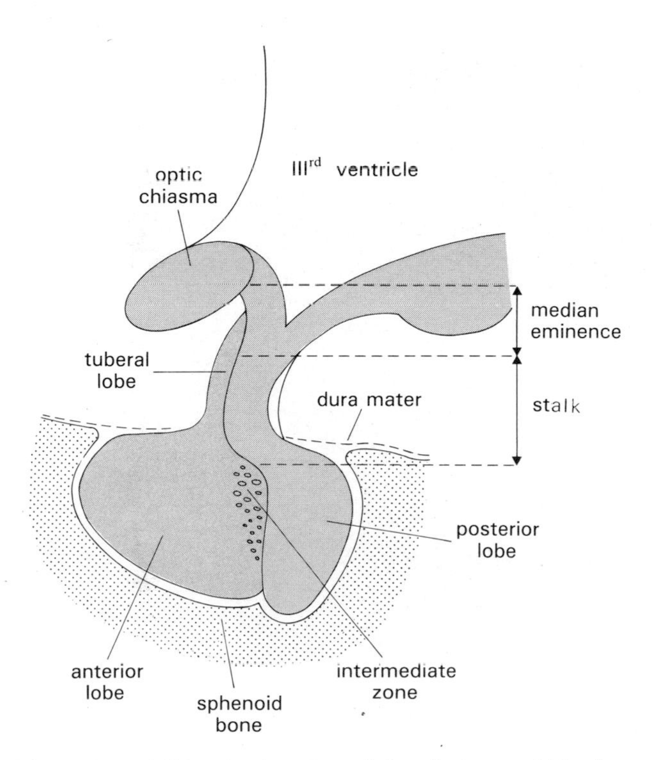

Figure II.2 Midsagittal section of the pituitary within the sella turcica of the sphenoid bone. The gland is divided into the anterior lobe, and the neural or posterior lobe. There is no intermediate lobe in the human, but the intermediate zone, with its numerous cysts, is a vestige of this structure. A superior extension, the tuberal lobe of the anterior pituitary, can be seen. Between the hypothalamus and pituitary is the pituitary stalk, or infundibulum.

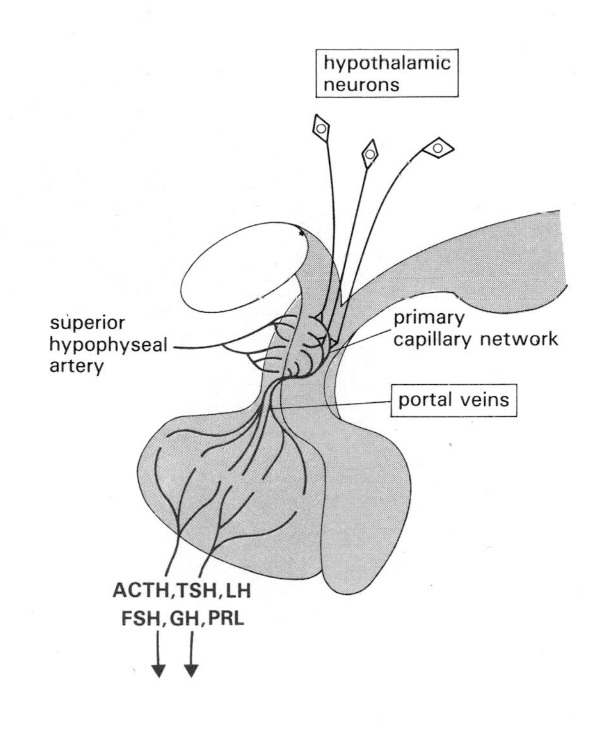

Figure II.3 Vascular connections between the hypothalamus and the anterior pituitary, involving the hypothalamus-pituitary portal system. This drawing, of what is in fact an extraordinarily complicated structure, has been simplified. It shows a series of anastomotic vessels, which make possible the physiological reversal of circulation in some areas.

hypothalamus is localized by immunohistochemical detection of the neurohormones they produce (Fig. II.4A and B). The function of these neurons, which form synapses with perycaryons or dendrites, is itself under the control of afferent fibers from near or distant regions. In addition, there is a distal system in the median eminence, near the capillary network, which involves various axonal fibers reaching the extremity of the neurosecretory axon. These axo-axonal synapses are able to modulate the release of neurohormones directly at the level of the capillary junction.

2. Organization of the Anterior Pituitary

The term *adenohypophysis* refers to three portions of a structure, found in almost all adult mammals, which is of ectodermal, pharyngeal origin (Rathke's pouch): the anterior lobe or *pars anterior*, the intermediate lobe or *pars intermedia*, and the tuberal lobe or *pars tuberalis* (Fig. II.2). In the adult human pituitary, there is no clear demarcation between the anterior and intermediate lobes, as they are fused. Only the presence of numerous cysts (residues of the pituitary cleft, which separates the two parts in the fetus) identify the caudal portion of the anterior lobe, generally known in the human as the *intermediate zone* or cystic zone. The anterior pituitary is situated in the sella turcica. However, the adenohypophysis also consists of the tuberal lobe, through which portal vessels pass. The pars tuberalis is essentially suprasellar, and completely surrounds the suprasellar portion of the stalk (stem) and the base of the median eminence. Although formed of cords as in the anterior lobe, this region has no standard pattern of cell type, and no clearly established secretory activity.

The *anterior lobe* is the largest portion of the pituitary (~400 mg). It is formed of anastomosed *cords* which are limited by a basal lamina and separated by capillaries and a perivascular space with variable

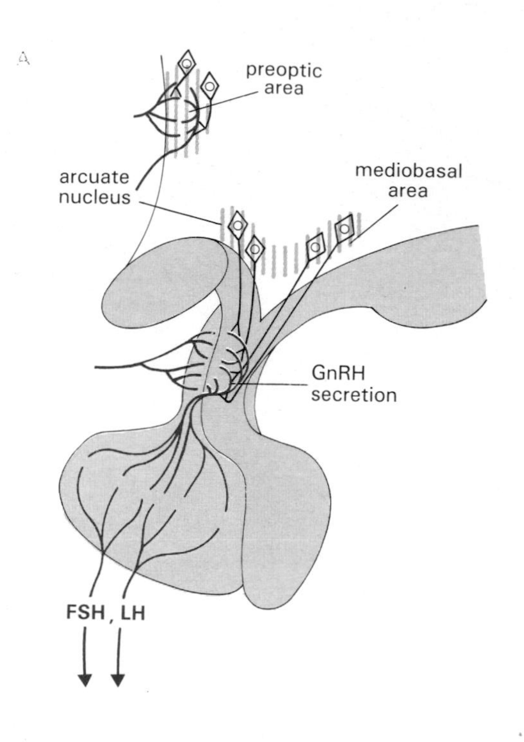

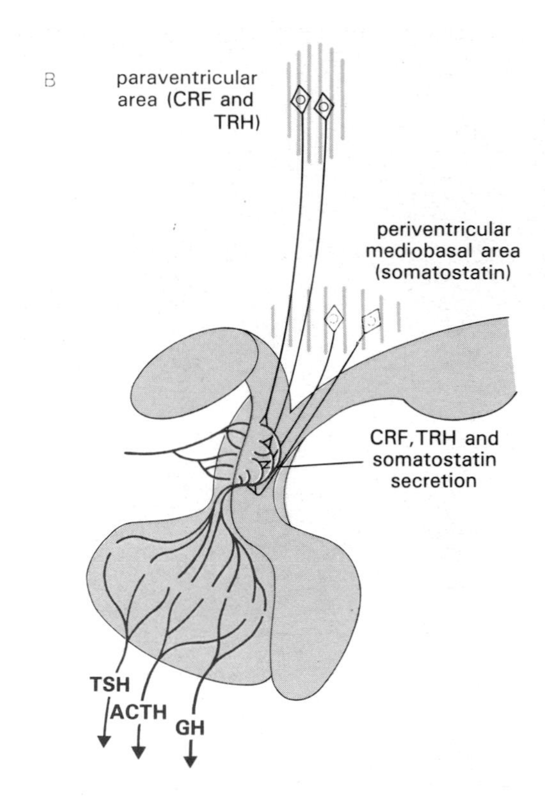

Figure II.4 **A.** The neurosecretory pathways of GnRH. The neurons of the mediobasal group reach the primary capillary network of the portal system, into which the neurohormone is released and carried to the anterior pituitary. There is also a group of GnRH neurons in the preoptic terminal area leading to capillaries in the lamina terminalis, however their functional significance is unknown. **B.** The neurosecretory pathway for TRH, CRF and somatostatin. TRH and CRF neurons are located mainly in the paraventricular nucleus, and somatostatin neurons are found in the mediobasal area.

amounts of connective tissue (Fig. II.5). These cords generally consist of several layers of cells, with several cellular types existing within them. Some are situated at the periphery, where there is a large area of contact with the basal lamina. Others, inside the cord, communicate only with the periphery, via extensions of varying size (Fig. II.5B).

Using optical microscopy, it is possible to distinguish several types of cells on the basis of the morphological and staining characteristics of their secretory granules (Fig. II.5A). Immunohistochemical identification of hormones is the best criterion by which these cells can be defined. *Five* categories of hormone-secreting cells have been identified: *somatotropic* cells for GH, *lactotropic* cells for prolactin, *corticotropic* cells for ACTH and its related peptides, *thyrotropic* cells for TSH, and *gonadotropic* cells for LH and FSH. There is also another category of cells, known as *follicular* cells, which apparently do not produce any hormone; they are capable of phagocytosis, and may play a role in ion exchange.

The developmental relationship between these cells is still poorly understood. It is generally agreed that, among gonadotropic cells, the majority produce and release both FSH and LH, while only a small number release just one of the two hormones. For somatotropic and lactotropic cells, it has been proposed that some are immunoreactive for both hormones in normal adult pituitary (this situation has been demonstrated more clearly in tumors secreting both GH and PRL).

The *release of hormones* from these cells occurs classically by the extrusion of secretory granules. The demonstration in several endocrine glands of the existence of two pools of hormones released under different conditions led to the concept that certain granules could be released preferentially. Undetected hormone may exist in cells in some form different from that stored in granules.

Most cords of the anterior pituitary contain several types of cells which are arranged in a nonrandom fashion, although their proportion may vary

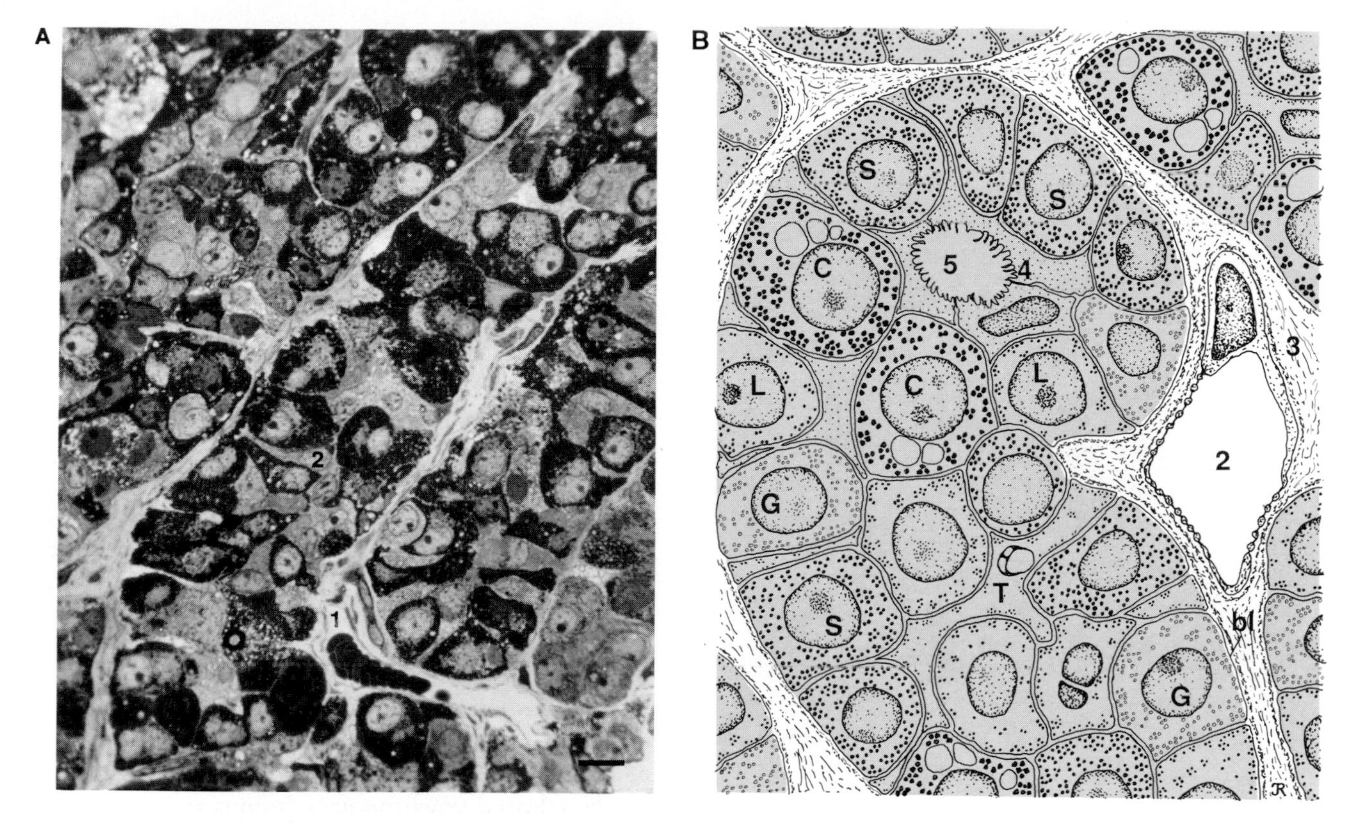

Figure II.5 A. Thin section of the anterior lobe of a human pituitary; 1: capillary with red blood cells. B. Schematic representation of an anterior pituitary cord; bl: basal lamina. 2: capillary with fenestrated endothelium. 3: pericapillary space. 4: follicular cell process. 5: lumen of a follicle limited by follicular cells. S: somatotropic cell. C: corticotropic cell. L: lactotropic cell. T: thyrotropic cell. G: gonadotropic cell. Note that some C and T cells contain vacuoles.

somewhat. Somatotropic and lactotropic cells are most numerous in the lateral wings of the pituitary, and corticotropic cells are concentrated mostly in the anterior and posterior portions of the mid-region, near the intermediate zone. The fact that *cells of different categories* coexist in the *same cord* suggests some functional interrelationship possibly involving paracrine activity; however this has not yet been documented in humans. Functional connections between gonadotropic and lactotropic cells have been shown in laboratory animals. It has also been shown that the α subunit of pituitary glycoprotein hormones may affect the secretion of lactotropic cells.

Stimulation of a secretory cell may lead, in addition to hormone synthesis and release, to hyperplasia. This process has been observed in lactotropic cells during pregnancy in mammals, including women. Their number increases greatly, resulting in a gain in pituitary weight of several hundred mg. Following pregnancy, these cells regress but do not disappear. The small cells form residual groups which are characteristic of women who have undergone several pregnancies. Hyperplasia of thyrotropic cells also occurs in congenital myxedema, leading to an increase in pituitary volume visible on X-rays of the skull.

3. The Hypothalamus-Postpituitary Neurosecretory Axis

The postpituitary neurosecretory system is similar to the hypothalamic-anterior pituitary system in certain respects, but there are also some differences.

A group of neurons is involved, recognized for a long time because of the large size of their cell bodies, which forms two pairs of *magnocellular* hypothalamic nuclei: the *supraoptic* and *paraventricular* nuclei (Fig. II.6). The axons of these neurons lead to the infundibulum. With the exception of a few axons which terminate near the capillaries of the median

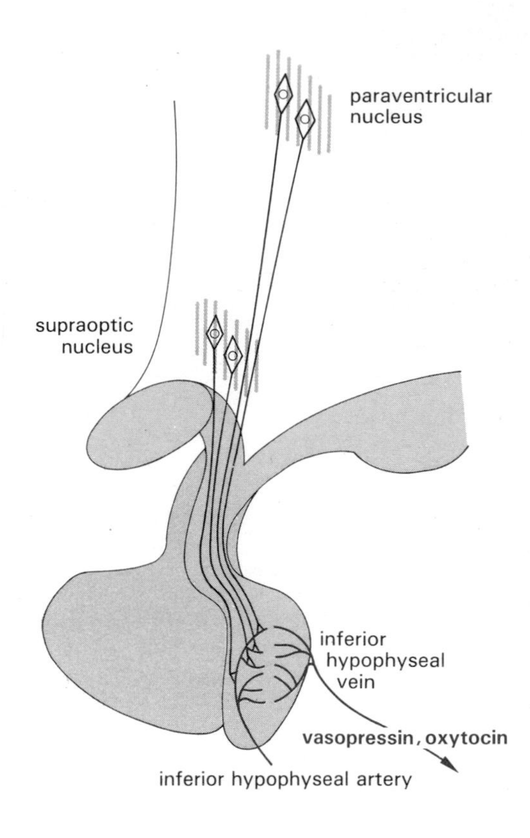

Figure II.6 Neural connections between the hypothalamus and posterior lobe of the pituitary. Vasopressin and oxytocin are synthesized in the magnocellular nuclei, transported, via long axons, to the posterior lobe, and are secreted into the general circulation.

eminence, most penetrate into the neurohypophysis. These fibers form the essential part of the pituitary stalk, terminating in the neural lobe or posterior pituitary, the most distal portion of the neurohypophysis, which is also located in the sella turcica (Fig. II.2).

The structure of the posterior pituitary is quite simple. Unlike the stalk, where nerve fibers are parallel, the axons of the neural lobe diverge in all directions, terminating near capillaries, thus forming a dense network. In addition to axons and capillaries, glial cells (pituicytes) are present, however their role is still unknown. There is little connective tissue around blood vessels. The arteries of the posterior pituitary are branches of the inferior pituitary arteries, themselves subsidiaries of the internal carotid. This vascularization essentially remains separate from that of the proximal part of the stalk and of the median eminence. Sometimes enlargements, called Herring's bodies (10 to 20 μm or greater), are found along axons (never at the extremities). They accumulate neurosecretory granules and function as a storage site. Lysosomes, which degrade secretory products, are also present.

In contrast to what occurs with the neurosecretory system controlling the anterior pituitary, neurohormones synthesized by the magnocellular nuclei have *remote targets*. As for other neuropeptides, the two neurohormones oxytocin and vasopressin pass from the axon into capillaries. However, since they are not associated with a portal system in the neural lobe, veins draining this area lead to the *general circulation*. The quantity of neurohormones released is relatively large. Altogether, the secretory cells form a mass which is clearly greater than that of the other neurosecretory nuclei. Posterior pituitary neurohormones circulating in the peripheral plasma at measurable levels act on distant organs (kidney for vasopressin, uterine myocytes and mammary cells for oxytocin). Direct stimulation by an action potential and *immediate release* of neurohormone at the axon terminal differentiate the secretion of neurohormones from that of glandular hormones. Given that groups of cells are connected by a rich synaptic network, neurohormone can be released synchronously from many cells (e.g., oxytocin during suckling).

General References

Bergland, R. M. and Page, R. B. (1978) Can the pituitary secrete directly to the brain? (Affirmative anatomical evidence). *Endocrinology* 102: 1325–1338.

Denef, C., Baes, M. and Schramme, C. (1986) Paracrine interactions in the anterior pituitary: role in the regulation of prolactin and growth hormone secretion. In: *Frontiers in Neuroendocrinology* (Ganong, W. F. and Martini, L., eds), vol. 9, Raven Press, New York, pp. 115–148.

Girod, C. (1983) Immunocytochemistry of the vertebrate adenohypophysis. In: *Handbook of Histochemistry* (Graumann, W. and Neumann, K, eds), vol. 8, suppl. V, Gustav Fischer, Stuttgart-New York.

Haymaker, W., Anderson, E. and Nauta, W. J. H., eds (1969) *The Hypothalamus*. Ch. Thomas publ., Springfield, Ill.

Knobil, E. and Sawyer, W. H., eds (1974) *Handbook of Physiology*, sect. 7, vol. IV. The pituitary gland and its neuroendocrine control. *Amer. Physiol. Soc.*, Washington D.C.

Kovacs, K., Horvath, E. and Ryan, N. (1981) Immunocytology of the human pituitary (De Lellis, ed) In: *Diagnostic Immunohistochemistry*. Masson USA Inc., New York, pp. 17–35.

Porter, J. C., Ondo, J. G. and Cramer, O. M. (1974) In: Knobil, E. and Sawyer, W. H. (*loc. cit.*), pp. 33–43.

Zimmerman, E. A. and Nilaver, G. (1984) The organization of neurosecretory pathways to the hypophysial portal system (Cammani, F. and Müller, E. E., eds) *Pituitary Hyperfunction: Physiopathology and Clinical Aspects*. Raven Press, New York, pp. 1–25.

Jacques Glowinski

Nerve Cells and Their Chemical Messages

In the vast domain of neurotransmission, major advances have been made in three complementary areas. Numerous *neurotransmitters* and their *receptors* have been discovered, a great *variety* of responses to these messages have been defined at the cellular level, and new information on *neuronal networks* has been obtained. As a result, there has been tremendous progress in the study of cerebral functions and dysfunctions, and in the understanding of the action of *psychotropic drugs*.

A Number of Neurotransmitters and Their Receptors

More than *50 molecules* which act as neurotransmitters or neuromodulators have been *identified*, and this number continues to grow. These molecules are grouped into *three families* (see Table 1): amino acids, amines and the flourishing group of neuropeptides, some of which are pituitary hormones or hypothalamic releasing factors.

Table 1 *The main neurotransmitters, or neuromodulators, divided into three classes.*

Amino Acids	*Amines*	*Neuropeptides*		
GABA	Acetylcholine	ACTH	Met-enkephalin	Prolactin
Glycine	Dopamine	Angiotensin II	Gastrin	Somatostatin
Glutamic acid	Norepinephrine	Bombesin	GnRH	Substance K
Aspartic acid	Epinephrine	Carnosine	Neuromedin K	Substance P
Proline	Serotonin	Cholecystokinin	Neurotensin	TRH
Taurine	Histamine	Dynorphin	Neuropeptide Y	VIP
		β-endorphin	Oxytocin	Vasopressin
		Leu-enkephalin		

Recently, it was found that *two or three* chemical mediators can exist *within the same neuron*. This observation has generated a great deal of excitement. Most frequently, neurons contain one classical chemical mediator (amino acid or monoamine) and one neuropeptide. The functional importance of the presence of two or three messengers within a neuron is not well understood. One suggestion is that *co-release* of a classical mediator and a neuropeptide depends on the frequency of stimulation of nerve fibers, and that the neuropeptide amplifies the response obtained by the classical neuromediator. On the other hand, the discovery of dendritic dopamine release has revealed that the transfer of information from neurons can be *bidirectional* (Fig. 1).

Among the *class of amino acids*, GABA (Fig. 2) and glutamic or aspartic acids are the chemical mediators released in most central synapses. GABA-ergic postsynaptic receptors (type A, p. 84), for which stimulation leads to an inhibition of target cells, have been studied in great detail since they are the sites of action of barbiturates and benzodiazepines which act on GABA-ergic transmission via different mechanisms. Presynaptic GABA-ergic receptors (type B), found most frequently on monoaminergic fibers (with a contrasting pharmacology to that of type A receptors), have been identified. Some derivatives of di- or tripeptides containing

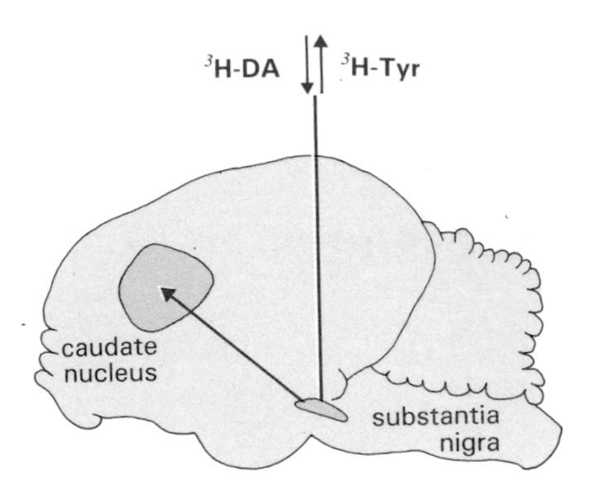

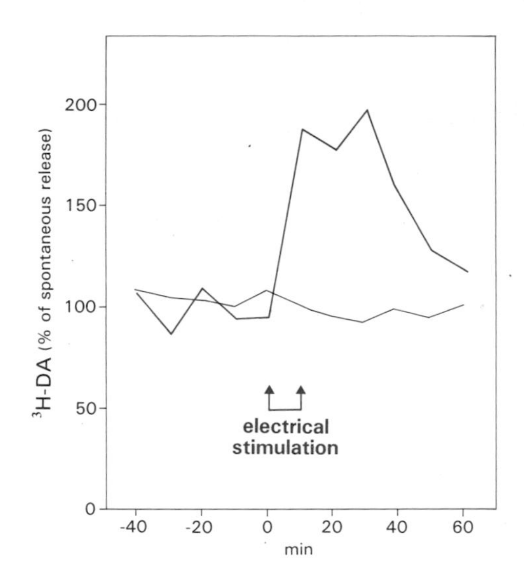

Figure 1 *Release of dopamine from dendrites of nigrostriatal dopaminergic neurons. The* dendritic release *of dopamine can be estimated in halothane-anesthetized cats implanted with a push-pull cannula in the substantia nigra by measuring the release of [^{3}H]dopamine (^{3}H-DA) continuously synthesized from [^{3}H]tyrosine (^{3}H-Tyr). The effect of electrical stimulation (10 min) of the pericruciate cortex on the dendritic release of newly synthesized [^{3}H]dopamine in the ipsilateral substantia nigra (LSN) is shown in stimulated animals.*

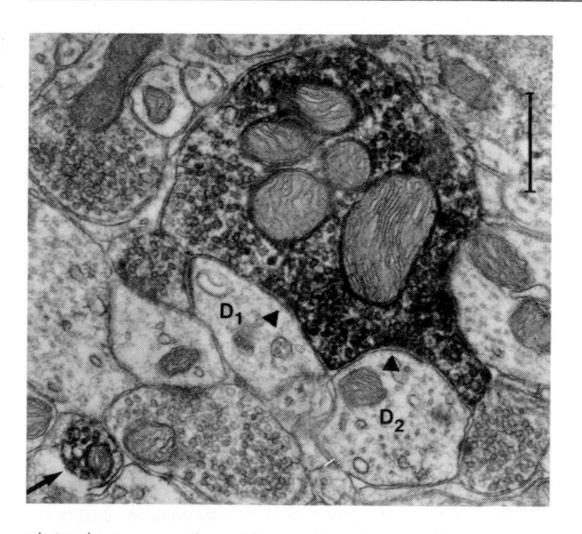

Figure 2 *GABA-ergic neurons can be identified by immunohistochemistry, using an antibody against glutamic acid decarboxylase (GAD). This electron micrograph of a GAD-immunoreactive axon terminal taken from the ventrolateral outgrowth of the medial accessory olive of the inferior olivary complex of the rat is in synaptic contact with two small dendritic profiles (D1 and D2). The synaptic complexes (arrowheads) are of the Gray type II, as is the case for about 80% of GAD-positive* boutons *in the inferior olive. The arrow points to an immunolabeled axonic profile which corresponds to a preterminal segment of a GABA axon. Note that the labeled bouton is filled with pleomorphic synaptic vesicles and abundant mitochondria. Bar=0.6 μm. (Courtesy of Dr. C. Sotelo, Hopital Pitié-Salpêtrière, Paris).*

glutamic or aspartic acids could act as excitatory neurotransmitters. The study of the mechanism of action of kainic acid, a neurotoxic agent which leads to the degeneration of different populations of neurons, and the discovery of several antagonists of excitatory amino acids have led to the description of several types of receptors for this class of neurotransmitters.

Many discoveries have been made in the field of *monoamines*. Diverse classes of dopaminergic receptors (Fig. 3) have been differentiated as a result of the use of new dopamine agonists and antagonists and in particular of novel neuroleptics in the benzamide series (e.g., sulpiride). The *first receptor* whose structure was determined was the *nicotinic acetylcholine receptor* (p. 84). More recently, the structure of the *β-adrenergic receptor* has been deduced from its cDNA (Fig. I.81). Endogenous peptides seem to regulate the specific *transport* of catecholamines or serotonin in nerve terminals, a mechanism which contributes to their inactivation. This could lead to the discovery of new antidepressants.

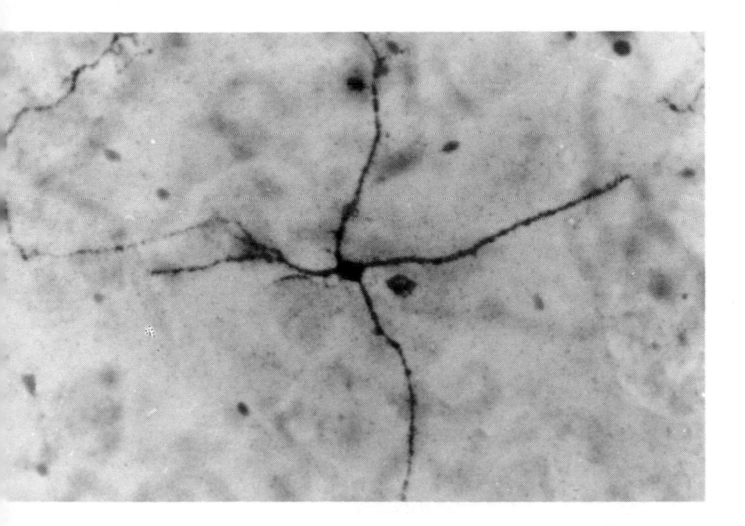

Figure 3 *Dopaminergic neurons can be identified and visualized, by autoradiography, following selective uptake and storage of tritiated dopamine in dissociated mesencephalic neurons of the mouse embryo grown in primary culture. (Courtesy of Dr. A. Prochiantz, Collège de France, Paris).*

In a relatively short period of time, spectacular results have been obtained with *neuropeptides*, especially with the *opiate* peptides and *tachykinines* (Fig. 4). These studies have led to a better comprehension of nociception and the monitoring of analgesia. However, the great diversity of brain structures innervated by neuronal systems containing these peptides suggests that they are involved in the regulation of multiple functions, including those of the limbic system. Molecular biological studies following the discovery of Leu- and Met-enkephalin have been very effective. *Three precursors* of opiate peptides are localized in *distinct neuronal systems* which have been mapped. These precursors are pro-opiomelanocortin, pro-enkephalin and pro-neoendorphin-dynorphin, which produce, respectively, β-endorphin, Met- and Leu-enkephalin and longer peptides, and finally α- and β-neoendorphin as well as dynorphin A and B (p. 186). These endogenous molecules, as well as morphine and its derivatives, exert their actions via one or more of the *four principal classes of opiate receptors.*

Substance P (Fig. 5), contained particularly in peripheral sensory fibers conducting nociceptive messages and in several central neuronal pathways, is one of the peptides for which strong evidence for its role as a neurotransmitter has been obtained. Thus, released by a calcium-dependent process, it exerts a potent excitatory influence on some spinal neurons. Substance P has also been shown to have neuromodulatory properties in the inferior mesenteric ganglion, since depolarization induced by substance P facilitates excitation of target cells provoked by afferent excitatory fibers, but cannot alone induce activation of these cells. Other peptides in the *tachykinin family* continue to be discovered, as evidenced by the recent identification of substance K and neuromedin K.

Variety of Responses

Classically, the transfer of information in the nervous system was only a matter of synaptic communication from one cell to another and included that of chemical neurotransmission.

Neurotransmitters are chemical mediators that generate a *brief* signal (order of ms); and act *within the synapse*, that is, in a well-defined space. However, as discussed already for substance P, in some cases chemical mediators may act as neuromodulators. Generally, *neuromodulators*, which modify the responses evoked by neurotransmitters, have a *longer* duration of action (seconds and even several minutes), and may act on receptors *far* from their site of production, to affect the properties of several populations of neurons or glial cells simultaneously. Indeed, recent studies have indicated that astrocytes, which exhibit several types of ion channels and receptors at their surface, are heterogeneous. Therefore, neuromodulators have properties intermediary between neurotransmitters and hormones, and participate in learning and memory processes at a neuronal level as well as in communications between neurons and glial cells. A *single substance* (e.g., monoamine or neuropeptide) may act as a neurotransmitter *or* neuromodulator, depending on the target cell and the receptor involved.

1	2	3	4	5	6
					NH_2
NH_2			pyro	pyro	Asp
Arg	NH_2	NH_2	Glu	Glu	Val
Pro	His	Asp	Pro	Ala	Pro
Lys	Lys	Met	Ser	Asp	Lys
Pro	Thr	His	Lys	Pro	Ser
Gln	Asp	Asp	Asp	Asn	Asp
Gln	Ser	Phe	Ala	Lys	Gln
Phe	Phe	Phe	Phe	Phe	Phe
Phe	Val	Val	Ile	Tyr	Val
Gly	Gly	Gly	Gly	Gly	Gly
Leu	Leu	Leu	Leu	Leu	Leu
Met	Met	Met	Met	Met	Met
$CONH_2$	$CONH_2$	$CONH_2$	$CONH_2$	$CONH_2$	$CONH_2$

Figure 4 *Neuropeptides of the tachykinin family. Some of these peptides (substance P, substance K and neuromedin K) are present in mammals. The tachykinins are characterized by their identical C-terminal sequence Phe-X-Gly-Leu-Met-NH_2. 1=substance P; 2=neurokinin A (substance K); 3=neurokinin B (neuromedin K); 4: eledoisin; 5: physalaemin; 6: kassinin. Substance P and substance K are encoded by the same gene, and neuromedin K by another, related gene. Their receptors are of the multiple membrane-spanning type, coupled to G protein (p. 79).*

The interaction of neurotransmitters or neuromodulators with their receptors results in the opening of *ion channels* or the formation (or mobilization) of *second messengers*. Various chemo-dependent ion channels intimately associated with, or constituted by receptor proteins have been demonstrated. Refined electrophysiological techniques allow single channel study. A number of second messenger molecules have now been described and include, as for hormones acting at the membrane level, cAMP, cGMP, diacylglycerol and calcium. These second messengers participate in the phosphorylation of diverse proteins, several of which are specific to certain populations of neurons. Such regulation may be important when one considers that the phosphorylated proteins could be receptors, ion channels, enzymes involved in the synthesis or release of chemical mediators, cytoskeletal proteins, proteins involved in the expression of genes, or endogenous inhibitors of phosphatases, such as DARPP 32 (a protein found exclusively in dopamine-nociceptive cells). Clearly, all these findings have led to more specific and quantitative approaches in neuropharmacology.

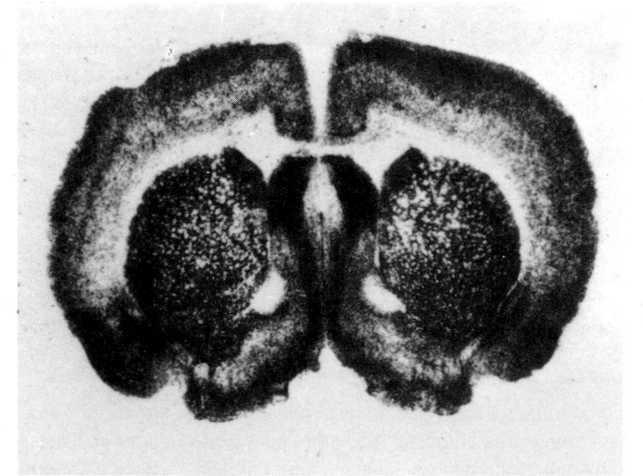

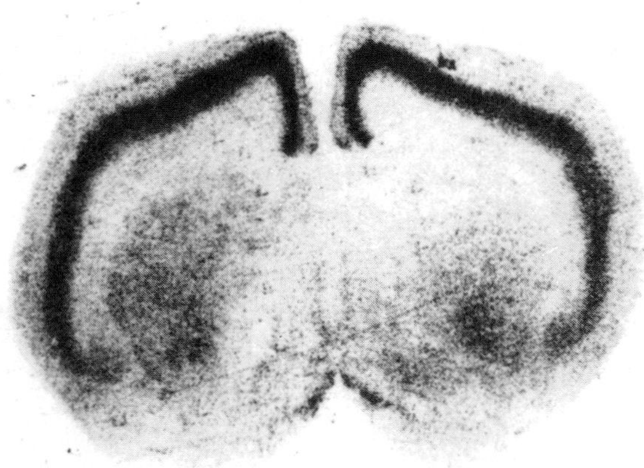

Figure 5 *Localization of substance P and neuromedin K receptors in the rat brain. The [^{125}I]Bolton-Hunter derivatives of substance P and Eledoisin can be used respectively to label substance P and neuromedin K receptors. Coronal slices of the rat brain (at the level of the striatum and the septum) were incubated with these ligands in conditions allowing their specific binding to the corresponding binding sites. Marked differences exist in the localization of [^{125}I]Bolton-Hunter substance P (left panel) and of [^{125}I]Bolton-Hunter Eledoisin (right panel) specific binding sites.*

Neuronal Networks and Their Regulation

Neurophysiology, neuropathology and psychiatry cannot be satisfied simply with the analysis of molecular or cellular events. An understanding of the *organization* of neuronal networks and their regulation is indispensable.

Biochemical *mapping* of the brain has recently become a very powerful technique. The anatomical distribution of various chemical mediators and neurotransmitters, as well as their receptors, in neuronal circuits of the brain has been described. For example, the discovery of dopaminergic innervation in the prefrontal cortex, a structure involved in cognitive functions, has furnished additional arguments in favor of the role of dopamine in *schizophrenia*. More recently, *Alzheimer's disease* has been attributed to the degeneration of cholinergic and somatostatinergic neurons innervating the cerebral cortex (p. 147).

Progressively, the principal interactions between neuronal systems have been identified. A method involving the *visualization* and quantification of *glucose utilization* in various brain structures has led to an overall vision of the neuronal network involved in some specific physiological and pharmacological situations. Finally, the discovery that *neurotoxins* have *selective* effects on certain neuronal populations has allowed the development of animal models which mimic some neurological disturbances in humans. Such models should help in the development of new therapeutic approaches. The best illustration is the observation that *Parkinson's disease* can be induced in monkeys by the administration of the neurotoxic compound MPTP, a contaminant of heroin, which selectively destroys neurons of the nigrostriatal dopaminergic pathway.

In conclusion, several discoveries have been made recently with the help of new pharmacological tools. One of the most fascinating developments is a new concept of the structural constituents of the brain. *A network* of neurons organized in a *rather diffuse and diversified* manner is *superimposed on* the precisely *"cabled" classical system*, with its large number of connections. The latter is essentially made of neurons rich in GABA or excitatory amino acids, which rapidly transfer important messages, fulfilling executive roles. The former network of neurons, which regulates the "cabled" system, is composed of aminergic and peptidergic neurons, exerting permissive or suppressive effects, and acting over a longer period of time. However, this dual classification system is too schematic; indeed, there are *often two or three distinct types of messengers per neuron* that have complementary functions. It is likely that both systems are expressed, but differentially, according to genetic and epigenetic factors.

References

Cheramy, A., Leviel, V. and Glowinski, J. (1981) Dendritic release of dopamine in the substantia nigra. *Nature* 289: 537–542

Florey, E. (1984) Synaptic and non-synaptic transmission: a historical perspective. *Neurochem. Res.* 9: 413–427

Hökfelt, T., Johansson, O. and Goldstein, M. (1984) Chemical anatomy of the brain. *Science* 225: 1326–1333

Iversen, L. L. (1979) The chemistry of the brain. *Sci. Amer.* 241: 118–129

Nestler, E. J., Vallas, S. I. and Greengard, P. (1984) Neuronal phosphoproteins. Physiological and clinical implications. *Science* 225: 1357–1364

Snyder, S. H. (1983) Molecular aspects of neurotransmitter receptors: an overview. In: *Handbook of Psychopharmacology*, vol. 17, (Iversen, L., Iversen, S. D. and Snyder, S. H., eds), Plenum Press, New York, 1–12

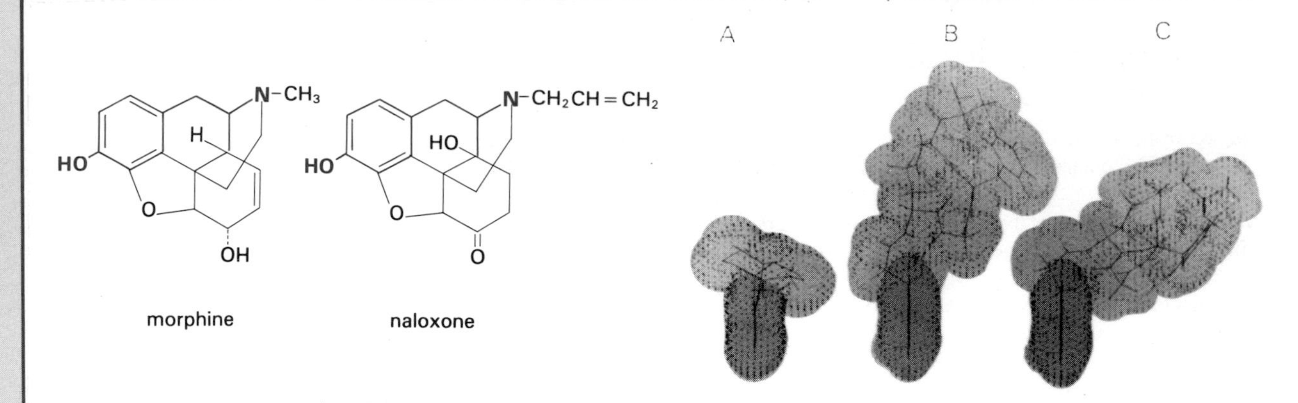

Figure 1 *Structure of morphine and naloxone. The tyrosine backbone is highlighted in red.*

Figure 2 *Three-dimensional representation of morphine and enkephalin. Visualization, on a monitor (program MANOSK), of the rigid structure of morphine (**A**), of the most stable conformation of enkephalin in solution (**B**), and of its postulated active conformation in the receptor (**C**). (Courtesy of Dr. J. P. Mornon, CNRS, Paris).*

Table 1 *The main opioid peptides*

Peptides derived from proenkephalin A or proenkephalin. (Four copies of Met-enk and one of each of the three other peptides per precursor molecule).	
	5
Met-enkephalin:	Tyr-Gly-Gly-Phe-Met
Leu-enkephalin:	Tyr-Gly-Gly-Phe-Leu
Heptapeptide:	Tyr-Gly-Gly-Phe-Met-Arg-Phe
Octapeptide:	Tyr-Gly-Gly-Phe-Met-Arg-Gly-Leu
Peptides derived from proenkephalin B or prodynorphin.	
	10
Dynorphin A:	Tyr-Gly-Gly-Phe-Leu-Arg-Arg-Ile-Pro-Lys-Leu-Lys-Trp-Asp-Asn-Gln
Dynorphin B‡:	Tyr-Gly-Gly-Phe-Leu-Arg-Arg-Gln-Phe-Lys-Val-Val-Thr
α-neoendorphin:	Tyr-Gly-Gly-Phe-Leu-Arg-Lys-Tyr-Pro-Lys
β-neoendorphin:	Tyr-Gly-Gly-Phe-Leu-Arg-Lys-Tyr-Pro
Peptide derived from POMC. (One copy of β-endorphin, which itself includes one sequence of Met-enk per molecule)	
β-endorphin:	Tyr-Gly-Gly-Phe-Met-Thr-Ser-Glu-Lys-Ser-Gln-Thr-Pro-Leu-Val-Thr-Leu-Phe-Lys-Asn-Ala-Ile-Ile-Lys-Asn-Ala-Tyr-Gly-Lys-Lys-Gly-Glu
	30

‡Also called rimorphin. Leu-morphine is C-terminally extended dynorphin B (29 aa).

Etienne-Emile Baulieu

From Morphine to Opioid Peptides

Pain, Stress, Reproduction

The discovery of opioid peptides represents a *landmark* in the biology of molecules that transfer information (reviewed in ref. 1). The search began for a CNS receptor for morphine and/or its antagonist naloxone[2–4] (Fig. 1). When a specific binding system became available, the endogenous ligands, *enkephalins*[5,6], were discovered (Table 1). Then followed the isolation of several molecules which have the same N–terminal structure but different C–terminal extensions. *Opioid peptides* have the same four amino acid N–terminal sequences, Tyr–Gly–Gly–Phe. The fifth amino acid is Leu or Met. The structure of *morphine* includes a phenol ring and an N–atom in a tyrosine–like arrangment (Fig. 1). This probably accounts for its ability to bind to opioid receptors (Fig. 2), which have been further diversified pharmacologically (Table 2).

All endogenous opioid peptides identified thus far are derived from *three precursors, POMC and proenkephalins A and B* (Fig. 1.23 and 1.24). These prohormones are of similar size and contain mutiple repeated units of different MSHs or enkephalins, segregated near the C–terminal and encoded by a single exon. There is considerable sequence homology between the two proenkephalins, which probably originate from a common ancestor by gene duplication. Radioimmunoassays must be verified thoroughly for specificity.

Pro-opioid peptides are distributed in many types of cells, especially in the brain and gastrointestinal tract, where prodynorphin derivatives are abundant (p. 583). POMC is found in high concentration in corticotropic cells of the anterior pituitary, and in many endocrine-related tissues (p. 231 and p. 477) (Fig. 3). Its differential processing is typical of a

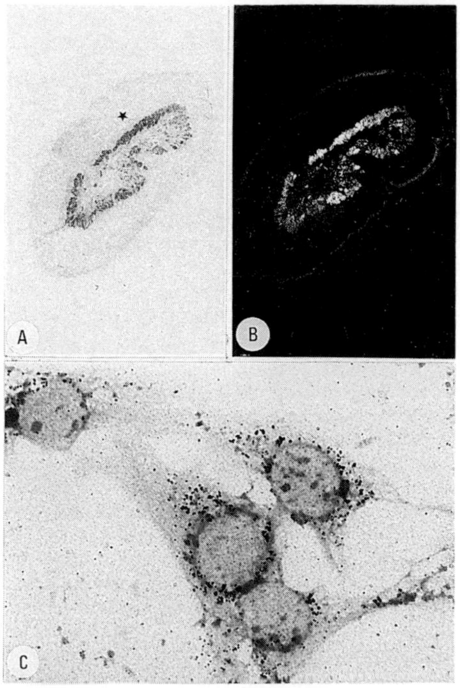

Figure 3 *Detection of enkephalin and enkephalin mRNA in cells of the adrenal medulla.* **A** *and* **B** *are adjacent sections of bovine adrenal gland.* **A**. *Met-enkephalin immunoreactivity, using immunoperoxidase in cell clusters located at the periphery of the adrenal medulla.* **B**. *Detection of preproenkephalin A mRNA (^{32}P cDNA, kindly provided by Dr. U. Gubler, Roche Institute for Molecular Biology, Nutley, N.J.). Dark-field illumination. Bar = 1 mm. Comparison between* **A** *and* **B** *shows enkephalin immunoreactivity and enkephalin mRNA in the same cells, indicating that enkephalin found in the adrenal is synthesized in situ, and that the cells in the center do not express the enkephalin gene. (The asterix indicates the adrenal cortex).* **C**. *Detection of preproenkephalin A mRNA in the cytoplasm of bovine adrenal medulla cells in culture. Bar = 1mm. (Bloch, B., Popovici, D., LeGuellec, E., Normand, S., Chouham, A. F. and Bihlen, P. (1986)* J. Neurosci. Res. *16: 183–200).*

Table 2 *The opioid receptors: relative affinities of morphine, antagonistic naloxone, and opioid peptides for the three most important pharmacologically defined opioid receptors.*

	M (μ)	Δ (δ)	K (κ)
Morphine	++	+	−
Enkephalins	+	++	−
Dynorphins		+	++
β-endorphin	++	+	−
Naloxone	++	+	−

++ means high affinity; + and − indicate low or no affinity. Natural peptides, with ++, are probably the endogenous ligands of the corresponding receptors. Other receptors have been partially characterized (e.g., the ε receptor, which preferentially binds endorphin gene products, and the specific haloperidol-sensitive σ receptor).

mechanism for generating diversity in different cell types (p. 27). The fundamental *raison d'être* of the wide and unequal distribution of POMC and proenkephalins in the nervous system is not fully understood. Its release after acupuncture has been related to beneficial effects of the procedure. Opioid peptides have *analgesic* properties, causing a decrease in motor activity and decreased responsiveness to noxious stimuli. They are often synthesized, in neurons, together with other neuropeptides and/or neurotransmitters. These peptides are abundant in various parts of the hypothalamus; they modify the function of the pulse generator oscillator, acting to decrease the frequency of LH pulses during the luteal phase[7], and modify the secretion of several hypothalamic-hypophyseal hormones (Table 3).

The *metabolism* of enkephalins, the most abundant of opioid peptides, is mainly due to enkephalinases, which are *metallopeptidases*. Specific inhibitors of these enzymes, for example kelatorphan, may produce physiological analgesia by decreasing inactivating metabolism[8,9]. The regulatory activities of opioid peptides are far from being understood, considering the mutiple chemical structures and sites of synthesis, the fact that they do not pass the blood-brain barrier, and the wide distribution of receptors.

Table 3 *Effect of opioid peptides on hormone secretion.*

ACTH ↓	PRL ↑
Vasopressin/oxytocin ↓	GH ↑
FSH/LH ↓	TSH ↑

References

1. Bertagna, X., ed (1986) Les peptides opioïdes. *Ann. Endocrinol.*47: 69–114.
2. Pert, C.B. and Snyder, S.H. (1973) Opiate receptor: demonstration in nervous tissue. *Science* 179: 1011–1014.
3. Simon, E.J., Hiller, J.M. and Edelman, I. (1973) Stereospecific binding of the potent narcotic analgesic ^{3}H-etorphine in rat brain homogenates. *Proc. Natl. Acad. Sci, USA* 69: 1835–1837.
4. Terenius, L. (1973) Characteristic of the "receptor" for narcotic analgesics in synaptic plasma membrane fraction from rat brain. *Acta Pharmacol. Toxicol.* 33: 377–384.
5. Hughes, J., Smith, T.W., Kosterlitz, H.W., Fothergill, L.A., Morgan, B.A. and Morris, H.R. (1975) Identification of two related pentapeptides from the brain with potent opiate agonist activity. *Nature* 258. 577–579.
6. Li, C.H. and Chung, D. (1976) Isolation and structure of an untriakontapeptide with opiate activity from camel pituitary glands. *Proc. Natl. Acad. Sci., USA* 73: 1145–1148.
7. Ferin, M., Van Vugt, D. and Wardlaw, S. (1984) The hypothalamic control of the menstrual cycle and the role of endogenous opioid peptides. *Rec. Prog. Horm. Res.* 40: 441–485.
8. Roques, B.P. (1985) Inhibiteurs d'enkephalinase et exploration moléculaire des differences entre sites actifs de l'enkephalinase et de l'enzyme de conversion de l'angiotensine. *J. Pharmacol.* 16, suppl. 1: 5–31.
9. Lecomte, J.M., Costentin, J., Vlaiculescu, A., Chaillet, P., Marcais-Collado, H., Llorens-Cortes, C., Leboyer, M., Gros, C. and Schwartz, J.C. (1986) Pharmacology of acetorphan and other "enkephalinase" inhibitors. In: *Innovative Approaches in Drug Research* (Harms, A.F., ed), Elsevier Science Publishers, Amsterdam, pp. 315–329.

CHAPTER III

Growth Hormone and Prolactin

Paul A. Kelly

Contents, Chapter III

Growth Hormone and Prolactin

Growth hormone and prolactin, in most animals including humans, together with placental lactogen (PL or chorionic somatomammotropin (CS)), are members of a family of polypeptide hormones which, based on amino acid sequence homology, were reported in 1971 to have arisen by duplication of an ancestral gene[1]. Recent observations using recombinant DNA approaches have in fact confirmed this theory.

Growth hormone and prolactin are produced by cells of the anterior pituitary. Numerous functions have been attributed to both hormones, although GH is most often associated with an overall stimulation of body growth and metabolism, and PRL with the induction and maintenance of lactation in mammals.

Although a great deal of information on the chemistry and molecular biology of these hormones has recently been gathered, *very little is known* about the actual *mechanisms* by which they induce their *actions* in their respective target cells.

1. Historical Background

The pituitary gland was originally thought to serve a function of lubrification for the nose cavities, and was later considered a vestigial structure. In 1886, Marie[2] first described and named *acromegaly* (p. 211) as a discrete disease. Although the importance of the pituitary was not appreciated as a causal factor in the disease at that time, the association of a pituitary tumor with the disease soon followed. At the end of the nineteenth century, it was recognized that *gigantism* (p. 211) and acromegaly represented the expression of the same disease at different ages. During the next two decades, the possible importance of the pituitary in *dwarfism* (p. 210) was recognized. Soon thereafter, a direct role of the pituitary was confirmed by cessation of growth in young hypophysectomized dogs.

Bovine GH was *first* isolated by Li and collaborators[3] in 1945. The isolation of GH from human pituitaries in the mid-1950s represented a major breakthrough, since the treatment of hypopituitary children with non-human GH preparations prior to this date had met with little success.

Early in the twentieth century, marked changes were observed in the histology of the anterior pituitary gland of women during pregnancy. The pregnancy cell was identified as a new cell type, which, during late pregnancy and postpartum periods, constitutes the most common pituitary cell type, later identified as *lactotropic cells* (p. 192). Prolactin was identified by Stricker and Grueter[4], in 1928, by the demonstration that bovine pituitary extracts could induce milk secretion in rabbits. Independently, in 1930, Corner observed milk secretion in rabbits treated with sheep pituitary extracts. Soon thereafter, a fraction of bovine pituitary extracts capable of stimulating growth of pigeon crop sacs was isolated and named *prolactin*[5]. Prolactin has since been identified in all

1. Niall, H.D., Hogan, M.L., Sayer, R., Rosenblum, I.Y. and Greenwood, F.C. (1971) Sequences of pituitary and placental lactogenic and growth hormones. Evolution from a primordial peptide by gene duplication. *Proc. Natl. Acad. Sci., USA* 68: 866–869.
2. Marie, P. (1886) Sur deux cas d'acromégalie: hypertrophie singulière non congénitale des extrémités supérieures, inférieures et céphaliques. *Rev. Med., Paris* 6: 297–333.
3. Li, C.H., Evans, H.M. and Simpson, M.E. (1945) Isolation and properties of the anterior hypophyseal growth hormone. *J. Biol. Chem.* 159: 353–366.
4. Stricker, P. and Grueter, F. (1928) Action du lobe antérieur de l'hypophyse sur la montée laiteuse. *C.R. Soc. Biol.* 99: 1978–1980.
5. Riddle, O., Bates, R.W. and Dykshorn, S.W. (1932) A new hormone of the anterior pituitary. *Proc. Soc. Exp. Biol. Med.* 29: 1211–1212.

mammals studied thus far, as well as in birds, reptiles, amphibians, and fish. This ubiquitous hormone appears to exist in all vertebrates, and over 85 biological functions have been attributed to PRL[6].

There was some question in the past whether or not human PRL was a discrete entity. Human GH preparations were lactogenic by conventional bioassays, and all early attempts to separate GH and PRL activities failed. There were, however, strong clinical, histological and immunological data suggesting that the two hormones were, in fact, separate entities. Finally, human PRL was successfully isolated and purified[7,8], which allowed the pathophysiology of PRL to be examined.

2. Morphology

2.1 Somatotropic Cells

Pituitary cells which secrete human GH are abundant; they contain many large round or oval membrane-bound secretory granules 350 to 500 nm in diameter (p.178). Somatotropic cells, which account for *35 to 45% of all pituitary cells*, are found in greater number in the lateral wings of the anterior pituitary. GH-secreting cells can be stained with orange G, but can be localized more specifically with an immunocytological technique[9].

2.2 Lactotropic Cells

Lactotropic cells are *less numerous* than somatotropic cells. They can be identified by either erythrosin or carmosin stains. Immunological staining reveals numerous secretory granules 275 to 350 nm in diameter which are round or oval[9]. Tumor cells may be somewhat different. In sections of a PRL adenoma (p. 213), the secretory granules vary from 150 to 700 nm. In the normal pituitary, lactotropic cells usually develop peripherally to somatotropic cells. The increase in the size of the pituitary that occurs during *pregnancy* is due principally to the *proliferation* of lactotropic cells. After delivery, the pituitary returns to normal size.

3. The Growth Hormone and Prolactin Gene Family

3.1 Structure of GH, PRL, and PL

Growth hormone, prolactin and placental lactogen form a *family* of polypeptide hormones with significant amino acid sequence homology, which also share some biological activities.

Human GH, which comprises as much as *10% of the dry weight of the pituitary*, is a single chain polypeptide of 191 aa, with two disulfide (S-S) bonds, between Cys-53 and Cys-165, and Cys-182 and Cys-189 (Fig. III.1). Human PL has a very similar structure, with the same number of amino acids and S-S bonds. Of the total 191 aa in hGH and hPL, 161 are identical, while 23 are highly or moderately compatible.

Human prolactin is related to both of these molecules, but, with an amino terminal extension, it is slightly longer, consisting of 199 aa (Fig. III.2). In addition to the two S-S bonds at the same relative positions, there is a third, at the amino terminal, between Cys-4 and Cys-11. *The high degree of structural homology* between GH and PL and PRL is shown in Fig. III.3.

3.2 Structure of GH, PRL, and PL Genes

It was in fact this high degree of structural similarity between GH, PRL and PL which first led to the concept that the corresponding genes were derived from a common precursor[1]. The sequence homology among the cDNAs of the mRNAs for human, rat, and bovine GH and PRL as well as human PL confirms this[10]. In addition, it has been observed that hGH[11], hPL-1 and

6. Nicoll, C. S. and Bern, H. A. (1972) On the actions of prolactin among the vertebrates: is there a common denominator? In: *Lactogenic Hormones* (Wolstenholme, G. E. W. and Knight, J., eds), Churchill-Livingstone, London, pp. 299–317.
7. Lewis, U. J., Singh, R. N. P., Sinha, Y. N. and Vanderlaan, W. P. (1971) Electrophoretic evidence for human prolactin. *J. Clin. Endocrinol. Metab.* 33: 153–156.
8. Hwang, P., Guyda, H. and Friesen, H. (1972) Purification of human prolactin. *J. Biol. Chem.* 247: 1955–1958.
9. Pelletier, G., Robert, F. and Hardy, J. (1978) Identification of human anterior pituitary cells by immunoelectron microscopy. *J. Clin Endocrinol. Metab.* 46: 534–542.
10. Miller, W. L. and Eberhardt, N. L. (1983) Structure and evolution of the growth hormone gene family. *Endocrine Rev.* 4: 97–130.
11. DeNoto, F. M., Moore, D. D. and Goodman, H. M. (1981) Human growth hormone DNA sequence and mRNA structure: possible alternative splicing. *Nucl. Acid Res.* 9: 3719–3730.

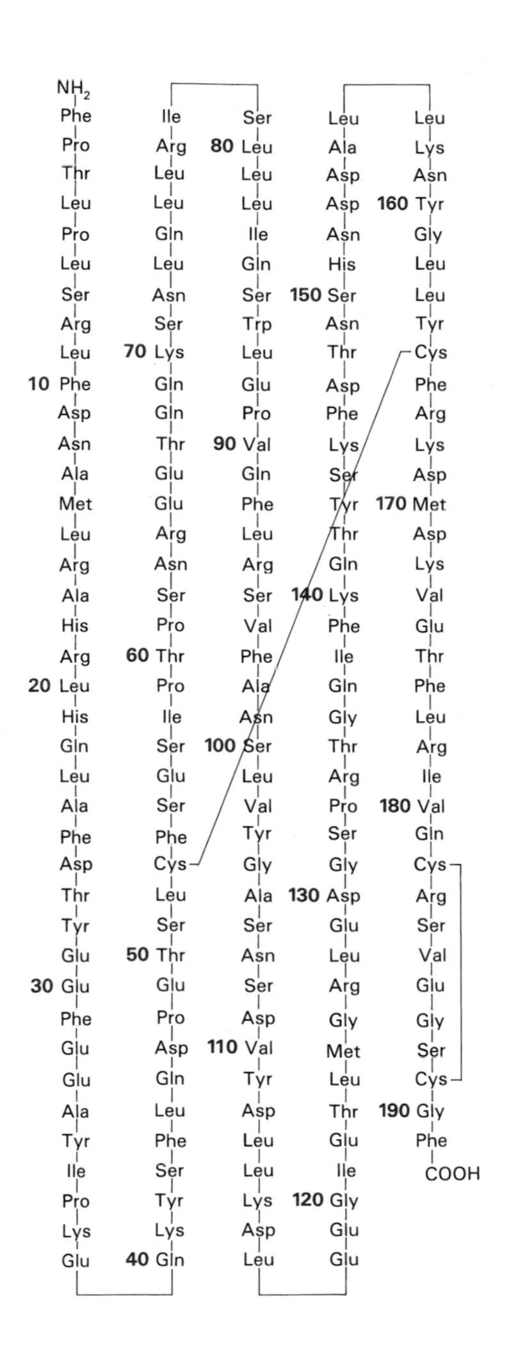

Figure III.1 Structure of human GH. It consists of 191 aa, and has two disulfide bridges.

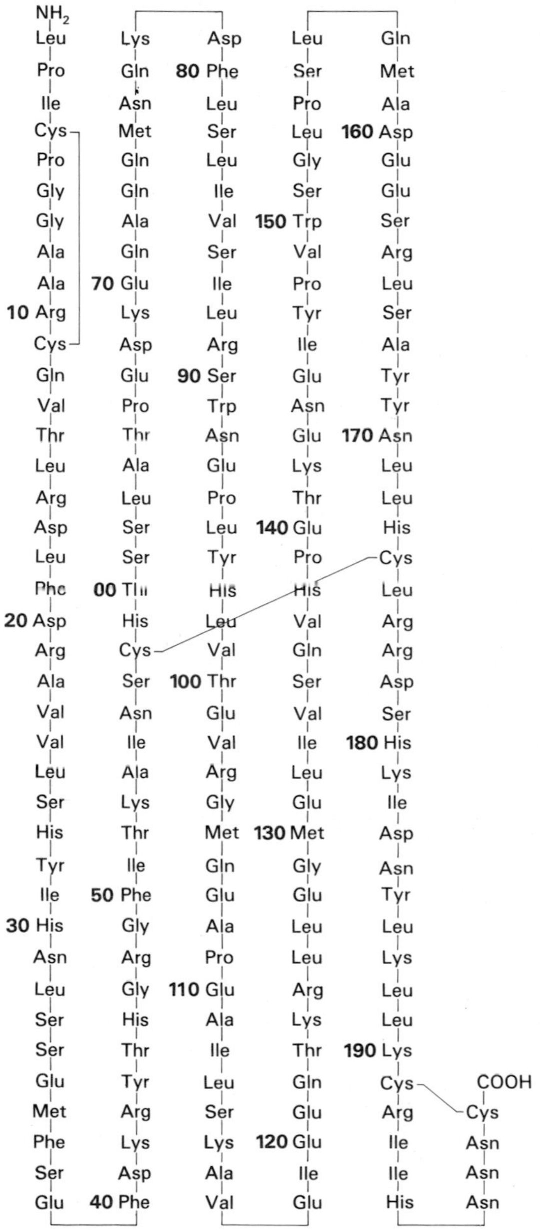

Figure III.2 Structure of human PRL. It consists of 199 aa, and has three disulfide bridges.

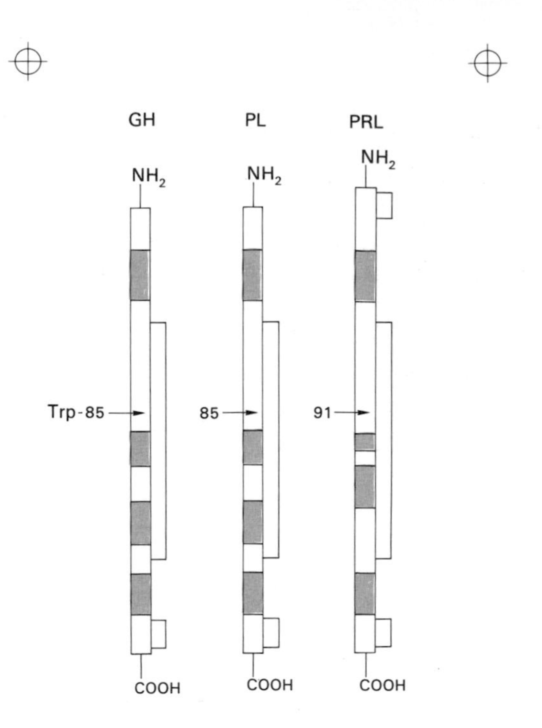

Figure III.3 Schematic representation of the structural homology between human GH, PRL, and PL. The colored regions represent the homologous sequences of amino acids seen in the three hormones. Note the similar location of both these homologous regions and of the disulfide bridges.

hPL-2[12] and hPRL[13] genes consist of five exons and four introns, and that they have highly conserved exon-intron boundaries (Fig. III.4). Note that the hPRL gene is much longer, spanning ~10 kb, due to the increased intron length. Human Alu middle repetitive sequences are found in both hGH and hPRL genes. For hGH, they only appear in the untranslated 3′ flanking region, whereas for hPRL, they are almost 10 times longer (~2 kb) and are found in both the 3′ flanking region and in intron D. There is a 92% sequence homology between the nucleotides encoding human GH and PL, while hPRL mRNA has only slightly over 40% homology with the nucleotides of the other two hormones.

Information derived from mRNA and DNA sequence analysis for human as well as animal *GHs and PRLs* suggests that gene duplication of a *common evolutionary ancestor* occured some *400 million years* ago, resulting in the two separate hormones (Fig.

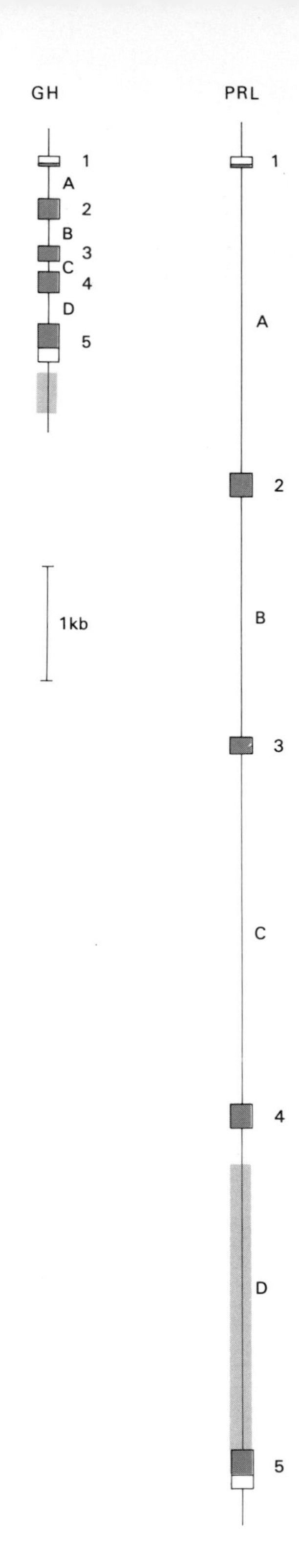

Figure III.4 Structure of the human growth hormone and prolactin genes. Note the difference in the length of the introns in two members of this gene family. Exons are identified by numbers, and introns by letters. (Moore, D. D., Walker, M. D., Diamond, D. J., Conkling, M. A. and Goodman, H. M. (1982) Structure, expression and evolution of growth hormone genes. *Rec. Prog. Horm. Res.* 38: 197–222; and ref. 13).

12. Selby, M. J., Barta, A., Baxter, J. D., Bell, G. I. and Eberhardt, N. L. (1984) Analysis of a major human chorionic somammotropin gene: evidence for two functional promoter elements. *J. Biol. Chem.* 259: 13131–13138.
13. Truong, A. T., Duez, C., Belayew, A. R., Pictet, R., Bell, G. I. and Martial, J. A. (1984) Isolation and characterization of the human prolactin gene. *EMBO. J.* 3: 429–437.

III.5). The ancestral genes for GH and PL appear to have duplicated much *more recently*, after mammalian speciation which occurred between 85 and 100 million years ago, although it is difficult to pinpoint

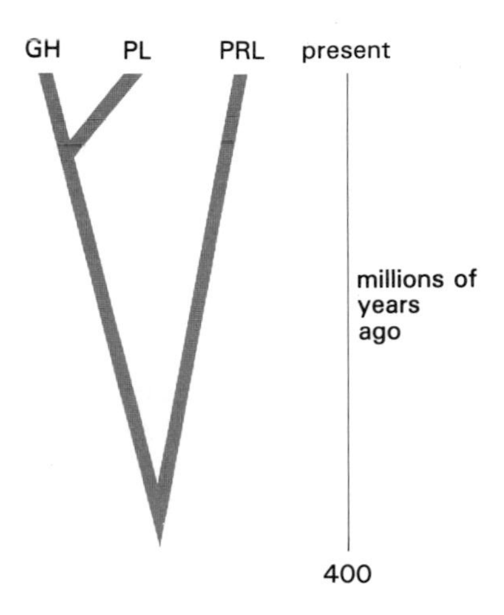

Figure III.5 Proposed evolution of GH, PRL and PL genes in humans, as determined by comparison of amino acid and nucleic acid sequences of the three hormones.

the exact time due to insufficient animal data and because gene conversion tends to invalidate the evolutionary clock[10]. The human *GH and PL genes* exist in *multiple* copies and are closely linked on *chromosome 17*, while the *PRL gene* is a *single* copy on *chromosome 6*. Nucleotide sequence data suggests that other mammalian PLs appear to be more closely related to PRLs rather than to GHs, and therefore hPL seems to be the exception to the rule.

The organization of the GH/PL/PRL gene family has suggested a possible model explaining the evolution of this family[10,12]. The ancestral gene contained an exon and a promoter region, possibly on a separate exon which was duplicated. Subsequently, an intron was removed between two of the original ancestral genes, accounting for two separate structural elements in exon 5. An exon different from the others was introduced and the gene was duplicated, resulting in the GH/PL and PRL precursor genes. Divergence of these precursor genes followed, and the final step involved the introduction of distinct regulatory sequences, which could explain the differential regulation of the three hormones.

Other Growth Hormone Related Genes

Mouse fibroblasts have been shown to produce a mRNA as a proliferative response to serum or PDGF. This mRNA is transcribed into a 25 kDa protein, *proliferin*, which shows a small but significant homology with prolactin and growth hormone[14]. This observation suggests that the GH/PL/PRL set of genes may in fact be evolutionarily related to other genes which together form a much larger family of growth-related genes. A similar but separate prolactin family of genes may also exist.

3.3 Different Forms of GH and PRL

Growth Hormone

There is a high degree of *heterogeneity* of the *hGH* molecule in pituitary extracts and plasma. The human pituitary contains from 5 to 10 mg of GH, produced exclusively in somatotropic cells. The 191 aa, 22 kDa form accounts for the majority of hGH activity. Some more acidic forms of hGH with increased biological activity have been observed in pituitary extracts, although they are not found in plasma.

One of these variants of potential molecular and clinical interest is 20 K hGH[15]. Interestingly, none of the variants of GH appear to be coded for by variant genes. The *20 K form* is not derived from 22 kDa hGH by post-translational modification, but rather is encoded by a separate species of mRNA derived from the same gene. Alternative processing and splicing of pre-mRNA to mRNA results in the removal of the nucleotides encoding amino acids 32–46 as well as those corresponding to the second intron. The frequency of occurrence is such that 10 to 20% of hGH in the pituitary is of the 20 K variant. None, however, appears in the peripheral circulation. The 20 K form of GH has the same weight gain and lactogenic activities as native hGH, however it does not possess the carbohydrate-regulating properties normally associated with hGH. Therefore, it *lacks the diabetogenic* properties of the intact hormone. This

14. Linzer, D. I. H. and Nathans, D. (1984) Nucleotide sequence of a growth-released mRNA encoding a member of the prolactin-growth hormone family. *Proc. Natl. Acad. Sci., USA* 81: 4255–4259.
15. Lewis, U. J., Dunn, J. T., Bonewald, L. F., Seavey, B. K. and Vanderlaan, W. P. (1978) A naturally occurring structural variant of human growth hormone. *J. Biol. Chem.* 253: 2679–2687.

localizes the carbohydrate activity of hGH to exon 3, which encodes the region of the molecule missing in the 20 K form.

Recently, a placental variant of hGH was identified by using a monoclonal antibody specific to the N-terminal region of the molecule[16]. *Placental hGH*, which may be coded by the hGH-variant gene (previously considered unexpressed), increases during the second half of pregnancy.

Prolactin

The human pituitary contains approximately 100 μg of prolactin. This hormone has also been reported to exist in different forms within the pituitary. A type of alternative processing similar to that seen for hGH exists for human prolactin, although it results only in a difference of a single amino acid in the leader peptide. In rats, a 16 kDa variant has been identified which has increased mitotic activity. This form apparently develops by an enzymatic cleavage of native 23 kDa prolactin. Very little is known about the secretion of this 16 kDa variant. *Both* animal and human PRLs appear in the pituitary in a *glycosylated* form. Glycosylated hPRL has a M_r of ~25,000. The carbohydrate unit is probably linked to Asn-31.

Growth hormone and prolactin exist in the peripheral circulation primarily in the monomeric form. However, "*big*" (dimer) and "*big-big*" (oligomeric) forms of each hormone are observed routinely. These are important because they are frequently measured by radioimmunoassay, but their biological activity is lower than the monomeric form of the hormone.

4. Metabolism

The half-life of both growth hormone and prolactin in plasma is 20 to 30 min. Human GH is cleared much faster than hPRL (125 compared to 45 ml/m^2 of body area per min). The metabolic clearance is similar in both normal men and women; however, GH clearance is reduced in patients with hypothyroidism or diabetes mellitus.

16. Hennen, G., Frankenne, F., Closset, J., Gomez, F., Pirens, G. and El Khayat, N. (1985) A human placental GH: increasing levels during second half of pregnancy with pituitary GH suppression as revealed by monoclonal antibody radioimmunoassays. *Int. J. Fert.* 30: 27–33.

5. Actions

The actions of growth hormone and prolactin are *numerous* and *varied*.

5.1 Receptors

The initial actions of both GH and PRL involve the binding of the polypeptide to specific receptors located in the *plasma membrane* of the target cell. Specific high affinity receptors for GH and PRL have been identified in all known target tissues for each hormone.

Receptors for *GH* have been detected in human *liver* and *lymphocytes*, as well as in adipose tissue, thymocytes, ovary and corpus luteum of subprimate mammals (Table III.1). It has been difficult to identify *PRL* receptors in human tissues, especially in normal *mammary* epithelial cells, although they have been clearly identified in human breast cancer biopsies as well as in continuous cell cultures of human breast cancers. In other mammals, such as rabbit, rat and mouse, PRL receptors are distributed in a *large number of tissues*, as illustrated in Table III.1[17].

Table III.1 Tissue distribution of GH and PRL receptors

GH Receptors	PRL Receptors
Human liver	Normal mammary gland
Adipose tissue	Mammary tumor
Lymphocytes	Liver
Thymocytes	Pancreas
Ovary	Kidney
Corpus luteum	Adrenal
	Placenta
	Ovary: granulosa cells and corpus luteum
	Testis: Leydig cells
	Epididymis
	Seminal vesicle
	Prostate
	Pancreas
	Lymphocytes
	Choroid plexus
	Hypothalamus

17. Hughes, J. P., Elsholtz, H. P. and Friesen, H. G. (1985) Growth hormone and prolactin receptors. In: *Polypeptide Hormone Receptors* (Posner, B. I., ed), Marcel Dekker, New York, pp. 157–199.

Structure

Both growth hormone and prolactin receptors are glycoproteins. Recent cloning and sequencing of their cDNAs have revealed that these receptors are monomeric, spanning the membrane once (Fig. III.6). The extracellular region contains seven cysteines and five potential *N*-linked glycosylation sites in the GH receptor, and five cysteines and three potential Asn sites in the PRL receptor. In the human, the overall size of both receptors is similar. There is strong sequence homology near the first two pairs of cysteines in the extracellular regions as well as in short spans in the cytoplasmic regions, Thus, GH and PRL receptors belong to a separate class of single membrane-spanning, non-tyrosine kinase hormone receptors. Their cDNAs encode proteins with a theoretical M_r of ~69,000 (GH) and ~67,000 (PRL). The glycosylated mature forms of the GH and PRL receptors have a M_r of ~130,000 and ~75,000, respectively. The large discrepancy between the apparent size of the GH receptor and that derived from the cDNA sequence cannot be explained entirely by glycosylation and the covalent attachment of ubiquitin. Smaller forms, probably derived by proteolysis, are observed for both receptors, although the cDNA of a second, shorter form (M_r ~40,000) of the PRL receptor is also found in the rat.

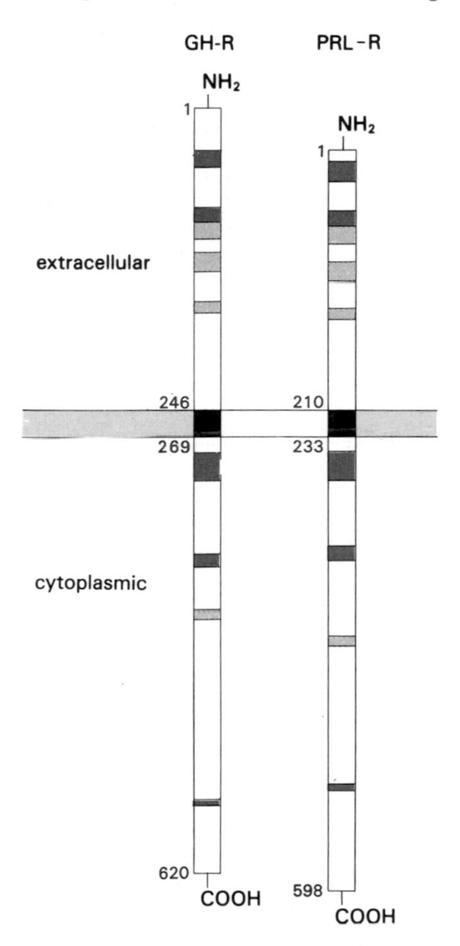

Figure III.6 Comparative structure of the human growth hormone and prolactin receptors. Both receptors span the membrane only once. Regions of high homology are shown in red and those of moderate homology are indicated in pink (see also Fig. I.78). (Leung, D. W., Spencer, S. A., Cachianes, G., Hammonds, R. G., Collins, C., Henzel, W. J., Barnard, R., Waters, M. J. and Wood, W. I. (1987) Growth hormone receptor and serum binding protein: purification, cloning and expression. *Nature* 330: 537–543; and Boutin, J. M., Edery, M., Shirota, M., Jolicoeur, C., Lesueur, L., Ali, S., Gould, D., Djiane, J. and Kelly, P. A. (1989) Identification of a cDNA encoding a long form of prolactin receptor in human hepatoma and breast cancer cells. *Mol. Endocrinol.* 3: 1455–1461).

The human and rabbit have a *serum GH-binding protein* (Table I.9 and p. 206). The N-terminal sequence for the rabbit binding protein is identical to that of the extracellular region of the GH receptor, suggesting that this serum protein may be produced by receptor proteolysis.

Regulation of Receptor Number

The regulation of GH and PRL receptors has been most extensively studied in the liver. There is a marked increase of both following puberty and during pregnancy and lactation[18]. The circulating concentration of peripheral factors, such as sex hormones, may be important regulators.

GH and PRL receptors appear to respond to the level of their own, endogenous hormone. Receptors can be *up- and/or down-regulated*, depending on the concentration of the hormone interacting with the receptors. Most of the studies on receptor regulation have been carried out in animals, although receptor regulation by the homologous hormone in humans has been demonstrated in cultures of lymphocytes for hGH and in breast cancer cells for hPRL.

Mechanism of Action Unknown

Although a great deal of information has been gathered on hormone receptor interaction for GH and PRL, very little is known regarding the mechanism by which these receptors mediate their action (Fig. III.7).

18. Kelly, P. A., Posner, B. I., Tsushima, T. and Friesen, H. G. (1974) Studies of insulin, growth hormone and prolactin binding: ontogenesis, effects of sex and pregnancy. *Endocrinology* 95: 532–539.

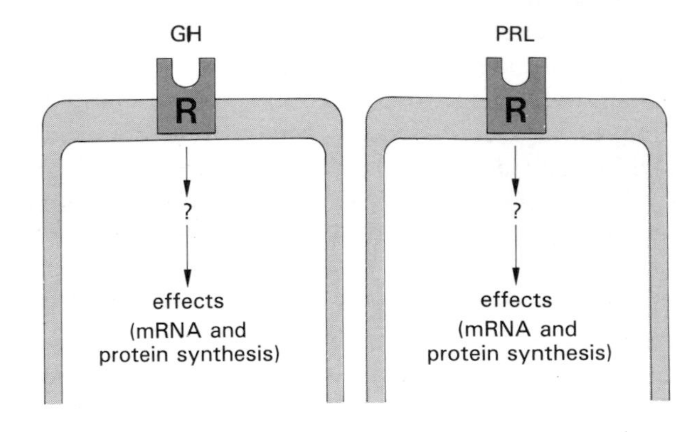

Figure III.7 Mechanism of action of GH and PRL. No intracellular mechanisms have been identified for either hormone.

It is clear that *neither the action of GH nor of PRL involves changes in the adenylate cyclase system*, as is true for many other hormone-receptor systems. Both GH and PRL provoke varied events within their respective target cells, including the stimulation of DNA synthesis and cell replication in addition to their specific effects on gene transcription. All of these are late events. *Neither* changes in *phosphoinositide* metabolism *nor* in *tyrosine kinase* activity appear to be involved in the mechanism of action of these hormones.

5.2 Growth Hormone

The most spectacular effect of GH is, as its name implies, the stimulation of *skeletal and soft tissue growth*. However, GH has a number of specific effects which can be divided into *two categories* (Table III.2). First are those effects related to *growth* which are mediated by *IGF-I (somatomedin)*; and second, those effects on carbohydrate and lipid *metabolism* which appear to be *direct* and mediated by receptors within the target tissues.

Table III.2 Principal actions of GH.

Action	Tissue
Direct ‡	
IGF-I production	Liver and fibroblasts in culture
Protein synthesis	Liver
Amino acid transport	Liver, muscle, and adipose tissue
Lipolysis	Adipose tissue
Carbohydrate metabolism (hyperglycemia)	Liver
Indirect (via IGFs)	
Chondrogenesis	Cartilage
Skeletal growth	Bone
Protein synthesis	Soft tissues
General cell growth	

‡Primate growth hormones are also lactogenic and possess all effects of prolactin on the mammary gland listed in Table III.3.

Indirect Effects (by IGFs)

One of the direct actions of GH is to stimulate the production of IGFs by the *liver* as well as by some other cells. This growth factor was originally discovered because the direct addition of GH to cartilage in vitro failed to stimulate growth and mitotic activity. Originally known as *sulfation factor* for its ability to stimulate sulfate incorporation into cartilage[19], these growth factors are now known as insulin-like growth factors or somatomedins. Two discrete peptides, each with a molecular weight of about 7500, have been identified, and their amino acid sequences have been determined. As can be seen in Figure III.8, these polypeptides have a structure similar to that of proinsulin; in consequence, the terms insulin-like growth factors (IGF) I and II have been accepted. The growth-related effects associated with GH are mediated by these two factors, and most specifically by IGF-I, which appears to be under direct GH regulation.

IGFs circulate in *plasma* bound to high molecular weight *binding proteins*. This appears to protect them from proteolysis, thus giving them a relatively long half-life. Two *specific receptors for IGFs* have been identified. The first binds IGF-I in preference to IGF-II, and only binds insulin weakly. A second receptor preferentially binds IGF-II over IGF-I, and has no affinity for insulin. The IGF-I receptor is structurally similar in both molecular weight and subunit composition to that of the insulin receptor, whereas the IGF-II receptor is a single polypeptide of high molecular weight (Fig. I.78).

IGFs are *strong mitogens*, and are essential for the growth of some cell lines in serum-free medium. Although many of the actions associated with IGFs are

19. Solomon, W. D. and Daughaday, W. H. (1957) A hormonally controlled serum factor which stimulates sulfate incorporation by cartilage in vitro. *J. Lab. Clin. Med.* 49: 845–836.

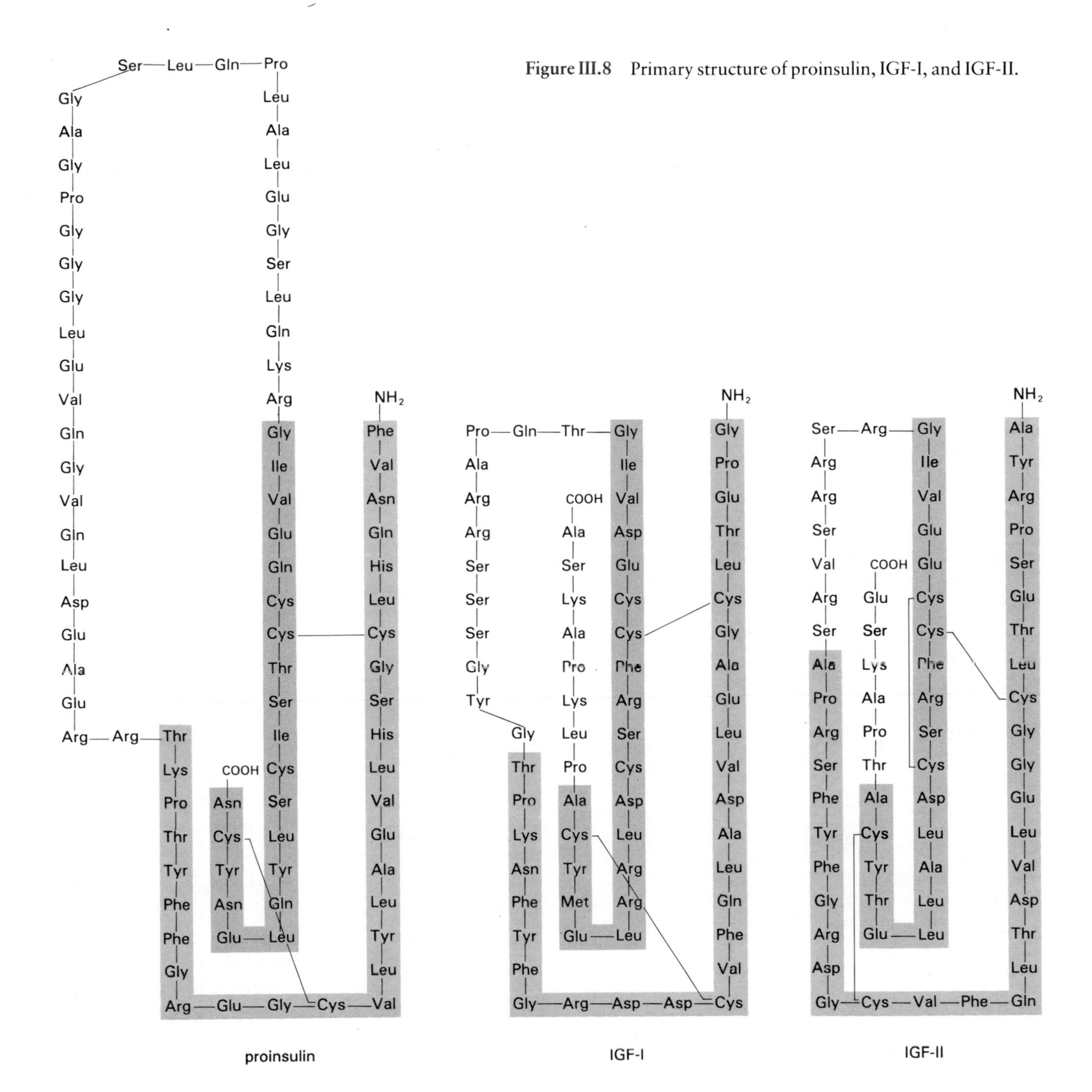

Figure III.8 Primary structure of proinsulin, IGF-I, and IGF-II.

in vitro effects, administration of IGF-I to hypophysectomized rats has been shown to have a direct effect on the stimulation of body weight, hypophyseal cartilage width, and DNA synthesis in cartilage[20].

The liver probably represents the major site of production of both IGFs and their specific binding proteins, although other cell types, such as pituitary cells and fibroblasts, are also capable of producing IGFs.

Serum concentrations of IGF-I, measured by RIA, are relatively constant over an extended period, and do not show the fluctuations associated with growth hormone concentrations. There are marked *changes* in the level of *IGF-I during development*, as shown in Figure III.9. Fetal levels are relatively low, but decrease even more just prior to parturition. Levels in

20. Schoenle, E., Zads, J., Humble, R.E. (1982) Insulin-like growth factor 1 stimulates growth in hypophysectomized rats. *Nature* 296: 252–253.

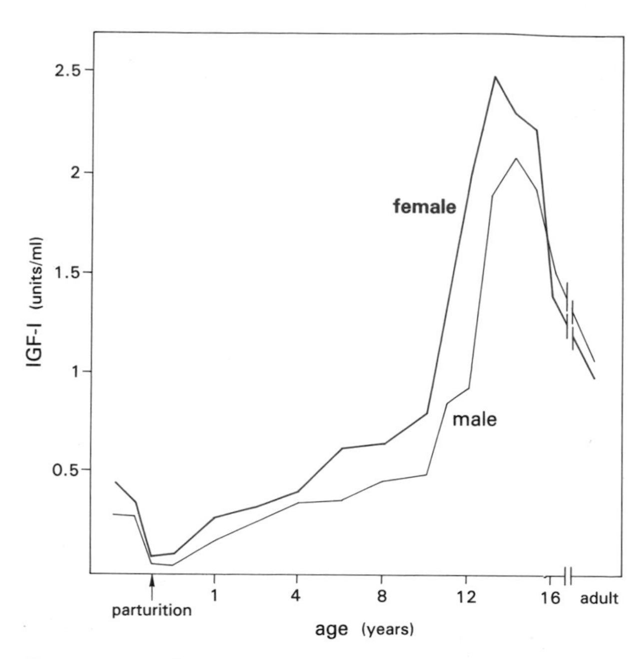

Figure III.9 Fluctuations in serum concentrations of IGF-I in males and females, as a function of age. (Adapted from Bala, R.M., Lopatka, J., Leung, L.A., McCoy, E. and McArthur, R.G. (1981) Serum immunoreactive somatomedin levels in normal adults, pregnant women at term, children at various ages and children with constitutionally delayed growth. *J. Clin. Endocrinol. Metab.* 52: 508–512).

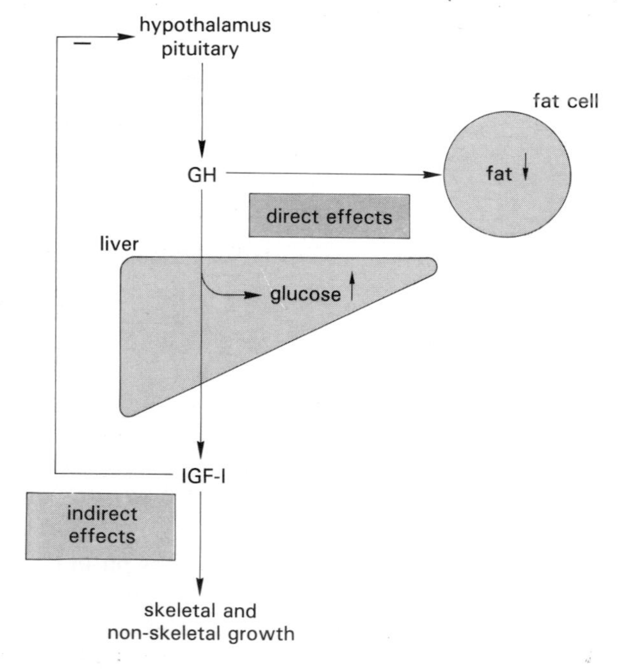

Figure III.10 Direct and indirect physiological actions of hGH. Direct actions, frequently antagonistic to insulin and synergistic to cortisol, include lipolytic and diabetogenic actions, as well as the stimulation of hepatic enzymes. Indirect actions, often insulin-like and antagonized by cortisol, but synergistic with thyroid hormones, include skeletal and non-skeletal growth mediated by IGF-I.

both males and females increase gradually but remain low up until about six to eight years of age, after which there is a sharp increase associated with puberty in both sexes. IGF-I levels decrease and remain relatively constant in adult years. *IGF-II* (not shown) may be under the control of hPL, and may play a role during fetal development.

Direct Effects

The effects of GH mediated by IGF-I on the growth of muscle and skeletal tissues are insulin-like, while the longer term, *direct effects of GH* on carbohydrate metabolism and lipolysis are opposed to those of insulin (Fig. III.10). Interestingly, GH and cortisol synergize in both their diabetogenic and lipolytic effects, but cortisol, which is catabolic, inhibits the indirect actions of GH on muscle and cartilage.

Generally, direct effects of GH can be demonstrated best in vitro or with isolated organs. In the liver, in addition to stimulating IGF-I production, GH directly stimulates cell replication and the processes involved in protein synthesis, including amino acid transport, RNA synthesis, and activation of the enzymes required for the translational process. In rat adipose tissue, GH leads to an increase in amino acid incorporation and lipolysis. Other direct effects have been reported in the diaphragm, heart, chondrocytes, fibroblasts, and hypothalamus.

5.3 Prolactin

The best known and most characterized action of PRL is on the *mammary gland*. In this organ, prolactin specifically stimulates *DNA* synthesis and epithelial cell proliferation, and the synthesis of *milk proteins* (casein, lactalbumin), free fatty acids, and lactose (Table III.3). PRL has been shown to stimulate specifically the rate of milk protein gene transcription as well as to have a very marked effect on the stabilization of the messenger RNA produced. The effect of PRL on DNA synthesis and casein production in mammary tissue in vitro is illustrated in Figure III.11.

Although PRL is the hormone which is primarily responsible for *mammary function*, *other hormones* are also involved. Ovarian steroids favor mammary growth. The specific mechanisms involved in the estrogen-induced stimulation of mammary mitogenesis is unknown, but it may involve IGF-I or other growth factors. Progesterone specifically inhibits milk protein (e.g., casein) mRNA accumulation, and

Table III.3 Principal actions of PRL.

Effect	Tissue
DNA synthesis Cell proliferation Milk protein synthesis FFA synthesis Lactose synthesis	Mammary gland
Prolactin-induced proteins	Mammary tumors
Corpus luteum: maintenance or regression	Ovary
Steroid biosynthesis	Ovary and testis
Immunostimulation	Lymphocytes
RNA synthesis	Liver
Ornithine decarboxylase stimulation	
Osmoregulation	Kidney, amnion, choroid plexus

inhibits PRL-induced translation of casein mRNA, which explains its potent inhibitory effect on lactogenesis.

Glucocorticosteroids are essential for normal secretory activity of the mammary cell, although their precise role is not completely understood. In all species studied, glucocorticosteroids potentiate PRL action on milk synthesis and accumulation of milk protein mRNA. In some species, they are capable of stimulating casein gene transcription directly. Progesterone

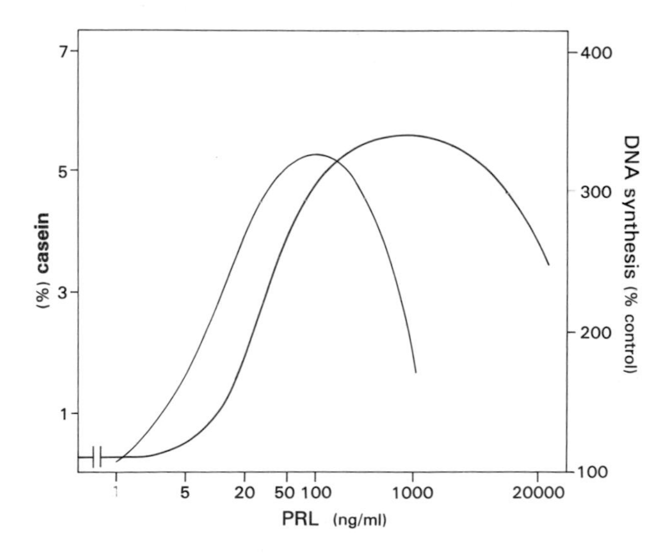

Figure III.11 Effect of PRL on DNA synthesis and casein production as a function of PRL concentration in cultured explants of pseudopregnant rabbit mammary gland.

may antagonize glucocorticosteroid action by competing for receptor binding or may act by competing for binding of steroid-receptor complexes with DNA.

Insulin is also important in mammary function. It favors *amino acid uptake* in the mammary cell and is *mitogenic*. Insulin and IGFs probably interact and potentiate the actions of prolactin.

Recently, prolactin has been shown to induce *specific proteins* (prolactin-inducible proteins) in human breast cancer cells[21]. These proteins have molecular weights of 11, 14 and 16,000, and may be involved in mediating the action of prolactin in these cells.

As shown in Table III.3, prolactin has other effects associated with the *reproductive system*, including effects on the corpus luteum and on the biosynthesis of steroids in both the ovary and testis, immunostimulation in lymphocytes, and the synthesis of RNA and ornithine decarboxylase in the liver, and osmoregulation in the kidney, amnion, and choroid plexus.

Table III.3 represents only a partial list of prolactin's actions. There are *approximately 100 actions* associated with this hormone, listed in a report that appeared a number of years ago[6], and the possible targets and actions of the hormone appear to be expanding continually.

6. Regulation of Secretion

Growth hormone and prolactin are two pituitary hormones which, until recently, were not considered to be regulated by the classical negative feedback control as exists for other anterior pituitary hormones (p. 32). However, for GH, at least, this concept must be reconsidered. Table III.4 lists a number of factors which have been shown to regulate the synthesis and secretion of these two hormones.

Hypothalamic neurohormones, having either stimulatory or inhibitory activity, are considered the *major* direct means of altering GH or PRL secretion.

6.1 Growth Hormone

The hypothalamus produces *two neurohormones* affecting GH secretion. *GH-releasing factor (GRF or somatocrinin)* is a peptide of 44 aa in humans (43 in rats) which specifically *stimulates* the secretion of GH. *Somatostatin* specifically *inhibits* GH secretion. It

21. Shiu, R. P. C. and Iwasiow, B. M. (1985) Prolactin-inducible proteins in human breast cancer cells. *J. Biol. Chem.* 260: 11307–11313.

Table III.4 Factors affecting GH secretion in humans.

Increase	Decrease
Sleep (stage III and IV)	Sleep (REM)
Stress	β-adrenergic agonists
α-adrenergic agonists	α-adrenergic antagonists
β-adrenergic antagonists	
Amino acids	HyperFFA
Hypoglycemia	Obesity
HypoFFA	Hyperglycemia
Some forms of diabetes	
GRF	Somatostatin
Glucagon	Glucocorticosteroids (high concentration)
Estrogens	Hypothyroidism
Androgens	IGF-I

exists as a 14 aa peptide and as a 28 aa N-terminally extended form. Therefore, the overall secretory rate of the hormone is influenced by a continual interplay of these two neuropeptides.

Growth Hormone Releasing Factor (GRF)

The primary structure of human pancreatic GRF as well as porcine and rat hypothalamic GRFs is shown in Figure III.12. The human peptide was first isolated from a pancreatic tumor from a patient who demonstrated clinical signs of excess growth hormone production[22,23]. The hypothalamic form of GRF appears to have bio- and immunoactivity similar to that of the pancreatic form.

The structure of the GRF gene and the mRNA encoding the protein is shown in Figure III.13. The gene spans over 9 kb and contains five exons and four introns[24]. The structure of the mRNA (from the cDNA) shows that the GRF peptide sequence is fol-

22. Rivier, J., Spiess, J., Thorner, M. and Vale, W. (1982) Characterization of a growth hormone releasing factor from a human pancreatic islet tumor. *Nature* 300: 276–278.
23. Guillemin, R., Brazeau, P., Bohlen, P., Esch, F., Ling, N. and Wehrenberg, W. B. (1982) Growth hormone-releasing factor from a human pancreatic tumor that caused acromegaly. *Science* 218: 585–587.
24. Mayo, K. E., Corelli, G. M., Rosenfeld, M. G. and Evans, R. M. (1985) Characterization of cDNA and genomic clones encoding the precursor to rat hypothalamic growth hormone-releasing factor. *Nature* 314: 464–467.

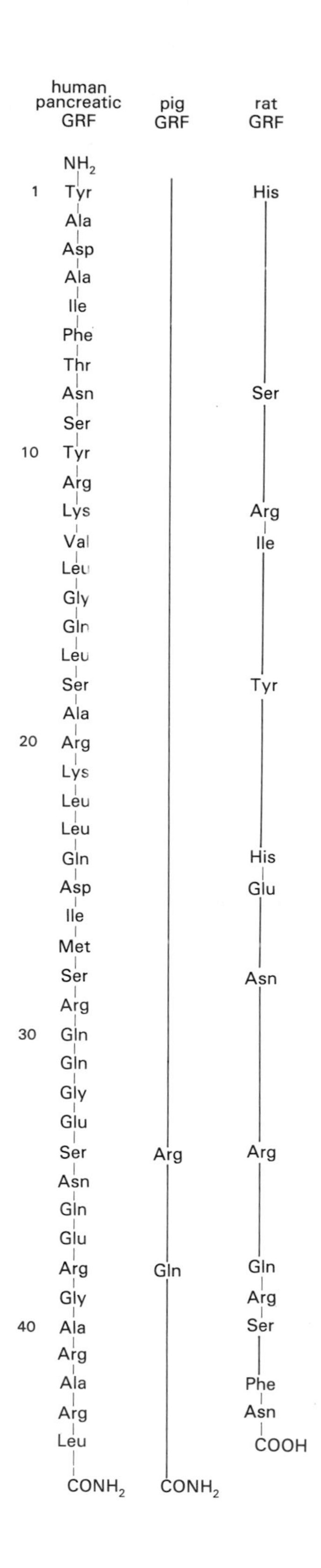

Figure III.12 Primary structure of human pancreatic and porcine and rat hypothalamic GRF.

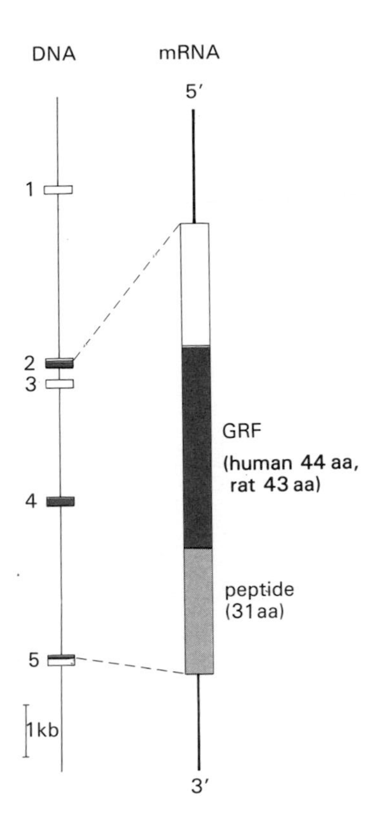

Figure III.13 Structure of the rat GRF gene and mRNA encoding the protein.

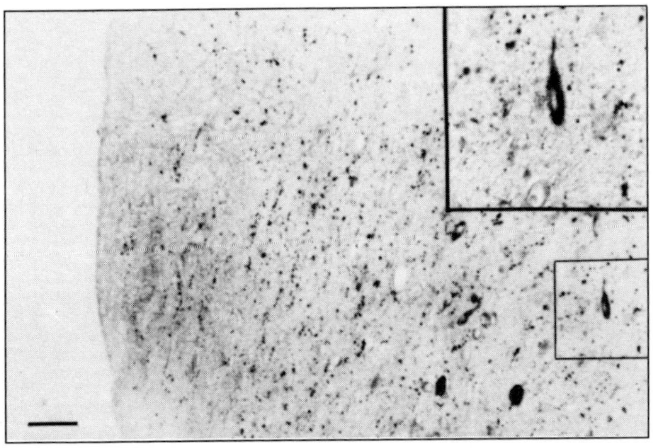

lowed by a single Arg residue, which probably serves as a signal for proteolytic processing. In the rat, this Arg is not preceded by a Gly residue, which serves as an amide donor in processed peptides that are C-terminally amidated. Rat GRF does have a free C-terminus, but all other GRFs appear to be C-terminally amidated. A 31 aa peptide of unknown function follows the Arg residue, followed by a 105 nucleotide untranslated sequence. In comparison to human cDNA sequences, GRF lacks the N-terminal portion of the precursor, which includes the signal peptide sequence and the 5′ non-translated sequences.

GRF is produced in the mediobasal hypothalamus. Immunohistochemical localization of human GRF cell bodies in the *arcuate nucleus* can be seen in Figure III.14.

The mechanism of action of GRF involves an *increase* in *cAMP* production, since its effects can be mimicked in pituitary cells in primary culture by the addition of phosphodiesterase or 8Br-cAMP (Fig. III.15). Highly specific receptors for the neurohormone have been located on somatotropic cells of the anterior pituitary. Other vectors mediating GRF action (Ca^{2+}, inositol lipids, and phosphorylation) have also been proposed.

Figure III.14 Immunohistochemical localization of GRF in the human arcuate nucleus. The inset is an enlarged view of a positive cell body. Bar=50 μm. (Courtesy of Dr. Georges Pelletier, Centre Hospitalier de l'Université Laval, Québec.)

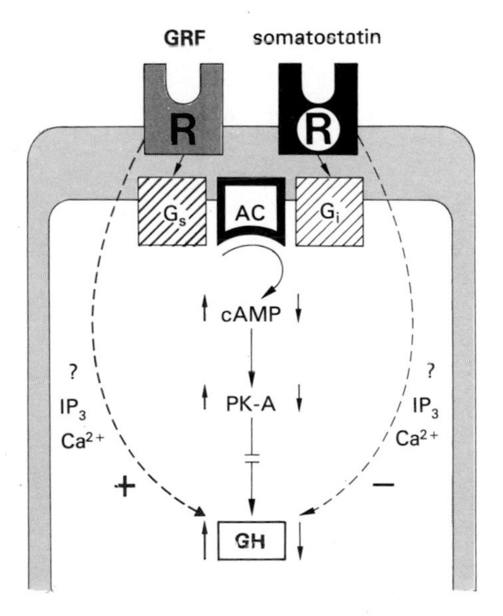

Figure III.15 Mechanism of action of GRF and somatostatin in a somatotropic cell. The most clearly established transducing system is adenylate cyclase, as indicated in the text. However, additional mechanisms implicating inositol phospholipids and Ca^{2+} are also involved.

Somatostatin

Figure III.16 shows the primary structure of somatostatin[25]. This neurohormone has a similar structure in all mammalian species tested thus far. The composition of somatostatin within the hypothalamus is heterogeneous; that is, it exists as the *normal 14 aa* form, as well as in a 14 aa N-terminally extended form known as *somatostatin-28*. Both of these peptides specifically inhibit the secretion of GH and PRL by a direct action on the corresponding pituitary cells.

The structure of the human somatostatin gene is shown in Figure III.17. In both the human and rat, the structure is similar[26], consisting of two exons and one intron. In exon 1, in addition to an untranslated region, the final portion corresponds to 46 aa of the (pre)prosequence of (pre)prosomatostatin. Exon 2 starts with the rest of the prosequence (42 aa), followed by the coding sequence corresponding to somatostatin-28, which is itself terminated by the 14 aa of

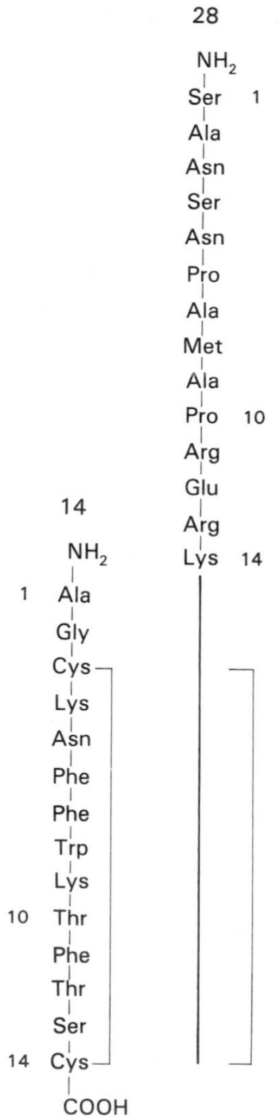

28
NH2
Ser 1
Ala
Asn
Ser
Asn
Pro
Ala
Met
Ala
Pro 10
Arg
Glu
Arg
Lys 14
14
NH2
1 Ala
Gly
Cys
Lys
Asn
Phe
Phe
Trp
Lys
10 Thr
Phe
Thr
Ser
14 Cys
COOH

Figure III.16 Primary structure of somatostatin-14 and somatostatin-28.

5′
untranslated
exon 1
prepro (46aa)
intron
pro (70aa)
exon 2
28
14
untranslated
100bp
3′

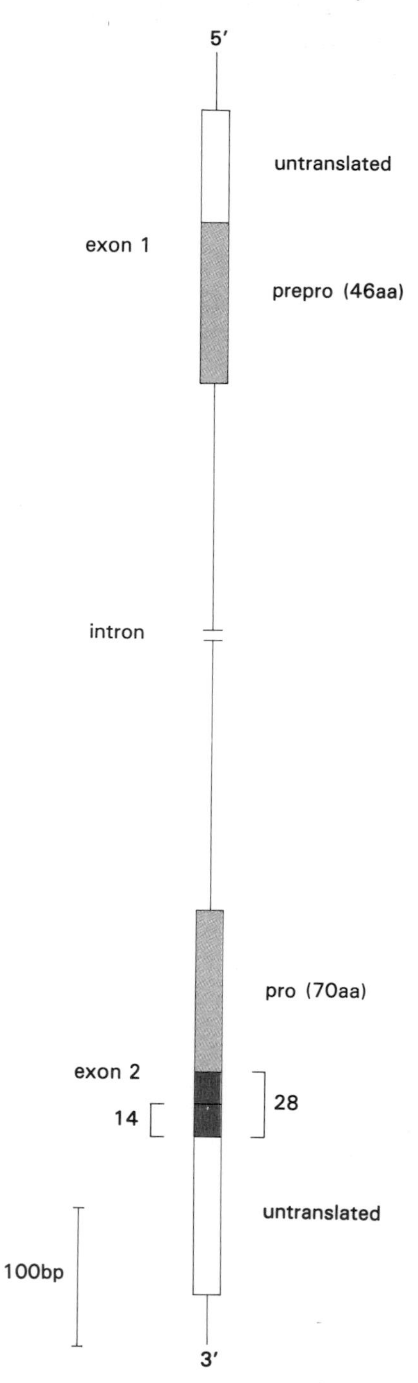

Figure III.17 Structure of the human somatostatin gene. Exon 1 codes for the prehormone and part of the prohormone. Exon 2 codes for the rest of the prohormone, as well as for somatostatin-14 and somatostatin-28.

somatostatin-14. The exon is terminated by another untranslated region.

Somatostatin is found in a *number of different tissues* throughout the body. As part of the nervous system, somatostatin neurons comprise a portion of the tuberoinfundibular system which terminates in the median eminence. Figure III.18 shows the localization of somatostatin within the arcuate nucleus of the human hypothalamus. Somatostatin is also located in a number of other regions within the central nervous system as well as in specific secretory cells of the *gastrointestinal* tract (p. 575).

Somatostatin acts on somatotropic cells by *inhibiting* the production of *cAMP*, and thus the secretion of GH (Fig. III.15). Somatostatin inhibits GRF-stimulated GH secretion as well as that stimulated by other agents. However, the fact that somatostatin is capable of inhibiting GH secretion induced by cAMP derivatives suggests that it does *not act exclusively* via the inhibition of cAMP synthesis.

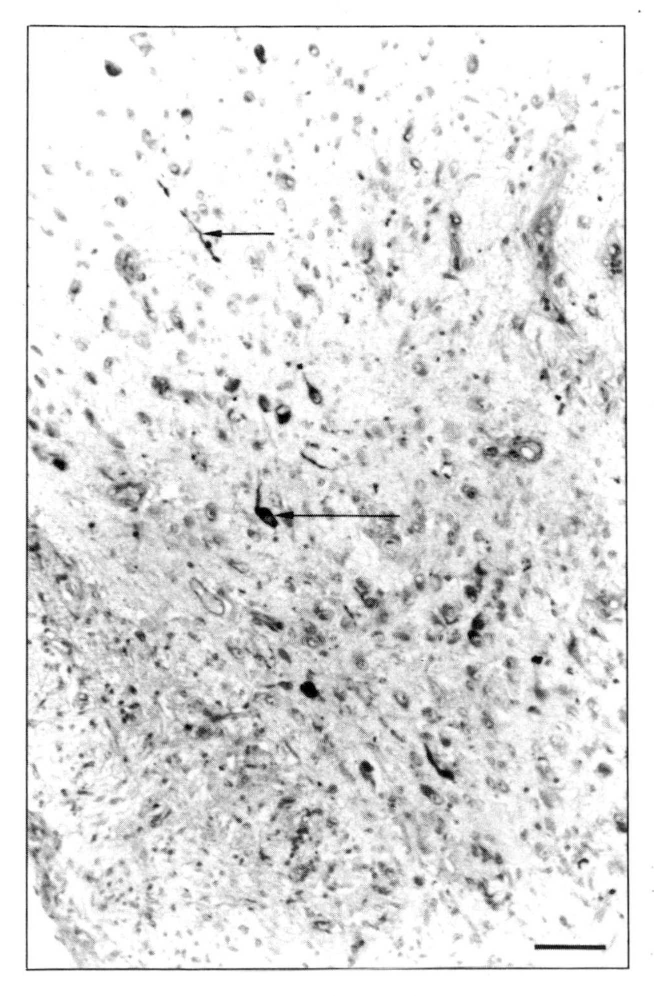

Figure III.18 Immunohistochemical localization of somatostatin in the human arcuate nucleus. A positive cell body (long arrow) and fiber (short arrow) can be seen. Bar=50 μm. (Courtesy of Dr. Georges Pelletier, Centre Hospitalier de l' Université Laval, Québec).

Other Factors (Table III.4)

Many other neuropeptides, monoamines or hormones are capable of modulating GH secretion. Endogenous opiate peptides or morphine administration increases GH secretion, probably via an α-adrenergic effect on GRF. VIP, which also stimulates GH, appears to inhibit somatostatin secretion or the effect of somatostatin on the somatotropic cell. Substance P and neurotensin reduce GH levels, probably via the stimulation of somatostatin. Bombesin and motilin both stimulate GH release by a direct action at the level of the anterior pituitary. The monoamines norepinephrine, epinephrine, dopamine and l-DOPA all stimulate GH release, probably by an effect on α_2 receptors.

In rat pituitary tumor cells in culture, both thyroid hormones and glucocorticosteroids increase the production of GH by enhancing transcription and stabilizing GH mRNA. Estrogens lead to an increase, and androgens to a decrease, of basal GH secretion. In humans, hypothyroidism inhibits GH secretion. In contrast to what is seen in rat pituitary tumors, large doses of glucocorticosteroids lead to a decrease in GH secretion.

Episodic or Pulsatile Secretion

There is an *ultradian* rhythm of GH secretion, which was first demonstrated in rats. Elevated peaks, exceeding 200 ng/ml, with a frequency of 3 to 4 h are routinely observed. In humans, a similar pulsatile secretion has been observed, although the peaks are less regular and are of lower magnitude[27].

Figure III.19 illustrates the peaks of GH by the interrelationship between somatostatin and GRF levels in the portal blood. As can be seen, GRF stimu-

25. Brazeau, P., Vale, W., Burgus, R., Ling, N., Butcher, M., Rivier, J. and Guillemin, R. (1973) Hypothalamic polypeptide that inhibits the secretion of immunoreactive pituitary growth hormone. *Science* 179: 77–79.

26. Shen, L. D. and Rutter, W. J. (1984) Sequence of the human somatostatin I gene. *Science* 224:168–171.

27. Vance, M. L., Kaiser, D. L., Evans, W. S., Furlanetto, R., Vale, W., Rivier, J. and Thorner, M. O. (1985) Pulsatile growth hormone secretion in normal man during a continuous 24-hour infusion of human growth hormone releasing factor (1-40): evidence for intermittent somatostatin secretion. *J. Clin. Invest.* 75: 1584–1590.

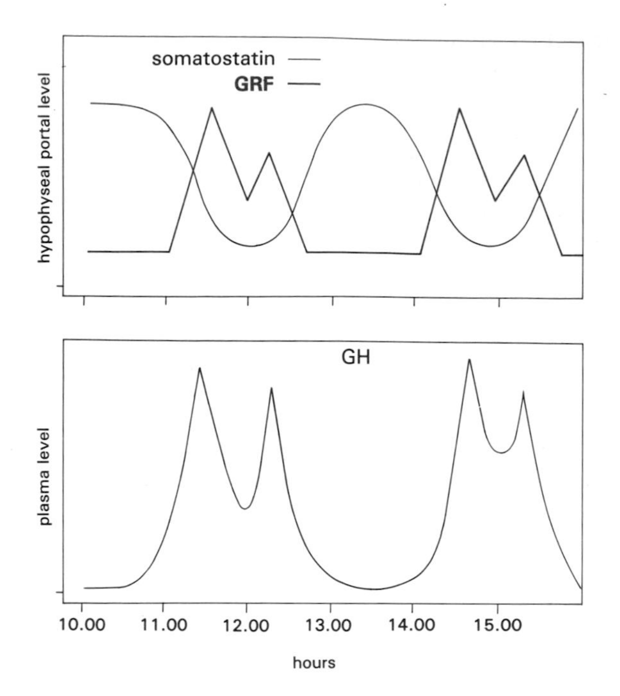

Figure III.19 Simplified representation of the rhythmic secretion of somatostatin and GRF, and the net result on plasma GH levels. Indirect evidence obtained by using anti-somatostatin antiserum permitted the correct alignment of the curves for the two neurohormones. Note that GH can rise when somatostatin levels are low, although the peaks of GH are directly correlated to GRF levels. (Adapted from Tannenbaum, G. S. and Ling, N. (1984) The interrelationship of growth hormone (GH)-releasing factor and somatostatin in the generation of the ultradian rhythm of GH secretion. *Endocrinology* 115: 1952–1957).

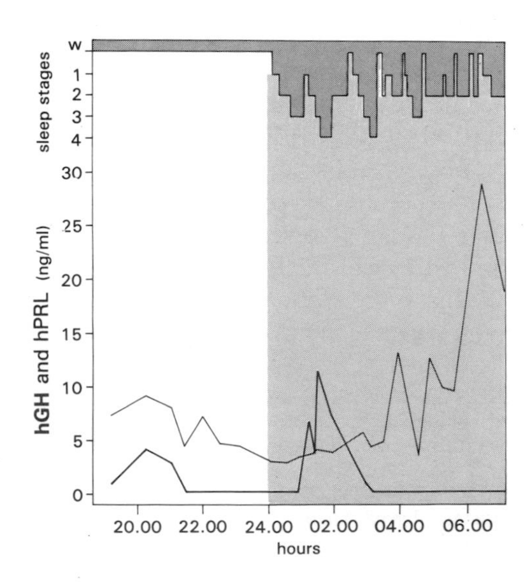

Figure III.20 Sleep-associated nocturnal increase in serum GH and PRL. Note that the GH peaks precede those of PRL.

lates the peaks at periods of low somatostatin concentrations, and when somatostatin levels increase, peripheral GH concentrations decrease.

Circulating Levels

Normal serum concentrations of GH average less than 5 ng/ml in humans. In both children and young adults, a *peak* of GH occurs approximately *1 h following the onset of deep sleep*, with the possibility of smaller peaks occurring at later periods in the sleep cycle (Fig. III.20). The function of this sleep-related GH is not certain, but it may be important in promoting skeletal growth, especially in children. *Fetal GH* levels appear in the first trimester of pregnancy, increase to a maximum of 150 ng/ml at 20 weeks, and fall to 30 ng/ml at birth. Although the presence of a M_r ~50,000–60,000 *plasma protein* specifically binding human growth hormone has been clearly established, its biological importance remains to be established.

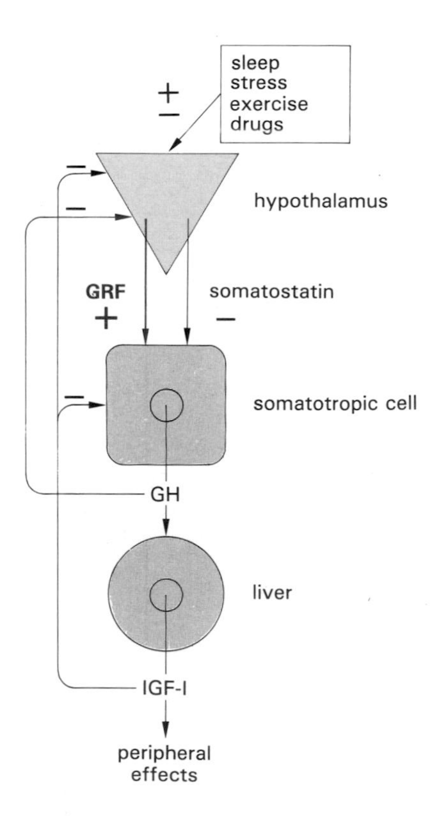

Figure III.21 Principal central and peripheral regulatory factors affecting GH secretion.

Feedback Control (Fig. III.21)

Insulin-like growth factor I, which is secreted by the liver in response to GH, has an *inhibitory* action on GH production at the level of the hypothalamus or anterior pituitary. In addition, *GH itself* may feed back on the hypothalamus and affect the level of its own secretion.

Stress and exercise, as well as the *emotional state* of the patient, can have marked effects on the circulating concentration of GH. Secretion of growth hormone is stimulated by α-adrenergic agents such as norepinephrine, clonidine, and l-DOPA, whereas α-adrenergic blockers inhibit secretion.

Nutritional status also has an important effect on the regulation of GH secretion. Hyperglycemia can prevent the GH increase which occurs during stress; in contrast, chronic hyperglycemia in diabetics does not suppress GH levels.

Increased circulating *free fatty acids* tend to *suppress* GH secretion, which may explain reduced sleep-induced GH secretion as well as the reduced response to stimulation tests observed in obese patients.

6.2 Prolactin

Prolactin is *under general negative control* by the hypothalamus. In rats, removal of the pituitary gland from its normal anatomical location and autotransplantation under the kidney capsule results in a marked elevation in serum prolactin levels. Similar increases are observed in experimental animals and humans following pituitary stalk section. For many years, the search for a definitive hypothalamic neuropeptide or *prolactin inhibiting factor* (PIF) has met with no success. Generally, it has been agreed that *dopamine* is the major factor responsible for PRL inhibition[28]. A large number of other hormones have also been shown to alter prolactin secretion, including estradiol, glucocorticosteroids, thyroid hormone, TRH, and somatostatin. A number of physiological and pharmacological factors can either inhibit or stimulate PRL levels specifically, in both animals and humans (Table III.5).

Table III.5 Factors affecting PRL secretion in humans.

Increase
Nipple stimulation
Stress (including psychogenic)
Sleep (stage I and II and REM)
Stalk section
Many pituitary and cerebral diseases
Prolactinoma
TRH
Pregnancy
Estrogens
Hypothyroidism
Adrenal insufficiency
Drugs affecting dopamine secretion or action (domperidone, reserpine, phenoghiazines, metoclopramide, chlorpromazine, cimetidine, opiates)
Decrease
Dopamine (analog such as bromocryptine, lisuride, pergolide, and mesulergine)
GAP or PIF

Dopamine

Dopamine has an *inhibitory effect* on PRL secretion. Dopamine, which affects the anterior pituitary, is secreted by neurons in the arcuate nucleus and travels in fibers of the tuberoinfundibular pathways to the median eminence. It then passes into the hypophyseal portal vessels and acts on the anterior pituitary to directly inhibit the secretion of prolactin. Dopamine

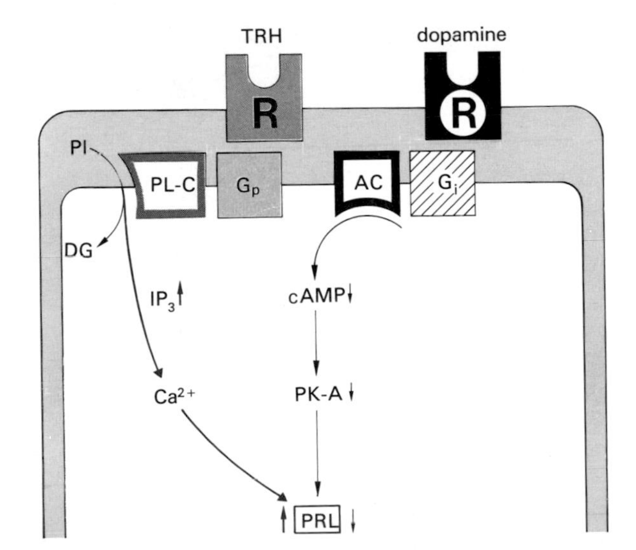

Figure III.22 Mechanisms of action of TRH and dopamine on PRL secretion.

28. Macleod, R. M. (1969) Influence of norepinephrine and catecholamine-depleting agents on the synthesis and release of prolactin and growth hormone. *Endocrinology* 85: 916–925.

binds to specific D_2 *receptors in lactotropic cells* of the anterior pituitary. This interaction induces an inhibition of adenylate cyclase and cAMP production. Since prolactin release can be mimicked by the addition of 8Br-cAMP or indomethacin, cAMP is thought to be a specific mediator of prolactin secretion. Therefore, dopamine inhibition of the adenylate cyclase system would offer a direct means of inhibiting PRL production (Fig. III.22).

GAP or PIF?

A 56 aa GnRH-associated peptide (GAP) which forms the C-terminal portion of the GnRH precursor (Fig. III.23) has been identified[29]. This peptide inhibits prolactin secretion, presumably by interacting with specific receptors on lactotropic cells. Whether GAP is the elusive PIF remains to be established.

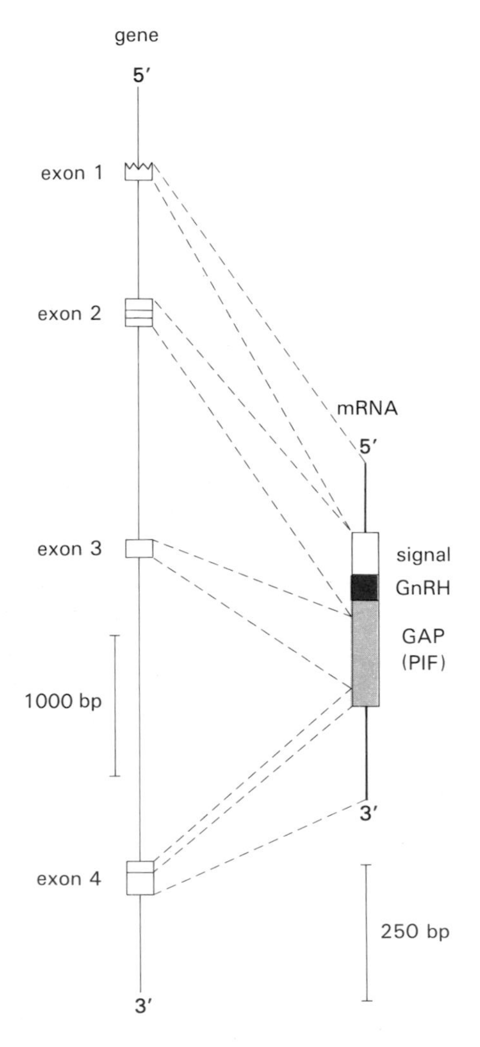

Figure III.23 The structure of the gene and mRNA for GAP (PIF), which forms the C-terminal portion of the GnRH precursor. It specifically inhibits PRL secretion[29].

Estradiol

Estrogens have long been known to have a *positive effect* on PRL secretion. In addition to inducing hypertrophy of the lactotropic cells, estradiol increases PRL production by directly stimulating prolactin gene transcription, leading to increased synthesis of PRL mRNA and prolactin[30]. Estradiol induces a several-fold increase in PRL gene transcription, occurring as early as 20 min after treatment with the steroid, suggesting a direct regulation of transcription by the estradiol-receptor complex.

Thyrotropin Releasing Hormone (TRH)

Thyrotropin releasing hormone, in addition to its effect on the release of TSH from thyrotropic cells, has been shown to specifically *increase* the release of PRL[31] in vivo as well as in pituitary cells in primary culture.

TRH neurons are widely distributed in the brain as well as in extra-CNS sites. Within the hypothalamus, TRH is localized within neurons in the median eminence, with cell bodies in the medial portion of the paraventricular nucleus (Fig. II.4B).

Thyrotropin releasing hormone binds to specific *receptors* which, in addition to their presence on thyrotropic cells, have been identified on lactotropic cells of the anterior pituitary. In animal models, TRH receptor numbers are increased by estrogens and decreased by thyroid hormones. Receptor number may be important in the overall control of prolactin secretion in response to TRH.

Originally, TRH was thought to induce prolactin release by increasing cAMP formation. VIP stimulates adenylate cyclase activity and the formation of cAMP, which is itself capable of inducing a several-fold increase in prolactin gene transcription and mRNA levels. The pattern of proteins phosphorylated following exposure of pituitary cells to VIP and TRH is quite

29. Nikilics, K., Mason, A. J., Szonyi, I., Ramachandran, J. and Seeburg, P. H. (1985) A prolactin-inhibiting factor within the precursor for human gonadotropin-releasing hormone. *Nature* 316: 511–517.
30. Maurer, R. A. (1982) Estradiol regulates the transcription of the prolactin gene. *J. Biol. Chem.* 257: 2133–2136.
31. Bowers, C. Y., Frisen, H. G. and Hwang, P. (1971) Prolactin and thyrotropin release in man by synthetic pyroglytamyl-hystidyl-prolinamide. *Biochem. Biophys. Res. Commun.* 45: 1033–1041.

different, suggesting a different mechanism of action for the two neurohormones. For TRH, it now appears that, following binding to its receptors, hydrolysis of inositol phospholipids occurs, resulting in the formation of intracellular IP_3 and DG. An increase in intracellular Ca^{2+} occurs as a result of the mobilization of intracellular pools and the stimulation of voltage-regulated channels. Secretion (exocytosis) may be stimulated directly by Ca^{2+}, or by subsequent phosphorylation of proteins, through a calmodulin-dependent protein kinase[32]. Studies using permeabilized GH_3 cells have shown that TRH-induced phospholipid hydrolysis involves activation of a phosphonositide phosphodiesterase. TRH regulation of this polyphosphoinoside hydrolysis is modulated by guanine nucleotides, however the GTP-binding protein involved (G_p) appears to be neither G_s nor G_i[33].

Other Factors

In addition, a number of other hypothalamic and peripheral factors have specific effects on prolactin secretion. GABA and, in some cases, somatostatin specifically inhibit PRL secretion. In experimental animals, neurotensin, VIP, EGF and bombesin are potent stimulators of secretion. Opiate peptides and morphine markedly stimulate secretion; they act by blocking the inhibition of prolactin secretion without affecting spontaneous PRL release.

Episodic or Pulsatile Secretion.

As is true for GH, PRL is secreted in a *pulsatile* fashion, although the frequency of the pulses is more irregular than for GH. The pulses are believed to result from the combined effect of hypothalamic inhibitory and stimulatory factors affecting lactotropic cells.

Circulating Levels (Fig. III.24 and Table III.5)

The human *pituitary* contains much *less PRL than GH*. To maintain a similar level of secretion, a higher rate of synthesis and turnover within the pituitary is required. Plasma concentration averages approximately 10 ng/ml in women (range of 1 to 20 ng/ml), and is slightly lower in men. There is a *sleep-induced increase* in plasma PRL concentration (Fig. III.20) which occurs later than the nocturnal increase in GH. Prolactin levels *increase during pregnancy* to a maximal concentration of approximately 200 to 300 ng/ml. Prolactin is also produced by the

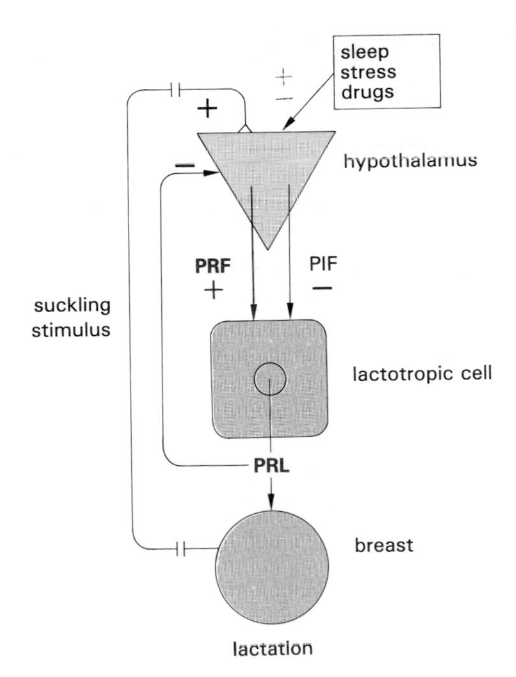

Figure III.24 Principal central and peripheral regulatory factors affecting PRL secretion. PIF and PRF refer to several compounds, as indicated in the text. While there is negative feedback of PRL at the hypothalamic level, PRL feedback at the pituitary level is still not clearly established. Note that the positive input of suckling involves an external stimulus indirectly connected to PRL secretion and action.

decidual cells of the placenta, which accounts for the increased PRL levels found in amniotic fluid (Fig. X.37). During *lactation*, nipple or breast stimulation induces a rapid increase in plasma PRL levels as a result of afferent *neural stimulation*. Pituitary adenoma (prolactinoma) is frequently the cause of elevated prolactin levels. In addition, drugs which affect endogenous dopamine secretion or action can also result in modifications of circulating prolactin. A number of other factors are able to increase prolactin, including *exercise, stress, surgery* (depending upon the anesthetic used), and sexual intercourse in women. Prolactin levels can be somewhat elevated in hypothyroidism, adrenal insufficiency, or renal failure.

32. Gershengorn, M. C., Geras, E., Spina Purrello, V. and Rebecchi, M. J. (1984) Inositol triphosphate mediates thyrotropin-releasing hormone mobilization of non-mitochondrial calcium in rat mammotropic pituitary cells. *J. Biol. Chem.* 259: 10675–10681.
33. Martin, T. F., Lucas, D. O., Bajjalieh, S. M. and Kowalchyk, J. A. (1986) Thyrotropin releasing hormone activates a Ca^{2+} dependent polyphosphoinositide phosphodiesterase in permeable GH_3 cells. GTPγS potentiation by a cholera and pertussis toxin-insensitive mechanism. *J. Biol. Chem.* 261: 2918–2927.

7. Pathophysiology

7.1 Dynamic Tests of Secretion

Since baseline levels of GH and PRL are known to vary as a function of the time of day or the medical treatment a patient might be receiving, it is important to perform provocative tests in order to establish the ability of the pituitary to respond, and also to attempt to distinguish between hypothalamic and pituitary disorders.

Growth Hormone

Administration of 0.05–0.1 U/kg body weight of *insulin* should produce at least a 50% decrease in blood sugar 20 to 30 min later, with a peak of hGH occurring between 45 and 75 min after injection. This test should be done under close surveillance, as severe symptoms of hypoglycemia may occur requiring the administration of iv glucose. Additional tests include the infusion of arginine (0.5 g/kg over 30 min), ornithine (20 g/1.73 m^2, iv), l-DOPA (0.5 g/1.73 m^2 administered orally), clonidine (4 μg orally) or glucagon (0.03 mg/kg, im or sc). GH peaks occur 60 to 180 min after administration of the stimulant. The oral administration of propranolol (30–40 mg or 0.75 mg/kg in children) 30 to 60 min prior to the provocative agent tends to potentiate the response.

Growth hormone releasing factor (GRF) may provide an important means of distinguishing GH deficiency due to a pituitary defect from a hypothalamic disorder. It may be important to prime the somatotropic cells by multiple injections with GRF if no response is observed following a single injection.

To be considered normal, growth hormone peaks should be in excess of 7 ng/ml following any of the provocative tests. Pretreatment with estrogen (DES) or propranolol increases the response, making interpretation easier.

The determination of *serum IGF-I* concentrations by RIA is also a useful technique in screening patients with growth-related problems.

Prolactin

Two general categories of tests are routinely performed, one stimulatory and the other inhibitory. The injection of *TRH* (100 μg iv or 1 μg/kg in children) induces a three- to five-fold *increase* in serum PRL levels. This is a result of direct stimulation at the level of the lactotropic cell. A central blockade of dopaminergic neurons by the administration of chlorpromazine (25 mg po, 0.4 mg/kg in children), or a direct blockade at the lactotropic cell with domperidone (4 mg) or metoclopramide (10 mg, iv), results in a more than two-fold increase in serum PRL.

Inhibitory tests, which include the administration of *l-DOPA* (500 mg) or *bromocriptine* (2.5 mg given po), should result in a 50% decrease in PRL concentrations. These compounds directly activate lactotropic cell dopamine receptors. Activation of the hypothalamic dopaminergic pathways by combining carbidopa (50 mg every 6 h, four times) followed by 35 mg carbidopa and 100 mg l-DOPA induces a >40% decrease in serum PRL. Carbidopa blocks the conversion of dihydroxyphenylalanine to dopamine in the pituitary. In patients with prolactinomas, only an insignificant decrease in prolactin occurs.

7.2 Growth Hormone Deficiency

A deficiency of GH production results in *dwarfism* (Fig. III.25). Three types of dwarfism of hormonal

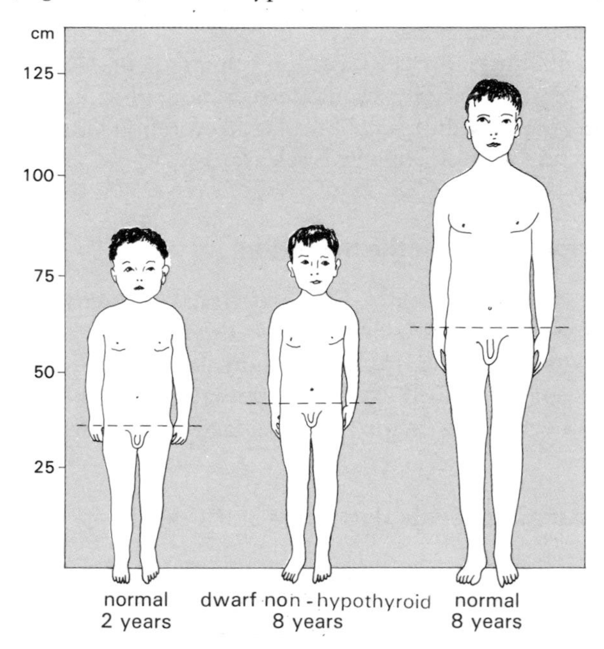

Figure III.25 The figures of normal boys, two and eight years of age, respectively, illustrate the change in the ratio of upper and lower skeletal segments measured from the symphysis pubis. Dwarfs of pituitary or primordial types attain the more mature proportions of their chronological age. (After Wilkins, L., ed (1953) Dwarfism. In: *The Diagnosis and Treatment of Endocrine Disorders in Childhood and Adolescence.* Charles C. Thomas, New York, pp. 119–137).

origin exist: those with a primary pituitary disease, those with a hypothalamic disorder, and those with normal hypothalamic-pituitary function which fail to respond to GH.

Primary Pituitary Dysfunction

There are several *genetic syndromes* which result in a deficiency of GH production, including pituitary hypoplasia, pituitary aplasia, familial panhypopituitarism, and familial isolated GH deficiency. This latter disease is transmitted by an autosomal recessive mode of inheritance. The genetic defect results in a deletion of 7.5 kb of DNA, including the gene that encodes normal GH, but not the variant GH gene[34]. In addition, some tumors or trauma may also induce pituitary destruction, resulting in GH deficiency. Some short children have been found to secrete a GH which is not biologically active, but which appears normal immunologically. Such patients have low IGF-I levels that increase following exogenous GH administration.

Hypothalamic Dysfunction

A hyposecretion of GH secondary to hypothalamic damage is classified most frequently under the heading of isolated GH deficiency of hypothalamic origin. Idiopathic hypopituitarism is generally the result of a perinatal insult occurring during abnormal delivery, and perinatal asphyxia. Various forms of infections and tumors can also be the cause of this disorder.

Growth Hormone Resistance

The best example of a syndrome of GH resistance is *Laron dwarfism*, a familial disorder transmitted in the autosomal recessive mode. This disease is most frequent in people from Middle-Eastern countries. Patients appear to have severe GH deficiency despite elevated serum GH levels. Serum IGF-I levels are quite low, similar to levels seen in hypopituitarism, and there is no increase following GH treatment. These observations suggest that Laron dwarfs have a defect in the GH receptor–effector system, with perhaps a specific reduction in GH binding by the liver.

African Pygmies also have a resistance to administered GH, although circulating GH levels are normal, with serum IGF-I levels low and non-responsive to GH administration. This disorder is also probably due to an inability to produce IGF-I.

Treatment

Humans do *not* respond to *non*-primate GH; therefore, *replacement with human GH* obtained from human pituitary glands removed at autopsy has been the only effective treatment in the past 30 years. The usual dosage is 0.1 U/kg body weight given im three times weekly. There is a direct dose–response relationship, and with the increased supply of human GH, larger doses may be indicated.

Recent results with hGH produced in bacteria, using *recombinant DNA* techniques, suggest that results are similar to those obtained by injection of hGH prepared from human pituitaries. Consequently, in the future, the supply problem might be alleviated. Another promising treatment is the use of *GRF* to stimulate the release of endogenous GH in children with GH deficiency.

7.3 Growth Hormone Excess

Hypersecretion of GH results in *acromegalo-gigantism*, when the excess occurs in children, and in *acromegaly*, when the excess occurs in adult life. These conditions usually result from an acidophil *pituitary adenoma*. Recently, some cases of GRF-producing pancreatic tumors[35] or ectopic GH-producing tumors[36] have been reported. Such tumors are rare, but should be looked for when evaluating for acromegaly. It is remarkable that GRF-secreting tumors can induce somatropic hyperplasia mimicking pituitary cell hyperplasia.

Acromegaly

Acromegaly consists of a coarsening of facial features and of the soft tissues of the feet and hands (Fig. III.26). There is an exaggerated growth of the mandible as well as of the frontal and nasal bones, resulting in characteristic facial development. Hypertrophy of connective tissue and of heart, liver

34. Phillips, J. A., Hjelle, B. L., Seeburg, P. H. and Zachmann, M. (1981) Molecular bases for familial isolated growth hormone deficiency. *Proc. Natl. Acad. Sci., USA* 78: 6372–6375.
35. Thorner, M. O., Perryman, R. L., Cronin, M. J., Rogul, A. D., Draznin, M., Johanson, A., Vale, W., Horvath, E. and Koracs, K. (1982) Somatotroph hyperplasia. Successful treatment of acromegaly by removal of a pancreatic islet tumor secreting a growth hormone-releasing factor. *J. Clin. Invest.* 70: 965–977.
36. Melmed, S., Ezrin, C., Kovacs, K., Goodman, R. S. and Frohman, L. A. (1985) Acromegaly due to secretion of growth hormone by an ectopic pancreatic islet cell tumor. *N. Engl. J. Med.* 312: 9–17.

A

B

C

Figure III.26 Progression of acromegaly. **A.** Age 16 years. **B.** Age 33 years. The disease is well established. **C.** Age 52 years. End-stage acromegaly with gross disfigurement: note the large head, the large nose and lips, the prognathism and the large hands. (Mendeloff, A. I., and Smith, D. E. (1956) Acromegaly, diabetes, hypermetabolism, proteinuria and heart failure. Clinical Pathological Conference. *Am. J. Med.* 20: 133–144).

and kidney also occurs. Glucose tolerance is reduced by approximately 50%, with actual diabetes being observed in approximately 10% of the patients. Growth hormone levels can vary from almost normal concentrations to several hundred ng/ml. Serum IGF-I levels are elevated in nearly all patients; IGF-I measurements are most valuable when GH levels are close to normal.

Treatment

The death rate of untreated acromegalics is twice that of the normal population. The primary objectives of treatment are to prevent complications from parasellar extension, further physical disfigurement, the progression of rheumatism, and pathophysiological complications such as diabetes mellitus and cardiovascular disease. The three major treatments for GH excess are neurosurgery, radiation, and medical administration of dopamine agonists.

Transsphenoidal *hypophysectomy* can be used for all but the largest tumors. Only approximately two-thirds of the patients with *microadenoma* (less than 10 mm) are cured by such surgery. A similar proportion of patients with macroadenomas are cured by this approach, although almost all of the pituitary must be removed. In an assessment of the long-term (5 to 11 years) success of selective adenomectomy, recurrence occurred in only one of eight patients with microadenoma, and in three of 13 with macroadenoma, for a 68% rate of success overall[37]. Therefore, transsphenoidal adenomectomy remains a valuable treatment of acromegaly.

Radiation may be used as a primary treatment, or it may be used in patients not cured by transsphenoidal surgery. Improvement is slow and may lead to general hypopituitarism. The delay associated with radiation treatment and the possibility of incompletely controlling hGH secretion are serious disadvantages of this form of treatment.

Medical treatment includes administration of bromocriptine or pergolide, to lower serum GH levels in acromegalics, although GH levels return to normal in only about 20% of the patients having received this treatment. Clinical improvement with bromocriptine can occur even though GH levels are only moderately affected. The decline in serum IGF-I may be greater

37. Serri, O., Somma, M., Comtois, R., Rasio, E., Beauregard, H., Jilwan, J. and Hardy, J. (1985) Acromegaly: biochemical assessment of cure after long term follow-up of transsphenoidal selective adenomectomy. *J. Clin. Endocrinol. Metab.* 61: 1185–1189.

than that of GH, explaining the improved clinical status. An alternative approach involves the use of long-acting analogs of somatostatin, such as SMS 201–995. Such treatment has the advantage of specifically affecting only GH secretion in the pituitary. In one study, elevated IGF-I levels returned to the near-normal range in acromegalic patients given 100 to 300 μg/day of the analog. A slight reduction in tumor size has also been reported with such treatment[38].

7.4 Decreased Prolactin Secretion

Reduced secretion of prolactin is usually associated with general pituitary hormone deficiency (Sheehan's syndrome). The only major consequence appears to be an inability to lactate following normal pregnancy.

7.5 Hyperprolactinemia

The clinical condition of hyperprolactinemia is the most common hypothalamic-hypophyseal disorder in humans. It is caused by *autonomous* PRL secretion by pituitary *adenomas*, reduced or absent dopamine effect on the lactotropic cell, or excessive stimulation of the lactotropic cell, overriding physiologic inhibition. Hypothyroidism, chronic renal failure and polycystic ovarian syndrome are also conditions associated with elevated circulating PRL concentrations.

Drug-Induced Prolactinemia

Drugs depleting the central monoaminergic neurons (reserpine, α-methylDOPA), blocking monoaminergic receptors (haloperidol, chloropromazine, metoclopramide) or blocking histamine-2 receptors (cimetidine), as well as opiates and sex steroids, are all capable of inducing hyperprolactinemia. Such drugs are commonly used in clinical practice; thus, before looking for a pituitary tumor, they should be ruled out as the cause of the elevated PRL levels.

Prolactinomas (Fig. III.27)

Prolactin-secreting tumors are the most common type of pituitary tumor, representing over one-third of all pituitary tumors. Many of the patients previously classified as having non-secretory chromophobe adenomas actually have prolactinomas. The diagnosis of prolactinomas has increased due to the use of specific RIAs as well as to greater clinical awareness.

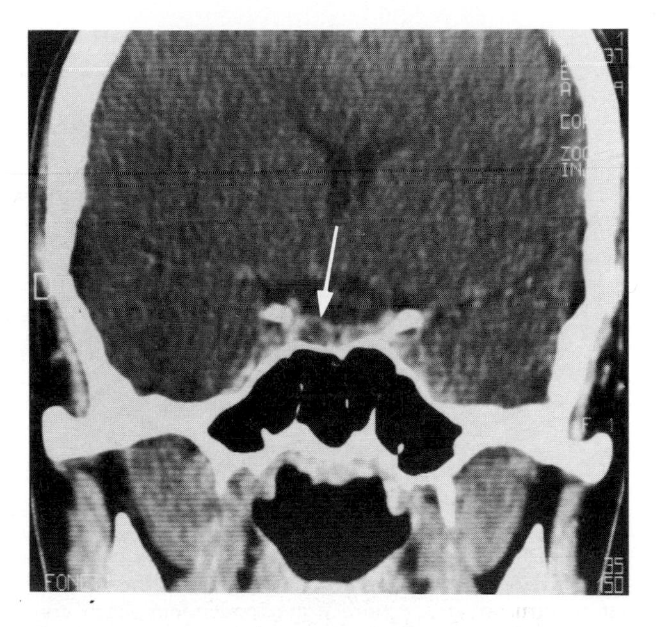

Figure III.27 Coronal CT scan of sella turcica demonstrating a prolactin-secreting pituitary adenoma (arrow). The adenoma is less dense than the remaining parenchyma. The upper margin of the pituitary is convex. (Courtesy of Dr. P. Bouchard, Hôpital de Bicêtre).

Prolactinomas occur five times more frequently in *women* than in men. In women, most prolactinomas are small (microadenomas), whereas, in men, macroadenomas are more common. The major clinical manifestations in women are frigidity, and primary or secondary *amenorrhea* due to inhibition of pituitary gonadotropin secretion. Galactorrhea (milk secretion), which was formerly thought to occur in all cases of prolactinoma, is only seen in about one-third of the patients with amenorrhea (p. 462). Hyperprolactinemia can be responsible for anovulation and/or infertility. However, in some women with elevated prolactin levels, normal ovulatory cycles can be observed. Most men with prolactinoma have disturbances of the visual field (bitemporal hemianopsia), due to suprasellar extension of a macroadenoma (Fig. III.28), and headache, due to parasellar involvement. Almost all men presenting with prolactinomas have some degree of decreased libido and potency, eventually associated with low serum testosterone levels, as well as azospermia or oligospermia.

38. Lamberts, S. W. J., Vitterlinden, P., Verschoor, L., VanDongen, K. J. and Del Pozo, E. (1985) Long-term treatment of acromegaly with the somatostatin analogue SMS 201-995. *N. Engl. J. Med.* 313: 1576–1580.

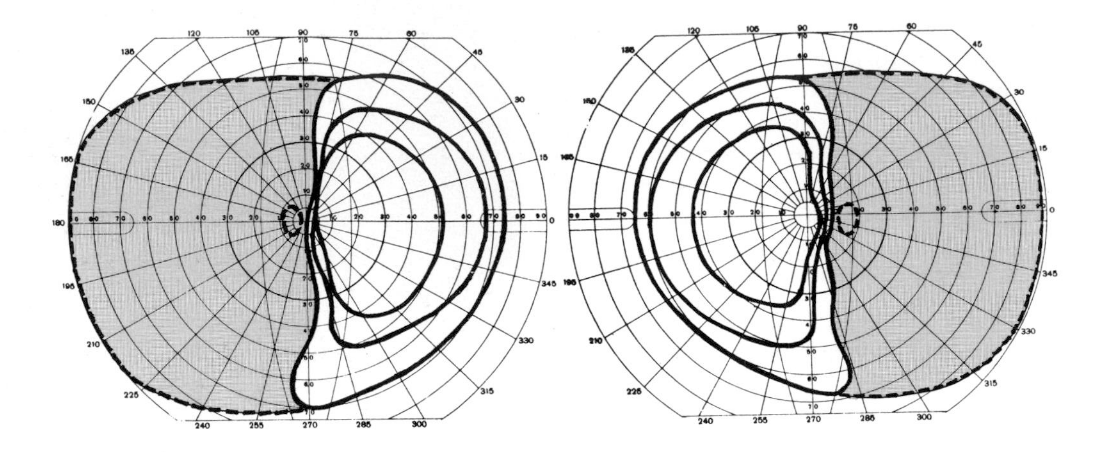

Figure III.28 Bitemporal hemianopsia due to the suprasellar protrusion of a pituitary macroadenoma. Each concentric line (isopter) passes through points of equal visual acuity. Hemianopsia means blindness (gray region) in one half of the visual field. Bitemporal hemianopsia indicates a lesion of the decussating nasal retinal fibers of the optic chiasma. The small dotted oval in the gray area is the blind spot. (Courtesy of Dr. P. Bouchard and Mrs. R. Barbaud, Hôpital de Bicêtre).

Treatment of Prolactin-Secreting Tumors

Over the last several years, there has been a great deal of controversy concerning the correct form of treatment for patients with prolactinomas. Since most prolactinomas occur in women, and since the vast majority of these are microadenomas which grow very slowly with little or no increase in secretion, those patients who do not want to become pregnant or who are not concerned by amenorrhea or galactorrhea can simply be followed at six- or 12-month intervals by RIA measurements. CT examinations should be performed when serum PRL increases with repeated sampling.

Prior to surgical, radiological or medical treatment, other endocrine disorders that affect PRL secretion, such as hypothyroidism, renal insufficiency or drug-induced hypoprolactinemia, must be ruled out.

Most clinicians agree that *treatment is necessary* for all women with hyperprolactinemia. The treatment reduces tumor volume, avoids estrogen deficiency and its bone risk, and allows the establishment of ovulatory cycles and, eventually, pregnancy. In men, treatment reverses impotence and decreases tumor volume. If hyperprolactinemia is associated with a radiologically *visible* pituitary adenoma, a good success rate (70 to 80%) can be obtained with transsphenoidal *resection* (Fig. III.29). One report has indicated that, despite the initial success of surgery, long-term follow-up (over five years) showed a high frequency of recurrence (12 out of 24 microadenomas and four out of five macroadenomas)[39]. In a contrasting report, in a group of 35 patients with prolactinomas, prolactin levels returned to normal in 26 patients. Hyperprolactinemia did not recur in any patient whose prolactin level had returned to normal six weeks after surgery, including 16 patients with macroadenomas[40]. Such data suggest that partial hypophysectomy offers an acceptable mode of treatment for some patients with prolactinomas.

When *no radiological* evidence of adenoma is seen, medical treatment with a *dopamine agonist* such as bromocryptine, pergolide, lisuride or mesulergine is appropriate (Fig. III.30). Treatment usually extends over a six-month period, and, following discontinuation of therapy for one month, PRL is measured. Treatment is resumed and the efficacy is tested again six months later, this time with a CT scan in addition to RIA. If hyperprolactemia persists, the patient may choose to continue medical treatment or proceed with transsphenoidal exploration. Frequently, surgery and medical treatment are combined (Fig. III.29).

39. Serri, O., Rasio, E., Beauregard, H., Hardy, J. and Somma, M. (1983) Recurrence of hyperprolactinemia after selective transsphenoidal adenomectomy in women with prolactinoma. *N. Engl. J. Med.* 309: 280–283.
40. Scanlon, M.F., Peters, J.R., Thomas, J.P., Richards, S.H., Morton, W.H., Howell, S., Williams, E.D., Hourihan, M. and Hall, R. (1985) Management of selected patients with hyperprolactinemia by partial hypophysectomy. *Brit. Med. J.* 291: pp. 1547–1550.

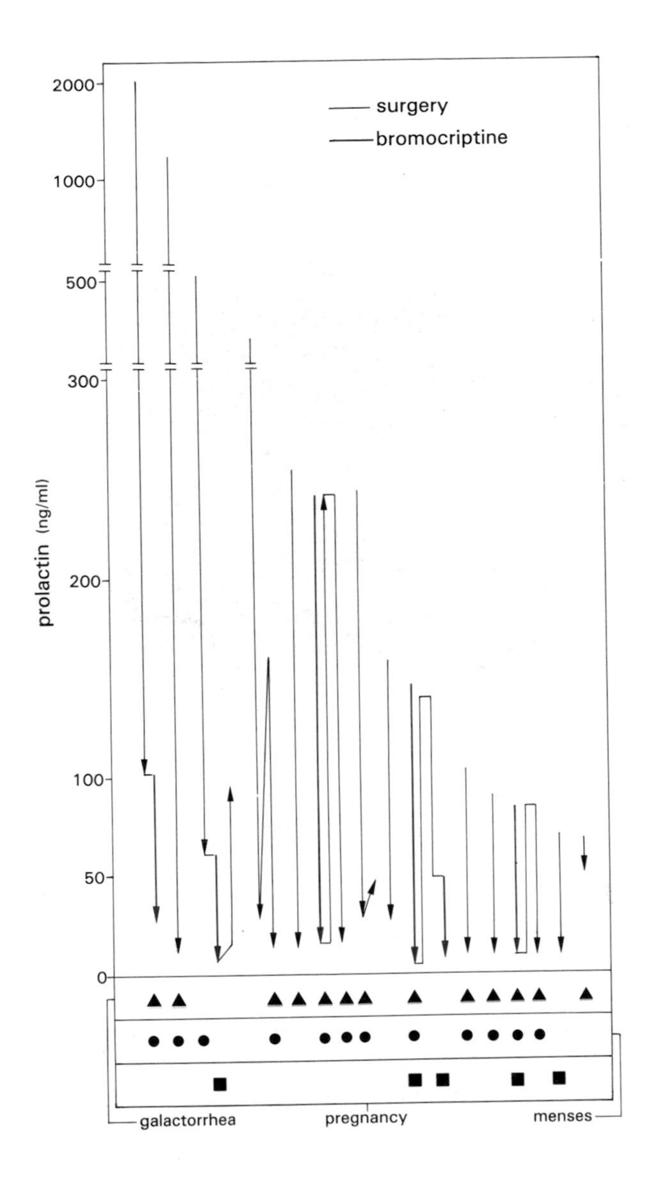

Figure III.29 Effects of transsphenoidal adenomectomy or bromocriptine on serum prolactin levels in female patients with prolactinomas. Note that, with one exception, serum PRL increased when bromocriptine was discontinued. (Adapted from Tolis, G. and Franks, S. (1979) Physiology and pathology of prolactin secretion. In: *Clinical Neuroendocrinology: a Pathophysiological Approach* (Tolis, G., Labrie, F., Martin, J. B. and Naftolin, F., eds), Raven Press, New York, pp. 291–317).

Radiation therapy is used for those patients who do not respond well to bromocriptine or related drugs or whose prolactin levels have increased again sometime after surgery.

In North America, transsphenoidal resection of the microadenoma is probably the preferred form of therapy. Regular menses return in the majority of cases, and conception is common. Failures generally occur in patients with the highest PRL levels. In such cases, treatment with a dopamine agonist is necessary.

In Britain and Europe, most patients with microadenomas are first treated with a dopamine agonist such as bromocriptine. Regular menses resume in >90% of the women within the first several weeks. If pregnancy occurs, treatment is discontinued until after delivery. Patients should be followed by visual field examinations, at regular intervals, in order to detect symptoms of growth of the microadenoma.

Treatment of *macroadenomas* is even more *controversial*. At best, surgery leads to a 50% cure rate. Initial treatment should be bromocriptine or another dopamine agonist. Over 80% of patients who respond actually show a decrease in tumor size on CT scan (Fig. III.30). This is probably the result of a reduction of cell volume rather than cell number. Serum PRL measurements are a good guide to the efficacy of treatment, and should be coupled with CT confirmation at regular intervals. Frequently, bromocriptine treatment is administered to reduce tumor size before surgery.

One of the disadvantages of long-term medical treatment of prolactinomas is the necessity of taking compounds such as bromocriptine two or three times a day. Bromocriptine therapy is occasionally associated with side effects in some patients. Efforts have been made to develop dopamine agonists with a longer duration of action and less adverse side effects. CV 205-502, an octahydrobenzo(g)-quinoline (Sandoz, Ltd.), is a new, long-acting, non-ergot dopamine agonist that has been shown to effectively inhibit prolactin secretion in all animal models. CV 205-502 is approximately 35 times as potent as bromocriptine (similar to pergolide (Lilly, Inc.)), with a prolonged duration of action. In a double blind study[41], a dose of 60 μg was sufficient to lower prolactin levels to normal in hyperprolactinemic women for at least 24 h. Side effects of CV 205-502 were mild and transient, and women who could not tolerate bromocriptine (or pergolide, which has the same side effects) had no problem accepting CV 205-502. Medical treatment of prolactin-secreting tumors with such a long-acting drug may become the preferred choice of therapy.

41. Rasmussen, L. C., Bergh, T., Wide, L. and Brownell, J. (1987) CV 205-502: a new long-acting drug for inhibition of prolactin hypersecretion. *Clin. Endocrinol.* 26: 321–326.

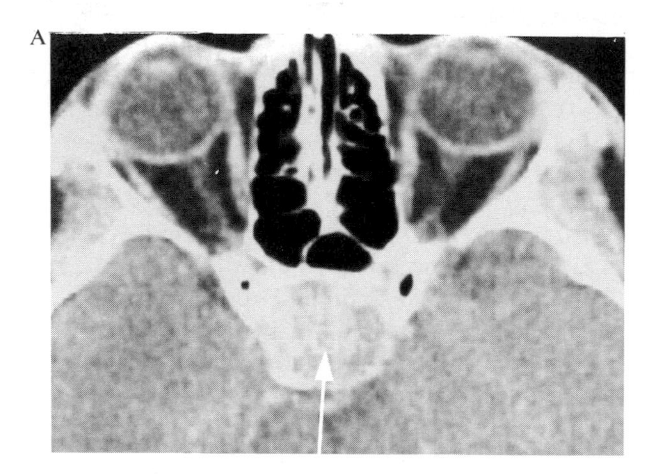

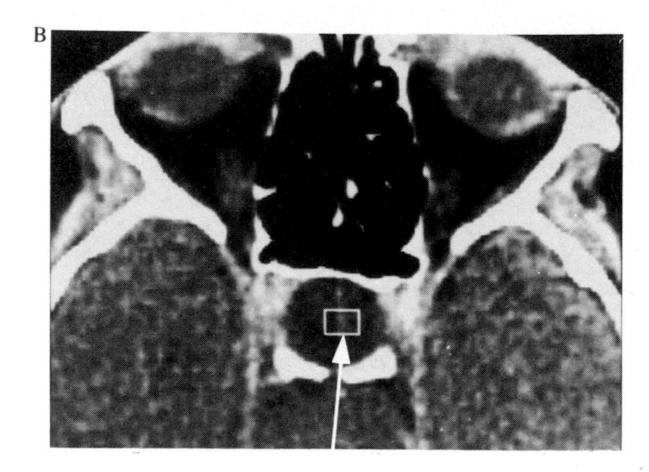

Figure III.30 CT head scans (postenhancement). Panel **A**: before therapy; panel **B**: three months after starting bromocryptine therapy. In **A**, the CT scan shows a large enhancing mass in the pituitary fossa. In **B**, the scan shows a dramatic reduction in tumor size, with a typical aspect of empty sella. (Courtesy of Dr. G. Schaison, Hôpital de Bicêtre).

8. Summary

Growth hormone and prolactin are pituitary hormones with *multiple actions*. GH is principally associated with body growth and metabolism, and PRL with lactation.

Growth hormone is produced by *somatotropic cells* of the anterior pituitary, while PRL is secreted by *lactotropic cells*. The increase in the size of the pituitary occurring during pregnancy is due to a proliferation of lactotropic cells. Human GH comprises as much as *10%* of the *dry weight* of the *pituitary*. It is a single polypeptide chain of 191 aa with two disulfide bonds. Human PRL is structurally similar to hGH, consisting of 199 aa with three disulfide bonds. It is much less abundant in the pituitary than GH. The placenta produces a hormone, *placental lactogen (PL)* or chorionic somatomammotropin (CS), structurally related to GH and PRL. Its physiological importance is not fully understood (p. 476).

The genes for hGH, hPL and hPRL all consist of five exons and four introns. There is 92% nucleotide sequence homology between the mRNAs coding for human GH and PL, while hPRL has slightly over 40% homology with the nucleotides of the other two hormones. Analysis of the amino acid and nucleotide sequences suggests that these three hormones evolved from duplication of a common evolutionary ancestor over 400 million years ago. *Different forms of hGH* exist in the pituitary. The most abundant is the 22 kDa monomer (191 aa), but a 20 kDa variant, which lacks the diabetogenic properties of GH, is also found. This form is not encoded by one of the variant GH genes, but is produced by alternative processing and splicing of the same pre-mRNA as 22 kDa GH, resulting in the removal of amino acids 32 to 46.

Growth hormone and prolactin act first by interacting with specific *receptors* in the plasma *membrane* of target cells. These receptors can be up- and/or down-regulated by the homologous hormone as well as by peripheral hormones. The molecular weight of the binding unit of the GH receptor is ~130,000, while that of PRL is ~75,000. The predicted amino acid sequences, deduced from their cDNAs, encode proteins of ~69,000 and ~67,000, respectively. Regions of strong sequence homology suggest that both receptors originated from a common ancestor. Neither GH nor PRL acts via changes in cAMP production, phosphoinositol metabolism, or tyrosine phosphorylation.

The *actions* of GH can be divided into two categories. 1. *Indirect* effects related to growth, mediated by *IGF-I*, and 2. effects on *metabolism* which are *direct*, mediated by specific receptors in target tissues. IGFs are factors that stimulate mitogenesis of many cells. IGF-I is much more responsive to GH than is IGF-II: The liver is the main site of production of both IGFs. Growth hormone is *regulated* by *two hypothalamic* factors, *GRF*, which stimulates secretion, and *somatostatin*, which inhibits it. In humans, GRF is a 44 aa peptide (43 in rats) coded for by a gene containing five exons and four introns. GRF acts primarily by binding to specific receptors on somatotropic cells, increasing the production of cAMP. Somatostatin, which exists as 14 and 28 aa forms, inhibits GH secretion. One gene, consisting of two exons and one intron, codes for both forms. Although somatostatin inhibits GRF-stimulated cAMP formation, other mechanisms, such as Ca^{2+} and the inositol phospholipid system, are also involved in its action. GH secretion is episodic. In humans, GH levels are usually <5 ng/ml. IGF-I can regulate GH secretion by negative feedback at the pituitary and hypothalamic level. The onset of *sleep* induces a characteristic increase in GH levels.

The best known action of PRL is on the mammary gland, where it stimulates DNA synthesis, epithelial cell proliferation, and synthesis of milk proteins, FFAs, and lactose. Other reproductive actions, as well as osmoregulatory effects, are also observed. The multiple actions of prolactin correlate well with the widespread distribution of its receptors. *Prolactin* is under *general negative control* by the hypothalamus. *Dopamine*, which binds to specific receptors on lactotropic cells, inhibiting cAMP levels, is thought to be the major factor regulating PRL secretion. *GAP*, a 56 aa C-terminal extension of GnRH, has been proposed as a PIF. Estrogens and TRH stimulate PRL secretion. PRL concentrations average ~10 ng/ml in women (levels are slightly lower in men). The nocturnal rise in serum PRL occurs much later than that of GH.

GH deficiency leads to *dwarfism*, which can be due to primary pituitary dysfunction, hypothalamic dysfunction, or peripheral resistance (Laron dwarfism). Treatment, except for the last category, generally involves injection of hGH. GRF therapy may be useful in the future. An excess of GH in children leads to *gigantism*, while in adults it results in *acromegaly*. Oversecretion is usually the result of a pituitary GH-secreting adenoma or an ectopic GRF-secreting tumor.

Hyperprolactinemia is the most common hypothalamic-hypophyseal pituitary disorder in humans. It also is caused by a pituitary adenoma. *Prolactinomas* are five times more frequent in women. Hypersecretion of PRL results in amenorrhea in women (sometimes accompanied by galactorrhea) and decreased libido and potency in men.

Treatment of both GH- and PRL-secreting adenomas involves transsphenoidal resection of the tumor. Success is good when the tumor is <10 mm in diameter (microadenoma). Medical therapy using bromocriptine and, more recently, a long-acting somatostatin analog, is used for GH tumors. Dopaminergic analogs, such as bromocriptine, are very effective therapy for prolactinomas. Sometimes medical treatment and surgery are combined.

General References

Daughaday, W. H. (1985) The anterior pituitary. In: *Textbook of Endocrinology* (Wilson, J. D. and Foster, D. W., eds), W.B. Saunders, Philadelphia, pp. 568–613.

Gershengorn, M. C. (1985) Thyrotropin-releasing hormone action: mechanism of calcium-mediated stimulation of prolactin secretion. *Rec. Prog. Horm. Res.* 41: 607–646.

Guillemin, R., Brazeau, P., Bohlen, P., Esch, F., Ling, N., Wehrenberg, W. B., Bloch, B., Mougin, C., Zeytin, F. and Band, A. (1984) Somatocrinin: the growth hormone releasing factor. *Rec. Prog. Horm. Res.* 40: 233–286.

Kelly, P. A., Djiane, J., Katoh, M., Ferland, L. H., Houdebine, L. M., Teyssot, B. and Dusanter-Fourt, I. (1984) The interaction of prolactin with its receptors in target tissues and its mechanism of action. *Rec. Prog. Horm. Res.* 40: 379–436.

Lewis, V. J., Singh, R. N. P., Tutwiler, G. F., Sigel, M. B., Vanderlaan, E. F. and Vanderlaan, W. P. (1980) Human growth hormone: a complex of proteins. *Rec. Prog. Horm. Res.* 36: 477–502.

MacLeod, R. M. (1976) Regulation of prolactin secretion. In: *Frontiers in Neuroendocrinology* (Martini, L. and Ganong, W. F., eds), Raven Press, New York, pp. 164–194.

Melmed, S., Braunstein, G. D., Horvath, E., Ezrin, C. and Kovacs, K. (1983) Pathophysiology of acromegaly. *Endocrine Rev.* 4: 271–290.

Moore, D. D., Walker, M. D., Diamond, D. J., Conkling, M. A. and Goodman, H. M. (1982) Structure, expression and evolution of growth hormone genes. *Rec. Prog. Horm. Res.* 38: 197–222.

Reichlin, S. (1983) Somatostatin. In: *Brain Peptides* (Krieger, D. T., Brownstein, M. and Martin, J. B., eds), Wiley, New York, pp. 711–752.

Seo, H. (1985) Growth hormone and prolactin: chemistry, gene organization, biosynthesis and regulation of gene expression. In: *The Pituitary Gland* (Imura, H., ed), Raven Press, New York, pp. 57–82.

Denis Gospodarowicz

Growth Factors

Growth factors are peptides which *stimulate cell proliferation*; they may also directly or indirectly stimulate *phenotypic transformation* of cells. All produce their effects by interaction with *specific membrane receptors*. They are involved not only in tumor evolution, but also in *normal development* (embryogenesis and hematopoiesis) and in some *pathological* or *repair* processes, such as atherosclerosis or wound healing. In most cases, the primary structure of these peptides is now known.

Epidermal Growth Factor (EGF) and the α and β Transforming Growth Factors (TGFs)[1–3]

EGF has been identified in a wide variety of tissues, but, with the exception of the submaxillary glands, all contain low concentrations. In the human, *urogastrone* (the original name of human EGF) is found in the urine, and has 70% amino acid homology with mouse EGF. In vitro, EGF is mitogenic for many ectodermal and mesodermal cells. In vivo, it increases the proliferation of the basal cell layer of many epithelia of ectodermal origin, and may act as a fetal growth hormone.

The action of EGF as an inducer of cell proliferation implicates binding to a specific *receptor* (p. 78). *Early metabolic effects* include increased nutrient uptake, ion fluxes, and phosphorylation of membrane proteins. *Subsequent* effects are the stimulation of RNA and protein synthesis, with an increase in ornithine decarboxylase and thymidine kinase activities, followed by the initiation of DNA synthesis.

Transforming growth factors *(TGFs)* induce a *transformed phenotype* in cultured cells, i.e. lack of density-dependent growth inhibition, growth in soft agar, and anchorage-independent growth. Characteristic morphological changes may occur. Todaro coined the initials TGF in 1978.

TGF-α, the first described, has homology with the N-terminal portion of EGF. It consists of one peptide chain. This factor acts via the EGF receptor, a ~170 kDa glycoprotein with Tyr-PK activity. TGF-α is mitogenic for epithelial and fibroblastic cells, and has effects that are synergistic with several transforming agents. TGF-α activity is found in the urine of all cancer patients, and, to a lesser extent, in the urine of non-cancer patients.

Mysterious TGF-β: (Anti) Growth Factor/Hormone?

TGF-β, with a MW of ~26,000, is a homodimer of two 112 aa chains connected by SS-bridges (nine cysteines in each chain). It does *not* interact with the *EGF receptor*, but only with its own receptor, formed of two glycosylated chains, each with a MW of ~300,000, and does not have Tyr-PK activity. TGF-β is *produced by many cells*, including platelets (in which it was discovered), and in bone matrix, in which, curiously, it is abundant. *Most synthesizing cells* also contain the *specific receptor*, and thus it is believed that TGF-β is involved in many *paracrine* and *autocrine* control activities related to growth. In fact, it is *often an inhibitor* of normal and tumoral cell growth, and can modulate, positively or negatively, the biological effect of other growth factors, such as EGF or FGF. In addition, it can promote growth of connective tissue (with increased collagen synthesis) and angiogenesis, two important factors for *wound repair*. Experiments have suggested that TGF-β, in the same cells and at the same concentration, *may alternatively* be a growth stimulator or growth inhibitor, depending on what other hormones and growth factors are present. The stimulatory activity of TGF-β on FSH secretion by gonadotropic cells and its homology with inhibin chains are reported on p. 453 and in Figure X.10. Moreover, TGF-β is an *inhibitor* of the *cholesterol* to *pregnenolone* stimulatory activity *of ACTH* in adrenocortical cells, independent of any change in cAMP level.

Platelet-Derived Growth Factor (PDGF)[4]

PDGF is stored in *α granules of platelets,* and is released, during platelet activation, when blood vessels are injured. It is a heterodimer composed of A and B chains with partially related structures. PDGF-B has been shown to have extensive sequence *homology with p24 sis,* the transforming protein of the simian sarcoma virus (p. 122). *PDGF* accounts for much of the *serum ability to stimulate the growth of cells in culture,* in particular that of fibroblasts, and glial and vascular smooth muscle cells. Since it is a chemoattractant for many cells, it may play a role in inflammation and tissue repair as well as in tumor cell growth, particularly glioma and osteosarcoma.

Fibroblast Growth Factors (FGFs)[5]

Also Known as Heparin Binding Growth Factors (HBGFs)

Fibroblast growth factor exists in two forms, a *basic form (bFGF), present in all organs, and an acidic form (aFGF),* present mostly in the brain and retina. Both forms have strong sequence homology. They interact with the *same receptors,* and, in vitro, trigger the proliferation of many mesoderm- and neuroectoderm-derived cells. Basic FGF has also been shown to induce differentiation of various cell types (chondrocytes, nerve cells, fibroblasts, and endothelial cells), as well as to delay their ultimate senescence. In vivo, both forms have angiogenic properties, and bFGF promotes limb regeneration in lower vertebrates. Recent studies have shown that bFGF is probably the vegetalizing factor responsible for mesodermal induction from animal pole cells in early embryonic development.

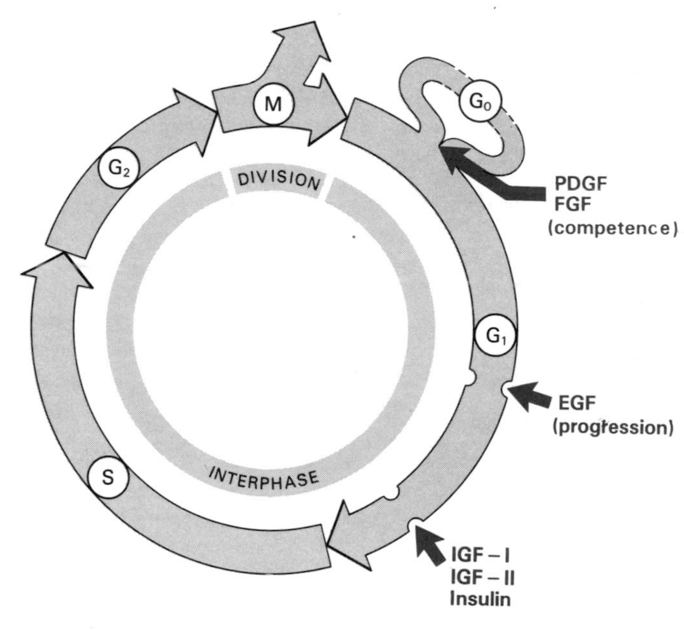

Figure 1 *Cell cycle and growth factors in a mammalian cell. In different growing cells, the S (synthesis), G_2 (gap) and M (mitotic) phases are approximately constant, lasting seven, three and one hour(s), respectively. G_1 is most variable, from two hours to several days. Tissue culture cells which, in the absence of appropriate hormone or nutrient, cease to grow are in G_0 (they are different from G_1 cells, since they are not preparing for DNA replication). The indicated growth factors stimulate the cycle at different restriction points in G_1.*

Insulin-like Growth Factors (IGFs)

Purified mostly from human plasma, and showing some insulin-like properties, *IGFs* have also been *known as somatomedins,* originally described as mediating growth hormone action on cartilage (p. 198). They have mitogenic effects, as does multiplication stimulating activity (MSA) originally isolated from an established line of rat liver cells (BRL-3A) and now termed IGF-II.

There is strong *amino acid homology between IGF-I, IGF-II, and proinsulin* (Fig. III.8). It is likely, therefore, that they originated from duplication of the insulin gene. However, while the *receptors for IGF-I and insulin* have a *similar* structure, the *IGF-II receptor* is *very different* (Fig. I.78). Since there is some interaction of all three peptides with the three corresponding receptors, there is some crossover in their activities. It is believed that the growth-promoting effect of insulin is exerted via IGF receptors (mostly IGF-I receptor), while the metabolic effects of IGFs are exerted mostly via the insulin receptor.

Table 1. *Growth factors.*

Factors	*Principal sources*	*Responsive cells or tissue*	$\sim M_r\ (10^{-3})$
Epidermal growth factors (EGFs) Urogastrone	Mouse submaxillary gland Human urine	Epidermal cells, fibroblasts	6
Transforming growth factor α (TGF-α)	Murine, rat, and human retrovirus cells	Frequently tested on rat fibroblast cells (NRK-49F)	5.6
Transforming growth factor β (TGF-β)	Most organs and transformed cells	Frequently tested on rat fibroblast cells (NRK-49F)	25 (homodimer, disulfide-linked)
Platelet-derived growth factor (PDGF)[a]	Human platelets (α granules)	Arterial smooth muscle cells, fibroblasts	32 (two partially identical subunits, disulfide-linked)
Basic fibroblast growth factor (bFGF)[b] (class II heparin-binding growth factor)	Bovine pituitary and brain also in most organs	Mesenchymal and neural crest-derived cells	16 and 14.5
Acidic fibroblast growth factor, (aFGF)[c] (class I heparin-binding growth factor)	Bovine brain and retina	Mesenchymal and neural crest-derived cells	14
Insulin	B cells of islets of Langerhans	Liver, muscle, etc.	5.7 (two non-identical chains)
Insulin-like growth factor I (IGF-I) or somatomedin C	Human plasma	Liver, adipose, muscle, cartilage, fibroblasts	7.6
Insulin-like growth factor II (IGF-II) or somatomedin A or multiplication stimulating activity (MSA)	Human plasma Rat: serum and liver	Liver, adipose, muscle, cartilage, fibroblasts	7.5 8
Nerve growth factor (NGF)	Mouse submaxillary gland and most organs	Sympathetic nervous system	13
Bone morphogenetic protein (BMP)	Bovine bone	Mesenchymal cells	18
Erythropoietin	Urine	Erythroblasts	34
Granulocyte-macrophage colony stimulating factor (GM-CSF)	Mouse lung (conditioned medium)	Granulocytes and macrophages	23
Granulocyte colony stimulating factor (G-CSF)	Mouse lung (conditioned medium)	Granulocytes	25
Multipotential colony stimulating factor (multi-CSF)	Leukemia cell	Granulocytes, macrophages, erythroid cells, eosinophils, megakaryocytes, mast cells, stem cells	15 to 28
Macrophage colony stimulating factor (M-CSF)	L cell	Macrophages	70
Interleukin-1 (IL-1)	Activated macrophages	T and B lymphocytes Fibroblast synovial cells	11
Interleukin-2 (IL-2)	T lymphocytes	T lymphocytes	15

[a]PDGF is also known as: osteosarcoma-derived growth factor, glioma-derived growth factor, fibroblast-derived growth factor, transforming protein of simian sarcoma virus (Fig. I.136).
[b]Basic FGF is also known as: eye-derived growth factor I, chondrosarcoma-derived growth factor, cartilage-derived growth factor, endothelial growth factor II, angiogenic growth factor, macrophage growth factor.
[c]Acidic FGF is also known as: retina-derived growth factor, eye-derived growth factor II, endothelial cell growth factor, endothelial growth factor I.

Bone Morphogenetic Protein (BMP)[6]

Bone morphogenetic protein is extracted from bone extracellular matrix. It favors adhesion, growth, and differentiation of bone cells. This factor induces the differentiation of pericytes (perivascular mesenchymal cells) into cartilage and bone cells.

Nerve Growth Factor (NGF)[7,8]

NGF has been identified in a wide variety of vertebrate tissues, but all contain very low concentrations, with exception of the *submaxillary* glands. NGF induces *differentiation* (neurite induction and elongation) of the sympathetic and portions of the embryonic nervous system. An additional important effect of this hormone is the *maintenance of the viability of responsive neurons*, in vivo and in vitro.

Table 2. *Growth factor receptors.*

Receptor	*Subunit Composition (kDa)*	*Activity*
EGF	One polypeptide chain (170)	Tyrosine kinase
TGF-α[a]	One polypeptide chain; two forms (170 and ~60)	Tyrosine kinase
TFG-β	Two identical disulfide- linked chains of ~300	No tyrosine kinase
PDGF	One polypeptide chain (185)	Tyrosine kinase
Basic FGF[b] or HBGF Class II	One polypeptide chain; two forms (145 and 125)	Tyrosine kinase
Acidic FGF[b] or HBGF Class I	One polypeptide chain; two forms (145 and 125)	Tyrosine kinase
Insulin	Two α chains (135) and two β chains (90)	Tyrosine kinase
IGF-I	Two α chains (135) and two β chains (90)	Tyrosine kinase
IGF-II	One polypeptide chain (250)	Not known
NGF	Two identical disulfide-linked chains of ~65	No tyrosine kinase
CSF or c-*fms*	One polypeptide chain (165)	Tyrosine kinase

[a]The 170 kDa receptor for TGF-α is the EGF receptor.
[b]Basic FGF interacts with both 145 kDa and 125 kDa polypeptides, while acidic FGF interacts preferentially with the 125 kDa.

Growth Factors for the Hematopoietic System[9–11]

Most of the red and white cells in circulating blood are short-lived and need to be replaced constantly throughout life. This process of blood cell formation, termed hematopoiesis, is regulated by a dynamic interaction between a set of progenitor cells and glycoprotein growth factors. Red cell formation is controlled by *erythropoietin*, a glycoprotein isolated from urine. The proliferation of white cells is under the control of the granulocyte-macrophage *colony stimulating factors*. These are well characterized, specific glycoproteins that interact to control the production, differentiation, and function of two related white cell populations of the blood, granulocytes and monocyte-macrophages (and red cells for IL-3). Widely produced in the body, these regulators probably play an important role in resistance to infection. The proliferation of myeloid leukemia cells remains dependent on stimulation by colony stimulating factors. One of them, however, also has the ability to suppress leukemic cell populations by inducing terminal differentiation.

Lymphoid Growth Factors: Interleukins[11–14]

The proliferation of both types of lymphocytes (B and T) is under the control of growth factors named interleukins (ILs) (p. 253). *IL-1,* released by macrophages undergoing an immune response, regulates *B lymphocyte differentiation.* It also controls the *release of IL-2 from T lymphocytes,* which is a potent *growth promoter* for that cell type. *IL-1* also controls the expression of receptors for IL-2 on T lymphocytes.

Growth Factors and Cellular Oncogenes

The remarkable multiple connections between growth factors and oncogenes are discussed on p. 122. Thus, the biology of growth factors is at the crossroad of hormonal mechanisms and cellular gene activation, with definite implications in virology and oncology.

References

1. Carpenter, G. and Cohen, S. (1979) Epidermal growth factor. *Ann. Rev. Biochem.* 48: 193–216.
2. Sporn, M. B. Roberts, A. B., Wakefield, L. M. and deCrombrugghe, B. (1987). Some recent advances in the chemistry and biology of transforming growth factor-beta. *J. Cell Biol.* 105: 1039–1045.
3. Sporn, M. B., Roberts, A. B., Wakefield, L. M. and Assoian, R. K. (1986) Transforming growth factor-β: biological function and chemical structure. *Science* 233: 523–534.
4. Ross, R., Raines, E. W. and Bowen-Pope, D. F. (1986) The biology of platelet-derived growth factor. *Cell* 46: 155–169.
5. Gospodarowicz, D., Ferrara, N., Schweigerer, L. and Neufield, G. (1987) Structural characterization and biological functions of fibroblast growth factor. *Endocrine Rev.* 8: 95–114.
6. Urist, M. R., Huo, Y. K., Brownell, A. G., Hohl, W. M., Buyske, J., Lietze, A., Tempst, P.,Hunkapiller, M. and De Lange, R. J. (1984) Purification of bone morphogenic protein by hydroxyapetite chromatography. *Proc. Natl. Acad. Sci., USA* 81: 371–375.
7. Yanker, B. A. and Shooter, E. M. (1982) The biology and mechanism of action of nerve growth factor. *Ann. Rev. Biochem.* 51: 841–868.
8. Levy-Montalcini, R. (1987) The Nerve Growth Factor: thirty-five years later. *EMBO. J.* 6: 1145–1154.
9. Goldwasser, E., Krantz, S. B. and Wang, F. F. (1985) Erythroprotein and erythroid differentiation. In: *Mediator in Cell Growth and Differentiation* (Ford, R. J. and Maizel, A. L., eds) Raven Press, New York, pp. 103–107.
10. Metcalf, D. (1985) Granulocytes-macrophages colony stimulating factors. *Science* 229: 16–22.
11. Sieff, C. A. (1987) Hematopoietic Growth Factors. *J. Clin. Invest.* 79: 1549–1557.
12. Lackman, L. B. (1985) The purification and biological properties of human interleukin 1. In: *Mediator in Cell Growth and Differentiation* (Ford, R.J. and Maizel, A.L., eds) Raven Press, New York, pp. 171–183.
13. Auron, P. E., Webb, A. C., Rosenwasser, L. J., Mucci, S. F., Rich A., Wolff, S.M. and Dinarello, C.A. (1984) Nucleotide sequence of human monocyte interleukin 1 precursor cDNA. *Proc. Natl. Acad. Sci., USA* 81: 7907–7911.
14. Smith, K. A. (1985) The determinants of T cell growth. In: *Mediator in Cell Growth and Proliferation* (Ford, R. J. and Maizel, A. L., eds) Raven Press, New York, pp. 185–192.

Ira Pastan and Mark C. Willingham

Receptor-Mediated Endocytosis

Peptide hormones, growth factors and many other biologically important molecules bind to specific receptors on the plasma membrane of cells. Each cell type has its characteristic array of receptors. Shortly after hormone binding, the hormone-receptor complexes are internalized in a process called receptor-mediated endocytosis. The structural elements involved in this type of endocytosis are: *coated pits*, associated with the cell surface, endocytic *vesicles*, termed *"receptosomes"* or *"endosomes"*, elements of the *Golgi* apparatus, and *lysosomes* (Fig. 1). Most receptors are transmembrane glycoproteins. Some receptors are also enzymes (e.g., the protein kinase activity associated with the insulin and EGF receptors). Others are in close association with enzymes (e.g., components of the adenylate cyclase system (p. 100 and p. 335)).

In most cases, the *hormone-receptor complexes* are not stationary in the plasma membrane but *diffuse laterally* with a diffusion coefficient of 3 to 8×10^{-10} cm^2/s. Randomly distributed in the plasma membrane are specialized structures, invaginations approximately 0.1μm in diameter, termed "bristle coated pits". The cytoplasmic face of each coated pit is covered with an unusual protein, termed *clathrin* (the major species has a MW of ~180,000). *A typical cell* has about *1000 pits*, and these occupy *1 to 2% of the area* of the cell surface. Based on studies with cultured cells, it has been estimated that each coated pit gives rise to a receptosome every 20 to 30 seconds. The details of how receptosomes form are still unclear, but in the process of formation, the contents of the coated pits are transferred into the vesicle. Coated pits capture and thereby concentrate receptors as they diffuse laterally in the plasma membrane. Based on the measured diffusion coefficients, each hormone-receptor complex will encounter a coated pit every three seconds. Some receptors are only captured in pits when occupied by a ligand (e.g., epidermal growth factor (EGF)); others are trapped even when unoccupied (e.g., transferrin, low density lipoprotein, asialoglycoprotein). Endocytosis has been studied mainly in epithelial and fibroblastic cells, where patching and capping are rarely observed. In these cells, the first site at which ligand-receptor complexes (or unoccupied receptors) have been found to be concentrated or clustered is in coated pits. Although changes in membrane fluidity could have a significant effect on the process of endocytosis, no direct evidence linking these two phenomena is available.

Receptosomes or endosomes are *uncoated transport* vesicles with a *low pH*; they move by saltatory motion along tracks of microtubules. Their acidic environment (pH 4.5 to 5.0) causes many hormone-receptor complexes to dissociate and may thereby arrest the actions of these hormones. Receptosomes do not contain appreciable amounts of degradative or processing enzymes. Thus, the ligand and receptor are not extensively modified in this compartment. *Ten to 15 min after* entry, these vesicles begin to fuse with tubular elements of the Golgi, termed the "*trans*-reticular Golgi" (*trans*-Golgi). As a result, the ligand and the receptor enter the tubules of the Golgi, where their ultimate destination is determined. *Three different types of sorting* have been observed. In the *first* instance, as with EGF, ligand and receptor are sent on to lysosomes, to be *destroyed*. This results in down-regulation of receptors (Fig. 2). In the *second* instance, as with transferrin, the ligand and the receptor are *returned* intact to the cell *surface*. This occurs via exocytic vesicular structures (Fig. 2). In the *third* instance, as with asialoglycoprotein, low density lipoprotein or α_2-macroglobulin‡, the *ligand* is transferred to lysosomes, where it is destroyed, and the *receptor* is returned to the cell *surface*, to be reused (Fig. 1). It requires about *10 to 20 min for a recycling receptor* to complete an entire cycle (i.e. to carry a ligand into the cell and then return to the surface). Such receptors carry out many cycles before they are degraded. Thus, they have a long half-life which is unaffected by receptor occupancy. When unoccupied, the EGF receptor also has a long half-life (of 6 h), which is reduced to 40 min when occupied.

The *trans*-Golgi is a membrane-limited compartment in which receptors and ligand-receptor complexes can diffuse laterally within the membrane. It contains small clathrin-coated pits about half the size of those at the cell surface. Their function is to collect and concentrate molecules destined to be transferred to lysosomes, and to carry out the transfer process.

‡α_2-macroglobulin is a protease inhibitor that circulates in the plasma until it combines with a protease. The complex is then endocytosed via a receptor found on fibroblasts and macrophages, and is destroyed in lysosomes.

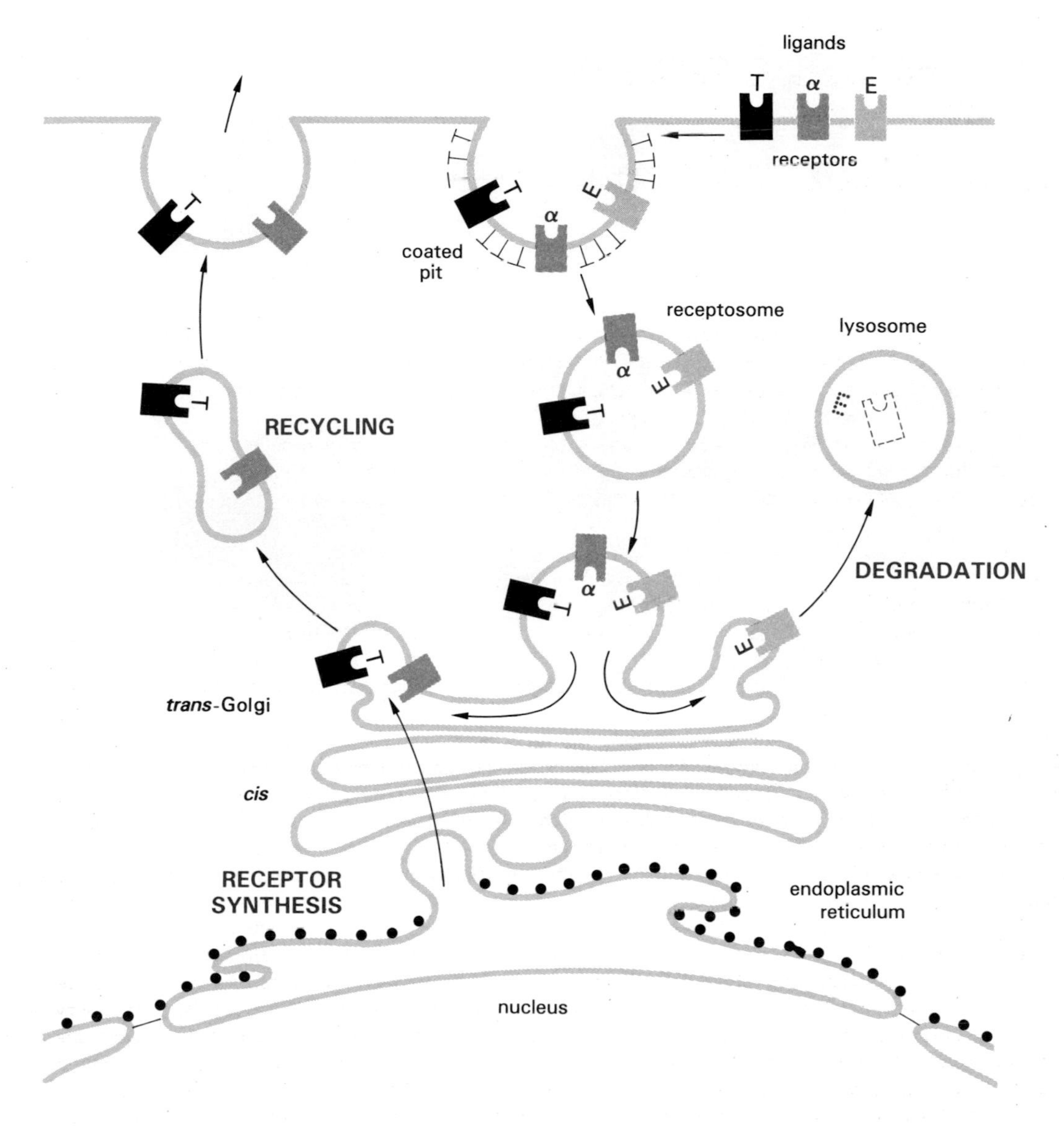

Figure 1 *Schematic representation of organelles that participate in endocytosis and recycling of materials to the cell surface. Note that each coated pit ordinarily internalizes several receptors and ligands, together. The Golgi stacks are functionally polarized.* "Cis" *refers to the side which receives material from the endoplasmic reticulum;* "trans" *refers to the exit side. Note that EGF and its receptor are degraded in lysosomes, as is* α_2*-macroglobulin, while the* α_2*-macroglobulin receptor, and transferrin and its receptor are recycled. (E, epidermal growth factor;* α, α_2*-macroglobulin; T, transferrin).*

With the morphological techniques employed thus far, peptide hormones have *not* been detected *in the nucleus*, even though many such hormones have actions on cell growth and gene expression. The failure to find peptide hormones in the nucleus has been interpreted as meaning that either the actions of such hormones are mediated by second messengers, or the amount of hormone which reaches the nucleus is too small for detection by current methods. Most investigators favor the second messenger hypothesis. Thus, a *direct* role of endocytosis in hormone action is *unproved*. However, endocytosis does have an *important* role in modulating the response to hormones, by clearing the hormone and receptor from the cell surface, and delivering the hormone, and often the receptor, to lysosomes, where destruction occurs.

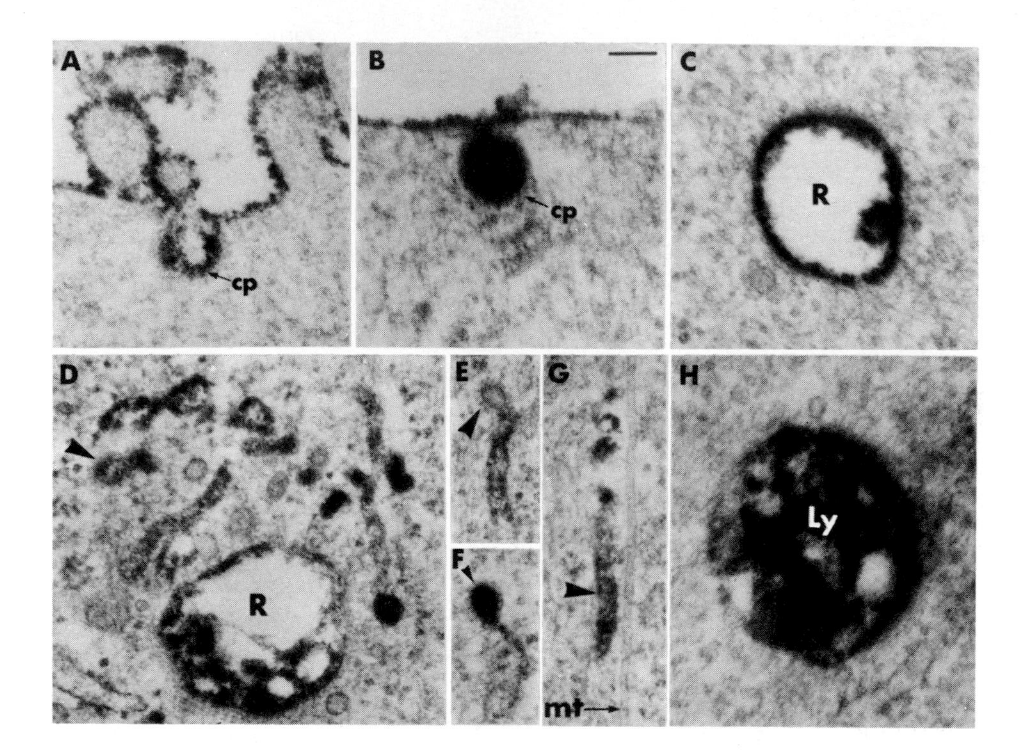

Figure 2 *Endocytosis of EGF and transferrin (T) visualized using ligand-peroxidase conjugates. KB human carcinoma cells were incubated at 4°C with EGF-HRP (**A,B,C,F,H**) or T-HRP (**D,E,G**). When fixed at 4°C without warming (**A**), EGF-HRP can be seen diffusely distributed on the cell surface, without significant concentration in coated pits (cp). However, when these cells were warmed to 37°C for 1 min (**B**), EGF-HRP rapidly concentrated in coated pits (cp). In contrast, T-HRP would appear concentrated in coated pits, even at 4°C (not shown). Five minutes after warming to 37°C (**C**), EGF-HRP can be seen in intracellular endocytic vesicles, termed receptosomes (R). These vesicles migrate by saltatory motion to the* trans-reticular *portion of the Golgi, with which they fuse. Labelled T-HRP, for example, can be seen at 12 min after warming to 37°C (**D**) in both the receptosome (R) and adjacent tubular elements of the* trans-Golgi, *some of which have the small coated pits characteristic of the Golgi apparatus (arrowhead). At this time, T-HRP can be found in these tubules without concentrating in these coated regions (**E**), but EGF-HRP is commonly found in a concentrated form in these small coated pits (**F**). A short time later, T-HRP can be found in elongated tubular elements (arrowhead, **G**) closely apposed to microtubules (mt), whereas EGF-HRP is not found in these structures. At a later time (15 to 30 min after warming), T-HRP is released back into the culture medium, whereas EGF-HRP (**H**) is found exclusively in lysosomes (LY). Bar=0.1 μm.*

References

Pastan, I. and Willingham, M. (1983) Receptor-mediated endocytosis: Coated pits, receptosomes and the Golgi. *Trends Biochem. Sci.* 8: 250–254.

Pastan, I. and Willingham, M. (1981) Journey to the centre of the cell: role of the receptosome. *Science* 214: 504–509.

Willingham, M. and Pastan, I. (1983) Formation of receptosomes from plasma membrane coated pits during endocytosis: Analysis by serial sections with improved membrane labeling and preservation techniques. *Proc. Natl. Acad. Sci., USA* 80: 5617–5621.

Beguinot, L., Lyall, R.M., Waterfield, M.D., Willingham, M.C. and Pastan, I. (1984) Down-regulation of the EGF receptor in KB cells is due to receptor internalization and subsequent degradation in lysosomes. *Proc. Natl. Acad. Sci., USA* 81: 2384–2388.

Brown, M.S., Anderson, G.W., and Goldstein, J.L. (1983) Recycling receptors: The round trip itinerary of migrant membrane proteins. *Cell* 32: 663–667.

Helenius, A., Millman, I., and Hubbard, A. (1983) Endosomes. *Trends Biochem. Sci.* 8: 245–250.

CHAPTER IV

ACTH and Related Peptides

Anthony S. Liotta and Dorothy T. Krieger‡

‡Editors' note: Dorothy Krieger died on April 2, 1985. She worked on this chapter until the end of her life.

Contents, Chapter IV

ACTH and Related Peptides

Introduction

Corticotropic cells of the anterior pituitary, in all species, and melanotropic cells of the intermediate lobe, in such animals as the rat, synthesize a glycoprotein *molecule containing* the *complete sequences of adrenocorticotropic hormone (ACTH), β-lipotropin, and a large glycopeptide* at the N-terminal region. This common precursor molecule has been named *proopiomelanocortin* (POMC) to indicate some of the biologically active peptides derived from it post-translationally[1,2].

ACTH stimulates adrenal steroid biosynthesis, especially that of glucocorticosteroids, i.e. mainly cortisol in humans. Glucocorticosteroids have a direct negative feedback effect on pituitary ACTH secretion, and also negatively control hypothalamic corticotropin releasing factor (CRF), which is involved in ACTH secretion. This CRF-ACTH-glucocorticosteroid sequence is one of the most important regulatory systems governing the reaction of the body to environmental stress and metabolic changes, and may be pathologically affected at each level. Besides ACTH, proopiomelanocortin encodes other important peptides, in particular MSH and β-endorphin.

1. Structure of POMC

Human, bovine, and rat POMC molecules exhibit a high degree of homology. POMC consists of *three* structural *domains*, preceded by a 26 aa *signal* sequence: 1. *ACTH*, which resides in the middle of the molecule; 2. *β-lipotropin*, which comprises the C-terminal sequence; and 3. the N-terminal sequence, variably referred to as the *N-terminal fragment* (named as such since a physiologically relevant bioactivity has not yet been conclusively demonstrated for it, nor has a peptide been derived from it), the *16 kDa fragment* (to indicate the apparent molecular weight exhibited by the mouse form in an SDS polyacrylamide gel system), or *pro-γ-MSH*. These segments of POMC are separated by the sequence Lys-Arg. This and other pairs of basic amino acids have been shown to be proteolytic cleavage sites in many precursor proteins containing biologically active peptides (p. 27). *Additional* pairs of basic amino acids in POMC thus represent other known *cleavage sites* (e.g., liberation of β-endorphin from β-lipotropin), and potential cleavage sites, which may be expressed in a *tissue-specific* manner (Fig. IV.1).

The three domains share sequence homology, corresponding to positions 2, 4, and 6 to 9 of ACTH (or *α-MSH*)[3]. The segment within the N-terminal fragment containing this sequence (bounded by pairs of basic amino acids) is referred to as γ-MSH.

Proopiomelanocortin exhibits size and charge heterogeneity, owing, at least in part, to several co-translational/post-translational covalent modifications, such as glycosylation, phosphorylation and

1. Chrétien, M. and Seidah, N. G. (1981) Chemistry and biosynthesis of propiomelanocortin. *Mol. Cell. Biochem.* 34: 101–127.
2. Krieger, D. T., Liotta, A. S., Brownstein, M. J. and Zimmerman, E. A. (1980) ACTH, β-lipotropin and related peptides in brain, pituitary and blood. *Rec. Prog. Horm. Res.* 36: 277–344.
3. Nakanishi, S., Ioue, A., Kita, T., Nakamura, M., Chang, A. C. Y., Cohen, S. N. and Numa, S. (1979) Nucleotide sequence of cloned cDNA for bovine corticotropin-β-lipotropin precursor. *Nature* 278: 423–427.

sulfation, which occur in corticotropic and melanotropic cells[4]. Available evidence indicates that such modifications are not responsible for the differential processing of POMC in the two pituitary lobes.

2. Structure of the POMC Gene

DNA complementary to bovine and mouse POMC mRNA and genomic DNA encoding most of rat, mouse and bovine POMC genes have been cloned and sequenced. POMC, POMC mRNA and POMC gene organization is very similar among the species studied to date.

The structure of the entire human POMC gene (Fig. IV.2) has been determined by nucleotide se-

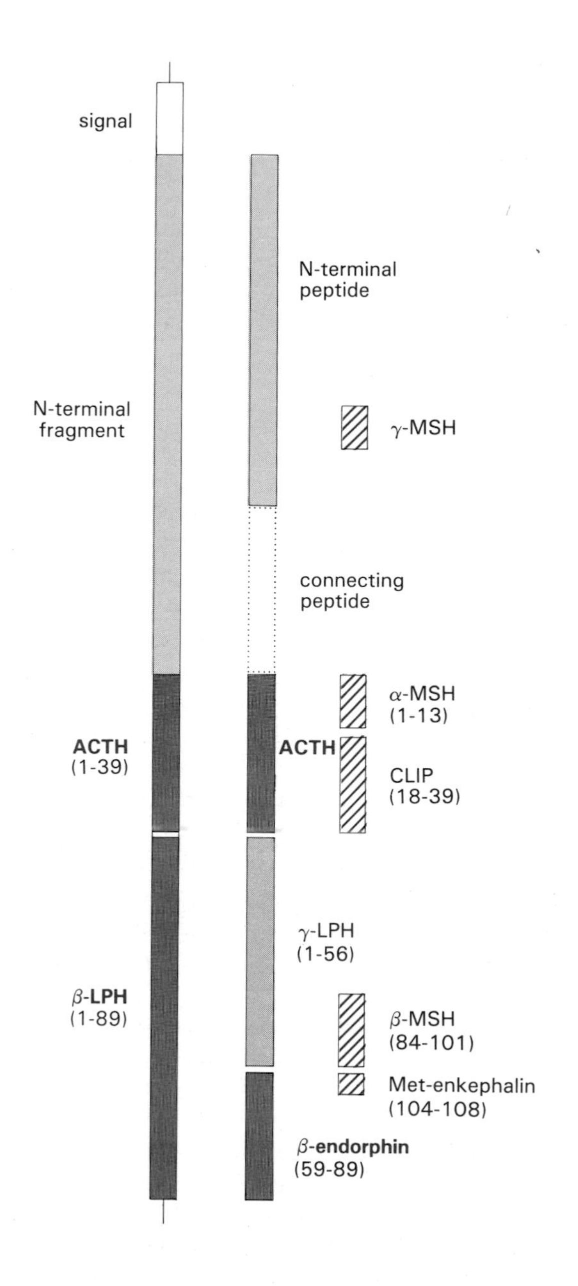

Figure IV.1 Human POMC molecule and post-translationally-derived peptides. γ-MSH has not been found in any mammalian anterior pituitary, nor has β-MSH, but it may be a final product in the intermediate lobe in some species. The connecting peptide has not been found in humans, is present in the rat, and has no recognized biological activity. The Met-enkephalin sequence is not a product of POMC processing in any tissue studied so far.

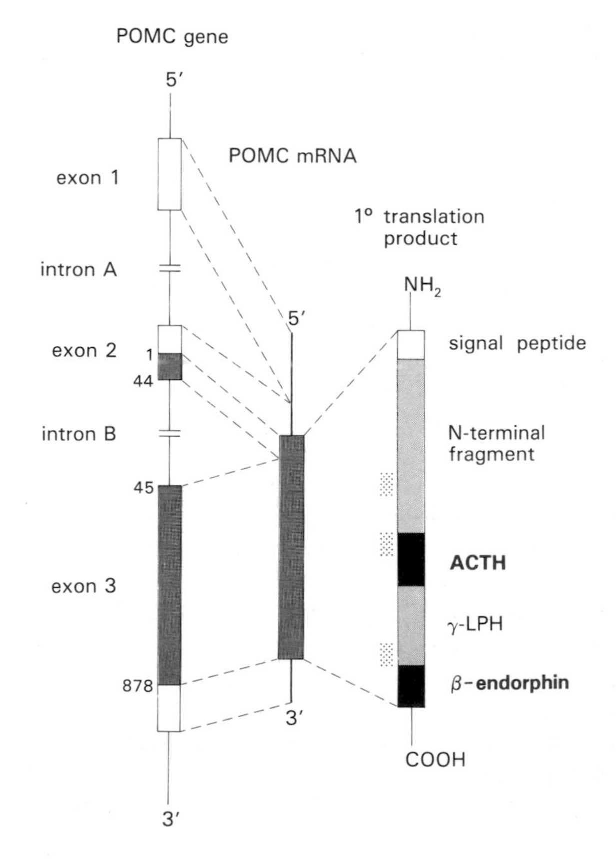

Figure IV.2 Organization of human POMC gene, mRNA, and POMC peptide. The stippled areas adjacent to POMC indicate internal homologies.

4. Eipper, B. A. and Mains, R. E. (1980) Structure and biosynthesis of proadrenocorticotropin/endorphin and related peptides. *Endocrine Rev.* 1: 1–17.

quence analysis of cloned human genomic DNA fragments[5]. The DNA encoding the mature messenger RNA (mRNA) is contained in three exons separated by two large intervening sequences. The first exon encodes 87 nucleotides of the 5′ untranslated region of the mRNA. An intron of approximately 3.6 kb in length interrupts this region 20 nucleotides upstream from the translation initiation site. The second exon (152 bp) encodes the remaining 20 nucleotides of the 5′ untranslated region of the mRNA, the 26 aa signal sequence of the pre-POMC molecule, and the first 18 aa of the N-terminus of the mature POMC molecule. The second intron, approximately 2.9 kb in length, interrupts the protein coding region between amino acids 18 and 19 of the mature POMC molecule (amino acids 44 and 45, counting from the initiation methionine of the signal sequence). The third exon (833 bp) contains all the DNA encoding the remainder of the POMC molecule (i.e. amino acids 19 to 241), which comprises the remainder of the N-terminal fragment peptide, ACTH, β-lipotropin, as well as the Lys-Arg pairs connecting these peptide domains and the 3′ untranslated segment of the mature POMC mRNA. Repetitive DNA sequences of the Alu family are present in the introns and in the 5′ flanking sequence of the human (and bovine) POMC gene.

Based on homology with DNA sequences of two other regulated genes (i.e. rat growth hormone and mouse mammary tumor virus), it has been postulated that a *21 bp DNA* segment, located *480 bp upstream* from the mRNA coding sequence and expressed in the anterior pituitary lobe[6], is involved in the *regulation* of the human POMC gene. The human genome contains a single POMC gene located on chromosome 2[7].

There are numerous examples of genes in which the biologically relevant protein-coding DNA sequences (i.e. functional or structural domains) are separated by introns (p. 376). On the basis of these findings, it has been proposed that exons represent separate functional or structural domains. In this regard, the *genomic organization* of the POMC gene is strikingly *different*, since *most* of the protein-*coding* sequence is contained in *one exon* (exon 3, as described above). The POMC gene organization also differs from many other genes in that an intron interrupts the sequence encoding the 5′ untranslated segment of the mRNA molecule. This region (believed to contain translation regulatory elements) is not split by introns in most other genes studied to date. Of interest is the finding that the two other known genes encoding opioid peptides, *proenkephalin A* and *proenkephalin B*, are similar to the POMC gene in organization, suggesting that all three genes evolved by similar mechanisms, and that they are *evolutionarily related*. In addition, the human CRF gene also has an intron inserted in the region encoding the 5′ untranslated region of the mRNA.

5. Whitfeld, P. L., Seeburg, P. H. and Shine, J. (1982) The human propiomelanocortin gene: organization, sequence and interspersion with repetitive DNA. *DNA* 1: 133–143.
6. Cochet, M., Chang, A. C. Y. and Cohen, S. N. (1982) Characterization of the structural gene and putative 5′-regulatory sequences for human proopiomelanocortin. *Nature* 297: 335–339.
7. Owerbach, D., Rutter, W. J., Roberts, J. L., Whitfeld, P., Shine, J., Seeburg, P. H. and Shows, T. B. (1981) The proopiocortin (adrenocorticotropin/β-lipotropin) gene is located on chromosome 2 in humans. *Somatic Cell Molec. Genet.* 3: 359–369.

3. POMC Processing Pathways in the Pituitary

Both anterior and intermediate pituitary lobes are present in most vertebrate species, including virtually all amphibians, reptiles and fish (with the exception of the simple Agnathan vertebrate, the hagfish), and most mammals. However, birds and some mammals lack an anatomically distinct intermediate lobe. In *humans*, an *intermediate lobe* is present during fetal life (comprising 4% of the total pituitary mass) but involutes shortly after birth; it is *not present in the adult*. It should be noted that, although some histochemical and immunocytochemical studies have suggested that a very small population of intermediate lobe-like cells may persist in the adult human, they are unlikely to have any physiological significance.

3.1 Anterior Pituitary

In the anterior pituitary *corticotropic cells*, ACTH, β-lipotropin, and a glycopeptide comprising approximately 75% of the N-terminal, non-ACTH, non-β-lipotropin region of POMC are major endproducts of POMC processing. In vitro pulse–chase studies have established the temporal order of cleavages producing these products. POMC is first cleaved to yield β-lipotropin and a biosynthetic ACTH intermediate consisting of ACTH still covalently bound to the N-terminal fragment. The second cleavage liberates *ACTH* and the intact N-terminal fragment of POMC. In some species (e.g., rat), some of the ACTH mol-

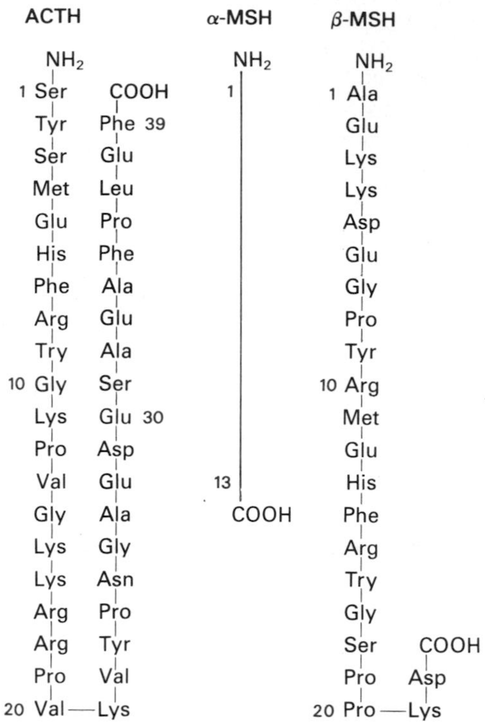

ACTH		α-MSH	β-MSH	
NH_2		NH_2	NH_2	
1 Ser	COOH	1	1 Ala	
Tyr	Phe 39		Glu	
Ser	Glu		Lys	
Met	Leu		Lys	
Glu	Pro		Asp	
His	Phe		Glu	
Phe	Ala		Gly	
Arg	Glu		Pro	
Try	Ala		Tyr	
10 Gly	Ser		10 Arg	
Lys	Glu 30		Met	
Pro	Asp		Glu	
Val	Glu	13	His	
Gly	Ala	COOH	Phe	
Lys	Gly		Arg	
Lys	Asn		Try	
Arg	Pro		Gly	
Arg	Tyr		Ser	COOH
Pro	Val		Pro	Asp
20 Val —	Lys		20 Pro —	Lys

Figure IV.3 Primary structure of human ACTH, α-MSH, and β-MSH.

ecules are cotranslationally glycosylated at Asn-29; *human* ACTH (Fig. IV.3) contains an Asp residue, at position 29, which is *not glycosylated.* In the human, but not in the rat pituitary, the N-terminal fragment (109 aa) appears to undergo an additional cleavage (removal of approximately 30 aa from the C-terminal), yielding a 76 aa N-terminal glycopeptide. Some of the β-lipotropin molecules are further processed to γ-lipotropin and *β-endorphin*, although *β-lipotropin* is the *predominant endproduct.* As noted above, all of these major cleavages occur at the Lys-Arg sequence (Fig. IV.1). In addition, desacetyl α-MSH, α- and γ-endorphin, and C-terminal fragments of β-endorphin appear to be minor products.

3.2 Intermediate Lobe

In the intermediate lobe, the initial proteolytic cleavages of POMC are the same as in anterior pituitary. ACTH, β-lipotropin, and the N-terminally-derived glycopeptide are all produced. However, in this tissue they serve as biosynthetic intermediates to *smaller molecular species.* β-lipotropin is efficiently cleaved to γ-lipotropin and β-endorphin (an active opioid peptide). β-endorphin is then α-*N*-acetylated, and on a slower time scale, C-terminally trimmed, yielding, successively, *N*-acetylated β-endorphin-(1–27) and β-endorphin-(1–26). The *N*-acetylated forms of β-endorphin, β-endorphin-(1–27), and β-endorphin-(1–26) are the major β-endorphin-related secretory products of the rat intermediate lobe. The extent of α-*N*-acetylation appears to be species-specific, the highest degree to date noted in the rat[8]. In contrast, little or no *N*-acetylated or C-terminally trimmed forms of β-endorphin are produced in the anterior lobe[9].

In analogy to β-lipotropin, ACTH is further processed in the anterior lobe: it is cleaved to ACTH-(1–13)-amide (possibly in two steps) and ACTH-(18–39)-related peptides. The ACTH-(1–13)-like peptide is then α-*N*-acetylated (yielding *α-MSH*) and subsequently *O*-acetylated (at the serine hydroxyl group at position 1). The majority of rat and porcine intermediate lobe α-MSH has been identified as α-*N*, *O*-diacetyl α-MSH[10]. The production of α-MSH, CLIP, and β-endorphin-sized peptides from POMC is characteristic of all vertebrate pituitary intermediate lobes studied to date (including several species of fish and amphibians). In the distinct *intermediate lobe* of the *human fetus*, processing of POMC is similar to that noted above, except for the absence of any appreciable acetylation of endproducts (e.g., desacetyl α-MSH is the predominant α-MSH-like peptide formed).

α- and γ-endorphin are also present in the intermediate lobe as minor products. It is not clear whether or not they represent specific endproducts in the POMC processing pathway, or, if they do, whether β-endorphin is an obligatory intermediary in their formation. Acetylation increases the melanocyte-stimulating activity of α-MSH, decreases the steroidogenic potency of ACTH, and abolishes the opiate activity of β-endorphin.

8. Eipper, B. A. and Mains, R. E. (1981) Further analysis of post-translational processing of β-endorphin in rat intermediate lobe. *J. Biol. Chem.* 256: 5689–5695.
9. Liotta, A. S., Yamaguchi, H. and Krieger, D. T. (1981) Biosynthesis and release of β-endorphin-, *N*-acetyl β-endorphin, β-endorphin-(1–27)-, and *N*-acetyl β-endorphin-(1–27)-like peptides by rat pituitary neurointermediate lobe: β-endorphin is not further processed by anterior pituitary. *J. Neurosci.* 1: 585–595.
10. Glembotski, C. C. (1982) Acetylation of α-melanotropin and β-endorphin in the rat intermediate pituitary. *J. Biol. Chem.* 257: 10493–10500.

It should be noted that, although the N-terminal pentapeptide sequence of β-endorphin comprises the methionine-enkephalin sequence (an opioid peptide), neither POMC nor β-endorphin serves as an intermediary in the biosynthetic pathway of its formation. Met-enkephalin is derived from the proenkephalin A gene, which also encodes other opioid peptides.

3.3 Secretion of POMC-Derived Peptides

All from one

In vivo studies measuring plasma levels of anterior pituitary POMC-derived peptides under basal conditions, and following endocrine manipulations associated with stimulation (e.g., CRF administration) or inhibition (e.g., administration of glucocorticosteroids) of ACTH secretion, reveal that under all circumstances the plasma peptide levels vary in parallel. These findings indicate that *all peptides* derived *from POMC* are *secreted concomitantly*. In species possessing an intermediate lobe, POMC-derived peptides of both anterior and intermediate lobes are detectable in plasma. POMC-derived peptides characteristic of pituitary intermediate lobe origin (e.g., α-MSH) are not normally detectable in human plasma, consistent with the lack of an anatomically distinct or functional intermediate lobe in normal humans (such peptides have been detected in the plasma of some human pituitaries containing ectopic POMC peptide producing tumors; analysis of the tumors revealed that some of them synthesized POMC and processed it in a fashion similar to that seen in the intermediate lobe).

In vitro studies using acutely dispersed, cultured, or superfused anterior pituitary or intermediate lobe cells have confirmed the in vivo observations, and have further demonstrated that POMC-derived peptides from both corticotropic and melanotropic cells are secreted in equimolar amounts[11]. These findings are consistent with the equimolar presence of these moieties within the POMC molecule, and with the fact that they are *cosequestered within secretory granules*.

11. Mains, R. E. and Eipper, B. A. (1981) Coordinate, equimolar secretion of smaller peptide products derived from pro-ACTH/endorphin by mouse pituitary tumor cells. *J. Cell Biol.* 89: 21–28.
12. Vale, W., Spiess, J., Rivier, C. and Rivier, J. (1981) Characterization of a 41-residue ovine hypothalamic peptide that stimulates secretion of corticotropin and β-endorphin. *Science* 213: 1394–1397.

4. Regulation of POMC Synthesis and Secretion of Derived Peptides

4.1 Anterior Pituitary POMC

Several regulators: CRF is the most important

Anterior pituitary corticotropic cells are regulated by substances of neural and peripheral origin. A variety of studies utilizing in vivo and in vitro model systems have shown that several naturally occurring substances possess the ability to stimulate the coordinated release of ACTH and other POMC-derived peptides from anterior pituitary corticotropic cells. The recently characterized CRF, AVP, the catecholamines epinephrine and norepinephrine, oxytocin, the neuropeptides VIP and peptide histidine-isoleucinamide, and angiotensin II all stimulate the release of POMC-derived peptides at the level of the anterior pituitary. Pituitary receptors for some of these compounds have been demonstrated. All of these substances are present in nerve terminals in the zona externa of the median eminence (most have been detected in hypophyseal portal blood), and hence can, potentially, reach anterior pituitary corticotropic cells. However, only CRF appears to fulfill the most important requirements of a bonafide hypothalamic anterior pituitary releasing factor: CRF is now recognized as the major agent regulating adenohypophyseal ACTH secretion.

Characterization of CRF

The first postulated hypothalamic releasing factor is now isolated. In 1955, Guillemin and Rosenberg, and Saffran and Schally described the presence of a hypothalamic factor that could stimulate the in vitro secretion of ACTH from the anterior pituitary. *Corticotropin releasing factor* (CRF) was the name given to this chemically unidentified substance. In 1981, Vale *et al.*[12] reported the isolation and characterization of a CRF isolated from ovine hypothalamic extracts. The complete amino acid sequence was determined; a synthetic replicate of this molecule was shown to be a very potent stimulator (<0.01 nM minimum effective dose) of ACTH release from anterior pituitary cells in vitro. Subsequently, the structure of rat hypothalamic CRF was determined by isolation and microsequencing of the peptide, and the

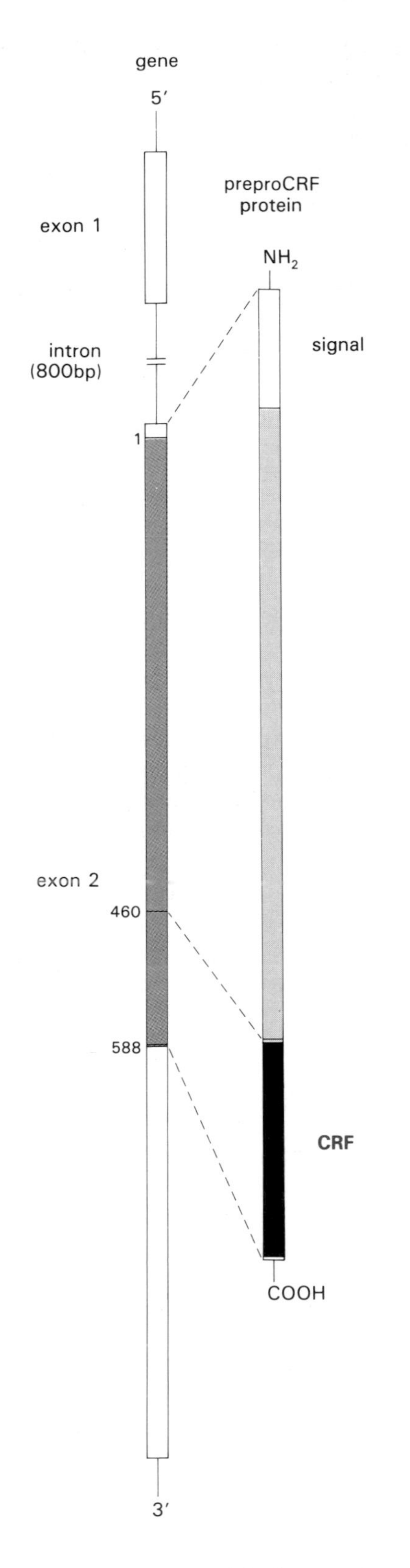

Figure IV.4 Organization of the human prepro-CRF gene and its corresponding protein[13].

amino acid sequence of a human prepro-CRF molecule and the genomic organization of the CRF gene have been predicted from the nucleotide sequence of a cloned human genomic DNA fragment (Fig. IV.4). The DNA encoding the mature mRNA is contained in two exons separated by an 800 bp intron. The first exon encodes the majority of the 5′ untranslated region of the mRNA. The second exon encodes the remainder of the 5′ untranslated region, the entire protein coding region and the 3′ untranslated region of the mRNA. Human prepro-CRF consists of 196 aa residues, including the putative signal sequence. The C-terminal end of the precursor protein containing the CRF sequence is preceded by the dipeptide Arg-Arg (putative signal for proteolytic cleavage), and is followed by the putative amidation consensus sequence Gly-Lys; the deduced structure of mature *human CRF* (*41 aa* with an amidated C-terminal) is identical to that of rat CRF (Fig. IV.5)[13]. Ovine CRF is 83% homologous to rat/human CRF, differing at seven of the 41 aa residues.

The CRF molecules are *structurally related to the non-mammalian peptide sauvagine*, isolated from frog skin, and *urotensin I*, isolated from the urophysis of sucker and carp fish (Fig. IV.5). All of these molecules possess the ability to cause the release of POMC-derived peptides from the anterior pituitary, and, as well, share other biological activities (e.g., inhibition of the release of pituitary growth hormone, and production of hypotension following systemic administration).

Distribution of CRF

Much broader than expected

Central nervous system. Extensive immunocytochemical mapping studies utilizing specific antisera raised to rat or ovine CRF have been conducted in the rat and, to a lesser extent, in the dog, sheep, monkey, and human (Fig. IV.6)[14]. Although immunoreactive CRF is widely distributed in the CNS, the *hypothalamus* contains the *greatest concentrations* (also established by specific radioimmunoassay of discrete brain areas). The *paraventricular nucleus*

13. Shibahara, S., Morimoto, Y., Furutani, Y., Notake, M., Takahashi, H., Shimizu, S., Horikawa, S. and Numa, S. (1983) Isolation and frequence analysis of the human corticotropin-releasing factor precursor gene. *EMBO J.* 2: 775–779.
14. Swanson, L. W., Sawchanko, P. E., Rivier, J. and Vale, W. W. (1983) Organization of ovine corticotropin-releasing factor immunoreactive cells and fibers in the rat brain: an immunohistochemical study. *Neuroendocrinology* 36: 165–186.

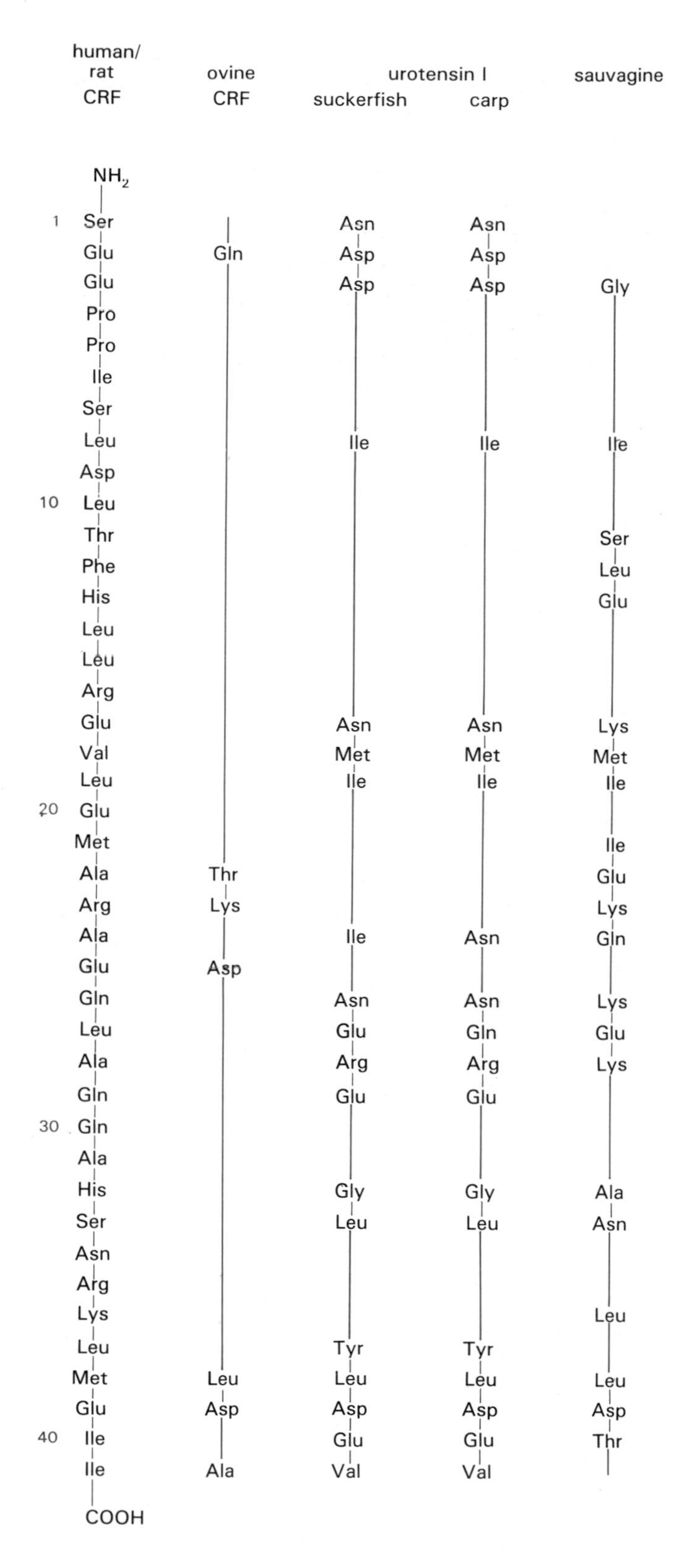

Figure IV.5 The CRF family.

contains the largest concentration of CRF-positive cell bodies in the hypothalamus. Within the paraventricular nucleus, the *parvocellular neurons* contain a majority of the CRF cells. A massive network of CRF-positive fibers projects from these neurons to the external zone of the *median eminence*. Adrenalectomy results in enhanced staining of these cells, consistent with the negative feedback effect of glucocorticosteroids on ACTH secretion. Adrenalectomy is also associated with the colocalization of vasopressin in most of the parvocellular CRF-positive neurons (not seen in brains of intact rats). This latter observation is consistent with the demonstration that vasopressin enhances the effect of CRF on ACTH release in vitro. In addition to the major parvocellular CRF-positive cell group, some CRF-stained cells reside within the magnocellular division of the paraventricular nucleus; a subgroup of these cells may contain both CRF and oxytocin. Some paraventricular CRF cells also project to the posterior pituitary.

Two additional groups of CRF-positive neurons are present in the *brain*: 1. *CRF cells* are present in several discrete nuclei in the *brain stem* and *basal forebrain*, areas known to be involved in the mediation of autonomic nervous system function (intraventricular administration of CRF is associated with multiple effects on sympathetic and parasympathetic nervous system function[15]; 2. CRF cells, many of which appear to be interneurons, are scattered in most regions of the cerebral cortex.

Peripheral distribution. Immunoreactive CRF has been reported in the adrenal, lung, placenta, pancreas and gastrointestinal tract, as well as in several types of endocrine tumors. This material exhibited chromatographic behavior identical to that of synthetic CRF (1–41) in molecular sieve and high pressure liquid chromatography systems, suggesting that it is a product of the expression of the authentic CRF gene in these tissues[16]. The physiological *significance* of this *widespread peripheral distribution* is *not known*, although it is a common finding for many other previously characterized neuropeptides. In the case of the *placenta*, CRF may be secreted into the maternal circulation, and be at least partly responsible

15. Brown, M. R., Fisher, L. A., Spiess, J., Rivier, C., Rivier, J. and Vale, W. (1982) Corticotropin-releasing factor: actions on the sympathetic nervous system and metabolism. *Endocrinology* 111: 928–931.
16. Suda, T., Tomori, N., Tozawa, F., Demura, H., Shizume, K., Mouri, T., Miura, Y. and Sasano, N. (1984) Immunoreactive corticotropin and corticotropin-releasing factor in human hypothalamus, adrenal, lung, cancer, and pheochromocytoma. *J. Clin. Endocrinol. Metab.* 58: 919–924.

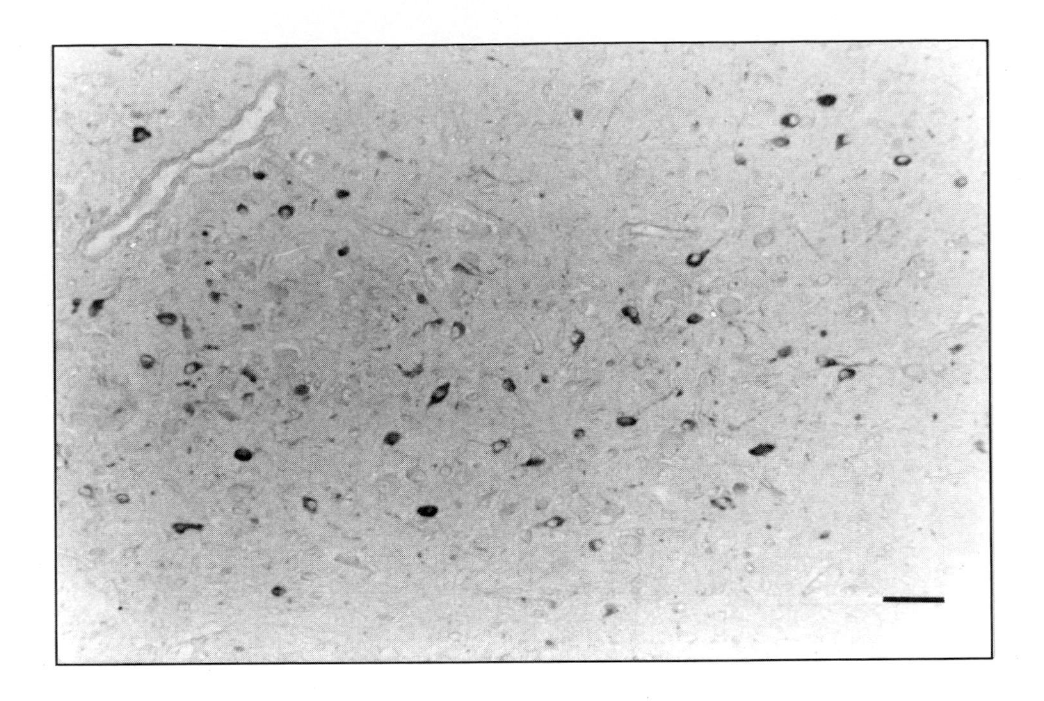

Figure IV.6 Immunohistochemical identification of CRF neurons in human hypothalamus: a section of the paraventricular nucleus; anti-ovine CRF has been used. Bar=50 µm. (Pelletier, G., Désy, L., Côté, J. and Vaudry, H. (1983) Immunocytochemical localization of corticotropin releasing factor-like immunoactivity in the human hypothalamus. *Neurosc. Lett.* 41: 259–263).

for the elevated plasma ACTH levels seen in the presence of high plasma free cortisol[17].

Actions of CRF

Regulation of Pituitary POMC by CRF

Corticotropin releasing factor exhibits the highest potency and intrinsic activity in the stimulation of the secretion of ACTH and other POMC-derived peptides in vitro. All the other ACTH secretagogues are much less potent and exhibit drastically reduced intrinsic activities (Fig. IV.7). Long-term exposure of cultured pituitary cells to CRF results in increased levels of medium and tissue concentrations of ACTH, indicating that CRF also increases the rate of synthesis of POMC. Recent data indicate that chronic administration of CRF to intact rats results in increased levels of POMC mRNA in the anterior pituitary, indicating that the effect on POMC synthesis is due to an *increased rate of transcription* of the POMC gene.

CRF is present in *hypophyseal portal blood* in *concentrations* shown to be effective in *stimulating ACTH* in vitro; this is not the case for the other secretagogues, with the possible exception of vasopressin. Immunoneutralization of endogenous rat CRF with specific anti-CRF serum results in a greatly diminished plasma ACTH response to ether stress, and greatly inhibits the release of ACTH following the administration of effective doses of vasopressin, oxytocin, or epinephrine to freely moving rats[18].

These in vivo and in vitro findings provide strong evidence that CRF plays a major role in the regulation of the adenohypophyseal POMC system. On the basis of physiological experiments, the paraventricular nucleus has long been implicated in the regulation of pituitary ACTH secretion. Destruction of the paraventricular nucleus blocks increases in plasma ACTH following stress. The biochemical, physiological, and anatomical data are thus consistent with the notion that CRF of paraventricular nucleus origin is transported to the median eminence; there, it is released into the pituitary portal circulation, to affect ACTH secretion.

17. Sasaki, A., Liotta, A.S., Luckey, M.M., Margioris, A.N., Suda, T. and Krieger, D.T. (1984) Immunoreactive corticotropin-releasing factor is present in human maternal plasma during the third trimester of pregnancy. *J. Clin. Endocrinol. Metab.* 59: 812–814.
18. Rivier, C., Rivier, J. and Vale, W. (1982) Inhibition of adrenocorticotropic hormone secretion in the rat by immunoneutralization of corticotropin-releasing factor (CRF). *Science* 218: 377–379.

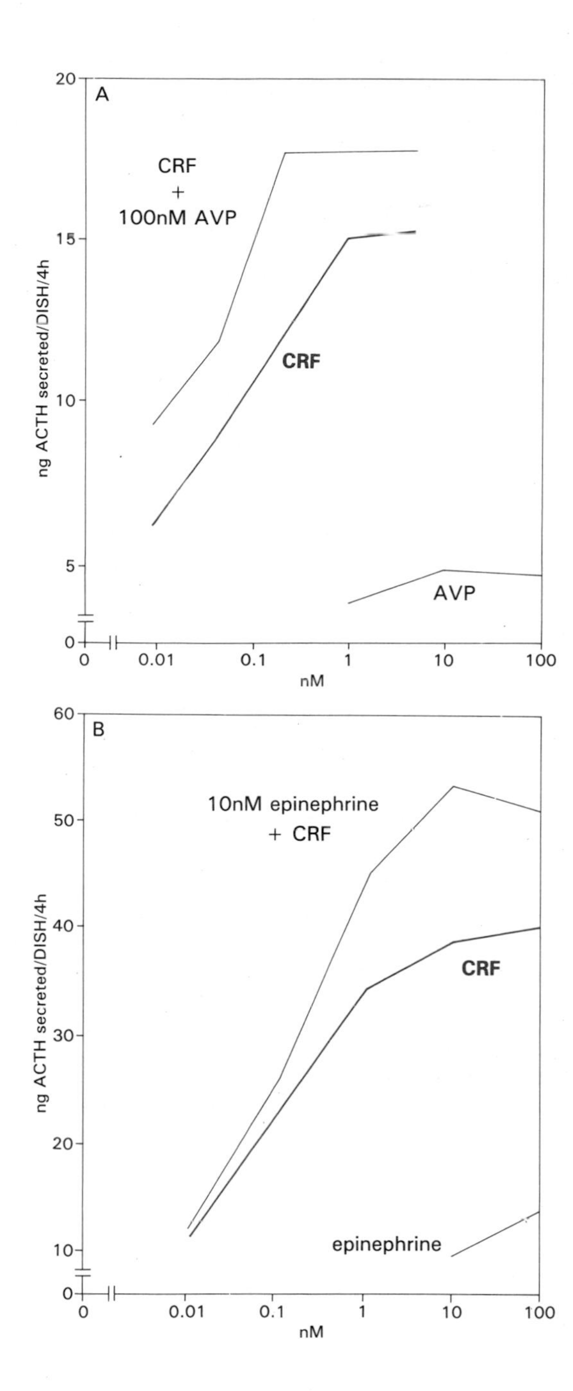

Figure IV.7 Effect of CRF and vasopressin (**A**) or epinephrine (**B**) on the secretion of ACTH by rat anterior pituitary cells in monolayer culture. (Vale, W., Vaughan, J., Smith, M., Yamamoto, G., Rivier, J. and Rivier, C. (1983) Effects of synthetic ovine corticotropin-releasing factor, glucocorticoids, catecholamines, neurohypophysial peptides, and other substances on cultured corticotropic cells. *Endocrinology* 113: 1121–1131).

CRF actions in the Central Nervous System

CRF acts *centrally* to *inhibit* the release of *pituitary LH and GH* (CRF has no direct effect at the pituitary level on LH and GH). The inhibition of LH does not involve the activation of opioid peptide neuronal pathways or the secretion of pituitary β-endorphin, as the effect is not blocked by potent opiate antagonists (central β-endorphin is known to inhibit LH release via modulation of dopaminergic activity). The inhibitory action on GH appears to be due to both the inhibition of GRF release and the stimulation of somatostatin. Infusion of CRF into the hypothalamic arcuate-ventromedial and mesencephalic central gray regions of the brain results in a dramatic inhibition of sexual receptivity in the female rat. Based on both the suppression of LH and the suppressive effect of CRF on sexual behavior, it has been suggested that centrally acting CRF may play a major role in the known *inhibitory action of stress on fertility*[19]. A variety of other experimental data provides evidence that centrally acting CRF can modulate a variety of behavioral and sympathetic and parasympathetic activities involved in an organism's response to stress.

Mechanism of Action of CRF

The presence of high affinity *receptors* (K_D~1 nM) for CRF has been reported on the *plasma membrane* of a subset of rat and human *anterior pituitary* cells, and also in discrete brain sites of rat *forebrain*. The putative receptors in the pituitary correlated well with the immunocytochemical localization of POMC-derived peptides. CRF *stimulates the production of cAMP* in anterior pituitary cells. Both ACTH secretion and cAMP formation increase in parallel in the presence of CRF. Increasing doses of CRF results in dose-related increases in cAMP-dependent protein kinase activity. The secretion of POMC-derived peptides is calcium-dependent. The other ACTH secretagogues do not act via activation of cAMP-dependent pathways (Fig. IV.8).

Interaction of CRF with other ACTH Secretagogues

Although vasopressin, oxytocin, and the other substances listed above are *weak ACTH secretagogues* in vitro, they *potentiate* the in vitro and in vivo

19. Sirinathsinghji, D. J. S., Rees, L. H., Rivier, J. and Vale, W. (1983) Corticotropin-releasing factor is a potent inhibitor of sexual receptivity in the female rat. *Nature* 305: 232–235.

effects of *CRF* at physiologically relevant concentrations (i.e. available in portal circulation from neural or peripheral routes) (Fig. IV.7). Immunoneutralization of endogenous CRF, or pharmacological blockage of endogenous CRF release, prior to exogenous administration of AVP, oxytocin, or catecholamines, abolishes the in vivo ACTH-releasing properties of these substances. Such results suggest that these ACTH secretagogues exert their actions by an interaction with CRF at the pituitary level. Recent findings have shown that, while the participation of CRF appears obligatory, the actual ACTH secretory response to a particular stress depends on the differential participation of some of these other modulating factors[20].

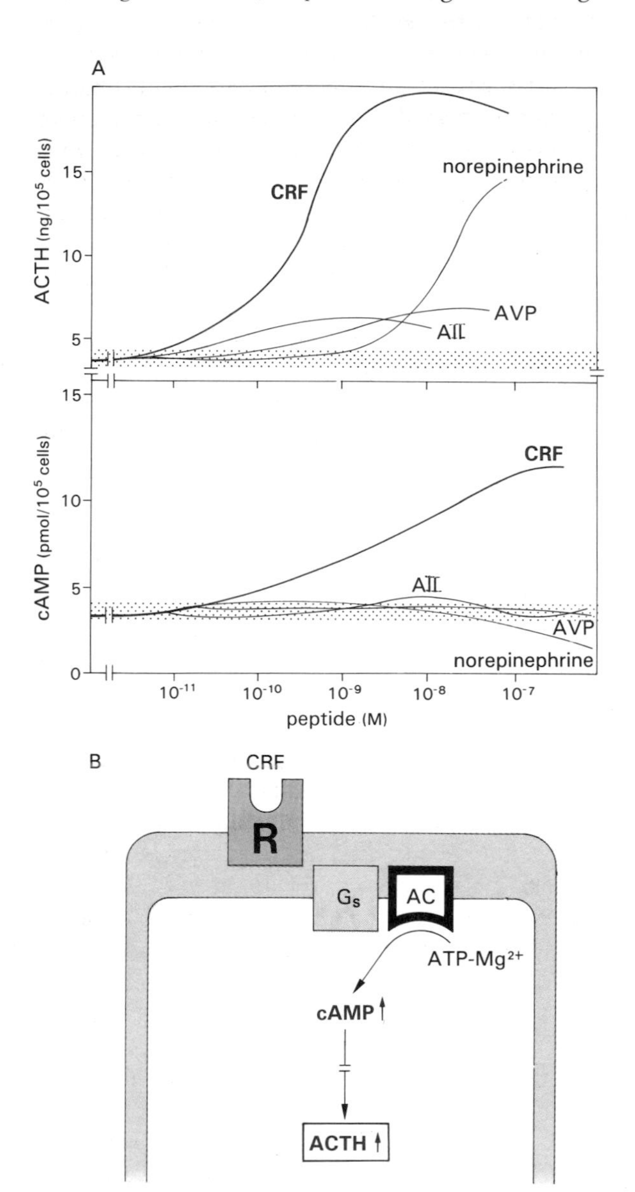

Figure IV.8 CRF acts via a cAMP-dependent mechanism. **A.** Stimulation of ACTH release (upper panel) and cAMP production (lower panel) by CRF, angiotensin II, vasopressin and norepinephrine in cultured rat anterior pituitary cells. **B.** Mechanism of action of CRF in a pituitary corticotropic cell. (Aguilera, G., Harwood, J. P., Wilson, J. X., Morell, J., Brown, J. H. and Catt, K. J. (1983) Mechanisms of action of corticotropin-releasing factor and other regulators of corticotropin release in rat pituitary cells. *J. Biol. Chem.* 258: 8039–8045).

In addition to the pituitary level of regulation, adenohypophyseal POMC synthesis and secretion of derived peptides is also influenced at hypothalamic and extrahypothalamic CNS levels. Serotonin is generally considered stimulatory, and norepinephrine and epinephrine inhibitory, presumably due to stimulation and inhibition of CRF release.

Apart from the classical neurotransmitter-mediated pathways involved, central participation by some putative pituitary modulators may also occur. For instance, intraventricular administration of vasopressin is associated with a dose-dependent reduction in portal blood CRF; a tonic inhibitory role of central vasopressin on CRF has been postulated[21]. Thus, a given regulatory factor may exert opposing actions on the secretion of POMC-derived peptides at different levels of organization of the regulatory network.

The *circadian periodicity* of both plasma ACTH and glucocorticosteroids is well documented. Recent studies have demonstrated that variations in the plasma levels of other anterior pituitary POMC-derived peptides correlate well with ACTH fluctuations, consistent with the coordinated synthesis and release of these peptides. This circadian variation is endogenously *driven at the CNS level.* In mammals, the major circadian pacemaker is thought to reside within the *suprachiasmatic nuclei*, and it is believed that its mechanism of action is related to neurotransmitter regulation of the release of hypothalamic CRF (imparting a periodicity to CRF release) and possibly other secretagogues. In humans, on a normal sleep-wake cycle (activity in the day, sleep at night), ACTH and cortisol exhibit a diurnal rhythm, with peak plasma concentrations occurring in the early

20. Plotsky, P. M., Bruhn, T. O. and Vale, W. (1985) Hypophysiotropic regulation of adrenocorticotropin secretion in response to insulin-induced hypoglycemia. *Endocrinology* 117: 323–329.
21. Plotsky, P. M., Bruhn, T. O. and Vale, W. (1984) Central modulation of immunoreactive corticotropin-releasing factor secretion by arginine vasopressin. *Endocrinology* 115: 1639–1641.

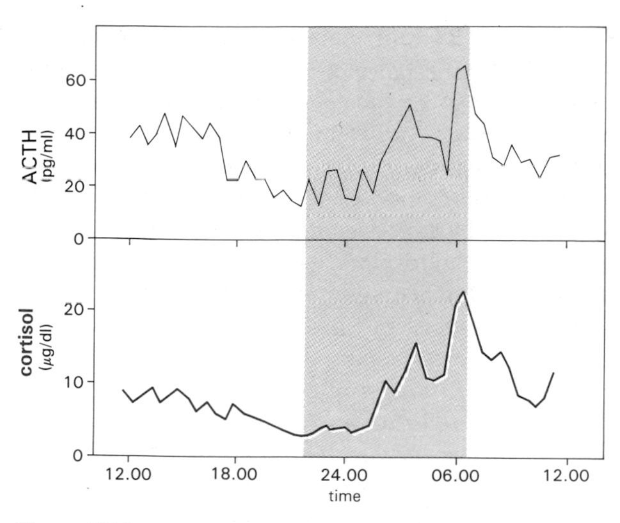

Figure IV.9 Changes in plasma ACTH and cortisol in a normal subject over a 24 h period. (Courtesy of Drs P. Linkowski and J. Mendlewiez, Department of Psychiatry, Erasmus Hospital, Brussels). (Tanaka, K., Nicholson, W. E. and Hort, D. N. (1978) Diurnal rhythm and disappearance half-time of endogenous plasma immunoreactive β-MSH (LPH) and ACTH in man. *J. Clin. Endocrinol. Metab.* 46: pp. 883–890.

morning hours, beginning prior to waking. The early morning period (03.00–09.00 h) is characterized by a series of *episodic peaks* in plasma hormone concentrations (reflecting bursts of episodic secretory activity) (Figs. IV.9 and IV.10).

Cortisol is secreted in a pulsatile fashion with approximately seven to nine spikes (i.e. short secretory episodes) normally occurring in a 24 h period. The initial phase of sleep is usually associated with an absence of cortisol secretory episodes. Some hours after the onset of sleep, secretory activity begins; plasma cortisol levels reach their highest concentration about the time of waking, as the result of relatively closely spaced secretory bursts. The frequency of cortisol spikes decreases dramatically during the remainder of the day, and activity essentially ceases prior to sleep. As a result of this decreasing pulse pattern, plasma cortisol levels drop during the remainder of the day and evening as the quiescent secretory period approaches (Figs IV.9 and IV.10).

The exact nature of the environmental entraining or synchronizing stimuli (i.e. *zeitgebers*) responsible for the maintenance of the circadian pattern of ACTH and cortisol release in humans is not known. However, the *light-dark, sleep-wake cycle* and routine *social interactions* (social cues) are all involved. The physiological importance of potential changes of circadian rhythms of ACTH and cortisol is yet to be evaluated.

Possible Involvement of Melatonin in Circadian Periodicity

Although the evidence is far from conclusive, melatonin may play a role in the modulation of *mammalian circadian rhythms* (including ACTH/glucocorticosteroid periodicities). In *primitive vertebrates* (most submammalian vertebrates), the *pineal gland* is *photosensory* in nature (light is the principal entrainer of rhythms in most of these species) and exhibits an endogenous circadian rhythm. In these species, the pineal gland is considered to be the site of the principal *pacemaker, or oscillator*, controlling other circadian rhythms. The pineal is the major site of synthesis of melatonin (p. 305). Melatonin exhibits a *nocturnal rise* (synthesis and release inhibited by light), and is thought to be the major messenger utilized by the pineal gland in the control of circadian rhythms.

The mammalian pineal gland also exhibits *nocturnal periodic* melatonin activity (in all species, irrespective of whether the glucocorticosteroid rhythm is diurnal or nocturnal), but this activity is not endogenous, and is *driven* by the principal pacemaker(s) or internal clock(s) *within* the *suprachiasmatic nuclei.* In mammals, it has been suggested that the pineal, by way of a "melatonin message," may function as a modulator of the suprachiasmatic nuclei pacemaker, especially in instances involving the entrainment of rhythms by light. The suprachiasmatic nuclei have been shown to accumulate melatonin preferentially, and melatonin is present in both *optic nerves*, and in the retinohypothalamic projections to the suprachiasmatic nuclei (the latter pathway is a major input to the suprachiasmatic nuclei). In addition, the mammalian *retina* synthesizes melatonin. Ocular melatonin exhibits nocturnal rhythmicity; the eyes may interact with central pacemakers via melatonin-secreting neural projections.

A shift in the phase of a (major) zeitgeber can initially disrupt the phase of the circadian rhythm and result in a phase-shift in the rhythm (i.e. reestablishment of a normal phase relationship to the shifted zeitgeber). For example, disruption of the circadian periodicity of ACTH and cortisol (as well as other rhythms) occurs in subjects who have rapidly crossed several time zones. Such individuals are out of phase with many potential zeitgebers. The symptoms attributed to *jet lag* may be a manifestation of the disruption of several biological rhythms involved in homeostasis. Studies of the effect of jet travel on the rhythms of urinary glucocorticosteroids, calcium, and

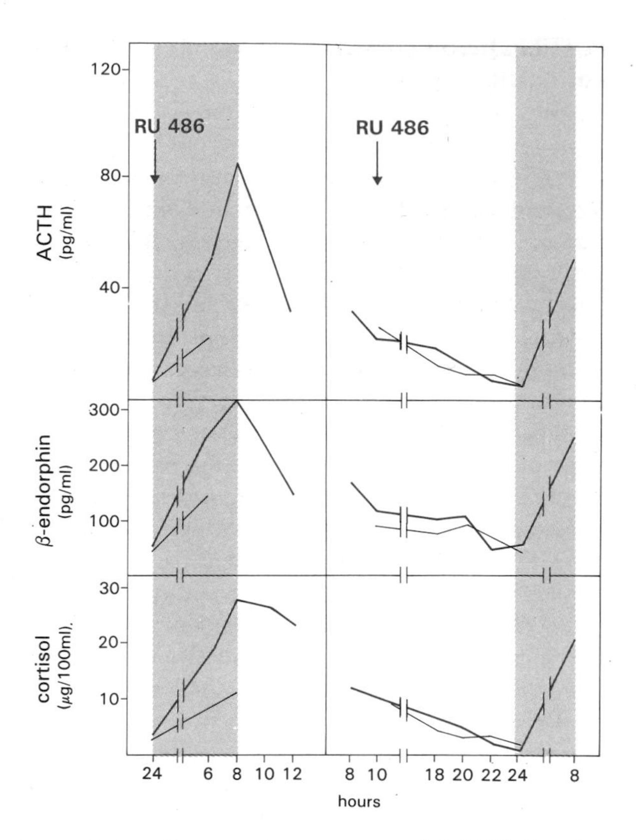

Figure IV.10 Comparative responses of ACTH, β-endorphin and cortisol to RU 486 administered either at midnight (left panel) or at 10.00 h (right panel) in normal young men. (Gaillard, R. C., Riondel, A., Muller, A. F., Hermann, W., Baulieu, E. E. (1984) RU 486: a steroid with antiglucocorticosteroid activity that only disinhibits the human pituitary-adrenal system at a specific time of day. Proc. Natl. Acad. Sci., USA 81: 3878–3882.

oral body temperature have shown a differential rate of re-entrainment.

Phase-shifts have been noted in some forms of psychiatric illnesses, and have even been implicated in the etiology of major affective disorders (endogenous depression) (section 8.1).

Regulation by Glucocorticosteroids

Glucocorticosteroids inhibit both the synthesis of POMC (acting at the transcription level) and the secretion of POMC-derived peptides. This negative feedback functions at the levels of the *pituitary* and *central* nervous system. Both *rate-sensitive "fast feedback"* and *level-sensitive "delayed feedback"* components of glucocorticosteroid suppression have been described[22].

Administration of dexamethasone or corticosterone to intact rats dramatically lowers plasma levels of ACTH and anterior lobe POMC mRNA levels, while adrenalectomy (removal of the endogenous source of glucocorticosteroids) markedly increases plasma ACTH and pituitary POMC mRNA levels[23]. Suppression of glucocorticosteroid activity by the administration of the antiglucocorticosteroid RU 486 in the late evening leads to an increase in ACTH and β-endorphin (with a consequent increase in cortisol). Interestingly, the same treatment in the late morning does not provoke the same response, confirming the circadian periodicity described above (Fig. IV.10). Administration of anti-CRF serum to adrenalectomized rats results in a rapid, statistically significant lowering of plasma ACTH levels. In primary cultures of anterior pituitary cells, glucocorticosteroids lower the apparent intrinsic activity (reduction in the maximum response) and shift the dose–response curve of CRF to the right; peptide secretion is never completely suppressed[24]. At the level of the CNS, glucocorticosteroids appear to inhibit the synthesis and release of hypothalamic CRF: adrenalectomy is associated with increased intensity of immunocytochemical staining of hypothalamic CRF neurons and modestly increased levels of hypothalamic pro-CRF mRNA[25].

Thus, the synthesis of POMC and the secretion of its derived peptides from the anterior pituitary appears to be *regulated by multiple factors at several levels* of organization. Each factor may be differentially controlled, by afferent neural pathways and peripheral input, in a stimulus-specific manner. CRF appears to be the major (possibly obligatory) component of this regulatory circuit. Pertinent to the multifactorial regulation of pituitary POMC is the finding that the pro-CRF molecule shares regions of

22. Dallman, M. F. and Yates, F. E. (1969) Dynamic asymmetries in the corticosteroid feedback path and distribution – metabolism – binding elements of the adrenocortical system. *Ann. N.Y. Acad. Sci.* 156: 696–721.
23. Schachter, B. S., Johnson, L. K., Baxter, J. D. and Roberts, J. L. (1982) Differential regulation by glucocorticoids of proopiomelanocortin mRNA levels in the anterior and intermediate lobes of the rat pituitary. *J. Clin. Endocrinol. Metab.* 110: 1442–1444.
24. Vale, W., Vaughan, J., Smith, M., Yamamoto, G., Rivier, J. and Rivier, C. (1983) Effects of synthetic ovine CRF, glucocorticoids, catecholamines, neurohypophysial peptides and other substances on cultured corticotropic cells. *Endocrinology* 113: 1121–1131.
25. Jingami, H., Matsukura, S., Numa, S. and Imura, H. (1985) Effects of adrenalectomy and dexamethasone administration on the level of prepro-corticotropin-releasing factor messenger ribonucleic acid (mRNAs) in the hypothalamus and adrenocorticotropin/β-lipotropin precursor mRNA in the pituitary of rats. *Endocrinology* 117: 1314–1320.

homology (either identical or chemically similar amino acids) with both POMC and the arginine vasopressin-neurophysin II precursor. Pro-CRF and arginine vasopressin-neurophysin precursor exhibit approximately 50% homology in their N-terminal segments (over a length of approximately 50 aa residues); the sequence of POMC encompassing the carboxyl terminus of the N-terminal fragment, ACTH, and γ-LPH, is approximately 50% homologous with the carboxyl two-thirds of pro-CRF. This structural relationship is of interest given the *functional* interactions between ACTH, CRF and vasopressin, and suggests an *evolutionary link between the three genes* encoding these precursor proteins, perhaps their derivation from a common ancestral gene.

4.2 Intermediate Lobe POMC

In the intermediate lobe, which is poorly vascularized in species possessing an anatomically distinct intermediate lobe (p. 232), regulation of POMC synthesis and the release of α-MSH and other POMC-derived peptides is believed to occur primarily via neurotransmitters released from the abundant catecholaminergic neurons (predominantly dopamine and norepinephrine) innervating this tissue, and epinephrine of peripheral origin is released from the adrenal gland following several types of stress[26].

The results of both in vitro and in vivo studies demonstrate that *dopamine* exerts a tonic *inhibitory* effect on *peptide secretion*, while *epinephrine* and *norepinephrine stimulate* secretion; mediation of these effects occurs at receptors present on the melanocorticotropic cells. In contrast to their effects on anterior lobe POMC release, catecholamines possess high potencies and intrinsic activities as intermediate lobe secretagogues, by direct interaction with β-adrenergic receptors present on melanotropic cells, in a CRF-independent manner.

Although CRF has been shown to stimulate peptide secretion in this tissue, it is much less potent on melanotropic cells than on corticotropic cells, and has weak intrinsic activity in comparison to epinephrine. In addition, very high doses of CRF only partially block the inhibitory effects of low doses of dopamine in vitro. Thus, epinephrine and norepinephrine (from both central and peripheral sources), and dopamine, appear to be the major positive and negative regulators of intermediate lobe POMC function; the physiological role of CRF is uncertain. Regulation of intermediate lobe POMC by catecholamines appears to be phylogenetically conserved: dopamine is a potent inhibitor of the secretion of POMC-derived peptides from amphibian intermediate lobe tissue in vitro.

5. Extrapituitary POMC

Although POMC was originally described in the anterior and intermediate lobes of the pituitary, still the major source of POMC-derived peptides, it is clear now that this precursor is expressed in a number of extrapituitary sites. POMC-derived peptides are present in the *central nervous system*, localized to neuronal cell bodies and fibers[27]. POMC-like mRNA has been detected in hypothalamic tissue extracts[28] and localized to hypothalamic neurons by in situ cDNA-mRNA hybridization[29]. POMC-derived peptides and POMC mRNA have also been identified in the *placenta* and in male and female *reproductive tracts*[30,31]. Evidence has been presented that POMC is also expressed in the *gastrointestinal tract*, *lung*, and some cell types of the *immune system*. As noted previously, there is evidence for only one POMC gene in the human, indicating that the same gene is expressed in multiple tissues. The finding of POMC-derived peptides and POMC-like mRNA in the testis, ovary, adrenal medulla‡ and placenta suggests that all steroid hormone secreting organs in mammals may utilize this peptidergic system.

26. Berkenbosch, F., Tilders, F. J. H. and Vermes, I. (1983) β-adrenoreceptor activation mediates stress-induced secretion of β-endorphin-related peptides from intermediate but not anterior pituitary. *Nature* 305: 237–239.
27. Finley, J. C. W., Lindstrom, P. and Petrusz, P. (1981) Immunocytochemical localization of β-endorphin-containing neurons in the rat brain. *Neuroendocrinology* 33: 28–42.
28. Civelli, O., Birnberg, N. and Herbert, E. (1982) Detection and quantitation of pro-opiomelanocortin mRNA in pituitary and brain tissues from different species. *J. Biol. Chem.* 257: 6783–6787.
29. Gee, C. E., Chen, C. L. C. and Roberts, J. L. (1983) Identification of proopiomelanocortin neurones in rat hypothalamus by in situ cDNA-mRNA hybridization. *Nature* 306: 374–376.
30. Margioris, A. N., Liotta, A. S., Vaudry, N., Boudin, G. W. and Krieger, D. T. (1982) Characterization of immunoreactive proopiomelanocortin-relatide peptides in rat testes. *Endocrinology* 113: 463–471.
31. Liotta, A. S., Houghten, R. and Krieger, D. T. (1982) Identification of a β-endorphin-like peptide in cultured human placental cells. *Nature* 295: 593–595.

‡ There are reports of immunoreactive ACTH and β-endorphin in adrenal cortical tissue.

The *POMC* gene may represent an *ancient gene* encoding basic messenger molecules. POMC-like peptides have been identified in a variety of invertebrate species, as well as in the unicellular organism Tetrahymena pyriformis (where material with the immunological and biological properties of ACTH has been detected). Complete characterization of the peptides detected in these lower phyla has not yet been performed; it is not yet clear whether these peptides represent products derived from (a) progenitor POMC gene(s).

Mechanism of Action of ACTH (Fig. IV.11)

Still poorly understood

Adrenocorticotropic hormone induced adrenal steroidogenesis is initiated with the binding of ACTH to high affinity *receptors* in the adrenocortical membrane, resulting in the *activation of adenylate cyclase*. (A plasma ACTH concentration of ~30 pg/ml may be sufficient to induce a maximal rate of steroidogenesis in the human and rat). The elevated intracellular levels of cAMP increase the transport of free cholesterol to a mitochondrial side-chain cleavage enzyme and activate cholesterol-ester hydrolase, thereby facilitating delivery of free cholesterol to the side-chain cleavage enzyme. These activities are associated with specific RNA and protein synthesis, phosphorylation and dephosphorylation of adrenal proteins (via both cAMP-dependent and cAMP-independent mechanisms), and turnover of adrenal polyphosphoinositides. Similar biochemical changes are associated with the actions of ACTH on brain function. The overall process is *calcium-dependent*.

Studies of the mechanism of action of β-*endorphin* in the *brain* indicate that similar intracellular biochemical events mediate its actions. While the net effect of ACTH appears to be an increase of the phosphorylation of specific proteins (in brain and adrenal gland), the action of β-endorphin appears to result in a net decrease in the phosphorylation of specific proteins. Specific phosphorylation-dephosphorylation reactions are known to represent a basic mechanism whereby both classical hormones and neurotransmitters control cellular events. With respect to brain function, differential effects on common intracellular kinases and phosphatases may represent at least one mechanism utilized by ACTH and β-endorphin-related peptides to differentially affect neurotransmitter function (see below).

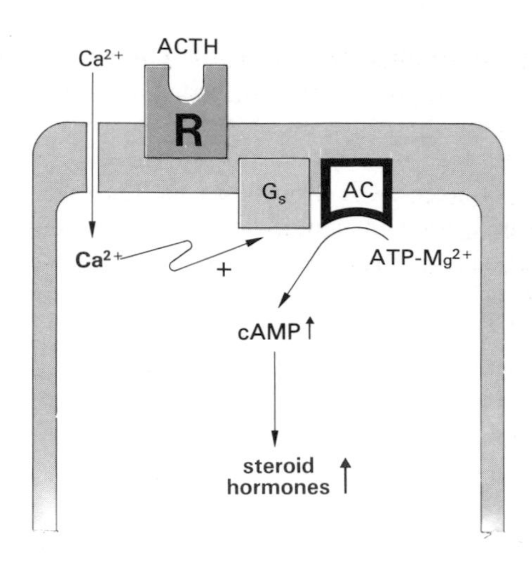

Figure IV.11 Mechanism of action of ACTH in adrenal cells. Ca^{2+} ions are involved in the cAMP response to ACTH, favoring the coupling of the occupied receptor to the adenylate cyclase system.

6. Functions of POMC-Derived Peptides

In addition to the well known action of ACTH in stimulating the synthesis and secretion of glucocorticosteroids from the adrenal cortex, *ACTH* has been shown to act at *extra-adrenal sites* as well. In vivo and in vitro studies have shown that ACTH stimulates *lipolysis* in adipose tissue and isolated fat cells. Specific ACTH receptors have been reported to be present on fat cells. This action of ACTH appears to be calcium-dependent, and is associated with increases in adipose adenylate cyclase activity. The ACTH-(4–10) sequence appears essential for the expression of lipolytic activity. This "core" sequence is also present in the γ-LPH segment of β-lipotropin and in the γ-MSH region of the N-terminal fragment of POMC (p. 229); all of these peptides possess weak lipolytic activity. However, ACTH is two to three orders of magnitude less potent in its action on fat cells than it is in stimulating adrenal steroidogenesis. The minimum effective concentration required to stimulate lipolysis is substantially above the plasma ACTH levels normally achieved. It is unlikely that this extra-adrenal effect of ACTH (and the other POMC-derived peptides) is of physiological relevance under normal circumstances.

The functions of the other pituitary POMC-derived peptides are not clear. The N-terminal fragment, or smaller fragments thereof (e.g., γ-*MSH*), has been reported to potentiate ACTH-mediated steroidogenesis in the adrenal cortex, to stimulate aldosterone secretion and adrenal mitogenesis (by a fragment not containing the γ-MSH sequence), and to possess natriuretic activity[32,33]. The biological significance of these observations remains to be determined. β-*lipotropin* possesses weak lipolytic activity, but this action does not appear to be physiologically relevant; it is generally believed that the intact molecule serves solely as a precursor of *β-endorphin*. The widespread distribution of peripheral and CNS opiate receptors suggests that pituitary β-endorphin may be the natural ligand at some of these peripheral sites (e.g., inhibition of uterine muscle contractions). However, the expression of the POMC gene at multiple tissue sites makes the source of the derived peptides less clear. *α-MSH* stimulates melanin dispersion in amphibia; adaptive color change can be correlated to changes in the concentrations of plasma α-MSH. The role of peripheral α-MSH in adult mammals is not known. In humans and other mammals that do not possess an intermediate lobe in adulthood, little or no α-MSH-like material has been detected in plasma (very low levels of desacetyl α-MSH are present in the anterior pituitary); a role for circulating levels of this peptide is unlikely. However, the existence of a distinct intermediate lobe in the human fetus suggests that α-MSH-like and β-endorphin-like peptides may play a role in the regulation of fetal metabolic pathways, and perhaps development. Consistent with this possibility, α-MSH has been shown to stimulate rat fetal growth and to reverse the glucocorticosteroid-mediated inhibition of body weight in chick embryos, in both cases exhibiting tropic activity. In addition, there have been reports that, in some species, α-MSH stimulates fetal adrenal steroidogenesis and DNA synthesis at a time when ACTH is ineffective. The functions of the pituitary intermediate lobe *N*-acetylated β-endorphin derivatives are not known; they are opiate inactive.

Central administration of ACTH and other POMC peptides brings about *effects* on the major mammalian *homeostatic systems*, such as blood pressure and temperature regulation[34]. These peptides produce dose-dependent effects on learning and memory processes. In many instances, β-*endorphin* and *ACTH* (or α-MSH) have *opposite actions*. For example, β-endorphin inhibits the turnover of dopamine in the tuberoinfundibular dopaminergic system, stimulates pituitary prolactin secretion, and inhibits pituitary LH release; α-MSH stimulates dopamine turnover, inhibits prolactin release, and stimulates LH release. β-endorphin inhibits, while α-MSH stimulates, sexual receptivity in the female rat (the effects on LH secretion and sexual activity appear to be mediated by differential effects of the peptides on GnRH release). Sufficiently high doses of β-endorphin induce analgesia, euphoria and sedation, while ACTH and α-MSH induce hyperalgesia and hyperactivity. ACTH and β-endorphin-related peptides have opposite effects on neurotransmitter turnover in selected brain regions; the overall actions of POMC-derived peptides in the central nervous system may be mediated by differential modulation of neurotransmission. These central actions are presumably mediated by POMC of central neuronal origin[35]. In

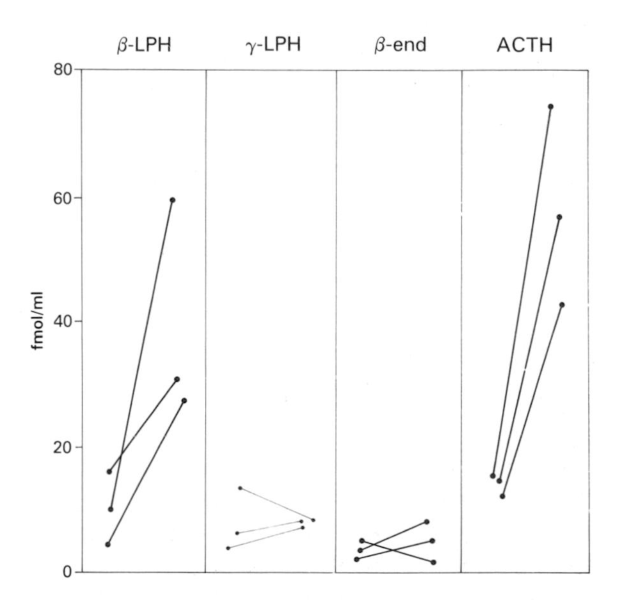

Figure IV.12 Immunoreactive plasma β-lipotropin, γ-lipotropin, β-endorphin and ACTH, under basal conditions and 30 to 45 min after insulin-induced hypoglycemia. (Yamaguchi, H., Liotta, A. S. and Krieger, D. T. (1980) Simultaneous determination of human plasma immunoreactive β-lipotropin, γ-lipotropin, and β-endorphin using immune-affinity chromatography. *J. Clin. Endocrinol. Metab.* 51: 1002–1008).

32. Lymangrover, J. R., Buckalew, V. M., Harris, J., Klein, M. C. and Gruber, K. A. (1985) Gamma-2 MSH is natriuretic in the rat. *Endocrinology* 116: 1227–1229.
33. Lowry, P. J., Silas, L., McLean, C., Linton, E. A. and Estivariz, F. E. (1983) Pro-γ-melanocyte-stimulating hormone in adrenal gland undergoing compensatory growth. *Nature* 306: 70–73.
34. Holaday, J. W., Wei, E., Loh, H. H. and Li, C. H. (1978) Endorphins may function in heat adaptation. *Proc. Natl. Acad. Sci., USA* 75: 2923–2927.
35. Liotta, A. S. and Krieger, D. T. (1983) Pro-opiomelanocortin-related and other pituitary hormones in the central nervous system. In: *Brain Peptides* (Krieger, D. T., Brownstein, M. J. and Martin, J. B., eds), Wiley, New York, pp. 615–652.

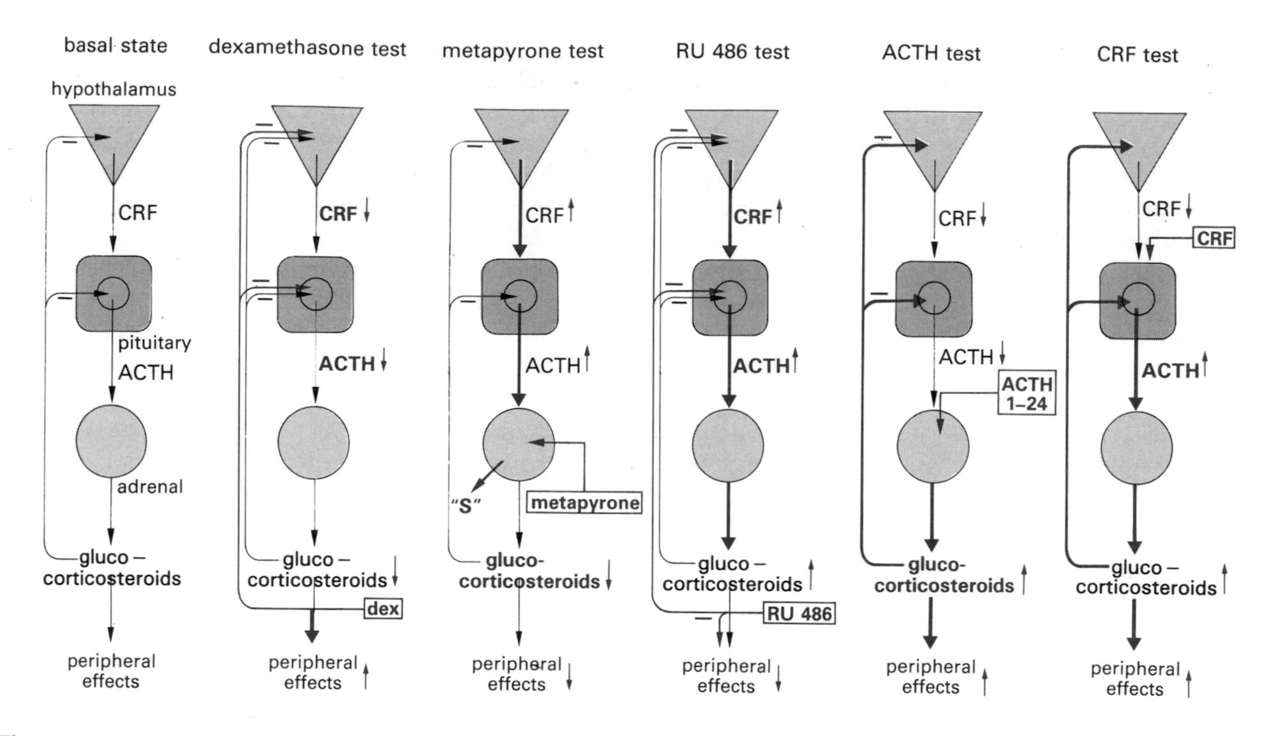

Figure IV.13 Stimulation and suppression test of the hypothalamic-hypophyseal-adrenal system. The basal state is indicated on the left. Represented successively are tests with a synthetic glucocorticosteroid (dexamethasone), a blocker of 11β-hydroxylation of glucocorticosteroids (metapyrone), the effects of an antiglucocorticosteroid (RU 486) acting at the receptor level, the stimulation test by ACTH, and the stimulation test with CRF.

the testis, ACTH and α-MSH stimulate growth and cAMP accumulation in Sertoli cells, while β-endorphin may inhibit Sertoli cell proliferation and secretion of androgen binding protein (ABP)[36].

Thus, the generation of multiple peptides from a single primary transcript provides a means of *coordinating* the synthesis and secretion of *functionally related peptides.*

7. Measurement of POMC Peptides in Plasma

Radioimmunoassay (*RIA*) is the most widely used method for estimating plasma levels of POMC-derived peptides. High affinity antisera provide adequate sensitivity for determining plasma levels of these peptides, which are normally present in femtomolar concentrations (Fig. IV.12). Because RIAs depend upon the interaction of an antibody with a limited amino acid sequence within the intact peptide, the assay alone cannot distinguish between the presence of a fragment of a peptide, the intact peptide, or a precursor of the peptide. For example, a specific β-endorphin antiserum may react with a fragment of β-endorphin (e.g., α-endorphin), the intact molecule, or β-lipotropin or POMC. For this reason, it is usually *necessary*, initially, to further *characterize* the nature of the *crossreacting species.* Molecular sieve chromatography, ion-exchange chromatography, and reverse phase high pressure liquid chromatography have all been successfully utilized for this purpose.

Bioassays for the measurement of ACTH provide additional information. Although ACTH is known to be metabolized to bio-inactive forms in the periphery (removal of N- and C-terminal amino acids in muscle and at other sites), reasonably good correlation between immunoassayable and bioassayable plasma levels in normal subjects has been reported by some groups. The most commonly used bioassay is based on the stimulation of steroidogenesis by isolated rat adrenal cortical cells (lowest detectable dose of ACTH is <1 femtomole). There is currently no useful bioassay or radioreceptor assay (based on specific binding to opiate receptor preparations) for the measurement of β-endorphin in plasma.

36. Gerendai, I., Shaha, C., Thau, R. and Bardin, W. C. (1985) Do testicular opiates regulate Leydig cell function? *Endocrinology* 115: 1645–1647.

Although POMC-derived peptides are released from the pituitary in a coordinated equimolar manner, relative plasma concentrations do not reflect this fact, due to *differences* in the *plasma half-life* of the individual peptides. The molar ratios of plasma ACTH, β-endorphin, and β-lipotropin have been shown to vary between individuals, and, as a function of time, within a given individual. *Episodic secretion* (as opposed to constant secretion) from the pituitary, and differences in the metabolic clearance rates of the individual peptides, may account for the differences observed in plasma concentrations. Figure IV.13 summarizes the *dynamic tests* used for the clinical evaluation of the hypothalamic-hypophyseal-adrenal axis.

8. Clinical Aspects of POMC Function

8.1 Stress and Depression

Stressful stimuli (physical and psychological) can elicit a variety of endocrine responses. These responses are presumably adaptive in nature and aid the organism to withstand stress and reestablish a homeostatic state. Many stresses result in the activation of adrenal cortical and adrenal medullary hormone synthesis, and in the subsequent secretion of corticosteroids and catecholamines, respectively.

Although the secretion of substances from the adrenal medulla was thought to be involved in responses to stress early on in "stress research", it was Selye who emphasized the *involvement of adrenal cortical activity* as a hallmark of the *response to stress*. He observed that stress induced a morphological change in the adrenal gland (enlargement), and showed that various noxious stimuli (e.g., exposure to cold, injury, spinal shock, injection of crude tissue extracts) resulted in a stereotypical series of responses, including adrenal enlargement, decrease in the size of the thymus, spleen, liver, loss of fat stores, drop in body temperature, and the appearance of gastric ulcers[37]. In an attempt to explain how an organism responds to such demands, Selye formulated the "*general adaptation syndrome*". In this view, stress could only be understood in relation to the physiological responses elicited, and those responses were regarded as stereotypical, non-specific, physiological reactions to the "message of stress".

Today, there is good evidence supporting the notion that *endocrine responses to stress* are specific responses *mediated by multiple* central nervous system level *afferent pathways* (some of which mediate the release of anterior pituitary hormone, hypothalamic releasing factors). Although Selye's basic concept of the stress syndrome, as originally formulated, did not account for stimulus-specific endocrine responses to stress, his continued focus on adrenal cortical responses as a measurement of stress undoubtedly stimulated research in this area. While there does not appear to be any general endocrine response to stress (such responses being stimulus-specific, in relation to the hormones involved and the magnitude of the stimulatory or inhibitory responses), hypophyseal-adrenal cortical (and adrenal medullary) responses are elicted by a great many diverse stressors, and this endocrine pathway serves as a good model for studying them.

37. Selye, H. (1975) Stress and distress. *Comp. Ther.* 1: 9.

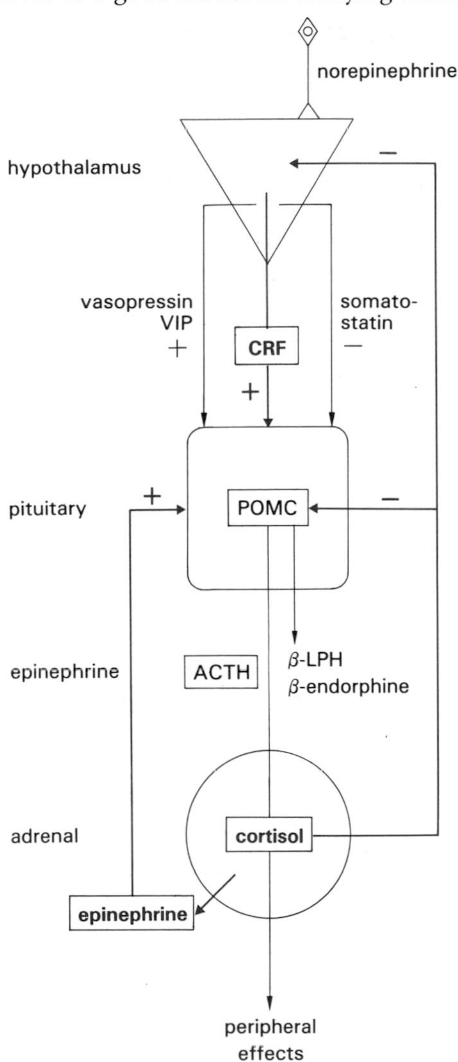

Figure IV.14 Multihormonal control of ACTH in corticotropic cells[38].

Catecholamines have long been known to be secreted preferentially from the adrenal medulla in response to certain types of emotional stimuli. In such an instance, peripheral catecholamines (in addition to effects on cardiac rate and output, mobilization of blood glucose, etc.) may serve as a major pituitary ACTH secretagogue (in the presence of a permissive amount of CRF) by direct action on the pituitary[38]. The synthesis of adrenal epinephrine is known to be stimulated by ACTH via glucorticosteroid-mediated increases in the activities of some of the enzymes involved in the biosynthesis of epinephrine. These interactions (Fig. IV.14) serve to illustrate the expression and interactions of two distinct endocrine systems during certain types of stress.

Pituitary-Adrenal Axis in Depression

Altered function of the pituitary-adrenal cortical axis is a classical symptom in most patients suffering from endogenous depression. *Hypercortisolism, reduced suppressive effect of dexamethasone*, and disturbances in the circadian rhythm of cortisol are typical features of this condition. The *rhythm* is both *blunted*, with respect to peak to nadir amplitude, and *phase-advanced*, with respect to the time of onset of the quiescent period or nadir in plasma cortisol levels. Whereas normal cortisol levels do not begin to rise until the second half of the sleep period (i.e. in the early morning hours), in depressed patients, cortisol secretion reaches its lowest point several hours before sleep, and secretory activity actually increases before the onset of sleep.

Based on the observation that hypercortisolism is a common response to a variety of stresses, some investigators have interpreted its association with psychiatric illnesses, according to the Selye model, as an endocrine response to stress of psychogenic etiology (e.g., anxiety or dysphoria). While this appears to be the case in some types of illness (e.g., schizophrenia), evidence strongly suggests that the endocrine abnormalities associated with endogenous depression are due to a dysfunction at the central nervous system level.

The phase-advance in circadian periodicity in depressives may be due to either a "fast running" circadian pacemaker or abnormal processing of environmental cues (i.e. a preferential entrainment, or abnormal responsiveness, to a zeitgeber), such as light. In this regard, it is of interest that nocturnal melatonin secretion in depressed patients is suppressed by an intensity of light insufficient to accomplish this in normal individuals. This finding may suggest that depressed patients are more sensitive to light. Thus, if melatonin has a role in the modulation of circadian periodicities, a hyperresponsiveness to light, mediated via melatonin, may be at least partially implicated in the circadian abnormality associated with depression. Pertinent to this hypothesis is the finding in experimental animals that monamine oxidase inhibitor-like antidepressive drugs blocked the suppressing effect of light on melatonin synthesis and caused a phase-delay in the circadian periodicity of melatonin activity.

If the same neurotransmitters are involved in the mediation of the symptoms characteristic of major depressive illness and in the regulation of at least some of the hypothalamic releasing factors controlling the release of anterior pituitary hormones, then a common defect at the level of the CNS may be responsible for both the affective disorder and the altered endocrine responses.

8.2 Hyposecretion of ACTH

Most cases of *hyposecretion* of ACTH and other POMC-derived peptides involve a general hypopituitary condition, with dysfunction of the secretion of most anterior pituitary hormones. Isolated ACTH deficiency is very rare and heterogeneous in nature (defects at steps in the synthesis of POMC and its processing to component peptides, as well as in the regulatory pathways, may be involved). In a few patients, vasopressin-stimulated ACTH release, demonstrating the ability of pituitary corticotrophs to synthesize and release ACTH, suggests a dysfunction at the hypothalamic level, perhaps in the ability to synthesize and release CRF[39].

8.3 Cushing's Syndrome and Disease

Over half a century ago, Harvey Cushing described the clinical *syndrome* (Fig. IV.15) associated *with bilateral adrenal hyperplasia*[40], now referred to as Cushing's syndrome. It was subsequently determined

38. Axelrod, J. and Reisine, T. (1984) Stress hormones: their interactions and regulation, *Science* 224: 452–459.
39. Stacpoole, P. W., Interlandi, J. N., Nicholson, W. E., Rabin, D. (1982) Isolated ACTH deficiency. A heterogeneous disorder. *Medicine* 61: 13–24.

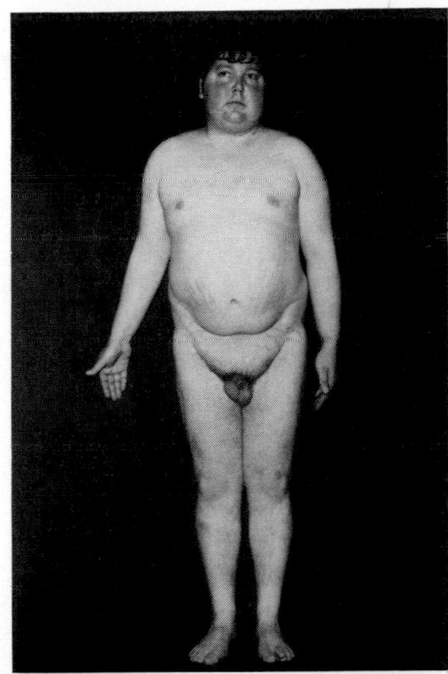

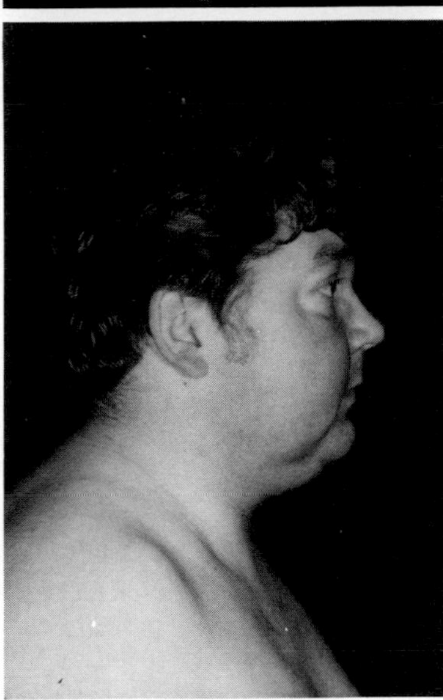

Figure IV.15 Cushing's disease. Note the supraclavicular and dorsal fat pads, the buffalo hump, the rounding of the face (moon face), and striae. (Courtesy of Dr. G. Schaison, Hôpital de Bicêtre).

that most of the clinical manifestations described were due to the excessive secretion of adrenal glucocorticosteroids (principally cortisol in humans). Hypercorticism may arise from multiple ACTH-dependent and ACTH-independent etiologies. However, the major disorder of the hypothalamic-hypophyseal POMC-axis is one of hypersecretion of anterior pituitary POMC-derived peptides and inappropriate secretion of ACTH, resulting in hypercorticism. *Pituitary-dependent* hypercorticism has been termed *Cushing's disease*, to distinguish it from *other causes of hypercorticism, or Cushing's syndrome*. Cushing's disease is responsible for the majority of cases of Cushing's syndrome in adults (see the 1982 monograph of D. Krieger, listed in the general references, for a clinical description).

Etiology and pathology

The majority of cases of Cushing's disease are caused by ACTH-secreting *pituitary adenomas* of the anterior lobe. Most of these corticotropic adenomas are classified as *microadenomas*, less than 1 cm in diameter (approximately one half of these are <0.5 cm in diameter). They are usually *basophilic* (some are chromophobes) and are unencapsulated. The cells of the adenomas usually display *Crooke's changes* (perinuclear hyalinization due to hypercorticism). Immunocytochemical studies have shown that adenoma cells stain positively for ACTH as well as N-terminal fragment β-lipotropin and β-endorphin. In some studies, not all cells staining for ACTH have been positive for the other POMC-derived peptides; this, however, appears to be due to technical difficulties, not to altered POMC processing. Electron microscopic studies have demonstrated that secretory granules (containing POMC and derived peptides) in some of these cells are larger (~ 200–700 μm) than those in normal corticotropic cells (200–450 μm); excess lysosomal activity has also been described[41].

Morphological changes in normal corticotropic cells surrounding the adenomas have been described; in general, these cells show signs of disintegration. In addition to the known long-acting suppressive effects of glucorticosteroids on normal corticotropic cells, the pathological changes seen in normal cells surrounding the adenoma provide an explanation for the post-operative ACTH-deficient state manifested in many patients following the surgical removal of the adenoma.

It has been suggested that Cushing's disease in humans may also arise from pituitary tumors of intermediate lobe origin. Such tumors contain neural fibers (based on positive argyrophilic staining), and in some instances "... it seemed as if nerves, ending around

40. Cushing, H. W. (1912) *The Pituitary Body and its Disorders: Clinical States Produced by Disorders of the Hypophysis Cerebri*. Lippincott, Philadelphia.
41. Pelletier, G., Robert, F. and Hardy, J. (1978) Identification of human anterior pituitary cells by immunoelectron microscopy. *J. Clin. Endocrinol. Metab*. 46: 534–542.

individual adenoma cells, could be seen"[42] (similar to the innervation of rat intermediate lobe). The presence of multiple microadenomas and/or multiple hyperplastic cell nests containing nerve fibers was also reported in two cases. Three of the six nerve-containing adenomas were located between the anterior and posterior lobes.

It has been proposed that remnants of the fetal lobe are represented in the adult by an incomplete layer of cells, in the colloid cyst region, between the anterior and posterior lobes; cord-like projections of these cells, called basophil invasion, extend to the posterior lobe. This group of adenomas may have originally been associated with a disorder at the level of the CNS, or with a neoplasm of neural origin, and not to intermediate lobe-like tumors.

Cushing's disease due to *corticotropic cell hyperplasia* in the apparent absence of adenoma has been reported (as well as a few cases in the absence of any apparent pituitary abnormality). However, it should be emphasized that the majority of all cases of Cushing's disease studied have been associated with pituitary adenomas. In addition, due to the very small size of many of these adenomas (they may not be detected by radiological procedures, including CT scan) and the inability to locate them surgically, the true incidence of pituitary adenomas as the cause of Cushing's disease is probably *underestimated.*

Pathophysiology and Differential Diagnosis

In Cushing's disease, POMC-producing pituitary adenomas appear to secrete ACTH episodically, in a manner similar to normal corticotropic cells. Such secretion is not completely autonomous. Patients with Cushing's disease *hyperrespond to stimuli causing ACTH release* in normal subjects (e.g., CRF (Fig. IV.16), metapyrone, arginine vasopressin), consistent with the increased adrenal cortical secretory rates associated with the syndrome. They exhibit an absent or *blunted circadian periodicity* of plasma ACTH and cortisol. Random ACTH levels (and other POMC-derived peptides) may be normal or modestly elevated. It is the persistent presence of such levels, not the maintenance of very high plasma ACTH levels, that probably accounts in large measure for the hypercorticism in Cushing's disease. In normal subjects, ACTH drops to a very low or undetectable level for several hours during a 24-hour period (onset depending on the sleep schedule); during this period, the adrenal cortical secretory rate drops significantly, resulting in very low plasma glucocorticosteroid levels. In contrast, in patients with Cushing's disease, ACTH levels during this quiescent period are in the low-to-normal physiological range, resulting in persistent adrenal stimulation.

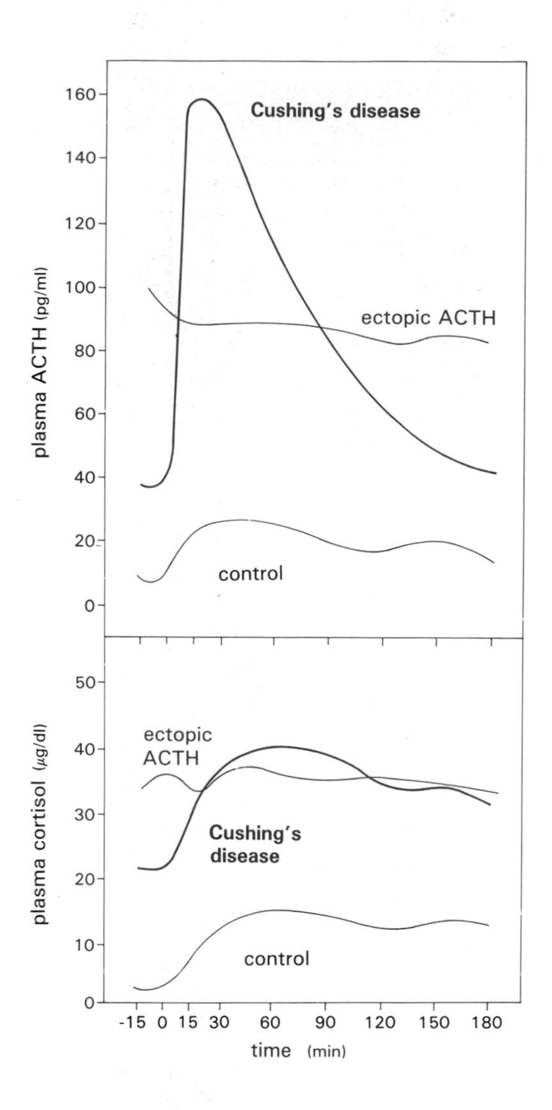

Figure IV.16 Stimulation test with CRF. ACTH and cortisol were measured in the plasma of patients with Cushing's disease (they show hyperresponsiveness) and with hypercorticism due to ectopic, tumoral ACTH secretion (they show no response).

Characterization of the forms of POMC-derived peptides present in adenomas and in plasma derived from patients with Cushing's disease and Nelson's syndrome (p. 250) indicates that processing of POMC

42. Lamberts, S. W. J., deLange, S. A. and Stefanko, S. Z. (1982) Adrenocorticotropin-secreting pituitary adenomas originate from anterior or intermediate lobe in Cushing's disease: differences in the regulation of hormone secretion. *J. Clin. Endocrinol. Metab.* 54: 286–291.

is similar to that noted in normal anterior pituitary corticotropic cells. The concentrations of such peptides, however, exceed those of the normal pituitary. In addition, the conversion of β-lipotropin to γ-lipotropin and β-endorphin appears to be more efficient[43]. To date, there has been no evidence of any such human adenoma processing POMC in an intermediate lobe-like manner. In a few cases where α-MSH was reported in adenomas, it was characterized as desacetyl α-MSH, and represented a minor peptide species in relation to ACTH, similar to findings in normal anterior pituitary. In contrast, Cushing's disease in the horse is associated with adenomas of the intermediate lobe, and the tumor appears to process POMC in a manner similar to normal intermediate lobe[44].

A hallmark of Cushing's disease is the relative *resistance* of pituitary ACTH secretion to the *normal negative feedback* inhibition by glucocorticosteroids. Compared to normal subjects, higher doses of exogenously administered glucocorticosteroids (usually dexamethasone) are required in order to obtain complete or partial supression of ACTH secretion. This altered responsiveness is manifested at least at the early, rate-sensitive phase of glucocorticosteroid inhibition. Some Cushing's patients have shown a paradoxical increase in ACTH secretion following administration of a low dose of dexamethasone, indicating the existence of possible abnormalities, beyond impaired inhibition, in glucocorticosteroid-mediated negative feedback mechanisms.

The diagnosis of Cushing's syndrome and the differential diagnosis of Cushing's disease can be difficult. This topic has been covered at length in many clinically oriented publications[45]. Briefly, the presence of endogenous hypercorticism must be established. Then, *pituitary-dependent hypercorticism* must be *differentiated from other* forms of *hypercorticism* (e.g., ectopic production of ACTH by non-pituitary tumors, or autonomous cortisol secretion by adrenal tumors). *No one* laboratory *test* has inherent diagnostic *specificity*. Several are performed, and the results of all of them are used in the clinical workup. The most important of these tests, at present, is based on the decreased sensitivity to glucocorticosteroid feedback inhibition in Cushing's disease. This test involves the administration of both low and high doses of the synthetic glucocorticosteroid dexamethasone (low: 0.5 mg orally every 6 h over 2 d; high: 2 mg every 6 h over 2 d). The lack of normal suppression of ACTH secretion (measurement of urinary free cortisol, or total urinary cortisol and its

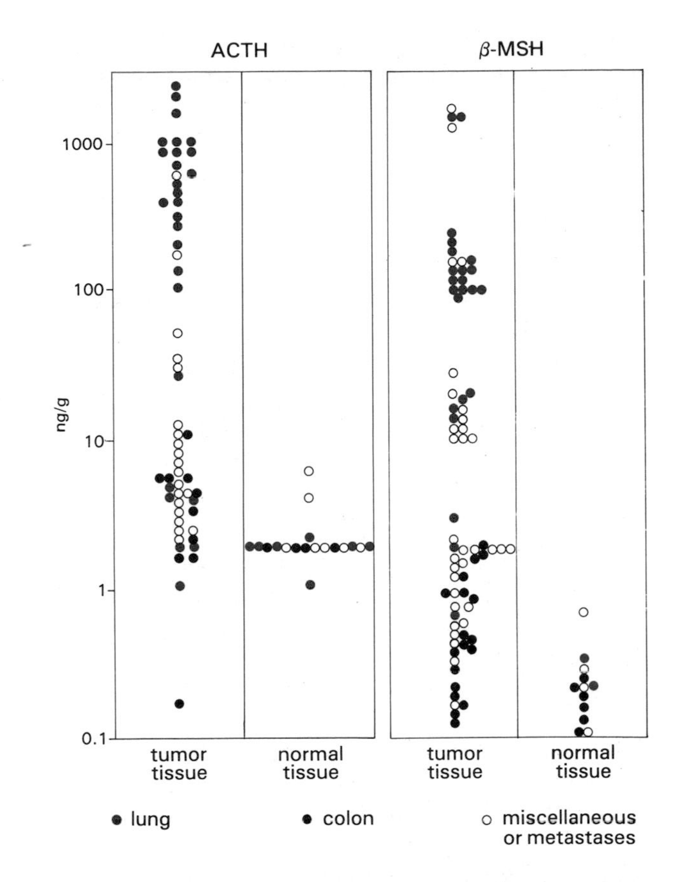

Figure IV.17 Immunoreactive ACTH and β-MSH in various carcinomas and corresponding normal tissues. (Odell, W. D., Wolfsen, A., Yoshimoto, Y. (1977) Ectopic peptide synthesis–a universal concomitant of neoplasia. *Trans. Assoc. Am. Physicians* 90: 204–227).

metabolites) following the low-dose test differentiates Cushing's syndrome from the eucorticosteroid state. The high dose test differentiates Cushing's disease, in which suppression or partial suppression of ACTH secretion occurs, from other causes of hypercorticism, such as glucocorticosteroid-secreting adrenal tumors (ACTH secretion is already suppressed in this condition, and basal plasma ACTH levels are usually undetectable) or non-pituitary, ACTH-secreting tumors (most such tumors do not respond to glucocorticosteroids) (Fig. IV.17), in which there is no suppres-

43. Suda, T., Liotta, A. S. and Krieger, D. T. (1978) β-endorphin is not detectable in the plasma from normal human subjects. *Science* 202: 221–223.
44. Wilson, M. G., Nicholson, W. E., Holscher, M. A., Sherrel, B. J., Mount, C. D. and Orth, D. N. (1982) Proopiomelanocortin peptides in normal pituitary, pituitary tumor, and plasma of normal and Cushing's horses. *Endocrinology* 110: 941–954.
45. Krieger, D. T. (1983) Physiopathology of Cushing's disease. *Endocrine Rev.* 4: 22–43.

sion of the synthesis and excretion of adrenal glucocorticosteroids. The administration of CRF has not been found useful in the differential diagnosis of Cushing's disease, as ACTH and cortisol responses to CRF in Cushing's disease are highly variable[46].

Treatment

The objective of the treatment of Cushing's disease is to *permanently remove the stimulus of increased ACTH* secretion and, thus, the consequent hypersecretion of cortisol which appears responsible for most of the clinical manifestations of the disease. When a pituitary adenoma is demonstrable, its removal by transsphenoidal microadenomectomy is usually the preferred treatment. Cure rates in more than 70% of cases followed postsurgically for several years have been reported. Some of the recurrences may have been due to the inability to find or remove the adenoma completely. In this regard, preoperative, bilateral, simultaneous blood sampling (to measure ACTH levels) of the inferior petrosal sinuses (which ultimately receive pituitary venous drainage) may indicate the side of the pituitary containing the adenoma (by detection of an ACTH gradient) and aid in the localization of the tumor (removal of half the pituitary ipsilateral to the sinus receiving a high ACTH gradient may be an acceptable option in cases where a tumor is not detectable[47]). Other therapeutic approaches include pituitary irradiation, the use of neuropharmacological agents to inhibit ACTH release at pituitary or CNS sites, and medical or surgical adrenalectomy.

8.4 Nelson's Syndrome

Ten to 20% of patients who have undergone *bilateral adrenalectomy* for the treatment of Cushing's disease have developed *pituitary tumors*. The clinical manifestation of this process was first described in 1958[48]. The tumors may be microadenomas, similar to those in Cushing's disease, or in some cases may be aggressive, rapidly growing neoplasia. In at least some cases, it is thought that such tumors derive from pre-existing microadenomas, after these small tumors are removed from the partial inhibitory influence of the pre-adrenalectomy hypercortisolism.

46. Orth, D.N. (1984) The old and the new in Cushing's disease. *N. Engl. J. Med.* 310: 649–651.
47. Oldfield, E.H., Chrousos, G.C., Schulte, H.M., Schaaf, M., McKeever, P.E., Krudy, A.G., Cutler, G.B., Loriaux, D.L. and Doppman, J.L. (1985) Preoperative localization of ACTH secreting microadenomas by bilateral and simultaneous inferior petrosal sinus sampling. *N. Engl. J. Med.* 312: 100–103.
48. Nelson, D.H., Meakin, J.W., Dealy, J.B., Matson, D.D., Emerson, K. and Thorn, G.W. (1958) ACTH-producing tumor of the pituitary gland, *N. Engl. J. Med.* 259: 161–164.

9. Summary

Adrenocortical hormone, or adrenocorticotropin (ACTH), a peptide of *39 aa* produced by the *corticotropic cells* of the pituitary, stimulates adrenal steroid biosynthesis (mainly cortisol in humans). ACTH is contained in a polyprotein encoded by the *proopiomelanocortin* (POMC) gene. This protein includes ACTH itself, *β-lipotropin*, which contains *β-endorphin*, and an N-terminal sequence including *β-MSH*. ACTH itself can be processed to α-MSH in the intermediate lobe of the pituitary. The processing of POMC in different cell types, to form several POMC-derived peptides, is a remarkable example of cell-specific processing of polyproteins.

ACTH is itself is under the control of *CRF*, a 41 aa peptide, which was the first hypophysiotropic factor to be postulated. CRF formation, like that of POMC in corticotropic cells, is under the *negative feedback control* of cortisol; it is mainly, but not exclusively, synthesized in the *paraventricular nucleus*. In addition, the synthesis and release of CRF and ACTH within their respective secretory cells is affected by a number of neurohumoral factors related to *stress*, including vasopressin and epinephrine. CRF, ACTH and cortisol demonstrate *circadian* periodicity, increasing nocturnally, with a maximum in the early morning; they are secreted in a *pulsatile* fashion, with approximatively 7 to 9 peaks/d. The mechanism of action of both CRF and ACTH involves a specific *membrane receptor* and, primarily, an increase of *cAMP* in the target cells. However, the mode of action of ACTH also includes an increase of intracellular Ca^{2+}, in particular in glomerulosa cells, where it increases aldosterone production.

The complexity of the POMC system is further demonstrated by evidence of *extrapituitary production* of ACTH and β-endorphin, especially in the placenta, endometrium, gonads, and several regions in the CNS. It is not established whether extra-adrenal activity of ACTH is physiologically significant. α-MSH increases melanin production in amphibian skin. A number of observations dealing with brain function suggest that the generation of several peptides from the single primary POMC translation product provides a coordinated network of information.

Although Selye's basic concept of "*general adaptation syndrome*" is questionable, involvement of the pituitary and adrenal cortex in diverse stress situations is definitively established, providing a fascinating model for the study of interactions between environment, brain processes, and general metabolism. In depression, the function of the hypophyseal-adrenal cortical axis is frequently altered. Cushing's syndrome involves clinical and metabolic symptoms which are due to excessive production of cortisol in humans. *Cushing's disease* is a Cushing's syndrome of pituitary origin. In most cases, a pituitary adenoma accounts for a blunted circadian periodicity of ACTH and cortisol, and for a relative resistance of ACTH secretion to the normal negative feedback inhibition exerted by glucocorticosteroids. Such a disease must be differentiated from other forms of hypercorticism, such as adrenal tumors and ectopic, non-pituitary tumors producing ACTH. In Cushing's disease, bilateral adrenalectomy is frequently performed. In 10 to 20% of patients, pituitary adenomas develop following adrenal surgery. The primary suppression of increased ACTH secretion may be obtained by removal of the pituitary microadenoma; other modes of treatment include pituitary irradiation or neurohormonal pharmacological agents. Most cases of hyposecretion of ACTH and other POMC-derived peptides involve general hypopituitarism.

General References

Herbert, E., Civelli, O., Birnberg, N., Rosa, P. and Uhler, M. (1982) Regulation of expression of proopiomelanocortin and related genes in various tissues: use of cell free systems and hybridization probes. In: *Molecular Genetics of Neuroscience* (Schmitt, F. O., Bird, S. J. and Bloom, F. E., eds), Raven Press, New York, pp. 219–230.

Vale, W., Rivier, C., Spiess, J., Brown, M. and Rivier, J. (1983) Corticotropin-releasing Factor. In: *Brain Peptides* (Krieger, D. T., Brownstein, M. J. and Martin, J. B., eds), Wiley, New York, pp. 961–974.

Vale, W., Rivier, C., Brown, M. R., Spiess, J., Koob, G., Swanson, L., Bilezikjian, L., Bloom, F. and Rivier, J. (1983) Chemical and biological characterization of corticotropin-releasing factor. *Rec. Prog. Horm. Res.* 39: 245–270.

Liotta, A. S. and Krieger, D. T. (1983) Pro-opiomelanocortin-related and other pituitary hormones in the central nervous system. In: *Brain Peptides* (Krieger, D. T., Brownstein, M. J. and Martin, J. B., eds), Wiley, New York, pp. 615–660.

Bardin, C. W., Shaha, C., Mather, J., Salomon, Y., Margioris, A. N., Liotta, A. S., Gerendai, I., Chen, C. L. and Krieger, D. T. (1984) Identification and possible function of pro-opiomelanocortin-derived peptides in the testis. *Ann. N.Y. Acad. Sci.* 438: 346–363.

Krieger, D. T. (1982) *Monographs on Endocrinology: Cushing's Syndrome*. Springer Verlag, New York.

Tyrrell, J. B. (1984) Diagnosis of Cushing's disease. In: *Secretory Tumors of the Pituitary Gland* (Black, P. McL., Zervas, N. T., Ridgeway, E. C. and Martin, J. B., eds), Raven Press, New York, pp. 263–272.

Jean-François Bach

Hormones of the Immune System

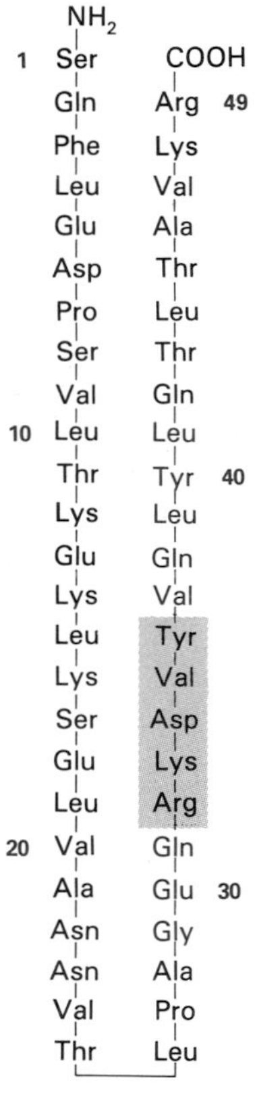

NH_2
1 Ser — COOH
Gln — Arg 49
Phe — Lys
Leu — Val
Glu — Ala
Asp — Thr
Pro — Leu
Ser — Thr
Val — Gln
10 Leu — Leu
Thr — Tyr 40
Lys — Leu
Glu — Gln
Lys — Val
Leu — Tyr
Lys — Val
Ser — Asp
Glu — Lys
Leu — Arg
20 Val — Gln
Ala — Glu 30
Asn — Gly
Asn — Ala
Val — Pro
Thr — Leu

pyro
1 Gln
Ala
Lys
Ser
Gln
Gly
Gly
Ser
9 Asn
COOH

Figure 1 (left) *Thymopoietin (Schlesinger, D. H. and Goldstein, G. (1975) The amino acid sequence of thymopoietin II.* Cell *5: 361–370).*

Figure 2 (right) *Thymulin (Bach, J. F., Dardenne, M., Pleau, J. M. and Rosa, J. (1977) Biochemical characterization of a serum thymic factor.* Nature *266: 55–57).*

Immune reactions perform the dual role of rejecting foreign invaders and maintaining the self-integrity of an organism. These functions are produced by the cooperative effect of *B cells*, which produce antibodies, and *T cells*. The latter, which differentiate under the regulation of the epithelial cells of the thymus, act either directly, as cytotoxic T lymphocytes, or via soluble factors, not specifically related to the antigen, known as *lymphokines*. These mediators come together at the reaction site, and then activate different types of cells, particularly monocytes and macrophages. The development of hormonal or cellular immune reactions involves numerous cellular interactions which amplify or restrain the intensity of the response. Although these interactions may involve direct contact between cells, they are more frequently produced by molecules which mediate communication between leukocytes, known as *interleukins*.

Principal Hormones of Immunity

The differentiation and action of the immune system involves several *hormones*. Those involved in *T cell differentiation* are produced by the epithelium of the *thymus*. Two of these have been well characterized, both biochemically and biologically. *Thymopoietin* is a peptide of 49 aa, of which a 5 aa fragment has most of the biological activity (Fig. 1) *Thymulin* (Fig. 2) is a metallopeptide of 9 aa. These two hormones, which do not share sequence homology, bind to different high affinity receptors on T cells and their precursors, in which they induce the appearance of differentiation antigens and the characteristic functions of the T cell lineage. Thymulin is produced exclusively by epithelial cells of the thymus, which are under feedback regulation by the hormone itself. Differentiation induced by thymic hormones is seen in the thymus before it is apparent in the periphery. In addition, precursor cells in the thymus recognize products of the major histocompatibility complex expressed by epithelial cells and macrophages. Other thymic hormones, less well defined chemically or biologically, have also been described. They include hormones of differentiation, such as those mentioned previously (*thymosins*, thymic humoral factor), or chemotactic hormones, attracting precursor cells into the thymus.

	NH_2			
1	Ala	Glu	Val	
	Pro	Glu	Leu	
	Thr	Leu	Asn	COOH
	Ser	Pro	Leu	Thr 133
	Ser	Lys	Ala	Leu
	Ser	Leu	Gln	Thr
	Thr	Glu	Ser	Ser 130
	Lys	Glu	Lys	Ile
	Lys	Glu	Asn	Ile
10	Thr	Leu	Phe	Ser
	Gln	Cys	His	Gln
	Leu	Gln	Leu	Cys
	Gln	Leu	Arg	Phe
	Leu	His	Pro	Thr
	Glu	Lys	Arg	Ile
	His	Leu	Asp	Trp
	Leu	Glu	Leu	Arg 120
	Leu	Thr	Ile	Asn
	Leu	Ala	Ser	Leu
20	Asp	Lys	Asn	Phe
	Leu	Lys	Ile	Glu
	Gln	Pro	Asn	Val
	Met	Met	Val	Ile
	Ile	Tyr	Ile	Thr
	Leu	Phe	Val	Ala
	Asn	Lys	Leu	Thr
	Gly	Phe	Glu	Glu 110
	Ile	Thr	Leu	Asp
	Asn	Leu	Lys	Ala
30	Asn	Met	Gly	Tyr
	Tyr	Arg	Ser	Glu
	Lys	Thr	Glu	Cys
	Asn	Leu	Thr	Met
	Pro	Lys	Thr	Phe

Figure 3 *Interleukin 2 (Taniguchi, T., Matsui, H., Fusita, T., Takaoka, C., Kashima, N., Yoshimoto, R. and Hamuro, J. (1983) Structure and expression of a cloned-cDNA for interleukin-2.* Nature *302: 305–310).*

Lymphokines are glycoproteins, produced by T lymphocytes, which bind to the different cell types they activate. Lymphokines may exert their very diverse functions by activating various cell types, particularly macrophages, or they may act in a totally independent fashion. Specific recognition of the antigen by the T cells is required for their physiological production, but lymphokines themselves are not specific to the antigen that was responsible for their synthesis. A number of distinct lymphokines for *various targets* have been identified; macrophages: migration inhibitory factor (MIF), macrophage arming factor (MAF); stem cells: colony stimulation factor (CSF), interleukin 3 (IL-3); cells producing histamines: granulocyte–macrophage colony stimulating factor (GM–CSF); bone cells: osteoclast activating factor (OAF); lymphocytes: blastogenic factor (BF). Some lymphokines, such as *γ interferon* (γ-IFN), have less clearly defined targets, but also act in a similar way following binding to a specific receptor.

Interleukins are responsible for the intracellular communication involved in the development of immune reactions. One of these, *interleukin 1* (previously known as lymphocyte activation factor, LAF), is produced by monocytes and macrophages stimulated by an antigen or a mitogen. The interleukin 1 gene has been cloned and its sequence determined. Interleukin 1 activates B lymphocytes and induces the production of interleukin 2 by T lymphocytes. Other known interleukins are produced by T lymphocytes, and form a part of the family of lymphokines already described. The best characterized of these is *interleukin 2* (IL-2, previously known as T cell growth factor, or TCGF) (Fig. 3), which acts on other T cells, stimulating growth and favoring proliferation induced by other signals. Although to a lesser degree, Il-2 also acts on B cells. Its amino acid sequence has been deduced from the sequence of the cloned gene. Il-2 binds to high affinity receptors expressed on activated T cells. Other factors interacting specifically with B cells have been described, such as B cell growth factor (BCGF), B cell differentiation factor (BCDF) and T cell replacing factor (TRF), which may be identical to BCDF. *Interleukin 4* is one of the B cell reactive interleukins. *Interleukin 3*, mentioned above as a CSF, is more similar to a lymphokine that stimulates cell lines than to an authentic interleukin involved in cellular cooperation.

Other Hormonal Controls

Although the substances described above are the principal hormones of immunity, the immune system also shares many hormonal peptides and receptors with the classical endocrine system. Additionally, other soluble mediators are involved in the function of the immune system (the proteins of complement, mediators of allergy such as histamine or serotonin, prostaglandins, etc.). However, these mediators should probably not be regarded as hormones.

The immune system is extremely sensitive to the action of a number of steroid or peptide hormones. *Glucocorticosteroids* suppress most immune responses by acting at several levels, especially at the level of interleukin production (IL-1 and IL-2). They may be involved, in an immunoregulatory feedback circuit, with immunological factors. *Androgens* reduce antibody production, e.g., autoantibodies, while estrogens, in certain cases, have the reverse effect (testosterone may also suppress thymus function throughout life, and castration accellerates autoimmunity). *Thyroid hormones* and *growth hormone* stimulate the function of the thymus and T cells. Finally, some data suggest the existence of a hypothalamic-hypophyseal control mechanism in some immune functions, e.g., the frequent appearance of Graves' disease (due to TSH-like autoantibodies) following psychic trauma.

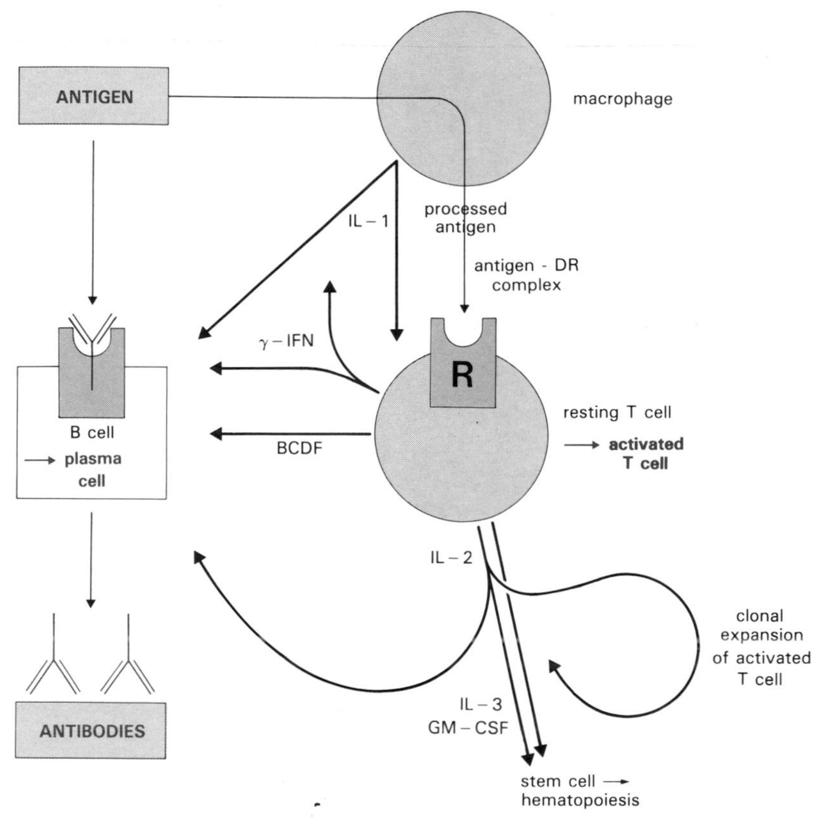

Figure 4 *Lymphokine production and action during the antigenic challenge. The macrophage takes up and processes the antigen. The processed antigen forms a complex with the histocompatibility molecule (DR) and, together with interleukin 1 (IL-1), stimulates the resting T cell. The activated T cell produces gamma-interferon γ-IFN), interleukin 2 (IL-2), interleukin 3 (IL-3), interleukin 4 (IL-4), and B cell differentiating factors (BCDF). The antigen also activates B cells, through surface antibodies, which become antibody-secreting plasma cells. Activated T cells also undergo clonal expansion, while IL-3 and granulocyte-macrophage colony stimulating factor (GM-CSF) stimulate marrow stem cells (adapted from Dinarello, C. A. and Mier, J. W. (1987) Current concepts. Lymphokines.* N. Engl. J. Med. *317: 940–945).*

References

Bach, J. F., ed (1983) Thymic hormones. In: *Clinics in Immunology and Allergy.* W. B. Saunders, New York, London, 3: 197–200.

Besedovsky, H., Del Rey, A., Sorkin, E. and Dinarello, C. A. (1986) Immunoregulatory feedback between interleukin-1 and glucocorticoid hormones. *Science* 233: 652–654.

Blalock, J. E., Harbour-McMenamin, D. and Smith, E. M. (1985) Peptide hormones shared by the neuroendocrine and immunologic systems. *J. Immunol.* 135: 858–861.

David, J. R., Remold, M. G., Liu, D. Y., Weiser, W. Y. and David, R. A. (1983) Lymphokins and macrophages. *Cell. Immunol.* 82: 75–81.

Jankovic, B. D., Markovic, B. M. and Spector, N. H., eds (1971) Neuro-immune Interactions. *Ann. N. Y. Acad. Sci.* vol. 496.

Nicola, N. A. and Vadas, M. (1984) Hemopoietic colony stimulating factors. *Immunology Today* 5: 76–80.

Robb, R. J. (1984) Interleukin 2: The molecule and its function. *Immunology Today* 5: 203–209.

Smith, K. A. (1988) Interleukin-2: Inception, Impact, and Implications. *Science* 240: 1169–1176.

CHAPTER V

Glycoprotein Hormones: Gonadotropins and Thyrotropin

Fernand Labrie

Contents, Chapter V

Glycoprotein Hormones: Gonadotropins and Thyrotropin

1. Pituitary Hormones

1.1 Secretory Cells

Thyrotropic Cells

Thyrotropic cells constitute approximately *10%* of all *pituitary cells*. They have an irregular form and are stained specifically by Alcian blue. At the electron microscope level, they can easily be identified by their secretory granules, which have a diameter of only 100 to 150 nm. They constitute the smallest of all anterior pituitary cells (Fig. II.5).

Gonadotropic Cells

Gonadotropic cells comprise *10 to 15%* of the total *pituitary cell* population. They can be identified by their round or oval form, their large volume, their almost uniform distribution throughout the pituitary, their basophilic staining, and the presence of two populations of secretory granules with respective diameters of ~275 and 375 nm. Although certain physiological observations suggest that LH and FSH are secreted by different cell types, no convincing morphological evidence has yet been presented to confirm such a theory.

1.2 Structure of TSH, LH, and FSH – Comparison with CG

The *glycoprotein hormones* of the pituitary: thyroid stimulating hormone (TSH), luteinizing hormone (LH) and follicle stimulating hormone (FSH), as well as human chorionic gonadotropin (hCG) secreted by the syncytiotrophoblastic cells of the placenta, have several common structural properties[1–6], and may be the product of common evolution, from the same primitive gene.

These four hormones consist of *two peptide chains*, known as α and β (Fig. V.1), and *15 to 30%* of the molecular weight is due to carbohydrates (Fig. V.2). The sugars which are present are fucose, mannose, galactose, *N*-acetylglucosamine and *N*-acetylgalactosamine. Sialic acid is also frequently present. *Microheterogeneity* of these glycoprotein hormones is largely due to variations in the carbohydrate composition of molecules with the same peptide structure. The *α chain, common* to all hormones, is formed of 92 aa, whereas the *β chain* consists of 112 to 145 aa, *depending* on the hormone (Table V.1). By

1. Sairam, M.R. and Li, C.H. (1973) Human pituitary thyrotropin: isolation and chemical characterization of its subunits. *Biochem. Biophys. Res. Commun.* 51: 336–342.
2. Shome, B. and Parlow, A. F. (1974) Human follicle-stimulating hormone (hFSH): first proposal for the amino acid sequence of the α-subunit (hFSH-α) and first demonstration of its identity with the α-subunit of human luteinizing hormone (hLH-α). *J. Clin. Endocrinol. Metab.* 39: 199–202.
3. Shome, B. and Parlow, A. F. (1974) Human follicle-stimulating hormone: first proposal for the amino acid sequence of the hormone-specific β-subunit (hFSH-β). *J. Clin. Endocrinol. Metab.* 39: 203–205.
4. Bellisario, R., Carlsen, R. B. and Bahl, O. P. (1973) Human chorionic gonadotropin. Linear amino acid sequence of the α-subunit. *J. Biol. Chem.* 248: 6796–6809.
5. Carlsen, R. B., Bahl, O. P. and Swaminathan, N. (1973) Human chorionic gonadotropin. Linear amino acid sequence of the β-subunit. *J. Biol. Chem.* 248: 6810–6827.
6. Pierce, J. G. and Parsons, I. F. (1981) Glycoprotein hormones: structure and function. *Annu. Rev. Biochem.* 50: 465–495.

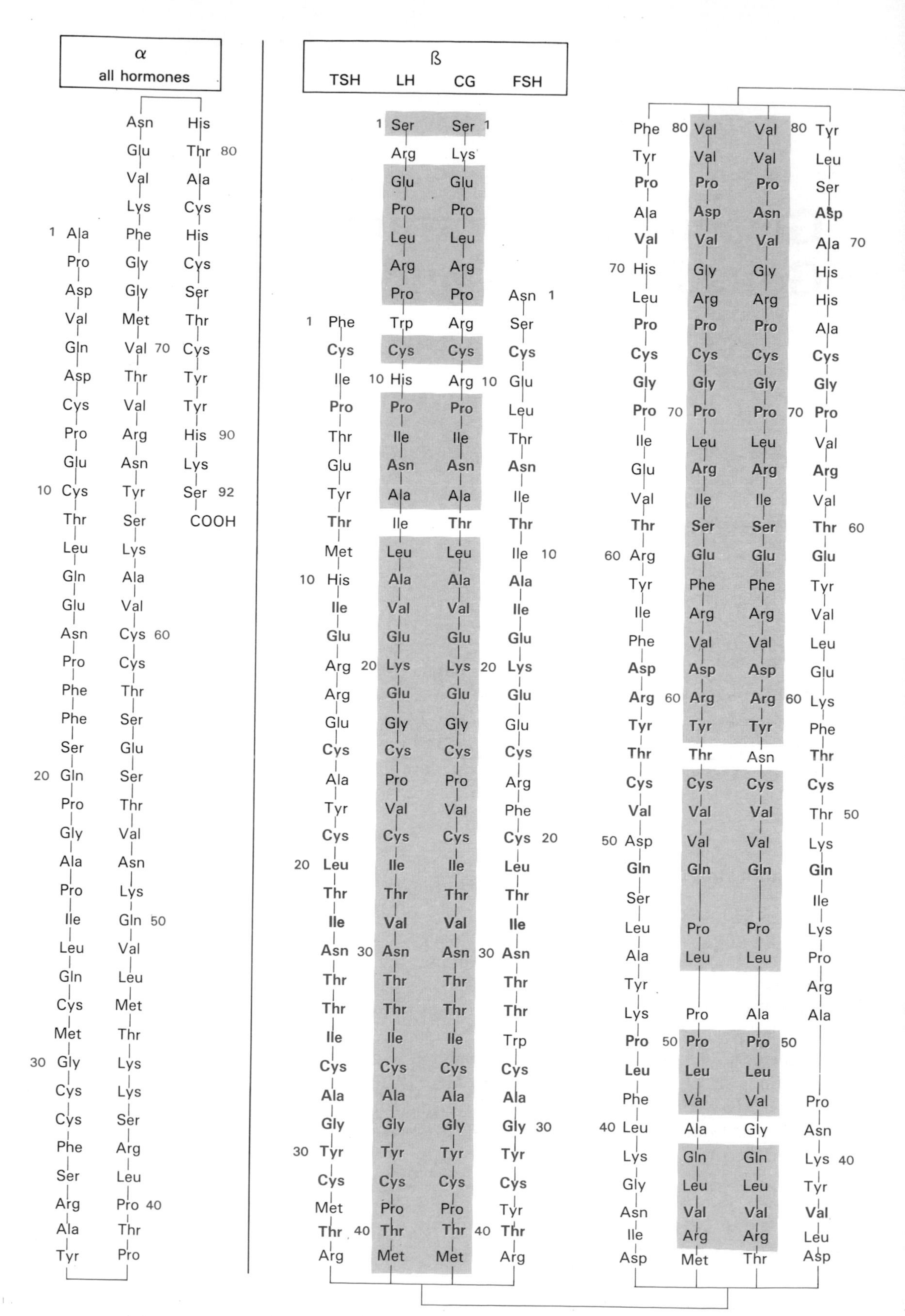

Figure V.1 Primary structure of the common α subunit and the β subunit of human LH, CG, FSH, and TSH.

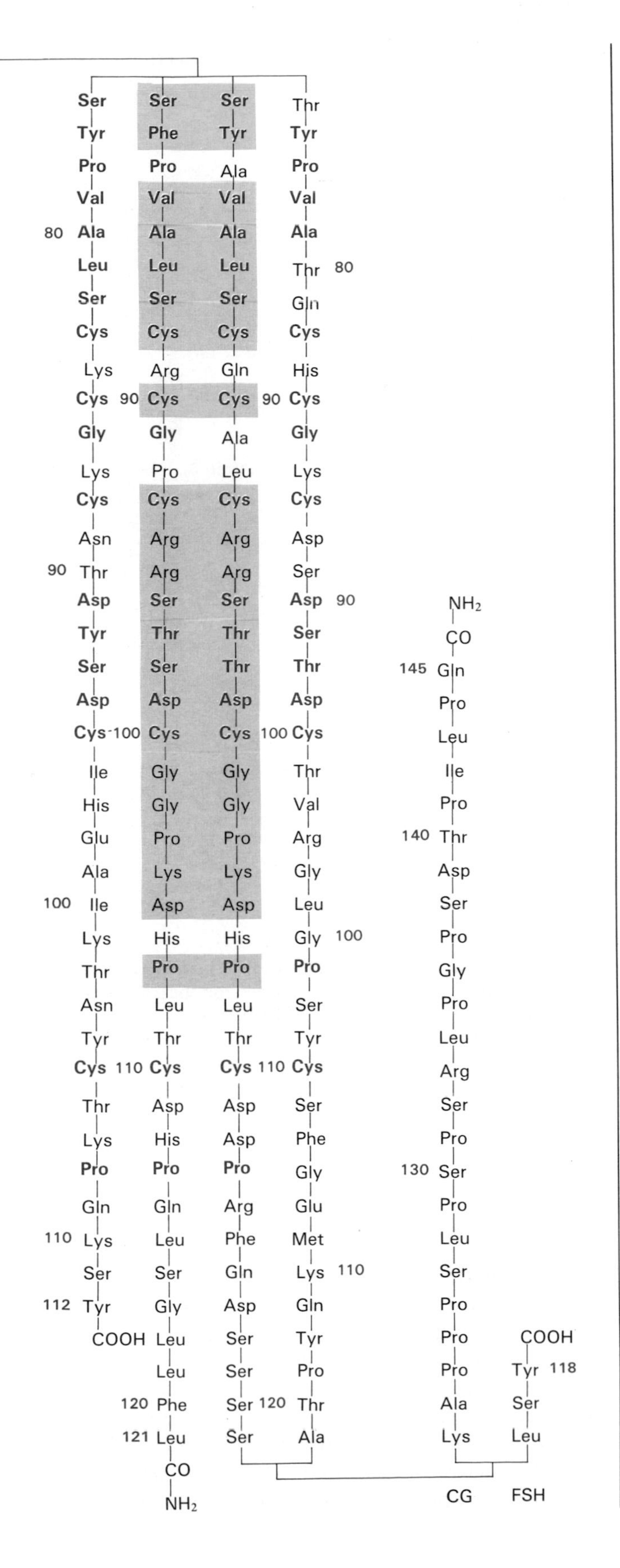

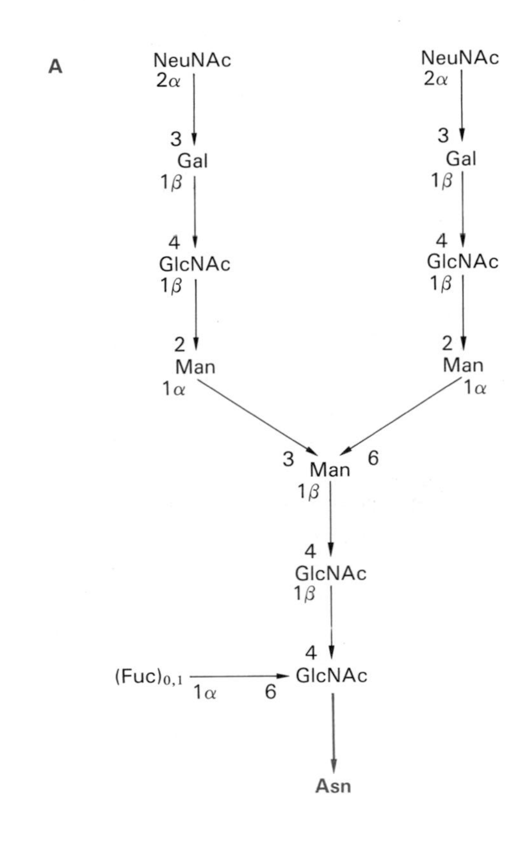

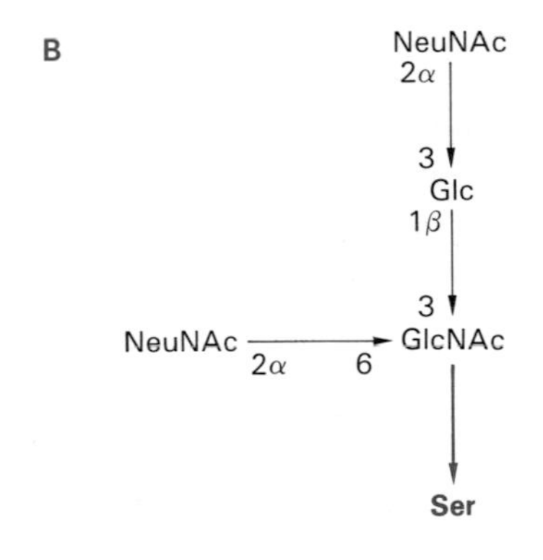

Figure V.2 *N*- and *O*-linked (**A** and **B**) oligosaccharide structure of hCG. The represented *N*-linked sugars are those of Asn-52 and -78 of the α subunit, and Asn-13 and -30 of the β subunit, whose structure also contains a fucose moiety. The α subunit does not have any *O*-linked sugars, but the β subunit has four sites of glycosylation, at Ser-121, -127, -132, and -138.

Table V.1 Number of amino acids in α and β subunits of human TSH, LH, FSH, and CG.

Hormone	α subunit	β subunit
TSH	92	112
LH	92	121
FSH	92	118
CG	92	145

comparison with human GH (which has only two disulfide bridges), human LH and TSH have 11 intramolecular disulfide bonds (five on the α subunit, and six on the β subunit). The α subunit contains two glycosidic chains, while the LH and hCG β subunits contain one and five glycosidic units, respectively.

The amino acid sequence of the α chain appears in Figure V.1. The *α chain* has *no biological activity by itself*, and the *specificity* of the *molecule* is *dependent on the β chain*, whose structure varies greatly among the four hormones. The proof of the specificity of the β chain can be demonstrated clearly by recombination experiments. Thus, while the isolated β chain has only weak biological activity, combination of an α subunit of TSH with a β subunit of LH forms a hybrid with exclusive LH activity, while the combination of a β subunit of TSH with an α subunit of LH produces a molecule with pure TSH activity.

The organization of glycoprotein hormone (α and β subunits) genes has been elucidated in humans and several other animals.

A single α subunit gene has been reported in the human, while separate β subunit genes are present[7]. The α subunit gene contains four exons and three introns (Fig. V.3). The corresponding mRNA encodes the precursor of the α subunit, a protein of 116 aa residues. This precursor contains the N-terminal signal leader and the α subunit itself.

There are at least seven genes for the β subunit of hCG, while a single gene encodes the β subunit of hLH. Each LH/CG β subunit gene examined so far contains three exons and two introns[8]. The *genes* for the *α* and the *β-LH/CG subunits* are located on *different chromosomes* in the human and mouse (Table V.2).

7. Chin, W.W. (1985) Organization and expression of glycoprotein hormone genes. In: *The Pituitary Gland* (Imura, H., ed), Raven Press, New York, pp. 103–125.
8. Talmadge, K., Vamvakopoulos, N. C. and Fiddes, J. C. (1984) Evolution of the genes for the β-subunits of human chorionic gonadotropin and luteinizing hormone. *Nature* 307: 37–40.

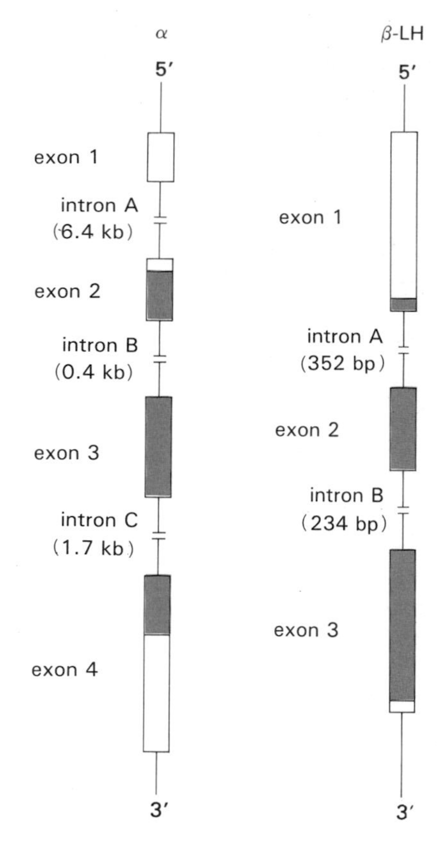

Figure V.3 Structure of the genes encoding the common α subunit of the glycoprotein hormones and the β subunit of rat LH and hCG.

Table V.2 Chromosomal localization of genes encoding the α subunit of glycoprotein hormones and the β subunit of LH, CG and TSH in the human and mouse.

Subunit	Human	Mouse
α	6	4
β-LH	19	7
β-CG	19	‡
β-TSH	1	3

‡Absent in this species.

The hCG β subunit contains an additional core of 24 amino acids at the C-terminal, as compared with the hLH β subunit. Study of the nucleotide sequence of the genes encoding the β subunits of hLH and hCG indicates that a mutation in the 3′ region of the β subunit gene is responsible for the additional 24 aa found in hCG (p. 155). Although hLH and hCG have the same activity at the level of testicular or ovarian receptors, it is possible, based on this minor difference in structure, to use a selective antibody as a specific pregnancy test, and even to attempt to neutralize hCG as soon as it appears, in order to "vaccinate" against pregnancy.

1.3 Synthesis and Intracellular Transport of TSH, LH, and FSH

TSH, LH and FSH, synthesized as other glycoprotein hormones (p. 28), are stored in *secretory granules.* The addition of carbohydrates to the polypeptide chains of TSH, LH and FSH occurs primarily in the Golgi apparatus, which is rich in glycosyltransferases[9]. The transfer of peptide hormones from the cisternae of the rough endoplasmic reticulum to the Golgi apparatus and from the vesicles of the Golgi to secretory granules is energy-dependent[9,10].

Secretory granules vary in dimension according to cell type. The anterior pituitary contains a *large store* of hormones, sufficient to respond to normal secretory needs for periods of 20 to 72 h. In response to the appropriate stimuli from the hypothalamus, the contents of the secretory granules is released into the intercellular space, by exocytosis, with a rapid dissolution of the contents of the granules before reaching the basal membrane, followed by a rapid diffusion (<1 min) of the hormones, across the fenestrated walls of the capillaries, into the general circulation.

1.4 Metabolism and Effects

Thyroid Stimulating Hormone

In humans, the half-life of TSH is approximately 60 min. The liver and kidney are the principal sites of inactivation of the hormone.

TSH is the main factor controlling the *formation* of *thyroid hormones* by the thyroid gland. The administration of TSH results in an increase in the size and vascularization of the gland, and in the synthesis and release of hormones. Following binding to specific membrane receptors (Fig. V.4), TSH provokes an *increase* in the intracellular concentration of *cAMP*, with a secondary *increase* in the uptake of *iodine*, in the synthesis of *thyroglobulin*, iodotyrosine and *iodothyronines*, in the proteolysis of thyroglobulin, and in the *release* of thyroxine and triiodothyronine by the thyroid gland (p. 355).

Although the physiological importance of such an *extrathyroid* effect has not been defined, TSH is capable of stimulating lipolysis in adipose tissue.

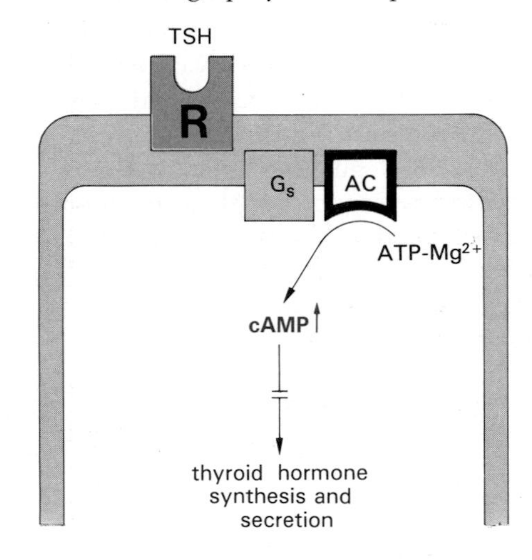

Figure V.4 Mechanism of action of TSH in a thyroid follicle cell.

Luteinizing and Follicle Stimulating Hormones

The plasma half-life of FSH is approximately 170 min, while that of LH is about 60 min.

In women, the primary actions of *FSH* involve the growth and maturation of the *ovarian follicle*, the final maturation of which is also under LH control. A more detailed description of the involvement of pituitary hormones in spermatogenesis, and the role of *LH* in ovulation and in the formation and maintenance of the *corpus luteum*, can be found on p. 451.

In men, LH and FSH act on the testis. The testis has two functions, which are performed by different components: secretion of androgens by the Leydig cells, and formation of sperm by the *seminiferous tubules*. The secretion of *androgens*, testosterone and dihydrotestosterone by the Leydig cells is under the direct control of *LH*. The key step in the stimulatory action of LH on androgen biosynthesis is the side-chain cleavage enzyme which converts cholesterol to pregnenolone. Androgens secreted by the testes are responsible for the development of the male genital tract and accessory sex organs (p. 439), as well as male

9. Pelletier, G. (1974) Autoradiographic studies of synthesis and intracellular migration of glycoproteins in the rat anterior pituitary gland. *J. Cell. Biol.* 62: 185–197.
10. Labrie, F., Pelletier, G., Lemay, A., Borgeat, P., Barden, N., Dupont, A., Savary, M., Côté, J. and Boucher, R. (1973) Control of protein synthesis in anterior pituitary gland. In: *Karolinska Symposium on Research Methods in Reproductive Endocrinology* (Diczfalusy, E., ed), *Acta Endocrinol.*, suppl. no. 180, pp. 301–339.

behavior. Some estradiol is also secreted by the testes, but its physiological importance is minor. Normal function of the *seminiferous tubules* requires *both androgens*, secreted by the Leydig cells, and pituitary *FSH*. *Sertoli cells* are the sites of binding and action of FSH. It is likely that normal spermatogenesis requires the concerted action of FSH and testosterone, although the precise requirements of each hormone are still under investigation.

In both males and females, in fact, all of the regulatory aspects of ovarian and testicular function are now being revised in view of recent findings on the structure of *inhibin* and related peptides (TGF-β, activin, p. 453). It is likely that both the feedback mechanisms regulating gonadotropin secretion and paracrine aspects of sex gland function will soon be greatly revised.

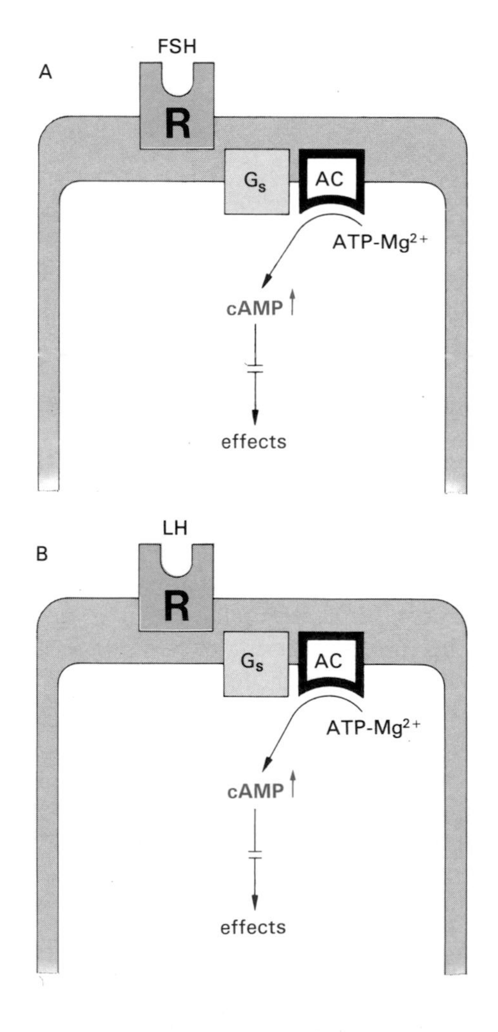

Figure V.5 Mechanism of action of FSH (**A**) and LH (**B**). Effects include an increase of respective steroid synthesis and secretion.

In terms of the mechanism of action of both FSH and LH, in both sexes, and in all relevant target cells, the glycoprotein first binds to a specific *membrane receptor*. An *increase* in *adenylate cyclase activity* and cAMP has been shown to be responsible for most of the observed effects of these hormones (Fig. V.5).

2. Hypothalamic Hormones

2.1 Structures

TRH and Its Prohormone

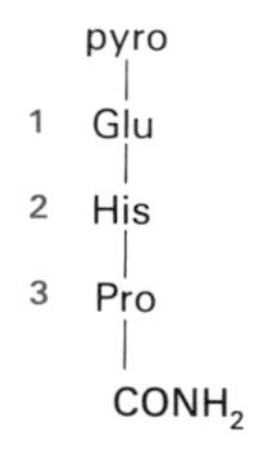

Figure V.6 Primary structure of TRH.

Thyrotropin releasing hormone (*TRH*) was the *first hypothalamic-hypophysiotropic hormone isolated*, and its primary structure was determined simultaneously, in 1969, by the groups of Schally[11] and Guillemin[12] (Fig. V.6). The fact that 500,000 pork or sheep hypothalami were required to produce approximately 1 mg of this material highlights the enormous difficulties encountered during the early purification of hypothalamic hormones.

The *tripeptide* structure of the *TRH precursor* (Fig. V.7) has been established in both the frog skin[13], a tissue with a relatively high abundance of TRH, and

11. Boler, J., Enzman, F., Folkers, K., Bowers, C. Y. and Schally, A. V. (1969) The identity of chemical and hormonal properties of the thyrotropin-releasing hormone and pyroglutamyl-histidyl-proline amide. *Biochem. Biophys. Res. Commun.* 37: 705–710.
12. Burgus, R., Dunn, T. F., Desiderio, D. and Guillemin, R. (1969) Structure moléculaire du facteur hypothalamique TRH d'origine ovine: mise en évidence par spectrométrie de masse de la séquence pGlu-His-Pro-NH$_2$. *C. R. Acad. Sci. Paris.* 269: 1870–1873.
13. Richter, K., Kawashima, E., Egger, R. and Kreil, G. (1984) Biosynthesis of thyrotropin releasing hormone in the skin of xenopus laevis: partial sequence of the precursor deduced from cloned cDNA. *EMBO J.* 3: 617–621.

in the rat hypothalamus[14], using recombinant DNA techniques. In both cases, a rather large protein was found which included repeated sequences of TRH, flanked by paired basic amino acids. Rat hypothalamic *preprohormone* contains 255 aa and can generate *five TRH* molecules. Seven other peptides of 10 to 49 aa in length can also be produced by post-translational processing, and some of these non-TRH peptides may be secreted. Interestingly, only the repeated TRH-coding units, unevenly dispersed throughout the precursor, are maintained between the amphibian and mammalian prohormones. Conservation of this pattern suggests an important mechanism in the amplification of hormone production.

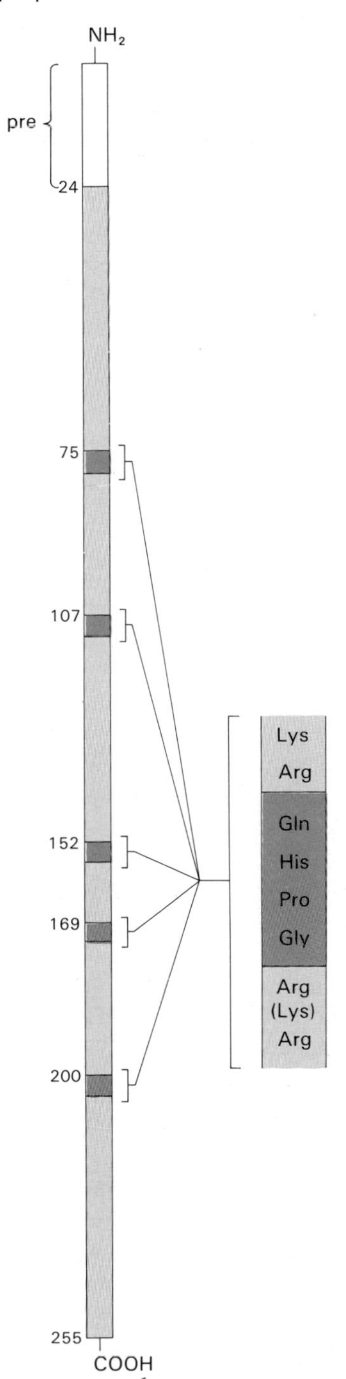

Figure V.7 Structure of prepro-TRH. Note that each repeated sequence includes a fourth amino acid residue, Gly, which is included in C-terminal amide formation during post-translational processing.

GnRH: Precursor, Localization, and Analogs

Two years after the elucidation of the structure of TRH in 1971, GnRH, the hypothalamic hormone which stimulates LH and FSH secretion, was purified from the hypothalamus of pigs, and its *decapeptide* structure was identified (Fig. V.8)[15]. Its structure was rapidly confirmed in sheep[16].

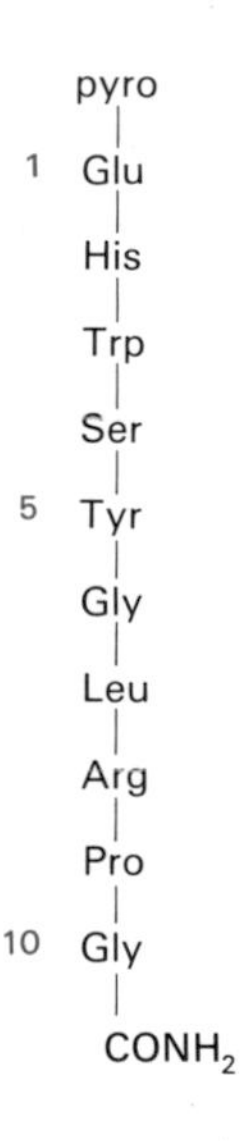

Figure V.8 Primary structure of GnRH.

14. Lechan, R. M., Wu, P., Jackson, I. M. D., Wolf, H., Cooperman, S., Mandel, G. and Goodman, R. H. (1986) Thyrotropin-releasing hormone precursors characterization in rat brain. *Science* 231: 159–161.
15. Matsuo, H., Baba, Y., Nair, R. M. G., Arimura, A. and Schally, A. V. (1971) Structure of the porcine LH- and FSH-releasing hormone. I. The proposed amino acid sequence. *Biochem. Biophys. Res. Commun.* 43: 1334–1339.
16. Burgus, R., Butcher, M., Ling, N., Monahan, M., Rivier, J., Fellows, R., Amoss, M., Blackwell, R., Vale, W. and Guillemin, R. (1971) Structure moléculaire du facteur hypothalamique (LRF) d'origine ovine contrôlant la sécrétion de l'hormone gonadotrope hypophysaire de lutéinisation. *C. R. Acad. Sci. Paris* 273: 1611–1613.

The cDNA of the gonadotropin releasing hormone *(GnRH) precursor* has been cloned[17]. It encodes the decapeptide preceded by a signal sequence of 23 aa, followed by the triplet Gly-Lys-Arg, necessary for enzymatic processing and C-terminal amidation of GnRH (Fig. V.9). The 56 aa immediately following form the *C-terminal* portion of the precursor and constitute the *GnRH-associated peptide (GAP)*[18], which was found to be an inhibitor of PRL secretion (p. 208). Active immunization against GAP, leading to enhanced PRL secretion, has been observed in rabbits. In addition, GAP stimulates the release of gonadotropin in rat pituitary cell culture. This latter property is also shown by a synthetic peptide comprising the first 13 aa of GAP (Fig. V.10)[19]. This peptide does not interact with pituitary GnRH receptors, and

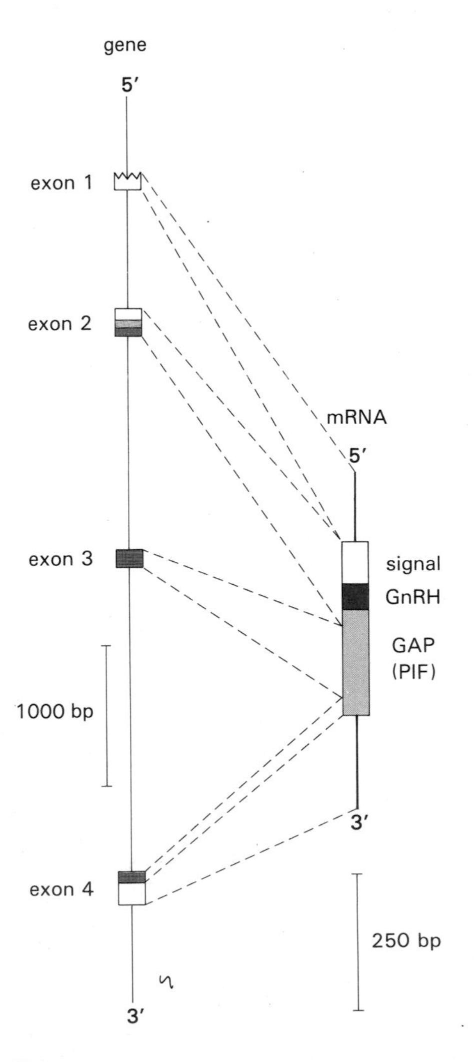

Figure V.9 Structure of human prepro-GnRH. Note that this drawing is similar to Fig. III.22, except that the synthesis of GnRH is highlighted.

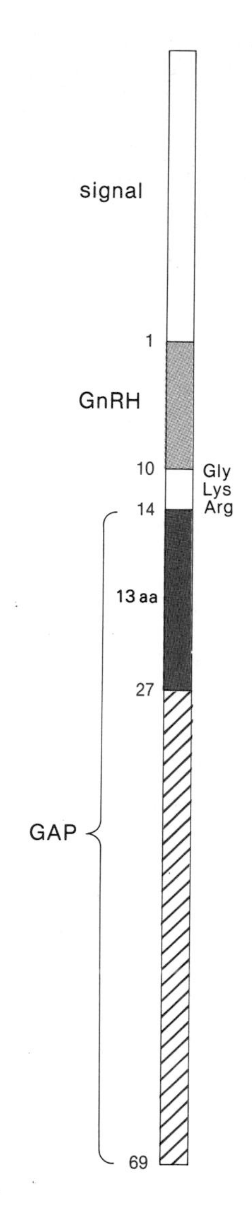

Figure V.10 Structure of human prepro-GnRH. The 13 aa, non-GnRH peptide with GnRH activity is shown, in red, as a portion of GAP.

17. Seeburg, P. H. and Adelman, J. P. (1984) Characterization of cDNA for precursor of human luteinizing hormone releasing hormone. *Nature* 311: 666–668.
18. Nikolics, K., Mason, A. J., Szonyi, E., Ramachandran, J. and Seeburg, P. H. (1985) A prolactin-inhibiting factor within the precursor for human gonadotropin-releasing hormone. *Nature* 316: 511–517.
19. Millar, R. P., Wormald, P. J. and de L. Milton, R. C. (1986) Stimulation of gonadotropin release by a non-GnRH peptide sequence of the GnRH precursor. *Science* 232: 68–70.
20. Pelletier, G., Labrie, F., Puviani, R., Arimura, A. and Schally, A. V. (1974) Immunohistochemical localization of luteinizing hormone-releasing hormone in the rat median eminence. *Endocrinology* 95: 314–317.

thus the GnRH prohormone curiously contains at least two distinct peptides capable of stimulating gonadotropin release. The 13 aa peptide does not affect TSH or PRL secretion. The same prohormone is synthesized in extrahypothalamic brain areas and gonadal tissues, suggesting a *complexity* in the central and probably paracrine functions of the hormonal peptides.

Localization

Using immunofluorescence and immunohistochemistry with specific antibodies, GnRH has been localized in the *hypothalamus*[20]. It is found principally in the lateral portion of the *external zone of the median eminence*. GnRH is contained exclusively in secretory granules which have a diameter of approximately 75 to 95 nm, present in nerve terminals near the basal membrane of fenestrated capillaries of the portal system (Fig. V.11). The neurohormone is stored in nerve terminals in the external region of the median eminence before it is released, through extru-

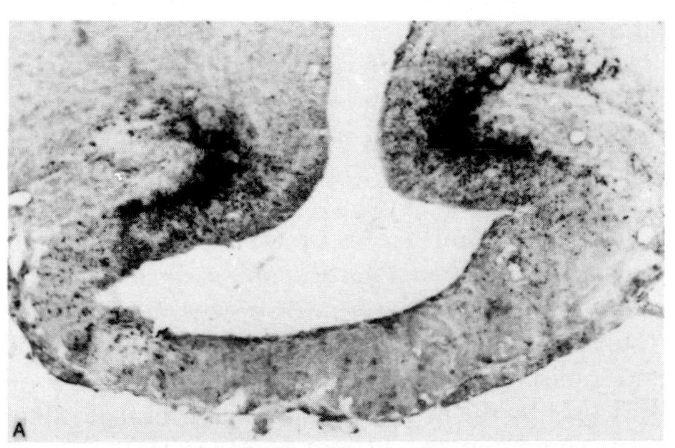

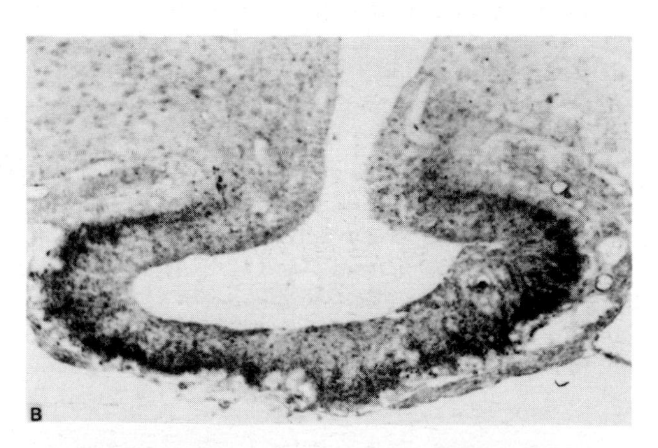

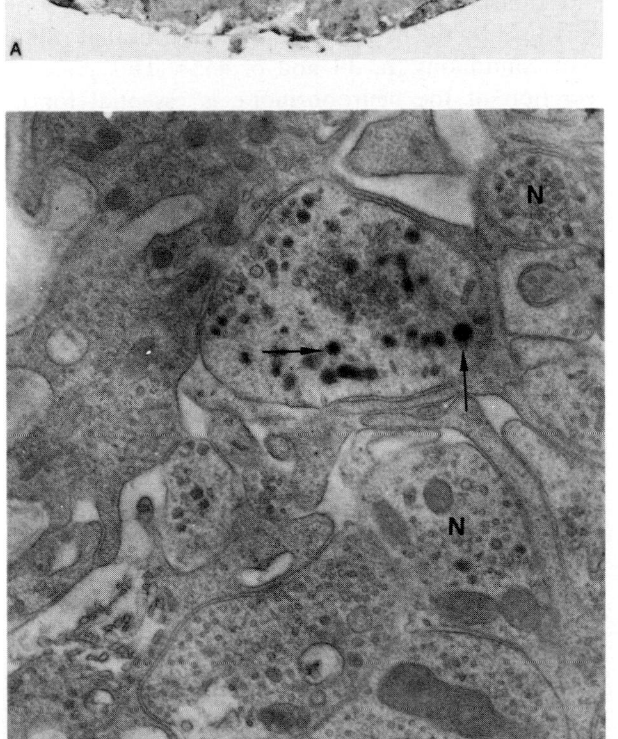

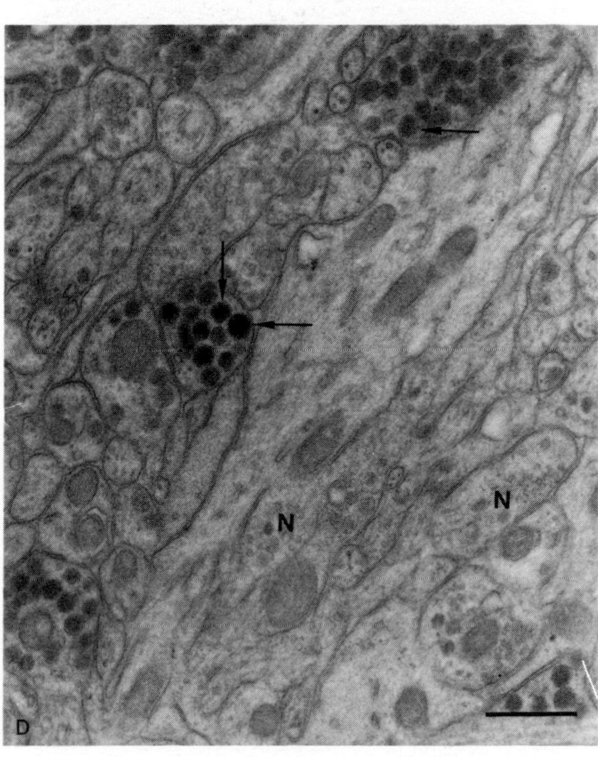

Figure V.11 Differential distribution of GnRH and somatostatin in rat hypothalamus. **A** and **B** show GnRH and somatostatin, respectively, on two consecutive thin sections of the median eminence. In panels **C** and **D**, the same two peptides are localized in vesicles of nerve terminals in the median eminence, as indicated by arrows. Negative nerve terminals are indicated by N. Bar=0.5 μm[20].

sion of the contents secretory granules, into capillaries of the portal system. Some pericaryons have also been identified dispersed in the zone extending *from* the *preoptic nucleus to* the caudal portion of the *tuber infundibulum*. GnRH has *also* been localized in nerve terminals of the vascular organ of the lamina termina and in ependymal and subependymal cells of the paraventricular organ, such as the subfornical organ, subcommissural organ, and the postrema center. The significance of the presence of GnRH in the periventricular organ remains unclear.

Analogs

Not only has the demonstration of the pulsatile administration of GnRH had important applications in the treatment of male and female infertility, but the field expanded rapidly following the synthesis of both superactive agonists and antagonists of the molecule. *Superactive* agonists of GnRH were first developed with the aim of improving fertility control. Figure V.12 shows the structure of three synthetic peptides for which the respective activities are three, 30 and 115 times that of natural GnRH[21]. Although these peptides are generally administered sc, the high activity of some of these compounds permit intranasal administration. In order to achieve inhibitory actions on the gonads during chronic administration of agonists, long-acting forms, resulting in prolonged release over a period of one to six months, have been developed. A great deal of effort has also been directed toward the development of synthetic *antagonists* with potential applications in the fields of contraception and hormone-dependent diseases (p. 120).

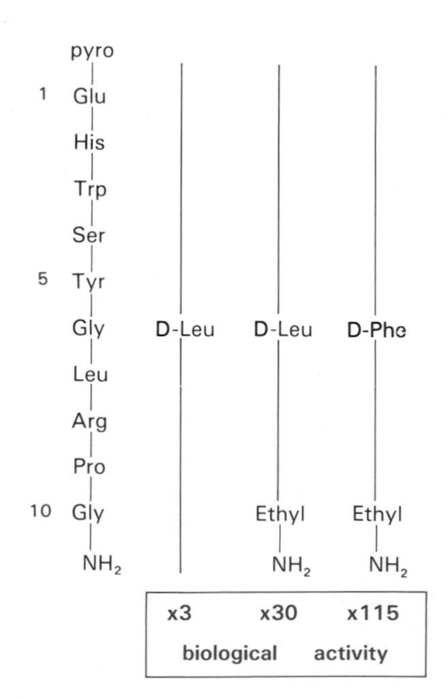

Figure V.12 Super-agonists of GnRH. Biological activity is compared to natural GnRH (activity=1).

2.2 Effects of TRH and GnRH

The tripeptide thyrotropin releasing hormone *(TRH) stimulates TSH and PRL* secretion in thyrotropic and lactotropic cells, while *GnRH stimulates* the secretion of *LH and FSH* from gonadotropic cells in the anterior pituitary.

The stimulatory effect of GnRH on the secretion of LH and FSH can be shown by the direct administration of the peptide, while its physiological importance can be demonstrated by the use of specific antibodies against the neuropeptide.

An important aspect of *GnRH* secretion, and essential for its *physiological action in* the *pituitary*, is its *intermittent release* from hypothalamic neurons. Thus, it has been clearly demonstrated in monkeys[22,23], and confirmed indirectly in humans, that the secretion of GnRH by the hypothalamus is intermittent rather than continuous (p. 34 and p. 455). This *pulsatile* secretion of the neurohormone is essential for its *stimulatory* action on gonadotropic cells. The *continuous* administration of the decapeptide causes a progressive *desensitization* of the pituitary mechanisms responsible for gonadotropin secretion.

Whether or not the recently discovered 13 aa "second" GnRH, or GAP, has any actual role in the secretion of FSH and LH remains conjectural.

2.3 Receptors

TRH Receptor

The *TRH receptor*[24] is located in the external *membrane* of *thyrotropic* and *lactotropic* cells. The properties of the TRH receptor appear identical in the two cell types. Receptor concentration is stimulated by estrogens and inhibited by thyroid hormones. Since changes in receptor levels are accompanied by parallel changes in the response of TSH and prolactin to TRH, it appears likely that the concentration of TRH receptor is an important factor in the control of thyrotropic and lactotropic cell activity.

GnRH Receptor

The characteristics of the pituitary *GnRH receptor* have been studied in detail using stable iodinated GnRH agonists[25,26]. It is located in the plasma membrane, and a high degree of specificity has been demonstrated with a large series of GnRH agonists and antagonists. Contrary to findings in many other receptor systems, the changes in responsiveness to GnRH observed after acute or chronic administration of GnRH itself, of its agonists, or after treatment with estrogens, androgens and progestins, are usually not accompanied by corresponding changes in the level of GnRH receptors. These data indicate that, although GnRH receptors are the first and obligatory site of interaction of GnRH with gonadotropic cells, most of the modulation of GnRH action occurs at steps subsequent to its binding to the receptor[27].

2.4 Mechanism of Action of GnRH and TRH

In cultures of anterior pituitary cells, when the concentration of intracellular *cAMP increases* (either by incubation with liposoluble cAMP analogs or by blocking phosphodiesterase activity by theophylline), the release of all pituitary hormones is increased. A direct stimulatory effect of GnRH on the accumulation of cAMP in the pituitary[28] is illustrated in Figure V.13. This change is accompanied by an increase in LH and FSH secretion. The parallelism between the level of cAMP and hormone release is observed as a function of time following the addition of the peptide, and also as a function of GnRH concentration (or of its analogs). The addition of *TRH* also leads to a parallel *increase* in the secretion of *cAMP* and the release of TSH. In addition to the probable role of cAMP, recent experimental evidence indicates that *calcium* and *inositol phospholipids* are involved in the mechanism of TRH and GnRH action (Fig. V.14 and p. 209). Thus, following binding to their specific membrane receptor, releasing factors probably induce responses via the intervention of several intracellular mediators, including cAMP, calcium, and phospholipids[29].

Since the pituitary is a storage gland, and since the initial effect of hypothalamic hormones is a stimulation or inhibition of the level of release of these hor-

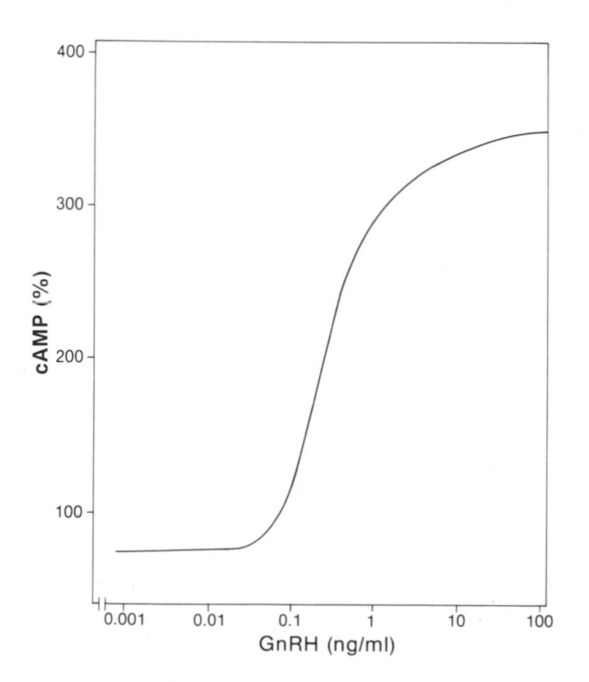

Figure V.13 Effect of increasing concentrations of GnRH on cAMP accumulation in the anterior pituitary (expressed in percent of control)[28].

21. Rivier, J., Rivier, C., Perrin, M., Porter, J. and Vale, W. (1981) GnRH analogs: structure-activity relationships. In: *LHRH Peptides as Female and Male Contraceptives* (Zatuchni, G. I., Shelton J. D., and Sciarra, J. J., eds), Harper and Row, Philadelphia, pp. 13–23.
22. Knobil, E. (1980) The neuroendocrine control of the menstrual cycle. *Rec. Prog. Horm. Res.* 36: 53–88.
23. Ferin, M., Van Vugt, D. and Wardlaw, S. (1984) The hypothalamic control of the menstrual cycle and the role of endogenous opioid peptides. *Rec. Prog. Horm. Res.* 40: 441–485.
24. Labrie, F., Barden, N., Poirier, G. and DeLéan, A. (1972) Characteristics of binding of [^{3}H] thyrotropin-releasing hormone to plasma membranes of bovine anterior pituitary gland. *Proc. Natl. Acad. Sci., USA* 9: 283–287.
25. Reeves, J. J., Séquin, C., Lefebvre, F. A., Kelly, P. A. and Labrie, F. (1980) Similar LHRH binding sites in the rat anterior pituitary gland and ovary. *Proc. Natl. Acad. Sci., USA* 77: 5567–5571.
26. Clayton, R. N. and Catt, K. H. (1980) Receptor binding affinity of gonadotropin-releasing hormone analogs: analysis of radioligand receptor assays. *Endocrinology* 106: 1154–1159.
27. Ferland, L., Marchetti, B., Séquin, C., Lefebvre, F. A., Reeves, J. J. and Labrie, F. (1981) Dissociated changes of pituitary luteinizing hormone releasing hormone (LHRH) receptors and responsiveness to the neurohormone induced by 17β-estradiol and LHRH in vivo in the rat. *Endocrinology* 109: 87–93.
28. Borgeat, P., Chavancy, G., Dupont, A., Labrie, F., Arimura, A. and Schally, A. V. (1972) Stimulation of adenosine 3′, 5′-cyclic monophosphate accumulation in anterior pituitary gland in vitro by synthetic luteinizing hormone-releasing hormone. *Proc. Natl. Acad. Sci., USA* 69: 2677–2681.
29. Raymond, V., Leung, P. C. K., Veilleux, R., Lefèvre, G. and Labrie, F. (1984) LHRH rapidly stimulates phosphatidylinositol metabolism in enriched gonadotrophes. *Mol. Cell. Endocrinol.* 36: 157–164.

A

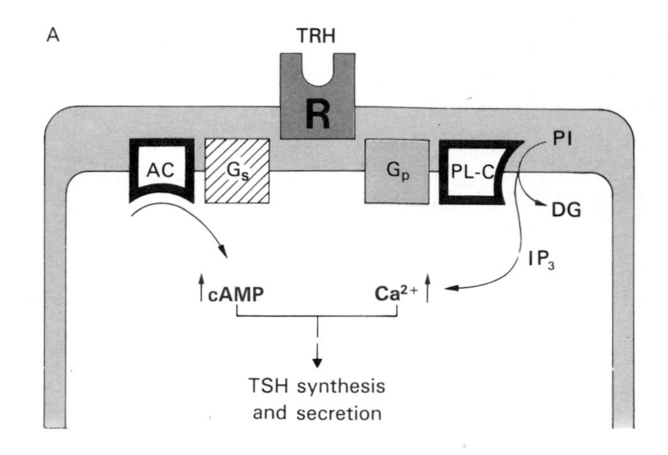

B

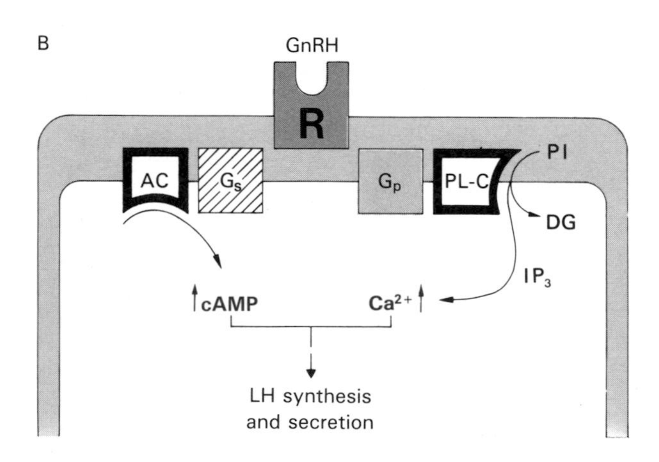

Figure V.14 Mechanism of action of TRH in a thyrotropic cell (**A**) and that of GnRH in a gonadotropic cell (**B**).

mones by exocytosis, changes in the phosphorylation of certain specific proteins of intracellular membranes, secretory granules and plasma membrane could be responsible for changes in the rate of exocytosis and release of hormones. It has been shown that cAMP stimulates the phosphorylation of specific proteins of the plasma membrane and secretory granule membrane. Phosphorylation of ribosomal proteins and chromatin has also been observed. These results suggest the possibility that the phosphorylation of some protein substrates may be a mechanism controlling the activity of the pituitary gland.

Prostaglandins of the E type have a marked stimulatory effect on the accumulation of cAMP in the pituitary, an effect which is accompanied by parallel changes in the release of growth hormone. These prostaglandins also have a small stimulatory effect on the secretion of TSH and the response of TSH to TRH. Although it seems possible that prostaglandins have a role in the control of TSH secretion, it does not appear that they affect the secretion of LH, FSH, prolactin, or ACTH, at least at the pituitary level.

3. Control of GnRH and TRH Secretion

Peptidergic neurons of the hypothalamus are capable of responding to classical *neurotransmitters*. They also appear to be sensitive to the action of *peripheral hormones* (Fig. V.15).

It has been suggested that dopamine is a negative regulator of GnRH secretion, based on immunohistochemical studies which demonstrated its high concentration in nerve terminals in the region rich in neurons containing GnRH; that is, in the lateral, external layer of the median eminence. Turnover studies indicate that the secretion of *GnRH* is *inhibited by dopamine* and *stimulated by norepi-*

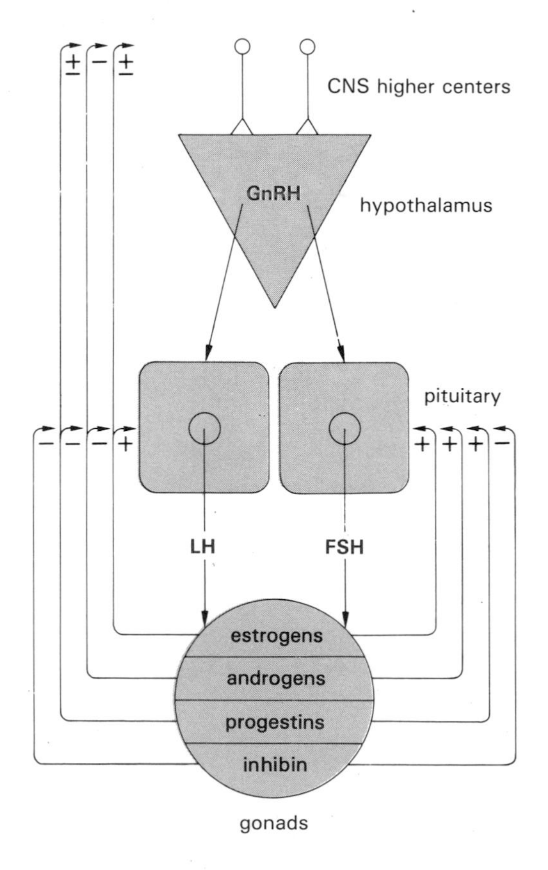

Figure V.15 The effect of peripheral hormones on GnRH and LH and FSH secretion.

nephrine. An *inhibitory* effect of *serotonin* on the secretion of GnRH has also been observed.

GnRH secretion is also under the control of hormones in the peripheral circulation. A stimulatory effect has been observed with *weak* doses of *estrogens*, whereas *elevated doses* have an *inhibitory* effect. *Androgens, progestins* and *inhibin* also have *inhibitory* actions on GnRH secretion (Fig. V.15).

Injection of prostaglandin of the E type stimulates the release of GnRH in rats, and entails a secondary increase in serum LH and FSH. This release of GnRH is accompanied by an increase in the concentration of cAMP in the median eminence, the probable site of action of intravenously injected prostaglandin. These results suggest that prostaglandin and cAMP play a role in the control of GnRH secretion.

Recently, it has been shown that *thyroxine*, in addition to its *inhibitory* effect on the secretion of *TSH* at the *pituitary* level, has an inhibitory effect, at the *hypothalamic* level, on the secretion of *TRH* (Fig. V.16).

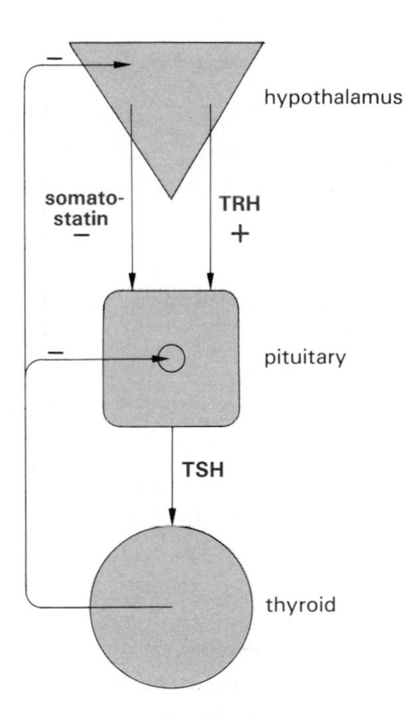

Figure V.16 The effect of thyroid hormones on TRH and TSH secretion.

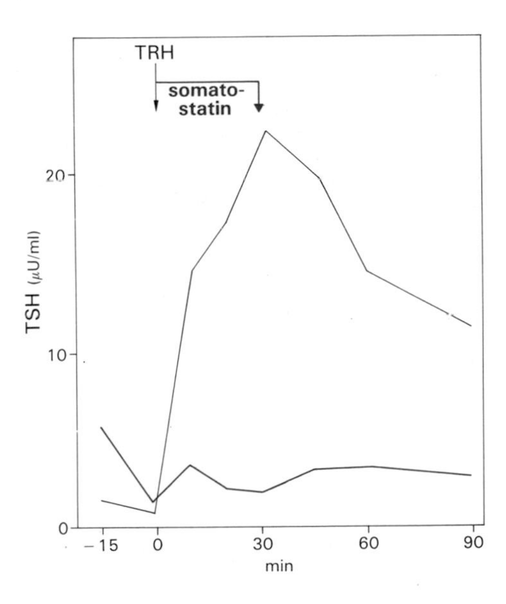

Figure V.17 Inhibitory effect of an infusion of somatostatin on the TSH response to a single injection of TRH in a normal subject.

3.1 Control of TSH

Role of the Hypothalamus

Lesions of the anterior portion of the hypothalamus or median eminence result in a reduced secretion of TSH and a reduction in thyroid gland function, which suggests an essential role of the hypothalamus in the control of TSH secretion. Electrical stimulation of the anterior or preoptic regions of the hypothalamus stimulates TSH secretion. The *overall influence* of the hypothalamus on TSH secretion, therefore, is highly *stimulatory*.

The isolation and elucidation of the structure of TRH, its availability, and the relative ease of synthesis of multiple analogs and derivatives has facilitated numerous studies on the control of TSH secretion. Recent studies have demonstrated that *somatostatin* can *inhibit TRH-induced TSH secretion* (Fig. V.16). The observation of a rapid increase in the baseline level of TSH as well as an increased response of TSH to TRH following the administration of antisomatostatin antibodies clearly indicates a role of somatostatin as a physiological inhibitor of TSH secretion. These observations suggest that the level of *TSH secretion* is the result of a *balance* between the two influences from the hypothalamus: stimulation by TRH, and inhibition by somatostatin (Figs V.16 and V.17).

Role of Thyroid Hormones

Although it has been known for a number of years that thyroid hormones have an inhibitory effect on the secretion of TSH, it has been difficult to demonstrate clearly their site of action. As shown in Figure V.16, thyroid hormones have an inhibitory action on TRH secretion in the hypothalamus as well as on the response of TSH to TRH at the pituitary level. This system of *negative feedback control* allows the maintenance of circulating levels of thyroid hormones within constant and precise limits. Various elements of the hypothalamus and pituitary respond independently to the concentration of plasma thyroid hormones, and adjust the secretion of TRH and TSH accordingly. A reduced concentration of thyroid hormones leads to hypersecretion of TRH and TSH, while inversely, increased thyroid hormone levels have a negative effect on TRH and TSH secretion.

Role of Estrogens

Estrogens have an important effect on TSH secretion at the pituitary level. Administration of these hormones in the rat is capable of completely reversing the inhibitory effect of thyroid hormones on the response of TSH to TRH. This effect of estrogens on the response of TSH is accompanied by a parallel increase in the number of TRH receptors in pituitary cells.

Response to Cold

In experimental animals, exposure to cold results in a rapid *increase* in *TSH* secretion, followed by a secondary increase in thyroid hormone level. Stimulation of TRH secretion by cold appears to be the major factor responsible for the increased secretion of TSH.

3.2 Control of LH and FSH

Sexual Differentiation of the Hypothalamus

While the secretion of *both LH* and *FSH* remains relatively *constant in the male*, the secretion of these hormones *varies cyclically in females*. Thus, in women, a secretory peak of LH and FSH occurs at the mid-point of the menstrual cycle, resulting in ovulation and in the subsequent formation of the corpus luteum.

Several experimental results obtained in rats suggest that the hypothalamus of both sexes has the tendency to evolve toward the female type (cyclic secretion of gonadotropins) unless the endocrine hypothalamus is exposed to male steroid hormones (androgens) at the time of birth[30,31]. This differentiation to female characteristics of gonadotropin secretion can occur in the male rat following castration just after birth or by hypothalamic implantation of an antiandrogen. Treatment with an antiestrogen in early life also causes a loss of cyclic gonadotropin secretion in female animals, a result confirming the necessary role of the aromatization of androgens to estrogens in male hypothalamic differentiation. Endocrine treatment also causes a disturbance of the secretion of gonadotropins in female rats. In the newborn female rat, a single injection of testosterone (or estradiol) leads to the absence of cyclic gonadotropin release and anovulation in adult life[30]. Androgens and estrogens thus play an essential role in the development of the hypothalamic control leading to the tonic (male) or cyclic (female) secretion of GnRH which controls the secretion of LH and FSH. In addition to this early effect definitively *imprinting* the type of gonadotropin secretion, sex steroids *also* exert positive or negative control, at the level of the hypothalamus, on the secretion of GnRH, and at the pituitary level, on the response of LH and FSH to GnRH (p. 446).

In contrast to findings in rodents, *masculinization* of the neurogenic mechanisms controlling cyclic gonadotropin secretion has *not* been *definitely established in humans*. While social influences and sex steroids play a major role in gender identity, there is also evidence for a role of early hormonal influences on sexual behavior[32].

Role of Hypothalamic GnRH for the Two Pituitary Hormones

Our knowledge of the control of ovulation and the secretion of LH and FSH has benefited enormously from the determination of the structure of GnRH. This hypothalamic hormone stimulates the secretion of both LH and FSH, as do the recently described GnRH-associated peptides (p. 264).

At present, the overall regulation of LH and FSH secretion is far from being clearly understood, since there are newly discovered peptides whose physiological significance remains unknown. Moreover, the mechanisms controlling the preferential stimulation of one or another gonadotropin are yet to be elucidated. *No hypothalamic factor with a negative action on gonadotropin* secretion has yet been discovered. Finally, the *actual implication* of inhibin and related peptides is still to be fully appreciated at the level of the hypothalamus and pituitary.

Effects of Estrogens and Androgens on the Selective Secretion of LH and FSH

Since GnRH stimulates the secretion of both LH and FSH, the dissociation of the secretion rate of these two hormones observed during the follicular phase of the menstrual cycle, as well as under different experimental conditions, suggests a *selective control* of the secretion of the two gonadotropins *by sex steroids.*

The use of pituitary cells in primary culture, combined with in vivo studies, has demonstrated specific actions of both estrogens and androgens on the secretion of gonadotropins, thus helping to distinguish between effects related to an alteration of GnRH secretion and those resulting from an alteration of the pituitary response to GnRH.

Figure V.15 illustrates the exclusively *stimulatory action* of *estrogens* on the secretion of LH at the *pituitary level*, while their effect on the secretion of GnRH in the *hypothalamus* is *stimulatory* at *physiological doses* but *inhibitory* at *pharmacological* concentrations. Recent studies have shown that, following castration, the levels of α and β LH mRNAs in the rat pituitary are increased in parallel with plasma LH levels. Conversely, the administration of elevated doses of estrogens, causing a reduction in plasma LH, reduces the concentration of α and β subunit mRNAs.

The action of *androgens* is exclusively *inhibitory* at the *hypothalamic* level, while, at the *pituitary* level, it *inhibits* the secretion of *LH* and is slightly stimulatory of that of FSH. This selective inhibitory action of androgens on the secretion of LH may partially explain the frequently observed dissociation of LH and FSH secretion in humans and other animals. Although the mechanism of action is still not understood, it is probable that a major portion of the activity of progestins commonly used as contraceptives is exerted through their androgenic properties. As illustrated in Figure V.15, *progestins* have an action similar to that of androgens.

The finding that sex steroids do not exert an inhibitory effect on FSH secretion at the pituitary level is compatible with the possibility that *inhibin* exerts a major negative control[33]. Indeed, many clinical observations support a role for a substance of testicular origin in the control of FSH secretion: high levels of circulating FSH are found in men with severe testicular damage of various causes, e.g., irradiation, cytotoxic agents, and cryptorchidism, mump orchitis,

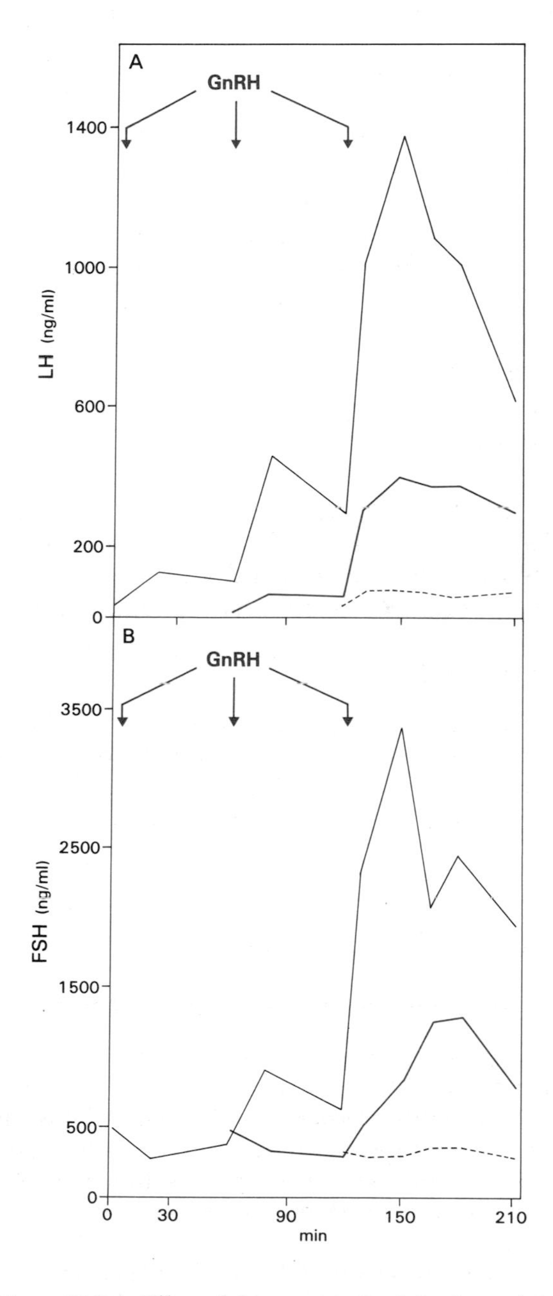

Figure V.18 Effect of three successive injections of GnRH given at 60 min intervals on the afternoon of proestrus in the rat; LH response (**A**), FSH response (**B**).

30. Barraclough, C. A. and Gorski, R. A. (1961) Evidence that the hypothalamus is responsible for androgen-induced sterility in the female rat. *Endocrinology* 68: 68–79.
31. McEwen B. S. (1981) Neural gonadal steroid actions. *Science* 211: 1303–1311.
32. Ehrhardt, A. A. and Meyer-Bahlburg, H. F. L. (1981) Effect of prenatal sex hormones on gender-related behavior. *Science* 211: 1312–1318.
33. McCullagh, D. R. (1932) Dual endocrine activity of the testis. *Science* 76: 19–20.

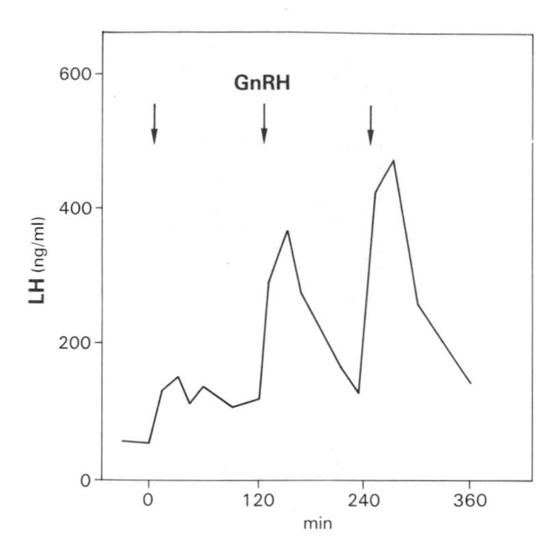

Figure V.19 Effect of GnRH: women with normal menstrual cycles were given 5 μg injections of GnRH, repeated three times, every two hours, 2 to 3 d before the presumed spontaneous LH peak.

and Klinefelter's syndrome. The site of production of inhibin in men is in Sertoli cells, and in women, in granulosa cells (p. 453). Follistatin, a specific inhibitor of FSH release, is also found in follicular fluid; it is structurally unrelated to inhibin.

Autopriming by GnRH

Prior injection of GnRH can increase the response of LH and FSH to subsequent GnRH injections. This priming effect is illustrated in Figure V.18, where it can be seen that the injection of 50 ng of GnRH in rats on the afternoon of proestrus increases the response of LH and FSH to a second injection of GnRH by a factor of four to six. This effect is even greater following the third injection of 50 ng of GnRH, the increase in the LH and FSH response being of the order of 30 to 50. Injection of low doses of GnRH in women during the two or three days preceding the preovulatory peak of LH may also suggest an autopriming effect of GnRH in humans (Fig. V.19).

Control of the Menstrual Cycle (p. 451)

Except during pregnancy and lactation, the secretion of LH and FSH in women goes through a series of complex variations which are repeated throughout the active reproductive period. The average cycle length is 28 d. There are multiple and subtle interactions between sex steroids, GnRH secretion and the secretory response of LH and FSH to GnRH, which allow the periodical preparation of the ovary and uterus for ovulation and pregnancy. The predominant

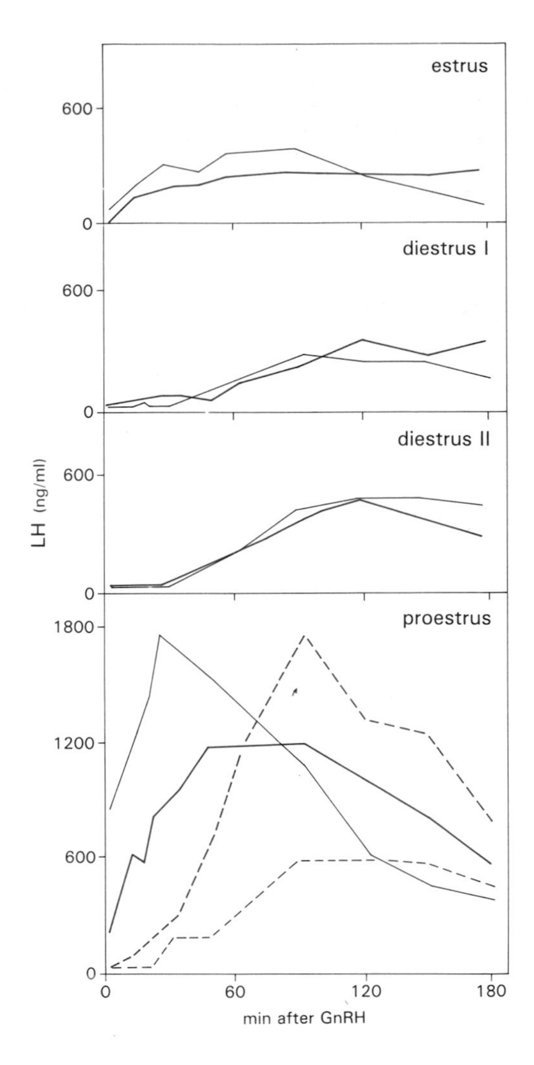

Figure V.20 LH response to a single injection of GnRH in the rat at different phases of the estrous cycle. Animals were injected at 10.00 h, 13.00 h, 15.30 h, and 18.00 h.

factor is, without doubt, the *preovulatory peak of LH and FSH* occurring at the mid-point of the menstrual cycle in women. Numerous data indicate that the LH peak is secondary to an increased secretion of estrogens occurring 24 to 48 h earlier in women. This increased secretion of *estrogens* leads to *increased responsiveness of the pituitary to GnRH*[34]. This phenomenon is illustrated in the rat (Fig. V.20) during the afternoon of proestrus (the period corresponding to the preovulatory period in women), where the response of plasma LH to the injection of a constant dose of GnRH is increased seven- to ten-fold.

The preovulatory peak of LH is probably the

34. Ferland, L., Borgeat, P., Labrie, F., Bernard, J., DeLéan, A. and Raynaud, J. P. (1975) Changes of pituitary sensitivity to LHRH during the rat estrous cycle. *Mol. Cell. Endocrinol.* 2: 107–115.

result of a direct stimulatory effect of estrogens on the sensitivity of the pituitary to GnRH, as well as the increased secretion of GnRH by the hypothalamus, which is probably also under the influence of sex steroids.

Control of Testicular Function

Androgens for LH and inhibin for FSH?

In men, the level of LH secretion appears to be controlled mainly by the inhibitory effect of circulating androgens of testicular origin. *Androgens*, as mentioned previously, have an *inhibitory* effect on the secretion of GnRH by the *hypothalamus* and on the response of LH to GnRH in the pituitary. As they do not appear to have a direct inhibitory effect on the secretion of FSH at the pituitary level, it makes sense that a part of the control of FSH secretion is exerted by inhibin (Fig. V.13).

Multiple Control Systems

LH production by pituitary cells and androgen/progestin formation in gonadal interstitial cells form a closed-loop feedback system. Similarly, FSH synthesis in gonadotropic cells and inhibin synthesized in granulosa and Sertoli cells also constitute a feedback system. In addition, although inhibin does not affect gonadal steroidogenesis, it enhances LH-stimulated androgen synthesis, and activin (p. 453) decreases it, thus providing a paracrine cross-connection between the two feedback loops.

4. Clinical Syndromes of Hypothalamic Origin

Many neuroendocrine diseases in the human are secondary to functional or organic lesions of the hypothalamus. It is quite rare that lesions of other regions of the brain result in neuroendocrine complications.

4.1 Dynamic Tests of LH and FSH

The administration of *GnRH* is used to evaluate the *capacity of the pituitary* to secrete LH and FSH. This test allows the identification of cases of hypogonadism of pituitary (deficiency of LH and FSH secretion) or hypothalamic (deficiency of GnRH secretion) origin.

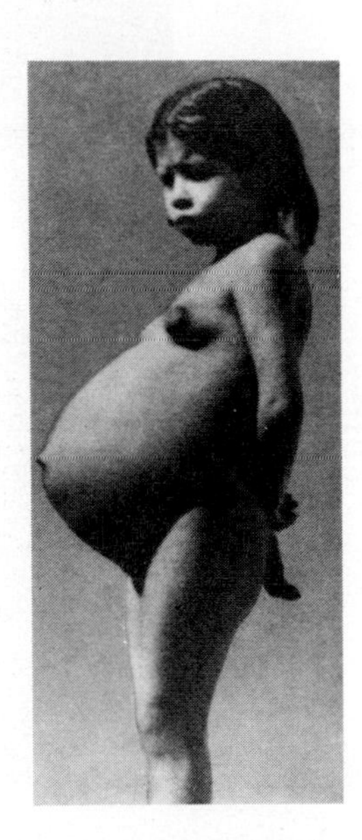

Figure V.21 Precocious puberty. The youngest known mother in the world. This Peruvian child started menstruating at three years of age and was pregnant at four years, ten months. A month after the photograph was taken, the child, who was 1.15 m tall at the time, underwent a cesarean section. The infant weighed 3.2 kg. (Escomel, E. (1939) *La Presse Médicale* 47: 744 and 875).

A normal response to GnRH is observed in hypothalamic deficiency, whereas a pituitary lesion is accompanied by a deficient response to GnRH.

4.2 Precocious Puberty

True precocious puberty is characterized by sexual maturation which provokes the premature activation of gonadotropin secretion. This can be *distinguished from pseudoprecocious* puberty, which refers to the premature development of secondary sexual characteristics following hypersecretion of estrogens or androgens of peripheral origin. Puberty is considered precocious when it occurs before the age of eight years in girls (Fig. V.21) and before ten years in boys (p. 457). True precocious puberty of central origin is most frequently idiopathic, or without a discernible organic cause. It is frequently accompanied by electroencephalographic changes, however. Precocious puberty can also be secondary to hypothalamic tumors or to compressions of this area by nearby tumors (craniopharyngioma, glioma, astrocytoma, or pinealoma). Although rare, hypothyroidism may also be associated with true precocious puberty. The most evident manifestations of precocious puberty are due to the action of sex steroids on secondary sexual

A 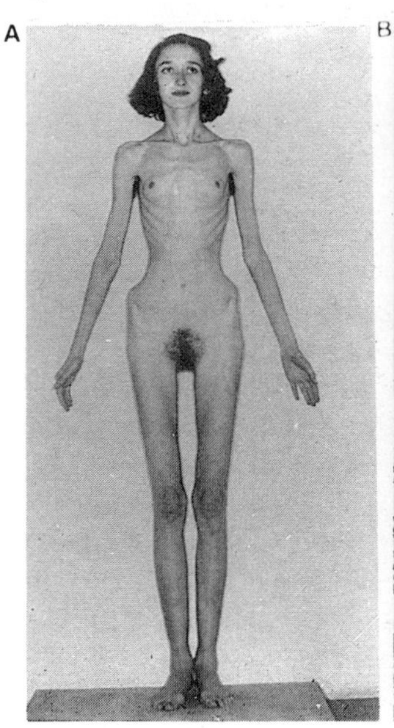 B 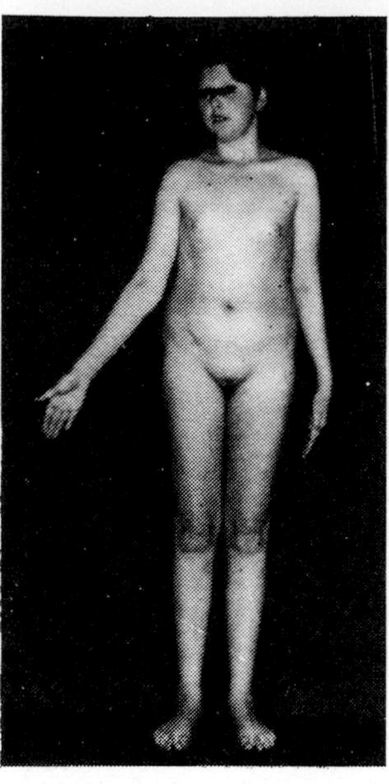

Figure V.22 **A.** Anorexia nervosa. (Courtesy of Dr. P. Bouchard, Hôpital de Bicêtre). **B.** Pituitary insufficiency. (Courtesy of Dr. F. Peillon, Hôpital Pitié-Salpêtrière, Paris).

organs (Fig. V.21). Because of the premature closing of the epiphyses, subjects who have undergone precocious puberty are generally shorter than normal.

4.3 Delayed Puberty

The main diagnostic problem in the male is distinguishing between subjects with constitutional delay of puberty and those with gonadotropin deficiency. In cases of constitutional delayed puberty, there is an important psychological component. Measurements of 24 h patterns of serum LH and testosterone, as well as a GnRH test, can assist the diagnosis. Although observation is the main recommendation, some situations require the short-term administration of androgens. *Pulsatile GnRH* administration is the *treatment* of choice *for true hypogonadotropic hypogonadism*.

Delayed puberty in girls can be treated with replacement estrogen therapy, for maturation of the secondary sexual characteristics. If pregnancy is desired, pulsatile administration of GnRH should be attempted[35] (p. 457).

35. Crowley, W. F. and McArthur, J. W. (1980) Stimulation of the normal menstrual cycle in Kallman's syndrome by pulsatile administration of LHRH. *J. Clin. Endocrinol. Metab.* 51: 173–182.

4.4 Psychogenic (Hypothalamic) Amenorrhea

Cessation of menstruation in the young, non-pregnant woman is frequently observed in patients with *pseudopregnancy*, anorexia nervosa (Fig. V.22A), and psychogenic amenorrhea (p. 459). The absence of ovulation may be all that is observed, or symptoms of marked estrogen deficiency may be present. Amenorrhea is always associated with psychological problems, which are normally temporary. Pseudopregnancy is associated with such signs as abdominal distension, changes in the breasts, and gastrointestinal symptoms. An elevated and continuous secretion of LH and an increase in prolactin secretion may explain the endocrine changes which accompany this syndrome. In cases of anorexia nervosa, amenorrhea frequently appears before anorexia and a reduction in body weight.

The diagnosis of this disease should first *exclude hypopituitarism*, which is generally accompanied by a more generalized hormonal deficiency (Fig. V.22B). Simple psychogenic amenorrhea or the simple reduction in menstrual frequency (oligomenorrhea) may be due to psychological stress of varied importance, e.g., entrance to university, travel, etc. This syndrome is probably due to abnormal functioning of the control system of GnRH secretion, which in turn causes suppression of the ovulatory peak of gonadotropins. Different degrees of estrogen deficiency may accompany psychogenic amenorrhea. In addition to psychological counselling, a normal menstrual cycle can be reestablished by administering progesterone alone or in association with estrogen, or by clomiphene (an antiestrogen with weak estrogenic activity, capable of inducing ovulation).

4.5 Decreased Secretion of Hypophysiotropic Hormones

Some patients with growth hormone deficiency also show signs of hypothyroidism and hypogonadism, while the pituitary response to the administration of TRH and GnRH is normal for TSH, LH, and FSH. These syndromes are probably secondary to a deficiency of TRH and GnRH secretion by hypothalamic neurons.

5. Summary

Three glycoprotein hormones, TSH, LH and FSH, are synthesized in the *anterior pituitary*, while the fourth, *chorionic gonadotropin* (in humans, *hCG*), is produced by the placenta. TSH is produced by *thyrotropic cells*, and *LH and FSH by gonadotropic* cells. Secretory granules of the pituitary secretory cells contain a significant stored pool of hormones. All glycoprotein hormones are formed by two subunits, α and β. There is *one common α subunit*, composed of 92 aa, for all four hormones. The *β subunits vary* in length, from 112 to 145 aa. The genes for the α and β subunits are located on different chromosomes. There is a high degree of homology between LH and CG, correlating these quasi-identical biological activities via a common receptor. There are also *significant homologies* between *all β chains*, and, even, although to a lesser degree, between them and the α subunit (see evolutionary aspects, p. 154). *Sugar* constituents of the four hormones constitute *15 to 30%* of their molecular weight.

Regulation of the synthesis and secretion of the pituitary hormones is primarily under the control of *hypothalamic releasing factors. TRH*, increasing TSH secretion, is a tripeptide synthesized in the form of a prohormone containing five times its sequence. *GnRH*, a decapeptide, simulates the secretion of both FSH and LH. The proprotein containing GnRH also includes a C-terminal extension (*GAP*), exhibiting, despite a lack of homology with GnRH itself, some stimulatory effects on gonadotropin secretion; however, its physiological significance is unknown. TRH and GnRH, produced in the hypothalamus, are found in nerve endings communicating with the portal system, through which they reach anterior pituitary cells. Both hypophysiotropic factors in thyrotropic and gonadotropic cells, respectively, act through a specific *membrane receptor* that activates, in turn, the membrane *phosphoinositide system* and *adenylate cyclase. GnRH* is secreted in a characteristically *pulsatile* fashion, which is a determining factor in its activity. Pulsatory secretion results in *stimulation* of LH and FSH, while *continuous* exposure of pituitary cells to GnRH leads to *desensitization*. Pulsatile administration of GnRH is used to stimulate gonadal activities in humans, while continuous administration results in "medical castration". The secretion of TSH, FSH and LH is *also controlled by peripheral hormones. Thyroid* hormones *regulate TSH negatively*, at both the hypothalamic and pituitary levels. The control of *gonadotropin* secretion is very *complex*. Physiologically, estrogens stimulate gonadotropin secretion, at the pituitary level, during the menstrual cycle. Larger amounts of estrogens are inhibitory on LH secretion, as are elevated concentrations of both progestins and androgens. Inhibin, a peptide of ovarian and testicular origin, inhibits FSH secretion. The overall control of both endocrine function and germ cell production in the gonads probably also involves as yet poorly defined paracrine steroid and peptide regulators. The effects of TSH, FSH, and LH are mediated by a specific membrane receptor for each hormone, and the only second messenger demonstrated so far is cAMP.

General References

Bhatnagar, A. S. (1983) *The Anterior Pituitary Gland*. Raven Press, New York.

Crowley, W. F. Jr., Filicori, M., Spratt, D. I. and Santoro, N. F. (1985) The physiology of Gonadotrophin-Releasing Hormone (GuRH) secretion in men and women. *Rec. Prog. Horm. Res.* 41: 473–532.

Fiddes, J. C. and Talmadge, K. (1984) Structure, expression, and evolution of the genes for the human glycoprotein hormones. *Rec. Prog. Horm. Res.* 40: 43–78.

Imura, H. (1985) *The Pituitary Gland*. Raven Press, New York.

Kourides, I. A., Gurr, J. A. and Wolf, O. (1984) The regulation and organization of Thyroid Stimulating Hormone genes. *Rec. Prog. Horm. Res.* 40: 79–120.

Mauvais-Jarvis, P., Sitruk-Ware, R. and Labrie, F., eds (1986) Médecine de la Reproduction. *Gynécologie endocrinienne*. Flammarion Médecine/Sciences, Paris.

Sheldon J. Segal

Contraceptive Choices

At the time of the publication of this book, the world population had grown to exceed *five billion people.* The demographic evolution is dramatic (Fig. 1). Given better hygiene, more vaccinations against infectious, parasitic and viral, diseases, and with improved lifestyle in developing countries, life expectancy will contiue to increase, and the number of inhabitants on earth will continue to grow at a rapid rate (Fig. 2). Today, 40% of the 800 million couples of reproductive age who use a modern form of contraception employ only a small number of methods. There are approximately 140 million men and women who are voluntarily sterilized, 70 million (of whom three-quarters are in China) use an IUD and about 40 million use condoms, while 55 million take oral contraceptives. Each year, 40 to 45 million women undergo a legal abortion, and about the same number in undeclared conditions.

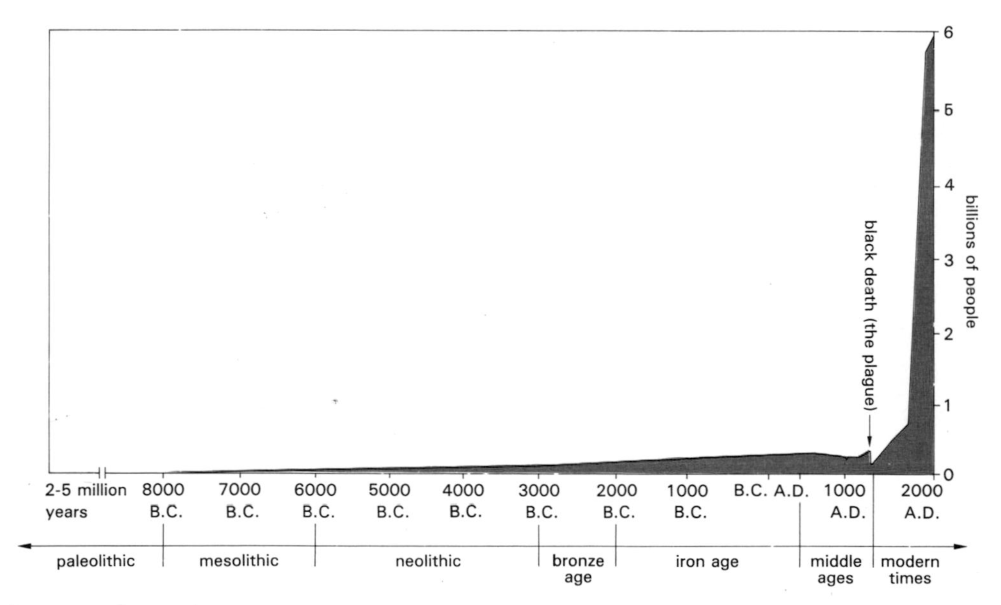

Figure 1 *Demographic evolution.*

It is *difficult* to imagine any one method of fertility regulation which would be *perfect and convenient for everyone.* It is known, however, that the more methods available, the greater the acceptance. This justifies the development of new techniques. However, *contraceptive research* is in a *critical state*; in addition to normal scientific problems, it is subject to variable ideological and political considerations, as well as to problems associated with the prohibitive cost of liability insurance.

Steroid Contraceptives

All currently available daily birth control pills and long-acting contraceptives are *based on steroids. Combined oral* contraceptives, known simply as *"the pill"*, contain synthetic forms of the naturally occuring female steroid hormones, estrogen and progesterone.

Their popularity is related to their high level of *effectiveness*, overall *safety*, and *accessibility*. The newer steroid contraceptives, of biphasic and triphasic formulations, vary the hormonal dose over the course of the cycle, reducing the total amount of ingested hormones. Despite continuing improvement, oral contraceptives are still associated with a number of minor side effects (nausea, weight gain, menstrual spotting). There are also some significant health benefits in addition to their effectiveness in preventing unwanted pregnancy, such as virtual elimination of ectopic pregnancy, lower rate of pelvic infections, reduced risk of ovarian and uterine cancers. However, oral contraceptives increase the risk of benign liver tumors and serious cardiovascular complications, including blood clots, heart attacks, and strokes. These problems are almost entirely confined, however, to women over the age of 35 with a predisposition to these diseases, because they are smokers, or because they have a high level of blood cholesterol. *Progestins* (progesterone-like drugs) have been proposed as oral contraceptives since they may have lower risk of cardiovascular side effects, but unfortunately their contraceptive failure rate is higher.

Injectable contraceptives, essentially progestins, are active for two to three months. They are more effective than birth control pills which must be taken every day. Suppressing menstruation, but frequently causing irregular bleeding, these injectable contraceptives are not well accepted in certain cultural contexts, but they do not carry the health risks that have been claimed by some of their critics.

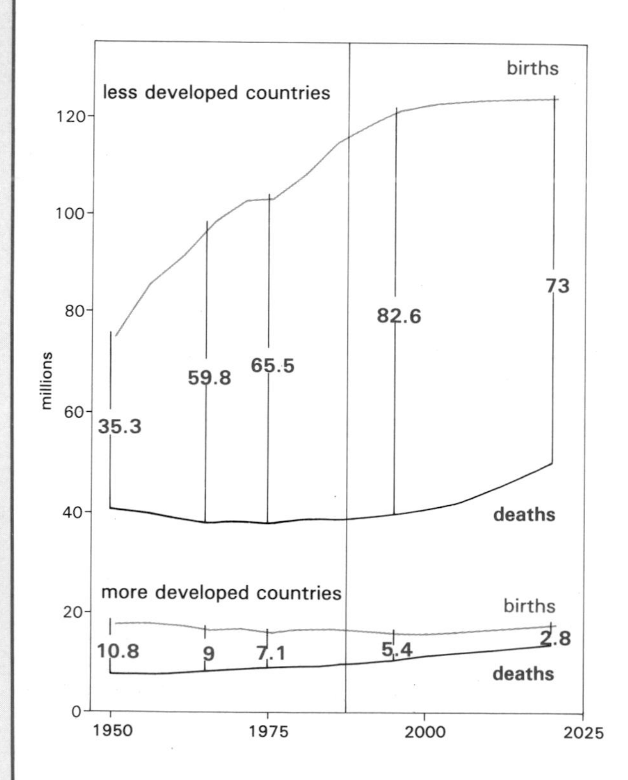

Figure 2 *The differential birth and death rates in more and less developed countries. The curves have been extrapolated for the next 40 years.*

Contraceptive subdermal *implants* have been developed, taking advantage of the tissue compatibility of Silastic®, a synthetic, rubber-like plastic. A medical-grade preparation of this polymer is used, and lipophilic steroids are soluble in it, and pass through its solid matrix. The most advanced version is called Norplant®. Six capsules placed under the skin of the upper arm deliver approximately 30–60 μg of levonorgestrel each day. A reversible method, implants exeed the effectiveness of the pill or IUD. The main undesirable side effect is irregular bleeding.

Oral contraceptive steroids pass first through the liver in high concentration before being distributed through the systemic circulation. Absorption of contraceptive steroids through the vaginal mucosa is very efficient, and the steroids are absorbed directly into the systemic blood, reaching the reproductive organs directly, without first passing through the liver. Contraceptive *vaginal rings* made of Silastic have been used, the most widely studied being based on release of levonorgestrel and 17β-estradiol, with a schedule of three weeks and removal for one week. Such a ring may be reused over several months. The acceptability of vaginal rings or vaginal pills by women is not yet known. As compared to vaginal suppositories or the diaphragm, such a method dissociates the use of the contraceptive technique from coitus.

Other Methods

The daily or protracted use of *GnRH* (p. 448) (or synthetic analogs) suppresses ovarian functions, including ovulation, resulting in a "medical castration". It may be difficult to use this peptide effectively, due to the greatly reduced estrogen and progesterone levels that occur.

Intrauterine devices (IUD) are, for many women, a very effective, long-acting, one-step method of contraception. Effectiveness is as great as that of the combined oral contraceptives. One to three pregnancies per 100 users occur during the first year, which is similar to the success rate of most combined oral contraceptives. However, effectiveness may be superior to oral contraceptives in countries where education is low. The IUD may not be a good method for women exposed to sexually transmitted diseases since it is associated with some risk of pelvic infection. However, over the years, *smaller* and more flexible IUDs and, importantly, *copper-releasing* and *progestin* (levonorgestrel)-*releasing* IUDs have been introduced. The latter is very efficient, and, by thickening the cervical mucus, it acts as a barrier to spermatozoa and bacteria.

A major *gap* in current birth control technology is the lack of a *postcoital method*, a *menstrual inducer*, which could be used a few days or a week after unprotected intercourse, to bring on a late menstrual period. The laws of a number of countries permit "menstrual therapy" even where they do not allow surgically induced abortion following the confirmation of pregnancy. Although any new method of postcoital contraception will certainly remain questionable to some segments of society, in nearly all parts of the world, an inexpensive, self-administered menstrual inducer could be expected to gain wide acceptance. *Already*, some methods of contraception can be used postcoitally. For example, a double dose of certain full strength combined oral contraceptives or estrogen preparations, taken within 72 h of unprotected intercourse and repeated in 12 h, can prevent pregnancy by interfering with fertilization or with the implantation of a fertilized egg. The insertion of an IUD following unprotected intercourse would have a similar result. However, neither technique is widely practiced.

A true menstrual inducer may become available within the next few years, and may be a future breakthrough in fertility control before the year 2000. Several *prostaglandins*, which cause contraction of the uterus, are presently being tested. However, to date, they have been associated with bothersome side effects such as nausea, diarrhea and occasionally incomplete evacuation of the uterus. The *antiprogesterone RU 486* is active on its own to interrupt the luteal phase of the menstrual cycle or early pregnancy. Alone or combined with a low dose of prostaglandin (thereby avoiding the previously mentioned side effects), this compound may become an effective method for bringing on a missed menstrual period.

Current research in immunology could lead to the development of a *reversible, long-term contraceptive vaccine*, which may have a dramatic impact on family planning. Anti-hCG vaccination, suppressing the role of the gonadotropin in maintaining implantation, can mimic the physiological failure of the development of pregnancy and result in menstruation. Thus, pregnancy is prevented in what would otherwise appear to be a normal menstrual cycle. A vaccine is now undergoing clinical testing in India. Another prototype vaccine consists of an oligopeptide corresponding to the amino acids 109–145 of the C-terminal extremity of the β subunit of hCG and a protein carrier (diphtheria toxoid) to which the polypeptide is conjugated. The World Health Organization (WHO) is conducting trials to assess tolerance, safety, and efficacy. Other vaccines aimed at immunization against constituents of sperm or eggs are presently at early stages of development.

Periodic abstinence, an efficient technique provided there is high level of motivation, is based on fundamental knowledge of the menstrual cycle and the period of ovulation. Two to four percent of married couples worldwide use some form of periodic abstinence, the method used most predominantly in Catholic countries. However, whatever variant of the method used, an irregular menstrual cycle or a lack of a clearly detectable pattern of change in basal body temperature, cervical mucus or other indicators make the failure rate as high as 10 to 30% per year.

Voluntary sterilization and *barrier methods* are used by a large number of couples, but a discussion of them is outside the scope of a book on hormones. Sterilization, an irreversible method (in both women and men), contrasts sharply with hormonal methods. The barrier methods often enhance protection against sexually transmitted diseases, and will continue to be used by a sizeable portion of the population that does not object to using a technique requiring a conscious action before each sexual act. Modern vaginal *sponges* may be used for up to 24 h, through more than one sexual act, but this method has an unacceptably high failure rate.

Contraceptives for men are not of hormonal nature. Besides condoms, withdrawal and vasectomy, male methods are non-existent. A major stumbling block has been the nature of the male reproductive system: sperm production is continuous, and many potential male contraceptives reduce testosterone, thereby resulting in a loss of libido and male secondary sexual characteristics. The reproductive capacity of men is therefore not as easy to interrupt, without unacceptable side effects, as that of women.

It has proved difficult to administer the proper regimen of *steroids to block spermatogenesis* without side effects. *GnRH* desensitization can be used, but necessitates some compensation by androgen. *Inhibin* may one day be the basis for a new male contraceptive, but there is growing doubt that the expected selective decrease of FSH will be achieved. The cotton-seed oil derivative, *Gossypol*, discovered in China, may not be relevant in a discussion about hormones. Because of imputed toxicity, it is presently at an early stage of development, awaiting clarification of the safety issue. Meanwhile, it has to be considered as promising because of its ability to suppress sperm production and/or motility without suppressing testicular hormone production.

Potential Impact of New Technology

Modern contraception was revolutionized by the work of Gregory Pincus and his colleagues, leading to the development of "the pill" by 1960. The last two decades of worldwide efforts to expand family planning services indicate that presently available contraceptive methods are acceptable to a significant number of people, in most countries, despite striking differences in cultural backgrounds and socioeconomic conditions. An array of new or improved methods, however, could make the job of reaching millions of non-users significantly easier. Certain types of new contraceptives would, for example, reduce the logistical and cost factors which limit contraceptive availability for very poor or very remote rural populations. Increased safety, reduced side effects and greater convenience would increase the attractiveness of contraception to potential users (side effects, although generally not serious for most women, are the principal cause of discontinuation and method-switching among users). The existence of a wider choice of methods, even if those methods were not significantly more attractive than existing ones, would broaden the consumer base, since individual needs and tastes vary greatly. The record of organized family planning efforts shows that sustained, high levels of contraceptive prevalence are more easily achieved if consumers are offered a wide range of birth control choices. Obviously, however, program costs and logistical problems increase with the number of choices offered, especially if the methods are inherently difficult to provide or to use. Although *no single method can be expected to cover the entire spectrum of desirable characteristics*, contraceptives must now go further towards meeting consumer demands for higher levels of reliability, safety and reversibility, and they should be made affordable and easily accessible.

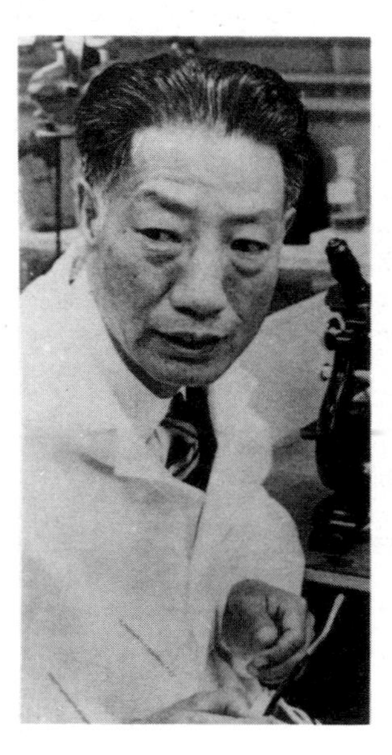

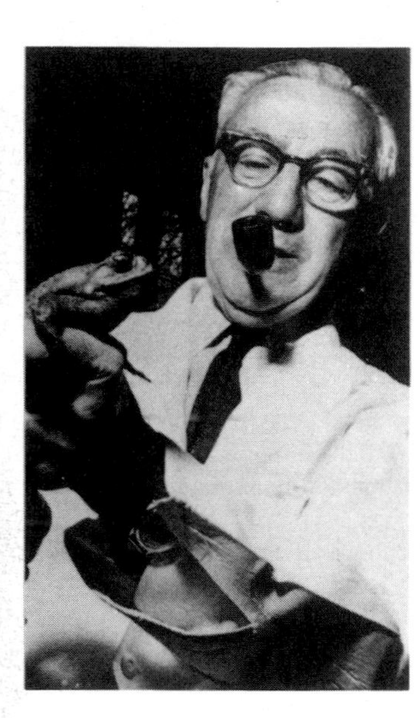

Figure 3. *Pioneers of oral contraception. Gregory Pincus, the pilot, between Min Chueh Chang, the biologist (left), and John Rock, the gynecologist (right).*

References

Baulieu, E. E. and Segal, S. J., eds (1985) *The Antiprogestin Steroid RU 486 and Human Fertility Control*. Plenum Press, New York.

Berquist, C., Nilliius, S. J. and Wide, L. (1969) Intranasal LHRH as a contraceptive agent. *Lancet* 2: 215–217.

Luukkainen, T., Lahteenmaki, R. and Toivonen, J. (1984) Long-term clinical trials with copper or progestin-releasing IUDs. In: *Fertility Control* (Sciarra, J. J., Pescetto, G., Marteni, L. and DeCecco, L., eds), Monduzzi editore, Bologna, pp. 151–162.

Pincus, G., Rock, J., Garcia, C. R., Rice-Wray, E., Paniagua, M. and Rodriguez, I. (1958) Fertility control with oral medication. *Amer. J. Obstet. Gynecol.* 75: 1333–1346.

Pincus, G., ed (1965) *The Control of Fertility*. Academic Press, New York.

Segal, S. J. (1984) Norplant subdermal contraceptive implants. In: *Fertility Control* (Sciarra, J. J., Pescettop, G., Martini, L., DeCecco, L., eds) Monduzzi editore, Bologna, pp. 5–12.

Segal, S. J., ed (1985) *Gossypol: a Potential Contraceptive for Men*. Plenum Press, New York.

CHAPTER VI

Vasopressin and Oxytocin

Serge Jard

Contents, Chapter VI

Vasopressin and Oxytocin

Introduction

Vasopressin and oxytocin were the *first* two *peptide hormones* for which the *chemical structure* was *defined* and for which the complete *synthesis* of active forms was carried out[1]. These hormones are *produced by nerve cells* whose cell bodies lie in the hypothalamus, their axons terminating in the neurohypophysis. They are released (by exocytosis) into the general circulation and transported to their peripheral target organs. Vasopressin and oxytocin are also produced in several other locations in the central nervous system, as well as in the adrenal medulla and gonads.

The major physiological action of *vasopressin* is the *control* of the *excretion of water* by the kidney. It therefore plays an important role in the regulation of the osmotic pressure of extracellular fluids, acting indirectly on their volume. A lack of vasopressin secretion is accompanied by the appearance of profuse diuresis and a reduction of the osmolarity of urine. Vasopressin is also one of the elements involved in the multifactorial regulation of a large number of other physiological functions. It participates, for example, in the regulation of blood pressure. It is from this vasopressor effect that vasopressin was named, but it is also frequently termed antidiuretic hormone (ADH).

The best known physiological actions of *oxytocin* are on *milk ejection* (contraction of the myoepithelial cells of the mammary gland) and on *uterotonus* (contraction of smooth muscle cells of the uterine myometrium).

1. Vigneaud, V. (1956) Hormones of the posterior pituitary gland: oxytocin and vasopressin. *Harvey Lecture*, series 50: 1–26.

1. Structure, Biosynthesis and Secretion of Vasopressin and Oxytocin

1.1 Structure

Vasopressin and oxytocin are *nonapeptides* with an intramolecular *disulfide bond* between cysteine residues in positions 1 and 6 (Fig. VI.1). The glycyl residue in the C-terminal position is amidated. These two peptides differ by only the amino acids present at positions 3 and 8. This structural homology may account for the fact that, at high concentrations, oxytocin can produce all the known effects of vasopressin, and vice-versa. Two types of vasopressin have been identified in mammals. In the majority of species studied, including *humans*, the natural vasopressin is

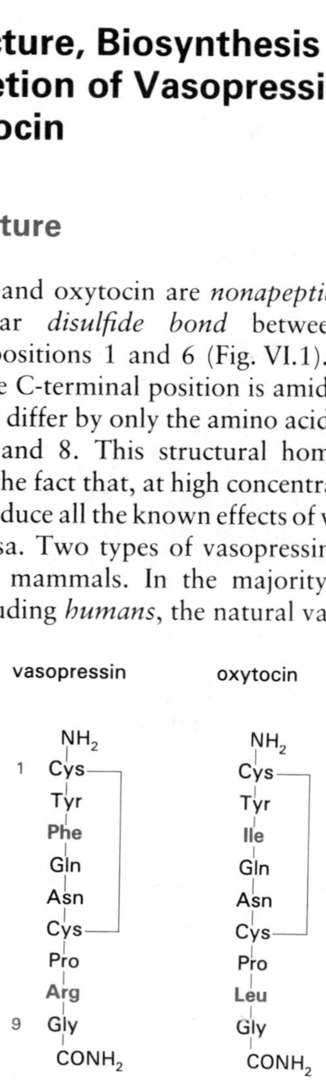

	vasopressin	oxytocin
	NH_2	NH_2
1	Cys	Cys
	Tyr	Tyr
	Phe	Ile
	Gln	Gln
	Asn	Asn
	Cys	Cys
	Pro	Pro
	Arg	Leu
9	Gly	Gly
	$CONH_2$	$CONH_2$

Figure VI.1 Structure of vasopressin and oxytocin.

arginine vasopressin (with an arginine residue at position 8). In the pig and other species of the swine family, the natural hormone is lysine vasopressin (with a lysine residue at position 8). The structure of neurohypophyseal peptides has undergone considerable evolution during phylogenesis; at least 10 peptides with different structures have been identified in the different classes of vertebrates. However, vasopressin has been identified only in mammals.

1.2 Biosynthesis

Sites of Production and Secretion

Vasopressin and oxytocin are synthesized by neurons whose cell bodies are located in several anatomically well defined nuclei in the central nervous system. Vasopressin and oxytocin are produced by different neurons. Both vasopressin and oxytocin neurons are present in two main sites of production, the *supraoptic* and *paraventricular nuclei*, also known as the *magnocellular nuclei*. The neurons which produce vasopressin and oxytocin are characterized by their large size. The cell bodies of these neurons possess all of the elements typical of *neurosecretory cells*: an abundant endoplasmic reticulum, a well developed Golgi apparatus, and numerous secretory granules with a diameter of approximately 150 to 200 nm. Magnocellular neurons are also present in several other hypothalamic sites, which differ from one species of mammal to another. Vasopressin neurons of a smaller size have been identified in the suprachiasmatic nucleus of the hypothalamus and in several other extrahypothalamic sites, such as the amygdala, the ventrolateral septum, and the bed of the stria terminalis.

The axons of the neurons of the supraoptic, paraventricular and annexed magnocellular nuclei pass through the internal portion of the median eminence, terminating in the neurohypophysis, in contact with *fenestrated capillaries* (Fig. VI.2). The hormones released in this region have direct access to the general circulation. The neurohypophysis also contains star-shaped glial cells, termed *pituicytes*, which are thought to be *supporting* cells (Fig. VI.3). Cholinergic and dopaminergic nerve fibers innervate the neurohypophysis.

In addition, *some* vasopressin *fibers* from the paraventricular nucleus terminate in the median eminence, in *contact* with capillaries of the hypophyseal *portal system*. The *hormone* released here is

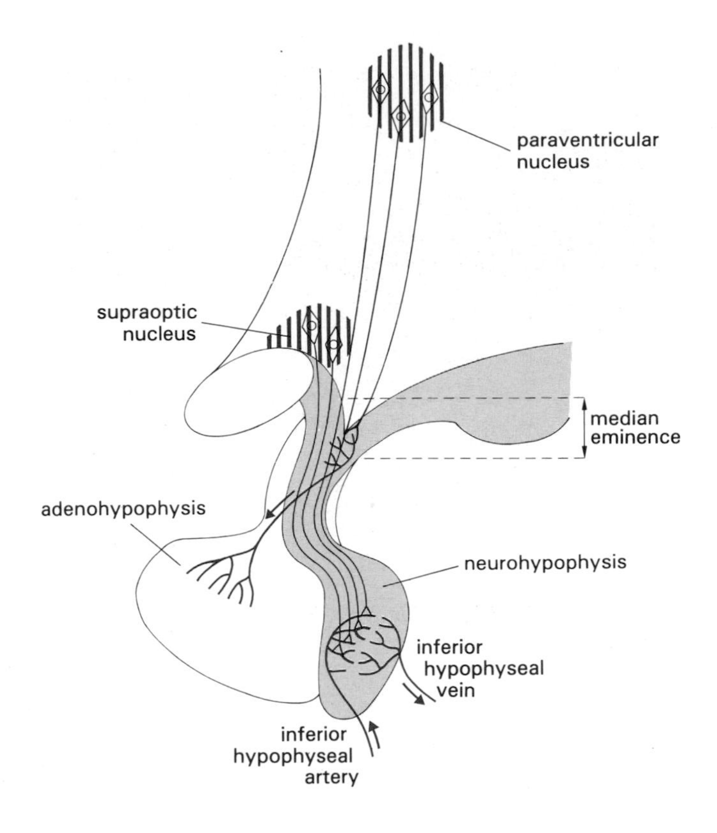

Figure VI.2 Anatomy and vascularization of the neurohypophysis. The diagram represents a sagittal section through the pituitary. Magnocellular oxytocin and vasopressin neurons originate in hypothalamic paraventricular and supraoptic nuclei, and send axonal projections to the posterior portion of the pituitary, the neurohypophysis. Their products are released directly into the general circulation. Axonal projections coming from cell bodies located in the paraventricular nucleus also terminate in the median eminence. Two capillary systems in series, one within the median eminence and the other in the anterior pituitary, allow the secretory products which are released by axon terminals in the median eminence to reach the anterior pituitary (p. 179).

first carried *to* the *anterior pituitary* before reaching the general circulation. Vasopressin and oxytocin fibers project into numerous zones of the brain and spinal cord. Depolarization induces the release of vasopressin and oxytocin from these terminals[2].

Vasopressin and oxytocin can therefore be secreted: 1. directly into the general circulation, reaching the peripheral target organs by transport in blood, thus acting as *hormones* in the classical sense; 2. into the hypophyseal portal system, like the majority of *hypophysiotrophic factors* of hypothalamic origin; 3.

2. Buijs, R. M. and Van Heerikhnige, J. J. (1982) Vasopressin and oxytocin release in the brain – a synaptic event. *Brain Res.* 252: 71–76.

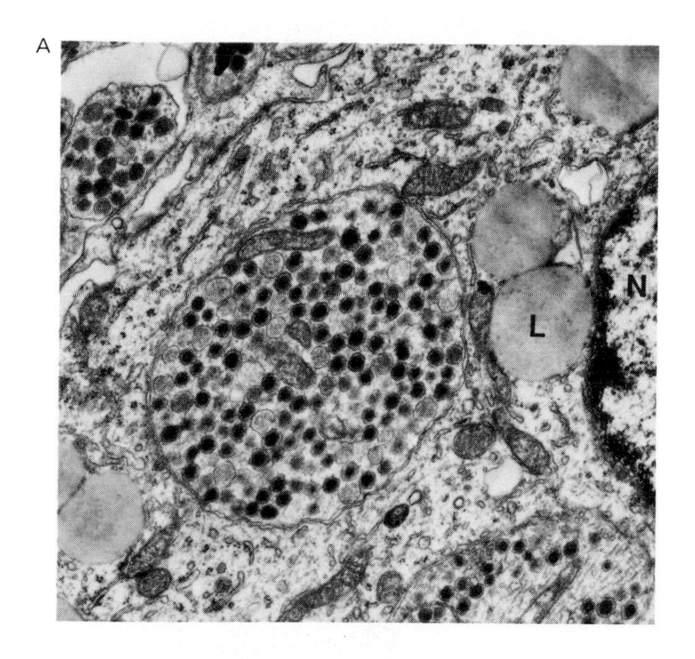

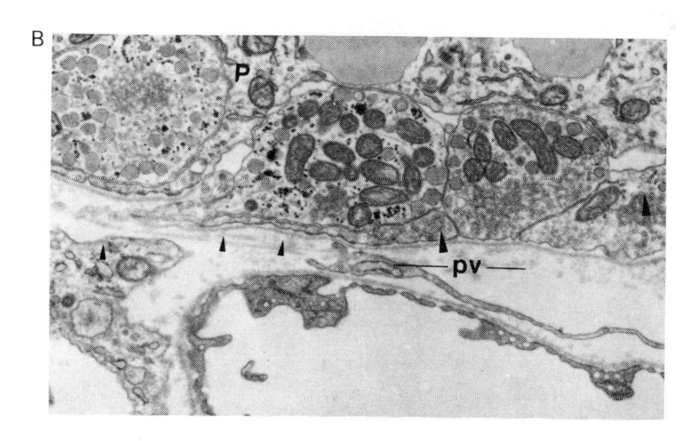

Figure VI.3 Ultrastructure of the neurohypophysis. **A** shows a transverse section of an axon of a magnocellular neuron, in which there is an abundance of neurosecretory granules. The fiber is surrounded by an adjoining cell, called a pituicyte, of which the nucleus (N) and a lipid inclusion (L) are apparent. The fiber and the pituicyte are separated by an extracellular space. Bar=1 μm. **B** shows zones of contact (small arrowheads) between neurosecretory terminals and the basal membrane of fenestrated capillaries (lower portion of the image). Pituicytes, or their prolongations, are also in contact with capillaries (large arrowheads). Bar=1 μm.

in the proximity of other neurons, thus acting as *neurotransmitters or neuromodulators.*

There is experimental evidence that vasopressin and oxytocin are *also produced peripherally.* The adrenal medulla, ovary and testis contain substances with immunological, physical and biological properties very similar to those of vasopressin and oxytocin[3]. Oxytocin secretion has been demonstrated in the ovary of several mammalian species. The physiological role of vasopressin and oxytocin produced peripherally has not been clearly established. Vasopressin-sensitive cells have been identified in the vicinity of putative peripheral sites of production. Thus, vasopressin receptors have been characterized in rat adrenal cortex and Leydig cells of the rat testis. Vasopressin was shown to have a mitogenic effect on cells from the zona glomerulosa of the adrenal cortex.

Cellular Aspects of Vasopressin and Oxytocin Biosynthesis

Vasopressin and oxytocin are synthesized in the rough endoplasmic reticulum of cell bodies, along with associated proteins known as *neurophysins.* The major role of neurophysins is to specifically bind oxytocin and vasopressin. The neurophysin associated with vasopressin (neurophysin II) is different from the neurophysin associated with oxytocin (neurophysin I).

Histologically, neurophysins can easily be detected by the Gomori reaction. This technique has been important in establishing the concept of neurosecretion. In the Golgi apparatus, vasopressin, oxytocin and their associated neurophysins are included in vesicles with a limiting membrane. These neurosecretory vesicles, or granules, contain approximately 10^4 molecules of vasopressin or oxytocin and/or the corresponding number of neurophysin molecules. They migrate along axons at a rate of 2 to 3 mm/h. The period of time separating the synthesis of vasopressin or oxytocin in hypothalamic nuclei and the arrival of these hormones in the posterior lobe of the pituitary is approximately 12 h in the human.

3. Adashi, E. Y. and Hsueh, A. J. W. (1981) Direct inhibition of testicular androgen biosynthesis revealing antigonadal activity of neurohypophyseal hormones. *Nature* 293: 650–652.

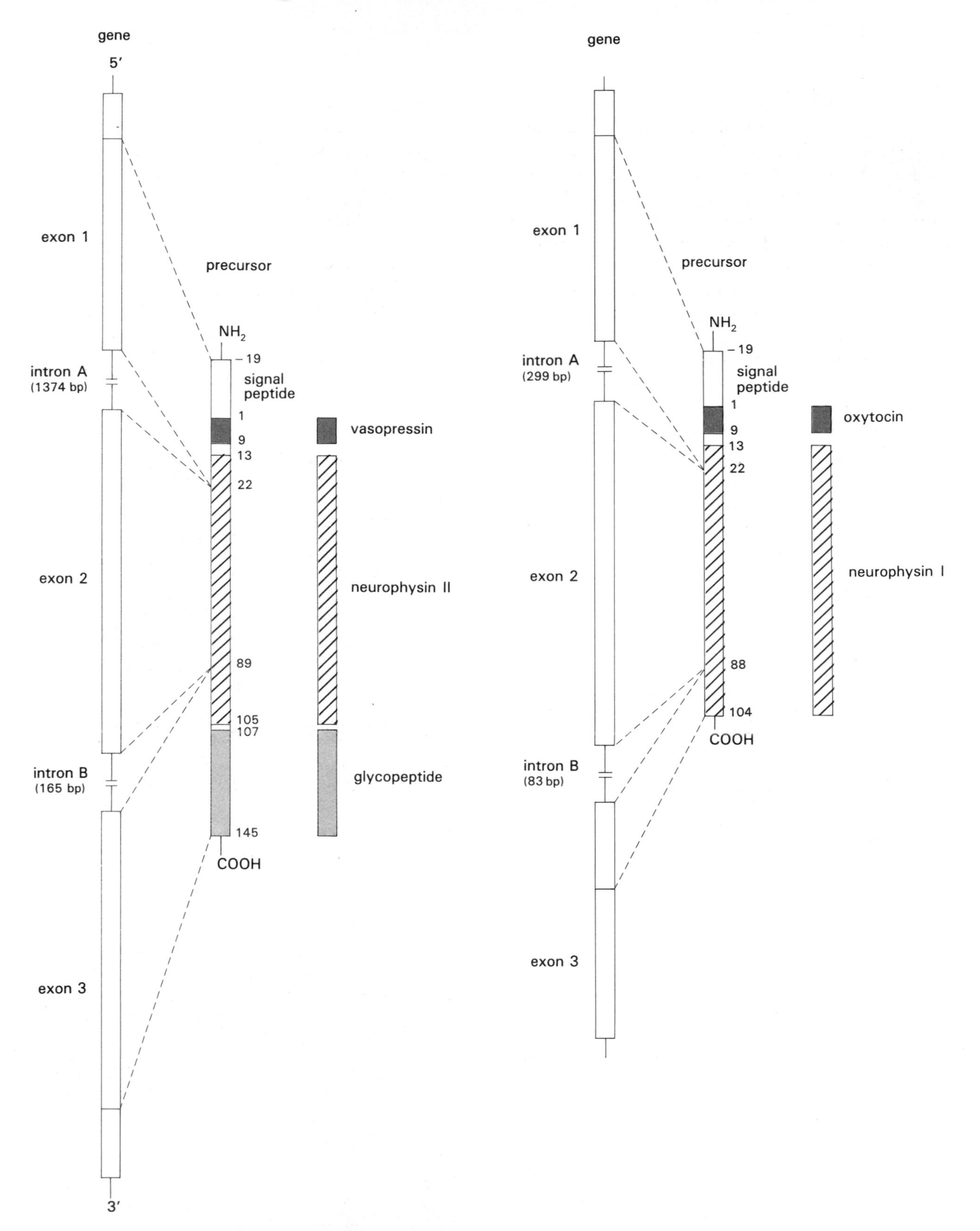

Figure VI.4 Structure of the human gene and precursor of vasopressin, prepropressophysin. The precursor contains four segments corresponding respectively to the signal peptide, vasopressin, neurophysin II, and a glycopeptide.

Figure VI.5 Structure of the human gene and precursor of oxytocin, preprooxyphysin. The precursor contains three segments corresponding respectively to the signal peptide, oxytocin, and neurophysin I.

Molecular Aspects

Vasopressin is synthesized in the form of a prepropeptide[4] consisting of 166 aa, of which 19 comprise the signal peptide (Fig. VI.4). The signal peptide is immediately followed by the sequence of arginine vasopressin, itself separated, by three amino acids, from the sequence of *neurophysin II*, the protein which is essential for the axonal transport of the hormone. The last 39 aa form a *glycopeptide* whose biological role has not been determined.

Oxytocin is synthesized in a fashion similar to vasopressin. Its precursor also contains the related transport protein, *neurophysin I*, which is very similar to neurophysin II, however there is *no glycopeptide* encoded in the 3′ extremity of the gene (Fig. VI.5). In fact, the genes for the two hormones are similar in their intron-exon structure, linked together with a 12 kb intervening sequence, and transcribed from opposite DNA strands[4]. Despite their proximity, the two genes are apparently always expressed in different cells, in discrete neurons.

A remarkable *mutation* involving a *single nucleotide deletion* within the second exon encoding the highly preserved region of neurophysin II in the *Brattleboro rat* leads to a complete *blockage* of *vasopressin production*[5]. The mRNA coded by the modified gene is correctly transcribed and spliced, but is not translated. The Brattleboro rat has been, and

4. Sausville, E., Carney, D. and Battey, J. (1985) The human vasopressin gene is linked to the oxytocin gene and is selectively expressed in a cultured lung cancer cell line. *J. Biol. Chem.* 260: 10236–10241.
5. Schmale, H. and Richter, D. (1984) Single base deletion in the vasopressin gene causes diabetes insipidus in Brattleboro rats. *Nature* 308: 705–709.

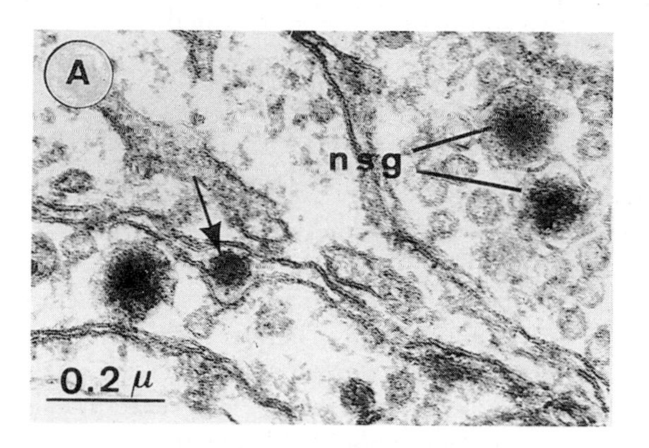

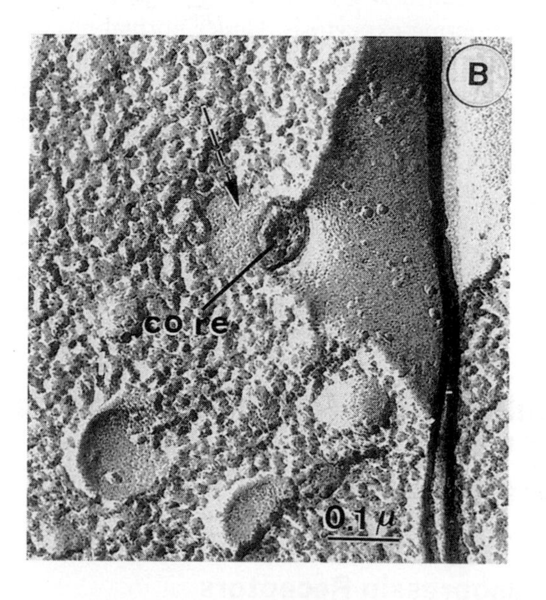

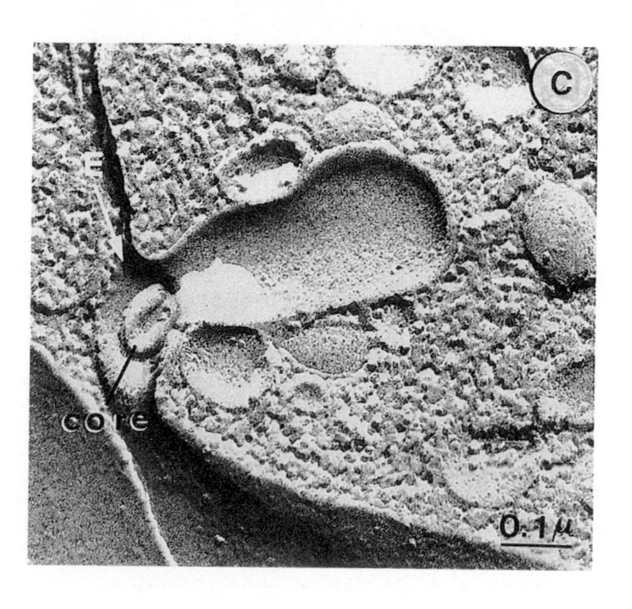

Figure VI.6 Exocytosis of neurosecretory granules at the extremities of magnocellular neurons. **A.** The round mass of electron-dense material designated by the arrow is comparable to intracytoplasmic neurosecretory granules (nsg). It fills an invagination of the plasma membrane in the nerve terminal. This invagination results from the fusion of the granule membrane with the plasma membrane. Bar=0.2 μm. **B.** Freeze-fracture micrograph of a nerve terminal, showing the internal (upper right) and cytoplasmic (left) layers of the plasma membrane. The fusion of the neurosecretory granule is indicated by the arrow. The core of the granule has been fractured. Bar=0.1 μm. **C.** Freeze-fracture micrograph of a nerve terminal, showing an open granule with core entering the extracellular space. Bar=0.1 μm. (Theodosis, D. T., Dreifuss, J. J. and Orci, L. (1978) A freeze-fracture study of membrane events during neurohypophysial secretion. *J. Cell. Biol.* 78: 5442–553).

continues to be, of great use in the study of many aspects of vasopressin action in mammals[6].

1.3 Secretion

Vasopressin and oxytocin are released from nerve terminals of the neurohypophysis by a mechanism of *exocytosis* (Fig. VI.6). The other major intragranular component liberated with vasopressin or oxytocin is an associated neurophysin. There is a close relationship between hormone release and a decrease in the number of neurosecretory granules in nerve terminals.

Current evidence indicates that an increase in *calcium* ion concentration in axon terminals is responsible for initiating exocytosis. Under normal physiological conditions, the secretion of vasopressin and oxytocin is induced by depolarization of the axon membrane, which leads to an opening of calcium channels and the transfer of extracellular calcium to the interior. In summary, the secretion of oxytocin and vasopressin secretion is controlled by alteration of the electrical activity of the neurons producing these hormones. The stimuli involved in this phenomenon are described below.

2. Cellular and Molecular Aspects of Vasopressin and Oxytocin Actions

2.1 Vasopressin Receptors

Vasopressin receptors have been localized on the *surface* of target cells. They have been characterized in several organs and cell types (kidney, liver, pituitary, blood vessels, platelets, testis, and several cell lines of tumoral origin). Receptors from these different tissues bind hormone with similar affinity, $K_D \sim 1$ nM.

A maximal biological response of vasopressin-sensitive cells is usually obtained for vasopressin concentrations much lower than those needed to achieve complete receptor occupancy. The ED_{50} value (hormonal concentration leading to half-maximal response) is much lower than the corresponding K_D value (hormonal concentration leading to half-maximal receptor occupancy). The ratio ED_{50}/K_D varies within large limits from one tissue to another. It is the lowest for antidiuretic action, which is undoubtedly the major physiological action of vasopressin. For several other peripheral actions of vasopressin, the observed ED_{50} values are at the upper limit of circulating vasopressin levels. For this reason, the physiological relevance of these actions has been questioned frequently.

Functionally, *two types of receptors (V_1 and V_2)* have been described (Fig. VI.7)[7]. V_2 receptors have been identified only in *renal tubules*. They are coupled to *adenylate cyclase*. Occupancy of V_2 receptors by vasopressin leads to activation of membrane adenylate cyclase, resulting in an increase in the intracellular

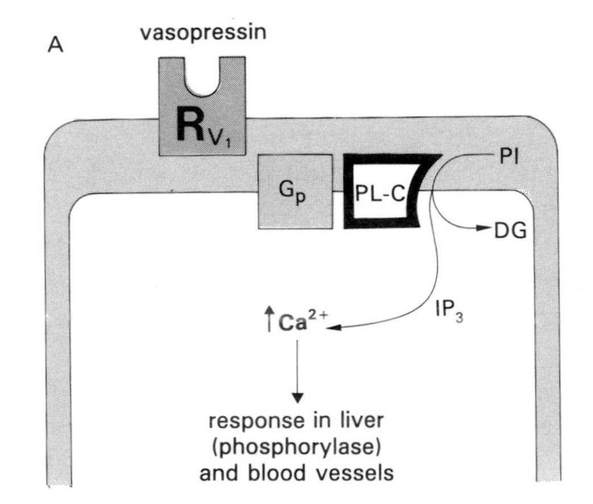

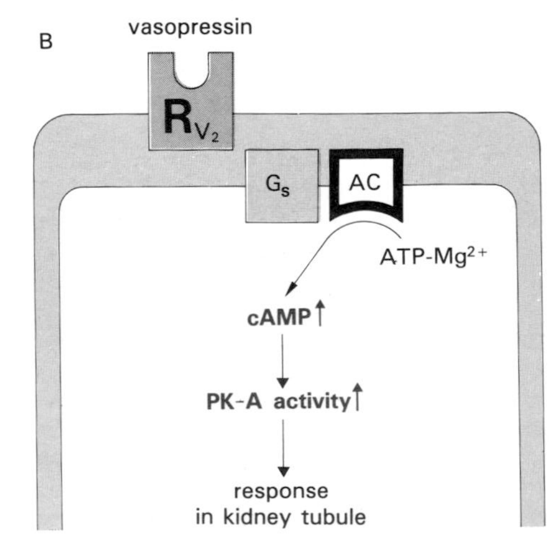

Figure VI.7 Mechanisms of action of vasopressin in extrarenal tissues, via the V_1 receptor (**A**), and in the kidney tubule, via the V_2 receptor (**B**).

6. Valtin, H. (1977) Genetic models for hypothalamic and nephrogenic diabetes insipidus. In: *Disturbances in Body Fluid Osmolality* (Andreoli, T. E., Grantham, J. J. and Rector, F. C. Jr., eds), American Physiological Society, Washington D. C., pp. 197–215.
7. Michell, B. H., Kirk, C. J. and Billah, M. M. (1979) Hormonal stimulation of phosphatidylinositol breakdown, with particular reference to the hepatic effect of vasopressin. *Biochem. Soc. Trans.* 7: 861–865.

Table VI.1 Pharmacology of vasopressin receptors.

Analog	Antidiuretic activity (1)	Vasopressor activity (2)	(1)/(2)
Arginine vasopressin	100	100	1
[Val-4] arginine vasopressin	228	9	25
Deamino [D-Arg-8] vasopressin	372	0.1	3520
[Val-4 D-Arg-8] vasopressin	202	0.01	20145
[Phe-2 Ile-3 Orn-8] vasopressin	0.17	34	0.005

Results in columns (1) and (2) are expressed in percent of arginine vasopressin activity. (From Sawyer et al. (1981) see general references).

concentration of cAMP. The molecular mechanisms involved in this activation are common to all hormone-sensitive adenylate cyclases (p. 335). V_1 receptors, first characterized in the liver and blood vessels, are not coupled to membrane adenylate cyclase. Their activation results in an increase in the cytosol concentration of *calcium ions*. Receptor-mediated hydrolysis of PIP_2, leading to the release of *DG* (diacylglycerol) *and* IP_3 (p. 102), probably represents an early step in the mechanism by which vasopressin increases cytosol calcium[8]. So far, *all extrarenal* effects of vasopressin appear to be mediated by vasopressin receptors of the V_1 type.

Pharmacology of Vasopressin Receptors

A vast number of vasopressin structural analogs have been synthesized, and their biological potencies determined, on the basis of two classical bioassays for vasopressin, namely, the *antidiuretic test* in water-loaded rats and the *vasopressor test*. These analogs are used extensively in studying the ligand specificities of vasopressin receptors. Several structural modifications affect vasopressin and antidiuretic potencies in a distinctly different manner (Table VI.1).

Deamino [D-arginine-8] vasopressin (dAVP) has elevated antidiuretic activity and is practically devoid of vasopressor activity. It has an increased metabolic stability in relation to vasopressin, and is active nasally. This vasopressin analog is currently used in the treatment of diabetes insipidus. Several antagonistic derivatives of vasopressin have been synthesized (Fig. VI.8).

8. Kirk, C. J., Creba, J. A. K., Hawkins, P. T. and Michell, R. H. (1983) Is vasopressin-stimulated inositol lipid breakdown intrinsic to the mechanism of Ca^{2+} mobilization at V_1 vasopressin receptors. In: *The Neurohypophysis: Structure, Function and Control. Progress in Brain Research.* (Cross, B. A. and Leng, G., eds), Elsevier Science Publishers, Amsterdam, vol. 60, pp. 405–411.

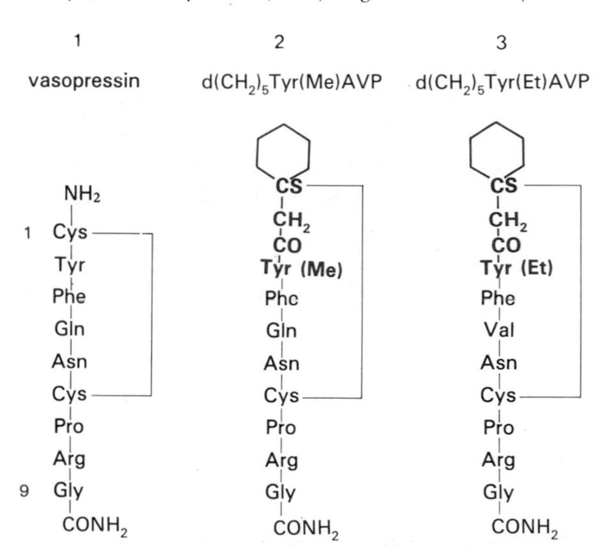

Figure VI.8 Structures of two antagonists of vasopressin. These antagonists are derivatives of vasopressin. The modifications are: 1. the replacement of a cysteinyl residue in position 1 by β-mercapto-β, β-cyclopentamethylene propionic acid; 2. methylation or ethylation of a tyrosyl residue; 3. replacement of glutamine by valine. The first peptide, $d(CH_2)_5Tyr(Me)AVP$, is an antagonist of the vasopressor effect of vasopressin. It has weak antidiuretic activity. The second peptide, $d(CH_2)_5Tyr(Et)VAVP$, is an antagonist of the vasopressor and antidiuretic effect of vasopressin in the rat. (Manning and Sawyer (1983). See general references, p. 302).

Vasopressin receptors can be distinguished on the basis of their ligand specificity. There is a fairly good correlation between the functional (see above) and pharmacological criteria which can be used to distinguish the different types of vasopressin receptors. Thus, several V_1 receptors from different tissues have very similar if not identical *ligand specificities*, which are markedly *different* from those of renal V_2 receptors. However, several *subclasses of* V_1 receptors probably exist, as suggested by studies showing that the specificity of pituitary receptors is clearly different from that of either vascular or renal receptors.

2.2 Oxytocin Receptors

Oxytocin receptors in uterine muscle and mammary gland have been characterized[9]. Oxytocin does not affect the concentration of cAMP in target tissues. *Calcium* ions are probably the second messenger for this hormone in the uterus and mammary gland. Functionally, oxytocin receptors are *homologous to type* V_1 vasopressin receptors (Fig. VI.9).

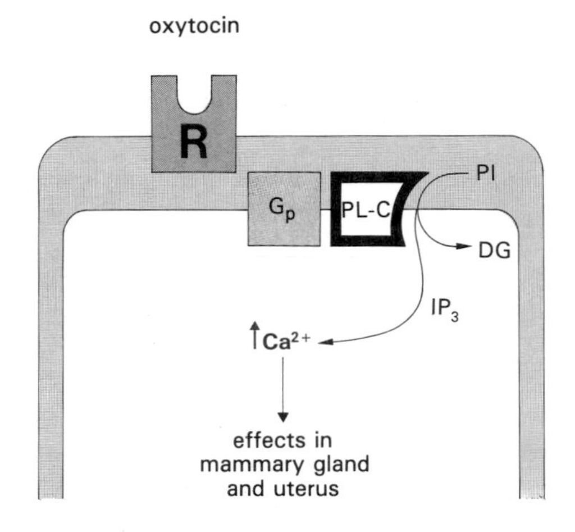

Figure VI.9 Mechanism of action of oxytocin.

Although structural modifications of the oxytocin molecule can have different effects on uterotonic and milk secretory activities, the existence of different classes of oxytocin receptors has not been clearly established. Potent antagonists of oxytocin have been synthesized.

Vasopressin binds to oxytocin receptors, with low affinity, and, reciprocally, oxytocin binds to vasopressin receptors, also with low affinity. At high doses, both oxytocin and vasopressin are capable of inducing all the effects of the other hormone. *Physiologically*, however, the *tissue-specific action of these two peptides* is probably always *maintained*.

9. Schroeder, B. T., Chakraborty, J. and Soloff, M. S. (1977) Binding of [^{3}H]oxytocin to cells isolated from mammary gland of the lactating rat. *J. Cell. Biol.* 74: 428–440.

3. Physiological Aspects of Vasopressin and Oxytocin Actions

3.1 Actions of Vasopressin on the Kidney: Its Role in Osmoregulation

Vasopressin has no significant effect on renal glomerular filtration rate (quantity of plasma filtered by the glomerulus), nor does it affect the total solutes eliminated in urine. Within broad limits, it does *regulate* the *amount* of *water in* which *solutes* are *excreted*. In the *absence* of vasopressin, the kidney produces a *large volume* of *urine* with an *osmotic pressure lower than* that of *plasma*. The osmotic pressure of urine may decrease to values of the order of 50 mOsm in humans, while normal osmotic pressure in extracellular fluids is 250 mOsm. Under such conditions, the kidney excretes *water in excess of solutes*. The production of hypo-osmotic urine tends to increase the osmotic presure of plasma and to correct for eventual hemodilution. Under *maximal stimulation* by vasopressin, the kidney produces a small volume of hyperosmotic urine (with an osmotic pressure greater than that of plasma). In humans, the maximal concentration attainable for urine is of the order of 1200 mOsm. Under these conditions, the kidney excretes an *excess* of *solutes in relation to water*, which tends to reduce the osmotic pressure of plasma and to correct for eventual hemoconcentration.

The ability of the mammalian kidney to concentrate urine is a function of the specific anatomical arrangement of the nephrons in the kidney and the water permeability and transport properties of the terminal portions of the nephron (Henle's loop and the collecting duct). In the deep regions of the kidney, "hair-pin" blood vessels (vasa recta) and Henle's loops are arranged in parallel. The blood and intratubular urine traverse the cortical regions, toward the extremity of the papilla, and then pass from the extremity of the papilla to the cortical regions. Such an organization of the deep regions of the kidney allows the majority of the vascular and urinary loops to function as a countercurrent concentration multiplier (Fig. VI.10). In the presence of vasopressin, *osmotic pressure* in the intravascular, interstitial and intratubular compartments *increases progressively* as it passes from the corticomedullary junction to the extremity of the papilla, where it reaches its maximum value. In the absence of vasopressin, the corticopapillary osmotic gradient is markedly reduced.

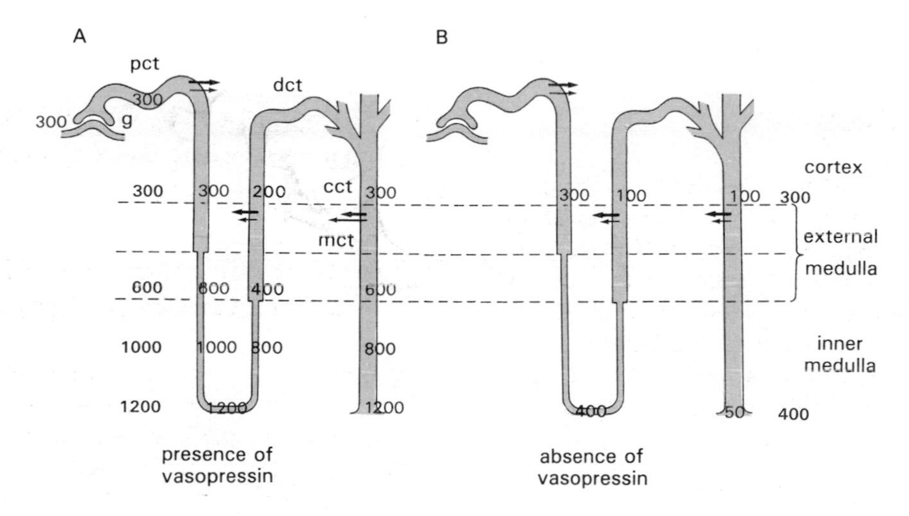

Figure VI.10 Evolution of intraluminal osmotic pressure along the nephron: effects of vasopressin. The abbreviations are: g, glomerulus; pct, proximal convoluted tubule; dct, distal convoluted tubule; cct, cortical and mct, medullary portions of the collecting duct. The comparison between **A** and **B** demonstrates the two effects of vasopressin on the kidney: 1. the hormone allows the establishment of a corticopapillary osmotic gradient; 2. it increases the permeability of the walls of the collecting duct to water. In the presence of vasopressin, an osmotic pressure equilibrium is established between intratubular urine and hypertonic interstitial fluid. (⇉ = isotonic reabsorption; → = reabsorption of water in excess of solute reabsorption; → = reabsorption of solute in excess of water reabsorption).

The final adjustment of the osmotic pressure of urine occurs during its movement along the collecting duct, and is a function of the osmotic permeability to water of the apical membrane of the cells limiting the lumen of the collecting duct. This permeability is controlled directly by vasopressin. In the absence of vasopressin, the permeability of the apical membrane of these tubular cells to water is low. The osmotic pressure of the intratubular fluid, lower than the osmotic pressure of plasma at the entrance of the collecting duct, remains constant during the corticopapillary transit, and may even diminish because of the reabsorption of solutes by the tubular walls. In the presence of *vasopressin*, the *osmotic permeability* of the apical membrane of tubular cells is markedly *increased*, thus allowing passive water reabsorption along the transtubular osmotic gradient. Consequently, the osmotic pressure of intratubular fluid equilibrates with that of the hypertonic interstitial fluid. The final osmotic pressure of urine is close to the osmotic pressure found in the renal interstitium at the extremity of the papilla (Fig. VI.10).

In summary, *vasopressin* exerts a *dual effect* on the kidney. It contributes to the build-up of the *corticopapillary osmotic gradient*, and controls *water reabsorption* by the collecting duct.

3.2 Localization of Sites of Vasopressin Action along the Nephron

Vasopressin stimulates adenylate cyclase in the cortical and medullary portions of the collecting duct (Fig. VI.11). In addition, the presence of a vasopres-

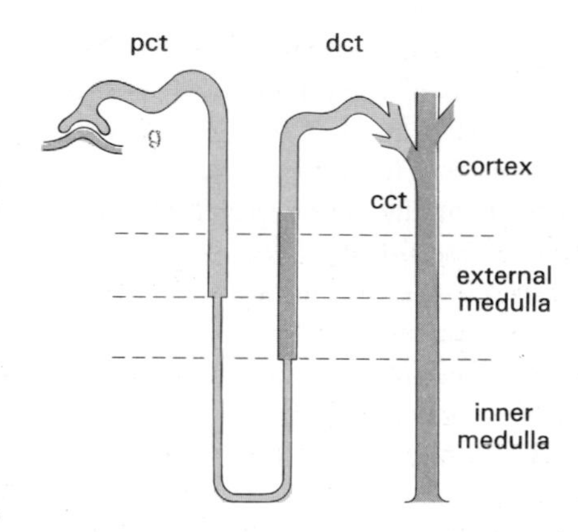

Figure VI.11 The segments of the nephron sensitive to vasopressin. This figure represents a diagram of the principal segments of the nephron (see legend of Fig. VI.10 for identification of the regions of the nephron). The red segments are sensitive to vasopressin. Sensitivity to vasopressin was measured by activation of adenylate cyclase-isolated segments. Receptors present in this area are of the V_2 type (functionally coupled to adenylate cyclase). This figure illustrates results obtained in the rat. Receptors for vasopressin are also present in the mesangial cells of the renal glomerulus (one of the cell types constituting the filtration barrier). Receptors present in this region are of the V_1 type.

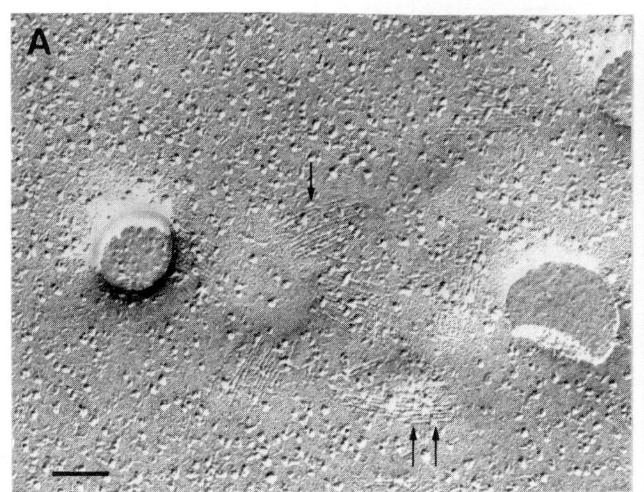

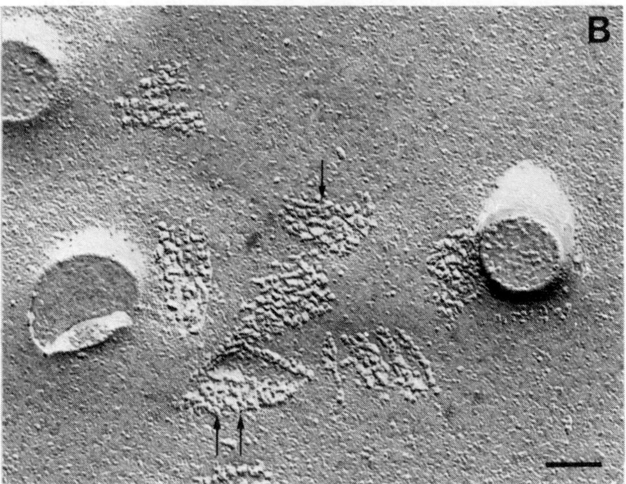

Figure VI.12 Freeze-fracture micrographs of the apical membrane zone of a frog bladder epithelial cell. Such cells are functionally homologous to tubular cells from the collecting ducts of the mammalian kidney. They respond to neurohypophyseal hormones by a marked increase in water permeability of their apical membrane. The figure shows a double replica of the fracture zone of the apical membrane of a vasopressin-stimulated cell. The transversal sections of microvilli indicate that the two replicas are complementary. The internal membrane leaflet in panel **B** demonstrates the aggregation of particles, resulting from hormonal stimulation. They contain the water channels, transferred to the apical membrane during the hormonal challenge. On panel **A**, striae seen on the external membrane leaflet are grooved areas which correspond to the aggregates (arrows). Bars=0.1 μm. (Courtesy of Dr. J. Bourguet, Centre d'Etudes Nucléaires, Saclay).

sin-sensitive adenylate cyclase in the ascending limb of Henle's loop has been demonstrated in several, but not all, mammalian species studied thus far[10]. In *Henle's loop*, vasopressin stimulates the reabsorption of solutes, in particular that of the of divalant cations $Ca^{2+}$and Mg^{2+}. This effect may be involved in the establishment of the corticopapillary osmotic gradient. In the *collecting duct*, vasopressin increases the permeability of the tubular wall to water. Hormone receptors are located in the basolateral membranes of tubular cells. The site of the ultimate action of the hormone is the apical membrane of the cells, which constitutes the limiting barrier in the transepithelial transfer of water.

The large variations in the permeability of the apical membrane to water result from a reversible incorporation of intracellular vesicles, rich in protein aggregates, into this membrane (Fig. VI.12). These aggregates probably represent preferential pathways for water reabsorption. The process of vesicle incorporation involves elements of the cytoskeleton. All the effects of vasopressin on the cells of the collecting duct can be mimicked by cAMP.

Independent of its tubular effects, vasopressin is a contracting agent for the mesangial cells of the renal glomerulus. These cells have the contractile property of smooth muscle cells, and are constitutive elements of the filtration barrier separating blood from the intratubular compartment. The degree of contraction of the mesangial cells probably affects the permeability of the filtration barrier to water.

3.3 Regulation of Vasopressin Secretion in Relation to Osmoregulation

The crucial experiment conducted by Verney[11] demonstrating the *role of plasma osmotic pressure* in the control of vasopressin secretion was performed in the dog undergoing water diuresis.

Water diuresis is a consequence of hemodilution resulting from the absorption of large quantities of water. In these circumstances, the circulating vasopressin level is reduced. As a consequence, urinary flow is increased, and the osmotic pressure of

10. Imbert-Teboul, M., Chabardes, D., Montegut, M., Clique, A. and Morel, F. (1978) Vasopressin dependent adenylate cyclase activities in the rat kidney medulla: evidence for two separate sites of action. *Endocrinology* 102: 1254–1261.
11. Verney, E. B. (1947) The antidiuretic hormone and the factors which determine its release. *Proc. Roy. Soc., London, B.* 135: 25–106.

urine is reduced. Water diuresis mimicks diabetes insipidus caused by a deficiency in the production or secretion of vasopressin. Administration of exogeneous vasopressin to the animal undergoing water diuresis results in a *dose-dependent antidiuretic response* which can be quantified by the reduction in urine flow or by the increase in urinary osmotic pressure. This represents the basis for the bioassay of vasopressin.

Verney's experiments (Fig. VI.13) consisted of: 1. exploring the effects of local modifications of blood osmotic pressure on the urinary output and osmotic pressure of urine. By adjusting the infusion rate and concentration of a hypertonic solution infused in the carotid artery, it is possible to increase, in a predictable manner, the osmotic pressure of blood reaching the brain, without affecting the osmotic pressure of peripheral blood due to the high ratio, total blood mass/volume of the vascular bed separating the carotid artery from the heart; 2. determining the amount of endogenously released vasopressin which could account for the observed antidiuretic responses elicited by osmotic stimuli, using the bioassay procedure mentioned above.

Experiments have established clearly that *vasopressin secretion* is *regulated by* osmosensitive structures called *osmoreceptors*. Osmoreceptors are located in the *brain*, but their precise anatomical location is not clearly established. The most important characteristics of osmoreceptors are: 1. their ability to regulate vasopressin secretion positively and negatively around a critical value of blood osmotic pressure corresponding to the normal physiological

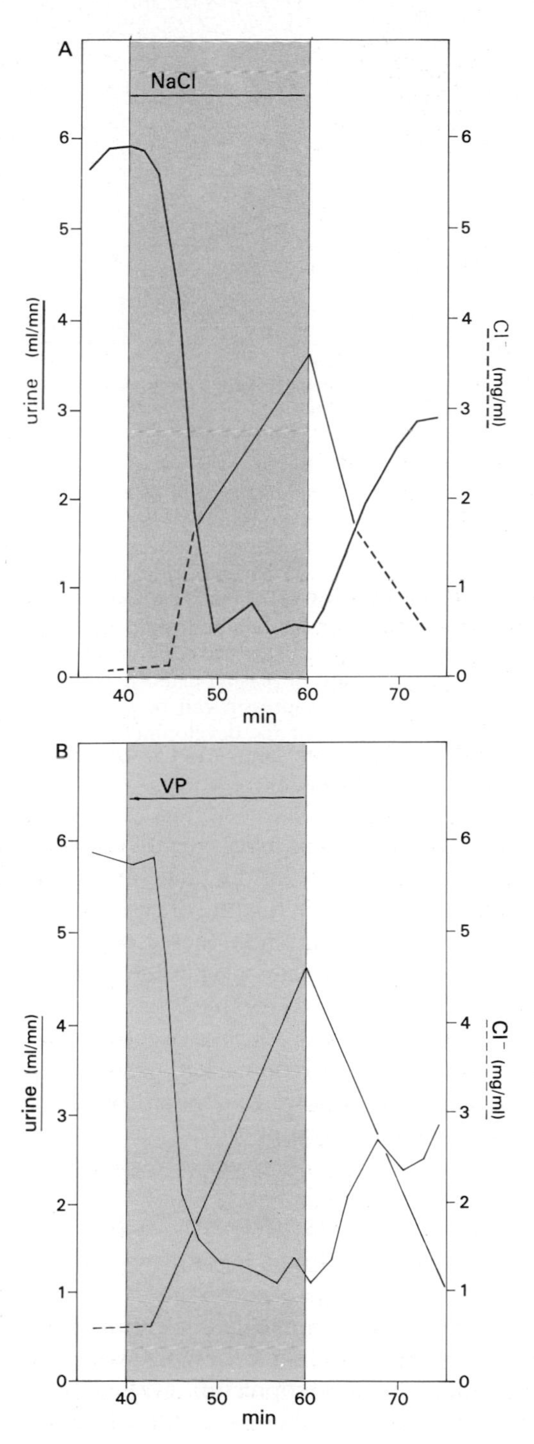

Figure VI.13 Role of plasma osmotic pressure in the control of vasopressin secretion. Verney's basic experiment was carried out in the dog undergoing water diuresis induced by gastric loading with a diluted saline solution. The secretion of vasopressin is reduced, resulting in an increased urinary output and a reduction in the osmotic pressure of urine (estimated here by its concentration in chloride ions). **A.** Effect of an osmotic stimulus. During the period indicated by the hatched zone, the animal received a continuous perfusion, via the carotid, of a hypertonic solution of NaCl (0.654 M) at a rate of 1.64 μl/min. This perfusion increased the osmotic pressure of the plasma irrigating the brain regions by 5.05%. It led to a reduction in urinary output and an increase in the osmotic pressure of urine. The same perfusion administered intravenously is without effect (not shown on this figure). Complementary experiments demonstrated that the stimulus is osmotic pressure (hypertonic solutions of the same osmotic pressure exert the same effects, independent of the nature of the solute used). Response did not occur following destruction of the neurohypophysis. **B.** Bioassay of vasopressin in the same animal. Intravenous perfusion of increasing doses of vasopressin allows the assessment of the quantity of vasopressin released in response to the osmotic stimulus by determining the amplitude of the diuretic response. This figure illustrates the effects of perfusion of 3.3 μU/sec of vasopressin.

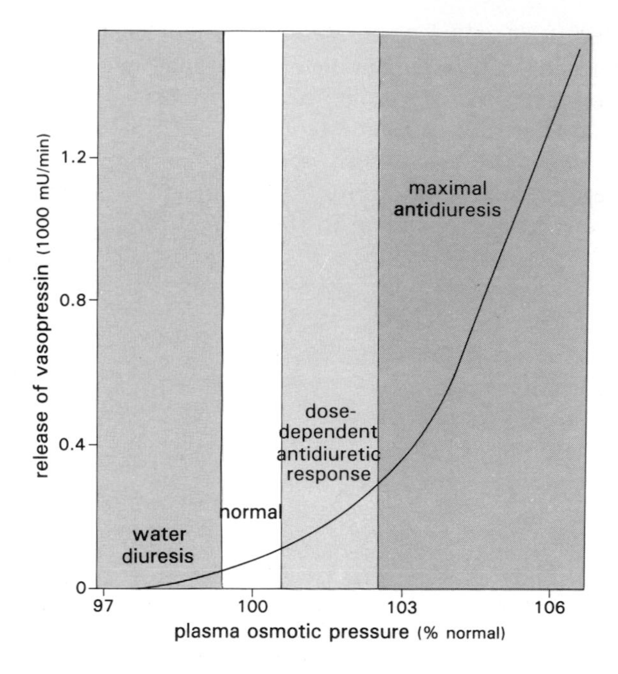

Figure VI.14 Relationship between plasma osmotic pressure and the circulating concentration of vasopressin. The secretion of vasopressin is modified by very small variations in plasma osmotic pressure (regulation operates from the point of reference 100%). The resulting modifications of renal excretion (elimination of excess water in relation to solutes, under conditions of hemodilution, and the inverse effect under conditions of relative hemoconcentration) allow an autoregulation of plasma osmotic pressure. Note that the secretion of vasopressin can be stimulated much more than is necessary for the development of a maximal antidiuretic response. (O'Connor, W. J. (1962) *Renal Function*. Edward Arnold, London).

value; 2. their high sensitivity. Relative variations in blood osmotic pressure (in the range of 1%) induced marked modifications in the rate of vasopressin secretion; 3. absence of adaptation. Increased or decreased vasopressin secretion rates persist until the osmotic pressure of blood has regained its normal value (Fig. VI.14). Stress and other nociceptive stimuli increase vasopressin secretion. Hemorrhage and ether anesthesia are most powerful stimuli for vasopressin release by the neurohypophysis.

3.4 Effects of Vasopressin in the Control of Blood Pressure

Because there are a number of systems controlling blood pressure (autonomic nervous system, renin-angiotensin system, vasopressin), evaluation of their respective in vivo contributions is difficult. Alterations of one of these systems can be compensated by the action of the others.

An *increase* in the plasma concentration of *vasopressin*, within normal physiological limits (1–20 pM), results in an *increase* in peripheral vascular *resistance*, but a *reduction in cardiac output*. These two effects are compensatory, with blood pressure remaining practically unchanged. Following denervation of baroreceptors, only the effect on peripheral resistance remains, resulting in a substantial increase in arterial pressure[12]. In circumstances where effects of the sympathetic nervous system and of the renin-angiotensin system are eliminated, the rapid effect of endogenous vasopressin on blood pressure can be

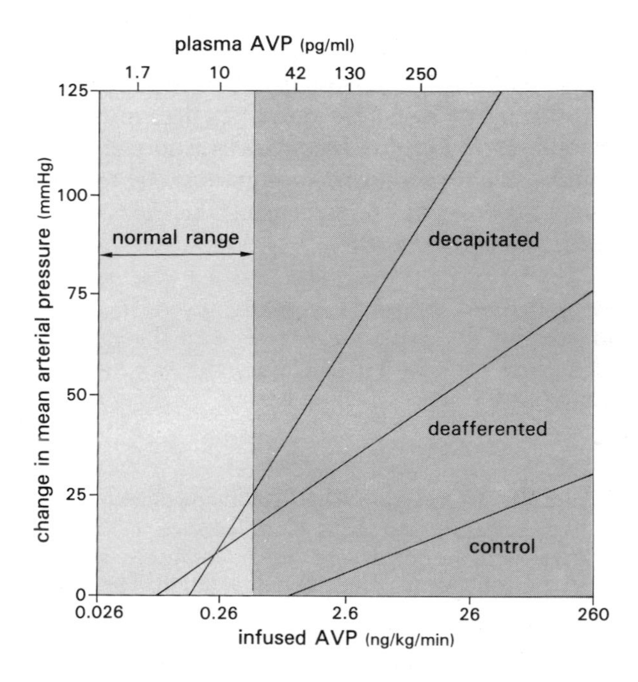

Figure VI.15 Vasopressor effect of vasopressin in the dog. An increase in blood pressure in response to an infusion of increasing doses of vasopressin has been measured in the dog. The effective blood vasopressin concentrations reached during the experiment were measured and are indicated on the abscissa. They can be compared to physiological concentrations of endogenous vasopressin (pink zone). In the awake animal, the response to vasopressin becomes significant only at concentrations of hormone which surpass the upper limit of physiological variation. The apparent sensitivity to vasopressin is increased following *denervation* of the atrial and aortic baroreceptors or suppression of the influence of the superior central nerves. As indicated in the text, these results are explained by a dual action of the hormone: a vasoconstrictor effect, and an increase in the efficiency of the reflex autoregulation of blood pressure by the autonomic nervous system. Under normal conditions, these two effects compensate each other[12].

demonstrated. In these experimental conditions, rapid withdrawal of blood produces a biphasic change in blood pressure: a rapid decrease followed by a compensatory phase. The latter is not observed after removal of the posterior pituitary or following the administration of antagonists to vasopressin. Within physiological limits of its concentration in plasma, vasopressin can therefore affect peripheral vascular resistance (Fig. VI.15).

In addition, vasopressin *increases* the efficiency of the *reflex regulation* of blood pressure by the autonomic nervous system. The stimulation of carotid and atrial baroreceptors (receptors for high and low pressure, respectively) reduces the electrical activity of magnocellular neurons of the supraoptic and paraventricular nuclei. Hemorrhagic stress, in contrast, is a potent stimulator of vasopressin secretion by the neurohypophysis. Finally, vasopressin neurons of the supraoptic and paraventricular nuclei project into the regions of the brain stem implicated in the control of the cardiovascular system.

The involvement of vasopressin in the control of blood pressure now appears clearly established. It results from a dual action of the hormone; one direct, on peripheral vascular resistance, and the other, on reflex regulation of the cardiovascular system by the autonomic nervous system.

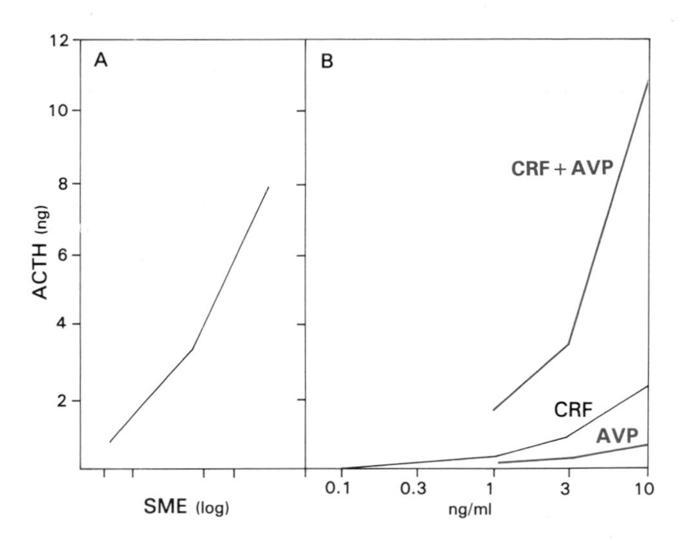

Figure VI.16 Vasopressin and ACTH secretion. Anterior pituitary cells were isolated, and ACTH released by these cells was monitored. **A.** Cells were stimulated with extracts of the median eminence (SME, stalk median eminence extract). Concentrations are expressed in arbitrary units: 1 unit = content of one median eminence in 1 ml of incubation medium. **B.** Cells were stimulated with vasopressin (AVP), corticotropin releasing factor (CRF), or with AVP+CRF. AVP, or CRF alone, is capable of stimulating ACTH release. A marked synergism can be seen between these two factors. The response of the combination of CRF and AVP is much greater than the sum of the responses of the two agents added alone. The response obtained by the mixture AVP+CRF is comparable to that induced by extracts of median eminence[13].

3.5 Effects of Vasopressin in the Control of ACTH Secretion

Vasopressin, which reaches the anterior pituitary via the portal system (see anatomical arrangement, Fig. VI.2), has a dual action on pituitary corticotropic cells: it *stimulates ACTH* secretion by these cells, and *potentiates* the action of hypothalamic *CRF* (Fig. VI.16)[13]. As mentioned previously, vasopressin fibers terminate in the median eminence, in contact with capillaries of the hypophyseal portal system. In the monkey and rat, it has been shown that blood from the hypophyseal portal system contains vasopressin in quantities sufficient to influence ACTH secretion by corticotropic cells. Adrenalectomy increases the synthesis of vasopressin in neurons of the paraventricular nucleus projecting into the median eminence. These neurons are not affected by chronic dehydration, which reduces the content of vasopressin in the neurohypophysis. It is now known that vasopressin, as well as CRF, is one of the elements of the multifactorial regulation of ACTH secretion. However, the relative participation of these elements under various physiological conditions has not been clarified.

12. Cowley, A.W. Jr. (1982) Vasopressin and cardiovascular regulation. In: *Cardiovascular Physiology IV. International Review of Physiology* (Guyton, A.C. and Hall, J.E., eds), University Park Press, Baltimore, pp. 189–242.
13. Gillies, G., Linton, E.A. and Lowry, P.J. (1982) The corticotrophin releasing activity of the new CRF is potentiated several fold by vasopressin. *Nature* 299: 355–356.

3.6 Effect of Oxytocin on Milk Ejection

The ejection of milk in the lactating female is initiated and maintained by *suction on the nipple*. Destruction of the motor innervation of the mammary gland does not eliminate reflex milk ejection. It does disappear, however, in hypophysectomized women (in which lactation is maintained by treatment with lactotropic hormones). In the cow, blood removed immediately following milking and perfused into the mammary gland of another animal results in immediate milk ejection. The administration of oxytocin to a lactating female also induces milk ejection. Finally, *suction* on the nipple causes the *release of oxytocin* from the *neurohypophysis*. These observations indicate clearly that oxytocin is one of the elements in the reflex control of milk ejection[14].

The intermittent character of milk ejection is coupled to the *pulsatile* nature of oxytocin release. In the rat, nipple suction leads to synchronized activation of hypothalamic oxytocin neurons, which generate a series of 70 to 80 action potentials within 3 to 4 s, resulting in the release of 0.5 to 1.0 mU of oxytocin (1–2 pmol). These waves of action potentials are separated by periods of rest, even if stimulation of the nipple is maintained. Other stimuli may substitute for nipple suction to induce milk ejection, as for example intercourse in women, or conditioned stimuli associated with nursing. Several neurotransmitters (e.g., acetylcholine, dopamine) participate in the activation of magnocellular oxytocin neurons. Oxytocin may feed back positively on the rate of its own secretion.

3.7 Role of Oxytocin in Parturition

Oxytocin induces *contraction* of smooth muscles of the *myometrium*. The sensitivity of the myometrium to oxytocin is largely influenced by hormonal status. In women, the sensitivity of the uterus to oxytocin is low during all phases of the menstrual cycle, however it is especially sensitive to vasopressin during the luteal phase. The response to oxytocin increases markedly during the last two weeks of pregnancy. Oxytocin is *released during labor* and parturition. It is not clearly established, however, that oxytocin is involved in the onset of either. Oxytocin release could be due to dilatation of the cervix and vagina, starting point of a reflex (Ferguson's reflex) resulting in the activation of magnocellular oxytocin neurons. Other stimuli, for example the violent abdominal contractions which accompany parturition in the rat, also participate in the regulation of oxytocin secretion. In addition, a progressive increase in the activity of oxytocin neurons is observed during pregnancy, with a marked increase in the minutes just preceding the onset of parturition. The same is true for the period just preceding the expulsion of the fetus.

3.8 Other Biological Effects of Vasopressin and Oxytocin

Other Peripheral Effects

Vasopressin, like a number of other hormones (glucagon, catecholamines acting via β- and α_1-adrenergic receptors, angiotensin), is a *glycogenolytic* agent (p. 510). The mechanism of action of vasopressin on the liver has been especially well studied; it is in this tissue that type V_1 receptors have been best characterized, both functionally and pharmacologically. It is clearly established that vasopressin increases the concentration of calcium ions within the *hepatocyte*[15]. The hormone thus *regulates phosphorylase* activity, which catalyzes the formation of glucose phosphate

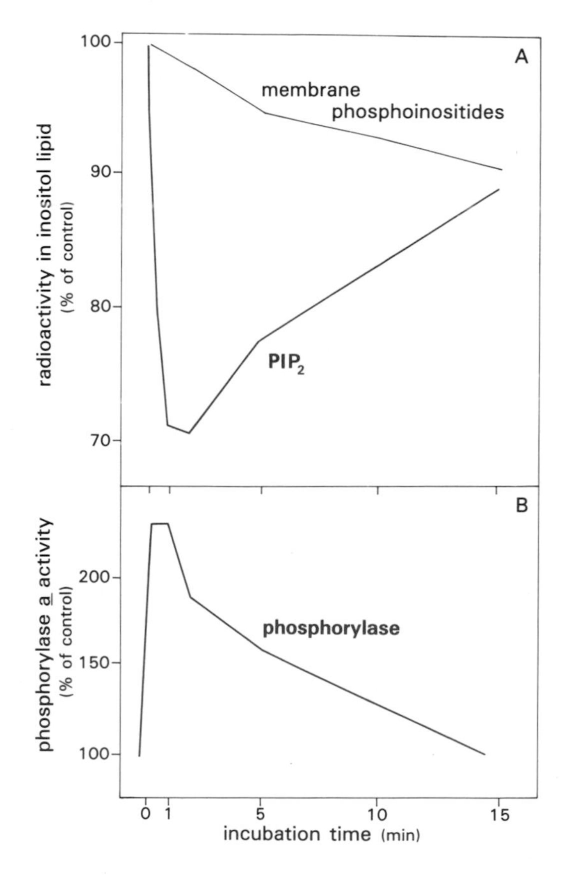

Figure VI.17 Activation of hepatic phosphorylase by vasopressin. There is a close correlation between the hydrolysis of PIP_2 (p. 102) and the increase of phosphorylase activity[8].

14. Bisset, G. W. (1968) The milk-ejection reflex and the actions of oxytocin, vasopressin and synthetic analogues on the mammary gland. *Neurohypophysial hormones and similar polypeptides*. In: *Handbook of Experimental Pharmacology* (Berde, B., ed), Springer Verlag, vol. 23, pp. 475–544.
15. De Wulf, H., Keppens, S., Vandenheede, J. R., Haustraete, F., Proost, E. and Carton, H. (1980) Cyclic-AMP independent regulation of liver glycogenolysis. In: *Hormone and Cell Regulation* (Nunez, J. and Dumont, J., eds), North Holland, Amsterdam, vol. 4, pp. 47–71.

from glycogen, and is rate-limiting in glycogenolysis (Fig. VI.17). The mechanism of activation of phosphorylase involves a phosphorylation of the enzyme, itself catalyzed by a phosphorylase kinase activated by calcium ions. Vasopressin, angiotensin and catecholamines, acting via α_1 receptors, activate this calcium-dependent regulatory pathway. Glucagon, and catecholamines acting via β receptors, activate a cAMP-dependent regulatory pathway. Hepatic glycogenolysis is activated by vasopressin at concentrations which are at the upper limit of circulating concentrations attained during maximal stimulation of the neurohypophysis, as seen, for example, during hemorrhagic stress.

In several species, including the human, vasopressin is one of the factors which causes *platelet aggregation* in vitro and the release of factor VIII. The involvement of vasopressin as an adjuvant factor involved in hemostasis in vivo has not actually been proved. The same is true for the *mitogenic* effects vasopressin has on several cell types.

Oxytocin is a *lipolytic* agent. It activates calcium-dependent regulation of lipolysis. In the cow, goat and sheep, oxytocin administered at high doses induces the *regression of the corpus luteum*. This effect is both direct (inhibition of the action of gonadotropic hormones on steroidogenesis in luteal cells) and indirect (release of $PGF_{2\alpha}$ from the uterine myometrium). These actions of oxytocin are only seen with concentrations greater than physiological levels in the circulation. However, the fact that in several species (including humans) the ovary is a major source of oxytocin suggests that it may act as a local hormone within the ovarian milieu. The role of vasopressin and oxytocin produced in the testis remains to be demonstrated. In vitro, they may inhibit gonadotropin-stimulated testosterone production by Leydig cells in culture[3].

Central Effects of Vasopressin and Oxytocin

Experimental data suggest that vasopressin and oxytocin can act as *neurotransmitters*; vasopressin and oxytocin neurons project into several regions of the central nervous system and medulla. In such regions, depolarization induced by elevated extracellular potassium levels results in the release of the two hormones. Specific binding sites for vasopressin and oxytocin have been detected in several regions of the central nervous system. Vasopressin and oxytocin introduced by iontophoresis modify electrical activity of several neuronal networks. Vasopressin administered centrally, into the cerebral ventricles, slows the spontaneous loss of conditioned avoidance behavior in the rat. Such data has lead to the hypothesis that vasopressin may be involved in the regulation of *memory*, although a direct mechanism has not yet been demonstrated.

4. Pathophysiology

4.1 Clinical Evaluation

Antidiuretic Hormone

Clinical evaluation of vasopressin depends, for the most part, on *indirect methods*. These tests involve the measurement of *urinary osmolality, in contrast to plasma osmolality*. In normal individuals in states of water diuresis, urinary osmolality may fall to 50 mOsm/kg. Under conditions of maximal water retention, urinary osmolality may reach 1200 mOsm/kg. The capacity of the kidney to concentrate or dilute the urine is measured by free water clearance. Free water clearance is defined as:

$$C = V - (V \times U_{osm})/P_{osm}$$

(V=urinary flow, U_{osm}=urinary osmolality, P_{osm}=plasma osmolality)

The quantity of solutes eliminated in the urine ($V \times U_{osm}$) in relation to plasma osmolality (P_{osm}) measures the apparent volume of isotonic liquid eliminated in urine; that is, the *osmotic clearance*, defined as:

$$C_{osm} = (V \times U_{osm})/P_{osm}$$

Therefore, the difference between the urinary output and osmotic clearance measures the quantity of water eliminated or reabsorbed in excess of solutes. *Positive free water clearance* occurs in states of diuresis where urinary volume surpasses osmotic clearance (water diuresis), and negative free water clearance occurs in states of diuresis where urinary volume is less than osmotic clearance (water restriction).

Free water clearance is not dependent on vasopressin secretion alone. It is also influenced by a number of *other factors*, renal (osmotic output, medullary blood flow, glomerular filtration, and the status of the corticopapillary osmotic gradient) and extrarenal (glucocorticosteroids, norepinephrine). Thus, the interpretation of the results of indirect clinical tests for vasopressin is made difficult by the numerous factors which affect urinary osmolality.

The *suppression test* consists of the oral intake of 20 ml/kg of water. Eighty percent of this water load is eliminated in 5 h: urinary osmolality should decrease below 100 mOsm/kg. The *stimulation test* involves perfusion with a hypertonic saline solution (3% salt) (maximum 300 ml) or a 5% saline solution in water-loaded subjects. The secretion of vasopressin occurs at a specific plasma osmolality (280 to 285 mOsm/kg). Secretion is demonstrated by the production of negative free water clearance. Renal sensitivity to vasopressin can be shown by the increase in urinary osmolality which follows the administration of vasopressin.

Whether measured in plasma or urine, the measurement of vasopressin can only be interpreted in relation to plasma osmolality. In addition, the conditions under which the measurement is taken must be strictly controlled, as the position of the subject being studied, the temperature of the room, and the subject's medication regimen can alter vasopressin levels. Ideally, stimulation–suppression tests should be coupled with measurement of the hormone. Figure VI.18 shows variations in urinary vasopressin as a function of plasma osmolality and natremia. These values are for normal subjects undergoing a suppression test, by 15 ml/kg of water given orally, and the stimulation test, by 7 ml/kg of hypertonic saline (3% salt) solution infused over 2 h. Using this technique, it is possible to diagnose syndromes of deficient or inappropriate vasopressin secretion.

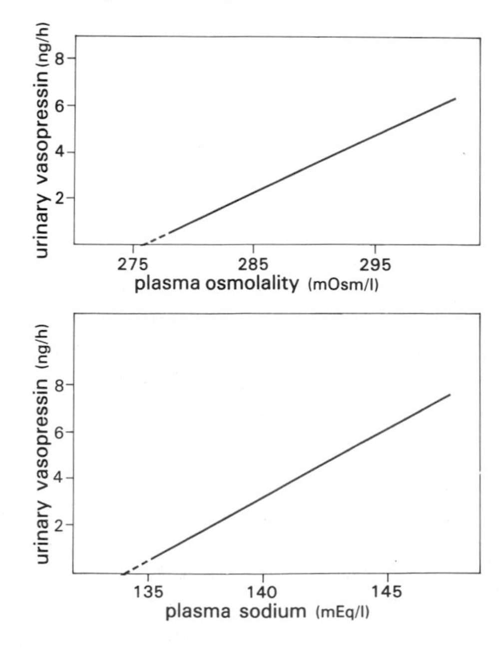

Figure VI.18 Correlation of urinary vasopressin excretion with plasma osmolality and plasma sodium concentration in normal human subjects.

4.2 Vasopressin Pathology

Exaggerated or Inappropriate Vasopressin Secretion

Schwartz–Bartter Syndrome

The *syndrome* of *inappropriate antidiuretic hormone secretion (SIADH)* consists of hyponatremia accompanied by normal or elevated natriuresis. In order to make this diagnosis, the biological abnormality should exist in the absence of renal, adrenal or thyroid insufficiency. A marked reduction in glomerular filtration hinders the elimination of free water and can cause hyponatremia. Cortisol deficiency reduces the elimination of free water by a dual mechanism operating both centrally (exaggerated release of vasopressin) and peripherally (increase in tubular permeability to water). In cases of thyroid insufficiency, the problem with free water elimination does not depend only on a possible excess of vasopressin, and SIADH disappears after administration of thyroxine. In this syndrome, contrary to what could be expected, blood volume is not decreased. Indeed, SIADH is associated with an expansion rather than a contraction of the extracellular fluid volume.

It is also important to determine whether hyponatremia is "true". The reduction in plasma water which accompanies an important hyperproteinemia (as in myeloma) could result in false hyponatremia. Plasma sodium is reduced, but the concentration of sodium in plasma water is normal.

SIADH can be clinically latent when natremia does not decrease to below 120 mEq. Below this value, signs of *water intoxication* are apparent: loss of appetite, nausea, vomiting, personality problems, insomnia, convulsive crises, muscle cramps, or positive Babinski reflex.

Hyponatremia is contrasted with a *conserved natriuresis*. The measurement of urinary osmolality theoretically shows values superior or equal to those of plasma osmolality (260 to 300 mOsm/kg). It should be remembered that low urinary osmolality (e.g., 200 mOsm/kg) may preclude the adjustment of plasma osmolality to a normal value, depending on intake of water and solutes. In addition, in patients with SIADH, creatinine and blood urea levels are normal, in contrast to those observed in renal insufficiency. Normal levels of urinary and plasma cortisol eliminate adrenal insufficiency. Plasma aldosterone is normal, and can be increased by low salt diet, angio-

Table VI.2 Causes of the Schwartz–Bartter syndrome, or inappropriate antidiuretic hormone secretion (SIADH).

1. Malignant tumors: lung, pancreas, duodenum, lymphosarcoma, thymoma.
2. Disorders of the central nervous system: subdural hematoma; meningeal hemorrhage; tuberculous, purulent, or viral meningitis; encephalitis; Guillain-Barré syndrome; skull fractures; cerebral stroke; cerebral tumors; intermittent acute porphyria; acute psychosis.
3. Pulmonary disorders: tuberculosis; viral and bacterial pneumonia; abscess; aspergillosis; chronic bronchopneumopathy.
4. Iatrogenic hyponatremia: administration of water and oxytocin during labor, chlorpropamide, vincristine, diuretics.

tensin, ACTH, and spironolactone. Plasma *renin* activity is *low*, however, almost undetectable.

The biological characteristics of SIADH, hyponatremia with hypernatriuresis, can be reproduced wholly in normal subjects who are given large quantities of water and treated with vasopressin injections. The biological and radioimmunological assays of vasopressin have confirmed that SIADH patients have an abnormal amount of vasopressin circulating in the blood, a quantity inappropriate for their plasma osmolality. This *inappropriate vasopressin secretion* could come either from the neurohypophysis or from malignant tumors which synthesize both neurophysin and vasopressin. The principal causes of SIADH are enumerated in Table VI.2. The amount of vasopressin present varies enormously from one individual to another. In some cases, there is very little circulating vasopressin not regulated by osmolality, while, in others, a large amount of circulating vasopressin is sensitive to variations in plasma osmolality.

An excess of vasopressin accounts for plasma hypo-osmolality. How can the natriuresis be explained? It is probably dependent on the hypotonic expansion of the extracellular space, which tends to increase glomerular filtration rate and, especially, to inhibit proximal tubular reabsorption of sodium. The natriuretic phenomenon would therefore be similar to the escape described following administration of mineralocorticosteroid, and may involve specific natriuretic factors (p. 619). That the proximal tubular rejection of sodium is responsible for the observed natriuresis is shown by the suppression of SIADH which occurs during thrombosis of the inferior vena cava, a clinical circumstance increasing proximal tubular reabsorption of sodium.

In addition to treatment of the causal disease, *corrective therapy* may include: 1. *water restriction*. This should be carefully conceived (300 ml of liquid daily, and total suppression of food rich in water). Water restriction is efficient but may be difficult to maintain over a long period, and its effects may be delayed; 2. *perfusion of hypertonic salt solution* in combination with the administration of *furosemide*. Furosemide, a diuretic acting on the concentrating segment (ascending branch of Henle's loop), inhibits the renal action of the inappropriate secretion of vasopressin. If this therapy is administered well, perfusion with a hypertonic salt solution maintains sodium equilibrium by equilibrating urinary loss of sodium. The loss of water is explained by the fact that urinary osmolality remains approximately 300 mOsm/kg under the influence of furosemide, while the liquid which is administered is hyperosmolar. Maintenance of potassium balance should be carefully controlled during such treatment; 3. administration of *lithium* is effective because it partially blocks the action of vasopressin on the kidney. Lithium leaves the kidney insensitive to the abnormal secretion of vasopressin, and consequently corrects hyponatremia. The inappropriate secretion of vasopressin is therefore masked by an iatrogenic nephrogenic diabetes insipidus. A careful surveillance of lithemia is necessary in order to maintain a level between 0.5 and 1.5 mEq/1; 4. *vasopressin antagonists* might be useful, in the future, for the treatment of SIADH.

Insufficiency of Vasopressin Secretion

Neurogenic Diabetes Insipidus

Diabetes insipidus designates a *permanent state* of *water polyuria*. Daily urinary volume is increased, and its osmolality varies between 50 and 200 mOsm/kg. It can be due to a deficit of vasopressin secretion (neurogenic diabetes insipidus), or to an insensitivity of the renal tubule to vasopressin (nephrogenic diabetes insipidus).

With the exception of the discomfort associated with polyuria and polydipsia, tending to compensate for renal loss of water, diabetes insipidus is a disease which is benign and *well tolerated*. Prognosis depends mainly on its cause.

Idiopathic, non-familial, neurogenic diabetes insipidus constitutes approximately 45% of the cases.

Familial forms have been described which are either autosomal dominant or sex-linked, thus differing from congenital diabetes insipidus of the rat (Brattleboro strain), which is transmitted through the autosomal recessive mode. Atrophy of the hypothalamic nuclei has been observed upon autopsy of such patients. The etiology of diabetes insipidus is diverse. Following hypophyseal surgery, the syndrome evolves in three phases. After initial polyuria lasting a few days, there is a second phase, with normal diuresis and a residual, poorly regulated secretion of vasopressin, hindering the excretion of a water load. A few days later, diabetes insipidus becomes definitive. Diabetes insipidus can also be a consequence of fractures of the base of the skull, sellar or suprasellar tumors, primitive or metastatic (especially from cancer of the kidney), various vascular lesions (thromboses, ruptured aneurisms), and meningioencephalitis.

The *diagnosis* of diabetes insipidus should demonstrate water polyuria. It is necessary, therefore, to begin with the elimination of osmotic polyuria as observed in diabetes mellitus (glucose), renal insufficiency (urea), and urinary loss of sodium (certain nephropathies, diuretics). It is then important to exclude other possible causes of water polyuria, e.g., potomania and renal diabetes insipidus.

Three dynamic tests are available, in which the measurement of free water clearance should be coupled with the measurement of antidiuretic hormone in plasma or urine: 1. the *water restriction test*. This necessitates constant medical presence for detection false cases, and, in bonafide cases, to stop the test if a sign of intolerance appears (faintness, hypotension). All drinking should stop at midnight, and restriction of liquid lasts for 12 to 18 h. Urinary osmolality remains below 300 mOsm/kg, even though weight loss attains or surpasses 3% of body weight. Very little vasopressin appears in either the urine or plasma, while plasma osmolality reaches or exceeds 290 mOsm/kg; 2. *perfusion of hypertonic* salt *solution* (e.g., 7 ml/kg, over 2 h, of a 3% salt solution). This does not increase plasma or urinary antidiuretic hormone levels, and does not result in negative free water clearance; 3. intravenous *infusion of vasopressin*. This makes free water clearance negative, thereby excluding a renal cause for diabetes insipidus. The cause of diabetes insipidus should be investigated radiologically.

Diabetes insipidus is moderate when associated with anterior pituitary insufficiency (Sheehan's syndrome). Hypernatremia, when associated with diabetes insipidus, is due to a thirst problem.

The *treatment* of diabetes insipidus involves the administration of long-acting, nasally active derivatives of vasopressin (deamino [D-Arg-8] vasopressin, see Table VI.1). Drugs that reduce free water clearance over a long period are also available. Thiazidic diuretics never inhibit free water clearance, but reduce it by increasing proximal reabsorption of sodium, coupled with a reduction of the extracellular space. They are effective in nephrogenic diabetes insipidus. Chlorpromide, clofibrate and carbamazepin are also active, and can inhibit clearance of free water, the last two being the better tolerated. They appear to act in the collecting tubule by potentiating the action of small quantities of vasopressin remaining in the circulation during diabetes insipidus.

Absence of Renal Response to Vasopressin

Nephrogenic Diabetes Insipidus

Nephrogenic diabetes insipidus is a sex-linked *hereditary* disease transmitted in the recessive mode. Most frequently, *boys* are affected, and the disease is transmitted by women who are not affected. Additional studies are needed in order that early detection of the disease may be made possible.

Severe forms occur in the first days of life, and are dangerous if dehydration is not correctly detected and treated. Less severe forms are characterized by thirst, polyuria, slow growth, and hyperuricemia.

Diagnosis is made on the basis of vasopressin-resistant water polyuria (urinary density 1.005, osmolality 80 to 150 mOsm/kg), and is observed in the absence of renal diseases such as uropathies related to malformation, nephronophtisis (medullary cystic disease), calciuria, chronic hypokalemia, depranocytosis, and cystinosis. Plasma or urinary vasopressin is normally adapted to plasma osmolality.

The mechanism of *renal resistance* to vasopressin is *unknown*. It has been demonstrated that vasopressin does not cause urinary excretion of cAMP in subjects with nephrogenic diabetes insipidus, while cAMP is increased in normal subjects or in those in which diabetes insipidus is a result of a deficiency of vasopressin. Glucagon and parathyroid hormone have a weak effect on urinary cAMP in these subjects. A *deficiency* of the renal tubular *adenylate cyclase* system is therefore a possible cause.

5. Summary

Vasopressin and oxytocin are neuropeptides of *nine amino acids*, with a disulfide bond connecting Cys-1 and Cys-6. Their large precursors are synthesized in several groups of neurons, mostly in the *magnocellular nuclei* of the hypothalamus. Each precursor includes, besides the hormone, the sequence for a specific *carrier protein* called *neurophysin* (II and I, respectively).

Vasopressin and oxytocin, released into the general circulation by the posterior pituitary, act as *peripheral hormones*. In the median eminence, vasopressin is hypophysiotropic, increasing ACTH synergistically with CRF. Vasopressin and oxytocin are also produced in several zones of the central and peripheral nervous system, where they act as *neuromodulators*. They may be involved in regulatory paracrine function in the adrenals and gonads.

Physiologically, *vasopressin* is the exclusive regulatory agent for *renal reabsorption of water*. In the *absence* of endogenous vasopressin secretion, or in the absence of a vasopressor effect on the kidneys, there is permanent *diabetes insipidus*, a syndrome characterized by excessive excretion of water (up to 15 l/d), in hypo-osmotic urine. The resulting hemoconcentration and the reduction of extracellular volume are compensated by excessive thirst and increased water intake. Vasopressin has a dual effect on the differential mechanism of water and solute excretion by the kidney. It establishes a corticopapillary osmotic pressure gradient and increases tubular permeability to water. Both of these effects lead to *passive reabsorption of water*. In the presence of vasopressin, there is reduced excretion of urine with an osmotic pressure higher than that of plasma. Vasopressin does not modify the amount of solutes excreted by the kidney, but rather the volume of excreted water. Vasopressin secretion is directly regulated by the osmotic pressure of extracellular fluids: it is increased by a small augmentation of blood osmotic pressure ($\sim 1\%$), thus providing a system of regulation. Vasopressin is also part of a multifactorial regulation of *a number of physiological functions*, such as the control of blood pressure, hepatic glycogenolysis, ACTH secretion, and aldosterone production.

Oxytocin provokes the *contraction* of the uterine *myometrium* and of myoepithelial cells of the *mammary gland*. Stimulation caused by uterine cervical dilatation or nipple suction triggers reflex pulsatile secretion, as observed in parturition and lactation.

In the *central nervous system*, vasopressin and oxytocin are probably involved in the control of *complex* behaviors. Injected centrally, vasopressin reinforces acquired behavior in the rat.

Membrane receptors for *vasopressin* have been characterized. The concentration of vasopressin in the blood (1 pM–100 pM) is much lower than the K_D of the receptor binding constant, and a maximal response is observed after occupancy of only a fraction of the receptors. In the *kidney*, vasopressin receptors *(V_2)* are located on tubule cells in the cortical and medullar segments of the collecting ducts, and in some cells of the ascending limb of Henle's loop. *Cyclic AMP* mediates the effect of the hormone on renal tubule water permeability. The mechanism involves the formation of protein aggregates in the apical membrane, which may function as micropores. In contrast, vasopressin receptors *(V_1)* implicated in *other effects* of the hormone (including in the nervous system) do not act via adenylate cyclase, but *increase calcium* in target cells, through the *phosphoinositide system*. Selective synthetic ligands, either agonist or antagonist, have been obtained for both classes of receptors. *Calcium* is also the second messenger mediating the effect of *oxytocin*.

Pathologically, a vasopressin-related defect is involved in diabetis insipidus as well as in the syndrome of inappropriate antidiuretic hormone secretion *(SIADH)*, also called *Schwartz–Bartter* syndrome. *Diabetes insipidus* may be due to a *secretory* defect (central origin) or to *renal insensitivity* (nephrogenic diabetis insipidus). The two forms may be differentiated on the basis of the measurement of vasopressin concentration and the evaluation of renal responsiveness to the administration vasopressin. In SIADH, there is *hyponatremia* associated with normal or increased natriuresis; when the concentration of sodium is lower than 120 mEq/1, signs of water intoxication are observed.

General References

Bargam, W. (1966) Neurosecretion. *Int. Rev. Cytol.* 19: 183–201.

Bartter, F. C. and Schwartz, W. B. (1967) The syndrome of inappropriate secretion of antidiuretic hormone. *Amer. J. Cardiol.* 42: 790–806.

Jard, S. (1983) Vasopressin isoreceptors in mammals: relation to cyclic AMP-dependent and cyclic AMP-independent transduction mechanisms. In: *Current Topics in Membrane and Transport* (Martin, R. and Kleizeller, A., eds), Academic Press, New York, vol. 18, pp. 257–285.

Manning, M. and Sawyer, W. H. (1983) Design of potent and selective in vivo antagonists of the neurohypophysial peptides. In: *The Neurohypophysis: Structure, Function and Control. Progress in Brain Research* (Cross, B. A. and Leng, G., eds), Elsevier Science Publishers, Amsterdam, vol. 60, pp. 367–382.

Morel, F. and Imbert-Teboul, M. (1982) Mécanismes de concentration de l'urine. In: *La fonction rénale: acquisitions et perspectives.* (Bonvallet, M., ed), INSERM-Flammarion Médecine, Paris.

O'Connor, W. J. (1962) *Renal function.* Edward Arnold, London.

Sawyer, W. H., Crzonka, Z., and Manning, M. (1981) Neurohypophysial peptides. Design of tissue specific agonists and antagonists. *Mol. Cell. Endocrinol.* 22: 117–134.

Tweedle, C. D. (1983) Ultrastructural manifestations of increased hormone release in neurohypophysis. In: *The Neurohypophysis: Structure, Function and Control. Progress in Brain Research* (Cross, B. A. and Leng, G., eds), Elsevier Science Publishers, Amsterdam, vol. 60, pp. 259–272.

Etienne-Emile Baulieu

Melatonin: A Neurohormonal Mystery

Sleep and Sex

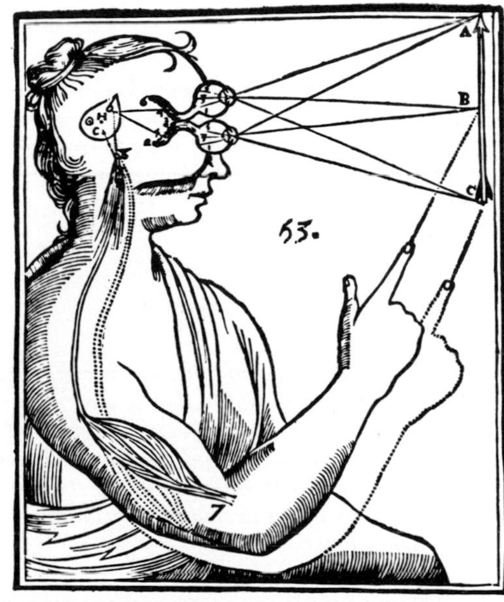

LXXVIII. Comment vne idée peut eſtre cõpoſée de pluſieurs; & d'où vient qu'alors il ne paroiſt qu'vn ſeul objet.

Et de plus, pour entendre icy par occaſion, comment, lors que les deux yeux de cette machine, & les organes de pluſieurs autres de ſes ſens ſont tournez vers vn meſme objet, il ne s'en forme pas pour cela pluſieurs idées dans ſon cerveau, mais vne ſeule, il faut penſer que c'eſt tousjours des meſmes points de cette ſuperficie de la glande H que ſortent les Eſprits, qui tendant vers divers tuyaux peuvent tourner divers membres vers les meſmes objets: Comme icy que c'eſt du ſeul point b que ſortent les Eſprits, qui tendant vers les tuyaux 4, 4, & 8, tournent en meſme temps les deux yeux & le bras droit vers l'objet B.

Figure 1 *An illustration published in René Descartes'* Traité de l'Homme *(1667). The pear-shaped cerebral structure labelled 'H' is the pineal gland, which Descartes thought to be the seat of the soul, involved in perception and in the initiation of effector commands.*

Melatonin is a neurohormonal intermediate involved in the timing of reproductive (and possibly other) events[1,2]. It is synthesized by the pineal gland, a small organ buried in the center of the brain. Descartes believed it was the seat of the soul! (Fig. 1). Later, it was more correctly compared to the third eye of low vertebrates, a photosensory epiphyseal organ. Heubner, a nineteenth-century French clinician, observed a relationship between precocious puberty and destruction of the pineal gland. Melatonin (Fig. 2), which has an anti-melanocyte stimulating hormone (MSH) effect in lower vertebrates, was discovered by the dermatologist Lerner.[3] It is abundant in the plasma of children between one and five years of age, decreases steadily until the end of puberty, and then remains low[4]. There is a daily rhythm in the production of melatonin in all mammalian species[5–7]. Secretion is stimulated by darkness and inhibited by light (Fig. 3). Changes in plasma melatonin level and in regulatory patterns have been observed in association with depressive disorders[8,9]. In mammals which breed seasonally, melatonin is a signal reflecting the changing environmental lighting, and is used to time reproductive cycles. The pineal gland thus senses seasons and regulates reproduction. Its activity may be related to a decrease in the amplitude and/or frequency of GnRH secretion. However, in strains of laboratory animals whose pineal gland does not make melatonin (contrary to the wild type), there are no obvious reproductive disorders.

The pathway from retina to pineal gland is indicated in Figure 4. Melatonin synthesis in the pineal gland is under α-adrenergic control[10]. The role of melatonin in humans remains a mystery. Administration in the evening does not provoke obvious endocrine changes, but it seems to make most individuals sleepy, a property which has been proposed for recuperation from jet lag.

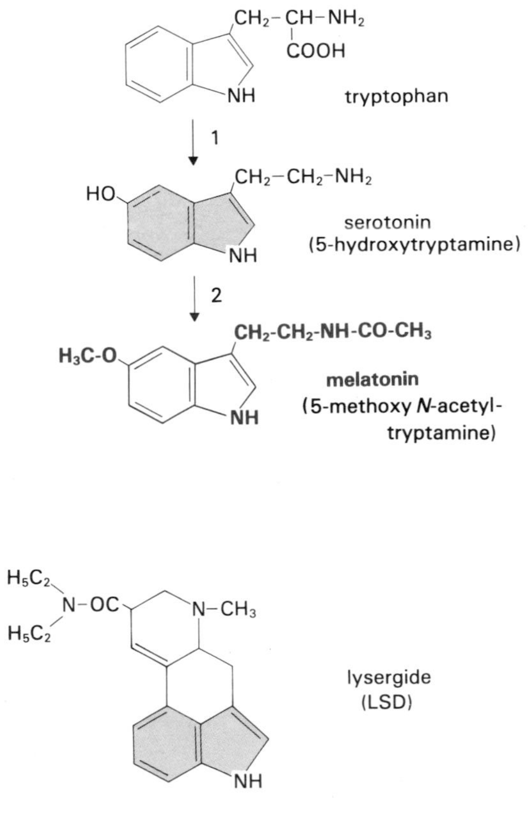

$CH_2-CH-NH_2$ / $COOH$ — NH — tryptophan

↓ 1

HO — $CH_2-CH_2-NH_2$ — NH — serotonin (5-hydroxytryptamine)

↓ 2

H_3C-O — **$CH_2-CH_2-NH-CO-CH_3$** — **NH** — **melatonin** **(5-methoxy *N*-acetyl-tryptamine)**

H_5C_2 / H_5C_2 > N-OC — N-CH_3 — NH — lysergide (LSD)

◁

Figure 2 *Biosynthesis of melatonin. Serotonin is also called enteramine or vasotonin. LSD (lysergide or lysergic acid diethylamide) is an antiserotonin. Reserpine, a tranquilizer, decreases serotonin in the brain.*

Figure 3 *Mean plasma melatonin concentration in four healthy subjects. On day 1, dusk was between 19.30 and 21.00 h, and dawn between 06.00 h and 07.30. Then, on day 2 and the days following, dusk was advanced to 16.00 h. Note the shift of the onset and the subsequent decrease of melatonin observed on the eighth day (red curve)[8].*

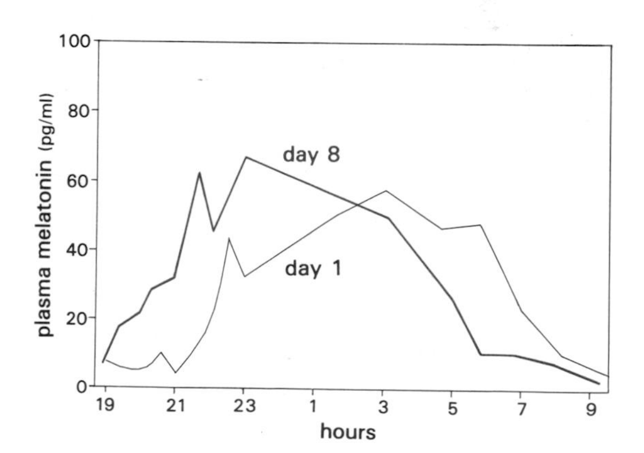

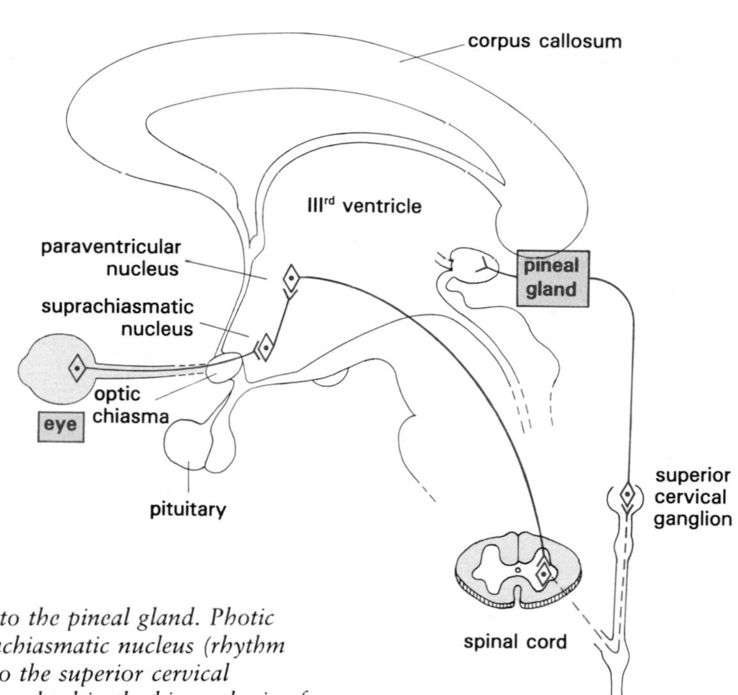

Figure 4 *The neural pathway from the eye to the pineal gland. Photic information is transferred, through the suprachiasmatic nucleus (rhythm generator) and the paraventricular nucleus, to the superior cervical ganglion. An adrenergic mechanism is then involved in the biosynthesis of melatonin. Melatonin may regulate, according to the season, the LH pulse generator (an ultradian oscillator with a frequency code) and its sensitivity to steroid negative feedback.*

References

1. Tamarkin, L., Baird, C.J. and Almeida, O.F.X. (1985) Melatonin a coordinating signal for mammalian reproduction. *Science* 227: 714–720.
2. Lincoln, G.A. (1984) The pineal gland. In: *Hormonal Control of Reproduction. Reproduction in mammals* (Austin, C.R. and Short, R.V., eds), Cambridge University Press, Cambridge, vol. 3, pp. 52–75.
3. Lerner, A.B., Case, J.D., Lee, T.H., Takahashi, Y. and Mori, W. (1958) Isolation of melatonin, the pineal factor that lightens melanocytes. *J. Am. Chem. Soc.* 80: 2587–2594.
4. Wurtman, R.J.. and Moskowitz, M.A. (1977) The pineal organ. *N. Engl. J. Med.* 296: 1329–33 and 1386–6.
5. Short, R.V., ed (1985) *Photoperiodism, melatonin and the pineal* Pitman Press, London.
6. Short, R.V. (1985) Photoperiodism, melotonin and the pineal: it's only a question of time. In: *Photoperiodism, Melatonin and the Pineal* (Short, R.V., ed), Pitman Press, London, pp. 1–8.
7. Cardinali, D.P. (1981) Melatonin. A mammalian pineal hormone. *End. Rev.* 2: 327–346.
8. Lewy, A.J., Sack, R.L., Miller, S. and Hoban, T.M. (1987) Antidepressant and circadian phase-shifting effects of light. *Science* 235: 352–354.
9. Beck-Friss, J., Ljunggren, J.G., Thoren, M., Von Rosen, D., Kjellman, B.F. and Wetterberg, L. (1985) Melatonin, cortisol and ACTH in patients with major depressive disorder and healthy humans with special reference to the outcome of the dexamethasone suppression test. *Psychoneuroendocrinology* 10: 173–186.
10. Axelrod, J. (1974) The pineal gland: a neurochemical transducer. *Science* 184: 1341–1348.

CHAPTER VII

The Adrenal Medulla

Jacques Hanoune

Contents, Chapter VII

The Adrenal Medulla

1. Introduction

1.1 Definitions

The term *catecholamine* designates *three* compounds which are derivatives of phenylethylamine, hydroxylated in positions 3 and 4 of the aromatic ring (Fig. VII.1). *Dopamine* is the biosynthetic precursor of epinephrine and norepinephrine; it acts essentially as a neurotransmitter in the central nervous system, particularly in the negrostriatal pathway. *Norepinephrine* is the neurotransmitter of sympathetic nerves. *Epinephrine* is the only catecholamine whose action is *essentially hormonal*; it is produced by the endocrine adrenal medulla, but is also found in the CNS.

1.2 Historical Background

Modern studies began in the second half of the nineteenth century. Henle first described the '*chromaffin reaction*', a red-brown precipitate of the granules of the medullary cells with potassium bichromate. The hypertensor effects of cellular extracts were shown in 1895, and soon thereafter, Abel and Crawford isolated epinephrine. In 1910, Langley observed that the effects of extracts of adrenal medulla and the secretory products of sympathetic nervous stimulation were identical. In 1933, Cannon and Rosenblueth attempted to explain the opposing physiological effects of epinephrine by the formation of 'sympathin E' or 'sympathin I', depending on whether the mediator was bound to a cellular component E (for excitation) or I (for inhibition), forming

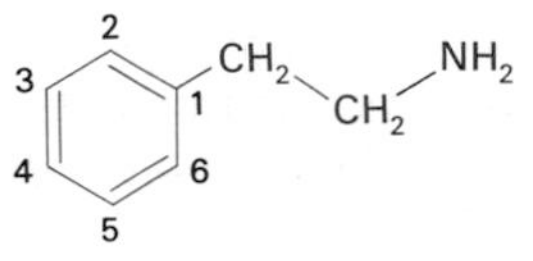

phenylethylamine

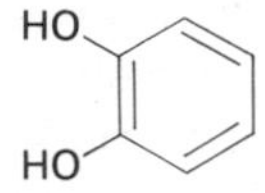

catechol

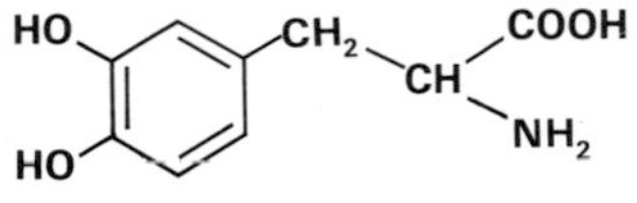

1,3,4-dihydroxyphenylalanine (DOPA)

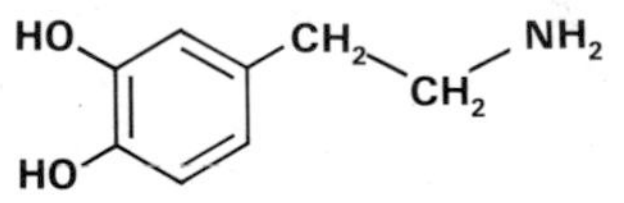

dopamine

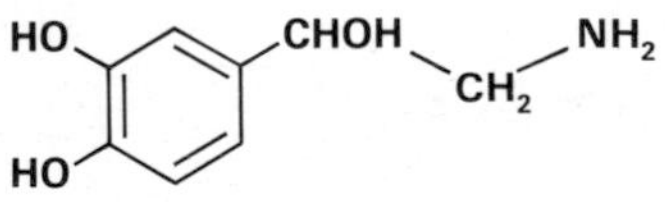

l-norepinephrine

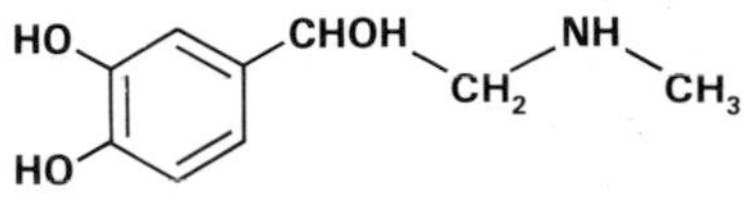

l-epinephrine

Figure VII.1 Structure of catecholamines. The basis for numbering carbons of phenylethylamine, a molecule structurally related to the natural hormones, is shown. Natural catecholamines are levorotary and are designated by the prefix l or (−). Their chemical configuration is of the d type (in comparison to mandelic acid) or of the R type (according to the rule of Cann, Ingold, and Prelog). In Europe, the terms adrenaline and noradrenaline are commonly used for epinephrine and norepinephrine.

a complex which was then liberated into the blood. It was only in 1948 that Ahlquist correctly interpreted the experimental data, in defining cellular *α and β receptors*. In 1946, von Euler identified norepinephrine as a sympathetic neurotransmitter, and its role as a neurotransmitter in the brain was demonstrated later. Blaschko and Axelrod defined the pathways of the synthesis and degradation of catecholamines.

1.3 Embryology

All neurons of the autonomic nervous system are derived from the *neural crest* (neuroectoderm). The superior portion of the neural crest is the origin of cholinergic neurons, and the median portion the origin of adrenergic neurons; the inferior portion is the origin of both cell types. *Nerve growth factor (NGF)* plays an important role in the overall development of the peripheral sympathetic nervous system.

Relatively early in embryonic development, primitive sympathetic cells, or *sympathogonia*, differentiate into two types of cells: *sympathoblasts*, which will become cells of the sympathetic ganglia, and *prechromoblasts*, from which the cells of the *adrenal medulla* are derived. Around the fiftieth day, the prechromoblasts invade the mesoblastic cells at the origin of the adrenal cortex. Between the second and fourth months, mesoblastic cells multiply and completely surround the medullary part. Primitive sympathetic cells predominate in the adrenal medulla at birth. They mature slowly into chromaffin cells and develop the capacity to secrete epinephrine and norepinephrine. During adult life, the principal product secreted is epinephrine. Outside the adrenal medulla, chromaffin cells of neuroectodermal origin can be found in the preaortic sympathetic plexus, forming, in particular, the organ of Zuckerkandl. Embryologically, the adrenal medulla appears analogous to postsynaptic cells of the peripheral nervous system.

1.4 Anatomy of the Adrenal Medulla

Situated above the kidneys in adults, the adrenal glands are triangular and weigh 5 to 7 g. They are

A

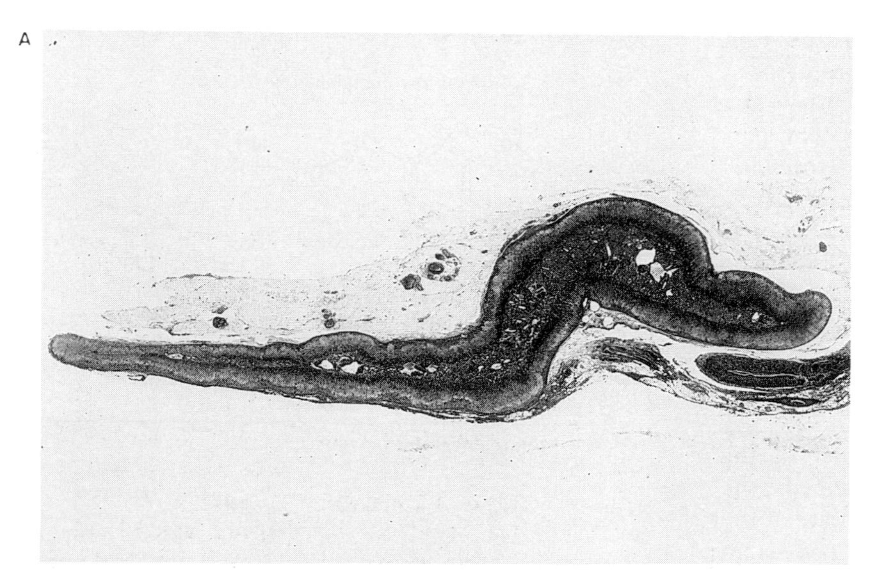

B

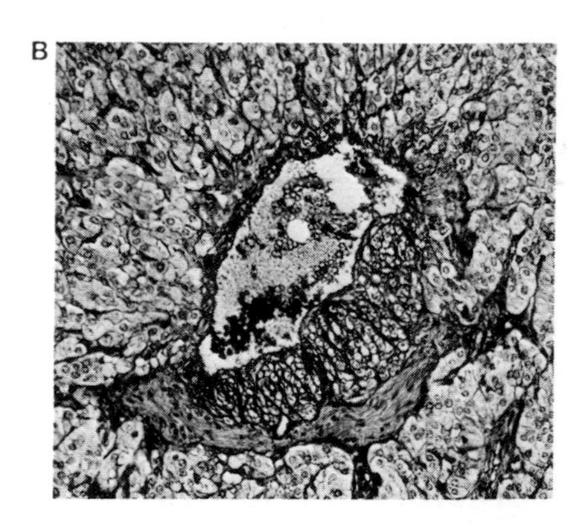

C

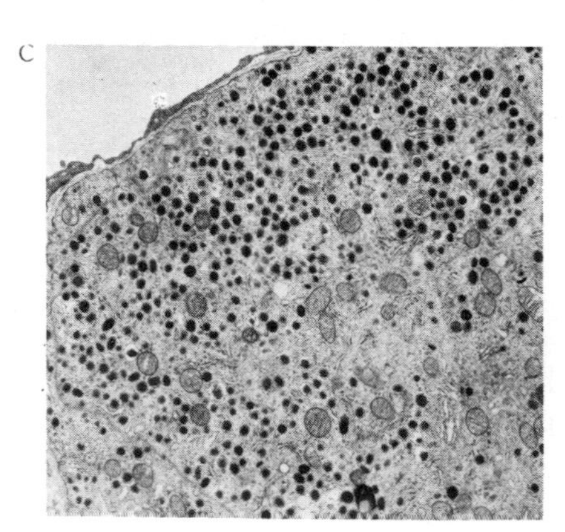

Figure VII.2 The adrenal gland. **A.** Section of a human gland. The cortex is of uniform thickness and surrounds the medulla, which is more variable in thickness. Enlarged two times. (Courtesy of Dr. J. A. Long, Department of Anatomy, University of California, San Francisco). **B.** Histological aspect of a normal adrenal medulla. It is made of polyhedral cells, and is highly vascularized. Bar = 100μm. (Courtesy of Dr. M. Forest, Hôpital Cochin, Paris). **C.** Electron micrograph of secretory granules. The cytosol of this hamster adrenal medulla cell is filled with dense secretory granules. Bar = 20 μm. (Courtesy of Dr. O. Grynszpan-Winograd, Laboratoire de Cytologie, Université Paris VI).

composed of a cortical zone, of mesoblastic origin, and a medullary zone, of ectodermal origin (Fig. VII.2). The adrenal medulla is made up of polyhedral cells characterized by *storage granules* which resemble storage vesicles in nerve terminals; these granules contain epinephrine or norepinephrine. The adrenal gland is vascularized by three arteries which come from the superior diaphragmatic artery, the aorta and the renal artery, reaching the medulla after having passed through the cortex. On the left, the adrenal vein merges with the renal vein, whereas, on the right, adrenal venous blood drains directly into the inferior vena cava. Finally, the adrenal medulla is innervated by preganglionic cholinergic neurons.

1.5 Comparison with Other Hormones

Catecholamines can be distinguished from other major hormones by a fundamental characteristic: they can be considered *either hormones or neurotransmitters*, according to their sites of action, release, or storage. This chapter will deal primarily with the endocrine aspects of catecholamines; that is, the action of epinephrine and, in part, that of norepinephrine. Certain recent developments concerning the central nervous system will also be reported.

2. Metabolism of Catecholamines

2.1 Biosynthetic Pathway

The biosynthesis of catecholamines *from tyrosine* is *simple* (Fig. VII.3). Four steps are involved: 1. hydroxylation of the phenolic ring; 2. decarboxylation of the lateral chain; 3. hydroxylation of the lateral chain; and 4. *N*-methylation. This pathway has been known for at least 30 years, and the different enzymes involved have been characterized and isolated. All of these enzymes are soluble and are present in the cytosol of the chromaffin cells, with the exception of dopamine β-hydroxylase, which is found only in granules. As most of these enzymes are relatively non-specific, other pathways of biosynthesis have been described; they are secondary, however, and will not be expounded here.

Tyrosine is the normal precursor of catecholamines. Its source is essentially in dietary intake, but it is also derived from the hydroxylation of

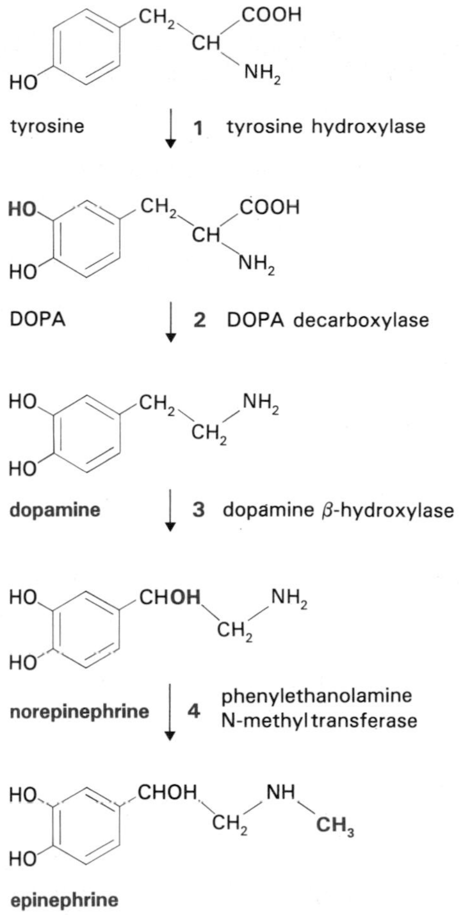

tyrosine ↓ 1 tyrosine hydroxylase
DOPA ↓ 2 DOPA decarboxylase
dopamine ↓ 3 dopamine β-hydroxylase
norepinephrine ↓ 4 phenylethanolamine N-methyltransferase
epinephrine

Figure VII.3 The biosynthetic pathway of catecholamines. The cofactors are: **1.** Fe^{2+}-reduced pteridin; **2.** pyridoxal phosphate; **3.** ascorbic acid; and **4.** *S*-adenosylmethionine.

phenylalanine. Tyrosine is transformed *into DOPA* (3, 4-dihydroxyphenylalanine) by the action of *tyrosine hydroxylase*; this enzyme uses as a cofactor a reduced pteridin, tetrahydrobiopterin (Fig. VII.4). Its activity is dependent on the presence of molecular oxygen and ferrous iron (Fe^{2+}), and involves an oxidoreduction of biopterin[1]. It is interesting to note that the concentration of biopterin was found to be greatly reduced in the caudate nucleus removed at autopsy from subjects having had Parkinson's disease. The transformation of tyrosine into DOPA is the essential *limiting step* in the biosynthesis of catecholamines, since the activity of tyrosine hydroxy-

1. Negatsu, T. (1983) Biopterin cofactor and monoamine-synthesizing mono-oxygenases. *Neurochem. Int.* 5: 27–38.

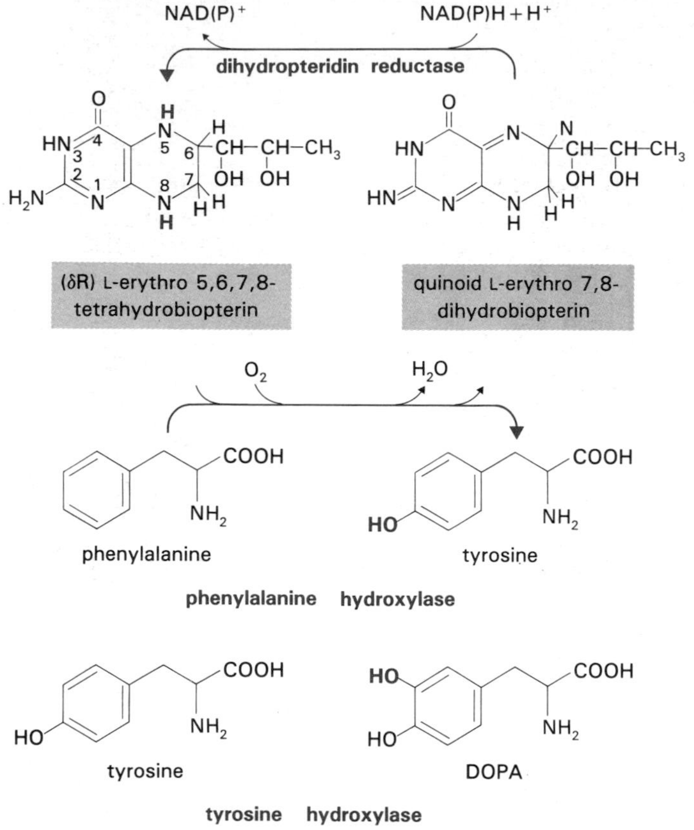

$NAD(P)^+$ $NAD(P)H + H^+$

dihydropteridin reductase

(δR) L-erythro 5,6,7,8-tetrahydrobiopterin

quinoid L-erythro 7,8-dihydrobiopterin

O_2 H_2O

phenylalanine

tyrosine

phenylalanine hydroxylase

tyrosine

DOPA

tyrosine hydroxylase

Figure VII.4 Synthesis of tyrosine and DOPA. These two compounds are derived, respectively, from phenylalanine and tyrosine, by hydroxylation, using the same biopterinic cofactor. Under the effect of the enzymes tyrosine or phenylalanine hydroxylases, 5, 6, 7, 8-tetrahydrobiopterin is transformed into 7, 8-dihydrobiopterin quinonoid. This product is then reduced back into tetrahydrobiopterin, in the presence of NADPH, by the action of a dihydropterin reductase, whose activity is sufficiently elevated as not to be rate-limiting. In contrast, the pathways of the biosynthesis of biopterin from GTP, and particularly dihydrofolate reductase which catalyzes the transformation of l-erythro-7,8-dihydrobiopterin into tetrahydrobiopterin (in the presence of NADPH), are most probably limiting; the biopterin cofactor, therefore, constitutes an important regulatory element in the synthesis of DOPA.

lase in the adrenal medulla is 200 times less than DOPA-decarboxylase and dopamine β-hydroxylase.

Because tyrosine hydroxylation is the key step in the synthesis of DOPA, it is not surprising that it is under *specific regulation.* Two forms of the enzyme exist, one of which is not phosphorylated and is only slightly active, and the other, phosphorylated and very active; the *phosphorylation* step is under the control of a cAMP-dependent protein kinase.

The activity of tyrosine hydroxylase is inhibited by a large variety of compounds, particularly derivatives of tyrosine, such as α methyltyrosine, which is efficient in the treatment of pheochromocytoma, and 3-iodotyrosine and its metabolites. Other inhibitors, such as dopamine, norepinephrine and epinephrine, exert a negative feedback on the activity of this enzyme, the first in the biosynthetic pathway. Tyrosine hydroxylase has been cloned[2].

The second enzyme, which *transforms* DOPA *into dopamine,* is *dopamine decarboxylase,* also called l-amino acid decarboxylase because of its lack of specificity for DOPA. It is widely distributed throughout the body, and requires pyridoxal phosphate in order to act. This enzyme has a high affinity for its substrate, with an elevated maximal rate of activity. It is therefore not limiting in the synthetic pathway of catecholamines. Dopamine decarboxylase is competitively inhibited by several analogs, one of which is α-methylDOPA. The use of other *inhibitors* of DOPA decarboxylase which do not penetrate the central nervous system is of importance in the treatment of *Parkinson's disease.* When administered simultaneously with l-DOPA, *carbidopa* or *benserazide* reduces the peripheral degradation of l-DOPA. Therefore, more l-DOPA is available to enter the brain and replenish the abnormally low levels of dopamine which characterize this disease. It is thus possible to reduce by 75% the amount of l-DOPA required, and thereby improve tolerance to the drug.

Hydroxylation of dopamine *to norepinephrine* is catalyzed by *dopamine β-hydroxylase*, in the presence of Ca^{2+}, ascorbic acid, and molecular oxygen; it is a mixed-function oxidase with little specificity for dopamine, and is present in secretory granules, associated with chromogranin A.

Finally, norepinephrine is methylated from S-adenosylmethionine by *phenylethanolamine* N-*methyltransferase*, a relatively non-specific enzyme. It is noteworthy that the three biosynthetic enzymes tyrosine hydroxylase, dopamine β-hydroxylase and phenylethanolamine *N*-methyltransferase might share similar protein domains in their primary structures and have common gene coding sequences[3].

2.2 Regulation of Catecholamine Biosynthesis

Hormonal Regulation: A Regional System

In 1965, Wurtman[4] and Axelrod demonstrated that the suppression of endogenous secretion of glucocorticosteroids by hypophysectomy, or by administration of dexamethasone, reduces the concentration of phenylethanolamine *N*-methyltransferase. This effect is reversed by the administration of ACTH or large doses of glucocorticosteroids. Physiologically, the *transfer of cortisol from the neighboring adrenal cortex* to the adrenal medulla by the *local portal system* provides the elevated concentration of cortisol necessary to sustain a normal concentration of enzyme. Tyrosine hydroxylase and dopamine β-hydroxylase are also increased by glucocorticosteroids, and are regulated by cAMP.

2. Lamouroux, A., Faucon Biguet, N., Samolyk, D., Privat, A., Salomon, J. C., Pujol, J. F. and Mallet, J. (1982) Identification of cDNA clones coding for rat tyrosine hydroxylase antigen. *Proc. Natl. Acad. Sci., USA* 79: 3881–3885.
3. Joh, T. H., Baetge, E. E., Ross, M. E., Lai, C. Y., Docherty, M., Bradford, H. and Reis, D. J. (1985) Genes for neurotransmitter synthesis, storage and uptake. *Fed. Proc.* 44: 2773–2779.
4. Wurtman, R. J., Pohorecky, L. A. and Baliga, B. S. (1972) Adrenocortical control of the biosynthesis of epinephrine and proteins in the adrenal medulla. *Pharmacol. Rev.* 24: 411–426.

Nervous Regulation

The nervous system is *not required for* the *biosynthesis* of epinephrine. However, it is an *essential* component of the *regulation* of enzyme biosynthesis. It has long been known that protracted stimulation of splanchnic nerves leads to an increase in the synthesis and secretion of catecholamines, which is coupled to an increase in the synthesis of tyrosine hydroxylase, dopamine β-hydroxylase, and phenylethanolamine *N*-methyltransferase. In this and other conditions, there is an increase in preganglionic cholinergic nervous activity.

Local Regulation

The activity of certain enzymes in the biosynthetic pathway is inhibited by their own metabolic by-products. This type of *negative feedback* regulation is difficult to demonstrate in vivo. This is the case for the inhibition of tyrosine hydroxylase by norepinephrine. Acute nervous stimulation of the adrenal medulla results in an increased secretion and a reduced intracellular concentration of catecholamines, thus removing the inhibition of tyrosine hydroxylase activity and leading to increased hormone synthesis.

2.3 Storage and Secretion of Catecholamines

In the Adrenal Medulla

Bilateral adrenalectomy leads to an almost complete disappearance of epinephrine production, whereas that of norepinephrine is practically unchanged. Almost *all circulating epinephrine* comes from the adrenal medulla, where the concentration of 1 mg/g of tissue ensures normal secretion.

Storage

Catecholamines are *stored* in cells of the adrenal medulla, in the form of *granules* surrounded by a thick membrane (3 to 10 nm). Their diameter varies between 60 and 200 nm. Hypo-osmotic lysis releases soluble components of the granular matrix. Intragranular composition is shown in Table VII.1. The components of low molecular weight are essentially ions, ATP and catecholamines, epinephrine or norepinephrine, according to the cell. The principal protein component is *chromogranin A*, an acidic protein with a M_r of ~77,000. Other structurally related pro-

Table VII.1 Composition of chromaffin granule matrix.

Catecholamines	600 mM
ATP	130 mM
Total nucleotides	180 mM
Ca^{2+}	18 mM
Mg^{2+}	5 mM
Ascorbate	24 mM
Proteins	200 mg/ml
pH	5.5

teins are also found, as well as dopamine β-hydroxylase and the precursors of Met- and Leu-enkephalin[5,6]. It is worth noting that chromogranin-like proteins are also present in a variety of endocrine cells that possess secretion granules, in the pituitary, the thyroid, or the parathyroid glands[7]. The molar ratio of catecholamine/ATP in the granules of the adrenal medulla is constant, at 4:1, and it is likely that the catecholamines are stored in granules, in complexes with ATP and magnesium. Chromogranins have been implicated in the formation of these complexes, but their exact role is unknown. The elevated concentration of catecholamines in granules is due to a mechanism of active transport, dependent on ATP and Mg^{2+}, and inhibited by reserpine. In fact, the concentration of catecholamines in cytosol is 20 μM, which establishes a concentration gradient of at least 10,000 between the granules (600 mM) and the cytosol. The entry of catecholamines into the granule is due to a transport system, indirectly coupled to the hydrolysis of ATP, which recognizes not only catecholamines, but also phenylethylamine and 5-hydroxytryptamine; it is specific for the natural, l-forms of the molecules. The affinity for substrates corresponds to a K_D of 20 μM, and binding is inhibited in a competitive manner by reserpine (K_D=0.2 μM). The amine transporter has been purified. It corresponds to a peptide with a M_r of ~45,000[8].

5. Lewis, J. W., Tordoff, M. G., Sherman, J. E. and Leibeskind, J. C. (1982) Adrenal medullary enkephalin-like peptides may mediate opioid stress analgesia. *Science* 217: 557–559.
6. Viveros, O. H. and Wilson, S. P. (1983) The adrenal chromaffin cell as a model to study the co-secretion of enkephalins and catecholamines. *J. Auton. Nerv. Syst.* 7: 41–58.
7. Landsberg, L. (1984) Chromogranin A. *N. Engl. J. Med.* 311: 794–795.
8. Gabizon, R. and Schuldiner, S. (1985) The amine transporter from bovine chromaffin granules. Partial purification. *J. Biol. Chem.* 260: 3001–3005.

Secretion

The secretion of catecholamines into the blood occurs by a process of exocytosis. The membrane of the granule becomes contiguous with the plasma membrane, and all of its content flows out into the blood, including catecholamines, dopamine β-hydroxylase, chromogranins, and ATP. The storage vesicle is not reutilized, but rather is probably degraded. The synthesis of new storage vesicles is necessary for an efficient synthesis of catecholamines. This process of exocytosis is started primarily by the release, from the adrenal medulla, of a preganglionic mediator, acetylcholine. Its action is due to an increase in Na^+ and Ca^{2+} in the chromaffin cell, and involves the microtubular system. The *secretion* of epinephrine is increased by *numerous* physiological *stimuli*: hypoglycemia, physical exercise, hypoxia, etc. These responses are blocked by the denervation of the adrenal medulla, and therefore involve the splanchnic nerves.

In Nerve Terminals

In 1956, von Euler and Hillarp demonstrated that *norepinephrine* was also *stored in granules* in *nerve terminals*. The catecholamine concentration of these granules is less than that of the granules of the adrenal medulla; the norepinephrine/ATP ratio is also different (12 to 18:1). The release of norepinephrine by the terminals of sympathetic neurons occurs in response to an action potential, along with an influx of calcium. *Numerous* substances *modulate* this *release* of norepinephrine. Cyclic nucleotides, β-adrenergic and nicotinergic agonists, as well as angiotensin, increase its release; α_2-adrenergic and muscarinic agonists, prostaglandins of the E series, opiates and enkephalins, dopamine and adenosine inhibit its release. The physiological and pharmacological importance of presynaptic inhibitors of α_2-adrenergic receptors will be discussed later. Sympathomimetic amines, such as tyramine and amphetamine, provoke the release norepinephrine from storage granules, leading indirectly to an increase in arterial pressure.

2.4 Catabolism of Catecholamines

The catabolism of catecholamines is *very rapid*: the half-life of exogenous epinephrine administered to animals is 10 to 20 seconds. In humans, the metabolic rate of plasma clearance of epinephrine at equilibrium

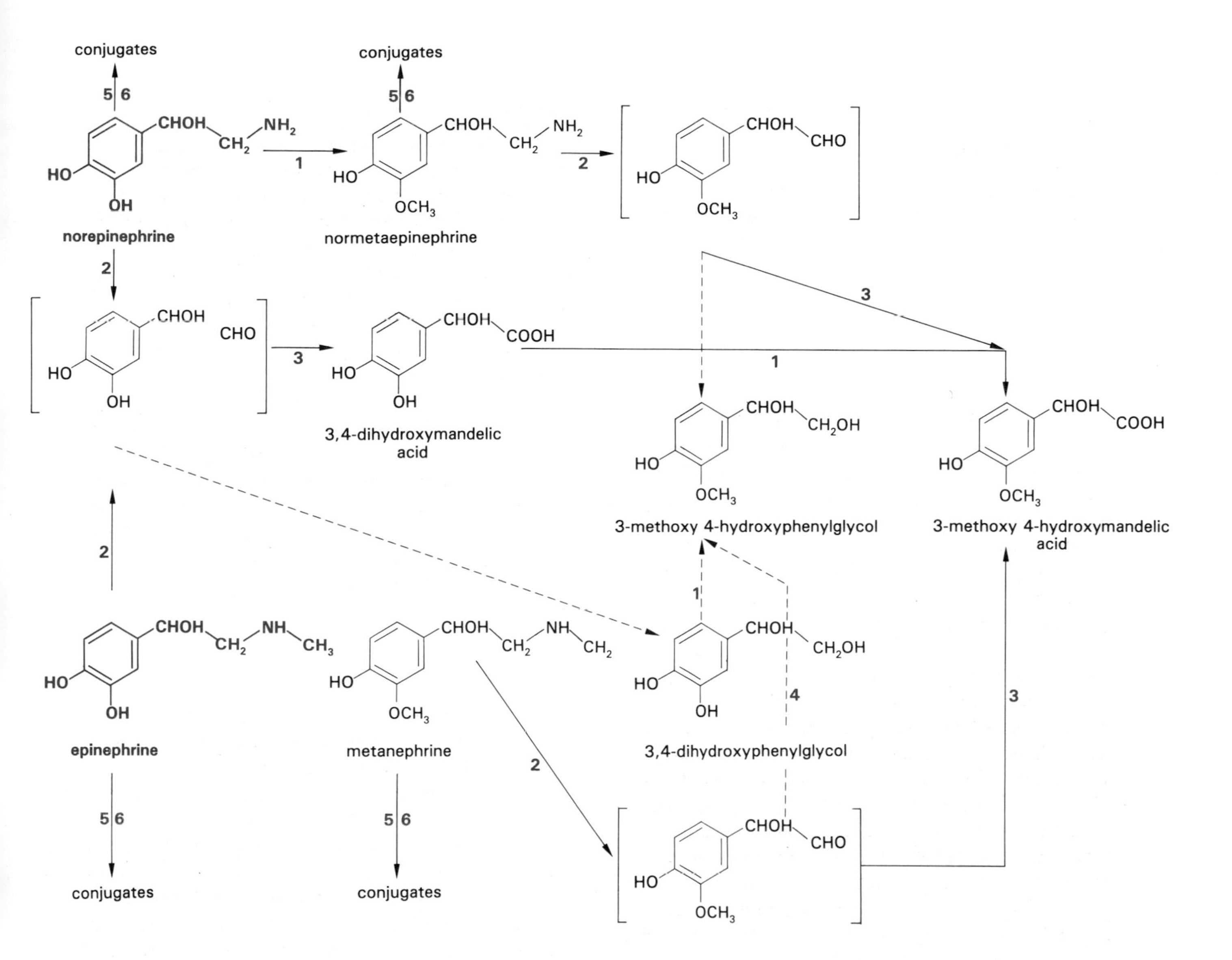

Figure VII.5 Degradation pathways of catecholamines **1.** catechol *O*-methyltransferase (COMT); **2.** monoamine oxidase (MAO); **3.** aldehyde dehydrogenase (Ad); **4.** aldehyde reductase (Ar); **5.** β-glucuronotransferase; **6.** sulfotransferase.

is 52 ml/kg/min (for a plasma concentration of epinephrine of 24 to 74 pg/ml). For norepinephrine, the clearance rate is 25 ml/kg/min (for a plasma concentration of 230 to 345 pg/ml). These values increase markedly when the plasma concentration of catecholamines increases.

General Pathways of Degradation

Degradation essentially involves *two enzymes: catechol O-methyltransferase*, which catalyzes the methylation of the hydroxyl in position 3, and *monoamine oxidase*, which catalyzes the oxidative deamination of the aliphatic chain. Both act on numerous substrates, and each is also active on the products of the other. In addition, their isolated or sequential action results in numerous urinary metabolites, because two other enzymes, an aldehyde reductase and an aldehyde dehydrogenase, transform unstable aldehydes produced under the action of monoamine oxidase into alcohol and carboxylic acid, respectively. Finally, these different products of catabolism can be excreted either *free* or *conjugated* to sulfates or glucuronides (Fig. VII.5). The major portion of catecholamines is excreted as deaminated metabolites. Only a small part is excreted unchanged or as 3-methoxylated derivatives (metepinephrine). The excretion of intact catecholamines or of metepinephrine is indicative of the activity of the adrenal medulla rather than of the excretion of deaminated metabolites, which essentially reflect norepinephrine metabolized in nervous tissue. In the human, vanylmandelic acid, an oxidized product, forms 40% of the total of urinary catecholamines, while in other species, 3-methoxy, 4-hydroxyphenylglycol, a reduced metabolite, is predominant.

Catechol *O*-methyltransferase

Catechol O-methyltransferase catalyzes the transfer of the methyl group of *S*-adenosylmethionine, in the presence of a divalent cation, to the hydroxyl in position 3 of epinephrine and norepinephrine, transforming them into metepinephrine and normetepinephrine. This enzyme is especially abundant in the *liver* and *kidney*, where the major portion of circulating catecholamines is degraded.

Monoamine Oxidase

Monoamine oxidase is an enzyme which deaminates catecholamines to aldehydes, and is distributed to *all tissues*, where it is found primarily in the external membrane of *mitochondria*. Its essential role seems to be to control the storage of norepinephrine in nerve endings. Pharmacologically, *two types* of monoamine oxidase, A and B, have been characterized. *Type A*, which is present in neurons, specifically deaminates serotonin, epinephrine, and norepinephrine; it is inhibited by clorgyline and harmaline. *Type B*, found in platelets, preferentially deaminates benzylamine and 2-phenylethylamine, and is inhibited by deprenyl. Increased levels of these forms have been found in association with *depression*.

2.5 Cellular Uptake

Catecholamines are taken up by the cells in which they are stored or degraded. This phenomenon plays *a major role* in the inactivation of adrenergic stimulation. There are two different types of uptake, depending on the tissue.

Type I uptake[9]*: in nerve terminals.* Type I uptake, in the adrenergic neurons of the central nervous system and of the peripheral autonomic nervous system, involves an *active* transport system. It is saturable, *stereospecific*, and has a greater affinity for *norepinephrine* than for epinephrine. It cannot transport isoproterenol. Many drugs (imipramine, amitryptiline, cocaine) potentiate nerve stimulation by inhibiting this process.

Type II uptake: in extraneural tissues. Type II uptake occurs in various extraneural peripheral tissues (smooth muscle, heart, certain glandular tissues). It is *not stereospecific*, and has a stronger affinity for *isoproterenol* and *epinephrine* than for norepinephrine. Its affinity for its substrates is much lower than that of type I uptake. In contrast, its maximal capacity is much greater. Physiologically, this uptake leads to a rapid intracellular degradation of catecholamines by catechol O-methyltransferase, in contrast with the "uptake-storage" of type I. Type II is inhibited by metepinephrine and normetepinephrine, estradiol and cortisol, and may be involved in the peripheral regulation of catecholamine action.

9. Iversen, L. L. (1973) Catecholamine uptake processes. *Brit. Med. Bull.* 29: 130–135.

3. Pharmacology of Catecholamines

Based on receptor concept

The receptor concept has played a major role in all recent physiological and pharmacological studies of catecholamines. It has allowed a classification based on simple biochemical properties, as well as the elucidation of the mechanism of action of catecholamines at the molecular level. It is now possible, therefore, to describe the action of catecholamines in terms of physiological and pharmacological effects and receptor molecule binding characteristics.

3.1 Ahlquist's Classification

Now classical

The physiological effects of catecholamines are *numerous* and varied, and are *often* of an *opposing nature*, depending on the tissue. Considerable progress in the understanding of the physiology of catecholamines was made relatively recently, in 1948, when Ahlquist proposed the now classical classification of *α- and β-adrenergic receptors*. This classification was based on the *comparison* between the relative *efficiency* of *five sympathomimetic amines* (norepinephrine, epinephrine, isoproterenol, and the methylated derivatives of epinephrine and norepinephrine) in different animals and on multiple physiological functions. Ahlquist found that the majority of physiological responses could be divided into two categories: for some, epinephrine was the most efficient compound, and isoprenaline the least efficient (type α), while for others the order was inversed (type β). Shortly thereafter, it was shown that the blocking drugs which existed at that time were capable of abolishing only the actions mediated by α receptors. Later, the discovery of substances which were specific β-blockers reinforced Ahlquist's conclusions. The use of these *specific agonists and antagonists* has allowed the classification of the actions of epinephrine according to the nature of the receptors involved. The molecular agonists and antagonists currently utilized physiologically or pharmacologically to characterize the type of response have thus permitted the establishment of four *categories* of receptors (Table VII.2).

α agonists: Norepinephrine and phenylephrine are the agonists most commonly used. Various imidazolines, such as oxymethazoline and naphazoline, also have α-type effects.

α antagonists: The α antagonists are classified according to their length of action. The effects of phentolamine are rapidly reversible. In contrast, phenoxybenzamine forms covalent bonds with the receptor and results in irreversible effects in vitro, and in long action in vivo. In contrast to β antagonists, α antagonists are not structurally similar to catecholamines.

β agonists: The β agonist most commonly utilized is isoproterenol.

β antagonists: The first β antagonist to be used was dichloroisoproterenol. It is also, however, a partial agonist. Later, pronethanol and propranolol were discovered, which are two very efficient β-blockers without any agonistic activity. Presently, numerous β antagonists are available, some of which are more specific than others for certain types of receptors.

3.2 Structure–Activity Relationships and Pharmacological Applications

Sympathomimetic amines are derived from a *simple* structure: a diorthohydroxylated phenyl moiety and an aminated side chain. This general structure is susceptible to *numerous structural modifications*, resulting in various analogs, from nearly pure α agonists, such as phenylephrine, to pure β agonists, such as isoproterenol.

Distance from the phenyl ring and the amino-group: The activity is maximal when two carbon atoms separate the phenyl ring from the nitrogen atom.

Substitution of the amino-group: All substitutions of the amine by an apolar alkyl group increase β, and especially β_2, activity. This effect is maximal for isoproterenol, in which the substitution is an isopropyl, or for protokylol, in which the substitution is even larger (Fig. VII.6). In contrast, α activity is reduced by all apolar groups larger than methyl. Drugs such as protokylol even act as α antagonists, although they are excellent β agonists. This probably explains the vasodilator effects of a series of drugs such as isoxuprine and nylidrine.

Substitution of the phenyl ring: α and β activity is maximal when the ring is substituted by two hydroxyls in positions 3 and 4. If one is missing, the total activity, and particularly that of β, is reduced. This is the case for phenylephrine (Fig. VII.6), in which the hydroxyl group in position 4 is missing. It has lost almost all its β activity and has become a pure, although weak, α agonist. The substitution of these hydroxyl groups also leads to a reduction in activity: metepinephrine, methoxylated in position 3, has lost all activity; salbutamol (Fig. VII.6) and mesuprine are weak agonists. Substitution of the phenyl ring, as for

Table VII.2 Adrenergic receptors.

	α receptors	β receptors
Physiological response	Vasoconstriction Uterine contraction Pupillary dilatation	Cardiac stimulation Lipolysis Bronchodilatation Vasodilatation
Order of efficiency of ligands	EPI≥NOR>ISO	ISO>EPI≥NOR
Selective agonists	Phenylephrine	Isoproterenol
Selective antagonists‡	Phenoxybenzamine Phentolamine	Propranolol

‡Labetalol is a mixed α and β antagonist. EPI: epinephrine, NOR: norepinephrine, ISO: isoproterenol.

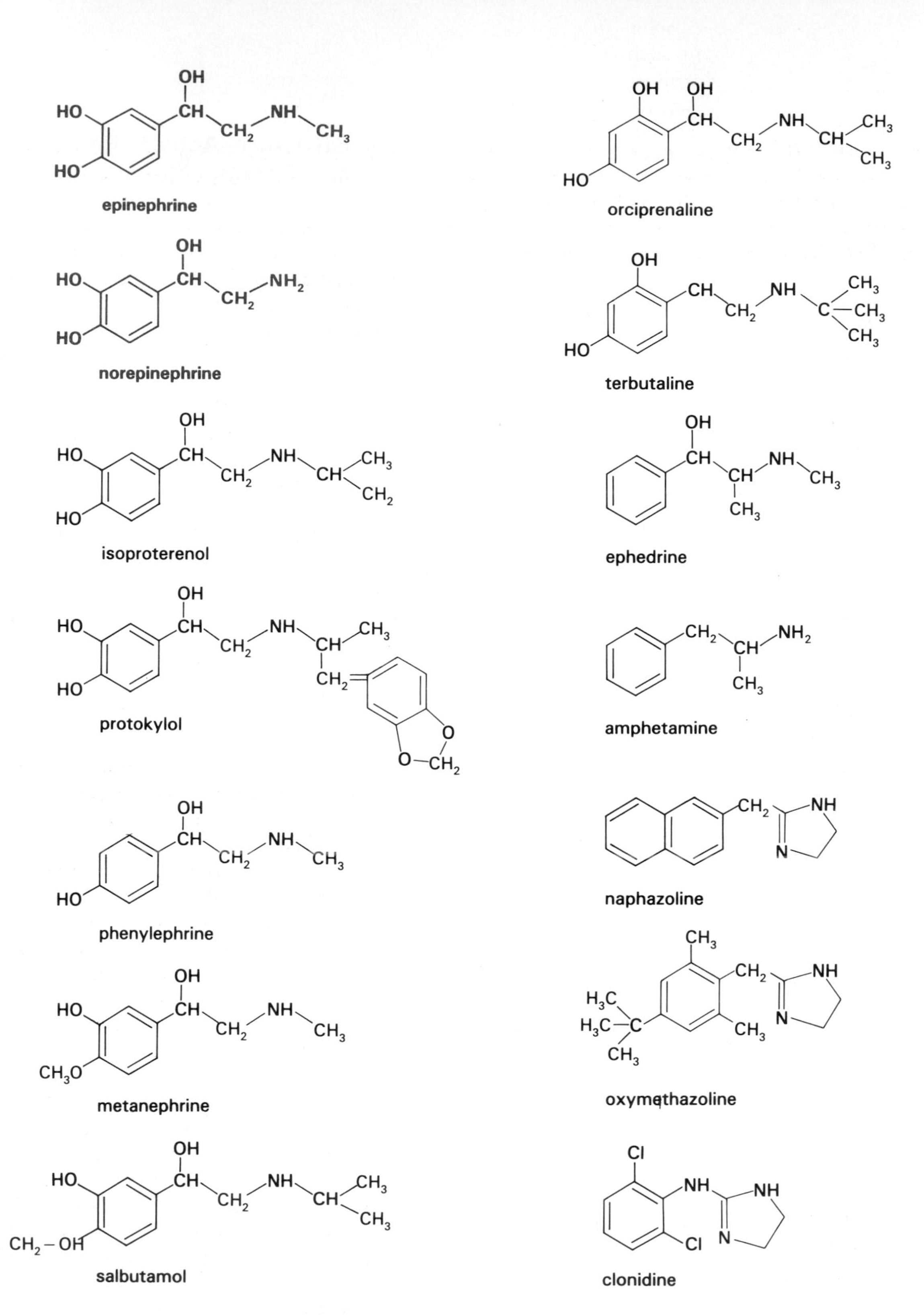

Figure VII.6 Structure of catecholamine agonists. As the substitution of the nitrogen of the amine increases, the affinity of the compound for β receptors increases; in contrast, its affinity for α receptors diminishes, and it may even become an antagonist. Depicted on the right hand side of the figure are structures of a few compounds which are not catecholamines, yet which possess α-adrenergic activity.

example in 3, 4-dichloroisoproterenol or *propranolol*, results in β antagonists (Fig. VII.7). Catecholamines, which are methoxylated by catechol O-methyltransferase, normally have a very short half-life. The absence of hydroxyl groups or the distance of one from the other, as for example in terbutaline or orciprenaline (Fig. VII.6) where they are found in positions 3 and 5, slows down the degradation of these products and increases their length of action in vivo. Derivatives which have neither of the two hydroxyls (Fig. VII.6) have a stimulatory effect on the central nervous system (ephedrine, amphetamine). Compounds which have a clear agonistic action on β-adrenergic receptors and which are not catecholamines are rare. Quinterenol, which has a β hydroxyquinoleine instead of a catechol group, and clenbuterol, in which this group is replaced by a 4-amino 3,5-dichlorobenzyl moiety, may be pharmacologically interesting. Certain α_2 agonists, such as imidazoline derivatives (oxymetazoline, naphazoline, tolazoline), also lack a catechol group.

Substitution of the α carbon of the side chain: This substitution impairs the activity of monoamine oxidase, thus increasing the pharmacological potency of non-catechol derivatives which are not degraded by catechol O-methyltransferase.

Substitution of the β carbon of the side chain: The presence of a hydroxyl group at the β position is indispensable for peripheral activity at both α- and β-adrenergic receptor sites.

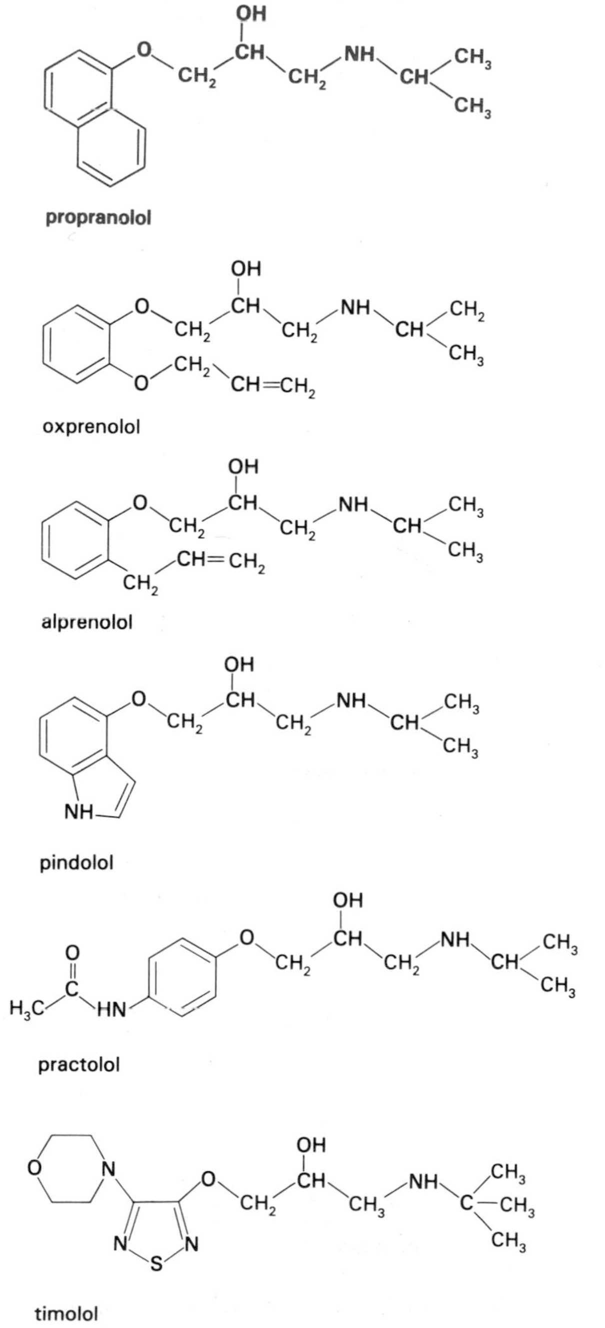

propranolol

oxprenolol

alprenolol

pindolol

practolol

timolol

Figure VII.7 Structure of some derivatives with β-blocking action. As can be seen by comparing their structures to that of epinephrine, they all differ by the absence of the catechol nucleus and by an extension of the lateral chain through the insertion of a OCH_2 group.

3.3 Subclassification of β-adrenergic Receptors

Lands and collaborators demonstrated the possible existence of *two subtypes of β receptors*[10]. They compared the activity of 15 sympathomimetic amines on bronchodilatation, vasodilatation, cardiac stimulation, and lipolysis. The relative efficiency of the different substances tested suggests that their effects on the heart and on lipolysis involve a receptor (β_1) different from the one involved in bronchodilatation and vasodilatation (β_2). However, there is apparently no difference in the mechanism of amplification of the hormonal signal by the two types of receptors. *Both of them activate adenylate cyclase.* Series of *agonists* and *antagonists* with a selectivity for β_1 or β_2 activities have been synthesized (Table VII.3). Practolol (Fig. VII.7) is a β_1 inhibitor, and butoxamine a β_2 inhibitor. Isoetharine, salbutamol, terbutaline and protokylol have β_2 agonist effects (Table VII.3). The therapeutic implications of this new classification are important. Salbutamol, in contrast to isoproterenol, causes bronchodilatation at low doses, and has very little effect on cardiac rhythm. Conversely, β_1-adrenergic antagonists can aggravate an asthmatic crisis. However, in several tissues the situation is more complex, since both receptor types are present in variable proportions. Eighty percent of the β receptors in

10. Lands, A. M., Arnold, A., McAuliff, J. P., Luduena, F. P. and Brown, T. G. Jr. (1967) Differentiation of receptor systems activated by sympathomimetic amines. *Nature* 214: 597–598.

Table VII.3 β_1- and β_2-adrenergic receptors.

	β_1 receptors	β_2 receptors
Site	Postsynaptic	Postsynaptic
Physiological response	Cardiac stimulation	Bronchodilatation
	Lipolysis	Glycogenolysis (muscle)
Order of efficiency of ligands	ISO>NOR≥EPI	ISO>EPI>NOR
Specific agonist	‡	Salbutamol
Specific antagonists	Practolol	Butoxamine

‡No pure β_1 agonist has been found.

the heart are of the β_1 type. The *lung* contains *25%* of β_1 and *75%* of β_2 receptors, and the *hypothalamus 30%* β_1 and *70%* β_2. More definitive studies in cell culture have shown that, even within the same cell, both types of receptors can exist. It has been proposed, but not demonstrated, that β_1 receptors are found mainly at sites where cells are innervated, for mediation of the responses of the neuromediator norepinephrine, whereas β_2 receptors would serve essentially as the binding sites of circulating epinephrine.

3.4 Subclassification of α-adrenergic Receptors

Anatomical Classification: Pre- and Postsynaptic α Receptors

Twenty years ago, it was demonstrated that α-blocking drugs increase, and α agonists reduce, the overflow of norepinephrine released from nerve terminals. This effect is coupled with the presence of presynaptic receptors specific for norepinephrine. Stimulation of these receptors leads to a reduction of the secretion of the neuromediator, thus permitting the regulation of the local concentration of norepinephrine in the synaptic cleft, by negative feedback (Fig. VII.8). In 1974, Langer[11] proposed the terms 'α_1' for postsynaptic receptors and 'α_2' for presynaptic receptors.

Pharmacological Classification

More recently, it has been shown by more detailed pharmacological studies that α_1- and α_2-adrenergic receptors were, in fact, general subtypes that could be found in numerous neuronal or extraneuronal tissues. At the present time, they should be defined not by their anatomical location, but only by the spectrum of their affinities for a series of drugs, especially adrenergic antagonists (Table VII.4).

11. Langer, S. Z. (1974) Presynaptic regulation of catecholamine release. *Biochem. Pharmacol.* 23: 1793–1800.

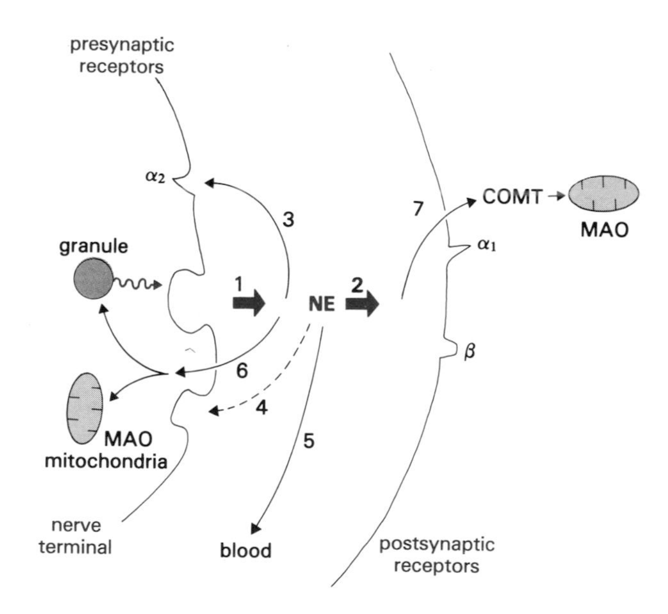

Figure VII.8 The fate of norepinephrine (NE) in the synaptic cleft. In the synaptic cleft, norepinephrine (NE) is released by exocytosis of granules (**1**). It then acts (**2**) on postsynaptic receptors of type β or α_1. It can also control its own release by the intermediary of presynaptic inhibitory α_2 (3), or activating β (4), receptors. Norepinephrine produced in excess (overflow) goes into the plasma (**5**), from which it can be taken up by peripheral cells. Intraneuronal reuptake (type I) (**6**) allows the storage of the neurotransmitter in vesicles or its metabolism by monoamine oxidase (MAO). Extraneuronal uptake (type II) (7) results essentially in its degradation by catechol O-methyltransferase (COMT).

Table VII.4 α_1- and α_2-adrenergic receptors.

	α_1 receptors	α_2 receptors
Site	Postsynaptic	Presynaptic (not exclusively)
Physiological effects	Vasoconstriction Hypertension Increase in blood glucose	Decrease in the release of norepinephrine by nerve terminals Decrease of lipolysis Platelet aggregation
Selective agonists	Methoxamine	Clonidine
Selective antagonists	Prazosin	Yohimbine

Phenylephrine is an α_1 agonist, weak but relatively pure; clonidine is an α_2 agonist which sometimes acts as an antagonist at α_1 receptors. As antagonists, prazosin is a good α_1 antagonist and yohimbine a good α_2 antagonist; they are frequently used to define the α receptor subtype (Fig. VII.9). In contrast to β_1 and β_2 receptors, which both activate adenylate cyclase, a major difference exists in the expression of activation of α_1 and α_2 receptors. Stimulation of *α_1 receptors* leads to a redistribution of intracellular *calcium*; stimulation of *α_2 receptors* leads to an *inhibition* of *adenylate cyclase*, at least in extraneural tissues. Blood platelets, even though they are not innervated, possess α_2-adrenergic receptors.

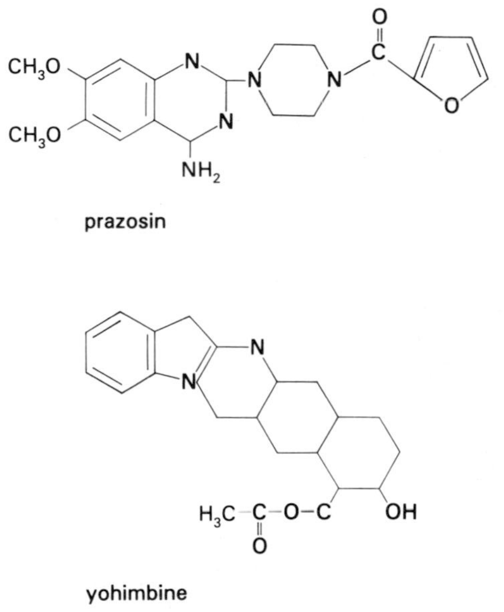

Figure VII.9 Structure of prazosin and yohimbine.

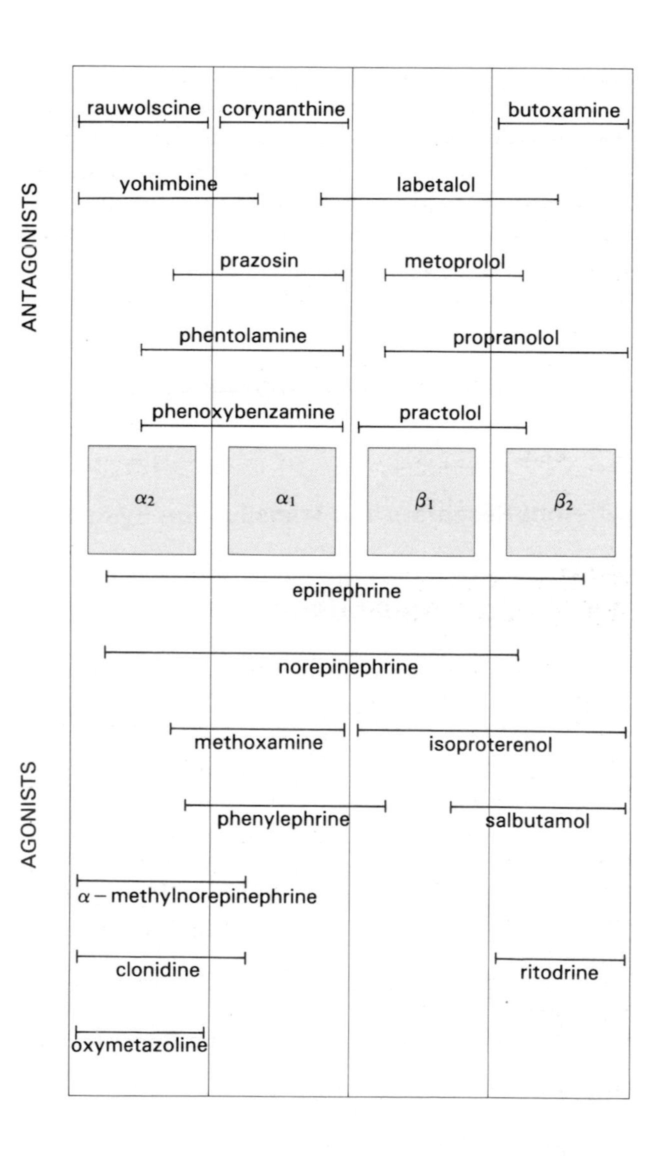

Figure VII.10 The spectrum of action of drugs acting on the different adrenergic receptors.

3.5 Individuality of Adrenergic Receptors

The definition of the various types of adrenergic receptors is primarily pharmacological. It is based on the order of efficiency of a number of agonists and antagonists, this order varying from one type to another (Fig. VII.10). It is important to remember, however, that these compounds are all synthetic, with the exception of epinephrine and norepinephrine, while all adrenergic receptors recognize both epinephrine and norepinephrine. Thus, it is possible to *question the biochemical individuality* of each of the receptors characterized pharmacologically. For example, in the past it was proposed that α and β receptors were different conformations of the same entity. Although this hypothesis has been abandoned, it drew attention, nevertheless, to the variation in ratio of receptors, α to β, in the same tissue. As a case in point, thyroid hormone administration increases the number of β-adrenergic receptors in the heart, and adrenalectomy has the same effect in liver.

4. Mechanisms of Adrenergic Effects (Fig. VII.11)

Different Receptors and Transduction Systems

4.1 β-adrenergic Effects

Early studies using radiolabelled catecholamines to identify β-adrenergic receptors in rat liver plasma membranes and other tissues met with only limited success due to the presence of non-specific catechol binding sites and to the rapid oxidation of the catechol moiety. More recently, it has become possible to study β-adrenergic receptors by direct ligand binding techniques using specific ligands such as [^{3}H]dihydroalprenolol and several [^{125}I]iodolabelled pindolol derivatives.

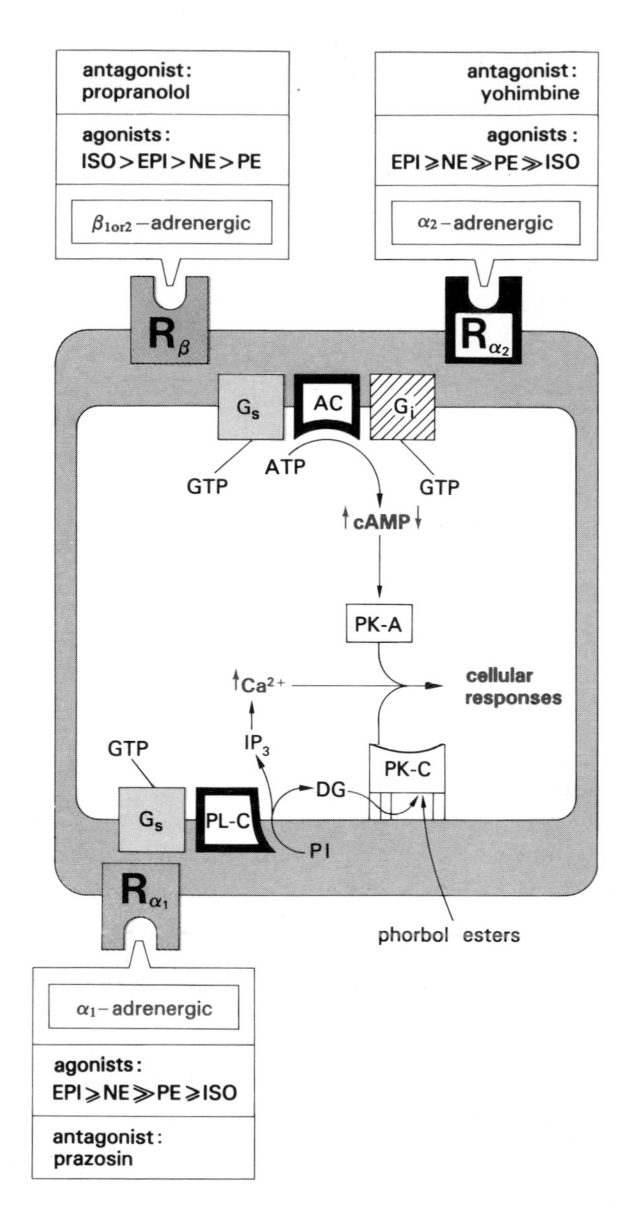

Figure VII.11 Mechanisms of action of adrenergic receptors. Epinephrine activates the β_2-adrenergic receptor and thus increases cAMP. By the intermediary of an α_1-adrenergic receptor, it augments cytosol calcium. It can also reduce the level of cAMP via an α_2-adrenergic receptor. (Courtesy of Dr. M. Caron, Duke University).

Structure and Gene of β-adrenergic Receptors

The β_2-adrenergic receptors from hamster and guinea-pig lung have been solubilized with digitonin and purified by sequential sepharose-alprenolol affinity and high performance steric exclusion liquid chromatography. After a 33,000-fold purification, a single receptor protein with high specific activity (14 nmol/mg protein) and a M_r of ~64,000 was obtained. Following insertion into phospholipid vesicles, functionality of the purified receptor was demonstrated by reconstitution with a pure G_s protein and with adenylate cyclase itself[12]. The M_r of the β_1-

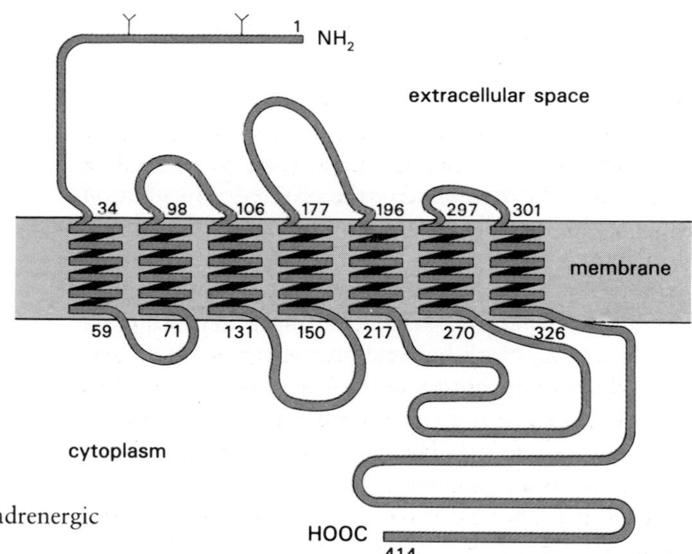

Figure VII.12 Proposed organization of the hamster β_2-adrenergic receptor, with seven membrane-spanning regions[13].

adrenergic receptor is similar to that of the β_2 receptor.

Recently[13], the *β-adrenergic* receptor from hamster lung was cloned. The coding sequence of the message corresponds to a polypeptide of 418 aa with a M_r of ~46,000, in close concordance with the apparent M_r~49,000 of the deglycosylated receptor. The *receptor sequence* encodes a largely *hydrophobic* polypeptide, with the *C-terminal* region of the molecule being *hydrophilic*. It does not contain a cleavable signal sequence, and may use an internal signal for the insertion of the protein in the membrane. Two potential sites for *N*-glycosylation are located near the N-terminal. Analysis of the amino acid *sequence* indicates *significant homology with* bovine *rhodopsin*, and suggests that, like rhodopsin, the β-adrenergic receptor possesses *multiple membrane-spanning regions* (Fig. VII.12, Fig. I.81 and p. 79). Noteworthy, also, is the *absence of introns* within the coding and 3′ untranslated regions of the receptor gene.

The β-adrenergic receptor gene has been localized to *chromosome 5*, on the same locus (5q31.2–q31.3) as the PDGF receptor.

Difference between Adrenergic Agonists and Antagonists

The two types of drugs, agonists and antagonists, bind equally well to β receptors, but only agonists lead to an efficient transformation of the receptor, capable of transmitting the hormonal information to the adenylate cyclase system. It has been shown[14] that the thermodynamic characteristics of binding for agonists and antagonists are very different: the affinity of agonists for the β receptor, but not that of antagonists, increases when temperature is reduced. *The binding of agonists* is associated with a *reduction in enthalpy*; that of *antagonists* is associated with *increased entropy*.

Role of Guanine Nucleotides

An important characteristic of *agonist binding*, but *not* of *antagonists*, is *sensitivity to guanine nucleotides*, such as GTP (p. 334). This nucleotide modifies the affinity of receptors for all agonists from a state of high affinity to a state of relatively low affinity.

12. Cerione, R. A., Sibley, D. R., Codina, J., Benovic, J. L., Winslow, J., Neer, E. J., Birnbaumer, L., Caron, M. G. and Lefkowitz, R. J. (1984) Reconstitution of a hormone sensitive adenylate cyclase system. *J. Biol. Chem.* 259: 9979–9982.
13. Dixon, R. A. F., Kobilka, B. K., Strader, D. J., Benovic, J. L., Dohlman, H. G., Frielle, T., Bolanowski, M. A., Bennet, C. D., Rands, E., Diehl, R. E., Mumford, R. A., Slater, E. E., Sigal, I. S., Caron, M. G., Lefkowitz, R. J. and Strader, C. D. (1986) Cloning of the gene and cDNA for mammalian β-adrenergic receptor and homology with rhodopsin. *Nature* 321: 75–79.
14. Weiland, G. A., Minneman, K. P. and Molinoff, P. B. (1979) Fundamental difference between the molecular interactions of agonists and antagonists with the β-adrenergic receptor. *Nature* 281: 114–117.

Desensitization

Of clinical importance

It has long been known that the administration of epinephrine to an animal or the prolonged incubation of a tissue or cellular preparation with epinephrine leads to a state which is refractory to subsequent stimulation. This phenomenon of desensitization is *complex*[15,16]; at the outset, it is simply an uncoupling of receptors, which become insensitive to GTP. This precedes the appearance of the receptor in cytoplasmic vesicles (agonist-induced internalization) by a very short period of time. Then, an actual reduction of the number of receptors is seen (down-regulation). This is known as homologous desensitization. *Heterologous desensitization* has also been described. In this case, desensitization due to treatment with a particular hormone is observed for all the hormonal effectors acting through the same adenylate cyclase system, and is related to the production of cAMP. In fact, it has been demonstrated that *receptor phosphorylation* is involved in homologous as well as heterogenous forms of desensitization. Studies of heterogenous desensitization carried out largely with avian erythrocytes have demonstrated that a cAMP-dependent protein kinase is involved in the process. Such phosphorylated receptors are functionally impaired with respect to their coupling to the adenylate cyclase system. In contrast, homologous desensitization of the β-adrenergic receptor seems not to be cAMP-mediated[17]. *Prolonged treatment* with adrenergic agonists leads to a *reduction* in the number of β-adrenergic receptors of circulating blood cells. This phenomenon explains *resistance to protracted treatment* with β-adrenergic agonists. Inversely, hypersensitivity to the action of β-adrenergic agonists occurs during prolonged treatment with β-blockers. This phenomenon, resulting in possible cardiac problems when propranolol treatment is stopped, is due to an increase (up-regulation) in the number of β receptors[18].

The Adenylate Cyclase System

The binding of epinephrine to the β-adrenergic receptor always leads to an *increase* in *adenylate cyclase activity* and, therefore, to an increase in the intracellular concentration of cAMP (Fig. VII.13). Detailed mechanisms involved in the transduction of adrenergic signals are archetypical of adenylate cyclase systems, as described on p. 100 and p. 334 Protein kinase A (PK-A) is involved in the activation of

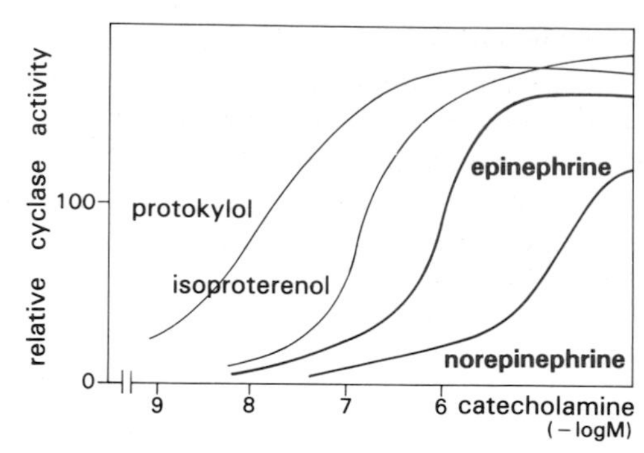

Figure VII.13 Action of catecholamines on liver adenylate cyclase. Results are expressed on a relative scale. The order of efficiency defines a β_2-adrenergic receptor.

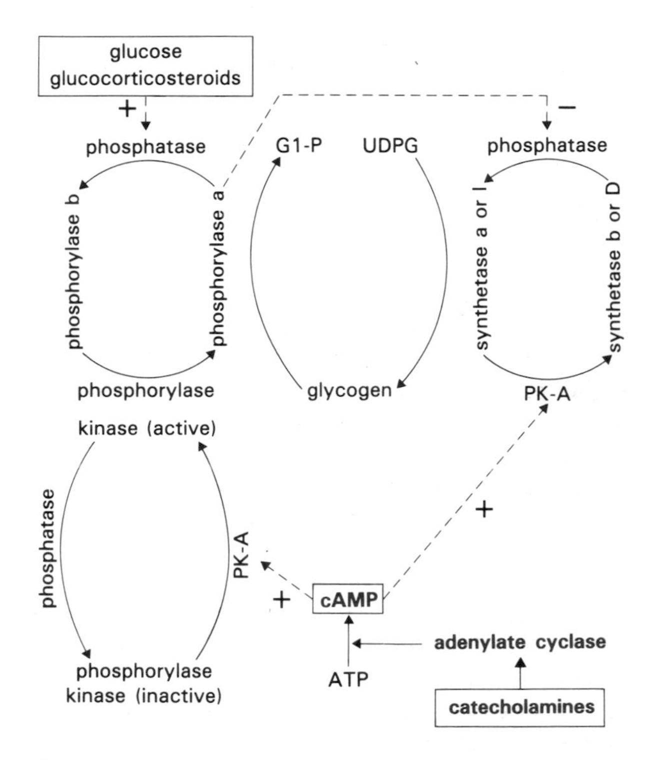

Figure VII.14 Hepatic metabolism of glycogen. UDPG: uridine diphosphoglucose, G1–P: glucose 1-phosphate.

15. Hertel, C. and Perkins, J. P. (1984) Receptor specific mechanisms of desensitization of β-adrenergic receptor function. *Mol. Cell. Endocrinol.* 37: 245–256.
16. Sibley, D. R. and Lefkowitz, R. J. (1985) Molecular mechanisms of receptor desensitization using the β-adrenergic receptor-coupled adenylate cyclase as a model. *Nature* 317: 124–129.
17. Benovic, J. L., Strasser, R. H., Caron, M. G. and Lefkowitz, R. J. (1986) β-adrenergic receptor kinase: identification of a novel protein kinase that phosphorylates the agonist occupied form of the receptor. *Proc. Natl. Acad. Sci., USA* 83: 2797–2801.
18. Aarons, R. D., Nies, A. S., Gal, J., Hegstrand, L. R. and Molinoff, P. B. (1980) Elevation of β-adrenergic receptor density in human lymphocytes after propranolol administration. *J. Clin. Invest.* 65: 949–957.

muscle and hepatic phosphorylases and adipose tissue lipases, as well as in the inactivation of glycogen synthase and a number of other enzymes (Fig. VII.14). Cyclic AMP also stimulates the gluconeogenic pathway by a mechanism which is not yet well understood. Together, these phenomena result in the metabolic effect of epinephrine (increase in glycemia, lactic acidemia, and plasma free fatty acids).

4.2 α_1-adrenergic Effects (Fig. VII.11)

Our understanding of α_1-adrenergic effects is much more recent than of cAMP-mediated mechanisms. Particularly important is the progress made in the field of calcium and phosphatidylinositol metabolism.

Characterization of the α_1-adrenergic Receptor

A number of radioligands have been used to identify α_1-adrenergic receptors. The most useful drugs for distinguishing between α_1- and α_2-adrenergic receptors are the *antagonists prazosin* and *yohimbine*. Of the two, prazosin has a considerably higher potency at α_1 than α_2 sites, while yohimbine is a selective α_2-antagonist. The concentration of α_1-adrenergic receptors measured in rat liver plasma membranes with prasozin is 800 fmol/mg protein, a value much higher than that of the β-adrenergic receptors in the same membrane preparation. As is true for the β-adrenergic receptor, the α_1-adrenergic receptor can be *desensitized by prolonged exposure* to the action of epinephrine. Also, as seen with the β receptor, it seems that two forms of the α_1-adrenergic receptor exist which are either of high or low affinity, and guanosine triphosphate (GTP) is involved in the equilibrium between them[19]. This regulation would appear to be directly related to receptor function, since a GTP effect is no longer observed in adrenalectomized animals exhibiting markedly decreased α_1-adrenergic receptor-mediated responses. These results suggest that guanine nucleotides are implicated in the coupling of α_1-adrenergic receptors to their effectors in a manner analogous to that described for adenylate cyclase-linked receptors.

Studies of the solubilization and purification of the α_1-adrenergic receptor are less advanced than those of the β-adrenergic receptor. The α_1-adrenergic receptor from liver or brain has been partially purified, following covalent labelling with tritiated phenoxybenzamine or with radioiodinated derivatives of prazosin. It appears to have a molecular weight of 80,000. Genomic and cDNA cloning experiments have demonstrated that the hamster α_1-adrenergic receptor is a glycoprotein consisting of 515 aa residues[20].

Coupling Mechanism

Unlike β-adrenergic responses, the action of catecholamines at α_1 receptors appears to be *independent of cAMP*, but is mediated by an *elevation* of cytosol *free Ca^{2+}*. Mobilization of intracellular Ca^{2+} by α_1-adrenergic agonists in liver cells[21] is extremely rapid. It can be detected within one to two seconds of binding, and is probably related to the activation of the phosphatidylinositol system. Although this hypothesis is not yet fully proved[22], the proposal that the second messenger of α_1-adrenergic effects is in fact *inositol triphosphate* is certainly very attractive[23] (p. 104). It is worth noting that phorbol esters, which are thought to act via protein kinase C, promote α_1-adrenergic receptor phosphorylation and inhibit α_1-adrenergic effects while enhancing adenylate cyclase.

4.3 α_2-adrenergic Effects

α_2-adrenergic receptors are the *least well known* of adrenergic receptors. Although their pharmacological characterization is relatively easy, it has been more difficult to study their function. Originally thought to be only presynaptic, they have been found mainly in extraneural tissues, such as platelets. Only a small portion of α_2 receptors in the brain are presynaptic.

19. Goodhardt, M., Ferry, N., Geynet, P. and Hanoune, J. (1982) Hepatic α_1-adrenergic receptors show agonist-specific regulation by guanine nucleotides. Loss of nucleotide effect after adrenalectomy. *J. Biol. Chem.* 257: 11577–11583.
20. Cotecchia, S., Schwinn, D.A., Randall, R. R., Lefkowitz, R. J., Caron, M. G. and Kobilka, B. K. (1988) Molecular cloning and expression of the cDNA for the hamster α_1-adrenergic receptor. *Proc. Natl. Acad. Sci. USA.* 85: 7159–7163.
21. Charest, R., Blackmore, P. F., Berthon, B. and Exton, J. H. (1983) Changes in free cytosolic Ca^{2+} in hepatocytes following α_1-adrenergic stimulation. *J. Biol. Chem.* 258: 8769–8773.
22. Ambler, S. K., Brown, R. D. and Taylor, P. (1984) The relationship between phosphatidyl inositol metabolism and mobilization of intracellular calcium elicited by α_1-adrenergic receptor stimulation of BC3H-I muscle cells. *Mol. Pharmacol.* 26: 405–413.
23. Bell, J. D., Buxton, I. L. O. and Brunton, L. L. (1985) Enhancement of adenylate cyclase activity in S_{49} lymphoma cells by phorbol esters. *J. Biol. Chem.* 260: 2625–2628.

Characterization of the α_2-adrenergic Receptor

The α_2-adrenergic receptor of circulating human platelets can be characterized using various labelled agonists, such as clonidine or paraamino-clonidine, or antagonists, such as yohimbine (Fig. VII.9) or rauwolcine. Yohimbine, in particular, binds with a high affinity (K_D=2.7 nM) to approximately 200 sites per platelet. Like the β-adrenergic receptor, the α_2-adrenergic receptor also exists in *two forms*, one of high and one of low affinity for agonists; *GTP*, in *synergy* with the *sodium ion*, leads to a *transformation* of the *high affinity* sites *to low* affinity sites. Labelling with tritiated phenoxybenzamine indicates a M_r of ~60,000, while radiation inactivation reveals a M_r of ~160,000. It has recently been purified, from human platelets, to apparent homogeneity (M_r~84,000)[24], functionally reconstituted with G protein in phospholipid vesicles[25], and completely cloned and sequenced[26]. Its structure is similar to that of other G protein-coupled receptors. It has two sites for N-linked glycosylation, near the N-terminal, and also possesses seven hydrophobic domains. Amino acid sequence homology is 45% with the β_1-adrenergic receptor, 39% with the human β_2-adrenergic receptor, and 31 to 28% with the M1–M2 muscarinic receptors (Fig. I.81). In situ hybridization localizes the α_2-adrenergic receptor to the long arm of chromosome 10 (q24–q26). Other, related genes have also been cloned, and they might represent additional α_2-adrenergic receptor subtypes.

Inhibition of the Adenylate Cyclase System

Little is known about the molecular mechanism by which the presynaptic α_2 receptor exerts its effect on nerve terminals. In contrast, a great deal is known about the mechanism of α_2-adrenergic effects in blood platelets[27]. Besides a possible acceleration of Na^+/H^+ exchange[28], it involves an *inhibition* of the *adenylate cyclase* system in the basal state or stimulated by prostaglandin E_1, fluoride, cholera toxin, or forskolin. This inhibition is blocked by yohimbine and can only be seen in the presence of GTP. The sodium ion also appears to be involved. It has been demonstrated that the inhibitory signal of adenylate cyclase activity requires the coupling protein G_i (p. 334). As α_2-adrenergic inhibition of adenylate cyclase is observed better in the presence of GTP than with analogs which are not metabolized (GTPγS, Gpp(NH)p), it is probably coupled to a stimulation of GTPase activity. Such α_2-adrenergic inhibition of adenylate cyclase has been described in other tissues where α_2 receptors are present (i.e. in adipose tissue, liver, thyroid, kidney, parotid gland). As with the other inhibitory systems, the effect of epinephrine on platelet adenylate cyclase can be prevented by islet-activating protein (i.e. Bordetella pertussis toxin, p. 335). The various ways in which epinephrine can interfere with its target cells are summarized in Figure VII.11.

5. Tissue Effects of Catecholamines

The respective roles of the different types of adrenergic receptors involved in integrated tissue effects of catecholamines remains difficult to elucidate since these receptors frequently transmit divergent messages and modulate the activity of the target tissue in opposite manners. The type of effect observed depends on the concentration of epinephrine or its duration of action. Only a limited number of examples having physiological or clinical importance will be described here (Table VII.5), even though catecholamines act upon almost all tissues.

5.1 Cardiovascular Effects

The cardiovascular effects of epinephrine result from a mixture of α- and β-adrenergic effects. It is possible to study them separately using specific agonists. *Phenylephrine*, an almost pure α_1 agonist, increases arterial systolic and diastolic pressure, and reduces

24. Regan, J. W., Nakata, H., de Marinis, R. M., Caron, M. G. and Lefkowitz, R. J. (1986) Purification and characterization of the human platelet α_2-adrenergic receptor. *J. Biol. Chem.* 261: 3899–3900.
25. Cerione, R. A., Regan, J. W., Nakata, H., Codina, J., Benovic, J. L., Gierschik, P., Somers, R. L., Spiegel, A. M., Birnbaumer, L., Lefkowitz, R. J. and Caron, M. G. (1986) Functional reconstitution of the α_2-adrenergic receptor with guanine nucleotide regulatory proteins in phospholipid vesicles. *J. Biol. Chem.* 261: 3901–3909.
26. Koblika, B. K., Matsui, H., Koblika, T. S., Yang–Feng, T. L., Francke, U., Caron, M. G., Lefkowitz, R. J. and Regan, J. W. (1987) Cloning, sequencing, and expression of the gene coding for the human platelet α_2-adrenergic receptor. *Science* 238: 650–656.
27. Jakobs, K. H. (1979) Inhibition of adenylate cyclase by hormones and neurotransmitters. *Mol. Cell. Endocrinol.* 16: 147–156.
28. Nunnari, J. M., Repaske, M. G., Brandon, S., Cragoe, E. J. and Limbird, L. E. (1987) Regulation of porcine brain α_2-adrenergic receptors by Na^+, H^+, and inhibitors of Na^+/H^+ exchange *J. Biol. Chem.* 262 12387–12392.

Table VII.5 Physiological effects of catecholamines.

Effector organ	Receptor	Physiological effect	Response
Cardiovascular effects:			
Heart:			
Atria and ventricles	β_1	↑ contractility	inotropic effect
Sinoatrial node	β_1	↑ conduction velocity	↑ heart rate
Arteries:			
Renal	α_1	constriction	↓ local blood flow
Splanchnic	α_1	constriction	↑ blood pressure
Coronary	β_1	dilatation	↑ local blood flow
Skin	α_1	constriction	
Skeletal muscle	β_2	dilatation	↑ local blood flow
Veins	α_2	constriction	↑ heart rate
Juxtaglomerular apparatus	β_2	renin secretion	↑ blood pressure
Metabolic effects:			
Liver	α_1 or β_2	↑ glycogenolysis ↑ gluconeogenesis	↑ blood glucose
Muscle	β_2	↑ glycogenolysis	↑ blood lactate
Pancreas	α_2	↓ insulin secretion	↑ blood glucose
	β_2	↑ insulin secretion	↓ blood glucose
Adipose tissue	β_1	↑ lipolysis	↑ free fatty acids
	α_2	↓ lipolysis	↓ free fatty acids
Gastrointestinal tract:			
Stomach glands	α, β‡	↓ secretion	↓ acidity
Stomach muscle	β_2	relaxation	
Intestinal muscle	β_2	relaxation	↓ motility
Intestinal sphincter	α_1	contraction	
Lungs:			
Bronchial muscles	β_2	relaxation	bronchodilatation
Urinary bladder:			
Detrusor	β‡	relaxation	inhibition of miction
Trigone and sphincter	α_1	contraction	
Eye:			
Radial muscle of iris	α_1	contraction	pupillary dilatation
Ciliary muscle	β‡	relaxation	accommodation for far vision
Uterus:			
Myometrium	α_1 and β_2	contraction	
	β_2	relaxation	
Vas deferens	α_1	contraction	
Platelets	α_2	aggregation	
Skin:			
Piloerector muscle	α‡	contraction	piloerection

‡Receptor subtypes have not been identified in all organs. (Bülbring, E. and Tomita, T. (1987) Catecholamine action on smooth muscle. *Pharmacol. Rev.* 39: 49–96).

cardiac rhythm. It leads to vasoconstriction in almost all vascular beds (kidney, skin, or conjunctival tissue). It does not, however, cause vasoconstriction of small coronary or cerebral vessels[29]. *Isoproterenol*, a pure β agonist, increases cardiac rate and output, as well as arterial systolic pressure, but reduces diastolic and mean blood pressure. It leads to vasodilatation of the vessels of skeletal muscles. The overall action of epinephrine and norepinephrine results from these two types of action. Contrary to classical data, α-adrenergic receptors exist in the mammalian heart. Their physiological importance is unknown, but probably involves a positive inotropic effect.

5.2 Metabolic Effects

The action of epinephrine results in hyperglycemia, hyperlactacidemia, and an increase in free fatty acids. It also increases basal metabolism.

Liver

Epinephrine increases the production of glucose in the liver by stimulating the two pathways of *gluconeogenesis* and *glycogenolysis*[30]. This action is mediated by β_2- or α_2-adrenergic receptors, depending on the species of the animal and on its stage in development. Similarly, the stimulation of phosphorylase kinase (which activates glycogen phosphorylase) occurs via an increase in cytosol calcium in rats, where α_1 receptors predominate, but via increased cAMP in dogs, where β_2 receptors predominate. Epinephrine also results in the release of hepatic potassium[31]; it increases ureogenesis, and inactivates acetyl-CoA carboxylase and glycogen synthetase. Except in periods of intense stress, the concentration of circulating epinephrine is not high enough to activate its receptors. In contrast, it is now recognized that the autonomic nervous system controls hepatic metabolism[32]. Stimulation of splanchnic nerves results in an increase in glycogenolysis, a phenomenon which concurs with the classical observation made by Claude Bernard in 1850 that puncturing the floor of the fourth ventricle results in transient glycosuria.

Skeletal Muscle

Catecholamines increase glycogenolysis and induce hyperlactacidemia by a mechanism which is purely β_2-adrenergic.

Adipose Tissue

The essential effect of catecholamines is to increase lipolysis, by a β_1-adrenergic mechanism, via the *phosphorylation* of a triglyceride *lipase*. This lipase causes an increase in glycerol and free fatty acids in the blood. When epinephrine binds to α_2-adrenergic receptors, it inhibits lipase. The proportion of β- and α-adrenergic receptors in adipose tissue varies not only according to the species, but also as a function of the type and location of the tissue. No α receptor has been found in rat adipose tissue. According to certain authors, α_2-adrenergic receptors could be more important than β receptors in controlling the total level of lipolysis within a given adipose tissue.

Kidney

Epinephrine increases renal gluconeogenesis by a dual mechanism involving calcium and cAMP.

Endocrine Pancreas

In pancreatic A cells, epinephrine acts via a β-adrenergic receptor to increase amino acid induced glucagon release[33]. In B cells, epinephrine acts via an α_2-adrenergic receptor to inhibit glucose-induced insulin release. In this system, it also appears that local catecholamines released following stimulation of the splanchnic nerves play a more important role than circulating catecholamines.

Metabolic Effects of Epinephrine in the Human

Administered in vivo, epinephrine increases the production of glucose, fatty acids, and ketone bodies. *Hyperglycemia* is explained essentially by direct β-adrenergic effects; the indirect effects which result from the inhibition of insulin secretion are mediated by α_2-adrenergic mechanisms. Epinephrine, with glucagon, is one of the major elements involved in the overall reaction normalizing glycemia following hypoglycemia. The plasma concentration of epi-

29. Moreland, R. S. and Bohr, D. F. (1984) Adrenergic control of coronary arteries. *Fed. Proc.* 43: 2857–2861.
30. Cryer, P. E. (1981) Glucose counterregulation in man. *Diabetes* 30: 261–264.
31. Epstein, F. H. and Rosa, R. M. (1983) Adrenergic control of serum potassium. *N. Engl. J. Med.* 309: 1450–1451.
32. Lautt, W. W. (1980) Hepatic Nerves: a review of their function and effects. *Can. J. Physiol. Pharmacol.* 58: 105–123.
33. Schuit, F. C. and Pipeleers, D. G. (1986) Differences in adrenergic recognition by pancreatic A and B cells. *Science* 232: 875–877.

nephrine required to produce these metabolic effects can be reached under normal physiological conditions. It varies, in fact, according to the effect observed (50 to 100 pg/ml for tachycardia; 75 to 125 pg/ml for lipolysis; 150 to 200 pg/ml for hyperglycemia; 400 pg/ml for the inhibition of insulin secretion).

5.3 Effects on Smooth Muscle

By the intermediary of a β_2 receptor, epinephrine leads to a *relaxation* of *bronchial* muscles (explaining the use of sympathomimetics in the treatment of asthma) as well as muscles of the uterus, intestine, and bladder. Via an α_1 receptor, epinephrine results in *contraction of sphincters* of the intestines and viscera, and of uterine, piloerector and splenic *muscles.*

5.4 Ocular Effects

Epinephrine, via an α receptor, leads to contraction of the radial muscle of the iris (mydriasis). Via a β receptor, epinephrine relaxes the ciliary muscle and allows accommodation for distant vision. The mechanism of its action on intraocular pressure is not well understood, but is most probably complex. One of the best treatments of chronic open angle *glaucoma* utilizes the local action of β-blockers, particularly *timolol*[34].

5.5 Effects on Blood Platelets

Blood platelets play a major role in hemostasis and its pathological expression, thrombosis, or platelet aggregation, a very complex phenomenon which can be stimulated by a broad spectrum of agents. Cyclic AMP is involved as an anti-aggregation factor, but its physiological importance is still debatable.

34. Potter, D. E. and Rowland, J. M. (1981) Adrenergic drugs and intraocular pressure. *Gen. Pharmacol.* 12: 1–13.
35. Bravo, E. L. and Gifford, R. W. (1984) Pheochromocytoma: diagnosis, localization and treatment. *N. Engl. J. Med.* 311: 1298–1303.

6. Catecholamines in Human Pathophysiology

6.1 Clinical Evaluation of the Adrenal Medulla

Clinical evaluation of the adrenal medulla is *limited* to the *measurement* of the level of either circulating or excreted *hormones*. The total secretion of norepinephrine and epinephrine normally attains 10 mg/d. The main derivatives, vanylmandelic acid and methoxylated catecholamines (in the urine), or free catecholamines (in the blood or urine), are measured routinely. The upper limit of normal excretion of vanylmandelic acid is 6.5 mg/d. Excretion of combined metepinephrine-norepinephrine ("metanephrine") is <1 mg/d. Non-metabolized hormones can be assayed, by fluorometry, radioimmunoassay or radioenzymeassay, in urine (150 μg/d for epinephrine and norepinephrine) and in blood (100 to 500 ng/l for total plasma catecholamines). The assay of blood is only of interest in the case of acute hypertension.

Among stimulation tests, the only one that is still used when diagnosis is doubtful is the intravenous administration of 1 mg of glucagon, which results in a well tolerated increase in arterial blood pressure. A suppression test with clonidine has been proposed.

6.2 Pheochromocytoma: A Rare Tumor

Pathophysiology of the adrenal medulla is restricted to one relatively *rare disease* known as pheochromocytoma[35] (which can also, even more rarely, develop from other chromaffin tissues. Fig. VII.15). This usually solitary tumor, developed from adrenal chromaffin tissue, is responsible for less than 0.5% of the cases of arterial hypertension. Approximately 1000 cases of pheochromocytoma are diagnosed each year. The classical clinical manifestations include a symptomatic triad (sweating attacks, tachycardia, and headaches) associated with a paroxystic arterial hypertension in 25% of the cases, and with chronic arterial hypertension in 60% of the cases. Diagnosis is based on *increased urinary vanylmandelic acid* (which can attain 10 to 600 mg/d), urinary "metanephrines" (1 to 150 mg/d) and free catecholamines (0.1 to 9 mg/d). The hematocrit is also elevated. Plasma catecholamine measurements seem less useful. When the diagnosis of

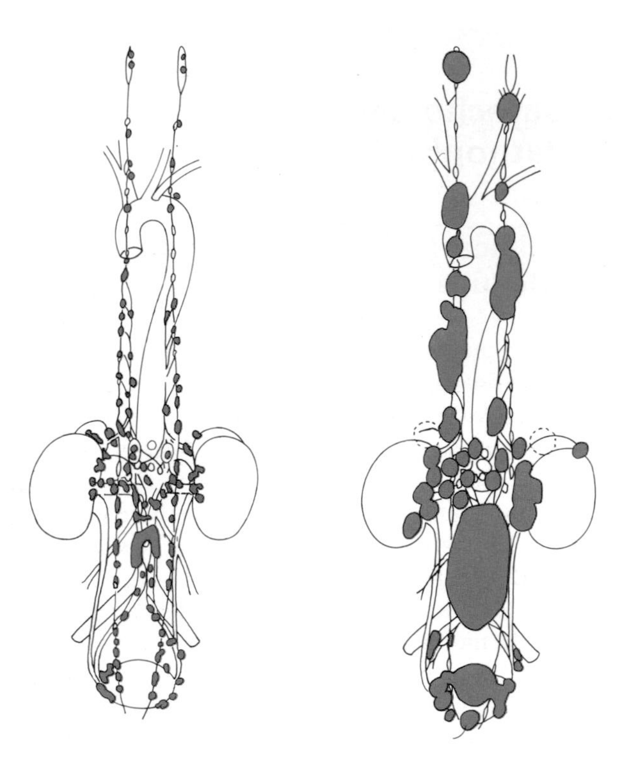

Figure VII.15 Distribution of chromaffin tissue in the newborn, as compared to the distribution of extra-adrenal pheochromocytomas. Extra-adrenal pheochromocytomas (right) occur in the newborn in sites containing chromaffin tissue (left). (Modified from Coupland, R. E. (1965) *The Natural History of the Chromaffin Cell.* Longmans and Green, London, pp. 192–194.

pheochromocytoma is not conclusive, e.g., because hypertension is not permanent, determination of platelet catecholamine content can be valuable[36]. For the localization of the tumor, the use of CT scanning is non-invasive and can detect tumors larger than 1.0 cm. Together with a scintigraphic procedure using [^{131}I]metaiodobenzyl guanidine, a compound taken up specifically by adrenergic cells[37,38], the need for

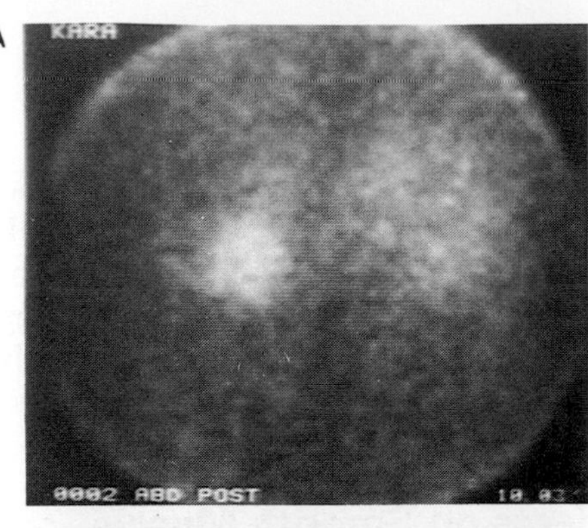

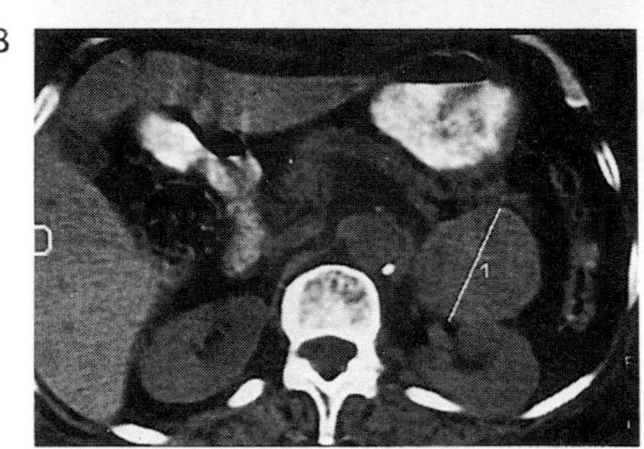

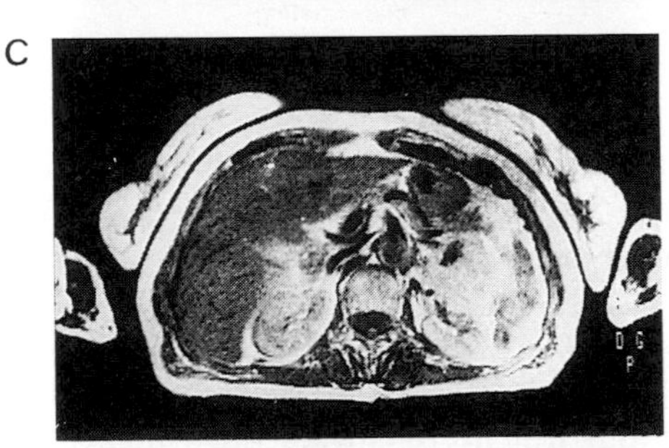

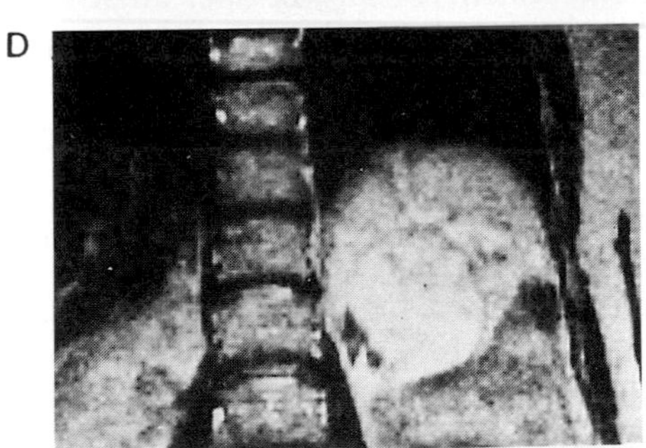

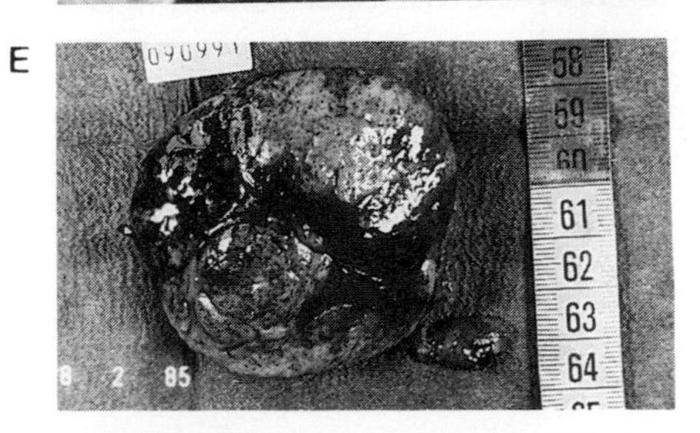

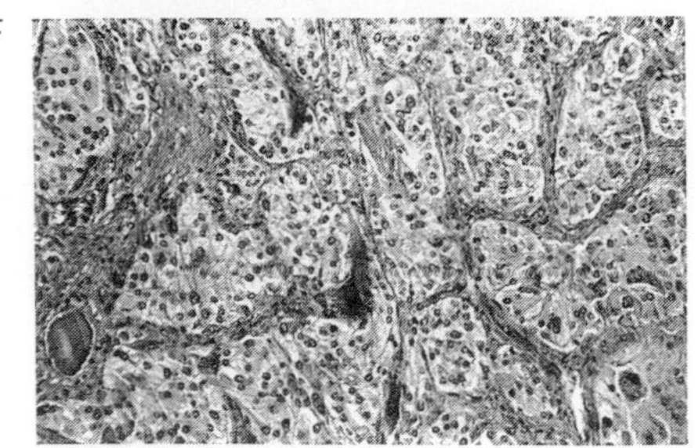

Figure VII.16 A case of pheochromocytoma. **A.** The posterior abdominal scintigraphic image with metaiodobenzyl guanidine reveals a unique sub-diaphragmatic site of fixation (right). **B.** Horizontal tomodensitometry shows that the tumor is anterenal, with a diameter similar to that of the upper pole of the kidney. **C.** and **D.** NMR imaging of the tumor (**C**, axial; **D**, frontal) allows a better contrast between liver, kidney, and tumor. The tumor has a heterogeneous structure (**C**). **E.** The operated tumor is encapsulated and hemorrhagic. **F.** Histologically, it is composed of large polyhedral cells within a well vascularized stroma. Bar=60 μm. (Courtesy of Drs. P.F. Plouin, A. Taieb and A. Baviera, Hôpital Saint-Joseph, Paris). ▷

other, more hazardous techniques, such as arteriography, is eliminated. The complete iconography of a patient suffering from a pheochromocytoma developed in the left adrenal medulla is presented in Figure VII.16. The only treatment is surgery (Fig. VII.17), and prognosis is good (only 8 to 10% of these tumors are malignant).

There has been *no report of syndromes of adrenal medullary insufficiency*[39]. Orthostatic hypotension can be seen in a series of primitive or secondary disorders of the autonomic nervous system, including Shy–Drager dysautonomia or familial dysautonomia, as well diabetic neuropathy.

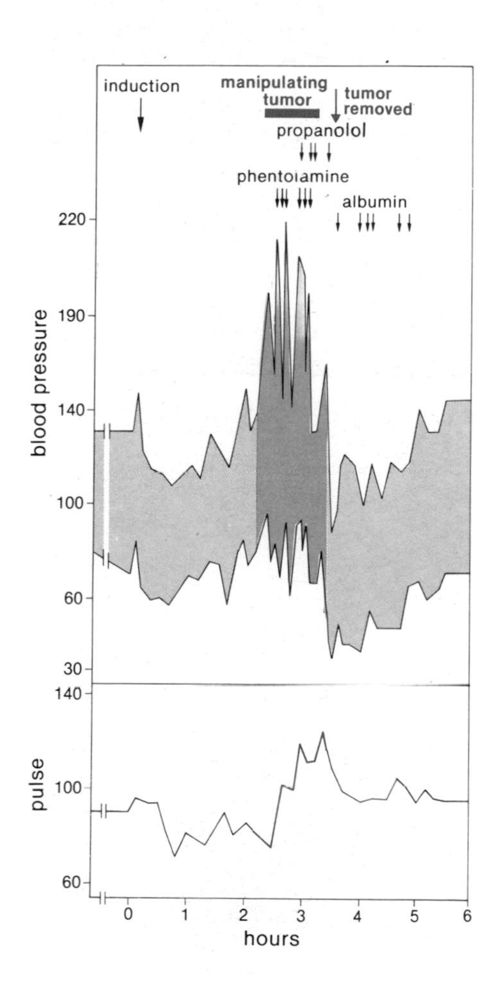

Figure VII.17 Pulse and blood pressure changes during the resection of a pheochromocytoma. Note the rise in blood pressure and pulse during induction and during manipulation of the tumor; blood pressure falls after tumor resection. (Bondy, P. L. and Rosenburg, L. E. (1980) *Metabolic Control and Disease.* 8th ed., W. B. Saunders, Philadelphia.

6.3 Hypertension

The model of pheochromocytoma led to the search for abnormalities of the autonomic nervous system in all forms of hypertension[40]. It is clear now that the *renin-angiotensin system* is *controlled by catecholamines.* The secretion of renin, the renal enzyme which catalyzes the formation of angiotensin, is increased by norepinephrine, released by renal nerves, or by circulating catecholamines. Its secretion is reduced by the administration of β-blockers. Measurements of catecholamines or adrenergic receptors do not indicate that the autonomic nervous system is involved in the early stage of essential hypertension. Nevertheless, it is quite probable that, in more than a quarter of hypertensive patients, an abnormally elevated sympathetic activity exists. Among the *drugs* currently used for the treatment of hypertension, the following *three classes* of compounds have a major role.

α_2-adrenergic agonists: Drugs such as *clonidine* are concentrated in the brain, and act negatively on the release of norepinephrine. α-methylnorepinephrine, the active metabolite of α-methylDOPA, acts in the same manner.

β-adrenergic antagonists: A drug of this type is propranolol. Several mechanisms have been proposed to explain its effect: a reduction in cardiac output, suppression of renin secretion, a central action, for example by the inhibition of presynaptic β receptors, an effect on the noradrenergic transmission of peripheral neurons, or an interaction with the production of prostaglandins.

α_1-adrenergic antagonists: A drug of this type is *prazosin*, which acts peripherally (vasodilatation).

36. Zweifler, A. J. and Julius, S. (1982) Increased platelet catecholamine content in pheochromocytoma. *N. Engl. J. Med.* 306: 890–894.
37. Sisson, J. C., Frager, M. S., Valk, T. W., Gross, M. D., Swanson, D. D., Wieland, D. M., Tobes, M. C., Beierwaltes, W. H. and Thomson, N. W. (1981) Scintigraphic localization of pheochromocytoma. *N. Engl. J. Med.* 305: 12–7.
38. Shapiro, B., Copp, J. E., Sisson, J. C., Eyre, P. L., Wallis, J. and Beierwaltes, W. H. (1985) Iodine-131 metaiodobenzylguanidine for the locating of suspected pheochromocytoma: experience in 400 cases. *J. Nucl. Med.* 26: 576–585.
39. Cryer, P. E. (1980) Physiology and pathophysiology of the human sympathoadrenal neuroendocrine system. *N. Engl. J. Med.* 303: 436–444.
40. Lake, C. R. (1984) Essential hypertension: are catecholamines involved? *Fed. Proc.* 43: 45–46.

6.4 Beginning of a Biochemical Approach to Psychiatric Diseases

The study of the biochemical bases of psychiatric diseases is just beginning. Progress in the area of biological diagnosis of mental diseases is most likely to develop very rapidly. Thus the few data which follow are preliminary. Results obtained from studies of certain mental diseases suggest an anomaly in the cerebral concentration of either neurotransmitter or receptor. Syndromes of *depression* may be associated with a partial or absolute *deficit of catecholamines*, especially norepinephrine, in functionally important zones of the brain. In contrast, in some manic conditions, an excess of catecholamines is found. It appears that the urinary concentration of 3-methoxy, 4-hydroxyphenylglycol (MHPG), the principal cerebral metabolite of norepinephrine, is low in depressed subjects. Certain assays performed in the blood may have diagnostic interest in some mental diseases. The *activity of monoamine oxidase* in platelets was found to be *lower* in certain *schizophrenics* and *increased* in some syndromes of *depression*. Drug-free patients with endogenous depression associated with psychomoter agitation have lower stimulated cAMP in lymphocytes than do control subjects[41]. In parallel with psychiatric disorders, catecholamines may also be involved in alcoholism and drug addiction. The mechanism of the euphoric effect of alcohol is possibly coupled with a modification of the metabolism of cerebral catecholamines: α-methyltyrosine, an inhibitor of catecholamine synthesis, blocks certain effects of alcohol, and the concentration of MHPG (Fig. VII.5) is increased in the cerebrospinal fluid during acute alcoholic intoxication, probably due to an increased activity of noradrenergic neurons.

41. Mann, J. J., Brown, R. P., Halper, J. P., Sweeney, J. A., Kocsis, J. H., Stokes, P. E. and Bilezikian, J. P. (1985) Reduced sensitivity of lymphocyte β-adrenergic receptors in patients with endogenous depression and psychomotor agitation. *N. Engl. J. Med.* 313: 715–720.

7. Summary

Epinephrine, the *only hormonal catecholamine*, is released from the *adrenal medulla. Norepinephrine*, also found in the gland, is mainly released as a *neurotransmitter*, from nerve terminals. *Dopamine*, the third catecholamine, is a *biosynthetic precursor* and a *neurotransmitter* in the central nervous system. Embryologically, the cells of the adrenal medulla are derived from sympathoblasts, and are termed *chromaffin* cells. Storage granules containing epinephrine and norepinephrine, associated with ATP, and specific proteins, such as chromogranins and enkephalins, are found in these cells. The *biosynthetic pathway* starts with tyrosine, which is first hydroxylated by the limiting enzyme *tyrosine hydroxylase*, and goes through 3, 4-dihydroxyphenylalanine (DOPA) and dopamine. Synthesis of norepinephrine from dopamine is catalyzed by dopamine β-hydroxylase. Norepinephrine is further methylated to epinephrine by a *N*-methyltransferase, an enzyme whose synthesis is controlled by the local concentrations of glucocorticosteroids from the adrenal cortex. Secretion occurs by exocytosis of granules, which is under preganglionic cholinergic control. Numerous stimuli result in the release of epinephrine, norepinephrine and their associated proteins into the blood. Catecholamines are degraded to a variety of compounds, which are deaminated, and the metabolic byproducts are excreted in the urine. *Monoamine oxidase (MAO)* transforms catecholamines into aldehydes, and plays an important role in the control of norepinephrine levels in nerve terminals. *Catechol O-methyltransferase (COMT)* is mainly involved in the degradation of epinephrine in peripheral tissues. *Catecholamines* are *taken up* by cells, either to be stored by a specific, high affinity process in the adrenergic *nerve* terminals (type I uptake), or to be degraded in *peripheral* tissues by a non-specific, low affinity system (type II uptake).

The physiological effects of catecholamines are *numerous*, and are classified pharmacologically in *two types*, *α and β* (effects of specific agonistic and antagonistic drugs). These effects were subsequently subclassified *into* β_1 and β_2 (both postsynaptic), and α_1 (postsynaptic) and α_2 (mainly but not exclusively presynaptic). They correspond to *specific receptors* located in the plasma membrane of the target cells.

α_1-adrenergic receptors cause contraction of smooth muscle and of the uterus, and lead to an increase of blood glucose. Specific agonists are norepinephrine and phenylephrine. Antagonists are phentolamine and prazosin. *α_2-adrenergic* receptors are responsible for the inhibition of the overflow of norepinephrine from nerve terminals, and for a few other peripheral effects. Agonists are norepinephrine and clonidine. The best antagonist is yohimbine. *β-adrenergic receptors* produce various effects in the cardiovascular system (vasodilatation and a positive inotropic effect) and in adipose tissue (lipolysis), via a β_1 receptor, and in bronchi (dilatation) and skeletal muscle (glycogenolysis), via a β_2 receptor. Membrane receptors for catecholamines have been characterized and purified, and the *β-adrenergic receptor* has been *cloned*. β-adrenergic effects, which are due to an *activation of adenylate cyclase*, lead to an increase in cellular cAMP and an activation of various protein kinases. *α_2-adrenergic* effects involve an *increase* in cytosol *calcium*, probably due to the release of *inositol triphosphate*. *α_2-adrenergic* effects are linked to a *decrease in adenylate cyclase* activity, at least in peripheral tissues.

Estimation of the various catecholamines and their metabolites in blood and urine is now easy. The only pathology of the adrenal medulla is *pheochromocytoma*, a tumor responsible for 0.5% cases of hypertension. Various derivatives and analogs of catecholamines with more or less specific agonist or antagonist action have been synthesized, and are now used routinely in a variety of diseases, from hypertension to glaucoma. Modifications of catecholamines are involved in several vascular and neurological diseases, and given an already abundant series of drugs, their elucidation may have important clinical implications.

General References

Blaschko, H. and Muscholl, E., eds (1972) *Catecholamines*. Springer Verlag, Berlin, pp. 1–1054.

Day, M. D. (1979) *Autonomic Pharmacology*. Churchill-Livingstone, Edinburgh, pp. 1–255.

Goodhardt, M., Ferry, N., Aggerbeck, M. and Hanoune, J. (1984) The hepatic α_1-adrenergic receptor. *Biochem. Pharmacol.* 33: 863–868.

Insel, P. A. (1984) Identification and regulation of adrenergic receptors in target cells. *Amer. J. Physiol.* 247: 53–58.

Kunos, G., ed (1981) *Adrenoceptors and Catecholamine Action*. Wiley, New York, pp. 1–343.

Lefkowitz, R. J., Caron, M. G. and Stiles, G. L. (1984) Mechanisms of membrane receptor regulation. *N. Engl. J. Med.* 310: 1570–1579.

Motulsky, H. J. and Insel, P. A. (1982) Adrenergic receptors in man. Direct identification, physiologic regulation, and clinical alterations. *N. Engl. J. Med.* 307: 18–29.

Ruffolo, R. R. (1984) Interactions of agonists with peripheral α-adrenergic receptors. *Fed. Proc.* 43: 2910–2916.

Weinshilboum, R. M. (1983) Biochemical genetics of catecholamines in man. *Mayo Clin. Proc.* 58: 319–330.

Lutz Birnbaumer

The Adenylate Cyclase System

The basic features of the adenylate cyclase system are described on p. 100. A complete adenylate cyclase system responds *to both stimulatory and inhibitory hormonal regulation* (Fig. 1). In addition, it is stimulated by non-hydrolyzable GTP analogs, $NaAlF_4$, forskolin, and cholera toxin. Inhibitory regulation by hormones, but not stimulatory regulation, is blocked by pertussis toxin. Figure 2 illustrates effects of regulatory ligands on a typical adenylate cyclase system. The following pages deal primarily with "*G proteins*", their regulatory functions, and their relationship with other GTP-binding proteins.[‡]

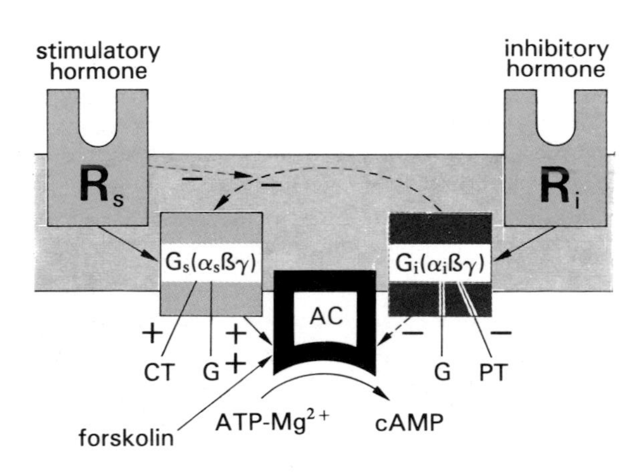

Figure 1 *Components currently known to constitute a dually regulated adenylate cyclase system, and scheme of regulatory features (C, catalytic unit of adenylate cyclase; G_s, stimulatory G protein; G_i, inhibitory G protein; G, guanine nucleotide (GTP or GDP); R_i, inhibitory receptor; R_s, stimulatory receptor; H_s, stimulatory hormone; H_i, inhibitory hormone; CT, cholera toxin; PT, pertussis toxin (islet-activating protein)). The system is shown embedded in the phospholipid bilayer of a plasma membrane,* with receptors oriented towards the outside *of the plasma membrane and* G and C proteins oriented towards its inner side. *Recent evidence suggests that C is a* glycoprotein, *and although not shown, it is likely that it traverses the plasma membrane. The $\alpha\beta\gamma$ subunit composition of G proteins is indicated. Regulatory influences are shown by plus and minus signs next to arrows.*

Regulation of Enzymatic Activity of the C Subunit by Guanine Nucleotide in the Absence of Hormone

In addition to receptors which recognize specific hormones, and to the catalytic unit C (M_r~150,000) which actually synthesizes cAMP, adenylate cyclase is composed of two GTP-binding G coupling proteins called G_s and G_i. On activation, G_s and G_i regulate C activity positively or negatively, respectively.[‡‡] The G proteins are true signal transducers, recognizing hormone receptor occupancy, and responding to this event. Both receptors and the C unit appear to be glycosylated, spanning the plasma membrane lipid bilayer. *G proteins are $\alpha\beta\gamma$ heterotrimeric oligomers* with a M_r of ~80–95,000. G proteins differ in their α subunits, but share common β and γ subunits (Table I). Both G proteins are *activated by GTP* and Mg^{2+}, and both are *GTPases.*

[‡]These proteins, because they bind guanine nucleotides, were also known as "N proteins".
[‡‡]A list of the hormones and corresponding receptors which activate G proteins is found on p. 101. *Activators of G_s* result in *stimulation* of adenylate cyclase activity. *Activators of G_i* result in *attenuation* of adenylate cyclase activity.

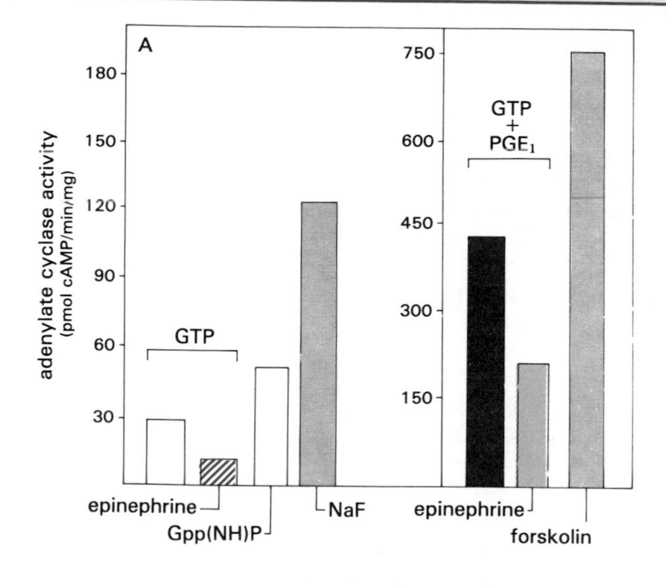

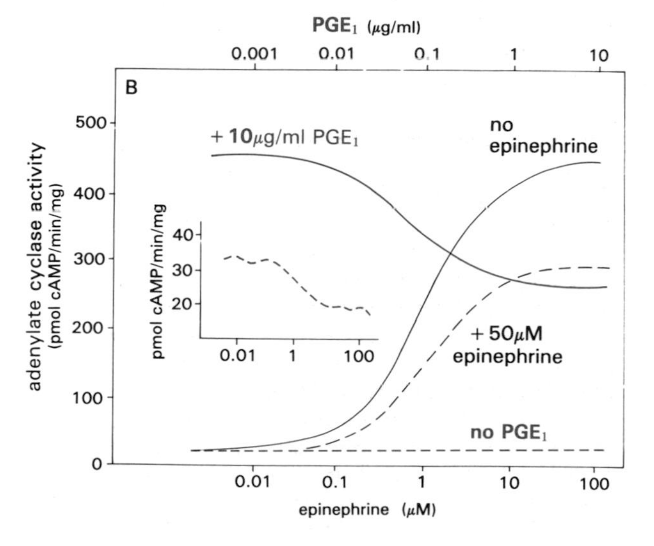

◁ Figure 2 *Regulatory responses of the adenylate cyclase system in human platelet membranes.* **A.** *Effects of saturating concentrations of ligands. Activity assayed in the absence of GTP (not shown) is about 50% that seen in its presence.* **B.** *Effect of varying the dose of the stimulatory hormone, PGE_1, in the absence or the presence of saturating inhibitory hormone, epinephrine, and the effect of varying the dose of the inhibitory hormone in the absence or the presence of saturating stimulatory hormone.* **Inset:** *Dose-response curve for epinephrine in the absence of PGE_1, plotted with an expanded y-axis. It is a general characteristic that positions of dose-response curves of stimulatory (or inhibitory) hormones are unaffected by simultaneous inhibitory (or stimulatory) regulation.*

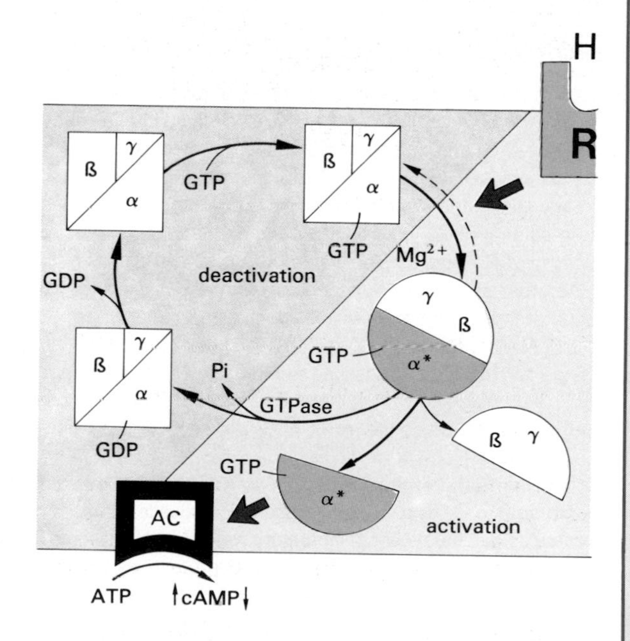

Figure 3 *Steps in the kinetic regulatory cycle of G proteins under the regulatory influence of GTP and Mg^{2+}, and built-in GTPase activity of these proteins. Inactive forms are represented in angular shapes, which, in the case of mammalian G protein (but not avian erythrocyte G protein), exchange nucleotides readily, without requiring intervention of regulatory receptors. Forms of G protein able to effect regulation of C protein activity are represented as rounded shapes. The α subunit in its active form tightly binds GTP (or GTP analogs), and then dissociates. Although only the non-dissociated form of a G protein is depicted as an active GTPase, this has not yet been substantiated experimentally.* ▷

This last property provides them with a built-in turn-off mechanism. Studies in detergent solution with pure G_s and G_i, using as regulatory ligands non-hydrolyzable GTP analogs such as GMP-P(NH)P and GTPγS, showed that the activation of a G protein involves both tight-binding of the nucleotide and a subunit dissociation reaction, to give an active α^{*G} complex plus a βγ complex. *Deactivation* of G proteins proceeds by two paths: 1. by reversal of the activation pathway, a reaction stimulated by the accumulation of βγ complexes, to give the deactivated oligomer, and 2. by hydrolysis of GTP to GDP, which is unable to maintain G proteins in their active conformation. These reactions give rise to a regulatory turnover cycle, as shown in Figure 3. For mammalian G proteins, the rate-limiting step is their activation by GTP and Mg^{2+}.

G_s and G_i stimulate and inhibit the catalyst C separately (Fig. 3) but not independently. Reconstitution experiments demonstrate that activated G_s stimulates C, and that activated G_i inhibits G_s-activated C but not G_s-free C. G_i-mediated inhibition may or may not involve the reversal of G_s activation.

Under *basal conditions*, i.e. in the absence of hormonal modulation, the cycling rate of G proteins is slow, and free βγ complexes appear to be sufficiently high that the overall proportion of G_s and G_i in their respective active forms is low, and the catalyst C is mostly, but not totally, in its unactivated form (Fig. 2).

Effects of Hormone Receptor Occupancy

Upon occupancy of receptors (R) by hormone (H), a *HR complex* is formed. HR complexes *regulate cAMP formation* by interacting with the regulatory G proteins and *facilitating their activation by GTP.* The resulting increased proportion of G proteins in their respective active states is then recognized by the catalytic unit C of the system, which either becomes activated, if the hormone is stimulatory and activates G_s, or attenuated, if the hormone is inhibitory and activates G_i. It has been shown that activation of adenylate cylase by H_sR_s complexes is the result of the formation of an *active* α^{*GTP} *complex* that stimulates C and an H_sR_s-mediated release of the system from inhibition by G_i. This release from inhibition is probably due to a block of the inhibitory effect of βγ complex on spontaneous receptor-independent G_s activation. Both G_s and G_i play a role in the fine tuning of hormonal stimulation of cAMP formation, and thus the term G_i may be a misnomer.

While receptor levels can be assessed by direct binding studies, levels of their C unit cannot be measured as yet, and, at present, the levels of G_s and G_i can only be approximated by studying binding of ^{35}S-labelled GTPγS and by quantifying ADP-ribosylation with cholera (G_s) and pertussis (G_i) toxins. Membranes from different cells contain different levels of G_s and G_i (with G_i usually being five to ten times higher than G_s).

Non-hormonal Regulators: Toxins, $NaAlF_4$, Forskolin

Cholera toxin and pertussis toxin: Cholera toxin transfers an *ADP-ribosyl* moiety from NAD^+ *to the α subunit of* G_s (an ADP-ribosyl factor (ARF), with a MW of ~ 21,000, is involved in ADP-ribosylation). In doing so, it *interrupts* G_s cycling and *GTP hydrolysis*, and converts GTP into a superactive ligand. In intact cells, this results in increased accumulation of cAMP and mimicry of hormonal stimulation. ADP-ribosylated G_s still interacts with receptors; the actions of stimulatory hormones are, however, obscured by an already stimulated system. *Pertussis toxin*, also called islet-activating protein (IAP), transfers an *ADP-ribosyl* moiety from NAD^+ to the α subunit of G_i. In so doing, it does not appear to affect GTP hydrolysis by G_i, but reduces its affinity for GTP, and *blocks* the *capacity of* G_i *to become activated.* As a result, cycling is also interrupted. ADP-ribosylated G_i does not interact with receptors and is therefore permanently uncoupled. In intact cells, this results in some elevation of cAMP levels, and, mainly, in a blockage of the action of those hormones which act via G_i.

Using [^{32}P]NAD^+, it is possible to radioactively label the α subunits of G_s and G_i in isolated membranes. This, as well as direct purification, has shown that there are isoforms of G_s and possibly also of G_i. The relative proportions of these isoforms of G_s vary from cell to cell, as do the levels of G_s per se.

$NaAlF_4$ (formerly referred to simply as *NaF*): Like non-hydrolyzable GTP analogs, the addition of NaF to G_s- and G_i- containing membranes or to pure G_s or G_i results in activation of both proteins. *Al^{3+}* is an obligatory co-factor for this action of NaF, being present as a contaminant of laboratory water and glassware. The *actual activator* is the *$NaAlF_4$* complex. As with non-hydrolyzable GTP analogs, activation of either of the G proteins by $NaAlF_4$ is associated with a reversible subunit dissociation reaction. It is not known whether $NaAlF_4$ affects G_s and G_i when added to intact cells. Its addition does not elevate cAMP levels, but if it acts on both G_s and G_i, G_s activation could be antagonized by the βγ complex from G_i. Also, as with a non-hydrolyzable GTP analog, the adenylate cyclase activities measured in the presence of $NaAlF_4$ are the result of a balance of its stimulatory and inhibitory effects.

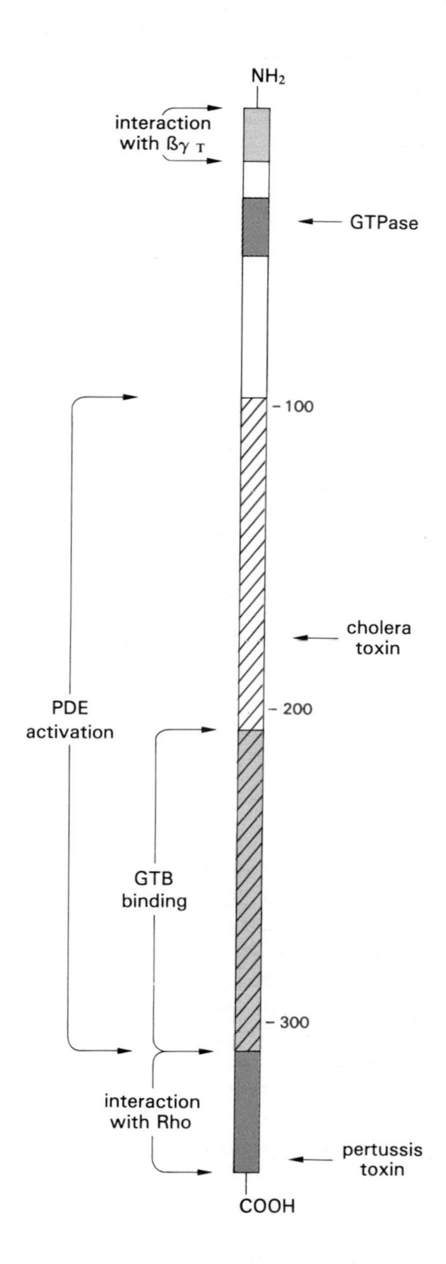

Figure 4 *Functional domains of the α subunit of transducin (α_t), as deduced from biochemical studies and its cDNA.*

Forskolin: This diterpene is a very potent stimulator of all somatic cell adenylate cyclase systems studied. It acts on adenylate cyclase activity in intact cells and in isolated membranes, and also after solubilization. Its effect is dependent neither on the presence of G_s nor on the presence of G_i, and appears to be due to a *direct* stimulation of the *catalytic* unit of the system.

GTP-binding Signal-transducing Proteins

In addition to G_s and G_i, two other proteins, termed *transducin (t)* and G_o, have an αβγ heterotrimeric structure related to that of G_s and G_i, and are also signal-transducing proteins. Synthesized only in the *photoreceptor cells* of the retina, *t* has been shown to be the GTPase/G protein that intervenes between light-activated rhodopsin and a cGMP-specific phosphodiesterase, acting, in a manner very similar to that of G proteins, to initiate the chain of events that ultimately leads to depolarization of the optic nerve. Its α subunit differs from those of the G proteins in molecular weight and in that it is ADP-ribosylated by both cholera and pertussis toxin. A tentative assignment of functions to different regions of the α_t polypeptide is shown on Figure 4. G_o purified from *brain* resembles a G protein in that it is a GTP-binding protein whose αβγ subunit structure has the same β and γ subunits as G_s and G_i, and in that it has GTPase activity. Among G proteins, it resembles G_i insofar as its α subunit is ADP-ribosylated only by pertussis toxin, but it has a smaller M_r (Table 1).
Yet other G proteins are currently being studied (Table I.16)

p21 ras Proteins and Their Congeners: Possible Relation to G Proteins

Recent biochemical and recombinant DNA studies have provided a molecular hypothesis for the tumorigenic properties of *p21 ras proteins* (p. 124). These studies showed that p21s are membrane-bound polypeptides which bind GTP, and, if coded for by a c-*ras proto-oncogene*, but not by a oncogenic c-*ras* gene, they readily hydrolyze GTP to GDP, in a manner similar to the hydrolysis of GTP by α subunits

Mutated ras oncogenes found in human tumors have a markedly *decreased* ability to *hydrolyze GTP*. This suggests that their transforming property depends on activation by GTP, with oncogenic p21s (Fig.5) being much more readily activated because of a lowered ability to become deactivated *via* their normal GTPase activity. In this sense, a mutated (oncogenic) p21 *ras* product appears to resemble G_s after permanent activation by a non-hydrolyzable GTP analog, or by GTP, following inactivation of its GTPase activity by ADP-ribosylation with cholera toxin.

In lower eucaryotes, such as yeast, *ras genes* and their products have also been found. Yeast *ras proteins* confer guanine nucleotide stimulation to a yeast adenylate cyclase obtained

Table 1. *Signal-transducing G proteins. General subunit structure: α, β, γ.*

	G_s	G_i
Found in mammals	all cells except spermatozoa	all cells except spermatozoa
Subunit compostion; Effect of activation; Rate-limiting step	$\alpha_s\beta\gamma$ (polymorphic in α_s) dissociates into:$\alpha_s^{*G}+\beta\gamma$ $\alpha_s\beta\gamma \xrightarrow[Mg^{2+}]{GTP} \alpha_s^{*GTP}\beta\gamma$	$\alpha_i\beta\gamma$ dissociates into:$\alpha_i^{*GTP}+\beta\gamma$ $\alpha_i\beta\gamma \xrightarrow[Mg^{2+}]{GTP} \alpha_i^{*GTP}\beta\gamma$
Properties of subunits	α_s (polymorphic): M_r=42–45,000 and 50–55,000; binds GTP; hydrolyzes GTP; ADP-ribosylated by cholera toxin. β: M_r=35,000. γ: M_r=5000. $\beta\gamma$ is the same as in G_i	α_i: M=40–41,000 binds GTP; hydrolyzes GTP; ADP-ribosylated by pertussis toxin (IAP). β: M_r=35,000. γ: M_r=5000. $\beta\gamma$ is the same as in G_s
Function	stimulation of activity of adenylate cyclase C	attenuation of stimulatory effect of G_s on adenylate cyclase C by direct interaction of α_i^{*GTP} with α_s^{*GTP}C and by $\beta\gamma$-mediated reversal of G_s activation
Regulated by	H_sR_s complexes	H_iR_i complexes
Effect of HR complexes	acceleration of rate of activation by GTP and analogs decrease in susceptibility to deactivation by $\beta\gamma$	acceleration of rate of activation by GTP and analogs

α_s, α_i, $\alpha_o=\alpha_{39}$, α_t, α subunits of G_s, G_i, G_o and G_t, respectively; γ_t, γ subunit of transducin; G_t transducin; Rho*, photoactivated rhodopsin; C, catalytic unit of adenylate cyclase. There are variants of β and γ subunits that have yet unknown significance.

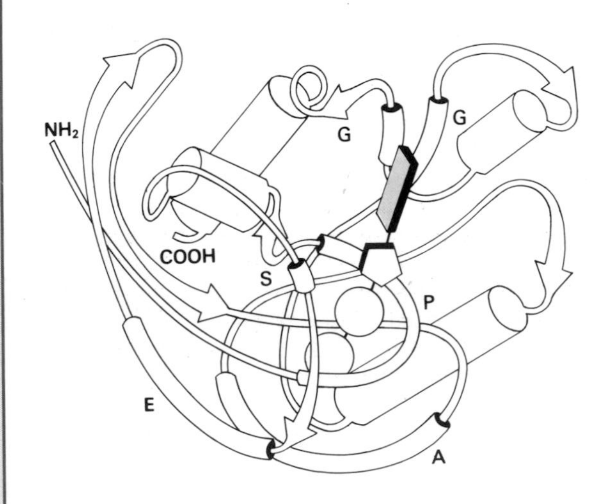

Figure 5 *Normal human c-H-*ras *p21 protein: backbone structure of the catalytic domain, looking into the GDP-binding pocket. This first three-dimensional structure of an oncongene product was obtained at 2.7 Å resolution with crystals of protein expressed in E. coli harboring a plasmid that contained a synthetic gene.*
*The flow of the backbone represented by a continuous flat ribbon, and α helices by a cylinder. The guanine base, the ribose and the phosphates are represented by a pink rectangular block, a pentagonal block, and spheres, respectively. The regions that bind the base (G), the sugar (S), the phosphate (P), a neutralizing antibody (A) and the effector region (E) are indicated by sleeved tubes. (De Vos, A. M., Tong, L., Milburn, M. V., Matias, P. M., Jancarik, J., Noguchi, S., Nishimura, S., Miura, K., Ohtsuka, E., and Kim, S. H. (1988) Three-dimensional structure of an oncogene protein: catalytic domain of human c-H-*ras *p21.* Science *239: 888–893.*

G_o	$\beta\gamma$("40 kDa protein")	G_t
nerve cells (brain)	all cells except spermatozoa	rod and cone cells of retina
$\alpha_o\beta\gamma$ (other properties as for G_s and G_i)	$\beta\gamma$ complex of G_s, G_i or G_o	$\alpha_t\beta\gamma_t$ dissociates into: α_t^{*GTP} rate-limiting step in activation (rhodopsin-dependent): $\alpha_t^{*GDP}\beta\gamma_t \xrightarrow[Rho^*]{} \beta_t\beta\gamma+GDP$
α_o: M_r=39,000; other properties as for G_i	M_r=40,000	α_t: M_r~38,000; binds GTP; hydrolyzes GTP; ADP-ribosylated by cholera and pertussis toxins. β: M_r=35,000. γ_t: M_r=8500. β in G_t is the same as β in other G proteins. $\beta\gamma_t$ functionally replaces $\beta\gamma$ of other G proteins.
?	inhibition of G_s and G_i activation by mass action reversal of subunit dissociation	activation of cGMP-specific phosphodiesterase
?		photo-activated rhodopsin (Rho*)
?		release of tightly bound GDP formed from GTP during the previous activation cycle, and concomitant facilitation of binding of new GTP

from a mutant lacking both *ras* proteins. This places them unequivocally in a class of proteins mediating signal transduction in this species. Normal human H-*ras* protein (Fig.5 p.124), albeit with lower efficiency, also reconstitutes nucleotide sensitivity to adenylate cyclase of ras^- yeast, but mammalian cyclase is not affected by yeast *ras*.

Although the biochemical function of *ras* proteins in mammalian cells is not yet known (they are clearly not identical to either α_s or α_i of G_s or G_i), there is structural evidence that relates *ras* proteins and α subunits of transducin and G proteins to a common ancestral gene. Homologies between sequences of N-terminal fragments of α_o, α_t and human H-*ras* and yeast *ras*-1 support this view.

The possible role, if any, of *ras* proteins in the regulation of mammalian adenylate cyclase systems is not known. In cultured cells, immunoneutralization of *ras* proto-oncogene products by micro injection of monoclonal anti-p21 antibody results in the inhibition of cell growth. This indicates a requirement for normal *ras* function in cell proliferation. The structural homologies between the α subunits of transducin and G_o and the c-*ras* proteins, the fact that all are membrane-associated proteins which both bind and hydrolyze GTP, and also the finding that, at least partially, mammalian c-*ras* protein (p21) can substitute for yeast *ras* protein in regulating yeast adenylate cyclase all suggest that *ras* proteins are involved in a signal-transduction event.

References

Birnbaumer, L., Codina, J., Mattera, R., Cerione, R. A., Hildebrandt, J. D., Sunyer, T., Rojas, F. J., Caron, M. G., Lefkowitz, R. J. and Iyengar, R. (1985) Regulation of hormone receptors and adenylyl cyclases by guanine nucleotide binding N proteins. *Rec. Prog. Horm. Res.* 41: 32–41.

Hurley, J. B., Simon, M. J., Teplow, D. B., Robishaw, J. D. and Gilman, A. G. (1984) Homologies between signal transducing G proteins and *ras* gene products. *Science* 226: 860–862.

Northup, J. K., Sternweis, P. C., Smigel, M. D., Schleifer, L. S., Ross, E. M. and Gilman, A. G. (1980) Purification of the regulatory component of adenylate cyclase. *Proc. Natl. Acad. Sci., USA* 77: 6516–6520.

Rodbell, M. (1975) On the mechanism of activation of fat cells adenylcyclase by guanine nucleotide. *J. Biol. Chem.* 250: 5826–5834.

Rodbell, M., Krans, H. M. J., Pohl, S. L. and Birnbaumer, L. (1971) The glucagon-sensitive adenyl cyclase system in plasma membrane. IV. binding of glucagon: effect of guanyl nucleotides. *J. Biol. Chem.* 246: 1872–1876.

CHAPTER VIII

Thyroid Hormones

Serge Lissitzky[‡]

‡Serge Lissitzky died on July 12, 1986. Dr. Y. Malthiéry, a colleague from his laboratory, kindly updated and proofread the manuscript.

Contents, Chapter VIII

Thyroid Hormones

The primary function of the *thyroid gland* is to synthesize and secrete thyroid hormones into the circulation. Hormonal production is under the *regulation of the anterior pituitary hormone thyrotropin (TSH)*. There is also a system of autoregulation within the thyroid gland itself that is still not well understood. The thyroid is differentiated, both morphologically and functionally, in a manner related teleologically to the function of its hormones. Thyroid hormones have major actions on almost *all tissues*, with the remarkable exception of the brain in the adult. In homeotherms, they control such essential functions as the regulation of *energy metabolism* and *protein synthesis*. Consequently, a constant supply of thyroid hormones to target tissues is necessary. Their synthesis is also dependent on the *exogenous supply of iodine*, a rare element in ingested food, whose content may vary greatly from day to day and from one region to another. It is not surprising that the *thyroid gland* possesses a series of elaborate mechanisms for the synthesis and storage of hormones capable of responding to metabolic needs and compensating for a temporary insufficiency in iodine intake. Thyroid homeostasis is provided by this buffering role of the gland, which gives it *unique characteristics* in terms of cellular organization and metabolism.

1. Microscopic and Macroscopic Anatomy

The thyroid is situated in the anterior portion of the neck. In humans, as well as numerous other mammals, the gland consists of two lobes. It weighs about 30 g in the adult, making it one of the largest endocrine glands. In addition, it is one of the tissues which is the *most highly vascularized*, with a blood flow of the order of 4 to 6 ml/min/g (in comparison, renal blood flow is 3 ml/min/g). The thyroid is innervated by both the sympathetic and parasympathetic nervous system.

The *functional unit* of the thyroid is the *follicle*, which has an irregular, spheroidal form. In the adult human gland, there are approximately *three million follicles*, each with a diameter varying between 50 and 500 μm. The follicle consists of a *monolayer of cuboidal cells* surrounding an amorphous and viscous mass, the *colloid* (Fig. VIII.1). This material consists essentially of an iodinated glycoprotein, *thyroglobulin (TG)*. The height of the follicular cells varies with the degree of glandular stimulation, being at their highest during periods of increased activity, and becoming flat when the gland is inactive. The epithelium rests on a basal membrane which separates it from the capillary network surrounding the follicle.

Ultrastructural study of the epithelial follicular cell (Fig. VIII.2A, B, C) permits the identification of microvilli (at the level of the apical plasma membrane which touches the colloid). They extend to the follicular lumen with protrusions of the apical pole of the cell, called *pseudopods*. These cells are classical *secretory cells* with a well developed rough endoplasmic reticulum and Golgi apparatus. A complex system of vesicles, either coated or with a soft surface, also exists, which is located on the apical portion of the cells. Some are termed colloid droplets (diameter of 0.5 to 3 μm). They contain TG-related material and are involved in pinocytosis.

In addition to follicular cells, the thyroid contains epithelial cells, which are larger, known as clear cells or *C cells*. They are situated in a parafollicular position, and are never associated with the follicular lumen. These C cells synthesize and secrete *calcitonin*

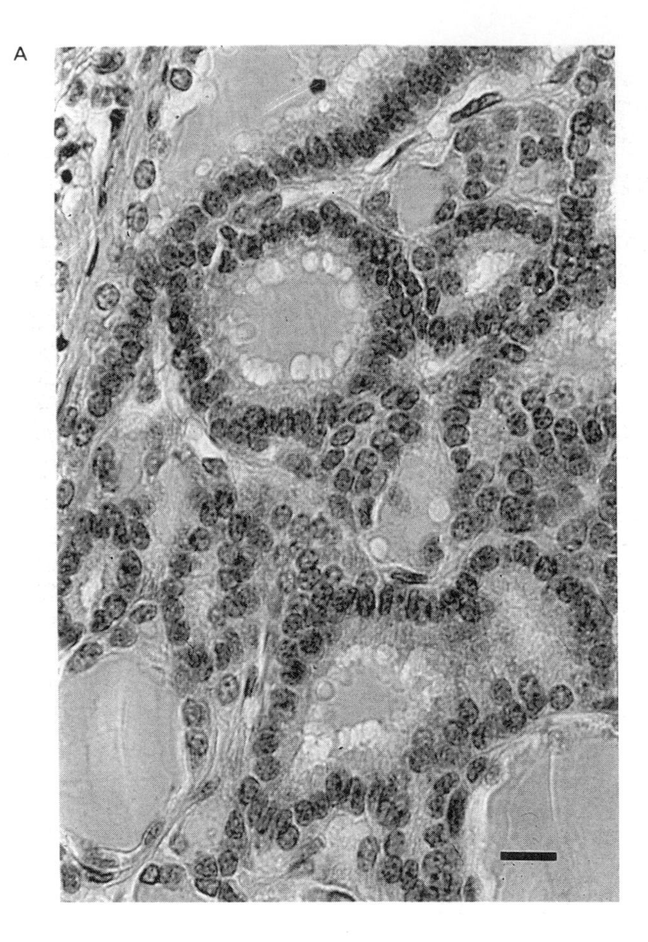

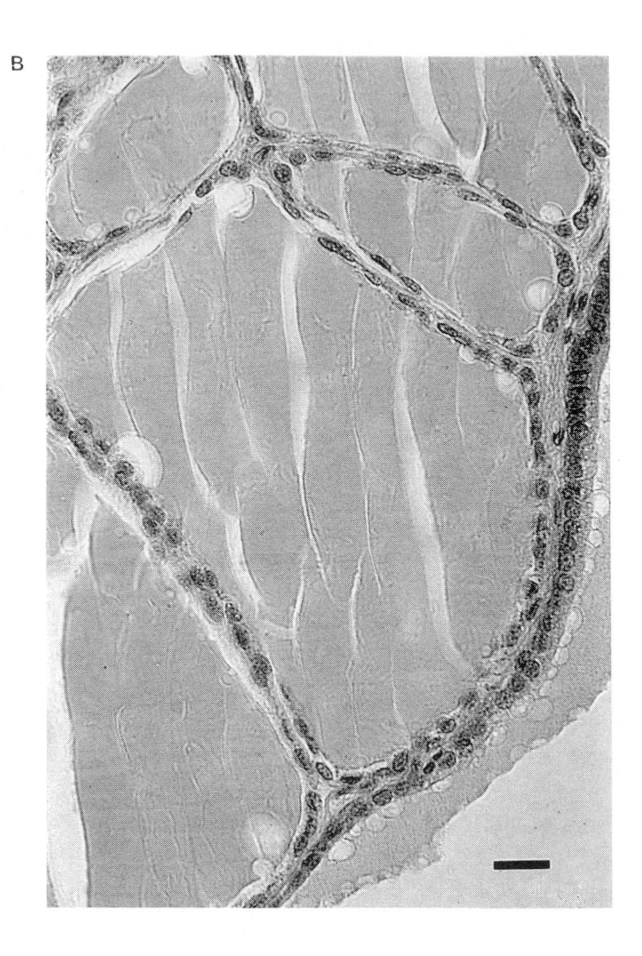

Figure VIII.1 Human thyroid gland. **A.** Patient with slight hyperthyroidism (note the height of the epithelial cells). **B.** Patient with goitrous hypothyroidism (note flattened epithelial cells and the distended follicle filled with colloid). Bar=25 μm. (Courtesy of Dr. Michel-Béchet, Faculté de Médecine Nord, Marseille).

(p. 650), and are derived from the fusion of the ultimobranchial body with the thyroid during embryonic development.

2. Embryology and Phylogenetic Development

The thyroid gland is derived from the *primitive gastrointestinal tract*. The thyroid outgrowth appears about the seventeenth day of human embryonic development as a thickening of the epithelium of the floor of the pharynx. The primitive stalk connecting the thyroid outgrowth to the floor of the pharynx elongates and becomes the thyroglossal duct, which later disappears around the second month of intrauterine life. A portion of the thyroid gland originates from the ultimobranchial portion of the IVth branchial pouches, and the cells which make up this tissue become the C cells of the thyroid gland.

Histologically, the thyroid is formed of a solid epithelial mass. Cellular cords develop, separated by vascularized connective tissue, and, during the *third month* of human fetal life, they become *follicles*. The capacity to synthesize TG appears shortly before the follicular structure becomes detectable. As soon as the follicles appear, the gland becomes capable of *concentrating iodide*, and simultaneously (at least in the human and rabbit) develops the ability to *oxidize iodide* and to form thyroid hormones (between the 70th and 75th days of fetal life in the human). Neither growth nor development of the thyroid gland appears to depend on TSH, at least for the first 10 weeks. However, when the fetal pituitary begins to secrete *TSH (at 10 to 12 weeks)*, it does appear to participate in the development of the gland.

A well developed thyroid gland with a follicular structure exists in *all vertebrates*. *Evolutionarily*, the thyroid develops from specialized cells in the digestive

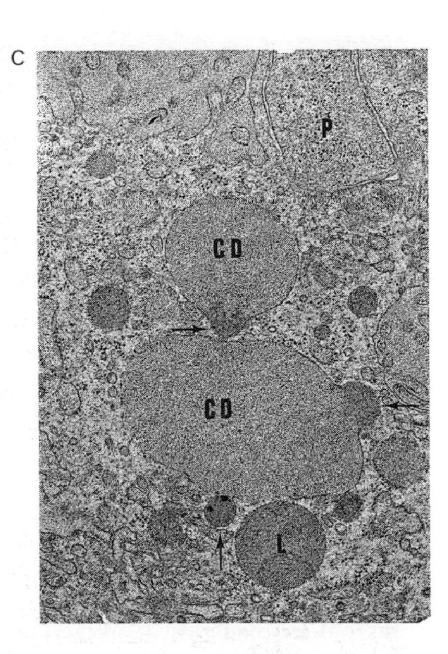

Figure VIII.2 The thyroid epithelial cell. **A.** Schematic representation of a follicular cell and adjacent follicular lumen (FL) showing the organelles involved in TG synthesis and transport into the follicle lumen: cisternae of the rough endoplasmic reticulum (ER), Golgi apparatus (G), and exocytotic vesicles (V). The right part illustrates the structures involved in absorption and degradation of TG: a pseudopod (P), colloid droplets (CD), and lysosomes (L). **B.** Electron micrograph (corresponding to the left part of (**A**) of rat thyroid. Bar = 0.4 μm. **C.** Electron micrograph corresponding to the right part of **A.** The rat was given TSH 20 min before sacrifice, to induce endocytosis. Several lysosomes (L) seem to have fixed with the colloid droplets (CD) (arrows). Bar=0.4 μm. (Ekholm, R. (1979) *Endocrinology* (De Groot, L. J., ed), Grune and Stratton, New York, vol. 1, pp. 305–307).

tract of protocords (tunicates and amphioxus), animals which are the closest relatives of vertebrates, intermediaries between vertebrates and invertebrates. In many invertebrates, iodinated scleroproteins are found in the fibrous structures of the cuticle and the pharyngeal teeth. These proteins do not have any functional importance. In protocords, the equivalent of thyroid cells secretes an iodoprotein, containing thyroid hormones, which is secreted directly into the digestive tract where it is hydrolyzed. The physiological role of thyroid hormones in these animals is not known, however.

The oldest *vertebrates*, the hagfish and the lamprey, are more highly evolved than the amphioxus. During the larval stage, the pharyngeal gland is also endocrine, but when the animal becomes an adult, cells capable of metabolizing iodine transform into follicles separated from the pharynx. It is from this point in evolution that a thyroid gland with an endocrine function differentiates.

It is interesting, therefore, that, in parallel with the establishment of endocrine thyroid follicles, secreted iodothyronines acquire the physiological role of hormone regulators.

3. Thyroid Metabolism and Synthesis of Thyroid Hormones

3.1 Iodinated Amino Acids and Radioactive Iodine

Thyroid hormones are *iodinated amino acids*. In the blood they are free molecules, while *in the thyroid gland* they constitute *an integral part of the peptide chain of TG*. Their biosynthesis depends on the metabolism of both thyroglobulin and iodine. The biosynthesis of thyroid hormones is comprised of the *following steps* (Fig. VIII.3): 1. the concentration of circulating iodide by the thyroid gland; 2. the oxidation of iodide and iodination of tyrosine residues of TG into residues of mono- and diiodotyrosine (Fig. VIII.4); 3. the coupling of the residues of iodotyrosine into residues of iodothyronines; 4. the

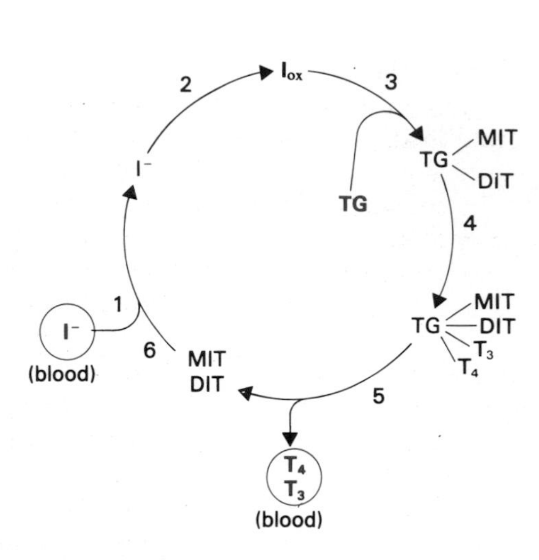

Figure VIII.3 Intrathyroidal cycle of iodine (I^-, iodide; I_{ox}, oxidized iodine (I^o or I^+); TG, non-iodinated thyroglobulin; TG (MIT, DIT) and TG (MIT, DIT, T_3, T_4), iodinated thyroglobulin). 1. Mechanism of iodine transport; 2 to 4, thyroid peroxidases; 5, intracellular proteases; 6, iodotyrosine dehalogenase.

release, by proteolysis, of iodotyrosines and free hormones; 5. the deiodination of free iodotyrosines and the recycling of the iodide thus produced. The availability, since 1945, of *radioisotopes of iodine* ^{131}I *and* ^{125}I proved to be a decisive step in thyroid research.

3.2 Thyroid Hormones

The first component with hormonal action contained iodine, and was isolated by Kendall in 1919. Synthesized in 1927[1], it was called *thyroxine* (T_4). It is the tetraiodinated derivative in 3, 5, 3′, 5′ of an unnatural amino acid, L-thyronine (T_o). In 1952, another thyroid hormone, 3, 5, 3′,-triiodo L-thyronine (*triiodothyronine*, T_3), was discovered simultaneously by two groups[2,3]. It differs from T_4 by only the absence of an iodine atom in the ortho-position of the phenol ring.

The *structure* of thyroid hormones is characterized by a diphenylether function. Phenolic-hydroxyl is found on one of the benzene rings, and an

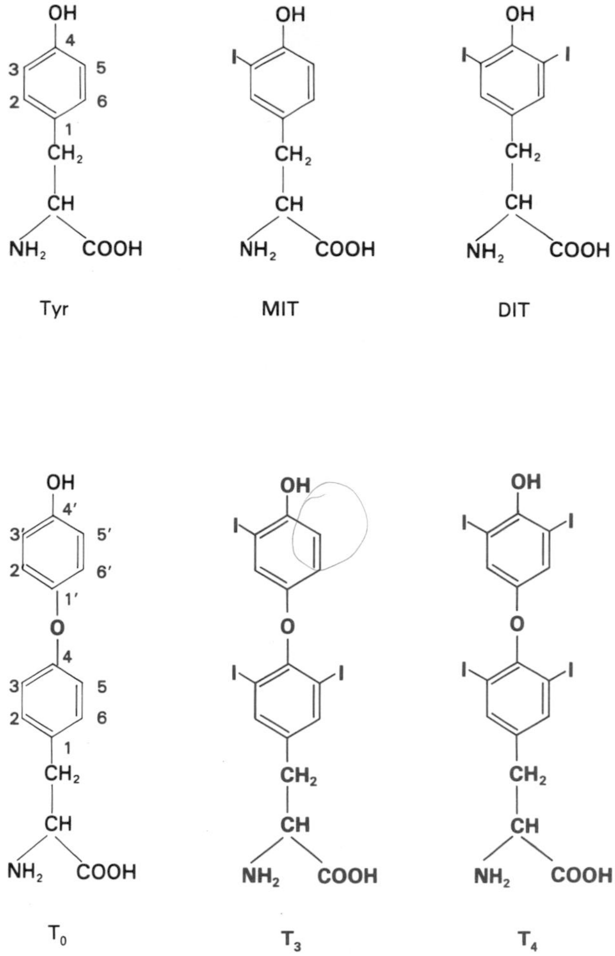

1. Harington, C. R. and Barger, G. (1927) Chemistry of thyroxine. III: Constitution and synthesis of thyroxine. *Biochem. J.* 21: 169–171.
2. Roche, J., Lissitzky, S. and Michel, R. (1953) Sur la triiodothyronine et sur sa présence dans les protéines thyroïdiennes. *Ann. Pharm. Franc.* II: 166–172.
3. Gross, J. and Pitt-Rivers, R. (1953) 3,5,3′-triiodothyronine. I. Isolation from the thyroid gland and synthesis. *Biochem. J.* 53: 645–650.

Figure VIII.4 Structures of tyrosine, thyronine (T_o), 3-iodotyrosine (MIT), 3,5-diiodotyrosine (DIT), thyroxine (T_4), and 3,5,3′-triiodothyronine (T_3).

alanyl chain on the other, in the para-position of the oxygen of the ether function (Fig. VIII.4). The aromatic rings form an angle of approximately 120°, and, in the most stable conformation, their planes are perpendicular (Fig. VIII.5). In T_4, the iodine atom in the 3′ position is probably farther from the non-phenolic ring than the iodine atom in the 5′ position.

The thyroid gland also contains *two other iodinated amino acids* which are quantitatively important: 3-iodo L-tyrosine (MIT) and 3,5-diiodo L-tyrosine (DIT) (Fig. VIII.4). These molecules, which are hormonally inactive, play a strategic role in the biosynthesis of thyroid hormones. In addition, other iodinated amino acids have been observed in very small quantities, such as 2-iodohistidine, 2,4-diiodohistidine, and several iodothyronines (3,3′-diiodothyronine, 3′,5′-diiodothyronine and 3,3′, 5′-triiodothyronine or reverse T_3, rT_3). None of these molecules has hormonal activity, and they are considered by-products of the deiodination of T_4 (p. 359).

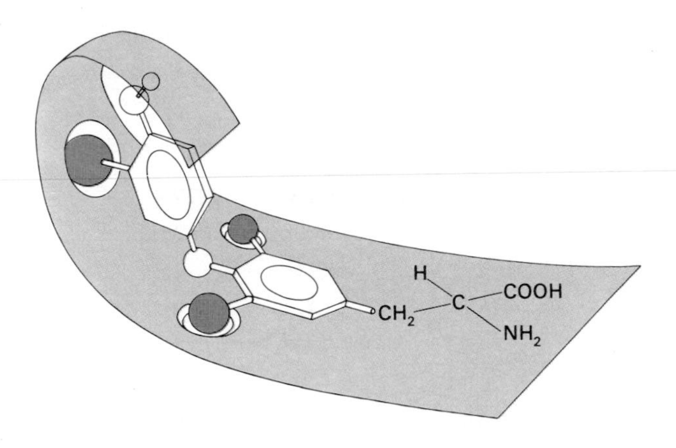

Figure VIII.5 Three-dimensional representation of steric and functional characteristics of a tri-substituted, hormonally active thyronine (T_3). (After Jorgensen, E. C. (1978) Thyroid hormones and analogs. I. Synthesis, physical properties and theoretical consideration. In: *Hormonal Proteins and Peptides* 6 (Li, C.H., ed), Academic Press, New York, pp. 107–204).

3.3 Origin and Absorption of Iodide

The major source of iodine is in the *diet*. Variations in iodine intake may occur on a daily basis, or may be chronic, related to geographical location. In areas of major deficiency, intake may be less than 10 µg/d, while in the United States it may reach 1 mg/d, although 200 µg are considered optimal. The major portion of iodine is ingested in the form of inorganic iodide; other forms are transformed to iodide in the intestine. However, iodine can also be introduced for medical reasons.

Following absorption, iodide remains essentially extracellular, with a distribution volume equivalent to 38% of the body weight. It is removed from plasma primarily by the kidney (clearance of 30 to 50 ml/h in normal humans). For an intake of approximately 150 µg/d of iodine, the thyroid clearance of iodide is around 17 ml/min, so that, in humans, the total metabolic clearance is 45 to 60 ml/min, corresponding to a half-life in the plasma of approximately 5 h (12%/h).

3.4 Thyroid Transport of Iodide

The thyroid gland has a very efficient mechanism for *concentrating* plasma iodide which is known as the *iodide pump*[4]. Under conditions of normal iodine intake, a concentration gradient of about 20- to 40-fold is established. This value may be multiplied by 20 when the gland is *stimulated by TSH*, when the diet is low in iodine, or when drugs interfere with the formation of thyroid hormones. Other tissues, such as the salivary glands and stomach, certain parts of the small intestine, mammary glands, or the choroid plexus, are also able to concentrate iodide. However, these tissues are unable to utilize iodide for incorporation into organic molecules. Iodide taken up by the thyroid is bound in an organic form within a few minutes. Under normal conditions, this transformation is so efficient that the iodide concentration in the gland is actually very low. It is very likely that the transport of iodide is the limiting step in the synthesis of organic iodide. The clearance of iodide in the thyroid, however, is greater than the combined effects of the organic transformation of iodine and its efflux, thus maintaining an iodide concentration in the thyroid greater than that in blood.

Active Transport of Iodide

The transport of iodide (I^-) occurs *against an electrochemical concentration gradient*. The concentration of iodide is higher in cells than in the follicular lumen or extracellular space. In addition, the intracellular potential across the basal membrane of the follicle is -40 to -50mV, while the potential between the outside of the follicle and its lumen is 0 to -10mV. This indicates that the potentials across the basal and the apical membranes are approximately identical, but if iodide is transported into the cell against an electric barrier, it moves along the electrical gradient of the cell, toward the lumen of the follicle. The structures making up the iodide pump are probably located in the basal membrane of the follicular cells.

The transport of iodide shows *saturation kinetics*. It can be studied quantitatively by postulating the existence of a carrier, and the K_M of active transport of iodide is of the order of 30 to 50 µM. It is approximately of the same order for other tissues which accumulate iodide. The thyroid has an accumulation capacity for I^- of 1 to 5 nmol/g of fresh tissue, a capacity approximately five times greater than that of the salivary and mammary glands.

4. Wolff, J. (1982) Iodide transport anion selectivity and the iodide 'trap'. In: *Diminished Thyroid Hormone Formation*. (Reinwein, D., and Klein, E., eds), Schattauer Verlag, Stuttgart, pp. 1–15.

Several *anions* are able to *inhibit* the uptake of I^-. The K_i of the *most active anion*, ClO^-_4, is 0.4 μM.

The transport of iodide requires *energy*. Ouabain, at low concentration, inhibits Na^+-K^+-dependent membrane ATPase (sodium pump), as well as the transport of iodide. Similar effects can be observed with inhibitors of aerobic metabolism (anaerobiosis, azide, hydrogen sulfide, and cyanide) and uncouplers of metabolic oxidation (2, 4-dinitrophenol).

The nature of the carrier involved in the mechanism by which I^- is concentrated is unknown.

Role of TSH

TSH is the *physiological stimulator of iodide transport*. Chronic hormonal stimulation caused, for example, by a low iodine diet, or by absorption of goitrogens (p. 357), increases the thyroid/serum (T/S) ratio of iodide. This occurs in relation to an increase of the capacity of the transport system, the K_M being unchanged. After a single injection in rats, the increase in T/S is preceded by reduction of this ratio. This biphasic effect has also been reproduced in vitro with either TSH or cAMP. TSH initially stimulates the efflux rate of I^-; in a second phase, it increases its influx, which depends on protein synthesis. In addition, self-regulating mechanisms of iodine transport may also be involved.

3.5 Thyroglobulin (TG)[5,6]

The use of iodide in thyroid hormone synthesis requires oxidation of iodide and its incorporation into the glycoprotein thyroglobulin (TG). These reactions are catalyzed by the membrane-bound enzyme thyroid peroxidase. TG, the *macromolecular support* of thyroid hormones synthesis, is the major protein component of the thyroid; 50 to 80 mg are extractable per g of fresh tissue. Small quantities of other iodoproteins are also present in thyroid extracts.

Physicochemical Properties[7–9]

Thyroglobulin has a sedimentation coefficient of 19 S, for a M_r of ~660,000. In the presence of agents which do not break covalent bonds, it dissociates into two 12 S subunits with a M_r of ~330,000. Table VIII.1 indicates some of the physicochemical properties of bovine TG. The stability and the form of TG in solution are dependent upon its iodination level. The non-iodinated dimer has a sedimentation coefficient of 17–18 S, as compared to 19 S for the iodinated molecule. Its molecular mass, however, is affected very little by iodination. Iodinated TG is more resistant to detergents and other dissociating agents. 19 S TG makes up approximately 95% of the proteins of the colloid, the remainder corresponding to 27 S tetramer (M_r~1,300,000) and 37 S hexamer (M_r~2,000,000).

Table VIII.1 Physicochemical properties of bovine thyroglobulin[7].

Property	12 S subunit	19 S thyroglobulin
$S^o_{20,w} \times 10^{13}\ s^{-1}$	12.1	19.4
MW	335,000	669,000
$d^o_{20,w} \times 10^7\ cm^2\ s^{-1}$	3.11	2.49
Axial ratio (a/b)	9.4	9.0

Composition

In addition to the peptide moiety, TG contains carbohydrates and iodine. There are 244 cysteine residues, and almost all are involved in disulfide bonds. Despite its importance in hormone synthesis, the *tyrosine content of TG is similar to that found in almost all proteins* (3%). Depending on the species, TG contains 8 to 10% sugars in the form of oligosaccharide units of two types, A and B, linked by a glycosylamine bond between the reducing carbon of the first *N*-acetylglucosamine and the amide group of special asparagine residues (Fig. VIII.6). Human TG also contains *N*-acetylgalactosamine-serine or- threonine groups and some units of chondroitin sulfate.

The iodine level of TG preparations varies with the dietary intake of iodine. In humans, for a TG iodine content of 0.5%, the protein contains 5,3,3 and 0.5 residues of MIT, DIT, T_4 and T_3, respectively. Of

5. Lissitzky, S. (1984) Thyroglobulin entering into molecular biology. *J. Endocrinol. Invest.* 7: 65–76.
6. Van Herle, J., Vassart, G. and Dumont, J. E. (1979) Control of thyroglobulin synthesis and secretion. *N. Engl. J. Med.* 301: 239–249 and 307–314.
7. Edelhoch, H. (1960) The properties of thyroglobulin. I. The effects of alkali. *J. Biol. Chem.* 235: 1326–1333.
8. Lissitzky, S., Mauchamp, J., Reynaud, J. and Rolland, M. (1975) The constituent polypeptide chain of porcine thyroglobulin. *FEBS Lett.* 60: 359–363.
9. Vassart, G., Brocas, H., Lecocq, R. and Dumont, J. E. (1975) Thyroglobulin mRNA: translation of a 33S mRNA into a peptide immunologically related to thyroglobulin. *Eur. J. Biochem.* 55: 15–22.

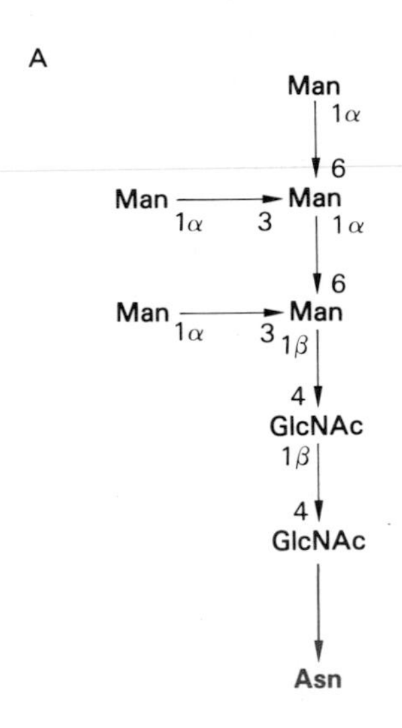

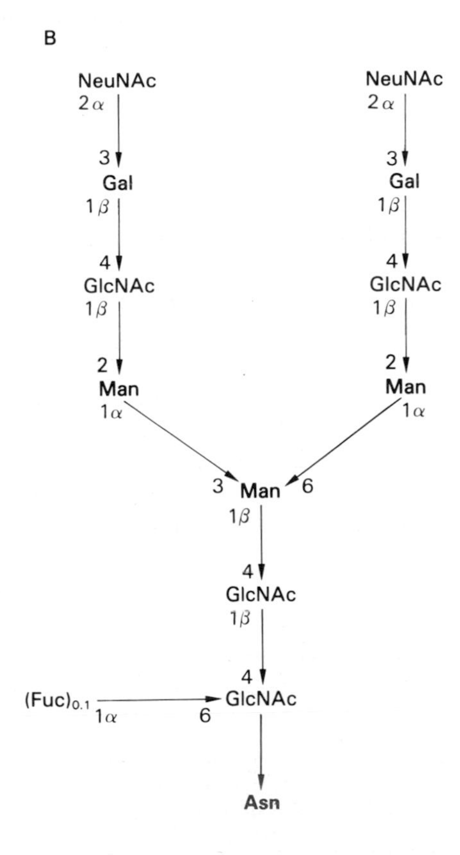

Figure VIII.6 *N*-oligosaccharide units of porcine thyroglobulin. Unit **A** contains five to nine mannoses. Unit **B**, which in addition to mannose (Man) and *N*-acetylglucosamine (GlcNAc) contains galactose (Gal), fucose (Fuc) and *N*-acetylneuraminic acid or sialic acid (NeuNAc), is represented in the form of its biantennary structure (triantennary structures are also present).

the 134 tyrosine residues within the protein, 25 to 30 are capable of being iodinated, and only six to eight of the resulting iodotyrosines can be coupled into hormone residues at *hormonogenic sites* (p. 353).

Structure

It is now well demonstrated that each 12 S subunit is formed from identical single-chain peptides, of M_r~330,000, with which the oligosaccharide units are associated. Electron microscopic examination after negative staining also revealed that the 19 S TG is present in the form of a 30×15 nm ovoid consisting of two symmetrical subunits[10].

The mRNA encoding human TG has been cloned (Fig. VIII.7), and its nucleotide sequence has been determined[11–14]. It is characterized by two short noncoding regions at both extremities (41 nucleotides at the 5′ end and 106 nucleotides at the 3′ end, respectively) and a coding region of 8301 nucleotides. The deduced amino acid sequence corresponds to a polypeptide of 2748 aa (mature, monomeric TG). The N-terminal Asn of the monomer is preceded by a signal peptide of 19 aa. Seventy percent of the molecule is composed of repeated structures. *Three types of internal homologies*, organized around Cys residues, are found successively from the N-terminal end. The first is the most characteristic, composed of 10 repeated units of about 50 aa, with conservation of Cys, Pro and aromatic residues in the same positions, suggesting duplication of an ancestral sequence. The second domain, repeated three times, is shorter; the third domain, with more limited homology, is repeated five times. The C-terminal part of the molecule includes a unique region containing a cluster of Tyr (representing up to 8% of the amino acids) which shows striking homology with acetylcholinesterase. This part of human TG has ~20% overall homology,

10. Berg, G. (1973) An electron microscopic study of the thyroglobulin molecule. *J. Ultrastructure Res.* 42: 324–336.
11. Bergé-Lefranc, J.-L., Cartouzou, G., Malthiéry, Y., Perrin, F., Jarry, B. and Lissitzky, S. (1981) Cloning of four DNA fragments complementary to human thyroglobulin mRNA. *Eur. J. Biochem.* 120: 1–7.
12. Malthiéry, Y., and Lissitzky, S. (1987) Primary structure of human thyroglobulin deduced from the sequence of its 8448-base complementary DNA. *Eur. J. Biochem.* 165: 491–498.
13. Mercken, L., Simons, M. J., Swillens, S., Massear, M. and Vassart, G. (1985) Primary structure of bovine thyroglobulin deduced from the sequence of its 8431-base complementary DNA. *Nature* 316: 647–651.
14. Baas, F., Ommen, G. J. van, Bikker, H., Arnberg, A. C. and Wijlder, J. J. M. de (1986) The human thyroglobulin gene is over 300 kb long and contains introns of up to 64 kb. *Nucl. Acids Res.* 14: 5171–5186.

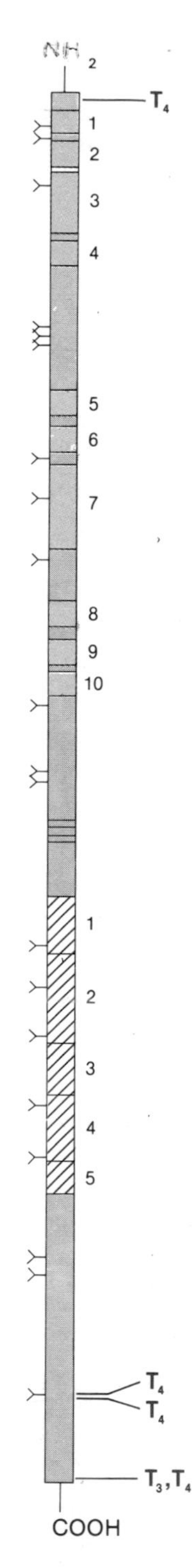

Figure VIII.7 Structural arrangement of the monomer of human TG. The N-terminal portion of the molecule contains 10 repeated sequences. The C-terminal portion consists of a region with five repetitive domains, differing from the preceding domains, and a non-repetitive region homologous to acetylcholinesterase. The hormonogenic sites for T_3 and T_4 are indicated at both ends of the molecule. Potential glycosylation sites are indicated, of which only half are actually glycosylated.

with acetylcholinesterase from Torpedo californica, and shorter regions have >50% homology. As yet, no function has been attributed to this region of the TG molecule. The structure of human TG suggests that it was derived from three different ancestral genes.

Determination of the amino acid sequences surrounding T_4 and T_3 residues in the mature protein have indicated the *tyrosine residues that are involved in hormone synthesis.* A remarkable feature is the presence of hormonogenic tyrosines, with a propensity for iodination and coupling, at both the N- and C-terminal ends of the chain. Other high affinity or low affinity sites are located more internally in the chain, near the C-terminal end. The conclusions that can be derived from the molecular structure of the hormone-forming sequences and from their position in the chain are as follows[5]: TG contains *several T_4 forming sites,* one of which (*in the fifth position* from the N-terminal) accounts for *50%* of T_4 synthesis. There is *only one T_3 forming unit,* near the C-terminal, which is highly susceptible to trypsin, possibly explaining preferential T_3 reaction. The *hormone-forming structures* of TG have been *highly conserved* throughout evolution.

There is a single gene for TG in the haploid genome, and its chromosomal location in the human is in the distal part of the long arm of *chromosome 8*[15]. It is the largest eucaryotic gene (~230 kb) identified thus far. The coding sequence is broken up into a large number of small exons (130 to 200 bp) separated by very large introns (15 to 70 kb). After processing and formation of the mature molecule, the 33 S TG mRNA (~8.5 kb) corresponding to the ~330,000 subunit is found in the cytoplasm associated with membrane-bound polyribosomes in the endoplasmic reticulum.

Post-translational Modifications

Post-translational modifications include glycosylation, iodination of certain tyrosine residues, and coupling of some of the resulting iodotyrosine into hormone residues. Following its synthesis by membrane-bound polyribosomes, TG peptides are released in the cisternae of the endoplasmic reticulum. *N*-glycosylation starts immediately. The first step in this series of complex reactions begins at the N-terminal end of the nascent chains. An oligosaccharide of the $(GlcNAc)_2(Man)_n$ $(Glc)_n$ type, preformed on a

15. Bergé-Lefranc, J.-L., Cartouzou, G., Mattei, M. G., Passage, E., Malezet-Desmoulins, C. and Lissitzky, S. (1985) Localization of the thyroglobulin gene by in situ hybridization of human thyroglobulin. *Human Genetics* 69: 28–31.

dolichol phosphate carrier, is bound to various sites of the nascent chain, to the asparagine of signal sequences (Asn-X-Ser or Thr), by a glycosylamide bond between the amide group of Asn and the reducing carbon 1 of the first GlcNAc[16] (Fig. VIII.6). This synthesis is catalyzed in the Golgi apparatus by monosaccharide transferases, the donor being a nucleotide sugar. The entire process of synthesis of the polypeptide chains and the sugar moities requires approximately three hours.

3.6 Thyroid Peroxidase[17,18]

The nature and *strategic role* of thyroid peroxidase in the synthesis of thyroid hormones has been determined only in the past ten years. This enzyme is strongly associated with the particles formed by homogenization of the thyroid, and can be solubilized by the action of detergents. It has been identified as the "microsomal antigen" detected in thyroid autoimmune diseases[19]. Its primary structure has been deduced from its cDNA sequence.[20]. In contrast to horseradish peroxidase, the prosthetic group does not appear to be ferriprotoporphyrin IX. In the presence of hydrogen peroxide, the enzyme oxidizes iodide and the tyrosine residues of the proteins.

The most potent inhibitors of the enzyme are cyanide, azide, and bisulfite. Other inhibitors act by competition (antithyroid drugs, p. 357).

3.7 Iodination of Thyroglobulin

In Vivo and In Vitro Experiments

The *iodination of thyroglobulin* occurs *independently of its biosynthesis*. The process is catalyzed by thyroid peroxidase, in the presence of iodide, and hydrogen peroxide. In vivo, it occurs very rapidly, as 50% of radioactive iodine injected into rats is organified within 2 min. Iodine is first incorporated into the protein in the form of iodotyrosine residues (MIT, DIT), before becoming iodothyronine residues (T_4, T_3). The latter process is much slower, requiring several hours or days.

All studies which have utilized radioactive iodine as a tracer have shown a *precursor–product relationship between MIT and DIT, and between DIT and T_4*. However, such a relationship is frequently obscured by the high degree of heterogeneity of the iodination of TG. Recent studies have shown that the relative distribution of iodinated amino acid residues is related to the iodine content of the preparations of human TG studied, not to their origin, i.e. from normal or pathological glands (Fig. VIII.8)[21]. These studies demonstrate the *great efficiency of TG* as a support for hormonal synthesis in the thyroid. In fact, it has been shown that, for 20 to 25 iodine atoms per mole of TG, 40 to 50% are included in the hormone residues, the yield being 20% when there are only one to five atoms of iodine per molecule[22] (Fig. VIII.9).

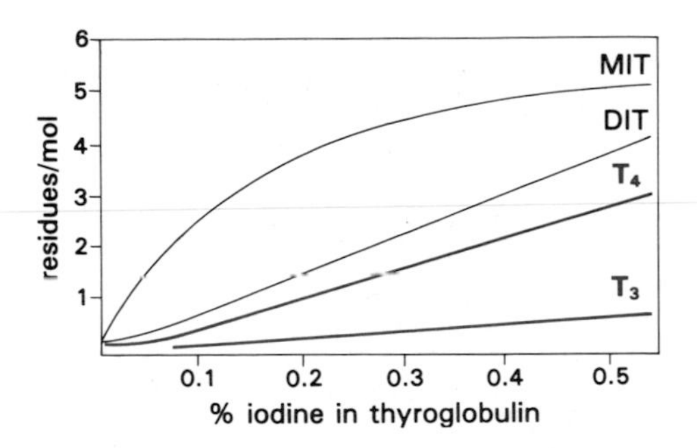

Figure VIII.8 Iodo-amino acid composition of thyroglobulin preparations from normal human subjects and patients with euthyroid or hypothyroid goiters, thyrotoxicosis, Hashimoto's thyroiditis or thyroid carcinomas[21].

In vitro, the formation of iodothyronine residues can be observed by allowing the progressive iodination (I_2+KI), chemically or enzymatically, at neutral pH, of very poorly iodinated TG preparations (0.02%), e.g., from an animal after prolonged

16. Ronin, C. and Bouchilloux, S. (1978) Cell-free labeling in thyroid rough microsomes of lipid-linked and protein-linked oligosaccharides. II. Glucosylated units. *Biochem. Biophys. Acta* 539: 481–488.
17. Nunez, J. (1980) Iodination and thyroid hormone synthesis. In: *The Thyroid Gland* (De Visscher, M., ed), Raven Press, New York, pp. 39–60.
18. Taurog, A. (1970) Thyroid peroxidase and thyroxine biosynthesis. *Rec. Prog. Horm. Res.* 26: 189–241.
19. Czarnocka, B., Ruf, J., Ferrand, M., Carayon, P. and Lissitzky, S. (1985) Purification of the human thyroid peroxidase and its identification as the microsomal antigen involved in autoimmune thyroid diseases. *FEBS Lett.* 190: 147–152.
20. Kimura, S., Kotani, T., McBride, O. W., Umeki, K., Hirai, K., Nakayama, T. and Sachiya, O. (1987) Human thyroid peroxidase: Complete cDNA and protein sequence, chromosome mapping, and identification of two alternately spliced mRNAs. *Proc. Natl. Acad. Sci., USA.* 84: 5555–5559.
21. Rolland, M., Montfort, M.-F., Valenta, L. and Lissitzky, S. (1972) Iodoaminoacid composition of the thyroglobulin of normal and diseased thyroid glands. Comparison with in vitro iodinated thyroglobulin. *Clin. Chim. Acta* 39: 95–108.
22. Rolland, M., Montfort, M.-F. and Lissitzky, S. (1973) Efficiency of thyroglobulin as a thyroid hormone-forming protein. *Biochim. Biophys. Acta* 303: 338–347.

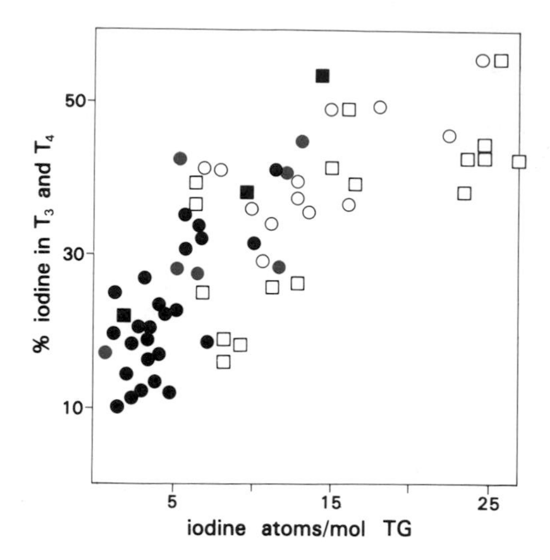

Figure VIII.9 Relationship between the percentage of iodine in thyroid hormones and the content of thyroglobulin of normal and pathological human thyroid glands; samples were obtained from normal gland (●), thyroid carcinoma (○), and different types of goiters[22].

administration of antithyroid drugs, or from certain human goiters. The sequential appearance of MIT, DIT and T_4 is observed. For identical iodine contents, however, the quantity of iodine incorporated into MIT and DIT (Fig. VIII.10) is greater in vitro than in vivo (Fig. VIII.11). This demonstrates that, in vitro,

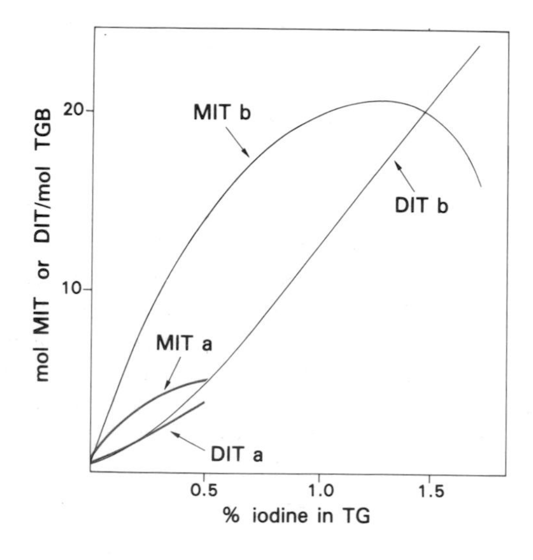

Figure VIII.10 Iodotyrosine composition of human thyroglobulin, iodinated in vivo (MIT a, DIT a) or in vitro (MIT b, DIT b), as a function of iodine protein content[21].

tyrosines are more easily iodinated and less easily coupled. All agents which induce a partial denaturation of TG reduce or abolish the formation of iodothyronines. It is quite clear that the conformation of TG that is found in the native protein is necessary for optimal synthesis of thyroid hormones.

Thyroid peroxidase is directly involved in the iodination of TG[18]. All antithyroid drugs inhibit protein iodination by the inhibition of thyroid peroxidase. This enzyme is also involved in the intramolecular coupling of iodotyrosine, producing iodothyronine residues. Propylthiouracil has an inhibitory effect on the formation of T_4, independent of its role in the iodination of TG.

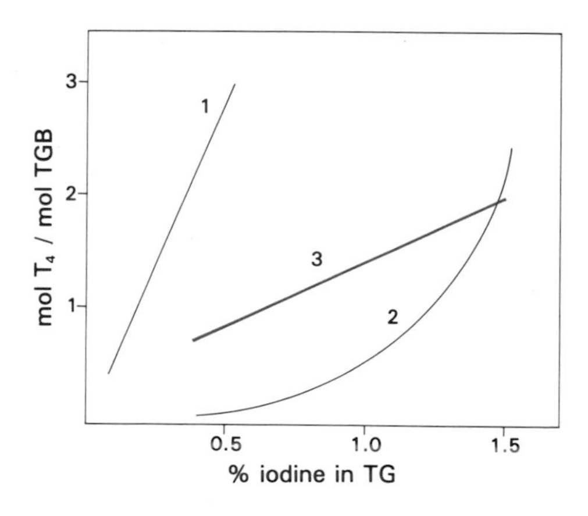

Figure VIII.11 Thyroxine content of human thyroglobulin iodinated in vivo (curve 1) and in vitro, either chemically (curve 2) or with thyroid peroxidase (curve 3)[18,22].

Physiological Role of Thyroglobulin

TG plays an essential role in the *hormonal homeostasis* of the thyroid: 1. because of its unique structure, it allows the *utilization of iodide*, resulting in a very high synthetic yield of hormones; 2. the gland allows the *storage of hormones* in a form which is not directly active, in the extracellular space (follicular lumen), away from any metabolic influences. The human thyroid also contains a *sufficiently high level of T_4* that, even without synthesis of new hormone, a euthyroid state is ensured *for two months*; thus, it can compensate for potential irregularity in dietary intake of iodine.

Nature of the Active Agent in Iodination

Although still not definite, it appears more and more probable that the *active group* in the iodination of TG is a *free radical of iodine bound to thyroid peroxidase*[17]. The H_2O_2-peroxidase complex has two active sites, one which oxidizes I^- to I^o or I^+, and the other, the phenol function of a tyrosine residue of the protein in the radical form, resulting in the formation of an intermediary quaternary complex. The iodination of tyrosine residues occurs by the reaction between these two radicals. The oxidation sites of the enzyme are not specific for either of the electron donors. In the presence of an iodide excess, these sites can be occupied by iodine radicals, which results in the formation of molecular iodine. This competition for the oxidation site of tyrosine residues would explain the inhibitory effect of high concentrations of iodide on the in vitro iodination of proteins by peroxidase, as well as the transitory inhibition of organic iodine formation in the thyroid of the rat by the administration of an excess of iodide (Wolff–Chaikoff effect). The nature of the chemical group of peroxidase with which the iodine radical could be associated is unknown.

Source of Hydrogen Peroxide

Hydrogen peroxide is probably produced by the oxidation of reduced pyridine nucleotides; the enzyme most probably involved in this process is microsomal NADPH-cytochrome *c* reductase[23]. Other sources, such as the transfer of electrons from the NADH to NADH-cytochrome *b* reductase and then to cytochrome b_5, or the production of H_2O_2 by monoamine oxidase, are also plausible alternatives.

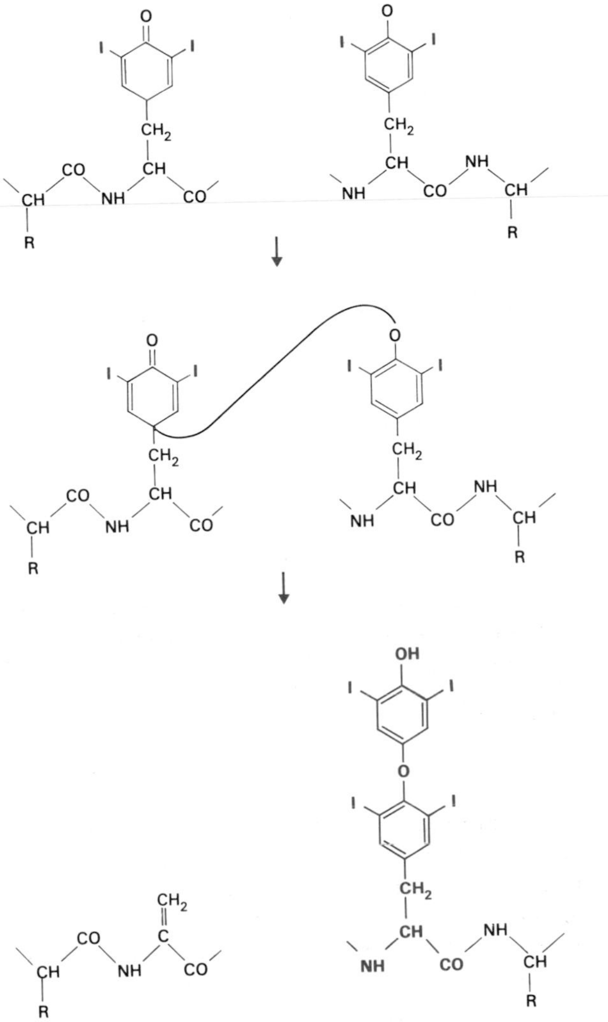

3.8 Mechanism of Synthesis of Thyroid Hormones

The *efficiency* of TG in the synthesis of thyroid hormones, even when available iodide is low, is *remarkable*. The hypothetical radical mechanism of *coupling*[24] (Fig. VIII.12) is compatible with the results of iodination of TG catalyzed by *thyroid peroxidase*. It implies the *oxidation of two residues of DIT of the protein*, their *coupling*, with the formation of a *quinol ether intermediate*, and the decomposition of the latter, with production of a T_4 radical and a residue of dehydroalanine or serine in place of the DIT residue having donated the phenolic cycle of T_4.

Figure VIII.12 Intramolecular coupling of diiodotyrosine residues, according to the radical mechanism. Note the formation of an intermediary quinol ether. DIT residues could be on the same or different peptide chains.

The coupling reaction described in Figure VIII.12 is *essentially intramolecular*, the role of the peroxidase being to catalyze the oxidation of the phenol groups of DIT *precursor residues*, which are *not* necessarily on the *same* peptide chain. Such a mechanism, implying a favorable conformation of native TG, is in accordance with studies carried out on the iodination of the protein in vitro.

23. Yamamoto, K. and DeGroot, L.J. (1974) Peroxidase and NADPH-cytochrome c reductase activity during thyroid hyperplasia and involution. *J. Clin. Endocrinol. Metab.* 39: 606–612.
24. Johnson, T.B. and Tewsbury, L.B. (1942) The oxidation of 3,5-diiodotyrosine. *Proc. Natl. Acad. Sci., USA* 28: 73–76.

3.9 Proteolysis of Thyroglobulin and Secretion of Thyroid Hormones

Before reaching the circulation, thyroid hormones stored in TG must be released from their peptide form. *First, TG penetrates* the cell by micro- or macropinocytosis, from the follicular lumen into the epithelial cell, this process appearing almost exclusively after a strong and acute stimulation by thyrotropin (TSH). Micropinocytosis results in the formation of small vesicles (0.05 to 0.3 μ in diameter) different from both the apical vesicles of exocytosis, which contain non-iodinated TG, and colloid vesicles, originating from pseudopod formations. The endocytotic vesicles fuse with lysosomes, forming phagolysosomes where the digestion of TG occurs. A coupling between the exocytosis of the newly formed TG molecules and the endocytosis of TG stored in the colloid has been demonstrated[25], suggesting an exchange of the different membrane components during the secretory process. The subsequent *proteolysis* occurs *in lysosomes*, and results in the *release of free iodinated amino acids* and iodide. Iodotyrosines are rapidly deiodinated, and the free iodide is recycled by the gland. *Free T_4 and T_3 are secreted* into the blood by a mechanism which remains unknown, T_3 appearing to be released more rapidly than T_4.

3.10 Deiodination of Iodotyrosines and Recycling of Iodide

In contrast to thyroid hormones, which undergo very limited deiodination in the thyroid, free iodotyrosines are *rapidly transformed into tyrosine* and iodide (Fig. VIII.13). This reaction plays an important role in the intrathyroid economy of iodine, and *permits* the *recycling* of it for the synthesis of hormones. This role has been demonstrated by the observation of patients with goiter and hypothyroidism due to a congenital defect in iodotyrosine-dehalogenase[26]. This condition, mimicking severe iodine deficiency, is caused by the loss of iodotyrosines into the blood which are then lost into the urine. In these patients, the administration of iodide allows, by compensation, the disappearance of the goiter and the maintenance of a euthyroid state.

25. Ericson, L. (1981) Exocytosis and endocytosis in the thyroid cell. *Mol. Cell. Endocrinol.* 22: 1–24.
26. Stanbury, J. B. (1972) Familial goiter. In: *The Metabolic Basis of Inherited Disease* (Stanbury, J. B., Wyngaarden, J. B. and Frederickson, D. S., eds), McGraw-Hill, New York, pp. 223–265.

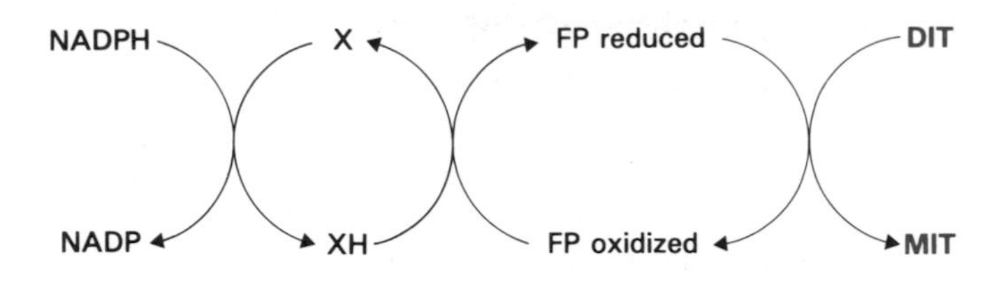

Figure VIII.13 Possible pathway of the enzymatic deiodination of DIT. In its native state, the NADPH-dependent reductase reduces the oxidized form of the flavoprotein deiodinase (oxidized FP), and the reduced enzyme acts on DIT to produce MIT and I^-.

Thyroid iodide is derived from two sources: uptake from the blood, and deiodination of free iodotyrosines. Studies performed at isotopic equilibrium have demonstrated the existence of a single pool of iodide, to which recycled and transported iodide contribute.

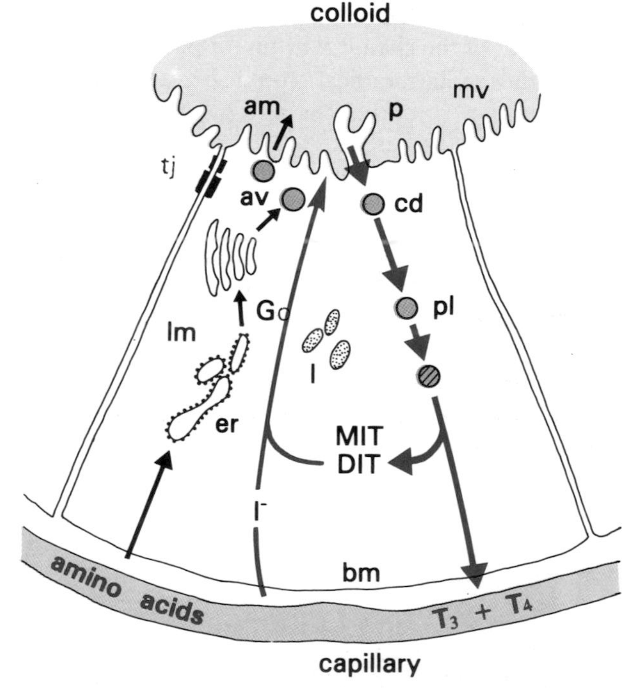

Figure VIII.14 Cytological localization of the metabolism of iodide. The black arrow shows synthesis, processing and secretion of non-iodinated thyroglobulin; the thin red arrow indicates uptake of iodide, iodination of thyroglobulin at the level of the apical membrane, and deiodination of the released iodotyrosines; the thicker red arrow indicates endocytosis, digestion of iodinated thyroglobulin, and secretion of thyroid hormones (bm, basal membrane; am, apical membrane; lm, lateral membrane; p, pseudopod; av, apical vesicle; cd, colloid droplet; l, lysosome; pl, phagolysosome; Go, Golgi apparatus; tj, tight junction; er, endoplasmic reticulum).

3.11 Localization of Iodine Metabolism in the Follicle

Thyroid cells have double secretory activity

Following its synthesis and glycosylation, non-iodinated *TG is transferred to the colloid within apical vesicles, and is iodinated in the apical membrane*[27] of follicular cells. The *iodination* of tyrosines of TG is a very rapid phenomenon, occurring within minutes. In contrast, the coupling of iodotyrosine residues requires hours, even days, to obtain a level of iodothyronines in equilibrium within TG. The present consensus is that *iodotyrosine residues are formed during the transfer of the protein across the apical membrane.* The system of microfilaments and microtubules, particularly developed in the apex of the cell and in the microvilli, is most probably the support for the movement of colloid required in this process.

The various steps in the secretion of hormones, as described above, demonstrate how the thyroid cell is involved in two seprate functions: the *exocrine secretion of TG*, from the cell to the follicular lumen, and the *endocrine secretion of thyroid hormones*, toward the blood, via the basolateral membrane.

4. Factors Influencing Thyroid Function

Various endogenous or exogenous factors are capable of modifying thyroid function. The major physiological factor is thyrotropin.

4.1 Thyrotropin or Thyroid Stimulating Hormone, (TSH)[28]

The function of the thyroid gland is under the *positive control of TSH* (p. 261). The secretion of TSH by the pituitary is controlled by the peripheral blood concentration of thyroid hormones, particularly their free forms, exerting a negative feedback control. The inhibitory action of T_3 on TSH secretion has been demonstrated, however the role of T_4 in this negative feedback is less clear (p. 270).

TSH is capable of *stimulating* a large *number of metabolic parameters* in the thyroid. Table VIII.2 summarizes some of the thyroid responses to an in vivo injection of TSH and the time required to obtain a

Table VIII.2 Thyroid responses to stimulation by TSH in vivo.

Response (increase)	Time after injection of TSH
Adenylate cyclase	2 min
Endocytosis of colloid	5 min
Oxidation of NADPH and NADH	5 min
Secretion of T_4 and T_3	10 min
Incorporation of [^{14}C]uridine in RNA	2 h
Uptake of iodide	10 h
Incorporation of [^{14}C]amino acids in proteins	15 h
DNA content	2 d
Mitotic activity	2 d

maximal effect. *Some* effects on thyroid function are *rapid*: those involved in the activation of adenylate cyclase and the increased concentration of cellular cAMP, the stimulation of endocytosis of the colloid, and general metabolic pathways such as the pentose-phosphate cycle and mitochondrial respiration. Biosynthetic effects, such as the incorporation of [^{32}P]orthophosphate into phospholipids and the increase of RNA and proteins, occur *much later*. The effect of TSH on the synthesis of proteins and the mitotic activity of thyroid cells (the latter is not agreed upon by all authors) could be the source of hypertrophy and hyperplasia of the thyroid (goiter). With the exception of aerobic oxidation of glucose and metabolism of phospholipids, all known effects of TSH can be reproduced in vitro or in vivo by cAMP or its dibutyryl analog. An interesting aspect is the time course of action of TSH on the *transport of iodide*. The immediate effect of the administration of TSH is *biphasic*: transport is reduced during the first few hours, after which time it is greatly increased.

Studies of isolated thyroid cells in culture or preparations of thyroid plasma membranes and iodinated TSH with a high specific activity have demonstrated that the hormone binds reversibly with *specific receptors* located on the external surface of the plasma membrane of thyroid cells and adipo-

27. Ekholm, R. and Wollman, S. H. (1975) Site of iodination in the rat thyroid gland deduced from electron microscopic autoradiographs. *Endocrinology* 97: 1432–1444.
28. Dumont J. E. (1971) The action of thyrotropin on thyroid metabolism. *Vitams. Horm.* 29: 287–412.

cytes[29,30]. A single class of glycoprotein *receptors* with high affinity (K_D~1 nM) and low capacity (about 1000 sites/cell) has been demonstrated. Recently, photoaffinity labelling of the detergent-solubilized human TSH receptor suggested the presence of two subunits, M_r~49,900 and 32,500, linked by disulfide bridges. The TSH receptor from different species and tissues seems similar but is not identical. After solubilization, the receptor can bind TSH and thyroid-stimulating antibodies found in the serum of Graves' patients. TSH is not inactivated following its binding and subsequent release from the receptor. The TSH-sensitive *adenylate cyclase* system has the same properties as shown for other hormone-sensitive adenylate cyclases (Fig. VIII.15) (p. 101).

The sympathetic nervous system has a direct, positive effect on the secretion of thyroid hormones in humans. In addition, intrathyroidal nerves containing VIP regulate thyroid activity. Carbamylcholine, via a muscarinic receptor, increases cGMP levels and negatively affects thyroid metabolism. These regulatory processes are of unknown physiological and pathological significance.

The complex circuits of regulation of the thyroid cell by TSH may involve the following components: 1. the cAMP system (positive regulation through an unknown mechanism); 2. a Ca^{2+} system (positive control of certain effects, and negative control for others); 3. the feedback control of iodide (negative control of the cAMP system)[31].

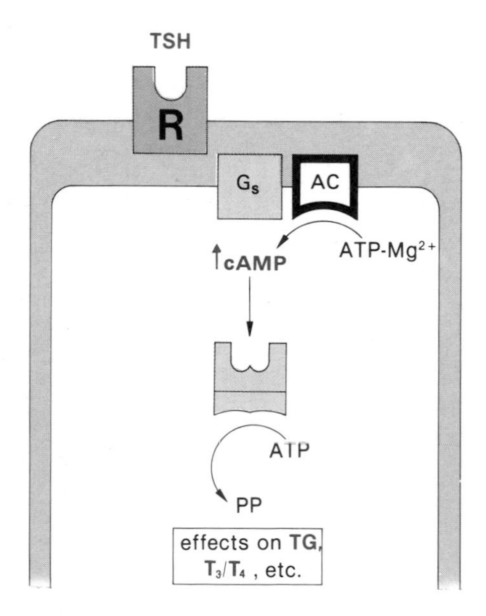

Figure VIII.15 Mechanism of action of TSH in the thyroid cell (Table VIII.2).

4.2 Iodine

Substrate in the synthesis of thyroid hormones, iodine *is also* involved as a *regulator* of certain metabolic reactions, in particular those which lead to the synthesis of hormones themselves. Administered *acutely*, low or moderate quantities of iodide (up to 2 mg in humans) have no influence on thyroid uptake of radioactive iodine given at the same time. In contrast, *a significant increase in plasma iodide provokes an inhibition* of iodine organification in the thyroid and a reduction of the synthesis of thyroid hormone (Wolff–Chaikoff effect)[32]. The small quantity of iodine incorporated into thyroglobulin is found essentially in the form of MIT, very little in the form of DIT, and practically none in the form of T_4. The thyroid is able to *resist* an *acute supply* of large doses of iodide, thus avoiding the excessive formation of hormones that could occur otherwise. An excess of iodide reduces the effects observed following an increase in cAMP concentration induced by TSH, but not when antithyroid drugs (MMI, PTU) are present, which block the oxidation of iodide.

The relative *inhibition* of organic iodination and of the synthesis of thyroid hormones is at least partially *reduced during chronic administration of iodide*. This phenomenon of *escape*, or adaptation, is due to a reduction in the transport activity of iodide, leading to a concentration of intracellular iodide insufficient to maintain the Wolff–Chaikoff effect at an optimal level. It constitutes a demonstration of the inhibition of iodide transport by autoregulation within the thyroid. In certain cases, this adaptation does not occur, and chronic inhibition of thyroid hormone synthesis is observed, with the appearance of a goiter, with or without hypothyroidism (iodide goiter).

29. Carayon, P., Guibout, M. and Lissitzky, S. (1979) The interaction of radioiodinated thyrotropin with human plasma membranes from normal and diseased thyroid glands. *Ann. Endocrinol, Paris* 40: 211–227.
30. Lissitzky, S. (1980) Thyrotropin receptor. In: *Autoimmune Aspects of Endocrine Disorders* (Pinchera, A., Doniach, D., Fenzi, G. F., Baschieri, L., eds), Academic Press, London, pp. 73–81.
31. Dumont, J. E. and Vassart, G. (1979) Thyroid gland metabolism and action of TSH. In: *Endocrinology* (DeGroot L. J. et al., eds), Grune and Stratton, New York, vol. 1, pp. 311–329.
32. Wolff, J. and Chaikoff, I. L. (1948) Plasma inorganic iodide as a homeostatic regulator of thyroid function. *J. Biol. Chem.* 174: 555–564.

4.3 Antithyroid Drugs[33]

Several Mechanisms

Antithyroid compounds block the synthesis of thyroid hormones and, since they lower the concentration of hormones in the circulation, lead to the formation of a goiter in response to the increase in TSH secretion. Antithyroid drugs are classified in *two groups*, according to whether they inhibit *thyroid transport of iodide* or the *transformation of iodide into organic iodine and the coupling of iodotyrosines.* In the *first group* are found monovalent anions, of which *thiocyanate* and *perchlorate* are the most active. Their action is reversed by high doses of iodide. A number of compounds appear in the *second group*. The most potent are characterized by the presence of the *thionamide* group within the molecule: S=C(N-)(X-) where X may be an atom of sulfur, oxygen or hydrogen. *6-n-propylthiouracil* (PTU) and 1-methyl 2-mercaptoimidazole (MMI) are the most active (Fig. VIII.16). Their action cannot be reversed by high doses of iodide. It is now clear that a large majority of these agents act by competitive inhibition of thyroid peroxidase[18], blocking the most sensitive coupling step.

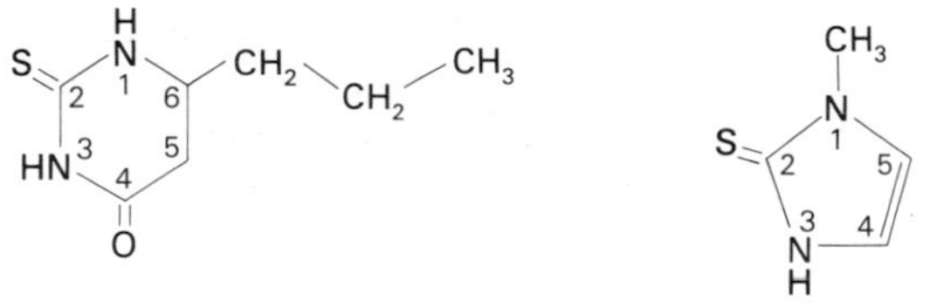

6-propyl 2-thiouracil

1-methyl 2-mercaptoimidazole

Figure VIII.16 Structure of two antithyroid drugs.

Natural antithyroid drugs are found primarily in cruciferous vegetables such as Brassica (cabbage, mustard greens). Goitrin, or L-vinyl 2-thiooxazolidone, isolated from rutabaga, is one of the most active. Glucosides which release thiocyanate have been found in certain vegetables. Cassava, a shrubby tree of the Euphorbiacae family which contains cyanogenic glucosides, is responsible for endemic goiter and cretinism in many regions in Africa.

5. Metabolism of Thyroid Hormones

5.1 Transport and Penetration into Cells[34]

Thyroid hormones are transported from the thyroid to tissues in a reversible association with certain plasma proteins called thyroid binding proteins (TBPs)[35]. In the human, at a pH of 7.4, 75% of T_4 is associated with a specific glycoprotein called *thyroxine binding globulin (TBG)*, 15% with a protein migrating before the serum albumin or *thyroxine binding prealbumin (TBPA)*, and the remainder with *serum albumin.* Various properties of these transport proteins are summarized in Table VIII.3.

TBG is an acidic glycoprotein possessing sialic acid residues. It binds T_4 with a high affinity (K_D ~50 pM), and T_3 with an affinity approximately 10 times lower. There is one binding site per molecule of protein.

TBPA is an oligomeric protein (four subunits of M_r~15,000). It binds thyroid hormones with an affinity lower than TBG, and it associates independently with a protein of lower molecular weight (21,000) which binds retinol (retinol binding protein). The binding of one ligand affects neither the interaction between the proteins nor the binding of the other ligand.

Interaction of T_3/T_4 with Plasma Proteins

As can be seen in Table VIII.3, the affinities of the three human TBPs for thyroid hormones are in inverse order with respect to their concentrations. For each, the affinity of TBP for T_3 is 10 times less than for T_4. The major result of the binding of thyroid hormones to TBPs is a considerable reduction in concentration of free hormones (p. 42). For T_4, the ratio of free hormone/total hormone is of the order of 0.025%, and for T_3, approximately 0.5%. However, the *function of TBPs* remains *unclear*. Subjects with congenital absence of TBG or serum albumin have normal use of

33. Merchant, B., Lees, F.H. and Alexander, W.K. (1979) Antithyroid drugs. In: *The Thyroid* (Hershman, J.A. and Bray, G.A., eds), Pergamon Press, Oxford, pp. 290–252.
34. Oppenheimer, J.H. (1968) Role of the plasma proteins in the binding, distribution and metabolism of the thyroid hormones. *N. Engl. J. Med.* 278: 1153–1662.
35. Gordon, A.H., Gross, J., O'Connor, D. and Pitt-Rivers, R. (1952) Nature of the circulating thyroid hormone plasma protein complex. *Nature* 169: 19–20.

Table VIII.3 Some properties of thyroid hormone binding proteins (TBPs).

Properties	TBG	TBPA	Albumin
MW	36,000	57,000	60,000
Concentration in serum:			
mg/ml	10	280	40,000
mol/l	2.7×10^{-4}	2×10^{-2}	0.6
K_D for T_4 (M)‡	0.5×10^{-10}	1×10^{-3}	1×10^{-5}
K_D for T_3 (M)‡	0.5×10^{-9}	1×10^{-8}	$<1\times10^{-4}$
Binding capacity (μgT_4/ml)	0.2	2.7	>450
Distribution of T_4 among TBP (%)	75	15	10

‡Each molecule of TBG or TBPA can maximally bind 1 mol T_4 or T_3 on its high affinity site. TBPA and albumin have low affinity sites capable of binding a higher number of T_4 (or T_3) molecules.

thyroid hormones. It is certain, however, that TBPs form a blood reservoir of hormones. Acting as buffers, they *regulate the concentration of free hormones*. It is believed that only the latter are able to leave the circulatory compartment and penetrate cells. The concentration of free hormone in plasma correlates better with the functional thyroid state than the level of total thyroid hormone in plasma. For example, during pregnancy, both the capacity of TBPs to bind T_4 and the quantity of T_4 bound increase. There is not, however, an increase in the concentration of free T_4, and no signs of thyrotoxicosis are observed, although total total T_4 is at a concentration normally associated with hyperthyroidism. Data on pathophysiological and pharmacological variations of the TBPs can be found in reference 36.

Free thyroid hormones penetrate cells by an *unknown* mechanism. This transport could occur at the plasma membrane level after the binding of T_4 and T_3 to a specific protein, resulting in their endocytosis and internalization. Low affinity binding proteins have been observed in the cytosol; they may play a buffering role in the regulation of the concentration of free hormone in the cell. Receptors are located in the nucleus (p. 70). The fact that T_4 is mostly extracellular, and T_3, intracellular, results from a reduced plasma binding and increased cellular binding of T_3.

5.2 Peripheral Metabolism and Excretion

In humans, the half-life of T_4 in serum is 6 to 7 d, and that of T_3, 1 d. The distribution volume of T_4 is 10 l, while that of T_3 is 43 l. The liver contains one third of extrathyroid T_4, and this pool is rapidly exchangeable with T_4 in the plasma.

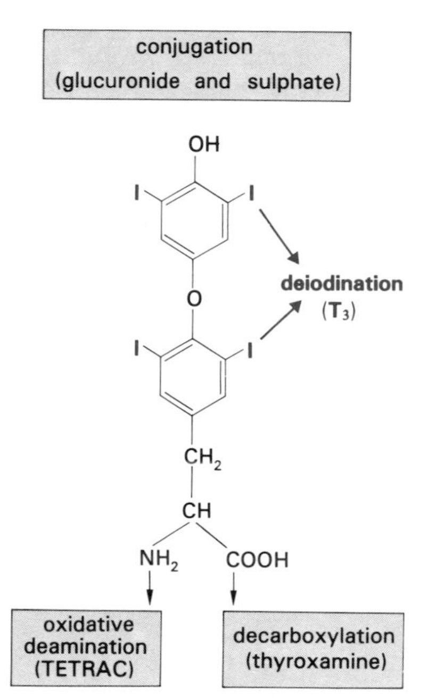

Figure VIII.17 Metabolic reactions of thyroxine in vivo. Some of the principal metabolites are indicated in parentheses.

Thyroid hormones are metabolized principally in the liver, kidney, brain, and muscle. The largest portion is deiodinated; the remainder is deaminated, decarboxylated, or excreted in the feces and urine (Fig. VIII.17).

36. Gershengorn, M. C., Glinoer, D. and Robbins, J. (1980) Transport and metabolism of thyroid hormones. In: *The Thyroid Gland* (De Visscher, M., ed), Raven Press, New York, pp. 81–121.

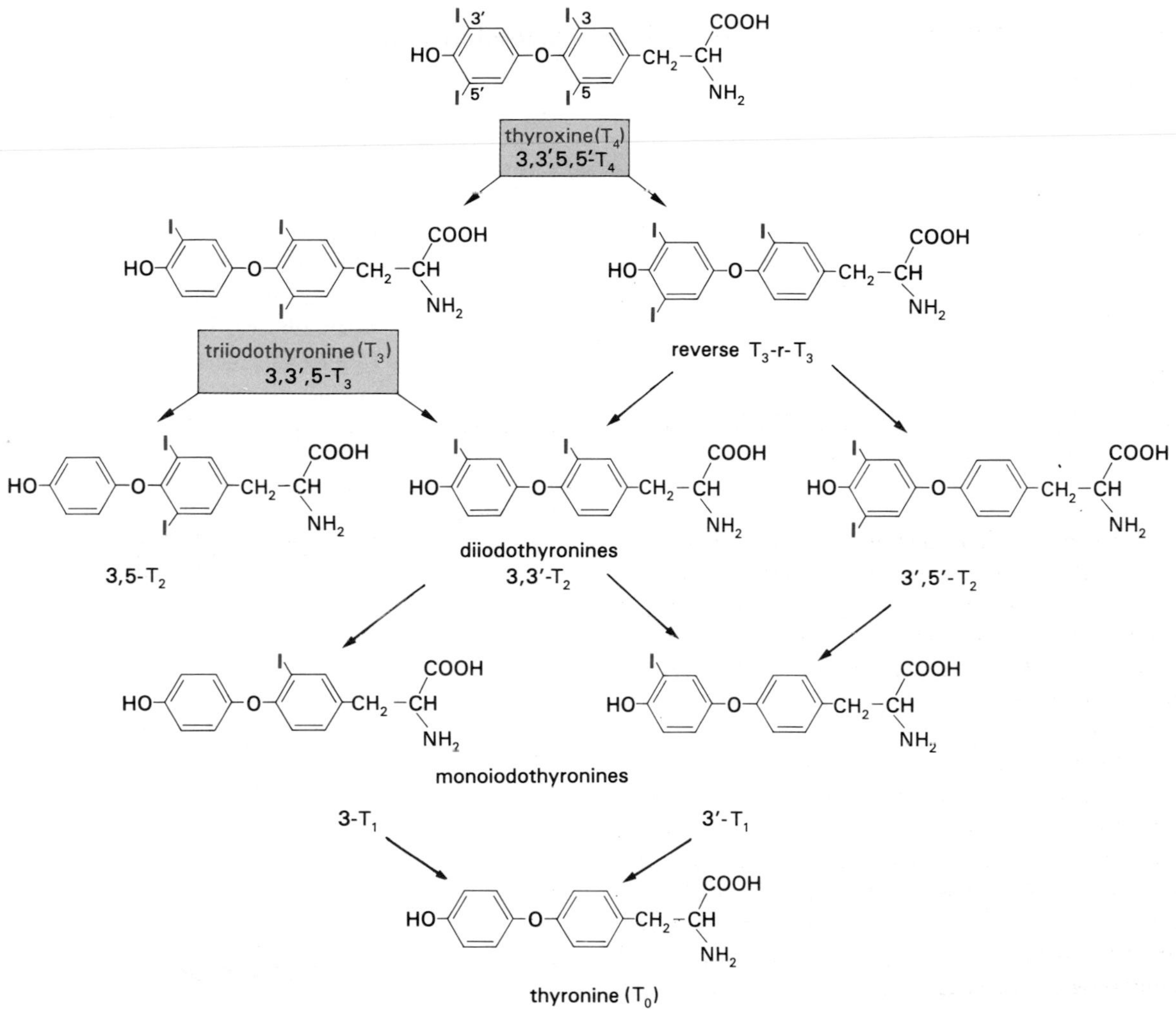

thyroxine (T_4)
3,3',5,5'-T_4

triiodothyronine (T_3)
3,3',5-T_3

reverse T_3-r-T_3

3,5-T_2

diiodothyronines
3,3'-T_2

3',5'-T_2

monoiodothyronines

3-T_1

3'-T_1

thyronine (T_0)

Figure VIII.18 Thyroxine and its products of deiodination.

Deiodination

Iodide which comes from the deiodination of thyroid hormones is *reutilized* for the synthesis of hormones in the thyroid, or is excreted. Most excretion products found in the urine and feces retain the diphenyl ether structure of the hormone.

The principal pathway for the metabolism of T_4 is *deiodination*[37]. It accounts for approximately 80% of its degradation. It was not until 1970 that a convincing demonstration of the conversion of T_4 to T_3 (5' monodeiodination) or rT_3 (reverse T_3) (5 monodeiodination) by extra-thyroid tissues was shown, following the development of RIAs specific for these compounds. The three diiodothyronines, the two monoiodothyronines, as well as thyronine, shown in Figure VIII.18, have been identified.

The *conversion of T_4 to T_3* is the essential pathway for the deiodination of T_4[37,38]. Approximately 80% of the circulating T_3, and 95% of the circulating rT_3, is derived from the peripheral monodeiodination of T_4. Figure VIII.19 shows the rate of production of the principal human iodothyronines.

In numerous test systems, the *biological activity of T_3 is the greatest*, while rT_3 is devoid of activity, as is probably also the case for T_4. Most, if not all, of the effects of T_4 can be explained by its conversion into T_3. This pathway can be considered an *activation*

37. Braverman, I. E., Ingbar, S. H. and Sterling, K. (1970) Conversion of thyroxine (T_4) to triiodothyronine in athyreotic human subjects. *J. Clin. Invest.* 49: 855–894.
38. Schwartz, H. L., Surks, M. I. and Oppenheimer, J. H. (1971) Quantitation of extrathyroidal conversion of L-thyroxine to 3,5,3' triiodothyronine. *J. Clin. Invest.* 50: 1124–1130.

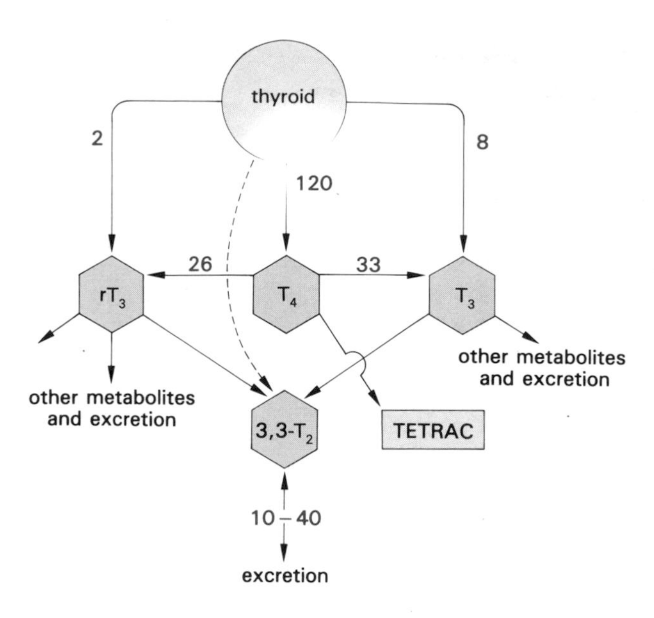

Figure VIII.19 Approximate thyroid and peripheral production rates (nmol/day) of some iodothyronines in humans. (Jansen, M., Krenning, E. P., Oostdijk, W., Docker, R., Brande, V. D. and Henneman, G. (1982) *Ann. Endocrinol.*, *43*, abst. n°144: 82A).

step, while the conversion of T_4 to rT_3 irrevesibly inactivates the hormone.

Deiodination enzymes have been localized in the liver, kidney, and brain. They probably exist in other tisues as well. They require the presence of thiol compounds, and reduced glutathione is probably the endogenous cofactor. In certain tissues, such as the pituitary and brain, the intracellular concentration of T_3 depends essentially upon the local production of T_3 from T_4. Generally, the regulation of T_3 production in these tissues is independent of that which occurs in the liver and kidney. However, almost nothing is known of the factors regulating the rate of degradation of T_3.

Thiouracil and its derivatives (propylthiouracil) are potent *inhibitors of iodothyronine-deiodinases.* It is possible that during the treatment of thyrotoxicosis this inhibition of conversion of T_4 to T_3 complements the effect of the drug on the synthesis of thyroid hormones in the gland.

The liver also contains non-dissociable (covalent) complexes formed by T_4, T_3, and proteins. At iodine equilibrium, they constitute approximately 10% of the total iodine in the tissue. Their half-life is approximately 4 to 5 d. The physiological role of these complexes is unknown.

The overall process of thyroid hormone synthesis suggests that the gland is regulated in such a way as to maintain the inactive precursor, or *prohormone (T_4)*, at a constant level. The quantity of *active hormone (T_3) produced is regulated at the level of the peripheral tissue*; namely, in target organs. Several clinical observations tend to indicate that the quantity of T_3 produced is adjusted to the needs created by specific circumstances. The modulation of the conversion of T_4 to T_3 by physiological alterations or pharmacologic agents has been observed during fasting, sugar intake, diabetes, and during the administration of PTU, dexamethasone, propranolol, diphenylhydantoin and various contrast media used in radiography.

Conjugation

The conjugation of the phenol group of thyroid hormones with *glucuronic acid* or *sulfuric acid* occurs in the *liver*. These conjugates are secreted with bile into the duodenum. About 0.6% of T_4 penetrating the liver is secreted in bile. T_3 is secreted at a rate 20 times greater than that of T_4. The conjugates are hydrolyzed in the intestine, and up to 30% of the free hormone produced is reabsorbed. These data indicate the importance of the enterohepatic cycle in the plasma T_3/T_4 ratio and the maintenance of a euthyroid state.

Oxidative Deamination and Decarboxylation

Some thyroid hormones or their deiodinated derivatives are transformed into *acetic derivatives* (triiodo- or tetraiodothyroacetic acid: *TRIAC or TETRAC*), and, by decarboxylation, into triiodo- and tetraiodothyronamines. These compounds are excreted into the urine and feces. It is rather unlikely that these derivatives are active forms of the hormone, although they do have a certain thyromimetic effect. It is possible, however, that TRIAC and TETRAC are produced in target cells and that they enter the nucleus.

Excretion

In addition to their excretion, in feces, in the conjugated form, a portion of thyroid hormones can be found in the urine. In humans, approximately 8 μg of T_4 and 3 μg of T_3 are eliminated in this manner each day.

6. Effects of Thyroid Hormones and their Mechanism of Action

Thyroid hormones exert regulatory effects that apparently have no direct relationship with one another, on *almost all organs and major functions* of the body (Table VIII.4). A *specific receptor* has been identified in the *cell nucleus* in all target tissues. The multiple effects of thyroid hormones can be classified in two categories: those which control *development*, and those which ensure the *regulation of metabolic activity* during and after development.

6.1 Effects on Development

Development is the sum of growth and differentiation.

Growth is defined as a permanent increase in the total mass of an organism. It could result from either an increase in cell size or increased cell division. *Differentiation* corresponds to all of the complex modifications that lead to diversification of cellular function and morphogenesis.

Early *fetal growth* appears to be controlled only by genes. There are no data to show that thyroid hormones are necessary for development before the appearance of the functional fetal thyroid. In fact, thyroid hormones do not cross the placental barrier. In several species, deprivation of maternal thyroid hormones does not interfere in any appreciable manner with growth before birth. However, these hormones are essential for the differentiation and maturation of fetal tissues, particularly the brain and skeleton, once the fetal thyroid gland has become functional. Recent studies suggest, however, that the young fetus constitutes an important site of maternal T_4 delivery and metabolism. The T_3/T_4 receptor is present in human fetal brain at mid-pregnancy.

Table VIII.4 General effects of thyroid hormones.

Effects on differentiation and growth
Brain maturation
Cartilage proliferation and epiphyseal maturation
Lung maturation (surfactant)
Endocrine development (growth hormone regulation)
Amphibian metamorphosis
Metabolic effects
Oxygen consumption and calorigenesis
Mineral balance
Carbohydrate, lipid and protein metabolism
Synergism or antagonism with other hormones (insulin, steroids, catecholamines)

In *postnatal* life, growth depends on the normal function of the thyroid, and a deficiency during the neonatal period leads to severe retardation of the growth of almost all organs. In addition, the growth of the organism is stopped by deprivation of thyroid hormone. A deficiency of growth hormone is possible, secondary to the primary deficiency of thyroid hormones. The reduced stature of hypothyroid infants is the result of a reduced cell population, a direct consequence of reduced mitotic activity. The effects of thyroid hormone deficiency on the differentiation of the skeleton and brain are most evident. In utero, growth of the skeleton is not affected, but differentiation is markedly retarded. After birth, both growth and differentiation are altered by hormonal insufficiency.

Thyroid hormones are especially necessary for the development of the *central nervous system*. Deficiency occurring during fetal life or at birth leads to the conservation of infantile characteristics of the brain, a reduced number of cortical neurons, retarded myelinization, and reduced vascularization. In postnatal life in rats and humans, numerous interneuronal connections are formed; growth of neurites as well as the assembly of tubulin in microtubules seems to be dependent, at least partially, on thyroid hormones[39]. In the absence of any corrective therapy, irreversible lesions characterized by a slowing of all intellectual functions (*cretinism*) are produced. After two years of age, however, hypothyroidism in the human has only very minor effects on mental development, and any problems which do appear are reversible.

One of the most spectacular effects of thyroid hormones is the induction *of tadpole metamorphosis*. A complex series of events leads to the resorption of the tail of the tadpole and the development of its adult physiognomy.

The action of thyroid hormones on development can be explored at a biochemical level; for example, in the liver, during the reinitiation of *growth* in young thyroidectomized *rats*, or during *metamorphosis of the tadpole*. A latent period of a few hours to several days precedes the primary responses. These include the activation of the synthesis of primary cellular ele-

39. Fellous, A., Lennon, A. M., Francon, J. and Nunez, J. (1979) Thyroid hormones and neurotubule assembly in vitro during brain development. *Eur. J. Biochem.* 10: 365–376.

Table VIII.5 Principal biochemical actions of thyroid hormones in the liver of thyroidectomized rats and in tadpoles.

Stimulation or increase of amount	Latent period (h)		Time for maximum effect (h)	
	rat	tadpole	rat	tadpole
Synthesis of rapidly labelled nuclear RNA	4–6	25–30	22	50–60
RNA polymerase I (rRNA)	10–12	—	40	—
RNA polymerase II (mRNA)	18–20	—	50	—
Incorporation of amino acids in proteins	18–24	—	40–45	—
Synthesis of phospholipids in microsomes	12–16	32–36	40	70–80
Mitochondrial respiratory enzymes	24–30	50–60	50–60	100
Serum albumin	—	90–100	—	250
Hemoglobin of the adult type	—	80–90	—	250
Enzymes of the urea cycle	—	60–70	—	150

Thyroidectomized rats received 15 to 25 μg T_3, and tadpoles of *Rana catesbeiana* 0.5 to 1 μg[44].

ments, which appear in a sequential manner (Table VIII.5)[40].

Biochemical studies of amphibian metamorphosis have shown that morphological changes due to the action of thyroid hormones are accompanied by specific changes in numerous organs; for example, in the appearance of enzymes involved in ureogenesis in the liver of the frog (these enzymes are absent in the liver of the tadpole). This does not appear to involve an activation of preexisting enzymes, but rather the induction of new enzymatic proteins. This is also the case for the induction, by T_3, of numerous enzymes and proteins, such as malic enzyme, $\alpha_{2\mu}$-globulin, Na^+-K^+ ATPase and carbamyl phosphate synthase, either in vivo or in hormone-sensitive cell lines in culture. It is therefore clear that thyroid hormones contribute to development and growth by stimulating the synthesis of specific proteins.

6.2 Metabolic Effects and Thermogenesis

Thyroid hormones affect the metabolism of proteins, sugars, lipids, water, mineral ions, and nitrogen (urea, creatine). One of their most remarkable effects is their capacity to increase *oxygen consumption and heat production*. As more than 90% of the total oxygen consumption of the organism takes place in *mitochondria*, these organelles were considered early on as the possible target of thyroid hormones.

Two Sources of Heat

Heat production in homeotherms is associated with respiration and concomitant oxidation of substrates of dietary origin. The quantity of energy released by the oxidation of food is progressive and controlled. A portion of the enthalpy of substrate oxidation (approximately 25%) is *conserved in ATP*. It is *not necessary, therefore, to uncouple phosphorylative oxidation and respiration* in order to generate relatively large amounts of heat during the oxidation of a substrate. Heat is produced immediately by the hydrolysis of ATP, and on a longer term involves the oxidation of substrates by the mitochondrial respiratory chain.

The action of thyroid hormones on *thermogenesis* is characterized by a *period of latency* of about 12 h and a maximal response 48 h after a single injection of the hormone. This action can be observed in adult mammals, in numerous tissues (skeletal muscle, heart, liver, and kidney). The stimulation of RNA and protein synthesis appears to be necessary for this action of thyroid hormones on mitochondrial activity. A continuous increase in mitochondrial respiration, however, requires an increase in the utilization of ATP (Fig. VIII.20). The process of the utilization of ATP stimulated by thyroid hormones[40] probably involves

40. Edelman, I. S. and Ismail-Beigi, F. (1974) Thyroid thermogenesis and active sodium transport. *Rec. Prog. Horm. Res.* 30: 235–254.

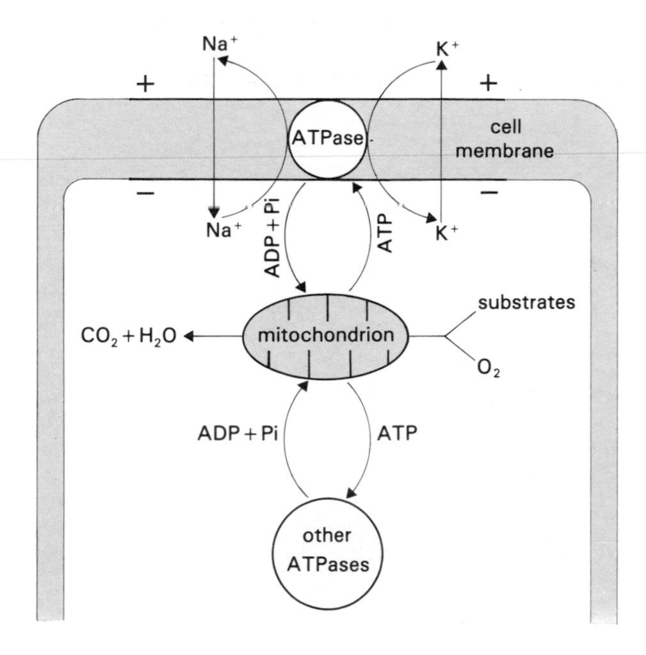

Figure VIII.20 Synthesis and utilization of ATP. Top, Na^+-K^+-dependent ATPase of the plasma membrane. Below, intracellular ATPases (other ATPases). Under conditions of adequate contributions of oxidizable substrates and oxygen, the ADP/ATP ratio controls the Qo_2. At equilibrium, the rate of hydrolysis of ATP is equal to its synthesis by rephosphorylation of ADP. The renewal of ATP depends on its consumption, provided that the mitochondrial capacity to synthesize ATP is not surpassed by the rate of hydrolysis. The synthesis of ATP from ADP+P_i by mitochondria is coupled to the oxidation of substrates. Two modes of ATP utilization have been shown by membrane and other intracellular ATPases[40].

the *sodium pump*, which consumes, in cells at rest, 20 to 45% of the total ATP synthesized. This membrane pump is controlled by Na^+-K^+-dependent ATPase, which stimulates the extrusion of intracellular Na^+, with a concomitant increase in extracellular K^+. In euthyroid rats, more than 90% of the increase in oxygen consumption (studied in vitro, in liver or skeletal muscles, after an injection of thyroid hormones in vivo) can be attributed to an increased utilization of ATP by the sodium pump and to a concomitant increase of Na^+-K^+ ATPase.

In fact, when the transmembrane transport of Na^+ is inhibited by ouabain, the effects of T_3 on respiration in liver slices are abolished, and those in kidney slices are reduced by 50%. However, this metabolic pathway might account only partially for the effect of thyroid hormones on thermogenesis. An important portion of the increase in O_2 consumption (50% according to certain authors) may be insensitive to ouabain or to Na^+ concentration.

The *initial studies* on the mechanism of action of thyroid hormones on *respiration* led to the conclusion that these hormones uncouple phosphorylative oxidation, since the synthesis of ATP per atom of oxygen consumed is reduced. Very high doses of hormone were used in these studies. These data, therefore, should be regarded only as evidence of pharmacological responses.

6.3 Molecular Mechanism of Action

Receptors and Diversity of Responses

In 1972, in vivo studies demonstrated saturable specific nuclear T_3 binding sites in rat liver and kidney[41]. Recently, the receptor has been identified by molecular cloning as *the normal product of the* c-*erb*-A *gene*. The T_3 nuclear receptor *belongs therefore, to a gene superfamily that includes related oncogenes and steroid receptor genes* (Fig. VIII.21A and p.70). There are small but significant tissue-specific differences in the T_3 receptor structure, e.g., between hepatic, cerebral or cardiac receptors. These differences could facilitate the development of specific T_3 analogs with important clinical applications. The human thyroid receptor has a molecular weight of ~45,000. The region most homologous to steroid receptors and *erb*-A proteins is the putative DNA binding site (p. 67 and Fig. 1.71). Table VIII.6 lists the biological activity and the nuclear receptor binding affinity of several T_3 analogs in rat liver.

Physiologically, 35 to 50% of nuclear receptors are saturated by T_3 in the rat. Table VIII.6 shows that the high affinity T_3 receptor ($K_D \sim 10^{-10}$) also binds T_4, however with an affinity 10-fold lower; *TRIAC* binds with high affinity. Other hormones, such as steroids, do not bind to the thyroid hormone receptor. A good argument in favor of an active nuclear receptor for T_3 is the relationship observed between the number of binding sites and the sensitivity of the tissue to thyroid hormones (~5000 sites per liver cell). There is also a good correlation between the thyromimetic activity of different structural analogs and their rela-

41. Oppenheimer, J. H., Koerner, D., Schwartz, H. L. and Surks, M. I. (1972) Specific nuclear triiodothyronine binding sites in rat liver and kidney. *J. Clin. Endocr. Metab.* 35: 330–333.

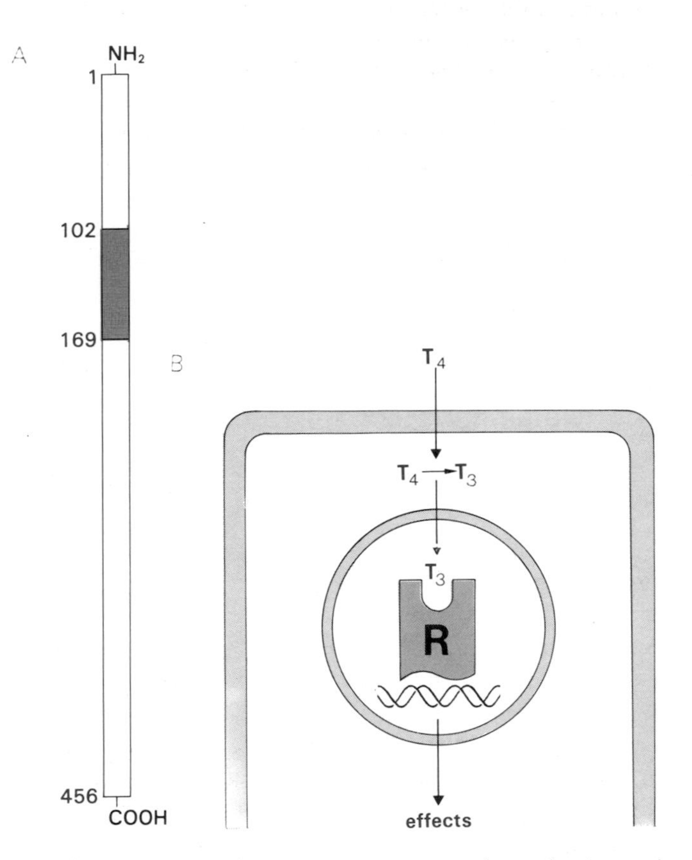

Figure VIII.21 Receptor structure and mechanism of action of thyroid hormones. **A.** The overall organization of the T_3/T_4 receptor protein (Fig. I.71). **B.** Mechanism of action of iodothyronines. T_4, shown entering the cell, is partially converted to T_3, which has higher affinity for the receptor.

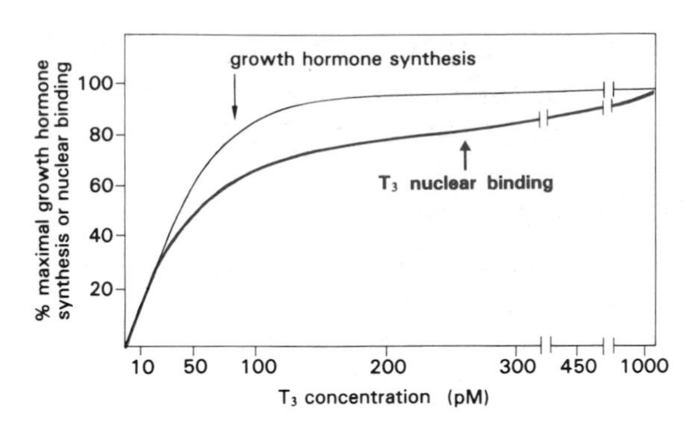

Figure VIII.22 Relationship of nuclear receptor occupancy to growth hormone synthesis in serum-free conditions in cultured GH_1 cells. For comparison, the results are expressed as a percent of maximal response or binding[42].

tive affinity for the nuclear T_3 receptor. In a line of pituitary cells in culture (GH_3), a quantitative relationship has been demonstrated between the occupation of the nuclear receptor by T_3, and the induction of growth hormone[42] synthesis (Fig. VIII.22). The nuclear receptor for T_3 ensures the initiation of the effect of T_3 (or possibly T_4) on gene transcription, leading to variations in the rate of mRNA synthesis (Fig. VIII.21B).

Such a *unified concept of the mechanism of thyroid hormone action* accounts for their anabolic effect and for the very small amount of hormone effective in vivo. The *diversity of cell responses depends on their state of differentiation.*

Whether other T_3 and/or T_4 binding sites described in various non-nuclear fractions of the cell[43] play a role in hormone response remains conjectural.

Table VIII.6 Biological activity and nuclear receptor binding affinity (in rat liver) of several T_3 analogs (% relative to T_3).

Analog	Biological activity (in vivo)	In vivo binding	Binding to isolated nuclei	Binding to solubilized receptor
T_3(L-T_3)	100	100	100	100
D-T_3‡	7.4	10	60	62
TRIAC	6	100	250	282
3′-isopropyl-T_2	100	100	125	92.5
T_4	18	10	12.5	14.4
rT_3	0.1	0	0.1	0.2
4′-NH_2-T_3	0.2	—	0.8	—
DIT	0	0	0	0
T_o	0.08	0	0	0

‡Refers to the optical isomer of natural T_3.

6.4 Relationship between the Chemical Structure and the Activity of Thyroid Hormone Analogs

Hundreds of analogs

The relatively simple structure of thyroid hormones has led to numerous attempts to establish the structural characteristics necessary for hormonal activity, and to the synthesis of analogs which are capable of mimicking or inhibiting the physiological effects of the hormones. Several hundred structural analogs of T_4 and T_3 have been synthesized[44]. Their activity has been determined in vivo for several aspects, such as oxygen consumption, antigoitrogenic effects in rats and mice receiving antithyroid drugs, and metamorphosis in tadpoles. They have allowed the definition of the essential *characteristics necessary for thyromimetic activity*, namely, a *lipophilic center*, furnished by thyronine, and *two specific anionic* groups sterically maintained at the opposite extremities of the molecule. A schematic representation of the steric and functional characteristics of an active hormonal molecule bound to its receptor is found in Figure VIII.5. Interaction with the receptor is likely to provoke a conformational change resulting in an effect on gene transcription. Studies using analogs may indirectly describe the chemical specificity of the hormone receptor. The 3, 5-diiodotyrosyl moiety bearing the alanine chain could be responsible for the transport and binding of thyroid hormones to the receptor, while the phenolic ring could represent a separate functional unit which would be the initiation point for hormonal action. The validity of this hypothesis will eventually be assessed once the three-dimensional structure of the nuclear receptor is known.

Analogs resulting from the replacement of iodine by alkyl groups were shown to have hormonal activity even greater than the hormones themselves. For instance, the 3,5-diiodo 3′-isopropyl derivative is 10 times more active than T_4. Analogs which bind to the receptor but which lack the anatomic and steric characteristics necessary for biological action were used as possible antagonists of T_4 and T_3. Large amounts of acetic rT_3 analogs block the stimulation of O_2 consumption by T_4 and T_3 in rat; their utility in the treatment of thyrotoxicosis remains to be established, since they may possess some thyromimetic activity.

42. Samuels, H. H., Stanley, F. and Shapiro, L. J. (1978) Dose-dependent depletion of nuclear receptor/re L-T_3: evidence for a role in induction of growth hormone synthesis in cultured GH_1 cells. *Proc. Natl. Acad. Sci., USA* 73: 3877–3881.
43. Menezes-Ferreira, M.-M. and Torresani, J. (1983) Mécanismes d'action des hormones au niveau cellulaire. *Ann. Endocrinol., Paris* 44: 205–216.
44. Jorgensen, E. C. (1978) Thyroid hormones and analogs. I. Synthesis, physical properties and theoretical consideration. In: *Hormonal Proteins and Peptides* 6 (Li, C. H., ed), Academic Press, New York, pp. 107–204.

7. Functional Investigation and Pathophysiology

Almost all symptoms of thyroid diseases result from an increased or decreased production of thyroid hormones. An enlarged thyroid gland is termed a *goiter*.

It is possible to study the *functional state of the thyroid* by measuring the blood concentration of hormones and other iodinated compounds, the *turnover of radioactive iodine* and its *distribution in the gland*, the *dynamics of thyroid function*, and the *peripheral effects* of thyroid hormones.

The radioisotopes of iodine most frequently used in research and for clinical investigation of the thyroid are ^{131}I (half-life=8 d, β and γ-emitter (0.606 MeV), principal photon energy (PPE) 364 KeV); ^{125}I (half-life=60 d, γ-emitter, PPE 28 KeV); ^{132}I (half-life=0.1 d, β-emitter (2.12 MeV), PPE 670 KeV); ^{123}I (half-life=0.55 d, γ-emitter, PPE 159 KeV).

7.1 Iodinated Compounds in Serum

The principal iodinated compounds in serum are *iodide* and *thyroid hormones*. In addition, small quantities of DIT and thyroglobulin are observed.

Table VIII.7 Iodinated components of normal human serum.

Compounds	Concentration‡ (M)	(/ml)
Iodide	3×10^{-4}	—
Total T_4	1×10^{-7}	78 ng
Free T_4	3×10^{-11}	23 pg
Total T_3	2×10^{-9}	1.3 ng
Free T_3	5×10^{-12}	3.2 pg
DIT	3×10^{-9}	1.2 ng
TG	3×10^{-11}	21 ng

‡These values should be considered approximate, since they vary with the method used, and since limits of physiological variations are not given.

Table VIII.7 indicates normal concentrations of these compounds, which are *mostly bound to TBP* (p. 43). Serum T_4 can be evaluated by a competitive protein binding assay with TBG, by radioimmunoassay (RIA), or by enzyme-linked immunosorbent assay (ELISA).

Measurement of the concentrations of free T_3 and free T_4 using column chromatography and RIA results in values of 2.4 to 4.9 and 5.8 to 18.2 pg/ml, respectively. The values observed in hyperthyroid and hypothyroid patients do not overlap the range of these normal values. Our results suggest that the most sensitive assays are those of free T_3 and free T_4 in hyperthyroidism, and TSH and free T_4 in hypothyroidism. RIAs for rT_3 and various diiodo- and monoiodothyronines have been developed recently, but they are of limited diagnostic interest. The determination of protein-bound iodine and butanol extractible iodine is now obsolete.

Thyroglobulin

TSH stimulates the secretion of TG into the circulation. In normal children older than one year, and in adults, plasma TG levels are 5 to 35 ng/ml. Fluctuations have been observed in thyroid diseases. In endemic goiter, an increase in plasma TG levels might be a sensitive sign of acute or chronic thyroid stimulation. In Graves' disease, variable results were observed (with, however, a tendency to an increase). This is probably due to the presence, *in the blood of Graves' patients, of autoimmune antibodies against TG* which interfere with the RIA. High levels of plasma TG were measured in patients with proved deficiencies of iodine metabolism, whereas undetectable levels were found in the plasma of patients with TG-deficient glands. In carcinoma, estimation of plasma TG levels is essential for the management of differentiated epitheliomas of the thyroid[45]. Radioisotope testing of patients receiving thyroid hormone therapy is useless when plasma TG levels are undetectable. A search for metastases should be undertaken when TG is detectable in the absence of residual thyroid tissue.

45. Schlossberg, A. H., Jacobson, J. C. and Ibbertson, H. K. (1979) Serum thyroglobulin in the diagnosis and management of thyroid carcinoma. *Clin. Endocrinol.* 10: 17–27.

7.2 Thyroid Turnover of Radioactive Iodine

The high iodine content of the thyroid and the existence of numerous radioisotopes of iodine with appropriate physical characteristics make an in vivo functional evaluation of the thyroid possible. Administered in the form of iodide, radioactive iodine mixes rapidly and uniformly with endogenous, stable iodide, allowing its metabolism to be followed.

Thyroid Uptake of Radioactive Iodine

After the oral administration of a *single tracer dose* of ^{131}I, and now of ^{123}I, it is possible to measure the *amount accumulated in the thyroid* after 2, 24 and sometimes 48 h. In these standard conditions, it is possible to express the value as a *percentage of the dose administered.* In normal subjects, maximal uptake occurs between 6 and 24 h (Fig. VIII.23), then

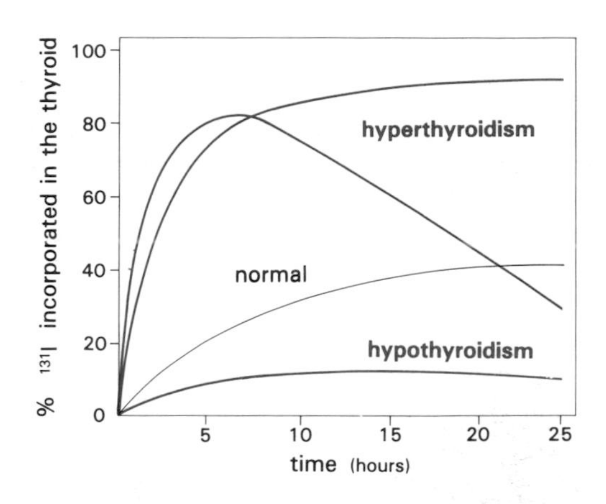

Figure VIII.23 Time-course of thyroid uptake of a tracer dose of [^{131}I]iodide after oral administration. Note the two types of response for hyperthyroid subjects. (De Visscher, M., and Barger, G. (1980) *The Thyroid Gland* (De Visscher, M., ed), Raven Press, New York).

the radioactivity associated with the thyroid decreases slowly. Normal values vary from one geographical region to another. They are inversely related to the dietary intake of iodine. In regions where the intake of iodine is sufficient, the uptake at 24 h is 25 to 30% of the doses administered. Maximal uptake is reduced in hypothyroid patients; in hyperthyroidism, it is increased and occurs sooner, and the disappearance of radioactivity associated with the thyroid is faster, thus reflecting the increase in hormone secretion.

The periods of 2 and 24 h have been chosen for practical reasons. The 2 h value is particularly increased in hyperthyroidism, while the 24 h value can be within normal limits. At 24 h the normal thyroid uptake reaches its maximum. In hypothyroidism, it is reduced. Because of its very short half-life (2 to 3 h), the use of ^{132}I is preferred in children. *Technetium* (^{99m}Tc) in the form of pertechnetate (TcO_4) is concentrated, as is iodide, in the thyroid, but is not incorporated into the organic form; after its uptake, it leaves the gland by simple diffusion. Its physical properties (half-life of 6 h, and weak γ emission without β emission) make it an ideal isotope for short-term studies of thyroid uptake.

Scintiscanning

Scintiscanning (Fig. VIII.24) allows the *localization* of radioactive iodine in the thyroid and the determination of the eventual presence of *zones that fix more or less* radioactive iodine. A scintigram is produced after the administration of a single dose of ^{131}I, either by linear scanning detectors, or now, by gamma-ray cameras which "see" at each instant the entire volume of the neck, reproducing it in the form of an image, with radioactive densities of the thyroid body in a single plane.

Figure VIII.24 Scintiscan of a normal thyroid gland after ^{131}I administration. Note the uniform distribution of the radioactivity in the lobes.

7.3 Radioimmunoassay of TSH

TSH is routinely measured in human serum by immunoassays, using monoclonal antibodies [46]. The detection limit is less than 0.1 μU/ml (~20 pg/ml) and allows the distinction to be made between patients with suppressed TSH (<0.1 μU/ml) and subjects with normal values (~0.3 to 3.0 μU/ml). Measurement of TSH is essential to the confirmation of primary hypothroidism (TSH levels up to 1000 μU/ml). In hypothalamus-pituitary-thyroid disorders, TSH varies from non-detectable to higher-than-normal levels[47].

7.4 Investigation of Peripheral Effects of Thyroid Hormones

A large number of tests

The measurement of *basal metabolic rate (BMR)*, defined as the quantity of heat produced by the organism under basal conditions, was previously one of the most useful measures for the evaluation of thyroid function. However, because of the difficulty of establishing basal conditions, and due to the large number of extrathyroid factors that influence measurement, it was abandoned. The same is true for the measurement of *cholesterol* in plasma, which is elevated in hypothyroidism. The study of the *Achilles' reflex time*, which examines the action of thyroid hormones on muscular contraction, is still used because of its simplicity. It consists of measuring the time passed between the moment of impact on the Achilles tendon and the half-relaxation of the muscle. Normal values are between 270 and 330 ms. In hypothyroidism, values are greater, whereas in hyperthyroidism they are lower.

7.5 Dynamic Tests

TRH Stimulation Test

The hypothalamic hormone *TRH stimulates the secretion of TSH*. The intravenous injection of TRH

46. Benkirane, M., Bon, D., Bellot, F., Princé, P., Hassoun, J., Carayon, P. and Delori, P. (1987) Characterization of monoclonal antibodies to human thyrotropin and use in an immuno-radiometric assay and immunohistochemistry. *J. Immunol. Methods* 98: 173.
47. Carayon, P., Martino, E., Bartalena, L., Mammoli, C., Costagliola, S. and Pinchera, A. (1987) Clinical usefulness and limitations of serum thyrotropin measurement by 'ultrasensitive' methods. *Hormone Res.* 26: 105.

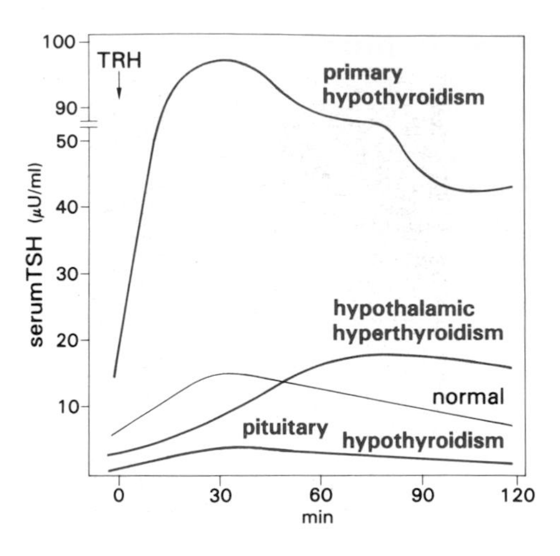

Figure VIII.25 TSH responses to TRH. (After De Visscher and Barger (1980) *The Thyroid Gland* (De Visscher, M., ed), Raven Press, New York).

leads to a rapid increase in the level of TSH, which is maximal at 30 to 40 min, returning to normal levels at 3 h (Fig. VIII.25). This test allows the evaluation of *pituitary reserves* of TSH in doubtful cases of hypopituitarism, and permits the differentiation between hypothalamic and pituitary lesions, particularly in the case of hypopituitary dwarfism.

TSH Stimulation Test

The TSH stimulation test allows the *distinction* to be made between *primary and secondary hypothyroidism*. In the former, the thyroid is maximally stimulated by endogenous TSH, and an injection of exogenous TSH does not cause any additional stimulation. In secondary hypothyroidism, the injection of TSH stimulates a significant increase in thyroid activity. This is measured by the evaluation of thyroid uptake of radioactive iodine before and after the injection of TSH.

Thyroid Suppression Test

In normal subjects, the administration of T_3 reduces thyroid activity by suppressing TSH secretion. In cases of Graves' disease and toxic adenoma where thyroid function is independent of TSH, T_3 does not affect thyroid function. Thyroid activity was evaluated by measuring thyroid uptake of iodide before and after injection of T_3. This test is no longer used. It has been replaced by the TRH test, which is much more rapid, and simpler, and more reliable.

7.6 Hypothyroidism

All anatomical or functional anomalies of the thyroid that lead to a deficiency of the synthesis and secretion of thyroid hormones reduce thyroid function. The *causes are multiple: congenital absence* or the *disappearance* of thyroid tissue due to a *destructive pathological process*, such as thyroiditis or therapeutic practices (treatment of thyrotoxicoses); insufficient stimulation of the gland by TSH due to *pituitary lesions* (for example, hypothyroidism secondary to a pituitary tumor); partial or total block of the synthesis of thyroid hormones by the gland, leading to an increased secretion of TSH and the formation of a *compensatory goiter*. If the compensatory response is sufficient, overcoming the problem causing the deficiency of hormonal synthesis, the patient will remain euthyroid with a goiter. In contrast, the patient will present with a more or less severe hypothyroid state, frequently with cretinism, if the hypothyroidism appeared in the first years of life. This last category of hypothyroidism can be *extrinsic in origin* (goiters related to an inadequate *dietary intake* of iodine or the absorption of an *antithyroid* substance) or *intrinsic* (congenital defects in thyroid metabolism of iodine).

The *clinical literature* describing hypothyroidism in the adult is extremely rich. Some of the *symptoms of hypothyroidism* are classified, by order of frequency, in Table VIII.8.

In typical hypothyroidism, there is an increase in the mucoprotein substances of the skin and sub-

Table VIII.8 Frequency of the principal symptoms observed in myxedema in the adult‡.

Symptoms	%
Weakness	99
Dry skin	97
Coarse skin	97
Lethargy	91
Slow speech	91
Edema of eyelids	90
Sensation of cold	89
Decreased ovulation	89
Edema of face	79
Memory impairment	66
Constipation	61
Weight gain	59

‡Means, J. H. (1948) *The Thyroid and its Diseases*. J. B. Lippincott Co., Philadelphia, p. 233.

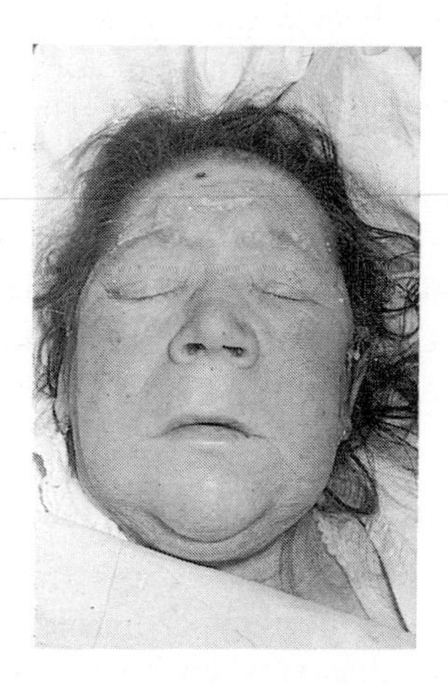

Figure VIII.26 Hypothyroidism (myxedema). (Courtesy of Dr. J. Vague, Hôpital de la Timone, Marseille).

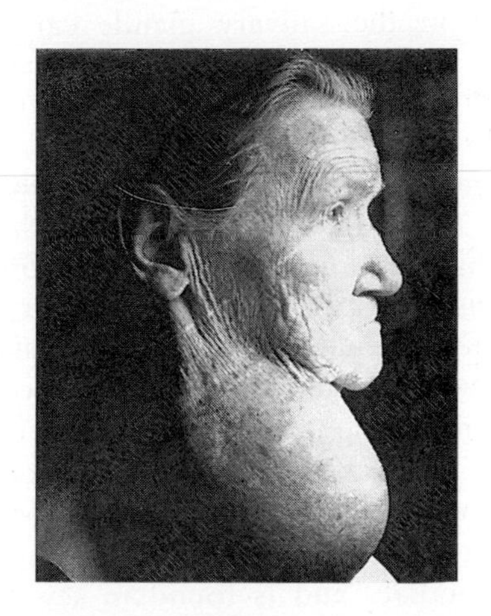

Figure VIII.27 Multinodular goiter caused by a congenital defect of thyroid hormone production.

cutaneous tissue. Mucosal infiltration leads to a thickening of features and a bloated aspect of the face. The term *myxedema* was given to this type of hypothyroidism (Fig. VIII.26). Other symptoms result from a slowing of the major metabolic functions.

Severe hypothyroidism which begins in childhood is called *cretinism*, characterized by severe intellectual problems and marked growth retardation (dwarfism) (Fig. VIII.27).

Neonatal hypothyroidism has become a public health concern due to the irreversible mental problems it causes and its relatively high frequency in the world (1/4000 to 1/8000 births in Europe). Systematic testing is currently being carried out in a number of countries, using TSH RIA from a single drop of the newborn's blood.

Goiters: A Series of Genetic Anomalies[48]

Although quite rare, *familial goiters* related to genetic defects[26] of thyroid hormone synthesis will be examined in further detail. The importance of goiters (Fig. VIII.27) depends on the degree of reduction of hormone levels in the circulation. In certain cases, the discovery of a metabolic disorder has led to a better understanding of normal thyroid physiology, and has clarified certain relationships between environment and heredity. For example, when the deficiency is in the mechanism for concentrating iodide and iodotyrosine dehalogenase, the metabolic deficiency can be *completely compensated* by increasing *iodide in the diet.* All disturbances affecting the enzymatic steps of thyroid hormone synthesis can lead to a reduction in secretion. Although the deficiency may vary, the clinical phenotype is frequently similar. The transmission of the genetic deficiency is via an autosomal recessive mode. Subjects with a goiter and hypothyroidism are homozygotic; the parents or brothers and sisters are probably heterozygotic. These conditions are observed in children and are of varying functional severity, ranging from severe hypothyroidism with cretinism to a near-normal state.

Although during the last 20 years the emphasis has been on congenital diseases related to metabolism of iodine within the thyroid, a new classification is emerging that includes more fundamental aspects of thyroid function, i.e. a better understanding of the structure of the TG gene and of TG itself. The possible genetic defects which result in deficiencies of TG synthesis, processing and transport will be considered. A few known genetic defects related to the thyroid are described below.

Failure of iodide transport is characterized by an absence or a considerable diminution of the capacity of the thyroid gland to concentrate iodide. This defect can be identified in vivo, by the measurement of thyroid uptake of radioactive iodide, or in vitro, utilizing fragments of thyroid obtained by biopsy. A paral-

48. Salvatore, G., Stanbury, J. B. and Rall, J. E. (1980) Inherited defects of thyroid hormone biosynthesis. In: *The Thyroid Gland* (De Visscher, M., ed), Raven Press, New York, pp. 443–487.

lel defect in the salivary glands can serve as a diagnostic criterion, i.e. the salivary/plasma I^- ratio. The specific nature of the problem involved in the transport of iodide is unknown.

A defect in organic iodination of iodide is characterized by an accumulation of iodide in the gland and the possibility of rapidly removing it from the blood by administering perchlorate or another antithyroid anion. This accumulation takes place quickly. It corresponds to an unusually large iodide space, indicating normal functioning of the mechanism of iodide transport, with a defect in the subsequent oxidation step. This biochemical anomaly appears to be due to the partial or total absence of a functional thyroid peroxidase for iodide, and is found in several types of goiters.

The defect of iodotyrosine-dehalogenase is the best characterized of the innate errors of thyroid metabolism. The problem is manifested biologically by a rapid and increased uptake of iodide by the thyroid, indicated by the presence of an increased quantity of MIT and DIT in blood and urine, and by the absence of deiodination of injected labelled DIT. The absence of deiodinating activity in the thyroid has been shown in many cases by incubating slices of goiter tissue with DIT labelled by radioactive iodine. The steps of intrathyroidal iodine metabolism are normal except for iodotyrosine deiodination. In these patients, urinary loss of a large quantity of iodine in the form of iodotyrosines leads to a state of iodine deficiency. Thus, stimulation of the thyroid and hyperplasia (goiter) create a vicious circle, accelerating the loss of hormonal precursors. In a specific case[49], it was possible to calculate that approximately 50% of the iodine ingested per day was excreted in the form of iodotyrosines or their derivatives. As in the case of defects of the mechanism of iodide transport, the administration of iodide to these patients allows a return to a euthyroid state and the disappearance of the goiter.

Cases of thyroid deficiency *without goiter* have been described. These are associated with a faulty response of the thyroid to TSH, or with a defect in the coupling between the TSH receptor and adenylate cyclase. Several cases of hypothyroidism interpreted as a peripheral resistance to the action of thyroid hormones may include deficiencies at the receptor level.

49. Lissitzky, S., Comar, D., Rivière, R. and Codaccioni, J. L. (1965) Etude quantitative du métabolisme de l'iode dans un cas d'hypothyroïdie avec goître due à un défaut d'iodotyrosine déshalogénase. *Rev. Franc. Etudes Clin. Biol.* 10: 631–640.

Hypothyroidism *justifies replacement by thyroid hormones*, which are administered orally, in the form of a natural or racemic derivative. The difficulty is in adapting the exogenous intake such that a constant level of hormones is maintained in the blood. Hypothyroid patients with congenital problems in the uptake of iodide or with a deficit in the iodotyrosine dehalogenase can be treated simply by the administration of iodide.

7.7 Hyperthyroidism

The syndrome of clinical hyperthyroidism, also called *thyrotoxicosis*, appears when tissues are subjected to excessive quantities of thyroid hormones. *Two main etiological varieties are* known: *toxic adenoma*, or multinodular toxic goiter, and *toxic goiter*, or *Graves' disease*.

The clinical symptoms of thyrotoxicosis are a reflection of the stimulation of metabolism induced by an excess of thyroid hormones in the circulation. Table VIII.9 lists several *principal functional and physical symptoms* observed.

Graves' disease is the *most frequent* of the thyrotoxicoses. It is a multisystemic disease characterized by one or several clinical entities, which include thyrotoxicosis associated with a diffuse goiter, exophthalmos (Fig. VIII.28), and a localized dermopathy of the leg with infiltration (pretibial myxedema). It manifests itself in populations with a genetic predilection for the disease, and is much more frequent in women than in men. Biologically, the level

Table VIII.9 Frequency of the principal physical and functional symptoms of hyperthyroidism.

Symptoms	%	Signs	%
Nervousness	99	Tachycardia	100
Increased sweating	89	Goiter	97
Hypersensitivity to heat	89	Skin changes	97
Palpitation	89	Tremor	97
Weight loss	85	Bruit over thyroid	77
Tachycardia	82	Eye signs	71
Weakness	70		

Williams, R. H. (1946) Thiouracil treatment of thyrotoxicosis, *J. Clin. Endocrinol.* 6:1–22; and Ingbar, S. H. (1985) The thyroid gland. In: Williams' *Textbook of Endocrinology*. 7th edit. (Wilson J. D., Foster D. W., ed), W. B. Saunders, Philadelphia, pp. 682–815.

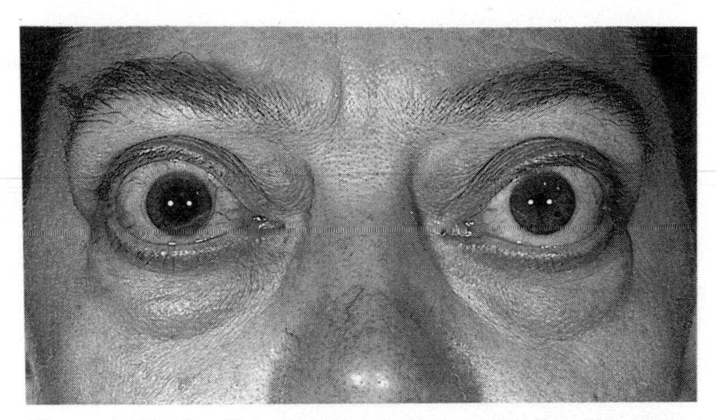

Figure VIII.28 Exophthalmos in a hyperthyroid patient (Courtesy of Dr. J. Decourt, Hôpital Pitié-Salpêtrière, Paris).

of circulating thyroid hormones is very high. TSH is undetectable. Another indication of the disease is the demonstration of a thyroid stimulating factor in the blood. Actually, Graves' disease is an *organ-specific autoimmune disorder* associated with the presence of thyroid stimulating antibodies directed against thyroid cell surface determinants closely related to the thyrotropin receptor.

Between 1956 and 1958, a factor stimulating the thyroid gland, called *long-acting thyroid stimulator (LATS)*, was discovered in the serum of a few patients with Graves' disease[50]. This finding directed research toward other potential factors involved in this disease. LATS was shown to differ from TSH by its prolonged biological action.

The *antibody* nature of LATS, demonstrated somewhat later, led to the supposition that Graves' disease could be classified among autoimmune disorders. However, it has not been possible to demonstrate the constant relationship between the levels of LATS and the intensity of hyperthyroidism, ophthalmopathy, or dermopathy. Another antibody, LATS-protector (*LATS-P*), was identified in 1967[51]. It competes with LATS for binding to human thyroid extracts in vitro, and also blocks the neutralization of LATS. Contrary to LATS, LATS-P is not active in mice, demonstrating species specificity, and there is good correlation between its concentration and the uptake of iodide by the human thyroid.

LATS and LATS-P are IgGs, and although they differ from TSH chemically, they have similar effects on the thyroid, both in vivo and in vitro. They are capable of stimulating the synthesis and secretion of thyroid hormones after binding to the plasma membrane of the thyroid cell, with subsequent activation of adenylate cyclase. In addition, the two stimulators induce the growth of the gland; antibodies to other receptors are also capable of mimicking the related hormonal effect. It is now clear that IgGs in Graves' disease, although often showing anti-human TSH thyroid receptor activity, are in fact heterogeneous, acting as activators of adenylate cyclase or as blocking antibodies[52–54].

The pathogenesis of Graves' disease involves an autoimmune mechanism, but the *underlying defects are still unknown*; genetic factors are implicated, as well as certain HLA alleles closely related to the immune response genes. They are associated with the disease, and are probably responsible for the impaired function of certain subsets of lymphocytes.

Treatment of thyrotoxicosis consists of reducing the glandular production of hormone in order to reestablish a euthyroid state. This goal can be attained by three types of treatment: 1. administration of *antithyroid compounds* which block the synthesis of thyroid hormones and possibly the peripheral conversion of T_4 to T_3; 2. local *radiotherapy* by ingestion of a high dose of radioactive iodine; and 3. subtotal *surgical* ablation of the gland. The indications of these different treatments are a function of age, sex, and the degree of clinical symptoms.

7.8 Endemic Goiter and Simple Goiter

In certain regions, and particularly in *mountainous regions*, there is an increased incidence of goiter characterized by a diffuse increase in the volume of the thyroid, a euthyroid state, a serum concentration of thyroid hormones at the lower limit of normal values, and an elevated level of serum TSH. This type of goiter is said to be *endemic* because it occurs almost invariably as a result of marked *insufficiency of the dietary intake of iodine*. This deficiency is the principal cause of endemic goiter.

50. McKenzie, J.M. (1974) Long-acting thyroid stimulator in Graves' disease. In: *Handbook of Physiology*, Endocrinology III, (Greer, M.A. and Solomon, D.H., eds), Amer. Physiol. Soc., Washington D.C., pp. 285–301.
51. Adams, D.D. and Kennedy, T.H. (1967) Occurrence in thyrotoxicosis of a gammaglobulin which protects LATS from neutralization by an extract of thyroid gland. *J. Clin. Endocrinol. Metab.* 27: 173–177.
52. Smith, B.R. and Hall, R. (1974) Thyroid stimulating immunoglobulins in Graves' disease. *Lancet* 2: 42.
53. Orgiazzi, J., Williams, D.E., Chopra, I.J. and Solomon, D.H. (1976) Human thyroid adenylcyclase-stimulating activity in immunoglobulin G of patients with Graves' disease. *J. Clin. Endocrinol. Metab.* 42: 341–354.
54. Drexhage, H.A., Botazzo, G.F., Bitensky, L., Chayen, J. and Doniach, D. (1981) Thyroid growth-blocking antibodies in primary myxedema. *Nature* 289: 594–596.

The formation of a goiter is the consequence of an increased secretion of TSH resulting from a reduced concentration of thyroid hormones in the blood. As long as there is dietary deficiency in iodine, the concentration of TSH in the blood remains higher than normal, and the goiter persists. These observations are in accordance with the concept of feedback control of thyroid function at the pituitary level. The effect of a disturbance of one of the factors involved in the regulatory loop, i.e. secretion of thyroid hormones, is compensated in large measure, but not completely, by an increase in TSH secretion; the tendency to reestablish a normal level of thyroid hormones is made at the cost of hypertrophy of the gland. Experience has demonstrated that only the reestablishment of an adequate iodine intake is sufficient to return the gland to normal size. *Prophylaxis*, by ingestion of iodinated salt, causes the disappearance or regression of a very large proportion of cases of endemic goiter.

Deficiency in iodine does not explain, however, *why* in a particular region *only certain* subjects are affected. The influence of natural goitrogens, the effects of certain mineral ions such as calcium and manganese, and an insufficient intake of proteins have been suggested.

Regions where dietary iodine deficiency is marked show an increased rate of *endemic cretinism*. This term describes mentally deficient subjects born in a zone where there is a high incidence of endemic goiters. The pathogenesis of this type of cretinism is still unknown. The hypothesis which is most reasonable is that the deficiency in iodine interferes with the function of the fetal thyroid. As thyroid hormones of the mother do not cross the placenta, the inability of the fetal thyroid to secrete hormones at a critical stage in the development of the brain could cause irreversible lesions in the central nervous system.

If the pathogenesis of goiters caused by iodine deficiency is well enough understood, such is not the case for simple goiters, observed despite the absence of iodine deficiency. Simple goiters correspond to a slow increase in the volume of the thyroid, *homogeneous or nodular*, resulting in the excessive formation of new follicles of different structures and functions. These nodules can be either "hot", i.e. having a highly effective iodination activity resulting sometimes in hyperthyroidism, or "cold", in the opposite case. The pathogenesis of these goiters has long remained a mystery. Clinical observations[55] suggest that epithelial cells of the follicles are of *polyclonal origin*, differing by metabolic properties such as iodination, peroxidase activity or endocytosis. It is clear, therefore, that, according to the particular capacity of follicle cells to respond to stimuli for growth (TSH, extrathyroid immunoglobulins or local factors which regulate cell growth), the new follicles could be morphologically and functionally different from one another, and give rise, by a slow process, to follicles that are either "hot" or "cold", resulting in a heterogeneous goiter. This appealing hypothesis should be substantiated by further observations and experimental findings.

55. Studer, H. and Ramelli, F. (1982) Simple goiter and its variants: euthyroid and hyperthyroid multinodular goiters. *Endocrine Rev.* 3: 40–61.

8. Summary

Thyroid hormones, *thyroxine (T_4)* and *3,5,3',-triiodothyronine (T_3)*, are iodinated amino acids which affect the activity of almost *all tissues* by controlling *protein synthesis* and *energy metabolism* (in homeotherms). In the evolution of vertebrates, thyroid hormones have assumed a dual role, being involved basically in developmental processes and metabolic regulation. While a definite level is required in target tissues, the *synthesis* of hormones in the thyroid gland is *critically dependent* upon the *exogeneous intake of iodine*. Various mechanisms of hormone synthesis and storage are capable of responding to metabolic requirements and to temporary deficiencies in iodine intake.

In the *thyroid gland*, epithelial secretory cells are organized in a closed monolayer structure, called the *follicle*, surrounding an acellular space filled with viscous *colloid*, essentially formed of *thyroglobulin (TG)*, an iodinated glycoprotein playing an *essential* role in *hormone synthesis*. The follicle is the functional morphologic unit of the thyroid gland. The thyroid is capable of *concentrating iodide* taken up from the plasma. This process, requiring *active transport*, is affected by competitor anions such as perchlorate and thiocyanate. Intrafollicular iodide is oxidized and bound to tyrosine residues of TG by a *peroxidase system* present at the apical membrane of follicular cells. TG is a glycoprotein with a molecular weight of approximately 660,000 and a sedimentation coefficient of 19 S. It is formed of two identical subunits (MW ~ 330,000, 12 S), each of which is made up of a single-chain peptide. The iodination of TG follows its synthesis. Residues of *iodotyrosines* (monoiodo- and diiodotyrosine) are formed rapidly. Oxidative coupling of iodotyrosine residues, although slower, allows the formation of thyroid hormones (iodothyronines). TG contains *precursors (iodotyrosines)* and *hormones in a molar ratio dependent* only on *the concentration* of *iodine* in the protein, and ultimately on the quantity of iodide present in the gland.

To reach the circulation, T_4 and T_3 are *released from TG* by lysosomal peptidases after phagocytosis or micropinocytosis at the apical pole of the thyroid cells. Free iodotyrosines are deiodinated by *iodotyrosine dehalogenase*, and the released iodide is reutilized for hormone synthesis. The thyroid thus has a dual function, of *exocrine secretion* (exocytosis of *TG* into the follicular lumen) *and endocrine secretion* (secretion of *hormones* into the blood).

In thyroid cells, *TSH stimulates* most of the metabolic processes of hormone synthesis and secretion. Its mechanism of action implies the occupation of a *receptor at the plasma membrane*, the activation of adenylate cyclase, and the increase of *cAMP*, the primary if not exclusive second messenger in the action of the hormone. As a substrate for the synthesis of hormones, iodine is also involved as a regulator of the implied reactions by mechanisms which are still not well understood (intrathyroid autoregulation).

More than 99.5% of thyroid hormones in the circulation are bound to *plasma transport* proteins, especially thyroxine binding globulin (TBG) and thyroxine binding prealbumin. Their physiological role is not well understood. Thyroid hormones appear to penetrate target tissues in their free form only.

The principal pathway of *extrathyroid metabolism* of thyroid hormones is *deiodination*. The 5′- monodeiodination of T_4 leads to T_3, *which is the most active* form of the hormone; approximately 80% of T_3 is formed by this process. The 5-monodeiodination of T_4 leads to an inactive product, 3,3′,5′-T_3 (reverse T_3 or rT_3), of which 95% is produced in this fashion. T_4 serves, therefore, as a prohormone of T_3, the only active hormone at the cellular level, with greater affinity for the *nuclear receptor*. It is possible, however, that T_4 has some hormonal activity itself. A portion of these hormones is also deaminated, decarboxylated and conjugated, forming inactive metabolites.

The *nuclear receptor* has been identified recently as the *product of the* c-erb-*A gene*. Therefore, it belongs surprisingly to the same family as steroid receptors.

The *effects* of thyroid hormones are *multiple*. In the human, from the eleventh week of fetal life, thyroid hormones are indispensable for the maturation and differentiation of numerous tissues, in particular the *brain* and skeleton. In *amphibians*, the stimulation of metamorphosis is spectacular. Thyroid hormones control many metabolic events, and, most remarkably, homeotherms increase oxygen consumption and *heat* production. This effect is coupled to an activation of the *sodium pump* (Na^+-K^+ ATPase). All effects of thyroid hormones involve increased RNA and protein synthesis. Responses depend upon the differentiation of the target tissue considered.

Studies of the synthesis, metabolism and mechanism of action of thyroid hormones owe a great deal to the use of *radioactive iodine*. The same can be said for the clinical investigation of the thyroid gland. In addition to the measurement, in blood, of total and free (non-protein bound) T_3 and T_4, and of TG, the *uptake of radioactive iodide* is particularly important for the evaluation of thyroid function. More or less active zones can be localized by *scintiscanning. Dynamic tests* of stimulation by TRH and TSH are frequently used.

Hypofunction of the gland can be due to a variety of causes, including those of pituitary origin. Familial *goiters* with *hypothyroidism* may be due to a genetic deficiency in the transport of iodine, defects of organic iodination or deiodination of iodotyrosines, or to other causes. Two etiological varieties of *hyperthyroidism* are frequent: toxic adenoma, and the diffuse toxic goiter of *Graves' disease*. This latter illness, the most frequent *thyrotoxicosis*, is an *organ-specific autoimmune disorder* characterized by the presence of thyroid stimulating antibodies directed against thyroid cell surface determinants closely related to the TSH receptor. *Endemic goiter* is due, principally, to a deficiency in dietary intake of iodine. Prophylaxis, by ingestion of iodinated salt, in affected regions has resulted in an almost complete disappearance of the disease. Propylthiouracil and methylmercaptoimidazole, inhibitors of peroxidase, are potent *antithyroid compounds*. Blocking hormone synthesis, they lower blood hormone levels and lead to the formation of a goiter by increasing the secretion of TSH, the pituitary hormone which stimulates thyroid function.

General References

Cavalieri, R. and Pitt-Rivers, R. (1981) The effects of drugs on the distribution and metabolism of thyroid hormones. *Pharmacol. Rev.* 33: 55–80.

Cody, V. (1980) Thyroid hormone interactions: molecular conformation, protein binding and hormone action. *Endocrine Rev.* 1: 140–166.

Greer, M.A. and Solomon, D.H., eds (1974) The thyroid. In: *Handbook of Physiology*, sect. 7: Endocrinology III. Amer. Physiol. Soc., Washington, D.C.

Lower, E.G., Medeiros-Neto, G.A. and DeGroot, L.J. (1983) Inherited disorders of thyroid metabolism. *Endocrine Rev.* 4: 213–239.

Oppenheimer, J. and Samuels, H.H. (1983) *Molecular Basis of Thyroid Hormone Action*. Academic Press, New York, p. 498.

Reed-Larsen, P., Silva, J.E. and Kaplan, M.M. (1981) Relationships between circulating and intracellular thyroid hormones: physiological and clinical implications. *Endocrine Rev.* 2: 87–103.

Robbins, J. and Rall, J.E. (1979) The iodine-containing hormones. In: *Hormones in Blood* (Gray, C.H. and Bacharach, A.I., eds), Academic Press, London, pp. 576–688.

Weetman, A.P. and McGregor, A.M. (1984) Autoimmune thyroid disease: developments in our understanding. *Endocrine Rev.* 5: 309–355.

Pierre Chambon

Gene Structure and Control of Transcription

How is gene expression controlled in eucaryotic cells? New approaches have become possible in the last few years thanks to discoveries in eucaryotic gene organization, polymorphism of RNA polymerases, and to the beginning of an understanding of chromatin structure. In addition, other important advances include the advent of in vitro genetics with gene cloning and site-directed mutagenesis, and, finally, DNA transfer into cells and organisms. The initiation of *transcription* of genes coding for proteins is one of the *crucial control mechanisms* operating in developing and in terminally differentiated eucaryotic cells. *Several*

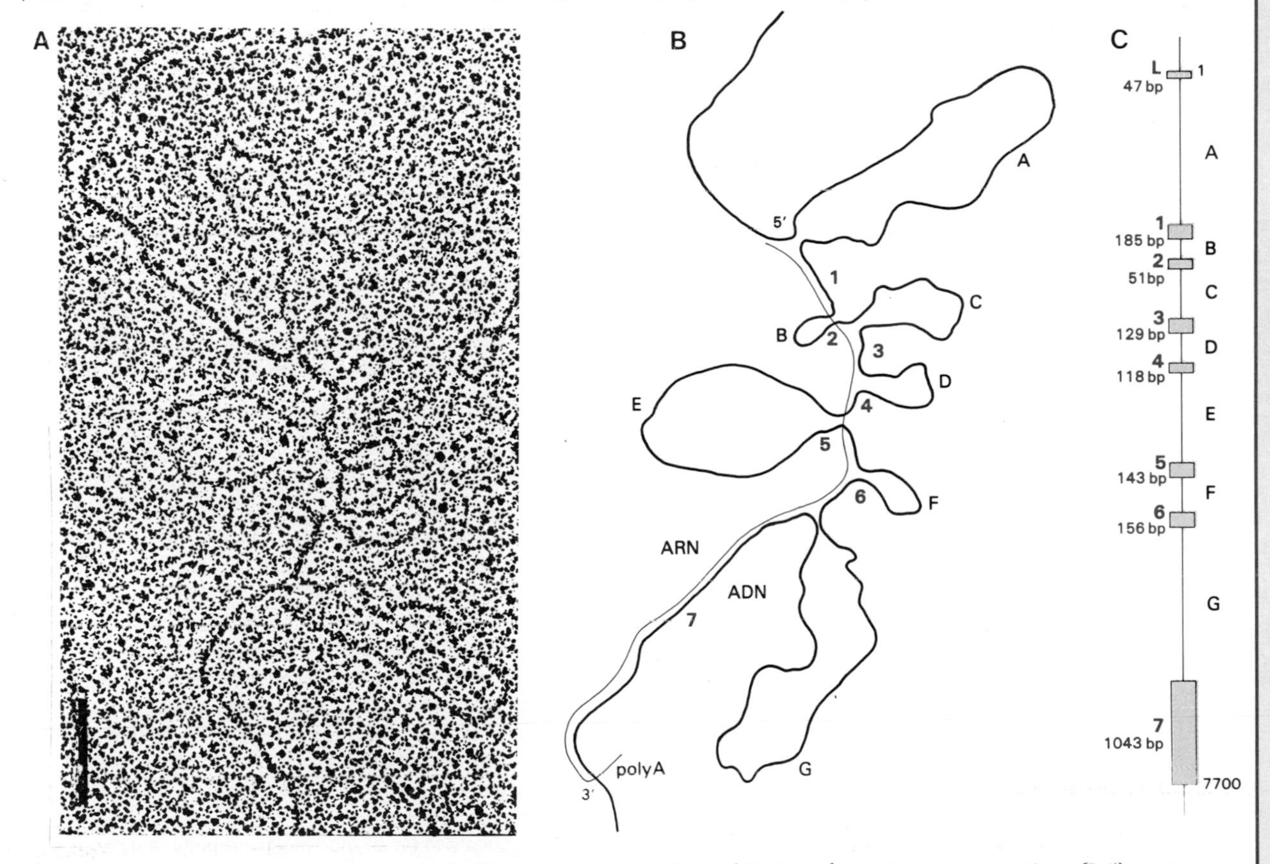

Figure 1 *Organization of a split gene.* **A.** *Electron micrograph, and* **B,** *its schematic representation.* **C.** *Structure of the ovalbumin gene. Single-stranded DNA containing the gene for ovalbumin was allowed to hybridize with ovalbumin RNA. The eight exons (labelled L and 1–7) of the gene anneal to the complementary regions of RNA, and the seven introns (A–G) loop out from the hybrid. The 5′ and 3′ ends of the messenger are indicated, as is the poly A tail. The number of bp of each intron is indicated. The size of introns varies between 251 bp (B) and ~ 1600 (G).*

strategies are used by eucaryotic organisms to control the initiation of transcription. Some of them are similar to those used by procaryotes, whereas others are specific to eucaryotes.

Structural Aspects

Higher eucaryotes contain approximately 1000 times more DNA than do procaryotes. Not only is eucaryotic DNA very large, but its structure is also far from being clearly understood. A large majority of the DNA is still of totally unknown function. *Genes* themselves are found *in pieces*, and *regulatory sequences* are *diverse* and *complex.*

During the past decade, perhaps the most unexpected of the discoveries that have challenged the view that "what is true for E. coli is true for the elephant" was the finding that *most genes* in higher cells were "*split*", in the sense that *mRNA-coding sequences* (called *exons* because their transcripts exit from the nucleus, to function in the cytoplasm) were *interrupted by apparently functionless* sequences (*introns*). The split organization of a gene is best visualized by electron microscopy of mRNA-DNA gene hybrids, as illustrated for the ovalbumin gene in Figure 1. During the nuclear processing of a *primary transcript*, which is *colinear* with the gene and which contains both exon and intron transcripts, intron transcripts are exactly excised, while exon transcripts are spliced together.

Introns exist in most mammalian and vertebrate genes, and also in genes of other eucaryotic organisms and even microorganisms (e.g., yeast), although sometimes with much lower frequency. The *length of introns* in vertebrate genes *varies* from approximately 40 to more than 10,000 base pairs with no obvious periodicity. Thus, the presence of introns accounts, at least to some extent, for the paradoxically large DNA content of eucaryotic genomes (C-value paradox, meaning that the amount of DNA per haploid cell does not correspond to the phylogenetic complexity of an organism).

Utilization of different promoters and/or polyadenylation sites results, for some genes, in the production of *multiple forms of mature mRNAs*. This can also be achieved by alternative splicing of a primary transcript, and in some cases both processes are combined. Thus, unique mRNAs, specific for certain terminally differentiated cells or different cells during development, can be generated from a single gene by using different 5′ or 3′ splice sites or even by eliminating whole exons. Several cases of such alternative splicing have been described (e.g., for the calcitonin gene, p. 651), but it is still unknown whether regulation of splicing is an important mechanism in the control of gene expression, nor is it known what mechanisms could be responsible for such regulation. In a few cases, transcriptional regulatory elements have been found in introns (such as the immunoglobulin heavy-chain gene enhancer (see below) and the enhancer-like glucocorticosteroid response element of the growth hormone gene). In one case, a coding sequence was found within one of the introns of another gene.

However, in most cases introns appear to be functionless, which raises the fascinating question of the *evolutionary origin of introns*. It is now widely accepted that the split genes of higher eucaryotes would reflect the structure of genes of primordial cells better than genes of contemporary procaryotic cells. This idea, which stems from Crick's view that *self-replicating RNA "machines"* were the original forms of life, has received strong support with the discovery of RNA self-splicing, in which RNA catalyzes its own cleavage and ligation. The early splicing process might have depended only on the sequence of introns, whereas the present-day, different splicing mechanisms would represent an evolutionary process refining the control of splicing with the help of proteins. It has been proposed by Gilbert that, if exons correspond to individual protein functions, *recombination within introns* could reassemble these functions, to generate new, larger, multifunctional proteins, thus giving an extraordinary evolutionary advantage to the organisms in which genes are split. However, it is now clear that single exons do not generally correspond to a protein domain, since most domains and single-domain proteins are encoded in a number of exons. Exons may correspond to sequential supersecondary structures ($\alpha\beta$ or $\beta\beta$ units), as proposed by Blake. It is possible that the large contemporary proteins have derived ultimately from a single small primordial exon through exon tandem duplication and shuffling, sometimes accompanied by intron deletions, thus leading in some cases to a functional unit encoded by a single exon.

Chromatin, as such, does not exist in procaryotes. The *packaging* of DNA appears to be a fundamental property of differentiated cells, *for the selective expression* of DNA in eucaryotic cells (Fig. 2).

The fundamental repeating structural subunit of chromatin is the *nucleosome*; in the central part, approximately 166 bp of DNA are wrapped, as a two-turn superhelix, around an *octameric histone* core [$(H2A)_2$, $(H2B)_2$, (H3), and $(H4)_2$], to constitute the nucleosome. Adjacent nucleosomes are connected by a segment of linker DNA, to form the basic level of organization of the chromatin, the *10 nm thick, "beads on a string"* nucleosomal fiber. This fiber is then coiled into a higher order structure to achieve

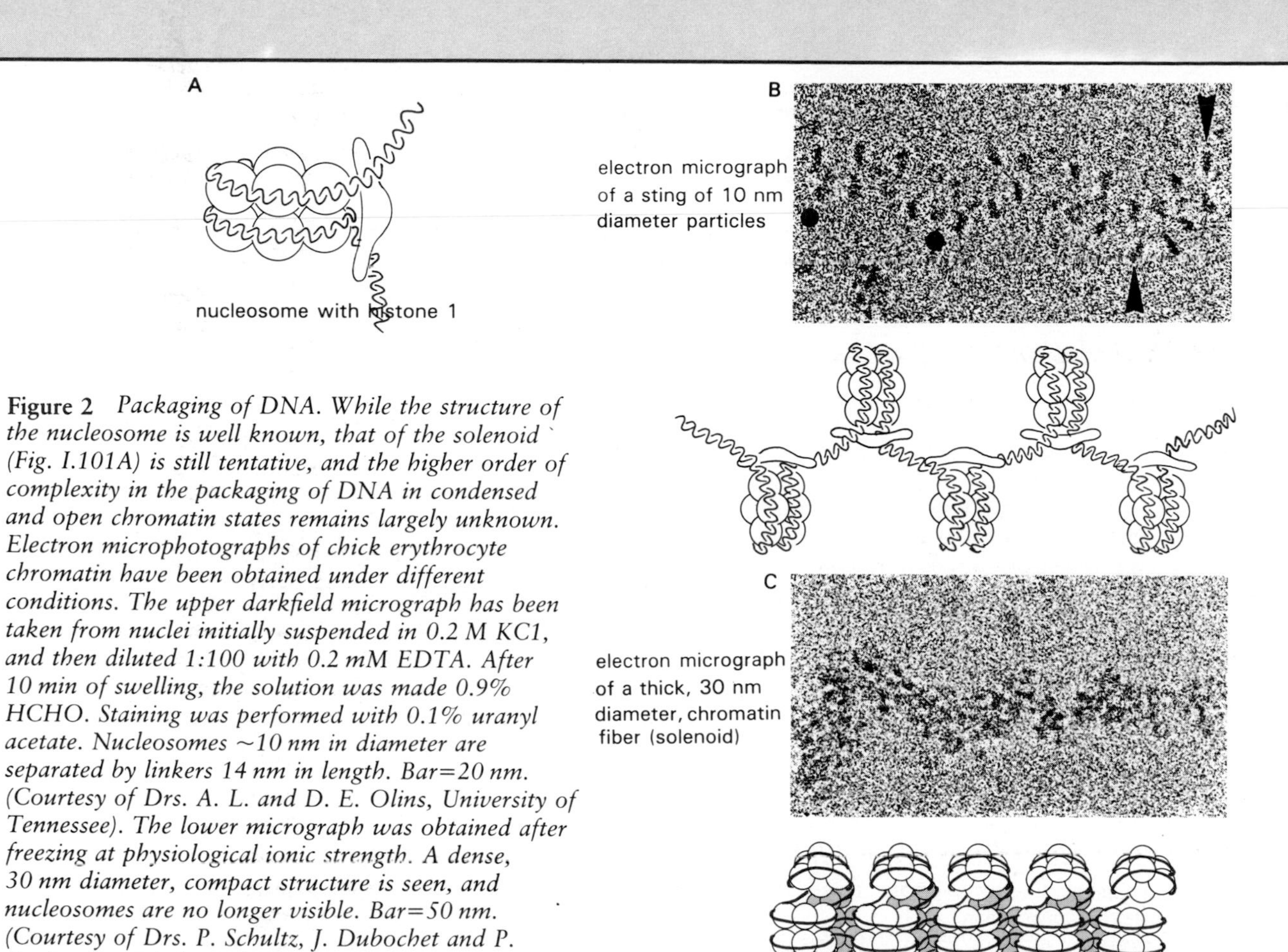

Figure 2 *Packaging of DNA. While the structure of the nucleosome is well known, that of the solenoid (Fig. I.101A) is still tentative, and the higher order of complexity in the packaging of DNA in condensed and open chromatin states remains largely unknown. Electron microphotographs of chick erythrocyte chromatin have been obtained under different conditions. The upper darkfield micrograph has been taken from nuclei initially suspended in 0.2 M KC1, and then diluted 1:100 with 0.2 mM EDTA. After 10 min of swelling, the solution was made 0.9% HCHO. Staining was performed with 0.1% uranyl acetate. Nucleosomes ~10 nm in diameter are separated by linkers 14 nm in length. Bar=20 nm. (Courtesy of Drs. A. L. and D. E. Olins, University of Tennessee). The lower micrograph was obtained after freezing at physiological ionic strength. A dense, 30 nm diameter, compact structure is seen, and nucleosomes are no longer visible. Bar=50 nm. (Courtesy of Drs. P. Schultz, J. Dubochet and P. Oudet, Faculté de Médecine, Strasbourg).*

the compaction of DNA found in interphase chromatin and in metaphase chromosomes. The next level of chromatin structure corresponds to the coiling of the 10 nm fiber into a *solenoid*, to form fibers approximately *30 nm thick. Histone Hl* is thought to be primarily responsible for organizing this higher-order structure of nucleosomes. It is widely believed that DNA packaged into such a higher-order structure is not available for transcription, and in particular, that it cannot interact with proteins controlling transcription. Such inaccessibility is thought to be responsible for the permanent repression of the genes (inactive genes) that will never be expressed in a given cell type. However, histones appear to be associated with most genes which are transcribed (active genes), to form nucleosomal structures. How, then, might the regulatory and transcription factors find their way to the promoter region of active genes?

In fact, the structure of *chromatin of actively transcribed genes* appears to *differ* from that of non-transcribed regions in many respects. First, it is probably not packed into a higher-order structure, which results in the so-called "*open*" chromatin state. This is characterized by an increased sensitivity to nuclease digestion (making it more accessible to any protein) and most likely by the absence of histone Hl or by a loosening of its binding to the nucleosomal fiber. The important question of how such an "open" state is established in the first place during differentiation, and how it is subsequently stably propagated through cell divisions, is still not resolved. *Chemical modification of* the *DNA* (methylation, see below), post-transcriptional *changes of histones*, possibly the absence of histone Hl, and interactions of DNA and/or histones with *"opening" proteins* (such as HMGs) have all been implicated in the generation of this "open" chromatin state.

Most interestingly, recent experiments suggest that nucleosomes are disrupted in the promoter regions of active genes. Such nucleosome-free regions, characterized by their hypersensitivity to digestion by nucleases, have been identified in the enhancer regions (most notably in SV40) for example, and in the upstream promoter region of the chicken β-globin gene. It is clear that the existence of such *nucleosomal gaps* in the promoter element regions would facilitate the reaction of regulatory and transcriptional factors with their cognate binding sites. It is not clear, however, how such "windows" are generated in chromatin. Whether the formation of a nucleosome is prevented by the binding of some of the factors involved in the initiation of transcription, and/or whether other specific regulatory proteins which do not interact directly with the transcription machinery prevent the formation of nucleosomes by binding to DNA sequences in the immediate vicinity of promoter element sites, remains to be established. The question of how the active chromatin state is generated and propagated represents a very challenging problem which will only be resolved by faithfully reconstructing the transcribing chromatin in vitro.

DNA Methylation

In contrast to procaryotes, *in many higher organisms* DNA is chemically modified by the addition of a methyl group at the carbon 5 position of cytosine. Almost exclusively, those cytosines that are followed by a guanosine are methylated (but not all CG sites are methylated), and the *pattern* of methylation is *cell-type specific* and practically stably inherited in any given cell type. Genes that are not expressed usually have a large fraction of methylated CG sites. In contrast, in a cell in which the same genes are active, there is a marked decrease in their level of methylation, most notably in their promoter regions. The mechanism by which the pattern of methylation of a given gene is specifically altered in various cell types during development is unknown. It is not yet clear whether the methylation of CG sites is *a cause or a consequence* of the initiation of transcription. The methylation of promoter regions of expressed genes could allow regulatory factors to interact with their cognate sites.

Control of the Initiation of Transcription

RNA polymerases have also shown structural and functional *complexity* in eucaryotes,and they are found in different subnuclear compartments. Polymerase I, or A, catalyzes the synthesis of ribosomal RNA, and polymerase III, or C, that of transfer RNA and 5 S RNA, whereas polymerase II, or B, transcribes principally the *genes coding for proteins*, and is thus involved in the synthesis of mRNAs.

Promoter Organization and Function

For eucaryotic genes, the term *promoter* is used in a broad sense to cover any *cis*-DNA sequence required for accurate and efficient initiation of transcription.

Capsite, or mRNA startsite, contains the bases coding for the 5′ terminal nucleotides of the mRNA. Its mutation is usually accompanied by a decrease in the amount of RNA initiated from the promoter region. The *TATA box* (p. 381) binds a ubiquitous, specific factor required for the formation of a *stable preinitiation complex* and subsequent specific initiation of transcription. Alteration of this structure results in the appearance of new heterogeneous initiation sites, and is usually accompanied by a reduction in the overall level of RNA synthesis. Initiation of transcription can be negatively regulated by the binding of *trans*-acting factors to the TATA box capsite region. Thus, the SV40 T-antigen regulates its own synthesis by binding to this region of the SV40 early promoter.

Upstream promoter elements (UE) are polymorphic in number and in position. Although these elements do not exhibit the clear evolutionary conservation seen for the TATA box, some upstream elements, which can be related to sequences such as the CAAT box or the CG-rich motif (5′-CCGCCC-3′ or the complementary sequence 3′-GGGCGG-5′), have been found in many different promoters, whereas others appear to be more restricted in their distribution, e.g., the heat-shock gene regulatory element. Upstream promoter elements are the *targets for specific transcription factors.* Some factors are *common* to many cell types (e.g., Spl), whereas others are species- or cell-*specific. Most* are involved in *positive* regulation, although some may involve negative regulation, which may be revealed by interaction with specific inducers (p. 381). The presence of multiple upstream promoter elements within a given promoter region is probably reflective of a multiplicity of regulatory factors. It is puzzling that at least some upstream elements function efficiently when their overall orientation in the promoter is inverted, even though their sequence is markedly asymmetrical. This observation has no precedent in procaryotes.

Enhancers are unique to eucaryotes (p. 381). They can *potentiate the initiation of transcription from* potential heterologous "natural" or "substitute" *promoter elements*, regardless of whether they contain a TATA box element, and *irrespective* of the presence of functional *upstream elements.* They can actually be located upstream, within, or downstream from a transcriptional unit. As upstream elements, they can also function *bidirectionally*, and some are active in a wide variety of organisms and cells, whereas others are species- or cell-specific. Enhancer elements contain several motifs which may cooperate. Each motif appears to bind a specific positive *trans*-acting factor, as for example the glucocorticosteroid receptor (p. 71 and p. 415). Furthermore, it appears that the activity of enhancers can also be negatively regulated by *trans*-acting protein factors. How enhancers stimulate the initiation of transcription is still not known.

It follows that a *global view* of a eucaryotic protein-coding gene would correspond to a *large nucleoprotein structure* assembled by the interaction between *specific factors bound at multiple DNA sites of upstream and enhancer elements*. Such a nucleoprotein complex may act as a specific "attractant" for the transcription machinery and thereby stimulate initiation of transcription. In any case, it is clear that the multiplicity of specific upstream elements and enhancer sequence motifs, and of their cognate factors, provides the *very large number of combinatorial possibilities* that are required for the fine regulation of transcription initiation in higher organisms. The recent discovery of *silencer* or blocker elements, which may be the negative counterparts of enhancer elements, further increases the number of these possibilities.

References

Gene organization and evolution. Split genes and RNA splicing

Blake, C. C. F. (1985) Exons and the evolution of proteins. In: *International Review of Cytology*. Genome evolution in prokaryotes and eukaryotes (Reanney, D. C. and Chambon, P., eds) Academic Press, New York, vol. 93, pp. 149–185.

Gilbert, W. (1978) Why genes in pieces? *Nature* 271: 501.

Gilbert, W. (1985) Genes-in-pieces revisited. *Science* 228: 823–824.

Padget, R. A., Grabowski, P. J., Konarska, M. M., Seiler, S. and Sharp, P. A. (1986) Splicing of messenger RNA precursors. *Ann. Rev. Biochem.* 55: 1119–1150.

Multiple classes of RNA polymerases in eucaryotes

Chambon, P. (1975) Eucaryotic RNA polymerases. *Ann. Rev. Biochem.* 44: 613–633.

Lewis, M. K. and Burgess, R. R. (1982) Eucaryotic RNA polymerases. In: *The Enzymes* (Boyer, P. D., ed), Academic Press, New York, vol. XV, part B, pp. 110–154.

Promoter organization and function

Chambon, P., Dierich, A., Gaub, M. P., Jakowlev, S., Jongstra, J., Krust, A., LePennec, J. P., Oudet, P. and Reudelhuber, T. (1984) Promoter elements of genes coding for proteins and modulation of transcription by oestrogens and progesterone. (Greep, R. O., ed), *Rec. Prog. Horm. Res.* 40: 1–42.

Dyan, W. S. and Tjian, R. (1985) Control of eucaryotic messenger RNA synthesis by sequence-specific DNA-binding proteins. *Nature* 316: 774–778.

North, G. (1984) Multiple levels of gene control in eucaryotic cells. *Nature* 312: 308–309.

Chromatin and DNA methylation

Doerfler, W. (1983) DNA methylation and gene activity. *Ann. Rev. Biochem.* 52: 93–124.

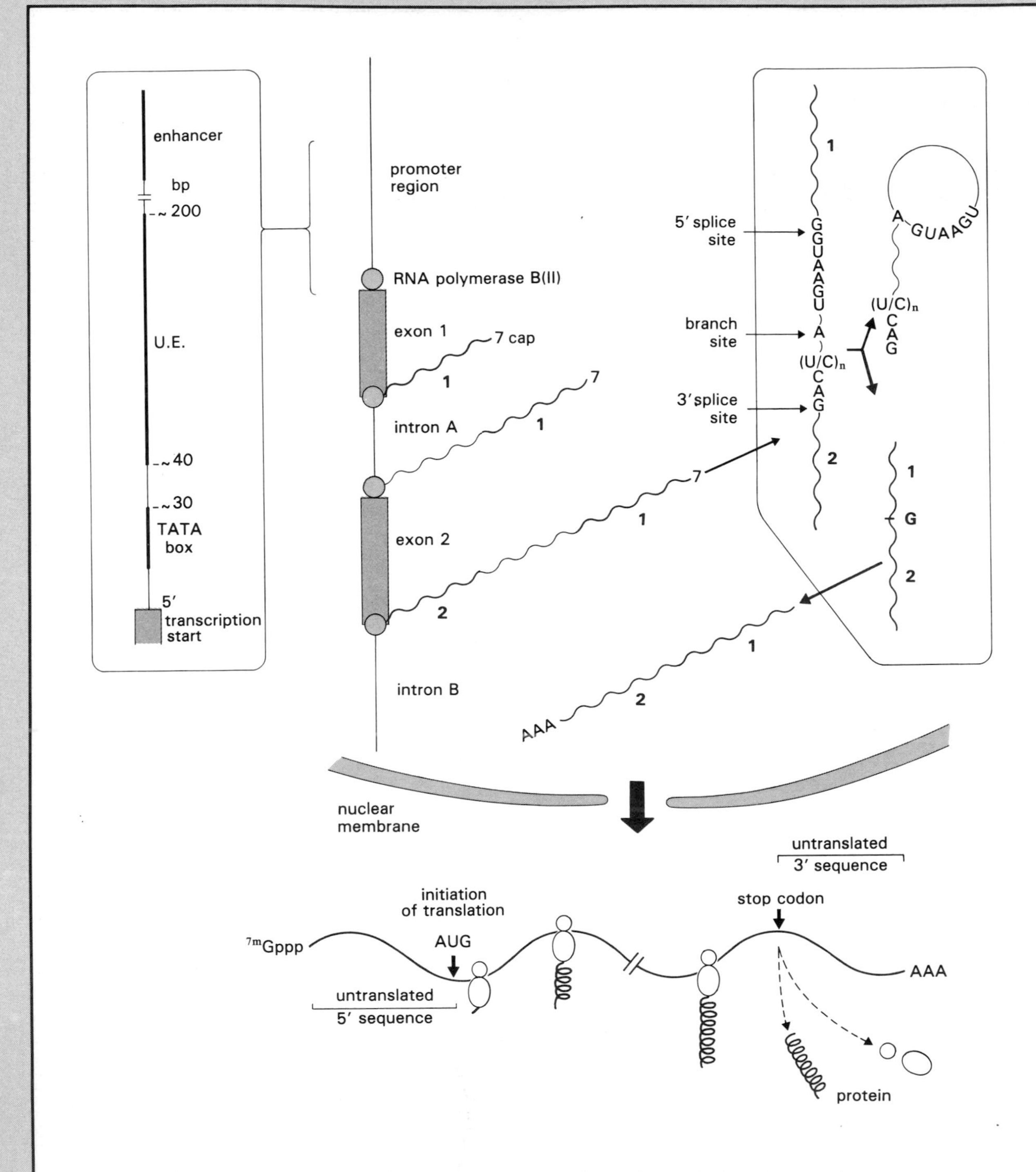

Figure 1 *Schematic representation of gene transcription and protein synthesis. The central part represents a gene being transcribed. Only two exons are indicated. The upper left, framed area deals with the promoter region. The upper right, framed area shows the basic mechanisms of RNA processing leading to mRNA, which includes the two sequences corresponding to exons 1 and 2. Below, the mRNA exported from the nucleus into the cytoplasm is translated. Different consensus sequences of the upstream elements (UE) have been identified, including GGGGCGC (GC box), CCAAT (CAT box), GCCACACCC, and ATGCAAAT.*

Phillip A. Sharp

Gene Transcription and Protein Synthesis

Protein synthesis involves *two essential stages: first*, the production of RNA in the nucleus that involves, successively, *transcription of genes* (Fig. 1) and *processing of primary RNA* products to form the definitive *messenger RNAs (mRNAs)* that will function in the cytoplasm, and *second*, the *translation* of these mRNAs in the cytoplasm, which directly determines the primary structures of the proteins. Transcription and processing of ribosomal RNAs and transfer RNAs, essential parts of the common machinery necessary for all protein synthesis, will not be dealt with here. The level of synthesis of a protein from a given gene within a cell can be regulated at a number of stages. The *primary stage* of *gene regulation* in mammalian cells is the control of the *rate of initiation of transcription by RNA polymerase II (or B)*. This enzyme, which synthesizes ribonucleic acids by transcribing a DNA template, is specific for eucaryotic mRNA synthesis. Initiation of transcription occurs (Fig. 1, left) with a nucleoside triphosphate (pppX) at either one site or a cluster of sites, and the RNA terminus is rapidly modified to produce a cap (7mGpppX), composed of a 7-methyl guanosine residue linked through a triphosphate group to the initiation nucleoside[1].

Promoter sequences are *DNA signals* which are recognized by factors that are essential to both accurate initiation and regulation of transcription. They can be divided into three classes on the basis of their physical location relative to the initiation site[2]. First, a majority of promoters contain a *"TATA" box*, related to the prototype TATAAA sequence, between 25 and 31 base pairs (bp) upstream of the initiation site. This element is probably responsible for positioning the RNA polymerase, as mutations in it often result in initiation at a number of nearby sites. Second, in most promoters, there are *upstream elements (UE)*, important for transcription, which can be identified by the effects of mutations. These elements lie typically between 50 and 200 bp upstream of the initiation site, and retain activity when moved by 50 or 100 bp in either direction. This class of elements can be distinguished from enhancers (see below), since UEs do not stimulate transcription when positioned downstream of an initiation site. Third, *enhancer sequences* stimulate the rate of initiation when positioned either 5′ (upstream) or 3′ (downstream) of the initiation site and over distances of thousands of base pairs[3]. Factors which recognize these enhancer elements can be cell-type specific, and, in some cases, may be hormone receptors[4]. Few factors which recognize sequences in enhancer, UE and TATA elements of cellular genes have been purified. However, it is widely accepted that alterations in the level of these factors, as a function of cell type and physiological state, are primarily responsible for specific gene regulation in mammalian cells.

All the RNA sequences that ultimately form a mRNA are typically synthesized from a transcription unit. Cleavage and polymerization of a tract of polyadenine (AA(A)n or "poly A") occurs rapidly after transcription, and generates the 3′ end of the nuclear precursor RNA. Such RNA is commonly referred to a *heterogeneous nuclear RNA (hnRNA)*. The RNA sequences derived from non-coding, intervening sequences (introns) are removed from the nuclear precursor RNA by splicing, leaving the exon-derived sequences as the *mature mRNA* ("processing"). Only the mature RNA is transported to the cytoplasm. The precursor RNA is probably organized into *complex ribonucleoprotein* structures containing a defined set of *proteins and small nuclear RNAs*. The predominant type of small nuclear RNA is U1 RNA, which is 165 nucleotides in length, and essential for splicing. Other small nuclear RNAs, such as U2, U4, U5 and U6, are also involved in splicing.

A complementary sequence at the 5′ end of the U1 RNA probably binds to sequences at the *5′ splice site*, specifying this site for *processing*[5] (Fig. 1, right). Introns have highly conserved dinucleotide sequences, GU and AG, at their 5′ and 3′ boundaries, respectively. A seven nucleotide consensus sequence, commonly found, is probably the major determinant specifying the 5′ splice site. The *3′ splice site* is often preceded by a long pyrimidine tract (20 to 30 nucleotides). It has been recognized recently that the intron-derived sequence is excised, as a lariat RNA, where the phosphate at the 5′ splice site is covalently bonded to the

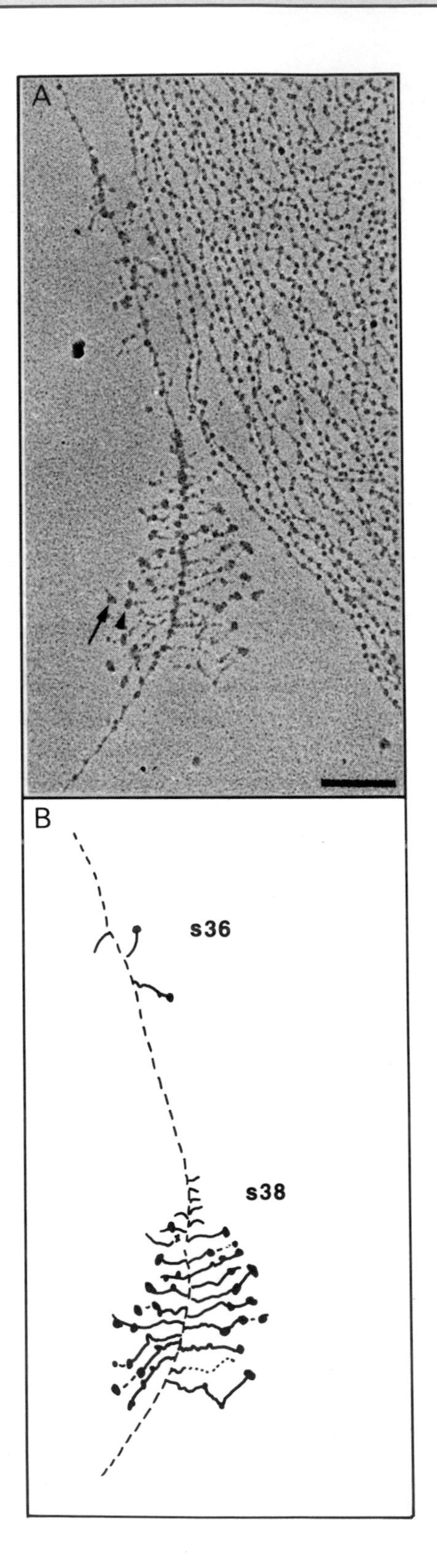

Figure 2 *Actively transcribed genes. Chromatin from* Drosophila melanogaster *stage 11 follicle cells was spread, revealing a tandem s36 and s38 gene pair separated by a non-transcribed spacer. Note the two ribonucleoprotein (RNP) particles, at the 5′ termini (arrow), at a more interior position (arrowhead) of the nascent transcript. They are associated with the 5′ and 3′ splice junction regions, respectively.* **A**. *Chorion genes s36 and s38.* **B**. *Interpretive tracing. Bar = 0.2 μm. (Courtesy of Drs. Y. N. Osheim, O. L. Miller and A. L. Beyer, with kind permission from MIT Press. (Osheim, Y. N., Miller, O. L. and Beyer, A. L. (1985)* RNP particles at splice junction sequences on drosophila chorion transcripts. Cell *43: 143–151)). On mature transcripts, the two particles frequently coalesce to give a ~ 40 nm particle, probably the spliceosome (on transcripts with longer introns, the intron is frequently observed as a loop, with a spliceosome present at the loop base).*

2′ hydroxyl of a ribose group[6]. This branch site is between 20 to 50 nucleotides upstream of the 3′ splice site, and the sequence in the vicinity of the branch site may have a role in specifying the splicing reaction. *Excised* intron-derived sequences are rapidly *degraded*, and probably have no functional activity independent of being a splicing substrate.

Regulation of mRNA synthesis has been documented at the stage of nuclear RNA processing. The *site of polyadenylation* of precursor RNA can vary in different cell types. Precursor RNAs that are polyadenylated at different sites frequently contain different sets of exons after splicing. The same set of DNA sequences can thus give rise to different but related proteins. *Different proteins* can be synthesized in *different cell types*, from the *same nuclear precursor RNA*, by skipping part, or all, of an internal exon during splicing (see the example of calcitonin and calcitonin gene related peptide, p. 651). The nature of the signals controlling this regulation during splicing has not been elucidated.

The synthesis of mRNA can probably also be regulated by the *rate of degradation* of nuclear precursor RNAs. Under particular conditions, some active promoters have been shown not to generate cytoplasmic mRNA[8].

Translation of mRNA into Proteins

Most mammalian cytoplasmic mRNAs have *5′ untranslated sequences* preceding the initiation AUG codon. In about 90% of the examples, the first AUG following the cap is utilized for *initiation of translation*. Recognition of the cap structure by a cap binding protein is probably important for initiation[7]. However, several additional factors are necessary to assemble the 40 S and 60 S subunits of the ribosome for initiation. These factors probably recognize other sequence elements in the mRNA in addition to the cap; however, the nature of these signals is not known. *Termination of translation* occurs at the first stop codon (UAG, UAA, or UGA) in the translational reading frame. At this site, a releasing factor frees the nascent polypeptide and ribosome. The 3′ untranslated sequence between the stop codon and poly A is typically between 50 and 150 bases in length. However, much longer regions are not uncommon.

A major determinant of the steady-state level of a particular *mRNA* is its *half-life* in the cytoplasm. This can vary from as short as five min to several weeks, a 1000-fold range. The cytoplasmic half-life might be influenced by the frequency of translation, and thus the regulation of the initiation of protein synthesis could effect mRNA levels. In addition, the stability of cytoplasmic mRNA is probably also regulated by recognition of the structure and sequence of the RNA.

References

1. Shatkin, A. J. (1976) Capping of eukaryotic RNAs. *Cell* 9: 645–653.
2. Breathnach, R. and Chambon, P. (1981) Organization and expression of eukaryotic split genes coding for proteins. *Ann. Rev. Biochem.* 50: 349–381.
3. Khoury, G. and Gruss, P. (1983) Enhancer elements. *Cell* 33: 313–314.
4. Chandler, V. L., Maler, B. A. and Yamamoto, K. R. (1983) DNA sequences bound specifically by glucocorticoid receptor in vitro render a heterologous promoter hormone responsive in vivo. *Cell* 33: 489–499.
5. Busch, H., Reddy, R., Rothblum, L. and Choi, Y. C. (1982) SnRNAs, SnRNPs and RNA processing. *Ann. Rev. Biochem.* 51: 617–654.
6. Konarska, M. M., Grabowski, P. J., Padgett, R. A. and Sharp, P. A. (1985) Characterization of the branch site in lariat RNAs produced by splicing of mRNA precursors. *Nature* 313: 552–557.
7. Kozak, M. (1983) Comparison of initiation of protein synthesis in procaryotes, eucaryotes and organelles. *Microbiol. Rev.* 47: 1–45.
8. Padgett, R. A., Grabowski, P. J., Konarska, S. S., Sharp, P. A. (1986) Splicing of messenger RNA precursors. *Ann. Rev. Biochem.* 55: 119–1150.

CHAPTER IX

Steroid Hormones

Edwin Milgrom

Contents, Chapter IX

Steroid Hormones

1. Historical Background

From the clinic to crystallized hormones

The *clinical symptoms* associated with a reduction of the secretion of steroid hormones have been *known* for a *long time*; indeed, well before the concept of a hormone was developed. Aristotle identified the effects of castration in humans and birds. In 1855, Addison[1] accurately described the symptoms of chronic adrenal insufficiency.

In 1889, Brown-Sequard, at the age of 72, prepared a testicular extract and tested it on himself. The enthusiastic results he described are viewed today as being due to a placebo effect. It is now understood that the aqueous extracts he used contained only minute amounts of androgens. However, his method, i.e. the preparation of an extract from a biological source and the evaluation of its compensating for hormonal deprivation, led eventually to the *isolation of active steroids*. The first success was reported in 1929, when Doisy and Butenandt crystallized estrone from the urine of pregnant women. Progesterone was later discovered by Corner. Reichstein, in the 1930s, identified corticosteroids[2], and the end of this period was marked, in 1953, by the isolation of aldosterone by Simpson and Tait[3].

1950s: Cortisone and contraception

Two discoveries had an important effect on the field of steroid research. In 1949, Hench[4] and his colleagues reported that cortisone treatment markedly improved patients suffering from *rheumatoid arthritis*; and, between 1950 and 1955, Pincus demonstrated the *contraceptive* possibilities of estrogen and progestin preparations[5]. These results established steroid hormones as potential drugs, beyond simple use in replacement therapy, and resulted in extensive chemical and pharmacological studies.

Shortly thereafter, the introduction of *radioactive steroids* of high specific activity enabled the first efficient studies of the mechanisms involved in hormone action. The discovery of steroid hormone *receptors* followed soon thereafter. The use of the same labelled steroids in *competitive binding assays* permitted the direct measurement of plasma concentrations of hormones. Clinical investigation of endocrine disorders was changed completely from that point on.

However, even though the fate of hormones within their specific target organs was better understood, the precise mechanism by which these hormones provoked cellular responses remained largely unknown. The direct transfer of results obtained from early molecular studies in bacteria and viruses to animal models proved impossible. It is only with contemporary developments in recombinant DNA technology that detailed studies of the mechanisms of hormone action began, bringing steroid research into a new era.

1. Addison, T. (1855) *On the constitutional and local effects of disease of the suprarenal capsules*. D. Highley, London.
2. Reichstein, T. and Shoppee, C. W. (1943) Hormones of the adrenal cortex. *Vitams. Horm.* 1: 345–413.
3. Simpson, S. A., Tait, J. F., Wettstein, A., Neher, R., v.Euw, J. and Reichstein, T. (1953) Isolierung eines neuen Kristallisierten Hormons aus Nebennieren mit besonders hoher Wirksamkeit auf den Mineralstoffwecksel. *Experientia* 9: 333–335.
4. Hench, P. S., Kendall, E. C., Slocumb, C. H. and Polley, H. F. (1949) The effect of a hormone of the adrenal cortex (17-hydroxy-11-dehydrocorticosterone: compound E) and of pituitary adrenocorticotrophic hormone on rheumatoid arthritis: preliminary report. *Proc. Staff. Meet. Mayo Clin.* 24: 181–197.
5. Pincus, G. (1965) *The Control of Fertility*. Academic Press, New York.

2. Embryology and Functional Anatomy of Steroid-Secreting Glands

2.1 Adrenal Cortex

The adrenal cortex is derived from the *mesoderm* and the medulla is of ectodermal origin. Between the fifth and sixth weeks of human embryonic development, the fetal adrenal cortex develops near the mesonephros, which is surrounded soon thereafter by cells of the definitive cortex. At the end of the eighth week, the gland begins to develop at the superior pole of the kidney, and actually becomes larger in size than the kidney.

At birth, the fetal cortex represents the major portion of the adrenal mass. It regresses very rapidly and actually disappears by the end of the first year of life. At the same time, the definitive or adult cortex develops, but its characteristic differentiation into three zones is completed only by the third year of life.

Fetal cortex and definitive cortex are very different

It has been possible to demonstrate that the *fetal cortex does not contain 3β-hydroxysteroid oxidoreductase*, and therefore that it cannot synthesize progesterone, glucocorticosteroids, or androstenedione. Functionally, the embryonic development of the adrenal cortex, which is initially found near the gonads and thereafter at the superior pole of the kidney, explains the possibility of finding accessory adrenals along this tract, and even occasional tumoral development in them (e.g., adrenal inclusion within the testis).

Three definitive zones

The definitive cortex consists of *three zones* (Figs. IX.1 and IX.2). The cells of the zona *glomerulosa* are arranged in circles and arcades; they secrete aldosterone and do not regress following hypophysectomy. In the zona *fasciculata*, which is the largest zone, cells are arranged longitudinally, in columns. The cells closest to the medulla form the zona *reticularis*. These latter two zones are under ACTH control, since they regress following hypophysectomy, an effect counteracted by the injection of ACTH.

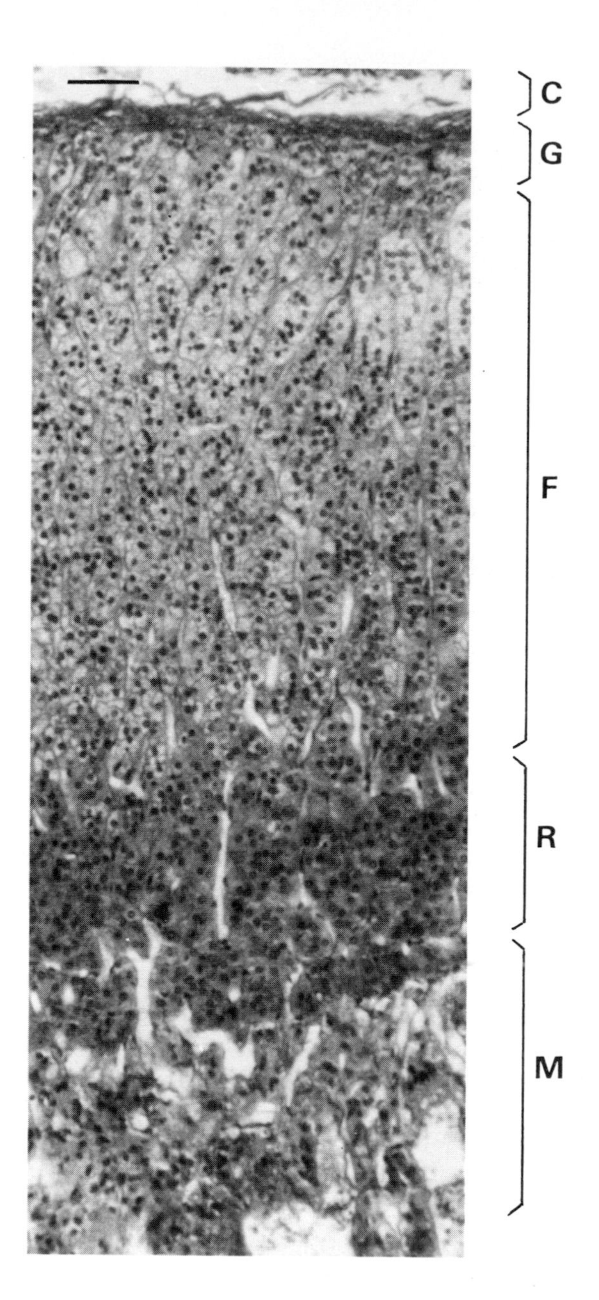

Figure IX.1 Human adrenal gland. Cross section showing division of the cortex into three concentric zones. (C, capsule; G, zona glomerulosa; F, zona fasciculata; R, zona reticularis; M, medulla. Bar = 100 μm). Long, J. A. Adrenal gland. In: Histology (Weiss, L., and Greep, R. O. ed.) 1977. McGraw-Hill, New York.

2.2 Testes

At six weeks in human pregnancy, the genital crest is well developed in the male fetus. Primary sexual cords are formed from the medullary region of this primitive gonad. These develop into *seminiferous tubules*, where spermatogonia of cortical origin will localize.

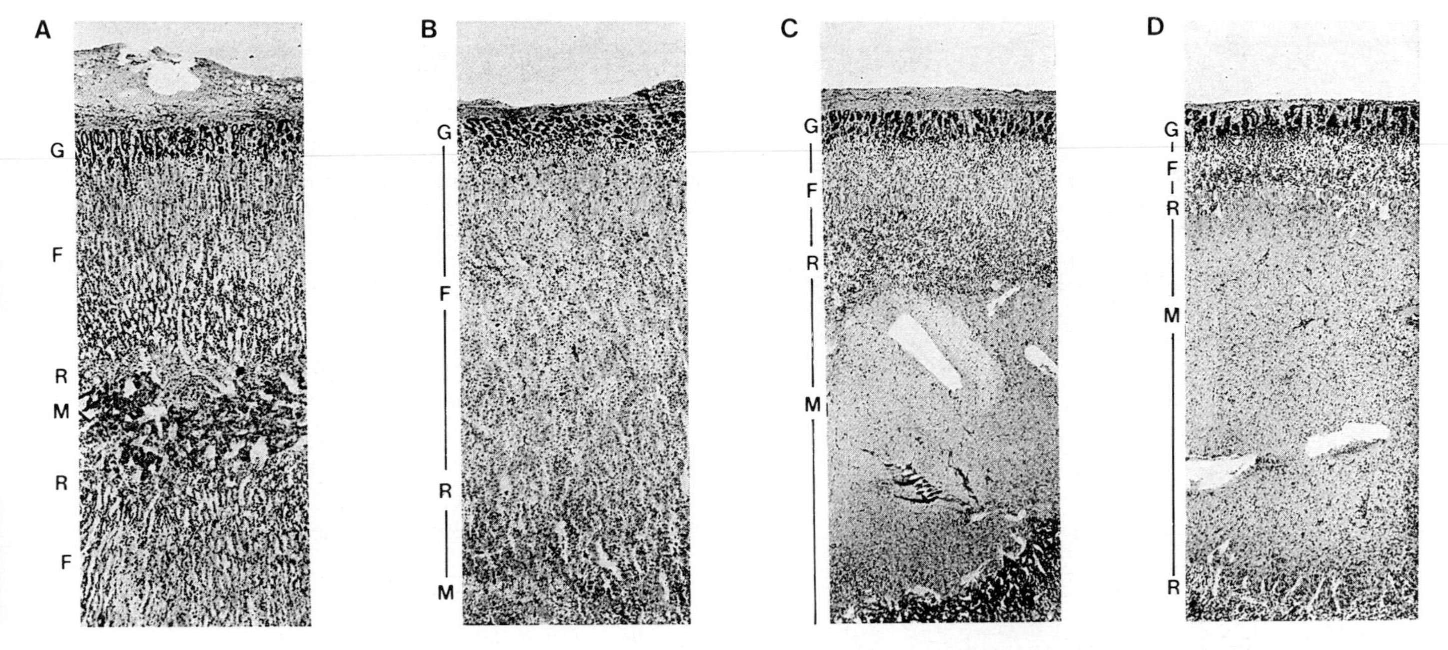

Figure IX.2 Rhesus monkey adrenal gland. **A.** Normal adrenal cortex. **B.** After ACTH. Note the increase in the width of zona fasciculata (F) and zona reticularis (R) (M, medulla). **C.** After treatment with cortisone. The inner zones are reduced in width. **D.** Eighty days after hypophysectomy. The inner zones are atrophied, but the width of the zona glomerulosa (G) remains the same. Bar = 0.3 mm. Knobil, E. et al. (1954). *Acta Endocrinol*, 17, 229.

Interstitial cells, or Leydig cells, appear in the eighth week and secrete androgens necessary for the development of the *external genital organs*. The Wolffian ducts become the vas deferens and seminal vesicles (p. 441). The Müllerian ducts involute under the influence of a testicular factor which is not a steroid but a protein (Müllerian inhibiting substance)[6]; its cDNA has recently been cloned (Fig. 5, p. 441).

Following a brief period of hormonal activity during the first months of life, the testes remain quiescent throughout the prepubertal period. At about 11 years of age, Leydig cells can again be identified in the interstitial tissue, and spermatogenesis begins between 11 and 14 years of age.

Two juxtaposed compartments

In adults, the testes consist of two juxtaposed entities (Fig. IX.3). In one, *interstitial cells* secrete testosterone. The other is composed of *seminiferous tubules*, which contain spermatic elements at all stages of maturation, and Sertoli cells, which are attached to one another and form a barrier at the periphery of the seminiferous tubes (blood/testes barrier)[7]. These cells control the flow of nutrients coming from the interstitial space, and are destined to become germinal cells. They also secrete numerous proteins: the best studied is androgen binding protein (ABP), the secretion of which is controlled, for the most part, by FSH and testosterone. Inhibin is also produced by *Sertoli cells*, which have several complex functions not yet clearly understood. Sertoli cells also aromatize a portion of testosterone to estradiol, which reaches them directly, by diffusion from Leydig cells.

2.3 Ovaries

Ovaries are also derived from the genital crest, but from the cortical portion. Their development is *retarded* by several weeks as *compared to* that of the *testes*. The secondary sexual cords that develop from the cortex penetrate into the medulla, bringing with them the primary germ cells. Fragments of the secondary cords form the primordial follicles. The medulla

6. Josso, N., Picard, J. S. and Tran, D. (1977) The antimüllerian hormone. *Rec. Prog. Horm. Res.* 33: 117–163.
7. Mather, J. P., Gunsalus, G. L., Musto, N. A., Cheng, C. Y., Margoris, A., Liotta, A., Berker, R., Krieger, D. T. and Bardin, C. W. (1983) The hormonal and cellular control of Sertoli cell secretion. *J. Steroid Biochem.* 19: 41–51.

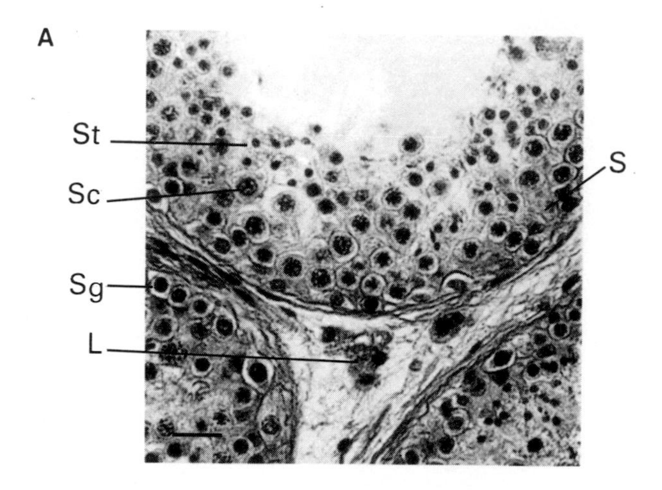

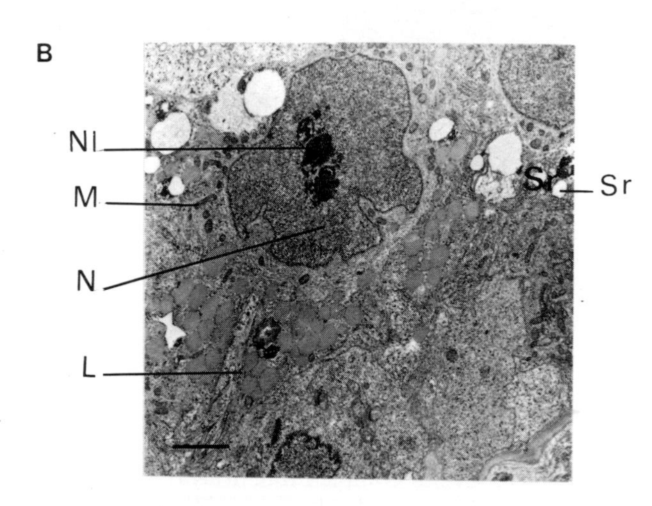

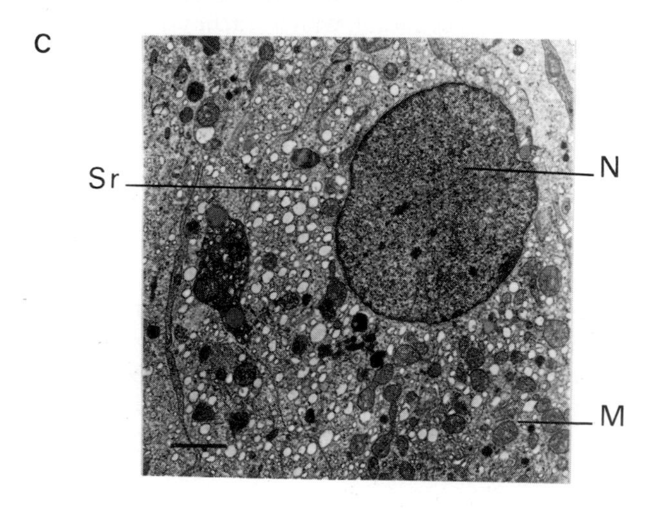

◁ **Figure IX.3** Human adult testis. **A.** Three seminiferous tubules with active spermatogenesis. (Bl, basal lamina; L, Leydig cells; S, Sertoli cells; Sg, spermatogonium; Sc, Primary spermatocyte; St, spermatid. Bar=10 μm). **B.** Sertoli cells. Note the special configuration of the nucleus, with deep indentation and a typical nucleolus. (L, lipid; M, mitochondria; N, nucleus; Nl, nucleolus; Bar=2 μm). **C.** Leydig cells. (M, mitochondria; N, nucleus; Sr, smooth endoplasmic reticulum. Bar=2 μm).

degenerates, and the proliferation of the cortex terminates at about the sixth month of age.

During the prepubertal period, certain follicles develop and may even form an antrum. Since gonadotropins are necessary for normal follicular development, these follicles degenerate.

The follicular cycle

After puberty, follicles pass through a complete cycle of development, during which they migrate from the ovarian cortex towards the interior. They are first surrounded by several layers of cells which form the granulosa. A growing cavity is formed in which follicular liquid accumulates. The peripheral cells form the theca. In a developed, or Graafian, follicle, the ovum is supported by cells of the granulosa (cumulus oophorus), and the antrum is surrounded by the granulosa cells and the external and internal thecal layers. At the time of *ovulation*, the follicle opens, the ovum is expelled, and the transformed granulosa cells organize into a corpus luteum, so named because of the presence of yellow lipids in the cells. The fate of the corpus luteum depends on the survival of the fertilized ovum. In the *absence of a conceptus*, it degenerates, whereas in the presence of a conceptus, it is transformed into the corpus luteum of pregnancy. Throughout life, only *a small fraction* of all follicles undergo this development (300 out of approximately 300,000 present at birth).

During *each cycle*, only *one follicle* undergoes maturation leading to ovulation. The mechanisms involved in the selection of a dominant follicle, as well as the process by which the dominant follicle blocks the development of ipsilateral and contralateral follicles, are the subject of a number of studies[8], but for the most part they remain poorly understood (p. 451).

8. Goodman, A. and Hodgen, G. D. (1983) The ovarian triad of the primate menstrual cycle. *Rec. Prog. Horm. Res.* 39: 1–74.

3. Biosynthesis of Steroid Hormones

3.1 Structure and Nomenclature

(Fig. IX.4 and Table IX.1)

Steroid hormones are derived from the cyclo-pentenophenanthrene ring, termed *gonane*, the *17* carbons of which are saturated by hydrogen. The addition of an 18th carbon atom results in the estrane ring *(C18)*, the addition of a 19th, in the *androstane* ring *(C19)*, and the addition of a two-carbon lateral chain produces the *pregnane* ring *(C21)*. The methyl groups at C18 and C19, and the lateral chain, project above the plane of the molecule (position β). The *structure* of all hormones and metabolites is the *same* with respect to the arrangement of rings B, C, and D. At carbon 5, hydrogen is found in one of two isomeric positions (Fig. IX.4). When it is in the α position (below the plane of the molecule), the A and B rings of the steroid are almost in the same plane. In contrast, when hydrogen is in the β position (above the plane of the molecule), the two rings form an angle. However, most of the active steroids have a double bond between C4 and C5 (and therefore no hydrogen at C5), and under such conditions the A and B rings are also essentially in the same plane. The common aspect of their form is such that the active hormones and their 5α metabolites frequently interact with the same proteins (plasma protein or receptor), whereas the 5β metabolites are unable to bind. The previously described rings bear hydroxyl groups and ketones. Secondary or tertiary alcohols are either α or β, according to their position below or above the plane of the molecule. The A ring of C18 steroids (*estrane* series) is usually phenolic, with the hydroxyl group at position 3. Table IX.1 lists the *common names*, their *abbreviations*, and the *chemical names*, of most of the steroids mentioned in this book.

A

C 27
cholestane

C 21
pregnane

C 19
androstane

C 18
estrane

B

5α–pregnane

5β–pregnane

Δ_5–pregnene

Δ_4–pregnene

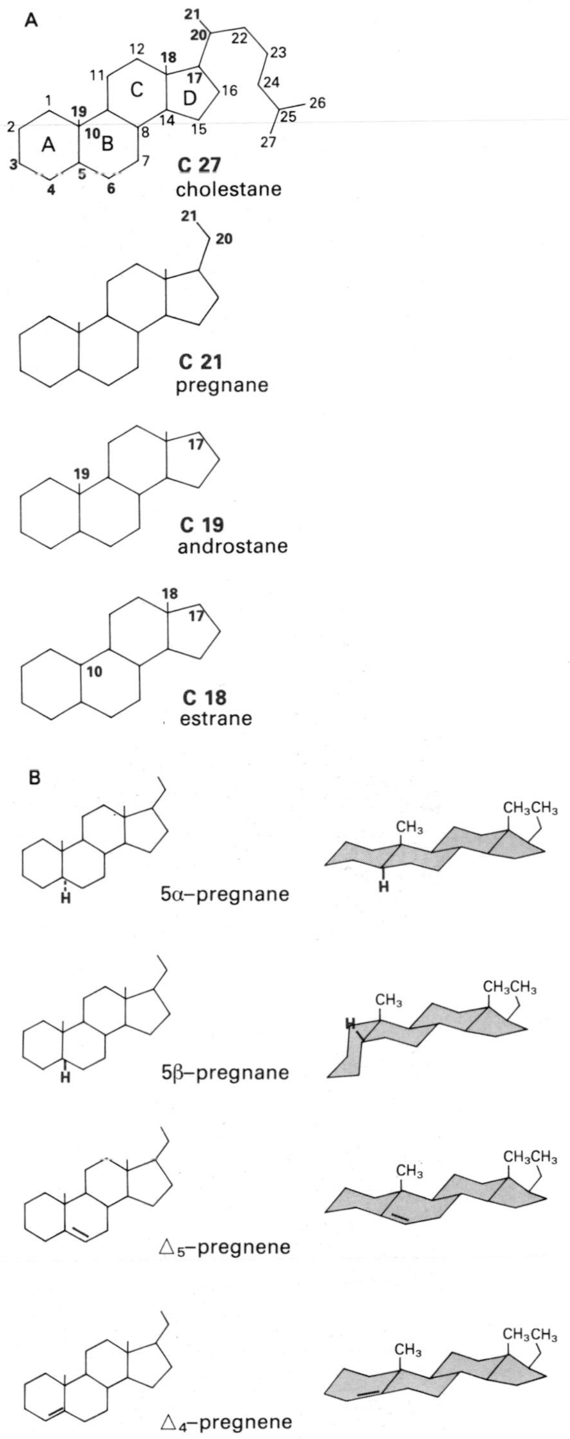

Figure IX.4 **A.** Principal ring systems forming the carbon skeleton of steroid hormones. **B.** Molecular structure of steroids and isomerism at carbon 5.

3.2 General Pathways of Biosynthesis

From Cholesterol to Pregnenolone

In steroid-forming glands, biosynthesis begins with cholesterol[9]. The *major portion of cholesterol* used comes from the *blood*. Its origin is in either hepatic synthesis or intestinal absorption. A very small

9. Hechter, O., Solomon, M. M., Zaffaroni, A. and Pincus, G. (1953) Transformation of cholesterol and acetate to adrenal cortical hormones. *Arch. Biochem. Biophys.* 46: 201–214.

Table IX.1 Common and chemical names and abbreviations of steroids.

Common name	Abbreviation	Chemical formula
Aldosterone		11β,21-dihydroxy Δ_4-pregnene 3,20-dione 18-al
3α-androstanediol	3α-diol	5α-androstane 3α, 17β-diol
3β-androstanediol	3β-diol	5α-androstane 3β, 17β-diol
Dihydrotestosterone (androstanolone)	DHT	17β-hydroxy 5α-androstane 3-one
Androstenedione	Δ_4	Δ_4-androstene 3, 17-dione
11β-hydroxyandrostenedione	—	11β-hydroxy Δ_4-androstene 3, 17-dione
Androsterone	—	3α-hydroxy 5α-androstane 17-one
Cholesterol	C	Δ_5-cholestene 3β-ol
Corticosterone	B	11β,21-dihydroxy Δ_4-pregnene 3,20-dione
18-hydroxycorticosterone	—	11β,18,21-trihydroxy Δ_4-pregnene 3,20-dione
Cortisol	F	11β,17,21-trihydroxy Δ_4-pregnene 3,20-dione
Dehydroepiandrosterone	DHEA	3β-hydroxy Δ_5-androstene 17-one
Dehydroepiandrosterone sulfate	DHEA-S	3β-hydroxy Δ_5-androstene 17-one-3-sulfate
Deoxycorticosterone	DOC	21-hydroxy Δ_4-pregnene 3,20-dione
11-deoxycortisol	S	17-21-dihydroxy Δ_4-pregnene 3,21-dione
Estradiol	E_2	$\Delta_{1,3,5}$-estratriene 3,17β-triol
Estriol	E_3	$\Delta_{1,3,5}$-estratriene 3,16α,17β-triol
Estrone	E_1	3-hydroxy $\Delta_{1,3,5}$-estratriene 17-one
Pregnandiol	—	5β-pregnane 3α,20α-diol
Pregnanetriol	PT	5β-pregnane 3α,17,20α-triol
Pregnenolone	Δ_5P	3β-hydroxy Δ_5-pregnene 20-one
17-hydroxypregnenolone	17 OH-Δ_5P	3β,17β-dihydroxy Δ_5-pregnene 20-one
Progesterone	P	Δ_4-pregnene 3,20-dione
17-hydroxyprogesterone	17 OH-P	17α-hydroxy Δ_4-pregnene 3,20-dione
Testosterone	T	17α-hydroxy Δ_4-androstene 3-one

portion of cholesterol is synthesized locally from acetate. The penetration of cholesterol into steroid-forming cells occurs in the form of complexes with low density lipoproteins (LDL) and high density lipoproteins (HDL). This penetration implies the interaction of LDL with specific receptors in the membrane (p. 533). After internalization, cholesterol is released and immediately utilized, or is stored in the form of esters, in cytoplasmic lipid droplets.

Side-chain cleavage of cholesterol occurs in *mitochondria*. It involves a complex enzymatic system of hydroxylation, at C20 and C22, followed by the action of a desmolase. The resulting compound is *pregnenolone*. It is at this step that *specific regulation* occurs either in the adrenals (by ACTH), or in the ovaries or testes (by LH).

Synthesis of Glucocorticosteroids (Cortisol and Corticosterone) (Fig. IX.5)

The synthesis of glucocorticosteroids occurs principally in cells of the *zona fasciculata*. Pregnenolone is *oxidized* at C3, resulting in a ketone function, and *isomerization of the double bond* occurs, forming the conjugated ketone of progesterone. A *series of hydroxylations* follows, at C17 (17-hydroxyprogesterone), at C21 (11-deoxycortisol or compound S), and then at C11β, terminating in cortisol. An alternative pathway exists which consists of hydroxylation at 17 (17-hydroxypregnenolone) before oxidation at C3, and isomerization (17-hydroxyprogesterone). *Cortisone* is not an adrenal secretory product (or at best it is secreted in very small quantities), and it is

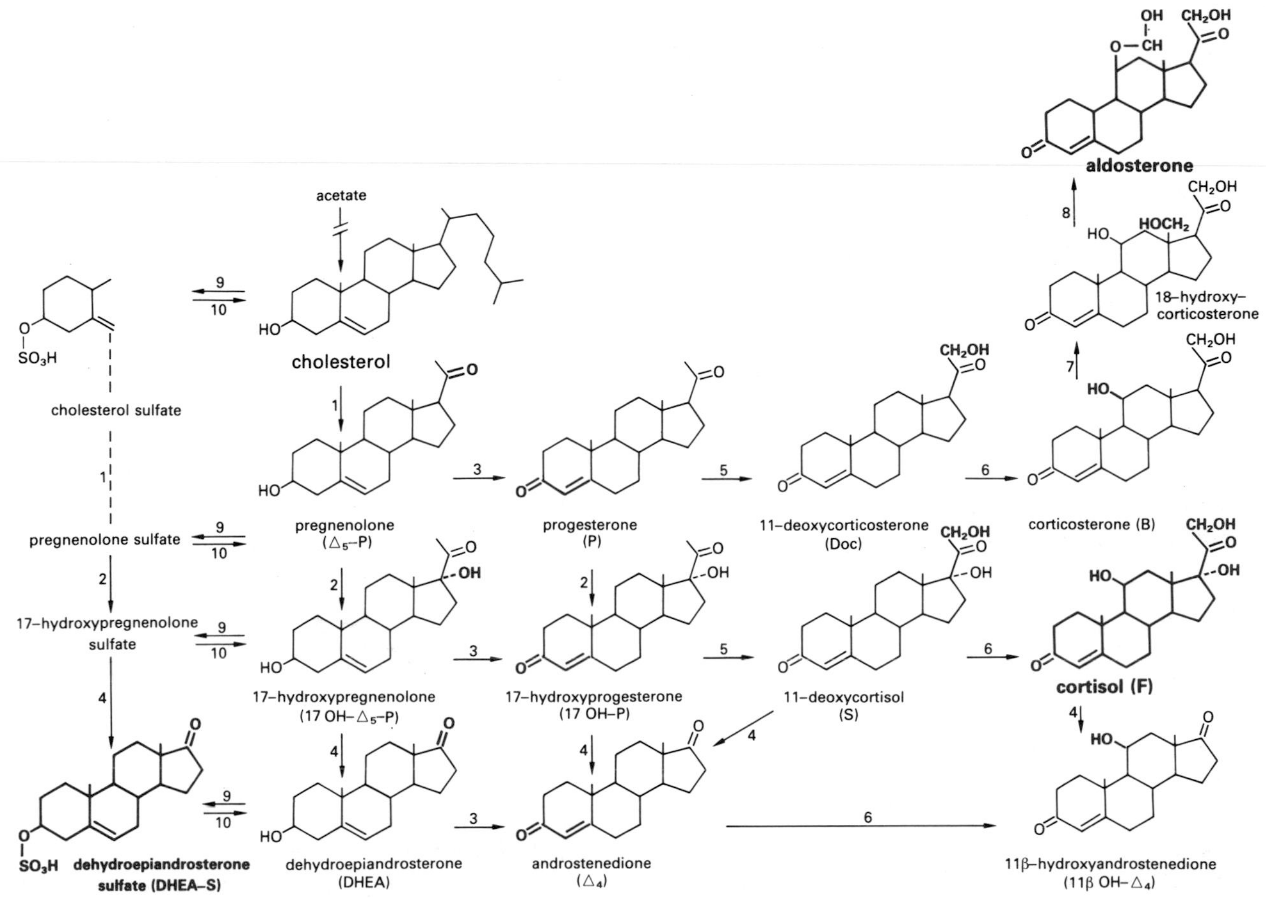

Figure IX.5 Biosynthesis of steroid hormones in adrenal glands: **1.** the desmolase system, consisting of 20- and 22-hydroxylases and of a C20-22-desmolase activity; **2.** 17-hydroxylase; **3.** 3β-hydroxysteroid oxidoreductase and $\Delta_{5\rightarrow4}$-3-oxosteroid isomerase; **4.** C17-20-desmolase; **5.** 21-hydroxylase; **6.** 11β-hydroxylase; **7.** 18-hydroxylase; **8.** 18-hydroxysteroid oxidoreductase; **9.** sulfotransferase; **10.** sulfatase.

derived essentially from hepatic metabolism of cortisol (p. 398).

Hydroxylation at C17 does not occur for *corticosterone*, the production of which is only 10% that of cortisol in the human. In rats and mice, 17-hydroxylase is practically absent, and therefore corticosterone is the only glucocorticosteroid in these species.

Synthesis of Mineralocorticosteroids (Aldosterone and Deoxycorticosterone)

(Fig. IX.5)

The synthesis of aldosterone and deoxycorticosterone occurs in the *zona glomerulosa* of the adrenal cortex, where there is no 17-hydroxylase. Progesterone is hydroxylated at 21 and 11β, producing corticosterone, then a hydroxylation at 18 (18-hydroxycorticosterone) takes place, followed by an oxidation of the alcohol function to aldehyde, resulting finally in *aldosterone* (which is present mostly in the hemiacetal form).

Synthesis of Androgens

Androgens can be classified in two categories: *adrenal* androgens (androstenedione and 11β-hydroxyandrostenedione, dehydroepiandrosterone and dehydroepiandrosterone sulfate) and *testicular* androgens (testosterone). A desmolase transforms 17-hydroxypregnenolone into *dehydroepiandrosterone* (DHEA), which is reversibly esterified, resulting in the ester *dehydroepiandrosterone sulfate* (DHEA-S). Alternatively, the 3β-hydroxyl group of DHEA can be oxidized, with isomerization of the double bond at C5-C4, which produces *androstenedione*. This product can also be obtained directly by removal of the lateral side chain of 17-hydroxyprogesterone. Hydroxylation at the 11β position, occurring only in the adrenal, produces 11β-hydroxyandrostenedione. The C17 reduction of androstenedione (quantitatively important only in the testes) produces *testosterone* (Fig. IX.6).

Cholesterol, pregnenolone, 17-hydroxypregnenolone, and dehydroepiandrosterone are in reversible equilibrium with their sulfate esters. A *sulfate pathway* has been described which consists of all reactions going from cholesterol sulfate to dehydroepiandrosterone sulfate, with the conservation of the acid group. However, this is a minor pathway, the sulfates probably constituting reserve forms for the principal biosynthetic pathway[10]. A similar sequence of biotransformation of *steroid fatty acid esters* (*"lipoidyl"*) has been described, but has not yet been evaluated physiologically.

Synthesis of Estrogens

Estrogens are derived from androstenedione and testosterone by *aromatization* of the A ring (Fig. IX.6). This reaction consists of a hydroxylation at C19, followed by an oxidation and removal of a molecule of formaldehyde. *Estrone* is obtained from androstenedione, and *estradiol* from testosterone. A large number of organs are capable of interconverting estrone and estradiol. This latter compound has approximately ten times greater estrogenic activity than estrone.

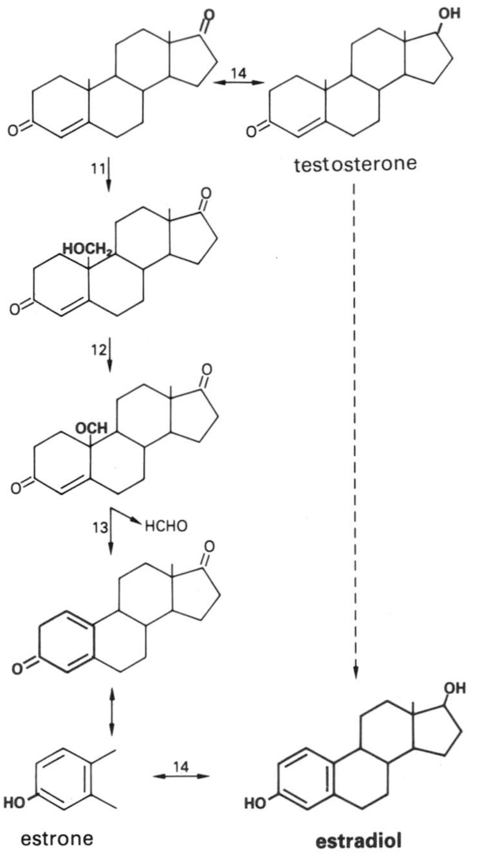

testosterone

estrone

estradiol

Figure IX.6 Biosynthesis of testosterone, estrone, and estradiol; **11.** 19-hydroxylase; **12.** 19-hydroxysteroid oxidoreductase; **13.** C10-19-desmolase; **14.** 17β-hydroxysteroid oxidoreductase.

3.3 Biosynthetic Enzymes

Hydroxylases catalyze reactions of the type:

$$RH+1/2\ O_2+NADPH \rightarrow ROH+NADP+H_2O$$

This is an irreversible reaction. Hydroxylases (or "oxygenases") are, in effect, complex enzymatic systems; they consist of a non-heme heteroprotein which contains iron, adrenoxine, adrenoxine reductase, which is a flavoprotein, and cytochrome P450 (Fig. IX.7). The latter compound is responsible for most of the known characteristics of these reactions, being auto-oxidizable (directly binding oxygen) and capable of binding carbon oxide, which inhibits its function. Hydroxylases are localized within mitochondria (C20, C22, C11, C18, C19) and in microsomes (endoplasmic reticulum) (C17, C21).

Desmolases break double bonds which have been destabilized by the introduction of hydroxyl (or ketone) groups on adjacent carbons.

3β-hydroxysteroid oxidoreductase and *$\Delta_{5\rightarrow4}$-3-oxosteroid isomerase* transform pregnenolone and DHEA into progesterone and androstenedione, respectively. They are microsomal enzymes, one being directly related to the other. The coenzyme is NAD^+.

17β-hydroxysteroid oxidoreductase is essentially testicular, ovarian, and placental. It has also been found in other organs (e.g., in liver, kidney, erythrocytes). The coenzymes are either NADH or NADPH. It can be either a soluble or microsomal enzyme.

Enzymes which transform androgens to estrogens are known as *"aromatases"*. They consist, in particular, of a *C19 hydroxylase* and a *hydroxysteroid oxidoreductase*.

10. Baulieu, E. E., Corpéchot, C., Dray, F., Emiliozzi, R., Lebeau, M. C., Mauvais-Jarvis, P. and Robel, P. (1965) An adrenal-secreted "androgen": dehydroisoandrosterone sulfate. Its metabolism and a tentative generalization on the metabolism of other steroid conjugates in man. *Rec. Prog. Horm. Res.* 21: 411–500.

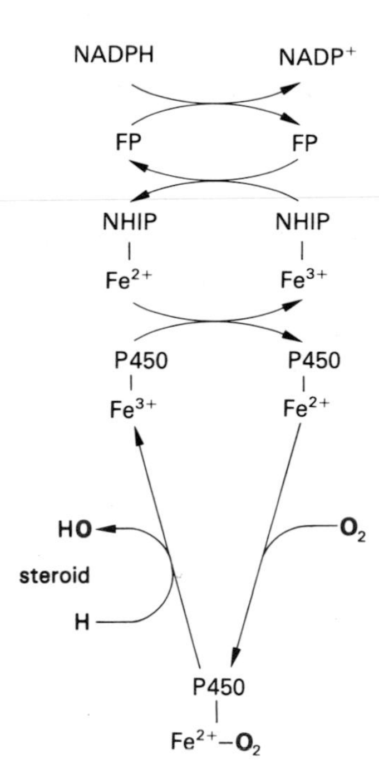

Figure IX.7 Mechanism of steroid hydroxylation. (FP, ferredoxin reductase; NHIP, non-heme iron protein (ferredoxin); P450, cytochrome P450).

Sulfotransferases use the sulfate contained in the coenzyme PAPS (3′-phosphate adenosine 5′-phosphosulfate), present principally in the zona reticularis. These enzymes are soluble. One (or several) steroid sulfase(s) can irreversibly hydrolyze steroid sulfates. They are present in low concentrations in the adrenals; higher amounts are found in the liver and placenta.

3.4 Biosynthesis in Different Endocrine Glands

Adrenal Cortex

Different zones

The zona fasciculata and zona reticularis possess all of the enzymatic equipment necessary for the synthesis of cortisol and adrenal androgens. The zona glomerulosa, deficient in 17-hydroxylase and possessing an 18-hydroxylase, synthesizes mineralocorticosteroids.

Testes

Testosterone, a local and peripheral hormone

Essentially, the testes secrete *testosterone* and *small* amounts of *estradiol*. Secreted testosterone is diluted in the peripheral circulation, but a fraction will go directly to the seminiferous tubules and genital tract, particularly the epididymis, perhaps in part via its binding to ABP (androgen binding protein)[11], which passes through the lumen of the tubules (Fig. I.47). It is now known that this local effect is indispensible for the maintenance of spermatogenesis; it is also involved in the activation of spermatozoa.

Ovaries

Periodic cellular cooperation

Ovaries secrete *primarily estrogens and progesterone*. The latter is an intermediary in the biosynthetic pathway of estrogens, and theoretically should not be secreted, since only terminal products of the biosynthetic chains are excreted into the circulation. This apparent paradox has been explained by the *two cell theory*, stating that progesterone is secreted by cells deprived of enzymes necessary for the transformation of progesterone to androgens and estrogens[12]. Cells of the *theca interna* of follicles secrete estrogens in increasing quantity as they develop during the follicular phase of the cycle. *Granulosa* cells, deficient in 17-hydroxylase and desmolase, synthesize progesterone, but since these cells are not vascularized at this stage, progesterone passes through the theca, where it is transformed into estrogens. Following ovulation, the cellular layers of the granulosa are penetrated by blood vessels, and progesterone reaches the circulation without undergoing further metabolism. The secretion of estrogens also continues during this phase of the cycle.

Finally, the *stromal portion* of the ovaries secretes *androstenedione*, normally of little importance but which can lead to virilization in diseases in which the ovarian stroma is overly developed (polycystic ovarian syndrome) (p. 464).

11. Hansson, V., Trygstad, O., French, F. S., McLean, W. S., Smith, A. A., Tindall, D. J., Weddington, S. C., Petrusz, P., Nayfeh, S. H. and Ritzen, E. (1974) Androgen transport and receptor mechanisms in testis and epididymis. *Nature* 250: 387–391.
12. Short, R. V. (1964) Ovarian steroid synthesis and secretion in vivo. *Rec. Prog. Horm. Res.* 20: 303–340.

4. Blood Transport

Binding Proteins in Plasma

Steroids are transported primarily in the *plasma*. Although low concentrations have been found in erythrocytes, their biological importance remains unknown. Plasma steroids, for the most part, are bound to proteins. The *free fraction*, in constant equilibrium with the bound fraction, is the only *active component*, at least for the most important actions.

4.1 Steroid Binding Plasma Proteins[13,14]

Both specific and non-specific proteins

Two proteins are known as specific (CBG and SBP), because they bind only certain steroids, with high affinity ($K_D 1 < 0.1$ μM) and low capacity. In contrast, binding to albumin is said to be non-specific, because this protein binds all steroids, with low affinity and high capacity.

CBG (Corticosteroid Binding Globulin)

Corticosteroid binding globulin, or *transcortin*, is an *α-glycoprotein* with a molecular weight of 52,000. It can be purified relatively easily by affinity chromatography. The protein binds *Δ_4-3-keto C21 steroids.* Cortisol, corticosterone, 11-desoxycortisol (compound S), progesterone and 17-hydroxyprogesterone are bound with high affinity. Among the synthetic steroids, prednisolone is bound, whereas dexamethasone and triamcinolone are not. Each protein molecule binds one steroid molecule. The plasma concentration of CBG is approximately 30 mg/l, which in molar terms represents a level two to three times higher than the plasma concentration of cortisol. In patients with a hypersecretion of cortisol, the binding capacity of transcortin is easily exceeded. The concentration of CBG increases during pregnancy and during treatment with estrogens (contraceptives, treatment of cancer of the prostate, etc.).

SBP (Sex Steroid Binding Plasma Protein)[15]

SBP, or *SHBG* (sex hormone binding globulin) or *TEBG* (testosterone estradiol binding globulin), is a β *glycoprotein* composed of two subunits with a M_r of ~42,000. This protein binds C19 or C18 steroids of planar structure (Δ_4-, 5α- and phenolic derivatives), and a C17β-hydroxyl. The metabolites of testosterone (DHT and 3α- or 3β-androstanediols), testosterone itself, and estradiol bind with a decreasing order of affinity. Neither estrone nor androstenedione is bound, nor are certain synthetic estrogenic compounds (diethylstilbestrol, DES) or androgens (R-1881). The normal plasma concentration of SBP is approximately 3 mg/l. It is slightly higher in older men, and is higher in women than in men. As is true for CBG, SBP *increases during pregnancy* (by a factor of five), or following administration of *estrogens*. Its concentration is also augmented by thyroid hormones (e.g., in Graves' disease).

Albumin

The concentration of albumin is approximately 1000 to 10,000 times greater (40 g/l) than that of the specific binding proteins. It binds all steroids, with a dissociation constant between 1 μM (estrogens) and 1 mM (cortisol). The affinities are inversely proportional to the polarity of the steroids.

Species Variations

Transport proteins vary from species to species. As such, *PBP* (progesterone binding plasma protein)[16] exists only in hystricomorphs (e.g., the guinea-pig) and appears in blood only during pregnancy (in maternal but not fetal blood). Estrogens are bound by *α-fetoprotein* in rats but not humans. In rodents, such as rats and mice, SBP is absent. In addition, the regulation of these proteins is also variable. Murine CBG increases under the influence of thyroid hormones, which appear to have no effect on human CBG.

13. Westphal, U. (1971) *Steroid-Protein Interactions.* Springer Verlag, Berlin.
14. Forest, M. G. and Pugeat, M., eds (1986) *Binding Proteins of Steroid Hormones.* vol. 149, INSERM/John Libbey Eurotext, J. Libbey and Son, London.
15. Mercier-Bodard, C., Alfsen, A. and Baulieu, E. E. (1970) Sex steroid binding plasma protein (SBP). *Acta Endocrinol.* suppl. 147: 204–224.
16. Milgrom, E., Allouch, P., Atger, M. and Baulieu, E. E. (1973) Progesterone-binding plasma protein of pregnant guinea pig. Purification and characterization. *J. Biol. Chem.* 248: 1106–1114.

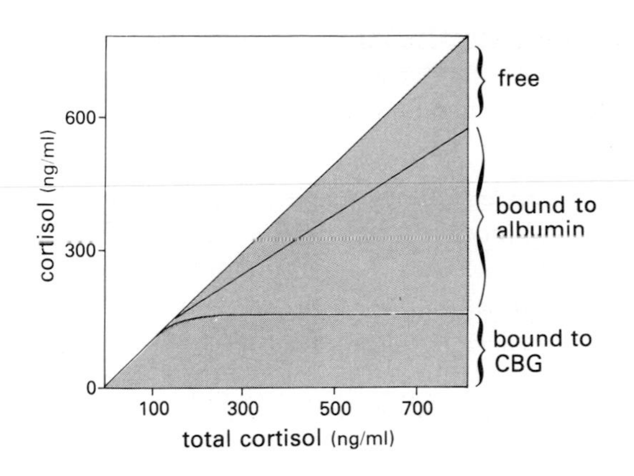

Figure IX.8 Distribution of cortisol, between different proteins and the free fraction, as a function of the total cortisol level.

4.2 Steroid Hormones in Plasma

Cortisol (Fig.IX.8)

The plasma concentration of cortisol is approximately 15 μg/dl at 08.00 h. Approximately 90% is bound to transcortin (K_D=1.5 nM at 2°C and 30 nM at 37°C), and the remainder is free, with practically no binding to albumin at normal plasma concentrations. When cortisol concentrations increase, CBG is rapidly saturated, and the free fraction, as well as that weakly bound to albumin, is greatly increased.

Testosterone

Testosterone is bound to three plasma proteins: SBP (K_D=0.8 nM at 4°C and 2 nM at 37°C), CBG (0.2 μM), and albumin (0.2 mM). The major portion of testosterone is bound by SBP in normal men. The free fraction represents approximately 2% of the total hormone in plasma.

Estradiol

Estradiol is bound by *SBP* (K_D=2 nM at 4°C) and albumin (K_D=1 μM).

Progesterone

Progesterone and cortisol, at 4°C, have similar affinities for transcortin (K_D=2 nM), but at 37°C the affinity of progesterone is higher (K_D=12 nM). It also binds to albumin (K_D=15 μM). The binding to orosomucoid (K_D=3 μM) is of unknown physiological importance. However, RU 486, an antiprogesterone, also binds this protein.

Aldosterone

Aldosterone is *weakly* bound to CBG and albumin, and circulates principally in the free form.

4.3 Physiological Significance of Plasma Binding of Steroids

A problem still unresolved

It seems that only the free fraction of hormones is active in target tissues. In addition, no obvious endocrine consequences are observed in patients with a congenital reduction in CBG concentration or those receiving treatment with estrogens (where binding proteins increase as well as total hormonal concentrations), or subjects suffering from Graves' disease (p. 370) (who have an increased testosterone concentration due to an increase in SBP). The direct injection of CBG counteracts the effect of an injection of cortisol (Fig. IX.9). Other observations led to the belief that binding proteins interact at the level of more precise regulation, such as differential induction of enzymes in various tissues. In any case, there is *no direct relationship between the affinity* of steroids for specific proteins in plasma and their *biological activity*. For example, the potent activity of synthetic steroids may be explained in part by a lack of binding to plasma protein.

The free fraction of a hormone controls the secretion of pituitary hormone via negative feedback mechanisms, mostly at the level of the hypothalamus and/or pituitary (p. 32 and p. 41). It is also preferen-

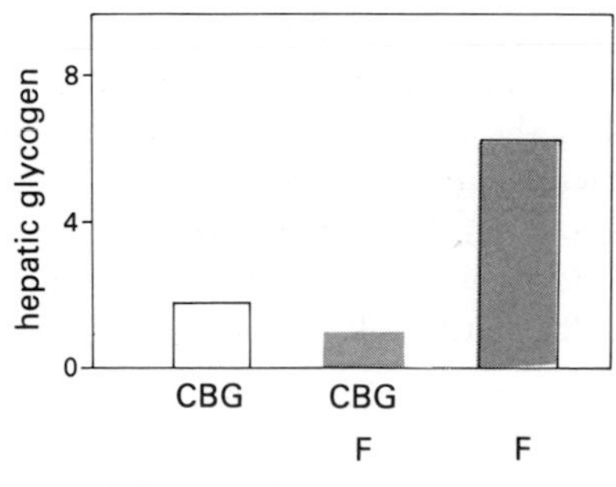

Figure IX.9 Inhibition of the biological effect of cortisol due to binding to CBG. Bars indicate liver glycogen levels (mg per 10 g of body weight) following injection of CBG, cortisol (F), or both, to adrenalectomized rats.

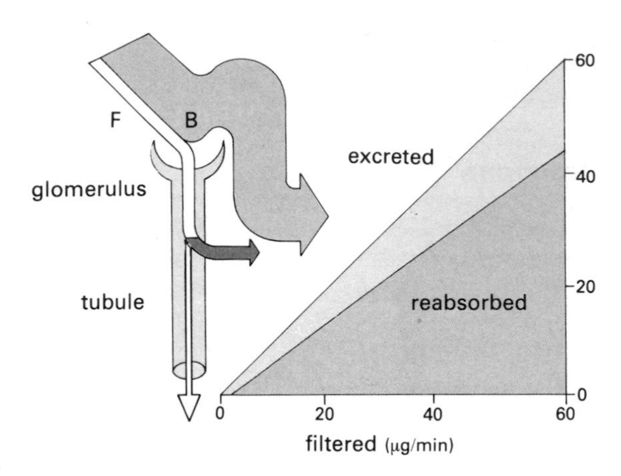

Figure IX.10 Renal filtration and reabsorption of non-protein-bound cortisol. The fraction of the hormone bound to protein (B) remains within the blood vessels. The free fraction (F) is filtered, but a large portion is reabsorbed. There is no saturation of this reabsorption phenomenon, so that excreted cortisol always remains proportional to filtered cortisol, and therefore also to the fraction not bound to proteins.

tially inactivated by hepatic metabolism, and undergoes ultrafiltration (and is partly reabsorbed in the case of cortisol) in the kidney (Fig. IX.10). The determination of free hormone in urine is therefore a good index of free plasma hormone concentration.

From these considerations, the *specific role* of binding proteins becomes *questionable*. A number of hypotheses have been put forward. The necessity of solubilizing steroids (which are lipid substances) by proteins appears to be excluded, because at physiological concentrations they are perfectly soluble in a simple aqueous milieu. The availability of a blood reservoir of hormones, rapidly mobilized when the concentration of free steroids is reduced, is an interesting hypothesis, in addition to the possibility that plasma binding allows an easier transfer of the hormone, from the cell which produces it, into the circulation. Finally, the presence of a transport protein in certain target organs suggests a role for these proteins that remains as yet undefined (CBG has been found in the uterus, kidney, lymphocytes and pituitary corticotropic cells, and SBP is found in the prostate) (p. 45).

5. Catabolism

Conservative and complex

The polycyclic carbon ring of steroid hormones is not degraded during metabolism. Following diverse chemical modifications, it is eliminated for the most part in urine, with a small quantity passing into the feces. These observations explain the absence of any danger in administering stable radioactive steroids, and thus radioactive isotopes can be used for the study of pathophysiology. The isotope is present within the body for only a very short time, and it cannot exchange with permanent components or those which are slowly turned over. Catabolism consists of a *series* of *reductions* and *hydroxylations*, and the steroid is finally conjugated with an acid (glucuronic or sulfuric). The hydrosoluble compounds thus formed are eliminated. *Biological activity* is irreversibly *lost* after the first (or one of the first) chemical transformations (frequently a reduction) of the hormone molecule.

5.1 Catabolism of Different Hormones

Cortisol (Fig. IX.11)

Cortisol is in reversible equilibrium with cortisone. These two compounds undergo an identical metabolism. The *reduction of the Δ_4 double bond* is the *decisive step* because it leads to compounds which have lost their hormonal activity. This results essentially in 5β-compounds. The dihydroderivatives formed can still be reduced to *tetrahydroderivatives* (reduction of the ketone at C3) and hexahydroderivatives (reduction of the ketone at C20), and these metabolites are glucurono-conjugated for the most part. The tetrahydroderivatives represent approximately 25% of the cortisol secreted, and hexahydroderivatives approximately 30 to 40%. A minor pathway (3%) consists of the removal of the lateral chain, producing *17-ketosteroids* which are *11-oxygenated*. *6β-hydroxylation*, very weak under normal conditions, increases markedly during pregnancy, following the administration of estrogens or Op′-DDD (mitotane, an antineoplastic agent), and in newborns. A small fraction of untransformed cortisol (<1%) is excreted *directly* into the urine, which is important for clinical testing.

Corticosterone undergoes the same modifications resulting in particular di-, tetra- and hexa-hydrogenated metabolites.

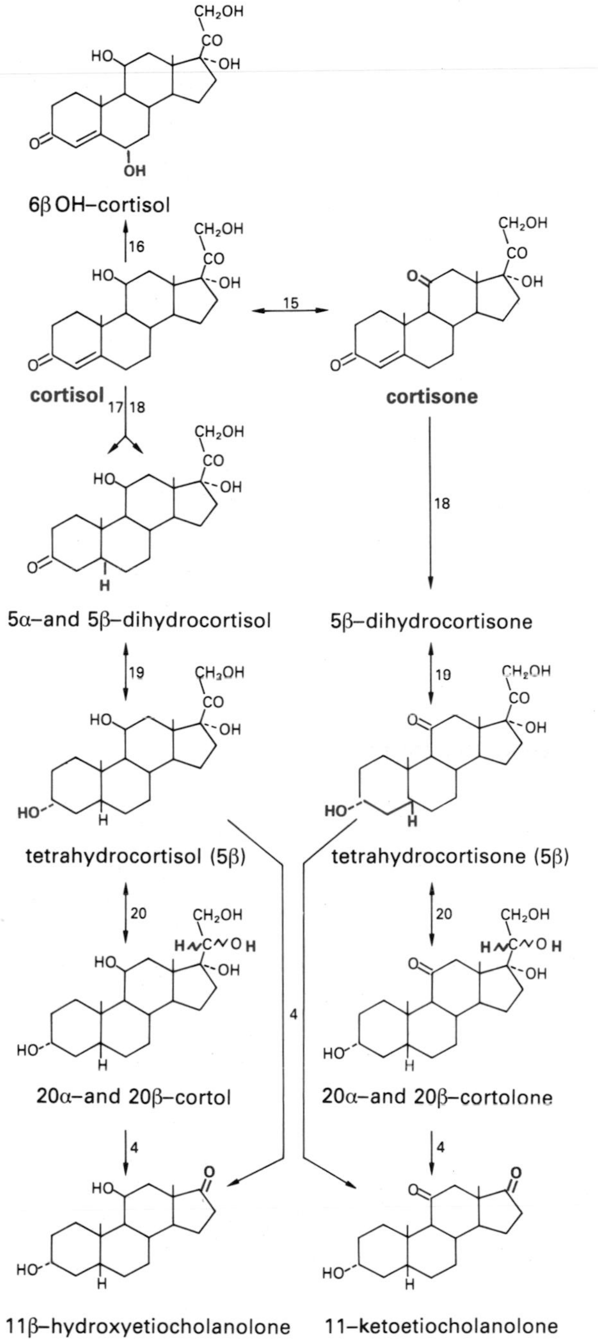

6β OH–cortisol

cortisol

cortisone

5α–and 5β–dihydrocortisol

5β–dihydrocortisone

tetrahydrocortisol (5β)

tetrahydrocortisone (5β)

20α–and 20β–cortol

20α–and 20β–cortolone

11β–hydroxyetiocholanolone

11–ketoetiocholanolone

Figure IX.11 Principal catabolites of cortisol. Abbreviations for the enzymes are: **15**=11β-hydroxysteroid oxidoreductase; **16**=6β-hydroxylase; **17**=Δ_4-3-oxosteroid 5α-reductase; **18**=Δ_4-3-oxosteroid 5β-reductase; **19**=3α-hydroxysteroid oxidoreductase; **20**=20α- or β-hydroxysteroid oxidoreductase; **4**=C17-20-desmolase.

Aldosterone

Aldosterone also undergoes a series of reductions. The *3α, 5β-tetrahydrogenated* derivative is glucurono-conjugated and represents approximately 30% of secreted aldosterone. *Hexahydrogenated* derivatives are also formed, resulting in bicyclic compounds, some of which are 21-methylated (loss of oxygen). From 5 to 10% of *aldosterone is directly glucorono-conjugated* at the level of the hemi-acetalic free hydroxyl group. In pathophysiological investigations, advantage is taken of the specific characteristic of this binding, for hydrolysis of the conjugate at room temperature, in an acid milieu (pH 1).

Androgens (Fig. IX.12)

DHEA is in reversible *equilibrium with* its *sulfate, DHEA-S,* due to the function of sulfotransferases and sulfatases, primarily in liver and kidney, and for the most part the conjugate is eliminated in the urine without further change[10]. A fraction of DHEA is metabolized, particularly in the liver, resulting in androstenedione. First, the double bond of the latter compound is reduced, resulting in 5α- and 5β-androstanedione. Reduction of the ketone at C3 produces various isomers. The two most important are *androsterone* (3α-hydroxy 5β-androstane 17-one) and *etiocholanolone* (or 5β-androsterone) (3α-hydroxy 5β-androstane 17-one). These metabolites are glucurono- and sulfo- conjugated.

The major portion of *testosterone* is oxidized to androstenedione, in the liver, before undergoing the previously described metabolism. Some testosterone follows a 17-hydroxyl metabolic pathway. It is reduced, producing *5α- and β-dihydrotesterone* (or *5α- and β-androstanolone) and then 5α- and β-androstanediols.*

The reduction of adrostenedione to testosterone may be important in women, but it is negligible in men, in comparison with testosterone production in the testes.

Estrogens (Fig. IX.13)

Estradiol and estrone are in reversible equilibrium due to hepatic and intestinal 17β-hydroxysteroid dehydrogenases. Estrone can undergo hydroxylation at C16α, leading to 16α-hydroxyestrone and *estriol*. Hydroxylation can also occur at the C15 position. Hydroxylation at the C2 and the C4 positions are also quantitatively important, especially that of C2. Hydroxylation is often followed by methylation.

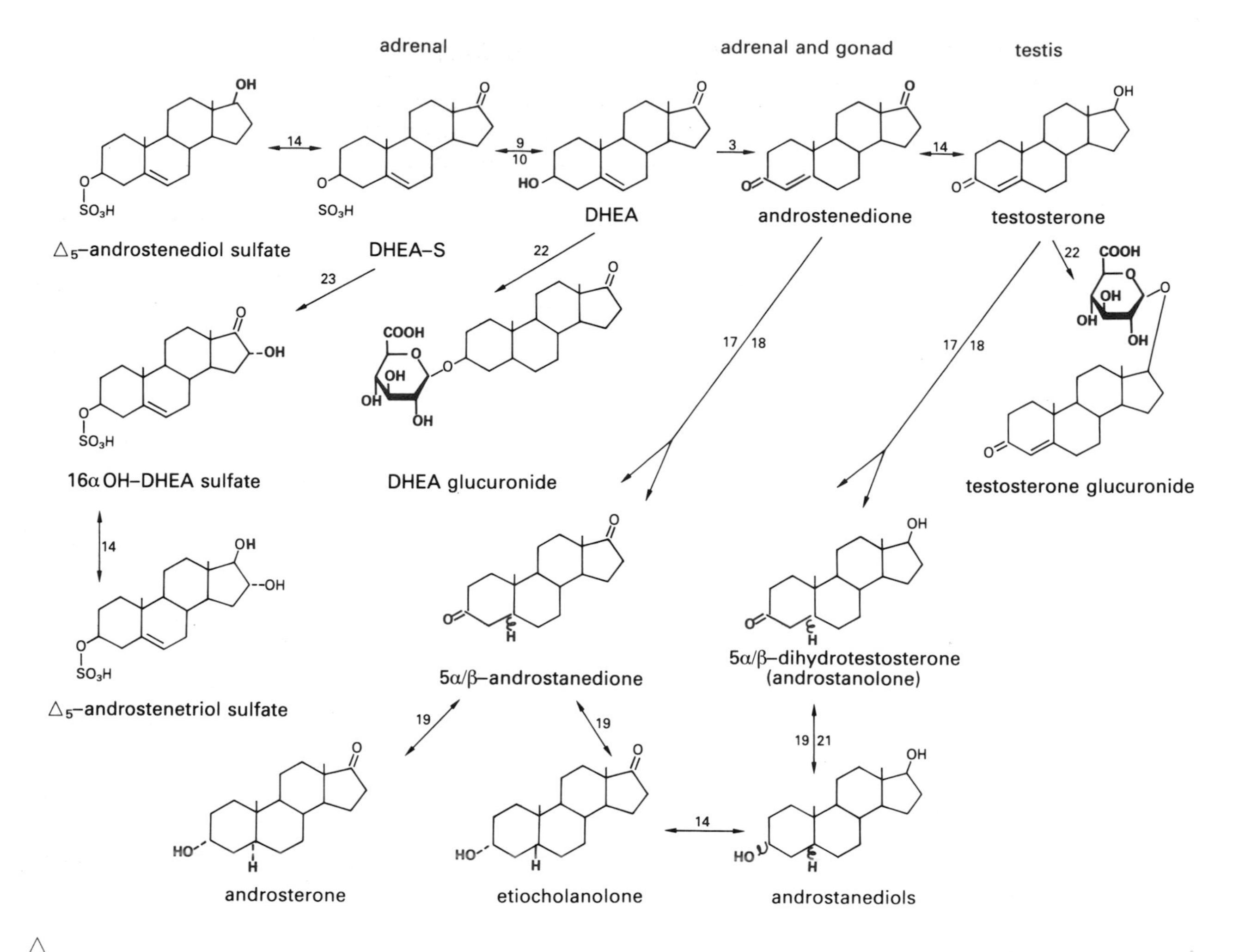

△
Figure IX.12 Catabolism of androgens. Abbreviations of the enzymes are: **22**=β-glucuronotransferase; **23**=16α-hydroxylase; **17**=Δ_4-3-oxosteroid 5α-oxidoreductase; **18**=Δ_4-3-oxosteroid 5β-oxidoreductase; **19**=3α-hydroxysteroid oxidoreductase; **21**=3β-hydroxysteroid oxidoreductase; **14**=17β-hydroxysteroid oxidoreductase.

◁
Figure IX.13 Catabolism of estrogens. Abbreviations of the enzymes are: **14**=17β-hydroxysteroid oxidoreductase; **23**=16α-hydroxylase; **24**=2-hydroxylase; **25**=O-methyltransferase.

These *catechol estrogens* are formed in relatively large amounts in the *brain*, in particular in the hypothalamus and pituitary. Their postulated activities include the inhibition of norepinephrine metabolism (by competition at the level of catechol O-methyltransferase), a local antiestrogenic effect (by competition with estradiol, at the receptor level), and direct interaction with dopaminergic receptors. All hydroxylated metabolites are excreted in rather large amounts in the *bile*, and undergo an enterohepatic cycle.

Progesterone and 17-hydroxyprogesterone

Progesterone undergoes *reductive metabolism*. In the dihydroderivatives, the 5β isomer predominates. Some tetrahydroderivatives (reduction of the ketone at C3) and some hexahydroderivates (reduction of the ketone at C20) are obtained. The glucuronide of *pregnanediol* (5β-pregnane 3α, 20α-diol) is the *major* metabolite (15 to 25% of progesterone secretion). It should be noted that a fraction of progesterone is reduced directly to dihydroderivatives 20α and 20β.

The principal metabolite of 17-hydroxyprogesterone, *pregnanetriol* (5β-pregnane 3α, 17α, 20α-triol), is eliminated in the form of a glucuronide.

5.2 Catabolic Enzymes

3-oxo Δ_4-steroid Reductases (or 5α-reductases)

5α-reductases catalyze the irreversible reduction of double-bond Δ_4. The *5α-reductases* are *membrane bound*, and are found essentially in microsomes and at the level of the external nuclear membrane of certain cells. *5β-reductases* are soluble. Their coenzyme is NADPH. Glucocorticosteroids and progesterone are principally 5β-reduced, while androgens are reduced to give 5α and 5β metabolites. The ratio of activity between the two categories of enzymes varies with sex, age, and thyroid state. These variations occur in humans and other animals, but they are not necessarily the same, quantitatively or qualitatively, throughout all species. The ratio between androgen metabolites 5α and 5β is higher in men than women, and can be reduced by the administration of estrogens.

Hydroxysteroid Oxidoreductases

Hydroxysteroid oxidoreductases catalyze *reversible* reactions.

17β-hydroxysteroid oxidoreductases are soluble or microsomal, and utilize NAD^+ or $NADP^+$ as a coenzyme. They have wide distribution (found essentially in liver, but also in kidney, skin, erythrocytes, etc.). The activity of these enzymes is biologically *important* since they transform active hormones (estradiol, testosterone) into metabolites with little activity (estrone, androstenedione).

3α- and 3β-hydroxysteroid oxidoreductases can utilize NAD^+ or $NADP^+$. They are soluble (3α-oxidoreductases), or primarily microsomal (3β-oxidoreductases). It is at the level of the C3 hydroxyl groups that conjugation with sulfuric acid or glucuronic acid occurs.

11β-hydroxysteroid oxidoreductase is essentially hepatic, but can also be found in lymphoid organs. It controls the equilibrium between cortisol and cortisone, the latter of which has reduced biological activity. Thyroid hormones increase the formation of cortisone at the expense of cortisol.

20α- and 20β-hydroxysteroid oxidoreductases are microsomal, and utilize NADPH as a coenzyme. They are found in the liver, but also in the ovaries and placenta. The biological activity of 20-reduced progesterone derivatives is still not defined.

Hydroxylases

Hydroxylases acting at C6, and at C16α and C2, are important in the metabolism of cortisol and estrogens, respectively.

Conjugating Enzymes

These are sulfuryl- and glucuronyltransferases. Only hydroxysteroids (including phenolsteroids) are conjugated, and there is no proof that conjugation of the enolic form of the ketone group takes place.

Sulfuryltransferases are soluble, and are found in the liver, intestine, skin, ovary, and placenta. They transfer sulfate from 3′-phosphoadenosyl 5′-phosphosulfate (PAPS).

Glucuronyltransferases are microsomal, and are found principally in the liver and kidney. The donor of glucuronic acid is uridine diphosphoglucuronic acid (UDPG), and the compound formed is a β-glucuronoside (hydrolyzable specifically by β-glucuronidases). It is the glucurono-conjugation which predominates for 5β derivatives and for hydroxyls at C3α, C16α, C17, C18, and C21. Inversely, 3β-hydroxylated derivatives of the Δ_5, 5α and 5β steroids are essentially sulfoconjugated.

Sulfatases are microsomal, and are found in high concentration in the placenta. They are also found in liver and testes. These enzymes hydrolyze sulfates of 3β-hydroxy 5α-steroids and 3β-hydroxy Δ_5-steroids, while androsterone sulfate and testosterone sulfate are not hydrolyzed. Actual in vivo *β-glucuronidase* activity is minimal.

6. Regulation of Plasma Hormone Concentration

Hormone secretion, concentration and catabolism are coupled in a regulated system

The concentration of hormones at the target cell level determines hormonal action. It depends on both hormone production and degradation. Regulatory factors affect the latter, as for example the administration of thyroid hormones, which increases the catabolism of glucocorticosteroids, whereas estrogens reduce it. However, in both cases, the plasma concentration of cortisol remains constant due to a regulatory system which adjusts hormone biosynthesis to the rate of degradation. Such a homeostatic system has been reported for *most steroid hormones*.

6.1 Adrenocortical Hormones

Regulation[17,18]

The biosynthesis of glucocorticosteroids and adrenal androgens is controlled by *ACTH*. The secretion of this pituitary hormone is regulated by corticotropin releasing factor (CRF) (p. 233). CRF is itself under dual control: a long loop feedback control regulated by the level of free cortisol, and regulation by the higher brain centers (a short loop feedback by the concentration of ACTH has also been proposed). This latter control consists of a "clock" regulating a *circadian rhythm*, with highest levels of cortisol and ACTH observed early in the morning, and lowest levels during the afternoon. In addition, this system is involved during periods of stress, causing the stimulation of adrenal cortical secretion. The secretion of ACTH occurs as spikes of short duration superimposed over circadian variations (Fig. IV.9). The feedback control of circulating cortisol occurs not only in the brain, but also at the pituitary level. The secretory spikes of ACTH lead to oscillations of plasma cortisol levels. It is remarkable that *ACTH stimulates* not only the synthesis of *cortisol*, but also that of adrenal *androgens and*, to a lesser degree, that of *aldosterone*, although the latter two hormones have *no* effect on the feedback control of ACTH. Discussion of the factors regulating the production of aldosterone can be found on p. 616.

Mechanism of ACTH Action[19] (Fig. IX.14)

Steroidogenesis and Tropic Effects

ACTH has two types of effects on adrenals. The *rapid* effect consists of the immediate release of adrenocortical hormones into the circulation. It is

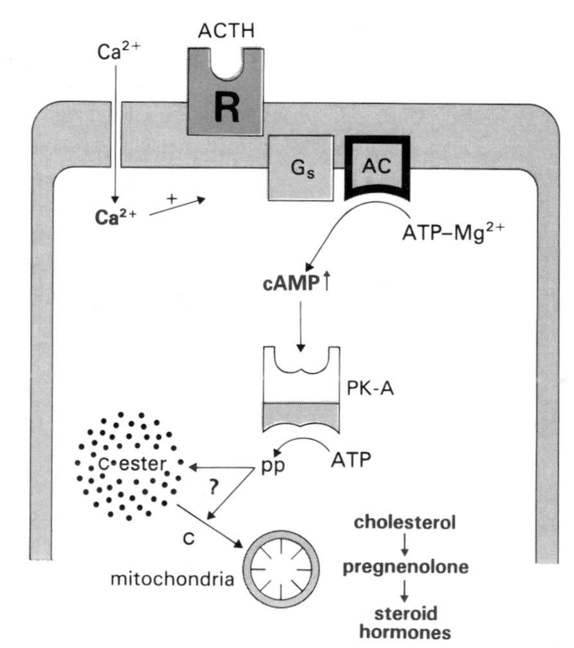

Figure IX.14 Mechanism of action of ACTH. ACTH binds to its membrane receptor and activates adenylate cyclase via G_s. Cyclic AMP then activates a PK-A. The latter phosphorylates a ribosomal protein, leading to increased protein synthesis, particularly of the regulatory protein which favors translocation of cholesterol (c), from lipid droplets to mitochondria. Concomitantly, the hydrolysis of cholesterol esters and the conversion of cholesterol to pregnenolone are increased[19].

17. Fortier, C. (1966) Nervous control of ACTH secretion. In: *The Pituitary Gland* (Harris, G. W. and Donovan, B. T., eds), Butterworths, London, pp. 195–234.
18. Yates, E. F. and Urquhart, J. (1962) Control of plasma concentrations of adrenocortical hormones. *Physiol. Rev.* 42: 359–421.
19. Garren, L. D., Gill, G. N., Masui, H. and Walton, G. M. (1971) On the mechanism of action of ACTH. *Rec. Prog. Horm. Res.* 27: 433–478.

dependent on protein synthesis, as shown by the inhibitory effect of puromycin or cycloheximide, but not on RNA synthesis, as demonstrated by the absence of any effect of actinomycin D. In addition, and *more slowly*, ACTH leads to *hypertrophy* of the adrenal cortex itself, an effect which requires the synthesis of both proteins and RNA.

As for other polypeptide hormones, ACTH interacts with a receptor at the level of the *plasma membrane* of target tissues, leading to an activation of *adenylate cyclase*. Cyclic AMP activates one or several PKs (protein kinases) of unknown substrate. Several possibilities have been put forward, one of which involves the activation of a phosphorylase, resulting in the formation of NADPH necessary for steroidogenesis. However, ACTH stimulates steroid production even under conditions in which NADPH is abundant and thus not rate-limiting. Other hypotheses involve a protein with rapid turnover which may favor the penetration of cholesterol into mitochondria, or an increase of the hydrolysis of cholesterol esters, or, yet, an increase of side-chain cleavage of cholesterol into pregnenolone, or, finally, the exit of pregnenolone from mitochondria.

Ca^{2+} also appears to *play a role* as an intracellular mediator for certain effects of ACTH, mostly synergistic to those of cAMP.

6.2 Testicular Hormones

The secretion of testosterone by interstitial cells is regulated by *luteinizing hormone* (LH). The mechanism of action of this hormone appears to be the same or very similar to that of ACTH. LH is secreted under the influence of hypothalamic gonadotropic releasing hormone (*GnRH*). Both LH and GnRH are negatively controlled by testosterone (Fig. V.15). The secretion of follicle stimulating hormone (*FSH*) is also regulated by GnRH, but it appears to be controlled additionally by inhibin, a peptide produced in seminiferous tubules. The complexity of this regulation is illustrated by the fact that spermatogenesis depends on FSH, and also that it is sensitive to testosterone produced locally by Leydig cells under the influence of LH.

6.3 Ovarian Hormones

The cyclical function of the ovary and the control thereof are complex phenomena consisting of both positive and negative feedback systems (Fig. V.15; p. 39 and p. 451).

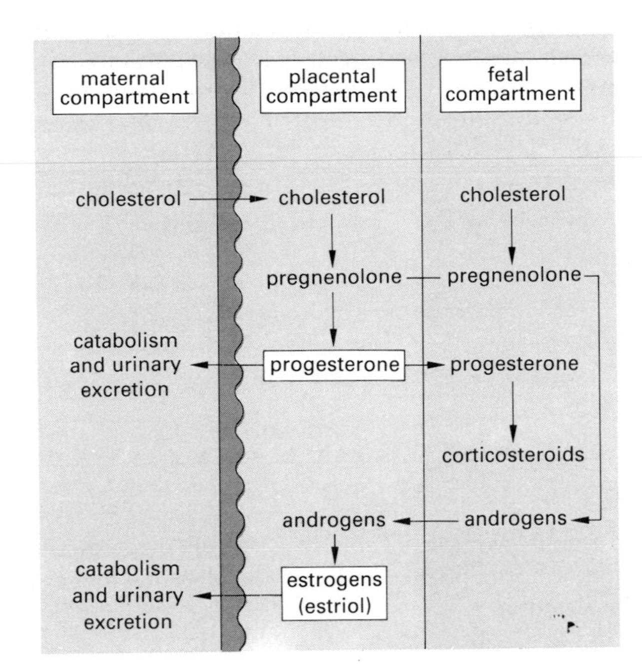

Figure IX.15 Biosynthesis of steroids in the feto-placental unit.

7. Steroid Hormones in Pregnancy

7.1 Estrogens and Progesterone

First the maternal ovary, then the feto-placental unit

During the *first trimester* of pregnancy in the human, the maternal ovary is the primary source of estrogens and progesterone. The corpus luteum is stimulated by chorionic gonadotropin (hCG), which increases progressively, attaining a peak at three months pregnancy, and decreases rapidly thereafter. From the end of the *second month*, the *placenta* begins to contribute to steroid production. In the second half of pregnancy, the role of the ovary is negligible, and the biosynthesis of estrogens and progesterone occurs in the *feto-placental unit* (Fig. IX.15). In effect, both the fetus and the placenta lack certain enzymes necessary for complete steroidogenesis; however, their *deficiencies are complementary* in such a way that the pathway of metabolites through these two compartments leads to complete

Table IX.2 Enzymes involved in the metabolism of steroids are not present in both the fetus and placenta.

Enzymes lacking in the placenta	Enzymes lacking in the fetus
C17–20 desmolase	3β-hydroxysteroid oxydoreductase
16α-hydroxylase	
Sulfotransferase	$\Delta_{5\rightarrow4}$-3-oxosteroid isomerase
	Sulfatase
	Aromatase

hormone synthesis (Table IX.2). A detailed description of hormone production in the placental unit can be found on p. 472.

7.2 Cortisol and Testosterone

Maternal corticosteroid binding globulin (CBG) and blood cortisol levels increase in parallel during pregnancy (Fig. IX.16). They begin to increase during the sixth week and stabilize at a maximal level (double or triple normal) in the third trimester. Not only does the bound fraction of cortisol increase, but there is also a doubling or tripling of the free fraction. This observation poses a double problem. First, the hypothalamic-hypophyseal feedback system maintains an increased plasma cortisol level. Secondly, there is no clinical sign of hypercortisolism, suggesting a diminution in sensitivity to active glucocorticosteroid hormones.

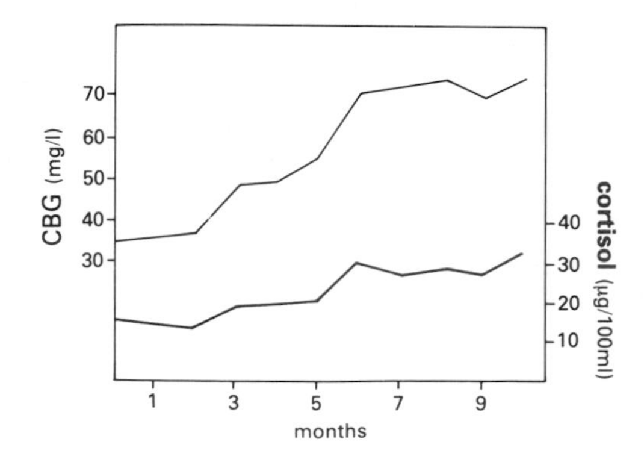

Figure IX.16 Variations of total cortisol and CBG concentrations during pregnancy.

SBP also increases during pregnancy, by a factor of two or three, with a resulting increase in plasma testosterone levels.

7.3 Steroids in Parturition

The best animal model

The role of steroids hormones involved in parturition has been particularly well studied in *sheep*[20]. The initial observation was made by injecting glucocorticosteroids or ACTH into the fetus, which led to premature delivery. Measurement of hormones in the blood of both the mother and fetus led Liggins to propose a precise scheme of events (Fig. IX.17). An increase in *fetal ACTH* (of unknown cause) is responsible for increased *cortisol* secretion. Cortisol then acts on the placenta, increasing the transformation of progesterone to 17,20α-dihydroxypregna-4-ene-3-one and the synthesis of *estrogens* and *prostaglandin* $PGF_{2\alpha}$. The *reduction* of circulating *progesterone* and the elevation of estrogens and prostaglandins increases myometrial contractility. This model applies equally well to the cow and the goat, but not to the human.

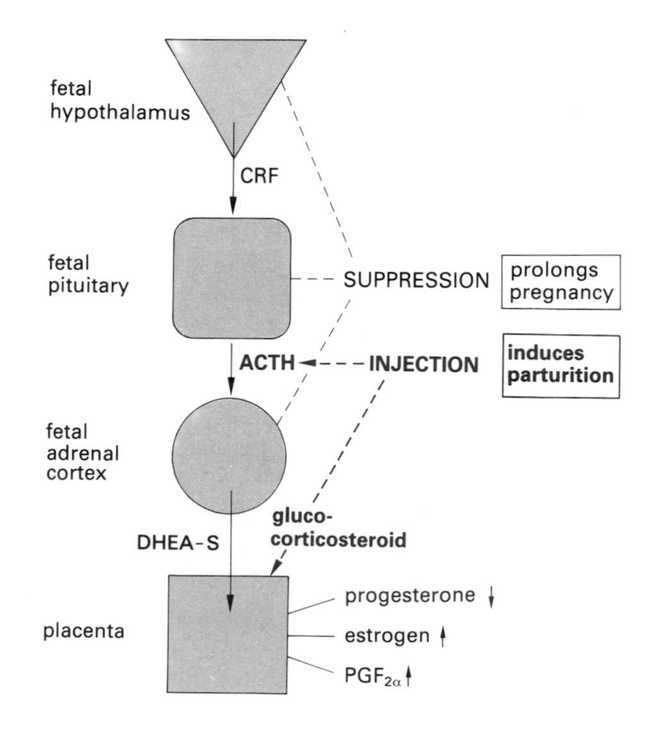

Figure IX.17 Mechanisms involved in the initiation of parturition in sheep. The main experimental arguments are represented.

7.4 Role of Steroids in Fetal and Neonatal Development

Glucocorticosteroids: Lungs, Liver

Although it remains uncertain whether increased fetal cortisol levels play a role parturition in humans, increased circulating cortisol is important in the maturation of fetal functions. In particular, in lung maturation, *alveolar surfactant* is cortisol-dependent[21]. It has been proposed that hyaline membrane disease in premature infants may be due to cortisol insufficiency, which can be prevented by the administration of glucocorticosteroids to women who are at risk of premature delivery.

Hepatic functions, in particular glycogenesis, also undergo maturation under the influence of glucocorticosteroids.

Androgens: Central Nervous System

An increase in circulating testosterone is observed in male rats just after birth. This hormonal change appears to result in an *irreversible differentiation of the nervous system*. The hypothalamus of a rat exposed to androgens during this period is "*androgenized*"; that is, the animal will have a tonic, constant secretion of LH and FSH after maturation[22]. Exposure to androgen is also necessary for normal *sexual behavior* in the adult male rat (copulation for example).

In boys[23], increases of plasma testosterone during the first months of life have also been observed. It is not known whether the role of these hormonal changes in the child are similar to those described in the rat. It is possible that a transitory alteration of testicular function in the newborn could be the cause of a central nervous system disorder and of permanent sexual anomalies first manifested at puberty.

20. Liggins, G. C., Fairclough, R. J., Grieves, S. A., Kendall, J. Z. and Knox, B. S. (1973) The mechanism of initiation of parturition in the ewe. *Rec. Prog. Horm. Res.* 29: 111–159.
21. Delemus, R. A., Shermata, D. W., Knelson, J. H., Kotas, R. V. and Avery, M. E. (1969) The induction of the pulmonary surfactant in the foetal lamb by the administration of corticosteroids. *Pediat. Res.* 3: 505–507.
22. Gorski, R. A. and Barraclough, C. H. (1963) Effects of low dosage of androgen on the differentiation of hypothalamic regulatory control of ovulation in the rat. *Endocrinology* 73: 210–216.
23. Forest, M. G., Sizonenko, P. C., Cathiard, A. M. and Bertrand, J. (1974) Hypophyso-gonadal function in humans during the first year of life. I. Evidence for testicular activity in early infancy. *J. Clin. Invest.* 53: 819–828.
24. Milgrom, E., Atger, M. and Baulieu, E. E. (1973) Studies on estrogen entry into uterine cells and on estradiol-receptor complex attachment to the nucleus. Is the entry of estrogen into uterine cells a protein-mediated process? *Biochim. Biophys. Acta* 320: 267–283.
25. Jensen, E. V. and Jacobson, H. I. (1962) Basic guides to the mechanism of estrogen action. *Rec. Prog. Horm. Res.* 18: 387–414.

8. Mechanism of Steroid Hormone Action

Diversity of actions and common mechanisms

Steroids exert *numerous and diverse actions* on their target tissues. For example, the selective changes of Na^+ transport controlled by aldosterone in kidney tubule cells do not resemble the massive growth-promoting effect of estrogens in the uterus. However, there is a great deal of *similarity* between the *molecular mechanisms* of action of these different hormones. For this reason, a *general scheme* of the *cellular mechanism* of steroid hormones will be presented (Fig. IX.18), after which the specific characteristics of each system will be examined.

One approach consists of describing modifications at tissular, cellular and finally molecular levels, in relation to the function of the target organ. Alternatively, the fate of the hormone within the target organs can be followed.

8.1 Hormone Penetration in Target Cells

Very few studies have dealt with the penetration of steroids *across the cell membrane*. These hormones, which are *lipophilic*, diffuse easily across membrane structures. However, in the case of estrogens[24] in the rat uterus, and glucocorticosteroids in cultured pituitary cells, indirect evidence has shown that *facilitated diffusion* into target cells could be involved. Within the target cells, the hormone is recognized and is bound by a specific protein structure, the receptor.

8.2 Receptors

Discovery of Receptors

The decisive instrument: tritiated estrogens[25]

Near the end of the 1950s, tritiated *estrogens* with *high specific activity* were synthesized, allowing

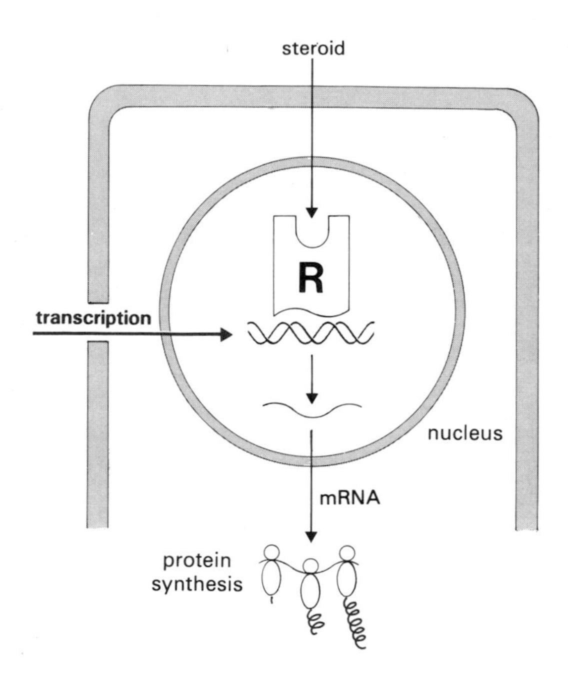

Figure IX.18 Mechanism of action of steroid hormones. A steroid enters the cell and binds to the nuclear receptor (Fig. I.70). The activated hormone-receptor complex acquires affinity for the DNA, modifying transcription in particular.

experiments to be carried out under physiological conditions. Jensen and Jacobson[25] injected [^{3}H]estradiol into rats, and studied the radioactivity present in various tissues. They observed that its *accumulation* in *target organs* was radically different from that in non-target organs (Fig. I.67). In non-target organs, radioactivity was present in much lower concentrations than in the blood, and persisted for only a short time. The steroid was highly concentrated in the uterus or vagina, as compared to plasma, and in addition was retained within these tissues. In the decade which followed, the existence of macromolecules responsible for these phenomena was established, and their characteristics were studied. Thereafter, similar receptors were described for other steroid hormones[26,27,28].

Characteristics of Soluble Receptors

In studies of sex hormones in hormone-deprived animals (for example, castrated or prepubertal animals), the receptor was found in the soluble fraction of cellular homogenates (*cytosol*). This observation was interpreted to mean that unoccupied receptor is present in the cytoplasm. Recent studies have indicated that for estrogen and progesterone *receptors* this does not seem to be the case, and that receptors are located *almost exclusively* in the *nucleus*, from which they are easily extractable by homogenization (p. 67). The number of receptors per target cell is of the order of 10^3 to 10^5.

For a number of years, hormone receptors were identifiable only by their hormone binding ability. Purification was almost impossible due to their very low concentration and instability.

Receptors are *oligomeric proteins* that *dissociate* into *subunits* upon an increase in ionic strength. The sedimentation coefficient of receptors is 8 to 10 S at low ionic stength (according to the hormonal specificity of the receptor), and 3 to 5 S at 0.3 to 0.5 M KCl or NaCl. The nature of the subunits is discussed below.

At equilibrium, the dissociation constants (K_D) of steroid-receptor complexes are of the order of 0.1 to 10 nM, lower for estrogens and androgens than for progesterone and corticosteroids. This dissociation constant is of the same order of magnitude as the concentration of hormones within the circulation, and thus a small variation in plasma hormone concentration leads to variation in receptor occupancy.

The *hormonal specificity* of receptors is remarkable. Receptors bind hormone agonists or antagonists with affinities which generally correlate with their biological potency. Of course, problems of distribution and metabolism must be considered. For instance, the ER binds estradiol, a potent estrogen, more strongly than estrone or estriol, which are weak estrogens. The estrogen receptor also binds tamoxifen, clomiphene, and other antiestrogens. There is, however, no cross-reaction with progesterone or cortisol. This receptor can be used pharmacologically to

26. Baulieu, E. E., Alberga, A., Jung, I., Lebeau, M.-Cl., Mercier-Bodard, C., Milgrom, E., Raynaud, J. P., Raynaud-Jammet, C., Rochefort, H., Truong, H. and Robel, P. (1971) Metabolism and protein binding of sex steroids in target organs: an approach to the mechanism of hormone action. *Rec. Prog. Horm. Res.* 27: 351–419.
27. Gorski, J. and Gannon, F. (1976) Current models of steroid hormone action: a critique. *Annu. Rev. Physiol.* 38: 425–450.
28. Gustafsson, J. A. and Eriksson, H., eds (1983) *Steroid Hormone Receptors: Structure and Function.* Nobel Symposium, no. 57, Elsevier, Amsterdam.

screen new molecules rapidly and economically.

Tissue specificity of receptors implies that they are found only in target organs. However, this notion has been somewhat modified with advances in detection techniques, and, in certain cases, it represents a quantitative distinction (more or less receptor) rather than a qualitative difference (receptor present or absent). In different cell types of the same organ, the concentration of receptor may differ.

Receptor Purification

One of the most efficient techniques of receptor purification is *hormonal affinity chromatography*. Two major problems arose during the development of this method for steroid receptors. It was necessary to find a region of the ligand molecule not involved in receptor binding which could be coupled to the support. For example, a lateral chain at C7 of estradiol, or derivatives at C21 for progesterone, does not preclude binding to the receptor. Secondly, considerable difficulties were encountered in the *recovery* of receptors from affinity columns. It was essential to reduce the concentration of the ligand attached to the chromatographic support and increase the specificity of receptor elution.

The availability of antireceptor antibodies has led to progress in *immunoaffinity purification*. Another method has utilized the property of transformed receptors in order to bind DNA with increased affinity. Non-transformed receptor is not retained by *columns of DNA-cellulose*, whereas after transformation (p. 409) it is strongly bound to this support. The first chromatographic step allows the removal of most proteins with an affinity for DNA from the cellular extract, and the second step, after transformation of receptor, results in an efficient purification of the receptor.

Three receptors have been purified in sufficient quantity to be studied physiochemically. The estradiol receptor is composed of a single subunit with a M_r of $\sim$66,000, and the glucocorticosteroid receptor has a molecular weight of $\sim$94,000. After a number of years of controversy, the hormone binding subunit of the progesterone receptor also appears to be a single peptide ($M_r \sim$110,000–120,000). Recent evidence suggests an M_r of $\sim$115,000 for the androgen receptor. These receptors are phosphoproteins, but the mechanism and the significance of their phosphorylation remains uncertain. All steroid receptors are "*native*" *7 to 9 S oligomers* with a M_r of $\sim$300,000 in absence of hormone[26,27,28]. They do not bind to DNA, and are therefore probably inactive. They all include a doublet of the same *hsp 90 molecule*[29,30,31], dissociating upon hormone binding and increased temperature and/or ionic strength. The released hormone-binding receptor (3 to 5 S) itself binds to specific DNA segments[32].

Recently, the cloning of the cDNA of human glucocorticosteroid, human and avian estradiol, and human, rabbit and avian progesterone, receptors has enabled the determination and comparison of their respective sequences[33–39]. All show homology with the sequence of *erb*-A (p. 123) (Fig. IX.19 and Fig.

29. Baulieu, E. E., Binart, N., Buchou, T., Catelli, M. G., Garcia, T., Gasc, J. M., Groyer, A., Joab, I., Moncharmont, B., Radanyi, C., Renoir, J. M., Tuohimaa, P. and Mester, J. (1983) Biochemical and immunological studies of the chick oviduct cytosol progesterone receptor. In: *Steroid Hormone Receptors: Structure and Function*. Nobel Symposium, no. 57 (Eriksson, H. and Gustafsson, J. A., eds), Elsevier, Amsterdam, pp. 45–72.
30. Catelli, M. G., Binart, N., Jung-Testas, I., Renoir, J. M., Baulieu, E. E., Feramisco, J. R. and Welsh, W. J. (1985) The common 90-kd protein component of non-transformed "8S" steroid receptors is a heat-shock protein. *EMBO J.* 4: 3131–3135.
31. Schuh, S., Yonemoto, W., Brugge, J., Bauer, V. J., Riehl, R. M., Sullivan, W. P. and Toft, D. O. (1985) A 90,000-dalton binding protein common to both steroid receptors and the Rous sarcoma virus transforming protein, pp60^{v-src}. *J. Biol. Chem.* 260: 14292–14296.
32. Von der Ahe, D., Janich, S., Scheidereit, C., Renkawitz, R., Schutz, G. and Beato, M. (1985) The glucocorticoid and the progesterone receptors bind to the same sites in two hormonally regulated promoters. *Nature* 313: 706–709.
33. Hollenberg, S. M., Weinberger, C., Ong, E. S., Cerelli, G., Oro, A., Lebo, R., Thompson, E. B., Rosenfeld, M. G. and Evans, R. M. (1985) Primary structure and expression of a functional glucocorticoid receptor cDNA. *Nature* 318: 635–641.
34. Weinberger, C., Hollenberg, S. M., Rosenfeld, M. G. and Evans, R. M. (1985) Domain structure of human glucocorticoid receptor and its relationship to the v-*erb*-A oncogene product. *Nature* 318: 670–672.
35. Green, S., Walter, Ph., Kumar, V., Krust, A., Bornert, J.-M., Argos, P. and Chambon, P. (1986) Human estrogen receptor cDNA: sequence, expression and homology to v-*erb*-A. *Nature* 320: 134–139.
36. Greene, G. L., Gilna, P., Waterfield, M., Baker, A., Hort, Y. and Shine, J. (1986) Sequence and expression of human estrogen receptor complementary DNA. *Science* 231: 1150–1154.
37. Krust, A., Green, S., Argos, P., Kumar, V., Walter, P., Bornert, J.-M. and Chambon, P. (1986) The chicken estrogen receptor sequence: homology with v-*erb*-A and the human estrogen and glucocorticoid receptors. *EMBO J.* 5: 891–897.
38. Loosfelt, H., Atger, M., Misrahi, M., Guiochon-Mantel, A., Meriel, C., Logeat, F., Benarous, R. and Milgrom, E. (1986) Cloning and sequence analysis of rabbit progesterone-receptor complementary DNA. *Proc. Natl. Acad. Sci.*, USA, 83: 9045–9049.
39. Misrahi, M., Atger, M., D'Auriol, L., Loosfelt, H., Meriel, C., Fridlansky, F., Guiochon-Mantel, A., Galibert, F. and Milgrom, E. (1987) Complete amino acid sequence of the human progesterone receptor deduced from cloned cDNA. *Biochem. Biophys. Res. Com.*, *143*: 740–748.

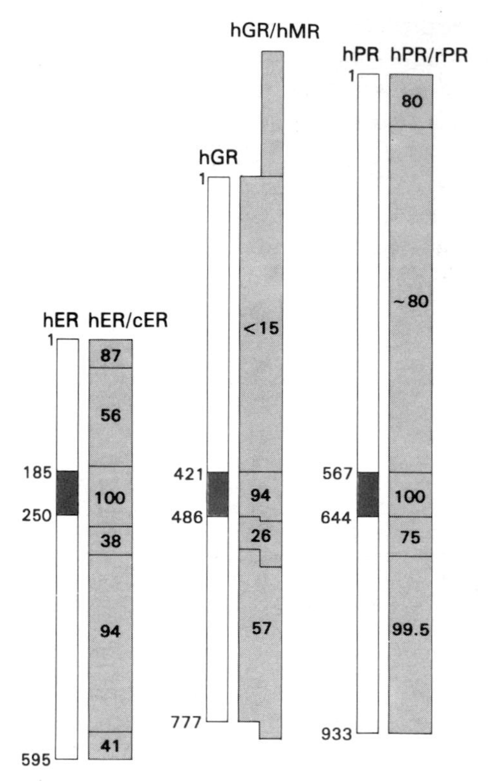

Figure IX.19 Structure of steroid receptors. The human ER, GR and PR are represented schematically. The DNA binding region is highlighted. Paired comparisons of homology between receptors indicated in the gray bars are expressed in percent. Note the remarkable homolgy between the human and avian (chick, c) ER, and between the human and rabbit (r) PR, in both the DNA and steroid binding regions. Differences between the human GR and MR are also indicated (see references 33–39 and Figs. I.73 and I.138).

I.71) in two zones: one forms two Zn fingers (p. 70) rich in cysteines and basic amino acids, and probably corresponds to the zone of interaction with DNA (C of Fig. I.71: homology ⩾50%); the other, rich in hydrophobic amino acids, is the putative hormone binding site (E of Fig. IX.18: homology is ~25%).

Polyclonal and then monoclonal antibodies have been produced from preparations of purified receptors. *Receptors for the same hormone*, prepared from different tissues of various species, in particular mammals, contain *common antigenic* determinants. For example, an antibody prepared against the estrogen receptor of calf uterus recognizes the human receptor, and vice-versa. Yet, even *within the same species*, there is *no common antigenic* determinant recognized from *one type* of receptor to *another* (estrogen and progesterone, for example) or with plasma proteins binding the same steroids. These results suggest the possibility of a common ancestral gene, and also that conservation of common antigenic determinants of receptors for the same hormone may have important phylogenetic significance.

Due to the difficulty of purifying receptors, it is very difficult to assess the monospecificity of polyclonal antibodies. This limitation has motivated the development of monoclonal antibodies (Fig. IX.20). Such antibodies were first prepared against the estrogen receptor[40], and then for other known steroid hormone receptors.

8.3 Nuclear Receptors

Transformation, Activation, and Interaction with Chromatin (Fig. IX.20)

Upon hormone binding, the *receptor binds to nuclear chromatin*[41]. In this way, information is transmitted to nuclear structures, thus ensuring gene transcription. This is a specific process, in which both the *steroid* and the *receptor* play a *precise role*. The hypothesis accepted currently is that a steroid molecule modifies the conformation of the receptor, and that, once modification has taken place, the hormone molecule is no longer necessary in order for the cell to respond. The steroid stabilizes the active form of the receptor. It is this modified receptor which is involved in altering gene expression. Three major questions concerning this mechanism can be asked: 1. Where is the receptor located within the cell?; 2. How do transformation and activation occur (modification of the receptor which gives it an increased affinity for chromatin)?; and 3. What is the nature of the nuclear acceptor, i.e. the structure which binds the receptor?

Immunocytochemical Studies of the Receptor

Immunocytochemical studies carried out using antibodies to the estrogen and progesterone receptor[42–44] (including monoclonal antibodies, at the ultrastructural level[45,46]) have demonstrated that, in the absence or presence of the hormone, the *receptor is intranuclear*. Another approach, involving the use of cytochalasin to enucleate cells, has confirmed these observations[47]. It seems, therefore, that,

40. Greene, G. L., Nolan, C., Engler, J. P. and Jensen, E. V. (1980) Monoclonal antibodies to human estrogen receptor. *Proc. Natl. Acad. Sci., USA* 77: 5115–5119.
41. Milgrom, E. (1981) Activation of steroid receptor complexes. In: *Biochemical Actions of Hormones* (Litwack, G., ed), vol. 8, Academic Press, New York, pp. 465–492.

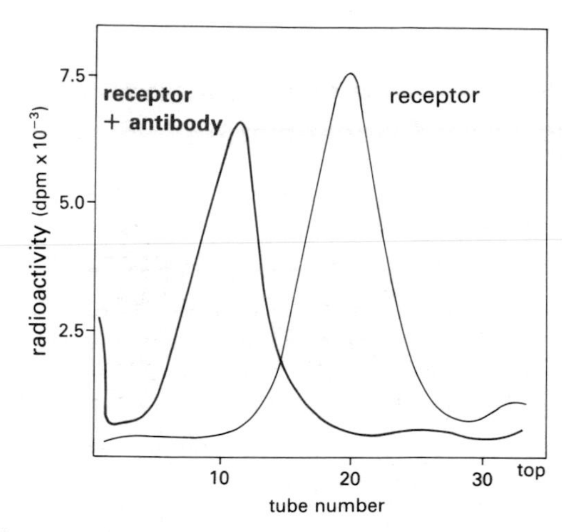

Figure IX.20 Demonstration, by ultracentrifugation on a sucrose density gradient, of the interaction between the rabbit progesterone receptor and monoclonal antibody. (Logeat, F., Hai, M.T.V., Fournier, A., Legrain, P., Buttin, G. and Milgrom, E. (1983) Monoclonal antibodies to rabbit progesterone receptor: cross-reaction with other mammalian progesterone receptors. *Proc. Natl. Acad. Sci., USA* 80: 6456-6459).

although the receptor is intranuclear in the absence of ligand, it is either only weakly or not at all bound to nuclear structures, and thus, following homogenization, the receptor is extracted into the soluble fraction. Once the receptor interacts with the steroid, the complex is strongly bound to chromatin.

It is still too early to generalize this notion of exclusive nuclear localization of receptors to all steroid hormones and target cells, however, since results may depend on differences in biological systems as well as technical developments (e.g., fixation).

Transformation and Activation of the Hormone-Receptor Complex (Fig. IX.20)

The hormone-receptor complex formed at low ionic strength and at low temperature is unable to bind to a nuclear acceptor. If exposed for a short time to an elevated temperature (15–37°C) or higher ionic strength, the complex is capable of binding to the nuclear acceptor, even at 0°C. The acquisition of this property is usually termed *transformation*, or *activation*. Even at low temperatures and weak ionic strength, a more precise analysis of the phenomenon has shown that the receptor begins transformation as soon as it binds the hormone. This reaction is extremely slow, and requires many hours for completion. An increase in temperature or ionic strength simply accelerates this reaction[48].

The actual mechanism involved in transformation is still controversial. For estrogen receptors, the activated form is a dimer with a sedimentation coefficient of ~5 S. Activation of glucocorticosteroid receptors could be due simply to a conformational change of the steroid-receptor complex, which would lead to the appearance, on the surface of the molecule, of a positively charged region with an increased affinity for natural or artificial polyanions. It has been proposed that the dissociation, from the receptor, of hsp 90, present in 8 to 10 S complexes, is involved (p. 69).

Activation is also accompanied by modification of ligand-receptor interaction, the rate of dissociation of the steroid being reduced in activated complexes[49].

Different methods have been proposed for measurement of activated steroid-receptor complexes. The one which approaches physiological conditions consists of employing an excess of nuclei from the same tissue as the receptor. Alternatively, polyanions (DNA-cellulose, phosphocellulose) or polycations (DEAE-cellulose) may be used[50]. Active complexes are retained by the former, whereas the latter retain only non-activated complexes. Finally, differences in the rates of dissociation of the steroid-receptor com-

42. Gasc, J.M., Ennis, B.W., Baulieu, E.E. and Stumpf, W.E. (1983) Récepteur de la progestérone dans l'oviducte de poulet: double révélation par immunohistochimie avec des anticorps antirécepteur et par autoradiographie à l'aide d'un progestagène tritié. *C.R. Acad. Sci.* 297: 477–482.
43. Perrot-Applanat, M., Logeat, F., Groyer-Picard, M.T. and Milgrom, E. (1985) Immunocytochemical study of mammalian progesterone receptor using monoclonal antibodies. *Endocrinology* 116; 1473–1484.
44. King, W.J. and Greene, G.L. (1984) Monoclonal antibodies localize estrogen receptor in the nuclei of target cells. *Nature* 307: 7454–7749.
45. Perrot-Applanat, M., Groyer-Picard, M.T., Logeat, F. and Milgrom, E. (1986) Ultrastructural localization of the progesterone receptor by an immunogold method: Effect of hormone administration. *J. Cell Biol.* 102: 1191–1199.
46. Press, M. F., Nousek-Goebl, N.A. and Greene, G.L. (1985) Immunoelectron microscopic localization of estrogen receptor with monoclonal estrophilin antibodies. *J. Histochem. Cytochem.* 33: 915–924.
47. Welshons, W.V., Krummel, B.M. and Gorski, J. (1985) Nuclear localization of unoccupied receptors for glucocorticoids, estrogens, and progesterone in GH_3 cells. *Endocrinology* 117: 2140–2147.
48. Jensen, E.V., Suzuki, T., Kawashima, T., Stumpf, W.E., Jungblut, P.W. and De Sombre, E.R. (1968) A two step mechanism for the interaction of estradiol with rat uterus. *Proc. Natl. Acad. Sci., USA* 59: 632–638.
49. Weichman, B. and Notides, A.C. (1979) Analysis of estrogen receptor activation by its [^{3}H] estradiol dissociation kinetics. *Biochemistry* 280: 220–225.
50. Milgrom, E., Atger, M. and Baulieu, E.E. (1973) Acidophilic activation of steroid hormone receptors. *Biochemistry* 12: 5198–5205.

plexes have been used to quantitate the fraction that is activated.

Some effectors of the activation reaction have been described. A low molecular weight inhibitor of unknown structure has been demonstrated in the cytosol of liver cells. Molybdate is also an inhibitor of the transformation reaction, stabilizing complexes of high molecular weight (8 to 10 S) even under conditions of increased ionic strength. It also stabilizes receptor binding activity[51].

Nature of the Nuclear Acceptor

The nuclear structure interacting with the hormone-receptor complex is termed the *acceptor*. In fact, the receptor binds to *chromatin*, primarily to DNA, with a possible involvement of *non-histone proteins* and a component(s) of the *nuclear matrix* (an insoluble fibrillar structure containing a very small fraction of DNA, for which a role in gene transcription and cell division has not been definitively established).

The binding of steroid-receptor complexes to DNA has been studied in two ways. It is possible to observe the formation of ternary, DNA-hormone-receptor, complexes by either chromatography or ultracentrifugation, or by utilizing DNA coupled to a solid phase (DNA-cellulose, for example). Secondly, the digestion of nuclei by DNase results in the solubilization of the major portion of the steroid-receptor complexes. However, difficulty in the interpretation of these experiments has developed due to the apparent non-specific character of this interaction. Receptors bind to all DNAs of either animal or bacterial origin. The binding of steroid receptors in nuclei is non-saturable (at least within the limits of the concentration of receptors existing in the cell). Consequently, the hypothesis was put forward that, in addition to a large number of non-specific sites present within DNA, a *small number* of *specific sites* involved in hormonal action are hidden. It was only recently that it was possible to verify this hypothesis by the cloning of hormone-dependent genes[32]. Other studies of the high affinity interactions of hormone-regulated genes and steroid receptor complexes are described on p. 92 and p. 416.

51. Baulieu, E. E. and Schorderet-Slatkine, S. (1983) Steroid and peptide control mechanisms in membrane of Xenopus laevis oocytes resuming meiotic division. In: *Molecular Biology of Egg Maturation*. Ciba Foundation Symposium, no. 98 (Porter, R. and Whelan, J., eds), Pitman Press, Bath, pp. 137–158.

8.4 Measurement of Variation in Receptor Concentration

Hormonal Sensitivity

It has long been known that, under certain well defined physiological or pathological conditions, the responses of target organs to hormones can vary. For these studies, it is necessary to *measure receptor concentrations* under different conditions. Although in the absence of endogenous hormone the situation is relatively simple (since the receptor site is free and is found entirely in the soluble extract of the cell), measurements are much more *complex* when *endogenous hormone* is present. In this case, a fraction of receptor sites is occupied by the hormone, and a large majority of measurable hormone-receptor complexes are found in the nuclear fraction.

Three different types of receptor measurements can be made. 1. *Unoccupied sites* are measured in the cytosol using radioactive hormone. 2. If endogenous steroids are present, an *exchange* of endogenous hormone with radioactive hormone added to the cytosol must be performed. Total cytosoluble receptors are thus measured. This is carried out either by prolonging the incubation sufficiently, at 0°C for certain receptors (progesterone), or by raising the temperature of incubation (estradiol). 3. *Nuclear-bound* receptors are also measured using an exchange assay (Fig. IX.21). It should be emphasized that, with all methods involving exchange in which incubations are long and temperatures elevated, the degradation of receptors should be rigorously controlled.

The use of *antireceptor antibodies* for the detection and *measurement* of receptors is *just beginning*, but certainly represents the future in this field. It will be possible to utilize either immunocytochemical methods, permitting the detection of receptors in histological sections, or immunometric techniques, in which antibodies are labelled by radioactivity or by coupling to an enzyme. A sandwich method has been proposed: a monoclonal antibody is attached to a solid support, and the receptor is bound by the antibody and revealed by a second monoclonal antibody labelled by radioactivity or by enzymatic activity. The two monoclonal antibodies must recognize two different antigenic determinants of the receptor.

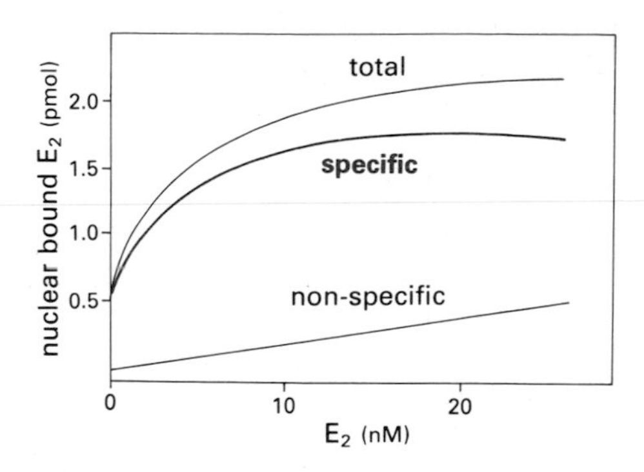

Figure IX.21 Exchange assay for the nuclear-bound estrogen receptor. Rats were injected 1 h before the experiment with non-radioactive estradiol. Nuclei of target tissues contain hormone-receptor complexes to be measured. They are incubated with increasing concentrations of radioactive hormones for 1 h at 37°C. The radioactive hormone is exchanged with unlabelled hormone of the hormone-receptor complexes, but it "sticks", in a non-specific fashion, to nuclei (total). In order to evaluate this non-specific binding, control studies are carried out in the presence of excess, non-radioactive hormone. This inhibits binding to the receptor, but does not modify non-specific binding (non-specific). The difference represents specific binding to the receptor (specific). (Anderson, J., Clark, J.H. and Peck, J. Jr. (1972) Estrogen and nuclear binding sites determination of specific sites by (^{3}H) estradiol exchange. *Biochem. J.* 126: 561–567).

8.5 Other Binding Proteins in Target Organs

In the preceding pages, a general scheme of receptors has been proposed, although many points remain unresolved. However, a certain number of observations have been made that are difficult to integrate into this scheme.

Membrane Receptor

Progesterone can act on *amphibian eggs* at the level of the plasma membrane (p. 82)[51]. It is possible, therefore, that in such a case the receptor is membrane-bound. A protein with a molecular weight of ~30,000, which could be the receptor, has been identified recently.

Plasma Proteins In Target Tissues

In the mammalian *uterus* (rat, rabbit, human, etc.), in addition to the progesterone receptor, a high concentration of protein is found which has many of the characteristics of *CBG*[52,53]. Numerous arguments have been put forward suggesting that this CBG-like protein is intracellular. Questions concerning its origin, whether it comes from the plasma or whether it is synthesized locally, and its function, are yet to be answered. The same protein has been found in the liver, kidney, lymphocytes, and pituitary corticotropic cells.

SBP has been found in the human *prostate* at concentrations which cannot be explained by simple contamination with plasma.

Steroid Binding Proteins within the Genital Tract

In the fluid of the retes testis and the epididymis, in rats, rabbits and human males, a protein which binds androgens (testosterone and dihydrotestosterone, DHT) has been found. *ABP*[11] is *synthesized* in *Sertoli cells* and is *under* the influence of *FSH*. ABP is absent in the inferior portion of the epididymis and in semen. Its proposed role is to transport androgens to cells of the seminiferous tubules and epididymis. It is similar to the corresponding plasma protein which binds androgens (SBP).

A protein called *prostatein*, or prostatic binding protein, is found in the prostate. It binds a large number of steroids, with low affinity[54]. *Uteroglobin*[55] is a small protein found in endometrial secretions of the rabbit; it binds progesterone and its metabolites with moderate affinity ($K_D \sim 10^{-7}$ M), and it has been suggested that this protein may be involved in implantation.

Antiestrogen Binding

When the binding of tamoxifen or 4-hydroxytamoxifen is studied in extracts of various target tis-

52. Leach, K. L., Dahmer, M. K., Hammond, N. D., Sando, J. J. and Pratt, W. B. (1979) Molybdate inhibition of glucocorticoid receptor in activation and transformation. *J. Biol. Chem.* 254: 11884–11890.
53. Milgrom, E. and Baulieu, E. E. (1970) Progesterone in the uterus and the plasma. II. The role of hormone availability and metabolism on selective binding to uterus protein. *Biochem. Biophys. Res. Commun.* 40: 723–730.
54. Heyns, W., Peeters, B., Mous, J., Rombauts, W. and DeMoor, P. (1978) Purification and characterization of prostatic binding protein and its subunits. *Eur. J. Biochem.* 89: 181–186.
55. Savouret, J. F. and Milgrom, E. (1983) Uteroglobin: a model for the study of progesterone action in mammals. *DNA* 2: 99–104.

sues of estrogens, a dual binding system is observed[56]. In addition to binding to estradiol receptors, these drugs bind to a non-estradiol binding protein, present mostly in the endoplasmic reticulum. The significance of such "*antiestrogen binding sites*" has not yet been defined. Most, if not all, of the antiestrogen activity of tamoxifen seems to be mediated by the estrogen receptor (p. 119).

Type II Binding Protein

Clark studied the estradiol binding system in cytosol and nuclei of rat uterus cells, under specific experimental conditions[57]. He used a wide range of hormone concentrations (up to 10^{-6} M), and his methods for separating bound and free hormone were much less stringent than those usually used to characterize receptors. Under such conditions, he observed, in addition to the classical receptor, a second system (Type II) with a weaker affinity (K_D about one order of magnitude higher), from which the ligand dissociates more rapidly. He proposed that this system may be important in certain delayed responses to estrogens, as for example cellular growth. However, this hypothesis remains to be confirmed. Similar observations have been made in relation to other target organs as well as other steroids.

8.6 Receptors and Hormone-Dependent Cancers

Breast Cancer

In 1896, Beatson[58] observed the *regression* of metastatic *breast cancer* after *ovariectomy*. Since that time, various types of endocrine therapy, surgical (hypophysectomy, adrenalectomy) or medical (progestins, androgens, antiestrogens, estrogens at high doses), have been employed. Unfortunately, only approximately one third of the patients treated benefit from such hormonal treatment. Consequently, a technique for the prediction of hormonal-dependency was needed in order to avoid inefficient and radical therapeutic procedures.

In 1961, Folca, Glasscock and Irvine[59] administered *tritiated hexestrol* to patients with metastatic breast cancer. They observed that the uptake of radioactivity by cancer tissues was much greater in those subjects whose metastases regressed following adrenalectomy. Jensen and collaborators, followed by numerous other laboratories, applied the measurement of estrogen receptors to human mammary tumors.

Receptor Measurements

Several techniques of receptor measurement have been developed. *Most involve* the *binding* of radioactive steroids to soluble proteins. Bound hormone can be separated by adsorption of the free hormone with activated charcoal, by ultracentrifugation, or by electrophoresis. However, the determination of bound hormone at a single steroid concentration is insufficient, since in certain tissue extracts, proteins such as albumin or specific plasma protein (SBP), which also bind estradiol in spite of their lower affinity, may influence the binding measurement. It is necessary, therefore, to measure the binding of a hormone at different concentrations. Using Scatchard analysis (p. 59), it is possible to distinguish different binding systems and thus accurately measure receptor concentration. The concentration of progesterone receptor is often measured using a synthetic progestin (for example R5020) which has high affinity for the receptor and weak affinity for CBG.

These binding methods are *laborious*, and necessitate a relatively large amount of tumor (~100 mg). Consequently, major efforts have been made to develop alternative techniques permitting the identification of receptors in histological sections. The use of either fluorescent steroid derivatives or antisteroid antibodies has been proposed. However, these methods have proved to be non-specific, and results have not correlated with biochemical determinations of receptors. The future of this area lies in the use of monoclonal antibodies that will enable the quantification of receptors by radioimmunology or immunoenzymology, or by immunocytochemistry.

Determination of Hormonodependence of Breast Cancer[60]

In most laboratories, the accepted lower limit of receptor concentration for defining *receptor-plus*

56. Sutherland, R.L. and Murphy, L.C. (1982) Mechanisms of estrogen antagonism by non-steroidal antioestrogens. *Mol. Cell. Endocrinol.* 25: 5–23.
57. Eriksson, H., Upchurch, S., Hardin, J.W., Peck, E.J. Jr. and Clark, J.H. (1978) Heterogeneity of estrogen receptors in the cytosol and nuclear fractions of the rat uterus. *Biochem. Biophys. Res. Commun.* 253: 7630–7639.
58. Beatson, G.T. (1896) On the treatment of inoperable cases of carcinoma of the mammary: suggestions for a new method of treatment with illustrative cases. *Lancet* 2: 104–107.
59. Folca, P.J., Glasscock, R.F. and Irvine, W.T. (1961) Studies with tritium-labeled hexoestrol in advanced breast cancer. *Lancet* 2: 796–798.
60. McGuire, W.L. (1980) Steroid receptors in breast cancer: treatment strategy. *Rec. Prog. Horm. Res.* 36: 135–136.

(R^+) samples is 10 fmol/mg of cytosol protein. Using this definition, 50 to 80% of breast cancers contain estrogen receptors. Numerous statistical studies have correlated the presence of estrogen receptor with various endocrine treatments, either surgical (ovariectomy, adrenalectomy, hypophysectomy) or medical (high doses of estrogens, androgens, and progestins, and especially antiestrogens, which in most cases is tamoxifen). If *receptors are absent* or the concentration is lower than 10 fmol/mg protein, remission is very unlikely (5 to 10% of the cases). When the estrogen receptor is present, a favorable response is observed in 50 to 60% of the cases. The probability is greater as the estrogen receptor level is higher. Therefore, the *presence* of *some estrogen receptor* appears *necessary*, but is *not sufficient*, in itself, to produce a response to endocrine treatment. Such a result is logical, since interaction with the receptor is only the first step in the action of the hormone, which may be interrupted at any further stage within the cell.

In addition to initial markers of estrogen action, a more *distal marker*, such as induced proteins or nucleic acids, is currently used. The first to be employed in this way was the *progesterone receptor*, which has been shown to be regulated by estrogens. On the whole, this hypothesis has received confirmation, since tumors containing both estrogen and progesterone receptors respond to endocrine therapy in approximately 75% of all cases (Table IX.3). Other markers of estrogenic effects are being studied. *Various proteins* have been described, the concentration or secretion of which is increased under the influence of estrogens. The mRNAs induced by these hormones have been characterized and corresponding cDNAs have been cloned; the clinical use of these probes is currently being analyzed (p. 148).

Table IX.3 Estrogen (ER) and progesterone (PR) receptor, and remission following hormone therapy in metastatic breast cancer‡.

Receptor status %		Objective remissions %
ER+:(65)	ER+, PR+(31)	77
	ER+, PR−(44)	17
ER−:(35)	ER−, PR+(4)	—
	ER−, PR−(31)	11

‡A study conducted in 358 patients

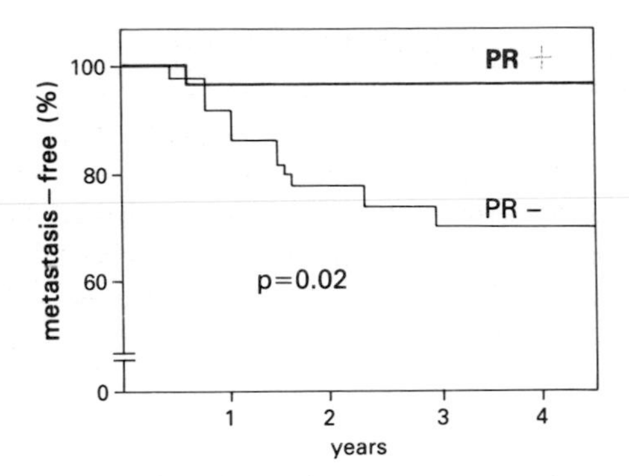

Figure IX.22 Breast cancer: correlation between the presence of progesterone receptors (PR) in primary tumors and the development of metastases.

Prognosis of Early Breast Cancer

Various studies have shown that adjuvant *chemotherapy* is able to *reduce* the frequency of *recurrence* in women operated for breast cancer. The deleterious effects associated with this treatment are well known. It is therefore very important to have a prognostic tool which can distinguish the different populations of patients. There are cases with good prognosis, and the side effects associated with chemotherapy do not warrant the minimal advantages that can be attained. There are other patients for whom chemotherapy is absolutely necessary in order to counteract a potentially disastrous prognosis. The presence and the number of affected lymph nodes, and their histological characteristics, have been used for the staging of the disease for some time. More recently, studies have compared the presence of receptors with the evolution of the cancer. A correlation between the presence of estradiol receptor and the absence of recurrence has been observed. However, the quantitative importance of this correlation, and therefore its clinical usefulness, is still being discussed by various groups carrying out these studies. Statistical differences suggest that the presence of the progesterone receptor is an important factor in predicting the evolution of this disease (Fig. IX.22). A strong correlation, even in patients with cancerous lymph nodes, is found between the presence of this receptor and a good prognosis. It is quite likely, therefore, that steroid hormone receptors, in particular progesterone receptors, are markers of cellular differentiation, and their presence may *help predict* a *less* dramatic *malignant evolution*[61].

61. Pichon, M. F., Pallud, C., Brunet, M. and Milgrom, E. (1980) Relationship of presence of progesterone receptors to prognosis in early breast cancer. *Cancer Res.* 40: 3357–3360.

Other Hormone-Dependent Tumors

The presence of sex steroid receptors has been described in cancers of the *endometrium*, *ovaries* and *prostate*, and *nervous system* (meningioma, glioma, etc.). Glucocorticosteroid receptors have been measured in *leukemia cells*. In these cases, clinical correlations appear less direct than in breast cancers, and receptor assays have not become a common medical test.

8.7 Effects of Hormones in Target Tissues

The effects of steroid hormones on their target organs can be classified in *four* different *categories*.

General Metabolic Activation and Cellular Growth

In the uterus and prostate, estrogens and androgens, respectively, are typical growth-promoting steriod hormones. A very *precise sequence of events* is observed in response to these hormones; an initial increase in RNA synthesis is followed by protein and, finally, DNA synthesis (Fig. I.93). Two hypotheses have been put forward to explain the general tropic effect of hormones. In the *cascade mechanism*, the effect of the hormone is on the synthesis of a limited number of messengers and on regulatory proteins. These then activate a large number of metabolic processes in the cell. In this mechanism, a specific and localized effect of the hormone (activation of some regulatory genes) leads to multiple metabolic effects in the cell. An alternative explanation is that the hormone acts *simultaneously* at the level of all genes for which it would have a direct regulatory role.

Metabolic Slow-Down and Killing Effect

In certain organs, steroid hormones provoke general suppression of macromolecular synthesis, leading eventually to *cell death*. This is the case, for example, with the *effect of glucocorticosteroids* on *lymphoid cells*[62,63]. Here again, it is not known whether the effect of the hormone is on a single, key component, or whether it is on the entire range of its metabolic effects.

Specific Effects

In contrast to the effects described previously in other systems, the action of steroids is much more *limited*. A function (*sodium transport* in *renal cells* under the effect of *aldosterone*) or an enzymatic activity (*tyrosine aminotransferase* induced by *glucocorticosteroids* in *hepatoma cells*) can be stimulated. It is not known, in opposition to the general stimulation cited previously, if the fundamental mechanisms implied in these precise activities follow the same general principles regarding mechanisms. However, the principle characteristics of receptors found in cells sensitive to steroid hormones are the same, independent of the type of response, and indeed there are always a number of proteins (domains) affected by the hormone.

Formation of New Cell Types

The fourth category of steroid hormone action is that of *differentiation*, with the appearance of new cell types. The example which has been studied in the greatest detail is the action of *estradiol* on the chicken *oviduct*. Under the influence of estrogens, cells divide, and the daughter cells are capable of synthesizing egg-white protein (ovalbumin) in response to appropriate stimulus. The original cells were incapable of producing ovalbumin, whereas the daughter cells remain in a

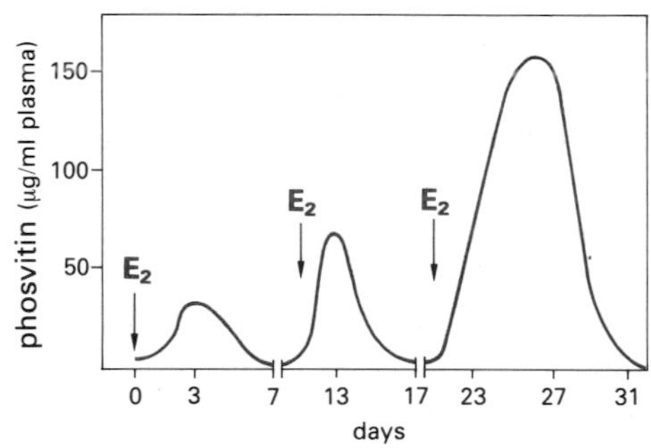

Figure IX.23 Amplification of the synthesis of a specific protein following successive injections of a hormone. The first injection of estradiol resulted in only a weak synthesis of phosvitin. The following injections lead to a greater and greater effect. Two mechanisms may be involved: each injection leads to the differentiation of new cells capable of synthesizing phosvitin under the influence of estrogens, or, alternatively, each cell becomes more responsive, probably due to an increased number of receptors. (Talwar, G.P., Jaikani, B.L., Narayana, P.R. and Narasimhan, C. (1973) Oestrogen induced synthesis of a protein in avian liver *Acta. Endocr.* suppl. 180: 341–356).

62. Sibley, C.H. and Tomkins, G.M. (1974) Isolation of lymphoma cell variants resistant to killing by glucocorticoids. *Cell* 2: 213–220.
63. Sibley, C.H. and Tomkins, G.M. (1974) Mechanisms of steroid resistance. *Cell* 2: 221–227.

stable, differentiated state. Differentiation, in general, is accompanied by specialization. Thus, ovalbumin represents approximately 60% of the secreted proteins of the oviduct. Cell differentiation may also be involved in the amplification of a specific hormonal effect in a given target organ (Fig. IX.23).

Effects at the Level of Transcription

The action of steroid hormones is usually mediated by an increase of the synthesis of several proteins (Fig. I.97).

The mechanism of this action has been studied in *model systems* in which a protein or a nucleic acid is synthesized in relatively abundant quantities in response to a given hormone. Initially, the model which was studied in greatest detail involved the induction, by estradiol and progesterone, of *egg-white proteins* (principally ovalbumin, but also conalbumin, lysosyme, and ovomucoid) in the *chicken oviduct*[64,65]. More recently, a great deal of important information has been provided by analysis of the induction of the *mouse mammary tumor virus* (MMTV) by glucocorticosteroids[66]. Other systems have also been studied: the regulation of the synthesis of *vitellogenin* by estrogens in the *liver of birds*[67] and *amphibians*, the induction of *uteroglobin* by progesterone in *rabbit endometrium*[55], and the induction of *protein* (prostatein), in the *ventral prostate* of the rat, by androgens[68].

In all of these studies, analysis has proceeded systematically from the level of protein to that of genes. Thus, in the case of ovalbumin, the initial observation that protein concentration increases following the administration of estradiol could be explained by either an increase in the rate of hormone synthesis or a reduction in degradation. In order to answer this question, fragments of oviducts were incubated in the presence of radioactive amino acids. The incorporation of the precursor into ovalbumin was measured by specific immunoprecipitation, while the incorporation in total proteins was measured by precipitation with TCA. The ratio between the two values permitted the calculation of the relative rate of *ovalbumin synthesis*, which was actually seen to *increase under* the influence of *estrogens*. This increased protein synthesis could then be due either to an increase in the concentration of mRNA, or to more efficient translation thereof. The answer to this question was obtained by first translating the message in an acellular system (reticulocyte lysate or wheat germ), and then immunoprecipitating the synthesized peptide. This method, low in sensitivity and difficult to perform, has been replaced by hybridization to a cDNA, in either liquid or solid phase (dot or Northern blot, Fig. I.96B). Rather than using cDNA obtained directly from mRNA (which almost always contains impurities), it is preferable to use a cloned cDNA which is homogeneous. The measurement of the concentration of mRNA does not permit the distinction between various possible mechanisms, i.e. change in transcription or in RNA turnover. This has been resolved by incubating nuclei in the presence of radioactive precursors. The specific mRNA is then isolated by hybridization to filters containing corresponding cloned cDNA. It was possible to ascertain from such studies that the *increase* in the concentration of *mRNA* was *due* principally *to* an increase in *transcription rate*, even if a simultaneous increase in the half-life of the mRNA was often also observed.

It is then logical to ask how the hormone-receptor complex is able to modify gene transcription. As a first experimental approach, the *direct interaction between the cloned gene and the purified receptor* can be analyzed. The retention, by nitrocellulose filters, of mixtures of receptors with fragments of hormone-regulated or non-hormone-regulated genes is studied. There is preferential binding of hormone-regulated genes to the filters, indicating the formation of receptor-DNA complexes. The use of smaller and smaller fragments obtained by digestion with restriction enzymes, or the use of the "footprinting" method, permits the delineation of precise regions recognized by receptors (Fig. IX.24). Such regions are found mostly upstream on the gene, beyond the TATA box. For example, in the case of the MMTV, these regions are situated between base pairs −72 and −124 and

64. Schimke, R.T., McKnight, G.S., Shapiro, D.J., Sullivan, D. and Palacios, R. (1975) Hormonal regulation of ovalbumin synthesis in the chick oviduct. *Rec. Prog. Horm. Res.* 31: 175–211.
65. O'Malley, B.W., McGuire, W.L., Kohler, P.O. and Korenman, S.G. (1969) Studies on the mechanism of steroid hormone regulation of synthesis of specific proteins. *Rec. Prog. Horm. Res.* 25: 105–160.
66. Groner, B., Ponta, H., Beato, M. and Haynes, N. (1983) The proviral DNA of mouse mammary tumor virus: its use in the study of the molecular details of steroid hormone action. *Mol. Cell. Endocrinol.* 32: 101–116.
67. Hayward, M.A., Brock, M.L. and Shapiro, D.J. (1982) Activation of vitellogenin gene transcription is a direct response to estrogen in Xenopus Laevis liver. *Nucl. Acid Res.* 10: 8273–8284.
68. Page, M.J. and Parker, M.G. (1983) Androgen-regulated expression of a cloned rat prostatic C_3 gene transfected into a mammary tumor cell. *Cell* 32: 495–502.

between −163 and −192. In certain cases (MMTV, uteroglobin gene), regions of high affinity are also found in the coding sequences of the gene (Fig. IX.25)[69]. In order to establish the *biological significance* of these *DNA sequences*, they can be used in *transfection experiments*[70,71] (Fig. I.72A). The most important studies have been carried out with the MMTV. The regulatory region is found in a DNA fragment known as the long terminal repeat (LTR), at the two extremities of the virus. By *in vitro mutagenesis*, it is possible to delete various regions of the LTR and then reintroduce the gene into cells containing glucocorticosteroid receptors. Hormonal induc-

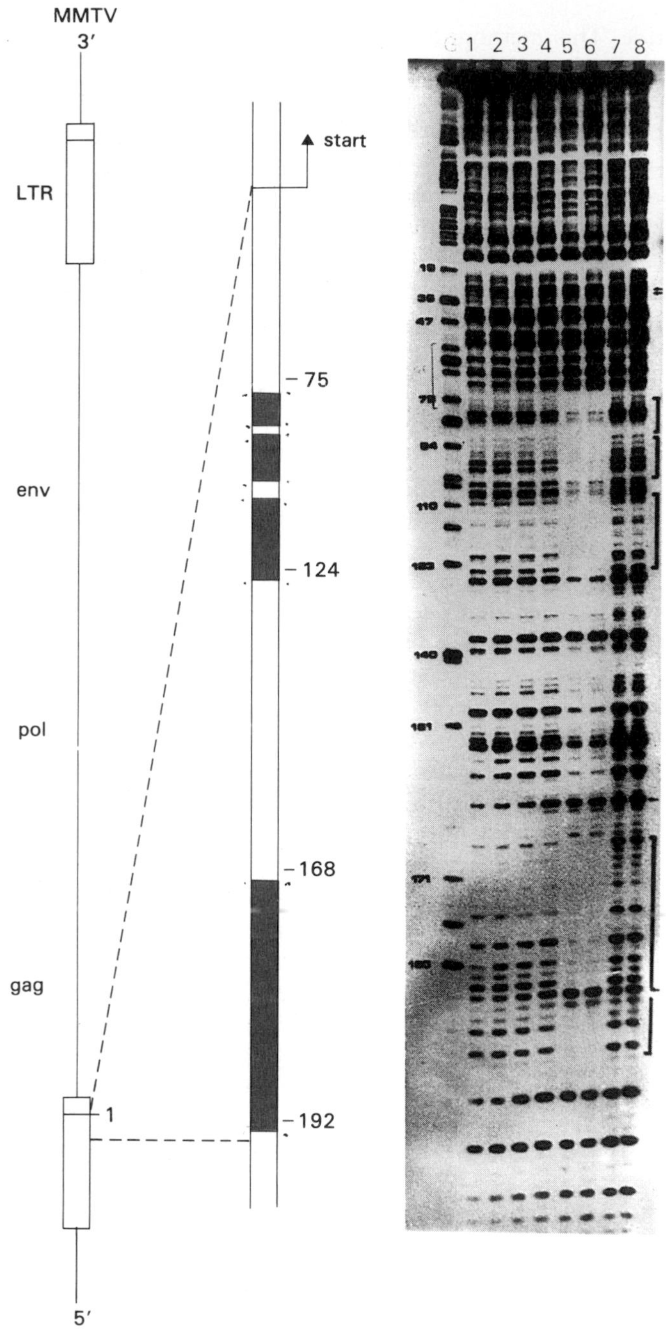

Figure IX.24 DNase I footprint in the LTR region the of MMTV: DNase protection experiment with the 94 kDa form of the rat liver glucocorticosteroid receptor and a 438 bp restriction fragment from the LTR region of MMTV (left). Numbers refer to the distance from the cap site. On the right, an autoradiogram of a 6% polyacrylamide sequencing gel is shown. Lane G represents a guanosine-specific sequence reaction. Lanes 1, 7 and 8 are DNase I digestions in the absence of added receptor. Lanes 2 to 6 are DNase I digestions in the presence of 50, 100, 180, 280 and 380 ng of receptor, respectively. The protected regions are indicated by the brackets to the right of the gel, and are shown in red in the drawing on the left.

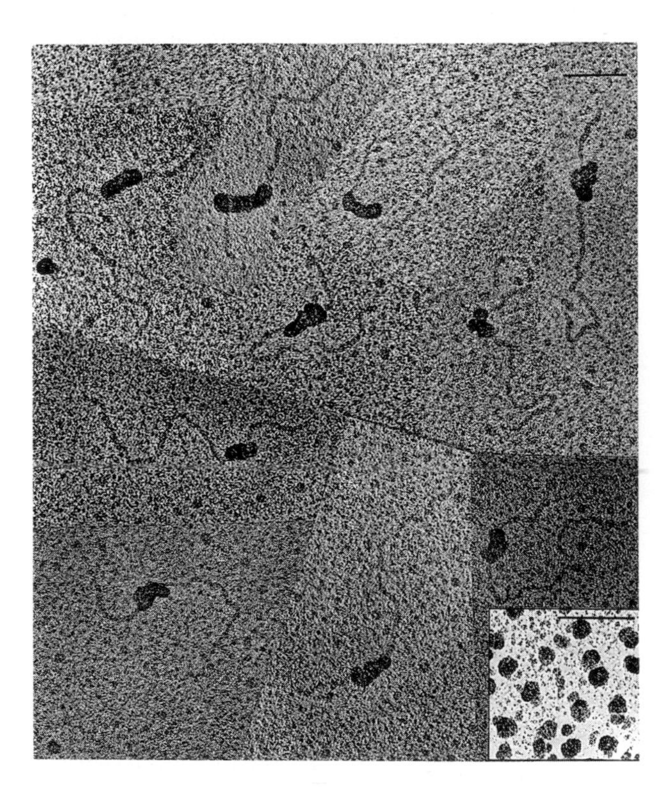

Figure IX.25 Electron micrograph of the binding of glucocorticosteroid receptor (purified from rat liver) to DNA of the MMTV long terminal repeat (LTR)[66]. Bars = 100 nm.

69. Payvar, F., Defranco, D., Firestone, G. L., Edgar, B., Wrange, O., Okret, S., Gustafsson, J. A. and Yamamoto, K. R. (1983) Sequence specific binding of glucocorticoid receptor to MMTV DNA and sites within and upstream of the transcribed region. *Cell* 35: 381–392.
70. Renkawitz, R., Beug, H., Graf, T., Matthias, P., Graz, M. and Schutz, G. (1982) Expression of a chicken lysozyme recombinant gene is regulated by progesterone and dexamethasone after microinjection into oviduct cells. *Cell* 31: 167–176.
71. Renkawitz, R., Schutz, G., Von der Ahe, D. and Beato, M. (1984) Sequences in the promoter region of the chicken lysozyme gene required for steroid regulation and receptor binding. *Cell* 37: 503–510.

tion is then studied. It is experimentally possible to construct a *genetic chimera* consisting of the more-or-less complete regulatory region of MMTV and the structural region of a marker gene (for example, that of thymidine kinase or chloramphenicol acetyl-transferase (CAT)). When the regulatory region is complete, it confers hormone regulation on a gene not normally under hormonal control (Fig. IX.26). It has been shown also that the regulatory region may be relatively distant from the promoter (up to 500 bp away), yet it retains properties of hormonal regulation. This regulation can occur regardless of its orientation ($5' \rightarrow 3'$ or $3' \rightarrow 5'$) on the same DNA strand, i.e. in *cis*. These properties are *similar to* those of *enhancers* (p. 375 and p. 380). Thus, the DNA regions which interact specifically with hormone-receptor complexes are probably a specific case of what belongs to a more general type of regulation. In the case of viral enhancers, the DNA regulatory sequences were described first, but the proteins with which they interact remained unknown. For steroid receptors, the regulatory proteins were the first to be described, while DNA regulatory regions were identified only later. The detailed mechanism by which these enhancers work remains unknown, but the hypothesis which is the most likely is that they act as regions of facilitated "entry" of RNA polymerase itself, or a protein factor modulating the activity of this enzyme. If one compares different DNA sequences recognized by the same receptor in the same gene or in different genes, it becomes clear that homologies exist. They have been termed "consensus sequences". Since this type of analysis is at a very early stage, it is difficult to define the structural characteristics of DNA which allow specific recognition by the receptor (p. 92).

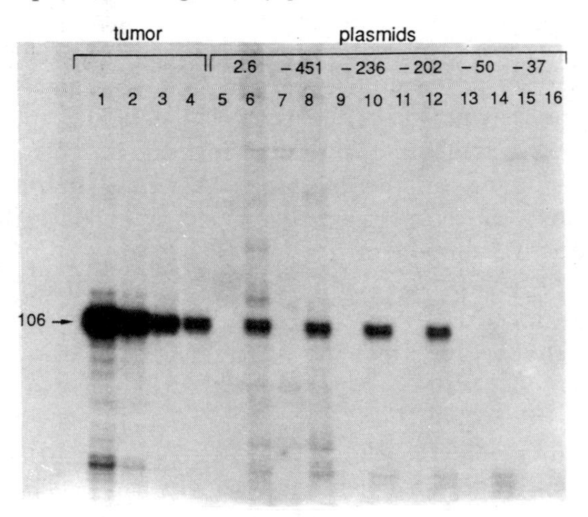

Figure IX.26 Steroid hormone response element. A chimeric gene was constructed containing the LTR of MMTV followed by the coding region of the gene for thymidine kinase (TK) of the herpes virus (2.6 kb plasmid). A series of derived genes was obtained by removing part of the LTR, the remaining portion terminating at −451, −236, −202, −50, and −37, respectively (bp before the start of the structural gene). A number of each of the chimeric genes was used to transform LTK^- cells (lacking thymidine kinase). Cells were either exposed (+) or not (−) to the action of glucocorticosteroids. Hybridization, by the technique of S_1-mapping, using a specific probe to the TK gene, allowed measurement of TK mRNA present in the cells. The arrow at 106 nucleotides indicates the size of the S_1 nuclease-resistant fragment. Induction of mRNA by glucocorticosteroids is seen as long as nucleotides up to position −202 are present[66].

8.8 Effects of Cortisol

Effects on Sugar and Protein Metabolism
(Fig. IX.27)

The most important effect of *cortisol* is on the *liver*, where it stimulates *glycogenogenesis*. Key enzymes, those involved in irreversible steps, are activated. The most important substrates of glycogen synthesis are metabolites of amino acids. Cortisol induces the synthesis of a whole series of enzymes involved in these metabolic transformations. Those most studied have been *tryptophan oxygenase* (or 'pyrrolase') *and tyrosine aminotransferase (TAT)*. The amino acids metabolized in this manner are derived from protein catabolism occurring in such organs as muscles, bone, skin, etc. In contrast, in the liver, protein anabolism is observed in the presence of glucocorticosteroids.

The overall effect of these metabolic modifications leads to a *negative nitrogen* balance, and tends to *increase glycemia*.

Other Effects

Numerous other effects of cortisol have been reported, which include specific aspects of *fat distribution* (buffalo neck of Cushing's syndrome), increased *gastric acid* secretion, an *antivitamin D effect* in the digestive tract, a *mineralocorticosteroid* effect (much less significant than that of aldosterone, which may be explained by the affinity of glucocorticosteroids for the aldosterone receptor), a *catabolic* effect on *lymphoid* tissue (with a reduction in antibodies), an *anti-inflammatory* effect and *anti-allergic* effects, an "excitatory" effect on the *central nervous system*, and a *permissive effect*, in particular on the vasoconstrictor action of *catecholamines*.

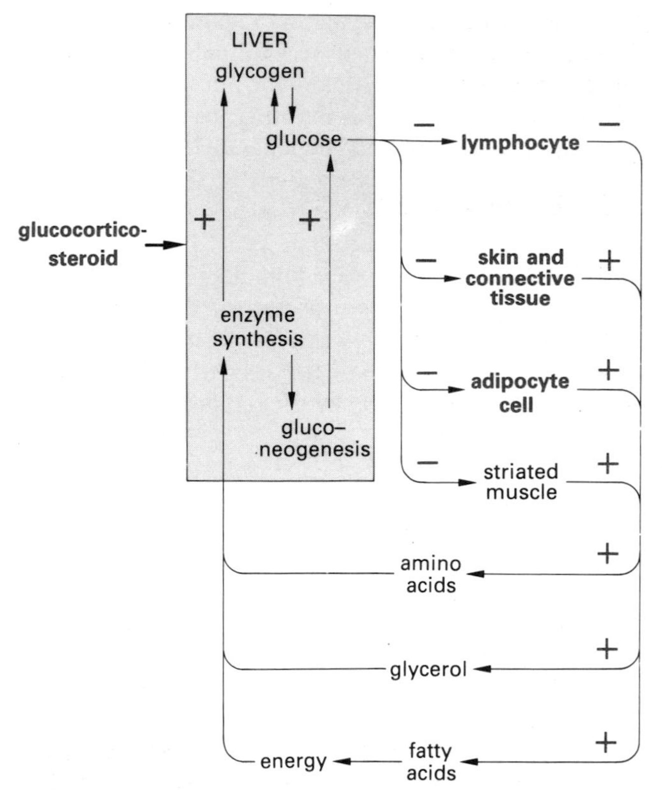

Figure IX.27 Effect of glucocorticosteroids on metabolism of carbohydrates, lipids, and proteins. Arrows indicate the flux of substrates under the influence of glucocorticosteroids. "+", stimulation; "−", inhibition. Not represented are: 1. permissive effects of glucocorticosteroids on other hormone actions; 2. the effects of hormones which are themselves secreted under the control of glucocorticosteroids; 3. gluconeogenesis in kidney. (Baxter, J. P. and Forsham, P. H. (1972) Tissue effects of glucocorticosteroids. *Amer. J. Med.* 53: 573–584).

This notion of a *permissive action*, defined by Ingle[72], is well exemplified in this case. Cortisol, alone, has no direct action on blood pressure; however, in order for catecholamines to exercise their hypertensor effects efficiently, the presence of cortisol is necessary. The molecular mechanisms involved in this action, and in permissive effects in general, remain unknown.

The diversity of the systems and organs on which glucocorticosteroids act is confirmed by the discovery of the corresponding *receptors in almost all cells* tested thus far.

8.9 Effects of Aldosterone

For a description of the effects of aldosterone, see p. 608.

8.10 Effects of Androgens

Androgens exert a tropic effect on *male secondary sexual organs*, e.g., the prostate and seminal vesicles, the rooster comb, antlers, etc.

In seminiferous tubules, testosterone has a stimulatory action on *spermatogenesis*. Locally, the concentrations of testosterone required for the maintenance of spermatogenesis are much higher than those of the general circulation which maintain secondary sexual characteristics.

Other examples of the *effects* of androgens include the development of the *Wolffian duct* during embryonic development, the functional *differentiation* of the *hypothalamus*, a *stimulatory* effect on *erythropoiesis*, and an *anabolic* effect on skeletal muscle and bone.

Metabolism of Androgens in Target Cells

The mechanism of cellular action of androgens frequently involves *local metabolism of testosterone in its target tissues* (Fig. IX.28)[73].

In certain organs (prostate, seminal vesicles, skin, etc.), *testosterone* undergoes a *5α reduction* catalyzed by an enzyme (5α-reductase) located in the endoplasmic reticulum. DHT, thus formed, is preferentially bound by the receptor. Therefore, in these organs testosterone is a *prohormone*, and DHT is the true active component.

In other organs (kidney, muscles), the quantity of DHT formed is very low, and the receptor binds *testosterone* (for which the affinity is less, however). It is in these organs, therefore, that testosterone itself *is* the *active* hormone.

In the hypothalamus, testosterone is partially aromatized to *estrogens*. Some are inclined to believe that estradiol, or a catechol estrogen derivative, is the active intermediary in certain effects of testosterone on brain structures.

The *regulation of the androgen receptor* has only been studied in the rat prostate[74]. The receptor itself is *androgen-dependent*. Its concentration decreases following castration, and is rapidly restored by treatment with androgens.

72. Ingle, D. J. (1954) Permissibility of hormone action. A review. *Acta Endocrinol.* 17: 172–186.
73. Robel, P., Lasnitzki, I. and Baulieu, E. E. (1971) Hormone metabolism and action: testosterone and metabolites in prostate organ culture. *Biochimie* 53: 81–96.
74. Blondeau, J. P., Baulieu, E. E. and Robel, P. (1982) Androgen dependent regulation of androgen nuclear receptor in the rat ventral prostate. *Endocrinology* 110: 1926–1932.

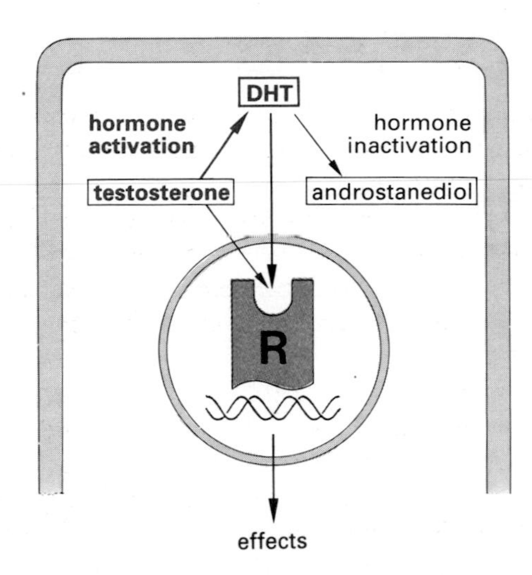

Figure IX.28 Relationship between intracellular metabolism of androgens and physiological androgen effects. Dihydrotestosterone has high affinity for the receptor, testosterone moderate affinity, and androstanediols have no affinity at all. The formation of DHT from testosterone by 5α-reductase is therefore a process of amplification, while the production of androstanediols by 3-hydroxysteroid oxidoreductase is a process of inactivation. Depending on enzymatic activities, either testosterone or DHT will be available to bind the receptor. The predominance of one or the other depends on the target organ involved and on endocrine status (e.g., stimulation of pituitary 5α-reductase by castration).

The genetically determined absence of active androgen receptors characterizes *testicular feminizing syndrome*, which has been observed in the rat, mouse, and human[75]. These subjects have functional testes, a normal or elevated concentration of testosterone in the blood, with, however, a female phenotype. In humans, in slightly more than half of the cases of complete androgen insensitivity, absence of the androgen receptor has been noted. In other cases, hormonal binding to the receptor is normal, which leads to the assumption that the abnormality is probably related to a post-receptor defect.

Patients with a *deficiency of 5α-reductase* suffer from sexual ambiguity[76]. During fetal life, differentiation of Wolffian ducts occurs almost normally, showing that testosterone acts directly in this target. However, external genitalia do not develop normally, demonstrating that DHT is the normally active hormone. At birth, these subjects are usually considered female. At puberty, the development of a penis makes the diagnosis clear.

75. Griffin, J. E. and Wilson, J. D. (1980) The syndrome of androgen resistance. *N. England J. Med.* 302: 198–209.
76. Imperato-MacGinley, J. and Peterson, R. E. (1976) Male pseudohermaphroditism: the complexities of male phenotypic development. *Amer. J. Med.* 61: 251–265.

8.11 Effects of Estrogens

Estrogens are responsible for the growth of *female secondary sexual organs.* The effects of estrogens on the metabolism of macromolecules in both the vagina and uterus have been described above. In the mammary gland, both estrogens and progesterone are necessary to preparation of the gland for lactogenesis, which will then be subject to control exerted by another hormonal system regulated primarily by prolactin, although glucocorticosteroids also play a major role.

Estrogens are active in the central nervous system, and *can modify sexual behavior.* These hormones are also responsible for the *modulation of LH and FSH secretion* by the hypothalamic-hypophyseal system. Depending on their concentration in plasma and the period of the cycle, they are involved in a negative feedback control, reducing the secretion of LH and FSH, or in a positive feedback control, stimulating gonadotropin levels (p. 271 and p. 446).

The *concentration* of *estrogen receptor* is *increased* by *estrogens* themselves. Progesterone inhibits this autoinduction of the receptor. These regulatory mechanisms may explain cyclic variation in estrogen receptor number in the uterus, with a maximal concentration observed at the time of the preovulatory estrogen peak, followed by a rapid decrease during the luteal phase.

8.12 Effects of Progesterone

Progesterone produces cellular *differentiation* in the *uterus*, a classical example of which is the production of "endometrial lace" (*dentelle utérine*) observed in the rabbit. Its action in the endometrium favors the appearance of decidual cells, and *prepares the uterus for* the *implantation* of a blastocyst. Progesterone exerts a relaxing effect on the myometrium. The disappearance of this "progesterone block" is one of the important factors leading to myometrial contraction and parturition or abortion.

The hormonal control of the *mammary gland* also involves progesterone. This control is complex, and differs among species[77]. In addition to participating in the differentiation of alveoli, progesterone also has an inhibitory effect on lactogenesis induced by prolactin.

Receptor Variation and Control

The *hormonal control of progesterone receptor level* is relatively well understood[78]. The receptor is under *dual regulation*, in which both estrogens and progesterone itself are involved (Fig. IX.29). Under the influence of *estrogens*, the *concentration* of progesterone receptor *increases*. This effect requires RNA and protein synthesis. It involves either the induction of the receptor or the induction of a hypothetical activator of the receptor. Following *administration of progesterone*, a *decrease in receptor* concentration is observed. This effect may be explained by an *increased receptor inactivation* rate under the influence of its own ligand. These studies show that *progesterone limits its own action.*

During the *estrous cycle* of the rat, guinea-pig, hamster, and in the menstrual cycle of women, the effects of ovarian hormones are interrelated, leading to cyclic variations of the progesterone receptor (Fig. IX.30). A peak concentration is observed at proestrus (slightly before ovulation), at a time when the level of circulating estrogen reaches a maximum. After ovulation, a rapid decrease in receptor level is observed.

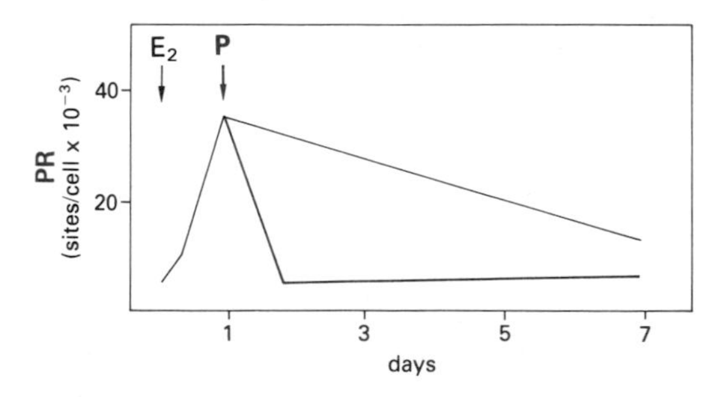

Figure IX.29 Effect of estrogen and progesterone on the concentration of progesterone receptors in the guinea-pig uterus. A single injection of E_2 results in an approximate 8-fold increase of the concentration of these receptors, in one day. If progesterone (P) is injected at this time, a marked reduction of the concentration of receptor is observed. In the absence of P, the concentration of receptors decreases slowly, with a half-life of approximately 5 d[77].

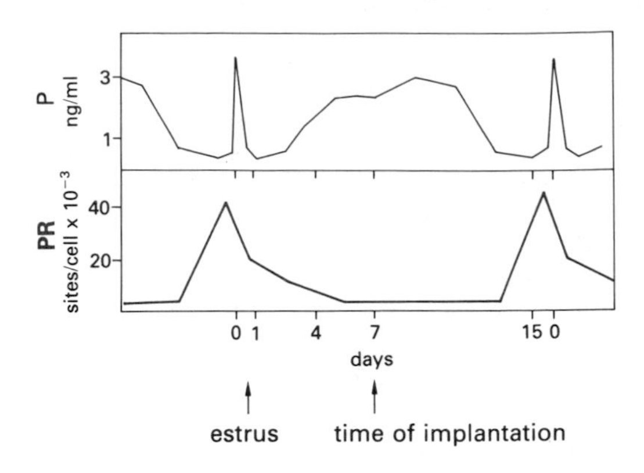

Figure IX.30 Variations in plasma progesterone concentration and in uterine progesterone receptor levels during the estrous cycle of the guinea-pig.

9. Pharmacology

Pharmacology of steroid hormones has undergone major development in the last 20 years. Only *three specific examples* will be described in detail.

9.1 Pharmacology of Glucocorticosteroids

Compounds having different effects on inflammation, salt retention, etc., are obtained by *various chemical substitutions* in the original structures of cortisol or corticosterone. The nature of a few of these compounds and their principal activities are described in Table IX.4 and illustrated in Figure IX.31.

The introduction of a *double bond between carbons 1 and 2* increases anti-inflammatory activity, while the effect on salt retention is unchanged or diminished. In contrast, *substitution* at C9α *by fluorine increases* both *activities*. The effect of the double bond at position 1 and fluorine at 9α is additive. 16α OH, 16α or β-CH_3 substitutions cause a marked reduction in salt retention.

77. Rosen, J. M., Matusik, R. J., Richards, D. A., Gupta, P. and Rodgers, J. R. (1980) Multihormonal regulation of casein gene expression at the transcriptional and posttranscriptional levels in the mammary gland. *Rec. Prog. Horm. Res.* 36: 157–193.
78. Milgrom, E., Luu Thi, M., Atger, M. and Baulieu, E. E. (1973) Mechanisms regulating the concentration and conformation of progesterone receptor(s) in the uterus. *J. Biol. Chem.* 248: 6366–6374.

Table IX.4 Synthetic glucocorticosteroids.

Steroid	Glucocortico-steroid and anti-inflammatory activity‡‡	Oral dose equiva-lence (mg)	Saline retention	Important chemical modifications			
				C_1-C_2 double bond	C6	C9	C16
Cortisol	1	20	++				
Prednisolone	4	5	+	+			
Prednisone‡	3.5	5	+	+			
Methylprednisolone	5	4		+	6α-methyl		
Triamcinolone	5	4		+		9α-fluoro	16α-hydroxy
Dexamethasone	30	0.8		+		9α-fluoro	16α-methyl
Betamethasone	35	0.6		+		9α-fluoro	16β-methyl

‡Ketone at C11 ‡‡Expressed as a factor of cortisol activity.

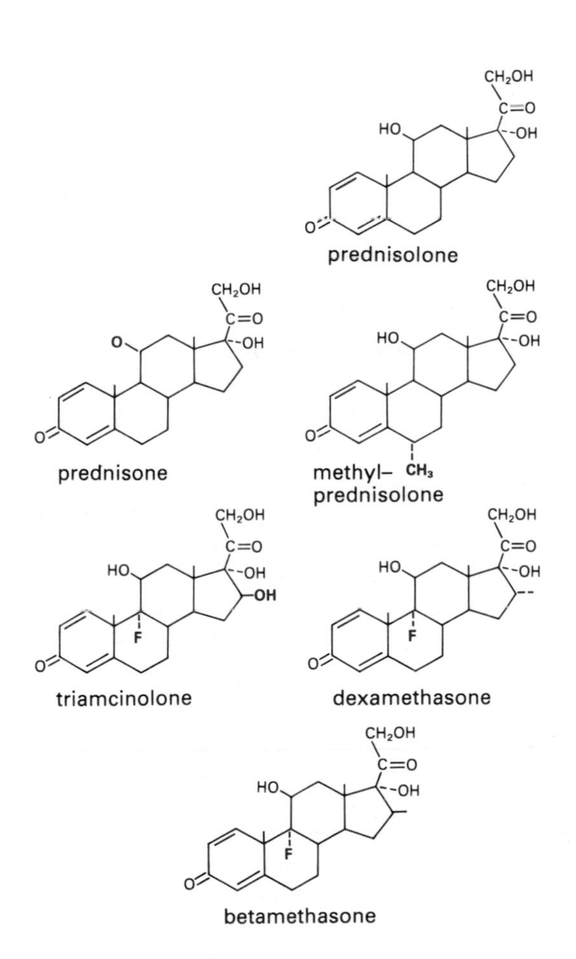

Figure IX.31 Structure of some synthetic glucocorticosteroids.

These synthetic steroids have notable *differences* with respect to their abilities to bind CBG. Prednisone and prednisolone are strongly bound, whereas dexamethasone or triamcinolone have only a very low affinity. Those glucocorticosteroids which are very active (such as dexamethasone or triamcinolone) have an affinity for the receptor greater than that of the natural steroids cortisol and corticosterone. Their affinity for mineralocorticosteroid receptors is weak, parallel to their feeble effect on salt retention. Conversely, derivatives of cortisol with mineralocorticosteroid activity, such as 9α-fluorocortisol, have an increased affinity for the aldosterone receptor.

9.2 Pharmacology of Progestins and Contraceptives

The activity of progestins is judged by a number of different criteria. The differentiation of rabbit endometrium following systemic administration of the hormone is frequently used (McPhail units). However, following uterine administration (McGinty), the same test may give different results. In rats, the induction of deciduoma, the inhibition of ovulation, and the maintenance of pregnancy in ovariectomized animals are also commonly tested.

Progestins frequently possess estrogenic, antiestrogenic or androgenic activities which are important to evaluate. Chemically, it is possible to distinguish *four families* of synthetic compounds: 1. derivatives of *17-hydroxyprogesterone*; 2. derivatives with an *estrane* ring (as 19-nortestosterone); 3. derivatives with an *androstane* ring; and 4. *isomers* of pro-

Table IX.5 Biological effects of progestins.

	McPhail test (rabbit)	McGinty test (rabbit)	Induction of deciduoma (rat)	Maintenance of pregnancy (rat)	Inhibition of ovulation (rat)	Estrogenic effect	Anti-estrogenic effect	Androgenic effect	Chemical series
Chlormadinone	++	++	+	±	±	−	+	−	17P
Dimethisterone	+		−	−		−	±	−	T
Lynestrenol	+		−	−	+	+		+	Nor
Megestrol	++	+		+	++	−	+	−	17P
Norethindrone acetate	+	−			++	+	+	+	Nor
Norethynodrel	+	−	−	−	+	+	−	−	Nor
Norgestrel	+	±		+		−	++	+	Nor
Medroxyprogesterone acetate (MPA)	++		±	+	++	+	+	±	17P

++ very active
+active
±weakly active
−inactive

17P=Derivative of 17-hydroxyprogesterone
Nor=Derivative of 19-nortestosterone
T=Testosterone analog

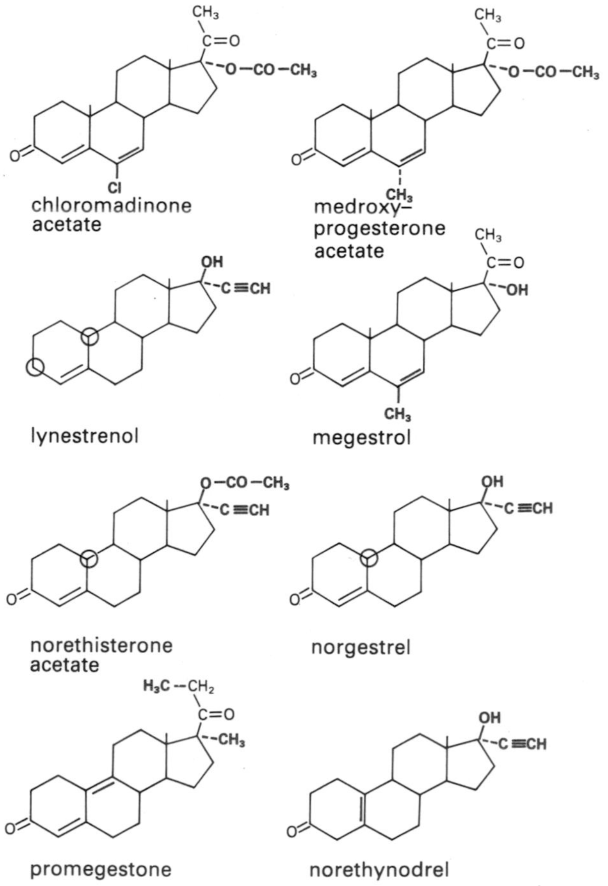

Figure IX.32 Structure of some synthetic progestins. The small red circles highlight important chemical features.

gesterone, such as retroprogesterone. The chemical structure of some of these compounds is shown in Figure IX.32, and their biological properties are summarized in Table IX.5. The first two groups of compounds are those used most frequently.

The derivatives of 17-hydroxyprogesterone in general have strong progestomimetic activity, as demonstrated by the McPhail and McGinty tests. They maintain pregnancy in ovariectomized rodents and are not estrogenic.

The derivatives of 19-nortestosterone are slightly less active in the McPhail test, and have weak estrogenic activity. They maintain pregnancy, induce deciduoma, and are very active in the McGinty test.

In a large number of *oral contraceptives*, a progestin is associated with an estrogen[5]. In most cases, the estrogen utilized is *ethinylestradiol* or *mestranol* (Fig. IX.33). The mechanisms of action of oral contraceptives include the prevention of ovulation, alteration of cervical mucus and tubular mobility, and endometrial modifications preventing implantation. In general, their effectiveness depends on the combined action of these parameters. At low doses (minipill), certain progestins, administered alone, are active contraceptives even though ovulation is often preserved[79].

79. Martinez-Manautou, J., Cortez, V., Giner, J., Aznar, R., Casasola, J. and Rudel, H. W. (1966) Low doses of progestogen as an approach to fertility control. *Fert. Steril.* 17: 49–57.

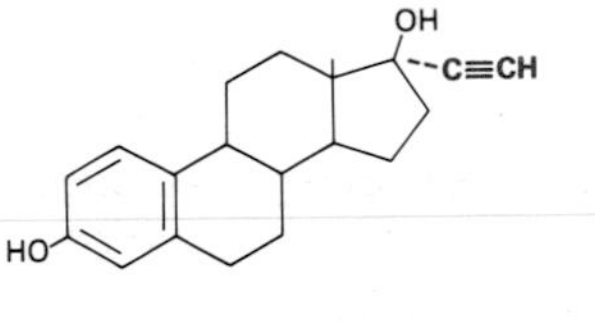

ethinylestradiol

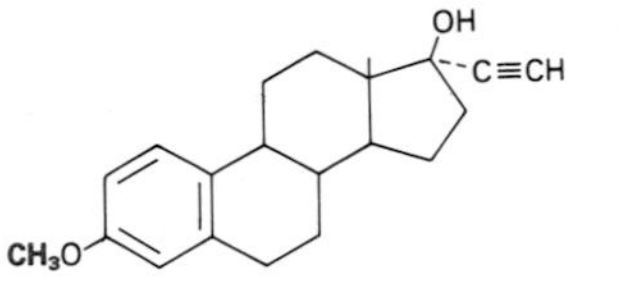

mestranol

Figure IX.33 Structure of the two principal synthetic estrogens used in oral contraceptive pills.

9.3 Antihormones (Fig. IX.34)

Numerous inhibitors of various steroid hormone actions have been described. Some of these compounds *inhibit hormone biosynthesis* (aminoglutethimide, metapyrone, Op'DDD). Others, *true antagonists*, interfere with the cellular action of steroids, and are thus true *antihormones*.

Antiandrogens

Cyproterone acetate[80] appears to act by displacing androgens from the cytosoluble receptor. This compound has been proposed for use in the treatment of prostate cancer and BPH (benign prostatic hypertrophy), precocious puberty (growth blockage), hirsutism, and acne. It has also been used to block the sexual drive of individuals convicted of sexual crimes.

A series of *non-steroidal* compounds has been proposed[81]. The most thoroughly studied are *flutamide* and *anandron*.

80. Neumann, F., Von Berswordt-Wallrabe, R., Elger, W., Steinbeck, H., Hahn, J. D. and Kramer, M. (1970) Aspects of androgen-dependent events as studied by antiandrogens. *Rec. Prog. Horm. Res.* 26: 337–410.
81. Neri, R. O. and Peets, E. A. (1975) Biological aspects of antiandrogens. *J. Steroid. Biochem.* 6: 815–819.

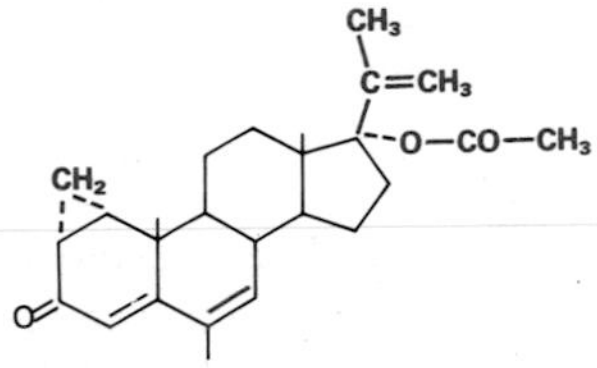

cyproterone acetate

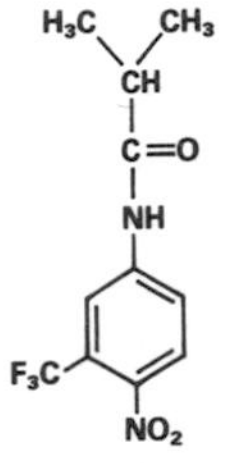

flutamide

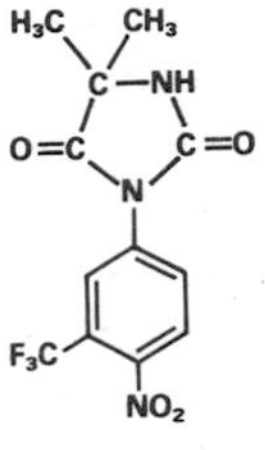

anandron

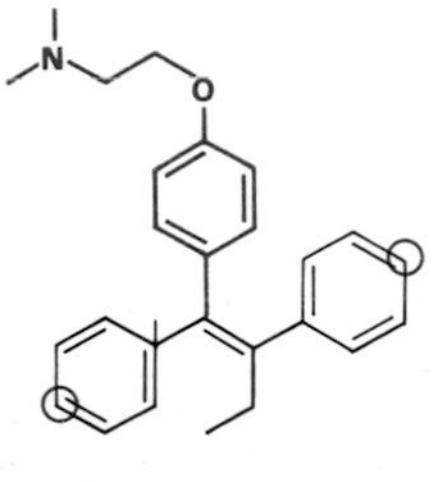

tamoxifen

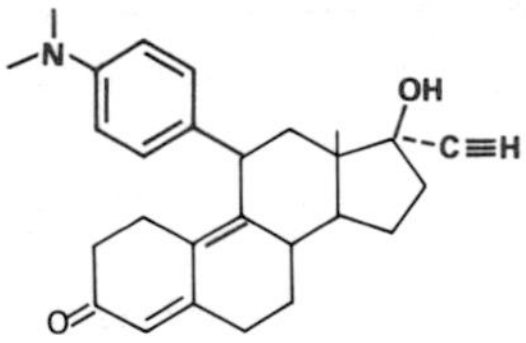

RU 486 (mifepristone)

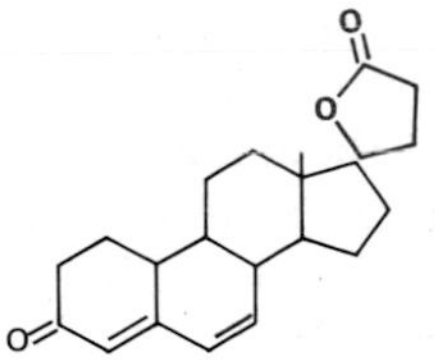

spironolactone 9376

Figure IX.34 Structure of some compounds with antiandrogen, antiestrogen, antimineralocorticosteroid and antiprogesterone activity.

Antiestrogens[82]

The two antiestrogens most often used are *clomiphene* and *tamoxifen*. The latter is metabolized to various components, in particular 4-hydroxy-tamoxifen, which is approximately 10 times more active than the former. The mechanism of tamoxifen and hydroxytamoxifen action, as well as that of other antiestrogens, remains poorly understood despite the large number of studies in this area. Antiestrogen-receptor complexes are transformed, and bind to chromatin of target cells. It is unclear why these compounds do not have agonistic effects. Antiestrogens can act as antagonists, as pure agonists, or as compounds with a combination of these two effects, according to the species and the target cell studied. For example, tamoxifen inhibits growth of human breast cancer cells (an antiestrogenic effect), but it induces the progesterone receptor (an estrogenic effect).

The natural androgens and progesterone also have antiestrogenic activity, but the mechanism of this action is not yet known.

Other Antihormones

Antialdosterone compounds such as *spironolactones* are widely used. Their mechanism of action is described on p. 614.

Recently, a compound with both *antiprogesterone* and *antiglucocorticosteroid* activities has been described: *RU 486* (mifepristone)[83]. It has been successfully used as a "contragestive" agent in women.

10. Clinical Investigation

Advances in hormone assays

While it is difficult to measure hormone levels in target cells, blood hormone levels and hormone metabolites in urine are readily determined. In the last 25 years, major methodological advances have been made since chemical methods which measure mg quantities have been replaced by methods which are sensitive at the pg level. This change of nine orders of magnitude has led to greatly improved pathophysiological investigation.

10.1 Measurement of Plasma Steroids

The method used initially in the measurement of plasma steroid concentration involved the formation of *derivatives doubly labelled* by radioactive atoms. It has a very great *specificity*, but requires a large amount of plasma, and involves multiple manipulations, thus making it *inapplicable for routine use.*

Gas–liquid chromatography and *mass spectrometry* are used in some cases.

Competitive Binding Assays

The most commonly used

Competitive binding assays take advantage of both the specificity and the high affinity (therefore sensitivity) of specific binding proteins.

Two kinds of proteins are used in competitive binding assays; 1. *plasma binding proteins* and *intracellular receptors*, which are rarely utilized; 2. *antibodies* with high affinity and specificity, which can be easily stored. *Steroids*, since they are small molecules incapable of provoking an immunologic reaction themselves, are coupled to immunogenic proteins (e.g., albumin, thyroglobulin). These antibodies recognize chemical features of a steroid in a position distant from the area of coupling, but not those close to this position. For example, when estradiol is coupled at position 17, the antibodies produced bind both estradiol (with an alcohol function at C17) and estrone (with a ketone function at C17). In contrast, a coupling at C6 or C7 allows the production of antibodies distinguishing these two steroids. In this clinically important field, monoclonal antibodies have not yet been shown to be superior.

Assays are *dependent* on the *specificity of binding proteins*. Therefore, if the protein specifically binds the hormone to be measured, it is possible to extract the steroid by an organic solvent, and to measure it directly by competitive binding. The assay is therefore very simple. If the specificity of the protein

82. Bhakoo, H. S., Ferguson, E. R., Lan, N. C., Tatte, T., Tsai, T. S. and Katzenellenbogen, J. A. (1979) Estrogen and antiestrogen action in reproductive tissues and tumors. *Rec. Prog. Horm. Res.* 35: 259–300.
83. Herrmann, W., Wyss, R., Riondel, A., Philibert, D., Teutsch, G., Sakiz, E. and Baulieu, E. E. (1982) Effet d'un stéroïde anti-progestérone chez la femme: interruption du cycle menstruel et de la grossesse au début. *C.R. Acad. Sci.* 294: 933–938.
84. Riad-Fahmy, D., Read, G. F., Walker, F. R. and Griffiths, K. (1982) Steroids in saliva for assessing endocrine function. *Endocrine Rev.* 3: 367–395.
85. Tait, J. F. and Burstein, S. (1964) In vivo studies of steroid dynamics in man. In: *The Hormones*, 5 (Pincus, G., Thimann, K. V. and Astwood, E. B., eds), Academic Press, New York, pp. 441–557.

is insufficient, chromatographic separation is required following extraction and before the binding assay. This necessitates the addition of a radioactive tracer in order to measure recovery obtained with the extraction method, which can vary from one sample to another. Under such conditions, the assay is less well adapted to multiple determinations.

The assay of steroids in *saliva* has been proposed as a means of circumventing repeated blood sampling and also as a way of measuring free hormone levels during hormone therapy; satisfactory results have been obtained[84].

10.2 Hormone Levels, Production Rate, and Metabolic Clearance Rate[85]

The blood hormone concentration is a reflection of both production and catabolism (p. 45). Values for production rates and MRCs have been established (Table IX.6).

10.3 Clinical Evaluation

Urinary Assays (Table IX.7)

The chemical determination of steroid metabolites in urine was the first method of clinical investigation applied on a large scale. The advantages include non-invasive multiple sampling, and urinary measurements integrate hormonal variations, for example circadian rhythms. A major inconvenience is *their lack of specificity*. For example, the assay of 17-ketosteroids in men does not provide a distinction between metabolites of testosterone and those of androstenedione, 11β-hydroxyandrostenedione, dehydroepiandrosterone, and dehydroepiandrosterone sulfate (androgens of adrenal origin with weak activity). Most assays measure metabolites of groups of hormones with very different physiological significance. Secondly, the *metabolites* studied most often represent only a *fraction of total hormonal secretion* (e.g., pregnanediol represents approximately 20% of

Table IX.6 Production rate, plasma concentration, and metabolic clearance rate of steroid hormones.

	Production (μg/d)	Plasma concentration (ng/ml)	MCR (l/d)
Cortisol	15,000	160–50‡‡	200
Aldosterone	100	0.07	1,500
Progesterone:			
Proliferative phase‡	4,200	0.3–1.5	—
Secretory phase‡	42,000	3–20	2,800
Testosterone:			
Men	7,000	7	1,000
Women	200	0.3	700
Androstenedione:			
Men	2,000	1.2	2,300
Women	4,000	1.6	2,000
Dehydroepiandrosterone	5,000	5	1,600
Dehydroepiandrosterone sulfate	15,000	1,000	15
Estradiol:			
Proliferative phase	40	0.06	600
24 h before ovulation	—	0.6	—
Secretory phase	200	0.2	800
Estrone:			
Proliferative phase	60	0.06	1,000
Secretory phase	120	0.1	1,200

‡Menstrual cycle
‡‡Nycthemeral cycle

Table IX.7 Urinary steroid metabolites (normal young adults, units/day).

	Men		Women
Creatininuria (g)	1.6–2.2		0.8–1.2
17-ketosteroids (mg)	9–17		5–14
DHEA-S (mg)	1–3		0.5–2
Testosterone (glucuronide) (μg)	50–150		10–30
17,21-dihydroxy 20-oxosteroids [Porter-Silber reaction] (mg)	5–8		3–6
17-hydroxycorticosteroids [Norymberski reaction] (mg)	8–15		3–6
Pregnanetriol (mg)	1–2		0.5–1.5
Pregnanediol (mg)	0.5–2	PP	0.5–1.5
		SP	4–8
Phenolsteroids [Kober reaction] (μg)	10–25	PP	5–25
		SP	20–45
Aldosterone (glucuronide) (μg)	5–15		5–15

PP=Proliferative phase of the menstrual cycle
SP=Secretory phase of the menstrual cycle

all progesterone secreted); it is known that catabolism is regulated independently of the plasma hormone concentration (for example, an increase in plasma cortisol metabolites associated with hyperthyroidism is without significant pathological importance). Practically, it is also *difficult to obtain complete urinary samples.*

Plasma Assays

Determination of blood hormone levels represents a *major advance* in hormone assays because none of the inconveniences described above for the measurement of urinary metabolites are encountered. However, *other problems* do *exist.* It is difficult to obtain multiple samples, and rapid fluctuations of the plasma concentration of most hormones sometimes make it difficult to interpret results. It may become useful, therefore, to challenge the endocrine system with tests involving stimulation or suppression, in order to identify pathological modifications clearly. On the other hand, abnormalities of hormone binding by plasma proteins can lead to pathophysiologically unimportant alterations of steroid levels.

Hormonal production rate is rarely measured. The only exception is the case of cortisol, for which the measurement of hypersecretion constitutes good evidence for the diagnosis of Cushing's syndrome.

11. Pathophysiology

11.1 Glucocorticosteroids

Urinary Assays

17-hydroxycorticosteroids. The *Porter–Silber reaction*, which detects 17,21-dihydro 20-oxosteroids, is the most commonly used. It measures the principal metabolites of cortisol and cortisone (~25%), and 11-desoxycortisol, produced during the metapyrone test and in certain cases of enzymatic blocks (p. 433). The 24 h values are a function of body weight, and are consequently augmented in obese individuals. Normal values are from 4 to 7 mg/d in the adult male, and from 3 to 5 mg/d in women.

Free cortisol. Free urinary cortisol represents <1% of the adrenal secretion of this hormone. It is a reflection of the fraction of plasma cortisol not tightly bound to proteins, i.e. the active fraction. The assay clearly distinguishes normal subjects from those with a hypersecretion of cortisol. However, this assay is less useful for the diagnosis of adrenal cortical insufficiency. The normal range is 50±30 μg/d.

17-ketosteroids. These are measured by the *Zimmerman reaction.* As already discussed, this colorimetric technique has little specificity, and its use is limited to cases of hypersecretion of adrenal androgens (adrenal tumor, enzymatic blocks). Normal values are 12 to 18 mg/d in men and 6 to 12 mg/d in women.

Plasma Assays

Cortisol

Colorimetric and fluorometric assays are no longer used. Radioimmunoassays have replaced other techniques, due to their specificity and sensitivity.

Cortisol levels undergo *circadian rhythm*: maximal levels are observed just before waking, with minimum levels reached at the end of the afternoon or early evening (p. 38 and p. 238). Normal cortisol levels are 160±70 ng/ml at 08.00 h. *Cortisol* levels are *elevated* in *pregnant women* and in subjects receiving estrogen treatment, but they have no important pathological repercussions. The assay of cortisol not bound to CBG (free cortisol) can therefore be of significance. Cortisol levels are also increased during stress and in patients with major illnesses.

ACTH

ACTH can be assayed radioimmunologically. *Rapid variations* of its concentration are observed, since ACTH secretion occurs in spikes, and since its half-life is short, resulting in a great range of normal values (20–75 pg/ml). The ACTH assay is most useful for determining the etiology of an insufficiency or hyperactivity of the adrenal cortex (p. 248).

Physiological Dynamic Tests

Variations in normal values necessitate challenging the system in order to establish various diagnoses.

Suppression

A single oral dose of 1 mg of *dexamethasone* at 24.00 h is the standard suppression test. The cortisol level measured the next *morning* at 08.00 h is markedly *decreased* in *normal* individuals (<30 ng/ml by RIA). If not, a prolonged test should be used. The administration of 3 mg/d of dexamethasone in four doses per day over two to three days reduces normal adrenal activity, but *not that of patients with Cushing's syndrome*[86]. The suppression test, using 8 mg of dexamethasone over two days, distinguishes syndromes due to adrenal hyperplasia, which are sensitive to the drug, from those provoked by tumors, which are insensitive (p. 249).

Stimulation

Stimulation by ACTH. Initially proposed by Thorn[87], and evaluated by the decrease of blood eosinophils, ACTH tests are now performed with synthetic ACTH (cosyntropin, i.e. the active fragment 1–24). It is administered at 08.00 h. The cortisol level is measured before the test and then 30 min and 1 h following administration. Normally, it should double. The slow-acting depot form results in a greater stimulation. Therefore, samples taken much later (8 and 24 h after injection) are added to those mentioned above. If the depot form is not available, eight hour infusions of ACTH can be administered over three consecutive days.

Metapyrone test[88]. This compound inhibits adrenal 11β-hydroxylation. As a result, 11-deoxycortisol (compound S) increases in the blood and is eliminated, in urine, in the form of tetrahydrodeoxycortisol (THS). This compound can be measured by the Porter–Silber reaction. The increase in ACTH due to reduced cortisol output also leads to an increased secretion of adrenal androgens, and consequently increases the elimination of urinary 17-ketosteroids.

The metapyrone test can be carried out in several ways. One involves the administration of metapyrone at a dose of 4.5 g per day, divided into six doses per day, over two days. 17-hydroxycorticosteroids and urinary 17-ketosteroids are measured over four days: before the test, during the two days of the test, and the day following. Hypothalamic-hypophyseal-adrenal activity is judged by the degree of stimulation. The two-day treatment with metapyrone is often poorly tolerated. In addition, the increase in 17-hydroxycorticosteroids may be difficult to estimate. When plasma samples are to be measured, metapyrone is administered over a 24 h period. The results are much more discriminating, since compound S, initially present at concentrations <10 ng/ml, increases to 120 ng/ml in normal subjects. A test involving a single administration of metapyrone at 24.00 h, with sam-

86. Liddle, G. W. (1960) Tests of pituitary-adrenal suppressibility in the diagnosis of Cushing's syndrome. *J. Clin. Endocrinol. Metab.* 20: 1529–1558.
87. Thorn, G. W., Jenkins, D., Laidlaw, J. C., Goetz, F. C., Dingman, J. F., Arons, W. L., Streeten, D. H. P. and MacRacken, B. H. (1953) Medical progress: pharmacologic aspects of adrenocortical steroids and ACTH in man. *N. Engl. J. Med.* 248: 232, 284, 323, 369, 588, 631.
88. Temple, T. E. and Liddle, G. W. (1970) Inhibitors of adrenal steroid biosynthesis. *Ann. Rev. Pharmacol.* 10: 199–218.

pling the next morning at 08.00 h, is used in pediatrics (tolerance is even better).

CRF[89], which has recently been synthesized, and *RU 486*[90], an antiglucocorticosteroid, may be utilized to assess the ACTH response of the pituitary gland.

Insulin test. The test involving insulin-induced hypoglycemia to obtain a response of the hypothalamic-hypophyseal-adrenal axis is rarely used now.

Insufficient Glucocorticosteroid Function

Addison's Syndrome or Chronic Adrenocortical Insufficiency

In Addison's disease, due to the destruction of the adrenal glands, glucocorticosteroid and mineralocorticosteroid insufficiency resulted in severe complications and was the main cause of death until the use of synthetic mineralocorticosteroids (approximately 30 years ago), such as *deoxycorticosterone acetate.* Tuberculosis remains a possible cause of the disease. Less frequent causes of adrenocortical insufficiency are metastatic cancers, hemochromatosis, amylosis, and candidosis, among others. Primitive adrenal insufficiency is due, in many cases, to autoimmunity. Clinically (Table IX.8), patients present with excessive *fatigue* and *melanodermia* due to the hypersecretion of ACTH and β-MSH resulting from a reduced level of plasma cortisol. This pigmentation predominates in wrinkles, scars, and mucosa. *Hypotension*, especially in the standing position, is provoked by both a reduction in the extracellular volume (deficit of mineralocorticosteroid) and a decrease in cortisol, with a decrease of the permissive effect on catecholamine action. Weight loss and digestive symptoms (anorexia, vomiting, and eventually diarrhea) are frequent. Spontaneous development can be marked by hypoglycemic episodes (see the role of cortisol in the control of glycemia, p. 417), and by acute adrenal crises with severe hypotension. *Sudden death* may occur at any time.

Table IX.8 Frequency of symptoms occurring with Addison's disease.

Sign or symptom	Incidence %
Weakness and fatigue	100
Weight loss	95
Anorexia	95
Hyperpigmentation	88
Vomiting and nausea	70
Diarrhea	13

Diagnosis is no longer dependent on indirect signs, which are often present only in severe forms of the disease (hyponatremia, hyperkalemia, hypoglycemia), or which lack specificity (e.g., results of the water test). The detection of *decreased elimination* of *17-hydroxycorticosteroids* and of very low blood cortisol makes the *diagnosis obvious* in most cases. It can be confirmed by the absence of a response to ACTH. More rarely, adrenal insufficiency is discovered before it is complete. Hormonal secretions are almost normal under basal conditions, but they do not increase following stimulation. The assays for urinary 17-ketosteroids and plasma ACTH, and the metapyrone test, are not useful in many cases. It remains very important to undertake a systematic etiological survey, paying particular attention to adrenal tuberculosis (visible calcification upon X-ray).

Treatment consists of the *administration of cortisol* (or an analog) at replacement dosage. In most patients, the administration of additional mineralocorticosteroids is necessary.

In the case of concomitant disease, stress, trauma, etc., the metabolic clearance rate of cortisol increases. Thus, treatment should be increased markedly, and the patient should be appropriately informed of this eventuality.

Adrenal Insufficiency of Hypothalamic-Hypophyseal Origin

In adrenal insufficiency of hypothalamic-hypophyseal origin, which includes a deficit in ACTH, all of the symptoms described previously for Addison's disease apply. However, hyperpigmentation, in contrast, is replaced by a characteristic paleness (deficiency of MSH). In addition, there may be clinical signs associated with an insufficiency of other pituitary and hypothalamic hormones, due to the local expansion of a pituitary tumor.

89. Gold, P. W., Loriaux, D. L., Roy, A., Kling, A. M., Calabrese, J. R., Kellner, Ch. H., Nieman, L. K., Post, R. M., Pickar, D., Gallucci, W., Augerinos, P., Paul, S., Oldfield, E. H., Cutler, G. B. and Chrousos, G. P. (1986) Responses to corticotropin-releasing hormone in the hypercortisolism of depression and Cushing's disease. Pathophysiologic and diagnostic implications. *N. Engl. J. Med.* 314: 1329–1335.
90. Gaillard, R. C., Riondel, A., Muller, M. F., Hermann, W., and Baulieu, E. E. (1984) RU 486: a steroid with antiglucocorticosteroid activity that only disinhibits the human pituitary-adrenal system at a specific time of day. *Proc. Natl. Acad. Sci., USA* 81: 3879–3882.

Biologically, a decrease in plasma cortisol and a reduction in urinary 17-hydroxycorticosteroids is observed as in peripheral adrenal insufficiency. However, *stimulatory tests* allow the origin of the deficiency to be defined. Indeed, the adrenals respond to ACTH; however, a prolonged secondary adrenal insufficiency may necessitate ACTH treatment, with long-acting cosyntropin, for several days, before obtaining a response. The assay of basal ACTH is of little interest. More important is the absence of an increase in ACTH during the administration of metapyrone.

Recently, the availability of CRF has made possible the clinical investigation of another regulatory aspect of ACTH secretion.

Iatrogenic Adrenal Insufficiency[91]

Iatrogenic adrenal insufficiency is a *frequently encountered* problem. It is manifested in patients treated with *large doses of glucocorticosteroids* for *prolonged* periods (rheumatism, allergy), in contrast to the substitution therapy used in cases of glandular insufficiency. When treatment is discontinued, normal function of the hypothalamic-hypophyseal-adrenal axis remains altered for periods of variable duration (often several weeks). The problem is to know at what moment the subject is *no longer* at *risk* of *acute* adrenal insufficiency, particularly in cases where there has been surgical intervention or in those complicated by concomitant disease.

The clearest case is when the basal level of cortisol is low. However, when cortisol level is normal, a negative test result to synthetic ACTH suggests atrophy, or at least hyporeactivity, of the adrenals. Normally, it is possible to reactivate the adrenal glands by a few injections of long-acting ACTH. The most difficult cases, however, are those where the adrenals are reactivated but the hypothalamic-hypophyseal axis remains inactive. Only the metapyrone test can detect such conditions. These patients are incapable of producing an increased cortisol level in the event of stress. A normal response to the metapyrone test should be obtained before concluding that a patient is not at risk of adrenal insufficiency.

91. Graber, A. L., Ney, R. L., Nicholson, W. E., Island, D. P. and Liddle, G. W. (1965) Natural history of pituitary adrenal recovery following long-term suppression with corticosteroids. *J. Clin. Endocrinol. Metab.* 25: 11–16.
92. Cushing, H. (1932) The basophil adenomas of the pituitary and their clinical manifestations (pituitary basophilism). *Bull. J. Hopkins Hosp.* 50: 137–195.

Hypercortisolism: Cushing's Syndrome[91,92]

Clinically (Fig. IX.35 and Table IX.9), *Cushing's syndrome is characterized by the redistribution of body fat* toward an excess in the facial and trunk regions, whereas the extremities are thin. Cortisol stimulates appetite, reduces the utilization of glucose, and favors the synthesis of fatty acids. This may explain the increase in body fat, but the reason for its particular distribution remains unclear. These same metabolic disturbances are responsible for alterations in the glucose tolerance test or even for *diabetes*. *Hypertension* is due, in part, to the mineralocorticosteroid effect of cortisol, but also to the potentiation

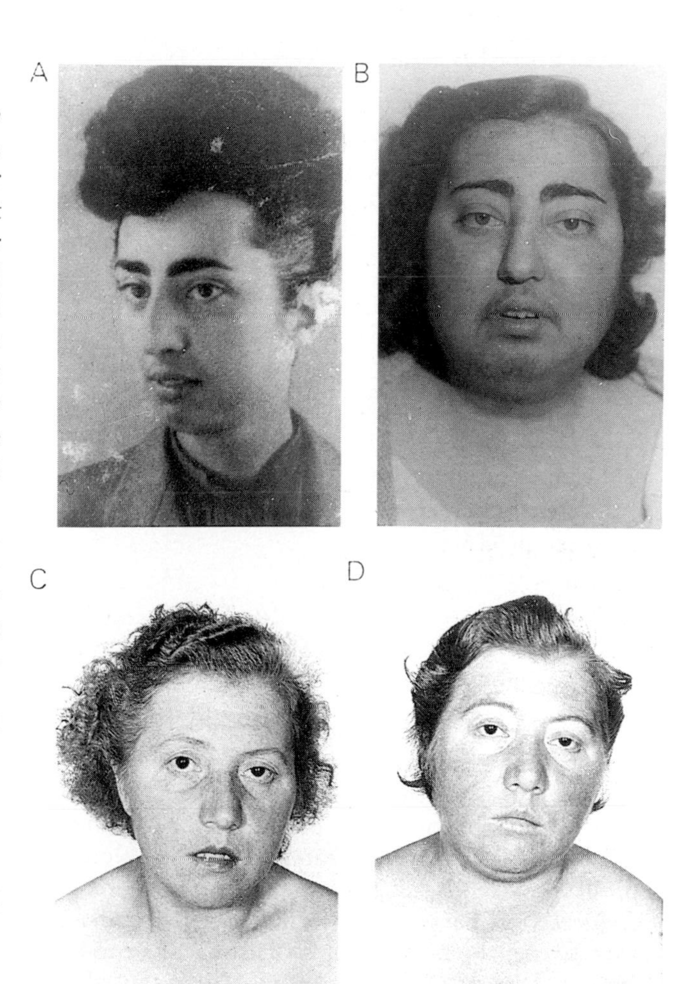

Figure IX.35 Cushing's syndrome patients. From a normal (**A**) to a diseased (**B**) state and from a diseased (**C**) to a treated (**D**) state (bilateral adrenalectomy). Note the characteristic facial features. (Courtesy of Dr. J. Decourt, Hôpital Pitié-Salpêtrière, Paris).

Table IX.9 Clinical symptoms of Cushing's disease.

Sign or symptom	Incidence %
Moon face	88
Obesity	86
Buffalo hump	54
Weakness	67
Fatigue	74
Hypertension	85
Plethora	77
Purple striae	60
Bruisibility	59
Ecchymoses	52
Mild polycythemia	20
Poor wound healing and leg ulcers	35
Ankle edema	57
Puffiness of eyes	26
Osteoporosis	58
Back pain and bone pain	54
Pathologic fractures	38
Kyphosis	25
Renal calculi	20
Frequent urination and nocturia	32
Polydipsia	28
Mental changes	46
Headache	40
Neurological symptoms	34
Menstrual disorders	77
Hirsutism (in females)	73
Acne	54
Exophthalmos	14

of the action of catecholamines. Protein catabolism leads to destruction of cutaneous elastic fibers, explaining the appearance of *striae* (Fig. IX.35A). Their *purplish* color is due partly to polycythemia, which is also responsible for the reddish, blotchy appearance of the face. The redistribution of body fat produces the characteristic buffalo neck and moon face of patients with Cushing's syndrome. Due to the fragility of vascular walls, bruising can be caused by the slightest trauma, and can occur during blood sampling. The destruction of the bone matrix results in *osteoporosis*, which is especially apparent in the vertebrae, and can lead to spontaneous fractures. *Muscular catabolism* is responsible for weakness, which can even reach a state of myopathy. Psychic repercussions associated with hypercortisolism are frequent. Patients may be agitated and particularly irritable. Hirsutism, acne and loss of hair are due primarily to the increased secretion of androgens by the adrenals.

Diagnosis relies on *elevated cortisol levels* in plasma which do not decrease in the afternoon or evening. Urinary 17-hydroxycorticosteroids are frankly elevated, especially free cortisol levels. The measurement of the rate of cortisol secretion is useful, if it is obtainable. However, in a large number of cases, basal values may remain within normal limits, and the best diagnostic evidence will therefore be the consequence of the loss of physiological feedback control[93] (Fig. IX.35). The administration of dexamethasone at moderate doses (p. 249), which suppresses the secretion of cortisol in normal individuals, has little or no effect in Cushing patients. There have even been some reports of paradoxical responses, with an increased secretion of cortisol in response to dexamethasone. Generally, these are due to tumors in which secretion is irregular, the variations observed, transitory, and which are, in fact, independent of the administration of dexamethasone.

It is important to consider the *etiology of the disease* (Table IX.10) in order to make a prognosis and define a therapeutic approach.

Hypercortisolism may be *primary*, i.e. in relation to an autonomous adrenal hyperactivity, or *secondary*, i.e. in response to a hypersecretion of ACTH.

Primary hypercortisolism is due to an adrenal *tumor* (25% of Cushing's cases), and may be a cancer, an adenoma, or microadenomatosis. This type of hypercortisolism is characterized by the autonomy of adrenal action. It can be identified by an increase in cortisol levels in the presence of very low plasma ACTH. That high doses of dexamethasone (8 mg for 2 d) have no effect on adrenal secretion can be accounted for by the fact that dexamethasone acts only on ACTH and that it has no direct adrenal action. For the same reason, the metapyrone test is negative. The *autonomy* of tumors defined by hyperactivity independent of ACTH should not be confused with an insensitivity to exogenous ACTH. Even if the majority of cancers do not respond to ACTH administration, approximately half of all adenomas are sensitive to ACTH. In addition, *malignant tumors* may also produce increased concentrations of *androgens* (elevated 17-ketosteroids), especially dehydroepiandrosterone

93. Aron, D. C., Findling, J. W., Fitzgerald, P. H., Forsham, P. H., Wilson, C. B. and Tyrrell, J. B. (1983) Cushing's syndrome: problem in management. *Endocrine Rev.* 3: 229–244.

Table IX.10 Etiological diagnosis of cases with signs of excess cortisol.

	Plasma Cortisol	Urinary 17 OH-CS	Urinary 17 KS	ACTH test	Metapyrone test	Dexamethasone test	Plasma ACTH
Hypersecretion of ACTH by the pituitary	↑	↑	↑	↑↑	↑	↓	↑ or N
Hypersecretion of ACTH of ectopic origin (tumor)	↑↑	↑↑	↑↑	↓ or N	↓ or N	↓↓	↑↑
Adrenal adenoma	↑	↑	↑ or N	↑ or ↓	↓	↓↓	↓
Adrenal cancer	↑↑	↑↑	↑↑	↓	↓	↓↓	↓
Iatrogenic	↑ or ↓	↑ or ↓	↓	↓ or N	↓		↓

↑=Increased ↓=Decreased N=Normal

sulfate. Testosterone levels are also frequently elevated. Enzymatic blocks can be observed, with an accumulation of intermediary compounds (17-hydroxyprogesterone, 11-deoxycortisol, etc.) and an increased secretion of their metabolites. The diagnosis of adrenal tumor may be facilitated by the evaluation of radiological signs (scanner), by the use of isotopes such as iodocholesterol, or by ultrasonography. Malignant tumors in which steroidogenesis is minimal are often detected by their large size, whereas, prior to visualization, hormone levels would have failed to suggest their presence. Benign tumors and microadenomatosis are more difficult to visualize.

Secondary hypercortisolism may be due to a hypersecretion of ACTH either by the *pituitary* or by an *ectopic tumor*. In the first case, *Cushing's disease* (~60% of the cases of hypercortisolism) caused by a pituitary tumor is associated with a bilateral hyperplasia of the adrenals. In spite of the pituitary origin, plasma ACTH is only slightly elevated or may even be normal. The adrenals respond normally, or even excessively, to the administration of ACTH. The metapyrone test is valuable for diagnosis, demonstrating an excessive response. Suppression of high cortisol levels can be obtained by large doses of dexamethasone (p. 249). Together, these results indicate that the control system is intact, but that it is regulated in such a way that supraphysiological levels of cortisol are secreted. It is important to look systematically for an eventual *tumor* at the *pituitary* level, by radiological and ophthalmological examinations. Such tumors often develop following bilateral adrenalectomy.

ACTH or an ACTH analog is sometimes produced by an *ectopic tumor* situated at a site distant from the hypophyseal-hypothalamic area (16% of patients with Cushing's syndrome). The tumor is frequently bronchial, pancreatic, or thymic. Since the production of ACTH is accompanied by that of MSH in many cases, an intense pigmentation is also observed. These malignant tumors result in emaciation, with a loss of muscle mass, an aspect quite different from symptoms of Cushing's disease. Biologically, the adrenal glands respond to ACTH. It is very difficult, however (except on rare occasions), to reduce the secretion of these tumors by dexamethasone. In general, plasma ACTH levels are quite elevated.

Treatment in the case of either adrenal or ectopic tumors consists of the removal of the tumor. *Bilateral adrenalectomy*, a difficult but very efficient intervention, was previously performed in cases of bilateral adrenal hyperplasia. However, it is now often replaced by *transsphenoidal microsurgery* of the pituitary (p. 250). Microadenomas are found in approximately 80% of patients with Cushing's syndrome. Op'-DDD and aminoglutethimide are inhibitors of adrenal function, and may be used to prepare patients for surgery and during critical periods. RU 486 is also useful in such cases.

11.2 Androgens

Clinical Evaluation

Plasma Assays

Testosterone is routinely assayed by radioimmunoassay. The antibodies used most frequently are formed from derivatives at position C3, which, in addition to testosterone, bind DHT and 3α- and 3β-androstanediols. The latter two compounds are normally present only at low concentrations in the blood, and therefore do not interfere with the assay. In men, *DHT* is present at a concentration of 0.4±0.1 ng/ml, representing approximately 6% of normal testosterone levels (7.9±1.8 ng/ml), whereas in women it represents almost 50% (DHT=0.27±0.06 ng/ml and testosterone=0.57±0.19 ng/ml). Consequently, a precise assay measuring both testosterone and DHT with the same antibody may require chromatographic separation prior to the assay. However, if it is thought that almost all of the DHT is derived from the metabolism of testosterone, it is then possible to measure the two compounds together, under conditions of clinical investigation. The measurement of other androgens, such as androstenedione, DHEA and DHEA-S, can also be carried out by radioimmunoassay. The measurement of the fraction of testosterone unbound to SBP may also be useful[14].

Urinary Assays

The measurement of *17-ketosteroids* by the Zimmerman reaction (p. 426) is only of value in *two specific cases*: congenital adrenal hyperplasia and adrenocortical tumors. In all other cases, the absence of assay specificity renders the interpretation of results difficult. Fractionation of 17-ketosteroids contributes little information. The assay of *urinary testosterone glucuronide* by gas chromatography or radioimmunoassay gives a reliable index of total testosterone production.

Testosterone Production in Women

The level of plasma testosterone production in women is of the order of 0.25 mg/d. Approximately 55% of this production is due to the metabolic transformation of androstenedione, and 15% to that of DHEA.

Virilism and Hirsutism

Virilism and hirsutism result in different clinical signs, depending on the age of onset.

Virilization of the female fetus results in a *masculine pseudohermaphrodite* (Fig. IX.36). In children, androgens stimulate a *precocious pseudopuberty* which is either iso- or heterosexual. In adult women, *virilization* may be of variable importance. Sometimes reduced to simple hirsutism (which is, or is not, accompanied by acne and seborrhea), virilization can also result in vocal modification as well as hypertrophy of the clitoris. Two points are particularly import-

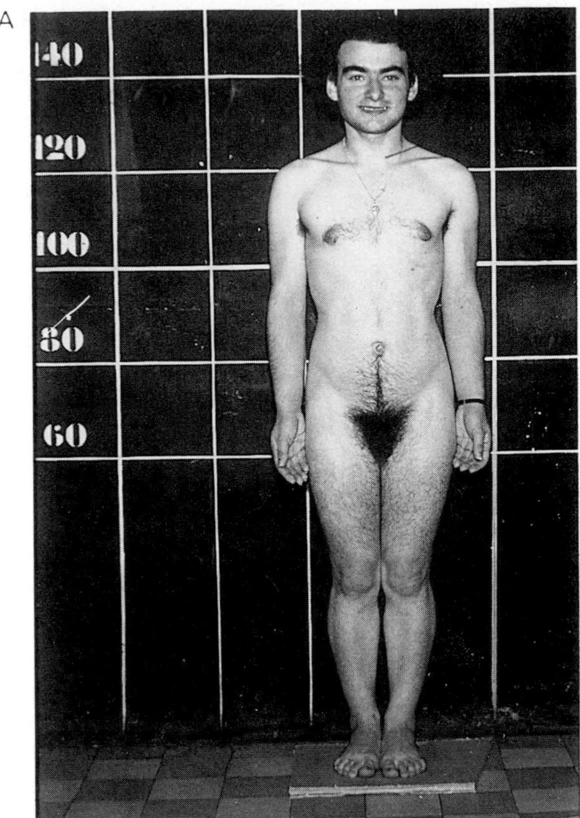

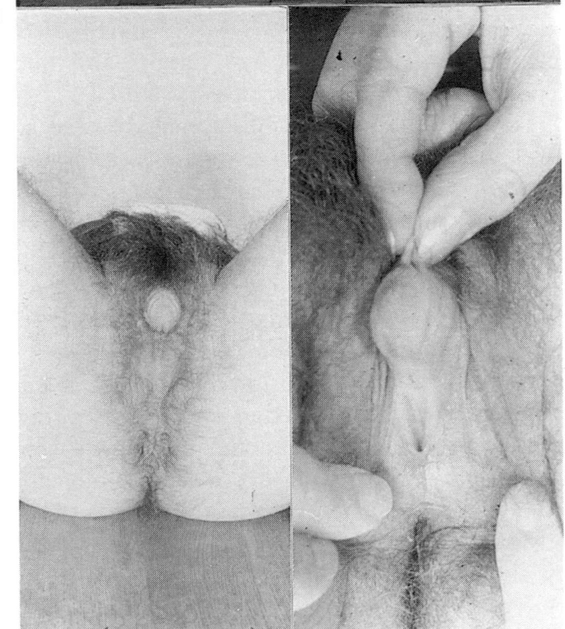

Figure IX.36 Congenital adrenal hyperplasia. Note the masculine look and the short stature (**A**), the peniform clitoris (**B**) and the opening of the vagina (**C**). (Courtesy of Dr. J. Decourt, Hôpital Pitié-Salpêtrière, Paris). Such cases of intensive masculinization are no longer observed, as patients are treated with glucocorticosteroids early in life.

ant in the determination of its etiology: is there an ethnic or familial explanation for hair growth, and are there modifications of the menstrual cycle? (p. 467).

Virilization of Adrenocortical Origin

There is a modest degree of virilization in almost all Cushing patients; however, this symptom is especially evident in virilizing adrenal tumors and in certain congenital enzymatic blocks.

Virilizing tumors[94] secrete large amounts of DHEA-S and androstenedione. The testosterone that masculinizes the patient is formed from these precursors rather than by more direct secretion. A significant elevation of urinary 17-ketosteroids and plasma testosterone, unalterable by synthetic glucocorticosteroids or by stimulation in the various tests described above, should result in the correct diagnosis and lead to the search for radiological signs of a tumor.

Specific Enzymatic Alteration[95,96]

The major forms of enzymatic alteration are detected in young patients, where related syndromes have been described. Attenuated forms detected in adults are also observed, as described on p. 469.

Congenital adrenal hyperplasia is characterized by enzymatic blocks of the synthetic chain of adrenal steroids (Fig. IX.37). Essentially, only two forms lead to virilization, with a block of the 21- or the 11β-hydroxylase. They are the most frequent forms of this disease. Other rare forms exist which are not normally accompanied by virilization. Enzymatic deficiencies are accompanied by either a reduction of cortisol secretion or a maintenance of normal levels (as the result of the hypersecretion of ACTH, with a stimulation of adrenal steroid precursors situated prior to the block). The hypersecretion of adrenal androgens occurs when there is no enzymatic deficiency in the chain of events leading to their synthesis. All alterations are exaggerated under the influence of ACTH. Inversely, androgens return to normal under the suppressive effect of dexamethasone. The clinical aspects of congenital adrenal hyperplasia have been well described[95].

A deficit of *21-hydroxylase* is accompanied by a hypersecretion of 17-hydroxyprogesterone and its urinary metabolite pregnanetriol. There is also an increased synthesis of androgens, and therefore an increased elimination of 17-ketosteroids. In more severe forms, a mineralocorticotropic insufficiency (loss of salt) is observed, which may be aggravated by the antialdosterone effect of progesterone and of 17-hydroxyprogesterone.

The gene corresponding to the 21-hydroxylase block is located on chromosome 6, near the HLA-B locus to which it is coupled. HLA typing is useful for the characterization of heterozygous carriers of the defect and also for prenatal diagnosis of enzyme deficiency. Late-onset forms of the disease, with minimal clinical symptoms, are relatively frequent (p. 469). Elevation of 17-hydroxyprogesterone following the administration of synthetic ACTH (cosyntropin) confirms the diagnosis.

11β-hydroxylase deficiency results in a similar condition, but in this case the metabolite which accumulates is deoxycortisol, or compound S. In urine, it is found in the form of tetrahydro derivatives, measured by the Porter–Silber reaction. Hypertension associated with 11β-hydroxylase deficiency (reversible by hormonal suppression) is attributable to a concomitant hypersecretion of deoxycorticosterone.

A deficit of *3β-hydroxysteroid dehydrogenase* results in sexual ambiguity in both sexes. There is an accumulation of pregnenolone and of its hydroxylated derivatives, and also of DHEA and DHEA-S. When the defect is complete, patients do not survive. The weak androgenic activity of DHEA and DHEA-S explains the discrete virilization of girls. The *17-hydroxylase* block prevents the synthesis of cortisol, but not that of corticosterone or deoxycorticosterone. The accumulation of the latter compounds accounts for the hypertension and hypokalemia observed in these patients. The block involves the gonads, preventing the synthesis of androgens and estrogens. Sexual problems observed in girls are manifested in the absence of development of secondary sexual characteristics, and amenorrhea, and in boys, in pseudohermaphroditism and gynecomastia.

Deficiencies of *18-hydroxylase* and *18-hydroxysteroid dehydrogenase* (which are involved only in aldosterone synthesis) result in symptoms involving mineralocorticotropic insufficiency.

The block of *desmolase* C20–22 leads to a lipid hyperplasia of the adrenal. The biosynthesis of cortisol, aldosterone and androgens is therefore impossible. All patients have a feminine phenotype and die shortly after birth.

94. Gallais, A. (1912) Le syndrome génito-surrénal. Thesis, Paris.
95. Wilkins, L. (1962) Adrenal disorders II. Congenital virilizing adrenal hyperplasia. *Arch. Dis. Child.* 37: 231–241.
96. Eberlein, W.R. and Bongiovanni, A.M. (1955) Congenital adrenal hyperplasia with hypertension: unusual steroid pattern in blood and urine. *J. Clin. Endocr. Metab.* 15: 1531–1533.

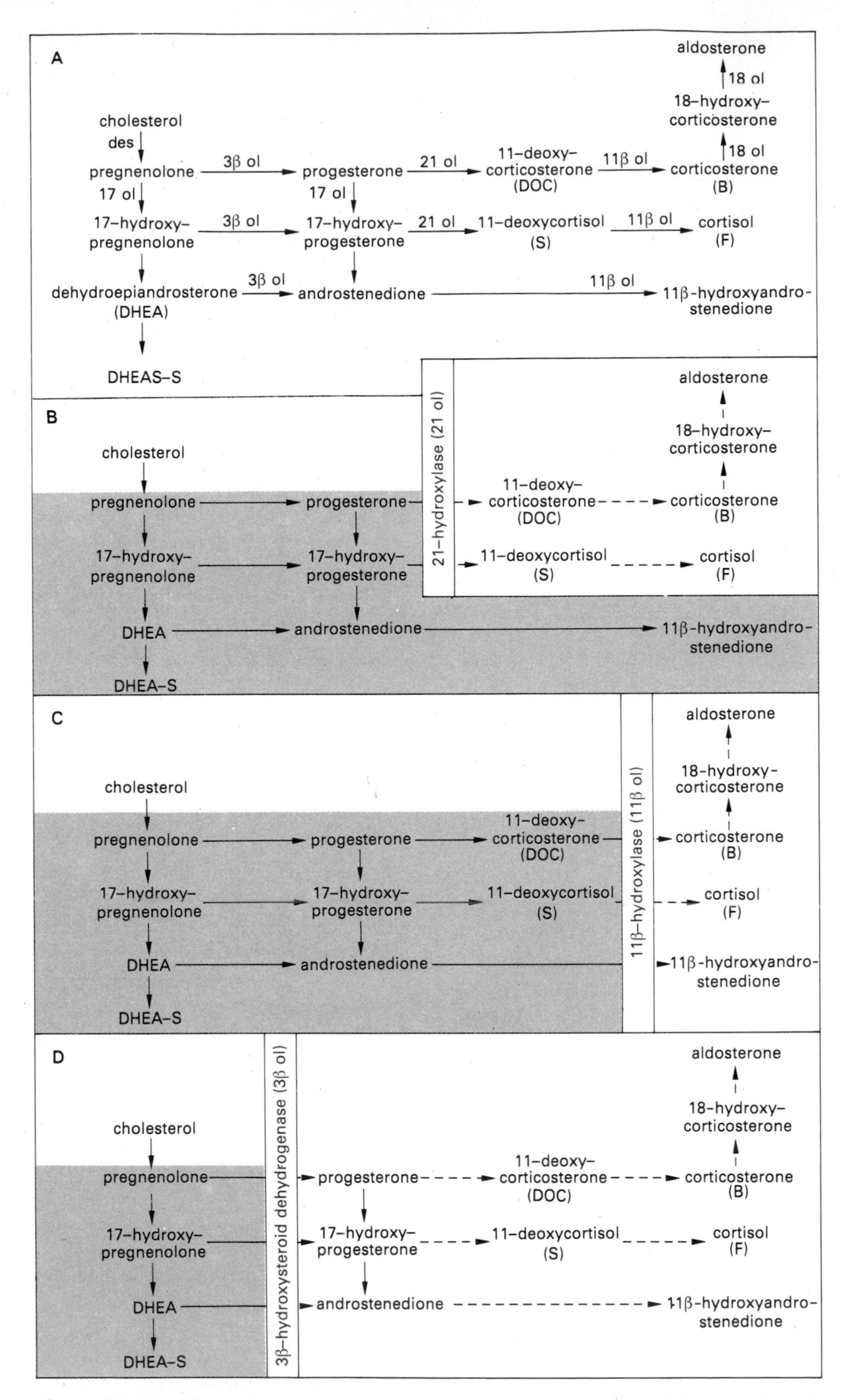

Figure IX.37 Schematic representation of three types of enzymatic defects in adrenal steroid biosynthesis. **A.** Normal biosynthetic pathway. In B, C, and D, solid arrows indicate pathways with increased synthesis preceding the enzymatic block. Dotted arrows denote pathways with residual or absent steroid synthesis. **B.** Defect of 21-hydroxylase; **C.** Defect of 11β-hydroxylase; **D.** Defect of 3β-hydroxysteroid oxidoreductase. Not represented are defects of C20–22 desmolase, 17-hydroxylase, 18-hydroxylase and 18-hydroxylase oxidoreductase.

Virilization of Ovarian Origin

Polycystic ovarian syndrome (PCO) is described in detail on p. 464. *Virilizing tumors* of the ovary are very rare (arrhenoblastoma, Leydig tumors, tumors of ectopic adrenal tissue, etc.). An elevation of both plasma and urinary androgens is observed which is alterable by neither dexamethasone nor the administration of a combination of estrogens and progesterone.

Iatrogenic Virilization

It is important to look carefully for a past history of treatment with androgens, or progestins with androgenic activity, which could account for signs of virilization.

Idiopathic Hirsutism

Hirsutism of *unknown cause* is the most *frequent*. Testosterone levels are within normal limits (however, there is a statistical difference between groups of hirsute women in comparison to normal women). Free testosterone is frequently elevated. In a certain number of cases, the production rate and the MCR of testosterone are increased. In any case, if the testosterone concentration is greater than 1.5 ng/ml, examination for a specific cause of elevated testosterone levels should be undertaken (e.g., a possible tumor). Treatment consists of cosmetic therapy (hair removal) and the administration of antiandrogens (cyproterone acetate) if necessary. The suppression of adrenal (by dexamethasone) or ovarian activity (by GnRH, Fig. X.4) is more controversial.

11.3 Other Steroids

Most pathophysiological aspects of female sex steroids are found in Chapter X, and those related to aldosterone, on p. 621.

12. Summary

Steroid hormones are involved in many processes affecting *metabolism, reproduction, and behavior.*

Steroid *biosynthesis* occurs in the *adrenal cortex, gonads*, and *placenta*. The precursor, *plasma cholesterol*, enters steroidogenic cells via the lipoprotein receptor system, and the early biosynthetic steps are identical for glucocorticosteroids and sex hormones. In *mitochondria*, a C21-steroid, *Δ_5-pregnenolone*, is produced initially. This is a complex, multienzymatic transformation stimulated by ACTH in the adrenal cortex, and by LH and FSH in the ovary and testis. Oxidation by 3β-hydroxysteroid dehydrogenase produces *progesterone*. The latter is a hormone secreted by luteal cells, and is an intermediate in other steroidogenic cells. In the zona fasciculata and the zona glomerulosa of the adrenal cortex, successive hydroxylations of progesterone result in the synthesis of the gluco- and mineralocorticosteroids *cortisol* (or corticosterone in some species) and *aldosterone*. In the zona reticularis, following hydroxylation at C17, C21-steroids are split to form C19-steroids: *dehydroepiandrosterone* (DHEA) and androstenedione, the former being largely *sulfoconjugated* before secretion. C19-steroids are also produced in gonadal cells, and, in testicular interstitial cells, androstenedione is reduced to *testosterone*. In the theca interna cells, C19-steroids are aromatized to form estrogens, mostly *estradiol* and estrone; androgens coming from theca cells are transformed into estrogens in luteal cells. *LH, FSH and ACTH stimulate* steroid *hormone* biosynthesis in their respective target cells, *via membrane receptors*. Adenylate cyclase activation, *increased cAMP* concentration and PK-A activity are clearly involved in the biological response. Cortisol, estradiol, progesterone and testosterone have *negative feedback control* on the secretion of the corresponding *pituitary hormones*, with the important exception of a *positive* preovulatory feedback mechanism exerted by estradiol on LH and FSH. ACTH stimulates aldosterone synthesis in the glomerulosa cells by both cAMP and a calcium-dependent mechanism. There is no feedback control of ACTH production by aldosterone, but, conversely, aldosterone production is under the influence of other hormonal and metabolic factors. The placenta synthesizes progesterone from cholesterol, but estrogens are produced from dehydroepiandrosterone sulfate and its 16-oxygenated derivatives which come primarily from fetal adrenals.

In *plasma*, steroid hormones are *bound* to *specific* high affinity *proteins*. In humans, *transcortin* (CBG) binds glucocorticosteroids and progesterone, while *sex steroid binding plasma protein* (SBP) binds testosterone and estradiol. *Highly* active synthetic hormones are only weakly bound by these proteins, as is also the case for aldosterone. These proteins increase under estrogen treatment and during pregnancy. Their respective concentration and regulation varies greatly according to species. In the *seminiferous tubules*, there is an *androgen binding protein,* very similar to the corresponding plasma protein synthesized in Sertoli cells, which transports testosterone to the head of the epididymis.

Catabolism of steroid hormones is *complex*, but *preserves* the steroid *nucleus*. The biochemical step responsible for the loss of biological activity is often the reduction of the C4–C5 double bond. Steroids are substrates for many specific hydroxyl oxidoreductases and hydroxylases, and the large majority of metabolites excreted in the urine are sulfo- and glucurono-conjugated.

Steroids *affect many cellular processes*. For instance, they *control* the preferential *synthesis* of secreted *proteins* (e.g., ovalbumin) or important intracellular proteins (such as the enzyme tyrosine aminotransferase, and sodium transport protein). Sex steroids have a *growth-promoting* effect on their target cells, and are sometimes *involved in* the process of *differentiation*. In contrast, glucocorticosteroids often have a negative regulatory effect, and are even able to *kill* certain cells (e.g., lymphoid cells). However, the early steps involved in the hormone–target cell interaction *seem to be the same in all cases*.

Steroids *enter the target cell* through the cell membrane, probably by passive diffusion. *Intracellular receptors* are proteins with high affinity for steroid ($K_D \sim 10–0.1$ mM), and their specificity is highly correlated with the hormonal activity of ligands. In the absence of hormone, the receptor is located primarily in the cell *nucleus,* where it is loosely bound. Upon binding of the hormone, the *receptor* is *transformed, or activated,* and binds with high affinity to chromatin, resulting in a *specific change of gene expression* in the cell. Steroid receptors have been cloned, and their sequences are known. Their molecular weights range from ~65,000 (estradiol receptor, 595 aa) to ~110,000 (mineralocorticosteroid receptor, 984 aa). Their primary structure demonstrates *two zones* of particular interest. The *first* one, ~70 aa in length, rich in cysteines, appears to be the potential *DNA binding site*; its structure probably includes "*Zn fingers*". The *other* region, ~220 aa in length and highly conserved between species for the same steroid, is located near the C-terminal extremity, and is involved in *hormone binding*. Other portions of the receptor (which show more variation, according to hormones and species) are also important for specific responses. In the absence of hormone, all steroid receptors are found in $M_r \sim 300,000$ "8 S" forms, and these *hetero-oligomeric, non-transformed complexes* are composed of receptor and heat shock protein (*hsp 90*), with a M_r of ~90,000. Consensus sequences of DNA, which resemble enhancer sequences, and which bind steroid hormone receptors with high-affinity, have been identified upstream of the transcription initiation site of steroid-regulated genes (HREs, hormone response elements). However, the mechanism of receptor transformation and activation, and the means by which transcription is modified, requires further investigation.

Steroid-receptor complexes act primarily on gene transcription; however, in the specific case of an *amphibian* oocyte, progesterone *receptors* have been identified at the *cell surface*, and protein synthesis is not dependent on a change in transcription.

Receptor concentrations in target cells vary physiologically and pathologically. Receptor measurements help define the hormone-dependency of certain cancers, and, thus, the selection of patients requiring endocrine therapy. For *breast cancer*, approximately 75% of patients with estrogen and progesterone receptors respond to some form of endocrine treatment.

Cortisol has effects in almost *all cells* of the body. In the liver, it increases glycogenogenesis and protein anabolism, but protein catabolism is augmented in both muscle and skin. Cortisol increases blood glucose and gastric acid secretion, and inhibits lymphoid tissue, resulting in a decrease of antibody formation. Cortisol also has anti-inflammatory and anti-allergic activities, and influences several CNS functions. *Permissive effects* suggest that cortisol, although not active by itself, is necessary for the action of other hormones.

Androgens are essential for the *differentiation* of male secondary sexual organs. They favor *protein anabolism* in muscle and bone as well as the maturation of sperm. In several tissues, testosterone is reduced to *5α-dihydrotestosterone*, which binds to the androgen receptor with higher affinity than does testosterone. A defect in

5α-reductase leads to incomplete *pseudohermaphrodism*, since testosterone can act directly in a number of target cells. A genetically defective androgen receptor is responsible for *testicular feminizing syndrome (tfm)*. Finally, *transformation to estrogens* is probably responsible for important differentiation and regulatory effects of testosterone in the hypothalamus.

Estrogens have important *growth-promoting* effects in female secondary sexual organs (uterus, vagina, oviduct). By increasing the concentration of *progesterone receptor*, estrogens *prime* the effect of progestins. They also have important effects in the CNS, as demonstrated by feedback control of FSH and LH secretion, and by changes in sexual behavior at the time of estrus. *Progesterone* is involved in the *differentiation* of certain cells (e.g., decidualization) and in specific protein synthesis (e.g., uteroglobin, ovalbumin), and has an inhibitory effect on myometrial contractility, which decreases with the diminution of the hormone at parturition. *Estrogen and progesterone effects are interrelated* physiologically and pharmacologically, often in a complex manner, differing as a function of the particular target organ. This includes changes of estrogen and progesterone receptor concentrations in the uterus and brain during the ovarian cycle.

Pharmacology and *therapeutics* related to steroid hormones involve a number of active synthetic *agonists and antagonists*. Among the most important are potent, orally effective glucocorticosteroids and progestins, antiestrogens and antiandrogens, antialdosterone and, as recent additions to this category, antiprogesterone and antiglucocorticosteroid.

Measurements of steroid hormones *in the blood* are useful *clinically*. To improve the significance of the results, several inhibition and stimulation tests, with appropriate hypothalamic and pituitary hormones, are also performed.

Addison's disease is characterized by chronic adrenocortical insufficiency due to the destruction of the adrenal glands. Functional insufficiency of the adrenal cortex may be of hypothalamic-hypophyseal origin. Iatrogenic adrenal insufficiency may occur after prolonged treatment with glucocorticosteroids.

Cushing's syndrome is due to hypercortisolism. It may be related to primary adrenal pathology, such as benign or malignant tumors. It can also be secondary to an increased production of ACTH, in most cases by the pituitary (often microadenoma), which characterizes *Cushing's disease*. The cause may be hypothalamic in origin, or it may be due to a pituitary tumor. Finally, increased ACTH may originate from an ectopic tumor (e.g., lung cancer).

Several *hyperandrogenic* syndromes have been described. Clinically, they differ as a function of the age at which excessive production begins. Virilizing *pseudohermaphroditism* results when the onset occurs during fetal life. Precocious pseudopuberty is observed in children, and varying degrees of *virilization* and hirsutism characterize the disorder in adults. The cause may be *adrenal* in origin: beside tumors, there are a number of enzymatic deficiencies of genetic origin. A defect of 21- or 11β-hydroxylase, the most frequent, leads to a decrease of cortisol, a secondary increase in ACTH, and, consequently, an overproduction of androgens. Treatment with corticosteroids is very efficient. Excess androgen may also be *ovarian* in origin, and the glands are often polycystic. The etiology of hirsutism is frequently unknown.

General References

Austin, C. R. and Short, R. V., eds (1984) Monoclonal control of reproduction. In: *Reproduction in Mammals*. vol. 3., Cambridge University Press, Mass.

Bardin, C. W., Milgrom, E. and Mauvais-Jarvis, P., eds (1983) *Progesterone and Progestins*. Raven Press, New York.

Baxter, J. D. and Rousseau, B. B. (1979) *Glucocorticoid Hormone Action*. Springer Verlag, New York.

Christy, N. (1971) *The Human Adrenal Cortex*. Harper and Row, New York.

Dorfman, R. I. and Ungar, F. (1965) *Metabolism of Steroid Hormones*. Academic Press, New York.

Felig, P., Baxter, J. D., Broadus, A. E. and Frohman, L. A. (1981) *Endocrinology and Metabolism*. McGraw-Hill, New York.

O'Malley, B. W. and Hardman, J. G., eds (1975) Hormone action. In: *Methods in Enzymology*. Academic Press, San Díego, vol. 37, 38, 39 and 40.

Sato, G. and Ross, R., eds (1979) *Hormones and Cell Culture*. Cold Spring Harbor Lab., Cold Spring Harbor, New York.

Wilkins, L. (1957) *The Diagnosis and Treatment of Endocrine Disorders in Childhood and Adolescence*. 2nd ed., C. C. Thomas, Springfield, Ill.

Alfred Jost

Hormonal Control of the Masculinization of the Body

During embryonic development, sexual differentiation of the body depends upon a chain of events (Fig. 1). The *genetic sex* first imposes the *sex of the gonads* (testes or ovaries), *which* in turn *control the later development of the body sex*. The gene TDF (testis determining factor), normally present on chromosome Y, determines the formation of the testes. Castration experiments performed on rabbit fetuses, and clinical evidence in humans, showed that maleness must be imposed upon the body by the fetal testes, against an inherent female trend (Fig. 2). In the absence of testes, whether ovaries are present or not, the feminine type of organogenesis prevails. Therefore, the endocrine story of a male begins with the conversion, into testes, of the common gonadal primordia located on the inner aspect of the fetal kidneys, the mesonephron (Fig. 3).

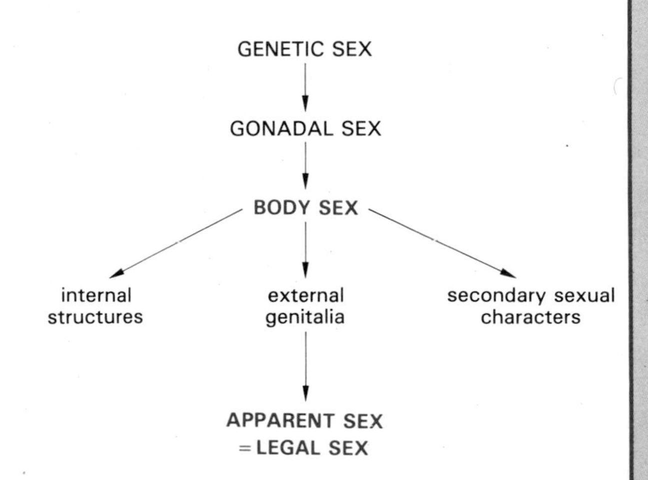

Figure 1 *Chain of events during sexual development. Gonadal sex and body sex are discussed in the text. It is noteworthy that apparent sex, or legal sex, is determined at birth, by only the conformation of the external genitalia. (Jost, A., in: Jones, H. W. and Scott, W. W. (1971)* Hermaphroditism. Genital Anomalies and Related Endocrine Disorders. *Williams and Wilkins Co., Baltimore).*

Testicular differentiation is initiated by the emergence of the Sertoli cell type in the gonadal primordium, and by the aggregation of these cells into seminiferous cords. Leydig cells appear somewhat later between the cords. The process can be followed in gonadal primordia cultured in vitro. Under experimental conditions, Sertoli cell aggregation and testicular organogenesis can be prevented without suppressing the differentiation in the unorganized gonad of the two kinds of endocrine cells of the fetal testis, i.e. the Sertoli cells and the Leydig cells[1]. It is conceivable that, if the same occurred in vivo, masculinization of the body would occur despite an atypical morphology of the testes.

The *differentiation of the inner and external genitalia*, from the common, indifferent condition, shown in Figure 3, *follows gonadal differentiation. In females*, the mesonephron and their ducts regress, whereas the Müllerian ducts develop into the fallopian tubes and uterus. The urogenital sinus contributes to the edification of the vagina, and the genital tubercle forms the clitoris. *In males*, the Müllerian ducts regress very early; actually, in humans and in most mammals, their involution establishes the male orientation of the genital tract. Moreover, parts of the mesonephric renal structures are incorporated into the genital tract, and the urogenital sinus and external genitalia are masculinized.

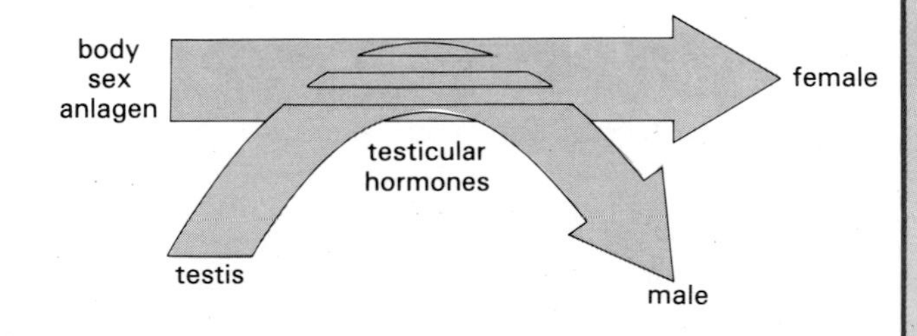

Figure 2 *Diagram showing that the differentiation of the body sex obeys a feminine program if maleness is not imposed by testicular hormones.*

Androgens administered to females or castrated male fetuses cause masculinization of the genitals, but they do not induce

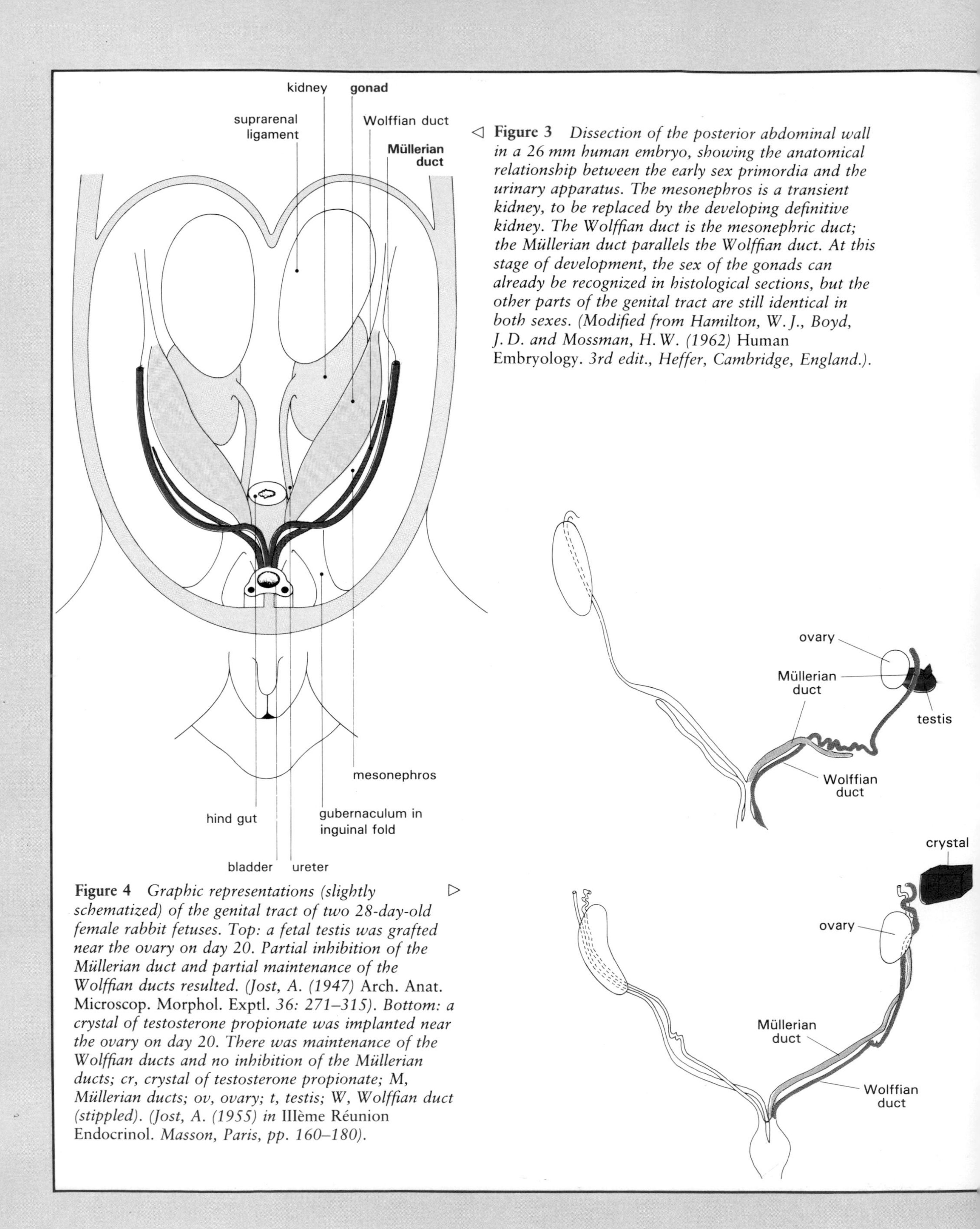

◁ **Figure 3** *Dissection of the posterior abdominal wall in a 26 mm human embryo, showing the anatomical relationship between the early sex primordia and the urinary apparatus. The mesonephros is a transient kidney, to be replaced by the developing definitive kidney. The Wolffian duct is the mesonephric duct; the Müllerian duct parallels the Wolffian duct. At this stage of development, the sex of the gonads can already be recognized in histological sections, but the other parts of the genital tract are still identical in both sexes. (Modified from Hamilton, W. J., Boyd, J. D. and Mossman, H. W. (1962)* Human Embryology. *3rd edit., Heffer, Cambridge, England.).*

Figure 4 *Graphic representations (slightly schematized) of the genital tract of two 28-day-old female rabbit fetuses. Top: a fetal testis was grafted near the ovary on day 20. Partial inhibition of the Müllerian duct and partial maintenance of the Wolffian ducts resulted. (Jost, A. (1947)* Arch. Anat. Microscop. Morphol. Exptl. *36: 271–315). Bottom: a crystal of testosterone propionate was implanted near the ovary on day 20. There was maintenance of the Wolffian ducts and no inhibition of the Müllerian ducts; cr, crystal of testosterone propionate; M, Müllerian ducts; ov, ovary; t, testis; W, Wolffian duct (stippled). (Jost, A. (1955) in* IIIème Réunion Endocrinol. *Masson, Paris, pp. 160–180).* ▷

regression of the Müllerian ducts (Fig. 4). A distinct Müllerian inhibitor, also called *anti-Müllerian hormone (AMH)* or *Müllerian inhibiting substance* (*MIS*)[2], is produced by *Sertoli cells*. It was recently isolated from the medium in which bovine fetal testes were cultured in vitro. Using monoclonal antibodies and immunoaffinity columns, the Müllerian inhibitor has been isolated; it is a multimer glycoprotein containing fucose (a monomer with a M_r of ~72,000)[3]. The antibodies obtained so far are species-specific for the bovine product. Recent cloning studies have indicated that the human gene has five exons that code for a protein of 560 aa (Fig. 5). Comparison of the bovine and human MIS proteins reveals a highly conserved C-terminal region that shows marked homology with human TGF-β and the β chain of porcine inhibin[4], and the decapentaplic complex of Drosophila.

The embryonic Müllerian duct is made of a monolayered epithelial duct surrounded by mesenchyme, with a basal membrane interposed. Müllerian regression has been attributed to the degeneration of epithelial cells by lysosomal enzymes. Recent studies emphasize the disappearance of the basal membrane and suggest dedifferentiation of epithelial cells, into mesenchyme-like cells, as an important factor in the regression of the Müllerian duct[5].

At an appropriate stage, the *fetal testes produces testosterone*, which governs the masculinization of the genital tract. The physiological significance of other fetal testicular steroids is still to be investigated. The mode of action of the testicular hormone varies as a function of the particular part of the genital tract[6].

The *Wolffian ducts* are prevented from degenerating and are converted into *vasa deferentia*; this means that the still unknown processes which result in the disappearance of the mesonephron and Wolffian ducts in females are opposed and suppressed in males. In addition, androgenic stimulation produces the outgrowth of seminal vesicles from the lower part of the Wolffian ducts and the growth and coiling into an epididymis of their upper part. The evidence obtained so far indicates that *testosterone* is the hormonal factor responsible for these effects on Wolffian ducts.

At the level of the external genitalia, androgens exert different actions. They control the cellular proliferation and possibly cellular death, and/or cellular differentiation, necessary for the morphogenic movements which result in the fusion of the borders of the urethral groove and in the inclusion of the male urethra into the penis. Growth of the whole set of involved tissues increases the ano-genital distance. In bulls and other ruminants, in which this process reaches an extreme degree, the penile shaft opens below the umbilicus. Masculinization of

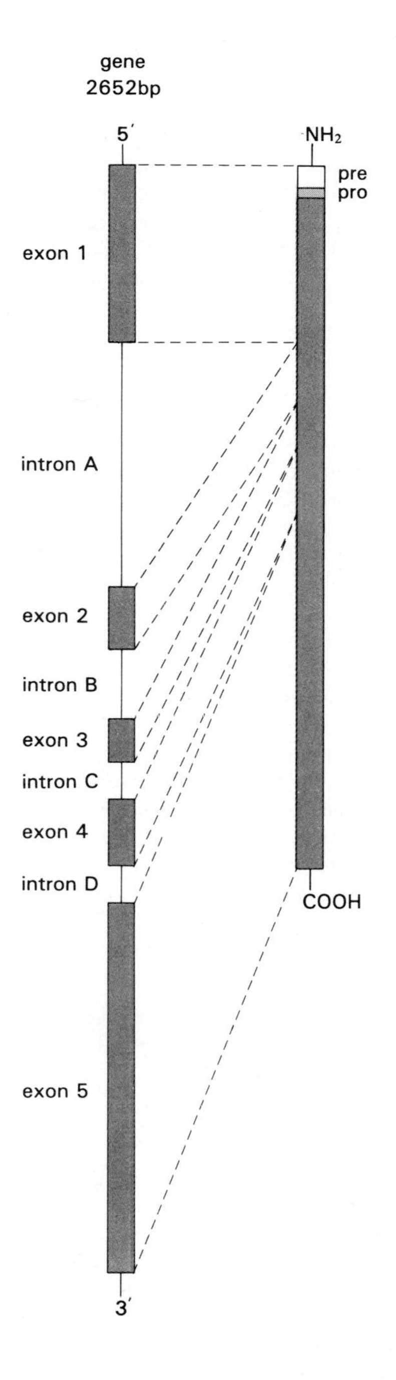

Figure 5 *Organization of the human gene and preproprotein of Müllerian inhibiting substance*[4].

the external genitalia is not produced by testosterone itself. Testosterone must be converted into *5α-dihydrotestosterone (DHT)* by the 5α-reductase enzymatic complex in cells of the target organ before it exerts its morphogenic action. Defects of this enzymatic system in human beings result in absent or incomplete masculinization. Fibroblasts in culture, obtained from skin of the sexual area of affected patients, reveal the defect. Complex mesenchymal-epithelial relations are probably involved in the development of the species-specific patterning of the genitalia.

At the level of the *urogenital sinus*, androgens impose two masculine features: 1. *vaginal* organogenesis is *suppressed* (the urogenital sinus becomes the male urethral canal); 2. *prostatic glands bud* from the wall of the urogenital sinus. In vitro studies of embryonic development of the prostate demonstrated that androgens exert their action mainly at the level of the *mesenchymal* cells. However, the mechanisms responsible for the development of male organs as opposed to the female counterparts are still not fully understood. Most experimental data suggest that *dihydrotestosterone* is the active hormonal trigger at the level of the urogenital sinus.

Finally, although different steroids (testosterone or dihydrotestosterone) affect different parts of the genital tract, a *single androgen receptor*, apparently controlled by a single gene, is operational. Its defect causes testicular feminization, including lack of stabilization of the Wolffian ducts, with absence of their derivatives, and lack of masculine organogenesis of the urogenital sinus and external genitalia. However, *the gene does not hinder normal Müllerian inhibition*, and no sex duct persists in affected individuals.

Clinical studies have revealed many variants in either androgen synthesis or cellular responsiveness to steroids. Varying degrees of defective masculinization of the body are known. Moreover, the testicular hormone exerts effects on some parts of the nervous system, at least in some animal species (e.g., the preoptic area of the rat hypothalamus). These important effects on nerve cells and the role of androgen *aromatization* in the *nervous system* are other aspects of the sexualization of the body (p. 48).

In conclusion, masculinization of the body is achieved through *pluri-hormonal control*, not only because *two different cell types* in the fetal testis (Sertoli and Leydig cells) (Fig. 6) produce *two* completely *different kinds of hormones* (Müllerian inhibiting substance and androgens), but also because intracellular conversion of testosterone to dihydrotestosterone takes place in some target organs and not in others. The same hormone receptor seems to be involved, but the cellular effects are multifarious. Androgens determine the characteristic male pattern of different parts of the body probably by stimulating, in one way or another, local developmental programs.

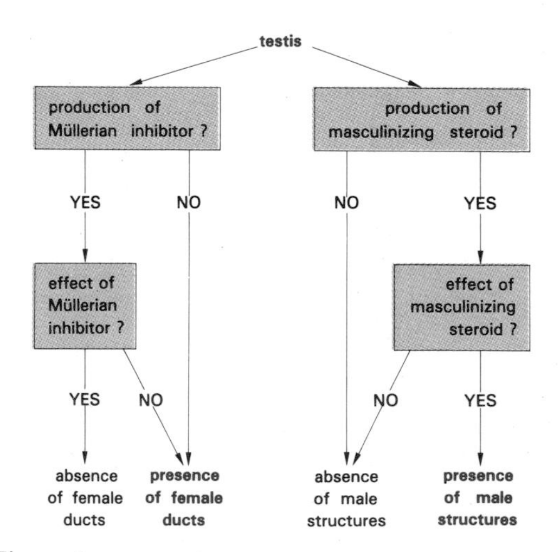

Figure 6 *General outline of reproductive physiology and its developmental background. (Modified from Figure 20 in Jost, A. (1970). In:* Mammalian Reproduction *(Gibian, H., Plotz, E. J., eds), Springer Verlag, Berlin, pp. 4–36).*

References

1. Donahoe, P. K., Budzik, G. P., Trelstad, R., Mudgett-Hunter, M., Fullter, A. Jr., Hutson, J. M., Ikawa, H., Hayashi, A. and McLaughlin, D. (1982) Müllerian-inhibiting substance: an update. *Rec. Prog. Horm. Res.* 38: 279–326.
2. Magre, S. and Jost, A. (1984) Dissociation between testicular organogenesis and endocrine cytodifferentiation of Sertoli cells. *Proc. Natl. Acad. Sci., USA* 81: 7831–7834.
3. Picard, J. Y. and Josso, N. (1984) Purification of testicular anti-Müllerian hormone allowing direct visualization of the pure glycoprotein and determination of yield and purification factor. *Mol. Cell. Endocr.* 34: 23–29.
4. Cate, R. L., Mattaliano, R. J., Hession, C., Tizard, R., Farber, N. M., Cheung, A., Ninfa, E. G., Frey, A. Z., Gash, D. J., Chow, E. P., Fisher, R. A., Bertonis, J. M., Torres, G., Wallner, B. P., Ramachandran, K. L., Ragin, R. C., Manganaro, T. D., McLaughlin, D. T. and Donahoe, P. K. (1986) Isolation of the bovine and human genes for Müllerian inhibiting substance and expression of the human gene in animal cells. *Cell* 45: 685–698.
5. Trelstad, R. L., Hayashi, A., Hayashi, K. and Donahoe, P. K. (1982) The epithelial-mesenchymal interface of the male rat Müllerian duct: loss of basement membrane integrity and ductal regression. *Develop. Biol.* 92: 27–40.
6. Wilson, J. D., Griffin, J. E., George, F. W. and Leshin, M. (1981) The role of gonadal steroids in sexual differentiation. *Rec. Prog. Horm. Res.* 37: 1–39.

CHAPTER X

Clinical Endocrinology of Reproduction

Samuel Yen

Contents, Chapter X

Clinical Endocrinology of Reproduction

1. Neuroendocrinology of Reproduction

The *reproductive system* requires integrated *control* by both *neurally* and *humorally conducted inputs*. In recent years, remarkable advances have been made in our knowledge and understanding of the manner in which neural signals participate in the control of endocrine activity. The central nervous system integrates internal and external stimuli, and sends temporally defined signals to its effector organ, the hypophysis. Thus, follicular maturation, ovulation and lactation are achieved via the appropriate release of gonadotropins and prolactin. The pivotal role of the central nervous system is also exemplified by the disruptive impact of external stresses on the normal rhythm of the reproductive process.

1.1 Neuroendocrine Control of Gonadotropin Secretion

Hypothalamic Gonadotropin Releasing Hormone (GnRH)

It is now established that the hypothalamus, particulary the *arcuate median eminence region* and *preoptic area*, produces and secretes the decapeptide GnRH (Fig. II.4). In the median eminence, GnRH nerve terminals secrete directly into portal capillaries; the hormone is then delivered by the *portal circulation* to the adenohypophyseal *gonadotropic cells* (Fig. X.1). The role of GnRH-associated peptide (GAP) (p. 264), which also stimulates the secretion of follicle stimulating hormone (FSH) and luteinizing hormone (LH), is unknown.

Biosynthesis and Secretion of GnRH

Biosynthesis of GnRH has been demonstrated in the hypothalamus, and the final product is derived

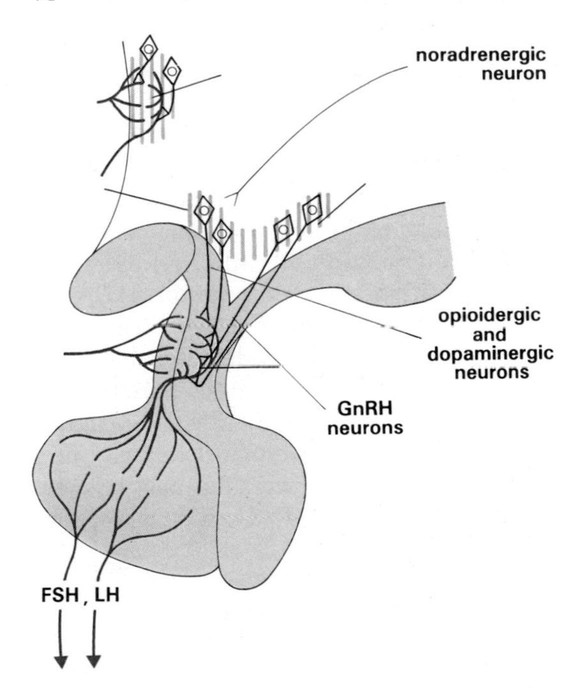

Figure X.1 Neuroanatomical relationship between noradrenergic, dopaminergic, opioidergic, and GnRH neurons within the arcuate-median eminence region and the preoptic, anterior hypothalamic area. Note GnRH and opioid peptides and dopamine axons terminating on the portal capillaries.

from a *large precursor* molecule (92 aa) by enzymatic processing (p. 264).

The *secretory* activity of *GnRH neurons* at nerve terminals is *pulsatile* in nature. The frequency and amplitude of secretory episodes are essential for the release as well as the synthesis of LH and FSH by gonadotropic cells[1].

Regulation of GnRH Secretion

Opioid-Catecholamine Systems

Within the mediobasal hypothalamus, GnRH-secreting neurons are connected with *noradrenergic neurons*, whose cell bodies are located in the locus ceruleus, and with *tuberoinfundibular dopaminergic and opiodergic neurons* (Fig. X.1). While conflicting data exist, evidence of an interacting system of opioid peptides and catecholamines in the control of GnRH secretion is emerging.

Compelling evidence indicates that in humans, as in experimental animals, endogenous *opioid peptides* have an essential *inhibitory* effect on GnRH secretion. Administration of morphine or opioid peptides induces a prompt decrease in pulsatile gonadotropin secretion, and treatment with naloxone, an opiate receptor antagonist, induces a prompt increase in pulsatile secretion of GnRH and LH.

Dopamine stimulates β-endorphin secretion, which in turn inhibits dopaminergic activity. This interaction may represent an ultra-short feedback system regulating the degree of opioidergic activity.

Opioid inhibition of GnRH release is probably due to a reduced influx of *excitatory adrenergic signals* in the vicinity of GnRH neurons. This view is supported by the finding that adrenergic blockage of the α_1 receptor promptly suppresses pulsatile LH secretion[2] and renders naloxone ineffective in inducing LH release. Thus, in a physiological context, the pattern and amount of GnRH secretion may be determined by a system involving opioid and catecholamine interaction (Fig. X.2).

Ovarian Steroid Modulation

The expression of the opioid-catecholamine system on GnRH neuronal activity is in the *regulation of the frequency and amplitude* of pulsatile *GnRH secretion.* That ovarian steroids modify endogenous opioid activity is suggested by the finding that naloxone induces an increased frequency and amplitude of pulsatile LH release in high estrogen and particularly

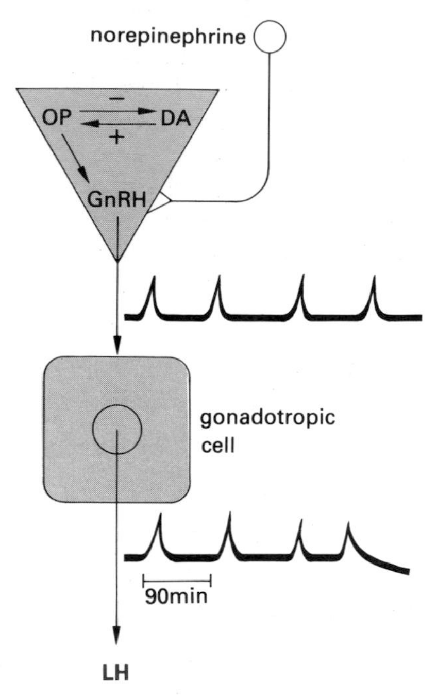

Figure X.2 Dopaminergic, opioidergic and α-adrenergic neurons represent an interacting system in the regulation of pulsatile GnRH secretion. The activity of this system is sex steroid dependent.

high estrogen and progesterone environments[3]. The disappearance of ovarian steroid feedback (i.e. in postmenopausal women) is associated with the uncoupling of this inhibitory influence, and could be functionally related to the high frequency of LH pulses and the unrestrained secretion of gonadotropin in postmenopausal women. Administration of estrogen and progesterone restores this inhibitory activity. Thus, the *negative feedback effect* of estrogen and especially the synergistic effect of estrogen and progesterone on gonadotropin secretion may be in part functionally coupled with increased activity of the opioidergic-catecholaminergic system in the inhibition of GnRH secretion (Fig. X.2).

1. Ferin, M., Van Gugt, D. and Wardlaw, S. (1984) The hypothalamic control of the menstrual cycle and the role of endogenous opioid peptides. *Rec. Prog. Horm. Res.* 40: 441–486.
2. Kaufmann, J.-M., Kesner, J. S., Wilson, R. C. and Knobil, E. (1985) Electrophysiological manifestation of luteinizing hormone-releasing hormone pulse generator activity in the rhesus monkey: influence of alpha-adrenergic and dopaminergic blocking agents. *Endocrinology* 116: 1327–1333.
3. Ropert, J. F., Quigley, M. E. and Yen, S. S. C. (1981) Endogenous opiates modulate pulsatile luteinizing hormone release in humans. *J. Clin. Endocrinol. Metab.* 52: 583–585.

Mode of Action of GnRH and Its Analogs

GnRH binds to specific *receptors* on *gonadotropic cells* and stimulates the synthesis and secretion of LH and FSH. This secretory activity of gonadotropins occurs intermittently, in apparent synchrony with, and in response to, the pulsatile discharge of GnRH from the hypothalamus.

GnRH induces it own receptors (*self-priming*), and changes in the number of GnRH receptors on the gonadotropic cells occur in different steroid milieu. In the absence of ovarian feedback (after castration), the increase in GnRH receptor number is induced by increased GnRH secretion (up-regulation). Thus, the number of GnRH receptors is reflective of changes in GnRH secretion by the hypothalamus.

At the level of the *pituitary*, gonadal *steroids* exert their *negative feedback* effect at a *post-GnRH receptor site* rather than by altering the number of GnRH receptors directly.

Continuous loading of GnRH receptors by constant delivery of GnRH or the administration of long-acting GnRH-agonist results in a *reduced synthesis* and *secretion* of gonadotropin, a phenomenon known as down-regulation, or desensitization.

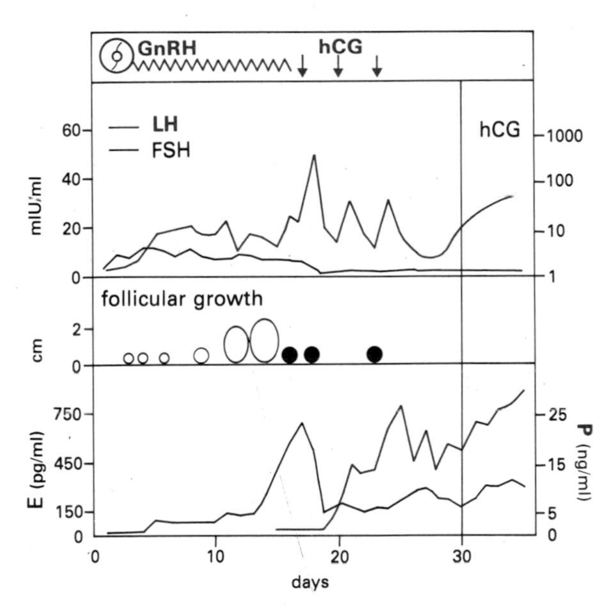

Figure X.3 Cyclic gonadotropin and ovarian steroid patterns, and follicular development (monitored by ultrasound). Black dots represent the corpus luteum during pulsatile GnRH stimulation (5 μg/pulse, every 96 min) in a patient with hypopituitarism due to pituitary stalk transection. Pregnancy resulted, as indicated by the rise of hCG. For comparison, in absence of GnRH treatment, the levels of FSH and LH were 2–3 mlU/ml; there was no follicular development with estradiol <10 pg/ml, and progesterone was not measurable.

Clinical Applications

The *pulsatile* nature of hypothalamic GnRH secretion in the control of episodic gonadotropin secretion provides a *physiological basis* for understanding the activation of the pituitary-gonad axis and the induction of ovulation in patients with endogenous GnRH deficiency. Pulsed administration of fixed amounts of GnRH (at 60 to 120 min intervals) *induces* appropriate *gonadotropin secretion*, a timely midcycle surge and ovulation, and, eventually, pregnancy may follow (Fig. X.3). This may be considered as up-regulation of the LH system, in contradistinction to the *down-regulation resulting from continuous infusion of GnRH or the administration of a long-acting GnRH agonist.* The loss of GnRH receptors in down-regulation provides an approach to the abolishment of the hypothalamic GnRH signal, and may offer new means of contraception as well as a

Table X.1 Clinical applications of GnRH and its analogs[4].

Activation of hypophyseal-gonadal function

- Physiologic hypogonadotropinism of puerperium
- Delayed puberty
- Cryptorchidism
- Induction of ovulation in hypogonadotropic amenorrhea
- Induction of multiple follicular maturation
- Male hypogonadotropic hypogonadism

Inhibition of hypophyseal-gonadal function

- Precocious puberty
- Endometriosis
- Hormone-dependent tumors
  - Uterine fibroids
  - Breast cancer
  - Prostate cancer
- Suppression of ovarian androgen excess
- Premenstrual syndrome
- Dysfunctional uterine bleeding, including clotting disorders
- Contraception
- Ovulation inhibition
- Induction of luteolysis

4. Yen, S.S.C. (1983) Clinical applications of gonadotropin-releasing hormone and gonadotropin-releasing hormone analogs. *Fert. Steril.* 39: 257–266.

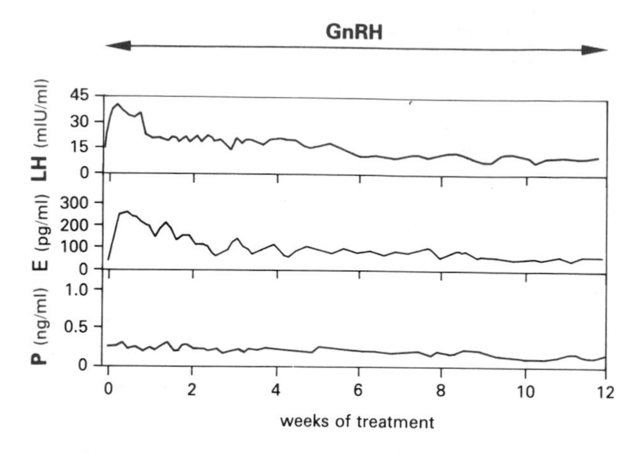

Figure X.4 An example of medical ovariectomy: the pattern of LH, estradiol (E) and progesterone (P) levels in five women with normal menstrual cycles, given daily subcutaneous injections of GnRH-agonist (2.5 µg/d,sc) over a period of three months. Ovulation was inhibited and amenorrhea was induced.

safe mode of inducing *reversible medical castration* (Fig. X.4) in a variety of sex steroid-dependent disorders (Table X.1).

1.2 Neuroendocrine Control of Prolactin Secretion

When viewed from an evolutionary perspective, prolactin (PRL), a lactogenic hormone, is *indispensable for the preservation of the species*. Survival of the young is dependent on a unique biological system which shifts the fuel delivery site from the uterus to the mammary gland, thus enabling the mother to continue providing nutrients after the birth of the child. Failure to do so would mean the extinction of humans and, indeed, of all mammalian species.

Among pituitary hormones, PRL apparently acts *without direct feedback* control by signals from peripheral target tissues, although it is under direct hypothalamic control. Despite the fact that the dominating role of hypothalamic inhibiting factors in the control of PRL release is well established now, substantial evidence suggests that stimulating factors also play an important role (Figs. X.5 and III.24).

Prolactin Inhibiting Factors: Dopamine

The *role* of *hypothalamic dopamine* as the *major PRL-inhibiting factor* (*PIF*) is firmly established. Dopamine is secreted into the portal vessels by dopamine-secreting neurons of the tuberoinfundibular dopamine system, whose cell bodies are located in the arcuate nucleus, with axons terminating in the external layer of the median eminence. The biosynthesis of dopamine occurs in the terminals of axons which abut the portal capillaries and deliver dopamine to the hypophyseal portal circulation. High-affinity dopamine receptors have been found on the surface of lactotropic cells (Fig. III.22).

The inhibitory effect of dopamine on PRL release can be shown by the administration of l-DOPA, a metabolic precursor of dopamine; l-DOPA readily crosses the blood-brain barrier and increases the rate of dopamine formation beyond the rate-limiting step, tyrosine hydroxylase. As a consequence, PRL levels decrease to a nadir 2.5 h following l-DOPA administration (0.5 g), and recovery includes a rebound to above basal levels.

Dopamine infusion, administered at different rates (from 0.02 to 8 µg/kg/min) for 3 to 4 h, produces a dose-related suppression of PRL levels in normal men and women, as well as in patients with hyperprolactinemia. *Dopamine does not cross* the *blood-brain barrier*, but since both the median eminence and

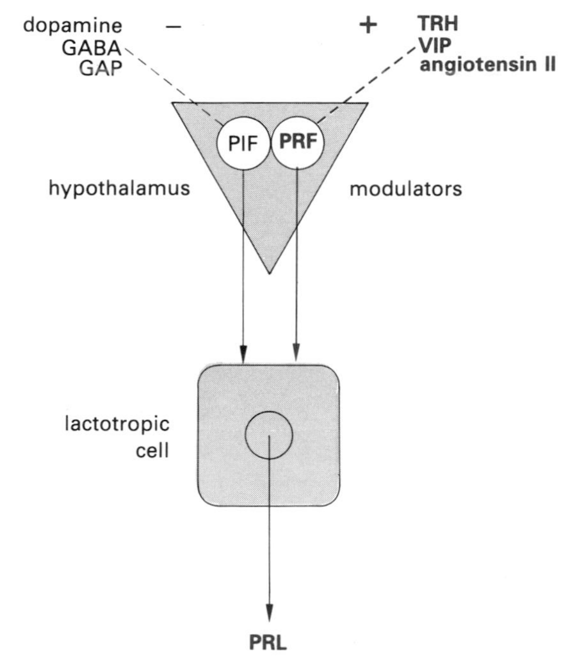

Figure X.5 Diagrammatic representation of neuroendocrine control of PRL secretion by PRL-inhibiting factors (PIF) and PRL-releasing factors (PRF). The process of PRL secretion is further modulated by a number of factors: serotonin, opioids, histamine, oxytocin, arginine, leucine, neurotensin, and substance P.

anterior pituitary are outside of this barrier, dopamine infused into the general circulation can be delivered to the median eminence pituitary system.

Dopamine receptor blockade by many pharmacological agents, such as metoclopramide, induces a rapid increase in PRL levels, and the effectiveness of metoclopramide is markedly attenuated by prior treatment with a dopamine agonist. Sequential dopamine and dopamine antagonist administration results in a corresponding reduction and augmentation of PRL secretion, indicating agonist and antagonist interaction at receptor sites.

A *peptide* sequence included in the prepro-GnRH molecule (p. 208) has recently been found to possess a *prolactin-inhibiting effect*. However, its physiological significance has not yet been assessed. Thus, the control of gonadotropin and prolactin secretion may reside, in large measure, in the same molecule within the arcuate nucleus.

Several lines of evidence derived from both in vivo and in vitro studies suggest that *GABA*, in addition to dopamine, *functions as a PIF*. First, GABA nerve terminals are present in the internal and external layers of the median eminence; they secrete GABA into the portal blood, where its concentration is increased several-fold, with a concomitant reduction in PRL secretion following pharmacological inhibition of GABA metabolism. Second, specific $GABA_A$ receptors are present on the pituitary lactotropic cell, and enhancement of endogenous GABAergic tone induced by sodium valproate (an inhibitor of GABA transaminase which degrades GABA at central and peripheral sites) reduces basal and breast-stimulated PRL release in women.

The inhibitory activity of dopamine is far greater than that of GABA. The marked rebound increase in PRL release seen both in vitro and in vivo following the cessation of dopamine infusion does not occur with GABA. This difference can be explained by the ability of dopamine, but not GABA, to induce the accumulation of newly synthesized PRL within lactotropic cells, which then appears to be released rapidly following withdrawal of dopamine inhibition. It has been proposed that, unlike dopamine, GABA as a PIF may function episodically in response to certain stimuli, as opposed to being secreted continuously into the portal blood.

Prolactin Releasing Factors (Fig. X.5)

While hypothalamic control of PRL secretion is dominated by a tonic inhibitory mechanism, *a functional role of PRL-releasing factor(s) (PRF)* appears necessary in order to account for acute secretory activities. For example, under certain conditions incremental PRL secretion is not accompanied by a measurable decrease in portal blood dopamine levels, and acute PRL release can occur under maximal inhibition by dopamine. Experimental evidence in several species, including humans, suggests that several other substances may be involved in the control of PRL secretion. Those which act as a neurohormone are secreted into the portal system and stimulate PRL release via specific receptors on the lactrotopic cell (thyrotropin releasing hormone (TRH), vasoactive intestinal peptide (VIP), and angiotensin II). Others function as neurotransmitters, as for example serotonin, opioid peptides, histamine, oxytocin, neurotensin, and substance P. Still others may act as a paracrine factor (GnRH).

PRL Secretion under Physiological Conditions (Fig. X.6)

Sleep-entrained PRL release. The *highest* plasma PRL concentration occurs during sleep. This nocturnal increase in PRL release is clearly sleep-entrained, as demonstrated by sleep reversal studies (Fig. III.20).

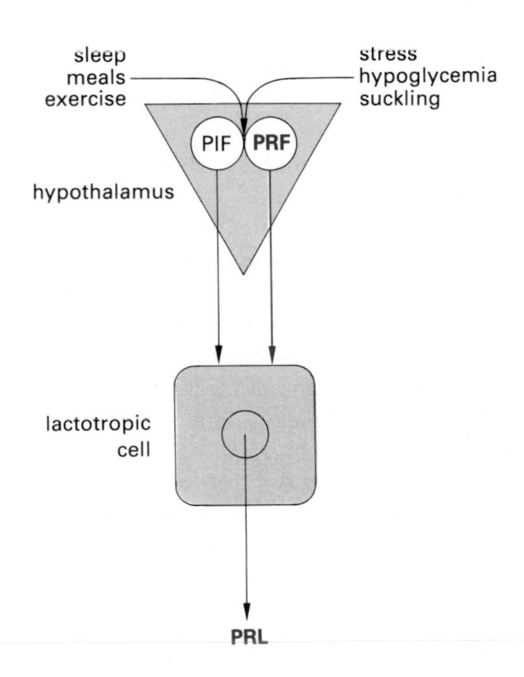

Figure X.6 Acute PRL secretion in response to a variety of physiological conditions.

Lunch-triggered acute PRL release. An acute release of PRL occurs following ingestion of a standardized mixed meal given at noon, but not after breakfast (08.00 h) or dinner (18.00 h). The composition of the midday meal has a clear effect: whereas carbohydrate meals induce no discernible effects, high protein and high fat meals cause a large increase in PRL secretion. Ingestion of neurotransmitter substrates l-tyrosine and l-tryptophan induces a remarkable increase in serum concentrations of PRL, suggesting that these essential amino acids may be active components of the high protein meal[5].

Stress and other stimuli. An elevated rate of PRL secretion is induced by a number of stress stimuli, including venipuncture, physical exercise, surgical stress, hypoglycemia, and general or conduction anesthesia in both men and women. Sexual intercourse is a potent stimulus for PRL release, and an even greater increase (10-fold) in PRL has been found in women experiencing orgasm. Limited studies indicate that coitus does not induce a PRL rise in males. Thus, PRL release is found in many different types of stress, which may be physical and/or emotional.

Short-loop feedback control (autoregulation). Since PRL release is not regulated by negative feedback signals from peripheral target sites, short-loop feedback, operating by way of hypothalamic regulation (via retrograde flow), assumes particular physiological significance. The presence of such an autoregulatory mechanism is suggested by a range of experimental evidence which includes the following observations: 1. the intraventricular injection of PRL results in an increase both in dopamine turnover in the median eminence and in the concentration of dopamine in portal blood; 2. the high turnover rate of dopamine in the median eminence found during lactation and pregnancy is reduced by hypophysectomy or by reducing PRL secretion, through direct pituitary inhibition, with the administration of bromocryptine; 3. autografts of prolactin-secreting tumors in rats cause a reduction in pituitary PRL content.

Thus, PRL controls its own secretion rate *through feedback regulation of hypothalamic dopamine.* Impairment of this autoregulatory mechanism may be a factor in the genesis of autonomous functional hyperprolactinemia. More generally, inappropriate PRL secretion is involved in many pathophysiological conditions (Table X.2).

5. Ishizuka, B., Quigley, M. E. and Yen, S. S. C. (1983) Pituitary hormone release in response to food ingestion: evidence for neuroendocrine signals from gut to brain. *J. Clin. Endocrinol. Metab.* 57: 1111–1116.

Table X.2 Conditions associated with inappropriate PRL secretion.

Pharmacological causes
Estrogen therapy
Anesthesia
Dopamine receptor blocking agents
Phenothiazones
Haloperidol
Metoclopromide
Domperidone
Pimozide
Sulpiride
Dopamine reuptake blocker
Nomifensine
CNS dopamine-depleting agents
Reserpine
α-methylDOPA
Monoamine oxidase inhibitor
Inhibition of dopamine turnover
Opiates
Stimulation of serotoninergic system
Amphetamines
Hallucinogens
Histamine H2-receptor antagonists
Cimetidine
Pathological causes
Hypothalamic or pituitary lesion
Craniopharyngioma
Glioma
Granuloma
Histoicytosis disease
Sarcoid
Tuberculosis
Stalk transection
Surgery (post-operation)
Head injury
Irradiation damage of the hypothalamus
Hypothalamic-hypophyseal disorders
Pseudocyesis
Cushing's disease
Nelson's syndrome
Acromegaly
Prolactinoma
Mixed adenomas secreting PRL and GH or ACTH
Hypothyroidism
Renal failure
Ectopic production
Lung cancer
Hypernephroma

Heterogeneity of PRL

In different physiological and pathological states, the ratio of biologically active PRL to immunoreactive PRL may vary considerably. Distinct forms of circulating PRL have been described (p. 196).

2. The Human Menstrual Cycle

(Figs. X.7 and X.8)

The menstrual cycle is the repeated expression of an integrated hypothalamic-hypophyseal-ovarian system, characterized by structural and functional changes in the target tissues of the reproductive tract (uterus, endometrium, vagina). Each cycle culminates in menstrual bleeding, and the first day of menstruation is an accepted reference point marking the beginning of a menstrual cycle.

In blood, *gonadotropins (LH and FSH), estradiol, estrone* and *progesterone* are measured by radioimmunoassays. In most cases, methods without a chromatographic step are used for the measurement of estrogens and progesterone. *17-hydroxyprogesterone* is also measured by radioimmunoassay; however, chromatographic separation is necessary in order to ensure specificity. In urine, the measurement of *pregnanediol* gives only a general idea of the secretion of progesterone. *Basal body temperature (BBT)*, which increases during the luteal phase, cervical mucus, vaginal smears and especially *endometrial biopsies* all permit an evaluation of hormone actions.

Women are endowed with a surplus of eggs (oocytes), arrested in the diplotene stage of the meiotic prophase, which are surrounded by a single layer of granulosa cells. The initial growth of these inactive primordial follicles to the early prenatal stage is independent of gonadotropins, and is a continuous process throughout approximately 30 years of ovarian function. Further growth and development

Figure X.7 Various phases of the human menstrual cycle. Day 1 represents the first day of menstrual flow (colored box). The follicular phase terminates with the onset of LH surge, and the luteal phase begins following ovulation.

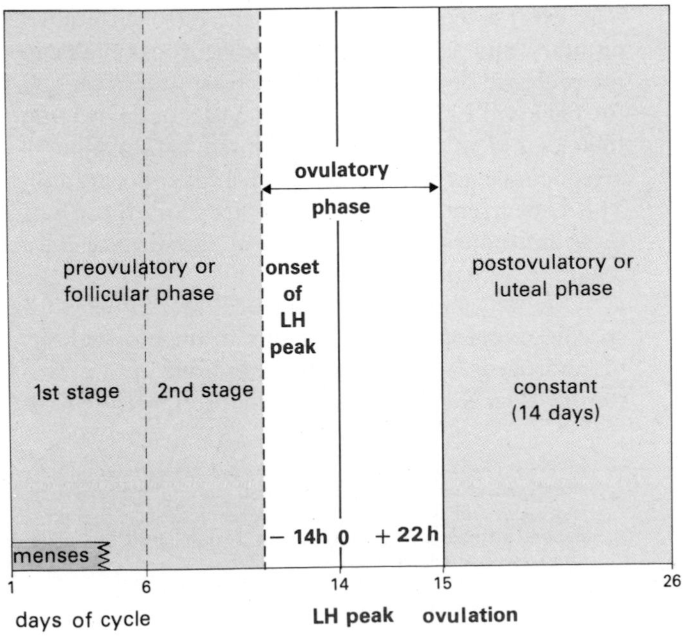

ovulatory phase
preovulatory or follicular phase
onset of LH peak
postovulatory or luteal phase
1st stage
2nd stage
constant (14 days)
– 14h 0 + 22h
menses
1
6
14
15
26
days of cycle
LH peak
ovulation

Figure X.8 The cyclic hormone pattern of the human reproductive cycle, centered on the day of LH peak (day 0). Basal body temperatures (BBT) are shown on top, and gonadotropin levels on the bottom. Ovarian steroids (estrogens, progestins, and androgens) are displayed in the center. Days of menstrual bleeding are indicated by colored boxes.

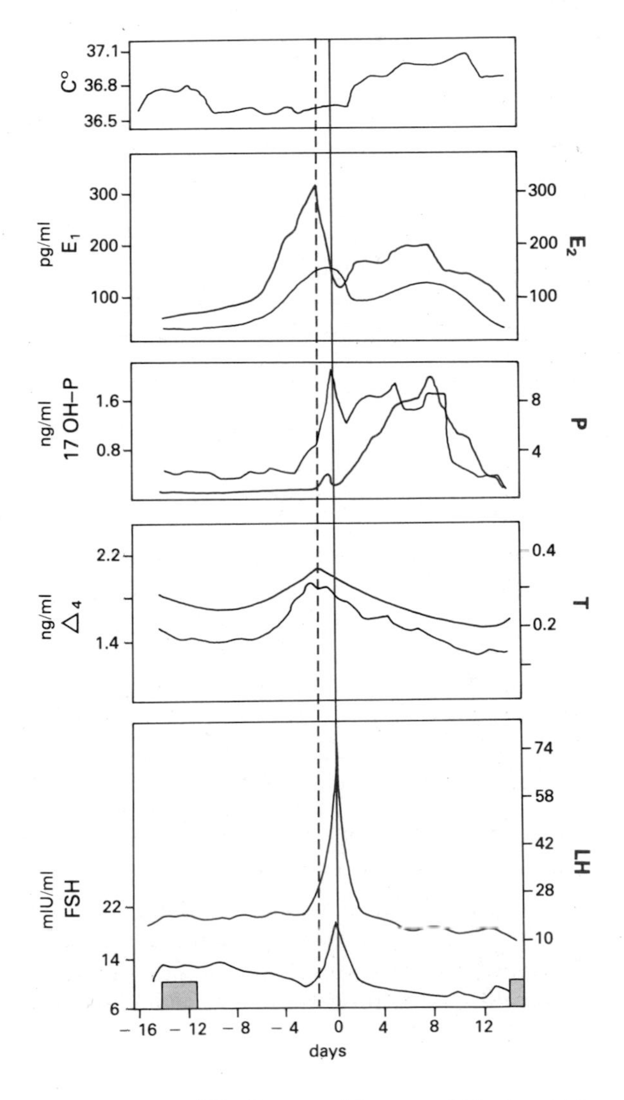

procedes with the recruitment of a cohort of follicles, the selection of a dominant follicle, and a phase of dominance by the preovulatory follicle. These events are associated with oocyte maturation and steroidogenesis, and are controlled by a coordinated interaction between gonadotropins and local ovarian factors.

The *target for FSH* is *exclusively granulosa cells*, while *LH* has *multiple target sites*, including theca, stroma, and luteal as well as granulosa cells. Ovarian follicular development involves synergistic interaction between gonadotropins (FSH-LH) and steroids (androgens-estrogens) operating on theca and granulosa cells through hormonally induced changes of specific receptors within a given crop of follicles. Figure X.9 illustrates the essential actions of the two gonadotropins (FSH and LH) on the two cell types (theca and granulosa) involved in the regulation of coordinated follicular growth and maturation, and in the provision of feedback signals for the modulation of gonadotropin output by the hypothalamo-pituitary unit.

FSH and granulosa cells. The granulosa cells of each of the follicles in the ovary possess receptors for FSH. Rising levels of FSH induce an increase in FSH receptor number, an early event in the orderly process of follicular growth. This increase in FSH receptor is reflective of an increased number of granulosa cells rather than an elevated number of receptors per granulosa cell. In addition, FSH induces granulosa cells to acquire *aromatizing enzymes* which enable the production of estradiol from androgens. Estradiol is able to increase the number of its own receptors, thereby exerting a direct mitogenic effect on granulosa cells, independent of that provided by FSH. Estradiol produces a paracrine effect which augments all the actions of FSH, including the generation of aromatase activity. This *synergistic* relationship between *FSH and estradiol* represents an intraovarian, autoregulatory, positive feedback mechanism responsible for stimulating rapid division of granulosa cells, thus promoting follicular growth (Fig. X.9).

Initially, FSH stimulates a progressive increase in FSH receptors, but has no effect on LH receptors. However, after exposure to increasing levels of endogenous estradiol, FSH receptor numbers increase more rapidly, followed by a delayed but pronounced induction of LH receptors. Thus, initially, estradiol promotes the ability of FSH to increase and maintain receptors for FSH, and, later, enhances the ability of FSH to increase and maintain LH receptors. The appearance of LH receptors in granulosa cells is responsible for progesterone production[6]. Pre-

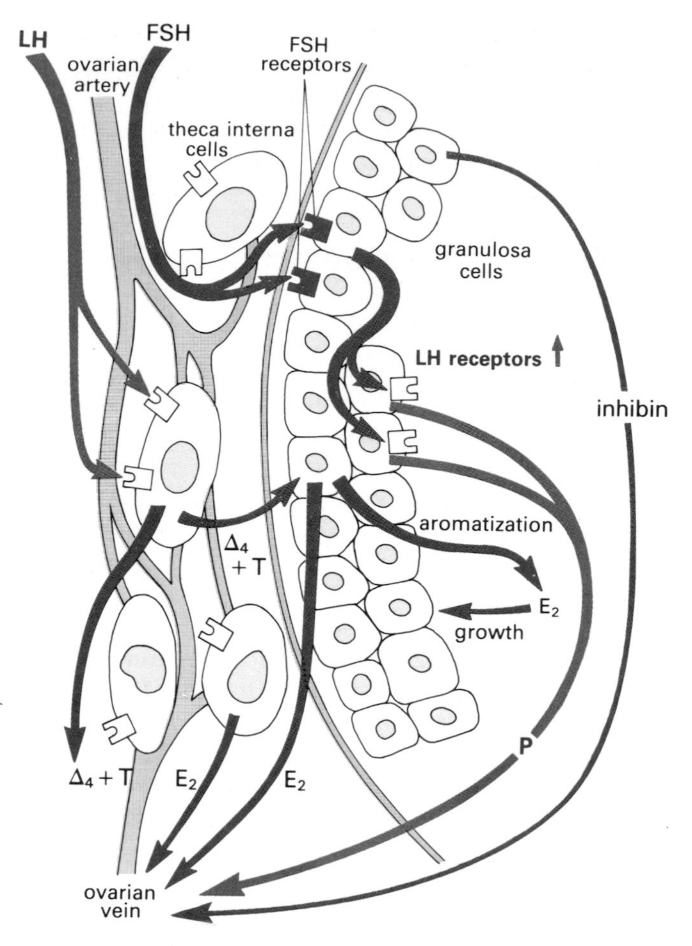

Figure X.9 Gonadotropin-ovarian interaction in the regulation of follicular maturation and steroidogenesis. E_2, estradiol; P, progesterone; T, testosterone; Δ_4, androstenedione.

ovulatory secretion of progesterone, although limited, may exert a feedback effect on the estrogen-primed pituitary and CNS-hypothalamic neurons. While in the presence of estradiol FSH increases the receptors for FSH and LH in the granulosa cells of developing follicles, LH, in association with luteinization, induces a reduction in receptors for FSH, LH, and estradiol. This concurrent decrease in receptors for all three of these hormones may have causal significance with regard to follicular atresia.

LH and theca cells. The theca interna develops specific receptors for LH, leading to the biosynthesis of androgens, mainly androstenedione and testosterone (Fig. X.9). Androgens produced by the thecal

6. McNatty, K. P., Makris, A., De Grazia, C., Osathanondh, R. and Ryan, K. H. (1979) The production of progesterone, androgens and estrogens by human granulosa cells, thecal tissue and stromal tissue from human ovaries in vitro. *J. Clin. Endocrinol. Metab.* 49: 687–699.

compartment diffuse into the follicular fluid and are aromatized to estrogens by granulosa cells. Formation of estrogens from androgens can also be achieved by theca cells, but the combination of theca and granulosa cells has been found to potentiate estrogen formation in vitro. Hence, the two-cell theory for estrogen formation has been sustained.

By the seventh day of the cycle, there is preferential uptake of gonadotropin only in the follicle emerging as morphologically dominant. By the ninth day, capillaries develop around the basal membrane, establishing a microcirculatory system which is twice as abundant in the dominant follicle as in other antral follicles. The increased amount of estrogen in the follicular fluid diffuses back to the perifollicular vascular network and into the systemic circulation. It is then delivered to multiple target sites, especially the hypothalamo-pituitary unit. The selected follicle also acquires the further advantage of an increased exposure to gonadotropic hormones provided by the enriched vascularity (Fig. X.9).

Ovarian inhibin. The presence of an inhibin-like substance has been demonstrated in follicular fluid of several species, including humans. The structure of inhibin from porcine follicular fluid has been determined from cDNA sequences (p. 29). It consists of two dissimilar subunits of $M_r \sim 18{,}000$, designated as the α subunit (134 aa) and the β subunit (116 aa) (Fig. X.10). There are two forms (A and B) of the β subunit, corresponding to the two forms, A and B, of inhibin. The β-type sequence has significant homology with human TGF-β, and it is tempting to speculate that inhibin may act as a growth regulator within gonadal tissue. Inhibin is a glycoprotein. Contrary to the pituitary glycoproteins that also contain one common and one variable chain, the α and β chains of inhibin A and B are coupled by disulfide linkage (p. 29). Granulosa cells, the ovarian counterpart of testicular Sertoli cells, are the sites of production of inhibin (as demonstrated in cultured human and monkey granulosa cells). Recent evidence suggests a progressive increase in intrafollicular inhibin during the course of follicular maturation, with levels higher than during the luteal phase. Intrafollicular progesterone appears to inhibit, whereas androgens seem to stimulate, the secretion of inhibin by granulosa cells.

Follicular inhibin diffuses into the surrounding microcirculation, thereby exerting a long-loop feedback effect at the pituitary level, *selectively inhibiting FSH secretion.* Thus, some of the existing questions concerning the differential regulation of LH and FSH secretion may be resolved. It seems physiologically appropriate for a specific target (granulosa cells) to provide a selective signal in the feedback modulation of its own tropic hormone (FSH) secretion. Consequently, a feedback control system of the granulosa cell-FSH axis may operate through both estrogen and follicular inhibin. In addition, intragonadal inhibin exerts paracrine action by way of enhancing the LH-mediated androgen production of theca interna cells. Thus, inhibin may link the granulosa (FSH-inhibin) and theca (inhibin-LH-androgen) systems.

A recent finding indicates that *at least one other inhibin-related peptide* has hormonal activity; it is also a S-S linked dimer consisting of two inhibin β

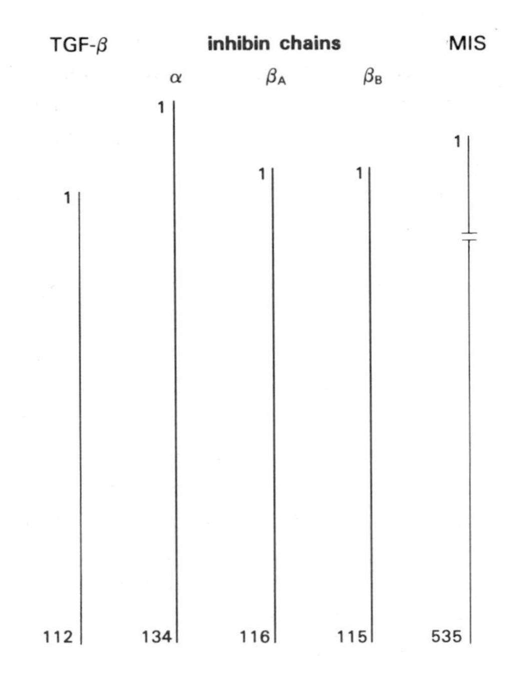

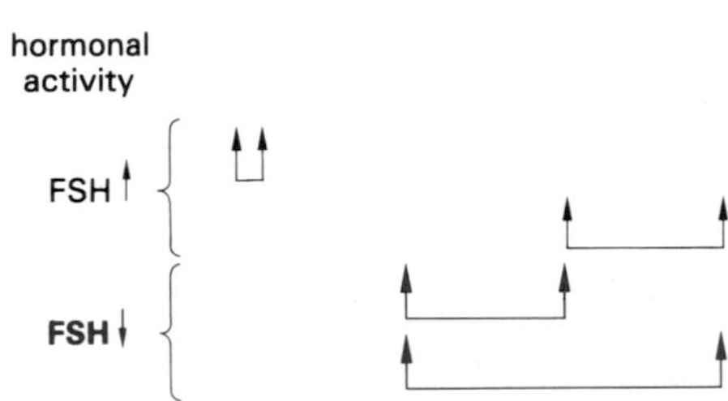

Figure X.10 Human proteins affecting pituitary FSH secretion. The inhibin chains and TGF-β are part of the same family, showing 70% homology in paired comparison. They are all disulfide-linked glycoproteins. Inhibin A is $\alpha+\beta_A$ and inhibin B is $\alpha+\beta_B$. Activin, or FRP (FSH releasing protein), is formed of two β chains. TGF-β (p. 219) is a homodimer with FSH-releasing activity. The C-terminal portion of MIS (p. 441) also shows homology with these proteins.

chains, and has potent FSH-stimulating activity (activin, or FSH-releasing peptide (FRP)). It does not stimulate LH or any other pituitary hormone. In fact, there are probably several dimeric proteins formed by combination of the β_A and β_B chains. The homodimer TGF-β itself (p. 219) has FSH-releasing activity.

2.1 Ovulation: The Preovulatory Follicle

Concomitant with an accelerated production of estrogen, the *preovulatory follicle* is *associated with dramatic changes*: 1. shifting of steroidogenesis in granulosa cells, from estrogen to progestin; 2. luteinization of granulosa cells; 3. resumption of meiosis (oocyte maturation); and 4. acquisition of biochemically differentiated granulosa-luteal cells with properties that participate in the mechanical aspects of ovulation. These interacting processes occur immediately before and during the 48 h span of the midcycle gonadotropin surge[7].

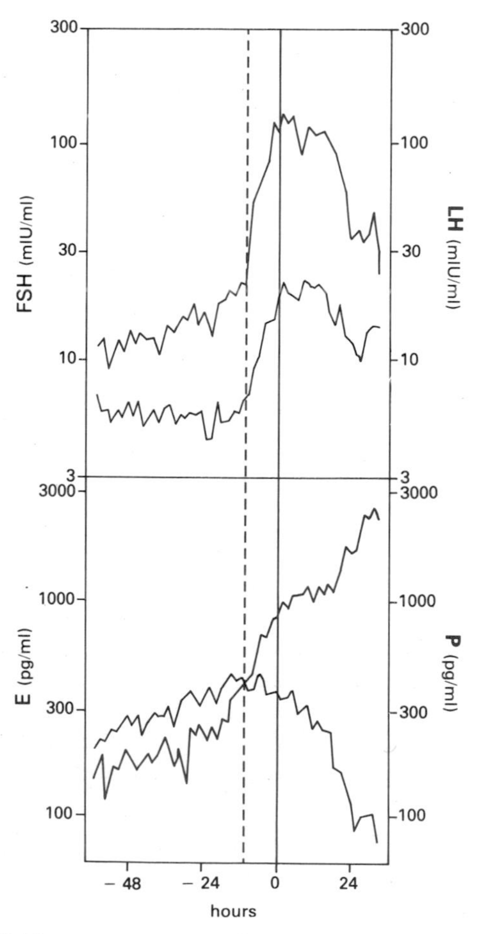

Figure X.11 LH, FSH, E, and P measured every 2 h for 5 d at midcycle in five women. Data are centered around the initiation of the gonadotropin surge. Note that the data are plotted on a logarithmic scale[6].

The *LH and FSH surges begin abruptly* (LH doubles within 2 h), and are *triggered by* the attainment of *peak estradiol levels* and the initiation of a rapid *rise of progesterone* 12 h earlier (Fig. X.11).

Ovulation (Fig. X.12)

Following the initiation of the LH surge, tissue concentrations of cAMP increase in the preovulatory follicle. A chain of events is activated by the LH-induced increase in cAMP; this includes oocyte maturation and luteinization of granulosa cells, with a parallel increase of the production of progesterone and prostaglandins[8]. Furthermore, cAMP and/or progesterone production may activate the proteolytic enzymes collagenase and plasmin, resulting in the digestion of collagen in the follicular wall, thus causing an increase in its distensibility.

The concentration of PGs of the F and E series increases markedly in preovulatory follicles or after exposure to high levels of LH, and are highest at ovulation. PGs appear to be involved in the induction of follicle rupture, in concert with actions of tissue plasminogen activator and collagenase on the follicle wall. Although the precise mechanism remains unclear, PGs and intrafollicular oxytocin may act synergistically to stimulate smooth muscle contraction, thereby facilitating the extrusion of the oocyte-cumulus mass[9].

2.2 The Corpus Luteum

When the luteinization of the newly ruptured follicle is complete, corpus luteum function is established. The *maximal activity* of the corpus luteum is attained by *seven to eight days* after the LH surge, and functional demise, that is, *luteolysis*, occurs two to three days before the onset of menses.

The most important endocrine feature of the luteal phase of the menstrual cycle is the marked

7. Hoff, J. D., Quigley, M. E. and Yen, S. S. C. (1983) Hormonal dynamics at midcycle: a reevaluation. *J. Clin. Endocrinol. Metab.* 57: 792–796.
8. LeMaire, W. H., Leidner, R. and Marsh, J. M. (1975) Pre- and post-ovulatory changes in the concentration of prostaglandins in rat Graafian follicles. *Prostaglandins* 9: 221–229.
9. Schaeffer, J. M., Liu, J., Hsueh, A. J. W. and Yen, S. S. C. (1984) Presence of oxytocin and arginine-vasopressin in human ovary, oviduct and follicular fluid. *J. Clin. Endocrinol. Metab.* 59: 970–973.

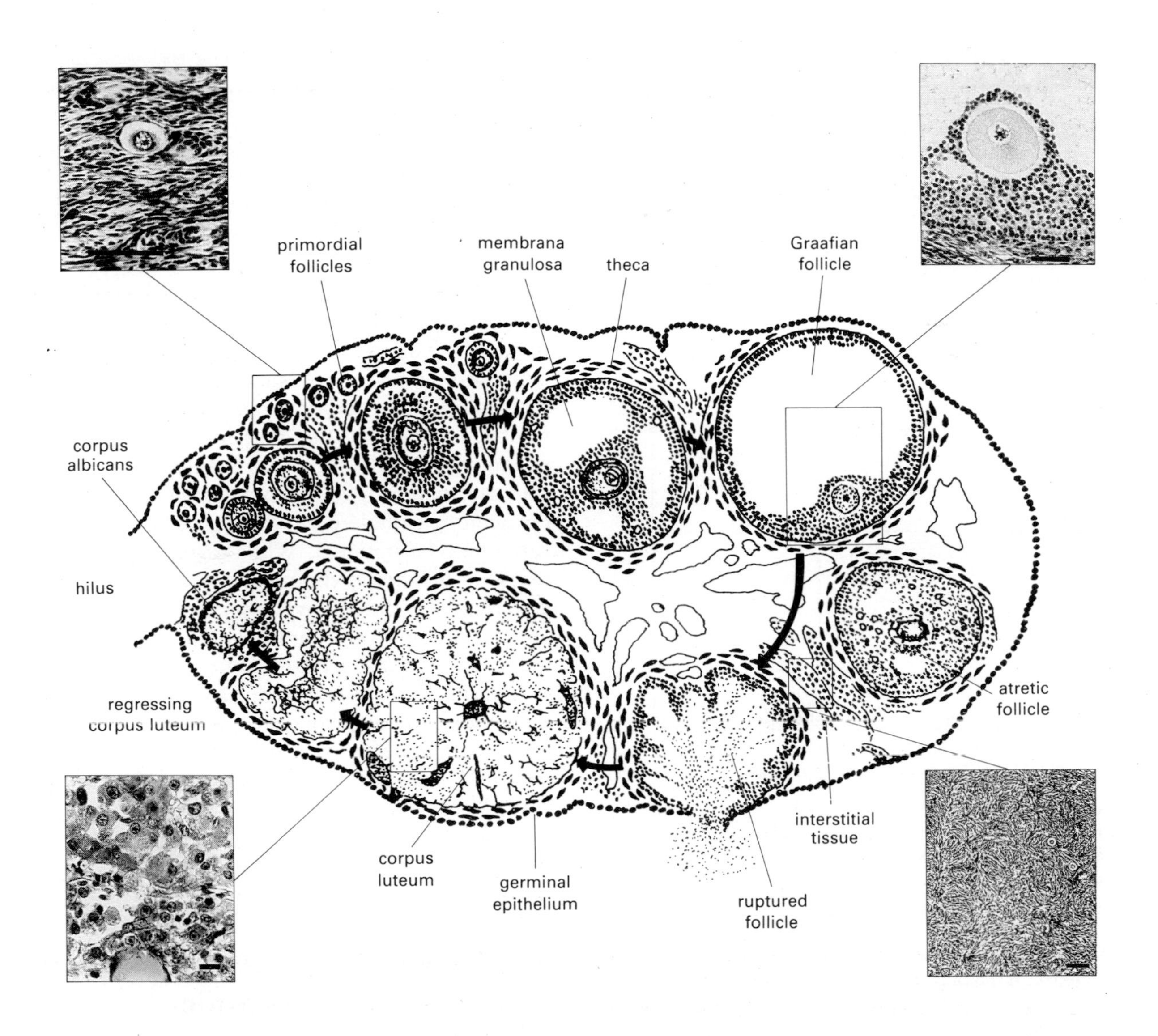

Figure X.12 Diagram of the human ovary, showing the development of a primordial follicle, to the stage of ovulation, after which a corpus luteum forms and ultimately regresses. On the right side of the ovary, the more common development of a follicle is indicated, terminating in atresia. In the four boxes, actual histological sections are shown. Bars=50 μm, except for interstitial tissue: bar=100 μm. (Courtesy of Dr. A. Gougeon, Hôpital de Clamart).

increase in *progesterone secretion* by the corpus luteum. The elevation of basal body temperature (BBT) is temporally related to a central effect of progesterone (Fig. X.8). *Progesterone* output reaches a maximum about eight days after the midcycle LH peak, with a *daily production rate of 25 mg*. There is a *parallel but smaller increase* in *17-hydroxyprogesterone, estradiol* and estrone levels. As *progestins* and *estrogens decrease at the end* of the luteal phase (luteolysis), plasma *LH* and *FSH* levels *begin* their *increase*, initiating follicular growth *for* the *next cycle* (Fig. X.13).

2.3 Integrated Neuroendocrine Control of the Menstrual Cycle

The *change in the pulsatile pattern* of gonadotropin secretion by the pituitary (Fig. X.14) is a direct result of *episodic GnRH release* from nerve endings in the *arcuate median eminence* region, in association with the effect of ovarian steroids. This intermittent delivery of endogenous GnRH appears to be a consequence of a yet undefined *signal oscillator* that triggers a periodic discharge of the neurohormone from GnRH-synthesizing neurons; neighboring opioid pep-

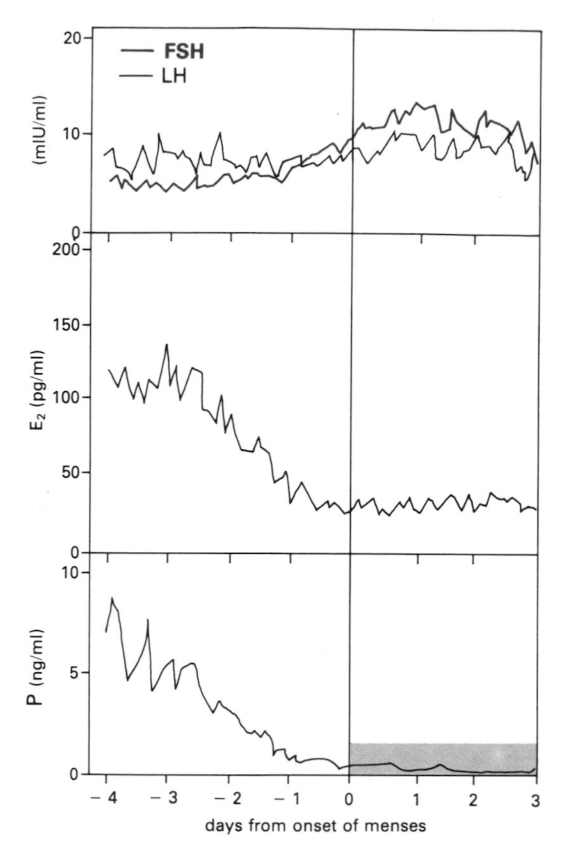

Figure X.13 Hormonal changes during luteolysis and the initiation of the next round of follicular development. Note the rise of FSH beginning one day before the onset of menstrual bleeding.

tide neurons appear to have a tonic inhibitory effect on this release. Dissociation of this interacting system occurs in a low ovarian steroid milieu. Thus, it is likely that *opioid inhibition of GnRH*, and consequently the low frequency of LH pulses seen during the luteal phase, becomes uncoupled following luteolysis; the result is an increased GnRH-gonadotropin pulse occa-

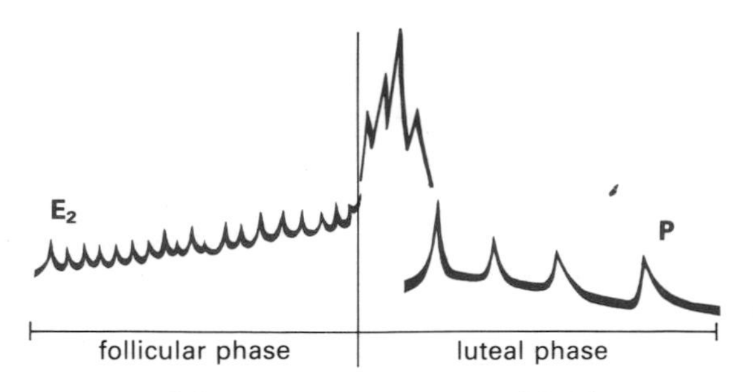

Figure X.14 Schematic representation of the changing pattern of pulsatile LH release during the follicular (estrogen-dominated) and luteal (progesterone-dominated) phases of the menstrual cycle.

sioned by the disinhibition of opioidergic influence, thereby initiating folliculogenesis.

The frequency and amplitude of GnRH pulses are crucial stimuli for the synthesis and secretion of gonadotropic hormones by the pituitary. The expression of such a stimulus-response is determined by *positive autoregulation* of *GnRH receptors by GnRH (self-priming)*. At the same time, the prevailing estradiol levels exert a negative feedback effect on the release of gonadotropin by the pituitary. In combination, they induce a remarkable increase in pituitary gonadotropin reserve. When estradiol levels exceed a threshold for a period of two to three days, a positive feedback effect on the functional activity of the gonadotropic cells occurs, as manifested by a rapid shift of gonadotropins, from the large reserve pool, to the acute releasable pool, through which a midcycle surge may be initiated. While *estradiol triggers the onset of the surge*, the increased secretion of progesterone by the preovulatory follicle (Fig. X.11) appears to amplify the duration of the surge. Although the site of *feedback action* is *principally* at the level of the *pituitary*, there is cogent argument for a hypothalamic site of estradiol action as well. *Progesterone*, in contrast, appears to *exert its feedback* effect *on* the *neuronal network* that *reduces* the *frequency* of GnRH *secretion*. However, the above observations have not yet led to a full understanding of the actual *physiological* significance of the acute GnRH surge in the induction of midcycle gonadotropin release.

3. Abnormalities of the Menstrual Cycle: Hypothalamic-Hypophyseal Dysfunction

There are functional and organic causes which account for dysfunction of the hypothalamic-hypophyseal-ovarian system and, thus, alteration of menstrual cyclicity.

Although the biochemical mechanisms of information exchange among neurons are far from clear, the influence of social and environmental stresses, lifestyles, and physiological factors on reproductive function is well recognized in lower animals as well as humans, and justifies the emerging field of *psychoneuroendocrinology*.

Neuroendocrine abnormalities resulting from organic disease can be traced to anatomic and functional derangements in a given region of the brain. Thus, hypothalamic lesions may cause varying degrees of hypopituitarism, depending upon the site and

Table X.3 Pathophysiological basis of chronic anovulation secondary to dysfunction of the CNS-hypothalamic-hypophyseal system.

Abnormalities in CNS-hypothalamus interaction
Physiological
Initiation of puberty
Postpartum amenorrhea
Neuropharmacological
Interference of neurotransmitters
Exogenous opiates
Psychoneuroendocrinological (psychogenic or functional amenorrhea)
Pseudocyesis
Persistent corpus luteum syndrome
Anorexia nervosa
Hypothalamic chronic anovulation (psychogenic, nutritional, exercise)
Defects within hypothalamo-pituitary unit
Galactorrhea-amenorrhea syndrome
Hypothalamic lesions, primary and secondary
Isolated hypothalamic GnRH deficiency
Interruption of vascular linkage (portal circulation) of the system (Sheehan's syndrome)
Cellular and anatomic defects of the gonadotropic cells
Isolated gonadotropin deficiency
Primary pituitary tumors
Empty sella syndrome

extent of the lesion. Regardless of the precise site or nature of the lesion, its ultimate expression is the *inadequate secretion of gonadotropin*, which results secondarily in chronic anovulation or amenorrhea. The term *chronic anovulation* implies the absence of intrinsic abnormalities of the ovary and the potential reversal to normal function, through appropriate gonadotropin stimulation (in both timing and quantity).

Based on these premises, chronic anovulation resulting from dysfunction of the hypothalamo-pituitary unit can be divided into two broad categories (Table X.3). The various sites of abnormalities in which several clinical syndromes are manifested are diagrammed in Figure X.15.

3.1 Physiological and Pharmacological Models

Puberty

The *pivotal role of the CNS* in the control of cyclic ovarian function is illustrated by two successive

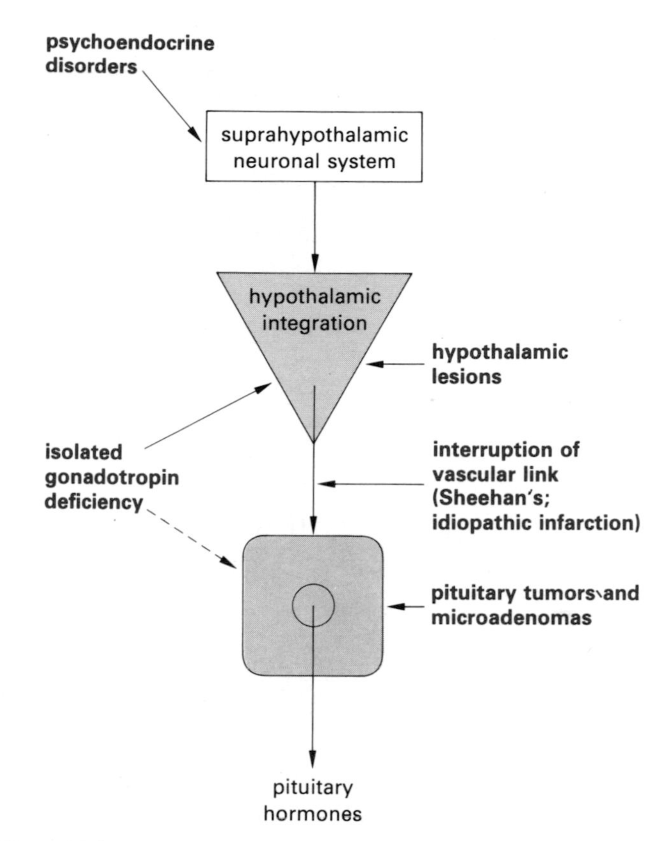

Figure X.15 Sites of abnormalities of chronic anovulation.

neuroendocrine patterns in the developing child. First, there is *central inhibition* of hypothalamic GnRH neuron activity, and secondly, gradual activation (disappearance of inhibition) of hypothalamic GnRH secretion, from complete *inactivity to sleep-entrained pulsatile release*[10]. The inhibition of the neural control system constitutes the *only limiting factor* in the initiation of puberty, since the *pituitary* and *ovaries* are *fully competent* in *prepubertal girls*, and since they can be activated by the premature onset of pulsatile GnRH secretion, as in central precocious puberty (Fig. X.16).

Postpartum Amenorrhea

Reinitiation of menstrual cyclicity after a full-term pregnancy appears to follow a neuroendocrine pattern *resembling* that of puberty[11]. Thus, the concept of "miniature puberty", with a time course of several

10. Boyar, R., Finkelstein, J., Roffwarg, H., Kapen, S., Weitzman, E. and Hallmann, L. (1972) Synchronization of augmented luteinizing hormone secretion with sleep during puberty. *N. Engl. J. Med.* 287: 582–586.
11. Liu, J.H., Rebar, R.W. and Yen, S.S.C. (1983) Neuroendocrine control of the postpartum period. *Clin. Perinatal.* 10: 723–736.

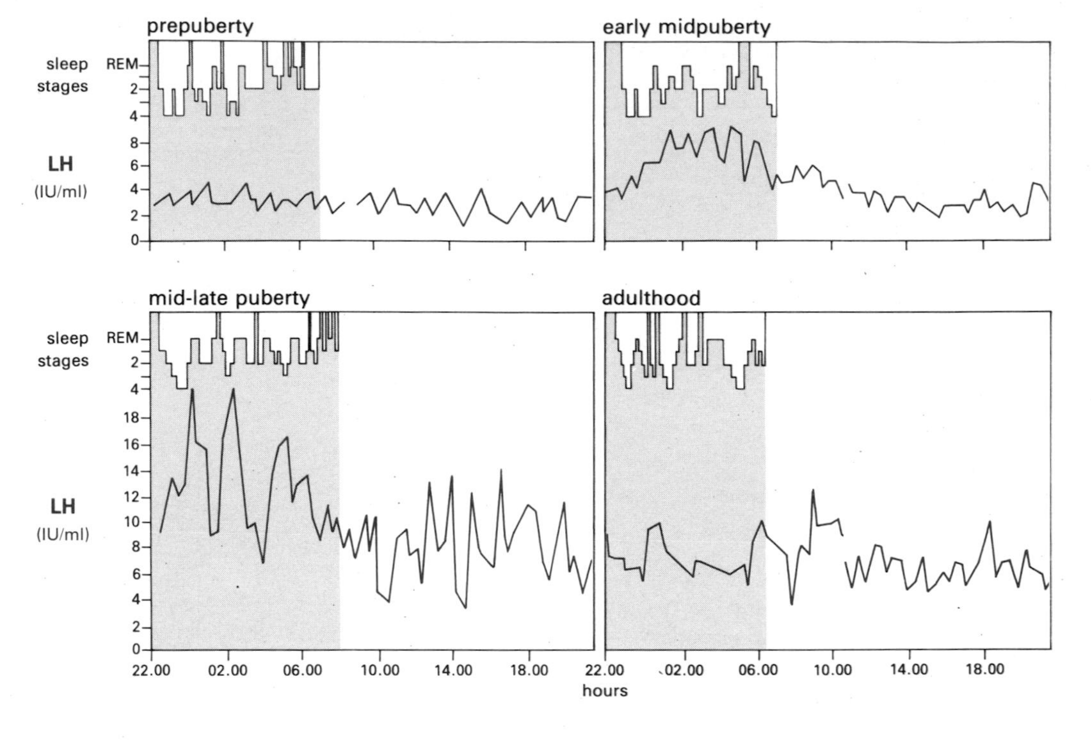

Figure X.16 Pulsatility of LH and sleep-entrained increments associated with pubertal development in a girl.

weeks instead of several years, has been advanced. In conditions in which pulsaltile GnRH secretion is disturbed, due either to a reduced stimulatory input or to inhibitory activity, hypogonadotropinism develops, resembling that of the prepubertal state.

Neuropharmacological Model

In view of the known involvement of catecholaminergic and opioidergic neuronal input in the hypothalamic GnRH secreting mechanism, one might anticipate that compounds interfering with catecholamine synthesis, metabolism and re-uptake, as well as with receptor agonists or antagonists, would result in disturbances of gonadotropin release and hence cause anovulation. *Reserpine*, commonly used as an antihypertensive agent, is frequently associated with amenorrhea or anovulatory cycles because of its catechol-depleting effect. The prolonged use of *tranquilizers*, especially phenothiazine, a dopamine-receptor blocker, represents another common cause of amenorrhea and chronic anovulation. *Opiate* addiction, also, is known to disturb menstrual cyclicity.

3.2 Amenorrhea Due to Hypothalamic-Hypophyseal Disorders

Pseudocyesis

Pseudocyesis, commonly known as "phantom pregnancy", is a classical example of the dominating role of mind, mood and behavior in the control of reproductive function.

The psychoneuroendocrine basis of this syndrome is supported by the association of an *increased pituitary secretion* of LH and prolactin sufficient to maintain corpus luteum function and lactation. The immediate fall of these hormones after the diagnosis is revealed to patients represents a most remarkable example of psychoneuroendocrine interaction.

Anorexia Nervosa

Anorexia nervosa is an extraordinary example of central causes of amenorrhea. These patients have *reduced gonadotropin* secretion, *similar* to that of *prepubertal children*. During remission, the sleep-associated pulsatile secretion of LH reappears. (Fig. X.17).

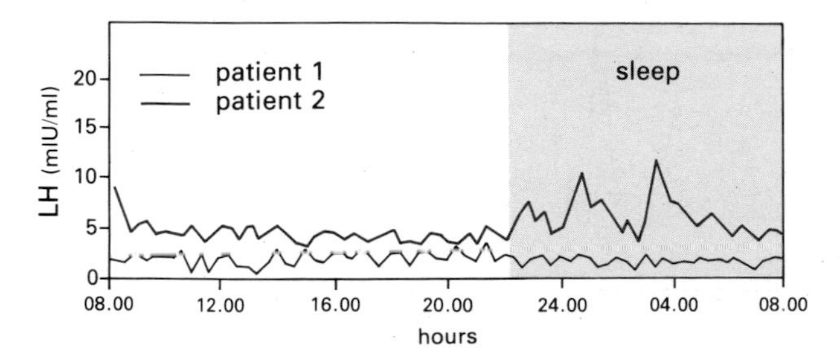

Figure X.17 Representative patterns of pulsatile LH secretion over a 24 h period in two patients with anorexia nervosa. Complete cessation of LH secretion and the absence of sleep-associated activation of GnRH pulses in one patient (top) is contrasted with the presence of small pulses with sleep-entrained amplification in another patient (bottom).

Neuroendocrine abnormalities, in addition to GnRH secretory dysfunction, include[12]: 1. *hypersecretion of cortisol* throughout the 24 h period, with maintenance of circadian rhythmicity. Elevated cortisol levels are due to increased secretory activity; following administration of dexamethasone, they are either incompletely suppressed or exhibit an early escape from suppression, a finding resembling that seen in patients with depression, and in Cushing's disease; 2. *low T_3 syndrome*: a selective decrease in serum T_3 levels, similar to that seen in association with starvation, occurs in anovulatory patients. This is due to impaired peripheral deiodination of T_4 to T_3, with a shift to the formation of metabolically inactive reverse T_3. The setting of malnutrition in association with a reduction of thyroid hormone action may serve a "protective hypometabolic" function, for survival in the face of a catabolic state; 3. *diabetes insipidus and abnormal thermoregulation* are additional features of hypothalamic dysfunction.

Thus, *anovulation* is a *multidimensional disorder* with a wide *variety* of *neuroendocrine-metabolic dysfunctions* and behavioral abnormalities (i.e. eating disorders and disruption of perceptual-cognitive processes).

Psychogenic (Hypothalamic) Amenorrhea

Cessation of the menstrual cycle in young women with no demonstrable organic endocrine abnormality represents one of the most common types of amenorrhea, the so-called '*college amenorrhea*'.

12. Meckelburg, R.S., Loriaux, D.L., Thompson, R.H., Anderson, A.E. and Lipsett, M.B. (1974) Hypothalamic dysfunction in patients with anorexia nervosa. *Medicine* 53: 147–159.

The underlying cause of this type of amenorrhea is social-environmental stress of sufficient intensity and duration. Vulnerability to the disorder is determined by predisposing factors and the adaptability of the individual. Failure to cope results in decompensation and subclinical depression. The cessation of menses represents nature's device to prevent the exposure of young to an unfavorable environment, when the potential mother is preoccupied with constant adjustment for her own survival. Indeed, the two silent features described below indicate that psychogenic amenorrhea is a stress syndrome.

Hypothalamic GnRH dysfunction. The degree of impairment of GnRH secretion varies widely, as evidenced by the diversity of pulsatile LH activity. The spectrum of abnormalities may very well reflect a pathophysiological continuum of this disorder. In severe cases, when few, quasi-pulses of LH are observed, ovarian activity virtually ceases, as demonstrated by markedly reduced estradiol, androstenedione and testosterone levels. Thus, the entire hypothalamic-hypophyseal-ovarian system regresses functionally to the prepubertal state (Fig. X.18). Under these conditions, progesterone-withdrawal bleeding will not occur, and treatment with clomiphene citrate would probably fail to induce cyclic ovarian function. In contrast, patients with substantial amplitudes but significantly reduced numbers of gonadotropin pulses, as compared to those of normal women during the follicular phases of the cycle, have a modest degree of ovarian estradiol secretion and normal levels of androgens (Fig. X.18). With time, spontaneous menstruation frequently ensues in these cases.

Hypercorticism. Abnormalities of cortisol secretion have been found recently in patients with psychogenic amenorrhea; a unique pattern of hypersecretion of cortisol, with a remarkable increase in the amplitude of secretory episodes, occurs only during the day hours (08.00–16.00 h) (Fig. X.19). Thus, the significant increase in 24 h mean cortisol secretion in these patients is due primarily to the selective daytime hypersecretion, without alteration in nocturnal secretory episodes. These observations indicate that an inappropriate activation of the CRF-ACTH-cortisol system occurs, which is consistent with the presence of identifiable stress and/or depression in patients with psychogenic amenorrhea. Further studies are needed for a full exploration of the significance of this abnormality.

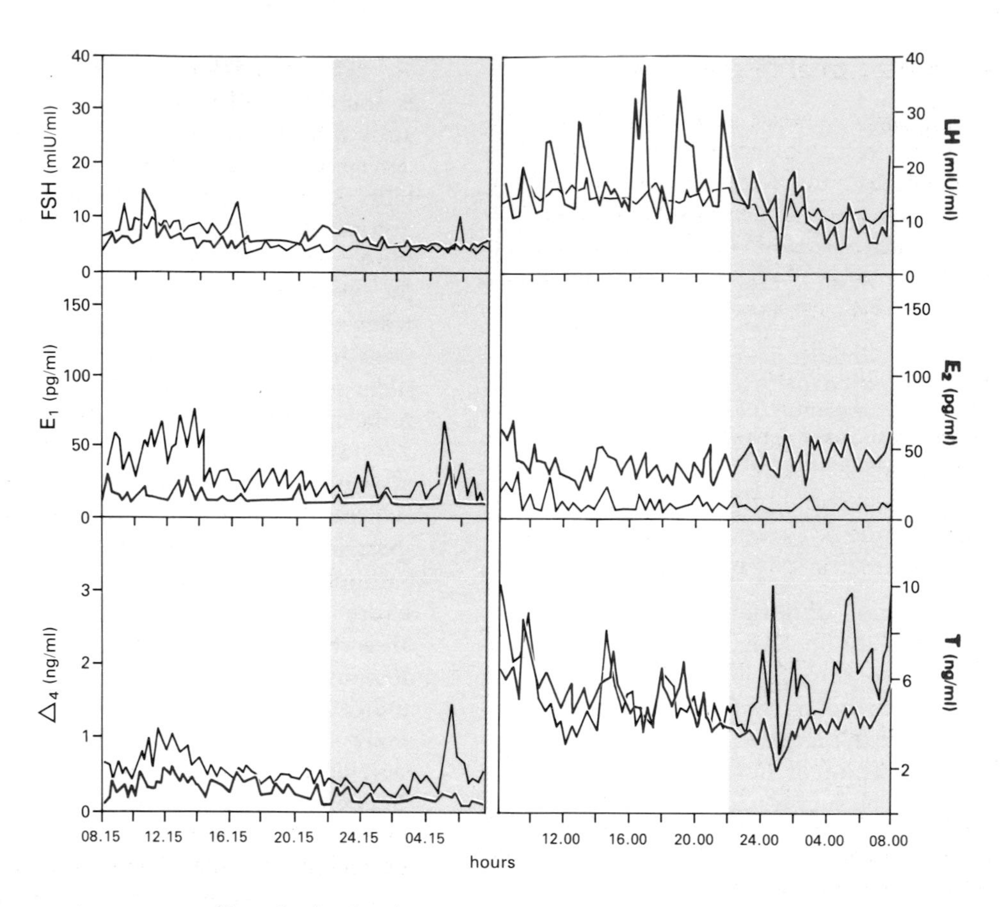

Figure X.18 Representative patterns of impaired pulsatile LH secretion and corresponding ovarian estrogen and androgen levels, determined at 15 min intervals over a 24 h period, in a severe and a mild case of psychogenic amenorrhea. (Drs. B. Suh, J. Liu, R. Kazer and S. S. C. Yen, University of California, San Diego).

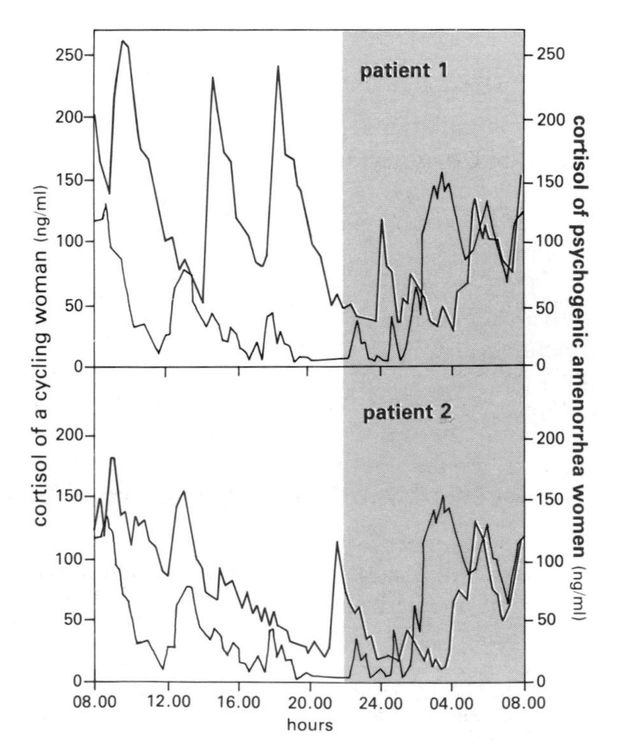

Figure X.19 Concentrations and episodic secretory activities of cortisol, studied over 24 h in two patients with psychogenic-amenorrhea, as compared to matched-normal women, during the early follicular phase of the cycle (same authors as Fig. X.18).

Exercise and Menstrual Dysfunction

Is jogging safe? Exercise appears to induce progressive dysfunction of ovarian cyclicity, including luteal phase defects, anovulatory cycles with amenorrhea[13], and delayed menarche in prepubertal girls.

Abnormalities of hypothalamic GnRH secretion, as reflected by pulsatile LH frequency and amplitude, have been observed in patients with amenorrhea caused by strenuous exercise. A spectrum of reduced frequency and/or amplitude of LH pulses is found, resembling those described in patients with luteal phase defects and hypogonadotropic amenorrhea. Defects in pulsatile LH release have also been observed in runners with normal menstrual cycles, in whom both pulsatile LH frequency and amplitude are significantly reduced in comparison with sedentary controls[14].

Thus, high intensity endurance exercise exerts a a central inhibitory effect on hypothalamic GnRH secretion; it is a reversible disorder, and recovery seems to follow a progressive and time-related increase in pulsatile GnRH activity. However, the neuroendocrine link between exercise-induced menstrual disorders and the inhibition of GnRH secretion remains to be determined. There is no evidence that endogenous hypothalamic opioid peptides are involved; changes in insulin secretion, related to fat redistribution, have been suggested as a possible cause.

3.3 Adrenal and Thyroid Dysfunction

Adrenal Dysfunction

Amenorrhea or oligomenorrhea (menstrual irregularity) is observed in the majority of premenopausal women with Cushing's syndrome. Not infrequently, these patients seek medical attention because of menstrual disorders and infertility. The precise causes of chronic anovulation and irregularity or absence of menstruation in Cushing's syndrome are not clear. Based on indirect evidence, several possible mechanisms may operate separately or together. When there is *dysfunction of GnRH secretion*, the frequency and amplitude of pulsatile LH secretion is reduced, and thus the secretion of GnRH appears to be inhibited. The central mechanism in the genesis of Cushing's disease appears to involve an increased CRF and serotoninergic input, which may impinge on the GnRH secretory function. Both in vivo and in vitro, CRF has been shown to exert an inhibition on hypothalamic GnRH secretion, and serotonin exerts an inhibitory effect on gonadotropin secretion. There is also increased peripheral production of estrogen. The varying degree of adrenal androgen excess in Cushing's syndrome of all types, together with a reduced SBP and increased fat reserves, may contribute significantly to the metabolism of sex steroids and to the excessive extraglandular production of estrogen from androgen.

Organ-sepecific autoantibodies directed against steroid-producing cells of the adrenal and gonads, as well as the placenta, have been found in Addison's disease, and may account for a high incidence of menstrual disorders (amenorrhea or anovulatory cycles). In addition to association with premature ovarian failure, a host of other endocrinopathies are frequently found: hyper- and hypothyroidism, Hashimoto's thyroiditis, hypoparathyroidism, and pernicious anemia. These combinations have been referred to as *autoimmune polyglandular failure* syndrome.

An autoimmune basis for *premature ovarian failure* includes the demonstration of circulating antibodies to ovarian tissue. In addition, the presence of lymphocytic infiltrates, with or without nests of plasma cells, is a feature denoting the autoimmune basis of ovarian failure. In ovarian dysfunction caused by autoimmunity, oocytes are either reduced in number or absent. Moreover, antigonadotropin receptor antibodies may also be present.

Thus, the presence of clinical and laboratory evidence of premature ovarian failure should alert the physician to search for other endocrine organ failures due to autoimmune disorders.

Thyroid Dysfunction

Both hyper- and hypothyroidism are associated with a variety of menstrual disorders, ranging from excessively prolonged uterine bleeding to complete cessation of menses. Both thyroid hormone deficiency and thyroid hormone excess induce significant changes in SBP (an increase in hyperthroidism and a decrease in hypothyroidism) and in the metabolism and interconversion of androgens and estrogens,

13. Bullen, B. A., Skrinar, G. S., Beitins, I. Z, von Mering, G., Turnbull, B. A. and McArthur, J. W. (1985) Induction of menstrual disorders in untrained women by strenuous exercise. *N. Engl. J. Med.* 312: 1349–1353.
14. Cumming, D. C., Vickovic, M. M., Wall, S. R. and Fluker, M. R. (1985) Defects in pulsatile LH release in normally menstruating runners. *J. Clin. Endocrinol. Metab.* 60: 810–812.

resulting in chronic anovulation with or without dysfunctional uterine bleeding.

3.4 Organic Hypothalamic-Hypophyseal Defects

Isolated Gonadotropin Deficiency

A syndrome involving hypogonadotropic hypogonadism associated *with anosmia* or hyposmia was described by Kallmann and associates in 1944. This syndrome occurs more frequently in males than in females. Anosmia is in fact very rare in women. However, many patients with isolated gonadotropin deficiency do not have anosmia or other distinguishable features, with the exception that responsiveness to GnRH is markedly impaired. The pathogenesis of the hypogonadotropinism in this syndrome has been the subject of much speculation. It is now recognized that there is a high degree of clinical, biochemical, developmental and genetic heterogeneity associated with it.

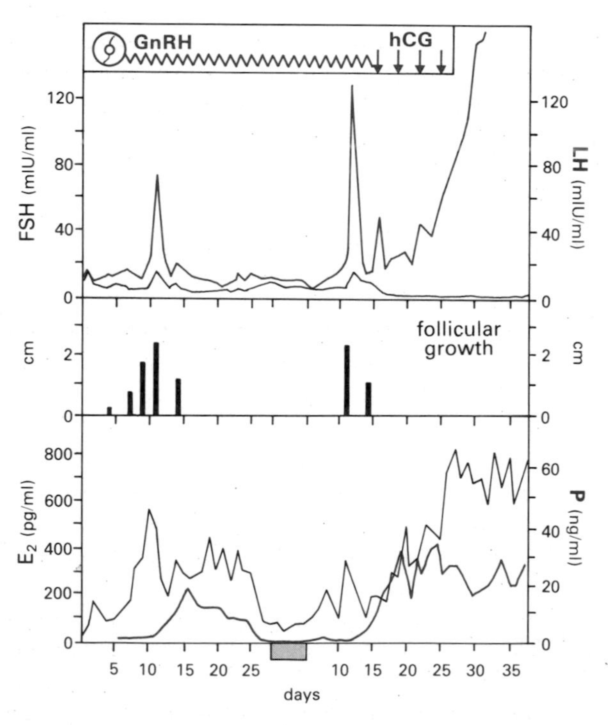

Figure X.20 Activation of cyclic gonadotropin-ovarian function in a patient with isolated gonadotropin deficiency by treatment with 1 μg/pulse of GnRH at 96 min intervals: ultrasonographic evidence of follicular development (measured in cm) and ovulation, which resulted in pregnancy during the second cycle of treatment; hCG, at a dose of 1500 IU, was injected, as indicated by arrows.

The primary site of the defect appears to be at the hypothalamic GnRH level; occasionally, it can occur at the level of gonadotropic cells. A full spectrum of GnRH secretory abnormalities have been observed. Activation of cyclic LH activity may be obtained by pulsatile administration of GnRH (Fig. X.20). There are also cases with nearly normal patterns of GnRH secretion, which may be explained by receptor defects and/or altered synthesis of gonadotropins in gonadotropic cells (e.g., preferential synthesis of an α subunit).

Hyperprolactinemia with or without Pituitary Adenoma

Hyperprolactinemia, a common condition, is frequently associated with amenorrhea (p. 213). The immediate cause of inactive hypophyseal-ovarian function is the absence of the reduced frequency of GnRH secretion. The mechanism accounting for the inhibition of GnRH neuron activity by hyperprolactinemia is illustrated in Figure X. 21.

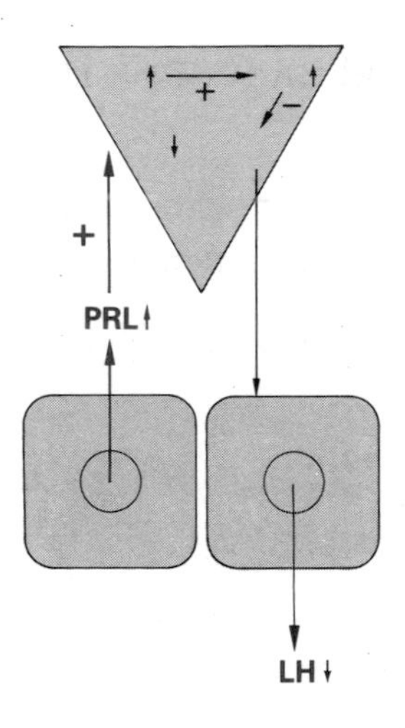

Figure X.21 An increased production of pituitary prolactin induces accelerated dopaminergic activity in the arcuate nucleus via short loop feedback. Dopamine, in turn, stimulates the activity of the opioidergic system, which inhibits GnRH and thus LH secretion.

Sheehan's Syndrome

Sheehan's syndrome is due to *infarction* of the anterior pituitary gland *secondary* to *postpartum hemorrhage* and hypotension, *resulting in panhypopituitarism*. The neurohypophysis is usually spared, but may be involved in severe cases, accompanied occasionally by diabetes insipidus. Clinical manifestations of hypopituitarism in patients surviving the period of postpartum shock are rapid and dramatic. Early mammary involution and failure to lactate are the earliest signs; fatigue, loss of vigor and hypotension are common findings during the puerperium, which are followed by loss of pubic and axillary hair and other features common to hypopituitarism. Failure to establish the diagnosis and promptly institute replacement therapy may be lethal.

The *pathophysiological basis* for pituitary necrosis in some patients with postpartum hemorrhage is *not* entirely *clear*. The anterior pituitary gland (but not the posterior gland) nearly doubles in size during the course of pregnancy (from 500 to 1000 mg), owing primarily to the hypertrophy and hyperplasia of the lactotropic cells. The enlarged gland may be vulnerable to ischemia resulting from postpartum hemorrhage, with attendant hypotension and shock. In addition, coagulation abnormalities encountered frequently during pregnancy may be important predisposing factors.

3.5 Peripheral Endocrine Disorders

Abnormal sex steroid signals to the hypothalamic-hypophyseal system result frequently in menstrual disturbances. The concentration and nature of steroids delivered to target organs depends in part on their metabolism and binding to plasma proteins.

For example, several abnormalities of reproductive function, including precocious puberty, amenorrhea, and irregular menstrual bleeding, can result from an *inappropriate extraglandular contribution* of *estrogens* derived through the peripheral conversion of androgens. Due to this additional estrogen production (not under direct control of pituitary gonadotropin), the cyclic feedback signal is masked, resulting in acyclic gonadotropin release and chronic anovulation, despite the absence of intrinsic abnormalities within the hypothalamic-hypophyseal ovarian system.

Sites of Extraglandular Estrogen Production

In both sexes, estrogens circulating in the blood do not originate exclusively from glandular secretion. In premenopausal adult women, most of the estrogens in the body are derived from ovarian secretion of estradiol, but a significant portion comes also from the extraglandular conversion of androstenedione and, to a lesser extent, testosterone. In males, the testes are the main source of estrogens, although peripheral conversion of androgens to estrogens plays an important role in maintaining the level of circulating estrogens.

Aromatization of androgens to estrogens has been demonstrated in several tissues. The *placenta* contains a large amount of aromatase, and serves as a unique model of aromatization of androgens to estrogens. *Muscle and adipose tissue* are major sites, muscles accounting for 25 to 30%, and adipose tissue, for 10 to 15%, of all extraglandular aromatization. The greater capacity of muscle for this process may explain the higher rate of aromatization observed in men than in women[15]. On the other hand, when there is an increase in fat cells, as in obese individuals, adipose tissue exhibits a greater rate of aromatization than does muscle. *Skin fibroblasts* and *hair follicles* are capable of aromatizing androgens to estrogens which may exert an effect in situ rather than contributing to a circulating pool of estrogens. The brain, especially the *hypothalamus*, contains an aromatase enzyme responsible for the in situ formation of estrogens in the hypothalamus. Although the *liver* is well known as the major site of metabolism of steroids, the aromatizing process in this organ is limited. In adults, it is responsible for less than 4% of peripheral aromatization. *Bone marrow* also exhibits aromatizing activity.

Conditions Associated with Increased Extraglandular Estrogen Formation

Under several pathophysiological conditions, the extraglandular production of estrogen may be excessive, resulting in a relatively constant and elevated level of circulating estrogen. Such acyclic extraglandular estrogen production may occur under the following circumstances: 1. *an increased precursor availability*, as in congenital adrenal hyperplasia, polycystic ovarian syndrome, androgen-producing tumors, and in some cases, Cushing's syndrome; 2. *ageing*, in men and women, which is associated with a two- to fourfold increase in extraglandular formation of estrone. This is due, in large measure, to the *increase in aromatase* enzyme activity in adipose tissue; 3. an increased total aromatizing enzyme activity in young men, resulting in feminization. Such an increase has not been reported in the female; 4. *increased extra-*

15. Longcope, C., Pratt, J. H., Schneider, S. H. and Fineberg, S. E. (1978) Aromatization of androgens by muscle and adipose tissue in vivo. *J. Clin. Endocrinol. Metab.* 46: 146–152.

glandular tissue for conversion, as in simple obesity or Cushing's syndrome. In contrast to ageing, this is a function of increased numbers of adipose cells rather than an increase in activity of aromatase per cell; and 5. *increased conversion* of *androgens to estrogens*, which also occurs in *hypo- and hyperthyroidism*, due to alterations of SBP and metabolic clearance rates of both precursor hormones and estrogens.

3.6 Polycystic Ovarian Syndrome (PCO)

The classical example of inappropriate feedback due to excessive extraglandular estrogen production is the PCO syndrome described by *Stein and Leventhal*[16].

Clinical Features

The combination of the following signs: failure to ovulate (and infertility), hirsutism (Fig. X.22), obesity, and bilateral polycystic ovaries, is now well recognized as clinical evidence of PCO syndrome (Table X.4). Certain types of historical information are found to be common to most patients, and may serve to distinguish PCO syndrome from chronic anovulation due to other causes.

Figure X.22 A woman with severe hirsutism due to polycystic ovarian disease. (Courtesy of Dr. G. Schaison, Hôpital de Bicêtre).

Table X.4 Frequency of clinical symptoms in polycystic ovarian syndrome.

	Frequency %
Hirsutism	95
Large ovaries	95
Sterility (primary or secondary)	75
Amenorrhea (primary or secondary)	55
Obesity	40
Dysmenorrhea	28
Persistent ovulation	20

The mean age at menarche (12.3 years) is close to the mean of 12.9 years found in the normal population. Primary amenorrhea occurs rarely. There is maintenance of postmenarchial *menstrual irregularity* and an onset of clinically discernible, excessive hair growth either before or around the time of menarche (a reliable history of this type can be obtained from the parents). Finally, patients are frequently considered overweight by their peers, even prior to menarche.

Thus, the clinical characteristics of PCO syndrome point to the existence of endocrine abnormalities *before puberty*, prior to the final establishment of cyclicity of the hypothalamic-hypophyseal-ovarian system.

Pathophysiology

Peripheral Hormone Levels (Fig. X.23)

Circulating concentrations of *several hormones* are significantly *elevated in PCO* patients, as compared to normal women, during the early follicular phase of the cycle: LH but not FSH; estrone but not estradiol; and both testosterone and androstenedione, as well as 17-hydroxyprogesterone. Approximately 50% of all PCO patients have elevated adrenal androgens, DHEA and its sulfate.

Amplification of Pulsatile Pituitary LH Release

High concentrations of LH found in PCO patients appear to be the result of *pulsatile LH* release of *greater* amplitude by the hypothalamo-pituitary unit. In contrast, FSH levels are low, with occasional small pulses (Fig. X.24).

16. Stein, I. F. and Leventhal, M. L. (1935) Amenorrhea associated with bilateral polycystic ovaries. *Am. J. Obstet. Gynec.* 29: 181–191.

Heightened sensitivity of the pituitary to exogenous GnRH in PCO patients accounts for the exaggerated pulsatile release of LH[17], which may be due to chronically elevated estrogen levels. Since GnRH induces its own receptor and thereby enhances its self-priming effect, especially in the presence of estrogens, an increase in the amplitude of pulsatile GnRH secretion may also be involved in the high amplitude of LH release in PCO syndrome.

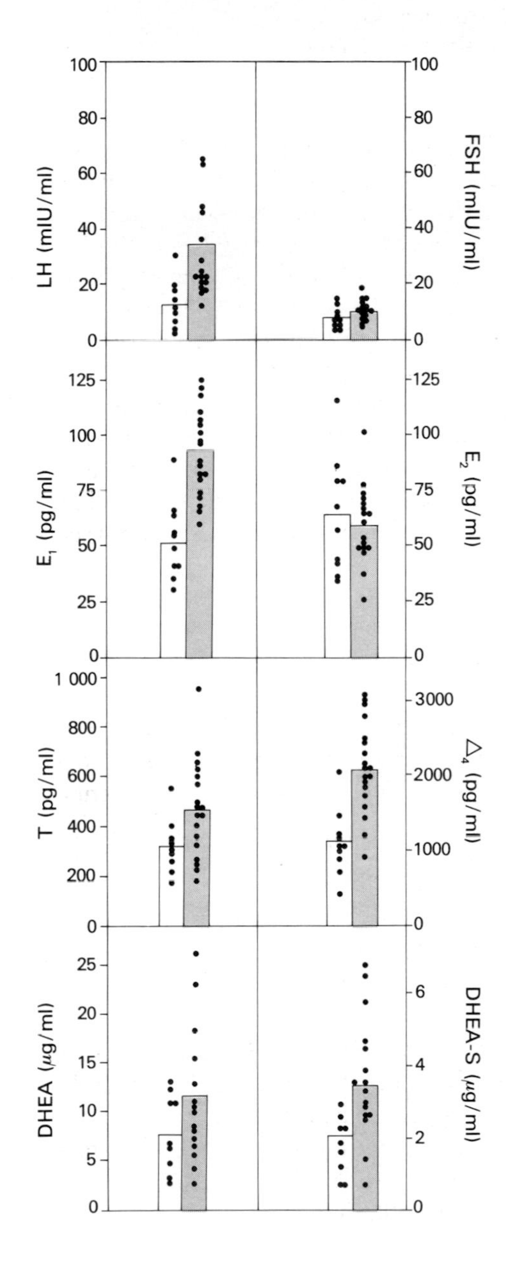

Figure X.23 Plasma hormone levels in PCO (red bars), as compared to those of normal women (open bars) (day 2 to 4 of the cycle).

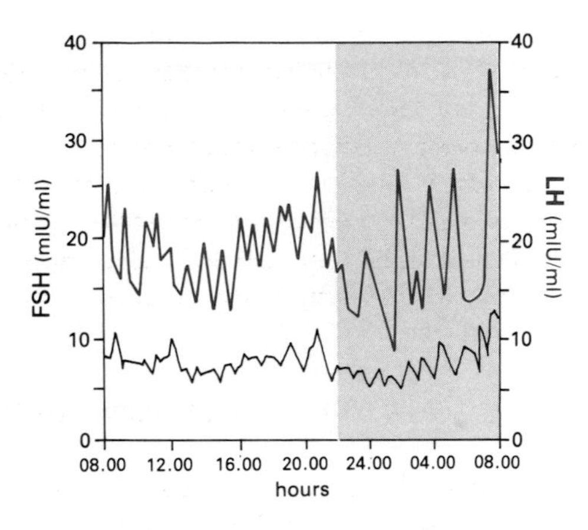

Figure X.24 Plasma gonadotropins in PCO. Exaggerated pulsatile LH and low FSH release over 24 h in one patient.

The *disparity* between *LH* and *FSH secretion* in patients with *PCO* can be explained by the following factors: 1. the inhibitory effect of both estradiol and estrone is greater on FSH than on LH; 2. FSH release is relatively insensitive to GnRH stimulation; and 3. the multicystic ovary in PCO patients may secrete increasing amounts of follicular inhibin, which would selectively inhibit the release of FSH.

Hypothalamic Dysfunction

In patients with hypothalamic dysfunction, in contrast to normal women, the administration of an opiate receptor antagonist (naloxone), of a dopamine receptor antagonist (metoclopramide), or of human synthetic β-endorphin, fails to alter LH secretory activity[18]. These findings, together with a greater LH sensitivity to the inhibitory action of dopamine in PCO patients, suggest that increased pulsatile GnRH release may result from the dysfunction of dopaminergic and opioidergic inhibitory mechanisms. The reversal of cyclic hypothalamic gonadotropin function in response to several forms of treatment argues against a primary hypothalamic defect in PCO syndrome. It seems likely that hypothalamic abnormalities in PCO syndromes are induced by inappropri-

17. Rebar, R., Judd, H. L., Yen, S. S. C., Rakoff, J., Vandenberg, G. and Naftolin, F. (1976) Characterization of the inappropriate gonadotropin secretion in polycystic ovary syndrome. *J. Clin. Invest.* 57: 1320–1329.
18. Cumming, D. C., Reid, R. L., Quigley, M. E., Renar, R. W. and Yen, S. S. C. (1984) Evidence for decreased endogenous dopamine and opioid inhibitory influences on LH secretions in polycystic ovary syndrome. *Clin. Endocrinol.* 20: 643–648.

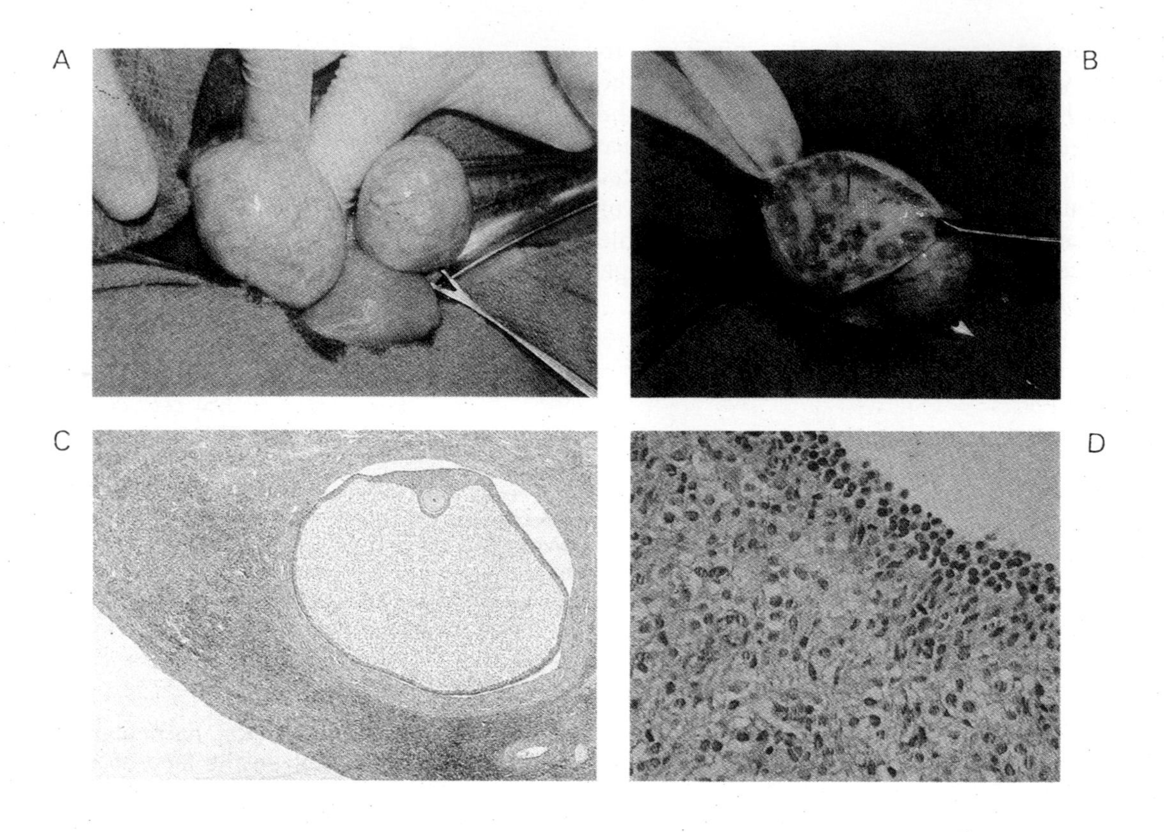

Figure X.25 The gross and microscopic characteristics of polycystic ovaries. **A.** Bilateral enlarged ovaries with a smooth and thickened capsule. **B.** On cut section, multiple follicular cysts surrounded by abundant ovarian stroma are found throughout the cortex of the ovary. **C.** The subcapsular cysts are lined with granulosa cells, with early stages of antrum formation. **D.** Hyperplasia of theca interna with luteinization.

ate ovarian steroid feedback, particularly chronic deprivation of progesterone.

Androgen Excess

Elevated levels of circulating *DHEA* and *DHEA-S* found in PCO syndrome are readily *suppressible by* the administration of *dexamethasone*. In *contrast*, elevated levels of *testosterone* and *androstenedione*, as well as of *17-hydroxyprogesterone*, are not significantly altered by dexamethasone, reflecting the ovarian contribution of these steroids.

Medical ovariectomy achieved by the administration of a GnRH agonist has been used in PCO patients as a probe for determining the relative contributions to the androgen pool[19]. Elevated levels of androstenedione, testosterone, and 17 OH-progesterone were reduced to castration levels, whereas in half of the patients, elevated adrenal androgens (DHEA and DHEA-S) remained unaltered. These findings indicate that androgen excess in PCO syndrome involves either the ovaries alone, or both the adrenals and the ovaries.

Extraglandular Formation of Estrogens and Chronic Anovulation

Increased androgen levels provide a substrate for peripheral conversion to estrogens. The relatively constant levels of estrogens that form the basis for acyclicity are reflected in the *chronically elevated levels of estrone* derived mainly from extraglandular formation of androstenedione. This creates acyclic feedback and inappropriate secretion of LH and FSH by the hypothalamic-hypophyseal system. It is presently thought to represent the *key factor* in the maintenance of chronic anovulation in PCO syndrome.

19. Chang, R.J., Laufer, L.R., Meldrum, D.R., DeDazio, J., Lu, J.K.H., Vale, W.W., Rivier, J.E. and Judd, H.L. (1983) Steroid secretion in polycystic ovarian disease after ovarian suppression by a long-acting gonadotropin-releasing hormone agonist. *J. Clin. Endocrinol. Metab.* 56: 897–903.

The Polycystic Ovary: Structure–Function Relationships

Morphologically, polycystic ovaries are enlarged bilaterally, and each has a smooth but *thickened capsule*. The capsule is avascular. On cut section, the ovary exhibits numerous subcapsular follicular cysts varying from 2 to 7 mm in diameter. The cysts are usually lined with a few layers of granulosa cells, but the most striking feature is the *hyperplasia of the theca internal* surrounding the many cystic follicles (Fig. X.25). An increased rate of follicular atresia also occurs. It should be emphasized that there is a great degree of *anatomical* and *clinical variability* in this syndrome.

The hyperplasia of theca cells is the result of chronic LH stimulation. The associated excess of intraovarian androgen production may be responsible for the increased rate of follicular atresia and, possibly, for the development of the thickened ovarian capsules.

Follicular cysts in the ovaries of PCO patients do not mature fully, and the absence of mature follicles results in low estradiol production. Granulosa cells, which are sparse and virtually devoid of aromatase activity in these follicles, can be increased by the administration of FSH. Since FSH plays a direct and specific role in stimulating aromatase activity in human granulosa cells, the absence of aromatase activity in PCO may be causally related to relatively low local concentrations of FSH rather than to an intrinsic abnormality.

Peripheral Effect of Androgen Excess: Hirsutism

In women, increased *hair growth* according to a pattern of distribution typical of the male, called *hirsutism*, is a manifestation of androgenic stimulation of the hair follicle. The degree of hirsutism in PCO syndrome does not always correlate with the magnitude of androgen excess. Severe hisutism may be the result of a slight elevation in the level of androgens, whereas a substantial elevation in androgens may not always be accompanied by hirsutism. This apparent dissociation between the level of androgens and the degree of hirsutism is due to differences in sensitivity of the hair follicle to androgen; the activity of 5α-reductase in the hair follicle is the major determinant of androgenic action on hair growth. This is because sexual hair is dependent upon dihydrotestosterone (DHT)[20]. It has been shown that the degree of 5α-reductase activity in PCO patients correlates with the severity of hirsutism[21].

In summary (Fig. X.26), *inappropriate gonadotropin* secretion with a *high LH/FSH ratio* is causally *related to* an *elevated* and non-linear *estrogen feedback* on the *hypothalamic-hypophyseal system*: an increased pulsatile secretion of GnRH occasioned by chronic progesterone deprivation and decreased opioidergic-dopaminergic inhibition (p. 446). Elevated free estradiol and estrone levels and ovarian inhibin preferentially inhibit FSH secretion. The LH-dependent hyperplasia of the theca cells and the associated hypersecretion of ovarian androgens are responsible for the development of hirsutism. The low levels of steroid binding protein (SBP) may facilitate the rapid tissue uptake of free androgens for peripheral formation of estrogen, and excess adipose tissue increases the conversion of androgen to estrogen, by virtue of an increase in tissue mass. The acyclic and elevated levels of estrogen result in self-perpetuating

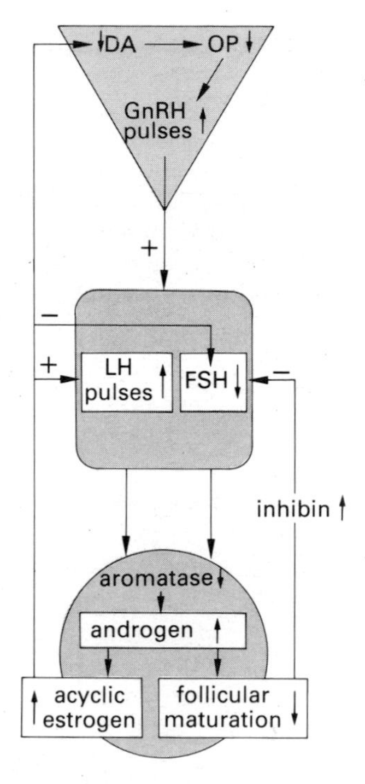

Figure X.26 The interdependent event of high LH-FSH ratio occasioned by an increased GnRH secretion as a consequence of reduced hypothalamic inhibition. This setting induces an increased ovarian androgen production by the theca cells and an acyclical estrogen feedback system in the maintenance of chronic anovulation in PCO syndrome.

20. Wilson, J.D. and Walker, J. (1969) The conversion of testosterone to 5 alpha-androstan-17β-ol-3-one (dihydrotestosterone) by skin slices of man. *J. Clin. Invest.* 48: 371–379.
21. Lobo, R.A., Goebelsman, U. and Horton, R. (1983) Evidence for the importance of peripheral tissue events in the development of hirsutism in polycystic ovary syndrome. *J. Clin. Endocrinol. Metab.* 57: 393–397.

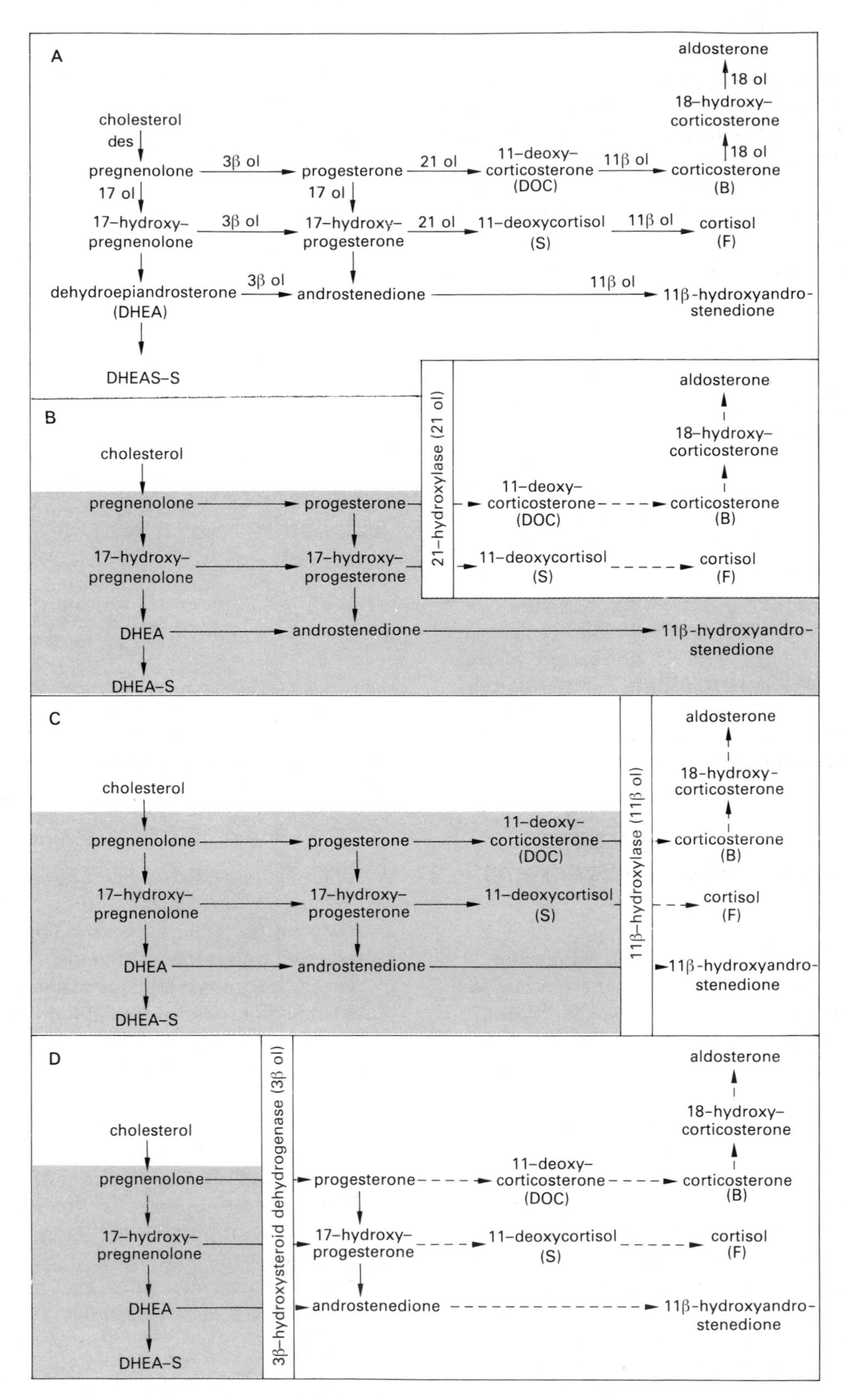

Figure X.27 Schematic representation of three types of enzymatic defects in adrenal steroid biosynthesis. **A.** Normal synthetic pathway. In **B**, **C**, and **D**, solid arrows indicate pathways with increased synthesis preceding the enzymatic block. Dotted arrows denote pathways with reduced or absent steroid synthesis.

acyclicity and chronic anovulation. Ovarian changes, such as inadequate follicular maturation and increased follicular atresia, are secondary events. The low estradiol production is associated with aromatase deficiency, which is causally related to inadequate FSH stimulation, not to an inherent defect in the granulosa cells.

Clinically, the major components of PCO are chronic anovulation resulting in infertility, androgen excess with hirsutism, and high incidence of endometrial hyperplasia due to unopposed estrogen. The specific aim in the *treatment* of PCO syndrome, apart from establishing fertility, is to reduce the probability of endometrial neoplasia and to suppress hirsutism by means of: 1. clomiphene citrate, an antiestrogen inducing an increase in FSH secretion, critical for the initiation of cyclic ovarian function; success rate is high; 2. combined clomiphene and dexamethasone, permitting the suppression of adrenal contribution of androgens, appears to facilitate the action of clomiphene; 3. purified human FSH of pituitary or urinary origin has been used successfully in inducing ovulation. 4. wedge resection of ovaries, although rarely used today, remains an effective therapeutic modality in cases where other medical treatment fails. The rate of success is approximately 80%.

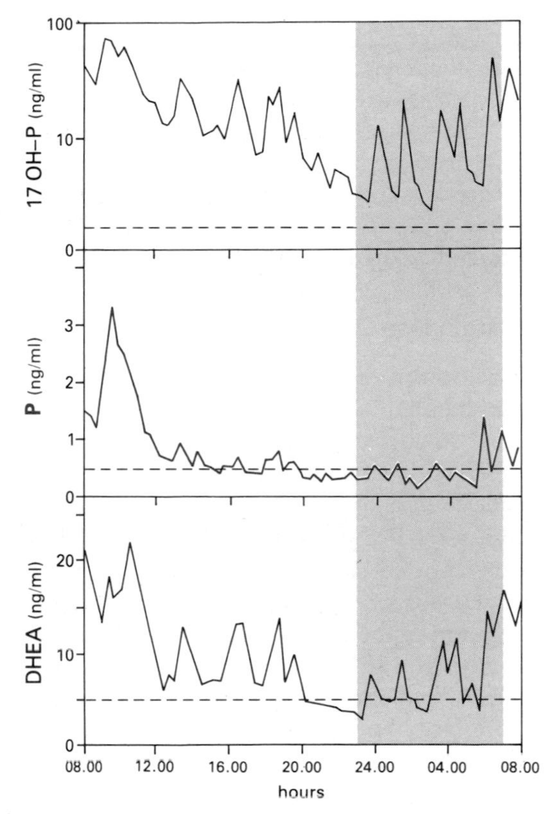

Figure X.28 Circulating levels of 17-hydroxyprogesterone, progesterone, and DHEA measured over a 24 h period in a patient with a late onset of congenital adrenal hyperplasia. The dotted lines represent upper-normal limits.

3.7 Partial Adrenal Enzymatic Deficiencies: PCO-like Syndrome

Late-onset 21-hydroxylase Deficiency

The *attenuated form* of 21-hydroxylase deficiency represents a mild enzymatic defect with late onset of manifestations mimicking idiopathic hirsutism of PCO syndrome. Although severe hirsutism, small stature, and family history of PCO-like syndrome may provide clinical clues, diagnosis can be made only by demonstrating elevated 17-hydroxyprogesterone levels. Because of a partial block at the 21-hydroxylation step in the biosynthesis of cortisol (Fig. X.27B and p. 433), 17-hydroxyprogesterone may be elevated, but is usually within the normal range during the day hours, becoming elevated, with episodic secretion, during the ACTH-cortisol circadian rhythm (Fig. X.28). Consequently, normal levels of 17-hydroxyprogesterone do not exclude the diagnosis. In suspected cases, an ACTH stimulation test is necessary for the assessment of the degree of rise in 17-hydroxyprogesterone.

Pubertal onset of hirsutism and menstrual irregularities may be explained by an increase in 17, 20-desmolase enzyme during adrenarche, thereby amplifying adrenal androgen biosynthesis in the face of a marked increase in substrate 17-hydroxyprogesterone. Nonetheless, once the diagnosis is made, this condition is highly amenable to dexamethasone suppression therapy. Incidence among hirsute women is estimated to be between 1.5 and 6%.

A *cryptic form* of 21-hydroxylase deficiency in family members of patients with classic congenital adrenal hyperplasia (CAH) has been described. Although both the basal and the ACTH-stimulated hormonal profiles are similar to those found in the attenuated form, unaccountably the cryptic form of CAH is asymptomatic[22]. Based on histocompatibility

22. Levine, L. S., Dupont, B., Lorenzen, F., Pang, S., Pollack, M. S., Oberfield, S. E., Kohn, B., Lerner, A., Cacciari, E., Mantero, F., Cassio, A., Scaroni, C., Chiumello, G., Rondanini, G. F., Gargantini, L., Giovanelli, G., Virdis, R., Barbolotta, E., Migliori, C., Pintor, C., Tato, L., Barboni, F. and New, M. I. (1981) Genetic and hormonal characterization of cryptic 21-hydroxylase deficiency. *J. Clin. Endocrinol. Metab.* 53: 1193–1198.

leukocyte antigen (HLA) *genotyping*, the cryptic form appears to have two genes for 21-hydroxylase deficiency: one classical and one cryptic (21 OH^{CAH}/21-$OH^{cryptic}$). It was proposed that allelic variants of the steroid 21-hydroxylase deficiency locus produce different degrees of enzymatic deficiency, resulting in the phenotypic diversity of classical, cryptic and attenuated forms of 21-hydroxylase deficiency[22].

11β-hydroxylase Deficiency

Deficiency in 11β-hydroxylase is the *second most common* adrenal biosynthetic defect. As seen on p. 433, 11β-hydroxylase deficiency disrupts the production of both cortisol and aldosterone, resulting in hypersecretion of ACTH and overproduction of DOC and adrenal androgen (Fig. X.27C). Adult onset has been reported, with manifestation of hirsutism, menstrual disorders, acne, and variable degrees of malformation of the external genitalia. Hypertension accompanying 11β-hydroxylase deficiency is less common. All patients have elevated plasma androgen levels, primarily androstenedione. Diagnosis is confirmed by demonstrating an elevation of plasma DOC, especially following ACTH stimulation.

3β-hydroxysteroid Dehydrogenase Deficiency

Although the condition is *rare*, reports of a defect in the conversion of pregnenolone and 17-hydroxypregnenolone to progesterone and 17-hydroxyprogesterone have appeared. The enzyme complex responsible for these conversions is the 3β-hydroxysteroid dehydrogenase isomerase system (Fig. X.27D). This defect affects both the adrenals and the gonads, resulting in an increase in Δ_5-progestins and Δ_5-androgens (DHEA and Δ_5-androstenediol), with normal Δ_4-androgens (androstenedione and testosterone). When the defect is partial, affected individuals may survive the neonatal period. In the adult, the female manifests hirsutism and menstrual disorders resembling the features of PCO syndrome. The diagnosis is made by the finding of an increased ratio of Δ_5- to Δ_4-steroids.

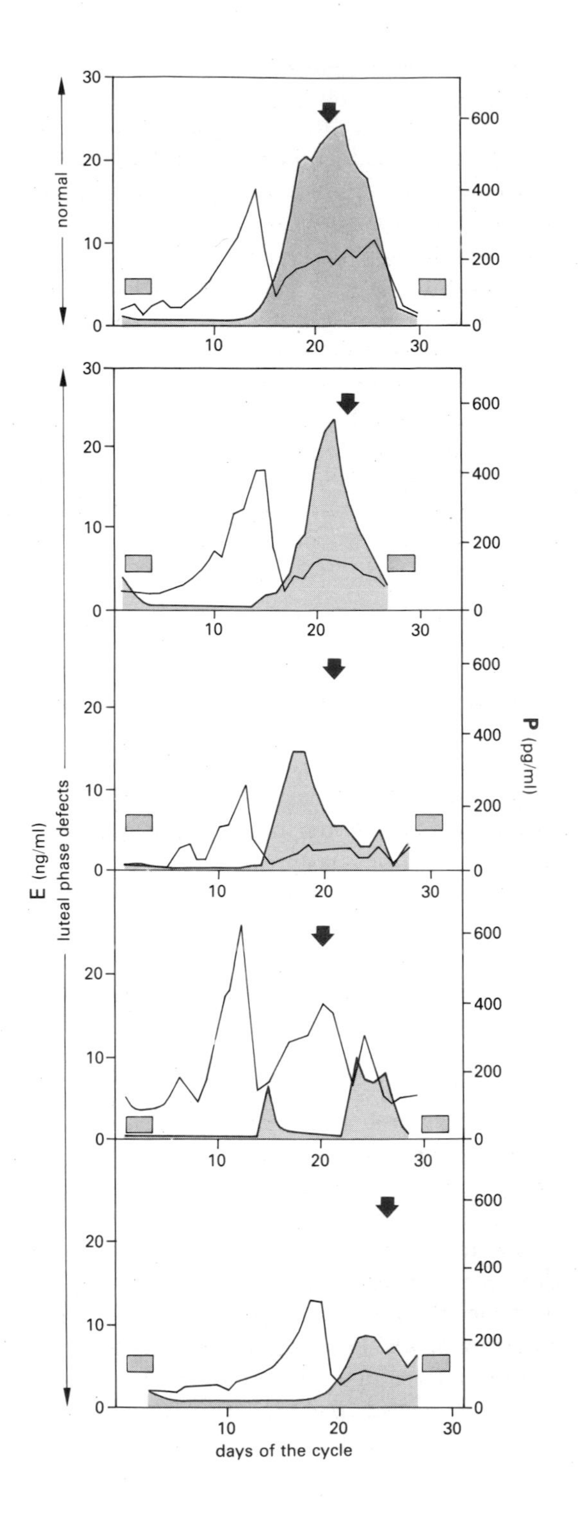

Figure X.29 Representative examples of luteal phase defects (LPD) as discerned by the duration, magnitude, and patterns of progesterone (P) secretion; concentrations in four infertile patients and in women with normal menstrual cycles. The red arrow indicates the estimated date of implantation. Colored boxes represent menstrual periods.

3.8 Inadequate Luteal Phase (Fig. X.29)

An inadequate luteal phase results from a relative *deficiency* in the *secretion* of *progesterone* by the *corpus luteum.* The term refers to *both* a *short interval* (<11 d) between ovulation and menstruation, with relatively normal peak values of progesterone, and, more commonly, to a luteal phase of *normal length, with lower than normal progesterone levels*. Both result in inadequate stimulation of the endometrium. A related condition may be caused by low levels of progesterone receptor in the endometrium. An inadequate luteal phase can be found in isolated cycles of normal women, and only if the defect is found repeatedly is it thought to be a significant factor in infertility. Approximately 3 to 4% of infertile women are diagnosed as having an inadequate luteal phase; the incidence may be higher in women with a history of habitual abortion.

It is disappointing to note that, despite the numerous articles written in the past 30 years about inadequate luteal phase, both its causes and its role in infertility remain elusive. Clinical studies do not substantiate the efficiency of reported treatments.

4. Endocrinology of Pregnancy

The initiation, maintenance, and termination of pregnancy are largely dependent on the interaction of hormonal and neural factors. Proper timing of the neuroendocrine events taking place within and between compartments (i.e. maternal, fetal placental, and amniotic fluids) is critical in directing proper fetal development and parturition.

Alteration of neuroendocrine functions during pregnancy represents a most remarkable adaptive phenomenon in biological systems. Within the pregnant uterus, the *fetal-placental-decidual unit* produces extraordinary *amounts of steroid* and *protein hormones, and neuropeptides*[23]. They appear to interact and function in a paracrine and autocrine fashion resembling that of a compressed hypothalamic-hypophyseal-ovarian system. These newly added endocrine units may conduct undirectional flow of nutrients from the mother to the fetus, provide a favorable environment within the uterus for cellular growth and maturation, and may convey signals when the fetus is ready for extra-uterine existence.

23. Simpson, E.R. and MacDonald, P.C. (1981). Endocrine physiology of the placenta. *Ann. Rev. Physiol.* 43: 163–188.

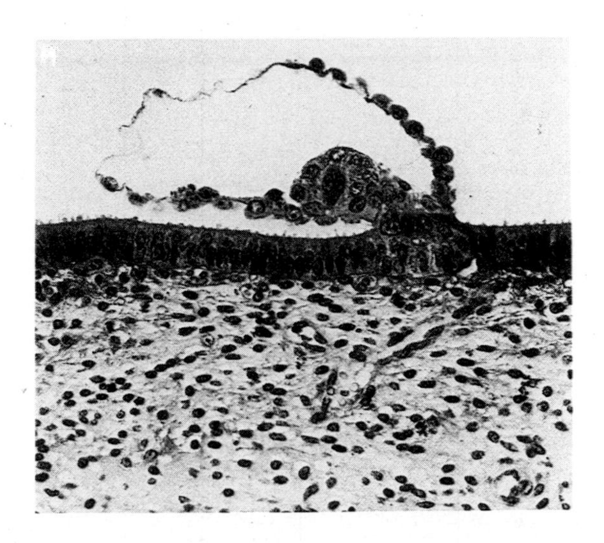

Figure X.30 Section showing the second stage of nidation of a monkey blastocyst. The entry of a single trophoblast cell through the endometrial epithelium is seen, occurring in the immediate vicinity of a maternal capillary. (Reproduced from: *Placental Vasculature and Circulation*. Ramsey, E. M. and Donner, M.W. (1980) W.B. Saunders Co., Philadelphia, p. 5).

4.1 Steroid Hormones: Formation and Secretion

Luteo-Placental Shift

The Corpus Luteum of Pregnancy

Within hours following implantation (Fig. X.30), the primitive trophoblast of the blastocyst secretes a luteotropic hormone, human chorionic gonadotropin (*hCG*), into the maternal circulation, thus rescuing *corpus luteum* function which would otherwise regress. Increased corpus luteum activity in response to *hCG stimulation* results in the biosynthesis and secretion of relaxin as well as an *increased rate of secretion of progesterone, 17-hydroxyprogesterone, and estradiol* (Fig. X.31).

A peptide hormone, *relaxin*, causes relaxation of the myometrium, and is detectable in serum between eight to ten days after conception (Fig. I.27). The appearance of ovarian relaxin in early pregnancy may function in conjunction with progesterone to reduce spontaneous uterine activity. Thus, relaxin plays a major role in the maintenance of early pregnancy.

The corpus luteum remains the major source of progestational steroids through approximately the *ninth week* of gestation, after which time the placental

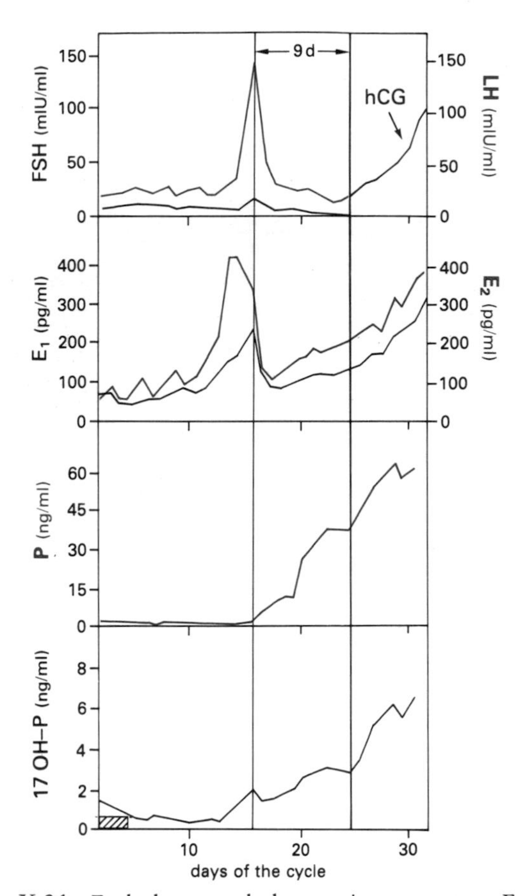

Figure X.31 Early hormonal changes in pregnancy. Following implantation on day 8 to 9 of the luteal phase, the function of the corpus luteum is rescued by hCG.

trophoblast (and decidua) assumes this role until the time of parturition. A *transitional* period of approximately one to two weeks may be required for this *luteal-placental shift*. Ovariectomy or corpus luteum removal before the eighth week of gestation invariably results in abortion, but has no influence on the course of gestation after the ninth week.

Maternal-Placental and Feto-Placental Units as Endocrine Organs

The placenta has evolved as that part of the reproductive equipment of mammals which serves to transmit nutrients to the fetus, to excrete waste products into the maternal blood, and, by means of its hormones, to modify maternal metabolism at various stages of pregnancy. The human placenta attains its mature structure by the end of the first trimester of pregnancy. The functional unit is the *chorionic villus* (Fig. X.32), which consists of a central core of loose connective tissue, with abundant capillaries connecting with the fetal circulation. Around this core are two layers of trophoblast, an outer syncytium (*syncytiotrophoblast*) and an inner layer of discrete cells (*cytotrophoblast*), the latter becoming discontinuous as the placenta matures.

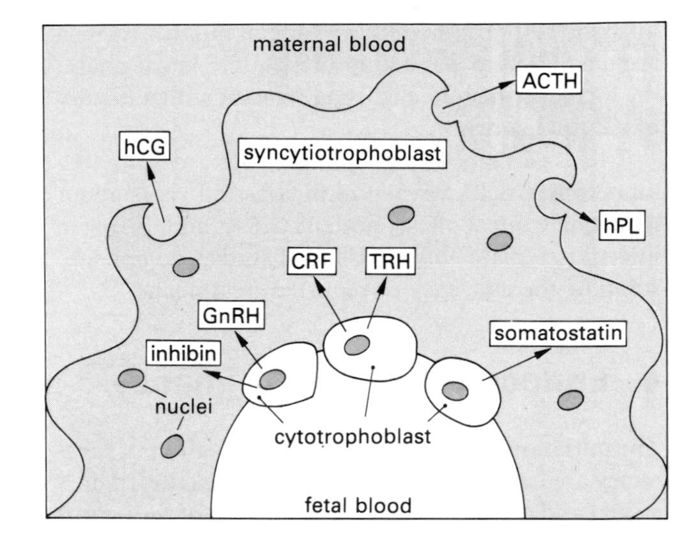

Figure X.32 Schematic representation of the anatomical relationship between the cytotrophoblast and the multinucleated syncytiotrophoblast, and their fine cellular structures. (Adapted from a cross-sectional electron micrograph. Fawcett, D.W., Long J.A. and Jones, A.L. (1969) The ultrastructure of endocrine glands. *Rec. Prog. Horm. Res.* 25:315–338.

The *developing fetus* and its *placenta* form an interdependent partnership in regulating the normal course of pregnancy. This functional relationship, commonly known as the *feto-placental unit*, is a unique endocrine system which produces a *substantial number of hormones*. Peptides, neuropeptides and steroid hormones are secreted, and many of these are identical to or at least mimic those produced by the hypothalamic-hypophyseal-gonadal system.

Formation of Progesterone: The Maternal-Placental Unit (Fig. X.33)

Progesterone is synthesized in the syncytiotrophoblast through hydroxylation and side-chain cleavage of cholesterol. *Cholesterol* and LDL-cholesterol are delivered to the placenta by the *maternal circulation*. The synthesis of progesterone follows the same pathway as described on p. 393 for other steroidogenic glands. This process is independent of a fetal precursor, as evidenced by the unchanged plasma progesterone levels following fetal death in utero. Thus, with an almost unlimited amount of maternal substrate, *placental progesterone production* represents an endocrine process exclusive

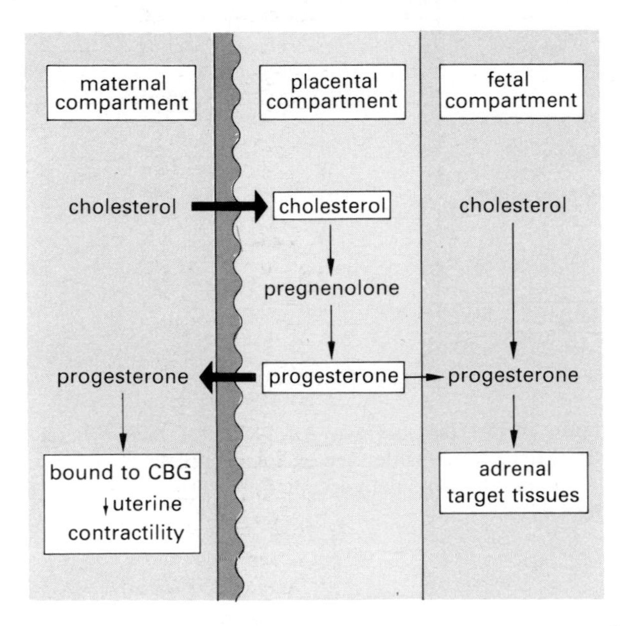

Figure X.33 The pathway of placental biosynthesis of progesterone. Only maternal-placental interactions have been represented.

to *maternal-placental interaction*, determined by the site and the perfusion of the placenta.

The maternal plasma progesterone concentration[24] increases from 40 to 160 ng/ml, from the first to the third trimester of pregnancy (Table X.5). At term, the placenta produces approximately 250 mg of progesterone per day, of which 90% is secreted into the maternal compartment and 10% into the fetal circulation. In human plasma, most progesterone is bound to CBG and albumin.

The myometrium contains progesterone receptor, and progesterone is known to decrease uterine sensitivity to oxytocin. Thus progesterone works synergistically with relaxin (see below) to reduce uterine motility and inhibit the propagation of uterine contractions.

Because of the absence of 17-hydroxylase, the placenta cannot produce 17-hydroxyprogesterone. The substantial increase in 17-hydroxyprogesterone

Table X.5 Plasma progesterone and urinary pregnanediol during pregnancy.

Weeks*	Plasma progesterone (ng/ml‡)	Urinary pregnanediol (mg/d)‡‡
Luteal phase	17	4
4	20	5
10	26	7.5
14	41	11
18	45	17
22	50	28
26	60	32
30	80	42
34	100	50
38	120	60
40	160	60

*Weeks following the first day of the last menstrual period.
‡Tulchinsky, D. and Okada, D. M. (1975) Hormones in human pregnancy. IV. Plasma progesterone. *Am. J. Obst. Gynec.* 121: 293–299.
‡‡Jayle, M. F. (1967) *Hormonologie de la grossesse.* Gauthier-Villars, Paris.

levels found in maternal blood after 32 weeks of gestation is largely derived from production by the fetal adrenal[24].

A marked increase in the plasma concentration of deoxycorticosterone (DOC), a potent mineralocorticosteroid, is due to steroid 21-hydroxylase activity in kidneys. The role of extra-adrenal production of DOC in water and salt retention during pregnancy is unknown.

The Placenta Converts Imported Androgens to Estrogens (Fig. X.34)

The placenta lacks 17-hydroxylase and 17,20-desmolase activity; hence, conversion of *C-21 to C-19* steroids (androgens) cannot occur. The source of androgens is the *maternal and fetal adrenal.* The placenta extracts DHEA-S from the maternal and fetal circulation, and, by hydrolyzing the sulfate (via sulfatase), converts it to free DHEA. The placenta is capable of converting *DHEA* to androstenedione and testosterone, which are readily aromatized *to form estrone* and *estradiol*[25], destined to be released mainly into the maternal circulation. Maternal serum testosterone and androstenedione levels increase two-

24. Tulchinsky, D., Hobel, C. J., Yeager, E. and Marshall, J. R. (1972) Plasma estrone, estradiol, estriol, progesterone and 17-hydroxyprogesterone in human pregnancy. I. Normal pregnancy. *Amer. J. Obstet. Gynecol.* 112: 1095–1100.
25. Ryan, K. J. (1959) Biological aromatization of steroids. *J. Biol. Chem.* 234: 268–272.

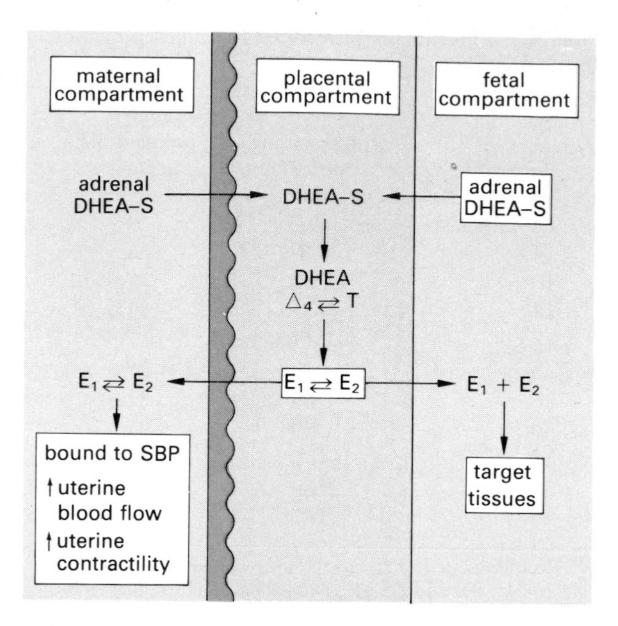

Figure X.34 The pathway of placental biosynthesis of estradiol (E_2) and estrone (E_1) from maternal and fetal androgen precursors: dehydroepiandrosterone (DHEA) and its sulfate (DHEA-S), testosterone (T) and androstenedione (Δ_4).

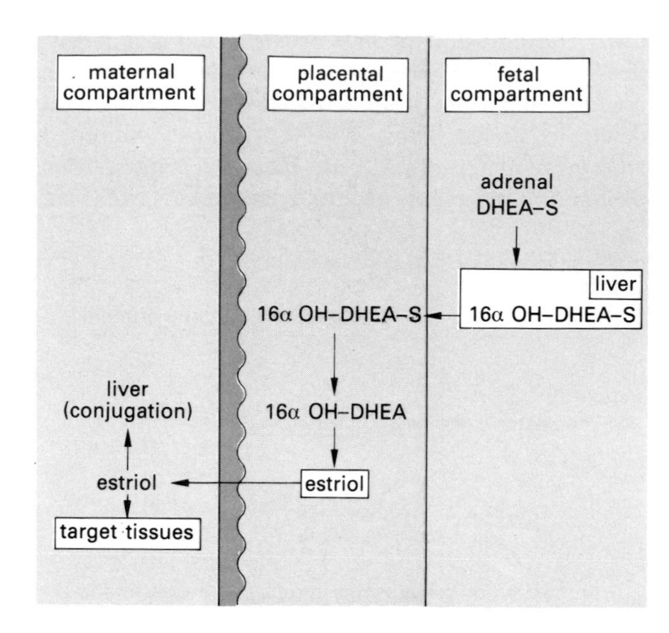

Figure X.35 The pathway of placental biosynthesis of estriol (E_3), which is derived exclusively from the fetal precursor 16α-hydroxydehydroepiandrosterone sulfate (16α OH-DHEA-S).

to three-fold during pregnancy, and most serum testosterone is bound to *SBP*, which is *markedly elevated* in the *maternal* compartment, but *very low in the fetal* compartment, independent of the sex of the fetus.

Thus, the placenta irreversibly aromatizes androgens to estrogens, a process of fundamental importance in pregnancy. Maternal serum estradiol levels increase throughout pregnancy, until term, reaching 20 to 30 ng/ml, a level 100 times greater than in non-pregnant women. SBP has a high affinity for E_2, and hence total serum E_2 levels are higher than those of E_1.

Formation of Estriol (E_3) and Estetrol (E_4) in the Feto-Placental Unit (Fig. X.35)

Because of its *lack* of *16-hydroxylase*, the *placenta* is unable to form E_3 from E_1 or E_2. The DHEA-S from the fetal adrenal is converted by the *fetal liver* to 16α-hydroxydehydroepiandrosterone sulfate (16 OH-DHEA-S). This is the principal fetal contribution to E_3 biosynthesis. In the placenta, 16 OH-DHEA-S is cleaved by sulfatase to 16 OH-DHEA. The placenta first converts 16 OH-DHEA to 16 OH-androstenedione and 16 OH-testosterone, and then aromatizes these compounds to E_3. Thus, the fetus and placenta play a joint and obligatory role in E_3 biosynthesis, independent of the mother (the fetal-placental unit). Estriol levels are frequently used as an index of fetal development.

In the absence of fetal ACTH stimulation, little or no DHEA-S is produced by the fetal adrenal, a condition found in association with congenital absence of the pituitary gland, in the anencephalic fetus, and when glucocorticosteroids administered to the mother cross the placenta, suppressing fetal ACTH secretion. Very rarely, a fetus may lack 16-hydroxylase activity. In these instances, there is relatively normal E_1 and E_2 but very low E_3 formation[24].

The placenta secretes E_3 into the maternal as well as the fetal circulation. In the fetus, E_3 is predominantly conjugated. Unconjugated E_3 crosses the placenta most readily. In the mother, E_3 metabolism consists almost entirely of conjugation, and E_3 is then excreted in the urine as glucuronosides. At term, urinary E_3 excretion averages 20 to 30 mg/d. E_3, E_1 and E_2 bind to estrogen receptors. However, because of its high MCR and rapid dissociation from receptors, the manifestation of biological activity requires sustained and larger amounts of E_3 delivered to the target cells. Because of the large quantity produced, E_3 is considered biologically important in estrogen-mediated events of pregnancy, i.e. in the increase of uterine blood flow.

Estetrol is 15α-hydroxyestriol, distinguishable from E_3 by a fourth hydroxyl group on C-15, introduced primarily in the fetal liver (15-hydroxylase is not present in the adult). This *uniquely fetal estrogen* binds to the estradiol receptor, but is devoid of estrogenic activity. Thus, E_4 may be an endogenous anti-estrogen serving to modulate the massive amount of biologically active estrogens on its target tissues.

Placental Sulfatase Deficiency

Placental sulfatase deficiency and an inability to cleave steroid sulfates result in very *low* E_3 formation. However, the aromatization of free DHEA to estrogens by the placenta remains intact. Thus, E_1 and E_2 levels in maternal circulation may be in the normal range, whereas E_3 and E_4 are markedly reduced. This lack of E_3 appears to cause a delay in the onset of labor, which may lead to postmaturity and intrauterine death unless cesarean section is performed. It is of interest that these patients are resistant to the induction of labor by oxytocin. These observations, as well as the fact that pregnancy is prolonged in the presence of an anencephalic fetus, support the theory that estrogens play an important role in determining the onset of labor. Placental sulfatase deficiency is X-linked. The affected offspring are male. Thus, family history may provide clues for prospective diagnostic exclusion (i.e. measurement of urinary E_3 in suspected cases).

4.2 Chorionic Peptide and Neuropeptide Hormones

Human Chorionic Gonadotropin (hCG)

(Fig. X.36)

The structure of hCG, the luteotropic hormone of pregnancy, is found in Figs. V.1 and V.2.

Secretion

The syncytiotrophoblast produces and secretes hCG into the intervillous space. This hormone is first detectable in maternal serum nine days following conception (Fig. X.31), corresponding closely to the time of penetration of the trophoblast-covered blastocyst into the endometrial stroma and its consequent direct apposition to the maternal circulation. Serum hCG levels increase rapidly over ten days following implantation, with a *doubling time of 1.7 d.* This exponential increase in hCG, reaching a peak of ~5 μg/ml (100 IU/ml) at the ninth week of gestation, is followed

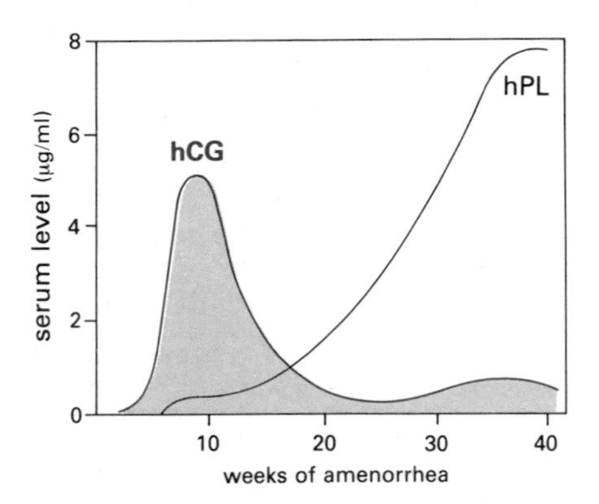

Figure X.36 Circulating concentrations of hCG and hPL during the course of pregnancy. Note the contrasting pattern and the shift of secretory activity of the two hormones by the syncytiotrophoblast at mid-gestation (data expressed as μg/ml; 1 μg=20 IU of hCG).

by a fall to a plateau of ~0.5 μg/ml for the remainder of pregnancy (Fig. X.36). In twin pregnancies, hCG levels at four to five weeks after the last menstrual period are at least twice that for singletons. During the last two-thirds of pregnancy, increasing amounts of hCG subunits are produced, with relatively little intact hCG. Relatively small but significant amounts of hCG appear in the fetal circulation.

Function

There are several functional roles of hCG in pregnancy. As indicated earlier, hCG provides stimulus for the continued production of progesterone and relaxin by the corpus luteum, which is essential for the maintenance of early pregnancy. Circulating hCG may also function to inhibit maternal gonadotropin secretion by the hypothalamo-pituitary unit. In addition to its luteotropic function, hCG also provides gonadotropic input to the fetus during the first trimester of pregnancy, which may be important in the sexual differentiation of internal genitalia in the male fetus. An understanding of the regulation of hCG secretion is just emerging, but recent evidence suggests that chorionic GnRH and inhibin are important regulators (see below).

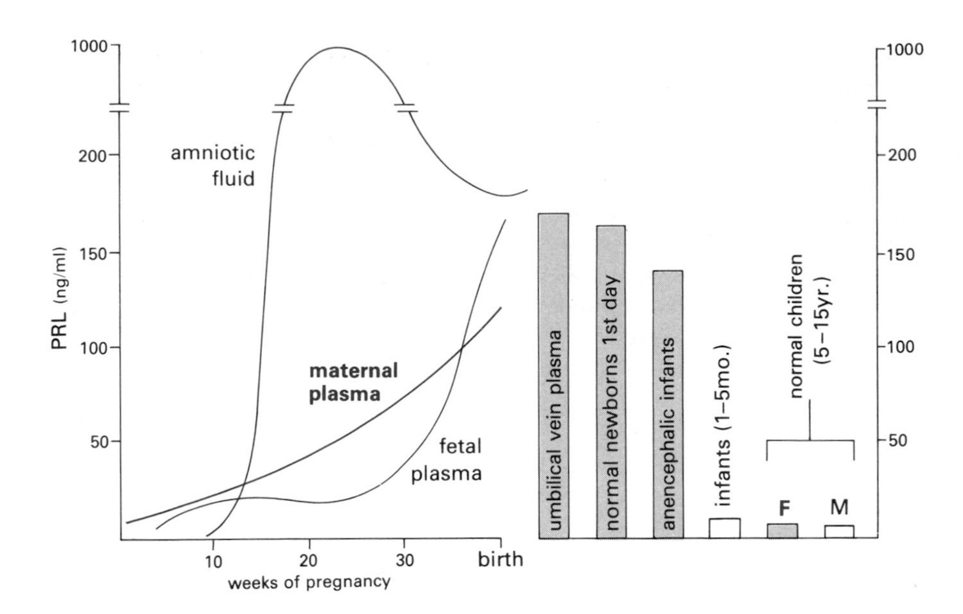

Figure X.37 Comparison of patterns of maternal, fetal and amniotic fluid prolactin levels during the course of human pregnancy. On the right, plasma levels of prolactin in normal and anencephalic newborns are compared with those of normal individuals. M, males; F, females (modified from refs. 32 and 33).

Human Placental Lactogen (hPL)

The structure of hPL, the metabolic hormone of pregnancy, is discussed on p. 192.

Nature and Secretion

The *syncytiotrophoblast* produces hPL. Unlike hCG, serum hPL levels increase concomitantly with placental growth (Fig. X.36), and daily hPL production averages 1 to 2 g at term. The metabolic clearance rate of hPL is 175 l/d, disappearing rapidly from the blood following delivery of the placenta (half-life 10 to 12 min)[26].

Function

Although hPL has lactogenic properties and may participate in the mammotropic effect of pregnancy, it apparently is *not involved* in *milk production* in humans. Very little hPL is found in the fetal compartment; *most* is secreted *into the maternal* circulation. It may function to inhibit maternal pituitary growth hormone secretion, which is known to be markedly attenuated during pregnancy. Importantly, hPL may function to invoke some of the metabolic changes characteristic of pregnancy. Most noteworthy is its glucose-sparing and lipolytic effects, which can be attributed to its anti-insulin action[26]. In this manner, the placenta directs maternal nutrients toward the fetus, assuring an uninterrupted supply of glucose and amino acids, which cross the placental barrier freely. The transplacental flux of free fatty acids, made available in abundance through the lipolytic action of hPL in the maternal compartment, is limited, and free fatty acids are thus preserved as metabolic fuel for the mother. The active transport of amino acids by the placenta, from the maternal circulation to the fetus, results in a decrease in amino acid concentration and in a limitation of substrates (especially alanine) required for gluconeogenesis. This, together with the hPL-induced lipolysis, accounts for the rapid development of hypoglycemia and ketonemia during fasting in pregnant women[27].

Neuropeptides and Gonadal Peptide of Chorionic Origin

During the past few years, several neuropeptides, first discovered in the brain, have been found in placental villi, in decidua, and in chorionic membranes. These include *GnRH, TRH, somatostatin*, cortico-

26. Grumbach, M. M., Kaplan, S. L., Sciarra, J. J. and Burr, I. M. (1968) Chorionic growth hormone-protein (CGP): secretion, disposition, biologic activity in man, and postulated function as the "growth hormone" of the second half of pregnancy. *Ann. N.Y. Acad. Sci.* 148: 501–531.
27. Felig, P. (1977) Body fuel metabolism and diabetes mellitus in pregnancy. *Med. Clin. of N. Amer.* 61: 43–66.

tropin releasing factor (*CRF*), and *proopiomelanocortin*[28]. More surprising is the recent discovery of the synthesis and secretion of a gonadal peptide, *inhibin*, by the placenta[29]. In contrast to the syncytiotrophoblastic origin of hCG and hPL, these neuropeptides and the gonadal peptide are localized exclusively in the *cytotrophoblast* of placental villi.

The immunocytochemical localization of GnRH-like peptide has been observed in the cytotrophoblast. Specific receptors for GnRH have been characterized, and the ability of GnRH to stimulate in vitro synthesis of bioactive hCG was clearly demonstrated[30]. In addition, there is a possibility that placental somatostatin regulates the secretion of hPL. It has been reported that placental CRF may stimulate ACTH secretion by the placenta[31]. The intimate association between neuropeptide-producing cells and those producing protein hormones suggests the intriguing possibility of a *trophoblastic control* system analogous to the neuroendocrine control of the hypothalamic-hypophyseal complex (Fig. X.32).

The demonstration that the cytotrophoblast of the human placenta produces and secretes an *inhibin*-like glycoprotein offers a new dimension in the understanding of the complex mechanisms of the endocrine function of the placenta. Inhibin appears to exert a tonic inhibitory effect on hCG secretion, and hCG stimulates the secretion of inhibin (cAMP-dependent). This interaction between inhibin and hCG appears to be mediated by GnRH[32]. Thus, a local feedback loop between inhibin, GnRH and hCG may be operative.

4.3 The Decidua as an Endocrine Compartment

Currently, the decidua, which lies between fetal membranes and the myometrium, is considered to be a specialized and complex endocrine structure as well as an endocrine target tissue. To date, the following endocrine features have been described: de novo biosynthesis of *prolactin* (Fig. X.37), de novo biosynthesis of *relaxin*, *prostaglandin* synthesis, the presence of *oxytocin* and *vitamin D receptors*, and *1 α-hydroxylase* activity.

The decidua denotes a membrane-like structure which is shed after termination of pregnancy. The decidual cell is a plump, glycogen-rich cell of stromal origin that appears during the late luteal phase of the endometrium, and proliferates in early gestation. Anatomically, three portions of decidua can be recognized; the d. basalis, at the site of implantation, constitutes the maternal portion of the placenta; the d. capsularis covers the gestation sac and disappears in late pregnancy; and the d. vera, lining the rest of the uterine cavity, becomes laminated to the chorion. Thus, the *decidua can communicate directly with the fetus*, via the amniotic fluid, *and* with the adjacent *myometrium*, by simple diffusion.

4.4 Diagnosis and Hormonal Monitoring of Pregnancy

Diagnosis of pregnancy is based essentially on *detection of elevated hCG*. Immunological kits specifically detecting the β subunit are available commercially. Unfortunately, there is *no technique* available, at present, to establish consistently the occurrence of *pregnancy before* the *expected* date of *menses*. Elevated levels of plasma progesterone and estradiol may be useful indications.

In the weeks following conception, hCG measurements are not clinically useful for the determination of those pregnancies which are in danger of termination. During the first two months, since estrogens and progesterone are produced by the corpus luteum of pregnancy, low values have suggested the idea that certain spontaneous abortions are due to luteal insufficiency. In fact, the *causative role of corpus luteum deficiency is controversial*, and abortions are due mostly to genetic abnormalities, detectable by caryotype. In such cases, hormonal rescue is not necessarily advisable.

Late in pregnancy, progesterone is essentially an index of placental function; it is now routinely measured in blood. Estriol (Fig. IX.13 and IX.15) is used mostly to monitor high-risk pregnancies (diabetes, hypertension, Rh factor incompatibility). Fetal death in utero is accompanied by a marked decrease in estriol (Table X.6). Whether the measurement of hPL is, or is not, a better index than that of estriol is still debated.

28. Liotta, A. S., Houghten, R. and Krieger, D. T. (1983) Identification of a β-endorphin-like peptide in cultured human placental cells. *Nature* 295: 593–595.
29. Petraglia, F., Sawchenko, P., Lim, A. T. W., Rivier, J. and Vale, W. (1987) Localization, secretion, and action of inhibin in human placenta. *Science* 237: 187–189.
30. Belisle, S., Guevin, J.-F., Bellabarba, D. and Lehoux, J.-G. (1984) Luteinizing hormone-releasing hormone binds to enriched human placental membranes and stimulates in vitro the synthesis of bioactive human chorionic gonadotropin. *J. Clin. Endocrinol. Metab.* 59: 119–136.
31. Petraglia, F., Sawchenko, P. E., Rivier, J. and Vale, W. (1987) Evidence for local stimulation of ACTH secretion by corticotropin-releasing factor in human placenta. *Nature* 328: 717–719.

Table X.6 Plasma and urinary estriol during pregnancy.

Weeks*	Plasma estriol (ng/ml)‡	Urinary estriol (mg/d)‡‡
20	30	4
24	42	6
28	65	9
30	75	10
32	123	12
34	126	13
36	148	15
38	197	18
40	230	19

*Weeks following the first day of the last menstrual period.
‡Goebelsmann, U., Kitagiri, H., Stanczyk, F. Z., Cetrulo, C. L. and Freeman, R. K. (1975) Estriol assays in obstetrics. *J. Ster. Biochem.* 6: 703–709.
‡‡Fransen, V. A. (1963) *The Excretion of Estriol in Normal Human Pregnancy.* Munksgaard, Copenhagen.

4.5 Prolactin and Lactation

The Mother

Maternal serum PRL begins to rise in the first trimester of pregnancy, and increases progressively to a level which, at term, is ten times higher than that of non-pregnant women. This increase follows an approximately linear pattern (Fig. X.37). The increase in serum PRL concentration is probably causally related to supramaximal estrogen stimulation during the course of human gestation, and is a functional reflection of hypertrophy and hyperplasia of pituitary lactotropic cells.

The *episodic* nature and *sleep-entrained* pattern of PRL release observed in non-pregnant women is maintained during pregnancy. The *midday surge* of *PRL* secretion (and cortisol) in response to lunch also persists in pregnancy, but is greater in magnitude.

Experimental data suggest that PRL may play a role in the regulation of the storage and mobilization of fat, and that PRL influence on fat metabolism is synergized by cortisol. In hyperprolactinemic patients, insulin secretion in response to glucose is exaggerated, as is the suppression of glucagon by glucose, a metabolic finding resembling normal pregnancy. These metabolic effects of PRL, together with the well recognized anti-insulin and catabolic actions of cortisol, may ultimately be shown to serve as components of an integrated endocrine control of metabolic homeostasis in pregnancy.

The Fetus

The pituitary gland of the human fetus is able to synthesize, store, and secrete PRL after the first 12 weeks of intrauterine life[32]. This is consistent with the first detection of pituitary lactotropic cells at 18 weeks of gestation and a sharp increase after 22 weeks. At term, the mean umbilical vein concentration of PRL is higher than that of the maternal plasma (Fig. X.37). This high PRL level decreases progressively to the range of normal children by the end of the first week of postnatal life.

The physiological role of fetal PRL is *unknown*. Experimental evidence suggests that, in the fetal compartment, PRL may participate in osmoregulation by the kidney and in lung maturation.

The Amniotic Fluid Compartment

The highest concentration of PRL is found in the amniotic fluid, where it is five- to ten-fold greater than in maternal serum[33]. Of special interest is the fact that this high PRL concentration in amiotic fluid peaks at the second trimester of pregnancy, when both the fetal and maternal PRL levels are relatively low (Fig. X.37). The human chorionic decidua is capable of de novo biosynthesis of PRL. The secretion of PRL by human decidual tissue in vitro is not influenced by dopamine agonists. PRL binding has been demonstrated in human chorionic laeve. A 50% decrease in amniotic fluid volume can be induced in rhesus monkeys by intra-amniotic injection of ovine PRL, an effect which persists for about 24 h. PRL has also been shown to stimulate uterine contractility. Thus, PRL produced locally may participate in the osmotic regulation of the amniotic fluid compartment, and decidual PRL may serve in concert with decidual relaxin (which inhibits myometrial activity) to modulate uterine contractility[33].

Lactogenesis and Lactation

PRL is the *key* hormone controlling *milk production*. The entire process of lactogenesis, however,

32. Aubert, M. L., Grumback, M. M. and Kaplan, S. L. (1985) The ontogenesis of human fetal hormones. III. Prolactin. *J. Clin. Invest.* 56: 155.
33. Schenker, K. G., Ben-David, M. and Polishuk, W. Z. (1975) Prolactin in normal pregnancy: relationship of maternal, fetal, and amniotic fluid levels. *Amer. J. Obst. Gynecol.* 123: 834–838.
34. Bigazzi, M. and Nardi, E. (1981) Prolactin and relaxin: antagonism on the spontaneous motility of the uterus. *J. Clin. Endocrinol. Metab.* 53: 665–667.

requires *multiple hormonal interaction*. Growth of the mammary duct system is dependent on estrogen synergized by the presence of growth hormone, PRL, and cortisol. The development of the lobulo-alveolar system requires both estrogen and progesterone in the presence of PRL. The synthesis of milk proteins and fat is reported on p. 200.

During pregnancy, the increasing levels of PRL, cortisol, placental lactogen, estrogens and progesterone combine to stimulate the development of the secretory apparatus of the breast, but lactogenesis is minimal, and lactation absent. While estrogen and progesterone act to stimulate mammary development, they inhibit the actual formation of milk; high levels of estrogen and progesterone block PRL action on mammary target cells through a negative regulatory effect on the PRL receptor. Following delivery, estrogen and progesterone levels fall rapidly, resulting in an unopposed autoregulatory mechanism and a prompt increase in PRL receptor number. When PRL receptors become abundant in glandular mammary tissue, lactogenesis and milk secretion are established soon after delivery.

During lactation, basal levels of PRL are not substantially elevated. In fact, PRL levels are near the upper-normal range by the third week postpartum. Lactation is maintained by periodic surges of PRL release stimulated by suckling, which acts to prime the breast for the next feeding. Optimal lactation is dependent on the frequency as well as the duration of suckling.

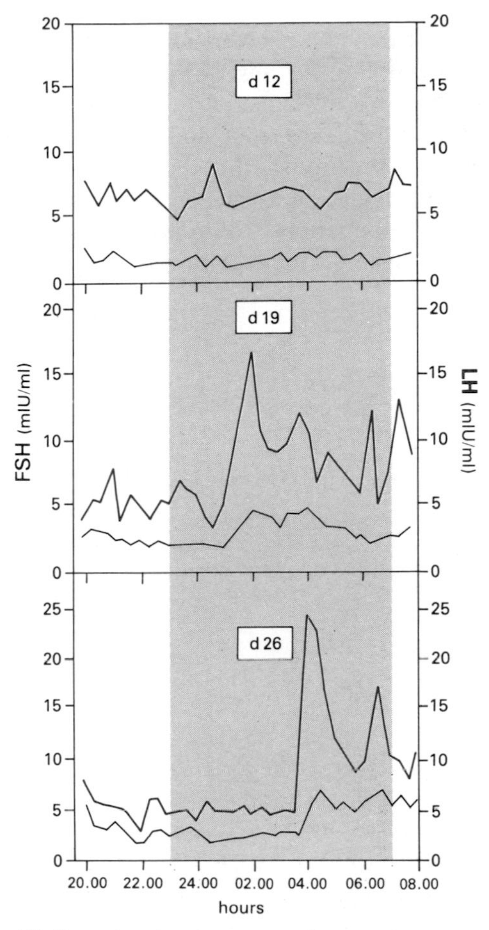

Figure X.38 The development of postpartum, episodic, nocturnal secretion of pituitary gonadotropins 12, 19 and 26 days after delivery.

Neuroendocrinology of the Suckling Reflex

The maintenance of lactation in the puerperium is dependent upon *mechanical stimulation* of the nipple by suckling. Sensory signals originating in the nipple during suckling are conveyed in an afferent pathway of the spinal cord, ultimately reaching the hypothalamus, where they induce an acute response of neuronal mechanisms controlling the release of oxytocin and PRL. Denervation of the nipple or lesions of the spinal cord and brain stem abolish the normal response to suckling.

Three interrelated neuroendocrine events result from *suckling: oxytocin release* is induced by afferent signals to both paraventricular and supraoptic nuclei, without concomitant release of vasopressin. The *myoepithelial cells* in the mammary alveoli and ducts are targets for oxytocin, which brings about contraction of these cells to induce the ejection of milk. An increase in *episodic oxytocin release* occurs when the mother plays with the infant or in anticipation of feeding. A milk let-down response may be observed. This phenomenon illustrates the involvement of psychic centers in the neuroendocrine control of oxytocin secretion.

When *nursing* is *initiated*, a prompt and *large release of PRL* occurs which is temporally associated with, but independent of, episodic oxytocin release. This transient increase of PRL secretion is sufficient to maintain lactogenesis and an adequate milk supply for the next feeding.

Gonadotropin secretion is *inhibited* during *pregnancy* and *lactation*. The inhibitory effect of endogenous opioids appears to be increased during the postpartum period, and diminishes progressively in non-lactating women; parallely, gonadotropin secretion increases, as does gonadotropic cell responsiveness to exogenous GnRH. This set of conditions may account for the transient hypogonadotropinism of puerperium, and for subsequent recovery. The latter is associated with a gradual increase in GnRH neuronal activity resembling the onset of puberty (Fig. X.38); dependency on GnRH activity is substantiated by experimentally induced reactivation of hypothalamic-gonadotropic function during the first 10 days postpartum, through appropriate administration of GnRH. Prolongation of postpartum hypogonadotropinism during lactation is probably due to episodic hyperprolactinemia induced by suckling. Lactational amenorrhea has had profound significance in birth spacing in ancient times. The same may be observed even today, among mothers of !Kung Hunter gathers[35].

35. Konner, M. and Worthman, C. (1980) Nursing frequency, gonadal function and birth spacing among !Kung Hunter gathers. *Science* 207: 788–791.

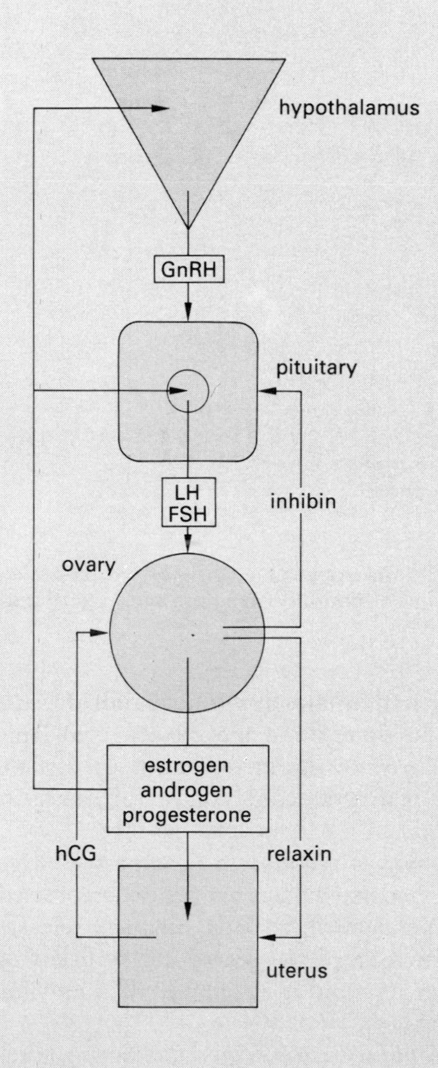

Figure X.39 Schematic and highly simplified representation of the hypothalamic-hypophyseal-ovarian-uterine axis in the control of menstrual cyclicity. In a conception cycle, the timely signal of human chorionic gonadotropin (hCG) is secreted immediately after implantation. Thus, the ovarian-uterine axis becomes an independent functional unit.

5. Summary

The reproductive system in women is composed of *interacting functional units* of the hypothalamic-hypophyseal-ovarian-uterine axis. They communicate, both *downstream* and *upstream*, by *hormonal signals*, in order *to achieve ovulation* and the synchronous development of the progestational endometrium for *conception* and *implantation* (Fig. X.39).

The hypothalamic gonadotropin releasing hormone, *GnRH*, plays a pivotal role in the control of the neuroendocrine axis. Secretory activity of the GnRH neuronal system is *pulsatile* in nature; the frequency and amplitude of its secretion appears to be modulated by neuropeptides, neurotransmitters, and ovarian steroids. During the menstrual cycle, appropriate frequency of GnRH stimulation of the target gonadotropic cells is required. An up-regulation of the functional capacity of the gonadotropic cell, occasioned by an increase in GnRH binding sites, is instrumental in the development of the *midcycle gonadotropin surge* and *ovulation*. When *pulsatile frequency* of GnRH secretion is altered, becoming either too fast or too slow, down-regulation of gonadotropin secretion occurs, and anovulation and/or amenorrhea ensues. Changes from pulsatile to constant GnRH stimulation also down-regulates gonadotropin secretion, resulting in "endocrine castration". Alterations of the menstrual cycle include pseudocyesis, anorexia nervosa, and psychogenic amenorrhea.

PRL, a *lactogenic hormone*, is an important component of successful reproduction. Its secretion is controlled by both hypothalamic inhibiting and releasing factors. When hyperprolactinemia occurs, gonadotropin secretion is impaired, resulting in menstrual disorders and infertility. Frequently, hypersecretion of PRL is due to a pituitary PRL-producing tumor; it can be effectively suppressed by the administration of dopamine or dopamine receptor agonist, with subsequent reversal of cyclic gonadotropin secretion.

Several abnormalities of reproductive function, including *precocious puberty, amenorrhea*, and *irregular* menstrual *bleeding*, can result from an *inappropriate extraglandular* contribution of *estrogens* derived through the peripheral conversion of androgens. Polycystic ovarian syndrome (PCO) constitutes a typical example of excessive extraglandular estrogen production. Consequent to this additional estrogen production, which is not under direct control of pituitary gonadotropin, the cyclic feedback signal is masked, resulting in acyclic gonadotropin release and chronic anovulation. Defects of thyroid and adrenal function (including late-onset deficiences of adrenal steroid biosynthesis) are also causative in the deregulation of reproductive processes.

Pregnancy represents a most remarkable adaptive phenomenon of biological systems. Within the pregnant uterus, the fetal-placental-decidual unit produces extraordinary amounts of steroid and protein hormones, and neuropeptides. They appear to interact and function in a paracrine fashion resembling that of a compressed hypothalamic-hypophyseal-ovarian system. These *supplementary endocrine units* provide regulatory mechanisms for the unidirectional flow of nutrients, from the mother to the fetus, and convey signals for the initiation of parturition when the fetus is ready for extrauterine existence.

General References

Crowley, W. F. Jr., Filicori, M., Spratt, D. and Santoro, N. (1985) The physiology of gonadotropin-releasing hormone (GnRH) secretion in men and women. *Rec. Prog. Horm. Res.* 41: 473–491.

Goodman, A. L. and Hodgen, G. D. (1983) The ovarian triad of the primate menstrual cycle. *Rec. Prog. Horm. Res.* 39: 1–32.

Knobil, E. (1980) The neuroendocrine control of the menstrual cycle. *Rec. Prog. Horm. Res.* 36: 53–67.

Krieger, D. T. (1982). Placenta as a source of "brain" and "pituitary" hormones. *Biol. Reprod.* 26: 55–71.

Siler-Khodr, T. M., Khodr, G. S., Valenzuela, G. and Rhode, J. (1986) Gonadotropin-releasing hormone effects on placental hormones during gestation: 1. Alpha-human chorionic gonadotropin, human chorionic gonadotropin and human chorionic somatomammotropin. *Biol. Reprod.* 34: 245–254.

Swanson, L. W. and Mogenson, G. J. (1981) Neural mechanisms for the functional coupling of autonomic, endocrine and somatomotor responses in adaptive behavior. *Brain Res. Rev.* 3: 1–267.

Tsonis, C. G. and Sharpe, R. M. (1986) Dual gonadal control of follicle-stimulating hormone. *Nature* 321: 724–728.

Van de Wiele, R. L., Boghumil, J., Dyrenfurth, I., Fevin, M., Jewelewicz, R., Warren, M., Rizkallah, T., Mikhail, G. (1970) Mechanisms regulating the menstrual cycle in women. *Rec. Prog. Horm. Res.* 26: 63–103.

Yen, S. S. C. (1983) Clinical applications of gonadotropin-releasing hormone and gonadotropin-releasing hormone analogs. *Fert. Steril.* 39: 257–266.

Yen, S. S. C. (1983) The endocrinology of pregnancy. In: *Maternal-Fetal Medicine: Principles and Practice.* (Creasy, R. K., Resnik, R., eds), W. B. Saunders, Philadelphia, pp. 331–360.

Yen, S. S. C. (1982) Neuroendocrine regulation of gonadotropin and prolactin secretion in women: disorders in reproduction. In: *Current Endocrinology, Clinical Reproductive Neuroendocrinology section.* (Vaitukaitis, J. L., ed), Elsevier, New York, pp. 137–176.

Robert G. Edwards, Colin M. Howles,
and Michael C. Macnamee

Hormones and In Vitro Fertilization

Foreword

The happiness of the individual and the family, and the evolution of societies–indeed, all human destiny–is related, to a large extent, to hormonal mechanisms which regulate fertility.

Hormones play a fundamental role in all aspects of fecundity and sexual reproduction, including sexual behavior. The formation of germ cells, implantation of fertilized ova, embryonic and fetal development, and parturition, are dependent on the concerted actions of hormones and growth factors.

As our knowledge of human reproduction becomes more complete, the individual is being provided the means of determining his or her fecundity, in the largest sense. Where once a woman over child-bearing age had no recourse to maternity, the technology of storing oocytes, and knowledge concerning the hormonal prerequisites for implantation, may give women greater options, and may ultimately alter the finality of menopause. Infertility caused by physical obstruction or absence of the fallopian tubes can now be treated by the transfer of oocytes fertilized in vitro (IVF).

Although IVF, in particular, among other recent developments in medicine, has provoked animated and discordant reactions concerning its application in humans, it cannot be denied that it has broadened the realm of possibilities, and that it complements those technologies presently at our disposal in the matter of birth control and fertility. Indeed, the success of the IVF procedure which led to the birth of Louise Brown, in 1978, represented a remarkable application of knowledge acquired in 50 years of research devoted to the understanding of human reproduction.

Daily Mail
8p
WORL
EXCLUS
And here she is...
THE LOVELY
LOUISE

It is primarily the broad implications of IVF that have aroused controversy. Some of the hypothetical uses of IVF are clearly beyond present technological means; others could potentially become reality. For example, the unprecedented notion of pregnancy in the male has been considered, although, among other problems, the exposure of the fetus to androgens inherent in the male makes the possibility remote – particularly in the case of a female conceptus.

What is not necessarily out of the question, however, is the development of a true "test tube" baby which would be incubated in vitro, under strictly controlled nutritive, environmental and hormonal conditions, until development is complete.

Nevertheless, with the controversial, moral aspects of fertility aside, IVF has already become very useful medically, not only in diseases related to infertility in women, as described above, but also in some cases of male infertility, and this is a considerable achievement.

E. E. B.

The Endocrinology of Ovarian Stimulation

The chances of pregnancy after IVF are maximized by a *superovulation strategy*. This is necessary in order to promote the synchronous development of several preovulatory follicles which will yield an adequate number of mature eggs for fertilization and replacement. The crux of such a strategy is the choice of effective forms of exogenous follicular stimulants that have no deleterious effects on the luteal phase or uterus.

Ovarian Stimulation with Clomiphene/Human Menopausal Gonadotropin

The most widely used stimulatory regime currently employed is a *combination of clomiphene citrate and human menopausal gonadotropin (hMG)*. Clomiphene citrate has both estrogenic and antiestrogenic properties which cause the release of gonadotropins from the pituitary gland[1]. Human MG is a partially purified and concentrated extract of the urine of postmenopausal women which contains bioactive LH and FSH. The standard stimulatory regime used at Bourn Hall (Cambridge) is comprised of the administration of clomiphene (100 mg), on days two to six, and hMG (equivalent to 150 IU of LH/FSH), from day five of the menstrual cycle.

Ovarian response is usually assessed by daily determination of *estrogen* concentrations in urine or blood, combined with frequent follicular *ultrasonography*, from day nine of the stimulated cycle.

At Bourn Hall, when the rising urinary estrogen level reaches more than 200 μg/24 h and the largest of the developing follicles exceeds 17 mm in diameter, ovulation is induced by an injection of hCG (5000 IU). Ova are surgically recovered from the ripening follicles just before they are due to ovulate (usually 34.5 h after hCG).

In the natural 28-day menstrual cycle, ovulation follows approximately 36 h after the endogenous LH surge which normally starts at 3.00 h sometime between day 12 and 16. In a small proportion (~18%) of IVF patients undergoing follicular stimulation, an attenuated endogenous LH surge occurs[2]. LH concentrations must be assayed, in urine or plasma, in all patients, in order to identify those who have an LH surge. In our clinic, three hourly urine samples are assayed for LH, using a rapid assay. LH monitoring is normally commenced when the leading follicles reach 12 mm in diameter. Human CG is administered immediately to sustain these weak surges, and laparoscopy for oocyte collection is timed for 25 to 28 h from the first rise in LH above basal values. The function of hCG in such cases is to support the abnormally attenuated LH surge and to achieve the degree of follicular maturity necessary for successful fertilization of the ova.

Plasma *progesterone* concentrations are also monitored in many IVF programs. Rising plasma progesterone levels can be indicative of approaching ovulation[3]. Persistently elevated plasma progesterone levels during the follicular phase (>1.2 ng/ml) of a stimulated cycle may be indicative of over-stimulation, which can result in follicular atresia and decreased fertilization rates.

Other Ovarian Stimulants

Although the vast majority of patients considered suitable for IVF can be successfully treated with clomiphene citrate/hMG on the regime outlined above, some recent innovations have opened up new possibilities for follicular stimulation. Preparations of *FSH-enriched,* postmenopausal urine may have advantages over conventional hMG, as a more physiological LH:FSH ratio may be achieved in stimulation cycles. This pretreatment with FSH might result in more synchronous follicular development than treatment with clomiphene citrate/hMG alone[4]. Nevertheless, a small trial carried out at various centers suggests that the number of follicles recruited, oocytes collected and the proportion fertilized, was similar in patients treated with FSH-enriched hMG and standard hMG. The possible benefits of more elaborate combinations of FSH/hMG and clomiphene citrate have not been explored.

GnRH and its analogs may also be of use in selected patients. Native *GnRH injected* in a *pulsatile* fashion has been used successfully to promote ovulation in patients with amenorrhea of diverse etiology[5,6]. A disadvantage of IVF is the relatively low incidence of multiple follicular development. An advantage is the virtual elimination of hyperstimulation of the ovaries.

Another variation of the pulsatile administration of hormones has been reported in which hMG was given iv, via a pulsatile pump, after clomiphene therapy[7].

Pituitary suppression with a GnRH analog has been used successfully to treat infertile women with abnormal hormone profiles[8]. The strategy of this treatment requires the administration of a high dose of GnRH, which causes a transient period of hypogonadotropism. *Ovarian stimulation is reinitiated* by the administration of *exogenous gonadotropins*, endogenous GnRH release remaining suppressed by a low dose of analog. With this regime, follicular growth can be controlled precisely, and the endogenous LH surge is blocked. This type of treatment may be useful for patients with a weak, undetected LH surge, premature luteinization, or high follicular progesterone concentrations in previous stimulatory cycles. Large quantities of hMG have to be administered in order to achieve an adequate response.

Implantation

Many *attempts* to *increase the chance* of implantation have proved *disappointing*. Even in the most successful centers, replacing three healthy-looking embryos gives only an incidence of 35% of clinical pregnancies, and 24% of all patients deliver one or more babies.

In all stimulated cycles, the uterus is exposed to high concentrations of estradiol, in the follicular phase, and progesterone, after oocyte recovery. These high levels of steroid hormones could compromise implantation[9]. The antiestrogenic effects of clomiphene might adversely influence the function of the uterus and corpus luteum[10].

Luteal phase support, by administering either exogenous progesterone or hCG, does not seem to be beneficial to the outcome of IVF[2]. Information is needed on the control of the implantation window, i.e. whether the period during which implantation can take place is highly limited.

After implantation, the incidence of abortion appears to be greater with IVF, implying that more fetuses are abnormal. Two factors influencing abortion are previous obstetric history[1] and age[11], both largely uncontrollable factors in an IVF program.

References

1. Adashi, E. Y. (1984) Clomiphene citrate: Mechanism(s) and site(s) of action – a hypothesis revisited. *Fert. and Steril.* 42: 331–341.
2. Edwards, R. G. (1984) The Ethical, Scientific and Medical Implications of Human Conception In Vitro. *Pontif. Acad. Sci. Scr. Varia* 51: 193–249.
3. Hoff, J. D., Quigley, M. E. and Yen, S. S. C. (1983) Hormonal dynamics at mid-cycle: a reevaluation. *J. Clin. Endocrinol. Metab.* 57: 792–796.
4. Hillier, S. G., Afnan, A. M., Margara, R. A. and Winston, R. M. L. (1984) Ovarian function before in-vitro fertilization. *J. Endocrinol.* 102 suppl, abstr. 81.
5. Hurley, D. M., Brian, R. J. and Burger, H. C. (1983) Ovulation induction with subcutaneous pulsatile gonadotrophin hormone-releasing hormone: singleton pregnancies with previous multiple pregnancies after gonadotrophin therapy. *Fert. Steril.* 40: 575–579.
6. Abdulwahid, N. A., Adams, J., Van der Spuy, Z. M. and Jacobs, H. S. (1985) Gonadotrophin control of follicular development. *Clin. Endocrinol.* 23: 613–626.
7. Afnan, A. M. M., Hillier, S. G., Margara, R. A., Franks, S. and Winston, R. M. L. (1984) Pulsatile gonadotrophin administration for in vitro fertilization. *Lancet* 8388: 1239.
8. Fleming, R., Adams, A. H. and Barlow, D. H. (1982) A new systematic treatment for infertile women with abnormal hormone profiles. *Brit. J. Obstet. Gynaecol.* 89: 80–83.
9. Edwards, R. G., Fishel, S. B., Cohen, J., Fehilly, C., Purdy, J. M., Slater, J. M., Steptoe, P. C. and Webster, J. (1984) Factors influencing the success of in vitro fertilization for alleviating human infertility. *J. In Vitro Fert. Embryo Transfer* 1: 3–23.
10. Clark, J. H. and Gutherie, S. C. (1981) Agonistic and antagonistic effects of clomiphene citrate and its isomers. *Biol. Reprod.* 25: 667–672.
11. Fishel, S. B., Edwards, R. G., Purdy, J. M., Steptoe, P. C., Webster, J., Walters, E., Cohen, J., Fehilly, C., Hewitt, J. and Rowland, G. (1985) Implantation, abortion and birth after IVF using the natural menstrual cycle or follicular stimulation with Clomiphene citrate and human menopausal gonadotrophin. *J. In Vitro Fert. Embryo Transfer* 2: 123–131.

General References

Trouson, A. and Wood, C. (1981) Extracorporal fertilization and embryo transfer. *Clin. Obstet. Gynecol.* 8: 681–713.
Seegar-Jones, G. (1984) Update on in vitro fertilization. *Endocrine Rev.* 5: 62–75.

Rosalyn S. Yalow

Radioimmunoassay of Peptide Hormones: Problems and Pitfalls

Radioimmunoassay (RIA) was first described for the measurement of plasma insulin, in humans, in 1959 (Fig. 1), and was subsequently applied quite rapidly to the measurement of other peptide hormones. The inhibition of binding of labelled peptide to specific antibody of unknowns is compared to that of known standards. Standards and unknowns do not need to be identical biologically or chemically. Although RIA is *simple in principle*, there are *problems* and *pitfalls* in its *practical application*.

Non-hormonal Effects

Non-specific factors, such as changes in pH, ionic strength and buffer, and a variety of anticoagulants or protective agents, can affect the antigen-antibody reaction. These non-specific effects do not behave in a predictable fashion, but depend on the particular antiserum and particular hormone employed. It is therefore important to ensure that standards and unknowns are in the same milieu, and that the assay is of sufficient sensitivity to permit dilution of plasma containing an unknown by a factor of 10 or more, in order to dilute out possible non-specific factors.

A simple test to confirm that non-specific factors do not interfere is to assay the unknown fluid before and after removal of the substance to be determined by a specific adsorbent. In the past we used precipitated silica (QUSO G32) for insulin, ACTH, etc. Affinity chromatography should prove to be equally satisfactory. Recently, we have employed C_{18} Sep-Pak cartridges (Water Associates) to extract peptides with a M_r smaller than 10–20,000. These cartridges are useful in extracting both acidic peptides such as gastrin and cholecystokinin (CCK) C-terminal octapeptide, and basic peptides such as glucagon, secretin and the 58 aa CCK precursor (p. 541). They also permit the concentration of hormones and the preparation of hormone-free plasma for standard curves.

Hormonal Effects

Unavailability of a reference preparation of the appropriate species, the presence of immunologically related but different hormones, and heterogeneity of hormonal forms are commonly encountered problems.

The validity of RIA procedure requires *superposition of a dilution curve of unknown*, along a *dilution curve of known*, standards. Some antisera produced in guinea-pigs by immunization with pig insulin may be highly sensitive (Fig. 2A). Note that bovine and human insulins behave almost identically in this assay. However, with an antiserum obtained from an insulin-resistant diabetic subject, the insulins are clearly distinguished (Fig. 2B). Therefore, antisera which do not distinguish between animal and human sources are useful for the measurement of insulin in humans. In fact, the assay of insulin poses no problem at present, since many species of purified insulins are currently available. However, for other peptides this may not be the case.

RIA procedures may be complicated because there are hormones which *share common amino acid sequences* but which have different biological activities, as for example gastrin and cholecystokinin (same C-terminal pentapeptide), and ACTH and MSH (similar N-terminal sequences) (Fig. IV.1). The placenta synthesizes several glycoproteins that are similar but not identical to pituitary hormones. This introduces problems in assays in pregnant women (p. 475).

Many if not all peptide hormones are found in *more than one form, in plasma*, glandular tissue, and other tissue extracts. These forms may represent either precursor(s) or metabolic product(s) of the biologically active hormone. It is fortunate that the first RIA was described for insulin, since it is the peptide with full biological activity (M_r=6000) which predominates in the circulation of virtually all subjects in the stimulated state. Only in patients with insulinoma, or those with a rare genetic abnormality that prevents

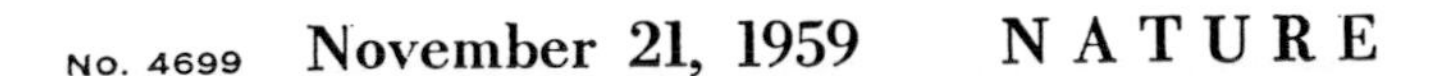

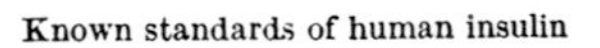

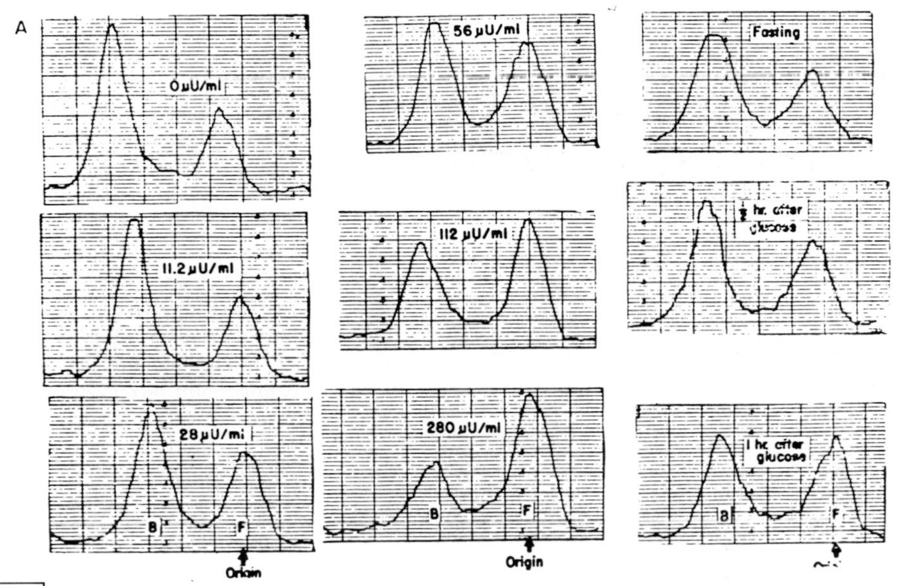

Figure 1 *The first assays of human insulin.* **A.** *Radiochromatoelectrophoretograms of mixtures containing [^{131}I]bovine insulin and anti-bovine insulin antiserum first used in RIA. The scans to the left and middle are of mixtures containing standard. Scans to the right are of mixtures containing a 1:4 dilution of plasmas obtained during a glucose tolerance test.* **B.** *Standard curve: experiment performed on January 7, 1960 with guinea-pig anti-insulin serum diluted 1:6000. [^{131}I]bovine insulin was used as ligand.*

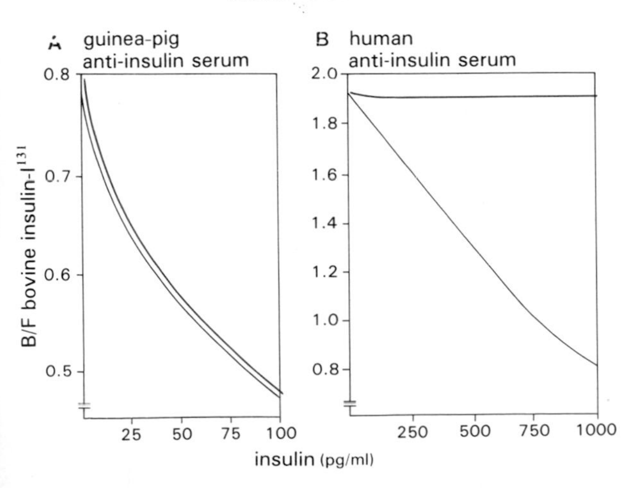

Figure 2 *Standard curves for insulin assay with serum from a guinea-pig* (**A**) *and an insulin-treated, diabetic, hypoglycemic patient* (**B**). *Black lines show standard curves for bovine insulin, and red lines indicate human insulin. Dilutions of serum from a non-diabetic, hypoglycemic patient (not shown) are parallel to the bovine insulin inhibition curves. Bovine insulin and human insulin cross-react almost identically in the guinea-pig antiserum assay* (**A**), *while human insulin hardly cross-reacts with human antiserum to insulin* (**B**), *permitting differential diagnosis between circulating endogenous and exogenous insulin. (Bauman, W.A. and Yalow, R.S. (1980) Differential diagnosis between endogenous and exogenous insulin reduced refractory hypoglycemia in a non-diabetic patient.* N. Engl. J. Med. *303: 198–199).*

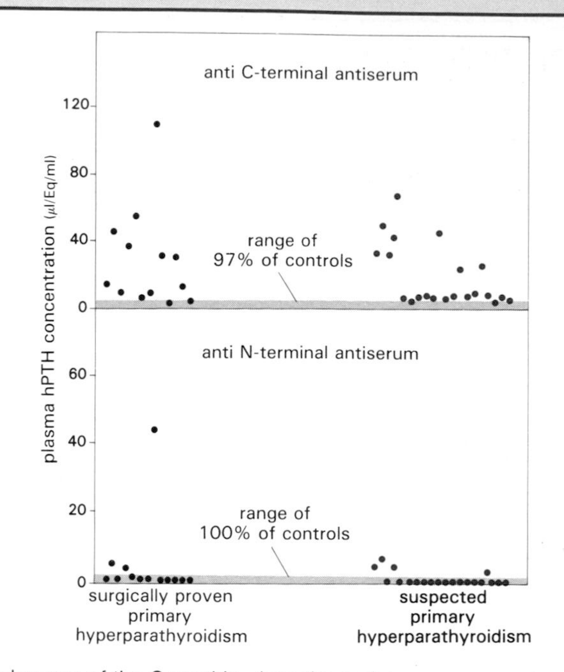

Figure 3 *Immunoreactive hPTH concentrations in plasma of patients with surgically proved or suspected primary hyperparathyroidisms, as measured with C-terminal (top) or N-terminal (bottom) antiserum. Gray areas show a range of normal values. Each patient is represented by a pair of points, one in the upper and one in the lower frame, equidistant from the vertical axis. (Silverman, R. and Yalow, R.S. (1971) Heterogeneity of parathyroid hormone: chemical and physiological implications.* J. Clin. Invest. *52: 1958–1971).*

cleavage of the C peptide, does the prohormone appear to predominate, and metabolic fragments are not immunoreactive with most antisera.

As an example, PTH assay is complicated by the presence of a biologically inactive C-terminal fragment with a long turnover time as compared to that of intact hormone. This fragment is removed primarily by the kidney, and is thus permanently elevated in uremic patients. The use of a C-terminal antiserum has an advantage in the diagnosis of primary hyperparathyroidism (Fig. 3), because, for a given secretory rate, its concentration in the plasma is amplified by the prolonged turnover time of the fragment. However, it has some disadvantages in uremia, since increased immunoreactivity may reflect decreased removal associated with deterioration of kidney function rather than increased secretion associated with secondary hyperparathyroidism. Other examples are described elsewhere, e.g., gastrin (p. 559) and ACTH (p. 244). It is interesting to note that ectopic ACTH, in the form of its proopiomelanocortin precursor, is found at the stage of squamous metaplasia in the lungs of heavy smokers.

Future Trends

Traditional RIAs have made use of heterogeneous animal antisera (polyclonal antibodies) and radioactive tracers.

Recently, *monoclonal antibodies* have come into use in RIA procedures. The development of such antibodies provides a virtually *unlimited amount* of homogeneous antibodies directed *against a specific antigenic site*. However, the affinity of these monoclonal antibodies for the antigen is usually lower than that of polyclonal antibodies. Thus, heterogeneous antibodies are more likely to provide highly sensitive assays than monoclonal antibodies. For research laboratories requiring only limited amounts of antisera, the simplicity of antibody production in animals, compared to the considerable effort required for the production of monoclonal antibodies, must be considered.

Increasingly, *non-isotopic labels* are employed in immunoassay procedures. In fact, immunological methods such as hemagglutination inhibition assays were described before RIA, but they did not have sufficient sensitivity to permit measurement of plasma insulin. Enzymatic reactions, fluorescent labelling and bioluminescence systems are potentially more powerful, but problems of non-specificity must be overcome. It should be appreciated that there is a very wide range in the plasma concentration of hormones. Thyroxine and cortisol concentrations are in the range of 100 to 500 pmol/ml; insulin, a thousand-fold lower, and basal secretin and cholecystokinin concentrations are about 1 fmol/ml. It seems likely that RIA will remain the method *of choice* for the assay of substances *in the fmol/ml concentration range*.

RIA and related methodologies are powerful tools for the measurement of a large number of clinical parameters of health and disease. Nonetheless, their use without regard to possible problems and pitfalls can detract from their very important role in clinical medicine.

CHAPTER XI

Pancreatic Hormones

Pierre Freychet

Contents, Chapter XI

Pancreatic Hormones

Diabetes mellitus has been known of for a long time. However, it was only in 1869 that Langerhans identified a new type of pancreatic cell, characteristically organized in a glandular parenchyma, in groups which Laguesse called *islets of Langerhans* in 1893. In 1889, von Mering and Minkowski demonstrated that diabetes mellitus, characterized in its most evident form by permanent hyperglycemia and glycosuria, could be induced experimentally in the dog by total pancreatectomy. This demonstrated the essential role of the pancreas in the regulation of glucose homeostasis. The hypoglycemic hormone responsible for this action was called *insulin* by De Meyer (1909), and was finally *extracted from the pancreas* by Banting and Best in 1922[1]. This discovery had a fundamental impact in various domains. It had an extremely beneficial effect on the prognosis and therapy of insulin-dependent diabetes, allowing a specific replacement treatment for endogenous insulin deficiency, which, untreated, leads to potentially fatal ketoacidosis. The arrival of the insulin era was also of major importance in the area of protein chemistry. Insulin was one of the first proteins to be *crystallized* (Abel 1926), and its *primary structure* was the *first to be elucidated*[2]; partial synthesis was carried out between 1964 and 1966, and total synthesis was completed in 1974. Human insulin, available commercially, is currently prepared by a modification of pork insulin or by a biosynthetic process using recombinant DNA technology.

The identification of a *hyperglycemic* pancreatic polypeptide called *glucagon* (a substance mobilizing glucose; Murlin, 1923) followed the discovery of insulin closely, but did not have the same clinical or physiological impact. Glucagon, however, has aroused increasing interest in the last 30 years. Its primary structure was identified in 1956, and its total synthesis was carried out between 1966 and 1968. The study of *the mechanism of glucagon action* led to the basic discovery of the "*second messenger*", *cAMP*, by Sutherland (p. 100). Recent findings in molecular biology have indicated that glucagon is part of an important family of regulatory peptides.

1. Anatomical, Histological, Embryological and Phylogenetic Aspects of the Pancreas

1.1 Anatomy and Histology

The *endocrine portion of the pancreas* (islets of Langerhans) accounts for *only* approximately *1% of the total weight* of the organ (70 g in the normal adult human). The average number of islets in the adult human pancreas is around 250,000; their diameters vary between 100 and 200 μm. In humans and other mammals, islets consist of at least *four cell types* (Fig. XI.1): *A* (or α2), *B* (or β), *D* (or α1 or δ) and PP (pancreatic polypeptide). *Glucagon is secreted by A cells*, and *insulin by B cells.* Two other polypeptides have been identified in normal pancreatic tissue: 1. *pancreatic polypeptide* (PP), a polypeptide whose physiological role is still poorly understood. This peptide, consisting of 36 aa in all species (with a greater interspecies variation), and for which the

1. Banting, F. G. and Best, C. H. (1921–1922) The internal secretion of the pancreas. *J. Lab. Clin. Med.* 7: 251–266.
2. Sanger, F. (1959) Chemistry of insulin. *Science* 129: 1340–1344.

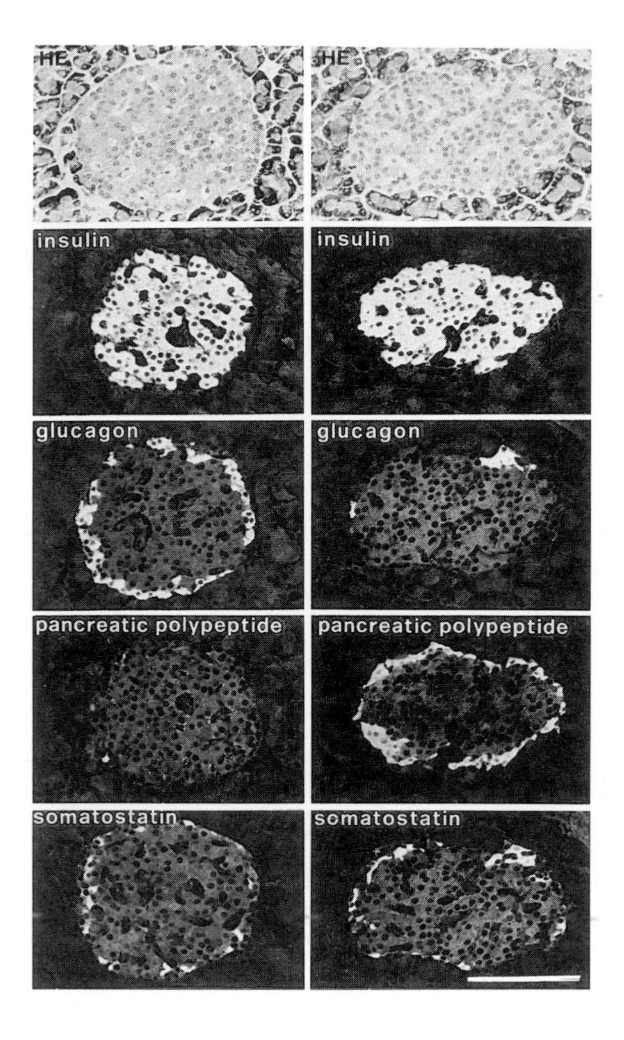

Figure XI.1 Immunohistochemical detection of four hormones in the islet of Langerhans: successive serial sections (3 μm) of two different islets in the rat (stained with hemalum eosin (HE) and with antisera to the indicated hormones, as revealed by indirect immunofluorescence). The series to the *left* shows a *glucagon-rich* or *dorsal type islet*, and the series to the *right* a *pancreatic polypeptide-rich* or *ventral type islet*. Bar=100 μm. (Orci, L. (1984) Patterns of cellular and subcellular organisation in the endocrine pancreas. *J. Endocrinol.* 102: 3–11).

human sequence is known, was identified in specific cells of the endocrine pancreas, present mainly in the dorsal portion of the head of the pancreas[3]; 2. *somatostatin*[4], a polypeptide isolated originally from the hypothalamus (p. 204). The cell type responsible for the pancreatic production of somatostatin has been identified as the D cell[5].

The morphology and *topological distribution* of the islets, and the relative importance of the *different cell types, vary* considerably between species. In the normal adult human, B, A, D and PP cells represent, respectively, 60, 15, 10 and 15% of the cellular population of islets. B cells are located in the center of the islet, the A cells at the periphery, and D cells are frequently interposed between A and B cells. In most species, *the three cell types A, B and D are often in intimate contact.* The possible physiological significance of this histological observation may involve an effect of glucagon on insulin secretion and an inhibitory effect of somatostatin on the secretion of insulin and glucagon. Quantities of endocrine and exocrine tissue vary according to species, but in most vertebrates, particularly mammals, endocrine islets are located in the middle of the exocrine tissue. Excessive production of peptides normally present in the digestive tract (of which some are also neuropeptides, p. 547), by pancreatic tumors, emphasizes the relationship between the endocrine pancreas and the gastrointestinal tract.

1.2 Embryology and Phylogeny

The pancreas and the liver are derived from a *common precursor* (the hepato-pancreatic ring), itself derived from the gastrointestinal tract (the anterior primitive intestine). From this precursor, evaginations penetrate the primitive mesenchyma. These diverticula form cellular chords which are first full, then hollow (tubules), giving rise to pancreatic canals, acini, and endocrine islets. During the development of the pancreas, at least one type of endocrine cell appears before the acini. In humans, A and B cells appear at the same time. In the fetal pancreas, A and D cells represent approximately 60% of the total cellular population of the islet, in contrast to 25% in the adult. The origin of pancreatic endocrine cells is controversial. It is generally recognized that the majority of insulin cells are derived from the endodermal epithelium of the gastrointestinal tract. However, the anterior primitive intestine and its derivatives (therefore the pancreas) are also "colonized" by cells of ectodermal origin belonging to the group of "APUD" cells (p. 549).

3. Orci, L., Malaisse-Lagae, F., Baetens, D. and Perrelet, A. (1978) Pancreatic polypeptide-rich regions in human pancreas. *Lancet* 2: 1200–1201.
4. Dubois, M.P. (1975) Presence of immunoreactive somatostatin in discrete cells of the endocrine pancreas. *Proc. Natl. Acad. Sci., USA* 72: 1340–1343.
5. Orci, L., Baetens, D., Dubois, M.P. and Rufener, C. (1975) Evidence for the D-cell of the pancreas secreting somatostatin. *Hormone Metab. Res.* 7: 400–402.

Insulin, or a larger molecule containing insulin, is present in the gastrointestinal tract of some *invertebrates*. It has been proposed that, during evolution, an ancestral molecule, represented by a digestive enzyme (proto-proinsulin), may have preceded the appearance of insulin[6]. The endocrine pancreas is separate in cyclostomes, even before the appearance of a differentiated exocrine tissue; it consists of type B cells and is apparently deficient in A cells, although glucagon (or a glucagon-like substance) is produced in the digestive tract. In vertebrates, islets are generally disseminated in exocrine tissue, although in some fish (teleosts) a large portion of the endocrine tissue is developed into a principal "islet". In birds, some islets contain an especially high concentration of A and D cells; glucagon is approximately 10 times more abundant than in the pancreas of mammals, in which endocrine tissue represents only a small portion of the pancreas. In mammals, at least in the adult, B cells are in the majority.

2. Insulin: Structure, Biosynthesis, Secretion, Circulation, and Metabolism

2.1 Structure

The insulin molecule consists of *two polypeptide chains, A(21 aa)* and *B (30 aa)*, connected by *two disulfide bridges*, A7–B7 and A20–B19; a *third* disulfide bridge links positions 6 and 11 of the *A* chain (Fig. XI.2A). This two-chain structure has been present throughout evolution, but major variations in the sequence are observed between species. The most frequent variations involve residues A8, 9 and 10, A12–14, and the N- and C-terminal portions of the B chain (Fig. XI.2B). These portions of the primary structure, more specifically residues A8–10, play an important role in the antigenic specificity of the insulin molecule. Pork insulin differs from human insulin by only a single residue (B30), whereas beef insulin differs by two residues (A8 and A10). Most of the mammalian insulin molecules have relatively similar sequences, with the exception of insulin from hystricomorph rodents (guinea-pig and coypu), which differs considerably (35–45% of the sequence) from human insulin. However, the *primary structure of insulin has been relatively stable throughout evolution*; human and hagfish (cyclostome) insulin sequences differ by 40%.

Most mammalian insulins (excepting guinea-pig insulin) have *similar biological potencies* in all mammals, including humans. Fish insulin has considerable potency in mammals, while immunologically it clearly differs from mammalian insulin. This evolutionary *stability of the biological activity* of insulin contrasts with the *variation in its antigenic specificity*. Its stability is associated with the three-dimensional structure of the insulin molecule, relatively constant throughout evolution, and this integrity is essential in order that insulin may bind its receptor and exert its biological effects.

The *three-dimensional structure* of the insulin molecule has been determined by X-ray crystallographic analysis[7]. The spatial configuration of the two chains is compact, the A chain resting on the central portion of the B chain, of which the N- and C-terminal extremities surround the A chain (Fig. XI.3A). Insulin can exist as a *monomer* (M_r=6000) or, by the aggregation of two monomers (association of two distinct molecules at the level of the B chains), as a *dimer* (M_r=12,000). Three dimers can aggregate in the presence of *two zinc atoms*, forming a *hexamer* (M_r=36,000) (Fig. XI.3B)[7]. Insulin circulates essentially in the form of a *monomer* because the physiological concentrations of the hormone are much lower (0.05–1.0 nmol/l) than the dissociation constant, dimer–monomer (approximately 1 μM at neutral pH). Certain residues of the A chain, Gly-A1 and Gln-A5, Tyr-A19 and Asn-A21, are on the surface of the molecule. They do not vary, and are not involved in the aggregation of the molecule. With residues of the C-terminal portion of the B chain (Gly-B23, Phe-B24, and Phe-B25), fully exposed in the monomer, they form a region, at the surface of the molecule, which is especially important for the integrity of the three-dimensional configuration and the biological activity of the hormone[8].

The degree of aggregation of insulin molecules is a function of various factors, including the concentration of hormone, pH, the presence of zinc, and temperature. Zinc, especially, favors aggregation, probably by binding to residue His-B10. The acid–

6. Steiner, D.F., Clark, J.L., Nolan, C., Ribenstein, A.H., Margoliash, E., Aten, B. and Over, P.E. (1969) Proinsulin and the biosynthesis of insulin. *Rec. Prog. Horm. Res.* 25: 207–282.
7. Blundell, T.L., Cutfield, J.F., Cutfield, S.M., Dodson, E.J., Dodson, G.G., Hodgkin, D.C. and Mercola, D.A. (1972) Three-dimensional atomic structure of insulin and its relationship to activity. *Diabetes 21*, suppl. 2: 492–505.
8. Gammeltoft, S. (1984) Insulin-receptor binding kinetics and structure-function relationships of insulin. *Physiol. Rev.* 64: 1321–1378.

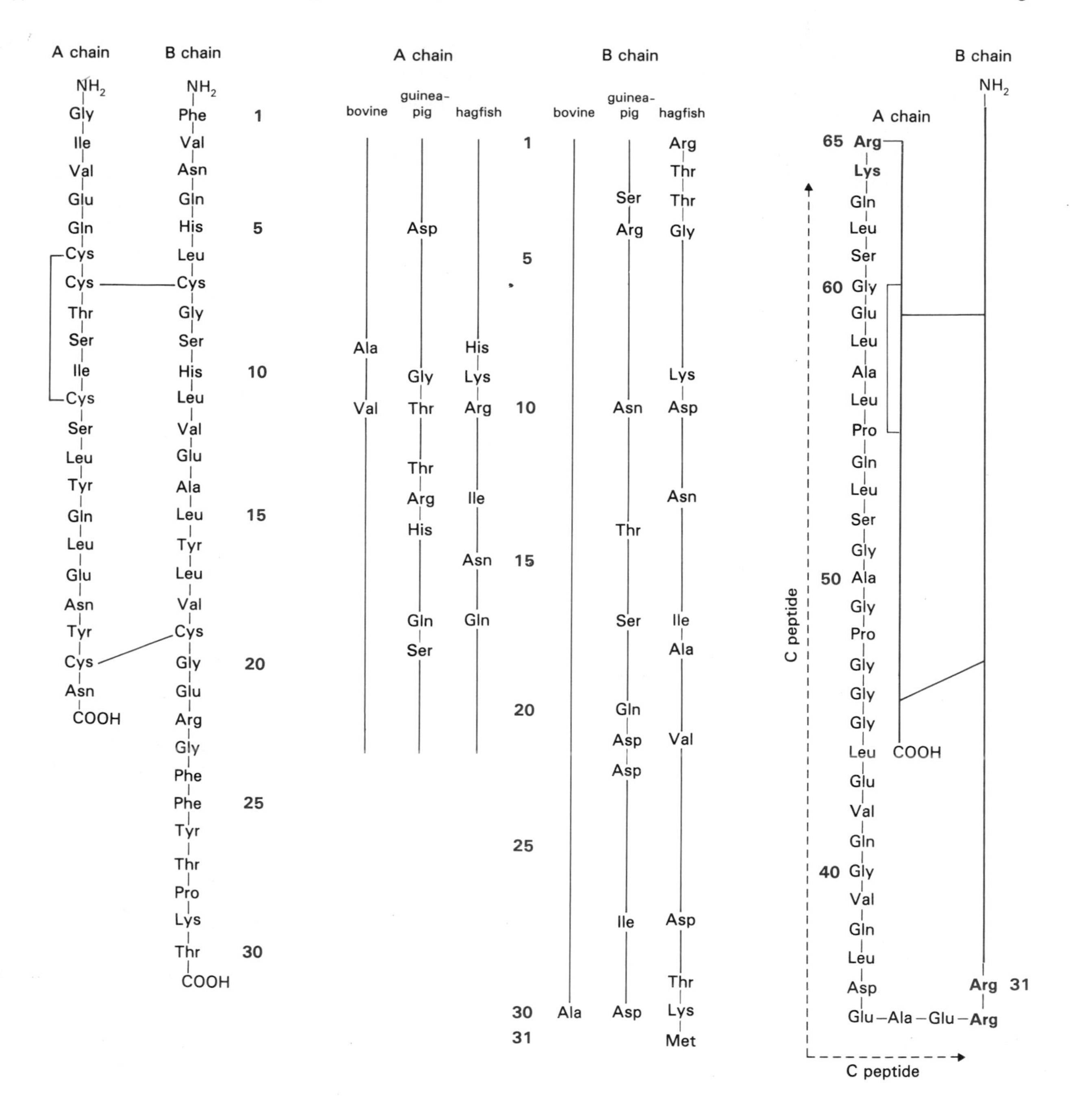

Figure XI.2 **A.** The primary structure of human insulin. **B.** Structure of insulin from cow, guinea-pig, and hagfish. Only the residues which differ from human insulin are indicated. Note the presence of a supplementary residue at the C-terminal portion of the B chain (Met-B31) in hagfish insulin. Porcine insulin, not represented, differs from human insulin by only a single residue (Ala-B30 in the place of Thr-B30). **C.** Structure of human proinsulin.

alcohol mixture used to extract insulin from the pancreas eliminates zinc. To crystallize insulin, the addition of zinc or another metallic ion is necessary (crystallized insulin normally contains 0.3 to 0.6% zinc). In the absence of zinc, insulin (M_r=6000) can dimerize (M_r=12,000) at acid pH. Insulin (pI=5.3) is relatively insoluble at pH between 4 and 7.

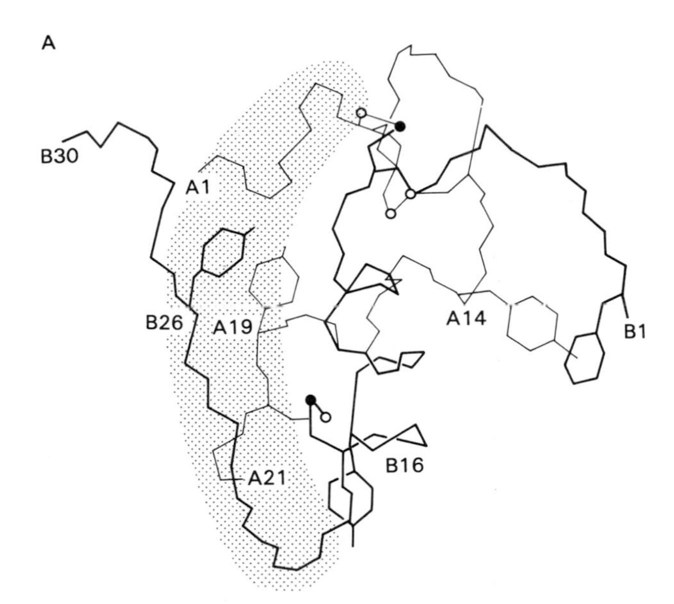

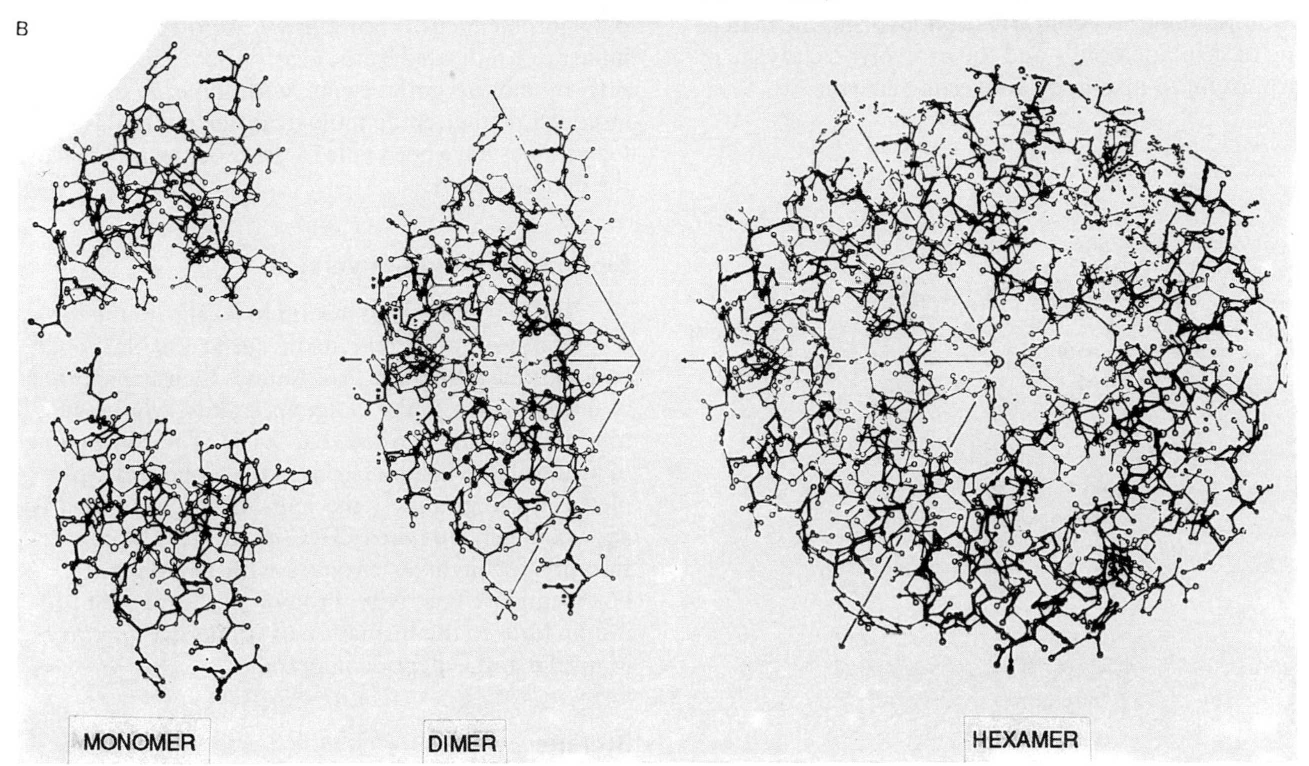

Figure XI.3 **A.** Structure of the insulin molecule (monomer). Arbitrarily, the B chain has been drawn thicker. The shaded area corresponds to the portion of the molecule involved in receptor binding . **B.** Insulin monomer, dimer and hexamer: three-dimensional atomic structure. The main contacts in the dimer are between non-polar B chain residues (e.g., Phe-B24 of one monomer and Tyr-B26 of the other monomer, and vice-versa), to form an antiparallel β-pleated sheet structure. (Blundell, T.L., Dodson, G., Hodgkin, D. and Mercola, D. (1977) Insulin: the structure in the crystal and its reflection in chemistry and biology. *Adv. Prot. Chem.* 26: 279–402).

2.2 Biosynthesis and Secretion

Proinsulin

Insulin is formed from a *biosynthetic precursor of a higher molecular weight, proinsulin.* Proinsulin (M_r=9000), like insulin, reacts with zinc to form a hexamer with a M_r of 55,000. It is also extracted from the pancreas by an acid–alcohol mixture, and can be obtained in a crystallized state. Proinsulin and intermediary fractions of a proinsulin type represented 3 to 6% of the total protein in previous commercial preparations of insulin. In current preparations (monocomponent or monopeak insulin), these contaminants are minimized. After complete reduction of proinsulin (which breaks the disulfide bridges), reoxidation regenerates proinsulin, with a recovery of 70 to 80%, whereas under the same conditions only ~1% of insulin is recovered. This fact emphasizes the importance of proinsulin in the biosynthetic processing of insulin, especially in the correct pairing of disulfide bridges.

Proinsulin consists of a continuous chain, beginning at the N-terminal extremity of the B chain and finishing at the C-terminal portion of the A chain, with a polypeptide segment (*connecting peptide or C peptide*) interposed between the C-terminal extremities of the B chain and the N-terminal portion of the A chain (Fig. XI.2C). There are important variations (more than 50%) in the sequence of the C peptide in different animals. The distribution of the residues in C peptide which do not vary suggests their involvement in the establishment of a conformation favoring the matching of disulfide bonds and the proteolytic cleavage of proinsulin to insulin by converting enzyme.

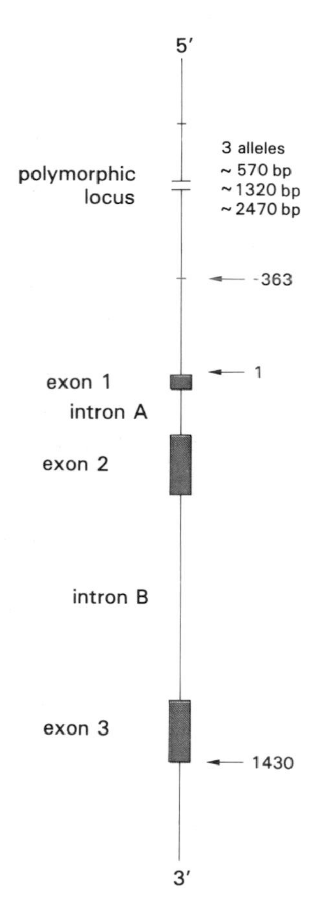

Figure XI.4 Organization of the insulin gene. Besides exons and introns of the proinsulin gene, the *polymorphic* locus of the 5′ region is represented. *Three classes of alleles* have been described, of 570 (the most frequent), 1320, and 2470 bp, respectively[11].

Proinsulin itself is derived from a preproinsulin[9] (M_r=11,500). The sequence of several preproinsulins in different animal species has been established based on the nucleotide sequence of cloned genes.

Gene Structure

The gene for insulin is located on the short arm of *chromosome 11* in humans[10]. It includes three exons encoding the mRNA of proinsulin (total 1430 bp) (Fig. XI.4). There is *polymorphism in the 5′ region* flanking the insulin gene[11], with a family of tandemly repeated sequences starting 363 bp upstream from the transcription initiation site. Various fragments have been described as occurring in higher frequency in type 1 and type 2 diabetes, but the function of the polymorphic locus is not known. A powerful genetic influence is indicated by the nearly 100% concordance rate in monozygotic twins with type 2 diabetes, although neither epidemiological nor molecular biology studies have been able to describe the genetics of the disease clearly.

Biosynthesis and Conversion

The biosynthesis of insulin in B cells (in the form of preproinsulin) occurs at the surface of the rough endoplasmic reticulum. Proinsulin is then transported in microvesicles to the Golgi apparatus, where secretory granules are formed (Fig. XI.5). The conversion of proinsulin to insulin begins in the Golgi and is completed in the granules; the half-life of conversion is approximately 60 min at 37°C in the rat. The conversion process involves enzymes with trypsin and carboxypeptidase B activity. Proteolytic cleavage of proinsulin leads to the formation of equimolar quantities of insulin and C peptide in granules.

Storage

Granules represent the morphological and functional unit of storage and secretion of the hormone (Fig. XI.6). Insulin granules (β granules) are quite large, and are less electron-dense than glucagon

9. Steiner, D. F. and Tager, H. S. (1979) Biosynthesis of insulin and glucagon. In: *Endocrinology*, vol. 2 (DeGroot, L. J., ed), Grune and Stratton, New York, pp. 921–929.
10. Owerbach, D., Bell, G., Rutter, W., Brown, J. and Shows, T. (1981) The insulin gene is located on the short arm of chromosome 11. *Diabetes* 30: 267–270.
11. Ullrich, A., Dull, T. J., Gray, A., Brosius, J. and Sures, I. (1980) Genetic variation in the human insulin gene. *Science* 209: 612–615.

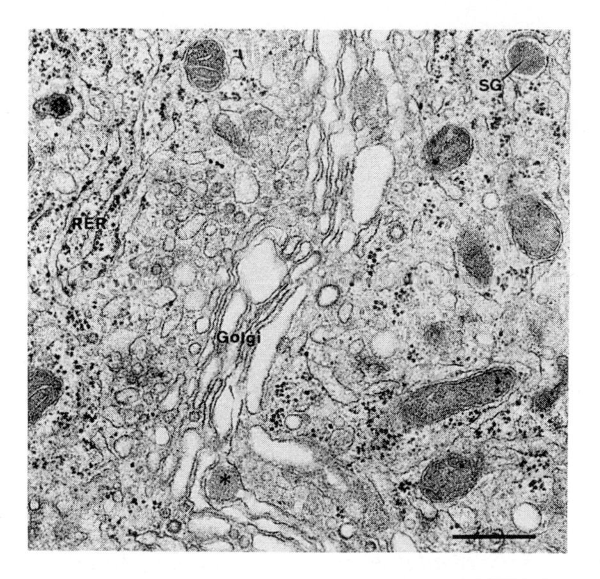

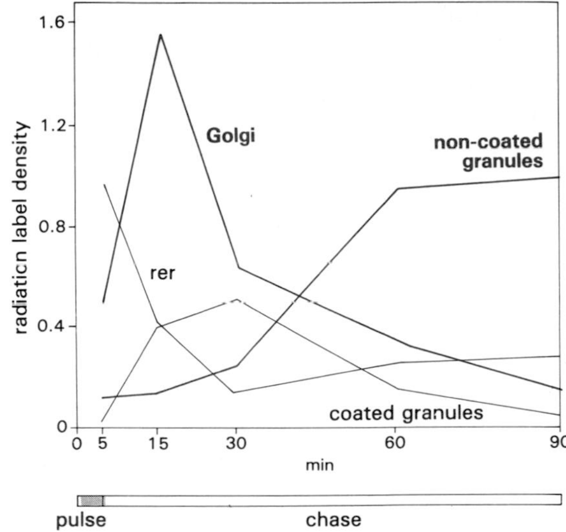

Figure XI.5 Insulin synthesis, packaging, and storage. The upper photograph represents a thin section of a pancreatic B cell, showing the three main intracellular membrane compartments involved in the processing of insulin: the rough endoplasmic reticulum (RER), site of synthesis; the Golgi apparatus, site where the conversion of proinsulin to insulin and the condensation (packaging) of the hormone is initiated; and the secretory granules (SG), site of insulin storage. Golgi cisternae with secretory material in the process of condensation are visible(*). The secretory granule is a coated granule (the coat is indicated by a dotted line). Coated granules are assumed to shed their coat and mature into pale-core and dense-core, granules (not shown, but see labelling in lower panel). The area between the RER and the Golgi shows numerous membrane vesicles which are assumed to carry newly synthesized proteins from the RER to the Golgi apparatus: transfer microvesicles. Bar=1 μm. Lower panel: isolated islets were pulse-chased with [^{3}H]leucine and leucine. The radiation-labelled density of the various intracellular membrane compartments is indicated. (Orci, L. (1984) Patterns of cellular and subcellular organisation in the endocrine pancreas. *J. Endocrinol.* 102: 3–11).

granules (α granules); in α and β granules, there is a clear space between the content and its limiting membrane. The form and variable density of β granules is a reflection of their maturation, varying as a function of pH, the presence of zinc and biogenic amines, and the ionic composition of the medium. The number of granules formed is related to the processes of synthesis and secretion. The adult human pancreas contains approximately 4 to 6 mg of insulin (or 100 to 400 U). The *stimulation* of the *synthesis and secretion of insulin by glucose* is accompanied by hypertrophy of the Golgi and the endoplasmic reticulum, and an increase in the number of ribosomes, whereas the number of granules in B cells is reduced because of the increased rate of secretion. Glucose (in vivo, the level of glycemia) is the *major stimulus* for both the synthesis and secretion of insulin. However, these two processes appear to be regulated independently.

Secretion

Intracellular Transport and Exocytosis

Insulin secretion *requires* the presence of Ca^{2+} in the extracellular medium. The microtubule system participates in the intracellular transport of granules in plasma membranes[12]. Ca^{2+} appears to initiate a contractile phenomenon by which *microtubules facilitate the displacement of granules towards* the *cell membrane.* Glucose favors the *uptake of calcium* by the β cells, while *cAMP* modifies the intracellular distribution of Ca^{2+} by increasing the cytosol pool at the expense of Ca^{2+} bound to intracellular organelles. *Protein kinase* C also appears to be involved in the secretion of insulin, as well as other secretory processes.

Exocytosis of the content of the β granule into the extracellular space represents the major process by which insulin is released into the circulation. It is characterized by the *fusion of the granule membrane with the plasma membrane* of the cell, and it appears to be accompanied by an inverse movement (endocytosis) of membrane material into the cytoplasm[13]. After fusion and subsequent rupture of the plasma membrane, the content of the granule is liberated into the extracellular space. Insulin and C peptide are liberated, in equimolar quantities, from mature

12. Malaisse, W. J. (1973) Insulin secretion: multifactorial regulation for a single process of release. *Diabetologia* 9: 167–173.
13. Orci, L. (1974) A portrait of the pancreatic B-cell. *Diabetologia* 10: 163–187.

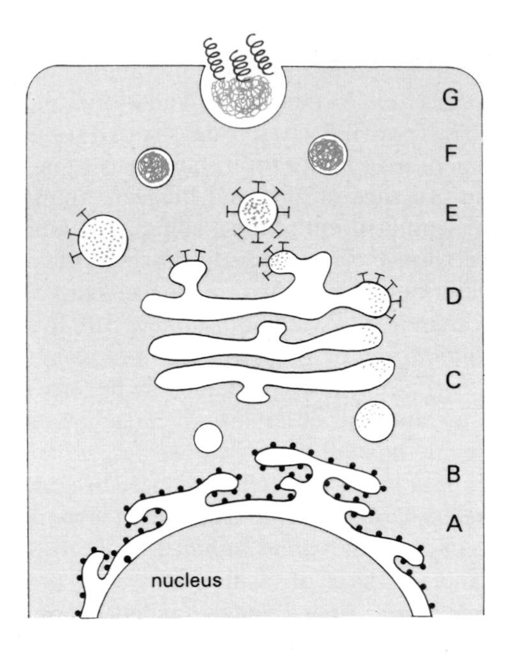

secretory granules (Fig. XI.6). Other forms of secretion may be involved in normal or pathological states: the secretion of insulin (and proinsulin) without prior incorporation into a granule; dissolution of the granule in the cytoplasm; rapid release of microvesicles or granules without maturation or previous intracellular storage.

Kinetics

In vivo, the *secretion of insulin is a complex* function which involves the interaction of different factors. There are *primary stimuli*, represented by various substrates, the *most important* being *glucose* and *secondary regulators* such as *hormones*. *Basal secretion* of insulin in the absence of any exogenous stimulation is an index of the secretory capacity of B cells (Fig. XI.7). The secretion of insulin in response to an *exogenous stimulus*, such as the sudden and sustained elevation of the extracellular concentration of glucose, results in a *biphasic response*: an immedi-

△

Figure XI.6 Pancreatic B cell, and the manufacture of insulin: a flow diagram of the main compartments involved (**A–G**). This drawing is based on data derived from chromatography and immunoprecipitation of secretory polypeptides, high resolution autoradiography, and immunocytochemistry with clathrin, proinsulin and insulin-specific antisera, or with probes, to reveal intracellular acidic compartments. **A.** Synthesis of preproinsulin by ribosomes attached to membranes of the rough endoplasmic reticulum (RER) and translation of preproinsulin mRNA. While synthesized, preproinsulin is discharged in the cavities of the RER, where a first proteolytic cleavage occurs, removing the pre-sequence; two disulfide bridges are established, to yield the characteristic proinsulin molecule. Other ribosomes on the RER membrane synthesize the proinsulin converting enzymes, which enter the cavities and which will be activated at a later stage in the manufacture of insulin (see **E**). **B.** Quanta of proinsulin (and probably of converting enzymes) enclosed within vesicular carriers pinch off from the RER membrane and are transported to the *cis* Golgi pole, where they are delivered to the *cis* Golgi cisternae. Whether proinsulin is delivered to *cis* Golgi cisternae mixed with other RER-derived proteins, or whether they are already sorted out by specific receptors on the RER membrane, is not known at present. **C.** Proinsulin within *cis* Golgi cisternae reaches the *trans* Golgi pole through intercisternal transport mediated by vesicular carriers budding from individual Golgi cisternae. Note that these vesicles are provided with a special type of coat (represented by spikes on the membrane), different from clathrin (see **D**). **D.** At the *trans* pole, proinsulin is concentrated, most probably through a receptor-mediated process, into the dilated extremities of the *trans*-most cisternae. These condensing areas are characterized by a clathrin coat (represented by spikes) on the outer aspect of the limiting membrane. The RER-*cis* transport of proinsulin, as well as the forward movement of proinsulin from the *cis* to the *trans* Golgi pole, are energy-dependent steps. **E.** The condensing extremities of clathrin-coated, *trans*-most Golgi cisternae pinch off and are released in the cytoplasm as clathrin-coated secretory vesicles rich in proinsulin. There is a progressive acidification (pH 6.5 to 5.5) of the coated secretory vesicle content, concomitant with activation of proinsulin converting enzyme, which results in the proteolysis of proinsulin into insulin and C peptide, which accumulate in vesicles. **F.** Coated secretory vesicles mature into secretory granules, rich in insulin, which lack a clathrin coat (= non-coated granules). Their pH is more acidic (pH 5.5 to 4.5) than clathrin-coated secretory vesicles. Non-coated granules are stored in the cytoplasm for prospective release, upon appropriate glucose stimulation. **G.** The limiting membrane of insulin-rich, non-coated secretory granules fuses with the plasma membrane, allowing the release of the granule contents, insulin and C peptide, into the extracellular space. Not shown on the drawing is the recapture of plasma membrane segments after the insertion of the secretory granule membrane, during sustained exocytosis, in order for the cell to maintain a constant surface (Fig. I.18). (Courtesy of Dr. L. Orci, University of Geneva; Orci, L. (1985) The insulin factory: a tour of the plant surroundings and a visit to the assembly line. *Diabetologia* 28:528–546; Orci, L., Ravazzola, M., Amherdt, M., Madsen, O., Perrelet, A., Vassalli, J.-D. and Anderson, R. G. W. (1986) Conversion of prosinsulin to insulin occurs coordinately with acidification of maturing secretory vesicles *J. Cell Biol.* 103:2273–2281.

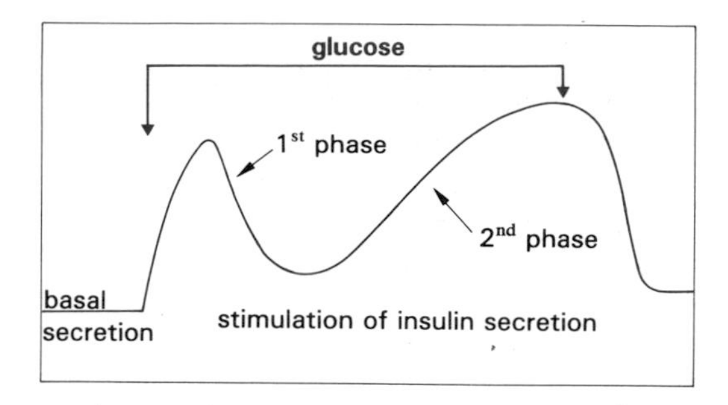

Figure XI.7 Biphasic time course (idealized) of the insulin secretory response to glucose stimulation.

ate increase (<1 min), reaching a first peak within a few minutes, is followed by a reduction, in spite of the persisting stimulus; then there is a second, slower increase, reaching a higher level of secretion. When the stimulus is stopped, a return to basal secretion occurs. The *first phase* (acute secretion) involves the *release of stored insulin*, and is reflective of the hormonal reserve contained in a quantitatively limited compartment which can be mobilized immediately. The *second phase* also represents the secretion of previously stored insulin, but, when this phase is *prolonged*, it *involves newly synthesized insulin*. The interaction of stimuli with the basal secretion and with the two phases of the stimulated secretion is variable. *Glucose influences all phases*, whereas other stimuli may affect one or both.

Substrates

The secretion of insulin is affected to varying degrees by *carbohydrates, proteins, and lipids*. Carbohydrates, primarily glucose, and to a lesser extent fructose, stimulate insulin secretion. Glucose, the major stimulus, acts via the intermediary of its intracellular metabolites and as an energy substrate. It also influences the biosynthesis of the hormone, which reconstitutes reserves, and leads, over the long term, to stationary levels of basal and stimulated secretion. It has been suggested that, in diabetes, the secretory anomaly could result, at least in part, from the improper recognition of glucose (or a signal it induces) within the B cell[14].

14. Cerasi, E. (1975) Mechanisms of glucose stimulated insulin secretion in health and diabetes: some re-evaluations and proposals. *Diabetologia* 11: 1–13.

Most amino acids stimulate insulin secretion. The stimulatory effect of leucine is clearly established, especially in pathological cases, but the effect of arginine is even more apparent. In addition to direct effects, some amino acids stimulate insulin secretion indirectly, by causing increased secretion of glucagon and growth hormone. The insulin secretory effect of amino acids is partially independent of their prior metabolism. The effects of glucose and amino acids (or proteins) on insulin secretion are synergistic.

Fatty acids and ketone bodies are also capable of stimulating insulin secretion, especially in the dog. In humans, the effect is much less pronounced.

Hormones and Neurotransmitters

The presence and absorption of substrates in the digestive tract stimulate the secretion of the *gastrointestinal hormones gastrin, secretin, CCK-PZ, and GIP* (p. 555). When administered exogenously, each of these peptides is capable of *increasing the secretion of insulin* (Table XI.1). However, for most of these peptides, a physiological role at the enteroinsular level (amplification of the insulin secretory effect of such primary stimuli as glucose and amino acids) is unlikely. Only *GIP*, the secretion of which is stimulated by glucose, amino acids and triglycerides, is capable of stimulating insulin secretion at physiological concentrations (p. 572).

Table XI.1 Effects of substrates, hormones and neurotransmitters on the secretion of insulin and glucagon.

	Insulin	Glucagon
Substrates:		
Glucose	↑	↓
Amino acids	↑	↑
Fatty acids	↑	↓
Hormones:		
Insulin	↓[1]	↓ or ↑[2]
Glucagon	↑	
GIP	↑	0 or ↑
Somatostatin	↓	↓
Neurotransmitters:		
Catecholamines	↓	↑
Cholinergics	↑	↑

[1]Direct inhibition of its own secretion by insulin is possible but remains controversial. In vivo, the administration of insulin inhibits endogenous insulin secretion via the hypoglycemia it induces.
[2]In vivo, the secretion of glucagon can be increased as a result of insulin-induced hypoglycemia.

Glucagon, both in vivo and in vitro, stimulates the secretion of insulin by a *direct effect on B cells.* Other hormones also stimulate insulin secretion. This may occur as a result of the administration (or physiological states that involve increased production) of GH, glucocorticosteroids, estrogens, and progesterone. It may involve a direct effect on B cells, or an indirect effect via an intermediary (glucose, amino acids).

Neurotransmitters affect insulin secretion. The stimulation of α-adrenergic receptors inhibits insulin secretion, whereas the stimulation of β-adrenergic receptors leads to an increase thereof. The net effect of catecholamines is to inhibit basal and stimulated insulin secretion[15]. The basal secretion of insulin is modulated by adrenergic tone; α-blockers increase this secretion, while β-blockers reduce it. The ventromedial area of the hypothalamus has an inhibitory effect on insulin secretion, while the ventrolateral region stimulates insulin secretion and increases vagal tone; acetylcholine also stimulates insulin secretion.

Somatostatin inhibits insulin and glucagon secretion in vivo and in vitro.

2.3 Circulation, Distribution, and Metabolism

Circulation

Insulin, proinsulin, proinsulin-like and insulin-like fractions (possible intermediaries in the conversion of proinsulin to insulin), and C peptide, are present in plasma in *free forms*, not bound to a protein. Insulin circulates as a monomer. With *most anti-insulin antisera*, the RIA of insulin gives an estimate of *total immunoreactive insulin*, including insulin, proinsulin, and the intermediary fractions. This heterogeneity of insulin, as normally measured in the circulation, is seen also for other polypeptide hormones, such as gastrin, growth hormone, parathormone, and ACTH. Various methods (p. 517) permit the separate measurement of insulin, and of proinsulin and proinsulin-like fractions which may account for as much as 20% of total insulin immunoreactivity in the basal state (i.e., after an overnight fast of 12 to 15 h). This percentage decreases after acute stimulation of insulin secretion; it can be considerably increased in certain pathological states.

The liver is the organ through which pancreatic hormones necessarily pass. It is also an important site of action and degradation of these hormones. Consequently, insulin levels are higher in the portal vein than peripherally. In the basal state, portal insulinemia is two to three times greater than peripheral insulin levels; following an acute secretory stimulus, such as a glucose load, portal insulin may be 10 to 15 times higher than peripheral insulin levels. The basal secretion of insulin at the site of exit from the pancreas and before it passes through the liver is estimated to be approximately 1 to 2 units/h (1–2 mg/d).

The level of total immunoreactive insulin (IRI) specifically reflects pancreatic insulin (and proinsulin), but it constitutes *only 5 to 10% of insulin-like activity (ILA)* in plasma. Thus, in the basal state, IRI represents only approximately 10 mU/l, for an ILA level of 100 to 200 mU/l, determined by biological assay of insulin activity in vitro (measured by the utilization of glucose in epididymal adipose tissue of the rat). The *difference (ILA–IRI)*, termed *non-suppressible insulin-like activity*[16,17] (NSILA), is *made up essentially of IGF-I and -II* (p. 198).

Distribution

The distribution of insulin within the organism is determined by the presence of the hormone in the extracellular compartment, its binding to receptor sites, and its degradation by various tissues and organs, particularly the liver and kidney. The amount of insulin removed by these different tissues is the sum of receptor-bound hormone and degraded hormone.

After crossing the capillary membrane, insulin *remains in the extravascular compartment much longer than in the vascular compartment*, where its half-life is short (~5 min). Insulin is filtered by the renal glomerulus, but it is almost completely reabsorbed by the proximal tubule, and is largely degraded by

15. Porte, D. Jr. (1966) A receptor mechanism for the inhibition of insulin release by epinephrine in man. *J. Clin. Invest.* 45: 228–236.
16. Froesch, E. R., Burgi, H., Muller, W. A., Humbel, R. E., Jakob, A. and Labhart, A. (1967) Nonsuppressible insulin-like activity of human serum: purification physicochemical and biological properties and its relation to total serum ILA. *Rec. Prog. Horm. Res.* 23: 565–616.
17. Zapf, J., Rinderknecht, E., Humbel, R. E. and Froesch, E. R. (1979) Non-suppressible insulin-like activity (NSILA) from human serum: recent accomplishments and their physiologic implications. *Metabolism* 27: 1803–1828.

the kidney. Bile contains insulin (or fragments of immunoreactive insulin). Almost 50% of hepatic portal insulin is removed by the liver in a single passage. The kidneys retain, in a single passage, nearly 40% of peripheral insulin. Reduced renal degradation explains, at least in part, the prolonged half-life of insulin in renal insufficiency.

The *half-life of proinsulin* in plasma (~20 min) is *longer than that of insulin*. The slower degradation of proinsulin by all tissues (except the kidneys), especially the liver, and the lower affinity of proinsulin for insulin receptors combine to prolong the plasma half-life of proinsulin as compared to that of insulin, and to increase the proinsulin/insulin ratio in the peripheral blood, in comparison with that of hepatic portal blood. *Proinsulin is not converted to insulin in the circulation, nor is it converted in target tissues.*

Degradation

Almost *all tissues* of the body degrade insulin, and none are able to resynthesize it from the degraded products; the *liver and kidneys* are major sites of degradation. Two enzymatic systems (either isolated or associated) are involved. Glutathione-insulin transhydrogenase catalyzes the reduction cleavage of disulfide bridges and thus separates the A and B chains. A protease, isolated and purified from striated muscle[18], is active at physiological pH, and its affinity for insulin (K_M=22 nM) is compatible with a physiological role. The presence, in *purified plasma membrane*, of degradation system(s) with similar properties[19] to that of the purified enzyme suggests that insulin may be degraded without entering cells. Another means of insulin degradation involves the interaction of the hormone with its receptor[20], followed by endocytosis and *intracellular degradation* of the hormone[21]. Degraded insulin is biologically inactive. Although degraded much more slowly than insulin, proinsulin is a competitive inhibitor of insulin degradation. The degradation of proinsulin is not associated with a conversion to insulin[19].

18. Duckworth, W. C., Heinemann, M. A. and Kitabchi, A. E. (1972) Purification of insulin-specific protease by affinity chromatography. *Proc. Natl. Acad. Sci., USA* 69: 3698–3702.
19. Freychet, P., Kahn, R., Roth, J. and Neville, D. M. Jr. (1972) Insulin interactions with liver plasma membranes. Independence of binding of the hormone and its degradation. *J. Biol. Chem.* 247: 3953–3961.
20. Terris, S. and Steiner, D. F. (1975) Binding and degradation of 125-insulin by rat hepatocytes. *J. Biol. Chem.* 250: 8389–8398.
21. Gorden, P., Carpentier, J.-L., Freychet, P. and Orci, L. (1980) Internalization of polypeptide hormones. Mechanism, intracellular localization and significance. *Diabetologia* 18: 263–274.
22. Bromer, W. W., Sinn, L. G. and Behrens, O. K. (1957) The amino-acid sequence of glucagon. V. Location of amide groups, acid degradation studies, and summary of sequential evidence. *J. Amer. Chem. Soc.* 79: 2807–2810.

3. Glucagon: Structure, Biosynthesis, Secretion, Circulation, and Metabolism

3.1 Structure

Glucagon is a polypeptide (M_r~3500)[22] consisting of 29 aa (Fig. XI.8). The sequences of human, porcine and bovine glucagon are identical; avian glucagons differ from mammalian glucagons by only a few amino acids. Thus, in contrast to insulin, the primary structure of glucagon does *not appear to have varied significantly throughout evolution*. In fact, the entire sequence of glucagon is necessary for the expression of hormonal activity, while the biological activity of insulin depends on the integrity of its three-dimensional configuration. This configuration does not appear to have been significantly affected by the variation in certain residues that occurred in the course of evolution. In an aqueous solution, and at physiological pH levels, ionic force and hormonal concentration, *glucagon, in contrast to insulin, does not have an orderly three-dimensional structure*. It is possible, however, that glucagon develops a secondary and a tertiary structure in the vicinity of its receptor. In contrast to insulin, *metals* such as zinc *do not form an integral part* of the glucagon crystal. In the crystallized state, glucagon has an α-helical structure, but it is easy to prepare glucagon-metal complexes, such as glucagon-zinc, which have prolonged biological activity.

The fragments 1–21 and 22–29 are devoid of activity. Deamidation of Gln residues at positions 3, 20 and 24 markedly reduces biological activity. The N-terminal histidine of glucagon, indispensable for biological activity, is one of the numerous structural *homologies that glucagon shares with secretin and VIP*. These three peptides also have some structural homology with GIP (Fig. XII.22).

glucagon

NH2
1 His
Ser
Gln
Gly
Thr
Phe
Thr
Ser
Asp
10 Tyr
Ser
Lys
Tyr
Leu
Asp
Ser
Arg
Arg
Ala
20 Gln
Asp
Phe
Val
Gln
Trp
Leu
Met
Asn
29 Thr
COOH

Figure XI.8 Primary structure of glucagon.

3.2 Biosynthesis and Storage

The biosynthesis of glucagon, like that of insulin, proceeds from a *larger precursor, preproglucagon*. The sequences of preproglucagons in different species have been established based on the nucleotide sequences of cloned cDNAs. The human preproglucagon gene[23] consists of at least three introns, which separate the polypeptide-coding portion of the gene into four exons (Fig. XI.9). The polypeptide regions of the precursor encoded by these exons correspond to the signal peptide and part of an N-terminal peptide, the remainder of the N-terminal peptide and glucagon, and two carboxyterminal glucagon-like peptides (GLP-1 and GLP-2). Thus, preproglucagon (179 aa in the human) is *a polyprotein precursor*. The N-terminal peptide is also referred to as *glicentin-related pancreatic peptide (GRPP)*. *Glicentin*[24], or *glucagon-like immunoreactivity I (GLI-I)*, is an *intestinal*

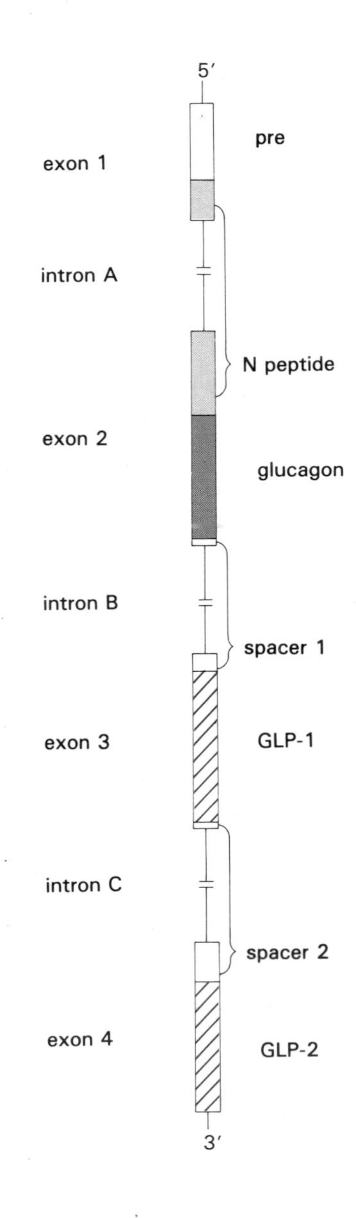

Figure XI.9 Structure of the human preproglucagon gene[23].

23. Bell, G. I., Sanchez-Pescador, R., Laybourn, P. J. and Najarian, R. C. (1983) Exon duplication and divergence in the human preproglucagon gene. *Nature* 304: 368–371.

glucagon-containing peptide which corresponds to pancreatic proglucagon (M_r=8100). Its 69 aa sequence contains the sequence 1–30 of GRPP and the integral sequence (33–61) of glucagon, the two sequences being linked by a pair of basic (Lys–Arg) residues (31, 32). The human preproglucagon gene[25] is located on *chromosome 2*.

The main fragments of the glicentin (or proglucagon) sequence, i.e. GRPP and glucagon, are *stored in secretory granules of the pancreatic A cell.* Glucagon immunoreactivity is restricted to the granule core, whereas glicentin-like material is found at the periphery of the granule[26]. Glucagon and GRPP are secreted synchronously and in approximately equimolar amounts by the pancreas, suggesting that they are derived from the same precursor stored in the glucagon-producing A cell. Although there are some discrepancies regarding the size of preproglucagon and proglucagon, and the smaller cleavage products, available data suggest that preproglucagon is processed into glucagon through stepwise proteolytic degradation (Fig. XI.10). *Glicentin* is also found in the secretory granules of the *intestinal L cells* (Table XII.3), but, in contrast to the pancreatic A cells, no conversion to glucagon appears to occur in the L cell. These observations suggest that tissue-specific processing of a common polyprotein precursor could generate the array of peptides of the glucagon family that is observed in the pancreas and intestine.

As for insulin in the B cell, granules in A cells represent the morphological and functional unit of glucagon storage and secretion. The glucagon content of the human adult pancreas ranges from 2 to 5 μg/g (wet weight).

Secretion and Regulation

The study of the secretion and regulation of glucagon has been limited by problems inherent in the RIA of glucagon in peripheral blood. These include the rapid inactivation of glucagon in plasma, the extrapancreatic (especially gastric) origin of glucagon of the pancreatic type, at least in some species, and

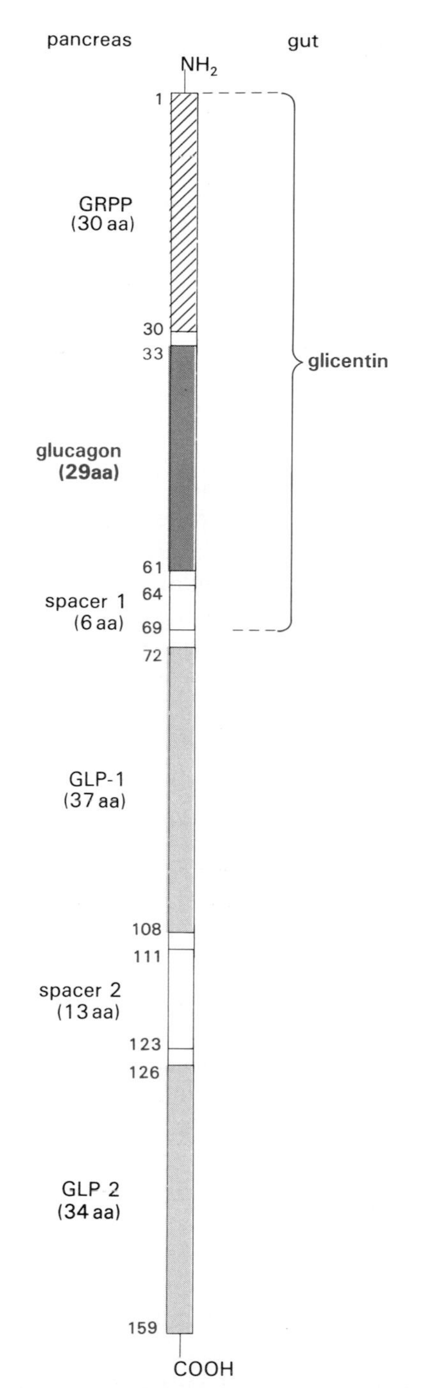

Figure XI.10 Tissue-specific processing of proglucagon. The common polyprotein precursor gives rise to different products. Processing in the pancreas (A cell) generates glicentin-related pancreatic peptide (GRPP) and glucagon, which are both released in the circulation. Processing in the gut (L cells) gives rise to glicentin, which is not cleaved into GRPP and glucagon. *Glicentin* (GLI) (or a glicentin-like molecule) is released, and may thereafter be cleaved to produce oxyntomodulin (Fig. XII.27). The two C-terminal glucagon-like peptides (GLP-1 and GLP-2) have not been isolated, and their possible biological function is unknown[23].

24. Moody, A. J., Holst, J. J., Thim, L. and Lindkaer-Jensen, S. (1981) Relationships of glicentin to proglucagon and glucagon in the porcine pancreas. *Nature* 289: 514–516.
25. Tricoli, J. V., Bell, G. I. and Shows, T. B. (1984) The human glucagon gene is located on chromosome 2. *Diabetes* 33: 200–202.
26. Ravazzola, M. and Orci, L. (1980) Glucagon and glicentin immunoreactivity are topologically segregated in the alpha-granule of the human pancreatic A-cell. *Nature* 284: 66–67.

especially the absence of specificity of the *RIA for pancreatic glucagon*, due to the interference of glucagon-like peptides with most antiglucagon antisera. The selection of antisera with greater specificity for pancreatic glucagon, the assay of glucagon in the pancreato-duodenal vein, and the development of several in vitro study systems have helped resolve some of these problems.

Substrates (Table XI.1)

Glucose. There is an inverse relation between the extracellular concentration of glucose and the pancreatic secretion of glucagon: in contrast to what is observed for insulin secretion, *hyperglycemia reduces, and hypoglycemia increases, glucagon secretion.* All glycemia levels ≤50 mg/dl are accompanied by an increased secretion of glucagon. In contrast, elevation of glycemia to levels >150 mg/dl reduces glucagon secretion. Variations in glucagon secretion in response to variations in glycemia are as rapid as those fluctuations seen in insulin secretion, and the range of glycemia (50–150 mg/dl) in which they are produced suggests that glucagon is effectively involved, as is insulin, in the rapid adjustment of glycemia homeostasis. In vitro, the same inverse relationship is observed between the concentration of glucose and the secretion of glucagon.

The *inhibitory effect of glucose* and, more generally, the inhibitory effects of energy substrates, such as fatty acids or ketone bodies, on the secretion of glucagon appear to be correlated with the level of energy produced by the metabolism of these substrates in A cells. When the intracellular metabolism of glucose is disturbed, an increased secretion of glucagon can occur together with hyperglycemia.

Amino acids. The iv and oral administration of amino acids (singly, as a mixture of several, or in the form of proteins) stimulates the secretion of both glucagon and insulin. *Arginine stimulates the secretion of the two hormones.* Alanine and other gluconeogenic amino acids especially stimulate the secretion of glucagon. This stimulation represents a physiological situation in which the secretion of glucagon and insulin is modified in the same direction. It has been suggested that the effect of amino acids on the secretion of glucagon is to prevent hypoglycemia resulting from their effect on insulin secretion[27]. Thus, an increase in insulin secretion may favor protein synthesis without disturbing glucose homeostasis.

Free fatty acids. An increased plasma concentration of free fatty acids (FFA) *reduces* the secretion of glucagon in most animal species, including humans. Inversely, the concentration of glucagon in the plasma increases with a decrease in FFA concentration[28].

Effect of Hormones

Insulin, within the islet, has a negative effect on the secretion of glucagon: an intrainsular negative feedback of a paracrine type probably exists between the A and B cells. In severe states of insulin deficiency, the interruption of this feedback leads to increased glucagon secretion which can be suppressed by insulin therapy. In normal subjects, exogenous administration of insulin may lead, in contrast to paracrine feedback, to an amplification rather than an inhibition of the secretory response of glucagon to a stimulus such as arginine[29].

Several gastrointestinal hormones affect the secretion of glucagon. CCK (cholecystokinin) increases glucagon secretion (and that of insulin) stimulated by infusion of amino acids. The administration of VIP into the pancreatic artery or to the isolated perfused pancreas also increases glucagon secretion. Secretin, in contrast, has an inhibitory effect, amplifying the depressive effect of glucose on glucagon secretion. However, the physiological role of these peptides on the secretion of glucagon (in particular their involvement in the enteroinsular mediation) has not been established[30].

Effects of the Nervous System

β-adrenergic agonists stimulate the secretion of glucagon. Stimulation of the ventromedial hypothalamus increases glucagon secretion. Activation of the sympathetic nervous system, especially during physical exercise, tends to increase glucagon secretion and reduce that of insulin. *Cholinergic stimulation* and *acetylcholine* also increase the secretion of glucagon (and of insulin). Somatostatin decreases the secretion of both glucagon and insulin; an intra-islet mechanism of the paracrine type is probably involved (at least in part).

27. Unger, R. H., Ohneda, A., Aguilar-Parada, E. and Eisentraut, A. M. (1969) The role of aminogenic glucagon secretion in blood glucose homeostasis. *J. Clin. Invest.* 48: 810–822.
28. Luycks, A. S. and Lefebvre, P.-J. (1983) Free fatty acids and glucagon secretion. In: *Glucagon II* (Lefebvre, P.-J., ed), Springer Verlag, Berlin, pp. 43–58.
29. Samols, E. (1983) Glucagon and insulin secretion. In: *Glucagon II* (Lefebvre, P.-J., ed), Springer Verlag, Berlin, pp. 485–518.
30. Pek, S. B. and Spangler, R. S. (1983) Hormones in the control of glucagon secretion. In: *Glucagon II* (Lefebvre, P.-J., ed), Springer Verlag, Berlin, pp. 99–111.

Table XI.2 Major metabolic effects of insulin and glucagon.

Metabolic effects	Insulin	Glucagon
Intracellular transport of:		
glucose	↑ adipose tissue, striated muscle	↑ liver
amino acids[1]	↑ striated muscle, adipose tissue	
Intracellular accumulation of K^+	↑ liver, adipose tissue, striated muscle	↓ liver
Glycogen synthesis	↑ liver, striated muscle	↓ liver
Glycogenolysis[2]	↓ liver	↑ liver, myocardium
Gluconeogenesis	↓ liver	↑ liver
Ketogenesis	↓ liver	↑ liver
Lipogenesis	↑ adipose tissue, liver	↓ liver
Lipolysis	↓ adipose tissue, liver	↑ adipose tissue[3], liver
Protein synthesis	↑ striated muscle, liver, adipose tissue	
Proteolysis	↓ liver, striated muscle, adipose tissue	↑ liver

[1]Insulin stimulates the transport of only some amino acids.
[2]Produces glucose in the liver (G 6-phosphatase) and in striated muscle (from lactate, in the glycolytic pathway).
[3]Variable lipolytic effects of glucagon depending on the animal species (p. 511).

3.3 Circulation, Distribution, and Metabolism

Circulation

Glucagon, like insulin, circulates in a *free form*, unbound to a protein. Multiple molecular forms of glucagon are present in the circulation: pancreatic glucagon (in some species, a glucagon identical to pancreatic glucagon may originate from fundic cells of the stomach), and glucagon-like molecules (on the basis of their immunoreactivity), of high molecular weight, of the GLI type (Fig. XI.10). Most antiglucagon antisera used in RIA are not completely specific for pancreatic glucagon, and thus the respective contributions of pancreatic glucagon and other molecular forms to the heterogeneity of circulating glucagon cannot be determined precisely. There is, as for insulin, a hepatic portal/peripheral plasma concentration gradient. For pancreatic glucagon, the concentrations in the efferent pancreatic veins are 15 to 30 times higher than in peripheral veins.

Distribution

In a period of 10 to 15 min, the apparent volume of glucagon distribution corresponds to body weight, which suggests *rapid degradation* of the hormone. The half-life of glucagon in the circulation is short (~5 min). Glucagon is filtered by the renal glomerulus and reabsorbed by the proximal tubule. Urine is practically devoid of glucagon when the proximal tubule functions normally. A portion of glucagon is excreted in bile.

Degradation

Glucagon is degraded by *several tissues*, especially the *liver and kidney*. An enzymatic activity of the dipeptidyl-aminopeptidase I type has been demonstrated with cleavage of the molecule, at the N-terminal extremity, between Ser-2 and Gln-3 (Fig. XI.8); in the absence of N-terminal histidine, the molecule is biologically inactive. However, this dipeptidase is not specific to glucagon. During its interaction with hepatic plasma *membranes*, glucagon binds to a specific receptor and also undergoes inactivation[31,32]; this process of inactivation is specific to glucagon and may play a regulatory role in hormonal action, reducing the active concentration of glucagon in contact with its receptor. As for insulin, a portion of the degradation of glucagon appears to be secondary to an intracellular transfer of the peptide[21]. After having been

31. Phol, S. L., Krans, H. M. J., Birnbaumer, L. and Rodbell, M. (1972) Inactivation of glucagon by plasma membrane of rat liver. *J. Biol. Chem.* 247: 2295–2301.
32. Desbuquois, B., Krug, F. and Cuatrecasas, P. (1974) Inhibition of glucagon inactivation. Effect on glucagon receptor interactions and glucagon-stimulated adenylate cyclase activity in liver cell membranes. *Biochim. Biophys. Acta* 343: 101–120.

absorbed by the proximal tubule, glucagon is degraded by the kidneys. The plasma half-life of glucagon is prolonged in renal insufficiency. Hepatic degradation particularly affects active glucagon of $M_r=3500$, while renal degradation also occurs for molecules of higher molecular weight. Plasma has an especially high capacity for degrading glucagon. This degradation may, in large part, be inhibited by protease inhibitors (trasylol) and benzamidine.

4. Effects of Insulin and Glucagon

4.1 Metabolism

Insulin has *multiple metabolic effects.* With their *opposing actions* (Table XI.2), especially in the liver, *insulin and glucagon* play a *major role in* the regulation of *glucose homeostasis. Insulin* stimulates *anabolic processes* and energy stores, whereas *glucagon* is *catabolic* and mobilizes energy substrates.

Carbohydrates

After entering the cell, *glucose is phosphorylated at position* 6 (glucose 6-phosphate, G 6-P) by ATP, in the presence of hexokinase and Mg^{2+}. This reaction, *irreversible in the absence of glucose 6-phosphatase*, present only in the liver and kidney, ensures the retention of sugar in the cell, since the cell membrane is impermeable to phosphoric acid esters. Four metabolic pathways are possible for G 6-P (Fig. XI.11). With the exception of the glucuronic acid pathway, all are stimulated by insulin to a varying degree. Glucagon essentially affects the metabolism of glycogen, and somewhat that of pyruvate, with opposite effects to those of insulin.

Glycolysis (Embden–Meyerhof). This reaction can be carried out in total anaerobiosis, resulting in lactate. Two moles of ATP are required for each mole of glucose metabolized. Phosphofructokinase, which catalyzes the conversion of fructose 6-phosphate (F 6-P) to fructose 1, 6-diphosphate (F 1,6-DP), is stimulated by insulin and inhibited by citrate and ATP.

Direct oxidative pathway (or pentose-phosphates). G 6-P is dehydrogenated, then hydrolyzed to 6-phosphogluconate, which is dehydrogenated and decarboxylated to ribulose 5-phosphate. This pathway involves the utilization of O_2 and produces CO_2. The NADPH formed is an important cofactor for lipogenesis and for the synthesis of cholesterol.

Glucuronic acid pathway. G 6-P is converted to G 1-P, which reacts with uridine triphosphate (UTP) to form uridine diphosphoglucose (UDPG). UDPG may undergo several transformations: formation of glycogen, formation of other polysaccharides, and oxidation of glucose, at C6, to form uridine diphosphoglucuronic acid (UDPGA) and NADPH. Glucuronic acid formed from UDPGA may be transformed to xylulose and enter the pentose-phosphate pathway. Insulin reduces the production of L-xylulose.

Synthesis and metabolism of glycogen (Fig. XI.12). Glycogen is the *major* hydrocarbon *storage form* of glucose. Its synthesis (glycogen synthesis) from UDPG and its degradation (glycogenolysis) to G 1-P are dependent on *glycogen synthetase and phosphorylase*, respectively. These two enzymes exist in two forms, active and inactive, interconvertible by phosphorylation and dephosphorylation. The *inac-*

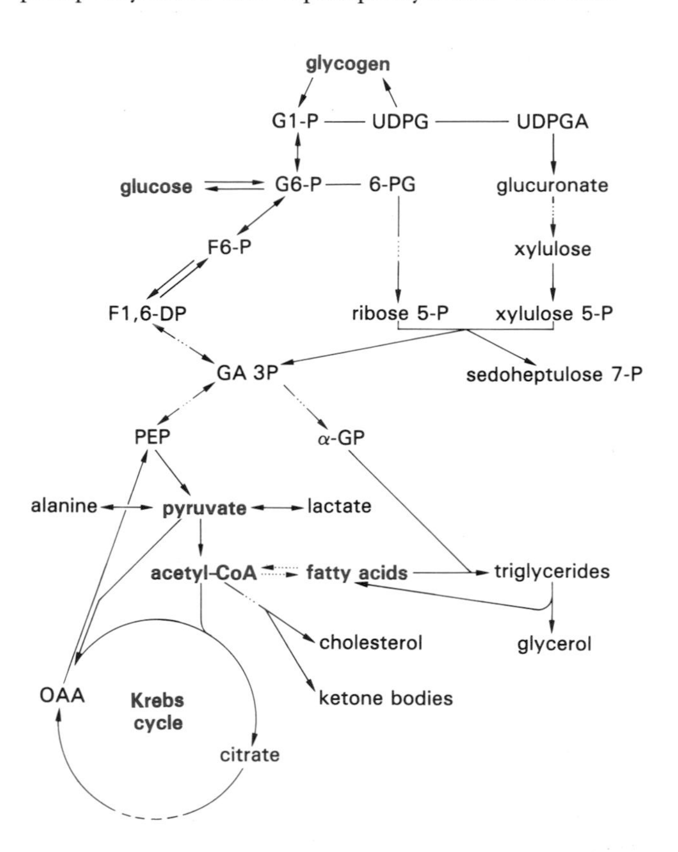

Figure XI.11 Metabolic pathway of glucose. G 6-P: glucose 6-phosphate; G 1-P: glucose 1-phosphate; UDPG: uridine diphosphoglucose; UDPGA: uridine diphosphoglucuronic acid; 6-PG: 6-phosphogluconate; F 6-P: fructose 6-diphosphate; F1, 6-DP: fructose 1,6-diphosphate; GA 3-P: glyceraldehyde 3-phosphate; α-GP: α-glycerol phosphate; OAA: oxaloacetic acid; PEP: phosphoenolpyruvate. Dotted lines indicate portions of the pathway that have not been detailed.

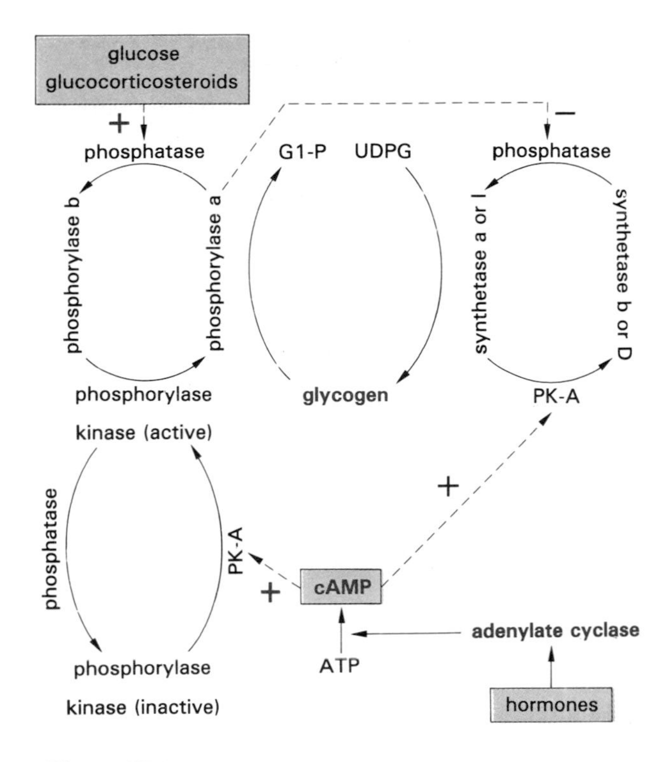

Figure XI.12 Hepatic metabolism of glycogen. UDPG: uridine diphosphoglucose; G 1-P: glucose 1-phosphate[34].

tive phosphorylase (dephosphorylated, b *form) is activated by phosphorylation (*a *form) by a* b *phosphorylase kinase*, which itself occurs in an inactive (dephosphorylated) or active (phosphorylated) form[33]. In contrast, the *active form I (or* a*) of glycogen synthetase is the dephosphorylated form*, while the *inactive form (D or* b*) is phosphorylated.* Phosphorylase *a* phosphatase and a glycogen synthetase D phosphatase catalyze the dephosphorylation which leads, respectively, to the inactive phosphorylase *b* and the active glycogen synthetase I.

Regulation occurs on different levels (Fig. XI.12): 1. *cAMP*, by the intermediary of a protein kinase, activates phosphorylase *b* kinase and phosphorylase, while it inactivates glycogen synthetase. This results in an increase in glycogenolysis and a reduction in glycogen synthesis; 2. the *phosphorylase a* itself inhibits glycogen synthase D phosphatase and, in consequence, glycogen synthetase and glycogen synthesis; 3. *glucose* inhibits phosphorylase *a* and stimulates phosphorylase *a* phosphatase, resulting in an inhibition of glycogenolysis and the stimulation of glycogen synthesis[34]; 4. *glucocorticosteroids* lead to the same effect by stimulating phosphorylase *a* phosphatase.

The effects of insulin and glucagon are opposed in the final expression of glycogen synthesis and glycogenolysis (Table XI.2), but the actions of these two hormones on the multienzyme system are complex. Intracellular cAMP concentration is increased by glucagon in the liver, but cAMP is not the sole mediator of glucagon activity[35].

Metabolism of pyruvate. Pyruvate occupies an *important position in intermediary metabolism* (Fig. XI.11). Oxidative decarboxylation leads to the formation of acetylcoenzyme A (acetyl-CoA). Pyruvate dehydrogenase is an enzymatic complex which, like glycogen synthetase, is inactivated, by phosphorylation, by a kinase, and activated following dephosphorylation by phosphatase. These two enzymes are dependent on Ca^{2+} and Mg^{2+} and the ratio of ATP to ADP. Acetyl-CoA and ATP inactivate

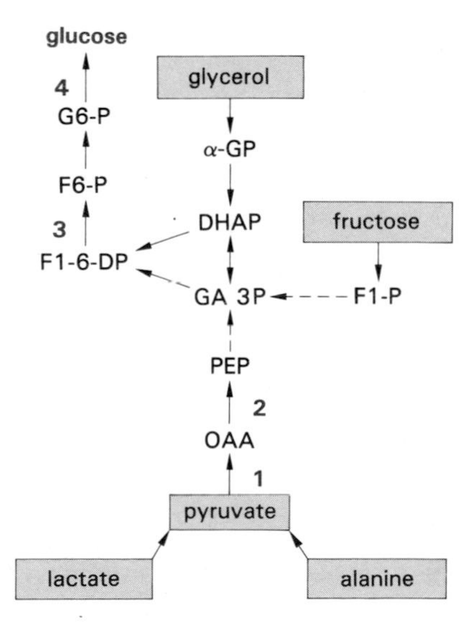

Figure XI.13 Gluconeogenesis and key enzymes. F 1-P: fructose 1-phosphate; DHAP: dihydroxyacetone phosphate; **1**: pyruvate carboxylase. **2**: PEP carboxykinase. **3**: F 1,6-diphosphatase. **4**: G 6-phosphatase. See previous figure for other abbreviations. Dotted line indicates a portion of the pathway that has not been detailed.

33. Krebs, E.G., de Lange, R.J., Kemp, R.G. and Riley, W.D. (1966) Activation of skeletal muscle phosphorylase. *Pharmacol. Rev.* 18: 163–171.
34. Stalmans, W., de Wulf, H., Hue, L. and Hers, H.G. (1974) The sequential inactivation of glycogen phosphorylase and activation of glycogen synthetase in liver after administration of glucose to mice and rats. *Eur. J. Biochem.* 41: 127–134.
35. Stalmans, W. (1983) Glucagon and liver glycogen metabolism. In: *Glucagon I* (Lefebvre, P.-J., ed), Springer Verlag, Berlin, pp. 291–314.

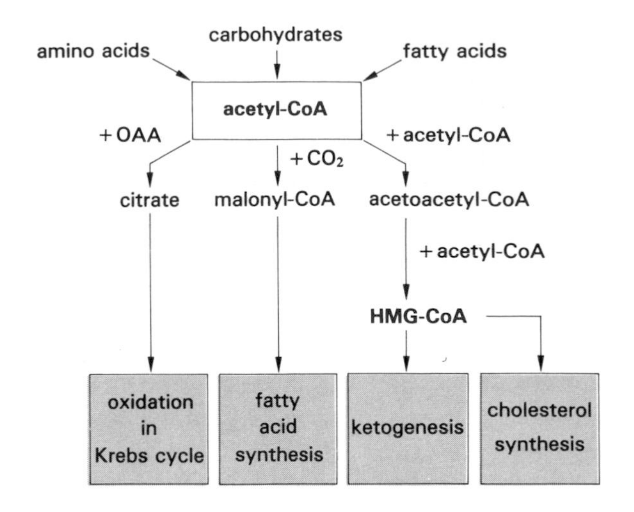

Figure XI.14 Metabolic pathways of acetyl coenzyme A. HMG-CoA: hydroxymethylglutaryl coenzyme A.

pyruvate dehydrogenase and activate pyruvate carboxylase, an enzyme involved in gluconeogenesis (Fig. XI.13). Insulin activates pyruvate dehydrogenase by promoting dephosphorylation of the subunit of this enzymatic complex[36]. *Acetyl-CoA* is used in *four principal metabolic pathways* (Fig. XI.14): 1. it enters the tricarboxylic (or citric) *Krebs cycle* by condensation with oxaloacetate, to form citrate. Approximately 90% of the energy produced by glucose is derived from the metabolism of pyruvate in the Krebs cycle; 2. it is incorporated *into long-chain fatty acids* and participates in *lipogenesis*; 3. it also participates in the *synthesis of cholesterol, and in ketogenesis.* The accumulation of acetyl-CoA favors ketogenesis in conditions where lipolysis is increased and lipogenesis reduced, as occurs with insulin deficiency; 4. it can also *favor gluconeogenesis*, from pyruvate, by activation of pyruvate carboxylase and inactivation of pyruvate dehydrogenase.

Lipids

Absence of insulin leads to a *reduction in lipogenesis* and an *increase in lipolysis* and ketogenesis. These changes account for a large portion of the energy stored from glucose, since approximately 90% of stored glucose is in the form of lipids. Insulin is indispensable for the maintenance and activity of the lipoprotein lipase; the absence of insulin leads to the accumulation of lipoproteins, rich in triglycerides, in the circulation. *Glucagon* exerts a variable *lipolytic* action according to the species; its ketogenic action requires elevated concentrations of the hormone.

Metabolism of fatty acids. The synthesis of fatty acids utilizes acetyl-CoA, which is carboxylated (acetyl-CoA carboxylase) to malonyl-CoA. Acetyl-CoA and malonyl-CoA undergo a series of condensations and reductions (by NADPH), catalyzed by an enzymatic complex (fatty acid synthetase), to form long-chain fatty acids. Acetyl-CoA carboxylase and fatty acid synthetase are activated by insulin. Glucagon reduces the activity of acetyl-CoA carboxylase.

The total *oxidation of fatty acids* into CO_2 and H_2O (via oxidation of acetyl-CoA in the Krebs cycle) is the *major source of energy.* Acetyl-CoA produced from the oxidation of fatty acids may also be used for ketogenesis (Fig. XI.14). The oxidation of fatty acids is increased in the absence of insulin.

Ketogenesis. The production of ketone bodies depends on the hepatic content of fatty acids. Ketone bodies (acetoacetic acid, acetone and β-hydroxybutyric acid) are oxidized mainly in peripheral tissues (Fig. XI.15). Under certain conditions, they are an *important energy source for muscle and even for the brain* (prolonged fast). Situations which accompany increased lipolysis, in particular insulin deficiency (insulin-dependent diabetes and fasting), are accompanied by an increased production of ketone bodies. Glucagon favors ketogenesis, probably due to an increase in hepatic lipolysis. Insulin opposes the ketogenic action of glucagon, and favors the peripheral utilization of ketone bodies. Ketone bodies inhibit oxidation of glucose and fatty acids, leading to their preferential utilization as a source of energy. When the production of ketone bodies exceeds utilization, they accumulate, leading to *ketoacidosis.* A *deficiency of insulin* therefore doubly favors the occurrence of ketoacidosis by increasing the production and reducing the utilization of ketone bodies.

Cholesterol synthesis (Fig. XI.14). Severe diabetic states, with major insulin deficiency, are accompanied by a reduction of cholesterol synthesis.

Proteins: Amino Acids and Protein Synthesis

Insulin stimulates the transport of certain amino acids in striated muscle, and, independently of this stimulation, increases the incorporation of amino acids into proteins; these two actions are independent of the effect of insulin on glucose transport. Insulin

36. Denton, R. M. Brownsey, R. W. and Belsham, G. J. (1981) A partial view of the mechanism of insulin action. *Diabetologia* 21: 347–362.

also accelerates the incorporation of acetate or pyruvate into proteins (by transamination, pyruvic acid can react with glutamic acid to produce alanine). The effects of insulin on protein synthesis are not identical in muscle and liver. Generally, insulin increases protein synthesis and reduces protein catabolism.

Insulin and growth. The anabolizing effect of growth hormone is reduced in the absence of insulin; in young animals, insulin deficiency can affect growth. However, it is difficult to demonstrate the direct effect of insulin on growth in normal animals, as can readily be shown with growth hormones. The relationship between insulin and other growth factors, particularly IGF-I and II, is shown in Figure III.8.

Insulin and glycoproteins. Hyperglycemia and *insulin deficiency* lead to an *accumulation* of *glycoproteins* in various tissues. Hyperglycemia, as such (independent of insulin deficiency), is responsible for a non-enzymatic glycosylation of proteins (p. 518). Experimental insulin deficiency is accompanied by an increased activity of glucosyl transferases, which are involved in the assembly of glycoprotein subunits in the basal membrane of capillaries.

4.2 Target Organs and Tissues

Liver

The liver plays a fundamental role in the storage of glucose, in the form of glycogen, and in the production of glucose. It is the *principal site of gluconeogenesis*, and the *exclusive site of ketogenesis*. It is also an important site for *protein and lipid synthesis*. Insulin and glucagon influence all of these processes. The *liver* is the major site of *glucagon* action.

Hepatic metabolism of glucose. Glucose penetrates hepatic cells by a reversible system of facilitated diffusion such that the intracellular and extracellular concentrations of glucose are almost always identical. The intracellular transport of glucose in the liver is therefore *not a limiting step* for its utilization. A glucokinase is present in the liver; it differs from hexokinase in that it has a much lower affinity ($K_M \sim 10$ to 20 mM, or 1.8 to 3.6 g/l) and a much greater specificity for glucose. The content of glucokinase in the liver is affected by nutritional and hormonal states. Glucokinase is reduced in the absence of insulin, whereas hexokinase remains unchanged. Insulin has an inductive effect on the synthesis of glucokinase.

Insulin favors the storage of glucose, in the form of glycogen, while glucagon mobilizes glycogen and increases the hepatic production of glucose. Insulin stimulates glycolysis by activating phosphofructokinase and pyruvate kinase, and the pentose phosphate pathway.

Glucagon plays an important role in the *stimulation of gluconeogenesis.* This stimulation occurs at different levels. The hepatic *uptake* of amino acids is increased, in particular the transport of alanine. The selective extraction of alanine by the liver, its conversion to glucose and its regeneration in muscle by transamidation of the pyruvate derived from glucose constitutes an important *glucose–alanine metabolic cycle*[37], similar to the glucose–lactate cycle of Cori. The gluconeogenic pathway is stimulated by the conversion of pyruvate into phosphoenolpyruvate (PEP), involving pyruvate carboxylase and PEP carboxykinase. Glucagon does not affect the portion of the gluconeogenic pathway beyond the triose phosphates (Fig. XI.13), in particular gluconeogenesis from glycerol and fructose. The *synthesis* of several hepatic *enzymes*, such as PEP carboxykinase, is stimulated by glucagon (and cAMP)[38]; glucocorticosteroids amplify this action. In contrast, insulin reduces the synthesis of the enzyme by inhibiting the transcription of its mRNA[39].

Substrates play an important role in the regulation of gluconeogenesis. In the absence of insulin, an increase in lipolysis, in adipose tissue and in the liver, and increased oxidation of fatty acids lead to an accumulation of acetyl-CoA, which inactivates pyruvate dehydrogenase, and activates pyruvate carboxylase. Oxidation of fatty acids furnishes the NADH necessary for gluconeogenesis. Glycerol produced during lipolysis can enter the gluconeogenic pathway, at the level of triose phosphates. The absence of insulin also favors protein catabolism and the hepatic uptake of gluconeogenic amino acids, particularly alanine and glutamine. During states of insulin deprivation (insulin-dependent diabetes and fasting), the hepatic uptake of alanine is markedly increased, while serum alanine levels are reduced. This phenomenon can be the result of an insulin deficiency, or an excess of glucagon, or both.

37. Felig, P. (1973) The glucose-alanine cycle. *Metabolism* 22: 179–207.
38. Salavert, A. and Iynedjian, P. B. (1982) Regulation of phosphoenolpyruvate carboxykinase (GTP) synthesis in rat liver cells. *J. Biol. Chem.* 257: 13404–13412.
39. Granner, D. K. and Andreone, T. L. (1985) Insulin modulation of gene expression. *Diabetes Metals. Rev.* 1: 32.

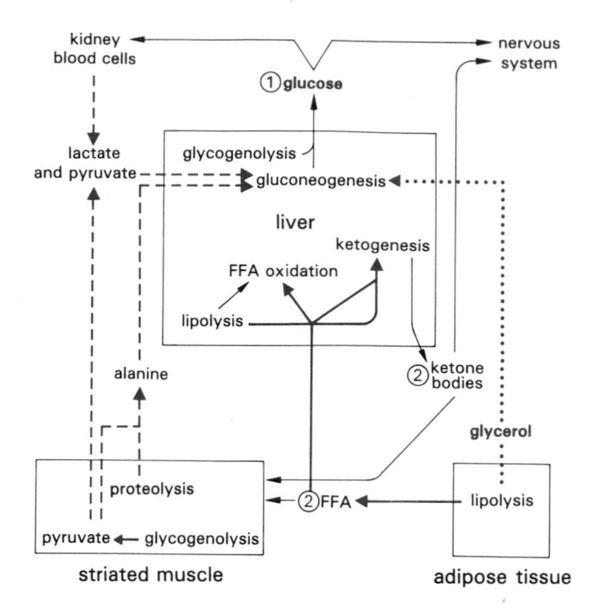

Figure XI.15 Metabolic pathways involved in the generation of substrates and in maintaining blood glucose level during fasting. Red arrows indicate substrates going to the liver. Black arrows indicate those going to various organs. **1**: glucose (produced by gluconeogenesis in the liver) goes to glucose-dependent tissue (in adult humans ~130–150 g/d to the CNS, particularly the brain; 30–50 g/d to red blood cells and medullary zones in kidneys). **2**: FFA and ketone bodies, which become important sources of energy in striated muscle and brain during prolonged fasting.

Hepatic metabolism of lipids. Insulin stimulates and glucagon inhibits hepatic lipogenesis. In the *absence of insulin*, the reduced activity of acetyl-CoA carboxylase and fatty acid synthetase limits the possibilities of synthesizing fatty acids from acetyl-CoA. In consequence, *ketogenesis* is increased; the concomitant *increase of lipolysis* in adipose tissue results in an increased afflux of FFA to the liver.

Glucagon has an *opposite action* to that of insulin, *inhibiting lipogenesis*, reducing the activity of phosphofructokinase and pyruvate kinase, and inhibiting acetyl-CoA carboxylase. These effects lead to an increase in ketogenesis. Thus, *insulin deficiency, and/or the excess of glucagon, inhibits hepatic lipogenesis and favors ketogenesis.* The depressive effect of glucagon on hepatic lipogenesis (and the increase in ketogenesis which follows) can be observed with physiological concentrations of glucagon; in such cases, its effects can be inhibited by insulin. The fine regulation of hepatic lipogenesis and ketogenesis appears to result from the antagonistic actions of insulin and glucagon on the hepatic metabolism of lipids.

In birds, glucagon favors the hepatic *synthesis of triglycerides* from FFA, the plasma concentrations of which are increased due to the marked lipolytic action of glucagon on adipose tissue in this species.

Hepatic metabolism of proteins. In the absence of insulin, the hepatic synthesis of proteins is altered. When insulin is administered to diabetic animals, one of the early effects (30 to 60 min) is an increase in RNA synthesis, which precedes the increase in protein. DNA content increases later (36 to 72 h), and an increased number of hepatic cells is also observed[40]. The *anabolic effect of insulin* is complemented by an *anticatabolic effect.* In contrast, glucagon has a catabolic effect coupled, at least in part, to lysosomal activation.

Glucagon and insulin stimulate the transport of *amino acids* in hepatocytes[41]. This effect can be seen in vivo when the secretion of insulin and glucagon is concomitantly increased by a diet rich in proteins. It probably contributes to both protein synthesis, stimulated by insulin, and the maintenance of glucose homeostasis by glucagon-stimulated gluconeogenesis.

Hepatic movement of K^+. Insulin increases the hepatic accumulation of K^+. This effect can be seen even at low concentrations of hormone, and is opposite to the release of K^+ by the liver, stimulated by glucagon and cAMP.

Adipose Tissue

Adipose tissue is the principal energy reservoir in the body. Its total mass is approximately 10 kg in a man of average weight, or the equivalent of about 90,000 kcal. Adipose tissue is the principal source of FFA; it also releases some amino acids, but not glucose. Glycerol liberated with FFA during lipolysis can serve as a substrate for gluconeogenesis. Insulin plays a major role in energy storage, in fat tissue, essentially in the form of triglycerides, which represent 90% of the mass of adipose tissue.

Intracellular transport of glucose. Within adipose tissue, in contrast to the liver, the *transport of glucose* into the cell represents an *essential limiting step* in the metabolism of glucose, at least in the absence of insulin and at normal or low concentrations of glucose. The absence of insulin can, in part, be

40. Steiner, D. R. (1966) Insulin and the regulation of hepatic biosynthetic activity. *Vitams. Horm.* 24: 1–61.
41. Fehlmann, M., Le Cam, A. and Freychet, P. (1979) Insulin and glucagon stimulation of amino acid transport in isolated rat hepatocytes. Synthesis of a high affinity component of transport. *J. Biol. Chem.* 254: 10431–10437.

compensated, at the level of transport, by an elevated extracellular concentration of glucose. In the presence of insulin, which stimulates glucose transport, hexokinase activity and the concentration of G 6-P in the cell become limiting factors. The accumulation of G 6-P inhibits hexokinase.

Effects of insulin. Many of the effects of insulin are the result of its *stimulatory effect on glucose transport.* Insulin also *stimulates the oxidative pathway of glucose (pentose-phosphates).* This stimulation, generating NADPH, is coupled to that of the synthesis of fatty acids. It results in a *marked stimulation of lipogenesis*, with an accumulation of glycerol 3-phosphate, derived from glucose. Insulin also stimulates the intracellular transport of certain amino acids, the accumulation of intracellular K^+, as well as the synthesis of proteins and glycogen. Insulin reduces lipolysis, glycogenolysis, and proteolysis. The *antilipolytic effect* of insulin is most important. It can be seen at very low concentrations of insulin (of the order of 0.01 nM, or 0.06 µg/l, or 1.5 mU/l). The persistence of some degree of antilipolytic effect, even under conditions of severe insulin deficiency, limits the importance of lipid catabolism and, to some extent, ketogenesis. The antilipolytic effect of insulin is independent of the stimulatory effect of the hormone on glucose transport and lipogenesis.

Effects of glucagon. The *physiological significance of the peripheral (extrahepatic) effects* of glucagon is *difficult to establish* because of the relative insensitivity of peripheral tissues to glucagon and the very low concentrations of glucagon in peripheral blood. Glucagon is considered a *lipolytic* hormone, but its effects on adipose tissue vary considerably between species. Avian adipose tissue is extremely sensitive to the lipolytic effect of glucagon, observed in vitro with a hormone concentration of 0.1 µg/l (or 0.03 nM). In humans, the lipolytic effect of glucagon is controversial. The stimulation of lipolysis by glucagon is accompanied by increased glycolysis in adipose tissue. The lipolytic effect of glucagon involves a *triglyceride lipase, activated by cAMP.* Except in birds, where the lipolytic effect of glucagon is especially strong, insulin inhibits lipolysis stimulated by glucagon when glucagon is present in relatively low concentration. The in vivo effect of glucagon on FFA in various species is counterbalanced by the insulin secretory and hyperglycemic effects of glucagon, which tend to reduce the plasma concentration of FFA. The inhibitory effect of FFA, like that of glucose on the secretion of glucagon, suggests that the lipolytic effect of the hormone is physiologically important. This negative feedback between the product of lipolysis (FFA) and the secretion of glucagon is *also* implicated in regulating the level of *energy substrates* other than glucose[42].

Striated Muscle

Striated muscle, either skeletal or cardiac, is a *major site of insulin action*; *cardiac muscle* is a *target organ of glucagon*, at least when pharmacological doses of this hormone are utilized.

Effects of insulin. In striated muscle, as in adipose tissue, insulin *stimulates glucose transport* into the cell, where hexokinase catalyzes the phosphorylation of glucose to G 6-P. In the absence of insulin, the transport of glucose into the cell represents the essential limiting step in its metabolism. In the presence of insulin, as in adipose tissue, the main limiting factor becomes the metabolism of G 6-P, the accumulation of which inactivates hexokinase. Insulin *stimulates glycogen synthesis,* the principal source of energy storage in muscle; however, it does not represent an important energy reserve, despite the large mass of muscle in the body. *Glycolysis stimulated by insulin* is the *predominant pathway of glucose utilization in muscle*, where dehydrogenases involved in the direct oxidative pathway are lacking. This lack of dehydrogenases is a limiting factor in the synthesis of NADPH, and, in consequence, causes a reduction in the lipogenic capacity of muscle. As is true for adipose tissue, muscle, lacking G 6-P, cannot produce glucose. Muscle is the principal source of alanine, pyruvate and lactate, the major gluconeogenic substrates (Figs. XI.13 and XI.15). *Insulin inhibits proteolysis and lipolysis in muscle.*

Conditions of insulin deficiency (insulin-dependent diabetes and fasting) lead, via the intermediary of an increased lipolysis in adipose tissue, to an increased level of circulating FFAs and ketone bodies. In muscle, FFAs and ketone bodies may become preferential energy substrates and enter into direct competition with glucose. An inverse relationship (glucose–fatty acid cycle) also seems to exist between the oxidation of fatty acids and the metabolism of glucose in muscle.

Insulin stimulates the active transport of certain amino acids across the plasma membrane of muscle

42. Lefebvre, P.-J. (1983) Glucagon and adipose tissue lipolysis. In: *Glucagon I* (Lefebvre, P.-J., ed), Springer Verlag, Berlin, pp. 419–440.
43. Le Marchand-Brustel, Y., Moutard, N. and Freychet, P. (1982) Amino-isobutyric acid transport in soleus muscles of lean and gold thioglucose-obese mice. *Amer. J. Physiol.* 243: 474–479.

cells[43]. Independent of its effects on transport, insulin stimulates the incorporation of all natural amino acids into muscle proteins. In the absence of insulin, both the number of active ribosomes and the initiation of the formation of peptide chains in muscle ribosomes are markedly reduced. The reduction of the number of ribosomes appears to be associated, at least in part, with reduced rRNA synthesis. The administration of insulin rapidly reestablishes ribosomal protein synthesis in muscles [44,45]. Insulin influences transmembrane ion fluxes, with an increase in intracellular K^+ and a reduction in Na^+. The effect of insulin on ion fluxes is associated with hyperpolarization of the membrane; thus, insulin may modify the transport function of the cell membrane.

Effects of glucagon. In *skeletal muscle*, glucagon does not stimulate adenylate cyclase, contrary to its effect in the myocardium. The positive inotropic and chronotropic effects exerted by glucagon are mediated, at least in part, by cAMP. They are accompanied by glycogenolysis and an increased production of lactate. These effects are observed with supraphysiological doses of the hormone, and should be considered pharmacological effects.

4.3 Other Effects

Insulin

Pregnancy and prolactin administration make the *mammary gland* sensitive to insulin, which then becomes a growth factor for the gland. The *placenta*, which insulin cannot cross, represents an important site of degradation, and in addition to the kidney, may also be a site of insulin action. Certain parts of the central nervous system, particularly the *hypothalamic* region involved in the sensation of satiety, appear to be directly sensitive to insulin. Insulin stimulates membrane ATPase in human *lymphocytes* in culture and the uptake of amino acids in thymocytes. The possible physiological significance of these effects has not been established. Insulin also appears to be involved in certain aspects of cellular immunity and in inflammatory processes.

Glucagon

Insulin secretory effect. This is a *direct effect*, increased by hyperglycemia (or a recent food intake), reduced by hypoglycemia (or fasting), and inhibited by catecholamines[29].

Pancreatectomy in birds. Partial pancreatectomy in birds *leads to hypoglycemia rather than hyperglycemia.* However, total pancreatectomy is also accompanied by a reduction in carbohydrate tolerance, with alternating hyperglycemia and hypoglycemia. Glucagon therefore plays a more important physiological role in birds than in mammals, whereas insulin appears to be equally important in both.

4.4 Insulin–Glucagon Interactions

In tissues where insulin and glucagon exert opposite effects (especially in the liver, and sometimes in adipose tissue), the molar ratio of insulin and glucagon required to counteract each of their effects is a useful index for the evaluation of the respective influence of the two hormones on various metabolic pathways. However, the *ratio of the peripheral plasma concentrations* of insulin and glucagon is not necessarily reflective of the actual ratio of the two hormones acting on the liver, because hepatic extraction of insulin and glucagon represents two distinct and independently regulated processes. The *ratio in portal venous blood* theoretically allows a better appreciation of the respective influence of the two hormones on the hepatic production of glucose, but this ratio (as such) is not the only determinant of the control of hepatic carbohydrate metabolism.

4.5 Pharmacological Effects of Glucagon

At *pharmacological* doses, glucagon exerts *numerous effects.* Some are directly related to the physiological action of the hormone, while others appear, or are first detected, at elevated doses.

Diagnostic Use

Stimulation of insulin secretion. This effect may be used to demonstrate hyperinsulinism in relation to an *insulin-secreting tumor.* Glucagon (1 mg iv) reveals hyperinsulinemia, and, due to its hyperglycemic action, minimizes the risk of concomitant hypoglycemia. The glucagon test is also useful in order to

44. Wool, I. G. and Cavicchi, P. (1967) Protein synthesis by skeletal muscle ribosomes: effect of diabetes and insulin. *Biochemistry* 6: 1231–1242.
45. Jefferson, L. S. (1980) Role of insulin in the regulation of protein synthesis. *Diabetes* 29: 487–496.

evaluate *residual endogenous insulin secretion* in diabetics.

Stimulation of catecholamine secretion. The iv administration of glucagon leads to a rapid *increase in plasma catecholamines* (in a few minutes), as well as to an increase in systolic and diastolic blood pressure. This effect can be used (*with caution*) as a provocative test for the diagnosis of pheochromocytoma.

Other diagnostic uses. The action of glucagon on the metabolism of glycogen is used as the *diagnostic test of glycogenosis*. In the disease of von Gierke (G 6-phosphatase deficiency), the hyperglycemic response normally observed folllowing administration of glucagon is abolished. The depressive effect of glucagon on gastrointestinal motility is utilized in radiology and in endoscopy.

Therapeutic Use

Hypoglycemia. Glucagon is a particularly useful therapeutic agent for the treatment of *insulin-dependent hypoglycemia*: 1 mg im is normally enough to correct hypoglycemia in a few minutes. Because of its stimulatory effect on insulin secretion, the use of glucagon is not always recommended for the treatment of *hypoglycemia due to hypoglycemic sulfonylureas.* Glucagon may be used efficiently for the symptomatic treatment of some hypoglycemias in adults and children. It is inefficient in acute alcoholic hypoglycemia.

Cardiac insufficiency. The positive inotropic and chronotropic effects of glucagon have led to its use, at pharmacological doses (1 to 5 mg), in the treatment of some cases of acute cardiac insufficiency, however results are inconsistent.

Other effects. Glucagon, at pharmacological doses, has a hypocalcemic effect which appears to involve the stimulation of calcitonin secretion. It has an inhibitory effect on gastrointestinal motility, as well as on that of the gall bladder and the sphincter of Oddi, and also has bronchodilator effects. It increases kalemia, in a transitory fashion, and is natriuretic.

5. Mechanism of Action of Insulin and Glucagon

5.1 Hormone—Receptor Interaction

The first step in the action of insulin and glucagon, like that of other polypeptide hormones, is the association of the hormone with a *specific receptor in the plasma membrane* of target cells (p. 74). The formation of hormone-receptor complexes can be measured using a radioactive ligand. The binding of the hormone to its receptor is specific. For insulin, it is strictly correlated to the active biological structure of the hormone. *Glucagon* receptors bind neither secretin nor VIP, even though these peptides belong to the same family (Fig. XII.22). The apparent dissociation constant (K_D) of insulin-receptor and glucagon-receptor complexes is if the order of 1 nM, and is therefore compatible with concentrations of hormones in vivo. A limited number of receptor sites are present. The total *number* of molecules of hormone which can be bound by a cell (~50,000-200,000 for insulin) *varies as a function of tissue type, species, physiological* (or pathophysiological) *state*, and experimental conditions[46].

5.2 Receptor Structure

Insulin

Three different approaches have been used to study the physicochemical structure of the insulin receptor: the biosynthetic labelling of the receptor, followed by a specific immunoprecipitation[47,48] by antireceptor antibodies; the covalent binding of insulin to its receptor using a photo-activated derivative of the hormone[49] or a cross-linking agent[50]; and cDNA cloning. The receptor consists of *two glycoprotein subunits*, of M_r~130,000 (α subunit) and 95,000 (β subunit), which combine to form a disulfide-linked heterodimer $(\alpha, \beta)_2$ (Fig. XI.16). The binding domain is restricted to the α subunit, and the function of the β subunit is discussed below. The subunits of the insulin receptor are *synthesized as a single polypeptide precursor* with a M_r of ~180,000.

46. Freychet, P. (1976) Interactions of polypeptide hormones with cell membrane specific receptors. Studies with insulin and glucagon. *Diabetologia* 12: 83–100.
47. Van Obberghen, E., Kasuga, M., Le Cam, A., Hedo, J. A., Itin, A. and Harrison, L. C. (1981) Biosynthetic labelling of insulin receptor: studies of subunits in cultured human IM-9 lymphocytes. *Proc. Natl. Acad. Sci., USA* 78: 1052–1056.
48. Kahn, C. R., Baird, K. L., Flier, J. S., Grunfeld, C., Harmon, J. T., Harrison, L. C., Karlsson, F. A., Kasuga, M., King, G. L., Lang, U. C., Podskalny, J. M. and Van Obberghen, E. (1981) Insulin receptors, receptor antibodies and the mechanism of insulin action. *Rec. Prog. Horm. Res.* 37: 477–533.
49. Fehlmann, M., Carpentier, J. L., Le Cam, A., Thamm, P., Sauders, D., Brandenburg, D., Orci, L. and Freychet, P. (1982) Biochemical and morphological evidence that the insulin receptor is internalized with insulin in hepatocytes. *J. Cell. Biol.* 93: 82–87.
50. Czech, M. P. (1980) Insulin action and the regulation of hexose transport. *Diabetes* 29: 399–409.

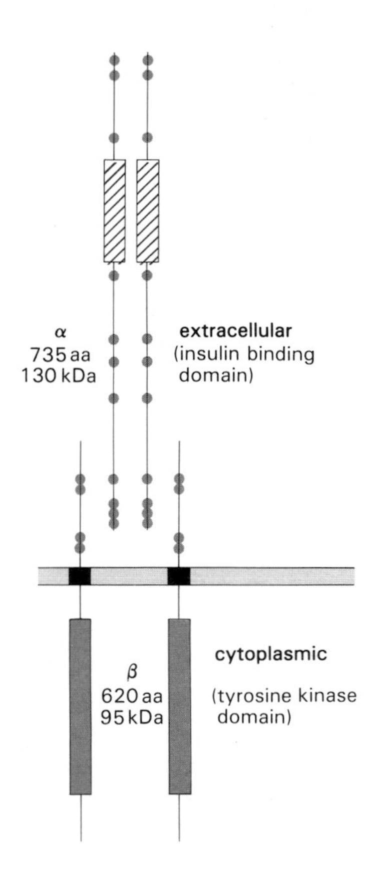

Figure XI.16 Schematic representation of the structure of the insulin receptor. Shaded boxes indicate cysteine-rich regions in the α subunits. Open circles on α and β subunits indicate single cysteine residues, possibly involved in the formation of $(\alpha, \beta)_2$ heterodimer. Since the insulin receptor is highly glycosylated (glycosylation sites are not shown), the M_r of both α and β subunits is greater than could be predicted from the amino acid composition of each subunit[51,52].

The primary structure of the human insulin receptor precursor has been deduced from its cDNA sequence[51,52] (Fig. XI.17). The precursor starts with a signal peptide sequence, followed by the α subunit, then the β subunit. Amino acid composition suggests that the α subunit is exclusively extracellular and that it contains the hormone binding site, whereas the β subunit contains a transmembrane sequence and the elements of a Tyr-PK in its cytoplasmic domain. The β subunit shows sequence homology with the epidermal growth factor receptor and some oncogene products (particularly the v-*ros* gene product)[51,53] (Fig. I.78). The human insulin receptor gene is located on chromosome 19[52].

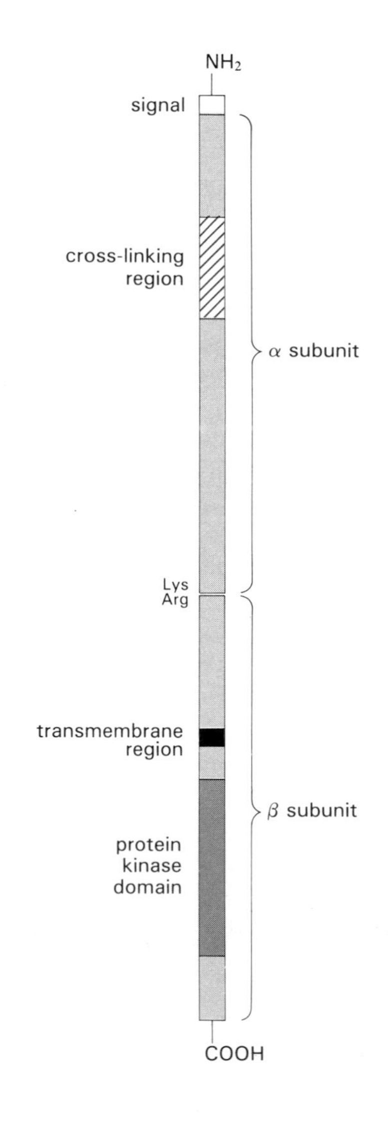

Figure XI.17 Schematic representation of the insulin receptor precursor[51,52].

51. Ullrich, A., Bell, J.R., Chen, E.Y., Herrera, R., Petruzelli, L.M., Dull, T.J., Gray, A., Coussens, L., Liao, Y.C., Tsubokawa, M., Mason, A., Seeburg, P.H., Grunfeld, C., Rosen, O.M. and Ramachandran, J. (1985) Human insulin receptor and its relationship to the tyrosine kinase family of oncogenes. *Nature* 313: 756–761.
52. Ebina, Y., Ellis, L., Jarnagin, K., Edery, M., Graf, L., Clauser, E., Ou, J.H., Masiarz, F., Kan, Y.W., Goldfine, I.D., Roth, R.A. and Rutter, W.J. (1985) The human insulin receptor cDNA: the structural basis for hormone-activated transmembrane signalling. *Cell* 40: 747–758.
53. Herberg, J.T., Codina, J., Rich, K.A., Rojas, F.J. and Iyengar, R. (1984) The hepatic glucagon receptor. Solubilization, characterization and development of affinity absorption assay for the soluble receptor. *J. Biol. Chem.* 259: 9285-9294.

Glucagon

A protein with a M_r of ~63,000 has been identified as the hormone-binding subunit of the hepatic glucagon receptor; the native receptor in the membrane appears to be a homodimer[53].

5.3 Mechanism of Action

Insulin (Fig. XI.18)

Paradoxically still a mystery

Recent studies have shown that insulin *binds to the α subunit* of its receptor, which results in the *autophosphorylation of the* β subunit and *activation of the* β subunit Tyr-PK[54–56]. This covalent modification of the receptor, modulated by the hormone, represents an early post-binding (but intraceptor) phenomenon in the sequence of events which follow the binding of insulin to the α subunit.

Three lines of evidence strongly suggest that the *phosphorylation of the receptor β subunit* and/or the *activation of its own kinase* are implicated in *transmembrane signalling* and *subsequent biological effects*. First, at physiological concentrations, and within one minute, insulin stimulates the phosphorylation and Tyr-PK of its receptor, a phenomenon which is rapidly reversible. Secondly, receptor kinase activity fluctuates, in parallel with insulin action, in states of resistance or increased sensitivity to insulin[57]. Thirdly, evidence that the PK activity of the insulin receptor is essential for insulin action has recently been documented with monoclonal antibodies to the receptor PK domain, and with site-directed mutagenesis of the receptor cDNA. In intact cells, insulin action results in covalent alterations (phosphorylation/dephosphorylation on Ser/Thr residues) of a number of proteins, including many enzymes (Fig. XI.18). One or several Ser-PKs, closely related to the receptor but not intrinsic to its structure, may be implicated in the *signalling process* between the receptor Tyr-PK and the subsequent cascade of phosphorylation/dephosphorylation reactions.

Many studies have involved a search for a *second messenger*, or an *intracellular mediator*, for insulin action. Most of the messengers or mediators potentially involved (cAMP and cGMP, ion fluxes, and alteration of membrane potential, calcium movements, small peptides) have not stood the test of time.

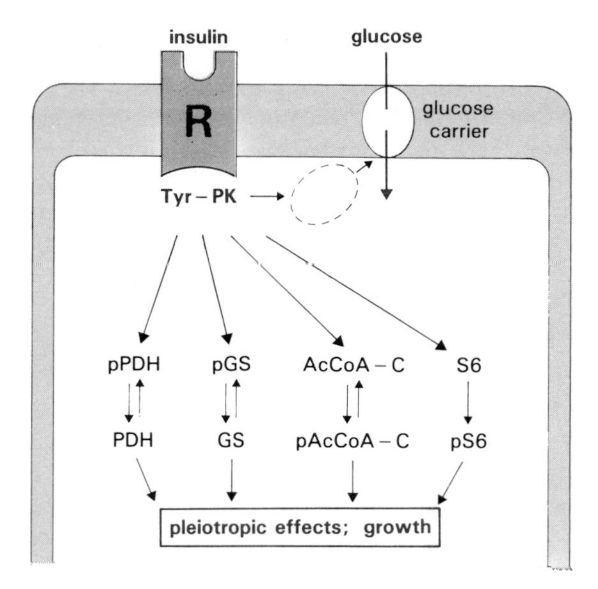

Figure XI.18 A proposed model of insulin action. Insulin binds to the α subunit of its receptor, and stimulates phosphorylation of the β subunit and activation of its tyrosine kinase. This results in a cascade of phosphorylation and dephosphorylation. Thus, ribosomal protein S6 and acetyl-CoA-carboxylase (AcCoAC) are phosphorylated, whereas pyruvate dehydrogenase (PDH) and glycogen synthase (GS) are dephosphorylated. Following binding to its receptor and receptor kinase activation, insulin also stimulates glucose transport by promoting the translocation of transport proteins, from an intracellular pool, to the plasma membrane. Insulin effects are also exerted at the gene level, on the transcription of specific mRNAs. This model does not exclude the possibility that a mediator is generated subsequent to receptor kinase activity; the mediator could modulate some of the effects of insulin within the cell.

Recently, *inositol phosphate-glycans,* released from the plasma membrane under the action of insulin, have been shown to modulate some of the hormone's effects[58]. Since inositol glycan generation is catalyzed by phospholipase C which is in turn regulated by pertussis toxin sensitive G proteins, it has been suggested that there is a functional linkage between the insulin receptor and the G protein regulatory system. Insulin has been shown to inhibit pertussis toxin-catalyzed ADP-ribosylation of G_i in plasma mem-

54. Kahn, C. R. (1985) The molecular mechanism of insulin action. *Annu. Rev. Med.* 36: 429–451.
55. Gammeltoft, S. and Van Obberghen, E. (1986) Protein kinase activity of the insulin receptor. *Biochem. J.* 235: 1–11.
56. Rosen, O. M. (1987) After insulin binds. *Science* 237: 1452–1458.
57. Le Marchand-Brustel, Y. (1987) Résistance à l'insuline dans l'obésité. *Médecine/Sciences* 3: 394–402.
58. Saltiel, A. R., Fox, J. A., Sherline, P. and Cuatrecasas, P. (1986) Insulin-stimulated hydrolysis of a novel glycolipid generates modulators of cAMP phosphodiesterase. *Science* 233: 967–972.

branes through a mechanism that appears to be independent of insulin receptor Tyr-PK-mediated G_i phosphorylation. Another aspect of insulin action deserves special mention. The stimulatory effect of the hormone on *glucose transport* across the cell membrane (one of the major insulin effects) involves the *recruitment of glucose transport proteins* that are transferred to the plasma membrane from intracellular sites[59]; in addition, insulin may cause a covalent modification of the glucose transport proteins, leading to their activation[50].

Following the binding to its receptor at the cell surface, *insulin is translocated into the cell*[21]. This process of *internalization*, which also *involves receptors*[49], appears essentially to serve a *degradative function*, especially for the hormone (Fig. XI.19). Receptors which are not degraded can be recycled to the plasma membrane[60]. It is unlikely that insulin thus internalized (or one of the products of its intracellular degradation) is involved in the mediation of the effects of the hormone. In fact, most of the available data indicate that the information necessary for the transfer of the hormonal signal is expressed at the level of the receptor itself, presumably by the intermediary of receptor kinase activation.

Glucagon

Glucagon acts *primarily* via *cAMP*; its intracellular concentration is increased following stimulation of adenylate cyclase by the hormone (Fig. XI.20). The nature of the coupling phenomenon which takes place between the occupancy of the receptor by glucagon and the activation of adenylate cyclase has largely been elucidated[61]. The glucagon-receptor complex acts by favoring the interaction of GTP with G proteins of the plasma membrane, which activate adenylate cyclase (p. 335).

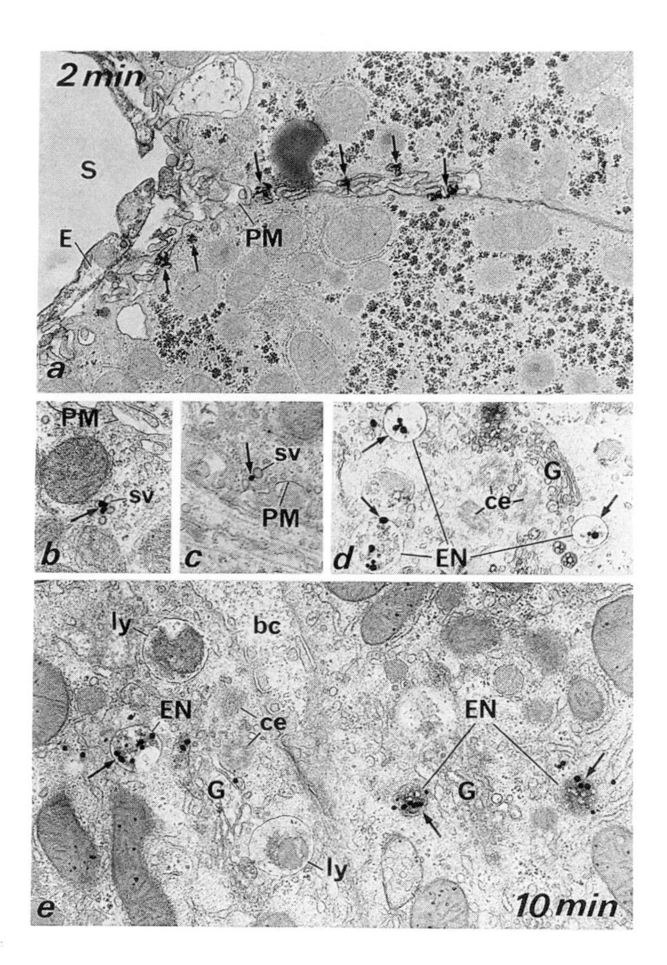

Figure XI.19 Insulin is internalized into target cells. The micrographs describe the initial binding (**a**) and progressive receptor-mediated internalization (**b–e**) of [^{125}I]labelled insulin into liver parenchymal cells in vivo. At first (**a**), radiolabelled insulin (the arrows mark silver grains which denote the radiation emanating from [^{125}I]insulin) is associated with the plasmalemma (PM) of hepatocytes. Endothelial cells (E) do not have appreciable levels of insulin receptors. 'S' denotes a liver sinusoidal space. In (**b**) and (**c**) are shown small (100–500 nm) subplasmalemmal vesicles (HSV) which appear to be clathrin-coated and marked with [^{125}I]insulin (arrow). In (**d**) and (**e**) are shown other components of the endosomal apparatus in which [^{125}I]insulin is subsequently found, i.e. after 10 min. The endosomes (EN) are 250 to 400 nm in diameter. They are in close proximity to the Golgi apparatus (G) and centrioles (CE). Liver endosomes often contain lipoprotein particles, most probably representing internalized lipoprotein. Their close proximity to centrioles, which function as microtubule organizing centers, is noteworthy, since endosomes migrate from the sinusoidal pole of the hepatocyte, to the bile canalicular pole (bc), along microtubules. Eventually, endosomes reside near the Golgi apparatus. As endosomes are rich in activated insulin receptor kinase activity, this new class of intracellular organelles is a candidate site for the regulation of insulin function. Bar=1 μm. (Courtesy of Drs. J.J.M. Bergeron and B.I. Posner, McGill University, Montreal).

59. Cushman, S. W. and Wardzala, L. J. (1980) Potential mechanism of insulin action on glucose transport in the isolated rat adipose cell: apparent translocation of intracellular transport systems to the plasma membrane. *J. Biol. Chem.* 255: 4758–4762.
60. Fehlmann, M., Carpentier, J. L., van Obberghen, E., Freychet, P., Thamm, P., Saunders, D., Brandenburg, D. and Orci, L. (1982) Internalized insulin receptors are recycled to the cell surface in rat hepatocytes. *Proc. Natl. Acad. Sci., USA* 79: 5921–5925.
61. Rodbell, M. (1983) The actions of glucagon at its receptor: regulation of adenylate cyclase. In: *Glucagon I* (Lefebvre, P.-J., ed), Springer Verlag, Berlin, pp. 263–290.

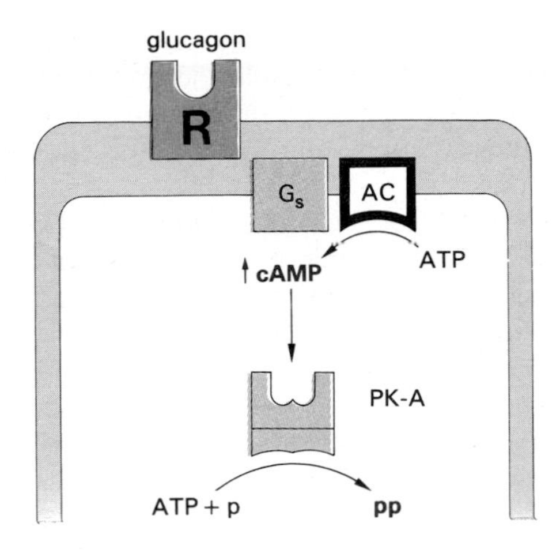

Figure XI.20 Mechanism of action of glucagon. Proteins phosphorylated by PK-A include enzymes of glycogen metabolism (Fig. XI.12).

6. Clinical Evaluation of the Endocrine Pancreas

6.1 Assay Principles in the Measurement of Insulin, Proinsulin, and Glucagon

Insulin, Proinsulin, and C Peptide

Plasma insulin in the peripheral blood, measured by RIA, represents *total insulin immunoreactivity (IRI)*; that is, insulin, proinsulin, and the intermediary circulating fractions. Insulin *normally constitutes the major portion (approximately 80%)* of the IRI, but the relative proportions of insulin and proinsulin, and the contribution of the latter to the IRI, vary according to the physiological or pathological circumstance[62].

Various techniques are used to measure proinsulin-like circulating fractions: 1. fractionation of plasma prior to RIA, with the separation of insulin and proinsulin based on their different molecular weights; 2. degradation of insulin by a specific protease; 3. use of *antisera directed against C peptide*. When the RIA of insulin is difficult or impossible due to the presence of anti-insulin antibodies in the circulation of patients treated with insulin, the RIA of C peptide in plasma[63] provides a valuable index of the residual biosynthetic and secretory activity of B cells.

The insulin receptor may also be used for measuring endogenous insulin (pancreatic or circulating) in a binding system which involves radiolabelled insulin and a preparation containing the receptor (plasma membrane or isolated cells). This receptor assay is more closely related to biological activity than RIA. It does not replace RIA, but it can be especially useful in clinical situations involving an increased proportion of proinsulin-like molecules, in relation to insulin, in the circulation.

Glucagon

There are difficulties involved in the RIA for glucagon. First, difficulties are encountered owing to the *absence of specificity for pancreatic glucagon* of most of the antiglucagon antisera; second, to the rapid inactivation of glucagon in plasma; and third, to the physiological level of the hormone in peripheral blood. However these difficulties *have been overcome* by the selection of antisera with greater specificity to pancreatic glucagon, by the fractionation of plasma, in order to separate the circulating forms of the hormones before RIA, by the use of inhibitors of glucagon degradation (p. 506), and, in certain experimental or clinical situations, by assessing venous pancreatic-duodenal or portal blood. A radioreceptor assay eliminates interference by glicentin and molecules, or fragments of molecules, which are immunologically but not biologically active. However, interference with unknown substances in plasma excludes the use of receptor assay for routine purposes[64].

6.2 Basal Measurements

Basal measurements are normally carried out following a 12 to 15 h fast.

Blood Glucose

Glycemia‡ values vary as a function of the *assay method* used. The difference between venous and

62. Gorden, P., Roth, J., Freychet, P. and Kahn, R. (1972) The circulating proinsulin-like components. *Diabetes 21,* suppl. 2: 673–677.
63. Heding, L. G. and Rasmussen, S. M. (1975) Human C-peptide in normal and diabetic subjects. *Diabetologia* 11: 201–306.
64. Holst, J. J. (1983) Radioreceptor assays for glucagon. In: *Glucagon II* (Lefebvre, P.-J., ed), Springer Verlag, Berlin, pp. 245–261.

‡ All values of normal glycemia reported in this chapter refer to assay in *venous plasma*; when the assays are carried out on total venous blood, the values are lower, by approximately 20 mg/dl (1.1 mmol/l).

capillary blood glucose is negligible during fasting but not after glucose load. The enzymatic assay method of glucose (glucose oxidase) is the most specific, and results in lower values: 80 to 90 mg/dl (4.5 to 5 mmol/l); values are slightly higher with the Hoffman method: 90 to 100 mg/dl, and with the Somogyi–Nelson method: 100 to 110 mg/dl.

In practice, the diagnostic interest of *fasting glycemia* is limited to two circumstances: 1. for the confirmation of *patent diabetes*. In this case, the values are usually greater than 140 mg/dl; 2. for the demonstration of *organic hypoglycemia,* with repeated determinations, during fasting, of less than 50 mg/dl. Hypoglycemia may become evident or may be accentuated by a diet of 50 g/d over 3 d, possibly followed by a complete fast (2 to 3 d), with close clinical and biological observation (determinations of glycemia several times per day). The *concomitant determination of insulin levels* in cases in which tumoral hyperinsulinemia is suspected (insulinoma, or insulin-secreting adenocarcinoma) permits diagnosis when an elevated or even normal insulin level is observed in association with reduced glycemia.

Glycosylated Hemoglobin

Glycosylated (or glycated) hemoglobin (HbA_{1c}, fast-migrating in electrophoresis) results from a post-translational, non-enzymatic glycosylation of the hemoglobin molecule, which is slow and continuous throughout the entire life of a red blood cell (120 d). In non-diabetics, the level of HbA_{1c} represents approximately 5% of total hemoglobin, whereas, in *diabetics* who are not treated or who are poorly controlled metabolically, this *level can reach 10 to 15% of total hemoglobin.* The level of HbA_{1c} thus allows a retrospective evaluation of the quality of glycemia control over the preceding two to three months; its determination is of value in the *follow-up of the quality of glycemia control* in treated diabetics.

Hormones

The level of *plasma insulin* in the peripheral blood, measured by radioimmunoassay following an overnight fast of 12 to 15 h, is normally between 2.5 and 15 mU/l (or 0.1 to 0.6 ng/ml, or 17 to 100 pmol/l). In this basal situation, and in normal subjects, proinsulin (and the proinsulin-like fractions) may contribute up to 20 to 30% of the IRI. This percentage is normally increased in conditions of hyperinsulinism (up to 80% of the IRI) due to pancreatic tumors (particularly malignant tumors), as well as in insulinopenic conditions (insulin-dependent diabetes, severe hypokalemia, pheochromocytoma); in obesity, the percentage is not modified, with respect to normal subjects[62].

The level of *plasma pancreatic glucagon* measured under the same nutritional conditions by RIA of peripheral blood is normally between 20 and 30 pg/ml (or 6 to 9 pmol/l).

Values of IRI and glucagon are higher in hepatic portal venous blood, even in the basal state, although the difference is clearly accentuated following the stimulation of insulin and glucagon secretion.

6.3 Stimulation Tests

Oral Glucose Tolerance Test (OGTT) (Fig. XI.21)

The oral glucose tolerance test represents the basis for the clinical evaluation of glucose regulation. However, it is *not useful for the diagnosis of diabetes* when fasting glycemia is elevated (>140 mg/dl, or 7.8 mmol/l). In normal adults, the level of fasting glycemia[65] is usually <115 mg/dl (6.4 mmol/l). There is therefore a rather large border zone between this value and the threshold value for the diagnosis of diabetes. Generally, this test is indicated when a reduction in carbohydrate tolerance (p. 523) or diabetes is suspected (based on an elevated postprandial level of glycemia), provided that fasting glycemia is less than <140 mg/dl; OGTT is also required for the diagnosis of gestational diabetes (p. 523), and for the substantiation of functional hyperglycemia as a reaction to the intake of glucose. OGTT is also of interest in the evaluation of diabetogenic endocrinopathies (when patent diabetes is not evident), and in organic hypoglycemia. This provocative, non-physiological test, because it involves the ingestion of a large quantity (75 g) of glucose in less than 5 min, has led to a reassessment of its value as a diagnostic tool, and has shown the need to regulate its use more strictly (Table XI.3).

A normal level of physical activity and a diet sufficiently rich in carbohydrates (300 g/d) for three days preceding the test is indispensable. Recent or concurrent acute diseases, physical inactivity, and a diet low in carbohydrates counterindicate the test and limit its usefulness in older and poorly nourished sub-

65. NIH National Diabetes Data Group (1979) Classification and diagnosis of diabetes mellitus and other categories of glucose intolerance. *Diabetes* 28: 1039–1057.

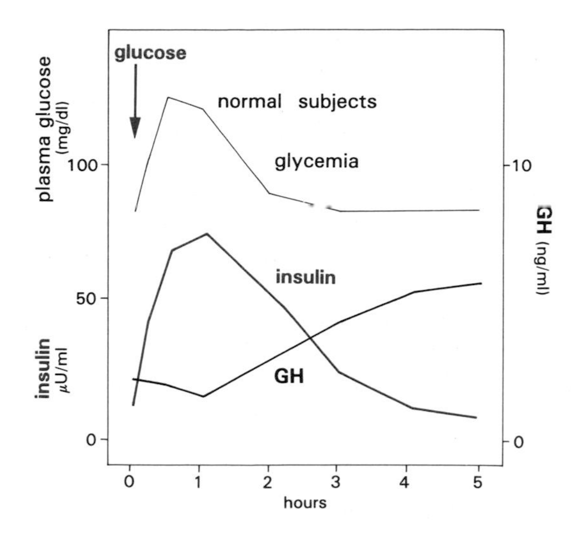

Figure XI.21 Oral glucose tolerance test in normal subjects. The insulin response precedes the increase of GH. The latter occurs concomitantly with the decrease of plasma glucose.

jects. A large number of therapeutic agents[65], glucocorticosteroids, diuretics, and hormonal contraceptives, for example, result in a usually diabetogenic dis-turbance. The test, carried out following an overnight fast, consists of the intake of a fixed quantity of glucose (75 g in adults; 1.75 g/kg of normal body weight in children, up to a maximum of 75 g). In pregnant women, the dose of 75 g is recommended by WHO[66], but the National Diabetes Data Group in the USA advocates a dose of 100 g.

For the diagnosis of diabetes, *WHO and most diabetes associations recommend the adoption of the plasma glucose values shown in Table XI. 3*, for both fasting and for two hours after the intake of 75 g of

Table XI.3 Diagnostic values, as established by the World Health Organization, for the oral glucose tolerance test administered under standard conditions. Glycemia levels indicated are for venous plasma. T_0 corresponds to fasting glycemia and T_{2h} to glycemia measured two hours after ingestion of 75 g of glucose.

	Diabetic mg/dl (mmol/l)	**Impaired glucose tolerance mg/dl (mmol/l)**
T_0	⩾140 (7.8)	<140 (7.8)
	and/or	and
T_{2h}	⩾200 (11.1)	>140 but <200 (7.8) (11.1)

glucose. The diagnosis of *impaired glucose tolerance* (intermediate between normal and diabetic states) and of *gestational diabetes* is based on specific criteria of the OGTT (p. 522).

The *RIA of insulin* during the OGTT allows the evaluation, in peripheral blood, of the secretory response of B cells to a large glucose load. The elevation of insulin levels occurs early, in a close temporal relationship with that of glycemia. The maximal level is attained 30 to 60 min following the ingestion of glucose, and normally remains lower than 100 mU/l (Figs. XI.21 and XI.22). Prolonged over a period of

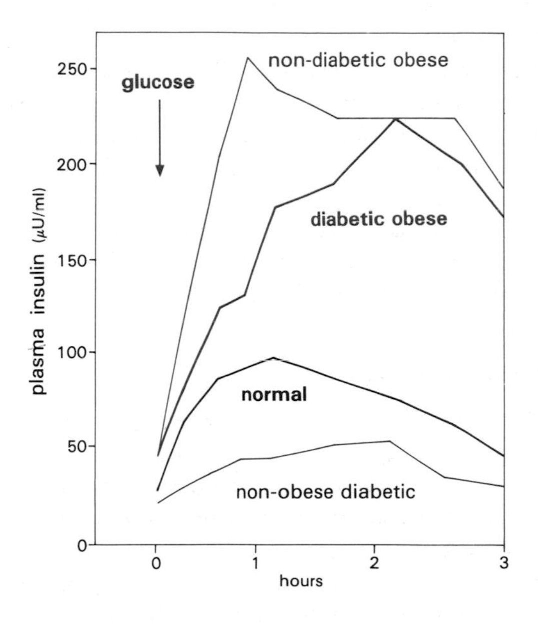

Figure XI.22 Plasma insulin during OGTT (in normal, non-diabetic obese, obese diabetic and non-obese diabetic patients). (Bagdade, J.D. Birman, E.L. and Porte, D.Jr. (1967). The significance of basal insulin levels in the evaluation of the insulin response to glucose in diabetic and non-diabetic subjects. *J. Clin. Invest.* 46: 1553).

five hours and combined with an insulin assay, this test allows evaluation of the return to normal levels of glucose and insulin. An increase in *GH* secretion normally accompanies the decline of glycemia in the late phase of the test, during a short fast (Fig. XI.21). In *non-obese diabetic patients*, the elevation of insulin levels is less apparent, despite values of glycemia which are much higher; the peak is frequently retarded and spread out in time (Figs XI.21 and XI.22). In

66. WHO (1980) *Diabetes Mellitus.* Technical Report Series 646, Geneva.

obese patients, the increase of insulin levels is frequently much greater; it is also delayed when diabetes is associated with obesity (Fig. XI.22). This relative or absolute hyperinsulinism (with respect to glycemia) is partly reflective of the insulin resistance usually observed in conjunction with obesity (p. 527).

The acute secretory stimulation brought on by an oral glucose load, or by any other stimulus, increases the plasma concentration of insulin without affecting that of proinsulin (or the proinsulin-like fractions). Thus, in the hour following stimulation, proinsulin normally represents a less important fraction of the IRI than in the basal state. The return to the proportion of proinsulin observed in the basal state takes place during the second hour after stimulation[62].

Modifications of plasma glucagon concentration during OGTT are not always easy to interpret, due to the heterogeneity of circulating glucagon and glucagon-like molecules of intestinal origin. The specific assay of pancreatic glucagon shows that its concentration in plasma decreases markedly, and can even become undetectable, following an intravenous glucose load, but the results are not as clear following an oral glucose load.

Intravenous Glucose Tolerance Test (IVGTT)

IVGTT, which involves a non-physiological administration of glucose, may have certain advantages when digestive tract disturbances make the oral test or its interpretation difficult. The IVGTT also allows the determination of the carbohydrate assimilation coefficient (K), which is reflective of the level of the peripheral effective insulin level and of the sensitivity of tissue to insulin; between 10 and 30 min after iv glucose administration, the plasma glucose level decreases linearly on a semilogarithmic plot. The K value is calculated from the half-life of this decrease (Fig. XI.23). The normal value of K, between 1.6 and 1.8×10^{-2}/min in young adults, diminishes with age, and ordinarily increases during the first trimester of pregnancy. Its value is usually less than 1×10^{-2}/min in patients with diabetes. The prolonged iv infusion of glucose is used less frequently. It is accompanied by an increase in insulinemia, while the plasma concentration of pancreatic glucagon is reduced and can even become undetectable. *Glucose clamp techniques*, consisting of iv infusion of glucose, producing various steady-state levels of glycemia (euglycemic and hyperglycemic clamps), are employed increasingly in the quantification of insulin secretion and insulin sensitivity (Fig. XI.25).

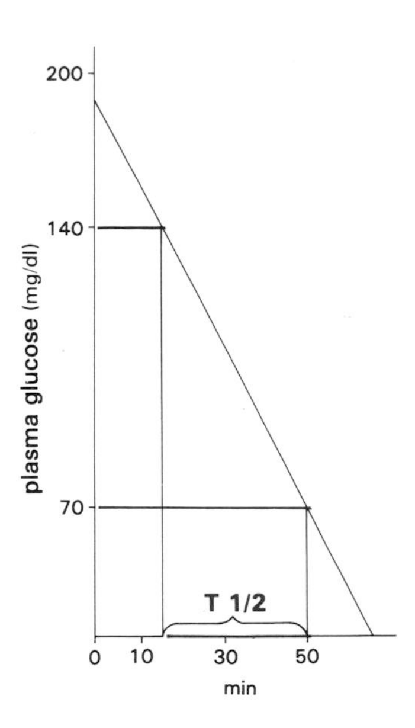

Figure XI.23. Intravenous glucose tolerance test (IVGTT). Semilog representation of the decrease of plasma glucose. The coefficient, K, of carbohydrate assimilation is calculated as: $K = \ln 2 / T_{1/2}$

Modification of Insulin and Glucagon Secretion by Agents Other Than Glucose

Tolbutamide Test

After iv administration of tolbutamide (1 g), the secretory response of insulin, which is directly stimulated by sulfonylurea and glycemia, is measured. Hypoglycemia is less accentuated and the onset is retarded in the diabetic as compared to a normal subject. The test is counterindicated in cases of insulin-secreting tumors, since an extreme insulin response can result in profound hypoglycemia.

Glucagon Test

The glucagon test offers the advantage of *stimulating insulin secretion* directly, and, because of its own hyperglycemic effect, minimizes the hypoglycemic risk in cases of insulin-secreting tumors. It is used in the evaluation of organic hypoglycemia and glycogenosis. The glucagon test is also used for evaluating the residual insulin secretory activity of B cells, in diabetes. Insulin levels are measured in non-insulin-dependent diabetics, and C peptide is measured in insulin-dependent diabetics, or in diabetics treated even transiently with insulin, because of the possible interference of exogenous insulin, and/or circulating anti-insulin antibodies, in the radioimmunoassay of insulin.

Leucine Test

L-leucine (200 mg/kg of body weight, per os) directly stimulates insulin secretion. In children, it allows the demonstration of hypersensitivity to leucine, characterized by a marked secretion of insulin, generating hypoglycemia, a response which is sometimes *severe*. Hypersensitivity to leucine is also observed in approximately 70% of patients with insulin-secreting tumors.

Arginine Test (Fig. XI.24)

The iv fusion of L-arginine (monochlorhydrate form, 25 g in 30 min in adults) directly stimulates insulin secretion. It also allows the evaluation of glucagon and growth hormone secretion. The arginine test offers the advantage of circumventing the risk of hypoglycemia due to insulin secretion; glycemia is slightly increased during this test.

Insulin-Induced Hypoglycemia

The iv administration of insulin (0.1 U/kg of body weight) is used mainly in two situations. First, in the assessment of suppressibility of endogenous insulin

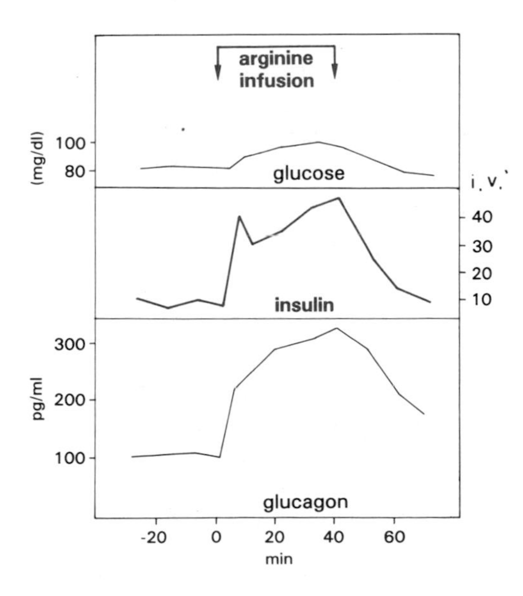

Figure XI.24 Effect of intravenous infusion of arginine on plasma glucose, insulin and glucagon in normal subjects. (Ungar, B.H. and Lefebvre, P.-J. (1972) In: *Glucagon*, Pergamon Press, Oxford, p. 232).

secretion (evaluated by the measurement of plasma C peptide levels), in the presence of hypoglycemia thought to be related to an insulin-secreting tumor. When autonomous secretion is due to a tumor, it is normally refractory to suppression. The test is also used to investigate endocrine reactions to hypoglycemia, including GH, ACTH and cortisol, epinephrine and glucagon. *In general, this test is counterindicated, or should only be used with extreme care*, as it can lead to severe hypoglycemia in cases where hypoglycemia is thought to be secondary to pituitary or adrenal insufficiency.

Meals of Variable Composition

Composed meals are rarely used in the clinical evaluation of insulin, due to the difficulty of standardizing the test. In normal subjects, the intake of a meal rich in carbohydrates is accompanied by an increase in insulinemia and a reduction in glucagonemia. As with the arginine stimulation test, the ingestion of a meal high in protein leads to an increase in insulin and glucagon levels.

6.4 Glucose Clamp Techniques in the Exploration of In Vivo Insulin Sensitivity

Insulin-induced hypoglycemia (provoked by the iv insulin test) does not permit a precise evaluation of insulin sensitivity because of the large dose of hormone used, and the secretion of hormones in response to hypoglycemia, counteracting insulin action. Glucose clamp techniques[67,68] consist of iv infusion of glucose, maintaining glycemia at a near physiological level (*euglycemic clamp*) in the face of varying levels of insulinemia achieved by increasing rates of iv insulin infusion. With this technique, the rate of overall glucose utilization is determined from the rate of glucose infusion required to maintain euglycemia at various steady-state insulin levels (Fig. XI.25). This allows the construction of *insulin dose-response*

67. De Fronzo, R., Tobin, J. D. and Andres, R. (1979) The glucose clamp techniques: A method for the quantification of beta cell sensitivity to glucose and of tissue sensitivity to insulin. *Am. J. Physiol.* 237: E214–E223.
68. Rizza, R. A., Mandarino, L. J. and Gerich, J. E. (1981) Dose-response characteristics for effects of insulin on production and utilization of glucose in man. *Am. J. Physiol.* 240: E630–E639.

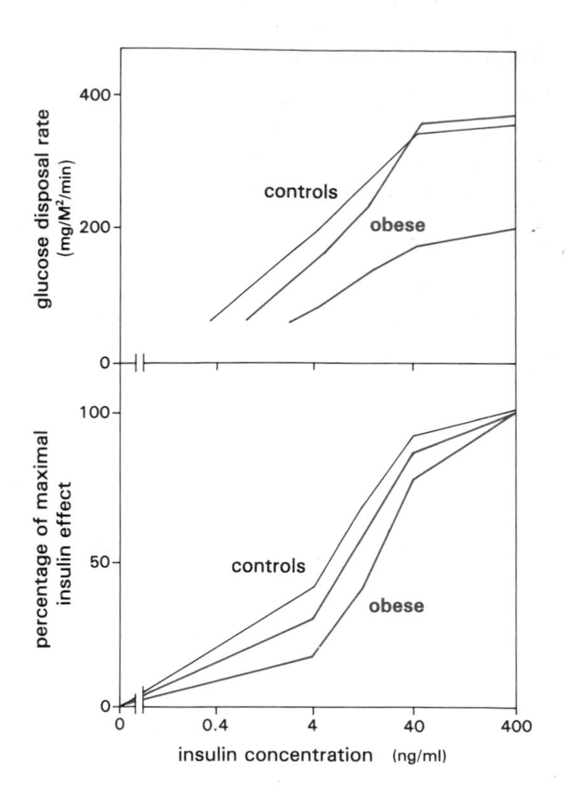

Figure XI.25 Dose-response curves: the effect of insulin on glucose utilization, by means of the glucose clamp technique, in control and obese subjects.[73]

curves of glucose utilization, and consequently the in vivo investigation of insulin sensitivity in pathological states.

6.5 Other Tests

In endocrine tumors of the pancreas, biselective celiac and superior mesenteric arteriography is of interest for localizing the tumor and evaluating the presence of hepatic metastases. The small size of pancreatic tumors frequently makes detection using ultrasonographic or tomodensitometric scanning difficult. The assay of blood hormone levels obtained by transhepatic catheterization of the portal vein and the pancreatoduodenal veins may be useful in the localization of the tumor.

69. WHO (1985) *Diabetes Mellitus*. Technical Report Series 727. Geneva.
70. Köbberling, J. and Tattersall, R., eds (1982) *The Genetics of Diabetes Mellitus*. Academic Press, New York.

7. Pathophysiology of the Endocrine Pancreas

7.1 Insulin

Classification of Diabetic States

The *classification* of *diabetic states* adopted by the *WHO* in 1980[66] (revised, with slight modifications, in 1985[69]) has largely been accepted and is in general agreement with that proposed by most diabetes associations throughout the world.

Clinical Classification

1. Patent Diabetes: Four Classes

Insulin-dependent diabetes mellitus (IDDM) is characterized by insulin deficiency and the patient's dependency on insulin for survival. *Type 1 diabetes*, an alternative denomination for IDDM, refers to specific pathogenetic mechanisms not always accessible to routine investigations, such as HLA typing and assay for islet-cell antibodies. Estimates of the prevalence (before 26 years of age) of IDDM range from 0.01 to 0.35%, with the highest rate found in Caucasian populations[69]. The proportion accounted for by IDDM among all diabetics (i.e. insulin and non-insulin-dependent) represents 10 to 15% (or more)‡. Most frequently, but not necessarily, IDDM affects subjects *younger than 40* years of age. The etiopathogenesis of IDDM (type 1 diabetes) is based on the combination of genetic predisposition and environmental factors. *Genetic susceptibility* is marked by certain alleles of the HLA system:B 8 and/or Bw 15 alleles in class I antigens; DR 3 and/or DR 4 alleles in class II antigens. The highest rate of IDDM is observed in individuals possessing both DR 3 and DR 4 alleles. The DR 3 allele is associated with IDDM in Caucasians and in some black populations[70], whereas DR 4 is associated with IDDM in all ethnic groups. A variety of *autoantibodies* directed against the islets of Langerhans *have been identified* in IDDM. Islet-cell cytoplasmic antibodies are found in the serum of 60 to

‡ Strictly speaking, IDDM refers to a condition where the patient needs insulin for survival, i.e. severe metabolic disturbances (ketoacidosis) occur in the absence of insulin treatment; however, a number of non-insulin-dependent diabetics may require insulin for proper metabolic control, and hence may be considered insulin-dependent diabetics.

80% of insulin-dependent diabetics with recent onset of the disease. The *acquired and environmental factors* presumably involve one or several *viral infections* which result in B cell alteration, a process possibly amplified by autoimmunity. The respective or combined roles of infection and autoantibodies in the pathogenesis of IDDM have not been fully established.

Non-insulin-dependent diabetes mellitus (NIDDM), also referred to as *type 2 diabetes* (former category of *mature-onset diabetes*), represents the *most common form* of the disease (85 to 90% of all diabetics). It is characterized pathophysiologically by the association of a *relative insulin deficiency* and an *insulin resistance* of variable importance. Its prevalence usually ranges from 2 to 7% of the general population, reaching more than 20% in certain American Indian and Micronesian populations; urban residents and migrants to urban areas have a higher prevalence of NIDDM than their rural counterparts of the same ethnic group[69]. A *family history* of diabetes is more frequent in non-insulin-dependent than in insulin-dependent diabetics. In contrast to IDDM, NIDDM is not associated with particular HLA profiles. Environmental factors are mostly linked to *nutrition and life style*, with *obesity and physical inactivity* being important risk factors for the development of NIDDM.

Malnutrition-related diabetes mellitus designates a category of diabetes which comprises at least *two subclasses*: *fibrocalculous pancreatic diabetes* and *protein-deficient pancreatic diabetes*[69]. The treatment of these forms of diabetes usually requires insulin. They occur mostly in developing tropical countries, where both protein malnutrition and the predominance of certain foods in the diet (such as those derived from cassava, of the fibrocalculous subtype) appear to play a major pathogenic role.

Other varieties of patent diabetes, which may or may not require insulin for treatment, include: diabetes *secondary* to a pancreatic disease (pancreatitis, hemochromatosis, cancer) or endocrinopathy (acromegaly, Cushing's syndrome or disease, pheochromocytoma, hyperaldosteronism), *iatrogenic* diabetes (especially caused by treatment with glucocorticosteroids) and rare forms of diabetes related to abnormalities of insulin or its receptors, or to certain genetic syndromes.

2. Impaired Glucose Tolerance (IGT)

GT is defined by the *criteria indicated in Table XI.3*, showing glycemia levels recorded after the administration of a glucose load. Certain subjects with IGT develop patent diabetes, but this evolution is also dependent on other factors, such as obesity, ageing, physical inactivity, and iatrogenic factors. Some individuals with IGT also have a greater risk of developing atherosclerosis, independent of the occurrence of diabetes.

3. Gestational Diabetes

Gestational diabetes is defined by an alteration of *glucose tolerance discovered during pregnancy*. The *WHO*[66,69] recommends that the diagnostic procedure (based on the OGTT) and criteria for pregnant women be the same as those proposed for adults. Thus, according to this recommendation, gestational diabetes is diagnosed following the administration of the OGTT (ingestion of 75 g of glucose) according to the same criteria as those defined for the diagnosis of impaired glucose tolerance (Table XI.3). However, the *National Diabetes Data Group* in the United States recommends that the OGTT (ingestion of 100 g of glucose) and the criteria indicated in Table XI.4 be applied for the diagnosis of gestational diabetes. Gestational diabetes involves a lesser risk of congenital malformations of the fetus than the risk associated with preexisting patent diabetes, but *fetal* and *perinatal morbidity* are the same (*macrosomia, hypoglycemia*) as for preexisting patent diabetes. Gestational diabetes also represents an increased statistical *risk* of subsequent occurrence of patent *diabetes* in the mother.

Table XI.4 Criteria of the oral glucose tolerance test in pregnancy[65]. Normal values are lower than those indicated, and gestational diabetes is defined by two (or more) of any of these values having been attained or surpassed. T_0 corresponds to fasting glycemia; T_{1h}, T_{2h} and T_{3h} to glycemia measured at 1, 2 and 3 hours following the ingestion of *100 g* of glucose.

	mg/dl	mmol/l
T_0	<105	5.8
T_{1h}	<190	10.6
T_{2h}	<165	9.2
T_{3h}	<145	8.1

Classification Based on Statistical Risk

The WHO distinguishes subjects at risk of developing NIDDM and those at risk of developing IDDM. The risk of NIDDM involves, in decreasing order, an identical twin whose sibling has NIDDM, subjects with first generation relatives who have NIDDM, obese patients, and women who have had one child (or several children) with a body weight at birth greater than 4 kg. In decreasing order, the risk of IDDM involves subjects with islet-cell antibodies, an identical twin whose sibling has IDDM, brothers or sisters with IDDM, and children of a parent with IDDM.

Patients with gestational diabetes and *impaired glucose tolerance are* also representative of individuals with an increased risk of becoming diabetic.

Insulin Deprivation Syndrome and Acute Metabolic Complications

Total pancreatectomy in mammals, and insulin-dependent diabetes in humans, results in an *insulinopenic syndrome,* the major symptoms of which are the direct consequences of a deficiency of insulin. The failure of intracellular penetration and utilization of glucose, and the increase in glycogenolysis and gluconeogenesis, creates *pronounced hyperglycemia.* When the level of maximal renal tubular reabsorption of glucose (300 mg/min) is attained, *glycosuria* appears. This leads to an *osmotic polyuria, generating polydipsia* resulting in dehydration. The *negative nitrogen balance* due to carbohydrate loss and protein catabolism results in *weight loss.* Lipolysis is increased, but a residual secretion of insulin, even a very low level, may be sufficient to suppress lypolysis and ketogenesis to a certain degree. The clinical signs of insulin-dependent diabetes, therefore, are polyuria, polydipsia, weight loss despite polyphagia, and a rapidly developing asthenia. The major metabolic complication (most frequently fatal before the insulin era) is *ketoacidosis.*

If insulin deficiency is aggravated, or if an external event increases protein and lipid catabolism, ketogenesis increases and surpasses the metabolic capacity to use ketone bodies, which accumulate and lead to ketonuria and *metabolic acidosis.* The hormonal profile of diabetic ketoacidosis does not involve only insulin deficiency. There is also an *associated hypersecretion of glucagon, GH, catecholamines and glucocorticosteroids*, which aggravates gluconeogenesis and ketogenesis. Clinical signs of ketoacidosis are frequently abdominal: nausea, vomiting, diffuse or localized abdominal pain (pseudoappendicitis). The amplitude and rhythm of respiration are modified (hyperventilation) in an attempt to compensate for metabolic acidosis. Hypothermia is sometimes observed; the existence of a fever, even moderate, indicates associated infection, which is frequently the factor responsible for metabolic decline. *Dehydration*, both cellular (dryness of the skin and mucosa, loss of weight) and extracellular (persistence of cutaneous fold, hypotonus of the ocular globes, oligouria, hypotension), is clinically evident. The patient is prostrated, with asthenia and adynamia which may be severe. *Emergency therapy* associates the administration of *insulin, and rehydration, with a correction of ionic disturbances.* Short-acting insulin is used, administered iv by a syringe (or pump) with an adjustable rate, delivering a dose equivalent to 0.15 to 0.3 U/kg of body weight/h; the administration of insulin by continuous iv infusion is usually preceded by an iv bolus of 20 U. Serious dehydration, both extra- and intracellular, which accompanies diabetic ketoacidosis provokes hypovolemia and alters renal function. This is as important to correct rapidly as the insulin deficiency. All cases of severe ketoacidosis are accompanied by a depletion of potassium, although the extracellular concentration of K^+ may be increased, at least transiently, as a result of acidosis. However, with the administration of insulin, which leads to an intracellular accumulation of K^+, and with the correction of acidosis, signs of severe hypokalemia may appear during treatment. The *factors contributing to the onset of coma* during ketoacidosis are not clearly established: hyperosmolarity due to hyperglycemia, dehydration, and disorders of the cerebral microcirculation, excess ketone bodies and acidosis itself (or, during treatment, an excessively rapid correction of systemic acidosis in relation to the correction of acidosis in the cerebrospinal fluid) are probably involved.

In patients in whom endogenous residual insulin secretion is still appreciable, the accentuation of hyperglycemia (frequently appearing as a result of treatment with a glucocorticosteroid, a diuretic, or diphenylhydantoin) associated with dehydration due to variable causes (fever, insufficient water intake, treatment with a diuretic) can lead to *non-ketonic hyperosmolar coma.* Treatment with biguanides can favor the occurrence of lactic acidosis in diabetics (which is rare but very severe), especially if complicated by renal insufficiency.

Hypoglycemia is a frequent complication of insulin therapy, and the patient should learn to *prevent it by proper dietetic and therapeutic education.*

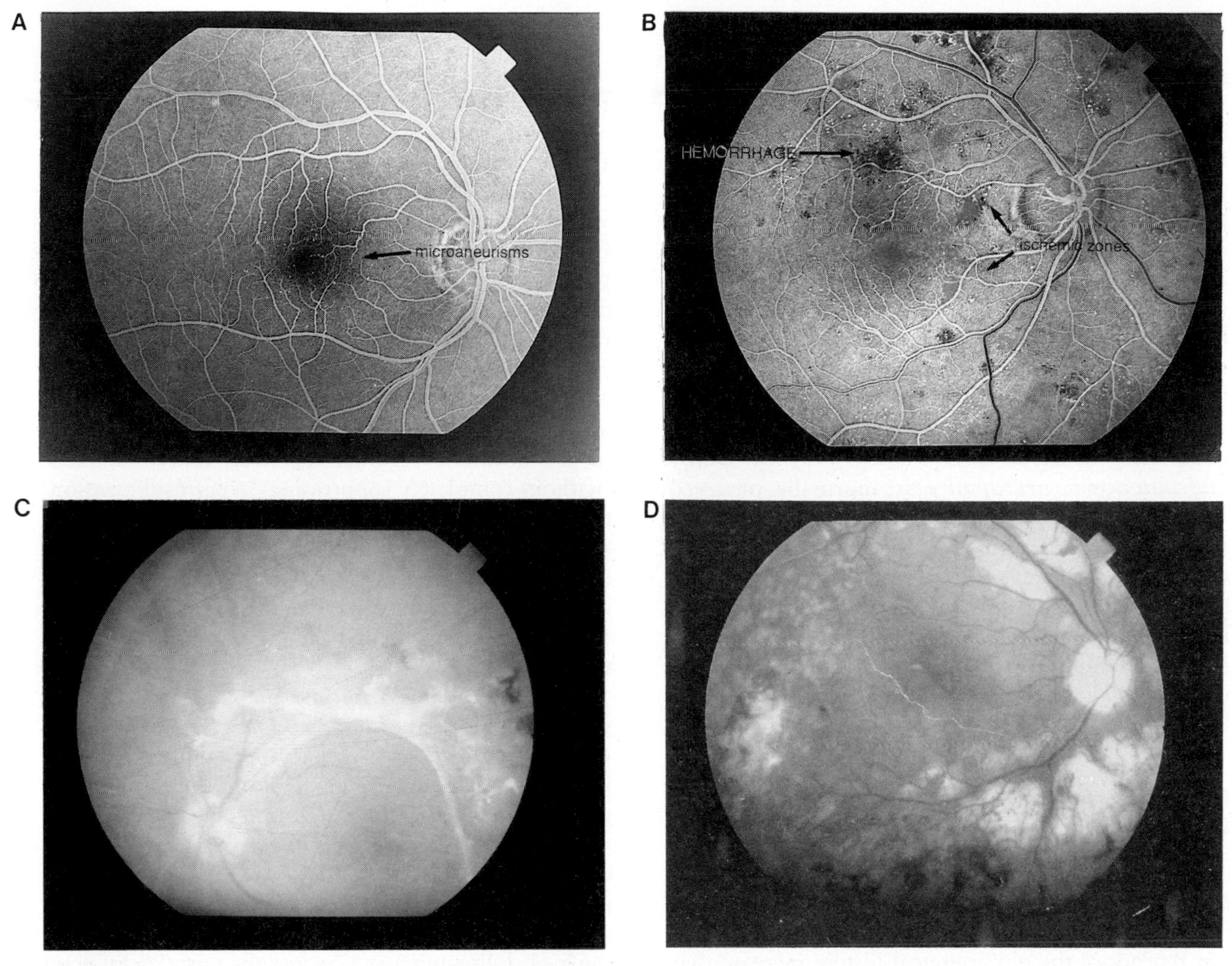

Figure XI.26 Retinal angiopathy **A.** Intravenous fluorescein angiography with retinal photography (early venous time of the procedure). Numerous microaneurisms, hemorrhages (darker spots) and ischemic zones (gray areas) can be seen. Ischemic zones (resulting from capillary closure) are regarded as the first step in neovascularization. **B.** Intravenous fluorescein angiography with retinal photography (venous time of the procedure). Several microaneurisms are seen in the macular area. **C.** Advanced proliferative retinopathy. Fibro-glial proliferation along the temporal vessels, from the optic papilla to the posterior pole of the eyeball. **D.** Argon laser panretinal photocoagulation. (Courtesy of Drs. P. Gastand and J. Darmon, University of Nice).

Non-insulin-Dependent Diabetes

The clinical symptoms of non-insulin-dependent diabetes are frequently less apparent (or even absent) than symptoms of insulin-dependent diabetes. Non-insulin-dependent diabetes may be revealed by a chronic complication affecting *both NIDD and IDD*, and thus medium or long term severity is conferred on the diabetic disease.

Chronic Complications of Diabetes

Macroangiopathy is characterized by vascular complications of *atherosclerosis* and, to a lesser degree for its clinical consequences, by lesions of *medial calcification* (calcification of the media in small and medium-sized distal arteries). Atherosclerosis and medial calcification are not specific to diabetes, but appear more frequently and are more precocious in diabetics than in non-diabetics. Atheromatous lesions are manifested primarily by coronary insufficiency and arteriopathy of the inferior limbs.

Diabetic microangiopathy is a specific complication. It is responsible in particular for *diabetic retinopathy* (Fig. XI.26) *and glomerulopathy.* It is potentially disabling (risk of blindness) and is life-threatening (risk of renal insufficiency). Diabetic microangiopathy is characterized morphologically by a thickening of the basal membrane of blood capillaries. Its incidence increases with the duration of the diabetic disease. It is generally agreed that the *principal pathogenic determinants* of microangiopathy are *chronic hyperglycemia and/or insulin deficiency.* These two factors are capable of inducing capillary lesions by increasing glycoproteins in the basal membrane. The possible pathogenic role of certain polypeptide growth factors is currently being investigated.

Mononeuropathy or multiplex neuropathy may occur early in the disease, whereas *polyneuropathy and autonomic neuropathy* are normally observed later. The sorbitol pathway, where sorbitol is formed from glucose under the effect of aldose reductase, is one of the pathogenic determinants of neuropathies and *diabetic cataract*; a deficiency in myoinositol, itself probably secondary to the accumulation of sorbitol, is probably also involved. The pathogenic role of excessive glycosylation of glycoproteins is possible, but has not yet been fully demonstrated.

Pathophysiological Bases for the Treatment of Diabetes

Insulin therapy, in its usual modes of administration, is not a perfect substitutive hormone therapy. The use of *long-acting insulin* is helpful, but it does not reproduce the physiological insulin increase that normally occurs after each meal (Fig. XI.21). Moreover, the mode of insulin administration, *necessarily parenteral* due to the destruction of insulin by digestive enzymes, means that only a small fraction of the injected hormone goes through the liver, in contrast to the entirety of insulin secreted under normal physiological conditions. It is generally thought that, to be efficient, insulin therapy *should be fractionated* (that is, divided into two or three daily injections or more), in order to approach physiological conditions of hormone secretion more closely, and to assure a more strict regulation of glycemia. Insulin therapy by *portable pump* (using short-acting insulin) is one of the best ways of administering this form of insulin therapy. Continuous sc infusion of insulin enables the establishment of basal and peak levels. Whether it is achieved by fractionated administration of insulin or by portable pump, optimal insulin therapy is of particular importance for the metabolic control of diabetes in pregnancy (particularly, if it is undertaken before conception, in order to minimize the incidence of congenital malformations), for the induction of remission of recent-onset diabetes, and for the treatment of disabling complications such as neuropathy. Optimal insulin therapy, whether it is achieved by several daily injections or by continuous sc infusion with a portable pump, leads to systemic hyperinsulinism, with risk of hypoglycemia in the short term, and possibly the risk of its own effects, of hyperinsulinization, over the longer term (induction of insulin resistance, possible atherogenic effect of insulin at pharmacological doses). The intraperitoneal (ip) administration of insulin, a more delicate type of administration, offers the advantage of a first passage of the hormone through the liver, and minimizes (without completely suppressing) hyperinsulinization observed with optimal insulin therapy administered subcutaneously. Technical failure of portable or implantable pumps results in the rapid development of diabetic ketoacidosis, owing to the very small reserve (sc or ip) of short-acting insulin used in this mode of therapy.

An *implantable artificial pancreas* is under study which is comprised of *both a miniaturized* system, to measure the level of glycemia continuously, and an instrument that can administer *insulin in response to glycemia levels. Biological protheses consisting of isolated islets of Langerhans in microspheres*, capable of secreting insulin, which are protected from immunological rejection, are at the experimental stage. *Grafts of islets of Langerhans*, especially within the liver, following injection into the portal vein, may open interesting possibilities of truly substitutive endogenous insulin therapy. *The segmentary transplantation of the pancreas* (coupled with a kidney transplant) is currently used for only a limited number of diabetics with chronic renal insufficiency; as is true for islet transplantation, the main problem is immunologic rejection.

Treatment, by *early immunosuppression*, using cyclosporin A in insulin-dependent diabetes (type 1) is presently under investigation.

The action of *hypoglycemic sulfonylureas* used in the treatment of non-insulin dependent diabetes is probably not limited to the stimulation of endogenous insulin secretion. These agents appear also to have extra-pancreatic effects; in particular, they amplify certain effects of insulin, at a postreceptor step. The possibility that biguanides potentiate insulin action at the receptor level is controversial.

Diet is an important basic element in the treatment of diabetes. It is especially important to avoid hyperglycemic and hypoglycemic periods, to *maintain normal body weight* in the non-obese diabetic, and to try to reduce the weight of the obese patient. *Physical exercise* is also important; it allows and accelerates utilization of glucose, and increases the efficiency of treatment.

The education of diabetic patients is a *fundamental* element in therapeutic effectiveness and in the prevention of complications. It is necessary for the patient to learn theoretical and practical information concerning diet, self-monitoring of metabolic control (urine and/or capillary blood glucose testing), use of insulin and adjustment of doses as a function of conditions of daily life (social and occupational activities, lifestyle), prevention of lesions of the feet, and the appropriate self-management (under medical control) of this chronic disease.

Insulin Resistance

Resistance to insulin is defined as a refractory state, variable in intensity, to the action of insulin. In the strict sense, insulin resistance defines states of *refractoriness to exogenous and endogenous insulin*, the latter being of normal biological activity. Because this section deals with the pathophysiological analysis of all states relevant to reduced insulin activity, the following passage also mentions rare situations that result from the production of abnormal or less efficient insulin. *Clinically, it is the combination of hyperinsulinism with normal or reduced carbohydrate tolerance which suggests insulin resistance.* This situation is frequently observed in obesity, acromegaly, Cushing's syndrome, and pregnancy. NIDD, even without obesity, associates a relative insulinopenia and a variable degree of insulin resistance. In IDD, insulin resistance is classically defined by an abnormally elevated need for insulin unexplained by intercurrent illness or by a major error in therapeutic management.

These *data* have been *refined* within the last 10 years, owing particularly to the development of techniques that facilitate the measurement of insulin receptors and sensitivity to insulin, which has enabled pathophysiological evaluation of the mechanism of cellular hormone action. Each of the successive steps in the production, transport and action of insulin can be the site of alteration(s) responsible for reduced action of insulin or for insulin resistance (Table XI.5). At the level of biosynthesis and secretion, an *abnormal insulin* with reduced biological activity can be produced[71]. There are also rare cases of *familial hyper-proinsulinemia*, characterized by failure to convert proinsulin to insulin[71]. Since proinsulin is biologically less active than insulin, this disorder is associated with an overall reduction in insulin action. At the level of transport, inactivation of insulin by *anti-insulin antibodies* is a classical but rare cause of insulin resistance. The degradation of insulin by the placenta participates in insulin resistance of pregnancy. It is especially knowledge concerning hormone action at the level of the cell that has elucidated the major elements and causes of insulin resistance in the last ten years. The study of various models of experimental or clinical insulin resistance has demonstrated the existence of *receptor defects (alteration in their number or affinity)* and *postreceptor* lesions.

Table XI.5 Main steps potentially involved in reduced insulin action.

Synthesis/secretion	Transport	Target cell
↓ Production	Blocking anti-insulin antibodies	Receptor defects
↓ Conversion of proinsulin to insulin	↑ Degradation	Postreceptor defect
Structural defect		

Insulin Resistance in States of Obesity and Diabetes

The binding of insulin to its receptors in different target tissues is reduced in various experimental models of *obesity* associated with insulin resistance[46]. Similar alterations in binding are observed in obese humans[72]. A reduction in insulin binding corresponds to a decrease in the number of receptors, while affinity is unaltered. Since the first step in insulin action involves binding of the hormone to the receptor, it can be expected that a decrease in receptor number could lead to insulin resistance. However, maximal insulin effects are usually achieved at a hormone concentration involving the occupation of less than the total number of available receptors. Those receptors in excess (relative to the maximal effect insulin) are termed *'spare' receptors*. In fact, this concept of spare

71. Tager, H. S. (1984) Abnormal products of the human insulin gene. *Diabetes* 33: 693–699.
72. Blecher, M. and Bar, R. S. (1981) Insulin receptors: obesity. In: *Receptors and Human Disease*. Williams and Wilkins, Baltimore, pp. 31–57.

receptors does not mean that some receptors are active in transmitting the hormone signal whereas others are inert, but rather that all receptors are fully functional, with receptor occupancy occuring as a random statistical event. Given this concept, the predictable biological consequence of a *decrease in receptor number (or affinity)* is a decrease in insulin action only at submaximally effective insulin levels. This leads to a shift to the right of the biological dose-response curve, without affecting the maximal response, and is termed *decreased sensitivity*[56,73,74]. If a *postreceptor* defect exists, one sees a proportionally decreased response at all insulin concentrations (including maximally effective). This is termed *decreased responsiveness. Combined receptor and postreceptor alterations result in both decreased sensitivity and decreased responsiveness.* Generally, experimental and clinical data concur with these predictions; obesity frequently involves a decrease in receptor number, which is responsible for a diminished *sensitivity* to insulin, with or without postreceptor alteration(s) responsible for a decreased *responsiveness* to the hormone. For example, studies conducted in obese subjects by means of the euglycemic insulin clamp technique have revealed a spectrum of response patterns[73] (Fig. XI.25). Some patients exhibit decreased sensitivity and normal responsiveness, compatible with a receptor defect as constituting the sole factor contributing to insulin resistance. Other obese subjects show both decreased sensitivity and decreased responsiveness, indicating that insulin resistance in these patients results from combined receptor and postreceptor alterations. In general, as insulin resistance progresses in obesity, postreceptor alterations combine with a receptor defect to decrease both the sensitivity and the response to insulin[74].

A reduction in caloric intake in obese patients corrects endogenous hyperinsulinism, the receptor deficit, and, to a large extent, insulin resistance. It is *possible, therefore, to modulate receptor number* on the basis of the inverse relationship between insulin receptor number and insulinemia.

The two components of insulin resistance observed in obesity are also found *in NIDDM patients* of normal weight. Reduced sensitivity to insulin explained by a diminution of the number of receptors and a diminution of the maximal response to the hormone suggests an alteration of the action of insulin at postreceptor steps. This *postreceptor alteration* appears to be the *major factor* in insulin resistance[73], at least in those diabetics who are most resistant. However, it can be corrected, for the most part, by insulin therapy. This suggests that an insulin deficit itself (even relative, as in NIDD) can be a factor in insulin resistance at postreceptor steps.

In IDDM patients, the number of insulin receptors is not reduced but may actually be increased before insulin therapy is initiated[75]. This observation is in concordance with the concept of an inverse relationship between insulinemia and the number of insulin receptors. However, the effect of insulin in insulin-dependent diabetes is sometimes reduced in vitro[75] and in vivo[76]. This alteration, due exclusively to postreceptor mechanisms, is amenable to some extent to insulin replacement. Insulin deficiency, whether relative (NIDDM) or more absolute (IDDM), leads to a state of insulin resistance, via postreceptor alteration.

The *mechanisms and sites of postreceptor alterations are largely unknown*. Recent studies have suggested that *defective activation* of the receptor kinase may be involved. This alteration would represent an early post-binding, intra-receptor defect. On the other hand, true postreceptor alterations, such as defects of the transport of glucose or of the activity of specific enzymes, are also involved, but their mechanisms remain to be elucidated.

Rare Syndromes of Extreme Insulin Resistance

Severe insulin resistance is observed in rare syndromes, and its frequency may be underestimated. These syndromes include the following elements: a marked resistance to exogenous and endogenous insulin, explaining major hyperinsulinemia (insulin being of normal biological activity), the frequent presence of the skin disorder acanthosis nigricans (although not in all cases), from which the term *insulin resistance with acanthosis nigricans* is derived, and the absence of other causes of insulin resistance: obesity, anti-insulin antibodies, associated endocrinopathy, or intercurrent infections. The study of insulin receptors in such syndromes has led to a div-

73. Olefsky, J. M. and Kolterman, O. G. (1981) Mechanisms of insulin resistance in obesity and non insulin-dependent (type II) diabetes. *Amer. J. Med.* 70: 151–168.
74. Le Marchand-Brustel, Y. and Freychet, P. (1978) Studies of insulin insensitivity in soleus muscles of obese mice. *Metabolism* 27, suppl. 2: 1982–1993.
75. Samson, M., Fehlmann, M., Morin, O., Dolais-Kitabgi, J. and Freychet, P. (1982) Insulin and glucagon binding and stimulation of aminoacid transport in isolated hepatocytes from streptozotocin diabetic rats. *Metabolism* 31: 766–772.
76. DeFronzo, R. A., Hendler, R. and Simonson, D. (1982) Insulin resistance is a prominent feature of insulin-dependent diabetes. *Diabetes* 31: 795–801.

ision into *three subgroups*[48]. In *type A*, the major defect is a *reduction* of the *number* of receptors; in *type B, antireceptor antibodies reduce* the *affinity* of receptors for insulin; *in type C*, the defects are *postreceptor*.

Type A normally affects adolescent or young women, usually with *polycystic ovaries*. Reduced insulin receptor binding in type A is not, or is only partially, reversible. This fact, demonstrated in cultures of fibroblasts from patients with type A syndrome and from infants with leprechaunism, suggests that such defects are responsible for these syndromes.

Type B generally occurs in older patients with *autoimmune* abnormalities[48]. This subgroup has circulating autoantibodies directed against the insulin receptor, capable of markedly altering receptor function. These antibodies are most frequently polyclonal IgGs; they reduce substantially the affinity of the receptor for insulin, and their presence probably accounts for extreme insulin resistance. Metabolic deterioration of these patients is sometimes associated with fluctuations of the level of insulin due to the insulino-mimetic properties of these antibodies, shown either in vitro or in vivo. The presence of anti-insulin receptor antibodies has also been reported in *ataxia telangiectasia*, a syndrome in which the anti-receptor antibodies belong to the low molecular weight, IgM class.

Some cases of insulin resistance with acanthosis nigricans involve neither alterations in receptor binding nor antireceptor antibodies; they represent a third subgroup, type C. *Lipoatrophic diabetes*, acquired or congenital, constitutes a heterogeneous group which can present as either type A or type C.

In some patients with the clinical phenotype of the type A syndrome, insulin receptor binding is normal, but *receptor kinase activity* is *defective*[56]. *This alteration points to a post-binding, intra-receptor defect* in *receptor tyrosine kinase* as a potential mechanism of insulin resistance.

States of Increased Sensitivity to Insulin

Pituitary and adrenal insufficiency and, to a less degree, thyroid insufficiency, lead to an increased sensitivity to insulin, which is in direct relation to a reduction, or absence, of hyperglycemic hormones. Insulinemia is low in these patients, who are extremely sensitive to exogenous insulin.

Hyperinsulinism

Obesity

Obesity is accompanied by *a varying degree of hyperinsulinism* characterized by an elevated insulinemia in the basal state and following stimulation (Fig. XI.22). This hyperinsulinism can coexist with a normal carbohydrate tolerance or with patent diabetes. The kinetics of insulin secretion may be altered, with a secretory delay following stimulation and a disappearance or modification of the early secretory phase. Hyperinsulinism may also be responsible, following dietary stimulation or a carbohydrate load, for *reactive or poststimulatory hypoglycemia*. The resistance to insulin is responsible, in large part, for the hyperinsulinism of obesity; hyperinsulinism, in turn, aggravates insulin resistance, and leads to a reduction in insulin receptor number (p. 526). However, the primary cause of hyperinsulinism of obesity has not been entirely elucidated; its pathogenesis probably involves, in addition to an excessively high caloric intake, one or several disturbances of the nervous regulation of insulin secretion[77].

Tumoral and Non-tumoral Hyperinsulinism

Pancreatic tumors secreting insulin (insulinomas) lead to organic hypoglycemia through primary hyperinsulinism. The main characteristic of the disorder is the association of fasting hypoglycemia with normal or elevated insulinemia (despite concomitant hypoglycemia). Clinical manifestations of hypoglycemia during fasting are mostly of the neuroglycopenic type (confusion, convulsion, coma). The persistence of normal or elevated insulinemia in the face of low glycemia differentiates insulinomas from other organic causes of hypoglycemia, e.g., endocrine (pituitary and adrenal) insufficiency or a tumor of non-pancreatic origin, both of which are characterized by a very low insulinemia. Neonatal hypoglycemia in children born to diabetic mothers is due to transient hyperinsulinism in conjunction with the hyperplasia of B cells, the fetus having been exposed to maternal hyperglycemia. Hypersensitivity of the B cell to leucine is also the cause of insulin hypoglycemia in certain children whose pancreas are shown to have B cell

77. Jeanrenaud, B. (1985) An hypothesis on the aetiology of obesity: dysfunction of the central nervous system as a primary cause. *Diabetologia* 28: 502–513.

hyperplasia. Leucine can also provoke an excessive insulin secretory response in cases of secreting pancreatic tumors. Aside from surgery, treatment of pancreatic tumoral hypoglycemia can involve the inhibition of insulin secretion by a non-diuretic, benzothiadiazine, diazoxide, or the destruction of B cells by the antibiotic streptozotocin. Insulin-secreting pancreatic tumors may coexist with other endocrine tumors, and thus be part of the so-called *multiple endocrine neoplasia* (MEN-1), a familial disease which can include insulinoma, gastrinoma (a pancreatic tumor producing gastrin, responsible for the Zollinger–Ellison syndrome), VIPoma (secreting VIP and responsible for the Verner–Morrison syndrome), glucagonoma, parathyroid hyperplasia, and even anterior pituitary, thyroid, or adrenal cortex adenoma.

7.2 Glucagon

Glucagon and Hypoglycemia

An insufficiency of glucagon secretion can be the cause, at least partial, of some cases of hypoglycemia in neonates and infants. The role of glucagon deficiency in reactive (poststimulatory) hypoglycemia is controversial. The instability of some insulin-dependent diabetics may imply an insufficiency of the secretory response of glucagon to hypoglycemia.

Glucagon and Endocrine Tumors

An excessive production of glucagon is observed in rare malignant tumors (glucagonoma) of the pancreas. These tumors may produce other peptides, in particular PP, in addition to glucagon.

Glucagon and Diabetes

In diabetic states, *glucagon secretion is often inappropriate*. This is characterized by a *normal* or even *increased* secretion of *glucagon* in the basal and stimulated states which persists despite hyperglycemia. The hypersecretion of glucagon is particularly clear-cut in diabetic ketoacidosis, but it also occurs in varieties of diabetes which do not involve a major insulin deficit. Insulin therapy only slowly or incompletely corrects this inappropriate secretion of glucagon. Thus, patent insensitivity of A cells to glucose in diabetes probably involves one or several factors other than insulin deficiency alone. The pathophysiological implication of this bihormonal profile in metabolic disorders of diabetic states is controversial. Diabetic states secondary to total pancreatectomy do not appear to be spared this bihormonal disorder, since glucagon of the pancreatic type can originate from extrapancreatic sources, particularly from A cells, identified in the gastric fundus of the dog and the human fetal stomach.

7.3 Somatostatin and Pancreatic Polypeptide

Rare cases of pancreatic tumors producing these peptides alone or associated with VIP and/or glucagon have been described; somatostatinomas have been reported, the clinical presentation of which consists of reduced carbohydrate tolerance or patent diabetes, steatorrhea, and achlorhydria. Excessive secretion of PP is also possible, either isolated, in cases of pancreatic tumor producing PP (PPoma), or associated with the tumoral production of VIP (VIPoma of the Verner–Morrison syndrome) or glucagon (glucagonoma).

7.4 Insulin and Glucagon: Interrelationship as a Function of Nutritional State

A very efficient responsive system

The *fasting state reproduces the hormonal profile of diabetes* since it associates, particulary during a prolonged fast, an *increased secretion of glucagon* with a *markedly reduced secretion of insulin*. However, in this pathophysiological situation, modifications of the secretory activity of the endocrine pancreas are physiologically appropriate, since maintenance of carbohydrate homeostasis, requiring an increased endogenous production of glucose, is achieved at the expense of lipid and protein stores (as in diabetic ketoacidosis). This hormonal profile is reversed in normal subjects when, immediately after a carbohydrate intake, the body can again restore energy sources. Finally, when dietary intake is exclusively (or predominantly) protein, the secretion of glucagon, concomitant with that of insulin, prevents insulin-induced hypoglycemia; insulin may then enhance protein synthesis and energy storage without compromising blood glucose. The *qualitative and quantitative regulation of the endocrine secretion of the pancreas, as a function of nutritional state*, is a remarkable example of adaptation to environmental conditions and integration of various regulatory processes.

8. Summary

The term *insulin* originates from the synthesis of hormone in the islets of Langerhans in the pancreas. Its isolation, in 1922, crystallization, the elucidation of its primary structure and conformation by X-ray crystallography, its radioimmunoassay, and, more recently, the detailed analysis of its receptor, are milestones in biology and medicine. However, its mechanism of action is still poorly understood. Whereas insulin is *hypoglycemic*, *glucagon*, the second pancreatic hormone, is a *hyperglycemic* hormone. Cyclic AMP was discovered during the study of the mechanism of glucagon action.

The endocrine portion of the pancreas accounts for only ~1% of the weight of the organ. The basic units are *islets*, composed of at least *four different cell types*: *B cells*(or β), the most numerous (~*60%*), which secrete *insulin*; *A cells* (or α_2) (15%), which make *glucagon*; *D cells* (or α_1), producing *somatostatin*; and *PP* cells, producing *pancreatic polypeptide*. Physiological interactions correspond to the *simultaneous presence of these cell types in the same islets*, as for example the insulin secretory effect of glucagon and the insulin- and glucagon-inhibitory effect of somatostatin.

Insulin is composed of *two peptide chains*, A, of 21 aa, and B, of 30 aa, which are linked by *two disulfide bridges*. Despite evolutionary differences in the primary structure, most insulins have similar biological activities, due probably to a relatively *fixed three-dimensional* configuration involved in the interaction with specific receptors. Insulin exists as a monomer of M_r~6000, a dimer of M_r~12,000, and three dimers form a hexamer, of M_r~36,000, in the presence of two zinc atoms. Indeed, metallic ions favor aggregation and crystallization of insulin.

Insulin is synthesized as *proinsulin* (M_r~9000), encoded by a gene located on the short arm of *chromosome 11* in humans. Proinsulin is formed by a *single-chain* polypeptide, starting with the N-terminal of the B chain and finishing, at the C-terminal of the A chain, with a connecting segment, or *C peptide*. The latter *allows appropriate alignment* of the disulfide bridges and the proteolytic cleavage of proinsulin to insulin. Insulin secretion, after selective proteolysis, occurs by Ca^{2+}-dependent exocytosis of the content of B cell granules. There is a *basal secretion* of insulin. *Primary stimuli* are metabolic substrates, the most important of which is *glucose*. When glucose concentration increases, a biphasic secretion response is observed, with the first phase representing the release of stored insulin, and the second, at least in part, newly synthesized insulin. *Amino acids* also *stimulate insulin* secretion, especially arginine, synergistically with glucose. *Glucagon* and other GI peptides stimulate insulin secretion, whereas catecholamines are inhibitors.

In the blood, insulin (and sometimes proinsulin) is measured by a specific radioimmunoassay. However, such a value accounts for only 5 to 10% of the total biological insulin-like activity (ILA). This difference is also called *NSILA* (*non-suppressible insulin-like activity*), and is made up essentially of *IGF-I and -II, growth factors* chemically related to insulin.

Glucagon is a 29 aa peptide which is evolutionarily highly conserved. It belongs to the same structural family as secretin and VIP. Glucagon is synthesized as proglucagon. The human proglucagon gene is located on *chromosome 2*. Processing of the translated product is responsible for the secretion of glucagon in the pancreas and *glicentin* in the gut (for glucagon-like immunoreactivity). *Increased blood glucose decreases*, and *decreased blood glucose stimulates*, glucagon secretion. However, amino acids increase glucagon secretion as they do insulin, resulting in a physiological situation in which the secretion of the two opposite hormones are modified parallely. *Insulin* has a negative effect, and *β-adrenergic* agonists, a stimulatory effect, on glucagon secretion.

Metabolic effects of insulin and glucagon are numerous. They play a major role in blood *glucose homeostasis. Insulin stimulates anabolic* processes and favors *energy storage*, whereas glucagon exerts a catabolic activity, consuming energy substrates. Thus, their effects are opposed on glycogenogenesis, glycogenolysis and gluconeogenesis, and their activities in relation to these multienzyme systems are complex. *Insulin favors fatty acid synthesis* and decreases lipolysis and ketogenesis, whereas *glucagon is lipolytic* and ketogenic. *Insulin stimulates the transport of glucose* and some amino acids, *in striated muscle and adipose tissue*. It also stimulates *protein synthesis* in the liver, striated muscle, and adipose tissue. Insulin, like growth factors, displays a *growth-promoting effect, and produces pleiotypic responses* in several experimental systems.

The liver is a major site of insulin and glucagon effects, primarily at the level of glucose, glycogen and lipid metabolism. In *adipose tissue*, insulin is essential for triglyceride storage. *Skeletal muscle* is also a target for insulin, as is *cardiac muscle,* in which glucagon is also active. The physiological role of *glucagon is more important in birds* than in mammals, since pancreatectomy leads to a decrease in blood glucose.

Insulin and glucagon act via specific *membrane receptors* in their target cells. Essentially, *glucagon effects are mediated by an increase in cAMP*, but the receptor itself is not yet chemically defined. In contrast, although the *mechanism of action of insulin* is *not clearly established*, its *receptor has been cloned,* and the corresponding amino acid sequence fully described. It is composed of two dimeric units, of which the two S-S linked chains of each dimer are derived from a single proprotein. The α chain, external to the cell, binds the hormone, while the β chain spans the membrane and includes a *tyrosine kinase* site which has been implicated in hormone action.

The concentration of insulin in the blood is of the order of 0.1 nM, and that of glucagon, 0.01 nM, in the basal state. Glucagon is more difficult to measure by radioimmunoassay than insulin because of its low level in the blood and because of its cross-reactivity with enteroglucagon. *Clinically*, the measurement of *fasting blood glucose* helps to confirm latent diabetes or to detect a hypoglycemic syndrome. In the absence of patent diabetes, glucose tolerance is best explored by *oral glucose administration. Insulin resistance,* observed frequently in conjunction with obesity, may be detected during this test, by insulin measurements. Many other challenge tests have been used to stimulate and then measure insulin levels.

There are *two forms of diabetes mellitus. Type 1 diabetes, or insulin-dependent diabetes mellitus (IDDM)*, is characterized by *insulin deficiency*, with permanently elevated blood glucose, glycosuria, polyuria, weight loss, polyphagia and polydipsia, which eventually leads to ketoacidosis. Treatment is based essentially on the correction of insulin deficiency as completely as possible, and includes the use of long-acting preparations and/or portable pumps. Active research deals with the concept of autoimmune disease, and with the development of different forms of artificial pancreas. *Type 2* diabetes, or non-insulin-dependent diabetes mellitus (*NIDDM*), is much *more common* (85 to 90% of all diabetics), occurs *later in life,* and is frequently associated with some degree of *obesity*. Complications of both type 1 and type 2 diabetes include *microangiopathy,* with thickening of the basal membrane of capillaries, most often in the retina and renal glomerulus. It includes relative insulin deficiency and insulin resistance of variable importance. This disease can be controlled largely by appropriate diet. An alteration in glucose tolerance discovered during pregnancy is called *gestational diabetes*. It is associated with some fetal and perinatal morbidity, and there is an increased risk of subsequent patent diabetes.

Glucagon-producing tumors have been reported. An increase in glucagon secretion occurs in diabetes, mostly during ketoacidosis.

General References

Czech, M. P., ed (1985) *Molecular Basis of Insulin Action*, vol. 1, Plenum Press, New York.

Kahn, C. R. (1985) The molecular mechanism of insulin action. *Annu. Rev. Med.* 36:429–451.

Köbberling, J. and Tattersall, R., eds (1982) *The Genetics of Diabetes Mellitus*. Academic Press, New York.

Lefebvre, P. J., ed (1983) *Glucagon I and Glucagon II.* Springer Verlag, Berlin.

Marble, A., Krall, L. P., Bradley, R. F., Chistlieb, A. R. and Soeldner, J. S., eds (1985) *Joslin's Diabetes Mellitus*, 12th edit, vol. 1, Lea and Febiger, Philadelphia.

Orci, L. (1974) A portrait of the pancreatic B-cell. *Diabetologia* 10: 163–187.

Rosen, O. M. (1987) After insulin binds. *Science* 237:1452–1458.

Unger, R. H. and Foster, D. W. (1985) Diabetes Mellitus. In: *William's Textbook of Endocrinology*, 7th edit. (Wilson, J. D. and Foster, D. W., eds), W. B. Saunders Co., Philadelphia, pp. 1018–1080.

WHO (1985) *Diabetes Mellitus*. Report of WHO Study Group. Technical Report Series 727, World Health Organization, Geneva.

Michael S. Brown and Joseph L. Goldstein

Lipoprotein Receptors

Lipoprotein receptors are *cell surface* proteins that *remove cholesterol-carrying lipoproteins from the circulation* through the process of receptor-mediated endocytosis (Fig. 1). Two classes of lipoprotein receptors have been identified: 1. those that bind lipoproteins containing exogenous cholesterol absorbed from the intestine, i.e. chylomicron remnant receptors, and 2. those which bind lipoproteins that carry endogenous cholesterol derived from the liver and other non-intestinal sources, i.e. *low density lipoprotein (LDL) receptors*.

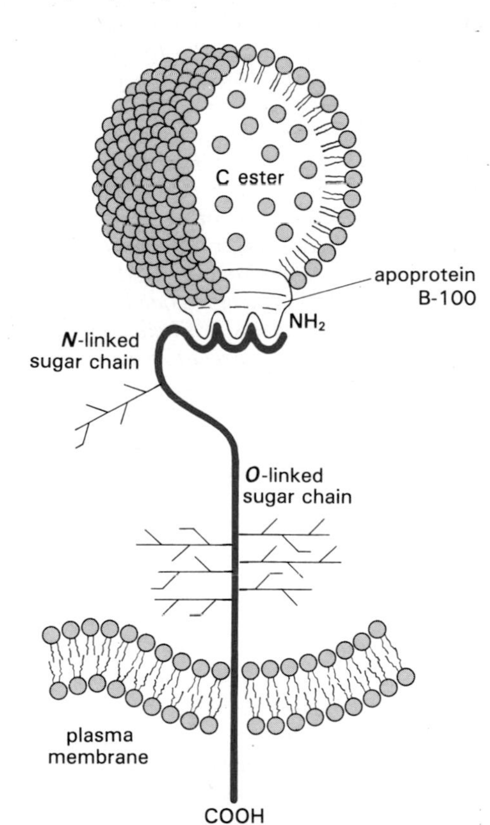

Figure 1 *LDL receptor, a glycoprotein embedded in the plasma membrane of most cells of the body. Note the site of attachment of* N- *and* O-*linked sugars. The spherical LDL particle (M_r~3×10^6, 22 nm diameter) consists of about 1,500 cholesterol ester molecules, and is surrounded by a coat of about 800 molecules of phospholipid, 500 molecules of cholesterol, and one large protein (apoprotein B-100).*

The *LDL receptors* were the first to be described; indeed, they are among the best characterized of all mammalian cell surface receptors. These receptors are present on the surface of essentially all cultured mammalian cells, where they mediate the uptake of plasma LDL, thereby providing cells with the cholesterol they need for growth. In the body, most LDL receptors are expressed in the *liver*, where they supply cholesterol for secretion into bile, for conversion to bile acids, and for resecretion into the plasma as newly synthesized lipoproteins. LDL receptors are also present in high concentrations in the *adrenal cortex* and ovarian *corpus luteum*, where they function to provide *cholesterol for steroid hormone formation*.

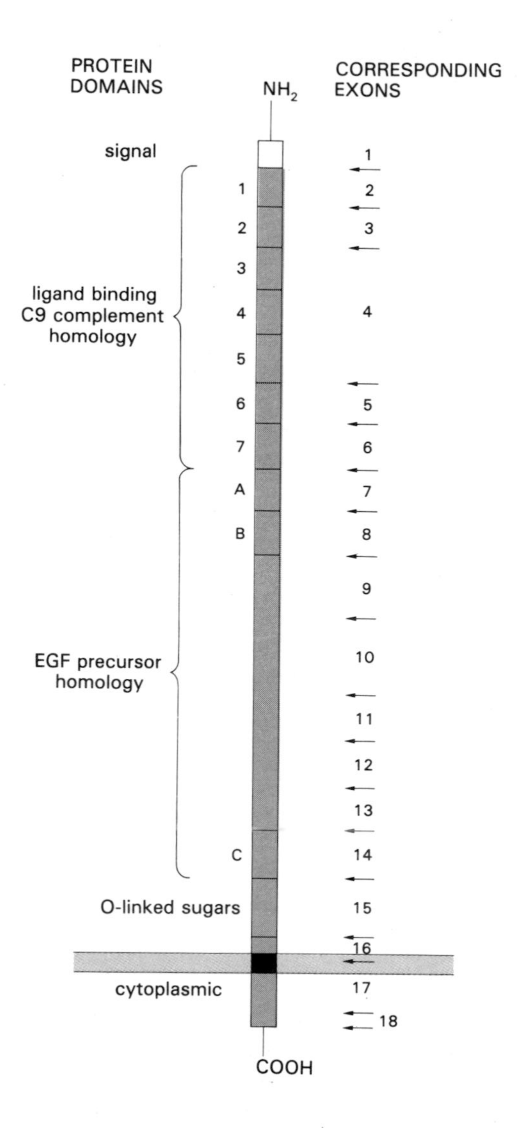

Figure 2 *Gene organization and protein domains of the human LDL receptor. Recent studies have determined the exon organization of the gene for the LDL receptor. As much as 45 kb in length, it contains 18 exons, most of which correlate with a functional domain. Thirteen of the 18 exons encode protein sequences that are homologous to sequences in other proteins: five similar to one in the C_9 component of complement; three similar to a repeat sequence in the precursor of EGF and in three proteins of the blood clotting system (factor IX, factor X, and C protein). Five other exons encode non-repeated sequences present only in the EGF precursor (Fig. I.78). Therefore, the LDL receptor appears to be a mosaic protein encoded by exons shared with different proteins and belonging to several "supergene families". The correlation between exon organization and the six protein domains of the human LDL receptor is indicated. Small arrows show where introns have been removed. A, B and C boxes correspond to cysteine-rich segments. There are also three copies of the Alu family of middle repetitive DNA sequences, which may be a site of recombination leading to pathological deletion.*

The human LDL receptor is a *glycoprotein* of 839 aa that *spans* the *plasma membrane* one time, with its *N-terminus* facing the *extracellular* environment and its *C-terminus* facing the *cytoplasm* (Fig. 2). The external domain contains a *cysteine-rich region* that bears the *binding site* for LDL. The high affinity of LDL receptor binding is Ca^{+}-dependent, and includes ionic interactions (negative charges are provided by the receptor). The external domain also contains amino acid sequences that serve as the sites of attachment for asparagine-linked and serine-threonine-linked *carbohydrate chains*. The cytoplasmic domain contains 50 aa residues that direct the receptor to its proper sites of function in the cell.

In order to carry its bound LDL into cells, the LDL receptor must first localize in *coated pits* on the cell surface. As described on p. 224, the pits *invaginate*, within minutes, to form coated endocytic vesicles which fuse to become endosomes. Within the endosome, *LDL dissociates* from its receptor, an event which allows the receptor to return to the surface, where it binds and internalizes another LDL particle, in a process known as *receptor recycling*. After it dissociates from the receptor in endosomes, LDL is carried to *lysosomes*, where its cholesterol ester component is hydrolyzed, and the *free cholesterol is liberated* for *metabolic and regulatory* purposes (Fig. 3).

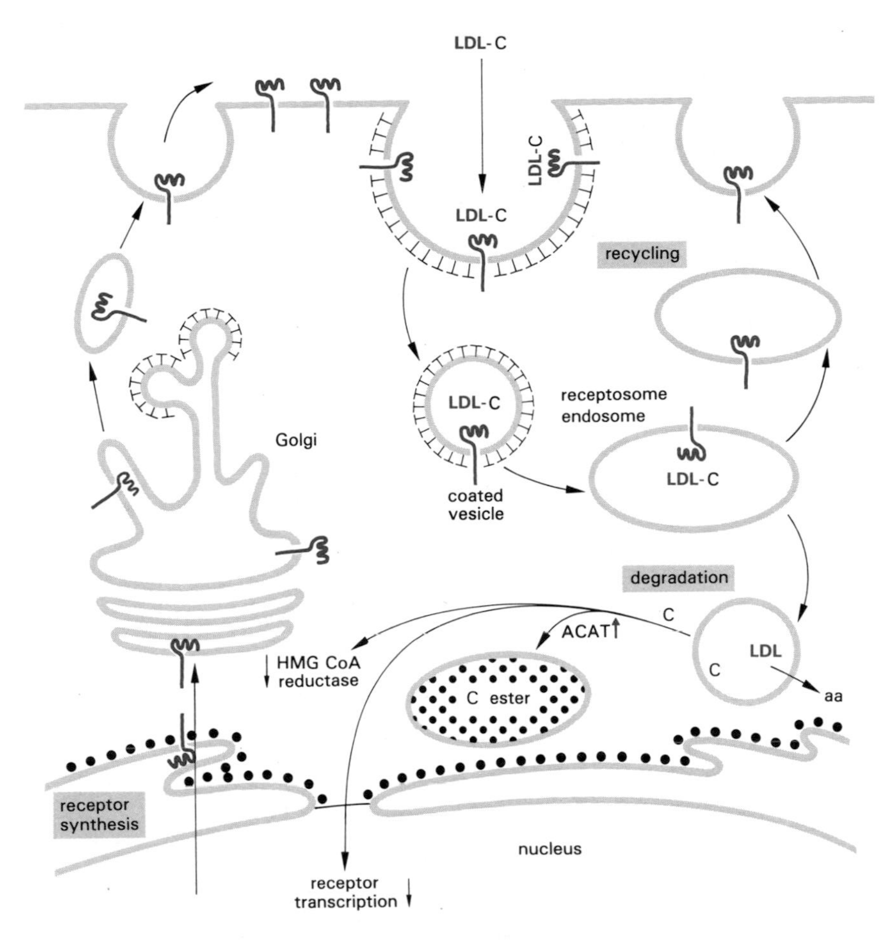

Figure 3 *Uptake of LDL by receptor-mediated endocytosis removes LDL from plasma and at the same time supplies cholesterol to cells for metabolic purposes. LDL receptor protein is synthesized on ribosomes attached to the rough endoplasmic reticulum. Next, it is carried in membrane vesicles to the Golgi apparatus, where the sugar chains are arranged to form mature receptors. The mature receptors are again incorporated into vesicles that carry them to the cell surface, where they are inserted into the plasma membrane, at random sites. On the surface, the receptors cluster in pits that are coated on their cytoplasmic surfaces with the protein clathrin. Circulating LDL binds to an LDL receptor in a coated pit. It is taken into the cell, together with the receptor, when the coated pit invaginates and pinches off to form a coated vesicle. The vesicle rapidly sheds its clathrin coat and fuses with other vesicles, to form an endosome. Proton pumps in the membrane of the endosome cause the internal fluid to become more acidic, and this causes LDL to dissociate from the receptor. The unoccupied LDL receptor then cycles back to the plasma membrane, where it again migrates to a coated pit, binds LDL, and delivers the LDL to endosomes. This receptor recycling is repeated every 12 min. After its deposition in endosomes, the LDL is delivered to lysosomes. In the lysosome, enzymes break down the apoprotein B-100 of LDL to amino acids and cleave the ester bond of the cholesterol esters, yielding long-chain fatty acids and free cholesterol. The free cholesterol is transported out of the lysosome and is used by the cell for the synthesis of membranes and for other purposes. The LDL-derived cholesterol also regulates three metabolic events, each of which helps stabilize the cellular level of cholesterol: 1. it suppresses the activity of 3-hydroxy 3-methylglutaryl coenzyme A reductase (HMG CoA reductase), which controls the rate of synthesis of cholesterol by cells; 2. it suppresses the production of new LDL receptors, a regulatory action that limits the total amount of LDL that enters the cell; and 3. it activates acyl-coenzyme A cholesterol acyltransferase (ACAT), which attaches a long-chain fatty acid to the incoming LDL-derived cholesterol, so that excess cholesterol can be stored in cholesterol ester droplets.*

Genetic Defects in the LDL Receptor

Genetic defects in the LDL receptor produce *familial hypercholesterolemia* (FH), a common cause of premature heart attacks in humans. Approximately one in every 500 persons has a single copy of a mutant LDL receptor gene, and thus suffers from *heterozygous FH*. The cells of these individuals produce about half the normal number of LDL receptors. As a result, LDL is removed from the circulation at half the normal rate: lipoprotein accumulates in blood to levels two times above normal (Fig. 4), and heart attacks occur typically in the fourth and fifth decades. Heterozygous FH causes about 5% of all heart attacks in people under age 60.

Rarely, two FH heterozygotes marry and produce a child with two mutant genes at the LDL receptor locus. These children are referred to as FH *homozygotes*, although they are often not true homozygotes in that they inherit different mutant LDL receptor genes from each of their parents. FH homozygotes have a much more severe clinical syndrome than do heterozygotes. Their cholesterol levels are six to ten times above normal, and they usually suffer heart attacks in early childhood.

Four classes of mutations in the LDL-receptor gene have been identified in patients with FH. One class of mutant genes produces no detectable receptors (so-called "null" alleles). The second class produces receptors, synthesized in the rough endoplasmic reticulum, which cannot be transported to the cell surface, and therefore cannot perform their normal function. The third class of mutations produces receptors that move to the cell surface normally but are unable to bind LDL owing to an abnormality in the binding site. The fourth class of mutations (Fig. I.155) produces receptors that are transported to the surface and bind LDL, but are unable to enter coated pits and therefore cannot carry LDL into cells (so-called "internalization defective" alleles). Studies of the molecular basis of mutations in the LDL receptor are beginning to reveal those parts of the protein that signal transport to the plasma membrane and those which signal incorporation into coated pits. Thus, these mutations are providing new insights in fundamental cell biology.

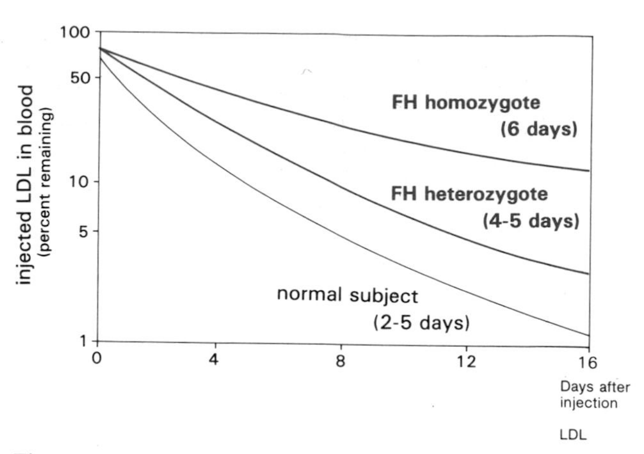

Figure 4 *Number of LDL receptors in the body. Labelled LDL is injected, and the amount of radioactivity in the blood is measured for several weeks: the decrease is reflective of the cellular uptake of LDL, and hence is an index of the number of LDL receptors.*

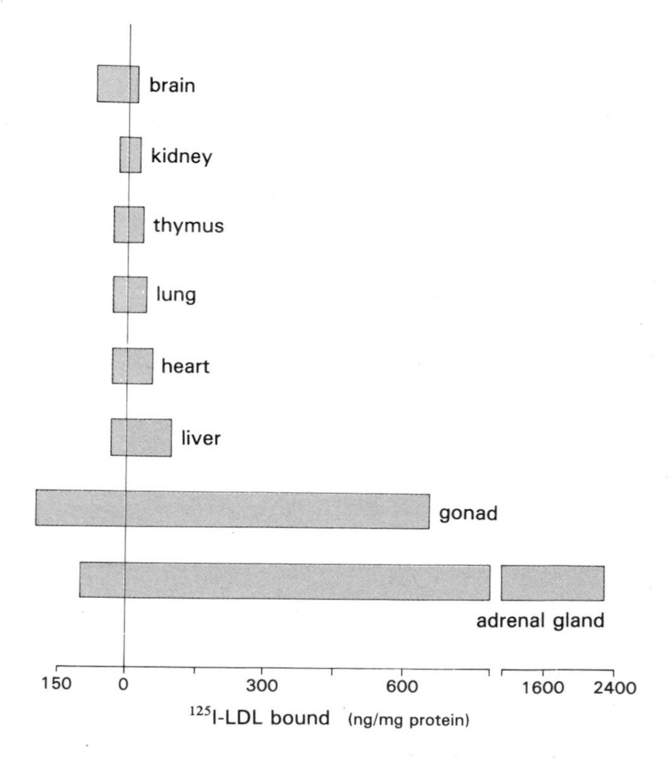

Figure 5 *Comparison of [^{125}I]LDL binding in human fetal membranes of various organs (non-specific binding is represented in gray, and specific binding in red).*

Understanding of the LDL receptor has also provided a rational *basis for treatment* of FH. Inasmuch as FH heterozygotes have a single copy of the normal receptor gene, it is possible to stimulate the normal gene to produce an increased number of receptor molecules, thereby overcoming the genetic deficiency. The possibility of such stimulation was raised when it was realized that production of *LDL receptors* is *under feedback regulation.* When cells accumulate excess cholesterol, they reduce their production of receptors. Conversely, when cells are deprived of cholesterol, they transcribe LDL receptor genes at a high rate, and produce increased amounts of the receptor. FH heterozygotes respond to agents that lower the content of cholesterol in liver cells, thereby stimulating the production of LDL receptors in the liver. Most commonly, this stimulation is achieved through the oral administration of resins that bind bile acids in the intestine, preventing their normal reabsorption. The liver responds by converting more cholesterol into bile acids, which depletes the liver of cholesterol, and causes the liver to produce more receptors. The effectiveness of bile acid resins can be enhanced by the simultaneous administration of drugs, such as compactin and mevinolin, that inhibit an enzyme in the cholesterol biosynthetic pathway. When hepatic cholesterol synthesis is inhibited, the liver develops an even larger increase in LDL receptors, in order to supply needed cholesterol. With a combination of bile acid binding resins and cholesterol synthesis inhibitors, it is possible to stimulate the normal LDL receptor gene sufficiently that LDL cholesterol levels are lowered to the normal range in FH heterozygotes.

Role of LDL Receptors in Steroid-Secreting Cells

LDL receptors are a *major* source of cholesterol in steroid-secreting cells of the adrenal and ovary. This role of the LDL receptor was originally established through study of cultured adrenocortical cells obtained from a steroid-secreting mouse adrenal tumor. In the adrenal, the LDL receptor is *regulated* so as to supply increased cholesterol at times when steroid secretion is accelerated. Thus, when adrenal steroid secretion is stimulated by *ACTH*, the *number of LDL receptors increases*. Conversely, when adrenal steroid secretion is suppressed, the number of LDL receptors in the adrenal cortex decreases.

Because of their large demand for cholesterol, the adrenal and ovary express the *highest* number of receptors per gram of tissue in comparison with all other organs. Figure 5 shows a survey of [^{125}I]LDL binding activity in membranes from human fetal tissues.

Because of the relative enrichment of LDL receptors in the adrenal cortex, bovine adrenal cortical tissue was used as a source for the initial isolation of the LDL receptor protein. Similarly, the human fetal adrenal served as the source of mRNA for the cloning of LDL receptor cDNA.

In a mutant strain of rabbits deficient in LDL receptors (Watanabe heritable, hyperlipidemic rabbits), the uptake of LDL in the adrenal gland is reduced, and the animals compensate by developing a high rate of de novo cholesterol synthesis within the adrenal cortex. This tissue is able to obtain cholesterol from at least two other sources. First, the adrenal is able to take up cholesterol from high density lipoproteins (HDL) by a mechanism that appears to be independent of the LDL receptor. The nature of the HDL uptake mechanism has not been defined. Second, the adrenal gland can synthesize its own cholesterol, and there is interplay between exogenous and endogenous sources of cholesterol in adrenal physiology.

References

Brown, M. S., Anderson, R. G. W. and Goldstein, J. L. (1983) Recycling receptors: the round-trip itinerary of migrant membrane proteins. *Cell* 32: 663–667.

Brown, M. S. and Goldstein, J. L. (1983) Lipoprotein receptors in the liver: control signals for plasma cholesterol traffic. *J. Clin. Invest.* 72: 743–747.

Brown, M. S. and Goldstein, J. L. (1984) How LDL receptors influence cholesterol and atherosclerosis. *Sci. Am.* 251: 58–66, 1984.

Brown, M. S., Kovanen, P. T. and Goldstein, J. L. (1979) Receptor-mediated uptake of lipoprotein-cholesterol and its utilization for steroid synthesis in the adrenal cortex *Rec. Prog. Horm. Res.* 35: 215–257.

Goldstein, J. L., Anderson, R. G. W., and Brown, M. S. (1979) Coated pits, coated vesicles, and receptor-mediated endocytosis. *Nature* 279: 679–685.

Goldstein, J. L., Kita, T. and Brown, M. S. (1983) Defective lipoprotein receptors and atherosclerosis: lessons from an animal counterpart of familial hypercholesterolemia. *N. Engl. J. Med.* 309: 288–295.

Mahley, R. W. and Innerarity, T. L. (1983) Lipoprotein receptors and cholesterol homeostasis. *Biochim. Biophys. Acta* 737: 197–222.

Tolleshaug, H., Hobgood, K. K., Brown, M. S. and Goldstein, J. L. (1983) The LDL receptor locus in familial hypercholesterolemia: multiple mutations disrupting the transport and processing of a membrane receptor. *Cell* 32: 941–951.

Yamamoto, T., Davis, C. G., Brown, M. S., Schneider, W. J., Casey, M. L., Goldstein, J. L. and Russell, D. W. (1984) The human LDL receptor: a cysteine-rich protein with multiple Alu sequences in its mRNA. *Cell* 39: 27–38.

Sudhof, T. C., Goldstein, J. L., Brown, M. S. and Russell, D. W. (1985) The LDL receptor gene: a mosaic of exons shared with different proteins. *Science* 228: 815–822.

CHAPTER XII

Gastrointestinal Hormones

Bernard Desbuquois

Contents, Chapter XII

Gastrointestinal Hormones

1. Historical Background

Traditionally, gastrointestinal (GI) hormones have been defined as peptides produced by *endocrine cells* located in the *GI mucosa*, released into the circulation under the influence of alimentary stimuli, which are involved in the regulation of secretion, motility and growth in the digestive system. It is now recognized that a number of GI peptides are *also* produced by *neurons* in the central and peripheral nervous system, particularly enteric neurons. They are delivered locally to their target cells, without entering the blood circulation, and/or exert biological effects outside the digestive system (Table XII.1).

Owing to the wide distribution of GI hormones throughout the tissues of the gut, and to the presence, in GI extracts, of non-hormonal substances acting as secretagogues, the development of gut endocrinology has been slow until recently. *Secretin* was described in 1902 by Bayliss and Starling[1] as an active product, present in an extract of intestinal mucosa, which, upon intravenous injection, stimulated the secretion of pancreatic juice. They postulated its release into the blood upon acidification of the upper intestine, a known stimulus of pancreatic secretion. The discovery of *secretin,* which *led to* the *concept of hormones*, was followed by that of *gastrin*[2], *cholecystokinin* (CCK)[3] and *pancreozymin*[4], described as substances which stimulated gastric acid secretion, gall bladder contraction, and pancreatic enzyme secretion. For years, these hormones were known only through their actions, and were used, by physiologists and pharmacologists, in the form of crude extracts.

In the 1960s, secretin[5], gastrin[6] and cholecystokinin-pancreozymin[7] were purified, and their amino acid sequences determined. *Vasoactive intestinal peptide* (VIP), *gastric inhibitor peptide* (GIP) and *motilin*, first identified through their actions, and *pancreatic polypeptide*, discovered fortuitously during the purification of insulin, were in turn isolated and characterized.

In the 1970s, immunochemical and immunocytochemical techniques made it possible to identify the cells that produce gastrin[8] and other GI hormones, and to quantitate and determine the physical characteristics of these hormones in blood and in tissue extracts. One major observation was the identification of *VIP and cholecystokinin in the brain*; *conversely*, *somatostatin*, *neurotensin*, *substance P*, *enkephalins* and *TRH*, which were first isolated from the brain, were shown to be present *in the GI tract as well.* Active research in cellular and molecular biology[9] has greatly extended our knowledge of the receptors, the mechanism of action, and the biosynthesis of GI hormones.

1. Bayliss, W. M. and Starling, E. H. (1902) The mechanism of pancreatic secretion. *J. Physiol. (London)* 28: 325–353.
2. Edkins, J.S. (1905) The chemical mechanism of gastric secretion. Proc. R. Soc. Lond. 69: 352–356.
3. Ivy, A. C. and Oldberg, E. (1928) A hormonal mechanism of gallbladder contraction and evacuation. *Amer. J. Physiol.* 86: 599–613.
4. Harper, A. A. and Raper, H. S. (1943) Pancreozymin, a stimulant of the secretion of pancreatic enzymes in the small intestine. *J. Physiol. (London)* 102: 115–125.
5. Jorpes, J. E., Mutt, V., Magnusson, S. and Steele, B. B. (1962) Amino acid composition and N-terminal sequence of porcine secretin. *Biochem. Biophys. Res. Commun.* 9: 275–279.
6. Gregory, R. A. and Tracy, H. J. (1964) The constitution and properties of two gastrins extracted from hog antral mucosa. *Gut* 5: 103–114.
7. Jorpes, E., Mutt, V. and Toczko, K. (1964) Further purification of cholecystokinin and pancreozymin. *Acta Chem. Scand.* 18: 2408–2410.
8. McGuigan, J. E. (1968) Immunochemical studies with synthetic human gastrin. *Gastroenterology* 54: 1005–1011.
9. Noyes, B. E., Mevarech, M., Stein, R. and Agarwal, K. L. (1979) Detection and partial sequence analysis of gastrin mRNA by using an oligodeoxynucleotide probe. *Proc. Natl. Acad. Sci., USA* 76: 1770–1774.

Table XII.1 Major peptides identified in mammalian gastrointestinal tract and pancreas, and their major cellular production sites.

Peptides	Endocrine cells	Neurons
ACTH	+	+
Calcitonin gene related peptide (CGRP)	−	+
Cholecystokinin (CCK)	+	+
Corticotropin releasing factor (CRF)[a]	+	−
Dynorphin	−	+
β-endorphin	+	+
Met- and Leu-enkephalins	+	+
Epidermal growth factor (EGF)	?	?
Galanin	−	+
Gastric inhibitory peptide (GIP)	+	−
Gastrin	+	−
Gastrin releasing peptide (GRP)[b]	−	+
Glucagon and enteroglucagon	+	−
Growth hormone releasing factor (GRF)	?	?
Insulin	+	−
Motilin	+	−
Neuropeptide Y (NPY)	−	+
Neurotensin	+	+
Pancreatic polypeptide (PP)	+	−
Peptide with N-terminal histidine and C-terminal isoleucine amide (PHI)[c]	−	+
Peptide YY (PYY)	+	−
Somatostatin	+	+
Secretin	+	−
Substance P[d]	+	+
Thyrotropin releasing hormone (TRH)	?	+
Vasoactive intestinal peptide (VIP)	−	+

[a]Structurally related to amphibian cerulein and phyllocerulein.
[b]Structurally related to amphibian bombesin.
[c]The human counterpart of porcine PHI is PHM (a peptide with N-terminal histidine and C-terminal methionine amide).
[d]Member of the tachykinin family of peptides, which includes additional mammalian, amphibian, and molluscan peptides (Fig. 4, p. 184).

In this chapter, the properties of GI peptides are described, as well as other hormonal activities related to GI functions.

2. General Aspects

2.1 Isolation

GI peptides were first isolated from the *GI tract of animals*, and some were later isolated from *human endocrine tumors* as well as from animal and human *brain*. Extraction methods were aimed at preventing proteolytic degradation; and isolation, first achieved by conventional methods, has been greatly facilitated by newer techniques such as HPLC.

2.2 Structure

With the exception of somatostatin, EGF and chymodenin, which have one or more disulfide bridge, all GI *peptides* are *straight-chained. Most consist of* <40 *aa* residues. Gastrins and CCK octapeptide, CCK-8, are acidic peptides, glicentin is neutral, and all other peptides are from weakly (motilin) to strongly (VIP) basic. Common (but non-exclusive) features include an N-terminal pyroglutamyl structure, as in gastrin and neurotensin, and a C-terminal amide structure, as in gastrin, CCK, secretin, VIP, and substance P. The only non-peptide constituent identified so far is sulfuric acid in CCK and one form of gastrin.

Many GI polypeptides are related to each other by structural similarities which enable the recognition of *several* polypeptide *families*. Thus, *gastrin and CCK* have the same C-terminal pentapeptide amide sequence; *glucagon, secretin, VIP, GIP,* and *PHI* (a peptide with N-terminal histidine and C-terminal isoleucine amide) share multiple amino acid identities when aligned from the N-terminal end; *pancreatic polypeptide* is related to *peptide YY* (a peptide with N-terminal tyrosine and C-terminal tyrosine amide) and *neuropeptide Y*; and *substance P* shows homology with *substance K* and *neuromedin K*. A certain degree of homology is also found between the C-terminal sequences of *CCK, VIP*, and *glucagon*; the N-terminal sequences of *CCK* and *GRP* (gastrin releasing peptide); between *peptide YY* and *neurotensin*; between the midsequence of *gastrin* and sequence 29–35 of GH and *somatomammotropin*; and between amino acids 16–23 of *CCK* and *calcitonin*. Structural homologies have been observed between *gastrin and* the *transforming protein of polyoma virus* (p. 122), and peptides of the *glucagon–secretin family* and *prealbumin*.

Another important feature of GI and pancreatic peptides is their *size heterogeneity*, the larger forms being extensions at either or both ends of shorter forms. Thus, two or more peptides which differ in chain length have been demonstrated chemically for gastrin, CCK, glucagon/enteroglucagon, somatostatin, GRP, substance P and neurotensin, and there is immunochemical evidence of heterogeneity for VIP,

GIP, motilin, pancreatic polypeptide, and enkephalins. Two lines of evidence support the contention that, at least with gastrin, CCK, somatostatin and glucagon/enteroglucagon, size heterogeneity reflects different degrees of post-translational processing of precursor peptides. First, pairs of basic amino acids generally precede, in the amino acid sequence of the large forms (as in gastrin or somatostatin), the sequence of the shorter forms. Secondly, short forms can be generated from larger peptides by proteolytic processing enzymes present in tissues.

Although it is now recognized that all forms of gastroenteropancreatic peptides are ultimately derived from larger precursors, few of these precursors have been isolated (with the exception of proinsulin), and none of them have been fully characterized using conventional peptide chemistry. However, the amino acid sequences of the precursors of gastrin, CCK, glucagon/enteroglucagon, VIP/PHM, pancreatic polypeptide, somatostatin, substance P/substance K and opioid peptides have been deduced recently from the nucleotide sequence of their corresponding cloned cDNAs. These studies have used, as a source of specific mRNAs, normal GI, pancreatic or neural tissues and peptide-secreting tumors, and, as hybridization probes to detect mRNAs, synthetic oligonucleotides corresponding to specific sequences of the mature peptides. Apart from those of opioid peptides, all precursors characterized so far consist of 95 to 180 aa, possess a signal sequence of about 20 aa, and include a single copy of the mature peptide, often located C-terminally. Several basic amino acids, often arranged in pairs, occur along their sequences; they represent potential cleavage sites for trypsin-like processing enzymes. One particular feature of the precursors for peptides possessing a C-terminal amide group, such as gastrin and CCK, is the presence of a Gly-Arg-Arg sequence immediately following the sequence of the mature peptide involved in the enzymatic amidation of the C-terminal amino acid. In general, a full concordance is observed between the actual sequences of the peptides isolated from tissue extracts and those of the peptides expected to result from processing of the precursor. In addition, molecular cloning techniques have allowed the identification and characterization of previously unknown sequences, such as that of PHM (the human counterpart of porcine PHI), located N-terminally to the sequence of VIP in the VIP precursor (p. 570), and those of two new glucagon-like polypeptides, located C-terminally to the sequence of glucagon in the glucagon precursor (p. 573).

2.3 Distribution

Concentration and Molecular Forms of GI Peptides in Tissues

The *regional distribution* of GI peptides in the human GI tract *depends on* the particular *peptide* (Table XII.2). The regions of *highest* concentration of *GI peptides* encountered are: *gastrin* in the *gastric antrum* and *proximal duodenum*; *somatostatin* in the *stomach*, *duodenum*, and *pancreas*; *CCK*, *secretin*, *GIP* and *motilin* in the *duodenum* and *jejunum*; *enteroglucagon* and *neurotensin* in the *ileum*; *VIP* in the *ileum* and *colon*; and *GRP* in the *gastric fundus* and *antrum*. For most peptides, the maximal concentrations found are in the range of 50 to 300 pmol/g tissue (wet weight), the highest being that of gastrin, but overall recoveries are probably highest for VIP and somatostatin, which have a more widespread distribution.

The *concentrations* of GI peptides in *individual layers* of the GI tract also *vary* as a function of the particular peptide (Fig. XII.1). The first group of peptides, which includes gastrin, secretin, GIP, enteroglucagon and neurotensin, is almost exclusively associated with the *epithelium*. The second group of peptides, which includes VIP, substance P, and bombesin/GRP, is

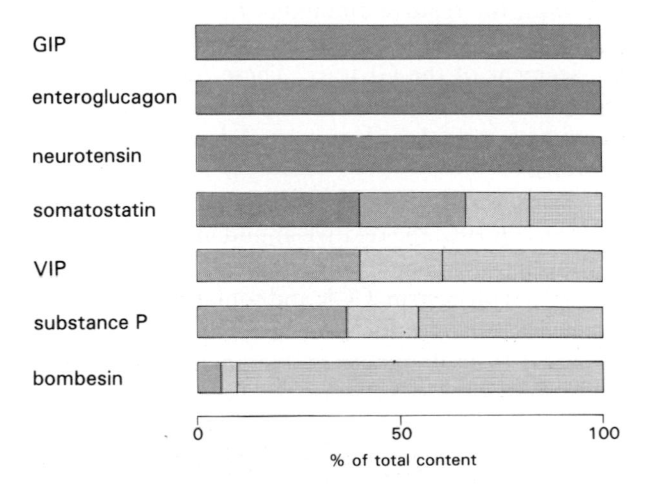

Figure XII.1 Tissue localization and relative distribution of regulatory peptides in separated layers from the human bowel. Red, epithelium; pink, lamina propia; gray, submucosa; white, muscle. (Ferri, G. L., Adrian, T. E., Ghatel, M. A., O'Shaughnessy, D. J., Probert, L., Lee, Y. C., Buchan, A. M. J., Polak, M. and Bloom, S. R. (1983) Tissue localization and relative distribution of regulatory peptides in separated layers from the human bowel. *Gastroenterology* 84: 777–786.)

Table XII.2 Concentrations of GI peptides in human gastrointestinal tract, brain, and peripheral blood plasma. Concentrations, measured by specific radioimmunoassays, are expressed in pmol/g wet weight for gastrointestinal tissues (whole thickness) and brain regions, and in pmol/l for blood plasma.

Peptide	Stomach		Duodenum				Colon			Brain					
	Fundus	Antrum	Proximal	Distal	Jejunum	Ileum	Ascending	Sigmoid	Rectum	Hypothalamus	Frontal cortex	Amygdala	Substantia nigra	Hippocampus	Blood plasma
Gastrin	5	820	420	150	3	1									5–25
CCK-8			214[a]			1									<0,2
CCK-33/39			203[b]		218[b]					49	52	64		50	
Secretin		5[c]	70[c]	130[c]	40[c]	5									1–2
VIP	290	125	580		190	215	428	290		23	17	21	2	11	1–7
GIP		5[c]	50[c]	70[c]	40[c]	7	1	2.5							30–60
Enteroglucagon	<1	<1	15		58	275	46	71	96						10–40
Somatostatin	390	363[c]	270[c]		188[c]	42	10	15		278	52	339	24	82	10–30
Pancreatic polypeptide															10–100
PYY	<1	<1	6		5	84	82	196	480						8–25
Motilin		3[c]	55[c]	75[c]	27[c]										30–60
Neurotensin					2	87	0.4	0.4	33	0.8	5	23	4		10–20
Bombesin/GRP	8	4	2		2	2	0.9	0.9							
Substance P			20		17	21	13	14		122	8.5	26	922	34	
Met-enkephalin		110	90		120	37				141	42	26	661	56	

[a]Concentration in neutral extracts. [b]Concentration in acid extracts. [c]Concentration in mucosa only.

associated with the *non-epithelial layers*. Variable amounts of these peptides are present in lamina propria, submucosa, and muscle. The third group, which includes somatostatin and CCK, shows a *dual distribution*, the relative amounts of peptide present in the epithelial and nonepithelial layers depending on the segment of the GI tract. These observations are reflective of differences in the distribution of individual peptides, between endocrine cells and neurons.

With those peptides exhibiting *two or more molecular forms*, the relative abundance of individual forms often varies along the length of the GI tract. For example, with gastrin, CCK and somatostatin, the low molecular size forms (G-17, CCK-8, and somatostatin-14) predominate at proximal sites, whereas the high molecular size forms (G-34, CCK-33/39, and somatostatin-28) are the most abundant distally. These findings suggest regional differences in the post-translational processing of the precursors.

GI polypeptides are associated with specific types of granules that can be identified by subcellular fractionation of tissue extracts. With secretin, VIP, GIP, enteroglucagon and somatostatin, a single population of granules is observed, but with gastrin, two distinct populations are found, differing both in their density and in the molecular form of gastrin they contain.

Although virtually all GI polypeptides have been immunochemically identified in brain, their *relative distributions* between the *GI tract* and *brain vary* widely among individual peptides. Thus, gastrin and possibly also GIP are found almost exclusively in the GI tract; VIP is more abundant in the GI tract and at other peripheral sites than it is in the brain; CCK, somatostatin and neurotensin are equally distributed between the GI tract and brain, although concentrations vary widely among individual regions; substance P, enkephalins and TRH are probably more abundant in the brain. Peptides identified immunochemically in the brain are associated with *synaptosomal fractions* and show unique regional distributions. Thus, VIP and CCK are concentrated particularly in the cerebral cortex and other telencephalic structures (amygdala and hippocampus), whereas other peptides are associated predominantly with the basal ganglia, pituitary stalk, median eminence, and hypothalamus.

Cells Producing GI Peptides

Two types of cells synthesize and secrete peptides in the GI tract: *endocrine* cells, found in GI mucosa and pancreatic islets, and *neurons*, located in all layers of the GI wall. *Outside* the digestive tract, *only neurons* produce GI peptides.

Gastrointestinal Endocrine Cells

GI endocrine cells are scattered in the epithelial lining of the gastric glands, intestinal crypts, and villi. When examined using light microscopy after conventional staining procedures, such as silver impregnation, masked metachromasia and lead hematoxylin, GI endocrine cells show certain characteristics that *distinguish* them *from* adjacent *epithelial* cells[10]. They are oval, pyramidal or columnar in shape, their cytoplasm is pale, and their nucleus is usually round and vesicular. *Secretory granules* are found in the basal region of the cell, while in the apical region endowed with a well developed brush border, cells usually reach the glandular lumen ("*open*" type). Some cells, however, located deep in the mucosa, lack luminal contacts ("*closed*" type). Examined using electron microscopy (Fig. XII.2), these features are more conspicuous, and the secretory granules vary in size (150–450 nm), shape, and electron density, depending on the cell type. A specific secretory product can generally be identified in the granules by immunohistochemistry. Additional ultrastructural features include the presence of a Golgi apparatus in a supra- or paranuclear position, smooth micropinocytic vesicles, microtubules, and microfilaments that occur as tonofilaments.

Certain GI endocrine cells, such as those which produce substance P and which contain 5-hydroxytryptamine, are identifiable by their ability to reduce silver salts (argentaffinity), or by the use of fluorescence histochemistry. Most GI endocrine cells, however, do not contain endogenous amines and do not show argentaffinity. Yet they can take up and decarboxylate exogenously applied amine precursors, a property which has led to the development of the *APUD cell* concept (p. 549). Another cytochemical property is the presence of *neuron-specific enolase*, an isoenzyme of the glycolytic enzyme enolase, proposed as a molecular marker for peripheral and central neuroendocrine cells.

To date, as many as *18 types of* endocrine *cells* have been identified in the human GI tract and pancreas on the basis of ultrastructural and immunohistochemical criteria. In general, a given cell type corresponds to one single secretory product, although occasionally the coexistence of two or more peptides in the same cell may occur. Each type of cell shows a specific regional distribution along the GI tract (Table

10. Grube, D. and Forssmann, W. G. (1979) Morphology and function of the entero-endocrine cells. *Hormone Metab. Res.* 11: 589–606.

A

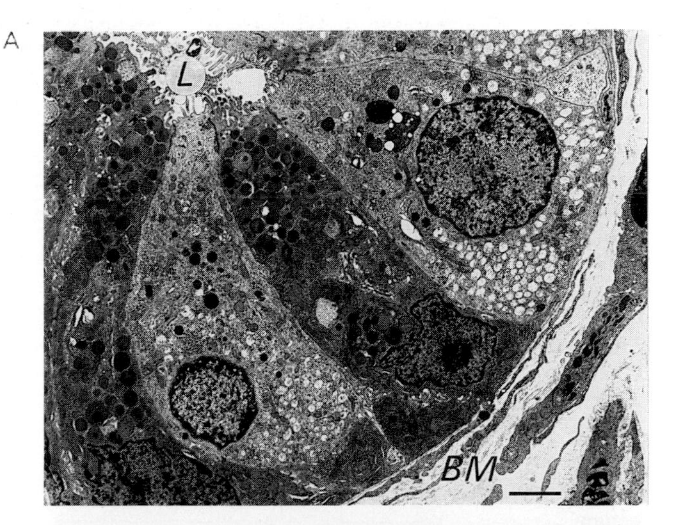

B

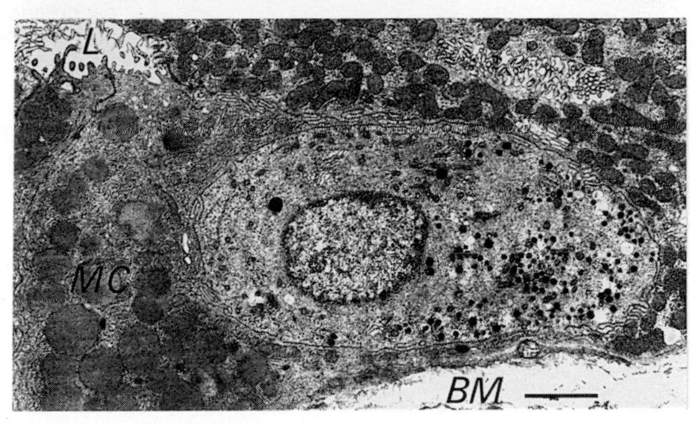

Figure XII.2 Ultrastructure of two typical endocrine cells of the human gastrointestinal mucosa. **A.** G cell ("open" type) of the antral mucosa. L, glandular lumen; BM, basal membrane. Note numerous lysosomes. This section was obtained from the stomach of a patient with an ulcer. The granules are partially depleted of gastrin, which has been excreted. **B.** Endocrine cell ("closed" type) of the fundic mucosa. L, glandular lumen; BM, basal membrane; PC, parietal cell; MC, mucosal cell. Bars = 0.2 μm. (Courtesy of Dr. T. Lehy, Hôpital Bichat, Paris).

XII.3 and Fig. XII.3). Enterochromaffin cells (EC) containing 5-hydroxytryptamine, which constitute the largest single endocrine cell population, are present in all regions of the GI tract; a few of these cells, which contain substance P, are found in ileum. Somatostatin cells (D) are also widely distributed throughout the GI tract, with highest densities in the stomach, proximal intestine, and pancreas. The majority of gastrin cells (G) are confined mostly to the gastric antrum; other gastrin cells, which differ from the G cells ultrastructurally (IG) or immunochemically (TG), are found in the small intestine. Cells containing CCK (I), secretin (S), GIP (K) and motilin (Mo) are associated predominantly with the proximal and

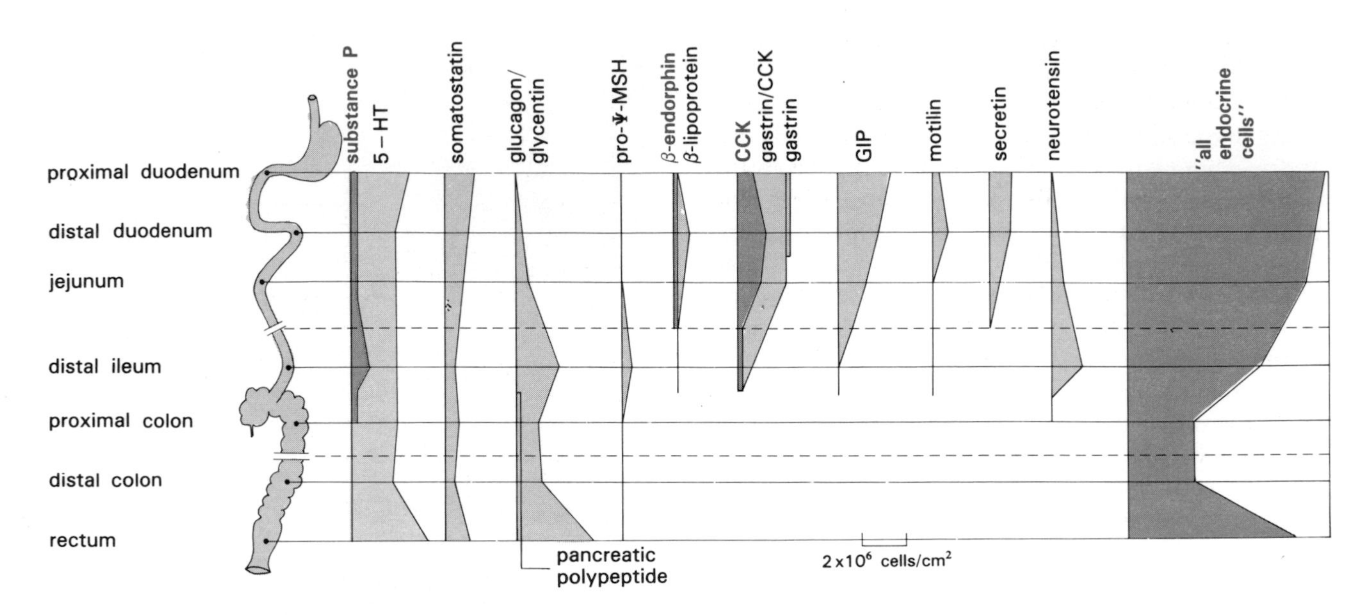

Figure XII.3 Regional distribution and frequency of individual endocrine cells in the human intestine. The cell number is expressed per cm^2 of mucosal surface. The bar represents 2×10^6 cells/cm^2 mucosal surface. Colored areas represent the number of specific subpopulations of cells (peptide underlined). (Sjölund, K., Sandén, G., Hakanson, R. and Sundler, F. (1983) Endocrine cells in human intestine: an immunocytochemical study. *Gastroenterology* 85: 1120–1130).

Table XII.3 Classification of the human gastroenteropancreatic endocrine/paracrine cells.

Cell type	Main product	Size of secretory granules (nm)	Pancreas	Stomach: Fundus	Stomach: Antrum	Small intestine: Upper	Small intestine: Lower	Large intestine
A	Glucagon	250	+	+[a,b]	–	–	–	–
B	Insulin	300	+	–	–	–	–	–
D	Somatostatin	350	+	+	+	+	+[c]	+[d]
D_1	Unknown	160	+[c]	+	+	+	+	+
EC	Serotonin, and uncharacterized peptides	200	+[d]	+	+	+	+	+
ECL	Unknown (histamine?)	450	–	+	–	–	–	–
G	Gastrin	300	+[a,b]	–	+	+	–	–
I	CCK	250	–	–	–	+	+[c]	–
IG	Gastrin	220	–	–	–	+	+	–
K	GIP	350	–	–	–	+	+[c]	–
L	Enteroglucagon	400	–	–	–	+[c]	+	+
Mo	Motilin	160	–	–	–	+	+[c]	–
N	Neurotensin	300	–	–	–	+[c]	+	+[d]
P	Unknown	120	+[b]	+	+	+	–	–
PP	Pancreatic polypeptide	180	+	–	+[a]	–	–	–
S	Secretin	200	–	–	–	+	+[c]	–
TG	C-terminal gastrin immunoreactivity	275	–	–	+[a]	+	+[a]	–
X	Unknown	300	–	+	+[a]	–	–	–

[a]Only exceptionally in humans. [b]In fetus or newborns; rare in adults. [c]Few. [d]Rare.
(From Solcia, E., Capella, C., Buffa, R., Usellini, L., Fiocca, R. and Sessa, F. (1981) *Physiology of the Gastrointestinal Tract* (Johnson, L. R., ed), Raven Press, New York, pp. 39–58; Grube, D. and Forssman, W. G. (1979) *Horm. Metab. Res.* 11: 589–606).

middle small intestine. Glucagon/glicentin cells (L) are found throughout the intestine, with highest densities in the distal intestine. Neurotensin cells (N) are most numerous in the ileum. Insulin (B), glucagon (A) and pancreatic polypeptide (PP) cells are confined to the pancreas, although A cells are also found in the gastric fundus of the dog.

The *contribution* of individual cell populations to the total population of endocrine cells in the *human intestine*[11] has been estimated as the following: serotonin cells, 38%; glucagon/glicentin cells, 18%; gastrin/CCK cells, 12%; GIP cells, 7%; somatostatin cells, 7%; neurotensin cells, 6%; specific CCK cells, 4%; secretin cells, 2%; and all remaining cell types, 6%.

Gastrointestinal Neurons

Neurons that contain classical transmitters, like most peptide-containing neurons in the digestive tract, belong to the *enteric nervous system*, a subdivision of the autonomic nervous system[12]. The latter is known to consist of five intramural plexuses; their interconnections and nerve fibers innervate the muscle, blood vessels and mucosa of the GI tract. The *mesenteric plexus* lies between the circular and the longitudinal layer of the muscularis externa, and the *submucosal plexus* lies within the connective tissue of the mucosa (Fig. XII.4).

Some peptide-containing neurons, whose cell bodies are located outside the GI tract, are *part of* the *extrinsic autonomic nervous system*. These include efferent sympathetic postganglionic fibres, which arise from prevertebral ganglia, vagal efferent preganglionic fibres, which originate in the dorsal motor nucleus of the vagus, and afferent neurons, whose cell bodies are located in the dorsal root ganglia.

The distribution and projections of peptide-containing neurons in the human and animal GI tract have been studied using immunohistochemical techniques applied to tissue sections or to whole mount preparations (Fig. XII.5); only in the latter can the full extent of the nerve processes within the different layers of the intestine be visualized (Fig. XII.6). The abundance of peptide-immunoreactive structures in the GI tract varies among individual peptides: VIP- and enkephalin-containing neurons are generally the most numerous,

11. Sjölund, K., Sanden, C., Hakanson, R. and Sundler, F. (1983) Endocrine cells in human intestine: an immunocytochemical study. *Gastroenterology* 85: 1120–1130.
12. Furness, J. B. and Costa, M. (1980) Types of nerves in the enteric nervous system. *Neuroscience* 5: 1–20.

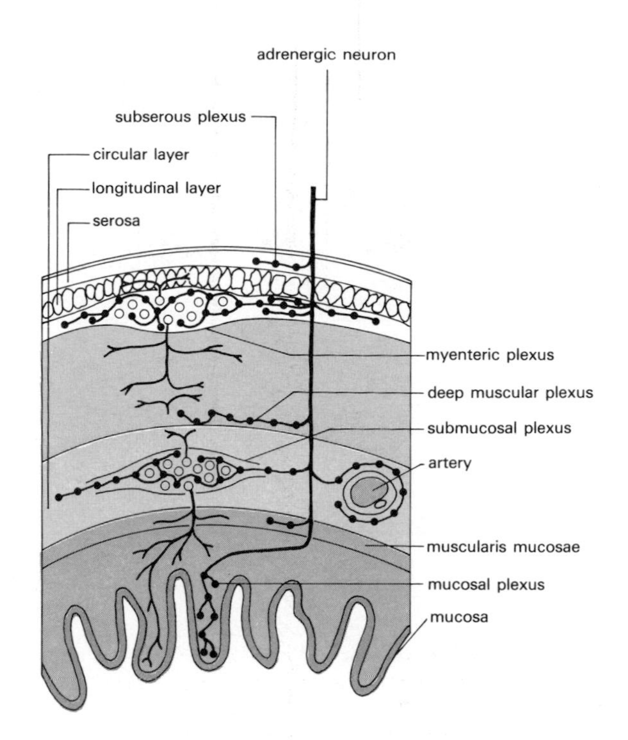

Figure XII.4 Innervation of the gut. (Goyal, R. K. (1983) Neurology of the gut. In: *Gastrointestinal disease*. (Sleisenger, M. H. and Fordtran, J. S., eds), W. B. Saunders Co., Philadelphia, pp. 97–115).

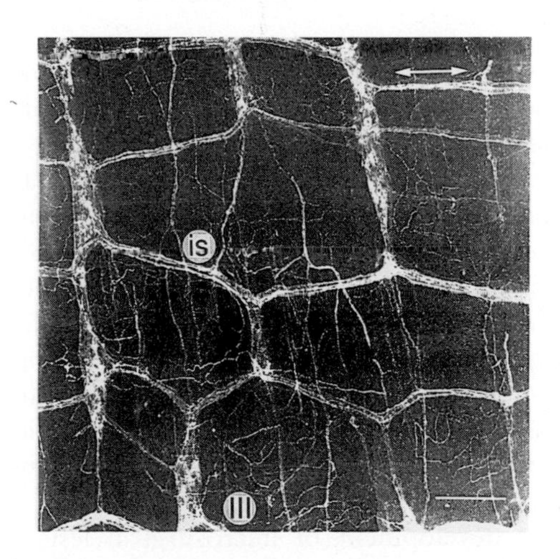

Figure XII.5 Enkephalin-containing nerves in a whole mount preparation of guinea-pig myenteric plexus. Enkephalin was localized using an indirect immunofluorescence technique. In this low power view, myenteric ganglia (g), internodal strands (is) and the tertiary plexus (III) are easily recognizable. The double-ended arrow is parallel to the longitudinal muscle. Bar=200 µm. (Costa, M. and Furness, J. B. (1982) Neuronal peptides in the intestine. *Brit. Med. Bull.* 38: 247–252.)

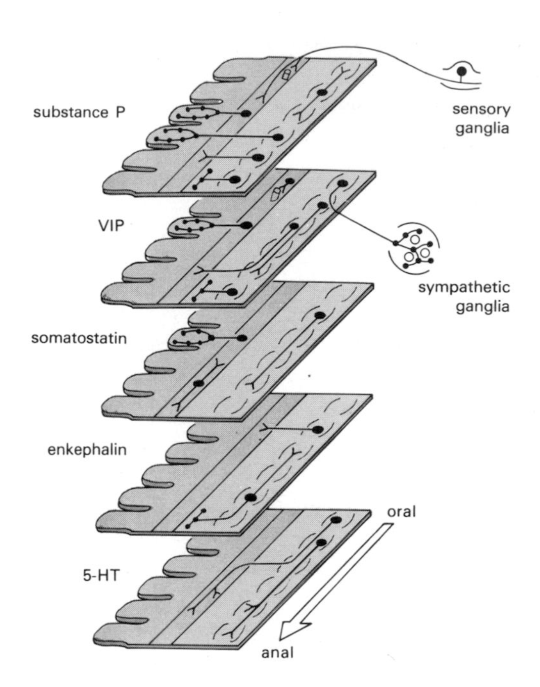

Figure XII.6 Diagrams of pathways of specific neuronal types in the small intestine, determined in guinea-pigs by examining the effects of microsurgical lesions. Note that each nerve type has specific patterns of projection. (Costa, M. and Furness, J.B. (1982) Neuronal peptides in the intestine. *Brit. Med. Bull.* 38: 247–252.)

and CCK-containing neurons the least abundant. *Variations* in the *regional distribution* of different peptide-containing neurons also occur; thus, enkephalin and VIP-containing neurons show a more widespread distribution than do somatostatin- and CCK-containing neurons, which are found mainly in the intestine. Finally, the relative proportions of nerve cell bodies in the two major ganglionated plexuses also vary among individual peptides. For example, in guinea-pig small intestine, VIP and somatostatin neurons are associated predominantly with the submucosal plexus, whereas enkephalin neurons are found exclusively in the myenteric plexus.

The development of microsurgical lesions has allowed the *identification* of *specific neural pathways* in guinea-pig intestine, each class of neurons having a well defined polarity, with a highly ordered set of projections[2] (Fig. XII.6). Thus, within the group of neurons with cell bodies in the myenteric ganglia, some neurons project to circular muscle (substance P, VIP, enkephalin) or to the mucosa (substance P); other neurons, acting as interneurons, project to adjacent myenteric (somatostatin, 5-hydroxytryptamine) or submucosal ganglia (VIP); still other neurons project to prevertebral sympathetic ganglia (VIP). Neurons with cell bodies in the ganglia of the submucosal plexus (substance P, VIP, somatostatin) project primarily to the mucosa. In most cases, although the projections of the neurons are well established, their nature, whether axonal or dendritic, is unknown.

Although some neuronal populations contain a single peptide, the *coexistence of several peptides*, or several classical transmitters, or both peptide(s) and transmitter(s), in the same population of neurons is quite common. Thus, in guinea-pig intestine, co-

A

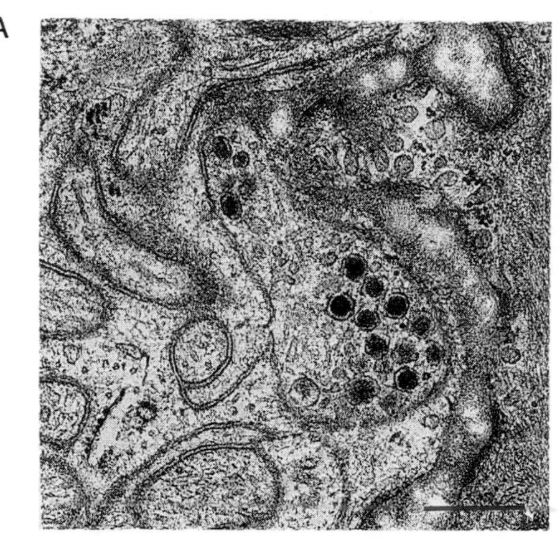

B

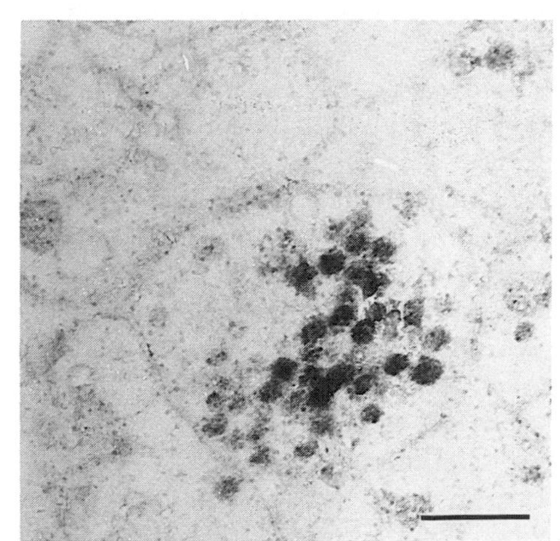

Figure XII.7 Ultrastructural properties of peptidergic nerves in the gut wall. **A.** Cat esophagus: a routinely stained section with a single, P-type nerve profile. **B.** Cat pylorus: a section stained for VIP, using the immunoperoxidase technique. Note the heavy deposit of reaction products over the neurovesicles only. Bar=500 nm. (Sundler, F., Häkanson, R. and Leander, S. (1980) Peptidergic nervous systems in the gut. *Clin. Gastroenterol.* 9: 517–543). (Courtesy of Dr. F. Sundler, University of Lund, Sweden).

localizations have been observed for VIP and dynorphin; for substance P and choline acetyltransferase; and for CCK, somatostatin, GRP, NPY, and choline acetyltransferase. Similarly, somatostatin and NPY have been identified in two separate subpopulations of extrinsic noradrenergic neurons that supply, respectively, the submucosal ganglia and the mucosa (somatostatin), and intestinal blood vessels (NPY). Some peptide-containing neurons (substance P, VIP, enkephalins) are characterized ultrastructurally by the presence of small round vesicles and large round vesicles with a core of varying electron density (Fig. XII.7). With regard to the neural control of intestinal motility, motor and sensory neurons containing substance P are excitatory, whereas motor neurons containing VIP, and sensory neurons containing enkephalin and somatostatin, are inhibitory.

2.4 Embryologic, Ontogenic, and Evolutionary Aspects

Embryology

Controversial

Several observations led to the initial proposal that GI endocrine cells were of *neuroectodermal origin*[13]: 1. the existence of cytochemical and ultrastructural similarities between certain endocrine and para-endocrine cells derived from the neural crest (adrenomedullary cells, calcitonin-producing cells, type I cells of the carotid body); one such similarity is their ability to synthesize bioactive amines from precursors, hence the acronym *APUD*[14] (amine precursor uptake and decarboxylation); 2. the existence of multiple and multihormonal endocrine tumors; 3. the identification of similar peptides in GI endocrine cells and neurons; and 4. the presence of tyrosine hydroxylase and neuron-specific enolase in GI endocrine cells. Initially thought to be from the neural crest, the postulated common embryonic origin of the APUD cells was later broadened to the neuroectoderm and to the "*neuroendocrine-programmed epiblast*".

Other observations, however, have recently supported an *endodermal* rather than ectodermal origin for GI endocrine cells[13]: 1. the lack of contribution of the neuroectoderm to the development of pancreatic and GI endocrine cells, as shown by interspecific graft experiments in embryos[15]; 2. the coexistence of endocrine and exocrine cells within certain GI tumors; and 3. the integrity of GI endocrine cells in Hirschsprung's disease (congenital megacolon), characterized by the absence of myenteric neurons in the colon, and caused by a defect in embryogenesis. In fact, the *functional similarities* between *GI endocrine cells and neurons no longer support the notion* that these cells are of *common embryological* origin.

Ontogeny

In the *human fetus*[16], most GI peptides are immunochemically detectable at *eight weeks*; concentrations and cell densities increase steadily, to attain adult patterns by 20 to 24 weeks, these changes being more rapid in the proximal than in the distal intestine (Fig. XII.8). With certain peptides (motilin, GIP), forms of high molecular weight predominate in early fetal life. In the pig[17], which is similar to the human in its fetal development and which gestates for 21 weeks, gastrin, somatostatin and pancreatic polypeptide cells appear earlier (4 to 6 weeks) than do CCK, secretin, motilin, GIP and neurotensin cells (6 to 8 weeks). These observations suggest that individual *hormones* play *specific roles* during the *development* of the fetal GI tract.

Evolutionary and Comparative Aspects

Peptides related to mammalian GI hormones chemically, immunochemically, and/or biologically have been identified in species representative of the main groups of vertebrates, and in some invertebrates, including simple multicellular organisms such as coelenterates[18]. These observations indicate that GI peptides were established *early in evolution*, presumably before the separation of deuterostomians from protostomians. Structural data show that the entire

13. Peranzi, G., Lehy, T. and Bonfils, S. (1984) L'origine embryologique des cellules endocrines du système digestif: une controverse toujours en cours. *Gastroent. Clin. Biol.* 8: 560–568.
14. Pearse, A. G. E. (1969) The cytochemistry and ultrastructure of polypeptide hormone-producing cells of the Apud series and the embryologic, physiologic and pathologic implications of the concept. *J. Histochem. Cytochem.* 17: 303–313.
15. Fontaine, J. and LeDouarin, N. (1977) Analysis of endoderm formation in the avian blastoderm by the use of quail-chick chimaeras. The problem of the neuroectodermal origin of the cells of the APUD series. *J. Embryol. Exp. Morphol.* 41: 209–222.
16. Bryant, M. G., Buchan, A. M. J., Gregor, M., Ghatei, M. A., Polak, J. M. and Bloom, S. R. (1982) Development of intestinal regulatory peptides in the human fetus. *Gastroenterology* 83: 47–54.
17. Alumets, J., Håkanson, R. and Sundler, F. (1983) Ontogeny of endocrine cells in porcine gut and pancreas. *Gastroenterology* 85: 1359–1372.
18. Dockray, G. J. (1979) Comparative biochemistry and physiology of gut hormones. *Ann. Rev. Physiol.* 41: 83–95.

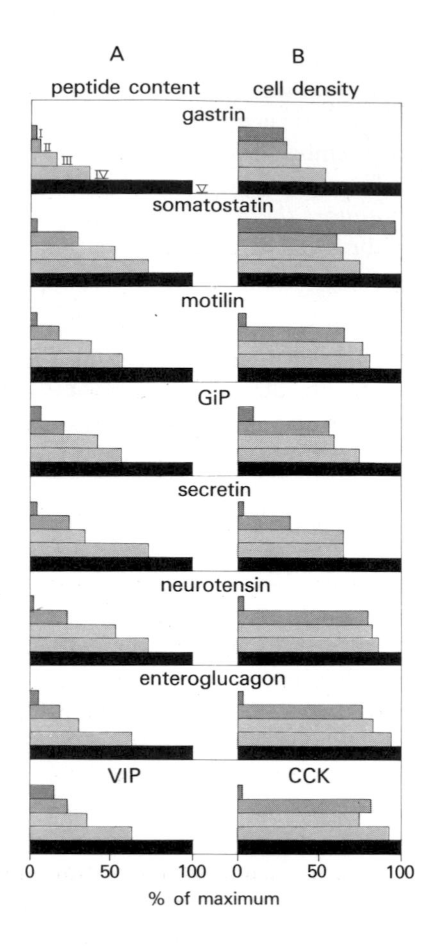

Figure XII.8 Developmental pattern of endocrine cells (**A**) and of the extractable peptide content (**B**) of human fetal GI tract. Intestines were removed from human fetuses aged 8 to 11 weeks (group I), 12 to 15 weeks (group II), 16 to 24 weeks (group III), 25 to 30 weeks (group IV), and 31 weeks to term (group V). Results of measurements of cell densities and peptide content are expressed as the percentage of values found in group V[16].

sequence of glucagon, VIP and somatostatin, as well as the C-terminal pentapeptide amide sequence of gastrin and CCK (also found in cerulein), has been strongly *conserved* throughout vertebrate history. Such data also suggest that structurally related polypeptides, such as secretin and *VIP*, or *gastrin* and *CCK*, are derived from *common ancestors*, presumably via the mechanism of *gene duplication*.

As in mammals, peptide-like activities identified in lower vertebrates and invertebrates are observed in both the alimentary tract and CNS; an *additional* site of *peptide-like activity* is found in *amphibian skin*, which contains peptides of the bombesin, tachykinin and endorphin/enkephalin families. In invertebrates, the CNS is the major site of localization. In primitive vertebrates, almost all endocrine cells, including insulin and glucagon-producing cells, are dispersed throughout the gut, a distribution well suited to responding directly to luminal stimuli. At later stages, insulin and glucagon cells migrate into pancreatic islets, suggesting that their secretion becomes preferentially linked to blood-born factors. Presumably, the increased regional specialization of the alimentary tract has favored the diversification of GI peptides and their sites of production during evolution.

2.5 Biological Actions

Multiple Effects

Together with the central and peripheral nervous systems, GI peptides *influence*, in a stimulatory or inhibitory manner, a *number of GI functions,* including *exocrine and endocrine secretions*, *motility*, *growth*, and *blood flow*. Some peptides also affect secretion and/or motility at extragastrointestinal sites, particularly in the circulatory, respiratory and urogenital systems. They control liver metabolism and exert various effects in the CNS, one of which is the *regulation* of *food intake*. Many of these effects result from a direct interaction of the peptides with epithelial, smooth muscle, or endocrine target cells, but some effects are hormonally and/or neurally mediated.

Gastrointestinal Exocrine Secretions

GI hormones *regulate* the secretion and/or absorption of *water*, *electrolytes*, *enzymes* and *mucus* by epithelial cells of the GI, pancreatic and biliary tracts. Thus, in glands of the *gastric fundus*, which consist of three morphologically and functionally distinct cell types (Fig. XII.9), GI hormones affect the secretion of H^+ ions by parietal (or oxyntic) cells[19], that of pepsinogen by chief cells, and the secretion of mucus by mucosal cells. Similarly, in the *exocrine pancreas*, which also shows cellular heterogeneity (Fig. XII.10), GI hormones affect the secretion of water and CO_3H^- ions by duct cells, and that of enzymes by acinar cells. In *intestinal epithelium*, they regulate secretion in the crypts and absorption in the villi.

The secretory responses of the GI tract to

19. Soll, A. H. and Walsh, J. H. (1979) Regulation of gastric acid secretion. *Ann. Rev. Physiol.* 41: 35–53.

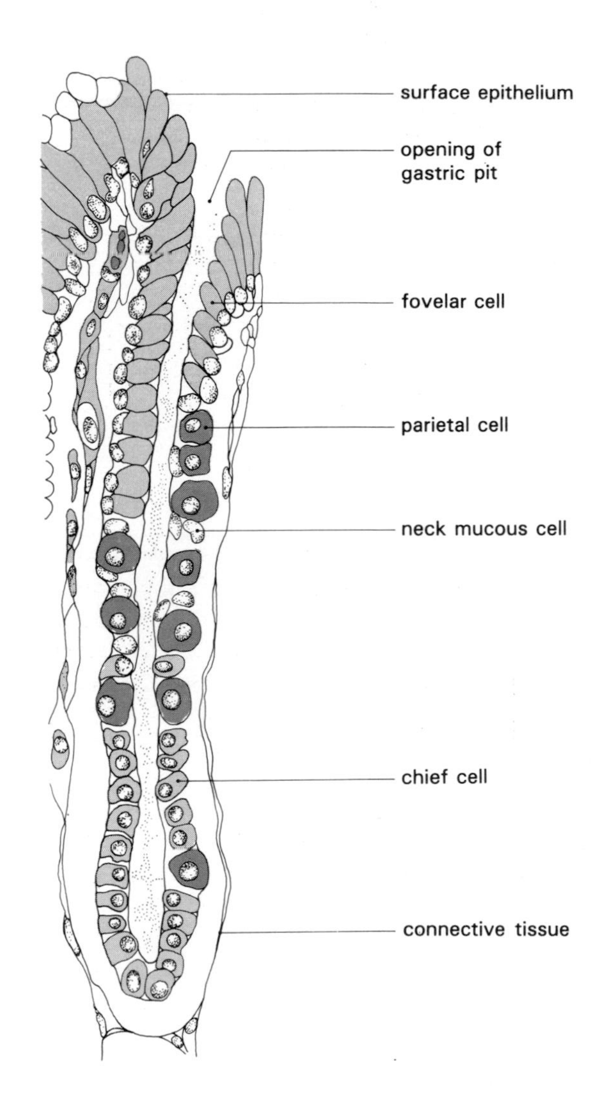

Figure XII.9 Fundic glands of human stomach. (Bloom, W. and Fawcett, D.W., eds (1968) *A Textbook of Histology*. W. B. Saunders Co., Philadelphia).

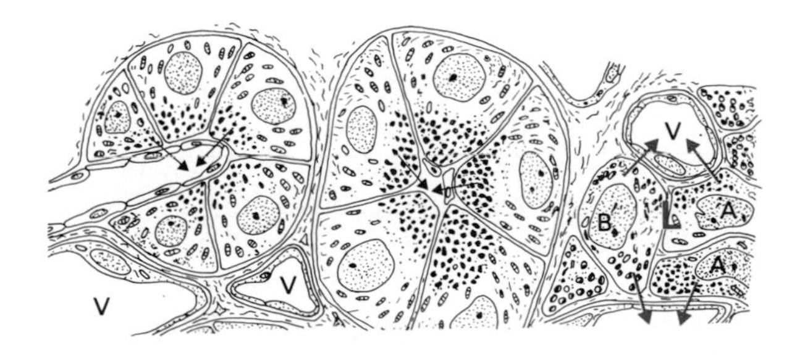

Figure XII.10 Schematic drawing of the human pancreas, showing two exocrine glands and part of an islet of Langerhans (L). Note that the exocrine cells accumulate secretory granules in the apex, ready to be excreted into the lumen. In the islet, the endocrine cells (e.g., B and A) have granules which release their content into blood vessels (V).

individual peptides are shown in Table XII.4. Gastric acid secretion is stimulated primarily by gastrin and GRP/bombesin; gastric pepsin secretion is stimulated by gastrin, CCK, secretin, and VIP; intestinal secretion of water and electrolytes by VIP; pancreatic secretion of water and CO_3H^- by secretin; pancreatic secretion of enzymes primarily by CCK but also by secretin, VIP, substance P and bombesin; and bile secretion by most peptides of the secretin/glucagon and gastrin/CCK families. Virtually *all* GI secretions in the basal and hormone-stimulated state are *inhibited by somatostatin*.

Gastrointestinal Motility

The effects of GI peptides on motor function of the GI tract[20], based on mechanical and/or myoelectrical activity in vivo, are summarized in Table XII.5. In general, gastrin, CCK, motilin, bombesin and neurotensin are primarily stimulatory, whereas glucagon, VIP, GIP, secretin and somatostatin are mostly inhibitory. Some peptides regulate the *migrating motor complex*, or *myoelectric complex*, a cyclical pattern of interdigestive motor activity that propagates from the stomach to the distal intestine[21]. Motilin initiates a fasting pattern, in contrast to gastrin, CCK, bombesin and neurotensin, which induce a fed pattern. The rate of gastric emptying and intestinal transit time are additional parameters affected by GI hormones in vivo, which, in part, reflect changes in motor activity. The effects of GI peptides on GI motility observed in vitro are generally consistent with those observed in vivo, although the type of smooth muscle affected and the mechanism of the response (direct and/or neurally mediated) varies depending on the particular peptide and GI region.

Gastrointestinal Growth

The effects of GI peptides on the growth[22] of epithelial cells in the GI tract and pancreas, based on cell number, DNA synthesis and DNA content, are shown in Table XII.6. Some peptides, such as gastrin, CCK and bombesin, are *stimulatory*, whereas other

20. Wienbeck, M. and Erckenbrecht, J. (1982) The control of gastrointestinal motility by GI hormones. *Clin. Gastroenterol.* 11: 523–543.
21. Itoh, Z., Aizawa, I. and Sekiguchi, T. (1982) The interdigestive migrating complex and its significance in man. *Clin. Gastroenterol.* 11: 497–521.
22. Johnson, L. R. (1981) Regulation of gastrointestinal growth. In: *Physiology of the Gastrointestinal Tract* (Johnson, L. R., ed), Raven Press, New York, pp. 169–196.

Table XII.4 Effects of GI peptides on gastrointestinal and biliary secretion in vivo.

Region and cell type	Secretion	Gastrin	CCK	Secretin	VIP	GIP	Glucagon	PP	Somatostatin	Neurotensin	GRP-Bombesin	Substance P	Enkephalins	TRH	EGF
Stomach:															
Parietal cells	Water, H^+	↑[a]	↑[b,c]	↓	↓	↓	↓	↑[b,c]	↓	↓	↑	↑[b,c]	↑	↓	↓
Chief cells	Pepsinogen	↑	↑	↑	↑	↓			↓	↓			↑	↓	
Duodenum	Water, electrolytes	↑	↑	↑			↑							↓	
Small intestine	Water, electrolytes	↑		↑	↑	↑	↑		↓			↑	↓		
Large intestine	Water, electrolytes				↑									↓	
Exocrine pancreas:															
Duct cells	Water, CO_3H^-	↑	↑[b]	↑	↑		↓	↓	↓			↑[b,d]	↓	↓	
Acinar cells	Enzymes	↑[b]	↑[a]	↑[a]	↑[a]		↓	↓	↓	↓	↑[a]	↑[a,b,d]	↓	↓	
Liver and biliary system	Water, electrolytes	↑		↑	↑		↑		↓			↓		↓	
Salivary glands	Water, enzymes											↑		↓	

[a]Effects occur in vitro also. [b]Weak stimulant. [c]Inhibits gastrin-stimulated secretion. [d]Inhibits secretin and CCK-stimulated secretion.

Table XII.5 Effects of GI peptides on the motor function of the gastrointestinal tract

Region	Gastrin	CCK	Secretin	VIP	GIP	Glucagon	Motilin[c]	Pancreatic polypeptide	Somatostatin[d]	Neurotensin[d]	GRP/Bombesin[d]	Substance P	Enkephalins	TRH
Lower esophageal sphincter	↑	↓	↓	↓	↓	↓	↑	↑		↓	↑	↑[e]		
Proximal stomach	↓[b]	↓[b]	↓	↓	↓	↓	↑	↑	↓	↓[b]	↓[b]	↑	↑	
Distal stomach	↑	↑	↓	↓	↓	↓	↑	↑	↓	↓	↑	↑	↑	↑
Pylorus	↓	↑	↑			↑	↑						↑[f]	
Duodenum			↓			↓	↑		↓	↓	↓	↑	↑	
Jejunum and ileum	↑	↑	↓	↓	↓	↓	↑	↑	↓	↓	↓	↑	↑	
Ileo-cecal sphincter	↓													
Colon	↑	↑								↑		↑	↑	
Gallbladder and bile ducts	↑	↑		↓		↓		↓[a]			↑		↑	
Sphincter of Oddi	↓	↓				↓		↑[a]						

[a]Weak effect.
[b]In vitro response is contraction.
[c]Initiates the migratory motor and myoelectric complexes.
[d]Disrupts the migratory motor and myoelectric complexes.
[e]Substance P mediates the esophageal response to acidification.
[f]Enkephalin mediates the pyloric response to duodenal acidification.

peptides, such as VIP and somatostatin, are *inhibitory*. Effects of GI hormones on the growth of GI epithelial cells can be functionally dissociated from the effects on secretion. GI hormones, especially gastrin, play an important role in the structural and functional changes of the GI mucosa which occur in response to feeding, in neonatal development, and post-operatively, all of which reduce the amount or alter the position of absorbing mucosa.

Gastrointestinal Endocrine Secretions

The effects of GI peptides on the secretion of other peptides in the GI tract and pancreas are summarized in Table XII.7. Most peptides of the gastrin/CCK and glucagon/secretin families stimulate the release of pancreatic hormones, especially insulin and somatostatin. Bombesin and somatostatin affect the secretion of virtually all peptides; the former is stimulatory, and the latter inhibitory. Many of these effects occur via *direct* interaction of the peptides *with endocrine* effector cells, although the inhibitory effects of glucagon, VIP, and secretin on gastrin release are mediated by an increase in the release of somatostatin.

Gastrointestinal Blood Flow

GI peptides regulate GI blood flow, in association with the autonomic nervous system, biogenic amines

Table XII.6 Effects of GI peptides on the growth of gastrointestinal mucosa and pancreas. Except when specified otherwise, arrows indicate changes in cell number or in DNA synthesis and tissue content of DNA. Such changes are generally accompanied by parallel changes in RNA and protein synthesis (or tissue content), although the latter may occur in the absence of changes in cell number.

Region	Gastrin	CCK	Secretin	VIP	Entero-glucagon	Insulin	Somato-statin	Bombesin	EGF
Stomach									
Fundus	↑		↓ [a]	↓ [a]		↑	↓	↑	
Antrum	0					↑	↓		
Small intestine	↑		↓ [a]		↑	↑	↓		
Large intestine	↑		↓ [a]		↑	↑ [e]			↑
Gallbladder	0	↑							
Exocrine pancreas	↑	↑	↑ [b]			↑ [d]	↓ [f]	↑	
Endocrine pancreas		↑ [c]			↑				

[a]Inhibition of the tropic effect of gastrin.
[b]Effect limited to an increase in weight and RNA content, especially marked in the presence of cerulein.
[d]Stimulation of pancreatic enzyme and protein synthesis.
[e]In cultured cells.
[f]Inhibition of the effect of cerulein.

Table XII.7 Effects of GI peptides (listed horizontally) on the secretion of other GI peptides in vivo (listed vertically).

Effector peptide	Stimulatory peptide: Gastrin	CCK	Secretin	VIP	GIP	Glucagon	Entero-glucagon	Somato statin	Pancreatic polypeptide	Neuro-tensin	Bombesin	Substance P	Enke-phalins
Gastrin			↓ [a]	↓ [a]	↓ [a]	↓ [a]		↓			↑		
CCK								↓			↑		
Secretin								↓			↑		
GIP								↓			↑		
Glucagon	↑	↑			↑			↓			↑	↑ ↓ [f]	
Insulin	↑	↑	↑	↑	↑ [b]	↑	↑	↓			↑	↑ ↓ [f]	
Pancreatic polypeptide	↑ [c]	↑ [c]	↑ [c]	↑ [c]	↑ [c]			↓		↑ [c]	↑ [d]		↑ ↓
Somatostatin	↑	↑	↑	↑	↑			↓			↑	↓ [e]	↓
Motilin		↑	↓					↓	↓		↑	↓	
Enteroglucagon								↓			↑		
Neurotensin											↑		

[a]Inhibition of secretion mediated by somatostatin.
[b]Stimulation of secretion occurs only under hyperglycemic conditions.
[c]Stimulation of secretion is mediated by cholinergic mechanisms.
[d]Inhibition of secretion occurs only in the human; in the dog, bombesin is stimulatory.
[e]Inhibition of secretion occurs only in the stomach; in the pancreas, substance P is stimulatory.
[f]In vitro response is inhibition of secretion.

(catecholamines, histamine), non-GI peptides (bradykinin, vasopressin, angiotensin), and intrinsic mechanisms (metabolic and myogenic)[23]. Thus, gastrin, CCK, secretin, glucagon and neurotensin increase gastric, intestinal and/or pancreatic blood flow, as do VIP and substance P, two vasodilator peptides which affect other vascular beds in the body. It has been suggested that endogenous CCK, secretin and GIP play a part in the phenomenon of *postprandial intestinal hyperemia*, and neurotensin has been implicated as a mediator of the increased *intestinal capillary permeability* seen with the absorption of fat.

Non-gastrointestinal Effects

A number of GI peptides, especially those produced by neurons, exert effects outside the GI tract. These include: contraction or relaxation of various

23. Granger, D. N., Richardson, P. D. I., Kvietys, P. R. and Mortillaro, N. A. (1980) Intestinal blood flow. *Gastroenterology* 78: 837–863.

Table XII.8 Main functions affected by GI peptides administered to the central nervous system.

	Gastrin	CCK	VIP	Somato-statin	GRP/ Bombesin	Neuro-tensin	Substance P	endorphins enkephalins	TRH	Calcitonin CGRP	CRF
Food intake		↓[a,b]	↓	↓	↓[a]	↓		↑	↓	↓	
Gastrointestinal functions:											
Gastric acid secretion	↑[a,b]	0	0	↑	↓	↓		↓	↑[b]	↓[a,b]	↓[a,b,c]
Gastric mucus secretion					↑[c]						
Prevention of stress-induced gastric ulcers					↑	↑					
Pancreatic secretion					↓			↓	↑[b]		
Gastrointestinal motor activity	↓	↓[d]		↑[d]	↓[b]	↓[b,e]			↑[b]	[e]	
Blood glucose concentration		↑[c]		↓	↑[c]			↑[c]	↑		
Adrenal epinephrine release					↑				↑	↑	
Body temperature		↓		↓	↓	↓[f]		↓	↑		
Blood pressure				↓	↑[c]					↑	
Heart rate				↓	↓			↑	↑	↑	
Locomotor activity		↓[b]		↑	↑		↑		↑		
Nociception		↓				↓[f]	↑	↓		↓[a,b]	
Pituitary hormone release:											
Growth hormone	↑	↑	↑	↓		↓	↑	↑			
Prolactin		↑	↑		↑	↓	↑	↑			
LH		↓	↑			↓	↑	↓		↓	
TSH	↓	↓						↓		↓	
ACTH		↑									

[a]Effect also occurring upon peripheral administration.
[b]Effect ablished by vagotomy.
[c]Effect ablished by adrenalectomy.
[d]Decreases the frequency of the migratory myoelectric complexes.
[e]Restores the fasting pattern of the migratory myoelectric activity.
[f]Effect antagonized by TRH.

types of *smooth muscle*, especially vascular smooth muscle (VIP, substance P, neurotensin), inotropic and chronotropic effects on heart (VIP), stimulation of *water and electrolyte secretion* in various exocrine glands (VIP), *metabolic* effects in liver and adipose tissue (VIP), and stimulation or inhibition of *pituitary hormone* release (CCK, VIP, somatostatin, substance P).

GI peptides produced by neurons also *elicit* a variety of *effects via the CNS* (Table XII.8). Most effects require central administration of the peptides, but some also occur upon peripheral administration; quite frequently, effects observed via these two routes of administration are opposite. Major functions, including appetite[24], GI secretion and motility, glucoregulation, thermoregulation, cardiovascular function, locomotor activity, nociception, and neuroendocrine secretion are affected. Some effects require the integrity of the adrenals and the vagus nerve. However, the specific brain regions and the mechanisms involved in most effects have only been partly identified.

24. Morley, J. E. and Levine, A. S. (1983) The central control of appetite. *Lancet* 1: 398–401.

2.6 Receptors and Mechanisms of Action

The localization of GI peptide receptors in the GI tract is consistent with their sites of action. For example, gastric parietal cells and fundic membranes specifically bind gastrin; intestinal epithelial cells bind VIP and insulin; pancreatic acinar cells, VIP, secretin, CCK, bombesin and substance P; gastric and intestinal smooth muscle, CCK, neurotensin and substance P; and gallbladder muscle cells bind CCK. In addition, synaptosomal fractions specifically bind peptides present in the CNS, and the regional distribution of such binding sites correlates well with that of the endogenous ligands. The apparent K_D of the ligands for receptors in the GI tract is ~0.1 to 1.0 nM, and the number of sites ~5000 to 100,000 molecules per cell, corresponding to 0.01 to 1.0 pmol/mg of membrane protein.

Two functionally distinct *mechanisms* by which GI peptides affect cell functions have been identified. For VIP, secretin and related peptides, the *adenylate cyclase* system is involved. For CCK, bombesin, substance P and TRH, an increase in intracellular Ca^{2+} subsequent to *phosphoinositol lipid* hydrolysis has been demonstrated. Other mechanisms, in particular those implicating protein kinases, calcium, calmodulin and others, have also been suggested in certain cases.

2.7 Structure–Activity Relationships

Studies with partial sequences and chemically modified analogs have shown that *determinants of receptor binding* and biological activity *vary among* individual *peptides*. In gastrin, CCK, neurotensin, substance P and bombesin/GRP, *only* the *C-terminal* region of the molecule is required, short C-terminal sequences being equally as potent or even more potent (as with CCK-8) than longer peptides. In VIP, secretin and glucagon, the *entire sequence* is involved, the N-terminal region being the most important functionally. Most peptide fragments (or analogs) are agonists of low potency, but some act as antagonists (such as secretin 5–27 and desphenylalanine amide CCK-8). In some cases, individual molecular forms differ in their relative abilities to bind to receptors of different target tissues (CCK) and/or to elicit biological effects in vivo and in vitro (gastrin).

The *structural similarities* between certain peptides, such as those which exist between gastrin and CCK or between secretin and VIP, account for the ability of one peptide to bind to receptors of its homologous counterpart and elicit biological responses. Thus, CCK, like gastrin, stimulates gastric acid secretion, and, conversely, like CCK, gastrin stimulates gallbladder contraction. Such low affinity effects are probably unimportant physiologically.

Presumably because of structural differences between peptides and/or receptors, *structure–activity* relationships in *lower vertebrate* species *differ from* those in *mammalian* species. Thus, in amphibians and birds, CCK-8 and cerulein are stronger stimulants of gastric acid secretion than mammalian gastrin. Similarly, it has been observed that mammalian CCK-8 does not cause gallbladder contraction in the hagfish (a primitive vertebrate), although hagfish intestinal extracts containing a CCK-like factor produce contraction in mammalian gallbladder.

2.8 Secretion

Most GI peptides produced by endocrine cells are secreted into the *blood circulation* and affect *distant* target cells in the GI tract. An alternative pathway has been suggested for somatostatin, with delivery to *adjacent* target cells in the GI mucosa and pancreas through direct diffusion into the extracellular space; other peptides may operate via this pathway as well. Yet *another* secretory *pathway* demonstrated for gastrin, somatostatin and secretin is that of secretion into the gastric or intestinal *lumen*; however, whether this pathway is important physiologically is as yet unknown.

Factors which regulate, directly or indirectly, the secretion of GI hormones in vivo act on GI endocrine cells via the GI lumen, the blood stream, or the autonomic nervous system.

Luminal Regulators

Luminal regulators affect primarily the secretion of peptides produced by endocrine cells; some act directly on these cells, whereas others act via nervous connections. Luminal regulators include the main classes of nutrients, the pH of the gastric or duodenal content, divalent cations such as Ca^{2+}, biliary and pancreatic secretions, and distension of the GI wall (Table XII.9). In general, *nutrients* are *stimulatory*, although they do not affect the secretion of all hormones in the same way. Small peptides and hydrophobic amino acids preferentially stimulate the release of gastrin and CCK, glucose stimulates the release of GIP and enteroglucagon, and fat that of CCK, enteroglucagon, and neurotensin. *Acidification* of the gastric content inhibits the release of gastrin, while duodenal acidification stimulates the release of peptides produced by the proximal intestine, particularly secretin. Gastrin and CCK release are stimulated by Ca^{2+} and Mg^{2+}. *Bile* stimulates secretin release, and *pancreaticobiliary secretions* that of motilin. *Gastric distension* stimulates gastrin release. Virtually all luminal regulators stimulate pancreatic polypeptide release, presumably via cholinergic entero-pancreatic reflexes.

Studies with regionally perfused nutrients have shown that the secretion sites of individual peptides in the GI tract correspond closely to their sites of localization. In addition, selective vein catheterization studies have shown that the response of GI endocrine cells to individual *stimuli* may *vary along the length* of the GI tract. Thus, in the dog, fat preferentially releases antral somatostatin, while amino acids favor the release of somatostatin by the fundus.

Circulatory Regulators

Circulatory substances affecting the release of GI peptides include *calcium*, various hormonal and

Table XII.9 Luminal factors affecting the release of GI peptides in vivo.

Stimulus	Gastrin	CCK	Secretin	VIP	GIP	Entero-glucagon	Motilin	Somato-statin	Pancreatic polypeptide	Neuro-tensin
Mixed meal	↑	↑	↑ [b]	→	↑	↑	↑	↑ [g]	↑ [h]	↑ [h]
Nutrients:										
Amino acids	↑ [a]	↑ [a]	→	→	↑ [c]			↑ [g]	↑ [h]	→
Amines	↑									
Peptides	↑	↑	→		↑				↑ [h]	
Glucose	→		→	→	↑	↑ [e]	↓	↑ [g]	↑ [h]	
Fat	→	↑	→	↑	↑ [d]	↑ [f]	↑	↑ [g]	↑ [h]	↑
Acid	↓		↑	↑	↑		↑	↑ [g]	→	
Alkali	↑		↓				↓			
Divalent cations	↑	↑								
Distension	↑						↑		↑ [h]	
Pancreatico-biliary juice				→			↑	↑	↑	

[a]Most potent releasers: phenylalanine, tryptophan, methionine, and valine.
[b]Acid accounts for the stimulation.
[c]Most potent releasers: arginine, histidine, isoleucine, leucine, lysine, and theronine.
[d]Long chain fatty acids are most effective.
[e]Secretion is also stimulated by fructose, xylose, and mannose.
[f]Triglycerides are most effective.
[g]Differential release occurs depending on the region where stimulus is applied (see text).
[h]Stimulation is mediated by cholinergic mechanisms.

non-hormonal *peptides*, *catecholamines* and, in some cases, absorbed *nutrients*. Intravenously administered Ca^{2+} resembles intraduodenally or orally administered Ca^{2+} in its stimulatory effect on gastrin, CCK and pancreatic polypeptide release; somewhat similar effects are elicited by Mg^{2+} and Zn^{2+}. Nearly all peptides affecting the secretion of GI peptides are themselves of GI and pancreatic origin (Table XII.7). The peptides which have the widest range of functions are bombesin and somatostatin; they stimulate and inhibit secretion, respectively. Physiologically, some of these peptides may act as hormones, but others may be delivered to secretory cells via paracrine (somatostatin) or neurocrine (bombesin) pathways. Nutrients administered intravenously are generally ineffective in stimulating the secretion of hormones produced by GI mucosal cells, nor do they stimulate the secretion of pancreatic polypeptide. However, they do stimulate the release of pancreatic somatostatin, glucose and amino acids being the most effective.

Neural Regulators

Neural regulators act largely via the *parasympathetic* and *sympathetic* nervous systems (especially the former), affecting mainly the secretion of gastrin, somatostatin, and pancreatic polypeptide. Activators of the parasympathetic system include *sham feeding* and *insulin-induced hypoglycemia*, which act centrally via vagal efferent fibers, and *intraluminal food*, which acts peripherally via long (vagovagal) or short (intramural) neural pathways. Exercise and gastric distension are examples of activators of the sympathetic system. Experimentally, neural regulation of secretion is assessed by section or electrical stimulation of the vagus or intramural nerves, and by injection of specific agonists or antagonists.

The *neurotransmitters* involved in the secretion of several GI hormones and the nature of the responses they evoke (stimulation or inhibition) are shown in Table XII.10. One major transmitter, *acetylcholine*, stimulates gastrin and pancreatic polypeptide release but inhibits somatostatin release; direct effects of this transmitter on isolated G and D cells have been demonstrated. Another important transmitter, *bombesin/GRP*, accounts for the non-cholinergic component of vagally-induced stimulation of gastrin secretion.

Secretion of Gastrointestinal Neuropeptides

Peptides produced by neurons are assumed to be *secreted locally*, to *act on neighboring target cells* (neurocrine secretion) *or other neurons* (neurotransmission or neuromodulation). In some circumstances, however, they may *enter the blood* stream and

Table XII.10 Neural regulators of GI peptide secretion.

Transmitter	Gastrin	Somatostatin	Pancreatic polypeptide	VIP	Motilin	Neurotensin
Acetylcholine	↑[a]	↓[b]	↑[c]	↑	↑	↑
Epinephrine:						
Acting at α receptors	↓	↓	↓	↓		
Acting at β receptors	↑	↑	↑		↑	
Dopamine	↑	↓	↓			
GABA	↑	↓				
Serotonin	↑	↓				
Bombesin/GRP[d]	↑	↑			↑	

[a]Stimulation mediated in part by inhibition of somatostatin release. There is also evidence for a cholinergic inhibitory pathway.
[b]There is also evidence for a cholinergic stimulatory pathway.
[c]Acetylcholine release mediates the effects of nutrients and hormones on PP release.
[d]Major candidate for the non-cholinergic stimulation of gastrin release and inhibition of somatostatin release by excitation of intramural gastric neurons.

reach distant targets by a neuroendocrine delivery system. Release of neuropeptides (VIP, bombesin, substance P, somatostatin) into the portal and/or peripheral circulation occurs upon electrical stimulation of the vagus and in response to mechanical or chemical stimulation of the intestinal mucosa.

2.9 Metabolism

Half-lives of circulating GI peptides vary from approximately *1 to 30 min.* In general, the larger the size, the longer the half-life. GI peptides are *removed* from the circulation mainly *in liver*, *kidney* and other *peripheral tissues*, where they undergo *degradation* by exopeptidases and endopeptidases. Some peptides (gastrin, secretin, and CCK), however, are relatively resistant to hepatic removal and to the action of exopeptidases, presumably due to the presence of N- and C-terminal blocking groups. *Peptidases in target tissues* may be important in terminating the action of GI peptides, although their contributions to the overall metabolism of circulating peptides remain to be assessed.

2.10 Clinical Aspects

Tumors

Endocrine *tumors*, although rare, are the *major cause* of *overproduction* of GI peptides in humans (Table XII.11, p. 581). Some tumors, such as insulinomas, glucagonomas and gastrinomas, arise from functionally well defined cell types, but other tumors, such as pancreatic VIPomas, originate from functionally undefined cells. *Most* tumors are located in the *pancreas*, but *some* are found in the *intestine* (substance P-producing carcinoid tumors) or outside of the pancreas and GI tract (VIP-producing ganglioneuroblastomas, bombesin-producing lung tumors). Major peptides produced by tumors include insulin, glucagon, pancreatic polypeptide, somatostatin, gastrin, VIP, enteroglucagon, substance P, neurotensin, and bombesin/GRP; although some tumors produce a single peptide, the production of more than one peptide is a common feature. Insulinomas, glucagonomas, gastrinomas and VIPomas are associated with specific clinical syndromes resulting from hypersecretion of the relevant peptide, but other tumors are silent, which may be explained by the secretion of biologically inactive molecular forms of peptides. The course of peptide-producing tumors varies widely among individual types; thus, 80% of insulinomas are benign, while *60%* of *gastrinomas* or VIPomas, although slow-growing, are *malignant.* Surgical removal may not always be possible; medical treatments include cytotoxic streptozotocin and long-acting somatostatin analogs.

Non-tumoral Conditions Affecting Gastrointestinal Endocrine Peptides

Non-tumoral conditions affecting gastrointestinal endocrine peptides include: 1. *surgical* intervention in the GI tract, such as gastric and intestinal resections, vagotomy, and bypass surgery; 2. *autoimmune* and *inflammatory* conditions, such as atrophic gastritis, celiac disease, tropical malabsorption, acute bowel infection, and cystic fibrosis; 3. *metabolic* conditions, such as insulin-dependent and non-insulin-dependent diabetes mellitus, obesity, hepatic failure, and chronic renal disease. In conditions affecting the upper small intestine (e.g., bypass surgery and celiac disease), plasma concentrations of hormones secreted by this region of the GI tract (CCK, secretin, GIP, and motilin) are decreased, whereas concentrations of hormones secreted by the distal intestine (enteroglucagon, neurotensin) are increased. This is presumably reflective of a reduced mass (or functional exclusion) and a compensatory hyperplasia of the relevant endocrine cells. Additional mechanisms responsible for altered plasma concentrations of GI hormones include removal of a physiological mechanism inhibiting secretion (as in hypergrastrinemia caused by atrophic gastritis), dysfunction of the autonomic nervous system, and reduced metabolism of the hormone.

Conditions Affecting Gastrointestinal Neuropeptides

Certain pathological conditions caused by dysfunction of the peptidergic innervation of the gut are characterized by a change in the concentration of neuropeptides in GI tissues. Thus, a marked decrease in local concentrations of VIP and in the density of VIP innervation is observed in *Hirschprung's disease*, or *congenital megacolon*, in *esophageal achalasia* (a congenital disease characterized by the absence of peristalsis in the body of the esophagus and failure of the lower sphincter to relax in response to swallowing), and in *Chagas' disease* (a parasitic disease leading to a denervation of the gut and subsequent megaesophagus, megaduodenum, and megacolon). In some cases, a decrease in somatostatin and substance P concentrations is observed as well. Conversely, increased VIP concentrations and hypertrophy of VIP nerves are found in association with *Crohn's disease* (a chronic form of intestinal inflammation which leads to the development of fibrosis), and both VIP and substance P concentrations in the mucosa are elevated in patients with celiac or duodenal *ulcers, presumably in reaction to chronic mucosal irritation.*

3. Gastrin

Gastrin was described in 1905[2] as a substance present in extracts of gastric antral mucosa, which, following intravenous injection, stimulated the secretion of gastric juice. For years, its existence as an entity distinct from histamine, another gastric secretagogue, was disputed, until gastrin was finally isolated and characterized[6].

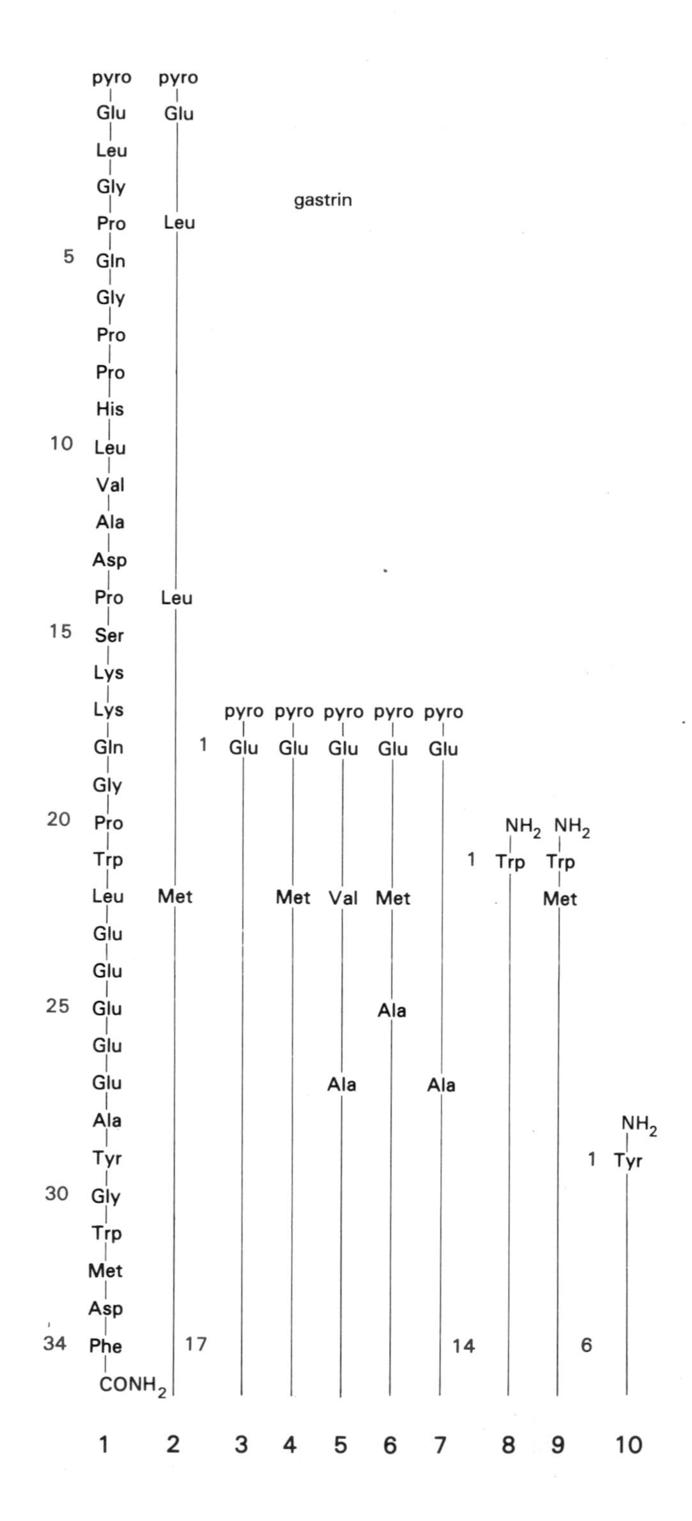

Figure XII.11 Primary structure of human (1) and porcine (2) gastrin-34, human (3), porcine (4), bovine (5), canine (6) and feline (7) gastrin-17, human (8) and porcine (9) gastrin-14, and mammalian gastrin-6 (10).

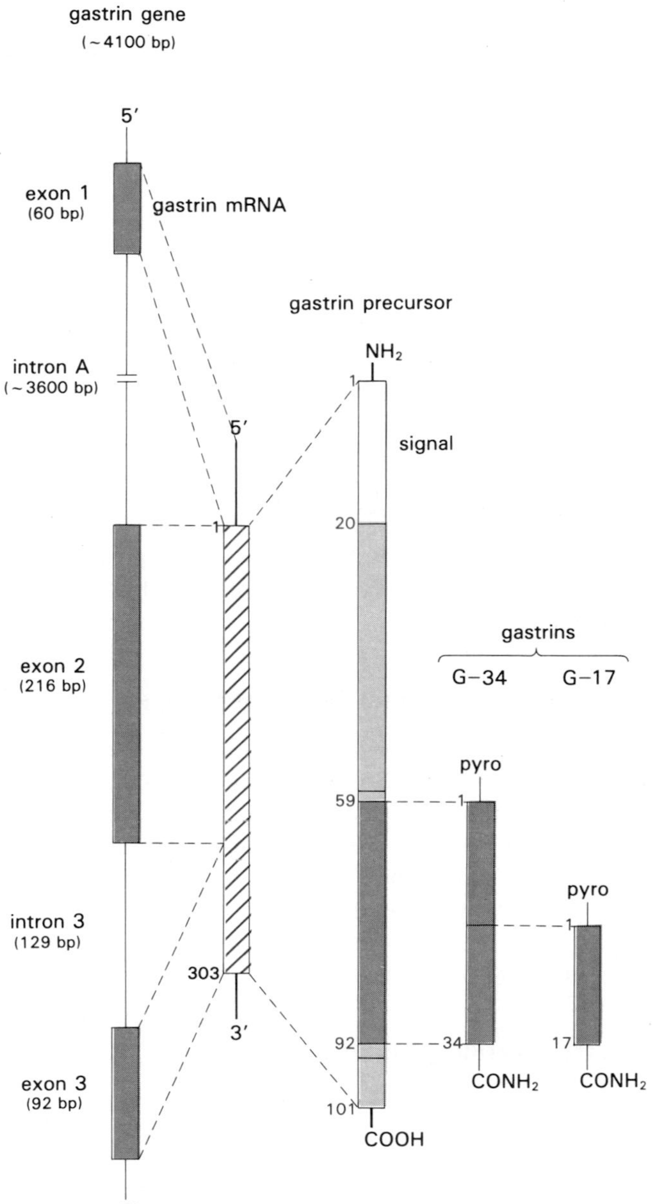

Figure XII.12 Structure of the human gastrin gene, mRNA precursor, and peptides[25,26].

Isolation and Structure

Gastrin exists in multiple molecular forms (Fig. XII.11). The *major* molecular forms are *gastrin-17* (G-17), a heptadecapeptide, and *gastrin-34* (G-34), a peptide of 34 aa which is an N-terminally extended form of G-17. Other molecular forms are *gastrins* 14, 6 and 4, corresponding to the C-terminal portion of the larger forms.

All mammalian gastrins are *characterized* by a *C-terminal pentapeptide amide* sequence *identical to* that of *CCK*, and by a single tyrosyl residue, at position 6 from the C-terminus, which may or may not be sulfated. The use of synthetic oligonucleotides corresponding to the C-terminal tetrapeptide amide sequence was instrumental in cloning gastrin cDNA (Fig. XII.12). Human preprogastrin consists of 101 aa[25]. The sequence of G-34 gastrin is located at the carboxyl end (positions 59–92). A striking homology is observed between amino acid sequences 41–54 and 74–87, and, to a lesser extent, between sequences 29–33 and 62–66, suggesting that the evolution of the gastrin gene could have involved gene duplication. The gene is ~4100 bp long and consists of three exons separated by two introns[26]. Intron B, 129 bp long, precedes the region encoding G-17 (codons 76–92), the principal hormonal form, thus separating it from the rest of the precursor.

Distribution

In mammals, virtually all extractable gastrin is present in the GI tract, with highest concentrations found in gastric antral mucosa. In humans, the duodenum contains nearly the same amount of gastrin as the antrum (Table XII.2). Small amounts have been identified in the vagus and sciatic nerves, in corticotropic and melanotropic pituitary cells, and in nerve terminals originating from cells in the hypothalamus.

In the gastric antrum, 90% of total gastrin is G-17. In the duodenum, however, G-17 and G-34 are present in equivalent amounts, and in the jejunum, G-34 is the predominant molecular form. In plasma, G-34 accounts for 70% of total gastrin in the basal state and 50% in the postprandial state.

25. Boel, E., Vuust, J., Norris, K., Wind, A., Rehfeld, J. F. and Marcker, K. A. (1983) Molecular cloning of human gastrin cDNA: evidence for evolution of gastrin by gene duplication. *Proc. Natl. Acad. Sci., USA* 80: 2866–2869.

26. Wiborg, O., Berglund, L., Boel, E., Norris, F., Norris, K., Rehfeld, J.-F., Marcker, K. A. and Vuust, J. (1984) Structure of a human gastrin gene. *Proc. Natl. Acad. Sci., USA* 81: 1067–1069.

In the GI tract, gastrin is produced *exclusively* by *endocrine* cells. The principal gastrin-containing cells, or *G cells*, are located in the antral glands of gastric mucosa. In addition to gastrin, they contain ACTH, Met-enkephalin and endorphin-like immunoreactivity. Additional gastrin-containing cells, the IG and TG cells, differing from the G cells with respect to the morphology of their secretory granules, have been identified in the mucosa of the duodenum and jejunum. They are revealed by antisera directed against the C-terminal tetrapeptide.

Biological Actions

The *major* physiological *effects* of gastrin are the *stimulation* of *acid secretion* by gastric parietal cells and the *regulation* of gastric *mucosal growth*.

Gastric Acid Secretion

In vivo, gastrin *increases* the secretion of H^+ and Cl^- ions, and does so by *increasing* their concentration in both gastric juice and in the *aqueous volume*. With increasing flow rate, the concentrations of H^+ and Cl^- ions increase hyperbolically, to attain maximal values of about 140 and 160 mEq/l, respectively, and, concomitantly, the concentration of Na^+ ions decreases (Fig. XII.13). In humans, the acid secretory response to a maximal dose of gastrin G-17 is about 30–35 mEq/h, and the dose of G-17 required for a half-maximal effect about 10–20 pmol/kg/h; under the latter conditions, plasma gastrin increases from about 15 to 30–40 pmol/l, a value comparable to that achieved physiologically.[27]

In isolated gastric parietal cells, gastrin stimulates oxygen consumption and aminopyrine accumulation, and causes expansion of the apical surface area.

Gastrin resembles *acetylcholine* (released by muscarinic nerve endings of parasympathetic postganglionic neurons) and *histamine* (produced endogenously in the gastric mucosa) in its stimulatory effect on acid secretion by gastric parietal cells. In vivo, there is an interdependence between the actions of the three secretagogues, which is demonstrated by the ability of *cimetidine* (a *histamine antagonist* acting at the H_2 *receptors*) and *atropine* (an *anticholinergic* agent) to *inhibit the action of gastrin*[19]. *In vitro*, however, the effects of gastrin are *unaffected* by antihistaminic and anticholinergic agents, suggesting that the effects of these agents in vivo are the result of the exposure of parietal cells to a constant background of histamine and acetylcholine[19].

27. Feldman, M., Walsh, J. H., Wong, H. C. and Richardson, C. T. (1978) Role of gastrin haptadecapeptide in the acid secretory response to amino acids in man. *J. Clin. Invest.* 61: 308–313.

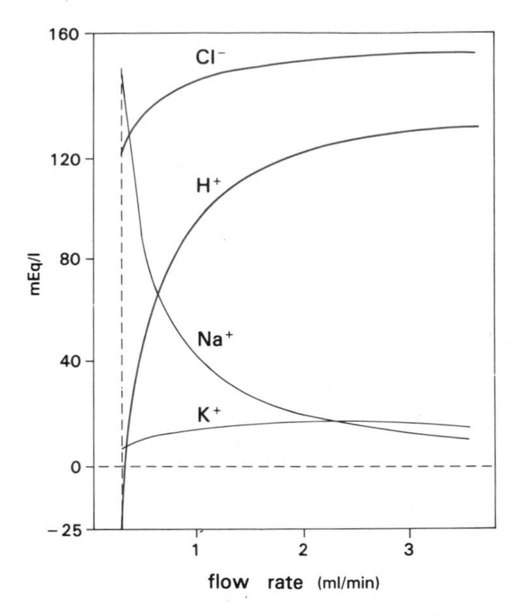

Figure XII.13 Relationship between ion concentrations in gastric juice and flow rate during stimulation of acid secretion in the human. (Makhlouf, G. M. (1981) Electrolyte composition of gastric secretion. In: *Physiology of the Gastrointestinal Tract* (Johnson, L. R., ed), Raven Press, New York, p. 553).

Other Gastrointestinal Secretory Effects

Gastrin stimulates water and electrolyte secretion by the pancreas, liver, Brunner's glands, and the small intestine. It stimulates gastric pepsin secretion in vivo, although, in isolated gastric glands, gastrin is less potent than CCK in eliciting this effect. Gastrin weakly stimulates pancreatic and intestinal enzyme secretion, and inhibits absorption of water, electrolytes and glucose by the small intestine.

Gastrointestinal Motility

Gastrin *stimulates* the motility of the gastric antrum, but, at least in the dog, relaxes the gastric fundus; it increases colon motility and weakly stimulates gallbladder contraction; it contracts the lower esophageal sphincter, but relaxes other sphincters (pyloric, ileal, Oddi). Effects of gastrin on GI smooth muscle in vitro occur both directly and via activation of cholinergic neurons.

Gastrointestinal Growth

A marked *increase* in *parietal cell* mass and in acid secretory capacity occurs in human subjects with *gastrinomas*. Similarly, in experimental animals, exogenous gastrin causes marked stimulation of mucosal growth in the oxyntic region of the stomach, and partially prevents the gastric atrophy which results from starvation or reduced food intake. Tropic effects of gastrin have also been described in the small intestine, colon, and pancreas. These effects are not affected by histamine antagonists or prostaglandins, suggesting that they can be functionally dissociated from the effects of this hormone on acid secretion.

Other Effects

Gastrin *increases blood flow* in the stomach, small intestine and pancreas, and stimulates the release of pancreatic hormones.

Receptors and Mechanism of Action

Specific gastrin *receptors*, the number of which is regulated by circulating gastrin, have been identified in fundic membranes and in isolated parietal cells. The *mechanism* of action of gastrin beyond the receptor is *unknown*. Although cAMP may mediate gastrin action in non-parietal cells, it has now been firmly established that, like cholinergic agents but unlike histamine, gastrin does not stimulate cAMP production in parietal cells.

The *transport of* H^+ ions across the luminal membrane of gastric parietal cells depends primarily on a K^+*-activated ATPase*; the transport reaction couples the electroneutral exchange of intracellular H^+ and extracellular K^+ to the hydrolysis of ATP. Whether gastrin (and other secretagogues) affects the activity of this enzyme or other biochemical events in parietal cells is unknown.

Structure–Activity Relationships

Mammalian G-17 gastrins are equipotent in their ability to stimulate acid secretion in vivo, as are their sulfated and unsulfated forms. Based on administered doses, G-34, G-17 and G-14 are equipotent, but based on concentrations attained in the blood, G-17 is six to eight times more potent than G-34.

Most of the information required for receptor binding and biological *activity* resides in the *C-terminal region*. Thus, in G-17, the C-terminal heptapeptide is equipotent to G-17, while G-5 and G-3 are, respectively, 10% and 0.1% as potent. *Removal* of the *C-terminal* amide group, oxidation of Met-15 and substitution of Asp-16 *abolishes activity*. The most *widely used* fragment in human studies is *pentagastrin*, a synthetic analog of G-5, in which the N-terminal Ala residue is blocked by a tertiary butyloxycarbonyl group.

Secretion and Metabolism

Luminal Regulators

The main *luminal stimulants* of gastrin secretion are *small peptides* and free hydrophobic amino acids; other stimulants include dietary amines and gastric distension. *Acidification* of gastric contents below pH 3 causes complete suppression of secretion. This provides a *feedback* mechanism whereby acid secreted under the influence of gastrin inhibits further hormone release.

Neural Regulators

Regulation of gastrin release by the *parasympathetic nervous system* is complex and involves both *cholinergic* (stimulatory and inhibitory) and *non-cholinergic* (stimulatory) pathways; these may be activated centrally or peripherally (p. 547). Studies using isolated perfused stomach have shown that *cholinergic* stimulation is mediated by the *inhibition of antral somatostatin release*[28] (Fig. XII.14), while *non-cholinergic* stimulation is mediated by the *stimulation of bombesin release*[29] (Fig. XII.15).

The sympathetic nervous system mediates the response of gastrin to gastric distension via β receptors.

Endocrine and Paracrine Regulators

The major endocrine regulator of gastrin secretion is *somatostatin*, which exerts direct *inhibitory*

28. Saffouri, B., Weir, G. C., Bitar, K. N. and Makhlouf, G. M. (1980) Gastrin and somatostatin secretion by perfused rat stomach: functional linkage of antral peptides. *Amer. J. Physiol.* 238: G495–G501.
29. DuVal, K. W., Saffouri, B., Weir, G. C., Walsh, J. H., Arimura, A. and Makhlouf, G. M. (1981) Stimulation of gastrin and somatostatin secretion from the isolated rat stomach by bombesin. *Amer. J. Physiol.* 241: G242–G247.

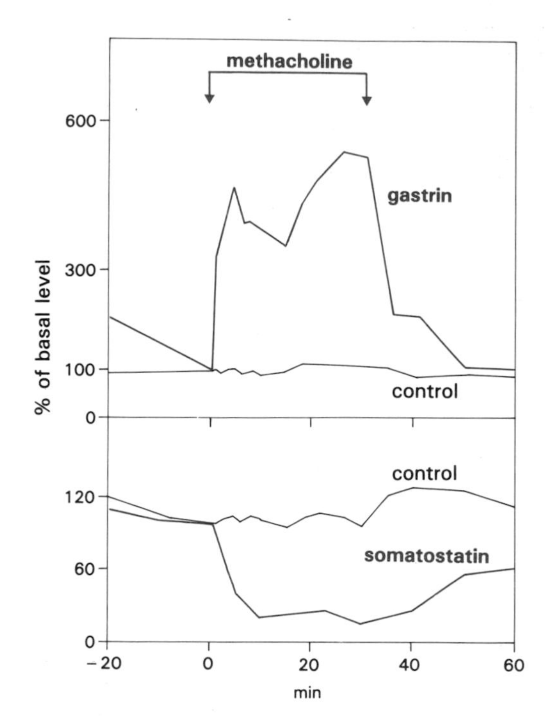

Figure XII.14 Effects of methacholine on gastrin (upper) and somatostatin (lower) release in isolated perfused rat stomach[28].

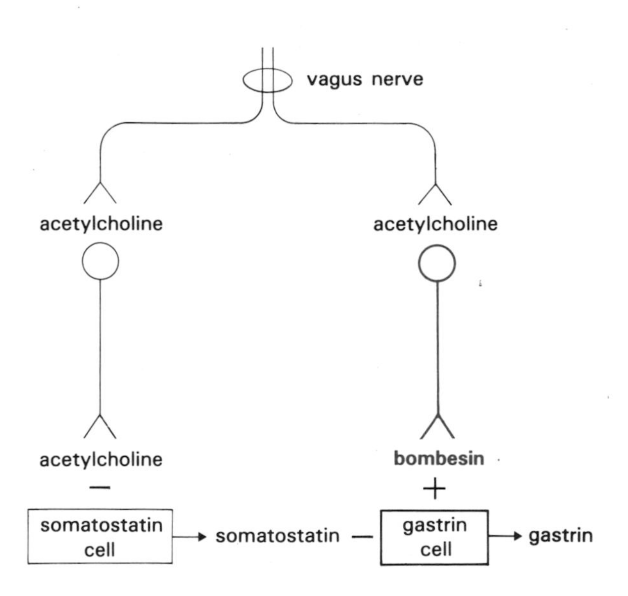

Figure XII.15 Model describing the neural control of gastrin secretion. Two intramural neural pathways are shown: a cholinergic postganglionic neuron innervating a somatostatin cell, and a peptidergic postganglionic (bombesin) neuron innervating a gastrin cell. Acetylcholine released from the cholinergic neuron inhibits somatostatin secretion, while bombesin released from the peptidergic neuron stimulates gastrin secretion. Both intramural neurons are under cholinergic (vagal) preganglionic control. Somatostatin inhibits gastrin secretion[39] via a direct effect on gastrin cells.

effects on G cells. Experimentally, antisera to somatostatin cause a marked stimulation of gastrin release, suggesting that somatostatin exerts a continuous restraint on gastrin secretion (Fig. XII.16). Additional inhibitors of gastrin secretion include secretin, glucagon, VIP, GIP and calcitonin, the effects of which are mediated by somatostatin.

Pathophysiology

Excess gastrin production may be associated with increased gastric acid secretion, as in gastrinomas, or with decreased gastric acid secretion, as in atrophic gastritis. Gastrin *deficiency* is always secondary to gastric resection.

Gastrinomas (Zollinger–Ellison Syndrome)

Gastrinomas are ectopic *gastrin-secreting tumors* which give rise to an *increase* in *gastric acid* secretion, hyperplasia of gastric parietal cells, multiple and recurrent duodenal ulcers, diarrhea, and steatorrhea (Fig. XII.17). Gastrinomas *generally* develop in the *pancreas* and, less commonly, in the stomach or duodenum. At least two-thirds undergo metastatic

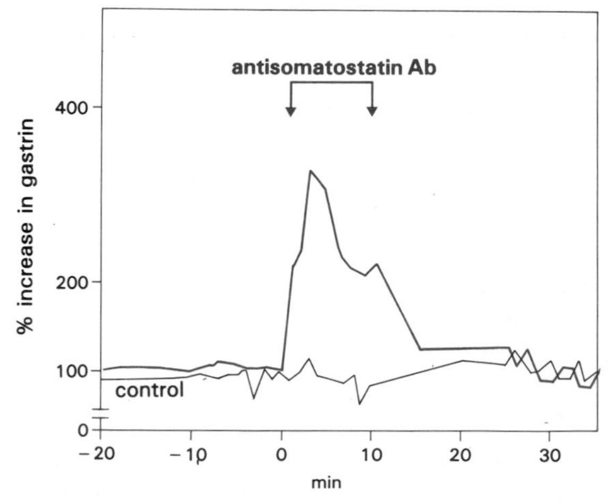

Figure XII.16 Enhancement of basal gastrin release from rat antrum perfused by an antibody to somatostatin. (Saffouri, B., Weir, G.C., Bitar, K.N. and Makhlouf, G.M. (1979) Stimulation of gastrin secretion from the vascularly perfused rat stomach by somatostatin antiserum. *Life Sci.* 20: 1749–1754).

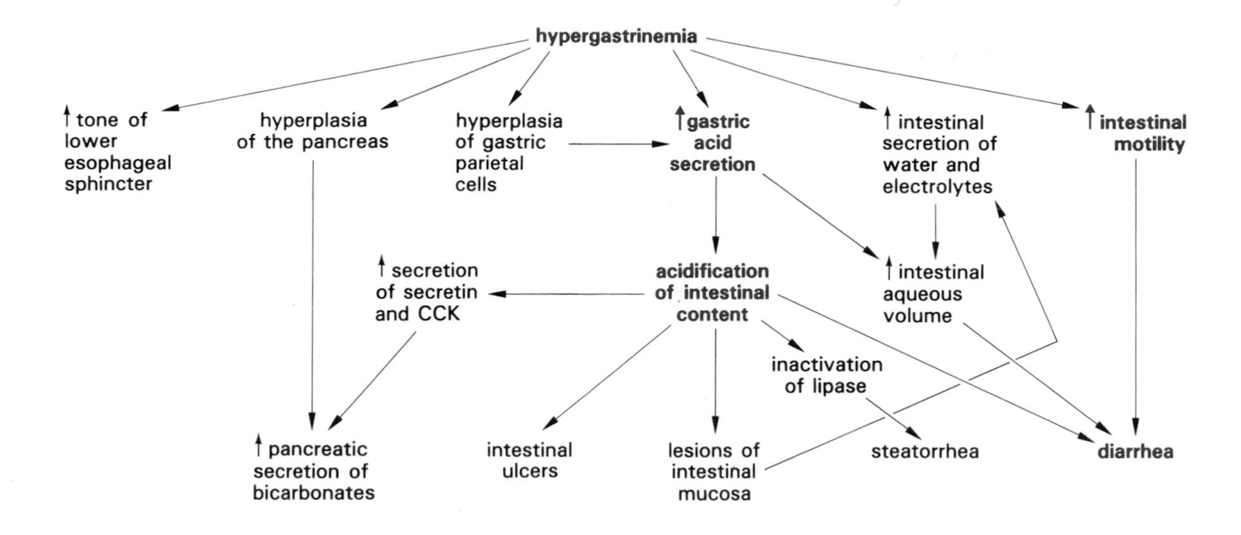

Figure XII.17 Pathophysiological symptoms of gastrinomas.

spread, and multiple tumors are common. Microscopically, they resemble carcinoid tumors, but it is difficult to distinguish benign from malignant tumors and to identify the type of endocrine cell involved. Fasting plasma gastrin concentration in patients with gastrinoma usually exceeds 200 pg/ml, and rises to abnormally high values following stimulation by protein meals, calcium infusion, and secretin infusion. This response of gastrin to secretin is particularly valuable, since secretin does not normally affect, or may actually decrease, serum gastrin.

Hyperparathyroidism is found in approximately *20%* of patients with gastrinomas; correction of hypercalcemia by parathyroidectomy may lead to a dramatic decrease in serum gastrin and gastric acid secretion. Other endocrine abnormalities which occur less commonly include pancreatic β cell, pituitary, adrenal, ovary, and thyroid tumors. This association, which may occur either spontaneously or in a familial form with autosomal dominant inheritance, is sometimes called *multiple endocrine neoplasia (MEN-1)* (p. 137).

As surgical resection of gastrinomas is seldom possible, the preferred surgical management is total gastrectomy. The latter produces good, immediate results, and ensures a survival rate of approximately 55% at five years. However, in a number of patients, treatment with cimetidine, provided it is continued indefinitely, ensures substantial reduction in gastric acid secretion and the disappearance of ulcers.

Other Causes of Hypergastrinemia with Acid Hypersecretion

Often associated with duodenal ulcers, the causes of hypergastrinemia with acid hypersecretion include: 1. antral *G cell* hyperplasia or *hyperfunction* in the absence of pancreatic tumor, in which antrectomy suppresses hypergastrinemia; 2. *pyloric obstruction* due to duodenal ulcer and hypertrophic pyloric stenosis, in which hypergastrinemia may be caused by gastric distension; 3. chronic *renal failure* and massive *intestinal resections*, in which hypergastrinemia results in part from alterations in gastrin metabolism.

Much attention has been given to the possible role of gastrin in gastric hypersecretion associated with the common *duodenal ulcer*. Although fasting serum gastrin is not increased in this condition, postprandial gastrin is usually greater than normal. Patients with duodenal ulcer are also more sensitive to stimulation of acid secretion by exogenous as well as endogenous gastrin released after a meal. However, it now appears that duodenal ulcer disease is a heterogeneous group of disorders, G cell dysfunction being *only one pathogenetic factor* among others.

Hypergastrinemias without Acid Hypersecretion

The major cause of hypergastrinemia is *atrophic gastritis*, especially that which spares the antrum and is associated with circulating parietal cell antibodies. Chronic hypochlorhydria leads to the hyperplasia of antral G cells, which in turn causes hypergastrinemia suppressible by intragastric administration of hydrochloric acid. In some patients, *pernicious anemia* develops as a result of the loss of secretion of intrinsic factor.

Increased serum gastrin concentrations with hypochlorhydria are also observed in carcinoma of the body of the stomach. This is presumably reflective of the atrophic gastritis and associated hypochlorhydria which pre-exist in some of these patients.

Diagnostic Use

Synthetic pentagastrin is widely used as a *stimulant* of *gastric acid* secretion. This peptide has also been used as a tool in the diagnosis of medullary carcinoma of the thyroid, a disease in which it induces the release of calcitonin.

4. Cholecystokinin (CCK)

Cholecystokinin and pancreozymin were described independently, in 1928[3] and 1943[4], as intestinal hormones which stimulated, respectively, gallbladder contraction and pancreatic enzyme secretion. They were isolated in 1964[7] and were shown to be a single polypeptide, which was called cholecystokinin (CCK). Initially thought to be confined to the GI tract, CCK was later identified in the CNS, where it elicits a variety of effects.

Isolation, Structure, and Biosynthesis

CCK, like gastrin, exists in *multiple molecular forms* of various length, the larger forms being N-terminally extended shorter forms. In the human, only the octapeptide, CCK-8, has been isolated, but in the pig, four major forms (CCK-58, -39, -33 and -8) have been isolated (Fig. XII.18). CCK-8 is preferentially extracted at neutral pH, whereas the larger, more basic forms are best extracted with acid.

The amino acid sequence of CCK-8 or of the C-terminal octapeptide of larger forms is identical in all species studied so far. This sequence is characterized by a *sulfated tyrosyl* residue at position 7 from the C-terminus, and by the *same C-terminal pentapeptide amide* as that of *gastrin* and the amphibian peptides

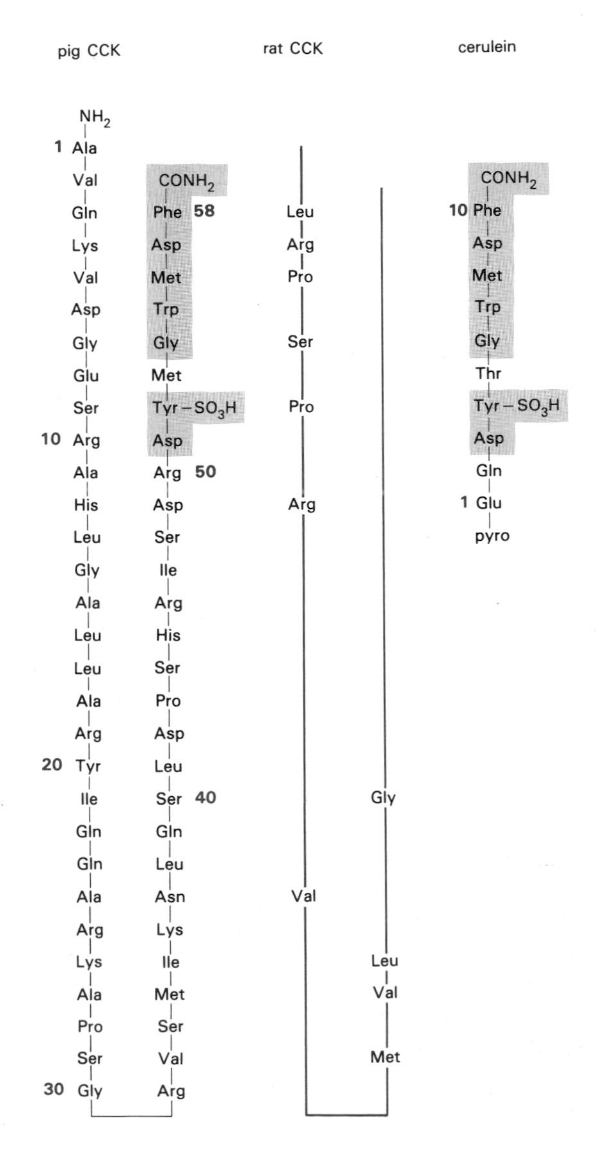

Figure XII.18 Amino acid sequences of pig and rat CCK, and of amphibian cerulein. In humans, only the C-terminal octapeptide has been isolated[30,31]. Homologies between CCK and cerulein are highlighted.

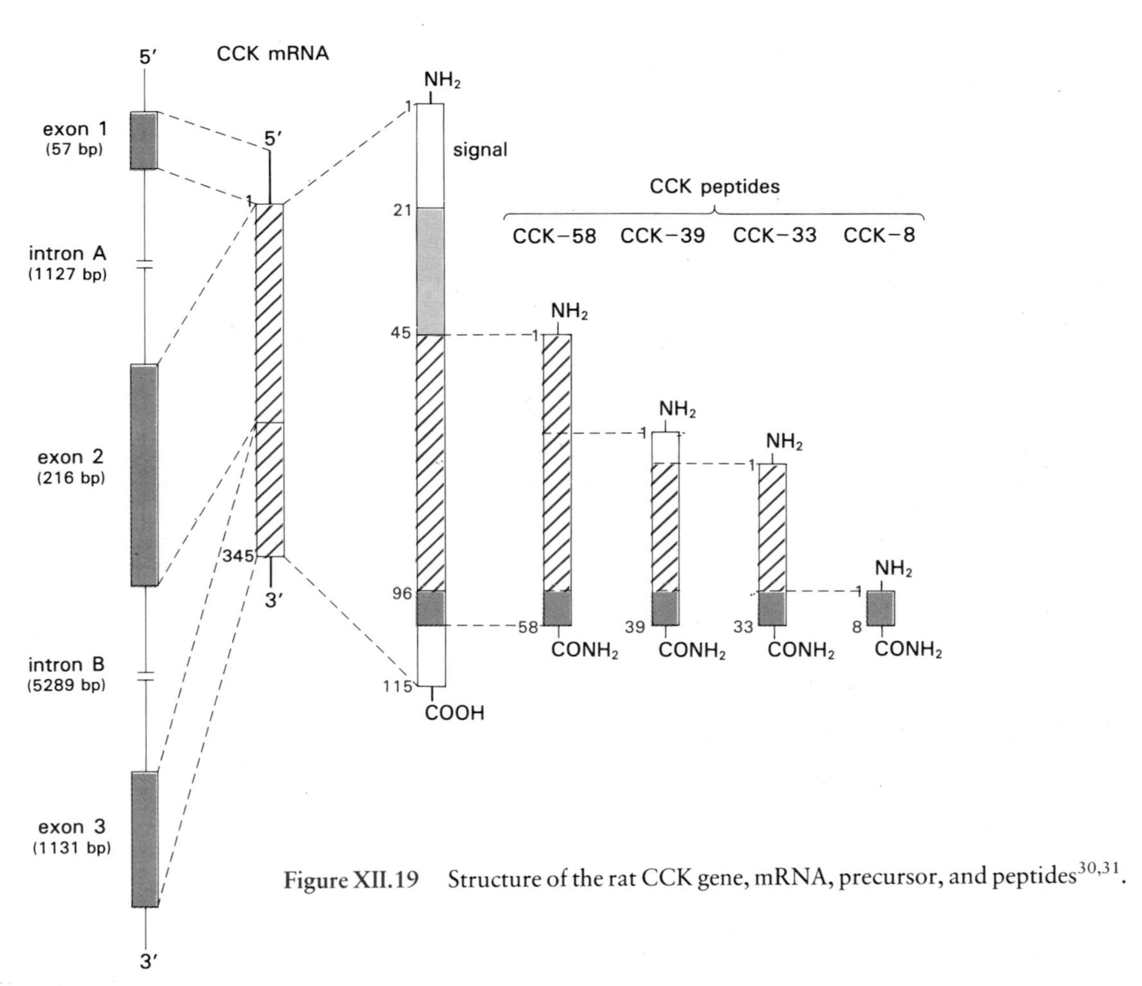

Figure XII.19 Structure of the rat CCK gene, mRNA, precursor, and peptides[30,31].

cerulein and phyllocerulein. Data suggest the processing of larger forms of CCK to CCK-8. From a medullary thyroid carcinoma, rat preprocholecystokinin cDNA has been cloned[30]. It predicts a peptide of 115 aa. Sequences corresponding to CCKs of various lengths are immediately preceded by a single arginine residue.

The organization of the rat CCK gene[31] is highly similar to that of the human gastrin gene (Fig. XII.12). The transcription unit spans 7 kb and consists of three exons interrupted by two introns. Intron B, of ~5500 bp, interrupts codon 72, thus separating the region coding for most of the sequence of CCK-33 from the region coding for the rest of the precursor (Fig. XII.19).

30. Deschenes, R. J., Lorenz, L. J., Haun, R. S., Roos, B. A., Collier, K. J. and Dixon, J. E. (1984) Cloning and sequence analysis of a cDNA encoding rat preprocholecystokinin. *Proc. Natl. Acad. Sci., USA* 81: 726–730.
31. Deschenes, R. J., Haun, R. S., Funckes, C. L. and Dixon, J. E. (1985) A gene encoding rat cholecystokinin. *J. Biol. Chem.* 260: 1280–1286.

Distribution

In human and porcine GI tracts, extractible CCK is found predominantly in the duodenum and proximal jejunum (highest concentrations are ~100–300 pmol/g tissue (wet weight) (Table XII.2)). At least 90% of this material is associated with the mucosa. CCK-8 is the most abundant form of the hormone found in the proximal intestine, whereas, in the mid-intestine, CCK-33 predominates. CCK is associated mostly with endocrine *I cells* in the *proximal intestine* and with neurons in the distal intestine. In the *brain*, CCK is found at higher concentrations than other neuropeptides, and is especially abundant in the frontal cortex, amygdala, hippocampus, and hypothalamus, where CCK-8 is the major molecular form.

Biological Actions

Pancreatic Exocrine Secretion

In vivo, CCK causes an *increase* in *hydrolase* concentration in *pancreatic juice*, with a moderate increase in fluid volume and bicarbonate concentration. In the presence of secretin, the dose of CCK-8 and CCK-33 required for a half-maximal stimulation of amylase and trypsin output (Fig. XII.20) is 40 pmol/kg/h, and the corresponding plasma CCK concentration is ~10 pM[32]. In isolated pancreatic acini, CCK peptides are also potent stimulators of amylase release[33], and their effects are potentiated by secretagogues, namely VIP and secretin, which act via cAMP production. Morphological and biochemical studies have shown that the main effect of CCK in acinar cells is to accelerate the fusion of zymogen granules with the luminal membrane, leading to the discharge of their content in the lumen.

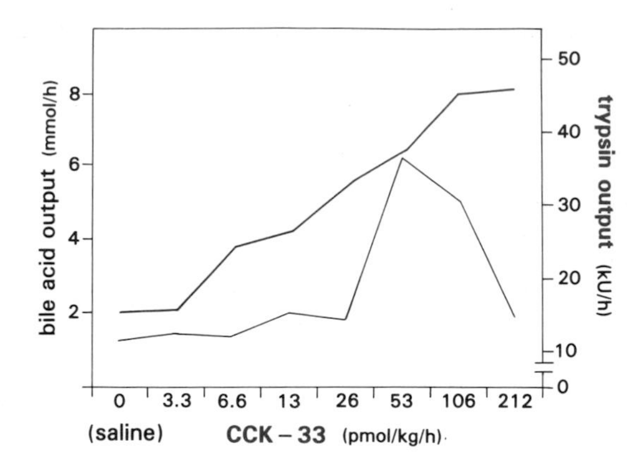

Figure XII.20 Effects of increasing doses of infused CCK-33 on duodenal trypsin output in the human. For comparison, the effect of CCK-33 on bile acid output (index of gallbladder contraction) is also shown. (Malagelada, J. R., Go, V. L. W. and Summerskill, W. H. J. (1973) Differing sensitivities of gallbladder and pancreas to cholecytokinin-pancreozymin in man. *Gastroenterology* 64: 950–954).

Other Gastrointestinal Secretory Effects

CCK *potentiates* secretin-induced *pancreatic secretion* of *fluid* and *bicarbonates*. It weakly stimulates gastric acid secretion and antagonizes gastrin-stimulated acid secretion. CCK increases intestinal lymph flow and releases intestinal peptidases.

Gastrointestinal motility

CCK is a *potent stimulant* of *gallbladder contraction*, and causes the *relaxation* of the *sphincter of Oddi*. This effect, not antagonized by atropine, leads to an increase in intraluminal gallbladder pressure, a decrease in gallbladder size, which can be demonstrated by X-ray examination or ultrasonography, and an increase in bile acid output (Fig. XII.20). In humans, the dose of CCK-8 required for a half-maximal decrease in gallbladder volume is ~10–20 pmol/kg/h, and the corresponding plasma CCK concentration is ~10–20 pM (Fig. XII.21). Additional effects of CCK include inhibition of gastric motility and delayed gastric emptying, stimulation of peristalsis and intestinal motility, relaxation of the lower esophageal sphincter, and contraction of the pylorus. *Effects* of

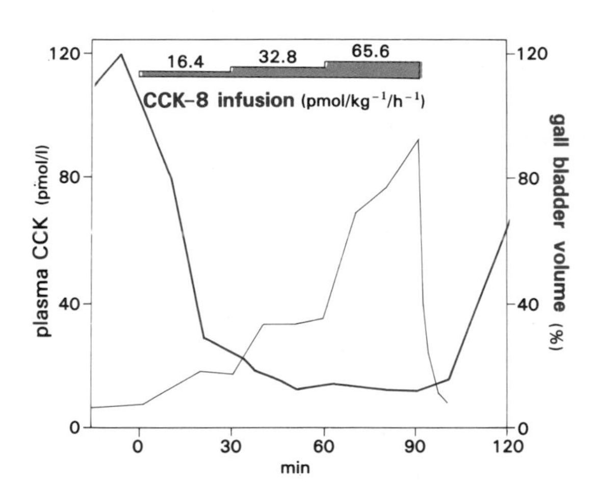

Figure XII.21 Plasma CCK concentration and gallbladder volume during infusion of CCK-8 in the human. (Byrnes, D. J., Barody, T., Daskalopoulos, G., Boyle, M. and Benn, T. (1981) Cholecystokinin and gallbladder contraction: effect of CCK infusion. *Peptides 2*, suppl. 2: 259–262).

CCK on gallbladder and gastric antral smooth muscle are *direct,* while effects on intestinal muscle are mediated by the release of acetylcholine from the myenteric plexus.

Other GI Effects

CCK exerts marked tropic effects in the exocrine pancreas and, to a lesser degree, in the endocrine pan-

32. Walsh, J. H., Lamers, C. B. and Valenzuela, J. E. (1982) Cholecystokinin-octapeptide-like immunoreactivity in human plasma. *Gastroenterology* 82: 438–444.
33. Jensen, R. T., Lemp, G. F. and Gardner, J. D. (1980) Interaction of cholecystokinin with specific membrane receptors on pancreatic acinar cells. *Proc. Natl. Acad. Sci., USA* 77: 2079–2083.

creas, gallbladder, gastric fundus, and duodenum (Table XII.6). It stimulates the secretion of pancreatic hormones (Table XII.7).

Central Nervous System

A major effect of CCK, when administered peripherally or centrally, is to *induce satiety*[34]. The latter can also be induced by the ingestion of phenylalanine, a stimulant of CCK secretion. Although CCK may induce satiety via a central effect, the lack of evidence that circulating CCK has access to central CCK receptors, and the demonstration that vagotomy abolishes the satiety effect, favors a peripheral site of action. Additional effects are indicated in Table XII.8.

Structure–Activity Relationships and Receptors

CCK *receptors* have been identified in exocrine pancreas[33], gallbladder and GI smooth muscle, and in the brain. Most, if not all, of the effects of CCK in pancreatic acinar cells are mediated by *calcium* mobilization.

Determinants for receptor binding and biological activity are contained in the *C-terminal* region[33], CCK-8 being approximately five times more potent than CCK-33, and CCK-4 0.1% as potent as CCK-8. In CCK-8 and CCK-7, the sulfate ester group of the tyrosyl residue, the tryptophan and aspartic acid residues and the C-terminal phenylalanine residue (the removal of which confers antagonistic properties) are essential for biological activity. Structure–activity relationships in exocrine pancreas and gallbladder are essentially identical, whereas in the stomach, desulfated CCK peptides retain biological activity. Brain receptors show much less specificity than pancreatic receptors for CCK peptides.

An interesting property of *CCK receptors* is that they *bind* various *compounds* chemically *unrelated to CCK*, such as dibutyryl cGMP, proglumide and benzotript (amino acid derivatives), and asperlicin (a non-peptide compound of fungal origin), which display antagonistic properties[35].

Secretion and Metabolism

The *secretion* of CCK into the blood is *stimulated by* luminal *lipids* (monoglycerides and long fatty acids) *and certain amino acids* (tryptophan and phenylalanine)[32]. The pancreatic response to fat and amino acids is fully inhibited by the CCK antagonist proglumide, suggesting that CCK is the major mediator of the intestinal phase of pancreatic secretion[36].

The half-lives of CCK-4 and CCK-8 in the blood have been estimated as 13 and 50 min, respectively.

Pathophysiology

CCK concentrations are diminished in the blood and duodenal mucosa of patients with *celiac disease*, a finding consistent with the impaired pancreatic and gallbladder responses to luminal nutrients observed in these patients. Conversely, plasma CCK concentrations may be increased in patients with exocrine *pancreatic insufficiency*, possibly because of the loss of the inhibitory influence of luminal trypsin and chymotrypsin on CCK secretion.

CCK peptides have been used to *stimulate gallbladder contraction* during cholecystography and to increase intestinal transit time during radiographic examination of the small intestine. They have also been used, in combination with secretin, to assess pancreatic function.

5. Secretin

Structure and Distribution

Porcine secretin[5] is a linear polypeptide of 27 aa which belongs to the secretin–glucagon family of peptides (Fig. XII.22). Its amino acid sequence is characterized by four arginine residues and three amidated carboxyl groups, including the terminal carboxyl group; this accounts for its strongly basic character. In the *human* GI tract, secretin is found *predominantly* in the *duodenum* and *proximal jejunum*, and is associated exclusively with endocrine *S cells*. In the rat and pig, secretin has also been detected in the CNS, particularly in the pituitary and pineal gland.

34. Smith, G. P., Gibbs, J., Jerome, C., Pi-Sunyer, F. X., Kissileff, H. R. and Thornton, J. (1981) The satiety effect of cholecystokinin: a progress report. *Peptides* 2, suppl. 2: 57–59.
35. Gardner, J. D. and Jensen, R. T. (1984) Cholecystokinin receptor antagonists. *Amer. J. Physiol.* 246: G471–G476.
36. Stubbs, R. S. and Stabile, B. E. (1985) Role of cholecystokinin in pancreatic exocrine response to intraluminal amino acids and fat. *Amer. J. Physiol.* 248: G347–G352.

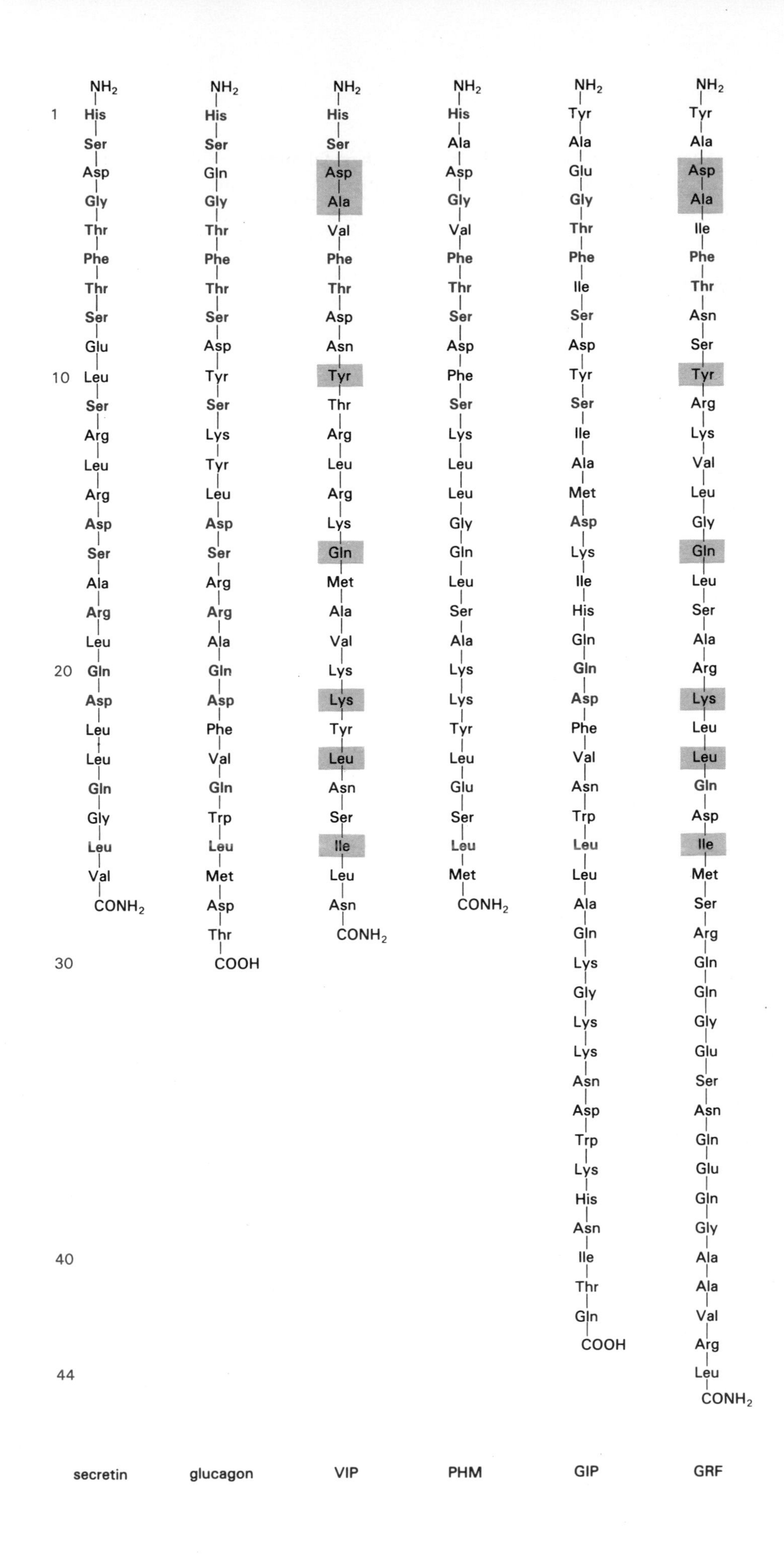
1
10
20
30
40
44
NH2 His Ser Asp Gly Thr Phe Thr Ser Glu Leu Ser Arg Leu Arg Asp Ser Ala Arg Leu Gln Asp Leu Leu Gln Gly Leu Val CONH2
NH2 His Ser Gln Gly Thr Phe Thr Ser Asp Tyr Ser Lys Tyr Leu Asp Ser Arg Arg Ala Gln Asp Phe Val Gln Trp Leu Met Asp Thr COOH
NH2 His Ser Asp Ala Val Phe Thr Asp Asn Tyr Thr Arg Leu Arg Lys Gln Met Ala Val Lys Lys Tyr Leu Asn Ser Ile Leu Asn CONH2
NH2 His Ala Asp Gly Val Phe Thr Ser Asp Phe Ser Lys Leu Leu Gly Gln Leu Ser Ala Lys Lys Tyr Leu Glu Ser Leu Met CONH2
NH2 Tyr Ala Glu Gly Thr Phe Ile Ser Asp Tyr Ser Ile Ala Met Asp Lys Ile His Gln Gln Asp Phe Val Asn Trp Leu Leu Ala Gln Lys Gly Lys Lys Asn Asp Trp Lys His Asn Ile Thr Gln COOH
NH2 Tyr Ala Asp Ala Ile Phe Thr Asn Ser Tyr Arg Lys Val Leu Gly Gln Leu Ser Ala Arg Lys Leu Leu Gln Asp Ile Met Ser Arg Gln Gln Gly Glu Ser Asn Gln Glu Gln Gly Ala Ala Val Arg Leu CONH2
secretin
glucagon
VIP
PHM
GIP
GRF

◁ **Figure XII.22** Amino acid sequences of peptides of the glucagon–secretin family. All structures represented are from the human, with the exception of secretin, which is from the pig. The structure of glucagon and VIP are identical in the human and pig, and PHM, GIP and GRF peptides differ only by two amino acids in the two species. Secretin–glucagon homologies are indicated in red, and identical VIP–GRF amino acid sequences are shaded.

Biological Actions

Pancreatic Exocrine Secretion

The major physiological effect of secretin is to stimulate the *secretion of water and bicarbonate* ions *by* the *duct cells* of the exocrine pancreas. With increasing secretory flow rates, the concentration of bicarbonate ions in the pancreatic juice increases hyperbolically, to attain a maximum of ~145 mEq/l, and, concomitantly, the concentration of Cl^- ion decreases (Fig. XII.23). In humans, the maximal output of bicarbonates is ~30 mEq/h, and the dose of secretin required for a half-maximal response is ~10–30 pmol/kg (bolus injection) or 2 to 8 pmol/kg/h (constant infusion)[37]. Under these conditions, plasma secretin concentrations attain physiological values (2–8 pM). In addition to its effects on duct cells, secretin weakly stimulates enzyme secretion and potentiates CCK-stimulated secretion in acinar cells.

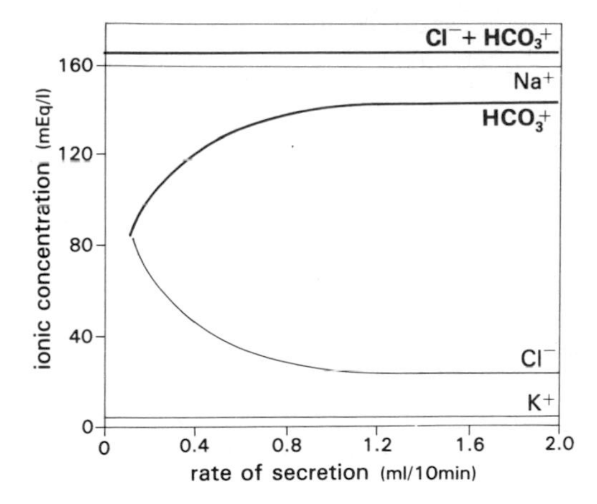

Figure XII.23 The electrolyte composition of pancreatic juice at different rates of secretion in the cat. (Jorpes, J. E. and Mutt, V. (1973) Secretin and cholecystokinin. In: *Secretin, Cholecystokinin, Pancreazymin and Gastrin* (Jorpes, J. E. and Mutt, V., eds), Springer Verlag, Berlin, pp. 1–179).

Other Effects

Secretin *inhibits* gastric *acid secretion*, *stimulates pepsin* secretion, and increases the biliary output of water, CO_3H^- ions and Cl^- ions (Table XII.4). It *inhibits GI motility* and potentiates the cholecystokinetic effect of CCK (Table XII.5). Secretion weakly stimulates pancreatic growth, and enhances the tropic effect of CCK on the pancreas (Table XII.6). It *inhibits gastrin release* (Table XII.7), causes redistribution of blood flow into the splanchnic circulation, and *stimulates diuresis* and *lipolysis*.

Structure–Activity Relationships and Receptors

Secretin *receptors*, functionally coupled to a secretin-sensitive adenylate cyclase, have been identified in *exocrine pancreas*, *gastric* epithelial *glands*, and *gallbladder* epithelial cells. Secretin is a more potent activator of adenylate cyclase in duct cells than in acinar cells.

Most of the *information* required for secretin activity is contained in the *N-terminal region*. However, while deletion of only 5 aa in this region abolishes biological activity, deletion of up to 13 aa partially preserves *receptor* binding ability, which accounts for the antagonistic properties of the resulting C-terminal fragments.

The *structural similarities* between *secretin* and *VIP* account for the ability of *secretin* to *bind* to *VIP receptors* and to stimulate *cAMP production* in a variety of tissues. However, the affinity of secretin for the VIP receptor is only 0.1% that of VIP, and more importantly, the chemical specificity of the secretin receptor differs from that of the VIP receptor.

Secretion and Metabolism

Acidification of the duodenal content is the major stimulant of secretin release into the blood, nutrients being ineffective[38]. Although an increase in plasma secretin does occur after a meal, this response requires a decrease in the duodenal pH to below 4, and is eliminated by the infusion of bicarbonates (Fig. XII.24). In dogs, the pancreatic bicarbonate

37. Schaffalitzky de Muckadell, O. B., Fahrenkrug, J., Watt-Boolsen, S. and Worming, G. H. (1978) Pancreatic response and plasma secretin concentration during infusion of low dose secretin in man. *Scand. J. Gastroenterol.* 15: 305–311.
38. Chey, W. Y., Lee, Y. M., Hendricks, J. G., Rhodes, R. A. and Tai, H. H. (1978) Plasma secretin in fasting and postprandial state in man. *Digest. Dis. Sci.* 23: 981–988.

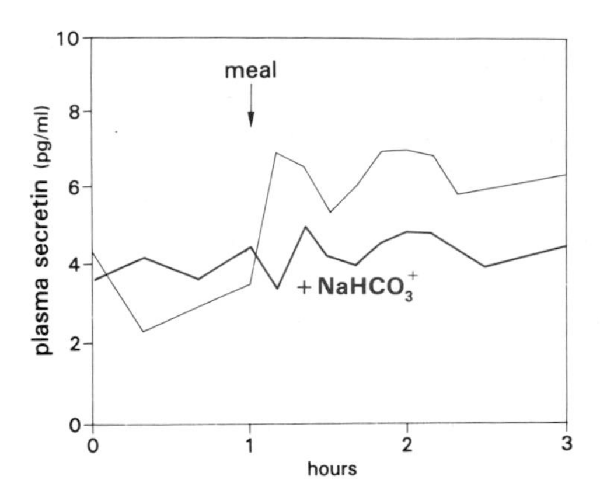

Figure XII.24 Normal plasma secretin response to a test meal in human subjects, and the effect of maintaining pH above pH 5.5 by intragastric infusion of bicarbonate[38].

response to a meal is suppressed by the administration of antibodies to secretin, indicating that the endogenous hormone plays a major physiological role.

The half-life of circulating secretin is 4 min. Kidneys are the major site of the degradation of the peptide.

Pathophysiology

Plasma concentrations of secretin are decreased in patients with *celiac disease* (a finding consistent with the impaired pancreatic secretory response to food seen in these patients), and fail to increase after a meal in *achlorhydric patients.* Conversely, plasma secretin concentrations are increased in patients with gastrinomas and other conditions with gastric acid hypersecretion, particularly duodenal ulcers.

Secretin, alone, or in combination with CCK, is widely used as a *stimulant* of *pancreatic secretion* in the assessment of exocrine pancreatic function. A decreased secretory response to secretin is found mainly in inflammatory and neoplastic diseases of the pancreas. Secretin has also been used as an effector of *gastrin secretion* in the evaluation of hypergastrinemia (p. 562); it causes a paradoxical increase in serum gastrin in patients with gastrinomas, but either suppresses or does not change serum gastrin in other hypergastrinemic states.

6. Vasoactive Intestinal Polypeptide (VIP), Peptide Histidine Isoleucine (PHI), and Peptide Histidine Methionine (PHM)

Identified by its potent vasodilator activity in peptide fractions from porcine duodenum, VIP was isolated in 1970[39], and was shown to exhibit a broad range of biological activities. Initially considered to be a gut hormone, it was shown to be widely distributed throughout the central and peripheral nervous system and to be *associated exclusively with neurons.*

Structure and Biosynthesis of VIP

Human VIP (structurally similar to porcine VIP) is a strongly basic polypeptide, with a C-terminal amide structure, which belongs to the *glucagon–secretin family* of peptides (Fig. XII.22).

The sequence of human prepro-VIP, deduced from cDNA studies, consists of 170 aa and includes the sequence of VIP at positions 125–152 (Fig. XII.25). An additional feature is the presence, at positions 81–107, of a 27 aa peptide, PMH-27, structurally and functionally related to VIP (p. 527)[40].

Distribution

VIP has been identified at *multiple sites* throughout the body, especially in the GI, respiratory, and urogenital tracts, as well as in the circulatory system and CNS. In most peripheral organs and tissues, VIP is associated with nerves supplying smooth muscle layers, exocrine glands, and blood vessels. In the GI tract, which accounts for the major part of VIP in the body, VIP is widely distributed, from the esophagus to the rectum, and is associated exclusively with neurons in non-epithelial layers. VIP neurons, which represent the largest population of enteric neurons, are especially abundant in the sphincters, the gallbladder wall, and the pancreas. In the CNS, VIP resembles CCK in that it is found predominantly in telencephalic areas.

39. Said, S. I. and Mutt, V. (1970) Polypeptide with broad biological activity: isolation from small intestine. *Science* 169: 1217–1218.
40. Itoh, N., Obata, K. I., Yanaihara, N. and Okamoto, H. (1983) Human preprovasoactive intestinal polypeptide contains a novel PHI-27-like peptide, PHM-27. *Nature* 304: 547–549.

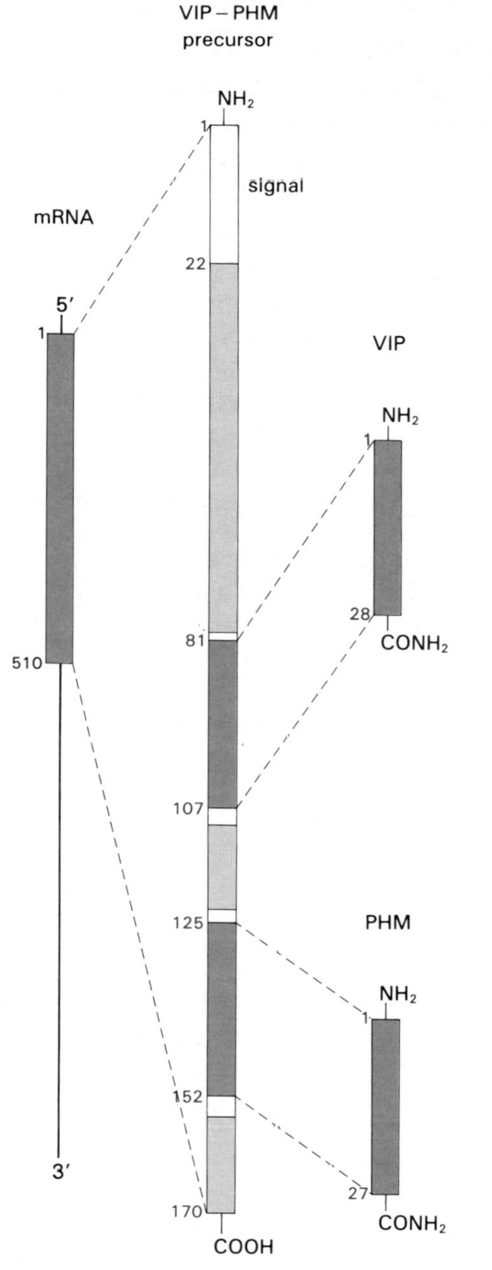

Figure XII.25 Structure of the mRNA and precursor of human VIP and PHM[40].

Biological Actions

The main effects of VIP include the *relaxation* of various types of smooth *muscle* and the *stimulation* of *exocrine* and *endocrine secretion* and *metabolism*; it also has effects on the *CNS*. VIP causes potent *vasodilatation* of peripheral, splanchnic, pulmonary, coronary, cerebral and urogenital vessels. It also exerts a moderate inotropic effect on heart. VIP augments ventilation, relaxes tracheobronchial and pulmonary vascular smooth muscle, and inhibits histamine release by lung mast cells. A major effect of VIP is to *inhibit absorption of water* and sodium in the small and large *intestine*[41]. This is of special interest in relation to the postulated role of VIP in the human watery diarrhea syndrome. Additional secretory effects of VIP in the GI tract are: to inhibit histamine and gastrin-stimulated gastric acid secretion, to stimulate pepsin secretion and pancreatic secretion of water, bicarbonates and enzymes, and to increase bile flow. VIP relaxes intestinal, urinary and uterine muscle, stimulates liver glycogenolysis and lipolysis, stimulates bone resorption, and inhibits platelet aggregation. It has widespread effects on many endocrine secretions (particularly insulin secretion). *In the CNS*, VIP causes neuronal *excitation*, stimulates glycogenolysis, and increases the activity of enzymes involved in the synthesis of serotonin and acetylcholine.

Structure–Activity Relationships and Receptors

The entire sequence of VIP, especially the N-terminal histidine residue, is required for full activity. VIP *receptors*, functionally coupled to a VIP-sensitive *adenylate cyclase*, have been identified in a wide range of tissues: intestine and gallbladder epithelia, pancreatic acini, liver, lung, myometrium, and brain. VIP receptors bind secretin, PHI/PHM and GRF with low affinity.

Pathophysiology

VIP-secreting *tumors* are the major cause of VIP overproduction in humans. Many are found in the pancreas, but some tumors occur elsewhere (e.g., neuroganglioblastomas). The main features of these tumors include *watery diarrhea* due to increased intestinal secretion, decreased gastric acid secretion, hypokalemia, hyperglycemia, hypercalcemia, dilatation of the gallbladder, and episodes of flushing (Verner–Morrison syndrome). Co-secretion of glucagon, somatostatin and pancreatic polypeptide may be observed. Enhanced plasma VIP concentrations have also been observed in terminal hepatic failure.

41. Krejs, G. J., Fordtran, J. S., Bloom, S. R., Fahrenkrug, J., Schaffalitzky de Muckadell, O., Fisher, J. E., Humphrey, C. J., O'Dorisio, T. M., Said, S. I., Walsh, J. H. and Shulkes, A. A. (1980) Effect of VIP infusion on water and ion transport in the human jejunum. *Gastroenterology* 78: 722–727.

Human diseases leading to defective VIP innervation in the gut have been described (p. 558).

Peptide Histidine Isoleucine and Peptide Histidine Methionine

PHI is a 27 aa peptide, structurally *related to VIP*, which has been isolated from porcine and bovine intestine and from the brain[42]. Its human counterpart, PHM, a peptide with N-terminal histidine and C-terminal methionine amide, was first identified as part of the sequence of the VIP precursor, and was later isolated from the colon[43] (Figs. XII.22 and XII.25).

The *distribution* of PHI/PHM closely *parallels* that of *VIP*. The concentration in the GI tract is approximately the same as that of VIP, except in the stomach, where it is about two to four times lower. PHI/PHM is associated exclusively with neurons, in which it coexists with VIP.

The biological *actions* of PHI *resemble* those of *VIP*, and include: 1. stimulation of intestinal secretion of water and electrolytes; 2. stimulation of pancreatic secretion of water and bicarbonate in vivo, and of amylase in isolated acini; 3. relaxation of tracheal, gastric and gallbladder smooth muscle and vasodilatation; 4. stimulation of insulin, glucagon and prolactin secretion. PHI, like VIP, stimulates adenylate cyclase activity, probably via the VIP receptors, which bind PHI with ~1/10 to 1/20 the affinity of native VIP. So far, no evidence for a physiological role of PHI/PHM has been established.

7. Gastric Inhibitory Peptide (GIP)

Isolated in 1970[44] from porcine intestine, GIP is a GI hormone whose major effects are to *inhibit* gastric *acid secretion* and to *enhance insulin* secretion in the presence of elevated glucose concentrations.

42. Tatemoto, K. and Mutt, V. (1981) Isolation of the intestinal peptide porcine PHI, a new member of the glucagon-secretin family. *Proc. Natl. Acad. Sci., USA* 78: 6603–6607.
43. Tatemoto, J., Jörnvall, H., McDonald, T. J., Carlquist, M., Go, V. L. W., Johansson, C. and Mutt, V. (1984) Isolation and primary structure of human PHI (peptide HI). *FEBS Lett.* 174: 258–261.
44. Brown, J. C., Mutt, V. and Pederson, R. A. (1970) Further purification of a polypeptide demonstrating enterogastrone activity. *J. Physiol.* 209: 57–64.
45. Andersen, D. K., Elahi, D., Brown, J. C., Tobin, J. D. and Andres, R. (1978) Oral glucose augmentation of insulin secretion. Interactions of gastric inhibitory polypeptide with ambiant glucose and insulin levels. *J. Clin. Invest.* 62: 152–161.

Structure and Distribution

Porcine GIP is a 42 aa peptide which belongs to the *glucagon–secretin family* (Fig. XII.22). In the human GI tract, GIP is present at highest concentration in the *duodenum* and *jejunum*, and is associated with endocrine K cells.

Biological Actions

One major effect of GIP is to *enhance insulin secretion under hyperglycemic* conditions[45]. Thus, when blood glucose is maintained above 1.25 g/l by a glucose clamp technique, the increase in plasma GIP concentration induced by oral glucose is accompanied by a simultaneous increase in plasma insulin concentration (Fig. XII.26). A similar enhancement of insulin release under hyperglycemic conditions occurs when GIP release is stimulated by the ingestion of fat. GIP may account for the enhanced insulin response induced by oral glucose as compared to intravenous glucose.

The *other GI actions* of GIP are: inhibition of gastric acid secretion, stimulation of intestinal secretion of water and electrolytes, and inhibition of gastric motor activity. It has been suggested that GIP mediates the inhibition of gastric acid secretion induced by fat ingestion.

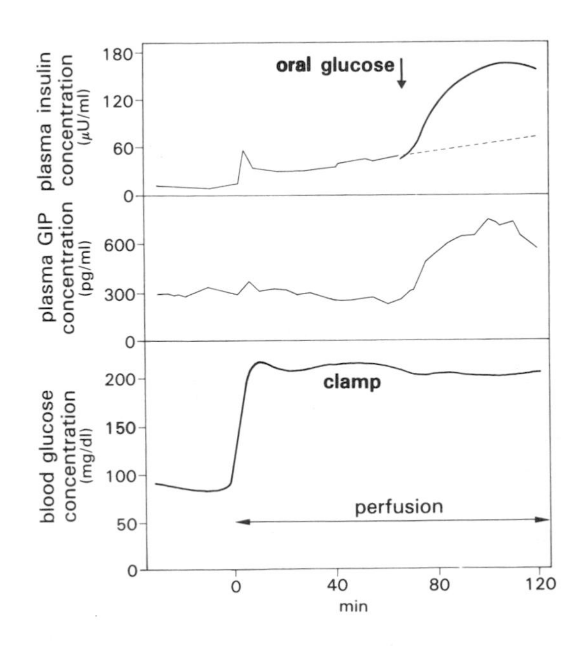

Figure XII.26 Simultaneous enhancement of plasma insulin and GIP concentrations caused by administration of oral glucose (40 g/m^2) in human subjects. Blood glucose is maintained at a constant hyperglycemic concentration[45].

Structure–Activity Relationships

Most of the N-terminal sequence is necessary for insulintropic activity, while most of the C-terminal sequence is required for the inhibition of gastric acid secretion. GIP receptors, functionally coupled to adenylate cyclase, have been identified in β cell tumors.

Secretion

GIP secretion is stimulated by the main classes of nutrients (Table XII.9), and is modulated by several hormones (Table XII.7).

Pathophysiology

Increased GIP levels in plasma and/or an exaggerated GIP response to glucose have been observed after vagotomy and duodenopancreatectomy, and in patients with diabetes mellitus, obesity, chronic pancreatitis, postprandial hypoglycemia, duodenal ulcer, and chronic renal failure. Decreased GIP levels have been observed in patients with celiac disease and with a jejuno-ileal bypass.

8. Enteroglucagon

Until recently, enteroglucagon was defined as *a group* of peptides; it is present in GI tissues and is immunochemically related to glucagon but heterogeneous as to molecular size, charge, and nature of the glucagon-like immunodeterminant. It is now recognized that these peptides are *extended forms of glucagon*, and that they are *derived* from the *same precursor* protein.

Isolation, Structure, and Biosynthesis

The two major forms of enteroglucagon in the pig are *glucagon-37, or oxyntomodulin*[46], and *glucagon-69, or glicentin*[47]; in both forms, the 29 aa sequence of *glucagon is extended* by a *basic octapeptide* at the C-terminus, and in *glicentin*, the sequence possesses an *additional* 32 aa *N-terminal extension* (Fig. XII.27). Apart from glucagon-29, additional molecular forms

46. Bataille, D. K., Tatemoto, C., Gespach, H., Jornvall, G., Rosselin, G. and Mutt, V. (1982) Isolation of glucagon-37 (bioactive enteroglucagon/oxyntomodulin) from porcine jejunoileum. Characterization of the peptide. *FEBS Lett.* 146: 79–86.
47. Thim, L. and Moody, A. J. (1981) The primary structure of porcine glicentin (proglucagon). *Regul. Peptides* 2: 139–151.

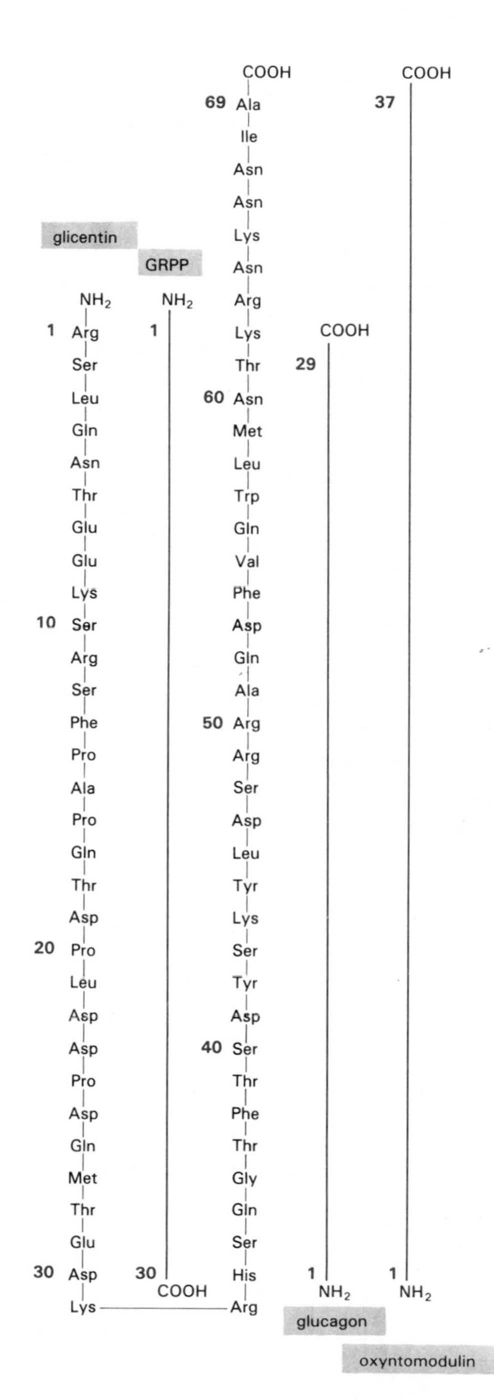

Figure XII.27 Amino acid sequences of porcine glicentin, glicentin-related pancreatic polypeptide (GRPP), glucagon, and oxyntomodulin. Continuous lines indicate a sequence identical to that of glicentin.

include two components which presumably correspond to the N-terminal 1–30 and C-terminal 13–69 sequences of glicentin. Glicentin 1–30, also called glicentin-related pancreatic polypeptide (GRPP), has been isolated from the pancreas, and is shown to be secreted synchronously and in equimolar amounts with glucagon.

The structure of the precursor protein encoding glucagon and glucagon-containing peptides has recently been determined by molecular cloning techniques. The human precursor, which consists of 180 aa, includes several functional regions: a signal peptide (sequence 1–20); GRPP (sequence 21–50); glucagon (sequence 53–81); the C-terminal basic octapeptide of enteroglucagon (sequence 82–89); and two glucagon-like peptides (sequences 92–128 and 146–180). Pairs of basic amino acid residues flank the sequences of glucagon and each of the glucagon-like peptides. The amino acid pair located C-terminally to the glucagon sequence belongs to the basic octapeptide. Thus, the structural differences between the pancreatic and intestinal peptides probably only reflect differences in post-translational processing, with pancreatic processing favoring cleavage of glicentin into GRPP and glucagon, and an intestinal processing favoring either the release of intact glicentin or the cleavage of glicentin into GRPP and oxyntomodulin.

Distribution

The *highest* concentrations of extractible enteroglucagon are found in the *ileum and colon*. Glicentin (G-69) accounts for about 60 to 80% of this material, and the rest is mostly oxyntomodulin (G-37)[48]. These extended forms of glucagon are associated with specific endocrine cells, the L cells, which differ morphologically from pancreatic A cells. It is only in dog fundic mucosa that glucagon-29, associated with authentic A cells, has been identified.

Biological Actions

Both oxyntomodulin and glicentin are *potent inhibitors* of *acid secretion* by gastric parietal cells. Additional effects of oxyntomodulin are to stimulate liver glycogenolysis and insulin release from the pancreas, via cAMP, albeit with lesser potency than pancreatic glucagon. The marked hypertrophy of intestinal mucosa observed in certain pathological conditions in which plasma enteroglucagon is elevated suggests that another important effect of enteroglucagon may be to stimulate intestinal growth.

In addition to its known glycogenolytic effect in liver, pancreatic glucagon has a wide range of effects in the GI tract, including the inhibition of gastric and pancreatic secretions, stimulation of intestinal secretion and bile flow, and inhibition of GI motility.

Secretion

Unlike pancreatic glucagon, most stimuli for the secretion of enteroglucagon are *luminal*; these include glucose, triglycerides, or the ingestion of a mixed meal. Secretion is also affected by bombesin and somatostatin, which are stimulatory and inhibitory, respectively.

Pathophysiology

Increased enteroglucagon levels in plasma and/or an exaggerated response of this peptide to ingested nutrients are seen in conditions leading to rapid gastric emptying or increased delivery of unabsorbed nutrients to the distal intestine, as is observed in conjunction with jejuno-ileal bypass, tropical sprue, and celiac disease. An increase in enteroglucagon concentration has also been reported in certain renal tumors.

9. Somatostatin

Initially isolated from the *hypothalamus* as a peptide inhibiting the secretion of growth hormone by the pituitary (p. 204), somatostatin has also been *identified subsequently in* the *GI* tract and in the *pancreas,* and has been shown to affect a variety of GI functions.

Isolation, Structure, and Biosynthesis

In mammals, GI and pancreatic somatostatin, which have the same amino acid sequence as their hypothalamic counterparts, are found in *two major* molecular *forms, somatostatin-14 and -28*. In the human, both forms are synthesized from a 116 aa precursor (Figs. III.16 and III.17).

48. Munck, A., Kervran, A., Marie, J. C., Bataille, D. and Rosselin, G. (1984) Glucagon-37 (oxyntomodulin) and glucagon-29 (pancreatic glucagon) in human bowel: analysis by HPLC and radioreceptorassay. *Peptides* 5: 553–561.

Distribution

In mammals, the *digestive system* accounts for at least *half* of the somatostatin present in the body, of which ~15% is found in the pancreas. Somatostatin in the human GI tract reaches highest concentrations in the stomach and duodenum[48], and is associated predominantly with the mucosa and submucosa (Fig. XII.1). Somatostatin-14 is the predominant molecular form found in the pancreas, gastric antrum and duodenum, but, in the gastric fundus, jejunum and colon, somatostatin-28 may represent from 50 to 70% of all somatostatin, with forms of larger molecular size also present in the jejunum. Somatostatin-14 is the predominant form of the hormone found in GI muscle.

Somatostatin is associated both with endocrine *D cells*, found in the *pancreas* and in the fundic, antral and duodenal *mucosa*, and with *neurons*, present in the *entire GI tract*. Fundic D cells of the closed type, and antral D cells of the open type, give off long cytoplasmic extensions that reach adjacent cells, and may be involved in the local delivery of somatostatin to these cells.

Biological Actions

Generalized inhibitory activity

The effects of somatostatin on GI functions are *exclusively inhibitory*, and are *opposite* to those of other *GI peptides. Somatostatin* affects *all* major *functions*, including gastric acid secretion, pancreatic exocrine secretion, intestinal absorption of sugars and amino acids, intestinal secretion of electrolytes, gastric and intestinal motility, GI and pancreatic growth, and hormone secretion. *Effects* of somatostatin on exocrine secretions and motility involve both a *direct* interaction of the peptide *with effector cells* and an *inhibition* of the release of *stimulatory hormones*.

Receptors and Mechanisms of Action

Specific somatostatin binding sites have been identified in endocrine and exocrine pancreas, gastric and intestinal mucosa, pituitary and brain. Effects of somatostatin on gastric glands, exocrine pancreas and pituitary are probably mediated by a decrease in *cAMP production*. In addition, it has been suggested that a decrease in cytosoluble Ca^{2+} and an increase in protein phosphatase activity may be involved in the action of somatostatin in certain target tissues.

Structure–Activity Relationships

Numerous analogs of somatostatin-14, characterized by amino acid deletions or substitutions, replacement of L-amino acids by their D-isomers, and creation of a second disulfide bridge, have been synthesized and characterized. Depending on the analog in question, biological activities may be globally reduced or enhanced, or even selectively affected[50].

Secretion

Somatostatin released from GI and pancreatic D cells is delivered to its target cells via both blood circulation and local diffusion. In humans, all major classes of *nutrients* stimulate somatostatin release. In contrast to *intraluminal* nutrients which affect the release of both *GI and pancreatic* somatostatin, circulating nutrients stimulate *only pancreatic* somatostatin release. Thus, while both types of somatostatin may regulate biological functions in the postprandial state, pancreatic somatostatin may be the only hormone that plays a regulatory role in postabsorption. Direct effects of glucose and amino acids on isolated pancreatic and fundic D cells have been demonstrated.

A number of GI peptides stimulate somatostatin release (Table XII.7). Inhibitors of secretion include endorphins and somatostatin itself.

The neural regulation of somatostatin release involves both a cholinergic inhibitory pathway and a non-cholinergic stimulatory pathway; the latter is mediated by bombesin/GRP. Additional neurotransmitters involved are listed on Table XII.10.

Pathophysiology

Somatostatin-producing *tumors* generally originate from the pancreas and give rise to predictable symptoms, which include mild diabetes, malabsorption, steatorrhea, cholelithiasis, weight loss, and reduced gastric acid secretion. Calcitonin has been identified as a secretory product of pancreatic somatostatinomas (and conversely, excess somatostatin may occur in medullary carcinomas of the thyroid).

49. Penman, E., Wass, J. A. H., Butler, M. G., Penny, E. S., Price, J., Wu, P. and Rees, L. H. (1983) Distribution and characterization of immunoreactive somatostatin in human gastrointestinal tract. *Regul. Peptides* 7: 53–65.

50. Rosenthal, L. E., Yamashiro, D. J., Rivier, J., Vale, W., Brown, M. and Dharmsathaphorn, K. (1983) Structure-activity relationships of somatostatin analogs in the rabbit ileum and rat colon. *J. Clin. Invest.* 71: 840–849.

A decrease in antral somatostatin content relative to gastrin has been observed in patients with duodenal ulcers, and it has been suggested that this may contribute to increased gastrin release and excessive acid secretion.

Somatostatin and somatostatin analogs have been used *therapeutically*. In patients with *type 1 diabetes* (insulin-dependent), somatostatin lowers blood glucose and markedly ameliorates hyperglycemia and ketosis after insulin withdrawal, possibly by reducing the hyperglucagonemia seen in these patients. In contrast, the *metabolic status* of *type 2* diabetics (non-insulin-dependent) is *worsened by* treatment with *somatostatin*, largely because of the loss of residual insulin secretion. Somatostatin has been used to treat a variety of hyperfunctioning endocrine tumors, severe bleeding from peptic ulcers, and hemorrhagic pancreatitis.

10. Motilin

Isolated from porcine intestine in 1970, motilin is a peptide which regulates the *interdigestive motility* of the stomach and small intestine. It has been identified as *both* a *neuroregulator* and a *hormone*.

Structure and Distribution

Motilin is a linear polypeptide of 22 aa (Fig. XII.28)[51] found in the GI tract and in the brain. In human and porcine duodenum and jejunum, it is associated with a population of non-argentaffin cells, or Mo cells, and also with a population of argentaffin cells distinct from those which contain substance P.

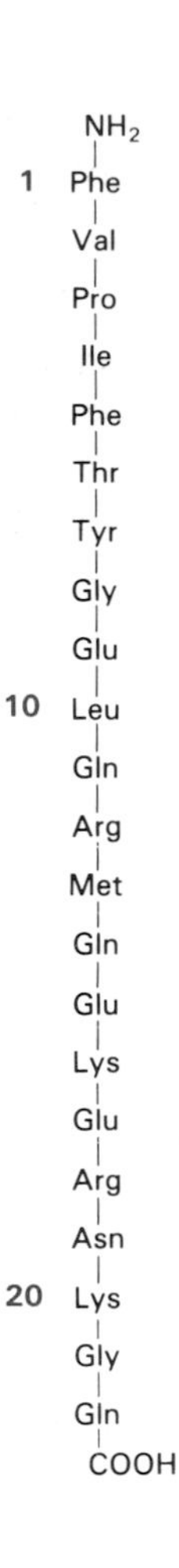

Figure XII.28 Amino acid sequence of porcine motilin.

Biological Actions

The major effect of motilin is to *initiate myoelectric discharges* in the *antroduodenal* region that propagate distally in the small intestine[52] (Fig. XII.29). Such myoelectric activity is identical to that which occurs spontaneously at 80 to 90 min intervals in the fasting state, and is presumably triggered by endo-

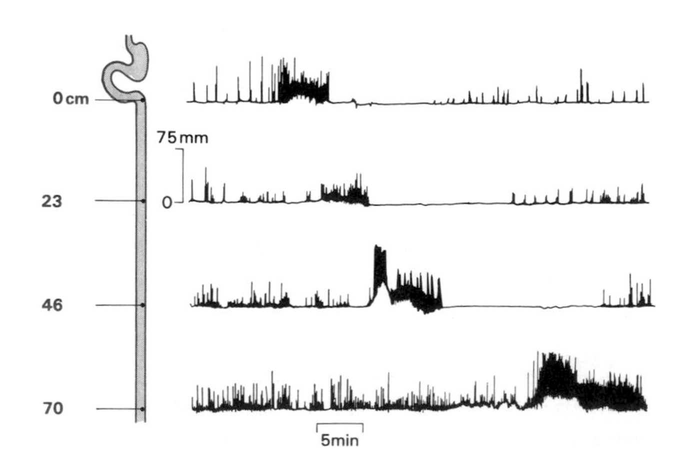

Figure XII.29 Effect of 13-norleucine (0.4 μg/kg/h) on the migrating interdigestive small intestinal motor complex (IMC) in the human. The IMC was recorded by continuously perfused (0.1 ml/min), open-tipped catheters placed at regular intervals, from proximal to distal, in the jejunum. The vertical scale on the left indicates the amplitude of pressure waves in mmHg[52].

51. Brown, J. C., Cook, M. A. and Dryburgh, J. R. (1973) Motilin, a gastric motor activity stimulating polypeptide: the complete amino acid sequence. *Can. J. Biochem.* 51: 533–537.
52. Vantrappen, C., Janssens, J., Peeters, T. L., Bloom, S. R., Christofides, N. D. and Hellemans, J. (1979) Motilin and the interdigestive migrating motor complex in man. *Digest Dis. Sci.* 24: 497–500.

genous motilin. Motilin also stimulates gastric pepsin secretion, but has little effect on other GI secretions.

Secretion and Metabolism

The concentration of motilin in plasma increases cyclically in the fasting state; this cyclical activity corresponds closely to and often coincides with the onset of the interdigestive myoelectric discharge in the duodenum. The release of motilin into the blood is also stimulated by luminal acid and pancreatico-biliary secretions, and by circulatory bombesin. Both cholinergic and vagal non-cholinergic mechanisms modulating plasma motilin concentration have been reported.

The half-life of circulating motilin is ~4.5 min.

Pathophysiology

A decrease in plasma motilin concentration has been observed in patients with inflammatory disorders of the GI tract, chronic pancreatitis, and diabetes.

11. Pancreatic Polypeptide (PP), Peptide YY, and Neuropeptide Y

Originally isolated from mammalian and avian[53] pancreas, pancreatic polypeptide (PP) is a hormone released through a *vagal cholinergic* mechanism which *affects* several *GI functions*. It is structurally related to two newly discovered peptides, peptide YY (PYY) and neuropeptide Y (NPY).

Isolation, Structure, and Biosynthesis

PP is a 36 aa peptide (Fig. XII.30). The sequence in mammals is more conserved than in birds and reptiles. PP has at least 17 aa in common with PYY and NPY.

The human prepro-PP has a M_r of ~11,000 and includes the sequence of PP at position 30–65. Another peptide of 20 aa occurs at position 69–88 in the precursor[54] (Fig. XII.31).

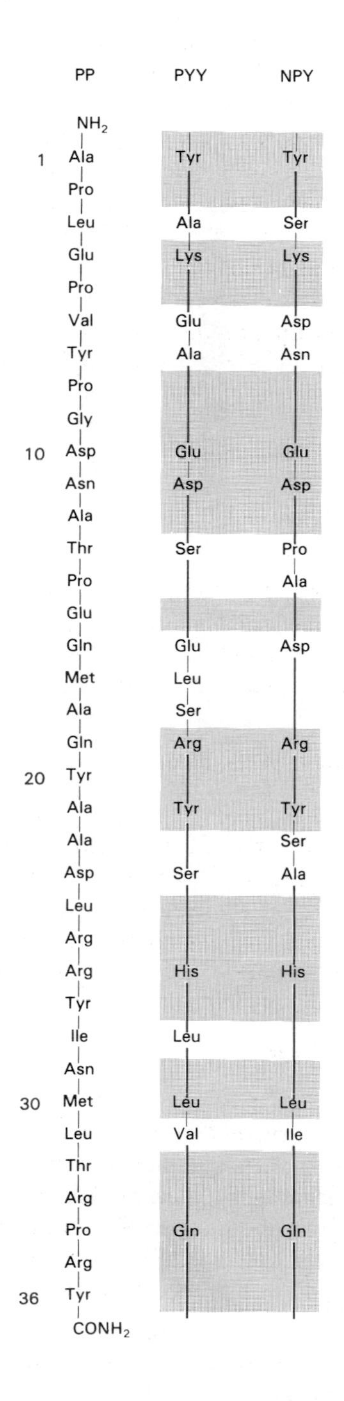

Figure XII.30 Amino acid sequences of human pancreatic polypeptide (PP), porcine peptide YY (PYY), and human neuropeptide Y (NPY). Identical aa between PPY and NPY are highlighted.

53. Kimmel, J. R., Hayden, L. J. and Pollock, H. G. (1975) Isolation and characterization of a new pancreatic polypeptide hormone. *J. Biol. Chem.* 24: 9369–9376.
54. Leiter, A. B., Keutmann, H. T., Goodman, R. H. (1984) Structure of a precursor to human pancreatic polypeptide. *J. Biol. Chem.* 259: 14702–14705.

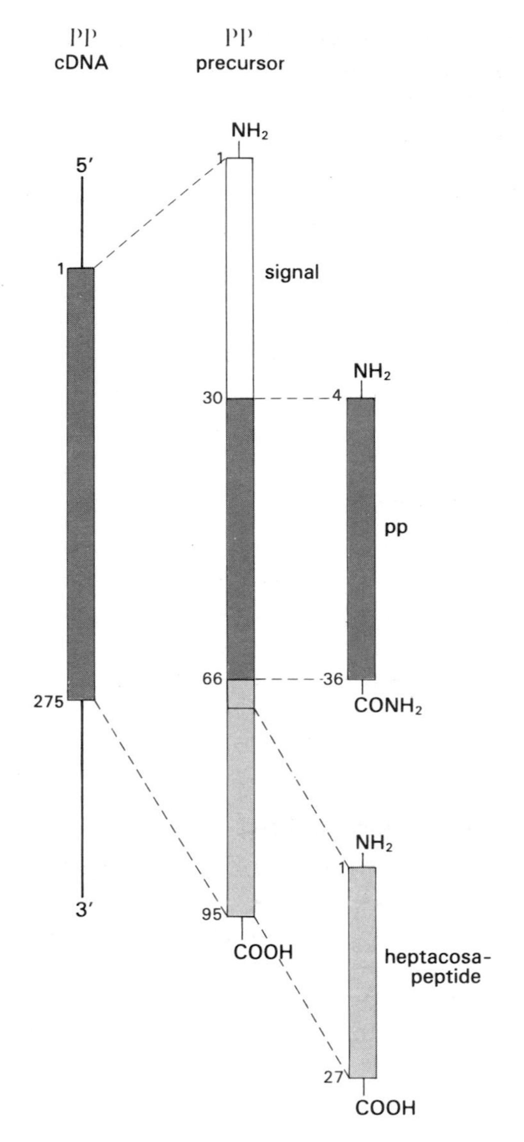

Figure XII.31 Structure of the human pancreatic polypeptide (PP) cDNA and precursor[54].

Distribution

PP is *produced by* specific *endocrine cells* in the *pancreas*, occuring in the islets as well as in the endocrine parenchyma, and is especially abundant near the duodenum (Table XII.3). PP-like *immunoreactivity* in the *intestine* and *brain* is mostly *attributable to PYY* and/*or NPY*.

Biological Actions

In mammals, PP is mostly an *inhibitor* of basal and stimulated *pancreatic exocrine* secretion and of *bile* secretion. PP also weakly stimulates gastric acid secretion, GI motility, and pancreatic growth. In birds, PP activity differs in that it stimulates acid and pepsin secretion in addition to causing changes in carbohydrate and lipid metabolism.

Secretion and Metabolism

The release of PP into the blood is stimulated by *luminal nutrients*, especially proteins and amino acids (Table XII.9), by GI peptides (Table XII.7) and, most importantly, by *vagal cholinergic* mechanisms; the latter appear to mediate the action of all other stimulants. In normal subjects, plasma PP concentrations show diurnal variations, and *increase with age*, almost quadrupling from the third to the seventh decade of life. The response of PP to food, typically biphasic, includes an initial or cephalic phase, with a maximum at 10 to 30 min, and a delayed, GI phase, which lasts for 6 to 8 h.

The half-life of circulating PP is 7 min.

Pathophysiology

Increased plasma concentrations of PP are often found in patients with endocrine pancreatic islet tumors. A reduced response of PP to food has been observed in chronic pancreatitis, cystic fibrosis, and diabetic autonomic neuropathy.

Peptide YY and Neuropeptide Y

PYY and NPY are 36 aa peptides which have been isolated using an assay based on the presence of an amidated C-terminal amino acid residue. *PYY* is found *only in* the *intestinal mucosa*, in which it is associated with endocrine cells, whereas *NPY* is found *both* in the *GI* tract and in the central and peripheral *nervous system*. Peripherally, NPY coexists with norepinephrine, in a distinct subpopulation of noradrenergic neurons, and with CCK and somatostatin, in enteric neurons. PYY inhibits gastric and pancreatic exocrine secretions, inhibits GI motility, and causes potent intestinal vasoconstriction. NPY, acting on various types of smooth muscle, reduces local blood flow and enhances adrenergic vasoconstriction.

The concentration of PYY in plasma increases markedly after a meal and after luminal introduction of fat and protein. PYY may mediate the inhibitory effect of oleic acid on gastric and pancreatic secretion, and may thus correspond to the so-called "enterogastrone", "pancreatone", and "anti-CCK hormone".

12. Neurotensin

Initially observed in the *brain*[55], neurotensin, which produces *vasodilatation* and *hypotension*, was *later isolated* from the *intestine*, where it affects several GI functions.

Structure and Distribution

Neurotensin is a tridecapeptide which is structurally related to xenopsin, an amphibian peptide (Fig. XII.32). In the GI tract, neurotensin predominates in the epithelium of distal jejunum and ileum, where it is synthesized in endocrine N cells (Table XII.3). In the CNS, neurotensin neurons are most abundant in the hypothalamus, substantia nigra, and brain stem.

Biological Actions and Secretion

In the GI tract, neurotensin *inhibits* basal and stimulated *acid* and *pepsin secretion*, *pancreatic exocrine secretion*, and *GI motility*. It causes intestinal and cutaneous *vasodilatation*, decreases blood pressure, and exerts inotropic and chronotropic effects on the heart. It stimulates glycogenolysis and causes hyperglycemia. Neurotensin causes the release of histamine from mast cells, and is probably involved in the inflammatory response. When administered in the CNS, neurotensin elicits a variety of effects (Table XII.8); these include inhibition of gastric acid secretion, inhibition of gastric mucosal blood flow, and prevention of stress-induced gastric ulcers.

Specific neurotensin *receptors* have been identified in *GI muscle*, colonic *carcinoma cells*, and in the *brain*. Determinants for biological activity are located C-terminally.

Neurotensin release is *stimulated by* intraluminal *lipids* but not by other nutrients. Its release following a meal is biphasic, with an early (20 min) and a late (90–120 min) increase, probably mediated by a cholinergic mechanism.

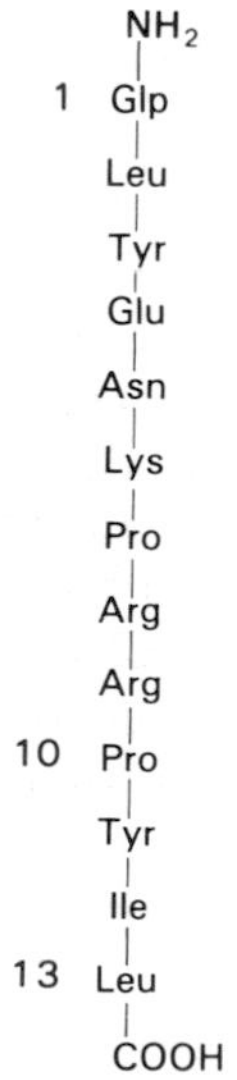

NH_2
1 Glp
Leu
Tyr
Glu
Asn
Lys
Pro
Arg
Arg
10 Pro
Tyr
Ile
13 Leu
COOH

Figure XII.32 Amino acid sequences of neurotensin (human and bovine).

Pathophysiology

Excess neurotensin production has been observed in pancreatic and gut tumors as well as bronchial carcinomas. Non-tumoral conditions leading to elevated plasma concentrations are the same as those which cause elevation of plasma enteroglucagon (p. 574).

13. Bombesin and Gastrin Releasing Peptide

Bombesin, isolated from the skin of the frog *Bombina bombina*[56], and *gastrin releasing peptide* (*GRP*), isolated from porcine GI tract[57], are structurally related peptides which have a *wide range of actions* in the mammalian GI tract and CNS.

Structure, Biosynthesis, and Distribution

Bombesin is a 14 aa peptide, and GRP is found in several different molecular forms, of various lengths, which have the same C-terminal octapeptide sequence as bombesin; the largest form consists of 27 aa (Fig. XII.33).

55. Carraway, R., Kitabgi, P. and Leeman, S. E. (1978) The amino acid sequence of radioimmunoassayable neurotensin from bovine intestine: identity to neurotensin from hypothalamus. *J. Biol. Chem.* 253: 7996–7998.
56. Anastasi, A., Erspamer, V. and Bucci, M. (1971) Isolation and structure of bombesin and alytesin, two analogous active peptides from the skin of European amphibians, Bombia and Alyts. *Experientia* 27: 166–167.
57. McDonald, T. J., Jornvall, H., Nilsson, G., Vagne, M., Ghatei, M., Bloom, S. R. and Mutt, V. (1979) Characterization of a gastrin releasing peptide from porcine non-antral gastric tissue. *Biochem. Biophys. Res. Commun.* 90: 227–233.

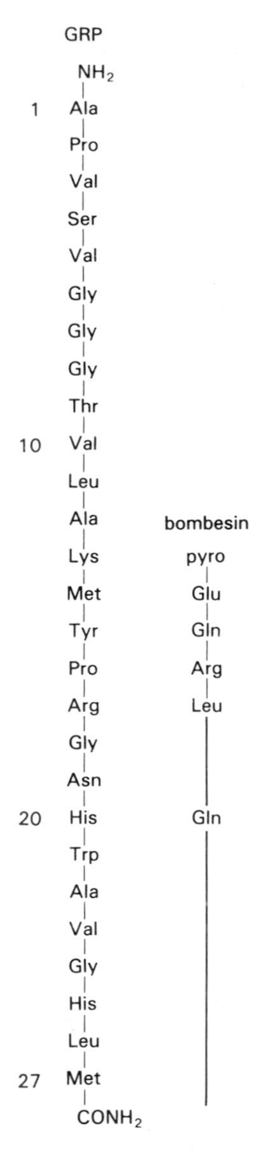

Figure XII.33 Amino acid sequence of human gastrin releasing peptide (GRP) and amphibian bombesin.

The precursor[58] of human GRP, characterized by cDNA studies, consists of 148 aa, and includes the sequence of GRP between positions 24 and 50 (Fig. XII.34).

Bombesin/GRP has been identified in the GI tract (where it is associated with neurons), CNS, lung, and milk. In the GI tract, it is found predominantly in the mucosa of the gastric fundus and the muscular layer of the antrum and intestine.

58. Spindel, E. R., Chin, W. W., Price, J., Rees, L. H., Besser, G. M. and Habener, F. (1984) Cloning and characterization of a cDNAs encoding human gastrin-releasing peptide. *Proc. Natl. Acad. Sci., USA* 81: 5699–5703.

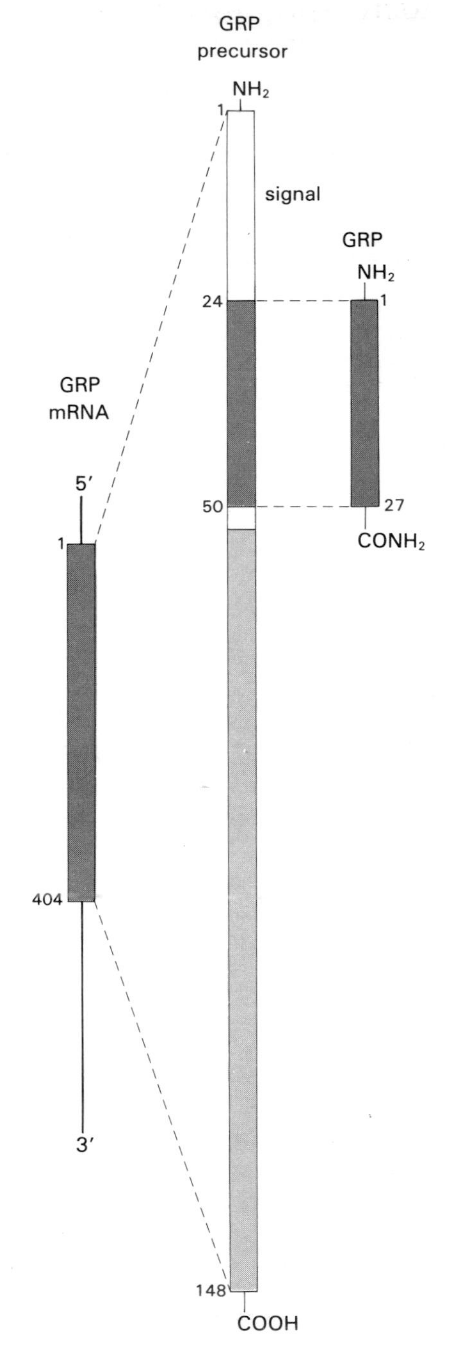

Figure XII.34 Structure of human GRP mRNA and its precursor[58].

Biological Actions

Bombesin and GRP *stimulate* the release of *gastrin* and a variety of other GI peptides (Table XII.7); endogenous bombesin appears to mediate the neural non-cholinergic stimulation of gastrin and

Table XII.11 Main features of human tumors producing GI and/or pancreatic peptides.

Tumor	Cell type	Peptide	Main location	Main features
Insulinomas	B	Insulin	Pancreas	Mental confusion or coma, adrenergic discharge, hypoglycemia
Glucagonomas	A	Glucagon	Pancreas	Weight loss, anemia, necrolytic migratory erythremia, diabetes, angular stomatitis, glossitis, abdominal discomfort, tendency to venous thrombosis
Gastrinomas	G	Gastrin	Pancreas; sometimes stomach or duodenum	Hypersecretion of gastric acid, hyperplasia of gastric parietal cells, multiple and recurrent duodenal ulcers, diarrhea and steatorrhea
VIPomas	Unknown	VIP	Pancreas (15–20% extrapancreatic ganglioneuro-blastomas)	Persistent severe watery diarrhea, hypokalemia, achlorhydria
Somatostatinomas	D	Somatostatin	Pancreas Intestine	Diabetes, malabsorption, steatorrhea, cholelithiasis, weight loss
PP-secreting tumors	PP	PP	Pancreas Intestine Bronchi	Unknown
Neurotensin-secreting tumors	Unknown	Neurotensin	Pancreas Bronchi	Unknown
Bombesin-secreting tumors	Unknown	Bombesin	Lung	Unknown

somatostatin release. Bombesin and GRP also stimulate gastric acid secretion[59], via gastrin release, and pancreatic enzyme secretion, via direct interaction with acinar cells. Bombesin exerts potent and complex effects on GI motility; it is mostly inhibitory in vivo, but is stimulatory in vitro. When administered in the CNS, bombesin elicits a variety of responses, some of which occur in the GI tract (Table XII.8).

Secretion and Pathophysiology

GRP increases in human plasma after the ingestion of a meal. It has been detected in some lung tumors (Table XII.11) as well as in fetal and neonatal lung. A decrease in the bombesin content of the respiratory tract has been observed in neonates with acute respiratory distress syndrome.

59. Varner, A.A., Modlin, I.M. and Walsh, J.H. (1981) High potency of bombesin for stimulation of human gastrin release and gastric acid secretion. *Regul. Pept.* 1: 289–296.
60. Chang, M.M. and Leeman, S.E. (1970) Isolation of sialogogic peptide from bovine hypothalamic tissue and its characterization as substance P. *J. Biol. Chem.* 245: 4784–4790.

14. Substance P

Originally identified in 1931 as a *vasodepressor* and a *gut-contacting* principle present in the brain and intestine, substance P was isolated in 1970[60], and was found to *stimulate salivary flow*. In the GI tract, substance P is *found predominantly in neurons*, and affects GI motility.

Structure and Distribution

Substance P is an 11 aa peptide which *belongs to* the *family of tachykinins*, characterized by the C-terminal pentapeptide amide sequence Phe-X-Gly-Leu-Met-NH_2 (Fig. 4, p. 184). Other members of this family include two peptides isolated from spinal cord, substance K (or neurokin α) and neuromedin K (or

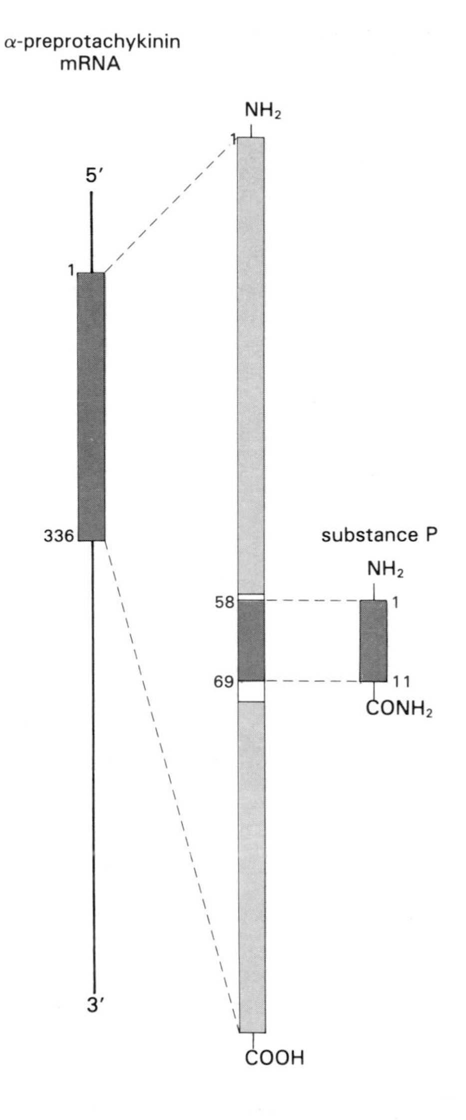

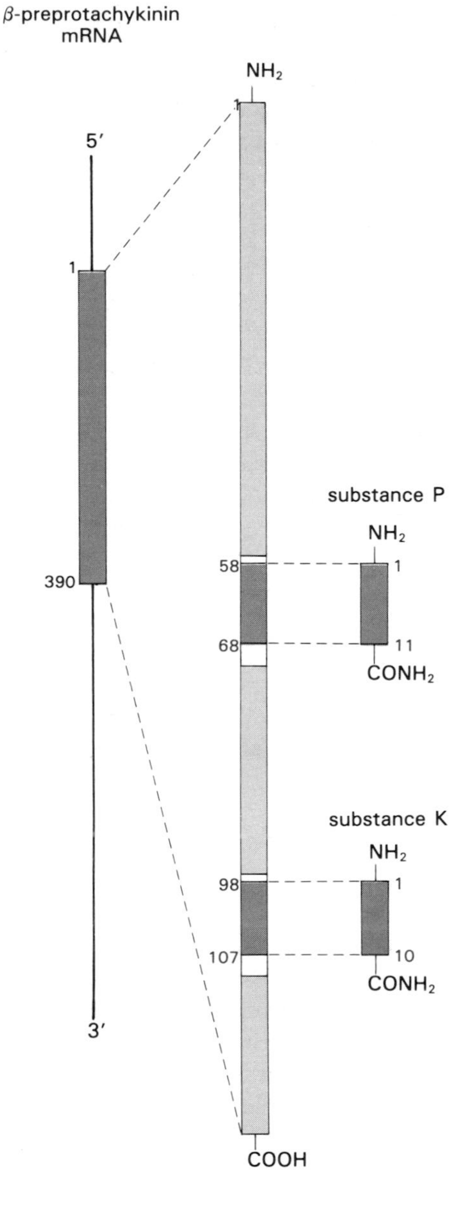

Figure XII.35 Structure of mRNAs and precursors of bovine substance P and substance K.[61]

neurokin β), peptides from amphibian skin (such as physalaemin and kassinin), and a peptide from the salivary gland of an octopod (eledoisin).

Substance P is synthesized from two precursors[61]: *preprotachykinin α*, which consists of 112 aa and includes the sequence of substance P at positions 58–68, and *preprotachykinin β*, which consists of 130 aa and includes the sequence of substance P at the same location as that of substance K, at positions 97–107 (Fig. XII.35).

Substance P is *widely distributed* throughout the body. *In the GI tract*, it is found *predominantly in* the *muscular layer* of the intestine, and is associated both with intrinsic neurons and with terminals of extrinsic, afferent neurons, the cell bodies of which lie in the dorsal root ganglia and in the central terminals of the dorsal horns of the spinal cord. Substance P is also present in a subpopulation of endocrine enterochromaffin cells. *Outside the GI tract*, substance P is present in *nerves* supplying skin, blood vessels, salivary glands

61. Nawa, H., Hirose, T., Takashima, H., Inayama, S. and Nakanishi, S. (1983) Nucleotide sequences of cloned cDNAs for two types of bovine brain substance P precursor. *Nature* 306: 32–36.

and sweat glands, mainly in afferent sensory neurons whose cell bodies lie in the dorsal root ganglia. *In the CNS*, substance P is especially abundant in *subcortical areas* of the telencephalon, hypothalamus and preoptic area, substantia nigra, lower brain stem and dorsal horns of the spinal cord, where it is associated with several neural pathways.

Biological Actions

Substance P is a potent stimulator of GI motility, and affects numerous other GI functions (Tables XII.4, XII.5, and XII.7). In the *circulatory system*, substance P causes potent *vasodilatation*. It is probably implicated as a mediator of *antidromic vasodilatation* which associates arteriolar vasodilatation with increased vascular permeability, and occurs upon excitation of cutaneous sensory nerve endings. Administered *centrally*, substance P provokes *complex* behavioral and sensory *responses*.

Receptors for substance P and other tachykinins have been identified in salivary glands, pancreatic acini, intestinal and bladder muscle, and brain. *Three receptor subtypes* have been described, referred to as SP-P (substance P), SP-E (eledoisin and neuromedin K), and SP-K (substance K)[62]. Determinants for binding and biological action of substance P are located C-terminally.

Secretion and Pathophysiology

Although circulating substance P is in part of intestinal origin, the mechanisms controlling the regulation of its secretion are practically unknown. Both substance P and substance K have been identified in extracts of intestinal carcinoid tumors. Conditions associated with defective substance P innervation have been described (p. 558).

15. Opioid Peptides

Originally identified in the brain and pituitary, opioid peptides (p. 187) were later shown to be *widely distributed* throughout the GI tract and pancreas, where they are associated primarily with *neurons*. Opioid peptides, like opiate drugs, *affect GI secretion*, *absorption* and *motility*, although their involvement as physiological regulators of these functions is not yet fully established.

Structure and Distribution

Opioid peptides identified in the GI tract include Met-enkephalin and short, C-terminally extended forms of Met-enkephalin, Leu-enkephalin, β-endorphin, and dynorphin (Table 1, p. 186). Opioid peptides are present in all *regions* of the GI tract, mainly in intrinsic *enteric neurons*. Enkephalinergic innervation is especially abundant in human esophagus. In the gastric antrum, duodenum and pancreas, Met- and Leu-enkephalins are found in endocrine G cells and enterochromaffin cells, respectively.

Biological Actions

Opioid peptides *increase* basal and stimulated *gastric acid* secretion, *inhibit* stimulated *pancreatic exocrine* secretion, and *stimulate* intestinal *absorption* of sodium and chloride[63]. Opioid peptides also exert potent actions on the *motility* of the gastric, intestinal and biliary tracts; in most species, with the exception of the guinea-pig, opioid peptides are *stimulatory*. Other effects of opioid peptides, beyond the scope of this chapter, include modulation of pancreatic hormone release and effects in the CNS.

Opioid peptides are known to exert their actions through *various* subtypes of *receptors* which are distinguishable by their affinities for individual ligands. In intestinal epithelial cells[63], they act via δ receptors, while in gastric[64] and intestinal smooth muscle, they act via both δ and μ receptors (p. 188). Receptors for opioid peptides in the GI tract appear to be located in *neural* structures rather than in epithelial or smooth muscle cells.

16. Thyrotropin Releasing Hormone (TRH)

TRH (p. 262) has been found *in* the *pancreas* and in the *mucosa* of the fundus, antrum, and duodenum. Pancreatic TRH is found in B cells, and shows a developmental pattern opposite to that of hypo-

62. Buck, S. H., Burcher, E., Shults, C. W., Lovenberg, W. and O'Donohue, T. L. (1984) Novel pharmacology of substance K-binding sites: a third type of tachykinin receptor. *Science* 226: 987–989.
63. Kachur, J. F., Miller, R. J. and Field, M. (1980) Control of guinea pig intestinal electrolyte secretion by a δ-opiate receptor. *Proc. Natl. Acad. Sci., USA* 77: 2753–2756.
64. Bitar, K. N. and Makhlouf, G. M. (1982) Specific opiate receptors in isolated mammalian gastric smooth muscle cells. *Nature* 297: 72–74.

thalamic TRH, being more abundant in the newborn than in the adult. The effects of TRH on GI functions are *mainly inhibitory*[65]. For example, it decreases basal and stimulated gastric acid secretion and pancreatic enzyme secretion, delays glucose absorption, and inhibits GI motility. The effects of centrally administered TRH on GI functions, however, are stimulatory (Table XII.8).

17. Epidermal Growth Factor (EGF) or Urogastrone

Epidermal growth factor (EGF) and urogastrone were *described independently* as factors present in mouse submaxillary gland and in human urine, that caused, respectively, premature opening of the eyelids and eruption of the incisors, and inhibition of gastric acid secretion and healing of experimental ulcers[66]. They were later shown to be a single peptide comprised of 53 aa (Fig. I.78).

EGF/urogastrone is *detected* in many GI exocrine glands, in saliva and gastric juice, and in plasma and urine. Besides its effect on gastric acid secretion, it stimulates mucosal growth in the GI tract and exerts a cytoprotective effect on the mucosa. However its *physiological role(s)* is (are) *not yet established*. Its receptor is described on p. 78.

18. Other Hormones and Mediators Affecting GI Functions

Insulin

Some effects of insulin on GI functions *result from hypoglycemia* induced by this hormone, and are vagally mediated. They include the stimulation of gastrin, gastric acid and pepsin secretion, and gastric motility. *Other* effects involve the *direct* interaction of insulin *with receptors* in GI epithelial cells; they include stimulation of several hydrolase activities in the intestinal epithelium, and stimulation of acinar cell function, in addition to the common pleiotropic effect associated with this hormone.

Calcitonin and CGRP (Calcitonin Gene Related Peptide)

Calcitonin and CGRP (p. 651) inhibit gastric acid and gastrin secretion[67], stimulate intestinal secretion of water and electrolytes, and cause satiety. Additional effects of CGRP are to cause vasodilatation, to stimulate pancreatic enzyme secretion, and to inhibit the release of certain GI hormones. Some effects of CGRP may be determined locally, since it is present in GI neurons. Excess calcitonin production may account for diarrhea observed in association with medullary carcinoma of the thyroid gland.

Vasopressin

Vasopressin *inhibits* the *absorption of water* and *electrolytes* by the human intestine, inhibits gastric secretion, and stimulates GI motility.

Angiotensin

Angiotensin exerts a dual action on the absorption of water and sodium by the intestinal epithelium; it is stimulatory at low doses and inhibitory at high doses[68]. These effects are mediated by the local release of norepinephrine and prostaglandins, respectively. Angiotensin also regulates colonic absorption, via aldosterone. The presence of angiotensin-like immunoreactivity in enteric nerves suggests a role for locally formed angiotensin. An additional effect of angiotensin in the GI tract is to stimulate motility.

Bradykinin

Via prostaglandins, bradykinin, a *hypotensive* peptide synthesized from plasma kininogen, simulates intestinal secretion of chloride[69] and causes contraction of intestinal smooth muscle.

65. Dolva, L. and Hanssen, K. F. (1982) Thyrotropin-releasing hormone: distribution and actions in the gastrointestinal tract. *Scand. J. Gastroenterol.* 17: 705–707.
66. Hollenberg, M. D. (1980) Epidermal growth factor-urogastrone, a polypeptide acquiring hormonal status. *Vitams. Horm.* 37: 69–106.
67. Kraenzlin, M. E., Ch'ng, J. L. C., Mulderry, P. K., Ghatei, M. A. and Bloom, S. R. (1985) Infusion of a novel peptide, calcitonin gene-related peptide (CGRP) in man. Pharmacokinetics and effects on gastric acid secretion and on gastrointestinal hormones. *Regul. Peptides* 10: 189–197.
68. Levens, N. R. (1985) Control of intestinal absorption by the renin-angiotensin system. *Amer. J. Physiol.* 249: G3–G15.
69. Manning, D. C., Snyder, S. H., Kachur, J. F., Miller, R. J. and Field, M. (1982) Bradykinin receptor-mediated chloride secretion in intestinal function. *Nature* 299: 256–259.

Pituitary Hormones

Hypophysectomy (including in humans) leads to atrophic changes in the stomach, small intestine and pancreas, accompanied by a corresponding decline in the function of these organs. Some of these effects are reversed by *growth hormone* which, like prolactin, stimulates intestinal absorption of water and electrolytes.

Adrenal Steroids

A variety of functions of the intestinal epithelium are influenced by *gluco-* and *mineralocorticosteroids*, the *latter* acting predominantly *in* the *colon*. They *stimulate sodium* and *water absorption*, inhibit calcium absorption, and increase the activity of certain enzymes. *Receptors* for both types of steroids have been identified in intestinal epithelial cells.

Acetylcholine

Widely distributed throughout the GI tract and associated with intrinsic enteric neurons, acetylcholine *regulates a variety* of *GI functions*, mainly via muscarinic receptors in epithelial and smooth muscle cells[70–72]. For example, it inhibits intestinal absorption (and/or stimulates secretion) of water and electrolytes, stimulates gastric acid and pancreatic exocrine secretion, regulates GI hormone secretion (Table XII.10), contracts GI muscle, and modulates norepinephrine release from adrenergic nerve terminals. Local release of acetylcholine is enhanced by nicotinic cholinergic agonists, and is inhibited by muscarinic cholinergic and α-adrenergic agonists, enkephalins, and somatostatin.

Norepinephrine

Also *widely distributed* throughout the GI tract but associated with extrinsic neurons, norepinephrine exerts *effects* on intestinal absorption of water and electrolytes and GI motility *opposite to* those of *acetylcholine*, via α_2 and β receptors[70,71]. It also causes the contraction of GI blood vessels and inhibits acetylcholine release from cholinergic nerves. Local release of norepinephrine is stimulated by nicotinic cholinergic agonists and by low concentrations of angiotensin, but is inhibited by muscarinic cholinergic and α-adrenergic agonists, prostaglandins, and high concentrations of angiotensin. *Physiologically, adrenergic tone predominates over cholinergic tone.*

Dopamine

Dopamine accounts for approximately *90%* of *all catecholamines* in the GI tract. Resembling norepinephrine, dopamine and its agonists are mainly inhibitory of GI motility[72]. In most cases, its activity is neurally mediated, via adrenergic receptors. Dopamine also exerts excitatory and inhibitory effects on GI secretions.

5-hydroxytryptamine (Serotonin)

Serotonin is *widely distributed* throughout the GI tract, as in the central and peripheral nervous system[70]. *GI* serotonin accounts for approximately *90%* of the hormone present in the body, with highest levels found in the proximal intestine. Most is found in the intestinal *mucosa*, associated with *enterochromaffin cells*. Elsewhere, it is present in enteric neurons. Serotonin inhibits gastric acid secretion, stimulates intestinal secretion of water and electrolytes, and affects GI motility and peristalsis as well as GI hormone release. Experimentally, GI serotonin can be released into the blood by mechanical, chemical, and electrical stimuli. In humans, its plasma concentration increases two-fold after a meal. It is the main secretory product of carcinoid tumors, where it may account for GI symptoms, particularly diarrhea.

Histamine

Histamine, *widely distributed* throughout the body, is abundant in the GI *mucosa*, particularly in the *oxyntic region* of the *stomach*[19]. It is located *in cells resembling mast cells*, and is found near parietal cells in the lamina propria. Its major effect in vitro is to *stimulate gastric acid secretion* and to affect parietal cell function; however, it has only 0.1% the potency of gastrin. The effect of histamine on gastric parietal cells is *mediated by cAMP* and *involves H_2 receptors*. It is inhibited by H_2 antagonists, such as metiamide and cimetidine, suggesting that endogenous histamine is a physiological *regulator* of *parietal cell function*[19]. An interdependence between the actions of histamine

70. Hubel, K.A. (1985) Intestinal nerves and ion transport: stimuli, reflexes and responses. *Amer. J. Physiol.* 248: G261–G271.
71. Tapper, E.J. (1983) Local modulation of intestinal ion transport by enteric neurons. *Amer. J. Physiol.* 244: G457–G468.
72. Willems, J.L., Buglaert, W.A., Lefebvre, R.A. and Bogaert, M.G. (1985) Neuronal dopamine receptors on autonomic ganglia and sympathetic nerves and dopamine receptors in the gastrointestinal system. *Pharmacol. Rev.* 37: 165–207.

and other gastric secretagogues has been described (p. 560).

Additional effects of histamine in the GI tract, mediated in part by H_1 receptors, include the stimulation of intestinal secretion of chloride, the stimulation of motility, and increased capillary permeability.

Prostaglandins

Prostaglandins (especially the E, F and I types) are *abundant* in the GI tract, where they influence secretion, motility, and cytotrophicity[73,74].

Prostaglandins of the *A*, *E* and *I* types are potent *inhibitors* of basal and stimulated *gastric acid secretion* when given parenterally, and some long-acting methylated analogs exert comparable effects when administered orally. Most prostaglandins (except PGI_2 and PGD_2) stimulate intestinal secretion of water and electrolytes, and cause diarrhea.

Prostaglandins *prevent the formation of gastric and duodenal ulcers* induced by a variety of experimental procedures. In addition, endogenous prostaglandins act as mediators in the phenomenon of adaptive cytoprotection. The cytoprotective effects of prostaglandins are unrelated to their antisecretory properties.

Effects of prostaglandins on GI motility vary depending on the chemical type of prostaglandin and on the animal species. For example, PGE_1 and methylated derivatives of PGE_2 inhibit antral gastric motility and increase intestinal motility in humans. In vitro, prostaglandins of the A, E and F types contract intestinal longitudinal muscle, whereas the E type relaxes it. F type prostaglandin causes the contraction of circular muscle.

Receptors for prostaglandins have been identified in gastric fundic mucosa, and cAMP has been proposed as a mediator of prostaglandin activity in gastric parietal cells and enterocytes.

Excessive prostaglandin production may account, in part, for the *diarrhea* observed in conjunction with tumors, such as *medullary carcinomas of* the *thyroid*, and with *Zollinger–Ellison syndrome*. Prostaglandins may be involved as mediators of diarrhea caused by infection or intestinal irritants. Patients with ulcers may have deficiencies in prostaglandin production, but this notion is yet to be confirmed.

73. Robert, A. (1981) Prostaglandins and the gastrointestinal tract. In: *Physiology of the Gastrointestinal Tract* (Johnson, L. R., ed), Raven Press, New York, pp. 1407–1434.
74. Miller, T. A. (1983) Protective effects of prostaglandins against gastric mucosal damage: current knowledge and proposed mechanisms. *Amer. J. Physiol.* 245: G601–G623.

19. Summary

Traditionally, gastrointestinal (GI) hormones have been defined as polypeptides *produced by endocrine cells* located *in* the *GI mucosa*. They are released into the *circulation* under the *influence of alimentary stimuli*, and are involved in the *regulation* of *GI functions*. It is *now* recognized that a number of GI peptides are *also produced by neurons inside and outside of the GI tract*; they are *delivered locally* to their target cells, and/or regulate non-GI functions. Initially restricted to gastrin, secretin, and cholecystokinin (CCK), the group of GI hormones or GI regulatory substances now includes *more than twenty polypeptides*. Most GI peptides are straight-chained and, with the exception of somatostatin, consist of less than 40 aa residues. Marked structural similarities define *families* of hormones of probable common evolutionary origin. *Size heterogeneity* is also a feature of some GI polypeptides (such as gastrin, CCK, and somatostatin), reflecting differential post-translational processing of precursors.

GI polypeptides are present in the GI tract at concentrations of 50 to 300 pmol/g of tissue (wet weight), and show *unique regional distributions*. Gastrin is found predominantly in the gastric antrum, whereas CCK, secretin, gastric inhibitory peptide (GIP) and motilin are found primarily in the proximal small intestine. Somatostatin and vasoactive intestinal polypeptide (VIP) show a more widespread distribution. GI peptides also show unique distributions within the GI wall. Gastrin, secretin, GIP and enteroglucagon are associated with the epithelium; VIP, substance P and gastrin releasing peptide (GRP), with non-epithelial layers; and somatostatin and CCK, with both layers. CCK, VIP, somatostatin, GRP, substance P and neurotensin are also present in the central nervous system, with specific regional distributions. GI peptides are *produced in specific endocrine cells and/or neurons*. Endocrine cells are widely scattered in the pancreas and *in the GI epithelium*, where they generally reach the glandular lumen. As many as *18 types of endocrine cells* have been identified in the human GI tract. Peptide-producing neurons, which are part of the enteric nervous system, are located in non-epithelial layers of the GI tract. The *coexistence* of several peptides and/or their association with non-peptide transmitters in the same population of neurons *is frequent*. Based, in part, on their ability to synthesize bioactive amines from precursors (hence the acronym APUD), GI endocrine cells were initially thought to originate from the neuroectoderm, but now appear to be of *endodermal origin*. In the human, most GI peptides are *detectable at eight weeks* of gestation and attain adult patterns by 20 to 24 weeks. Peptides and/or peptide-like materials resembling mammalian hormones have been detected in the main groups of vertebrates and in some invertebrates, indicating their appearance *early in evolution*. In vertebrates, these substances are present in both the alimentary tract and central nervous system, but in invertebrates they predominate in the latter.

Together with the central and peripheral nervous system, GI peptides influence *a variety of GI functions*, in a stimulatory or inhibitory manner, including water, electrolyte, enzyme and mucus secretions, hormone secretion, motility, growth, and blood flow. Some peptides also affect secretion and/or motility at non-GI sites, particularly in the circulatory, respiratory and urogenital systems. They regulate metabolism and exert several effects in the central nervous system, one of which is the regulation of food intake. Most effects exerted by peptides in the GI tract result from a direct interaction of the peptides with epithelial, smooth muscle or endocrine cells, but some are hormonally and/or neurally mediated. *Membrane receptors* and *second messengers* (cAMP, Ca^{2+} ions) have been identified for a number of GI peptides. With some peptides (gastrin, CCK, neurotensin, substance P, and bombesin), determinants for biological activity are located C-terminally, but the function of other peptides (VIP, secretin) depends on the entire sequence.

The *secretion* of GI hormones into the blood is *regulated by* ingested *nutrients and other luminal factors* (pH of the gastric or duodenal content, Ca^{2+}, distension), *by circulating* GI and/or pancreatic *hormones*, and *by* locally released neurotransmitters *(acetylcholine, bombesin/GRP). GI peptides have a circulatory half-life of 1 to 30 min; the larger the size, the longer the half-life. Liver, kidney and other tissue-specific peptidases are presumably involved in their overall metabolism.*

Endocrine tumors are the major *cause of overproduction* of GI peptides in humans. Most tumors are located in the pancreas, but some may occur in the intestine or at non-GI sites. Insulinomas, glucagonomas, gastrinomas and VIPomas are associated with specific clinical syndromes resulting from hypersecretion of the relevant peptide, but other tumors are often silent. *Non-tumoral conditions* affecting GI endocrine peptides include, mainly, *surgical* intervention (gastric and intestinal resections, bypass surgery), atrophic gastritis, celiac disease, tropical malabsorption, acute bowel injection, diabetes, and chronic renal disease. Conditions affecting GI neuropeptides, particularly VIP, include Hirschsprung's disease, esophageal achalasia, Chagas' disease, and Crohn's disease.

Gastrin occurs mainly as 17 and 34 aa peptides, the latter being an N-terminally extend form of the former. The precursor, preprogastrin, is made up of 101 aa, and the gene, including three exons separated by two introns, is 4.1 kb long. Gastrin is *produced* almost exclusively by *endocrine* G cells in the *gastric antrum* and *proximal duodenum*. Its major physiological effect is to *stimulate acid secretion* by gastric parietal cells; it also regulates the *growth* of the gastric *mucosa*. Most information required for the biological activity resides in the *terminal pentapeptide*. The release of gastrin into the blood is stimulated by luminal peptides and amino acids, and by locally released acetycholine and bombesin/GRP; it is inhibited by acidification of the gastric contents and by somatostatin. The major causes of gastrin overproduction in humans are: 1. pancreatic tumors or gastrinomas, often associated with multiple and recurrent duodenal ulcers (Zollinger–Ellison syndrome); and 2. atrophic gastritis, in which hypergastrinemia is the consequence of a decrease in gastric secretion. Synthetic pentagastrin is widely used as a stimulant in the assessment of gastric acid secretory response.

CCK occurs as *various molecular forms*, of 8, 33, 39, and 58 aa, the largest forms being N-terminally extended forms of the shortest. CCK is produced by endocrine cells localized in the *proximal intestine* (I cells) *and* by *neurons* in the central and peripheral nervous system, including enteric neurons.

The two major physiological effects of CCK in the GI tract are the *stimulation* of *gallbladder* contraction and *pancreatic enzyme* secretion; an important effect in the CNS, vagally mediated, is the *induction of satiety*. Determinants for biological activity are in the C-terminal region, the octapeptide being the most potent molecular form. The release of CCK into the blood is stimulated by luminal lipids and some amino acids. CCK is the major mediator of the intestinal phase of pancreatic exocrine secretion.

Secretin consists of 27 aa and is produced by *endocrine S* cells located in the *proximal intestine*. Its main effect is to stimulate the secretion of *water* and *bicarbonate* ions by *pancreatic duct* cells. Additionally, it stimulates pepsin secretion, inhibits secretion of gastric acid, and inhibits GI motility. The N-terminal region confers biological activity, and its deletion creates antagonistic peptides. Acidification of the duodenal contents is the major physiological stimulant of secretin release into the blood. Secretin, alone or in combination with CCK, is widely used as a stimulant, in the assessment of exocrine pancreatic function.

VIP consists of 28 aa and is structurally related to PHM-27. Originally isolated from the intestine, VIP is *widely* distributed throughout the body and is exclusively associated with *neurons* in the central and peripheral nervous system. VIP relaxes various types of smooth muscle (vascular bed, respiratory and GI tracts), increases myocardial contractility, stimulates intestinal secretion of water and electrolytes as well as glycogenolysis and lipolysis in the liver, and stimulates endocrine secretion in the pancreas, pituitary, and adrenal cortex. Although detectable in plasma, VIP appears to act *mainly* as a *neurotransmitter* and *neuroparacrine* regulator. VIP-secreting tumors, generally located in the pancreas, are the major cause of VIP overproduction in humans; one of the major symptoms of the disorder is watery diarrhea (Verner–Morrison syndrome).

GIP consists of 42 aa and is produced by K cells of the proximal intestine. Its major physiological effects are to enhance insulin secretion under hyperglycemic conditions and to inhibit gastric acid secretion. GIP release into the blood is stimulated by the main classes of nutrients and by several hormones.

Enteroglucagons, extended forms of pancreatic glucagon, are derived from the same precursor protein. The two major forms are *oxyntomodulin* (29 aa) and *glicentin* (with an N-terminal extension of 32 aa). Enteroglucagon peptides are produced by L cells located in the ileum and colon. Both oxyntomodulin and glicentin are potent inhibitors of gastric secretion; additional effects of oxyntomodulin are the stimulation of liver glycogenolysis and insulin release. Secretion is mainly stimulated by carbohydrate and fat.

Initially isolated from the hypothalamus but subsequently identified in the GI tract and pancreas, *somatostatin* occurs as two major molecular forms, of 14 and 28 aa, the latter being an N-terminally extended form of the former. Somatostatin is produced by endocrine D cells, found mainly in stomach, duodenum and pancreas, and by neurons in the GI tract. Its effects are exclusively *inhibitory*; all major functions, including endocrine secretion, motility and growth, are affected. Synthetic somatostatin analogs selectively inhibit insulin and/or glucagon secretion, intestinal ion transport, and/or gastric acid secretion. Somatostatin reaches its target cells via the blood circulation and by local diffusion; secretion is stimulated by luminal nutrients, circulating peptides, and locally released bombesin/GRP. In humans, pancreatic tumors give rise to expected symptoms of hypersomatostatinemia. Somatostatin is used in the management of type 1 (insulin-dependent) diabetes, hyperfunctioning endocrine tumors, and severe GI bleeding.

Motilin (22 aa) is produced in the proximal intestine. Its major effect is to initiate myoelectric complexes in the antroduodenal region that propagate distally in the small intestine. Plasma motilin concentration increases cyclically in the fasting state, and this coincides with the onset of the interdigestive myoelectric complexes in the duodenum.

Pancreatic polypeptide (PP) (36 aa) is produced by PP cells in the pancreas. The major effects are inhibition of pancreatic exocrine secretion and bile secretion.

Increased plasma concentrations of PP are often found in patients with pancreatic tumors.

PYY and NPY (36 aa) have about 50% homology with PP. PYY, associated with endocrine cells in the intestinal mucosa, inhibits GI secretion and motility, while NPY, found strictly in neurons, acts on various types of smooth muscle. Release of PYY into the blood is stimulated by luminal fat.

Initially isolated from brain, *neurotensin*, a tridecapeptide, is also found in the epithelium of the distal intestine (in N cells); it inhibits gastric and pancreatic exocrine secretions and motility. Release of neurotensin into the blood is stimulated by luminal fat.

Bombesin and *gastrin releasing peptide* (GRP) are related peptides which have been isolated from amphibian skin and mammalian GI tract, respectively. Bombesin consists of 14 aa; GRP possesses the same C-terminal octapeptide sequence as bombesin. Bombesin/GRP immunoreactivity is present in the GI tract, lung and CNS, and is associated strictly with neurons. Bombesin and GRP exert a wide range of actions in the GI tract and CNS; one of their major effects is the stimulation of the release of gastrin and other GI peptides. There is evidence that bombesin mediates neural non-cholinergic stimulation of gastrin and somatostatin release from gastric G and D cells, respectively.

Substance P, isolated from mammalian GI tract and brain, consists of 11 aa and belongs to a group of peptides known as *tachykinins*. It is synthesized as preprotachkinins α and β, the latter also including the sequence of substance K. Substance P is widely distributed throughout the central and peripheral nervous system; in the GI tract, it is associated mainly with neurons, but some is also present in endocrine cells.

In the GI tract, substance P causes contraction of intestinal smooth muscle and stimulates salivary, pancreatic and intestinal secretions. Although substance P has been identified in the blood, it appears to act mainly as a neurotransmitter.

Finally, the involvement of several other hormones, peptides and neurotransmitters is briefly reviewed.

General References

Bloom, S. R. and Polak, J. M., eds (1981) *Gut hormones*. Churchill-Livingstone, Edinburgh.

Creutzfeldt, W., ed (1980) Gastrointestinal hormones. In *Clinical Gastroenterology*. W. B. Saunders Co., London, Philadelphia, Toronto, vol. 9: 483–803.

Desbuquois, B. (1985) Receptors for gastrointestinal polypeptides. In: *Polypeptide Hormone Receptors* (Posner, B. I., ed), Marcel Dekker, New York, Basel, pp. 419–480.

Gregory, R. A., ed (1982) Regulatory peptides of gut and brain. *Brit. Med. Bull.* 38: 219–238.

Grossman, M. I. (1950) *Gastrointestinal hormones. Physiol. Rev.* 30: 33–90.

Jerzy-Glass, B. D., ed (1980) *Gastrointestinal Hormones*. Raven Press, New York.

Johnson, L. R., ed (1981) *Physiology of the Gastrointestinal Tract*. Raven Press, New York.

Mutt, V. (1982) Chemistry of the gastrointestinal hormones and hormone-like peptides and a sketch of their physiology and pharmacology. *Vitams. Horm.* 39: 231–427.

Walsh, J. H. (1983) Gastrointestinal hormones. In: *Gastrointestinal Disease* (Sleisenger, M. H. and Fordtran, J. S., eds), W. B. Saunders Co., Philadelphia, pp. 54–96.

Gastrin, Cholecystokinin, Secretin

Rehfeld, J. F. (1981) Four basic characteristics of the gastrin-cholecystokinin system. *Amer. J. Physiol.* 24C: G255–G266.

Hubel, K. A. (1972) Secretin: a long progress note. *Gastroenterology* 62: 318–341.

Morley, J. E. (1982) The ascent of cholecystokinin (CCK). From gut to brain. *Life Sci.* 30: 479–493.

Walsh, J. H. and Grossman, M. I. (1975) Gastrin. *N. Engl. J. Med.* 292: 1324–1334, 1377–1384.

VIP

Rosselin, G., Maletti, M., Besson, J. and Rostene, W. (1982) A new neuroregulator: the vasoactive intestinal peptide or VIP. *Mol. Cell. Endocrinol.* 27: 243–262.

Said, S. I. (1984) Vasoactive intestinal polypeptide (VIP): current status. *Peptides* 5: 143–150.

Enteroglucagon

Holst, J. J. (1983) Gut glucagon, enteroglucagon, gut glucagon-like immunoreactivity, glicentin. Current status. *Gastroenterology* 84: 1602–1613.

Somatostatin

McIntosh, C. H. S. (1985) Gastrointestinal somatostatin: distribution, secretion and physiological significance. *Life Sci.* 37: 2043–2058.

Reichlin, S. (1983) Somatostatin. *N. Engl. J. Med.* 309: 1495–1501 and 1556–1563.

Motilin

Fox, J. E. T. (1985) Motilin: an update. *Life Sci.* 37: 2043–2058.

Pancreatic Polypeptide, Peptide YY, and Neuropeptide Y

Schwartz, T. W. (1983) Pancreatic polypeptide: a hormone under vagal control. *Gastroenterology* 85: 1411–1425.

Solomon, T. E. (1985) Pancreatic polypeptide, peptide YY and neuropeptide Y family of regulatory peptides. *Gastroenterology* 88: 838–844.

Neurotensin and Substance P

Leeman, S. E., Aronin, N. and Ferris, C. (1982) Substance P and neurotensin. *Rec. Prog. Horm. Res.* 38: 93–132.

Pernow, B. (1983) Substance P. *Pharmacol. Rev.* 34: 85–141.

Opioid Peptides

Ambinder, R. F. and Schuster, M. M. (1979) Endorphins: new gut peptides with a familiar face. *Gastroenterology* 77: 1132–1140.

Hughes, J., ed (1983) Opioid peptides. *Brit. Med. Bull.* 39: 1–106.

Elisabeth Granström, Pär Westlund, and Bengt Samuelsson

Prostaglandins and Related Compounds

Certain polyunsaturated fatty acids can be metabolized into a large family of compounds, the "prostanoids", or eicosanoids. They include the *prostaglandins, thromboxanes, prostacyclins, leukotrienes, lipoxins* and a large number of related mono-, di- and trihydroxy fatty acids. Most of these compounds are biologically very active and display a wide variety of effects in many different biological systems. The most important compounds originate in *arachidonic acid.*

Investigations in this field began in the 1930s with the discovery of the prostaglandins by von Euler[1]. They were first found in seminal plasma and accessory genital glands, such as the prostate; hence their name. Later, it was discovered that prostaglandins are found almost everywhere in mammalian tissues, and they have also been discovered in many lower species.

The *structure and biosynthesis* of prostaglandins (PGs) was elucidated in the early 1960s[2]. *Prostaglandins have a unique fatty acid skeleton* called *prostanoic acid* (Fig. 1). This compound, however, does not exist in nature; all naturally occurring PGs have various oxygen-containing substituents in the molecule. The nature and position of these *substituents* determine the biological activity of the compound; the various types are indicated by the *letters*: A, B . . . I. The most important positions in the PG molecule are carbons 9, 11, and 15 (Fig. 1). Prostaglandins are further subgrouped into *three dif-*

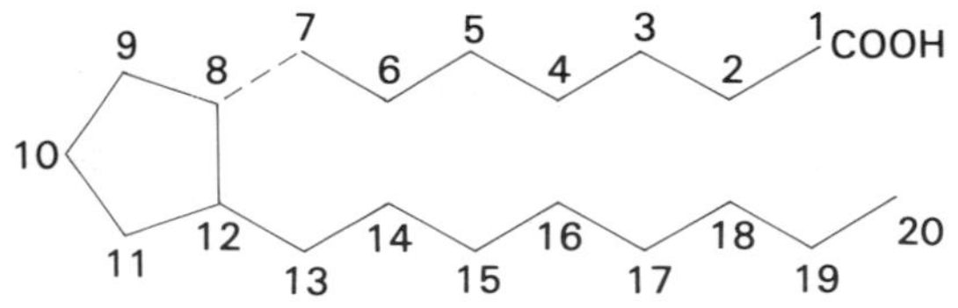

Figure 1 *Structure of the prostaglandin skeleton, "prostanoic acid", with the numbers of the carbon atoms indicated.*

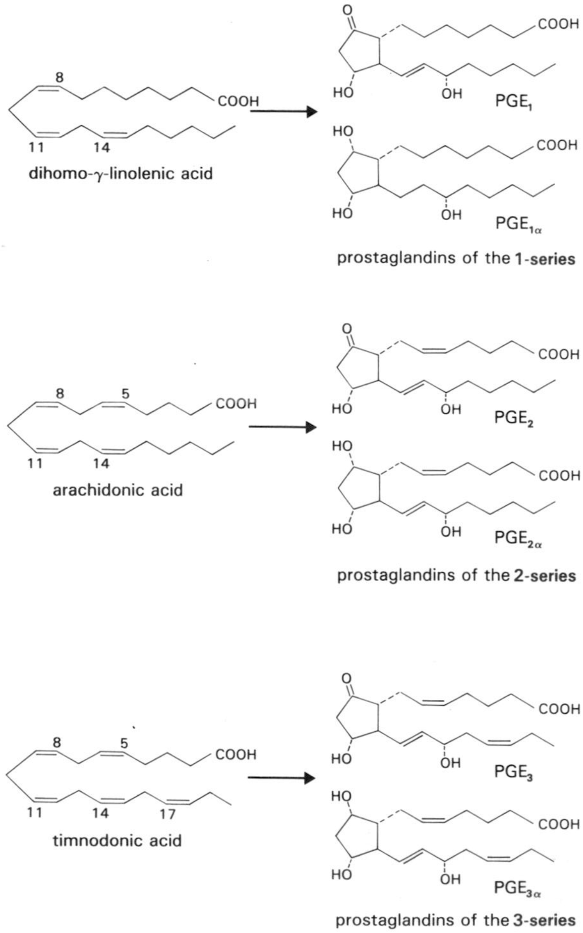

Figure 2 *Biosynthesis of some prostaglandins of the 1-, 2- and 3-series from dihomo γ-linolenic acid, arachidonic acid, and an eicosapentaenoic acid, timnodonic acid.*

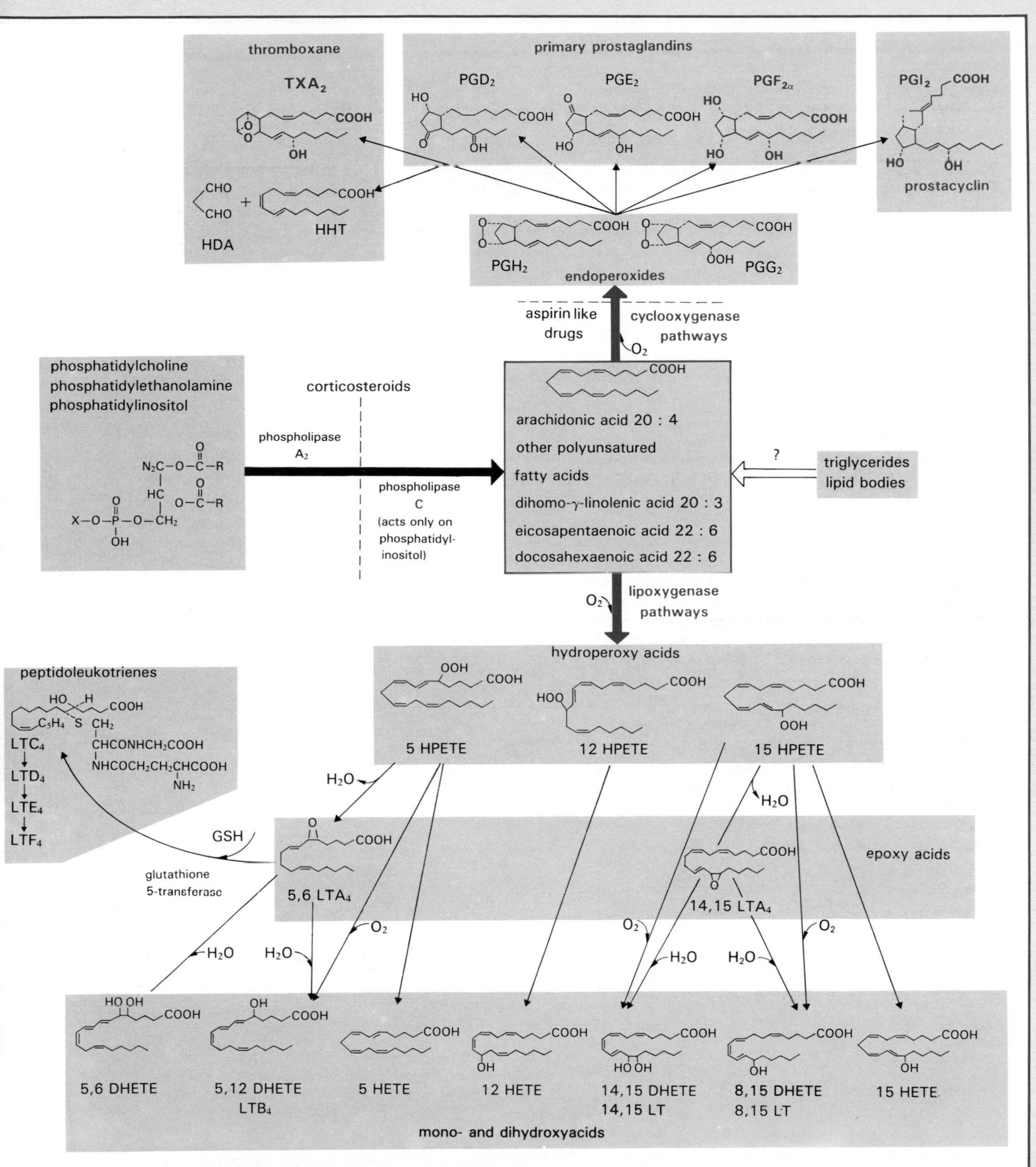

Figure 3 *Simplified diagram of oxygenations of polyunsaturated fatty acids.*

ferent series, the 1-, 2- and 3-series, differing in their *degree of unsaturation*, and originating in three different polyunsaturated fatty acids (Fig. 2). The number indicating the series is added as an index after the prostaglandin name, e.g., PGE_1, PGD_2. Arachidonic acid gives rise to the 2-series.

During conversion of the fatty acid into the end products, two important intermediates appear. These are the potent prostaglandin *endoperoxides* PGG and PGH, in which a peroxide bridge has been formed between positions 9 and 11 (Fig. 3). Their formation is catalyzed by an enzyme called fatty acid *cyclo-oxygenase*. In 1971, it was found that *non-steroidal anti-inflammatory drugs*, such as aspirin and indomethacin, inhibit this enzyme and probably exert at least part of their action via this mechanism[3]. The well known clinical effects of aspirin suggested the involvement of cyclo-oxygenase products in many physiological and pathological processes, such as thrombosis or bleeding disorders, dysmenorrhea, and inflammation.

A few years later, it was discovered that the endoperoxides can be metabolized not only into the classical prostaglandins, but also into other potent compounds with somewhat different structures, namely, *thromboxanes* and *prostacyclins*[4,5]. Prostacyclins are derivatives of prostanoic acid; hence, they are commonly referred to by their systemic prostaglandin name, PGI. Thromboxanes (TXs), on the other hand, do not possess the prostanoic acid skeleton, but have a different ring structure. At present, two types of thromboxanes are known, TXA and TXB.

All of these compounds are products of the cyclo-oxygenase, and thus their biosynthesis is inhibited by substances such as aspirin. *Oxygenation* of the *precursor* fatty acid can *also* take place *at other positions* in the molecule and by different mechanisms. For example, a number of *lipoxygenase enzymes* have been demonstrated which introduce oxygen at carbons 5, 8, 9, 11, 12 or 15 in arachidonic acid. The product formed initially is a *hydroperoxy fatty acid* (Fig. 3), which can then be converted into a monohydroxy fatty acid or undergo other transformations, for example conversion into the *leukotrienes* (LTs)[6]. Some of these, such as the LTs of the B type, are relatively simple dihydroxy fatty acids. However, the LTs of the C, D, E and F type have been formed by introduction of small peptides or amino acids. The most complex of these, LTC, has incorporated a whole molecule of glutathione, and the others are formed by step-wise hydrolysis of this peptide chain.

COMPOUNDS: **Trivial name** (Abbreviation)	**Formula** (Systemic name)	**BIOLOGICAL EFFECTS:** (note: pronounced species and tissue differences may occur) ↑: induction, stimulation, increase; ↓: inhibition, decrease.
Precursor fatty acids: Dihomo γ-linolenic acid (20:3ω6)	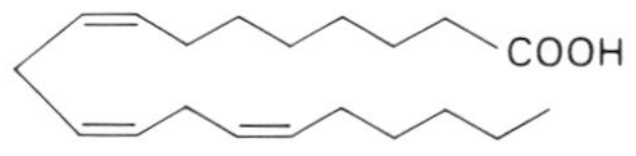 COOH (8,11,14-eicosatrienoic acid)	Precursor of 1-series of prostaglandins and one 3-series of leukotrienes
Arachidonic acid (20:4ω6)	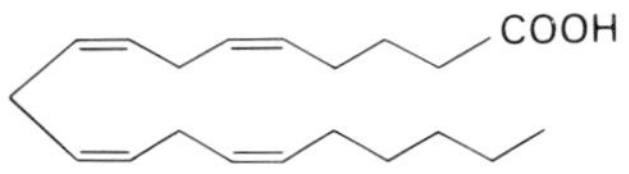 COOH (5,8,11,14-eicosatetraenoic acid)	Precursor of 2-series of prostaglandins and 4-series of leukotrienes
Timnodonic acid (20:5ω3)	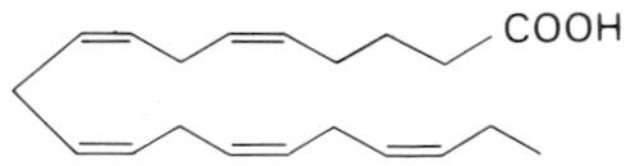 COOH (5,8,11,14,17-eicosapentaenoic acid, "EPA")	Precursor of 3-series of prostaglandins and 5-series of leukotrienes
Dihomo γ-linolenic acid metabolite: Prostaglandin E_1 (PGE_1)	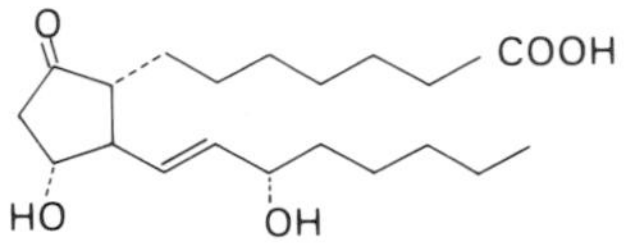 O, COOH, HO, OH	Vessel tone (general) ↓, renin release ↑, systemic blood pressure ↓, platelet aggregation ↓, bronchial tone ↓, myometrial activity ↑ or ↓, smooth muscle of GI tract: longitudinal ↑, circular ↓, gastric secretion ↓, pancreatic secretion ↓, renal blood flow ↑, diuresis ↑, body temperature ↑, hyperalgesia, bone resorption ↑, cyclic AMP ↑, lipolysis ↑

COMPOUNDS: Trivial name (Abbreviation)	**Formula**	**BIOLOGICAL EFFECTS:** (note: pronounced species and tissue differences may occur) ↑ : induction, stimulation, increase; ↓ : inhibition, decrease.
Arachidonic acid metabolites: Cyclo-oxygenase pathway:		
Prostaglandin E_2 (PGE_2)	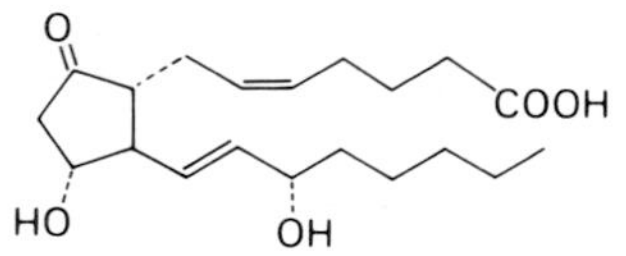 O, COOH, HO, OH	Vessel tone (general) ↓, renin release ↑, systemic blood pressure ↓, platelet aggregation ↑ or ↓, bronchial tone ↓, myometrial activity ↑, GI smooth muscle: longitudinal ↑, circular ↓, gastric secretion ↓, renal blood flow ↑, body temperature ↑, hyperalgesia, lipolysis ↑, cyclic AMP ↑ (or ↓), norepinephrine output (adrenergic nerve endings) ↓, cytoprotection of GI mucosa and other tissues.
Prostaglandin $F_{2\alpha}$ ($PGF_{2\alpha}$).	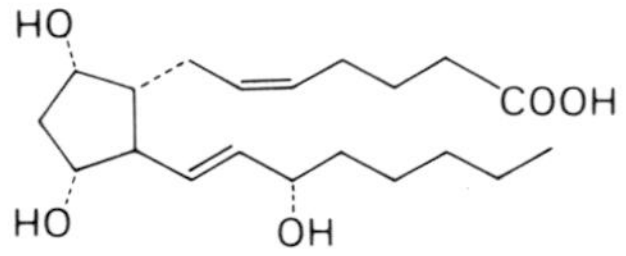 HO, COOH, HO, OH	Vessel tone ↑ or ↓, cardiac output ↑, bronchial tone ↑, myometrial activity ↑, GI smooth muscle ↑, luteolysis ↑
Prostaglandin D_2 (PGD_2)	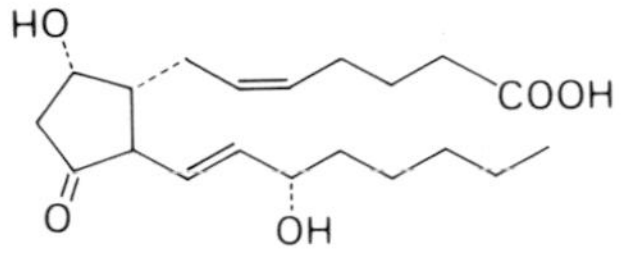 HO, COOH, O, OH	Vessel tone (general) ↓, smooth muscle activity (general) ↑, platelet aggregation ↓, renal blood flow ↑, GI smooth muscle ↑, enteropooling ↓, bronchial tone ↑
Prostaglandin endoperoxides: Prostaglandins G_2 and H_2 (PGG_2, PGH_2)	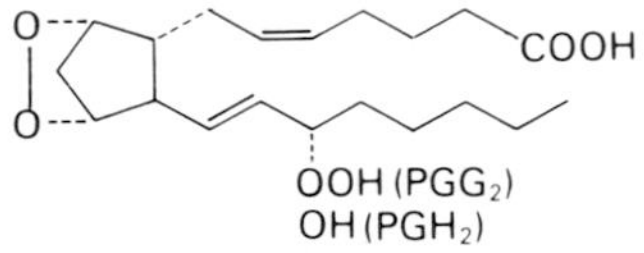 O, O, COOH, OOH (PGG_2), OH (PGH_2)	Platelet aggregation ↑, vessel tone ↑, bronchial tone ↑ (the effects may to some extent be caused by conversion into other cyclo-oxygenase products by the studied tissues.
Thromboxane A_2 (TXA_2)	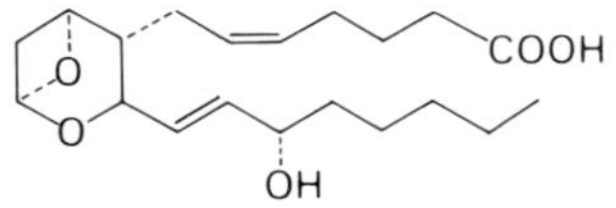 O, O, COOH, OH	Platelet aggregation ↑, vessel tone ↑, bronchial tone ↑, gastric ulcerations ↑
Prostaglandin I_2 (prostacyclin, PGI_2)	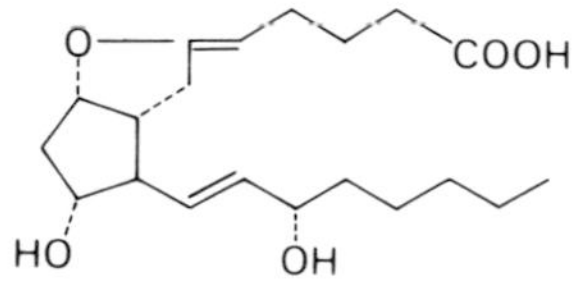 O, COOH, HO, OH	Vessel tone ↓, platelet aggregation ↓, systemic and pulmonary blood pressure ↓, cytoprotection of GI mucosa, gastric secretion ↓, renin release ↑, GI smooth muscle activity ↓, enteropooling ↓
Lipoxygenase pathways:		
Leukotriene B_4 (LTB_4)	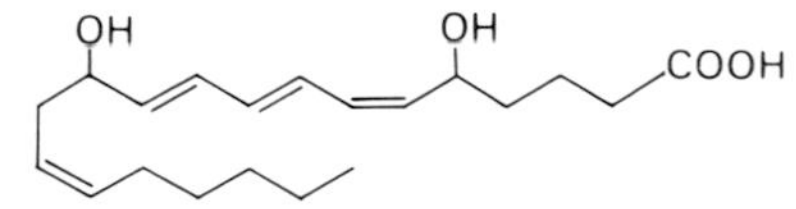 OH, OH, COOH	Effects on polymorphonuclear neutrophil leukocytes: chemokinesis ↑, chemotaxis ↑, aggregation ↑, adherence ↑, release of lytic enzymes ↑, release of activated oxygen species ↑. Other effects: vascular leakage (indirect) ↑, airway tone ↑ (indirect)
Leukotriene C_4 (LTC_4)	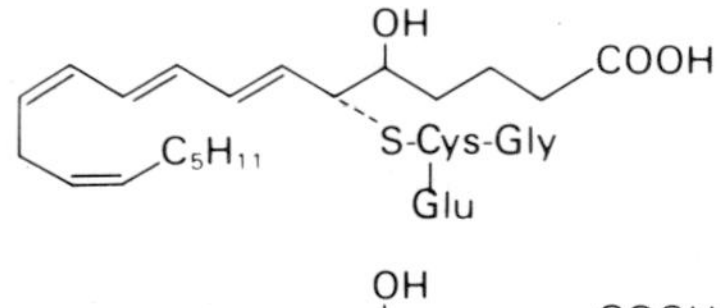 OH, COOH, C_5H_{11}, S-Cys-Gly, Glu	Bronchial tone ↑, vascular permeability ↑, negative inotropism, vessel tone ↑ or ↓, mucus secretion in airways ↑, smooth muscle tone (GI, urinary, myometrium, etc.) ↑ (applies to LTC_4, LTD_4 and LTE_4)
Leukotriene D_4 (LTD_4)	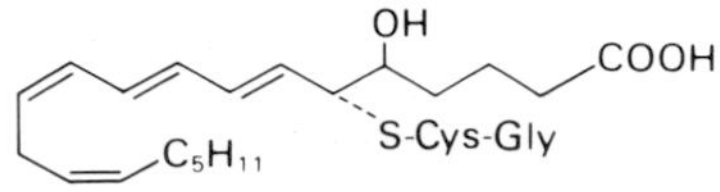 OH, COOH, C_5H_{11}, S-Cys-Gly	
Leukotriene E_4 (LTE_4)	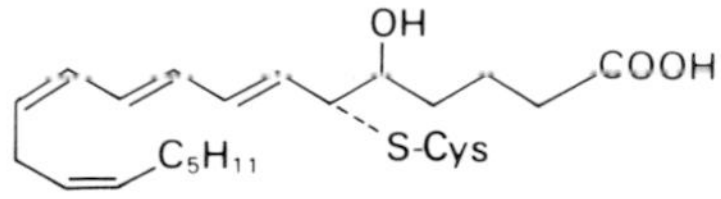 OH, COOH, C_5H_{11}, S-Cys	

The lipoxygenase-catalyzed pathways are generally not inhibited by aspirin or by other non-steroidal anti-inflammatory drugs. There are some specific *lipoxygenase inhibitors*. However, the most potent inhibitors of leukotriene biosynthesis, *glucocorticosteroids*, act at an earlier step, i.e. in the release of the precursor fatty acid from phospholipids; they are thus inhibitors of the total metabolism of the fatty acid, including the cyclo-oxygenase pathway[7]. The potent clinical effects of glucocorticosteroids, as well as their pronounced side effects, compared with aspirin for example, indicate that leukotrienes, or other lipoxygenase products, may be involved in a large number of biological phenomena, such as asthma and other anaphylactic or allergic responses, inflammation, gastric ulcer, among others.

Biochemical pathways in the eicosanoid field and the biological effects of most metabolites are found in Figure 3 and in the Table.

References

1. von Euler, U. S. (1935) Über die spezifische blutdrückssenkende Substanz des menschlichen Prostata- und Samenblasensekretes. *Klin. Wochschr.* 33: 1182–1183.
2. Bergström, S., Carlsson, L. A. and Weeks, J. R. (1968) The prostaglandins: a family of biologically active lipids. *Pharmacol. Rev.* 20: 1–48.
3. Vane, J. R. (1971) Inhibition of prostaglandin synthesis as a mechanism of action for aspirin-like drugs. *Nature* 231: 232–235.
4. Hamberg, M., Svensson, J. and Samuelsson, B. (1975) Thromboxanes: a new group of biologically active compounds derived from prostaglandin endoperoxides. *Proc. Natl. Acad. Sci., USA* 72: 2994–2998.
5. Johnson, R. A., Morton, D. R., Kinner, J. H., Gorman, R. R., McGuire, J. C., Sun, F. F., Whittaker, N., Bunting, S., Salmon, J., Moncada, S. and Vane, J. R. (1976) The chemical structure of prostaglandin X (prostacyclin). *Prostaglandins* 12: 915–928.
6. Samuelsson, B., Dahlén, S.-E., Lindgren, J., Rouzer, C. A. and Serhan, C. N. (1987) Lenkotrienes and Lipoxins: Structures, Biosynthesis, and Biological Effects. *Science* 237: 1171–1176.
7. Blackwell, G. J. and Flower, R. J. (1983) Inhibition of phospholipase. *Brit. Med. Bull.* 39: 260–264.

CHAPTER XIII

The Renin-Angiotensin-Aldosterone System and Atrial Natriuretic Factor

Pierre Corvol and Joël Ménard

Contents, Chapter XIII

The Renin-Angiotensin-Aldosterone System and Atrial Natriuretic Factor

The *regulation* of *water* and *ion* metabolism, and the *control* of tissue *blood flow* and of arterial *pressure*, depends on numerous *hormonal systems*. The *renin-angiotensin-aldosterone system* plays a key role in the control of blood pressure, blood volume, and the metabolism of sodium and potassium. Recently, a factor with *natriuretic* and *vasodilator* activity *(atrial natriuretic factor, ANF)*, produced in the atria of the heart, has been discovered, which may play an important role in salt and water homeostasis and in the control of blood pressure. *Vasopressin* is the hormone which controls water metabolism, but it has only a minor role (at least in humans) in the maintenance of blood pressure (p. 294). In addition, there are also several *tissue mediators* (bradykinin and prostaglandins) which modulate renal blood flow and renal excretion of electrolytes.

Since the *kidney* is the principal organ which controls the excretion of water and electrolytes in response to the needs of the body, it is not surprising that it is the *target organ* for the hormones regulating water and electrolyte metabolism. These hormones function to maintain a constant homeostasis in extra- and intracellular compartments. The kidney is also a hormone-secreting organ, producing renin in response to discrete variations in blood pressure and salt and water metabolism. Thus, this chapter is devoted to a new branch of endocrinology, *renal endocrinology*.

1. The Renin-Angiotensin System

The renin-angiotensin system has an important role in the homeostasis of *two major functions* of the body: *water* and *electrolyte balance* and the *regulation of arterial pressure*. It is the site of interaction of a number of different systems involved in bioregulation. The renin-angiotensin system can be considered *both* an *intra-* and *extrarenal endocrine* system. It is *intrarenal* in that renin is secreted into the kidney in situ and produces angiotensins which have local vascular and cellular actions. It is *extrarenal*, because renin secreted by the kidney can induce the production of angiotensins, at some distant site, which then act on a number of other target organs.

The renin-angiotensin system can have primary or secondary actions in diseases involving abnormal blood pressure and edematous syndromes. The importance of the renin-angiotensin system becomes apparent when the multiple actions of angiotensin are considered. However, in contrast with what can be shown for other endocrine organs (ablation and hormonal replacement), it is not possible to remove the secretory function of the kidney without suppressing its excretory functions. The recent development of *specific inhibitors* of the renin-angiotensin system, used experimentally and therapeutically in humans, has confirmed the key role of this system in the control of normal and pathological blood pressure, as well as in water and electrolyte metabolism.

1.1 Fundamental Elements

Anatomical Aspects

Renin is synthesized and secreted by the *juxtaglomerular apparatus* of the kidney, where the three fundamental intrarenal structures come together[1] (Fig. XIII.1).

1. Cantin, M. (1983) Morphopathology of the renin-angiotensin system. In: *Hypertension* (Genest, J., ed), McGraw-Hill, New York, pp. 280–308.

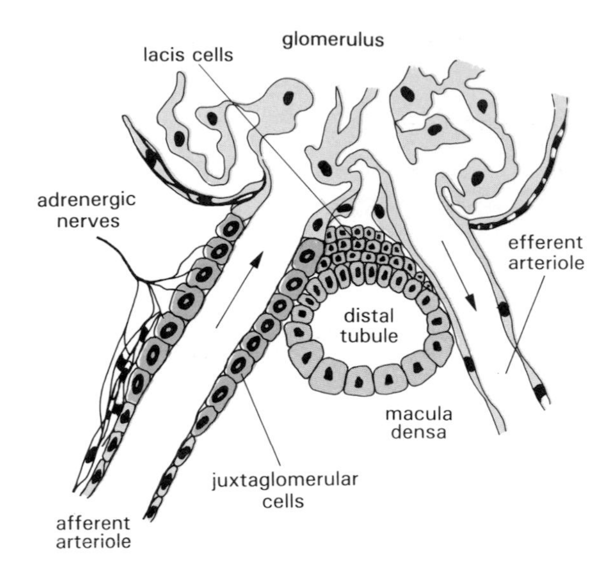

Figure XIII.1 Structure of the juxtaglomerular apparatus. Renin is synthesized and stored in the vascular wall of the afferent arteriole.

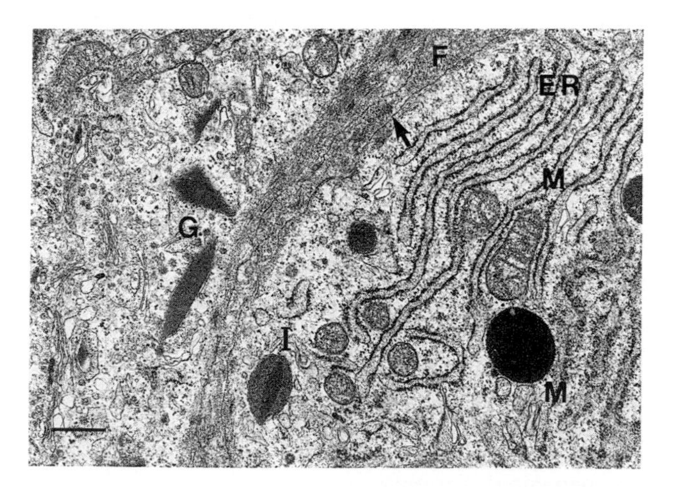

Figure XIII.2 Electron micrograph of a renin-secreting cell. All types of renin secretory granules are visible (P, protogranules; M, mature granules; I, intermediate forms; G, rhomboid crystalloid protogranules). The rough endoplasmic reticulum (ER) is well developed. Myofilaments (F) and attachment bodies (arrows) are present and show the myoepithelial origin of the cell. The renin-secreting cell, therefore, has the unique features of both an endocrine cell and a vascular smooth muscle cell[2]. Bar=1 μm.

The afferent arteriole of the glomerulus. Immediately before the capillaries form the glomerulus, the afferent arteriolar wall contains a specific structure which exists nowhere else. *Myocytes* (vascular smooth muscle cells) *differentiate* into *endocrine cells* and *secrete renin*. Granules of renin are visible with certain stains (e.g., Bowie) and can be demonstrated particularly well by immunofluorescence, with specific anti-renin antibodies[2] (Table XIII.1). Using electron microscopy, secretory granules are observed, as in almost all endocrine cells. The presence of myosin filaments in these cells is a sign of their origin in smooth muscle (Fig. XIII.2).

On the surface of the *afferent arteriole* are baroreceptors, sensitive to variations in pressure and flow rate.

The macula densa. The initial portion of the distal tubule is marked by the junction between the proximal portion of the nephron, where maximal hydrodynamic effects are exerted (e.g., glomerular filtration, proximal tubular reabsorption), and the terminal portion, where fine hormonal regulation of sodium, potassium and water takes place. The cells of the macula densa constitute, for the afferent arteriole, a kind of chemoreceptor, sensitive to variations in the concentration of urinary ions (Na^+, Cl^-, Ca^{2+}) at the beginning of the distal tubule.

The lacis cells are localized in the triangle formed by the macula densa and the afferent arteriole. This lacis is continuous with the glomerular mesangium. These cells, and their multiple cytoplasmic projections, are surrounded by a basal membrane.

The existence of *adrenergic and cholinergic innervation*[3] of the juxtaglomerular apparatus was first shown by silver staining and confirmed later at the electron microscope level. The physiological significance of this adrenergic system is now well known. However, little is known about the cholinergic system. This type of innervation essentially affects myoepithelial cells. Nerve terminals have not been identified in the macula densa.

Biochemical Elements

The description of enzymatic reactions of the renin-angiotensin system should be completed by the

2. Camilleri, J. P., Phat, V. N., Bariéty, J., Corvol, P. and Ménard, J. (1980) Use of a specific antiserum for renin detection in human kidney. *J. Histochem. Cytochem.* 28: 1343–1346.
3. Barajas, L. (1978) Innervation of the renal cortex. *Fed. Proc.* 37: 1192–1201.

Table XIII.1 Detection, by immunohistochemistry, of the three components of the renin angiotensin system in the human anterior pituitary gland, using monospecific antisera.

Antibody	Human pituitary cells				
	Lactotropic	Corticotropic	Thyrotropic	Gonadotropic	Somatotropic
Antirenin	+	–	–	–	–
Antiangiotensinogen	+	–	–	–	–
Anticonverting enzyme	+	–	–	–	–

(Courtesy of Dr. J. P. Saint-André, Laboratoire d'Anatomo-pathologie, CHU, Angers). The entire renin-angiotensin system is present only in the lactotropic cells. The presence of angiotensin II has also been detected in these cells by immunohistochemistry. This suggests a paracrine action of angiotensin II on prolactin secretion.

description of different inhibitors capable of blocking the system at one of the several steps which lead to the formation of two active peptides, angiotensin II and III.

Renin

Renin (M_r~40,000) is an enzyme which cleaves *angiotensinogen*, to produce *angiotensin I*. Recently, it has been completely purified in several species, including humans[4]. The primary structure of renin from the *submaxillary gland of mice* has been determined by conventional sequence analysis[5], and that of its precursor by nucleotide sequencing of its cDNA[6]. A model of the processing of preprorenin to renin has been proposed (Fig. XIII.3). Recently, *human* renin has been cloned, and its amino acid structure (deduced from the cDNA sequence) is very close to that of mouse renin (68% homology). As is the case for other prohormones, prorenin is initially inactive and is activated following cleavage of the profragment, the cleavage occurring at a site following a pair of basic amino acids (Lys-Arg). Mouse submaxillary renin consist of two chains, A and B, linked by a disulfide bridge. Human renin probably undergoes the same processing as mouse submaxillary renin, but it consists of only a single polypeptide chain. In contrast to mouse submaxillary renin, human renin has two potential sites of *N*-glycosylation.

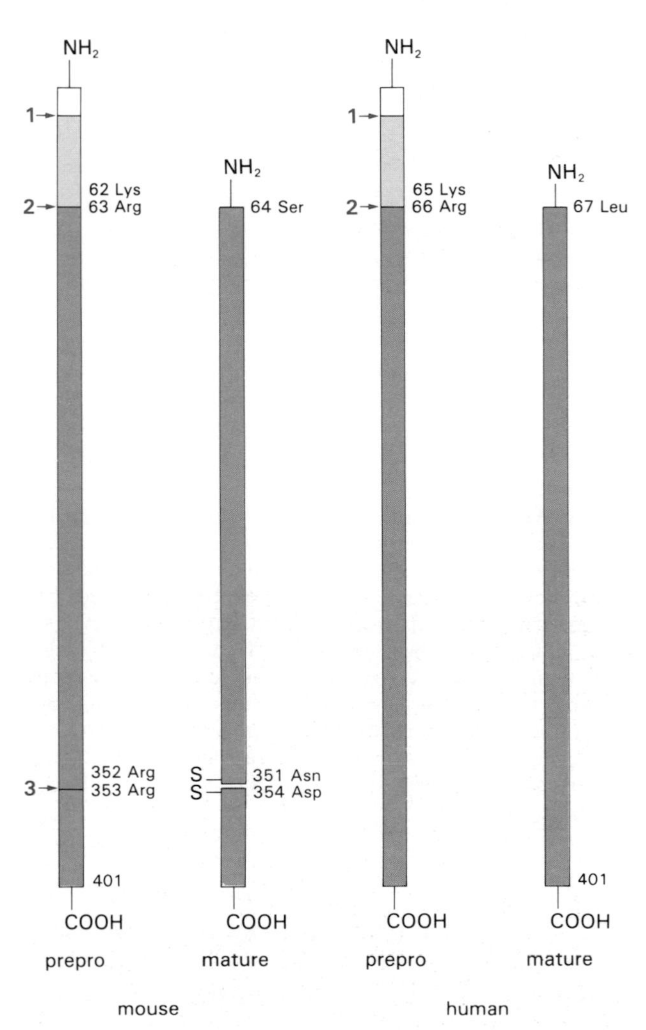

Figure XIII.3 Structure of human renal and mouse submaxillary renin. Renin is synthesized as a preprorenin precursor. It is then processed into a fully active enzyme. Two sequential steps are involved: hydrolysis of the prepeptide (1), and hydrolysis of prorenin (inactive zymogen) into active renin (2). This processing occurs after a dibasic peptide (Lys-Arg). In the case of *submaxillary renin*, a *third* cleavage occurs (3), leading to two renin chains linked by a disulfide bridge.

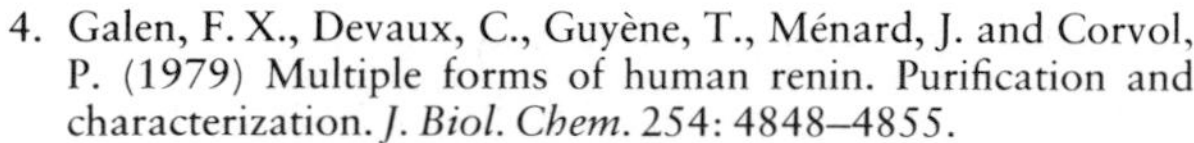

4. Galen, F. X., Devaux, C., Guyène, T., Ménard, J. and Corvol, P. (1979) Multiple forms of human renin. Purification and characterization. *J. Biol. Chem.* 254: 4848–4855.
5. Misono, K. S., Chang, J. J. and Inagami, T. (1982) Amino acid sequence of mouse submaxillary gland renin. *Proc. Natl Acad. Sci., USA* 79: 4858–4862.
6. Panthier, J. J., Foote, S., Chambraud, B., Strosberg, A. D., Corvol, P. and Rougeon, F. (1982) Complete amino acid sequence and maturation of the mouse submaxillary gland renin precursor. *Nature* 298: 90–92.

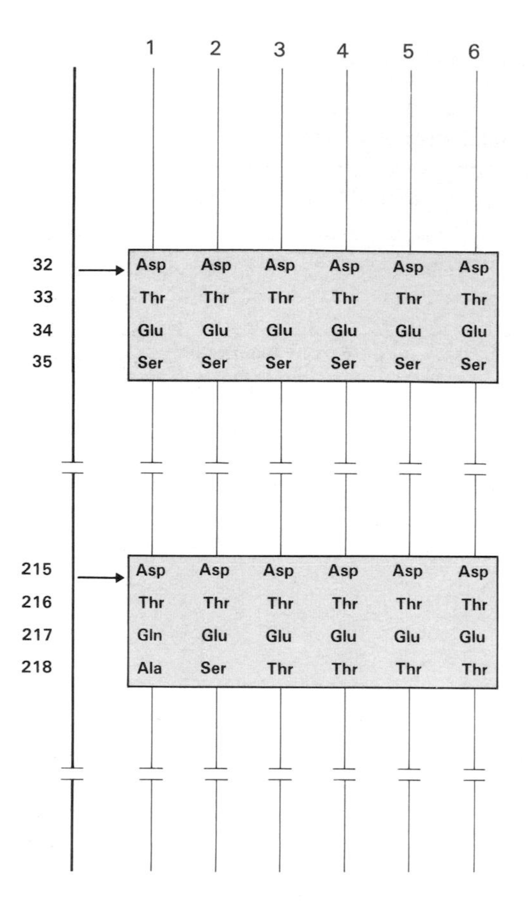

Figure XIII.4 General structure of renin and acid proteases. All acid proteases possess two aspartic acid residues which are essential for catalytic activity (Asp-32 and Asp-215). Note the sequence homology of the amino acids following these two Asp residues. Each acid protease evolved from a single gene, bearing a single aspartic residue, which was then duplicated, explaining the sequence homology found in the two lobes of each acid protease. 1, human renin; 2, mouse renin; 3, endothia pepsin (this enzyme, extracted from microorganisms, was crystallized and used as a three-dimensional model for renin); 4, chymosin; 5, penicillopepsin; 6, pepsin.

A study of the amino acid sequence of mouse and human renin shows striking *homology* with that of *acid proteases*. Pepsin, cathepsin D, chymosin (the gastric enzyme involved in milk coagulation) and other enzymes, such as penicillopepsin, form a *family of enzymes*. All these enzymes, including renin, have the following points in common: 1. two aspartic acid residues involved in the catalytic site (Fig. XIII.4); 2. disulfide bridges localized at the same sites; 3. a potent competitive inhibitor, pepstatin; 4. a similar crystalline structure. Renin has been crystallized, but its three-dimensional structure is not known. The study of the crystal structure of other acid proteases has shown the existence of two separate lobes, with a deep cleft in which the active site is found; 5. finally, these proteins have an N-terminal polypeptide extension of 45 to 46 aa, responsible for the inactive form of the enzyme (pepsinogen, prochymosin, prorenin)[7].

The existence of a bilobar structure in this family of acid proteases is due to the duplication of an ancestral gene which encoded a single acid protease; dimerization ensured enzymatic activity. Gene duplication explains the existence of these two similar zones, each containing an aspartic acid involved in the catalytic activity, which together act as an acid protease. Analysis of the gene structure of submaxillary renin of mice has confirmed this hypothesis recently[8]. Thus, aspartyl proteases have undergone a long evolution, starting from a common gene and terminating as a variety of proteins, one of which is renin, a highly specialized enzyme, responsible for a single function only.

Renin is *synthesized and secreted by the juxtaglomerular cells of the afferent arteriole*, which have all the enzymatic and granular equipment, for hormonal processing and storage, of an endocrine cell. It is probable, as shown in Figure XIII.5, that prorenin is converted to active renin and stored in secretory granules. However, another pathway possibly involves the release of inactive prorenin into the plasma, without prior storage in granules[9]. This would explain the presence of a relatively *high proportion* of *prorenin* (50 to 80% of total renin) in *plasma*. Variations in blood pressure (via baroreceptors) as well as in β-adrenergic hormones, ionic factors and angiotensin II probably all act on these secretory cells. They are responsible for the rapid release of renin which has been previously synthesized and stored, as well as for the initiation of de novo hormone synthesis. Under conditions of intense renin stimulation (sodium depletion, renal ischemia, etc.), there is not only an abundance of renin in the juxtaglomerular cells, but also a recruitment of new secretory cells along the afferent arteriole. Remarkably, vascular smooth mus-

7. Evin, G., Devin, J., Castro, B., Ménard, J. and Corvol, P. (1984) Synthesis of peptides related to the prosegment of mouse submaxillary gland renin precursor: an approach to renin inhibitors. *Proc. Natl. Acad. Sci., USA* 81: 48–52.
8. Holm, I., Ollo, R., Panthier, J. J. and Rougeon, F. (1984) Evolution of aspartyl proteases by gene duplication: the mouse renin gene is organized in two homologous clusters of four exons. *EMBO. J.* 3: 557–562.
9. Galen, F. X., Devaux, C., Gubler, M. C., Mounier, F., Camilleri, J. P., Houot, A. M., Ménard, J. and Corvol, P. (1984) Renin biosynthesis by human tumoral juxtaglomerular cells. Evidence for a renin precursor. *J. Clin. Invest.* 73: 1144–1155.

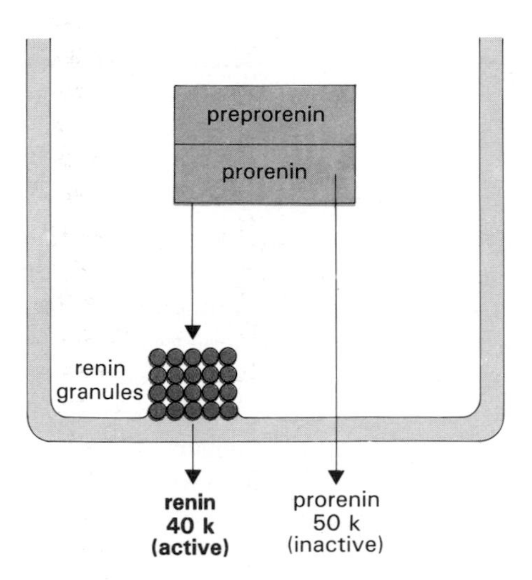

Figure XIII.5 Model of renin biosynthesis and release in the juxtaglomerular cell. Preprorenin is synthesized in the endoplasmic reticulum, and the prefragment is rapidly released. Prorenin could then follow two pathways: *pathway 1* (regulated pathway, leading to a mature form of renin stored in the secretory granules): renin could be released from these granules by acute stimulation (β-agonists, baroreceptors); *pathway 2* (constitutive pathway): renin is not processed or stored in granules, but is excreted in the plasma as prorenin, explaining the relative abundance of this protein in plasma.

cle cells can then be transformed into endocrine cells when necessary, and this differentiation is reversible.

Extrarenal Renin

Plasma renin activity is almost completely *suppressed* in *nephrectomized patients*. It is clear, therefore, that renal secretion contributes primarily to the circulating concentration of this hormone. Plasma renin (mainly in the form of prorenin) which remains following nephrectomy could come either from the arterial walls or from the adrenals. A *small amount* of renin has been demonstrated in a number of *other tissues*, such as the uterus, ovary, testis, placenta, and brain. It is theoretically possible that a local hormonal system exists in such tissues; the circulating concentrations would therefore not be reflective of the real concentration within tissues.

Renin in vascular walls. The presence of renin activity in vascular walls has been suggested by the in vitro observation that tissue extracts can produce angiotensin from a plasma substrate, and also by the persistence of vascular renin activity following nephrectomy[10]. Moreover, aortic myocytes in culture synthesize renin. However, in vivo models have not been developed for defining the physiological or pathological role of renin produced by vascular walls.

Renin in the brain. Components of a renin-angiotensin system have been found in the brain, but it is not universally accepted that the brain is a site of actual synthesis. Although angiotensinogen converting enzyme and angiotensin II are certainly present in the brain, the nature of the enzyme which cleaves angiotensinogen remains controversial. Most of the "renin activity" measured could be related to cathepsin D; however, using antibodies specific to renal renin, the existence of a *small* amount of *authentic renin* different from cathepsin D has been confirmed. It has been localized especially in the region of the *hypothalamus* (supraoptic and paraventricular nuclei) and in the *anterior pituitary* (Table XIII.1).

The *central administration* of renin results in *polydipsia*, an increase of arterial pressure, and in vasopressin secretion. These effects are accompanied by an augmentation in the concentration of angiotensin II in cerebrospinal fluid. They can be reproduced by the intracisternal administration of angiotensin II.

Enzymes with an *activity similar* to that of renal renin have been found in human adrenals, placenta, uterus, and testes. The role of an amniotic enzyme in the regulation of utero-placental blood flow remains uncertain. The possibility that renin-angiotensin systems work locally, in a paracrine fashion, opens a number of possibilities. For instance, angiotensin may act as an intracerebral neuromediator or as a modulator of tissue blood flow.

Angiotensinogen

In situ, renin acts on a *privileged substrate*, angiotensinogen, in the kidney and plasma (Fig. XIII.6). Angiotensinogen is a *globulin* with a MW of 60,000 which is *synthesized by the liver* and immediately released *into the circulation*. Its primary structure has been deduced from the nucleotide sequence of its cDNA in both the rat[11] and human[12]. Human angio-

10. Swales, J. D., London, M., Bing, R. F. and Thurston, H. (1983) Renin in the arterial wall. *Clin. Exp. Hypertens.* A5: 1127–1136.
11. Ohkubo, H., Kageyama, R., Ujihara, M., Hirose, T., Inayama, S. and Nakanishi, S. (1983) Cloning and sequence analysis of cDNA for rat angiotensinogen. *Proc. Natl. Acad. Sci., USA* 80: 2196–2200.
12. Kageyama, R., Ohkubo, H. and Nakanishi, S. (1984) Primary structure of human preangiotensinogen deduced from the cloned cDNA sequence. *Biochemistry, USA* 23: 3603–3609.

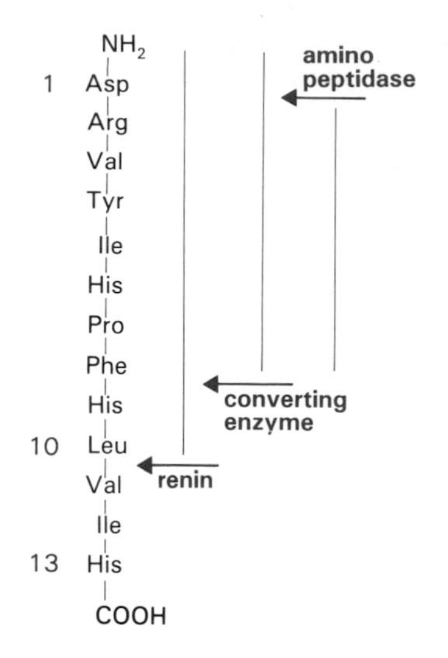

Figure XIII.6 Angiotensin formation in the human. Renin cleaves the Leu-Val bond of angiotensinogen, producing angiotensin I, the N-terminal tetradecapeptide of angiotensinogen. *Angiotensin I* is converted into angiotensin II, the biologically active angiotensin, by converting enzyme (a carboxydipeptidase). *Angiotensin II* is then converted into angiotensin III by an aminopeptidase. Finally, *angiotensin III* is degraded into inactive products by several exo- and endopeptidases.

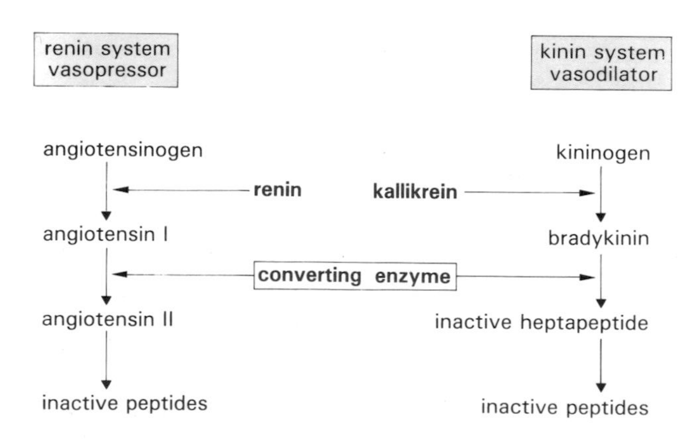

Figure XIII.7 Action of converting enzyme on the renin angiotensin and kallikrein-kinin systems. The same enzyme cleaves the inactive angiotensin I into an active vasopressor peptide and inactivates bradykinin. The inhibition of converting enzyme will lead to the blockade of a vasopressor system and to the potentiation of a natriuretic and vasodilator system.

tensinogen has four potential sites of glycosylation, which accounts for its microheterogeneity.

Converting Enzyme

Angiotensin I only acquires biological activity following its transformation *into* an *octapeptide, angiotensin II*, under the influence of *converting enzyme*. This enzyme is present in plasma, vascular endothelia, and numerous tissues such as kidney and lung[13]. It releases a dipeptide, histidyl-leucine, from the C-terminal end of angiotensin I. This converting enzyme also hydrolyzes bradykinin, a vasodilator and natriuretic peptide, to an inactive heptapeptide (Fig. XIII.7). It is therefore a *key enzyme* in the regulation of arterial pressure and water and salt metabolism, since it *controls, simultaneously, a vasoconstrictor system* (angiotensin II), resulting in retention of water and salt (by the intermediary of aldosterone secretion), *and a vasodilator and natriuretic system*, which functions via the regulation of bradykinin levels.

Angiotensins

The structures of angiotensin I, II and III are shown in Figure XIII.6. The *octapeptide angiotensin II is the strongest known vasoconstrictor*. It is broken down rapidly in peripheral capillary systems by several enzymes, called angiotensinases. Among the metabolites produced, desaspartyl-1 angiotensin II has physiological activity: this *heptapeptide (angiotensin III)* is abundant in the adrenal; recently, its physiological importance in the secretion of aldosterone has been shown. Angiotensin III can also be formed directly, by the action of converting enzyme on desaspartyl-1 angiotensin I. The *hexapeptide* desaspartyl-1-arginyl-2 angiotensin II also has specific biological activity, i.e. suppression of renin secretion and stimulation of aldosterone secretion.

Angiotensinases

The term *angiotensinase* is used for all proteolytic enzymes capable of hydrolyzing a peptide bond of angiotensin II, although *none* appear to be *specific*. Their enzymatic activity operates in numerous tissues as well as in plasma. They explain, in part, the *rapid degradation* of *angiotensin II*, since, following injection of an iv bolus, its pressor effect lasts for only three minutes.

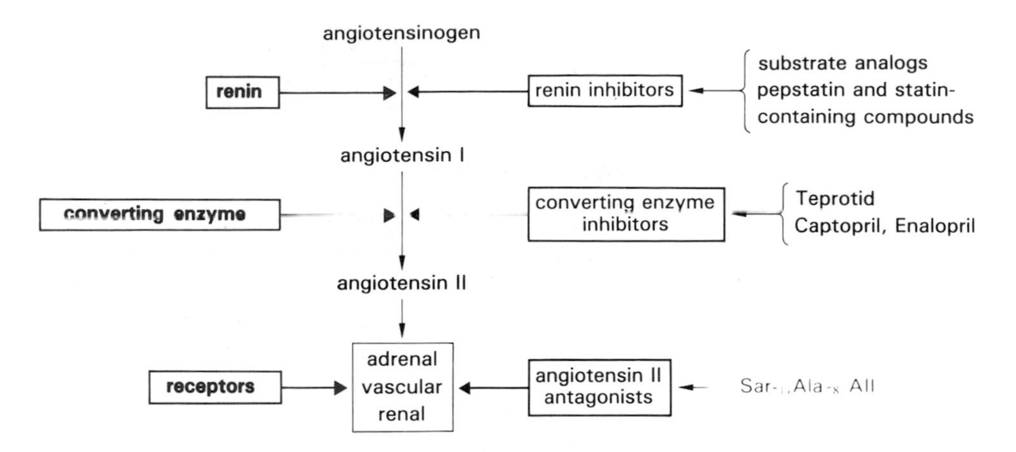

Figure XIII.8 Inhibitors acting on the renin angiotensin system at different levels. Orally active converting enzyme inhibitors (captopril, enalapril) are currently used in the treatment of essential hypertension.

Inhibitors of the Renin-Angiotensin System

The *cascade* of *enzymatic reactions* (renin, converting enzyme, aminopeptidases), and biochemical and physiological events of the renin system (synthesis and secretion of renin by the kidney, synthesis and secretion of angiotensinogen by the liver, interaction of angiotensin II and III with their receptors), *can be blocked at different levels* (Fig. XIII.8). These inhibitors allow the demonstration of the exact role of the renin system in the control of blood pressure and in the maintenance of water and electrolyte balance.

A reduction of the secretion of renin by the kidney is caused by β-blockers. More *specific inhibitors* have been synthesized to block the specific effect of angiotensin II, or to *inhibit converting enzyme or renin competitively* (for example, Saralasin[14]). Angiotensin III can also be inhibited by competitive peptides. Converting enzyme is inhibited by *site-specific analogs,* such as orally active captopril or enalapril, which are used in the treatment of hypertension and heart failure[15]. Competitive peptide derivatives can impede renin action on its substrate. Some are highly specific and active, and are currently under study for their potential clinical applications[16].

13. Ryan, U. S., Ryan, J. W., Whitacker, C. and Chin, A. (1976) Localization of angiotensin converting enzyme (Kininase II). II. Immunocytochemistry and immunofluorescence. *Tissue Cell* 8: 125–145.
14. *Kidney Intern.* (1979) 15: special issue.
15. Symposium on the renin angiotensin system: treatment of hypertension and heart failure. (1983) *Suppl. J. of Hypertension* 1.
16. Corvol, P., Michel, J. B., Evin, G., Gardes, J., Castro, B. and Ménard, J. (1983) Inhibiteurs du système rénine au niveau de la réaction rénine-substrat. *Ann. Endocr., (Paris)* 44: 339–342.
17. Reid, I. A., Morris, B. J. and Ganong, W. F. (1978) The renin angiotensin system. *Annu. Rev. Physiol.* 40: 340–377.
18. Keeton, T. K. and Campbell, W. B. (1980) The pharmacological alteration of renin release. *Pharmacol. Rev.* 32: 82–202.
19. Davis, J. O. and Freeman, R. H. (1976) Mechanisms regulating renin release. *Physiol. Rev.* 56: 1–56.

1.2 Regulation of the Renin-Angiotensin System

At different levels

Secretion of Renin by the Juxtaglomerular Apparatus of the Kidney

Active renin is an essential factor in the regulation of the renin-angiotensin systems, and is the one which is most studied[17–19].

Five Essential Variables

Pressure in the Afferent Arteriole

When *pressure* is *reduced* in the renal artery (constriction of the artery, hypovolemia or orthostatism in the human), the secretion of *renin increases* independently of any reduction in glomerular filtration or renal blood flow, and in the absence of any renal hypoxia. Davis' model of the non-filtering kidney has allowed the demonstration that variations in pressure or tension in the afferent arteriole can modulate the secretion of renin independently of the macula densa and of sympathetic innervation of the juxtaglomerular apparatus.

The precise mediator(s) involved in the release of renin under the influence of stretch receptors, or baroreceptors, is not known, but the involvement of calcium ions or prostaglandins has been suggested.

Modifications of Urine Composition at the Level of the Macula Densa

The release of renin is inversely related to the level of sodium and the transport of sodium across the macula densa[20]. Thus, when the secretion of renin is stimulated by a reduction in renal arterial pressure, the injection of a diuretic *increases sodium uptake* across the macula densa, and *reduces the secretion of renin.* The ion responsible for this regulation may, in fact, be *chloride*[21].

Stimulation of the Adrenergic System

Electrical stimulation of sympathetic perirenal nerves *increases renin* secretion whether the kidney is filtering or not. A release of *catecholamines* (norepinephrine) induced by carotid occlusion or administration of a central anticholinesterase (physostigmin) increases renin and provokes a release of epinephrine induced by hypoglycemia. Epinephrine, norepinephrine and isoproterenol stimulate renin secretion (Fig. XIII.9)[22], which can be blocked by β-blockers[18]. *β-adrenergic receptors* (β_1 and β_2) are present in the juxtaglomerular apparatus, however it is less certain that α-adrenergic receptors play an important role in the regulation of renin secretion.

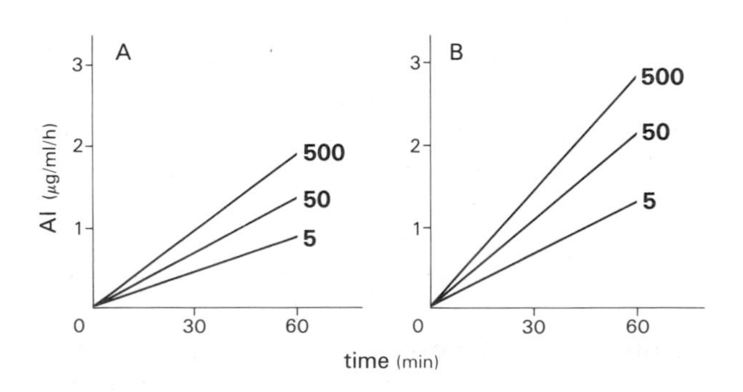

Figure XIII.9 Effects of β-adrenergic agonists on renin secretion in an isolated perfused kidney. Two β-adrenergic agonists, expressed by their effects on angiotensin I (AI), were studied at different doses, 5 nM, 50 nM, and 500 nM. **A.** Isoproterenol, a non-selective β agonist. **B.** Salbutamol, a selective β_2 agonist. Both stimulate renin secretion in a similar dose-response curve[22].

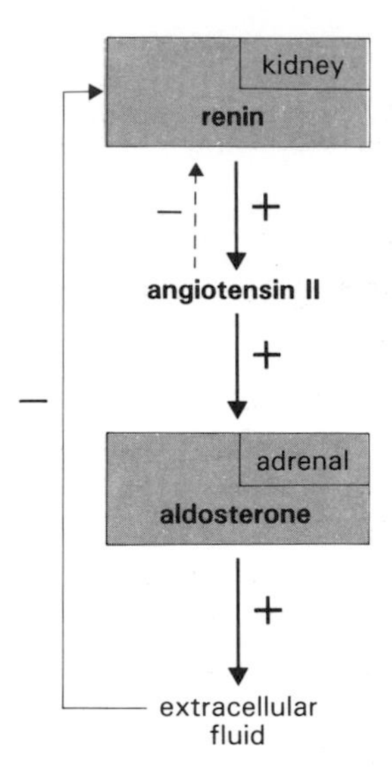

Figure XIII.10 Feedback control of renin secretion. Two negative feedback mechanisms are present: a long feedback loop, via extracellular fluid variations, which controls renin secretion through the baroreceptor, and a short and potent feedback loop, through a direct, negative action of angiotensin II (and III) on renin secretion.

Angiotensin II

Angiotensin II, III and the hexapeptide suppress the secretion of renin independently of a positive sodium balance, and of the three, angiotensin II has the greatest activity (Fig. XIII.10). The importance of this *short feedback mechanism* on the secretion of renin has been demonstrated by the administration of substances blocking the renin-angiotensin system. Antagonists of angiotensin II (Sar-1 Ala-8 angiotensin II) and inhibitors of converting enzyme (captopril) increase the secretion of renin and suppress the negative feedback effect of angiotensin II on the juxtaglomerular cell.

Calcium

Calcium may play the role of a chemical mediator *coupling* the *stimulus* of the *intrarenal baroreceptor both to secretion and to β-adrenergic stimulation.* Several direct experiments suggest that the release of

20. Vander, A. J. (1967) Control of renin release. *Physiol. Rev.* 47: 359–382.
21. Kirchner, K. A., Kotchen, T. A., Galla, J. H. and Luke, R. G. (1978) Importance of chloride for inhibition of renin by sodium chloride. *Amer. J. Physiol.* 235: F444–F450.
22. Nakane, H., Nakane, Y., Roux, A., Corvol, P. and Ménard, J. (1981) Effect of selective and nonselective beta adrenergic agents on renin secretion in isolated perfused rat kidney. *J. Pharmacol. Exp. Ther.* 212: 34–38.

renin and the intracellular level of calcium are inversely related, the increase in intracellular calcium having opposite effects on juxtaglomerular and endocrine cells.

Availability of Angiotensinogen and Angiotensins

Angiotensinogen is a limiting factor in the production of angiotensin II. Thus, to a certain degree, its availability contributes to the stimulation or inhibition of the renin-angiotensin system. In contrast to renin in the kidney, angiotensin is not stored in the liver cell in which it is synthesized. An increase in *thyroid* hormones (hyperthyroidism), *estrogens* (oral contraceptives, pregnancy) as well as *glucocorticosteroids* (Cushing's syndrome, corticotherapy), leads to *increased levels.* In contrast, adrenalectomy and hepatic insufficiency reduce the renin substrate. A reduction of the level of angiotensin II during chronic administration of captopril also reduces the level and the production of angiotensinogen. It is possible that this hormone has a direct, positive effect on the secretion of angiotensinogen. Adrenalectomy and treatment with an inhibitor of converting enzyme are accompanied by an increase in the peripheral consumption of angiotensinogen.

1.3 Biological Properties of Angiotensins

Whereas angiotensin I has a minimal physiological role, *angiotensin II* is the main effector of the renin-angiotensin system, maintaining circulatory homeostasis and the *stimulation of aldosterone* via its direct effect on vascular smooth muscle (Table XIII.2).

Table XIII.2 Biological and pharmacological properties of angiotensin.

Modulation of renal hemodynamics: Action on renal blood flow Action on glomerular filtration rate Direct tubular natriuretic effect
Stimulation of vascular smooth muscle
Stimulation of aldosterone secretion
Negative feedback on renin secretion
Positive myocardial inotropic effect
Maintenance of plasma osmolality
Stimulation of thirst
Catecholamines, stimulation and potentiation
Hypertension, by a central effect
Stimulation of vasopressin release
Stimulation of prostaglandin production
Increase in uterine contractility

Effect on Vascular Smooth Muscle

Angiotensin II exerts its action *in 10 to 15 s following iv administration.* Its effect varies depending on the region: vasoconstriction occurs in mesenteric and cutaneous circulation without affecting blood flow in muscle.

Angiotensin II Directly Stimulates Adrenal Steroidogenesis

Angiotensin II has a specific *stimulatory* effect on the secretion of *aldosterone* by the zona glomerulosa of the adrenal. Infusion of angiotensin II and III increases the secretion of aldosterone, even at doses which are subthreshold for blood pressure increase. The increase in adrenal aldosterone secretion is due to stimulation at an early stage in biosynthesis (between cholesterol and pregnenolone) (Fig. IX.5). Prolonged stimulation produces a tropic effect on the zona glomerulosa as well as stimulation of a later stage in the biosynthesis of aldosterone: the *conversion of corticosterone to 18-hydroxycorticosterone.*

The determining factors in the control of aldosterone secretion are the end-products of the renin-angiotensin system. Angiotensin II and to a lesser extent angiotensin III are the essential elements of the regulation of aldosterone in the face of variations in sodium balance. Sodium restriction or diuretics can increase the production of renin, angiotensin II, and aldosterone. The stimulation of the renin-angiotensin system can be detected slightly before an increase in aldosterone. An acute or chronic increase in the intake of sodium results in a reduced level of activity of plasma renin, angiotensin II, and aldosterone. Finally, the inhibition of the renin-angiotensin system by an inhibitor of converting enzyme is accompanied by a fall in the production of aldosterone (p. 393 and Fig. XIII.8).

It is possible, therefore, to demonstrate that the renin-angiotensin system is the primary cause of variations in aldosterone levels observed with changes of sodium balance. This, however, does not mean that it is the only factor responsible for modification of aldosterone levels.

Angiotensin II Directly Affects Renal Hemodynamics

At low doses, angiotensin II leads to a *reduction in glomerular filtration and renal blood flow,* as well as a reduction in the secretion of sodium and water. The reduction in renal blood flow is generally greater than that of glomerular filtration, resulting in an increase in the filtered fraction (ratio of glomerular filtration to renal blood flow)[23].

This increase in the filtered fraction can be interpreted as an action of angiotensin II on the efferent arteriole of the glomerulus. At high doses, an increase in arterial pressure leads directly to natriuresis, which masks the antinatriuretic effect of angiotensin.

It has been suggested that angiotensin II reduces external cortical blood flow at the expense of that of the deep cortex and medulla. The fact that the long-loop juxtaglomerular nephrons are better able to reabsorb sodium may explain the antinatriuretic action of angiotensin II.

Other Pharmacological Effects

Angiotensin II has a positive *inotropic* effect on the myocardium and an indirect pressor effect via the intermediary of the central nervous system. The *central* action of angiotensin II also involves a *stimulation of thirst* and a net *stimulation of vasopressin* secretion. It may also play a role in the *maintenance of plasma osmolality* (Fig. XIII.11). Angiotensin II stimulates the secretion of catecholamines by the adrenal medulla and sympathetic nerve terminals.

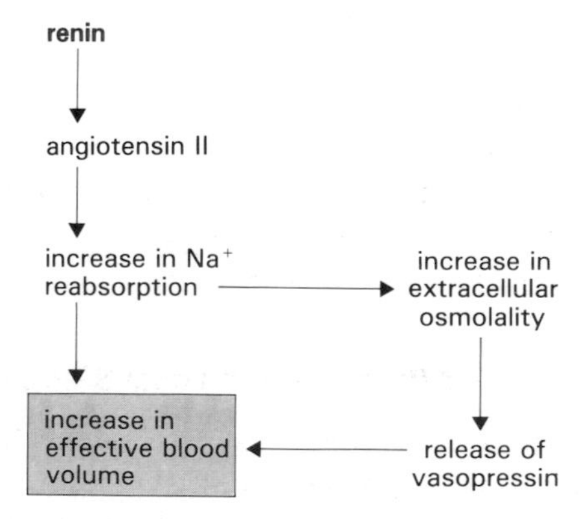

Figure XIII.11 Interaction between the renin-angiotensin system and vasopressin in the maintenance of plasma volume and plasma osmolality. In addition to its indirect effect on the secretion of vasopressin, angiotensin II has been shown to control vasopressin release directly.

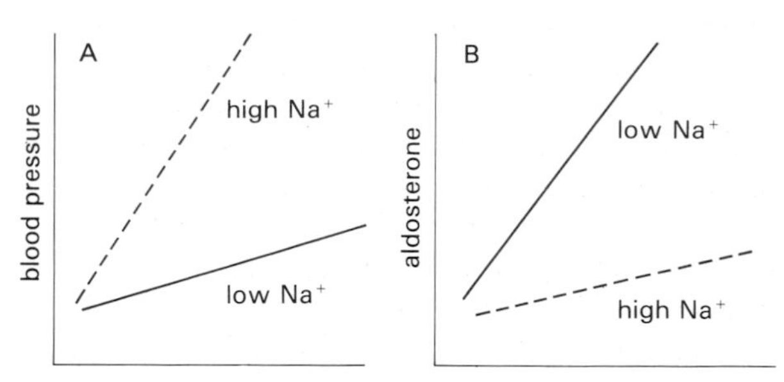

Figure XIII.12 In vivo effects of angiotensin II on blood pressure (**A**) and aldosterone secretion (**B**). The magnitude of the effects *depends* on the status of the *sodium balance*, and varies with the pressor and aldosterone-stimulating effect.

Angiotensin II Activities

Figure XIII.12 shows that the same doses of angiotensin II lead to an effect which *depends on the sodium balance* of the animal or patient: the same dose of angiotensin II has a more pronounced pressor effect and is less active in aldosterone stimulation given a sodium repleted state than one of sodium depletion. Thus, with a reduced sodium diet, infusion of angiotensin II has only a weak pressor effect, while it leads to a large increase in the secretion of aldosterone.

Angiotensin III

It is possible that at least a portion of the adrenal action of angiotensin II could be due to its partial conversion to angiotensin III.

Angiotensin Receptors

Angiotensin receptors have been demonstrated by observing the correlation between the kinetics of high affinity binding of radioactive angiotensin and the biological response (contraction of smooth muscle)[24]. These receptors are located on the *plasma mem-*

23. Keyssac, P. P. (1967) Intrarenal function of angiotensin. *Fed. Proc.* 26: 55–59.
24. Devynck, M. A. and Meyer, P. (1978) Angiotensin receptors. *Biochem. Pharmacol.* 27: 1–5.

brane of *vascular smooth muscle* cells and *adrenal cells*. There are *also* receptors in the liver, brain and pituitary, and on the basal membrane of the glomerulus. Recently, the coupling mechanism responsible for the biological effects of angiotensin II (e.g., stimulation of gluconeogenesis in the liver and prolactin secretion in the pituitary) have been partly elucidated. Activation of the adenylate cyclase system is not involved. Intracellular free *calcium* is *increased*, as well as the *metabolism* of *membrane inositol lipids*. It is not yet clear whether angiotensin receptors differ from one tissue to another. For example, receptors in the adrenal zona glomerulosa have a greater affinity for angiotensin III than for angiotensin II, and the reverse is true in vascular tissue. Such an observation is potentially interesting in that it may lead to the selectivity of a receptor for an agonist or antagonist.

Receptor concentration is not fixed. There is a reduction in the pressor effect of exogenous angiotensin II in the presence of elevated plasma concentration of endogenous angiotensin II, which may be related to down-regulation of receptors (p. 85). The potent action of angiotensin II on the secretion of aldosterone in low sodium conditions is correlated with an increase in angiotensin II receptor concentration in the adrenal.

1.4 Effects of the Renin-Angiotensin System

The renin-angiotensin system is *unique in the regulation of blood pressure* in that it alone is capable of *simultaneously* influencing the state of vasoconstriction or vasodilatation of the *arterial* bed and *extracellular* blood volume (Fig. XIII.13).

Vasoconstrictor Action

Angiotensin II, a vasoconstrictor, acts *directly* on vascular sensitivity to angiotensin II, which is diminished when dietary intake of potassium is reduced.

This direct pressor action of angiotensin II is controlled by sodium intake, as described above. In addition, intake of potassium has a positive effect on vascular sensitivity to angiotensin II.

Action on Blood Volume

The quantity of sodium excreted into the urine represents the small difference between the relatively large amount filtered by the glomerulus and that reab-

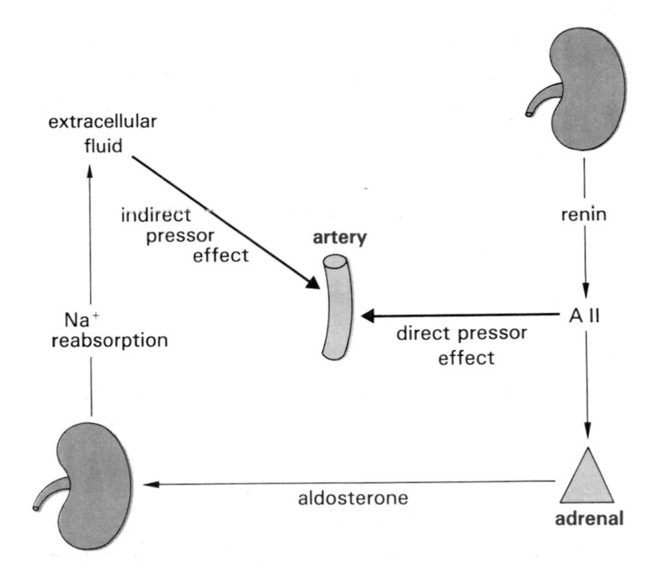

Figure XIII.13 Actions of the different hormones of the renin-angiotensin-aldosterone system on blood pressure and extracellular fluid volume.

sorbed by the tubule. Variation in the quantity filtered by the glomerulus has only a minimal influence on the quantity excreted; thus, the factors controlling the *reabsorption of sodium* are *determinant*. Among these, aldosterone plays a major role. The secretion of *aldosterone* in vivo depends essentially on the renin system. This system has a major role in the control of aldosterone secretion in relation to sodium balance. Parallel to an increase in total sodium under the influence of aldosterone is an increase in the secretion of vasopressin, which is dependent upon angiotensin II, thus maintaining plasma osmolality (Fig. XIII.11).

In conclusion, angiotensin II is an *essential element in the control of blood pressure, even in moderate sodium depletion*. It is also the major stimulus for aldosterone to restore sodium balance, thus exerting actions on both the tonicity of the vessel wall and the sodium content of the vessel. Its action is amplified by interaction with the nervous system, and is modulated by the intake of sodium (Fig. XIII.12).

Action on Blood Pressure and Local Blood Flow

The renin-angiotensin system is activated by factors reducing blood pressure or renal perfusion pressure (such as hemorrhage, shock, or heart failure), or by factors which reduce the uptake of sodium in the distal tubule (such as sodium depletion). This regulation is under both short- and long-loop feedback control (Fig. XIII.10).

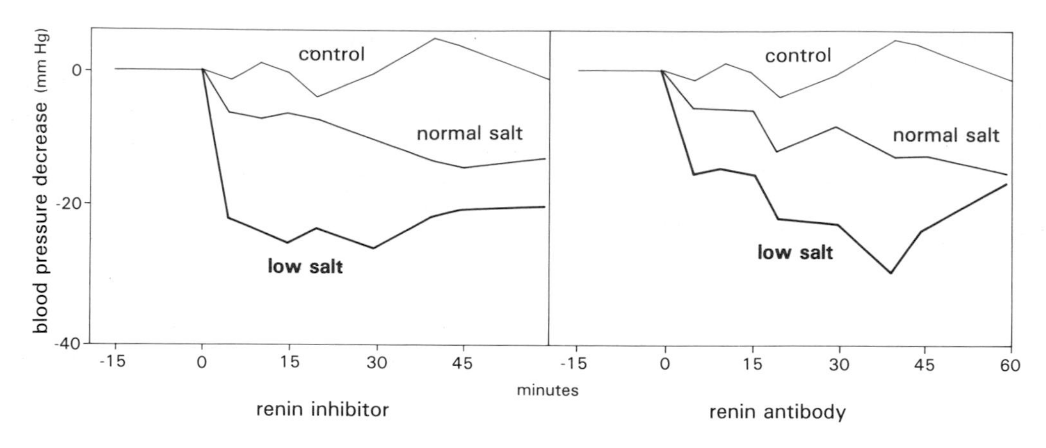

Figure XIII.14 Effect of an inhibitor (teprotide) of the converting enzyme and of human renin antibodies on blood pressure in the marmoset. Both substances, acting at two different levels of the renin system, produce a similar decrease in blood pressure. Their effects depend on the status of the animal's sodium balance, the effect being more pronounced in salt depletion[25].

The *renin system* therefore plays a major physiological role in the control of blood pressure. This role becomes *more important as sodium depletion is increased*, as demonstrated by specific inhibitors of the renin system. Under conditions of *elevated sodium intake*, the administration of angiotensinogen antibodies in the rat, or captopril in humans, leads to a slight but significant reduction in arterial pressure. For a *normal sodium diet*, teprotide and captopril in humans, as well as inhibitors of the renin–substrate reaction in animals, leads to about a 10 to 15% reduction in blood pressure (Fig. XIII.14)[25–28]. Given *low sodium intake*, the activity of the renin-angiotensin system is underscored by the elevated plasma renin and aldosterone levels and the significant decrease in blood pressure (15 to 40%) observed under the influence of inhibitors of the renin system.

25. Michel, J. B., Wood, J., Hofbauer, K., Corvol, P. and Ménard, J. (1984) Blood pressure effects of renin inhibition by human renin antiserum in normotensive marmosets. *Amer. J. Physiol.* 246: F309–F316.
26. Szelke, M., Leckie, B., Hallett, A., Jones, D. M., Suiras, J., Atrash, B. and Lever, A. F. (1982) Potent new inhibitors of human renin. *Nature* 299: 555–557.
27. MacGregor, G. A., Markandy, N. D., Roulston, J. E., Jones, J. C. and Morton, J. J. (1981) Maintenance of blood pressure by the renin angiotensin system in normal man. *Nature* 291: 329–331.
28. Haber, E. (1984) Which inhibitors will give us true insight into what renin really does? *J. Hypertension* 2: 223–230.

2. Aldosterone

Aldosterone was isolated from urine and adrenals by Simpson and Tait. It is *the physiological mineralocorticosteroid* hormone (Fig. XIII.15) (p. 393).

Its role is essential in the *conservation of sodium* and in the *excretion of potassium* from the body. Because of this, under physiological conditions it cooperates simultaneously in the regulation of blood volume and blood pressure. In pathological states, excessive production of aldosterone causes edema and some forms of arterial hypertension. Aldosterone deficiency poses a major problem for the body due to loss of sodium and retention of potassium.

2.1 Renal and Extrarenal Actions of Aldosterone on Excretion of Electrolytes

Excretion of Sodium by the Nephron

Before describing the hormonal control of sodium excretion, a description of the major sites of the nephron which control sodium excretion is warranted (Fig. XIII.16).

Glomerular Filtration

Glomerular filtration regulates the quantity of sodium reaching the renal tubules. The level is *very high*, 20,000 mEq Na^+/d in humans, in comparison with a daily urinary excretion of 50 to 200 mEq Na^+. Thus, a variation, even minimal, of filtration could, theoretically, have a major influence on sodium excretion.

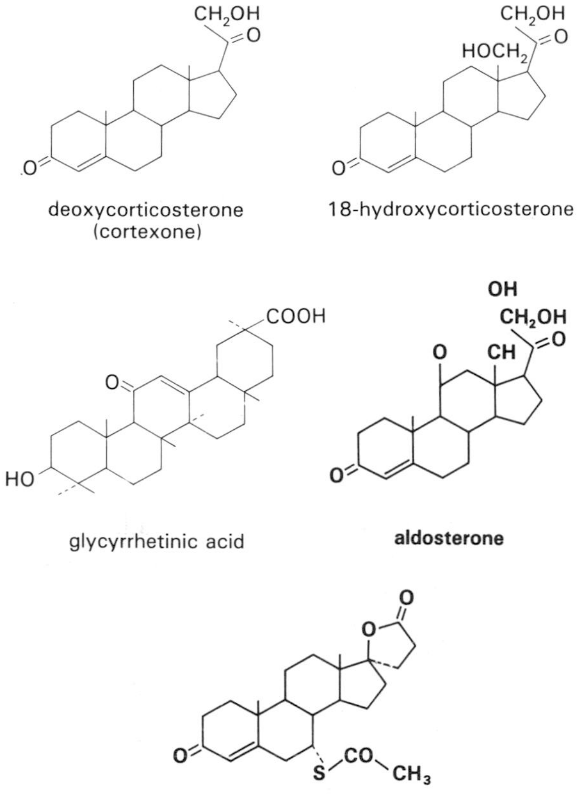

deoxycorticosterone (cortexone)

18-hydroxycorticosterone

glycyrrhetinic acid

aldosterone

spironolactone (aldactone®)

Figure XIII.15 Structure of several mineralocorticosteroids: deoxycorticosterone (DOC), 18-hydroxydeoxycorticosterone, glycyrrhetinic acid, and aldosterone. The most potent natural mineralocorticosteroid is aldosterone. DOC, 18-hydroxy-DOC and glycyrrhetinic acid (contained in licorice) possess, respectively, about 10%, 1% and 0.1% of the mineralocorticosteroid activity of aldosterone. Also shown is the structure of spironolactone, a potent antimineralocorticosteroid agent.

Proximal Tubule

Sixty to 80% of the sodium filtered is reabsorbed in the proximal tubule. There is an equilibrium between glomerular filtration and the reabsorption of sodium in the proximal tubule, which is probably dependent on intrarenal physical phenomena. *Oncotic pressure* and the *hydrostatic pressure* in proximal peritubular capillaries play an important role in sodium reabsorption. Reduction of oncotic pressure and increase of hydrostatic pressure following an expansion of blood volume are accompanied by a reduction in proximal tubule sodium reabsorption.

Henle's Loop

In the *ascending branch of Henle's loop*, there is an *active reabsorption of solutes* without reabsorption of water. A counter-current system is developed such that high osmolalities are found at the lowest level of the loop.

Distal and Collecting Tubules

When urine reaches the distal tubule, the major portion of sodium, and all potassium, has been reabsorbed. Only the *distal tubule* is involved in the *final regulation of sodium and potassium excretion.* It is a privileged site of exchange between cations, that of Na^+ for K^+ or H^+. This is the level at which mineralocorticosteroids have their primary action. The cortical collecting tubule, at the point where it merges with the distal tubule, has the same properties (Fig. XIII.16).

Finally, there are *two categories of nephrons*: one is *cortical*, characterized by a short Henle's loop, and the other *juxtamedullary*, with a long Henle's loop. The latter has a greater ability to retain sodium than the former. Variations in the distribution of renal blood flow may favor the perfusion of one or another type of nephron. In summary, all variations in sodium balance activate mechanisms affecting the retention or elimination of sodium by the kidney. The *most important* site of *ionic exchange* is situated *upstream* of the *distal convoluted tubule.* The only clearly established *hormonal control* of sodium excretion occurs at the level of the *distal convoluted and collecting tubules,* and is under the influence of aldosterone.

Renal Effects of Aldosterone

Effect of a Single Administration of Aldosterone in Animals

The *suppression of aldosterone* production by adrenalectomy results in urinary loss of sodium, a *negative sodium balance*, and retention of potassium. The *administration of aldosterone* in adrenalectomized animals *restores sodium* and potassium *equilibrium*, and *prolongs life*.

However, the doses of aldosterone used in early studies did not take into account the fact that *aldosterone has dual activity*: at low and physiological doses, aldosterone exerts a mineralocorticosteroid effect, whereas at high and pharmacological doses, it

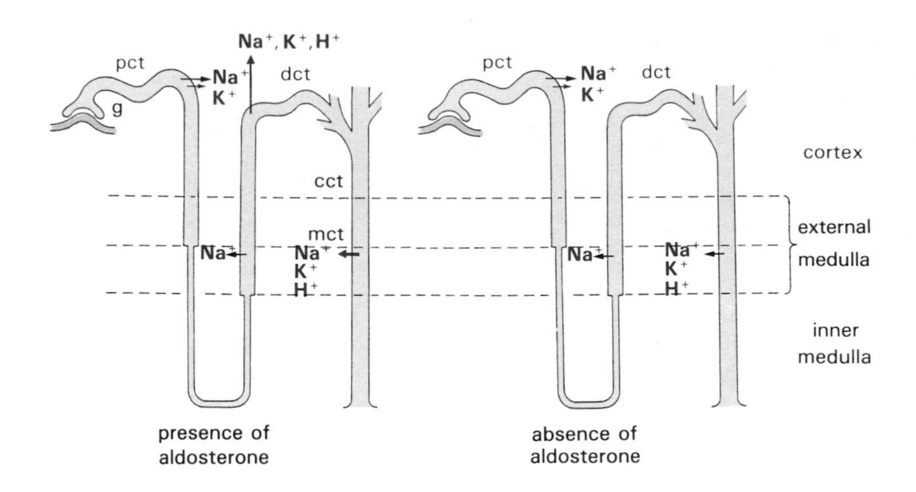

Figure XIII.16 Metabolism and effect of aldosterone on the kidney. Sites of excretion and reabsorption of sodium along the nephron (see also Fig. VI.10) (g, glomerulus; cct, cortical collecting tubule; dct, distal convoluted tubule; pct, proximal convoluted tubule; mct, medullary collecting tubule). Aldosterone and the other mineralocorticosteroids act primarily on the cct and dct, but the action of aldosterone on the mct remains controversial.

also has a glucocorticosteroid effect. There are indeed mineralocorticosteroid and glucocorticosteroid receptors in the kidney. The antinatriuretic effect of aldosterone is always observed, whereas its kaliuretic effect may be less pronounced, depending on several factors, such as sodium delivery at the distal tubule, and potassium balance. Under conditions of normal sodium and potassium intake, aldosterone exerts a strong antinatriuretic effect, within 30 to 60 min, which lasts for several hours (Fig. XIII.17). Under the same conditions, the kaliuretic effect of aldosterone appears at the same time as its antinatriuretic effect. Similarly, aldosterone seems to increase H^+ ion excretion, but its mechanism is uncertain. Aldosterone has no direct action on the excretion of calcium, magnesium, or phosphorus. At physiological doses, aldosterone has no effect on the excretion of free water by the kidney.

The effects of aldosterone on the metabolism of sodium and potassium are *shared by other natural mineralocorticosteroids* (deoxycorticosterone, 18-hydroxydeoxycorticosterone) *and synthetic steroids*, such as 9α-fluorocortisol. Since deoxycorticosterone was isolated before aldosterone, the mineralocorticoid activity of this steroid, or its acetate (*DOCA*), frequently serves as a reference (Fig. XIII.15).

The action of all mineralocorticosteroids can be specifically *antagonized by spironolactone* (Fig.

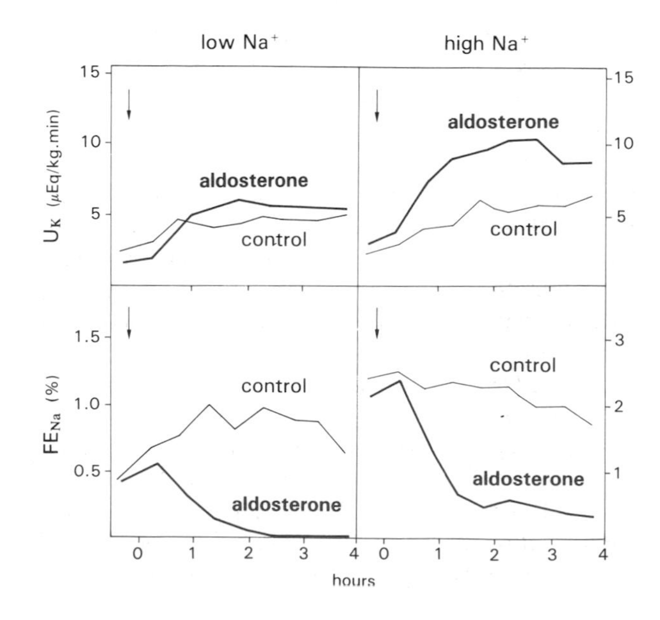

Figure XIII.17 Time course of effects of aldosterone on urinary potassium excretion (upper panel) and fractional sodium excretion (lower panel). Each point represents one 30 min clearance period. Arrows indicate the time of onset of aldosterone administration (1 μg/kg followed by an infusion at a 1 μg/kg/h). **A.** Rats receiving a low Na^+ solution (22.0 mM) and 4.0 mM K^+. **B.** Rats receiving a 70 mM Na^+ and 4.0 mM K^+ solution. Note that the kaliuresis is not increased in low Na^+-loaded rats and is highly increased by aldosterone in the salt-loaded rats. This illustrates the importance of Na^+ delivery at the level of the distal tubule.

XIII.15). This antimineralocorticosteroid inhibits the binding of aldosterone to its receptor; the spironolactone–receptor complex has only a weak affinity for chromatin, and has no agonistic effect. Spironolactone is both natriuretic and antikaliuretic; the magnitude of its effect depends on the circulating level of aldosterone and the quantity of sodium which reaches the distal tubule.

Effects of Aldosterone in Humans

The effect of aldosterone in normal individuals, or in patients deprived of glucocorticosteroids (Addison's disease), is *similar to* effects observed in animals. A single injection of aldosterone leads to a retention of sodium and an excretion of potassium.

Delayed Action of Aldosterone

In humans and other animals, there is *always a delay* between the administration of aldosterone and evidence of its effects on the excretion of electrolytes. This delay varies between 60 and 180 min, and is also observed with other mineralocorticosteroids. It probably corresponds to the time required for the *synthesis of protein* necessary for the active transport of sodium.

Site of Action

Clearance and stop-flow studies have indicated that the site of aldosterone action is most probably the *collecting tubule (cortical and medullary, cct and mct)*. Aldosterone allows the *exchange* between Na^+ *ions* in the *tubular region*, which are *reabsorbed,* and H^+ *and* K^+ *ions* in the *plasma, which* are *excreted.* The importance of these exchanges depends on the quantity of sodium ions available in the distal tubule and on the concentration of aldosterone: cation exchange is more important when the sodium load in the distal tubule increases and when the secretion of aldosterone is elevated (Fig. XIII.18). The results of micropuncture studies in adrenalectomized rats treated with aldosterone have confirmed that aldosterone acts mainly on the cortical collecting tubule and the distal convoluted tubule.

More recent studies on the localization of aldosterone receptors, by autoradiography, on different segments of the isolated rabbit nephron[29], as well as the study of Na^+-K^+ ATPase activity in these segments following administration of aldosterone[30], have shown that aldosterone acts not only on the collecting tubule, but *also* on the *connecting segment* following dct (Fig. XIII.16). It is possible that these *different segments* correspond to *separate actions* of aldosterone; thus, the transport of sodium occurs essentially in the cct, whereas the transport of protons takes place in the mct.

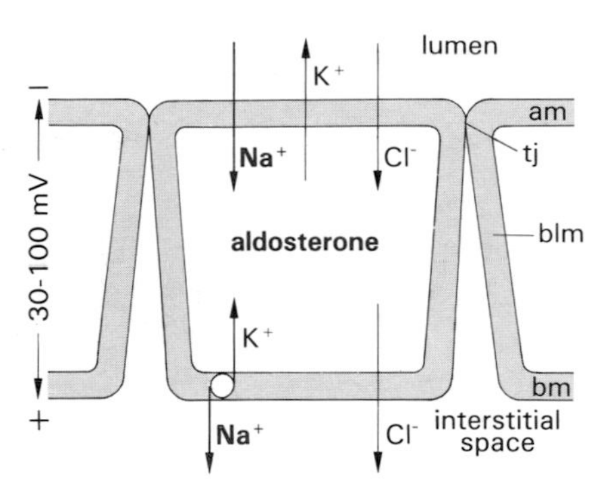

Figure XIII.18 Cation exchanges within the distal tubule. The importance of these exchanges is dependent upon two factors: the concentration of Na^+ at the entrance of the distal convoluted tubule and the intracellular concentration of aldosterone (apical membrane (am); basolateral membrane (blm); basal membrane (bm); tight junction (tj)).

Prolonged Administration of Aldosterone: The Phenomenon of Renal Escape

In normal animals, including humans, the *effect* of aldosterone on sodium retention is *lost after a few days*[31]. The daily injection of 3 to 6 mg of aldosterone over 5 to 8 d leads to a positive sodium balance, with an increase in body weight, and an increased secretion of potassium into urine and feces (Fig. XIII.19). Progressively, even though aldosterone is continually injected, there is an *escape* from mineralocorticosteroid effects; *excretion of sodium increases* and returns to control levels, the urinary loss of potassium is reduced, and a *new* state of *equilibrium* is established. Nevertheless, the net balance is an increase in body weight of 1 to 2 kg, corresponding to an increase in plasma volume and extracellular fluid.

29. Vandewalle, A., Farman, N., Bensacth, P. and Bonvalet, J. P. (1981) Aldosterone binding along the rabbit nephron: an autoradiographic study on isolated tubules. *Amer. J. Physiol.* 240: F172–F179.
30. El Mernissi, G. and Doucet, A. (1983) Short-term effects of aldosterone and dexamethasone on Na^+-K^+-ATPase along the rabbit nephron. *Pflugger's Archiv.*, 399: 147–151.
31. August, T. J., Nelson, D. M. and Thorn, G. W. (1958) Response of normal subjects to large amounts of aldosterone. *J. Clin. Invest.* 37: 1549–1555.

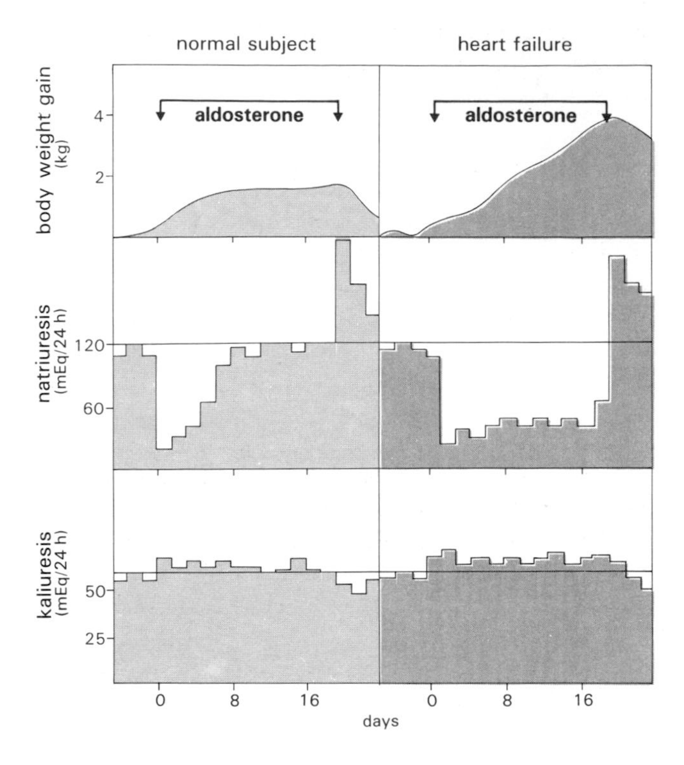

Figure XIII.19 The phenomenon of escape from the effects of aldosterone in a normal subject (left). During prolonged administration of aldosterone, sodium excretion becomes equilibrated after several days, and body weight stabilizes. A urinary loss of potassium follows. In patients with heart failure (or cirrhosis), continual sodium retention and weight gain occurs[31].

Moreover, there is *persistent hypokalemia* due to an inappropriately elevated urinary excretion of potassium. Finally, blood pressure is slightly increased during this experiment.

The phenomenon of escape is also observed with other mineralocorticosteroids, in humans and other animals. On the other hand, it is not seen in animals or patients with heart failure paired with edema. In such cases, the administration of aldosterone leads to a continual retention of sodium and water, and a gain in weight of several kg.

The phenomenon of renal escape in response to aldosterone is produced in normal subjects when a certain degree of filling of the plasma compartment has occurred. *Numerous hypotheses* have been put forward to explain this mechanism. There is no modification of glomerular filtration. Intrarenal physical phenomena may be responsible for proximal tubular rejection of sodium, because it has been shown that acute blood volume expansion in the rat leads to a reduction in oncotic pressure in the proximal peritubular capillaries, with an increase in hydrostatic pressure. When there is an in situ reduction in the concentration of proteins in the fluid of the peritubular capillaries, the same proximal tubular excretion of sodium is seen. Another appealing hypothesis is that an increase in blood volume leads to an elevation of right atrial filling pressure, provoking the secretion of ANF (p. 619). A final hypothesis involves a hypothalamic factor inhibiting Na^+-K^+ ATPase[32]. This would have a digitalis-like effect on the Na^+ pump[33].

Other Renal Effects of Aldosterone

Tubular secretion of H^+ ions. Aldosterone leads to an active reabsorption of sodium across the epithelial cell of the distal tubule, from the apical surface to the basal surface. It also creates a difference in electrical potential across the cell, the apical surface being negative with respect to the basal surface. This movement of sodium leads to a quantitively equivalent, passive transport of Cl^- ions. The secretion of H^+ or K^+ ions represents another possible means of restoring transcellular electroneutrality. An increase in urinary ammonium ions (NH_4^+) and phosphates (NaH_2PO_4) results from ionic exchange along the distal tubule.

At physiological doses, aldosterone has *no effect* on the *excretion* of *free water* by the kidney, but favors the osmotic water flux response to vasopressin in the cct.

Extrarenal Effects of Aldosterone

The essential action of aldosterone is to control the metabolism of electrolytes by the kidney. Organs other than the kidney are involved in the transfer of electrolytes, and are under the influence of aldosterone. Their role, and that of aldosterone, in the final adjustment of electrolytes is of minor importance, however.

Action on the Digestive Tract

In the colon, aldosterone leads to an *active reabsorption* of *sodium* in exchange of potassium[34]. The flux of sodium occurs from the intestinal lumen to the serosa. This effect of aldosterone is observed in animals, including humans, and can be quantitated by the measurement of the Na^+/K^+ ratio in the feces; administration of aldosterone (or other mineralocorticosteroids) reduces this ratio. It is also possible to measure this effect by observing variations in the electrical potential between the rectum and the pelvic colon, which is increased by the effects of aldosterone. Mineralocorticosteroid receptors have recently been demonstrated in rat and human colonocytes.

Action on Salivary Glands

Submaxillary and parotid salivary glands respond to the administration of aldosterone by a reduction in the Na^+/K^+ ratio in salivary secretions. It is not known, however, if aldosterone increases the reabsorption of sodium in the salivary ducts, or if it reduces the secretion of sodium. One of the best physiological models for the study of aldosterone is the sheep parotid gland, used by the group at the Florey Institute in Melbourne[35].

Action on Sweat Glands

Aldosterone can reduce the sodium content of sweat, and thus *plays a role* in the prevention of sodium depletion in sweat *during adaptation to heat*. This effect on the excretion of sodium is abolished by spironolactones.

Actions on the Heart

Aldosterone is capable of *increasing* the *contractile* force of the myocardium, but elevated doses are required to demonstrate this.

Actions on Salt Appetite

Adrenalectomy in rats stimulates salt appetite, a phenomenon which can be corrected by low doses of deoxycorticosterone acetate. The site and exact mechanism of this action is unknown.

Actions on Vessels

Aldosterone has a *direct role* in the *transfer* of *Na^+ and K^+* ions in vascular smooth muscle fibers[36], as demonstrated in the isolated caudal artery of the rat. Aldosterone acts rapidly on passive transport of Na^+, and in addition has a retarded action on active Na^+ transport (ouabain-dependent and independent), thus regulating vascular contractility.

Actions on the Brain

Aldosterone does *not* have an effect on the *control of ACTH* secretion. However, it is possible that it has some action in the *central nervous system*, since specific aldosterone receptors have been localized in the brain[37].

Summary of Effects on Electrolytes

In conclusion, aldosterone acts on tubular epithelial cells of the kidney, salivary glands and digestive tract, *stimulating sodium reabsorption* and increasing extracellular sodium, plasma volume, and blood pressure. This retention of sodium is *not indefinite*, because the kidneys in normal individuals can *"escape"* from this effect of aldosterone. Such is not the case in other target organs. *Aldosterone is not involved only in sodium retention*; its *role in the excretion of potassium is important* and is not subject to renal escape. Finally, aldosterone can play an important role in the regulation of the *intracellular concentration* of *free sodium* in the *arterial wall*. Thus, *aldosterone acts directly* and *indirectly* (by control of blood volume) *in the physiological modulation of vascular tone and arterial pressure*.

2.2 Mechanism of Action of Aldosterone

Induction of Protein Synthesis

Aldosterone stimulates the active transport of sodium in the skin and in amphibian bladder, confirming the major functional analogies between these epithelial cells and the cells limiting the distal portion of the mammalian nephron. The threshold of the effect occurs at a concentration of approximately 0.5 nM,

32. De Wardener, H. E. and MacGregor, G. A. (1980) The possible role of a circulating sodium transport inhibitor in the aetiology of essential hypertension. *Kidney Int.* 18: 1–9.
33. Hamlyn, J. M., Ringel, R. and Schaeffer, J. A. (1982) A circulating inhibitor of (Na^+K^+)ATPase associated with essential hypertension. *Nature* 300: 650–652.
34. Edmonds, C. J. (1967) Transport of sodium and secretion of potassium and bicarbonate by the colon of normal and sodium depleted rats. *J. Physiol. (London)* 193: 589–602.
35. Blair-West, J. R., Coghlan, J. P., Cran, E., Denton, D. A., Funder, J. W. and Scoggins, B. A. (1979) Increased aldosterone secretion during sodium depletion with inhibition of renin release. *Amer. J. Physiol.* 224: 1409–1414.
36. Moura, A. M. and Worcel, M. (1984) Direct action of aldosterone on transmembrane ^{22}Na efflux from arterial smooth muscle. *Hypertension* 6: 425–430.
37. Krozowski, Z. S. and Funder, J. W. (1983) Renal mineralocorticoid receptor and hippocampal corticosterone binding species have identical intrinsic steroid specificity. *Proc. Natl. Acad. Sci., USA* 80: 6056–6060.

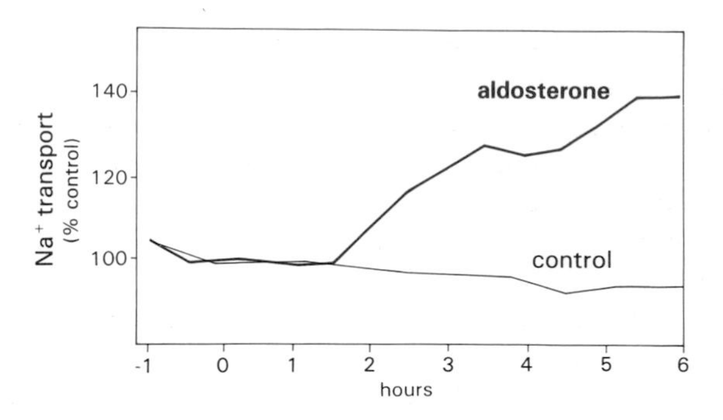

Figure XIII.20 The action of aldosterone on the active transport of sodium in the toad bladder. Two fragments of the same bladder were placed, as a diaphragm, between two chambers filled with Ringer's solution (similar in composition to plasma). The net flow of sodium across the epithelium is measured at regular intervals. One of the preparations is stimulated by aldosterone added in the internal compartment at a concentration of 0.7 μM; the other preparation serves as a control. Measurement of net flow of sodium is expressed as a percent of the value obtained at time 0 of the experiment (N=19). It should be noted that, following a latency of approximately 1 h, aldosterone produces an increase in the active transport of sodium. (Edelman, I. S. and Fimognari, G. M. (1968) On the mechanism of action of aldosterone. *Rec. Prog. Horm. Res.* 24: 1–44; Edelman, I. S. and Marver, D. (1980) Mediating events in the action of aldosterone. *J. Steroid Biochem.* 12: 219–224.).

and the maximal response at 0.1 μM. Handler and co-workers developed a renal cell line of the toad which could respond to aldosterone administration by an increase in the transport of Na^+, measurable in cell culture[38].

Edelman and co-workers studied the steps involved in the action of aldosterone on the active transport of sodium in the *amphibian bladder*, and demonstrated its dependency on *protein synthesis*. The addition of aldosterone to the incubation medium bathing the internal surface of the preparation leads to an increase in the active transport of sodium after a latency of 40 to 120 min (Fig. XIII.20). The effect reaches its maximal level at about 3 h. Removal of the hormone from the incubation medium following the latency period results in a full biological response, suggesting an indirect action of aldosterone on the system of active sodium transport. The inhibition of protein synthesis at the transcriptional level abolishes the response to aldosterone. Together, these observations suggest that one of the early effects of aldosterone is to induce the production of mRNA encoding several proteins responsible for the stimulation of active sodium transport.

Mineralocorticosteroid Receptor

Aldosterone binds to an *intracellular receptor* of high affinity (K_D=0.5 nM) and high specificity for mineralocorticosteroids and steroidal antimineralocorticosteroids. This receptor is regulated by the level of circulating aldosterone. Aldosterone reduces the level of receptors, whereas adrenalectomy increases it[39]. There is a parallel between maximal occupation of mineralocorticosteroid receptors and maximal physiological response[40], although it is not clear that there is a linear correlation between a given level of receptor saturation and the hormonal response. The human *mineralocorticosteroid receptor* (MR) has recently been cloned[41] (Figs. I.71 and IX.18). Aldosterone *also binds with low affinity to glucocorticosteroid receptors*, which explains the glucocorticosteroid effect of this hormone when administered at high doses. Glucocorticosteroids can also bind to mineralocorticosteroid receptors, and thus have some mineralocorticoid effect on water and electrolyte metabolism. Spironolactones exert this effect by preventing the binding of aldosterone to the mineralocorticosteroid receptor.

Role of Proteins Synthesized in Response to Aldosterone

The transepithelial reabsorption of Na^+ implies passage across two permeability barriers:. 1. the apical membrane in contact with the exterior milieu, and 2. the basal membrane in contact with the interior milieu; these two membranes delimit the intracellular compartment (Fig. XIII.21).

The intracellular *penetration of external sodium* occurs by a process of *facilitated diffusion*, which is *probably passive* in nature. The transport of *sodium, from the cell to the interior milieu*, is *active*; it is accomplished *by a Na^+/K^+ pump* using energy

38. Handler, J. S., Steele, R. E., Sahib, S. K., Wade, J. B., Preston, A. S., Lawson, N. L. and Johnson, J. P. (1979) Toad urinary bladder epithelial cells in culture: maintenance of epithelial structure, sodium transport, and responses to hormones. *Proc. Natl. Acad. Sci., USA* 76: 4151–4155.
39. Claire, M., Oblin, M. E., Steimer, J. L., Nakane, J., Misumi, J., Michaud, A. and Corvol, P. (1981) Effect of adrenalectomy and aldosterone on the modulation of mineralocorticoid receptors in rat kidney. *J. Biol. Chem.* 256: 142–147.
40. Farman, N., Kusch, M. and Edelman, I. S. (1978) Aldosterone receptor occupancy and sodium transport in the urinary bladder of Bufo Marinus. *Amer. J. Physiol.* 235: C90–C96.
41. Arriza, J. L., Weinberger, C., Cerelli, G., Glaser, T. M., Handelin, B. L., Housman, D. E. and Evans, R. M. (1987) Cloning of human mineralocorticoid receptor complementary DNA: structure and functional kindship with the glucocorticoid receptor. *Science* 237: 268–275.

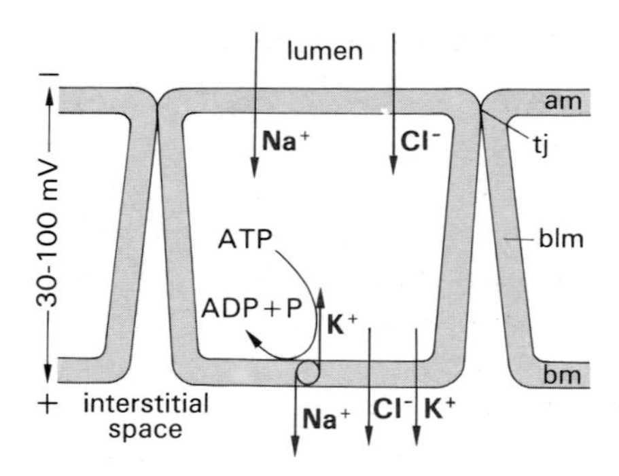

Figure XIII.21 Schematic representation of the mechanisms of transepithelial transport of sodium and chloride ions by amphibian bladder. The basal membrane (bm) and basolateral membrane (blm) of epithelial cells contain a pump, which exchanges Na^+ for K^+, shown in red in the figure. This pump uses energy released by the hydrolysis of ATP to transfer sodium ions from the intracellular medium to the interior medium and potassium ions in the opposite direction. A concentration gradient develops, for Na^+ and K^+ ions, between the intracellular medium and the two compartments separated by the epithelium. The intracellular concentration of sodium is low compared to its concentration in the exterior medium (bladder urine) or in the interior medium of the animal. The reverse situation occurs for potassium ions. Sodium is transferred from the exterior medium to the intracellular medium by facilitated diffusion, which is probably passive in nature (net flow in the direction of the concentration gradient). The low permeability of sodium for the basal and basolateral membranes prevents any appreciable diffusion of this ion from the interior medium to the intracellular compartment. The balance of *sodium* exchange is a *transepithelial transfer* of the exterior medium *towards the interior*. The apical membrane (am) is only slightly permeable to potassium, whereas basal and basolateral membranes are freely permeable to this ion. The result is that potassium actively transported by the pump returns to the interior medium by passive diffusion across the basal membrane. The system of active basal transport, therefore, does not create a net transepithelial flow of potassium ions. The diffusion of sodium at the level of the apical membrane and of potassium at the basal membrane results in the development of an electric potential between the two cells. The electrical gradient permits a passive transepithelial transfer of chloride ions. Tight junctions (tj) ensure the cohesion of the epithelium, and eliminate the possibility of direct exchange between the exterior and interior medium via the intercellular space.

42. Crabbé, J. (1961) Stimulation of active sodium transport by the isolated toad bladder with aldosterone in vitro. *J. Clin. Invest.* 40: 2103–2110.
43. Law, P. Y. and Edelman, I. S. (1978) Induction of citrate synthase by aldosterone in the rat kidney. *J. Membrane Biol.* 41: 41–48.

furnished by the *hydrolysis of ATP* to transfer Na^+ ions against an electrochemical gradient. Aldosterone may stimulate the active transport of sodium by increasing the efficiency of the apical facilitated diffusion process. It could act, for example, by inducing the recruitment of a permease for Na^+ ions, or by an increase in intracellular ATP (resulting in an increased activity of the Na^+-K^+ ATPase), or by the stimulation of basal active transport (increase in the number of Na^+-K^+ ATPase molecules).

Active Sodium Transport

Three hypotheses (Fig. XIII.22)

Crabbé has suggested that the *primary action* of aldosterone in the toad bladder is to *increase apical permeability to Na^+ ions*[42]. The transitory increase in intracellular Na^+ would result in an activation of the basolateral Na^+ pump.

Another possibility is that aldosterone acts by *increasing the synthesis of enzyme(s) of the Krebs cycle*, thus permitting an increased supply of oxidizable substrates and ATP[43]. In fact, the modifications of mitochondrial enzymes may be secondary to the

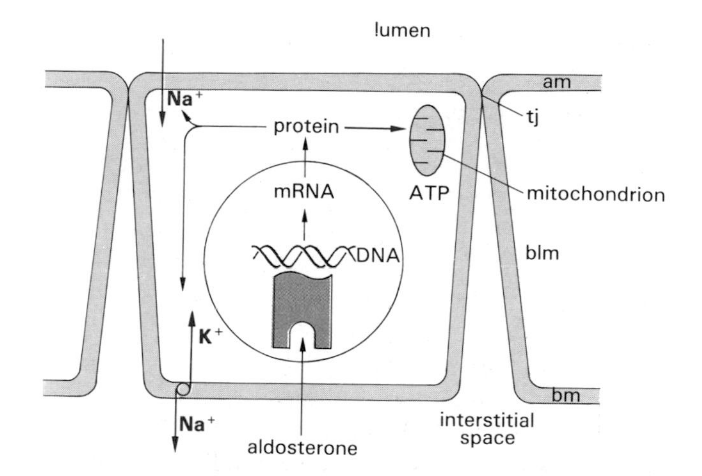

Figure XIII.22 Mechanism of action of aldosterone at the molecular level. Aldosterone binds to the mineralocorticosteroid receptor and induces the transcription of mRNA. Several proteins are then synthesized. Three hypotheses could account for the active transport of sodium from the lumen to the interstitial space: 1. an increase in the number of *Na^+ channels* at the apical membrane; 2. an increase in *ATP* content within the target cell; 3. an increase in *Na^+-K^+ATPase* molecules. None of these theories is exclusive, and it is likely that aldosterone exerts a pleiotropic effect. The sequence of events mediated by aldosterone is not yet known.

increased activity of Na^+-K^+ ATPase rather than the cause of active sodium transport. Another alternative is an *initial increase in synthesis of* Na^+-K^+ *ATPase*[44].

In summary, it is not currently possible to determine what key protein is induced. However, an early effect of aldosterone is an increase in apical membrane permeability to sodium as a result of an increased permeability of the active basolateral transport of Na^+ induced by Na^+-K^+ ATPase. This results in a continuous flow of Na^+ across the target cell, without intracellular accumulation. In parallel, an increase in the activity and synthesis of mitochondrial enzymes allows an energy build-up in the cell.

2.3 Control of Aldosterone Secretion

Several factors regulate the secretion of aldosterone, the *most important* of which are *angiotensin II*, *kalemia*, and *ACTH*. *Natremia* and *other modulators also* participate in this regulation. The study of these different factors is complicated by the fact that the capacity of the adrenal cortex to secrete aldosterone is greatly influenced by the duration of exposure to these factors and by their interaction. The culture of adrenal zona glomerulosa cells has made possible a detailed analysis of different stimuli affecting aldosterone secretion.

Role of Renin-Angiotensin

The role of the renin-angiotensin system in the control of aldosterone secretion has been controversial. As described previously, the renin system plays a major role in aldosterone secretion. It is responsible for aldosterone secretion during periods of reduced blood volume caused by hemorrhage, sodium depletion, diuretics, etc. Inversely, suppression of the renin system by an overload of sodium results in a reduction of aldosterone secretion. The key role of this system has been confirmed by the use of *inhibitors* of *converting enzyme* (teprotide, captopril, lopril). These compounds *inhibit* the *formation* of *angiotensin II* and *reduce* the plasma and urinary level of *aldosterone*, in the rat and human. Prolonged treatment with these inhibitors results in approximately a 50% reduction in urinary excretion of aldosterone.

Role of Kalemia

Potassium is the *second stimulus*, in the order of *importance*, in aldosterone secretion. Intravenous

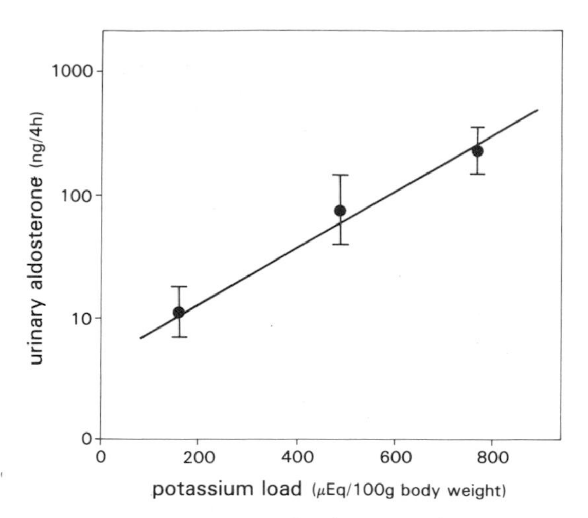

Figure XIII.23 Relationship between the excretion of urinary aldosterone and three different potassium loads in the rat (N=7, mean±SD). There is a strong, positive correlation between the amount of oral potassium loading and the secretion of aldosterone (as deduced from measurement in urine). (Corvol, P., Oblin, M. E., Degoulet, P., Fressinaud, Ph. and Ménard, J. (1977) Effect of acute potassium loading on plasma renin and on urinary aldosterone in rats. *Endocrinology* 100: 1008–1013).

infusion of potassium (Fig. XIII. 23) to intact animals, or direct administration of potassium into the adrenal artery, increases aldosterone secretion. The addition of potassium to a preparation of adrenal zona glomerulosa cells also increases aldosterone secretion[45]. Relatively *minimal fluctuations* of *kalemia* (on the order of 0.2 mEq) can *increase aldosterone secretion*. In contrast, potassium depletion reduces the secretion of aldosterone.

There is another possible involvement of potassium in the entire renin-angiotensin-aldosterone system. *Independently* of all variations in sodium balance, the administration of potassium can block the secretion of renin, probably by an intrarenal effect. This action of potassium on renin is opposed to the direct action of potassium on aldosterone, and tends to reduce this steroid by inhibiting angiotensin II. The physiological significance and importance of this effect remain unknown.

Aldosterone secretion, in response to an infusion of potassium, is affected by sodium balance. A nega-

44. Geering, K., Girardet, M., Bron, C., Kraechenbuhl, J. P. and Rossier, B. C. (1982) Hormonal regulation of (Na^+K^+)ATPase biosynthesis in the toad bladder. Effect of aldosterone and 3,5,3′-triodo-L-thyronine. *J. Biol. Chem.* 257: 10338–10343.
45. McKenna, T. J., Island, D. P., Nicholson, W. E. and Liddle, G. W. (1978) The effects of potassium on early and late steps in aldosterone biosynthesis in cells of the zona glomerulosa. *Endocrinology* 103: 1411–1416.

tive sodium balance increases aldosterone secretion induced by an elevation of kalemia, whereas a positive sodium balance reduces this response. In other words, as is true for angiotensin II, at the same level of kalemia, aldosterone secretion increases with an increasingly negative sodium balance.

The mechanism of action of potassium on the secretion of aldosterone is *still not known* with certainty. It is not clear whether the stimulation of aldosterone by potassium depends on the intra- or the extracellular content of potassium. Potassium could influence the extracellular level of calcium and thus exert its effect on aldosterone secretion, but it does not influence the secretion of glucocorticosteroids. It is not known if potassium acts on the synthesis of aldosterone in zona glomerulosa cells at an early (cholesterol → pregnenolone) or late phase (corticosterone or DOC → aldosterone) in the biosynthetic pathway.

Proximo-Distal Cooperation and Maintenance of Sodium-Potassium Equilibrium (Fig. XIII.24)

Due to the dual action of *aldosterone*, all variations in its secretion have *concomitant, opposing effects on sodium and potassium.* An excess of aldosterone leads to retention of sodium and loss of potassium. Reduced aldosterone results in retention of potassium and loss of sodium. When an abnormality in sodium metabolism occurs, variation in the secretion of aldosterone tends to compensate, but there is a risk of a concomitant modification of potassium equilibrium. When a disturbance of potassium metabolism occurs, variations in aldosterone secretion tend to compensate for it, but sodium equilibrium is also affected.

The concept of an *exchange of sodium for potassium*, although not applicable to all circumstances, is an acceptable explanation of the complex phenomena regulating the level of these two cations. Thus, the effect of aldosterone is a function of both the quantity of aldosterone interacting with its receptor and the amount of sodium available to the cells of the distal tubule. This second variable is proportional to the quantity of sodium filtered by the glomerulus, and is inversely proportional to the quantity of sodium reabsorbed by the proximal cells of the renal tubule.

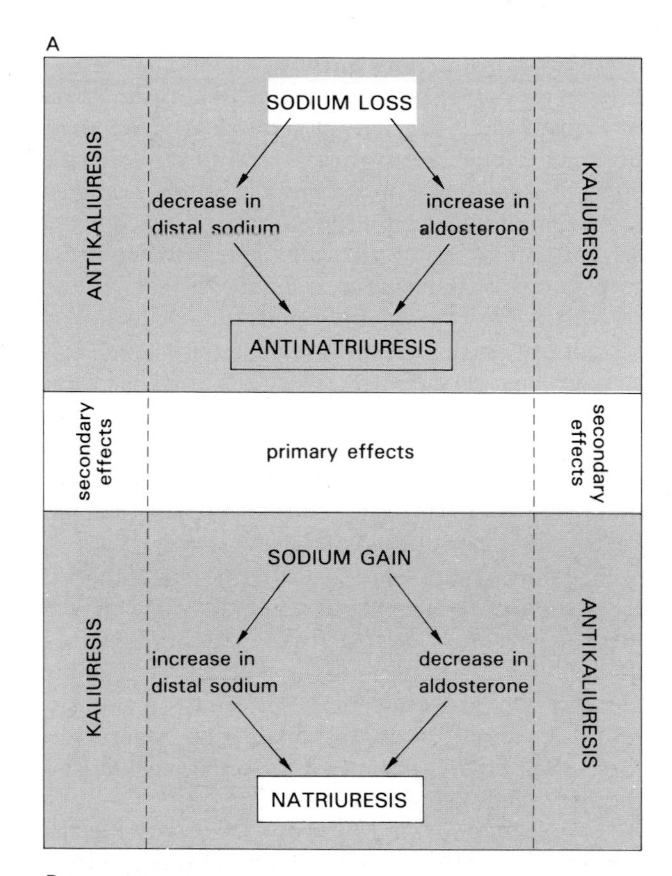

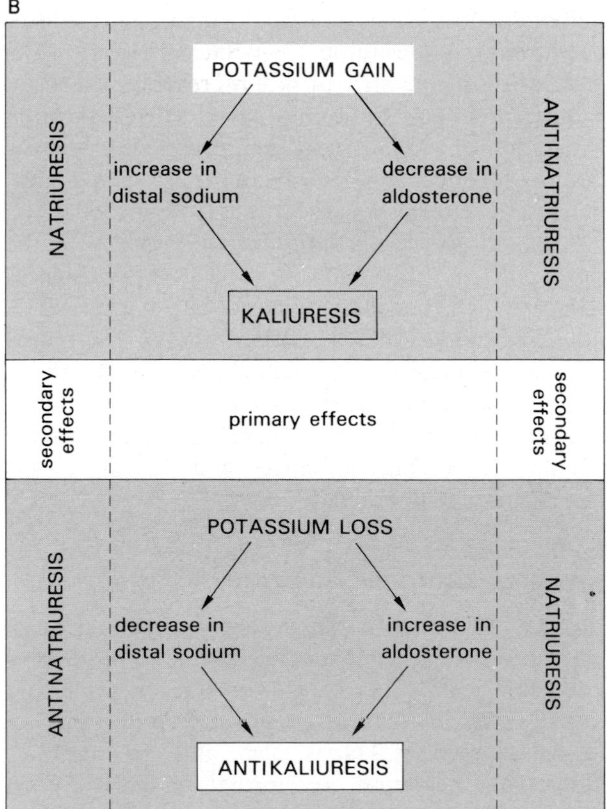

▷

Figure XIII.24 Proximo-distal nephron cooperation in the regulation of urinary sodium excretion (**A**), and urinary potassium excretion (**B**).

Variations in the quantity of sodium which reaches the distal tube can therefore modify the action of aldosterone. Variations in glomerular filtration or proximal tubular reabsorption tend to reinforce the effect of aldosterone on the initially poorly regulated cation, and to neutralize the effect of aldosterone on the cation that is not involved. This proximo-distal cooperation is summarized in Figure XIII.24.

Loss of sodium stimulates aldosterone. In the distal portion of the nephron, aldosterone suppresses sodium loss (first beneficial effect), but increases kaliuresis (undesirable associated effect). The reduction in the quantity of sodium reaching the distal tubule due to an augmentation of proximal reabsorption reinforces the primary (antinatriuretic) effect and neutralizes the associated (kaliuretic) effect.

Sodium uptake suppresses aldosterone. In the distal portion of the nephron, sodium loss is increased (first beneficial effect), but kaliuresis is reduced (undesirable associated effect). The increase in the quantity of sodium reaching the distal tubule as a result of a reduction in proximal reabsorption reinforces the primary (natriuretic) effect and neutralizes the associated (kaliuretic) effect.

Potassium uptake stimulates aldosterone. In the distal portion of the nephron, aldosterone increases potassium loss (first beneficial effect), but reduces natriuresis (undesirable associated effect). The increase in the quantity of sodium reaching the distal tubule due to an inhibition of proximal reabsorption of sodium by potassium reinforces the primary (kaliuretic) effect and neutralizes the associated (antinatriuretic) effect.

Loss of potassium suppresses aldosterone. In the distal portion of the nephron, the loss of potassium is reduced (first beneficial effect), but natriuresis is increased (undesirable associated effect). The reduction in the quantity of sodium reaching the distal tubule as a result of the reduction of glomerular filtration induced by potassium depletion reinforces the primary (antikaliuretic) effect and neutralizes the associated (natriuretic) effect.

Adrenocorticotropic Hormone (ACTH)

ACTH *increases* aldosterone secretion at pharmacological doses, but it is impossible to prolong its effect longer than 24 h: aldosterone levels, which are initially stimulated, return to normal. At near physiological doses, ACTH is also able to stimulate aldosterone secretion. In normal patients, ACTH plays an important role in the nycthemeral variation of plasma aldosterone concentrations[46]. In rats, the secretion of aldosterone, and its increase due to sodium depletion, is reduced following hypophysectomy. In humans, panhypopituitarism is associated with a reduction in aldosterone secretion.

Atrial Natriuretic Factor

Atrial natriuretic factor (ANF) has recently been shown to inhibit in vitro aldosterone production in adrenal cell cultures. In vivo, ANF also inhibits aldosterone section, but the magnitude of its effect depends on the concomitant stimulation of the renin system.

Aldosterone Stimulating Factor

Several studies carried out in anephric and decapitated dogs have suggested that *other pituitary factors* could control aldosterone secretion. An aldosterone stimulating factor (ASF), isolated from human pituitary, has been described[47].

Role of Natremia

Variation in natremia influences aldosterone secretion primarily *via* the *renin-angiotensin system*. However, important modifications of natremia can also *directly* regulate aldosterone secretion, hyponatremia leading to an increase in aldosterone secretion.

Other Factors

Pharmacological studies have suggested a *negative dopaminergic regulation* of aldosterone secretion[48]. *Serotonin* and numerous *prostaglandins* are capable of *increasing* aldosterone *production* under certain experimental conditions. The role of these factors in either the physiology or pathology of aldosterone secretion is unknown.

46. Katz, F.H., Romfh, P. and Smith, J.A. (1975) Diurnal variation of plasma aldosterone, cortisol and renin activity in supine man. *J. Clin. Endocrinol. Metab.* 40: 125–134.
47. Sen, S., Sharinoff, J.R., Bravo, E.L. and Bumpus, F.M. (1981) Isolation of aldosterone stimulating factor and its effect on rat adrenal glomerulosa cells in vitro. *Hypertension* 3: 4–10.
48. Carey, R.M., Thorner, M.O. and Ortt, E.M. (1980) Dopaminergic inhibition of metoclopramide induced aldosterone in man. Dissociation of responses to dopamine and bromocriptine. *J. Clin. Invest.* 66: 10–18.

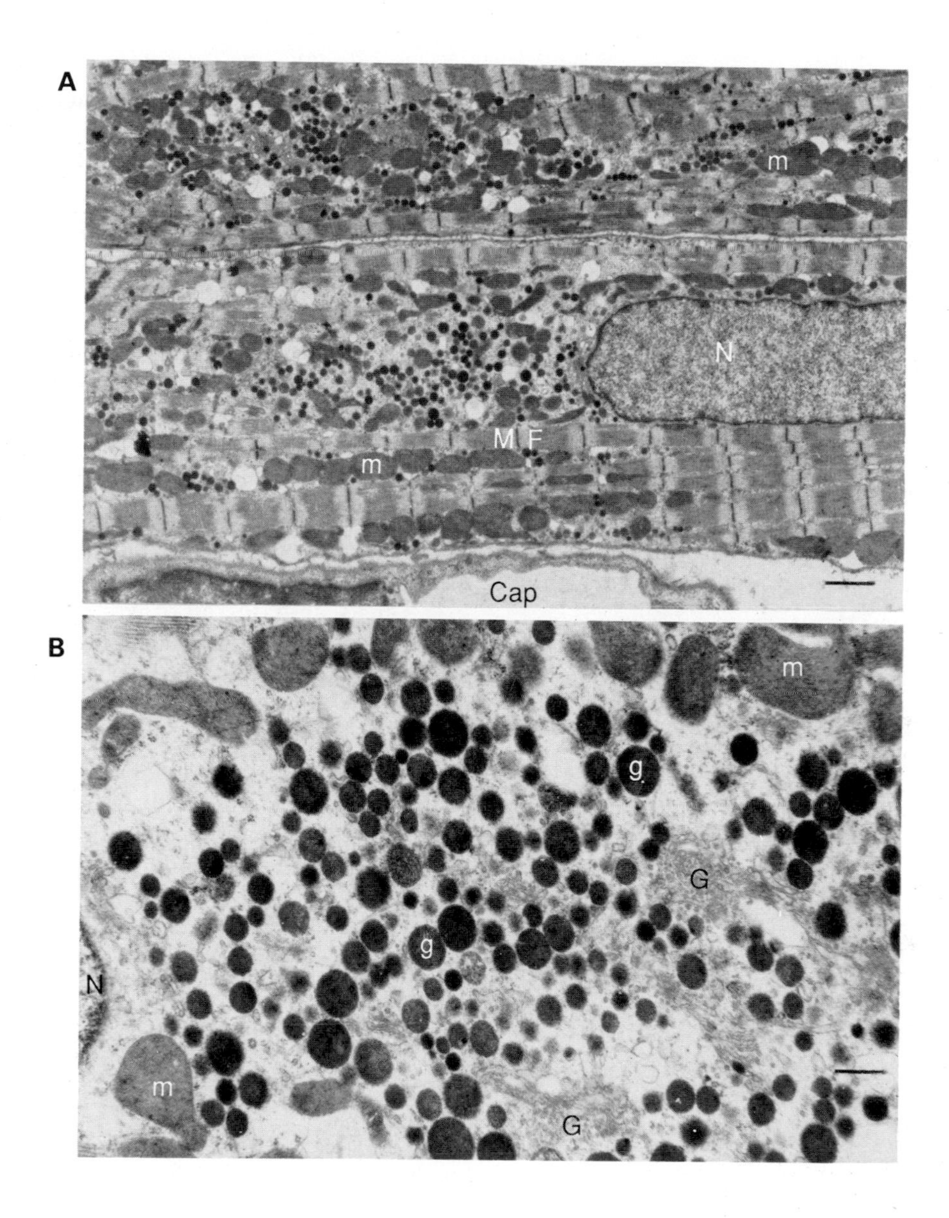

Figure XIII.25 **A.** General aspect of atrial cardiocytes; nucleus (n), mitochondria (m), and myofibrils (mf). The specific grains are particularly abundant in the perikaryon. Cap, capillary. The rat was sacrificed after five days of water restriction and the right atrium was removed. **B.** Specific grains of the Golgi apparatus in a normal rat. The following structures can be seen: nucleus (n), mitochondria (m), Golgi apparatus (G), specific grains (g). Bars=1 μm.[69].

3. Atrial Natriuretic Factor (ANF)

Heart atria contain a potent natriuretic factor, *ANF*. It has been known for a long time that certain cells in the *right atrium* contain *secretory granules*[49] whose number vary quantitatively as a function of water and sodium balance (Fig. XIII.25)[50]. It was demonstrated that *extracts* of atria have *diuretic and natriuretic activity* when injected into rats[51].

Recently, ANF has been sequenced and synthesized, and its cDNA has been cloned. It circulates in plasma and acts, therefore, as an atrial hormone (also

49. Jamieson, J. D. and Palade, G. E. (1964) Specific granule in atrial muscle cells. *J. Cell. Biol.* 23: 151–172.
50. Marie, J. P., Guillemot, H. and Hatt, P. Y. (1976) Le degré de granulation des cardiocytes auriculaires. Etudes planimétriques au cours des différents apports d'eau et de sodium chez le rat. *Pathol. Biol.* 24: 549–554.
51. DeBold, A. J., Borenstein, H. B., Veress, A. T. and Sonnenberg, H. (1981) A rapid and potent natriuretic response to intravenous injection of atrial myocardial extract in rats. *Life Sci.* 28: 89–94.

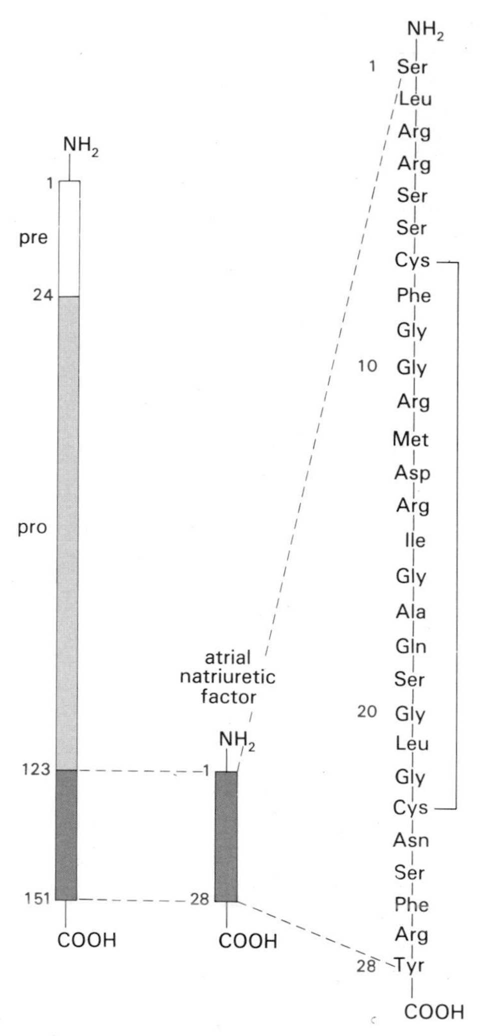

Figure XIII.26 Representation of preproatrial natriuretic factor and its processing to ANF.

known as ANP: *atrial natriuretic peptide*, *auriculin*, or *atriopeptin*). The preprohormone precursor is 152 aa long, and active ANF itself is a *28 aa peptide* (Fig. XIII.26). Rat and human ANF are identical, except for Met in the position of Ile-12 in the rat[52–55].

ANF and synthetic analogs have a potent and abrupt diuretic and natriuretic effect, but provoke moderate kaliuresis. Furosemide provokes a similar diuretic response, but, as compared to ANF-induced effects, diuresis is longer lasting, and its onset is slightly delayed. ANF also increases chloride, phosphate and calcium excretion and free water clearance. It does not alter the effective renal plasma flow, but increases glomerular filtration and, thereby, the filtration fraction. ANF possibly redistributes the renal plasma flow to the middle and outer cortex. It is not yet clear whether the diuretic and natriuretic effects result only from changes in renal hemodynamics, or if ANF exerts a direct effect on sodium transport at the tubule level[56–58]. In addition, ANF has a potent *vasodilator* effect on isolated arteries preconstricted by angiotensin II and norepinephrine, and on isolated veins. In vivo, ANF has a hypotensor effect. Its iv infusion in the human results in a decrease in systolic and diastolic blood pressure, a decrease in cardiac output (at high doses), a decrease in peripheral vascular resistance, and an increase in heart rate. These effects are independent of the rapid loss of water and salt, since they are also observed in binephrectomized animals.

Renal and vascular effects appear to be mediated by *membrane receptors* coupled to *guanylate cyclase*, and cGMP is a second messenger of ANF activities.

Unlike diuretics, all of which increase renin and aldosterone secretion as a consequence of sodium depletion, *ANF* paradoxically *decreases renin and aldosterone secretion*. There is a direct inhibitory effect on renin and aldosterone production in isolated juxtaglomerular cells and aldosterone-producing cells, respectively[59,60]. Thus, ANF is a factor which antagonizes the renin-angiotensin system at different levels: renin and aldosterone production, angiotensin

52. Thibaut, G., Garcia, R., Cantin, M. and Genest, J. (1983) Atrial natriuretic factor. Characterization and partial purification. *Hypertension* 5 (suppl. 1): 175–180.
53. Oikawa, S., Imai, M., Ueno, A., Tanaka, S., Noguchi, T., Nakazato, H., Hangawa, K., Fukuta, A. and Matsuo, H. (1984) Cloning and sequence analysis of cDNA encoding a precursor for human atrial natriuretic polypeptide. *Nature* 309: 724–726.
54. Atlas, S. A., Kleinert, M. D., Camargo, M. J., Januszcwicz, A., Sealey, J. E., Laragh, J. H., Schilling, J. W., Lewicki, J. A., Johnson, L. K. and Maack, T. (1984) Purification sequencing and synthesis of natriuretic vasoactive rat atrial peptide. *Nature* 309: 717–719.
55. Currie, M. G., Geller, D. M., Cole, B. R., Siegel, N. R., Fok, K. F., Adams, S. P., Enbanks, S. R., Gallerpi, S. R. and Needleman, P. (1984) Purification and sequence analysis of bioactive atrial peptides. *Peptides* 223: 67–69.
56. Cody, R. J., Atlas, S. A., Laragh, J. H., Kubo, S. H., Covit, A. B., Ryman, K. S., Shaknovich, A. S., Pondolfino, K., Clark, M., Camargo, M. J. F., Scarborough, R. M., Lewick, J. A. (1986) Atrial natriuretic factor in normal subjects and heart failure patients. *J. Clin. Invest.* 78: 1362–1374.
57. Ballerman, B. J. and Brenner, B. M. (1985) Biologically active atrial peptides. *J. Clin. Invest.* 76: 2041–2048.
58. Needleman, P. (1986) The expanding physiological roles of atrial natriuretic factor. *Nature* 321: 199–200.
59. Ataradhi, K., Multow, P. J., Franco-Saenz, R., Snajdar, R. and Rapp, J. (1984) Inhibition of aldosterone production by an atrial extract. *Science* 224: 992–994.
60. Laragh, J. H. (1985) Atrial natriuretic hormone, the renin-aldosterone axis, and blood pressure electrolyte homeostasis. *N. Engl. J. Med.* 21: 1330–1340.

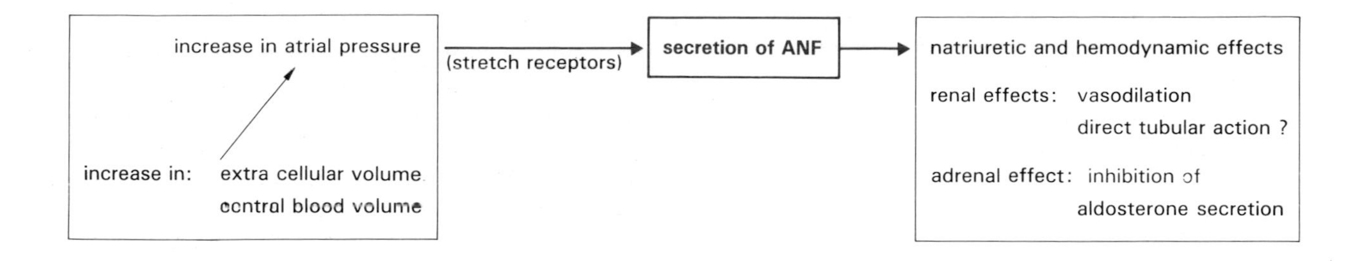

Figure XIII.27 Possible regulation of ANF secretion and its mechanism of action.

II, vasoconstriction, and aldosterone-induced sodium retention. Finally, ANF probably increases capillary bed permeability, since the hematocrit is elevated independently of volume depletion (which is persistent in binephrectomized rats).

Therefore, the overall effects of ANF *improve the hemodynamics* of the *heart* in case of sudden overload: natriuresis, vasorelaxation, increase in capillary bed permeability, and inhibition of the renin-aldosterone system. Blood volume expansion provokes the release of ANF stored previously in cardiocytes (Fig. XIII.27). In different models of acute volume overload, ANF levels increase markedly, as does natriuresis. Under DOCA-salt treatment or cardiac insufficiency, ANF is also produced in heart ventricles, indicating a *recruitment* of all of the *myocardium for* its *gene expression*[61]. However, it has not yet been demonstrated that the aldosterone escape phenomenon (p. 611) is due to ANF secretion.

Thus, ANF may be considered at least as an *emergency hormone* for improving heart hemodynamics by decreasing right atrial filling pressure and peripheral vascular resistance. Its role in semichronic or chronic conditions of volume expansion is still debatable. In fact, very little is known of its possible involvement in *pathophysiology*. There are no gross abnormalities in ANF levels in human hypertension. When *administered* to patients with congestive *heart failure*, it *improves cardiac output*, but does *not produce* the expected *diuresis*. The first possible use of ANF might be in cases of acute renal insufficiency. In practical terms, its therapeutic future will depend on the synthesis of orally active analogs, or, alternatively, the discovery of compounds able to inhibit its degradation or stimulate its release[58].

61. Lattion, A.L., Michel, J.B., Arnauld, E., Corvol, P. and Soubrier, F. (1986) Recruitment of all the myocardium for ANF mRNA increase during volume overload in the rat. *Amer. J. Physiol.* 251: H890–H896.

4. Pathophysiology

4.1 Clinical Evaluation of the Renin-Angiotensin System

Because of the numerous factors which influence the secretion of renin and aldosterone, the *evaluation* of these hormones, in order to be interpretable, should be done *under strictly controlled conditions*. In other words, as is true for other hormones, a single measurement may not be significant. Studies involving inhibition and/or stimulation should be carried out in addition to basal determinations, in order to estimate the dynamic response of the adrenal glands. Finally, studies should include not only the measurement of aldosterone, but also that of angiotensin II or plasma renin, and ACTH or cortisol (as a measure of corticotropic activity), as well as plasma sodium and potassium.

Conditions for Clinical Evaluation of Aldosterone

Conditions for clinical evaluation of aldosterone are summarized in Table XIII.3.

Age

The *basal* secretion of aldosterone *decreases* progressively *with age*. This reduction is probably secondary to a decrease in plasma renin activity, since there is a highly significant, negative correlation between renin activity and age. Values should therefore be compared to normal values according to age; sodium depletion leads to a less intense stimulation of the renin-angiotensin-aldosterone system in older than in younger patients.

Table XIII.3 Factors influencing renin measurement.

Factors	*Conditions of measurement*
Age (physiological decrease with age)	Age-matched normals
Nycthemeral rhythm	Sampling at 08.00 h
Posture	Supine position (2 h); upright position (1 h or 2 h of active orthostatism)
Sodium balance	Control of 24 h natriuresis. If possible, under strictly controlled sodium balance
Menstrual cycle (increase during the luteal phase of the menstrual cycle); prorenin increases markedly during ovulation, as a result of LH stimulation	If possible, during the follicular phase of the menstrual cycle

Nycthemeral Rhythm in Humans

As is true for glucocorticosteroids, plasma aldosterone *increases* between *04.00 h and 06.00 h*, and declines during the day, with the minimal level reached at about 18.00 h. This rhythm has been determined in supine subjects, in order to eliminate any variation due to *posture*.

Menstrual Cycle

Aldosterone secretion and its concentration in plasma are *increased* two-fold during the *luteal phase of* the menstrual *cycle*. At the same time, there is an *increase in plasma renin* activity. Increased progesterone secretion during the luteal phase probably induces natriuresis due to its antialdosterone effect, resulting in turn in secondary hyperaldosteronism.

Pregnancy

The secretory level of aldosterone *increases progressively* during pregnancy[62]. Metabolic clearance of the hormone is not altered, however its metabolism is altered, since the percentage of aldosterone excreted in the form of tetrahydroaldosterone is reduced, in favor of the glucuronide of aldosterone. Aldosteronemia is increased three- to four-fold and decreases progressively at term. This hyperaldosteronism is *secondary* to an *elevation of plasma renin* activity (Table XIII.4). The activation of the renin-angiotensin system could be the result of two phenomena: first, the *natriuretic effect of progesterone*; second, *increased estrogens*, which augment the level of plasma *angiotensinogen*, followed by increased renin activity and angiotensin II. It is clear that the sodium balance is positive during pregnancy, but there is no hypokalemia, despite secondary hyperaldosteronism.

Sodium Balance and Posture

The clinical evaluation of renin and aldosterone can only be carried out under strict conditions of sodium balance, and with controlled posture, as *supine* patients produce *less aldosterone* and *angiotensin II*.

Stimulation Tests

Stimulation tests involve aspects of the mechanisms involved in aldosterone secretion.

Table XIII.4 Influence of pregnancy on the renin-angiotensin system.

	Weeks‡			
	7–17	**18–28**	**29–39**	**Postpartum**
Active renin (ngAI/ml/h)	4.7±0.7	5.7±0.8	6.9±1.2	0.8±0.1
Total renin (ngAI/ml/h)	34.5±3.2	29.9±3.3	30.6±4.0	4.9±0.8
Angiotensinogen (ng/ml)	146±14	161±12	191±16	73±10
Angiotensin I converting enzyme (ng/ml)	337±32	319±30	322±33	423±44

Measurements of plasma renin activity, total renin (prorenin and active renin), angiotensinogen and converting enzyme were performed, in the supine position, in 11 normotensive women (mean±SEM). Plasma renin increases during the first trimester of pregnancy and remains stable until parturition.

‡ Weeks following the first day of the last menstrual period.

Renin-Angiotensin System

The postural stimulus (1 h of active *orthostatism*) represents an efficient and simple means of increasing aldosterone production. *Sodium depletion*, induced by the administration of diuretics or provoked by a low sodium diet, of 10 mEq of Na^+, for five days, also increases aldosterone secretion, via the renin-angiotensin system. Finally, it is possible to stimulate aldosterone secretion directly, by *the infusion of angiotensin* II or III.

Adrenocorticotropic Hormone

Plasma aldosterone concentration *doubles* following a single injection, or an infusion, of ACTH over a 3 h period.

Potassium

Potassium is no longer used in clinical evaluation of the renin-angiotensin system.

Suppression Tests

The renin-angiotensin system can be suppressed by *acutely increasing plasma volume*, by an infusion of two liters of isotonic physiological saline solution over 2 to 3 h[63], *or chronically*, by the administration, over a period of 5 d, of a regimen of 200 mEq of sodium. In the same way, volume expansion can be achieved by the injection of mineralocorticosteroids other than aldosterone (9α-fluorocortisol, deoxycorticosterone). Finally, *aldosterone* secretion can be *suppressed* by acute administration of an *inhibitor of converting enzyme*, such as captopril, at 1 mg/kg per os[64]. Plasma aldosterone is reduced by approximately

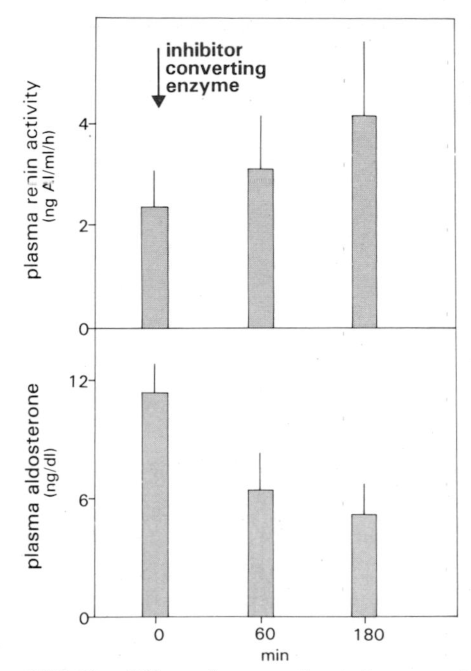

Figure XIII.28 Effect of captopril on plasma renin activity and plasma aldosterone in 26 patients with essential hypertension (n=26, mean±SEM). Measurements were made before and 60 and 180 min after administration of 1 mg/kg of captopril per os[61].

50% 2 to 3 h following administration of the inhibitor (Fig. XIII. 28), whereas at the same time *plasma renin activity increases* about two-fold due to the suppression of the negative feedback of angiotensin II on its secretion.

4.2 Hyperaldosteronism

Hypersecretion of aldosterone due to a *tumor* or to *bilateral hyperplasia* of the adrenals is known as *primary* hyperaldosteronism. Hyperproduction of aldosterone due to *an excess* of physiological secretory *stimuli* is called *secondary hyperaldosteronism.*

Primary Hyperaldosteronism and Other Hypermineralocorticisms

Primary Hyperaldosteronism[65–67]

Two years after the isolation and identification of aldosterone, Conn described the first case of hyperaldosteronism due to an *adrenal adenoma*[66]. The syndrome of primary hyperaldosteronism allows a better

62. Wilson, M., Morganti, A. A. and Zervondakis, I. (1980) Blood pressure, the renin aldosterone system and sex steroids throughout normal pregnancy. *Amer. J. Med.* 68: 97–104.
63. Holland, B., Brown, H., Khunert, L., Fairchild, C., Risk, M. and Gomez Sanchez, C. (1984) Further evaluation of saline infusion for the diagnosis of primary aldosteronism. *Hypertension* 6: 717–723.
64. Guillevin, L., Lardoux, M. D. and Corvol, P. (1981) Effects of captopril on blood pressure, electrolytes and certain hormones in hypertension. *Clin. Pharmacol. Ther.* 29: 699–704.
65. Weinberger, M. H., Grim, C. E., Hollifield, J. W., Ken, D. C., Ganguly, A., Kraner, N. J., Yune, H. Y., Wellman, H. and Donohne, P. J. (1979) Primary aldosteronism: diagnosis, localization and treatment. *Ann. Intern. Med.* 90: 386–395.
66. Conn, J. W. (1955) Primary aldosteronism. A new clinical syndrome. *J. Lab. Clin. Med.* 45: 3–6.
67. Streeten, D. H., Tomycz, N. and Anderson, G. H. (1979) Reliability of screening methods for the diagnosis of primary aldosteronism. *Amer. J. Med.* 67: 403–413.

Table XIII.5 Plasma and urinary electrolyte abnormalities in patients with primary hyperaldosteronism, as compared with essential hypertensive patients.

		Essential hypertension (n=16)	Primary hyperaldosteronism	
			Tumoral (n=20)	Non-tumoral (n=4)
Plasma electrolytes (mEq/l)	Na^+	141±2	144±3	144±2
	K^+	4.1±0.4	2.8±0.5	3.2±0.6
Urinary electrolytes (mEq/24 h)	Na^+	106±24	86±32	84
	K^+	55±17	51±18	33

Corvol, P., Houde, M., Ménard, J. and Milliez, P. (1977) Le système rénine-angiotensine-aldostérone chez les sujets hypertendus. I. Etude analytique chez 124 patients. *Nouv. Presse Med.* 6: 2569–2572.

understanding of the effect of aldosterone in the long term. The phenomenon of renal *escape* accounts for the equilibration of water and sodium balance and the *absence of edema*, whereas the sodium/potassium ratio in the saliva, sweat and feces is low. *Potassium depletion* and metabolic alkalosis are due to abnormal excretion of K^+ and Na^+ by kidney tubules (p.611). The mechanism responsible for the *elevation of blood pressure* always accompanying this syndrome is not well understood. In the initial stage of the disease, cardiac output is elevated, and peripheral resistance is normal (probably secondary to an increase in plasma volume).

Clinically, there is moderate hypertension associated with signs of potassium depletion (cramps, muscle fatigue, sometimes with transitory hypokalemic paralysis, tetany, etc.). The major biological symptoms are summarized in Table XIII.5. With respect to hormones, *plasma renin activity is low* and does not respond to stimuli. The presence of an increase in circulatory and extracellular volume accounts for the reduction in plasma renin activity. The renin-angiotensin system is not stimulated by sodium depletion (diuretics or salt-free diet). Aldosterone secretion is increased and is not suppressed by blood volume expansion (Fig. XIII.29), nor is it suppressed by the administration of an inhibitor of converting enzyme, demonstrating the autonomous secretion of the tumor. An accentuation of hypokalemia and kaliuresis is observed after a sodium load. With an increased accumulation of sodium in the distal tubule and the secretion of aldosterone which is not suppressible, there is an increased exchange in the distal tubules, of Na^+ for K^+ and H^+. The *tumor* can be localized *radiographically*.

Spironolactone, at doses of 2 to 5 mg/kg for four to eight weeks, completely *corrects all abnormalities*

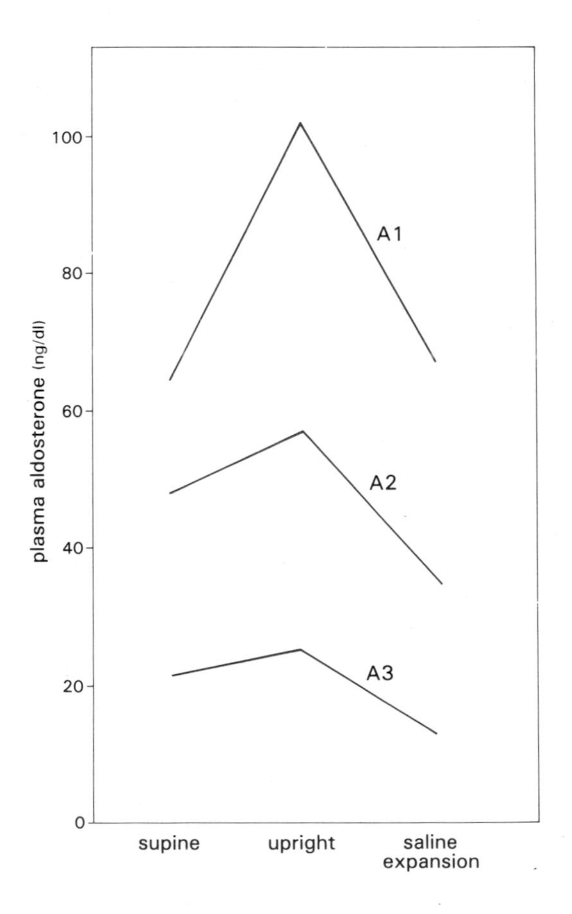

Figure XIII.29 Plasma aldosterone in supine and upright positions, and after volume expansion (two liters) Physiological saline solution was infused over 2 h in patients with primary aldosteronism (A1, A2) and essential hypertension (A3). After saline infusion, there is greater reduction in plasma aldosterone in bilateral adrenal hyperplasia (A1) than in adrenal adenomas (A2).

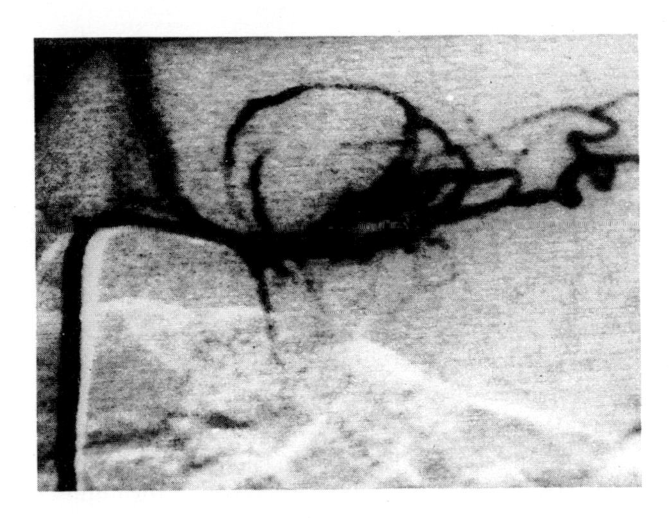

Figure XIII.30 Aldosterone-producing adenoma from the left adrenal. The tumor is indirectly seen by the displacement of the adrenal vein. Digitalized angiography allows insertion of the catheter in the adrenal vein and facilitates separated sampling for the assay of cortisol and aldosterone.

associated with primary hyperaldosteronism. There is a good correlation between medical treatment and the efficiency of possible surgical treatment. However, the use of antialdosterone is not a definitive test, since some forms of essential hypertension are corrected by spironolactone. Sodium depletion induced by diuretics which are not antagonists of aldosterone can also correct hypertension of primary hyperaldosteronism, demonstrating that positive sodium balance is responsible for the hypertension accompanying this syndrome.

Primary hyperaldosteronism is *most frequently due to a single adenoma* of the adrenal zona glomerulosa (Fig. XIII.30). Electron microscopy shows that the tumor consists of clear cells which resemble those of the zona fasciculata. In approximately *one-third* of the cases, there is *bilateral adrenal hyperplasia*, occasionally with some nodules. Without visualization of the adrenal gland (phlebography, adrenal scintigraphy with iodocholesterol, CT scan), preoperatory detection is difficult. In bilateral hyperplasia, hypokalemia is less accentuated, renin activity is not as low, and aldosterone is less increased than in adrenal adenoma. After 4 h of orthostatism, plasma aldosterone remains elevated in cases of bilateral hyperplasia of the adrenals, whereas it decreases in cases of adrenal adenoma. Aldosterone is more markedly reduced during acute blood volume expansion than in cases of adenoma (Fig. XIII.29) or following the administration of converting enzyme inhibitors.

Other Hypermineralocorticisms

An *excess* of *mineralocorticosteroids other than aldosterone* (Fig. XIII.16) results in hypertension associated with electrolyte problems identical to those of primary hyperaldosteronism. There is an increase in exchangeable sodium and a decrease in exchangeable potassium. The positive sodium balance leads to a reduction in plasma renin activity. The involvement of a mineralocorticosteroid other than aldosterone in this syndrome is well demonstrated by the existence of a *very low basal aldosterone level* which is not increased during sodium depletion.

Deoxycorticosterone leads to hypermineralocorticism with arterial hypertension when congenital adrenal hyperplasia is due to 11β-hydroxylase deficiency (p. 433 and Fig. IX.37).

Rare cases of deoxycorticosterone-producing tumors have been reported. *Glycyrrhetinic acid*, the active principal of licorice, has weak mineralocorticosteroid activity. The excessive intake of licorice can lead to hypertension, with signs of electrolyte imbalance.

In all of these cases, hypertension and abnormalities of electrolyte metabolism are corrected by spironolactone.

Secondary Hyperaldosteronism

Secondary hyperaldosteronism may be due to extra-adrenal *stimulation* resulting from of an excess of angiotensin, potassium, or ACTH (Table XIII.6).

Table XIII.6 Classification of secondary hyperaldosteronism according to the stimulus of aldosterone secretion.

Aldosterone stimulus	Secondary hyperaldosteronism
Angiotensin II (and III)	↓ Plasma volume (sodium depletion, hemorrhage) ↓ Effective plasma volume (cirrhosis with ascites, congestive heart failure, nephrotic syndrome) ↓ Perfusion pressure in renal artery (renal artery lesion, aortic coarctation)
Potassium	↑ Kalemia and potassium pool (renal insufficiency)
ACTH	↑ Aldosterone (short-term)

Secondary Hyperaldosteronisms Due to Hypersecretion of Angiotensin

Normally, the adrenal response to stimuli such as hemorrhage or sodium depletion induced by salt restriction, diuretics, or elevated ambient temperature, is adapted to the circumstances. Hyperaldosteronism is secondary to the normal response of the renin-angiotensin system, which is itself stimulated by the juxtaglomerular baro- and adenoreceptors.

In other cases, secondary hyperaldosteronism is pathological because it is *inappropriate in response to the physiological situation*, being secondary to an inappropriate stimulation of renin. This hypersecretion *persists* even though there is a positive sodium balance (edema) or hypertension (renovascular hypertension). Manifestations of secondary hyperaldosteronism can be similar to those of primary hyperaldosteronism, or completely different, depending on whether or not the escape phenomenon occurs.

Hypersecretion of Aldosterone and Edema

In edema caused by *hepatic cirrhosis, nephrotic syndrome* and, to a lesser degree, heart failure, there is a reduction in the effective circulatory volume. This results in an increased proximal tubule reabsorption of sodium chloride, and small quantities of sodium ions arrive at the location of cation exchange in the distal tubule. Hyperaldosteronism is secondary to a reduction in effective blood volume, but because of the low quantity of sodium ions in the distal tubule, it does not lead to potassium depletion, as it does in primary hyperaldosteronism. *Edematous states* are characterized by the *absence* of *escape* from mineralocorticosteroid action, evidenced by the continuous retention of sodium in dogs with experimental heart failure treated with aldosterone.

Elevated plasma aldosterone concentration in hepatic cirrhosis and heart failure is due not only to adrenal hypersecretion associated with excessive angiotensin secretion, but also to a reduction in metabolic clearance of the hormone, due, in turn, to a reduced hepatic blood flow and a reduction in the hepatic metabolism of aldosterone.

Secondary Hyperaldosteronism and Arterial Hypertension

Primary Reninism

Primary reninism is a syndrome in which *hypertension* is due to *renin* produced by a *tumor* of the kidney or of other tissues.

Several *juxtaglomerular cell tumors* have been described[68]. The three main clinical features of patients with these tumors are severe hypertension, hypokalemia, and high plasma renin activity. In addition, no renal artery abnormality or renal lesion can account for the high renin secretion.

Once the diagnosis is suspected, *localization* of the tumor is *difficult*. Several methods can be used: renin measurement in renal veins is one method, but separate renin measurements and segmental renal vein samplings do not always provide a clear answer. Renal tumors are small (0.5 to 3.0 cm), and are detected in the cortex by CT scan. The diagnosis of juxtaglomerular cell tumors is difficult to establish due to their small size and location, and it is likely that the number of such tumors is underestimated. However, this diagnosis should be considered more often, since ablation of the tumor enables the cure of hypertension.

Some other renal tumors (*Wilms' tumors*) can synthesize renin. Ectopic renin secretion[69] has also been observed in several cases (pulmonary, pancreatic and paraovarian cancers). Again, it is likely that the number of these cases is underestimated. When sudden hypertension with hypokalemia is observed with malignant disease, two diagnoses should be discussed: an ectopic ACTH-secreting tumor (with low plasma renin activity), and an ectopic renin-secreting tumor (with high plasma renin activity).

Renal Hypertension[70]

Experimentally, renal ischemia leads rapidly to arterial hypertension with secondary hyperaldosteronism (Fig. XIII.31A). Reduced perfusion pressure of the renal artery provokes an increase in renin secretion and an elevated circulating concentration of angiotensin II. The secretion of renin results in elevated arterial pressure and secondary hyperaldosteronism involving hypokalemia with metabolic alkalosis. The administration of antiangiotensin II or inhibitors of converting enzyme normalizes arterial pressure and reduces aldosterone secretion. Correc-

68. Baruch, D., Corvol, P., Alhenc-Gelas, F., Dufloux, M. A., Guyenne, T. T., Gaux, G., C., Raynaud, A., Brisset, J. M., Duclos, J. M. and Ménard, J. (1984) Diagnosis and treatment of renin secreting tumors. Report of three cases. *Hypertension* 6: 760–766.
69. Corvol, P. (1984) Tumor-dependent hypertension. *Hypertension* 6: 593–596.
70. Genest, J., Cartier, P., Roy, P., Lefebvre, R., Kuchel, O., Cantin, M. and Hamet, P. (1983) Renovascular hypertension. In: *Hypertension* (Genest, J., ed), McGraw-Hill, New York, pp. 1007–1034.

A

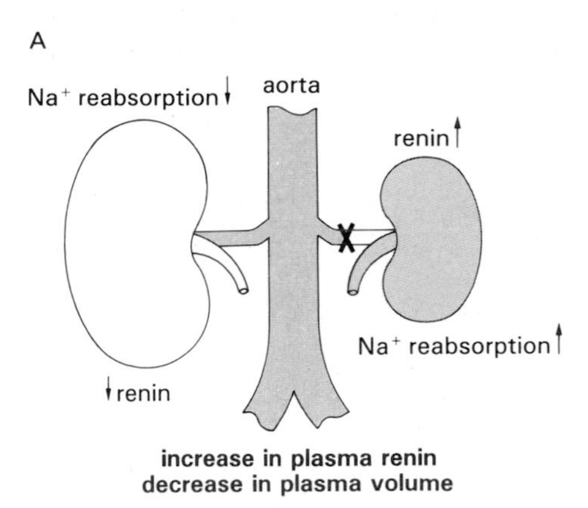

B

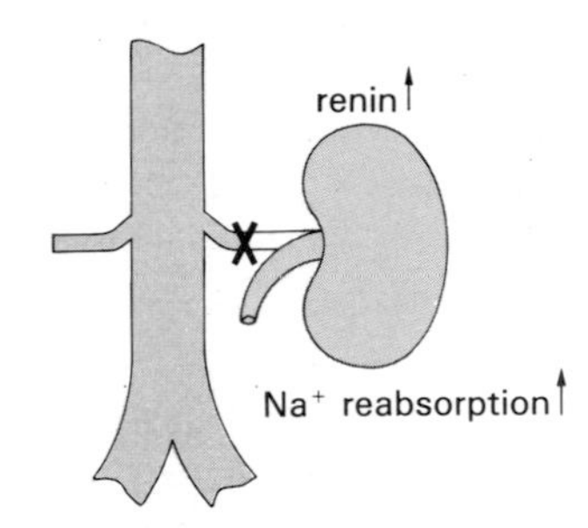

Figure XIII.31 Renovascular experimental hypertension in the rat. **A.** *One clip, two kidneys.* In this model, renin is increased in the clipped kidney, resulting in hypertension and hyperaldosteronism. Renin secretion is decreased in the contralateral kidney. The stenotic kidney tends to retain sodium and water, whereas there is increased natriuresis by the unaffected kidney, producing a net decrease in plasma volume. **B.** *One clip, one kidney.* A kidney is removed (or clipped). Sodium reabsorption by the stenotic kidney creates a slightly positive sodium balance, which in turn tends to decrease renin secretion. The two models have been investigated to show the predominance of renin in the genesis of hypertension (model **A**) and the role of renin *and* volume expansion in the second model (**B**).

tion of the renal arterial stenosis reestablishes normal arterial pressure and suppresses hypersecretion of renin and secondary hyperaldosteronism.

When the contralateral kidney is stenotic or absent (Fig. XIII.31B), compensatory elimination of sodium is no longer possible, and blood volume increases as a result. Renin and aldosterone are still abnormally elevated with respect to the positive sodium balance. Nevertheless, in this case arterial hypertension is not exclusively dependent on renin, as it has been found that administration of antiangiotensin II, or inhibitors of converting enzyme alone, does not correct elevated blood pressure. In order to correct hypertension dependent on both renin and sodium balance, it is necessary to reduce sodium in addition to administering inhibitors of the renin system.

Secondary Hyperaldosteronism and Elevated Kalemia

Hyperaldosteronism can, theoretically, be a response to hyperkalemia. In animals and patients which have undergone bilateral nephrectomy, potassium retention is observed between periods of extracorporal dialysis, and there is secondary hyperaldosteronism.

Secondary Hyperaldosteronism Due to an Increased Secretion of ACTH

In humans, the infusion of ACTH over a few hours results in an increased plasma concentration of aldosterone, but, if it is maintained over several days, a reduction in plasma concentration and urinary excretion of aldosterone is seen, with concentrations reduced to below basal levels. Thus, aldosterone levels are often normal or low in *Cushing's syndrome* and in congenital enzymatic abnormalities (p. 429 and p. 433) associated with hypersecretion of ACTH.

4.3 Hypoaldosteronism

Hypoaldosteronism is much rarer than hyperaldosteronism.

Hypoaldosteronism of Adrenal Origin

Hypoaldosteronism due to *Addison's disease, hypopituitarism*, or adrenal *enzymatic* defects is described on p. 428 and p. 433.

Hypoaldosteronism Secondary to Hyporeninism

A marked hyperkalemia due to insufficient aldosterone secretion can develop in *diabetics* with partial renal insufficiency[71]. The aldosterone response to ACTH or angiotensin II is normal. It seems that this

71. Schambelan, M., Sebastian, A. and Biglieri, E.G. (1980) Studies of the prevalence, pathogenesis and functional significance of aldosterone deficiency in hyperkaliemic patients with renal insufficiency. *Kidney Int.* 17: 89–101.

hypoaldosteronism is secondary to diminished renin secretion, as suggested by the absence of any increase in the renin-angiotensin system during sodium depletion. This hyporeninism could itself be secondary to a diabetic neuropathy affecting the mechanisms of β-adrenergic renin stimulation. Another hypothesis suggests an abnormality in the conversion of prorenin (inactive) to active renin, given the abnormally elevated proportion of inactive renin in the plasma of these patients.

Pseudohypoaldosteronism Secondary to Target Cell Abnormality

Rare cases of pseudohypoaldosteronism involving a target cell abnormality have been reported in young patients. Clinically, they show signs of salt depletion, dehydration, hypotension, and hyperkalemia[72]. There is no renal involvement, and the production of aldosterone is elevated, with a concomitant increase in renin secretion. These symptoms are due to a lack of response of the distal tubule to aldosterone. Either a receptor or a postreceptor defect may be involved.

72. Oberfield, S. E., Levine, L. J., Carey, R. M., Bejar, R. and New, M. I. (1979) Pseudohyperaldosteronism: multiple target organ unresponsiveness to mineralocorticoid hormones. *J. Clin. Endocrinol. Metab.* 48: 228–234.

5. Summary

The *renin-angiotensin system* plays a key role in the control of blood pressure and sodium and water metabolism. Inhibitors of the renin system and of aldosterone have been developed, and are now widely used in the treatment of human hypertension and electrolyte disorders. *Renin* is synthesized as a precursor, preprorenin, in the *specialized cells of the afferent arteriole* of the glomerulus. Preprorenin is converted to prorenin (inactive), and then to active renin. In blood, 50 to 80% circulates as prorenin, and 20 to 50% as active renin. Renin has also been found in extrarenal tissues such as brain, testes, ovaries, and walls of blood vessels. Renin acts on angiotensinogen, a glycoprotein synthesized by the liver, *producing angiotensin I*, an inactive peptide which is *hydrolyzed by converting enzyme* to the important octapeptide *angiotensin II*. This compound has *numerous actions*, all of which cooperate to *maintain blood pressure* by acting both on the fluid content of vessels (increase in the volume of vascular filling) and on the vascular system (direct vasoconstrictor effect on the arterial wall).

The *regulation of* renin secretion is dependent on *several stimuli*. Cells secreting renin are *sensitive to variations in blood pressure* in the *afferent arteriole (baroreceptor)*, and to *ion concentrations* at the *beginning* of the *distal tubule (chemoreceptors)*, at the level of the macula densa. β-adrenergic nerves also control renin secretion; angiotensin II has a direct, negative feedback effect on it.

Any reduction in sodium balance or in blood pressure stimulates renin secretion. *Angiotensin* therefore provokes the *contraction* of vascular smooth *muscle fibers* and increases aldosterone secretion without increasing blood volume. The secretion of renin is controlled by both long (restoration of blood pressure and plasma volume) and short (direct inhibitory effect of angiotensin II) feedback mechanisms.

Aldosterone is the physiological mineralocorticosteroid. It is fundamentally responsible for the *retention* of *sodium* and the excretion of K^+ and H^+ ions. Following glomerular filtration, most of the sodium is reabsorbed in the proximal tubule of Henle's loop. Aldosterone acts on the distal portion of the *nephron* by *reabsorbing* Na^+ ions in exchange of K^+ *and* H^+, which are *excreted*. After several days of hormonal stimulation, there is an "escape" from the effect of sodium retention by aldosterone, the mechanism of which is not clearly defined. Aldosterone also increases the reabsorption of sodium in the digestive tract and reduces its excretion by the salivary glands. In target cells, aldosterone binds to the *mineralocorticosteroid receptor*, and controls the synthesis of one or several proteins involved in the *active transport of sodium*. Aldosterone has an early effect on the synthesis of Na^+-K^+ *ATPase* in cells of the distal tubules of the *mammalian* nephron and in *amphibian bladder* cells.

The regulation of aldosterone secretion is essentially dependent upon the renin-angiotensin system, and any *loss* in sodium stimulates aldosterone secretion. Variations in kalemia also regulate aldosterone secretion, hyperkalemia resulting in an increase in aldosterone secretion and, therefore, kaliuresis, and hypokalemia having the opposite effect. Thus, aldosterone is the hormone responsible for the conservation of sodium, but it is also involved in potassium metabolism. *ACTH* has only a *minimal* role in the control of aldosterone secretion.

The importance of the renin-angiotensin-aldosterone system has been confirmed using *specific inhibitors acting on different levels* of the system; that is, on the actions of renin, converting enzyme, and aldosterone. These inhibitors have shown that renin and sodium balance control blood pressure, the effect of the renin system being more pronounced when *natremia* is reduced. These inhibitors are primarily used for the treatment of hypertension.

Atrial natriuretic factor (ANF) is a 28 aa peptide in humans, synthesized in myocytes of the right and left *atria*. It exerts important *vasodilator* effects, especially in the kidney, and has potent diuretic effects. Its *secretion* is *stimulated* by *increased blood pressure* in the atria, but its exact role in cardiovascular physiology remains to be defined.

Measurements of renin and aldosterone in plasma should be performed under clearly defined conditions, since the renin-angiotensin-aldosterone system is affected by age, sodium balance, posture, nervous state, and by the time of day. Clinically, *primary hyperaldosteronism* consists of hypertension accompanied by potassium depletion. Plasma renin activity is low, and responds poorly to stimuli; the level of aldosterone secretion is increased, and is not suppressible by volume expansion. Both blood pressure and abnormalities in ion balance are corrected by antialdosterone compounds (e.g., spironolactone). In two-thirds of all cases, the disease is due to an adrenocortical adenoma, and, in the remainder, to non-tumoral, bilateral hyperplasia of the adrenals. *Hyperaldosteronism secondary* to an excessive secretion of renin is sometimes observed in conjunction with certain types of hypertension, especially renovascular, and is associated with most *edematous* syndromes. These observations have led to the use of inhibitors of the renin-angiotensin-aldosterone system. The excessive secretion of renin in *primary reninism* is responsible for severe hypertension with secondary hyperaldosteronism. It is due, most frequently, to a renal tumor, but sometimes to the secretion of an ectopic tumor. *Hypoaldosteronism* is primarily associated with severe adrenal insufficiency (Addison's syndrome). Rare cases of isolated aldosterone deficiency are either of adrenal origin or are secondary to hyporeninism.

General References

Anderson, J. V. and Bloom, S. R. (1986) Atrial natriuretic peptide: what is the excitement all about? *J. Endocrinol.* 110: 7–17.

De Bold, A. J. (1986) Atrial natriuretic factor: an overview. *Fed. Proc.* 45: 2081–2085.

Corvol, P., Claire, M., Oblin, M. E., Geering, K. and Rossier, B. (1981) Mechanism of the antimineralocorticoid effects of spirolactones. *Kidney Int.* 20: 1–6.

Corvol, P. and Menard, J. (1987) From the renin gene to the renin inhibitors. *Adv. Nephrol.* 16: 17–36.

Edelman, I. S. and Marver, D. (1980) Mediating events in the action of aldosterone. *J. Steroid Biochem.* 12: 219–224.

Ganong, W. F. (1984) Control of aldosterone secretion. *Res. Steroids* 11: 111–122.

Genest, J. G., Kuchel, O., Hamet, P. and Cantin, M., eds (1983) *Hypertension.* McGraw-Hill Co., New York.

Gibbons, G. H., Dzau, V. J., Farhi, E. R. and Barger, A. C. (1984) Interaction of signals influencing renin release. *Annu. Rev. Physiol.* 46: 291–308.

Inagami, T. (1989) Atrial Natriuretic Factor. *J. Biol. Chem.* 264: 3043–3046.

Keeton, T. K. and Campbell, W. B. (1980) The pharmacologic alteration of renin release. *Pharmacol Rev.* 32: 81–202.

Mantero, F., Biglieri, E. G., Ganten, D., Hackenthal, E., Haufbauer, K. G., Inagami, T. and Edwards, C. R., eds (1982) *Endocrinology of Hypertension.* Serono Symp., 50, Academic Press, London.

Proceedings of the symposium on renin angiotensin system: biochemistry, pharmacology, clinical aspect. (1983) *Clin. and Exp. Hypertens.* A5: 721–1427.

Rossier, B. C. (1989) Mechanisms of action of mineralocorticoid hormones. *Endocr. Res.* 15: 203–226.

Anthony R. Means

Calmodulin

Table 1. *Ca^{2+} binding proteins of eucaryotic cells.*

Calmodulin
Troponin C
Parvalbumin
Calregulon
Calcimedins
S-100
Vitamin D induced intestinal Ca^{2+} binding protein
Ca^{2+} activated protease
Myosin light chains
Aequorin

Calmodulin (CaM) is *one member* of a *family* of relatively acidic *calcium binding proteins* (Table 1). Many of these proteins, such as parvalbumin, troponin C, intestinal calcium binding protein and S-100, are found only in vertebrates, and are restricted to a small number of tissues or cells. In addition, the function, when known, is specific and limited to one metabolic process or pathway. *Calmodulin* is the unique member of this protein class, as it is present *in all eucaryotic cells* and serves many functions. It is a single 148 aa polypeptide containing *four Ca^{2+} binding sites*, and is evolutionary highly conserved.

A schematic representation of the amino acid sequence of vertebrate calmodulin is presented in Figure 1. The sequence is arranged to illustrate the topology of the molecule, with squares representing the residues in the seven α helices. The other portion of the figure shows a drawing of the α carbon backbone of the Ca^{2+}-calmodulin complex, determined by X-ray crystallography[1]. The two pairs of Ca^{2+} binding sites are separated by an eight-turn α helix. This helix serves to separate the two highly homologous halves of the protein. Separation of the molecule between amino acids 79 and 80 allows the halves to be rotated so that each shows the same spatial orientation. The two Ca^{2+} binding sites in the N-terminal half of the molecule demonstrate one order of magnitude greater affinity for Ca^{2+} than do those in the C-terminal half[2]. Even in the absence of Ca^{2+}, calmodulin is 40 to 45% α-helical. In the Ca^{2+}-free state, the N-terminal and C-terminal domains wrap around the first and last parts of the central helix, respectively. Thus, in the absence of Ca^{2+}, the central helix is buried, and consequently the protein is unable to interact with calmodulin-dependent enzymes[1]. Binding of Ca^{2+} increases α helicity to 63%, forms two hydrophobic pockets (each between the Ca^{2+} binding loops at each end of the molecule), and exposes the central helix. The calmodulin binding drugs, such as the phenothiazines and naphthalenesulfonamides, associate with the hydrophobic pockets, whereas the enzymes apparently interact predominantly with the α helix.

The *primary mechanism of activation* of calmodulin is the *Ca^{2+}-dependent conformational change*. The drugs mentioned above bind to the Ca^{2+} calmodulin complex and are *competitive inhibitors of the association of calmodulin with calmodulin-dependent enzymes*[3]. Whereas the specific enzyme interactions are characterized by K_D values in the nanomolar range, the drugs exhibit micromolar affinities. These drugs are all hydrophobic, and as such do *not* represent calmodulin-*specific* antagonists. Each effective drug contains a multi-ring component that interacts with hydrophobic pockets formed between adjacent Ca^{2+} binding loops and a side chain that interacts with the central helix of calmodulin. It is this latter association that seems to interfere with enzyme binding. In addition to Ca^{2+} binding, calmodulin can be modified *post-translationally*[2]. Vertebrate calmodulins are *N*-acetylated and trimethylated on Lys-115. There is no evidence that these modifications decrease its activity in vitro. Finally, it is unlikely that calmodulin is phosphorylated to any significant degree. Thus, the Ca^{2+} calmodulin interaction apparently controls the activity of this multifunctional protein.

Most of the proteins *bound* specifically by calmodulin are *enzymes*[4]. In every case, the interaction between Ca^{2+} and calmodulin results in an activation of the enzyme. It is unusual that the same mechanisms exists for activation of such a diverse array of cellular enzymes. Table 2 lists enzymes which have been reported to be stimulated by calmodulin. The major enzyme classes include kinases, phosphodiesterases, phosphatases, cyclases, and transferases. The listed enzymes are involved in cyclic nucleotide, carbohydrate and lipid metabolism, ion transport, as well as contractility. These enzyme systems are generally characterized by their *high substrate specificity* and their *utilization of calmodulin as a positive regulatory subunit.* The primary exception is the enzyme called multisubstrate protein kinase, or calmodulin kinase II. This enzyme phosphorylates a number of enzymes involved in the synthesis of neurotransmitters. These enzymes are also listed in Table 2. In addition, calmodulin kinase II phosphorylates structural proteins, such as those involved in membrane specializations and cytoskeletal formation.

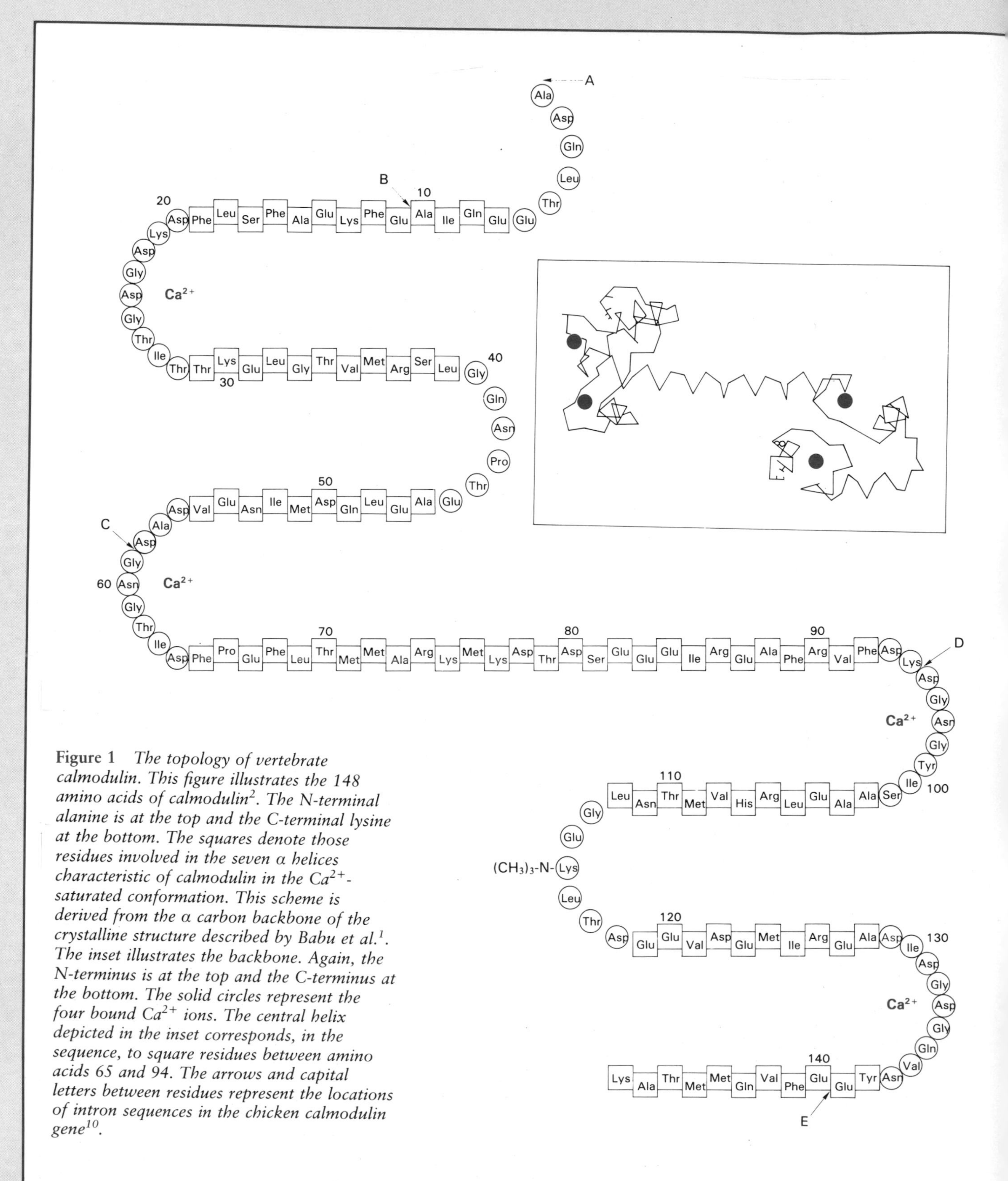

Figure 1 *The topology of vertebrate calmodulin. This figure illustrates the 148 amino acids of calmodulin[2]. The N-terminal alanine is at the top and the C-terminal lysine at the bottom. The squares denote those residues involved in the seven α helices characteristic of calmodulin in the Ca^{2+}-saturated conformation. This scheme is derived from the α carbon backbone of the crystalline structure described by Babu et al.[1]. The inset illustrates the backbone. Again, the N-terminus is at the top and the C-terminus at the bottom. The solid circles represent the four bound Ca^{2+} ions. The central helix depicted in the inset corresponds, in the sequence, to square residues between amino acids 65 and 94. The arrows and capital letters between residues represent the locations of intron sequences in the chicken calmodulin gene[10].*

Both Ca^{2+} and cAMP serve as intracellular messages that mediate the action of a large number of peptide hormones[4].

The general similarities between cAMP and Ca^{2+} are striking, and are depicted in Table 3. In each cascade of reactions, the messenger binds to a protein, which may be seen as a type of receptor (calmodulin or the regulatory subunit of cAMP-dependent protein kinase, R), and the complex formed activates an enzyme which in turn phosphorylates protein substrates. In the case of cAMP, there is a single enzyme (C subunit) that can phosphorylate multiple substrates, whereas calmodulin activates a variety of enzymes, each of which exhibit considerable substrate specificity (with the exception of calmodulin kinase II). Therefore a *cell's response to cAMP* is *determined by* its complement of *PK-A substrates* (p. 106), whereas the response to *Ca^{2+}* is predicted on the basis of its complement of *calmodulin-dependent enzymes*.

Table 3 reveals how Ca^{2+} and cAMP can act either as agonists or as antagonists. Peptide hormones that activate adenylate cyclase result in increased cellular levels of cAMP. However, Ca^{2+}-calmodulin is not only capable of activating adenylate cyclase, but in addition controls the activity of phosphodiesterase. Calcium regulates cell contractility by activating the calmodulin-dependent myosin light chain kinase. The subsequent phosphorylation of the light chain of myosin (M_r=20,000) allows stimulation of its ATPase and energy production. However, myosin kinase can also be phosphorylated by cAMP-dependent protein kinase. This phosphorylation inactivates the enzyme by preventing the binding of calmodulin. Many other examples exist which illustrate the *interrelationships between these two intracellular messenger systems*[4,5].

Even though some hormones, such as β agonists, act primarily through the cAMP system, and others, such as α agonists, utilize the Ca^{2+} signal, calmodulin plays a role in some of the mechanisms of action of nearly all peptide hormones. This is particularly true with respect to those hormones which stimulate the secretion of steroids, biogenic amines, peptides, or proteins. Whereas FSH, LH, ACTH and TSH all stimulate adenylate cyclase and increase cAMP in their respective target cells, they also exert influences upon Ca^{2+} flux, and calmodulin action is therefore required for secretion of the target cell products. Part of this effect can be explained by the fact that calmodulin is a component of secretory granules. Peptide hormones (or growth factors) may provoke an increased Ca^{2+} flux into target cell or a release of Ca^{2+} from intracellular compartment(s). The latter effect can be due to the action of IP_3 generated from the metabolism of phosphoinositide in the plasma membrane. The increased intracellular Ca^{2+} binds to calmodulin and, in turn, the complex binds to secretory granules. In addition, the Ca^{2+}-calmodulin complex stimulates contractility of the microfilaments via activation of myosin light chain kinase. Together, these events result in the movement of secretory granules to the cell periphery and the release of their contents.

The involvement of calmodulin in hormone action generally *results in altered Ca^{2+} levels*, affecting the activity of the binding protein *rather than changing* in the *intracellular concentra-*

Table 2. *Enzymes activated by Ca^{2+}-calmodulin.*

Substrate-specific enzymes

Cyclic nucleotide metabolism
- Cyclic AMP phosphodiesterase
- Cyclic GMP phosphodiesterase
- Guanylate cyclase
- Adenylate cyclase
- Cyclic GMP-dependent protein kinase
- Lipid metabolism
- Dolichol kinase
- Phospholipase A_2

Carbohydrate metabolism
- Phosphorylase kinase
- Glucose 1,6-diphosphatase
- Phosphofructokinase
- Phosphoprotein phosphatase II*b*
- NAD^+ kinase
- Ion transport
- Ca^{2+}-activated ATPase

Multisubstrate calmodulin-dependent protein kinases
- Tyrosine hydroxylase
- Phenylalanine hydroxylase
- Tryptophan 5′-monooxygenase
- Tyrosine 3′-monooxygenase
- Succinate dehydrogenase
- Acetyl-CoA carboxylase
- Tubulin
- Microtubule-associated proteins
- Estrogen receptor
- Vimentin
- Synapsin
- Myosin light chain

Table 3. *The relationship between the Ca^{2+} and cAMP messenger systems.*

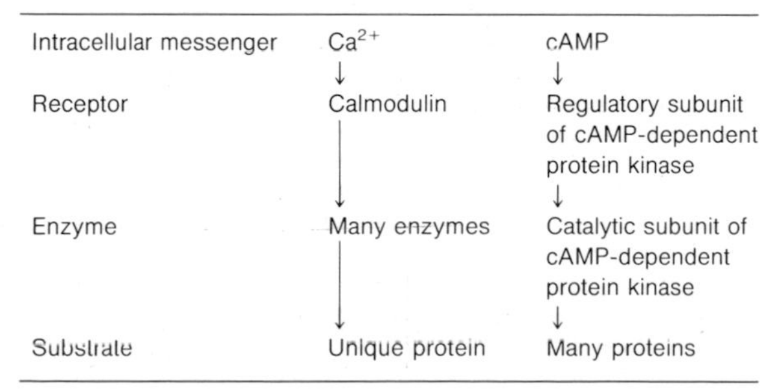

Intracellular messenger	Ca^{2+}	cAMP
	↓	↓
Receptor	Calmodulin	Regulatory subunit of cAMP-dependent protein kinase
	↓	↓
Enzyme	Many enzymes	Catalytic subunit of cAMP-dependent protein kinase
	↓	↓
Substrate	Unique protein	Many proteins

tion of calmodulin. Indeed calmodulin is constitutively expressed in target cells for both peptide and steroid hormones[4]. Primary regulation of cellular calmodulin occurs *during the cell cycle*: calmodulin synthesis is restricted to the late part of G_I, when the cellular content is doubled. Anticalmodulin drugs added to a proliferating cell population specifically and reversibly block cells at the G_I/S boundary (Fig. 1, p. 220). Mitogenic stimulation of a quiescent population of cells by growth factors results in a rapid decrease, and then increase, in calmodulin mRNA and calmodulin. Both of these changes precede the initiation of DNA synthesis. Again, anticalmodulin drugs prevent the return of these cells to the proliferative pathway. The role of the increased calmodulin prior to DNA synthesis seems to be due to a requirement in DNA repair[5]. Indeed the antiproliferative effect of interferon is associated with the inhibition of the synthesis of calmodulin that normally occurs at the G_I/S boundary of the cell cycle.

Calmodulin content of cells or tissues can be *assayed* by stimulation of a calmodulin-dependent enzyme. The most popular enzyme assay involves activation of a calmodulin-deficient cyclic nucleotide phosphodiesterase. It is in fact the subunit characteristics of this enzyme[6] that led to the discovery of a Ca^{2+}-dependent protein activator by Kakiuchi and colleagues[7]. Drawbacks of the enzyme assays include the requirement of Ca^{2+} and the fact that other calmodulin binding proteins frequently contaminate both calmodulin-containing extracts and the enzyme preparations. To obviate these difficulties, affinity purified polyclonal antibodies were prepared and utilized, to develop a highly specific and sensitive radioimmunoassay[8]. This assay is performed routinely in the presence of metal chelators, and thus is not affected by the presence of other calmodulin binding proteins. Subsequently, many other types of assays have been developed for the quantitation of calmodulin and calmodulin binding proteins[9].

Antibodies have also been utilized to localize calmodulin in cells and tissues, at both light and electron microscope levels[4]. It is a component of virtually all intracellular membranes and a variety of non-membranous organelles, such as microfilaments and microtubules. Calmodulin is a prominent protein in the mitotic apparatus, and seems to be involved in the regulation of chromosome movement. Antibodies were also used to develop a translation assay for calmodulin mRNA, which led to the molecular cloning of the cDNA.

The entire transcription unit of the chicken calmodulin gene has now been elucidated[10]. This gene is composed of six exons divided by five introns; four of the introns are in the amino acid coding region (Fig. 1). The DNA that encodes the central helix is contained on a single exon, suggesting that this region may be a functional unit. Three of the four Ca^{2+} binding loops are interrupted by introns. One approach that can now be utilized to study structure/function relationships for calmodulin is to alter appropriate regions of the molecule, to see how these modifications affect enzyme binding and activation as well as protein conformation. This is feasible, as the three-dimensional structure is now known[1], and calmodulin cDNA has been introduced into bacterial expression vectors, to yield large amounts of easily isolatable protein[11]. Site-directed mutagenesis can now be used to alter strategic portions of this receptor, such as the Ca^{2+} binding loops and the central helix. This will allow the eventual determination of the mechanism of action of calmodulin, in precise molecular terms.

References

1. Babu, Y. S., Sack, J. S., Greenhough, T. J., Bugg, C. E., Means, A. R. and Cook, W. J. (1985) Three-dimensional structure of calmodulin. *Nature* 315: 37–40.
2. Klee, C. B. and Vanaman, T. C. (1982) Calmodulin. *Adv. Protein Chem.* 35: 212–321.
3. Inagaki, M. and Hidaka, H. (1984) Two types of calmodulin antagonists: a structurally related interaction. *Mol. Pharmacology* 29: 75–84.
4. Means, A. R., Tash, J. S. and Chafouleas, J. G. (1982) Physiological implications of the presence, distribution and regulation of calmodulin in eukaryotic cells. *Physiol. Rev.* 62: 1–39.
5. Means, A. R., Slaughter, G. R. and Putkey, J. A. (1984). Postreceptor signal transduction by cyclic adenosine monophosphate and the Ca^{2+}-calmodulin complex. *J. Cell Biol.* 99: 226s–231s.
6. Cheung, W. Y. (1970) Cyclic 3′,5′-nucleotide phosphodiesterase, demonstration of an activator. *Biochem. Biophys. Res. Commun.* 38: 533–538.
7. Kakiuchi, S. and Yamazaki, R. (1970) Calcium dependent phosphodiesterase activity and its activating factor (PAF) from brain. *Biochem. Biophys. Res. Commun.* 41: 1104–1110.
8. Chafouleas, J. G., Dedman, J. R., Munjaal, R. P. and Means, A. R. (1979) Calmodulin: development and application of a sensitive radio-immunoassay. *J. Biol. Chem.* 254: 10262–10267.
9. Means, A. R. and O'Malley, B. W., eds (1983) Calmodulin and calcium binding proteins. In: *Methods in Enzymology 102*, Academic Press, New York.
10. Simmen, R. C. M., Tanaka, T., Ts'ui, K. F., Putkey, J. A., Scott, M. J., Lai, E. C. and Means, A. R. (1985) The structural organization of the chicken calmodulin gene. *J. Biol. Chem.* 260: 907–912.
11. Putkey, J. A., Slaughter, G. R. and Means, A. R. (1985) Bacterial expression and characterization of proteins derived from the chicken calmodulin cDNA and a calmodulin processed gene. *J. Biol. Chem.* 260: 4704–4712.

CHAPTER XIV

Calcium Regulatory Hormones

Pierre Royer and Byron Kemper

Contents, Chapter XIV

Calcium Regulatory Hormones

1. Regulation of Calcium Metabolism

In vertebrates, calcium and phosphorus have been shown to play an important role in *bone mineralization*. *Diet* is the *main source* of calcium, which enters the body primarily through intestinal absorption. The level of *blood calcium* is determined mainly by bone metabolism and urinary excretion. Regulation of calcium metabolism is *under endocrine control*, largely by parathyroid hormone and calcitriol. The physiological importance of plasma calcium concentration is not completely understood. The regulation of bone metabolism is probably under the control of several other factors acting locally via paracrine or autocrine mechanisms. *Calcium* acts not only at the level of bone, intestine and kidney, but *also participates in the function of most, if not all, cells*. Examples of physiological functions in which calcium plays a major role include myocontraction, nerve transmission, coagulation of blood and milk, the activation or inhibition of various enzymes, the function of several hormones, exocytosis, cell-cell interaction, and cell duplication. In addition to cAMP, calcium is often considered a *fundamental intracellular regulatory component*.

In this chapter, both endocrine and local control of calcium will be considered, and it is important not to confuse them. It should be kept in mind that, in bone cells under hormonal control, calcium can be both a second messenger of hormones and the final target of endocrine activity.

1.1 Physiological Aspects

The regulation of *calcium equilibrium* occurs principally in the *digestive tract* and *kidney*, and throughout the body, by an exchange between different compartments, of which *bone* and *extracellular space* are the most important.

The dietary intake of calcium varies, between 300 and 2000 mg/d, as a function of geographical considerations. In the Western world, it averages approximately 1000 mg/d in normal adults. A double flux of calcium is observed in the intestine: approximately 325 mg/d of calcium are absorbed from nutritional intake, and about 150 mg/d are excreted. Thus, the sum of *net intestinal absorption* represents the difference between dietary intake and fecal excretion of calcium, which averages about 175 mg/d. The percentage of dietary calcium involved in net intestinal absorption represents the *coefficient of net digestive uptake*. Urinary excretion of calcium is approximately 150 mg/d. This sum represents the difference between the quantity of calcium filtered by the glomerulus (approximately 10,000 mg/d) and the quantity of calcium reabsorbed by the renal tubules, which constitutes the majority of all calcium passing into the glomerulus. In addition, there is a minor tubular secretion (up to 100 mg/d) of calcium. The overall sum of the net absorption of calcium and urinary excretion is the *calcium balance*, a figure which does not take into consideration calcium lost in sweat and altered requirements associated with pregnancy and lactation. Normal adults are at equilibrium, whereas calcium balance is very positive in growing children, averaging between 100 and 200 mg/d; they retain about 1 kg of calcium over a period of 18 years.

Because of the considerable mass of calcium in bone, in contrast to extracellular calcium, an important factor in the regulation of the level of blood calcium is the exchange which occurs between these two compartments, known as *accretion* and resorption (*osteolysis*), each of which is equilibrated under normal circumstances in adults. *Three principal hormones* are involved in overall calcium regulation: *parathyroid hormone* (PTH), or parathormone, *calcitonin*, and *calcitriol*. Bony tissue and the kidneys are affected by all three hormones, whereas calcitriol acts fundamentally in the intestine.

Use of radioactive isotopes of calcium and strontium has provided important information regarding true intestinal absorption, the endogenous excretion of calcium and the rate of bony accretion, and has contributed to the definition of two compartments, the *rapidly and slowly exchangeable pools*[1].

1.2 Distribution and Principal Roles

Plasma and Extracellular Calcium

The total quantity of plasma and extracellular calcium is approximately 1 g in adults.

Calcium is present in human blood at a concentration of *100 mg/l* (or 2.5 mmol/l). It circulates as *free* calcium ions, *and* as phosphates, citrates and salts of other organic acids, as well as in a form *bound* to albumin. In contrast to the former fractions which are freely *diffusible*, bound calcium is not. The concentration of *serum free calcium ions*, which are *physiologically active*, is approximately *42 to 46 mg/l* (or 1.1 mmol/l).

What are the respective roles of the intestine, bone and kidney in the maintenance of calcium equilibrium? All three organs are involved, but the relative importance of their actions under physiological conditions has not been clearly established; it changes during fetal development, at birth, in the adult, and in old age. The intestine is important because it can adapt to dietary intake, but the net flux appears to remain moderate, as it is in the kidney. However, a possible importance of urinary calcium excretion, particularly in adults and older people, has been postulated. Because of the major exchanges which occur within bones, this tissue is of fundamental importance, especially during childhood.

Calcium in Bone

Calcium, in the form of *hydroxyapatite crystals*, is present in the *collagen* matrix of bone. *Ionic exchange* with extracellular fluid is controlled in part by a layer of *osteoblasts*, bordering the bony trabeculae, in addition to physiological changes governed by *osteocytes* and *osteoclasts* located deep in the bone. These exchanges lead to the continual remodelling of the skeleton and the adaptation of bone structure to mechanical forces, resulting in the production and resorption of bony tissue. At birth, approximately 30 g of calcium can be found in bone, whereas in adults this level attains or exceeds 1000 g.

What is the origin and the role of bone cells? Osteoblasts are derived from proosteoblasts or from osteoprecursor cells (OPC), and develop into osteocytes, or osteoblasts at rest, after having produced collagen and other proteins of the bone matrix, such as osteocalcin (or GLA protein), which is vitamin K-dependent and stimulated by calcitriol and osteonectin. Osteocalcin binds calcium, and osteonectin the calcium to collagen. Monocytes are considered precursors of osteoclasts.

The dynamic aspects of bone remodelling led Frost to put forth the concept of *bone modelling units* (BMU). These units associate osteoclasts and osteoblasts, and constitute the numerous structural groups which function sequentially: activation, resorption, and formation. The *life span* of a BMU in normal adult men is *four months*, but can be as long as 10 years in osteomalacia. Endocrine influences occur at the level of osteoblasts and osteoclasts (Fig. XIV.1), and BMUs are often identified by *quantitative histomorphometry*.

Intracellular Calcium

The *total cellular concentration of calcium* is of the same order of magnitude *as extracellular* calcium; that is, approximately *1 mmol/l*. The total quantity of cellular calcium in humans is quite low, approximately 2 g. Almost all *intracellular calcium* is *bound* to specific *proteins*, or is *sequestered* in *mitochondria* or in the *endoplasmic reticulum*. Free intracellular calcium is *<1/1000* of the total cellular concentration, and is approximately 10^{-7} *M*. Thus, there is an *enormous concentration gradient* between free calcium and calcium in the extracellular and intracellular spaces. *Movement of calcium across biomembranes* is the major factor responsible for the very low concentration of soluble free calcium.

A *dual system of regulation* controls ionic influx and efflux across the plasma membrane, and determines the ratio of intracellular to extracellular

1. Bronner, F. (1973) Kinetic and cybernetic analysis of calcium metabolism. In: *Calcium and Phosphorus Metabolism* (Irving, J. T., ed), Academic Press, New York, pp. 159–186.

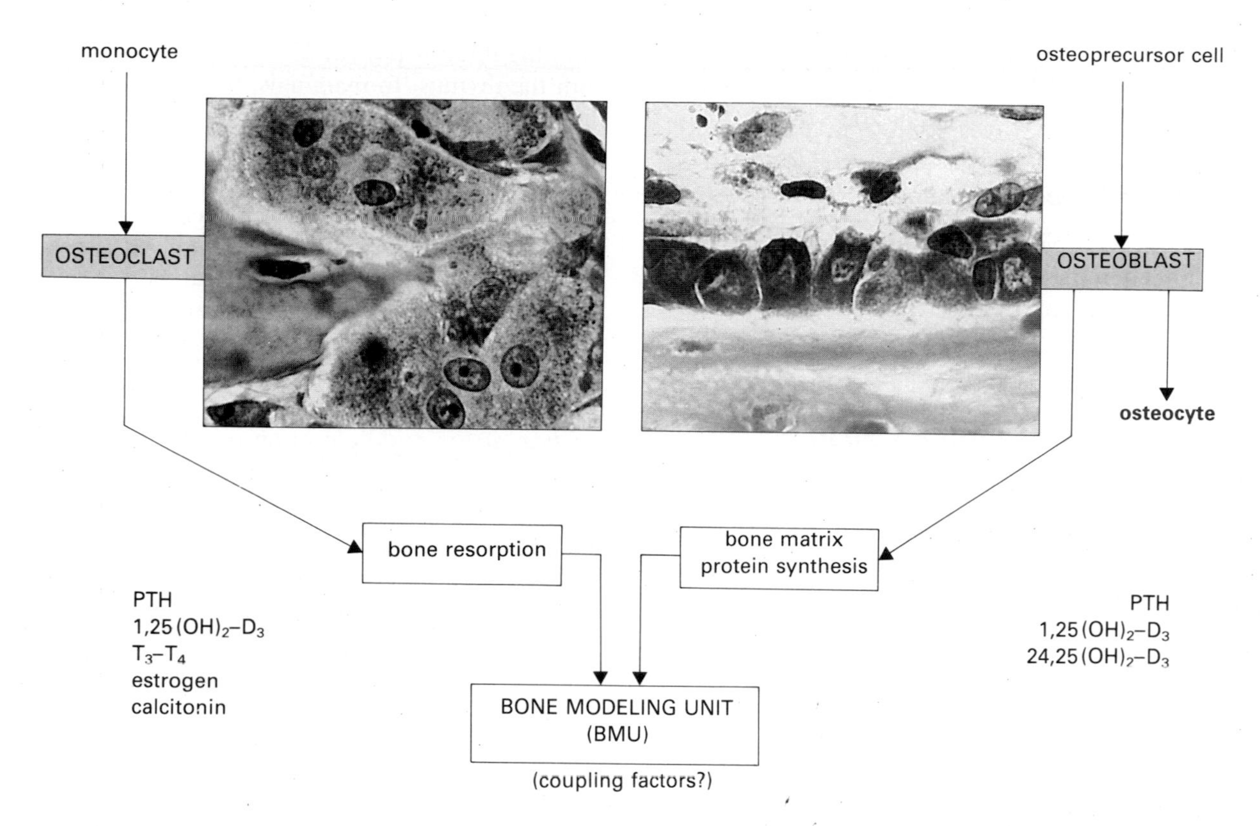

Figure XIV.1 Action and cooperation of bone cells involved in bone differentiation and renewal.

calcium. 1. Calcium *channels* allow the movement of calcium into the cell. Surface receptors are coupled to such channels, which open upon activation. This opening is brief, and the signal transmitted is transitory. These calcium channels can be blocked by various inorganic cations, such as cobalt, lanthanum and manganese, or by specific *blockers*, the three most important of which are nifedepine, diltiazem and especially verapamil, of interest for its therapeutic applications in cardiology[2]. In contrast, other molecules have been found to facilitate membrane permeability to ions, and thus form channels: these are *ionophores*, such as valinomycin, gramicidine A, and especially the ionophore A 23187, which transports divalent cations, such as Ca^{2+} and Mg^{2+}. 2. The plasma membrane plays a fundamental role in the release of calcium, against the concentration gradient, into the extracellular space. This occurs as the result of a *calcium pump (Ca^{2+}-ATPase)* bound to the membrane, which uses energy released from the hydrolysis of ATP. In some cells, other calcium pumps couple calcium efflux and sodium influx.

The concentration of intracellular calcium is also controlled by fluxes across *membranes of intracellular organelles* that sequester calcium. The surface of these biomembranes is from 10 to 100 times greater than that of the plasma membrane. Within the *endoplasmic reticulum*, the flux depends on *Ca^{2+}-ATPases* bound to the membrane. In *mitochondria*, the influx of calcium follows an *electrochemical gradient* generated by the transfer of electrons associated with oxidative phosphorylation.

Free Calcium and Calcium Binding Proteins

The concentration of *free* calcium in the cytoplasm, apart from the flux across biomembranes, is regulated by the level of calcium *phosphates* and by macromolecules such as *calcium binding proteins (CaBP)* (Table 1, p. 631). The major calcium binding proteins identified in cytoplasm are: *troponin C* in striated muscle; *parvalbumin*, also found in muscle; *S-100 protein* in the nervous system, melanocytes and Langerhans cells of the skin; intestinal *CaBP*, which is *vitamin D-dependent*, CaBP of cartilage, bone and teeth, and also vitamin D-dependent CaBP of basal cells of the skin and malpighian mucosa. Osteoblasts and bone also contain osteonectin and vitamin K-

2. Braunwald, E. (1982) Mechanism of action of calcium-channel-blocking agents. *N. Engl. J. Med.* 307: 1618–1627.

dependent GLA (γ-carboxyglutamic acid) protein, or osteocalcin[3]. *Calmodulin* is a *ubiquitous CaBP* found in all animal and plant cells (p. 631).

Calcium as a Second Messenger

The varied aspects of calcium as an intracellular regulatory component are described, for each relevant hormone, in the appropriate chapter, and are summarized on p. 104.

1.3 Hormones Affecting Calcium Regulation

The regulation of both serum and cellular calcium is under the control of *several hormones*, primarily parathyroid hormone (PTH), vitamin D_3 (or cholecalciferol), and calcitonin. They exert their actions principally within their *main target organs*: in intestine, kidney, and bone.

It it is actually necessary to treat two complementary aspects of calcium regulation: the metabolism of calcium itself, and its effects on various cellular events.

2. Parathyroid Hormone

2.1 Biology

Parathyroid hormone (PTH) is produced and secreted by the *parathyroid gland*. These glands are present in all terrestrial vertebrates. In 1909, MacCallum and Voetglin demonstrated that surgical removal of parathyroid glands in dogs reduced the calcium level in blood and led to tetany. In 1925, Collip and his associates prepared the first parathyroid extract capable of reversing hypocalcemia in a parathyroidectomized dog. Since 1970, the structures of bovine, porcine, human and rat PTH have been elucidated (Fig. XIV.2), and several fragments, including the active 1–34 aa portion, have been synthesized.

Morphology and Development of the Parathyroid Glands

The parathyroids are derived from the IIIrd and IVth pharyngeal pouches, the inferior glands developing from the superior pouch, and the superior glands from the inferior pouch. The inferior glands migrate with the thymus. In mammals, the dorsal portion of the pouches develops into the parathyroids, whereas the ventral portion forms the thymus.

The parathyroids vary in length, from 3 to 15 mm, with a mass of approximately 40 to 150 mg. They are brown or yellow in color, and vary in number from two to six, the most frequent number being four. There is an inconsistency in the position of the glands, and these anatomical differences were studied in detail once parathyroidectomy became a frequent surgical procedure. Their most common position is *near the trachea*, at the middle portion of the thyroid gland, with one pair above and one pair below the inferior thyroid artery. However, they can be found more laterally, or even in the thyroid, thymus, mediastinum, and pericardium. Usually, the inferior thyroid artery supplies the parathyroids.

Histological examination revealed that parathyroid secretory cells are arranged in a series of cords grouped into lobes separated by connective tissue. Half of the gland consists of adipocytes. The secretory cells have several specific histological and ultrastructural characteristics. *Chief cells* are round or polygonal, ranging in diameter from 5 to 12 μm. Their cytoplasm is slightly granular and dark, containing glycogen, which, when dissolved, leaves small vacuoles. The nucleus is large and contains one or several nucleoli. PTH has been demonstrated in these cells, by immunoflorescence. Occasionally, vacuoles invade the cell, causing the cell to increase in size. In this way, what are known as *clear cells* are formed. Another type of cell is represented by *oxyphil cells*, with a diameter from 7 to 18 μm, a small nucleus, and a cytoplasm rich in mitochondria containing acidophilic granules. The function of these cells is unknown. Not abundant in children, their number increases after puberty.

Biosynthesis of PTH

Parathyroid hormone was the *first polypeptide hormone* for which the entire biosynthetic pathway was elucidated, from the initial translational product to the mature form containing 84 aa (Fig. XIV.2). PTH is initially synthesized with an extra 31 aa at the N-terminus (prepro-PTH), of which 25 are proteolytically removed to produce pro-PTH, which is in turn cleaved to produce PTH[4] (Fig. XIV.3).

Pro-PTH was discovered during biosynthesis experiments with radioactive amino acids in bovine parathyroid slices. A radioactive protein larger than

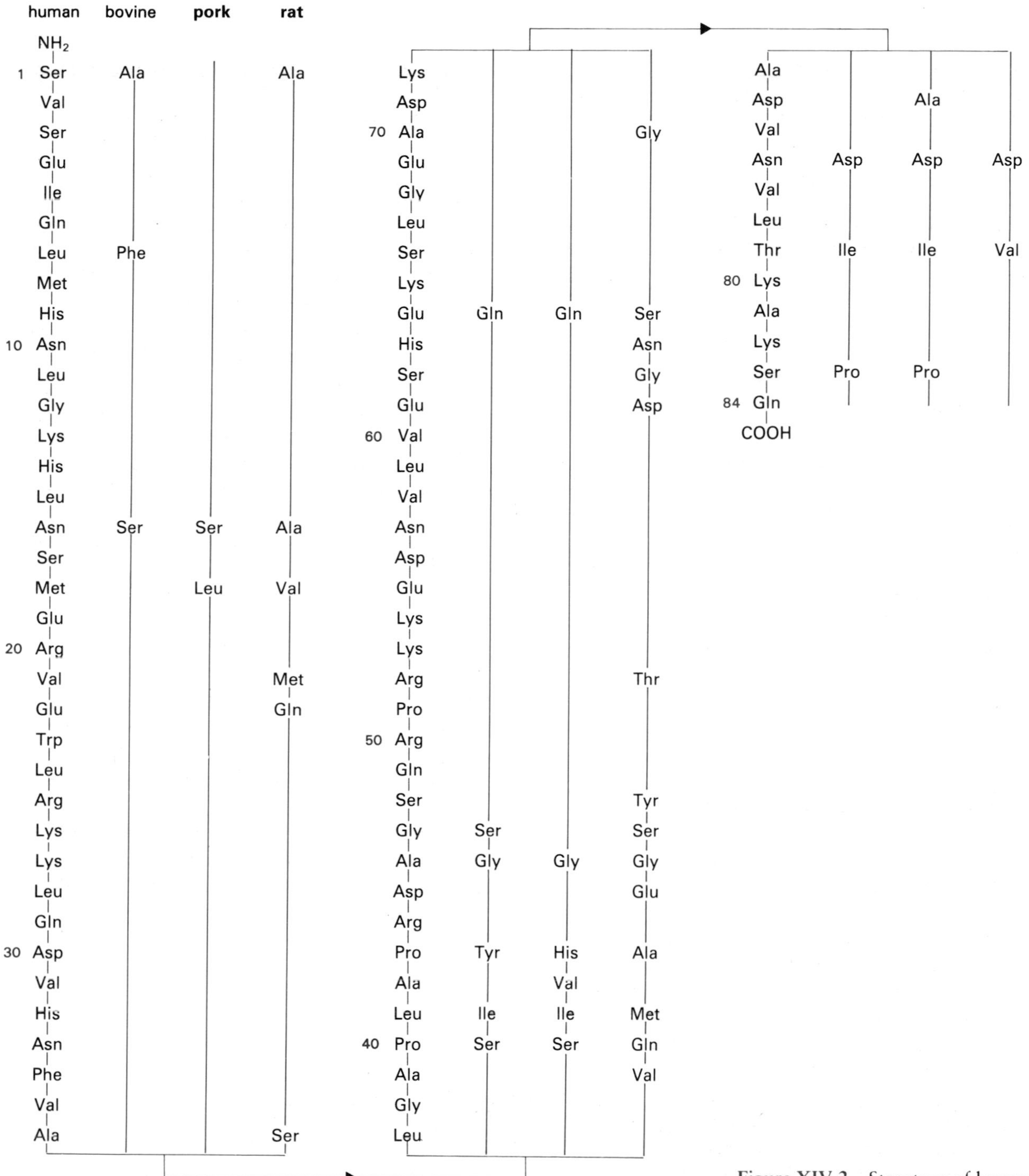

Figure XIV.2 Structure of human, bovine, porcine and rat PTH.

3. Gallop, P. M., Lian, J. B. and Hauschka, P. V. (1980) Carboxylated calcium-binding proteins and vitamin K. *N. Engl. J. Med.* 302: 1460–1466.
4. Cohn, D. V., MacGregor, R. R., Chu, L. L. H., Kimmel, J. R. and Hamilton, J. W. (1972) Calcemic fraction-A: biosynthetic peptide precursor of parathyroid hormone. *Proc. Natl. Acad. Sci., USA* 69: 1521–1525.

PTH was obtained and precipitated by antisera to PTH. Analysis of the amino acid sequence demonstrated conclusively that this larger protein was PTH, with the sequence Lys-Ser-Val-Lys-Lys-Arg at the N-terminal. An identical pro-sequence is present in human pro-PTH, and similar sequences are found in

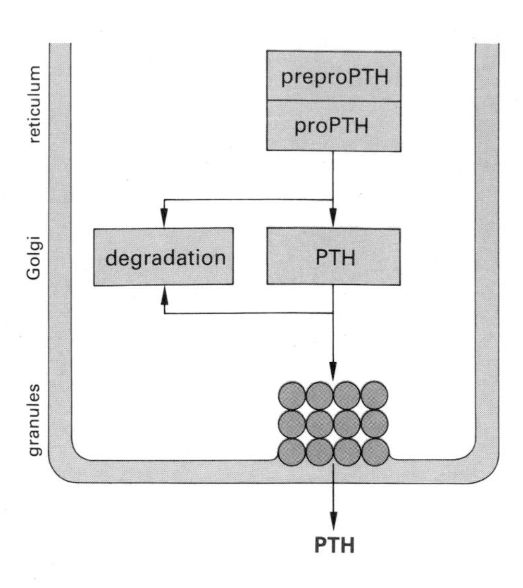

Figure XIV.3 Synthesis, storage and secretion of PTH.

rat, pig, dog and chicken pro-PTH molecules. The dibasic amino acids, the site where the molecule is cleaved to produce PTH, are similar to those at cleavage sites involved in the processing of most pro-hormones into their mature form. The cleavage of pro-PTH to PTH occurs 15 to 20 min after the synthesis of pro-PTH, at about the time the precursor reaches the Golgi apparatus in the cell. The function of the pro-sequence of pro-PTH remains unknown.

Prepro-PTH was observed when PTH mRNA was translated in cell-free systems[5]. Amino acid sequence analysis showed 25 extra amino acids at the N-terminus of pro-PTH. This pre-sequence contains a core of 12 hydrophobic amino acids, and is cleaved, from prepro-PTH, to produce pro-PTH, by an enzyme in microsomal membranes. These properties are consistent with the belief that the pre-sequence serves the function of the *signal* sequence. Human and rat pre-sequences have been deduced from cloned cDNAs and cloned genes, respectively, and share considerable homology with the bovine sequence determined by sequencing the cell-free translational product. While these sequences are similar, in their hydrophobicity, to other signal sequences, they differ in that they have few leucine residues and an unusually large number of methionine residues.

PTH fragments smaller than 1–84 aa have been identified in blood. C-terminal fragments are stable but have little or no biological activity, although they cross-react in radioimmunoassays. The *heterogeneity* of the *circulating forms* of immunoreactive PTH complicates the interpretation of certain physiological and pathological conditions.

PTH Genes

The mRNA for bovine PTH was shown to contain about 700 nucleotides, or about twice that required to code for prepro-PTH[6]. The sequence of bovine and human mRNAs has been derived from the sequence analysis of cloned cDNA[7], while the rat sequence has been derived from the sequence of the cloned gene[8]. In the regions coding for protein, the bovine and human genes are nearly 90% homologous, while there is about 80% homology between rat and both human and bovine cDNAs. Based on the derived protein sequences, the *PTH family* required about 5×10^6 years to undergo one amino acid change per 100 residues. These proteins, therefore, have been relatively *well conserved* throughout evolution. The conservation of sequences along the protein molecule is variable, with strongly conserved regions from amino acids 6 to 15, 23 to 38, and 49 to 84. The first and second regions are required for PTH activity, and the second is probably involved in receptor binding. The *conservation of the C-terminal* amino acids (49 to 84) was *unexpected*, since this region is not required for the biological activity of the hormone and has no known function.

The genes for bovine[9] (Fig. XIV.4), human[10] and rat[8] PTH have been isolated, and their structures have been characterized. Each of the genes has two introns which interrupt the coding sequence at exactly the

5. Kemper, B., Habener, J. R., Mulligan, R. C., Potts, J. T. and Rich, A. (1974) Pre-proparathyroid hormone: a direct translation product of parathyroid messenger RNA. *Proc. Natl. Acad. Sci., USA* 71: 3731–3735.
6. Stolarsky, L. and Kemper, B. (1978) Characterization and partial purification of parathyroid hormone messenger RNA. *J. Biol. Chem.* 253: 7194–7201.
7. Hendy, G. N., Kronenberg, H. M., Potts, J. T. and Rich, A. (1981) Nucleotide sequence of cloned cDNAs encoding human preproparathyroid hormone. *Proc. Natl. Acad. Sci., USA* 78: 7365–7369.
8. Heinrich, G., Kronenberg, H. M., Potts, J. T. Jr. and Habener, J. F. (1984) Gene encoding parathyroid hormone. Nucleotide sequence of the rat gene and deduced amino acid sequence of rat preproparathyroid hormone. *J. Biol. Chem.* 259: 3320–3329.
9. Weaver, C. A., Gordon, D. F., Kissil, M. S., Mead, D. A. and Kemper, B. (1984) Isolation and complete nucleotide sequence of the gene for bovine parathyroid hormone. *Gene* 28: 411–422.
10. Vasicek, T. J., McDevitt, B. E., Freeman, M. W., Fennick, B. J., Hendy, G. N., Potts, J. T. Jr, Rich, A. and Kronenberg, H. M. (1983) Nucleotide sequence of the human parathyroid hormone gene. *Proc. Natl. Acad. Sci., USA* 80: 2127–2131.

region of the mRNA, the second exon codes primarily for the pre- or signal sequence, and the third exon encodes PTH and the 3′ untranslated region of its mRNA. In contrast to many other mammalian genes, only a single PTH gene per haploid genome has been observed in the three species studied. *No pseudogenes* or other *related genes* have been detected.

The 5′ flanking regions, which presumably contain the transcriptional control regions for the PTH gene, also show considerable homology among the three species. In the bovine system, PTH mRNA is heterogeneous in size at the 5′ terminus, and transcription is initiated from two locations, about 30 bp apart on the gene. In both human and bovine genes, *two TATA* sequences, which are thought to participate in the initiation of transcription in eucaryotes, are present, separated by about 30 bp, in the proper location to direct the synthesis of the two size classes of PTH mRNA. The presence of two TATA sequences, although conserved in the two species, is not required for PTH gene transcription, since in the rat gene only the upstream TATA sequence is conserved. Consistent with this observation, only the larger size class of PTH mRNA is observed in the rat.

Analysis of a panel of hybrid human–rodent cells revealed that the gene for human PTH is located on the short arm of *chromosome 11* (Fig. XIV.5). Polymor-

Figure XIV.4 Structure of the bovine PTH gene.

same location in each species. The size of intron A is about the same in the rat and bovine genes (1600 to 1700 bp), but is twice as large in the human, while the smaller intron B contains about 100 bp in all three species. The three exons encode the functional regions of the protein. The first exon encodes the 5′ untranslated

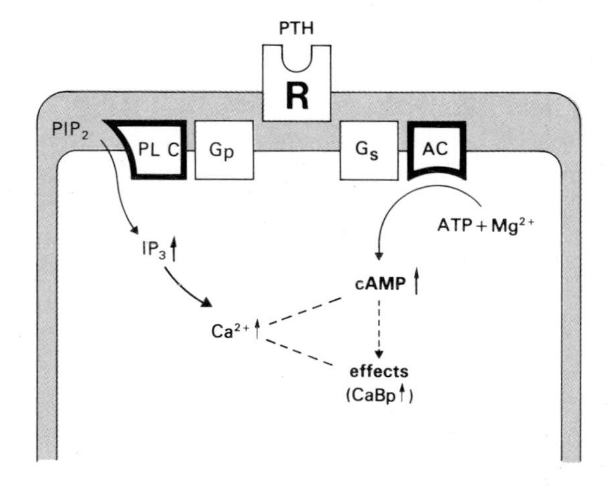

Figure XIV.5 Chromosomal localization of the PTH gene on the short arm of human chromosome 11. (Zabel, B. U., Kronenberg, H. M., Bell, G. I., and Shows, T. B. (1985) Chromosome mapping of genes on the short arm of human chromosome 11: parathyroid hormone gene is at 11p15 together with the genes for insulin, o-Harvey-ras 1, and β-hemoglobin. *Cytogenet. Cell. Genet.* 39: 200–205).

phism has been observed in a cleavage site for the restriction enzyme Pst I, 3′ to the PTH gene, in all ethnic groups examined thus far. This polymorphism and other polymorphisms near the genes for β-globin and insulin were used to determine the order of these genes on chromosome 11. The PTH gene was closely linked to the β-globin gene cluster, and weakly linked to the insulin gene. The proposed order is: the centromere, PTH and β-globin genes, and the insulin gene. Interestingly, the gene for *calcitonin* has *also* been localized to the short arm of chromosome 11. Thus, the genes for both of the major peptide hormones involved in calcium metabolism are present on chromosome 11. The localization of the PTH gene should permit the delineation of other polymorphisms which may be correlated with various abnormalities of PTH production. Such correlations should be useful in the diagnosis of abnormalities and in understanding their etiology.

Regulation of PTH Gene Expression

The biosynthesis of PTH is regulated by *changes in calcium* concentration. Chronic decrease in calcium levels increases the production of PTH in vivo, but whether this is related to a specific increase in the biosynthesis of PTH or to an overall increase in protein synthesis by the parathyroid gland has not been determined. In addition, chronic hypocalcemia leads consistently to hyperplasia of the gland.

Finally, parathyroid glands *also synthesize* another protein, *secretory protein I*, or SP-I[11]. This protein is similar or identical to chromogranin A found in the adrenal medulla. SP-I is found along with PTH in the secretory granules, and it may be involved in the process of secretory exocytosis.

Secretion[12]

Parathyroid hormone is secreted by exocytosis. The membrane of the secretory granule containing the hormone fuses with the plasma membrane, which is subsequently lysed, releasing PTH into the blood stream. *Interestingly, a reduction in Ca^{2+} level increases exocytosis of PTH*, whereas elevated Ca^{2+} generally has a stimulatory effect on endocrine or exocrine glands. Cyclic AMP may be involved in this Ca^{2+}-controlled exocytosis of PTH.

Control of Secretion

The major regulatory *role of Ca^{2+}* concentration is *at the level of hormone secretion*. In vitro, parathyroids incubated in a low calcium medium secrete abundant quantities of PTH. The perfusion of an isolated thyroid-parathyroid system with hypocalcemic blood leads to the secretion of immunoreactive PTH. In cows, calcium infusion decreases, whereas EDTA markedly increases, the concentration of immunoreactive PTH in blood. The kinetics of this regulation are precise. The decrease in immunoreactive PTH following iv injection of calcium is almost immediate, reaching a maximum after two hours. After injection of EDTA, the increase in immunoreactive PTH is reached much more slowly; its maximum is observed after three to four hours. There is an inverse relationship between the serum concentration of immunoreactive PTH and that of calcium, as long as the calcium level remains between 40 and 120 mg/l. Orthophosphate ions appear to act only indirectly, by modification of blood calcium levels. Variations in phosphatemia at constant calcium levels have no effect on the concentration of immunoreactive PTH in serum. The physiological role of magnesium ions is still uncertain. Epinephrine may play a role as modulator of PTH secretion in extreme situations. Calcitriol *receptors* are found *in parathyroids*, and a *vitamin D-dependent calcium binding protein* of low molecular weight has been identified in the glands. Calcitonin, cortisol and growth hormone may have a stimulatory effect on the parathyroid glands, whereas somatostatin may be inhibitory. There is no clear proof, to date, that any factor other than the concentration of extracellular calcium is involved in the normal regulation of PTH biosynthesis and secretion. However, circadian rhythms and variations during sleep cycles do exist which are apparently not regulated by calcium levels.

The content of PTH in the thyroid is approximately 11 pmol, and the *secretion rate* of PTH under normal physiological conditions is about *1 pmol of PTH/min/kg* of body weight.

Distribution and Peripheral Metabolism

Parathyroid hormone *circulates* in the form of the *intact 1–84 aa hormone*, and *C-terminal* fragments, with a long half-life, approaching several hours, are also found. Fragmentation of the hormone

11. Cohn, D. V. and Elting, J. (1983) Biosynthesis, processing and secretion of parathormone and secretory protein I. *Rec. Prog. Horm. Res.* 39: 181–203.
12. Habener, J. F. (1981) Regulation of parathyroid hormone secretion and biosynthesis. *Annu. Rev. Physiol.* 43: 211–223.

is favored by elevated circulating Ca^{2+}. Under stable conditions, RIA utilizing an antiserum directed against the C-terminal provides the best indication of parathyroid function. *Chronic renal insufficiency* results particularly in an increase in carboxyterminal fragments. The liver reduces the 1–84 aa PTH to smaller fragments. The kidney also intervenes in PTH metabolism[13].

Physiological Effects

Parathyroid hormone acts on the bone to *liberate* not only calcium, but also *orthophosphates, citrate, magnesium, osteocalcin, and hydroxyproline*. Following injection of PTH, the intensity of the response observed depends on the pre-existing level of calcium; it is weaker under conditions in which calcemia is low. The intensity and kinetics of the hypercalcemic response to PTH *vary considerably according to the species and the developmental stage* of the animal. PTH has little action in adult rats, whereas there is a clear response to PTH in neonatal rats. Hypercalcemia induced by PTH is characterized by a rapid and a slow component: the slow component is abolished by actinomycin D, whereas the rapid component, initiated after 10 minutes, is not affected by actinomycin. *PTH* probably activates bone resorption by *direct action* on *osteoblasts* which *secondarily* stimulate *osteoclasts*.

Parathyroid hormone has *several effects* on the *kidney*. A rapid increase in *phosphodiuresis* appears five to eight minutes following injection, and disappears after 60 to 90 min. This property prevents hyperphosphatemia secondary to the effect of PTH on bone. Micropuncture techniques have established that phosphodiuresis is the result of an inhibition of phosphorus reabsorption in the proximal tubule. PTH also inhibits the reabsorption of sodium at this level. In molar terms, inhibition is greater for sodium than for phosphorus, and they are both dependent on cAMP. PTH also causes proximal *inhibition* of the *reabsorption* of certain amino acids, and distal inhibition of the exchange between Na^+ and H^+ and K^+ ions. PTH *reduces* the *renal excretion* of *calcium*, an effect secondary to an increase in the reabsorption of calcium in the distal tubule, and *stimulates 25 OH-D_3 1α-hydroxylase activity* (p. 657).

Parathyroid hormone acts on organs other than kidney and bone; teeth, lens, mammary and salivary glands, and hepatocytes are also affected by PTH. In addition, it has an indirect effect on the small intestine, favoring intestinal absorption of calcium. The intermediate mechanism is an increase in renal biosynthesis of calcitriol, which itself acts on calcium absorption.

Mechanism of Action

Specific *membrane receptors* for PTH have been identified in bone, kidney and circulating lymphocytes[14]. In bone, they are found only in osteoblasts. Their number is regulated by the circulating level of PTH itself, such that, under conditions of spontaneous or experimental hyperparathyroidism, there is a reduction in the number of cellular receptors and desensitization to the action of PTH[15].

In kidney and bone cells, PTH acts via the *cAMP system*, and, although the molecular intermediates are

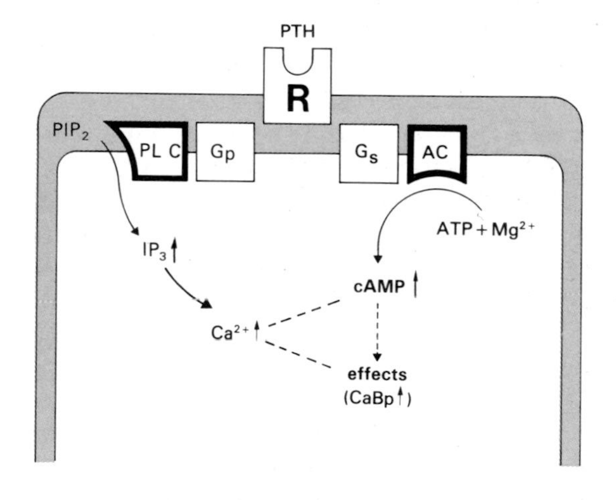

Figure XIV.6 Mechanism of action of PTH. PTH interacts at the plasma membrane, activates adenylate cyclase and the phosphoinasitide system, and increases calcium in the cytosol. This diagram is valid for osteoblasts (no receptor has been found in osteoclasts) and renal tubule cells. In intestinal cells, the effect of PTH on calcium is indirect, via the formation of calcitriol. In several types of target cells, synthesis of a calcium transport protein (CaBP) occurs.

13. Hruska, K. A., Korkor, A., Martin, K. and Slatopolsky, D. (1981) Peripheral metabolism of intact parathyroid hormone. *J. Clin. Invest.* 67: 885–892.
14. Mahoney, C. A. and Nissenson, R. A. (1983) Canine renal receptors for parathyroid hormone. *J. Clin. Invest.* 72: 411–421.
15. Yamamoto, I., Potts, J. T. and Segre, G. V. (1983) Circulating bovine lymphocytes contain receptors for parathyroid hormone. *J. Clin. Invest.* 71: 404–407.

unknown, an increase in cAMP is responsible for the elevation of cytosol Ca^{2+} (Fig. XIV.6). The biochemical events which, under the influence of PTH, lead to *resorption* of *bone* are still *unclear*. In the presence of PTH, osteoclasts undergo a rapid increase in RNA synthesis, retain glucosamine and leucine, and secrete collagenase, phosphatidic acid, hyaluronidase, and cathepsin.

Parathyroid hormone controls several renal functions, including excretion of phosphates, reabsorption of Ca^{2+}, hydroxylation of 25 OH-D_3, and the transport of amino acids.

2.2 Clinical Evaluation

Most of the diagnostic tests of PTH function involve simple biochemical techniques: assays of calcemia, phosphatemia, calciuria and phosphaturia, calculation of the percentage of tubular reabsorption of phosphorus (TRP), and determination of serum alkaline phosphatases and urinary hydroxyproline cAMP.

Assay of urinary and plasma cAMP can be carried out after PTH administration. The cAMP produced by the kidney and excreted into urine is known as nephrogenic cAMP. Urinary secretion of cAMP is increased in primary hyperparathyroidism and reduced in cases of hypercalcemia of non-parathyroid origin.

The assay of PTH is carried out by *radioimmunoassay*. Obtaining an adequate antiserum in the presence of several circulating forms of immunoreactive PTH has proved difficult. The level of total immunoreactive PTH (fragment 1–84) in normal subjects is less than 1 ng/ml of plasma; hypocalcemia induced by EDTA rapidly increases this level. Most frequently, the antiserum directed against the synthetic carboxyterminal extremity of human or bovine PTH (fragment 53–84) is used[16]. Normal levels are from 0.3 to 0.5 ng/ml, and are lower (0 to 0.3 ng/ml) when an N-terminal fragment has been used as the antigen. A biological test may still be of interest[17]. *Biopsy* of the iliac *bone* is sometimes useful in diagnosing parathyroid disease.

2.3 Hyperparathyroidism

A distinction is made between primary hyperparathyroidism, a disease initially of parathyroid origin, ectopic hyperparathyroidism, and secondary and tertiary hyperparathyroidism.

Primary Hyperparathyroidism

Primary hyperparathyroidism is *frequent*. Patients present with hypercalcemia (thirst, polyuria, mental confusion, pruritus, headache, calcified cornea, and metastatic calcifications), radiological signs of demineralization, renal disorders such as nephrocalcinosis and lithiasis, gastric ulcers, and pancreatitis; observed much more rarely are: bone pain, skeletal deformations and fractures, cysts, resorption of the phalanges and of subperiostial zones or of the lamina dura of dental alveolae, asthenia, bone anomalies, chondrocalcinosis, and arterial hypertension. Asymptomatic cases are frequently discovered due to a fortuitous assay of blood calcium.

The diagnosis is based on *hypercalcemia, hypophosphatemia* with a reduction in the percentage of tubular reabsorption of phosphorus, occasional elevation of alkaline phosphatases and hypercitraturia, and *increased urinary hydroxyproline* and amino acids. *Nephrogenic cAMP* is usually *elevated*. In addition, immunoreactive *PTH in serum* may be elevated. However, in certain cases, values obtained with antisera against the C-terminal fragments are normal. Reduced renal function may result in alkalosis.

The *cause* of primary hyperparathyroidism is either a parathyroid adenoma (and much more rarely parathyroid carcinoma), or a hereditary anomaly, which is much less frequent.

Parathyroid adenoma is a disease which afflicts adults. The incidence of parathyroid adenoma in the general population is 25 per 100,000 per year. It occurs rarely in children, and is hardly ever found in newborns. Diagnosis is based on the measurements cited above, and on scanning techniques.

The clinical evolution of parathyroid adenoma is variable. Certain forms are asymptomatic during long periods; others evolve rapidly and are frequently accompanied by a marked increase in alkaline phosphatases, resulting in skeletal effects. Postsurgically, calcemia should be monitored closely.

Parathyroid carcinoma. Secreting parathyroid cancers are very rare, and prognosis is poor.

Hereditary hyperparathyroidism. There are *several forms* of hereditary hyperparathyroidism: hyper-

16. Malette, L. E., Wilson, D. P. and Kirkland, J. L. (1983) Evaluation of hypocalcemia with a highly sensitive homologous radioimmunoassay for the midregion of parathyroid hormone. *Pediatrics* 71: 64–69.
17. Nissenson, R. A., Abbott, S. R., Teitelbaum, A. P., Clark, O. H. and Arnaud, C. D. (1981) Endogenous biologically active human parathyroid hormone: measurement by a guanyl nucleotide amplified renal adenyalte cyclase assay. *J. Clin. Endocrinol. Metab.* 52: 840–846.

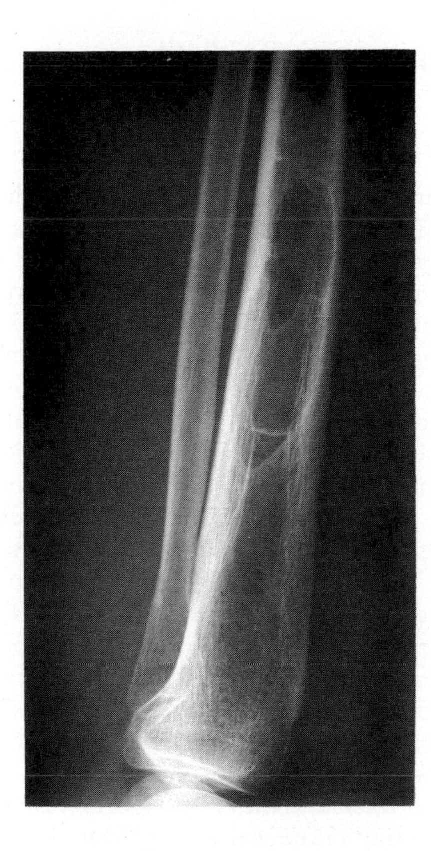

Figure XIV.7 Multiple parathyroid adenomas. Radiological image of cysts in the tibia. This is the typical radiological aspect of hyperparathyroidism.

plasia or simple adenomatosis (Fig. XIV.7), endocrine polyadenomatosis, multiple endocrine neoplasia (MEN 1), and hereditary parathyroid hyperplasia (MEN 2) (p. 137). The relationship between hereditary hyperparathyroidism and a similar disease, hereditary benign hypercalcemia or familial hypercalcemic hypocalciuria, should be considered[18,19,20].

Ectopic Hyperparathyroidism

Hypercalcemia of varied origin is observed in patients with *malignant hemopathy* and solid tumors, and is a condition seen with *high frequency* in *adults*. Some are related to a metastatic destruction of bone; others, to a secretion of hypercalcemic substances by malignant cells. When PTH or a PTH fragment is produced, the disorder is known as ectopic hyperparathyroidism. In other cases, the substance or substances other than PTH which cause hypercalcemia may be prostaglandin, or may remain unknown.

Secondary and Tertiary Hyperparathyroidism

Secondary or tertiary hyperparathyroidism evolves with chronic disease involving hypocalcemia. Immunoreactive PTH is increased in serum. This increase may be suppressed by the intravenous infusion of calcium (secondary hyperparathyroidism). However, an elevated PTH concentration can also be refractory to the administration of calcium (tertiary hyperparathyroidism). The degree of PTH suppression necessary to eliminate hyperparathyroidism is independent of the anatomical state of the parathyroids (normal, hyperplastic, or adenomatous). Secondary and tertiary hyperparathyroidism is observed in avitaminosis D, characterized by digestive malabsorption of calcium and chronic renal insufficiency, and can be provoked by protracted treatment with furosemide and prolonged bed rest.

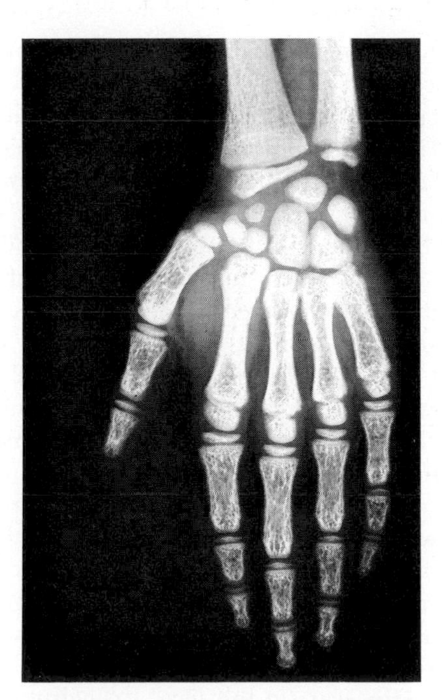

Figure XIV.8 Renal osteodystrophia. Radiography of a hand. Note the generalized microcystic aspect of the bone, especially the metacarpals and phalanges, and the subperiostal resorption at the inferior extremity of the cubitus, the fifth metacarpal, and the first phalanx.

18. Foley, T.P., Harrisson, H.G., Arnaud, C.D. and Harrison, H.E. (1972) Familial benign hypercalcemia. *J. Pediat.* 81: 1060–1067.
19. Marx, S.J. (1980) Familial hypocalciuric hypercalcemia. *N. Engl. J. Med.* 303: 810–811.
20. Lillquist, K., Illum, N., Brockjacobsen, B. and Lockwood, K. (1983) Primary hyperparathyroidism in infancy associated with familial hypocalciuric hypercalcemia. *Acta Pediat. Scand.* 72: 625–629.

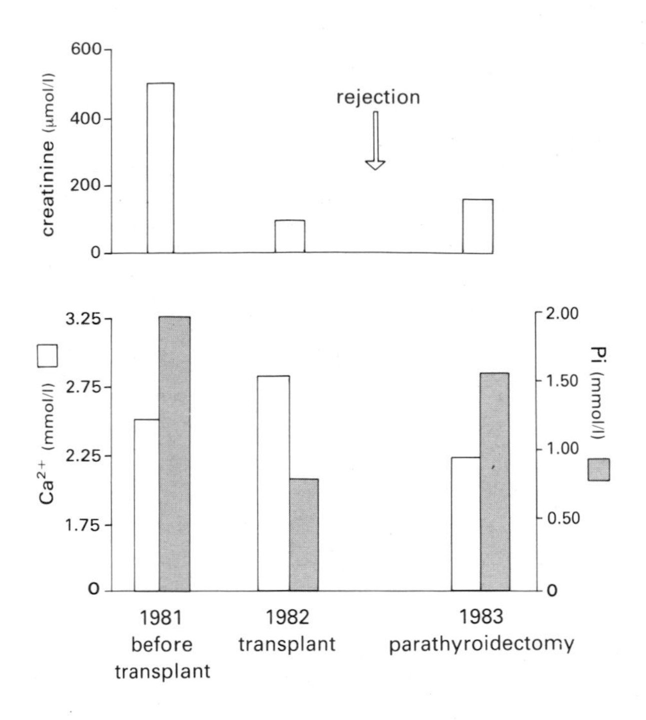

Figure XIV.9 Hyperparathyroidism secondary to renal insufficiency. After renal transplantation, hypercalcemia appears, which is corrected only after parathyroidectomy. Rejection of the grafted organ is indicated.

Chronic uremia is frequently accompanied by hypocalcemia. The known causes of hypocalcemia are elevated phosphatemia, and insufficient calcitriol synthesis. Hypocalcemia produces a permanent stimulatory effect on the parathyroids. The elevation of immunoreactive PTH in serum occurs as soon as glomerular filtration is reduced to less than 80 ml/min per 1.73 m^2. In severe uremia, the level of PTH is greater than that observed in adenomas, since PTH is no longer metabolized by the kidneys. Calcium infusion reduces or has no effect on serum PTH. The symptomatology of hyperparathyroidism in uremic patients is *fibrocystic osteitis* (Fig. XIV.8), metastatic *calcifications*, particularly of the cornea and blood vessels, and intense *pruritus*. The most effective treatment is with calcitriol or 1α $OH\text{-}D_3$, at daily doses of a few μg. In cases of tertiary hyperparathyroidism, this treatment leads to hypercalcemia, and justifies subtotal parathyroidectomy. A successful kidney transplantation should normalize calcium homeostasis in uremics. However, in the days following the graft, severe hypercalcemia is sometimes observed, which may be due to tertiary hyperparathyroidism. Subtotal parathyroidectomy may be indicated in this case (Fig. XIV.9).

2.4 Hypoparathyroidism

Parathyroid Insufficiency Syndrome

Parathyroid insufficiency, regardless of the age of onset, leads to *hypocalcemia*, hypocalciuria, sometimes hypomagnesemia, *hyperphosphatemia* with an increase in the percentage of tubular reabsorption of phosphorus (TRP), a lowered citratemia and citraturia, and a normal or low level of alkaline phosphatases. A regimen rich in phosphorus and low in calcium, or a test with EDTA, is of diagnostic value. The limited sensitivity of PTH assays rarely allows the measurement of low PTH levels. In cases of hypocalcemia with hyperphosphatemia, it is possible to observe either levels of PTH which are not increased, defining parathyroid insufficiency due to hypoparathyroidism, or increased levels of PTH, which define *pseudohypoparathyroidism*. Parathyroid insufficiency is accompanied by a fall in plasma calcitriol, explained by the role played by PTH in the activation of renal 25-hydroxy 1α-hydroxylase, and is confirmed by results of therapeutic use of calcitriol.

Hypocalcemia is the *most important* clinical sign of parathyroid insufficiency. *Convulsions*, which are either generalized or localized, tetany, laryngospasm, intracranial hypertension, tendency to moniliasis, changes in teeth and nails, cataract, and depressive or even psychotic states are the principal manifestations. In newborns and infants, tremors, convulsions and convulsive states, as well as cardiomegaly with cardiac insufficiency, are the most severe problems observed in relation to parathyroid insufficiency.

True Hypoparathyroidism

Complete or incomplete forms are observed *in adults and in older children*. Symptoms may include intracranial hypertension due to a cerebral tumor, chronic untreatable diarrhea, or, in other cases, osseous anomalies such as osteomalacia may be encountered (now explained by a deficiency in the activation of vitamin D in the absence of PTH). The *causes* of hypoparathyroidism are *varied*. The destruction of the parathyroids by a metastatic tumor is rare. More frequently, trauma of the parathyroids (or thyroid) is involved, often as the result of neck injury in automobile accidents. The *hereditary absence of parathyroids* (with a retarded manifestation of parathyroid insufficiency) is also *rare*. It is transmitted in an autosomal recessive or an autosomal dominant mode, or yet, in a recessive mode, coupled to the X chromosome. *Autoimmune hereditary hypoparathyroidism* is often associated with Addison's disease. *Hypoparathyroidism* with associated *deafness and nephrotic syndrome* is observed in children after three years of age.

Hypoparathyroidism in the Newborn

During the first two days of life, a state known as *precocious neonatal hypocalcemia* occurs frequently, and complicates toxemia associated with diabetes of pregnancy, fetal malnutrition, and premature birth. The major cause of hypocalcemia is an intense uptake of calcium into the skeleton and possibly an elevated concentration of blood calcitonin. Parathyroid insufficiency is not a determining factor. On the other hand, several varieties of hypoparathyroidism are characteristic of the neonate.

Non-hereditary, transitory hypoparathyroidism appears in the first week of life, but normal function is recovered one to eight months thereafter. It is observed especially in neonates born to patients with high risk pregnancies, and in retarded or premature newborns. The treatment of transitory hypoparathyroidism should be active, and can be very efficient.

Chronic hypoparathyroidism in the neonate (DiGeorge's syndrome) occurs in the fifth and sixth weeks of intrauterine life, and associates specific features (micrognathia, hypertelorism, and abnormal ear position) with an absence of parathyroids and an immunodeficiency syndrome. Survival rarely exceeds one year of age.

Pseudohypoparathyroidism

In 1942, Albright described a disease associating dysmorphia, mental retardation, and a biological syndrome of parathyroid insufficiency untreatable by the administration of PTH. In 1952, the same author described patients with the same clinical syndrome but without parathyroid insufficiency. The first group was termed *pseudohypoparathyroidism (PHP)*, and the second, *pseudo-pseudohypoparathyroidism (PPHP)*[21]. There is vertical familial distribution involving several successive generations; girls are affected twice as frequently as boys; transmission from mother to daughter, mother to son, father to daughter has been observed, but never father to son. This disease appears to be dominantly X-linked. It is possible, within the same subject or the same family, to observe aspects associated with PHP and PPHP[22].

Symptoms appear very early, sometimes in the first year of life, but are never seen at birth. The *major anomalies* observed are: small stature, due to precocious closing of the growth cartilages, moon features and obesity, brachymetacarpia, brachymetatarsia, periarticular and muscular calcification, as well as calcification of the central gray nuclei, cataracts, and mental deficiency.

The *parathyroids* of these patients are frequently *hyperplastic*, and the plasma concentration of immunoreactive *PTH* is abnormally *elevated*. The current interpretation of the etiology of pseudohypoparathyroidism implicates *peripheral resistance to PTH*. Variants include: isolated renal resistance, isolated bone resistance, and mixed renal and bone resistance[23]. Isolated renal resistance corresponds to the symptoms described in relation to "hypohyperparathyroidism" associated with signs of hypoparathyroidism, which include lesions of fibrocystic osteitis and the presence or absence of dysmorphia characteristic of Albright's dystrophy[24]. More recently, a study of responses to PTH has revealed two forms of PHP. *Type II* is very *rare*: after injection of exogenous PTH, urinary cAMP is elevated, but phosphodiuresis is unaffected. *Type I* is more *frequent*: calcemia, phosphaturia and hydroxyprolinuria are observed, and urinary cAMP is not increased following the injection of PTH, although injection of dibutyril-cAMP leads to urinary phosphate levels identical to those of normal subjects. Type I PHP is also sometimes associated with endocrine problems: diabetes mellitus, prolactin deficiency, insufficient thyroid response to TSH. It has been suggested that a generalized defect in the activation of adenylate cyclase may be involved in pseudohypoparathyroid syndromes. A partial *deficiency* of the *G_s subunit* (p. 334) is found in membranes of lymphocytes, platelets, fibroblasts and renal cells in these patients[25,26]

21. Albright, F., Burnett, C. H., Smith, P. H. and Parson, W. (1942) Pseudohypoparathyroidism, an example of "Seabright-Bantam syndrome". *Endocrinology* 30: 922–932.
22. Van Dop, C. and Bourne, H. R. (1983) Pseudohypoparathyroidism *Annu. Rev. Med.* 34: 259–266.
23. Frame, B., Hanson, C. A., Frost, H. M., Block, M. and Arstein, A. R. (1972) Renal resistance to parathyroid hormone with osteitis fibrosa. *Amer. J. Med.* 52: 311–321.
24. Costello, J. M. and Dent, C. E. (1963) Hypohyperparathyroidism. *Arch. Dis. Child.* 38: 397–407.
25. Downs, R. W., Levine, M. A., Drezner, M. K., Burch, W. M. and Spiegel, A. M. (1983) Deficient adenylate cyclase regulatory protein in renal membranes from a patient with pseudohypoparathyroidism. *J. Clin. Invest.* 71: 231–235.
26. Editorial (1983) Pseudohypoparathyroidism: continuing paradox. *Lancet* ii 8347: 439–440.

Treatment

The treatment of hypoparathyroidism is *very effective*[27]. It consists of the oral administration of 500 to 1000 mg *of calcium* with calcitriol, and 0.5 to 2.0 μg of *calcitriol* or 1 to 4 μg of 1α $OH\text{-}D_3$ per day, with a *follow-up of calcemia and calciuria measurements*. An interesting observation made during treatment of pseudohypoparathyroidism with vitamin D or its derivatives is the *reappearance* of *normal sensitivity to PTH*, either transitory or permanent. This has led to the uncertain concept of "transitory PHP"[28].

3. Calcitonin

In 1962, Copp and associates[29] discovered a hypocalcemic humoral factor which they thought to be of parathyroid origin, and called it calcitonin. In 1964, Foster[30] established that its true origin was the thyroid. Its cellular origin was defined in later years. The structure of the porcine hormone was first established in 1968.

3.1 Biology

Calcitonin is synthesized and secreted by a specific cellular system. The cells belonging to this system are derived by precocious embryonic migration of the neural crest[31], and form a part of the APUD system (p. 549). They then migrate to the level of the last branchial pouch. In birds and fish, they form an individualized organ, the ultimobranchial body. In mammals, these cells, rich in mitochondria and secretory granules, are concentrated *in the thyroid gland*, and form the *parafollicular or C cells*, in which calcitonin has been identified by histological techniques of immunofluorescence and immunochemistry. In addition, C cells are also found in parathyroids and thymus, but in lower concentrations. During evolution and adaptation to a terrestrial habitat, the ultimobranchial body appears before the parathyroids in elasmobranches, whereas the two organs are found simultaneously in amphibians. The evolutionary *development of calcitonin before* that of *parathyroid hormone* necessary for the regulation of calcium *in living organisms* is substantiated by the high concentration of calcium in sea water and its relatively low level in the terrestrial environment.

C cells are the primary source of calcitonin in humans. In the absence of the thyroid gland, calcitonin levels are undetectable in the blood. However, hypocalcemic activity has been found in parathyroids, thymus, and adrenal medulla. Substances similar to calcitonin have been identified immunohistochemically in the pituitary, brain, lung, digestive tract, liver, and bladder.

Biosynthesis of Calcitonin

Early studies of calcitonin biosynthesis were more difficult than those of PTH, as mammalian tissue was not easily obtainable for in vitro studies. In vitro studies with fish and chicken ultimobranchial bodies suggested that calcitonin was synthesized as a larger precursor with a MW of ~15,000. In mammals, information about the biosynthesis of calcitonin was derived initially from studies of calcitonin mRNA isolated from rat medullary tumor tissue or thyroid tissue.

The sequence of the *calcitonin precursor* was derived from cloned cDNA[32]. It is comprised of *136 aa*, and the 34 aa peptide corresponding to calcitonin has flanking amino acid sequences on both the amino and carboxyl ends. The 25 amino acids located at the N-terminus are hydrophobic and resemble the signal sequence of secreted proteins. These are followed by 59 aa, terminating with dibasic amino acids, followed by calcitonin and three basic amino acids, and finally 16 aa at the C-terminus. Thus, these studies suggested that, in addition to calcitonin, two peptides, of 59 and 16 aa, could be formed from the precursor. While the function of these proteins remains unknown, peptides immunoreactive to antisera against the 16 aa peptide are present in thyroid C cells and rat medullary carcinoma cells. Thus, the biosynthesis of calcitonin involves *extensive post-transla-*

27. Markowitz, M. E., Rosen, J. F., Smith, C. and DeLuca, H. F. (1982) 1,25-dihydroxyvitamin D_3-treated hypoparathyroidism: 35 patients years in 10 children. *J. Clin. Endocrinol. Metab.* 55: 727–733.
28. Stogmann, W. and Fischer, J. A. (1983) Pseudohypoparathyroidism. Disappearance of the resistance to parathyroid extract during treatment with vitamin D. *Amer. J. Med.* 59: 140–144.
29. Copp, D. H. and Cameron, E. C. (1961) Demonstration of a hypocalcemic factor (calcitonin) in commercial parathyroid extract. *Science* 134: 2038–2039.
30. Foster, G. V., MacIntyre, I. and Pearse, A. G. E. (1964) Calcitonin production and the mitochondrion-rich cells of the dog thyroid. *Nature* 203: 1029–1030.
31. Le Douarin, N. and Lelievre, C. (1971) Sur l'origine des cellules à calcitonine du corps ultimo-branchial de l'embryon d'oiseau. *Bull. Assoc. Anatomistes* 152: 558–568.
32. Jacobs, J. W., Goodman, R. H., Chin, W. W., Dee, P. C., Habener, J. F., Bell, N. H. and Potts, J. T. Jr. (1981) Calcitonin messenger RNA encodes multiple polypeptides in a single precursor. *Science* 213: 457–459.

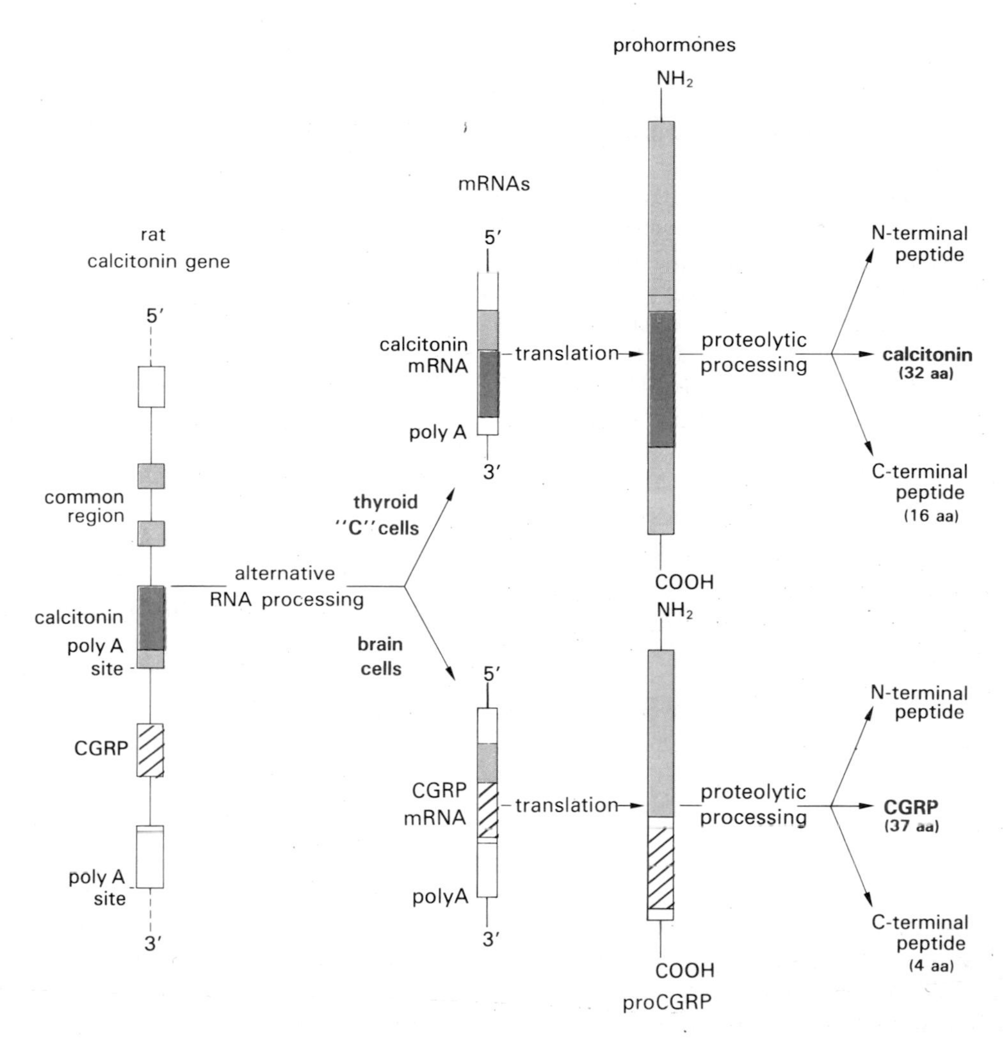

Figure XIV.10 Alternative RNA processing pathways in the expression of the calcitonin gene which predict the synthesis of a novel neuropeptide, CGRP, in the brain[34].

tional modifications, which include: 1. the cotranslational cleavage of the signal sequence; 2. two trypsin-like cleavages to remove the amino and C-terminal extension; 3. carboxypeptidase-like activity to trim the remaining basic amino acids from the C-terminus; and 4. conversion of pro-Gly, at the C-terminal end, to prolinamide.

Calcitonin Genes

Calcitonin mRNA was estimated by Northern blot analysis to be about 1000 nucleotides long, approximately 600 nucleotides longer than necessary to encode the calcitonin precursor. In studies using serially transplanted rat medullary thyroid tumor, it was noted that the size of the mRNA differed between some tumor lines. The size difference correlated with the permanent switching of the tumor line from high to low production of calcitonin. To elucidate the basis for this change, the corresponding cDNA was cloned and analyzed[33]. It contained a single open reading frame that encoded a 128 aa protein, which was termed *calcitonin gene related peptide (CGRP)*. The 5′, 227 nucleotides of CGRP and calcitonin cDNAs extending into the proteins were found to be identical, but the remainder of the sequences differed. Analysis of mRNA levels in various tissues demonstrated that calcitonin mRNA was present primarily in C cells or medullary tumor cells of the thyroid, whereas CGRP mRNA was localized in neural tissue.

To determine whether the two mRNAs were produced by two genes, or whether selective processing of the transcripts from a single gene produced the two mRNAs, the genomic DNA of rat calcitonin was

33. Amara, S. G., Jonas, V., Rosenfeld, M. G., Ong, E. and Evans, R. M. (1982) Alternate RNA processing in calcitonin gene expression generates mRNAs encoding different polypeptide products. *Nature* 298: 240–244.

cloned and analyzed[34]. From these gene studies, the remarkable discovery was made that the calcitonin gene actually encodes both calcitonin and CGRP (Fig. XIV.10). Comparison of the gene structure of both calcitonin and CGRP cDNA sequences revealed that the gene contains six exons. Exons 2 and 3, which basically code for the amino terminal peptide of the calcitonin precursor, are common to both calcitonin and CGRP cDNAs. Exon 1, which encodes the 5′ untranslated region of the mRNA, is probably also common to both, although the sequence in this region was not present in the cloned CGRP cDNA that was analyzed. Exon 4 is specific for calcitonin mRNA, and contains the regions encoding calcitonin, the 16 aa peptide at the C-terminus, and the 3′ untranslated region of the mRNA. Exons 5 and 6 are specific for CGRP mRNA, containing the coding region for CGRP and the 3′ untranslated region of the mRNA, respectively. Since high molecular weight precursors of the mRNAs contain both calcitonin and CGRP-specific sequences, the most plausible explanation is that both mRNAs are formed, by differential processing, from the same initial transcript. The *processing pathway* is *tissue-dependent*, with *calcitonin mRNA* predominating in *thyroid C cells*, and *CGRP mRNA* preponderant in *nerve tissue*.

The gene for calcitonin was localized to the short arm of *chromosome 11* by analyzing a panel of rodent–human hybrid cell lines. In situ hybridization experiments suggest that the gene is present between bands 13 and 15 on the short arm, and is thus probably near the insulin gene (Fig. XIV.5).

The discovery of CGRP generated a flurry of research activity, the *goal* of which was to define the *function of the protein*. Oligopeptides containing the CGRP sequence were synthesized and an antiserum to the peptides was produced. This antiserum was used to confirm the presence of CGRP in neural tissue[34] and to show that CGRP is secreted from rat trigeminal ganglion cells grown in vitro. A human peptide with antigenic determinants to rat CGRP has also been isolated, and was shown to have 89% homology with the rat protein. CGRP appears to have significant biological activity, causing an increase in mean arterial pressure and heart rate when injected into the cerebral ventricles, and reducing the mean arterial pressure when injected intravenously.

34. Rosenfeld, M. G., Mermod, J. J., Amara, S. G., Swanson, L. W., Sawchenko, P. E., Rivier, J., Vale, W. W. and Evans, R. M. (1983) Production of a novel neuropeptide encoded by the calcitonin gene via tissue-specific RNA processing. *Nature* 304: 129–135.

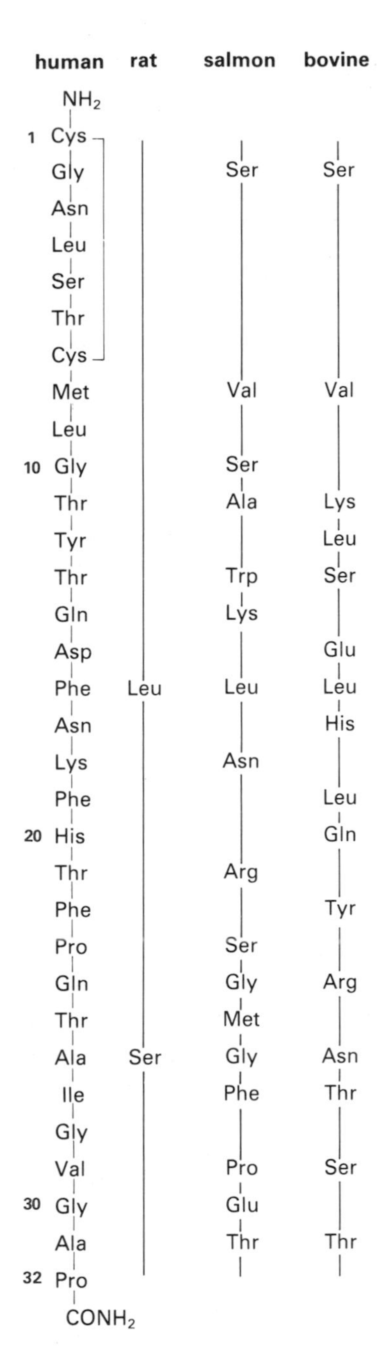

Figure XIV.11 Structure of human, rat, salmon and bovine calcitonin.

Thus, studies of calcitonin biosynthesis have revealed a complex processing of both the initial RNA transcript and the initial translation product. The *calcitonin gene* encodes two quite distinct mRNAs, and the calcitonin protein precursor can, potentially, produce *two proteins* in addition to calcitonin. Three new proteins have been discovered, and, although the significance of these proteins has not yet been determined, the power of methods of molecular biology in opening new areas of research has been demonstrated by the study of calcitonin.

The 32 aa of calcitonin are different in the human, rat, cow, sheep, pig, eel, and salmon. Common to the amino acid sequences of all of these species, however, is a disulfide bridge of half-cystine residues, at positions 1 and 7, and a proline-amide at the C-terminal end. Rat and human calcitonins (Fig. XIV.11) differ by only two amino acids. Salmon calcitonin differs from that of the human by 16 aa. Immunological studies have shown that antisera to calcitonin from different species have little or no cross-reactivity. Calcitonin from the ultimobranchial body, such as that of salmon, has greater activity than hormones from C cells of the thyroid. Analogs have been prepared with increased activity, for example by deleting the Ser-2 of salmon calcitonin.

Regulation of Secretion

Calcemia, or more probably the concentration of ionized calcium, is the determining factor in the regulation of the synthesis and secretion of calcitonin. At normal calcium levels, the endogenous rate of calcitonin secretion is low in normal subjects, averaging 125 μg/d[35]. A slight elevation in blood calcium increases this secretion. *Perfusion with calcium increases the concentration of calcitonin* by a factor of three to 15, a response which occurs very rapidly. Magnesium deficiency decreases the secretion of calcitonin in rats, and magnesium infusion lowers immunoreactive calcitonin in patients with medullary carcinoma of the thyroid.

No hypothalamic or pituitary factor appears to be involved in the *regulation* of *calcitonin* secretion. Gastrin increases blood calcitonin, and is used clinically in the diagnosis of medullary carcinoma of the thyroid. Secretin, cholecystokinin and cerulein are less active[36]. Glucagon also stimulates calcitonin secretion. Drugs which block α- and β-adrenergic activity inhibit this stimulation, whereas they have no effect on the stimulatory action of calcium. Finally, somatostatin inhibits calcitonin secretion in humans; it has, in fact, been localized in some C cells.

35. Kanis, J. A., Heynen, G., Cundy, T., Cornet, F., Paterson, A. and Russel, R. G. G. (1982) An estimate of the endogenous secretion rate of calcitonin in man. *Clin. Sci.* 63: 145–152.
36. Sethi, R., Kukreja, S. C., Bowser, E. N., Hargis, G. K. and Williams, G. A. (1981) Effect of secretin on parathyroid hormone and calcitonin secretion. *J. Clin. Endocrinol. Metab.* 53: 153–157.

Metabolism and Excretion

The half-life of human and porcine calcitonin has been determined to be less than 15 min. The metabolic clearance of human calcitonin in normal subjects is approximately 8 ml/kg per min, and is less in renal insufficiency. Following injection of a labelled hormone, the two organs which concentrate the hormone to the greatest extent are the liver and kidney. In the human, dog and rat, the *kidney* appears to be the principal site of binding and *degradation* of calcitonin.

Effects

The two principal target organs of calcitonin are bone and kidney.

The effect exerted by calcitonin on *bone* has been defined both in vivo and in vitro. Calcitonin is potently *inhibitory* of *bone resorption* in animals, including humans. This action is responsible for hypocalcemia and a reduction in urinary excretion of hydroxyproline. Calcitonin has no effect on bone formation, although contradictory results have been reported. A reduction in calcium resorption by osteoclasts occurs 15 to 30 min following the administration of calcitonin.

The effect of calcitonin on the *renal cortex* is clear. Membranes of *tubular* cells have specific receptors for calcitonin. The hormone activates adenylate cyclase, but there is no change in urinary cAMP levels. Calcitonin also *increases urinary excretion of calcium*, phosphate, sodium and potassium, and reduces urinary excretion of magnesium.

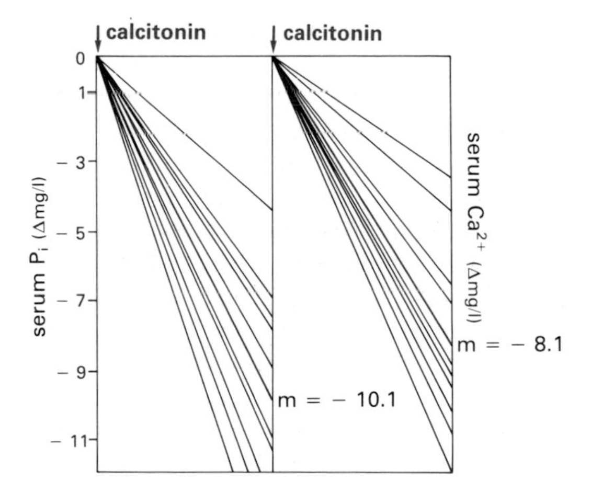

Figure XIV.12 Effects of calcitonin on calcium and phosphorus levels in children.

The action of calcitonin on *calcemia* is the result of its *combined effects* on bone and kidney. A *fall* in blood *calcium* (Fig. XIV.12), more important when the levels are initially elevated, is obtained rapidly, reaching a maximum in 90 min, and returning to normal levels 3 h after a single intravenous or subcutaneous injection. This decrease is accompanied by a *fall* in blood *phosphate* levels. The effect of this hormone on the reduction of calcium levels is dependent upon the type of calcitonin used; porcine calcitonin has effects almost equal to those of human calcitonin but less than those of salmon calcitonin. Moreover, calcitonin exerts greater effects in young rather than fully grown animals, due, most probably, to the elevated rate of bone remodelling during growth.

Calcitonin appears to have a fundamental role in the *protection of bone mass*, by limiting resorption and possibly stimulating bone formation. It may be involved in the physiological regulation of calcemia in mammals.

In addition to its function in bone and kidney, calcitonin also affects some other organs, such as digestive glands, gallbladder, lung, heart, brain, and pituitary.

Mechanism of Action

Specific *receptors* for calcitonin have been identified in the plasma membrane of *kidney* and *bone* cells[37]. The cellular mechanism of action of calcitonin is *still unclear*. It does not act by inhibiting PTH activity, as its effects are observed even in parathyroidectomized animals, nor is its cellular action dependent upon vitamin D. Although contradictory results have been reported, it appears, generally, that calcitonin results in a decrease in intracellular calcium concentrations, which may be responsible for its different effects on bone and kidney. In these two organs, calcitonin *increases cAMP levels* (Fig. XIV.13). In bone, only osteoclasts are targets for calcitonin, which has an inhibitory effect on these cells.

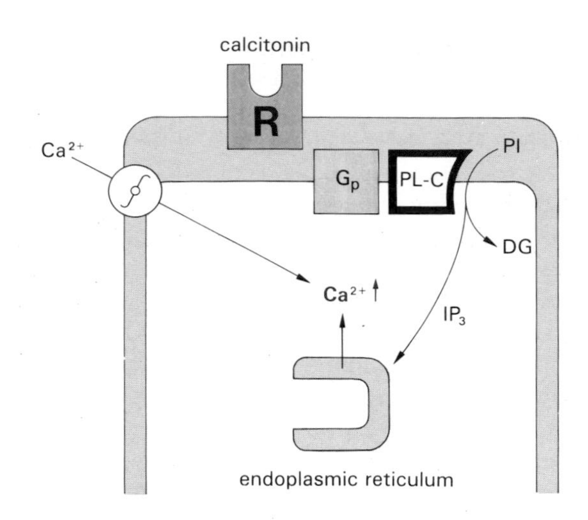

Figure XIV.13 Mechanism of the cellular action of calcitonin. It interacts with a plasma membrane receptor, activates adenylate cyclase, and decreases calcium concentration in the cytoplasm. This scheme is valid for osteoclasts (calcitonin has no effect on osteoblasts and renal tubule cells).

3.2 Clinical Evaluation

Assay Techniques[38]

Following the initial use of an in vivo bioassay, the availability of synthetic human calcitonin has allowed the development of different *radioimmunoassays*.

The radioimmunoassay of calcitonin is difficult because of low circulating concentrations in normal subjects and heterogeneity of the circulating forms. (Normal fasting levels are ~60 pg/ml). In patients with medullary carcinoma of the thyroid, a large immunoreactive form (M_r~10,000 to 20,000) is often found in plasma, urine, and tumoral tissue; it is biologically inactive.

Radioimmunoassays have shown an elevated concentration of calcitonin in the first days of life; it remains low during childhood and increases once again in adulthood. Calcitonin levels are slightly higher in men than in women, decreasing in old age more markedly in women than in men. Finally, calcitonin levels are increased during lactation.

37. Marx, S.J., Woodard, C.J. and Aurbach, G.D. (1972) Calcitonin receptors of kidney and bone. *Science* 178: 999–1001.
38. Austin, L.A. and Heath, H. (1981) Calcitonin. Physiology and Pathology. *N. Engl. J. Med.* 304: 269–278.

3.3 Pathology and Pharmacology

Hypercalcitoninemia

Tumors

A large number of *thyroid cancers* produce calcitonin. Three general types of tumors are accompanied by an excessive production of calcitonin. The first is benign, such as *C cell adenomas*. Two are malignant: trabecular cancer, rich in mucopolysaccharids, and especially medullary carcinoma of the amyloid stroma, which represents 6 to 8% of all thyroid cancers. Adenopathies exist in 80% of the cases, and frequently bone metastases are also observed. They are sometimes associated with diarrhea and episodes of flushing. Hypercalcintoninemia may be either aquired or hereditary. In the latter case, transmission is via the autosomal dominant mode, and it is classified as one of the multiple endocrine neoplasia (MEN 2, p. 137). There are several diseases associated with thyroid tumors, e.g., pheochromocytoma (Sipple's syndrome), mucous neuroma of the tongue and lips (Gorlin's syndrome), and a form which causes arachnodactylia, or Marfan's disease. Treatment is surgical.

Renal Insufficiency

An elevated level of calcitonin can be found in patients with either accute or chronic renal insufficiency. These increased levels persist even during hemodialysis.

Other diseases

Marked elevations of immunoreactive calcitonin in serum have been observed in pancreatitis, in acute hepatitis with coma, and in fulminating infantile meningococcemia. Elevated calcitonin concentrations have also been identified in the thyroid gland of patients with pseudohypoparathyroidism. However, hypercalcemia of primary hyperparathyroidism is not associated with hypercalcitoninemia.

Reduced Production of Calcitonin

A reduction or complete disappearance of calcitonin production is sometimes observed in congenital athyreosis and in patients whose thyroid has been destroyed, either surgically or isotopically. However, the analysis of hypercalcemia stimulated by the administration of thyroxine appears to show that T_4 does not completely counteract the slow return to initial levels of blood calcium. The hypocalcemic effect of calcitonin is more important in young children[39]. Immunoreactive calcitonin in plasma is reduced in infants with non-goitrous cretinism.

Pharmacology[40]

The *hypocalcemic* effect of calcitonin is more significant when calcium levels are elevated; bone renewal is more rapid, and the rapidly exchangeable calcium pool is smaller. Its therapeutic effectiveness is greater in infants and young children than in adults.

Calcitonin is *used in the treatment of hypercalcemia*. It has been shown to be useful in the treatment of hyperparathyroidism, bone metastases, vitamin D intoxication, sarcoidosis, and idiopathic hypercalcemia in children. In most cases, other forms of therapy are equally as active. Some syndromes do not react to calcitonin, such as hypercalcemia of hereditary hypophosphatasia. The doses utilized are approximately 4 MRC U/kg or 20 mg/kg. Calcitonin, administered intravenously, particularly synthetic salmon or human calcitonin, can be used for treating severe hypercalcemia, and is followed by corticotherapy or administration of phosphate; a diet low in calcium can be associated with, or may even replace, calcitonin treatment.

The *protective effect of calcitonin on bone* which has undergone an exaggerated osteolysis is also a desired effect[41]. Calcitonin is used in chronic treatment of alveolar paradontosis, idiopathic algodystrophia and osteoporosis, but the results are still questionable. Its main use remains in the treatment of *Paget's disease*, where it leads to a disappearance of pain, the sensation of cutaneous heat, and to an improvement of various skeletal abnormalities. Resistance or habituation to this treatment has been observed, the side effects of which include nausea, reddening of the face, diarrhea, and hives. Successful treatment with calcitonin is followed by a reduction in serum alkaline phosphatase, osteocalcin, and urinary hydroxyproline.

39. Carey, D. E., Leejones, K., Parthemore, J. and Deftos, L. J. (1980) Calcitonin secretion in congenital nongoitrous cretinism. *J. Clin. Invest.* 65: 892–895.
40. Deftos, L. J. and First, B. P. (1981) Calcitonin as a drug. *Ann. Intern. Med.* 95: 192–197.
41. Talmage, R. V., Grubb, S. A., Norimatsu, H. and Vanderwiehl, C. S. (1980) Evidence for an important physiological role for calcitonin. *Proc. Natl. Acad. Sci., USA* 77: 609–613.

Remarkable effects have been observed in children with *idiopathic hyperphosphatasemia* with macrocrania, a disease very similar to Paget's disease. The therapeutic effects of calcitonin have been carefully examined in osteogenesis imperfecta, idiopathic juvenile osteoporosis and progressive forms of fibrous bone displasia, but unfortunately it has been found to have limited or no effect.

4. Cholecalciferol or Vitamin D_3

4.1 Biology of Cholecalciferol

Vitamin D_3 has an important role in the regulation of calcemia. The steps necessary to produce active hormone include hydroxylation, at carbon 25, in the liver, followed by a second hydroxylation in the kidney, at carbon 1. Calcitriol and other di- or trihydroxylated derivatives are active steroid hormones. Vitamin D_3 is thus a prohormone.

Chemical Structure

Vitamin D compounds are sterols which have undergone photochemical modification by ultraviolet rays opening the B ring between C9–C10 of the phenanthrenic ring. These are *secosteroids* (Fig. XIV.14). Cholecalciferol, or vitamin D_3, and ergocalciferol, or *vitamin D_2*, are natural D vitamins. Synthetic substances such as tachysterol also have vitamin D like activity. Vitamin D_2 and D_3 have the same ring structure, with a different lateral side chain. Tachysterol has the same lateral side chain as vitamin D_2, but within the ring structure, a methyl group replaces methylene at carbon 10.

Four specific features are *required* for biological activity. The first is the opening of the ring at C9–C10. The second is the presence of a free hydroxyl group at C3. The third is the existence of an intact lateral side chain, mostly of the type found in cholesterol; however, ergocalciferol, or vitamin D_2, for which the side chain is different, retains 100% of activity in both the rat and human; 22, 23-dihydroergocalciferol retains 50% of the activity. The fourth chemical feature is the presence of three conjugated double bonds, at positions C5–C6, C7–C8, and C10–C19. Structural isomers such as isovitamin D, or geometric isomers such as 5, 6-transvitamin D, have only partial or no activity with respect to the natural compound.

Biosynthesis

Fish are capable of synthesizing cholecalciferol without the aid of ultraviolet radiation. Amphibians, reptiles, birds and mammals have zones which are receptive to ultraviolet radiation, such as the ears of

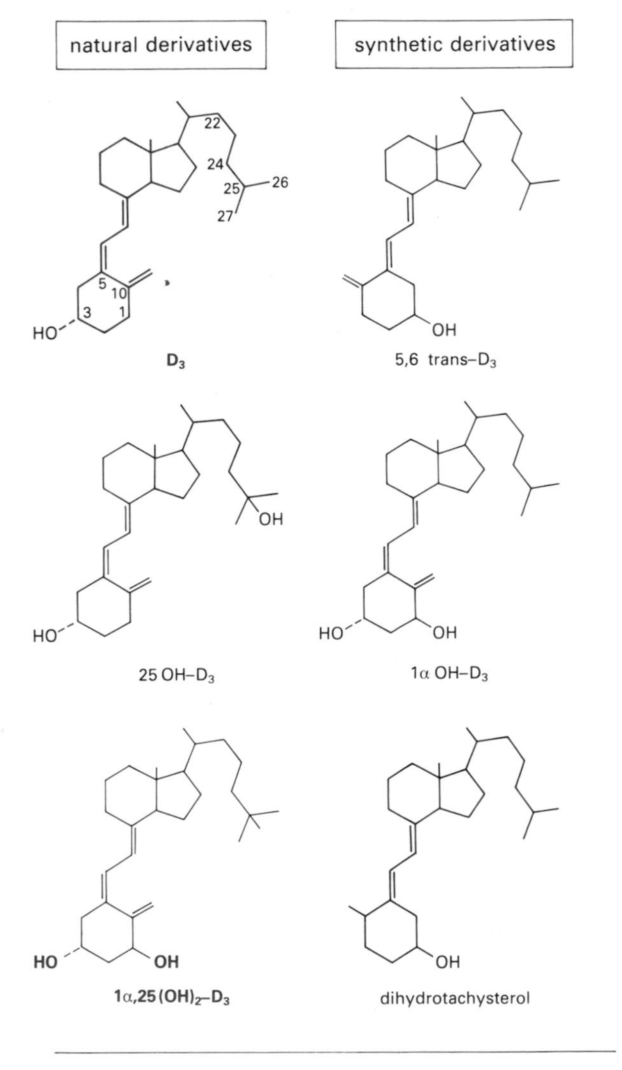

Figure XIV.14 Natural and synthetic derivatives of vitamin D. The vitamin D_2 series differs from the D_3 derivatives by only a double bond at C22–C23 and a methyl at C24.

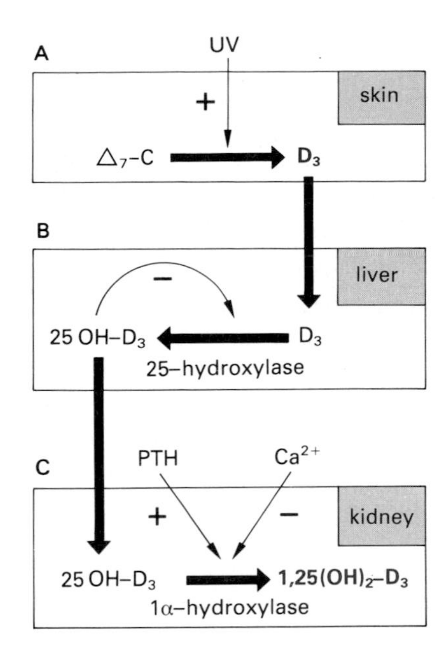

Figure XIV.15 Synthesis and activation of vitamin D_3. Δ_7-C: 7-dehydrocholesterol; **A.** 25-hydroxylase (soluble); **B.** 1α-hydroxylase (mitochondria); **C.** 24-hydroxylase. Not represented: 1. dietary intake of vitamin D; 2. overall storage in adipocytes; 3. enterohepatic cycle of hydroxylated metabolites; 4. target tissues.

rabbits, the feet of birds, and the skin of humans. In humans, the normal dietary intake of cholecalciferol is low, this substance appearing as an *ultraviolet-dependent prohormone* synthesized in a deep region of the *skin*, beneath the pigmented layer (Fig. XIV.15). Cholecalciferol (vitamin D_3) is produced *in two steps*, beginning with photochemical *activation* of 7-dehydrocholesterol, photochemically transformed to previtamin D_3, which is then isomerized to cholecalciferol. This compound is taken up by a *transport protein* and passes into the general circulation. The regulation of cholecalciferol synthesis within the skin is poorly understood, and the actual amount produced is unknown; thickness, temperature and pigmentation of the skin are all important factors. Vitamin D_3, and especially vitamin D_2, can be absorbed in the digestive tract. In normal adults, approximately *85%* of circulating D vitamins and derivatives are of the *D_3 series*, and the remainder is in the D_2 form.

Activation

In 1952, Carlson established that there was a lag time of eight to ten hours between the intravenous injection of vitamin D and the initiation of its effect on the intestinal transport of calcium. This suggested the existence of a series of biochemical events necessary for the activation of the vitamin.

The first activation involves a hydroxylation, at C25, in the liver: *25 OH-D_3* is formed, called *calcifediol* or calcidiol (Fig. XIV.15). The hepatic 25-hydroxylase is microsomal and mitochondrial. This enzyme is a cytochrome P450 mono-oxygenase.

The second step in vitamin activation is the formation of di- or trihydroxylated derivatives. The best known metabolite is *1,25 $(OH)_2$-D_3*, or calcitriol. This metabolite is *the most active* in the stimulation of intestinal absorption of calcium and in metabolism of bone calcium. Its synthesis is dependent upon the renal enzyme 25 OH-D_3 1α-hydroxylase[42]. Using isolated nephrons, it has been shown that the principal and perhaps exclusive location of this enzyme is in the proximal tubule. Its activity is increased by dietary restriction of calcium and phosphate, by hypocalcemia and hypophosphatemia, as well as by PTH. Enzymatic activity is suppressed by the administration of calcitriol, by hypercalcemia, calcitonin, or parathyroidectomy. Prolactin, growth hormone and estrogens stimulate the synthesis of calcitriol. An extrarenal synthesis of calcitriol has been demonstrated in bone and in the placenta. A *second metabolite, 24,25 $(OH)_2$-D_3*, is formed in the kidney when calcitriol synthesis is blocked. This metabolite is less active in the intestinal transfer of calcium than the 1,25 form, and is inexistent in anephric animals. It is assumed that it only acts on calcium transfer following transformation to 1,24,25 $(OH)_2$-D_3. Similarly, 24,25 $(OH)_2$-D_3 has no direct effect on the mobilization of calcium in bone. 24-hydroxylation of 25 OH-D_3 occurs outside the kidney. Chondrocytes from developing cartilage and bony tissue can easily transform 25 OH-D_3 to 24,25 $(OH)_2$-D_3, which favors mineralization of new bone.

Other derivatives of vitamin D have been studied, including 25,26 $(OH)_2$-D_3, and a glucoside of calcitriol, from plants, capable of inducing endemic hypercalcemia in sheep.

Metabolism[43]

Circulating vitamin D and its various derivatives are derived from two sources: mostly endogenous (D_3), and also from the diet (D_2 and D_3).

42. Kumar, R. (1980) The metabolism of 1,25-dihydroxyvitamin D_3. *Endocrine Rev.* 1: 258–267.
43. De Luca, H. F. (1981) Recent advances in the metabolism of vitamin D. *Annu. Rev. Physiol.* 43: 199–209.

In blood *serum*, a *specific protein* binds these substances. This protein *(DBP)* differs among species. The genetic control is complex (Table I.9), and a number of alleles are known. It binds vitamin D derivatives, with high affinity for 25 OH-D_3, 24,25 and 25,26 $(OH)_2$-D_3 and 25 OH-D_3-23,26-lactone; binding affinity is lower for calcitriol and vitamin D itself. This binding protein represents approximately 0.5% of serum proteins, and is low in patients with nephrotic syndrome and hepatic insufficiency. Its molar concentration is 100 times greater than that of 25 OH-D_3 (p. 42)[44].

Storage of vitamin D occurs in serum, muscle, and fat tissue. The primary form is 25 OH-D_3. There is a bile secretion and an enterohepatic circulation of the 1,25 $(OH)_2$-D_3 derivative.

Inactivation of vitamin D and its principal circulating metabolite, 25 OH-D_3, appears to be under the regulation of a microsomal enzymatic system of the *liver*. A number of compounds, such as antiepileptics, and in particular phenobarbital, are activators of this system. The principal pathway of the elimination of vitamin D and its hydroxylated and sulfate derivatives is in bile, and at least 2% appears in urine. The half-life of 25 OH-D_3 is approximately 10 to 20 d, whereas calcitriol has a shorter half-life, of 12 h.

Target Organs of Cholecalciferol

The most important target organs of cholecalciferol are the *intestine, bone, kidney, muscle, and parathyroids*. Intestine and kidney are the sites of transcellular calcium transfer; bone and muscle utilize calcium for such specific functions as mineralization and contraction. Specific *receptors* for calcitriol have been found in the small intestine, kidney, bone, cartilage, parathyroids, pancreas, pituitary, skin, parotids, and yolk sack.

The following effects have been determined in the *intestinal mucosa*: increased permeability of the entire small intestine to calcium (not involving the active transport system); increase in the capacity of the duodenum and superior portion of the jejunum to establish a system of active transport (requiring energy); stimulation of intestinal transport of phosphorus. These are all dependent on calcitriol.

In bone, the only definite effect of this dihydroxylated metabolite is the *mobilization* of calcium *from mature bone*. This action on bone resorption is slower than that of prostaglandin, osteoclast activating factors (OAF), and parathormone; only osteoblasts and osteoprogenetic cells have calcitriol receptors. The calcium thus mobilized is available to the extracellular space as well as to bone and osteoid tissue undergoing mineralization. Vitamin D probably has a direct role on development and mineralization of the skeleton, since a decrease in calcium levels is observed at the beginning of treatment of rickets with this vitamin. This, of course, implies the existence of a direct action of calcitriol on developing cartilage and bone, particularly on the stimulation of osteoblasts. The derivative 24,25 $(OH)_2$-D also appears to have a direct action on bone mineralization[45]. Parvalbumin and another calcium binding protein which is vitamin-dependent exist in calcified cartilage of bone and ameloblasts of teeth. They elevate the concentration of calcium at the calcification front. In addition, vitamin D elevates an action on bone collagen, inhibiting synthesis and increasing destruction[46].

In the *kidney*, calcitriol, at physiological doses, has a direct, positive effect on the reabsorption of phosphorus and possibly of calcium.

Muscle contains an abundant supply of vitamin D receptors. This vitamin is involved in muscle tone and contraction, probably via the intracellular calcium flux and the synthesis of ATP which it controls.

Parathyroids contain a calcitriol-dependent *calcium binding protein (CaBP)* which may play a *role in* the suppressive action of calcitriol on *PTH secretion*. 24,25 $(OH)_2$-D_3 has the opposite action.

In conclusion, the effects of vitamin D or its active derivatives have *one major consequence*: the *regulation of calcemia and phosphatemia*.

Other effects of vitamin D and its metabolites have been studied. For example, in *skin* it increases tyrosinase activity during exposure to ultraviolet radiation, and it controls the synthesis of a specific CaBP in stratified epithelium involved in cell proliferation and differentiation[47]. The relationship between vitamin D and the *hematopoietic* system is complex. In particular, human circulating monocytes have receptors for calcitriol which induce differentiation of monocytes to multinucleated macrophages involved in bone resorption. Vitamin D has complex and possibly important effects on the immune system[48].

Mechanism of Action (Fig. XIV.16)

The action of active vitamin D derivatives is mediated by a *nuclear receptor* very similar to that of other steroid hormones (Fig. I.71), *calcitriol* having the *highest affinity* for it. The mechanism of action, although not yet studied in detail, is similar to that of

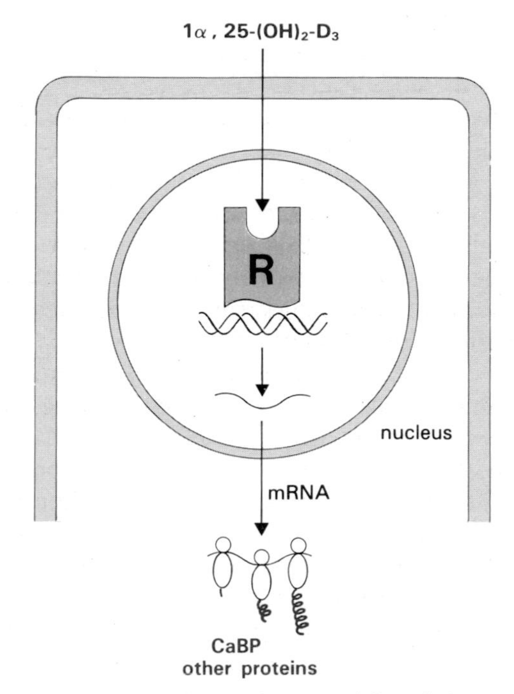

Figure XIV.16 Possible mechanism of the cellular action of calcitriol. Other proteins are alkaline phosphatase, Ca^{2+}-dependent ATPase, etc.

other steroid hormones (p. 51 and p. 405). In the *intestine*, calcitriol results principally in an increase in the synthesis of calcium transport protein (CaBP). An *increase in* a *calcium-dependent ATPase* in the brush border of intestinal microvilli, modifications of fatty acid components of membrane lipoproteins of intestinal cells, and an increase in *alkaline phosphatase* synthesis are effects observed following administration of vitamin D or its derivatives. *CaBP* was first discovered in the intestinal mucosa of the chicken, and then in mammals, including humans; two different proteins were identified, with a M_r of 9000 and 28,000.

4.2 Clinical Evaluation

An *insufficiency* of vitamin D provokes hypocalcemia, hypophosphatemia, elevation of serum alkaline phosphatases, reduction in calciuria and citraturia, and hyperaminoaciduria. Three steps should be taken in its clinical evaluation. The first is to recognize the existence of *rickets* or *osteomalacia*. The clinical and radiological signs are often obvious. In certain cases, bone biopsy is useful, demonstrating characteristic osteoidosis. Secondly, *sensitivity* to vitamin D or a derivative should be determined. Finally, the *plasma concentration* of vitamin D and its derivatives[48] should be measured. Under physiological conditions, blood concentrations of vitamin D are 20 to 40 ng/ml; for 25 OH-D_3, they are 10 to 30 ng/ml; for 24,25 $(OH)_2$-D_3, they are between 0.8 and 2.3 ng/ml; and for calcitriol, they average 30 pg/ml. The serum concentration of 25 OH-D_3 is a reflection of the vitamin D reserves in the body; it increases slightly with age, and is greatly influenced by sunshine[49]. The serum concentration of calcitriol is high in the infant during the first year of life, in the adolescent, and during pregnancy[50]. The assay of DBP may be important during pregnancy, in order to have an estimate of the free forms of calcitriol[51]. To assess excessive vitamin activity, derivatives of vitamin D should be measured in serum. Intoxication by massive doses of vitamin D is accompanied by an elevated level of metabolites.

4.3 Pathology Associated with Vitamin D Hypoactivity or Deficiency

The clinical signs associated with reduced vitamin D activity include *a reduction in growth* and, especially, the development of *osteoid tissue* in bone (Fig. XIV.17). Osteoidosis predominates in the metaphyseal extremity, in the infant and child, and develops into rickets; osteoidosis of the endostium and subperiostium in adults is the determining feature of

44. Daiger, S. P., Schonfield, M. S. and Cavalli-Sforza, L. L. (1975) Group-specific component (Gc) proteins bind vitamin D and 25-hydroxyvitamin D. *Proc. Natl. Acad. Sci., USA* 72: 2076–2080.
45. Dickson, I. R., Hall, A. K. and Jande, S. S. (1984) The influence of dihydroxylated vitamin D metabolites on bone formation in the chick. *Calcified Tissue Int.* 36: 114–122.
46. McCarthy, D. M., Hibbin, J. A. and Goldman, J. M. (1984) A role for 1,25-dihydroxyvitamin D_3 in control of bone marrow collagen deposition. *Lancet* i: 78–80.
47. Louari, D., Pavlovitch, H., Deceneux, G. and Balsan, S. (1980) A vitamin D dependent calcium-binding-protein in rat skin. *FEBS Lett.* 111: 285–289.
48. Chesney, R. M., Rosen, J. F., Hamstra, A. J. and DeLuca, H. F. (1980) Serum 1,25-dihydroxyvitamin D levels in normal children and in vitamin D disorders. *Amer. J. Dis. Child* 134: 135–139.
49. Chesney, R. M., Rosen, J. F., Hamstra, A. J., Smith, C., Mahaffey, K. and De Luca, H. F. (1981) Absence of seasonal variations in serum concentrations of 1,25 $(OH)_2D$ despite a rise in 25 $(OH)_2D$ in summer. *J. Clin. Endocrinol. Metab.* 53: 139–142.
50. Rosen, J. E. and Chesney, R. W. (1983) Circulating calcitriol concentrations in health and disease. *J. Pediat.* 103: 1–17.
51. Bouillon, R., Van Assche, F. A., Vanbaelen H., Heyns, W. and Demoor, P. (1981) Influence of the vitamin D binding-protein on the serum concentration of 1,25 dihydroxyvitamin D_3. *J. Clin. Invest.* 67: 589–596.

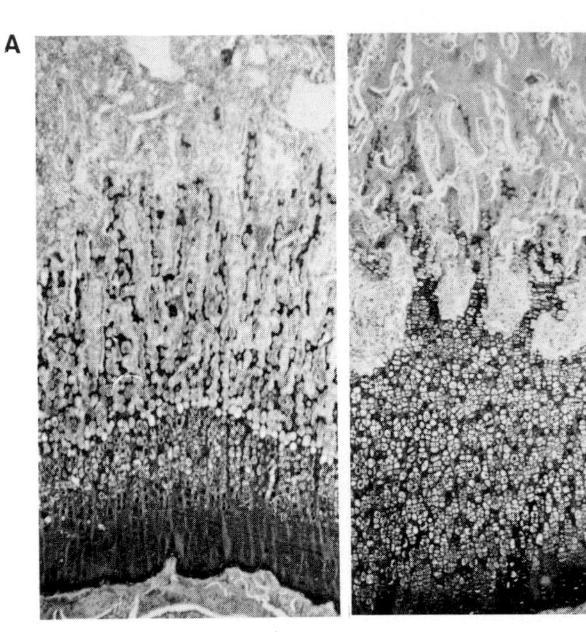

Figure XIV.17 **A.** Epiphyseal plates in children. Normal cartilage with orderly parallel columns of hypertrophic cartilage, and the calcification front. **B.** Rachitic cartilage with absence of orderly parallel columns; note the increased number of cells in irregular rows and the absence of calcification of the osteoid.

osteomalacia. Disorders may be either acquired or hereditary, and either sensitive or resistant to vitamin D.

Avitaminosis D Sensitive to Vitamin D

The daily requirement of vitamin D in adults is 5 to 10 μg/d, and about 1 μg/kg body weight per day in young children. Normally, it is assumed that, under conditions of moderate sunshine, a dose of 100 to 400 IU (1 μg=40 IU) per day protects the infant from rickets and the adult from osteomalacia.

Vitamin D-Sensitive Deficiency (Common Rickets)

Deficiency of vitamin D is not necessarily of *dietary* origin, since normal dietary intake is insufficient to meet physiological needs (the portion contributed by diet varies between 0.5 and 1.0 μg/d). It is, rather, a *deficiency* of *exposure of the skin to sunlight*, in unfavorable climates, or resulting from dressing habits, housing conditions, and industrial smoke, which filters ultraviolet rays. Individuals with pigmented skin who move to intemperate climates and to industrial areas are particularly affected.

The clinical signs of avitaminosis D associated with accidents duc to hypocalcemia include convulsions, tetany, spasm of the glottis, cardiomegaly, and behavioral modification. Hypocalcemia leads to secondary hyperparathyroidism and bone alterations. In young children, the disease is known as *vitamin D deficiency rickets*, still frequent, unfortunately, despite easy prevention. The principal signs include craniotabes, rachitic rosary, metaphyseal widening in long bones, submammary contraction of the thorax, curvature of the long bones, coxa-vara, and deformation of the spine (Figs. XIV.18 and XIV.19). *Radiological examination* of the skeleton demonstrates flattening and deformation of the bones as a result of poor mineralization. Metaphyses are irregular and enlarged. Diaphyses are bowed, with a dense portion, sometimes thin, sometimes thick, and sometimes modified by lamellar periostosis (Fig. XIV.20). Looser's striae and actual fractures can be seen. Regions of epiphyseal ossification are fuzzy and irregular, and show signs of growth retardation. The distribution of skeletal defects changes with age, affecting bones of the skull, thorax, and limbs of infants, knees and hips (coxa-vara) of older children, and vertebrae of adolescents.

Calcemia is low or normal, with reduced phosphatemia and elevated serum alkaline phosphatases. Levels of both calcium and phosphorus are elevated in feces. Calciuria is low. Frequently, there is generalized hyperaminoaciduria, hypocitraturia, and a reduction in plasma bicarbonates. Serum 25 OH-D_3 is low, but serum calcitriol is usually normal or subnormal[52].

Anemia (microcytic) is frequently observed. The effect of calcitriol on collagen synthesis explains the fibrous invasion of the bone marrow and the development of extramedullary hemopoiesis with considerable splenomegaly and myelocytosis (Von Jacks–Luzet syndrome), all of which are cured by vitamin D.

Hypotonia and muscular weakness are frequently observed. The association of tetany may constitute a serious complication, and can lead to death. Growth retardation is seen in severe cases of rickets.

The *clinical evolution* of rickets is divided into three stages characterized by different signs; stage I: hypocalcemia, normal phosphoremia, tubular phosphate reabsorption and amino aciduria, normal PTH, and very discrete bone lesions; stage II: normal calcemia, reduced phosphoremia and TRP (tubular reab-

52. Garabedian, M., Vainsel, M., Mallet, E., Guillozo, H., Toppet, M., Grimbert, R., Nguyen, T.M. and Balsan, S. (1983) Circulating vitamin D metabolite concentrations in children with nutritional rickets. *J. Pediat.* 103: 381–386.

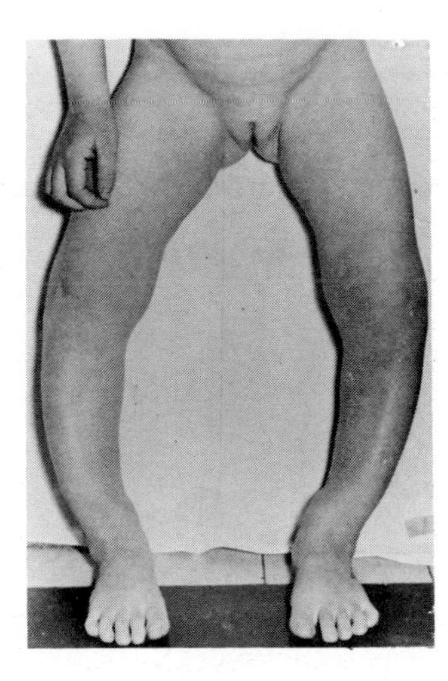

Figure XIV.18 Vitamin D deficiency rickets. Note the malformation with outward curvature of the lower limbs.

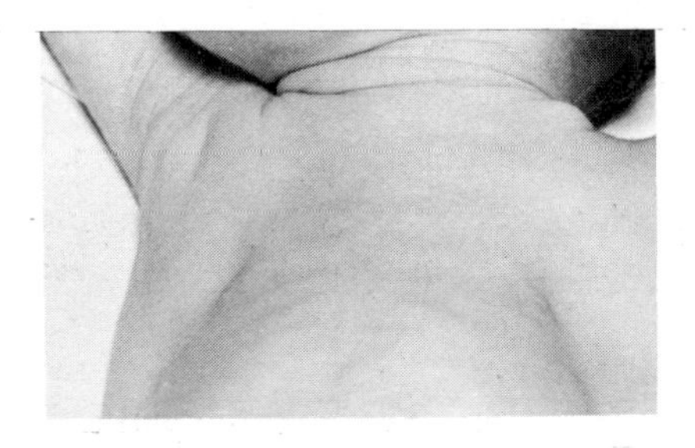

Figure XIV.19 Severe vitamin D deficiency rickets. Note the malformation of the clavicles and ribs.

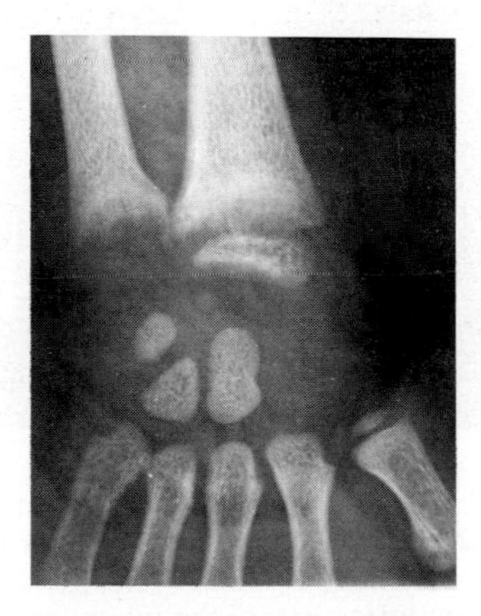

Figure XIV.20 Vitamin D deficiency rickets: radiography of the wrist. Note the cupular aspect with irregular fringed edges of the inferior extremity of the cubitus and radius.

sorption of phosphate), hyperaminoaciduria, elevated PTH with retention of PTH sensitivity, and radiological lesions; stage III: hypocalcemia and hypophosphoremia, elevated PTH and non-responsiveness to exogenous PTH, and severe radiological lesions.

Treatment of rickets caused by vitamin D deficiency involves the use of vitamin D_2 or D_3, but not of hydroxylated derivatives. Curative treatment consists of a single dose of 2 mg of vitamin D or, preferably, 50 μg (2000 IU) per day over six weeks. Preventive treatment consists of a daily intake of 10 to 30 μg of vitamin D during the first and second years of life (double for a transplanted individual with pigmented skin). In some countries, vitamin D is systematically added to milk. Rickets caused by deficiency in vitamin D is rare in breast-fed children[53].

Avitaminosis D of the Newborn (Congenital Rickets).

Avitaminosis D in newborns of mothers who themselves have avitaminosis D has been described. Congenital rickets due to a maternal deficiency of vitamin D can be distinguished from severe osteopenia of the premature infant, which is primarily related to an insufficient intake of phosphorus from the mother's milk.

Vitamin D-Sensitive Osteomalacia in the Adult

This condition is sometimes seen in *women* subjected to a confined life or to a non-equilibrated diet, and is especially prevalent in women with pigmented skin who have moved to industrialized societies.

Defects in the Intestinal Absorption of Vitamin D

Various digestive disorders may lead to poor intestinal absorption of vitamin D and to rickets sensitive to parenteral injection of 2 to 5 mg of vitamin D_2 or D_3 two or three times each year, as in celiac disease or in mucoviscidosis.

Vitamin D-Resistant Rickets

Many types

Many diseases in children and adults are accompanied by *osteopathy* with clinical, anatomical, radiological and biochemical characteristics which resemble (fully or partially) rickets and osteomalacia due to avitaminosis. However, these diseases differ by

53. Brooke, O. G. (1983) Supplementary vitamin D in infancy and childhood. *Arch. Dis. Child.* 58: 573–574.

the absence of an effect of physiological doses of vitamin D, and symptoms relapse when treatment with very high doses of vitamin is stopped.

Several *hepatic or biliary disorders* lead to a defect in the 25-hydroxylation of vitamin D and its clinical consequences[54]. It is interesting that 10% of epileptics receiving treatment with phenobarbital have hypocalcemia. The one to two percent who suffer from rickets or osteomalacia resistant to normal doses of vitamin D can be cured by the elimination of barbiturates or by increasing the dose of vitamin D. The explanation for this is that *anticonvulsive drugs* stimulate microsomal hydroxylases of the liver, thus favoring the metabolism of vitamin D to inactive polar compounds and reducing the amount of 25 $OH\text{-}D_3$ formed.

Hypoparathyroidism has been treated by vitamin D for a long time; favorable results are due to renal hydroxylation of vitamin D. It follows that treatment with calcitriol is very efficient.

In chronic *renal insufficiency*, osteopathies may occur; they are sometimes associated with osteopenia, fibro-osteoclasia, and osteoidosis. Hypocalcemia, hyperphosphatemia, and an elevated level of PTH are frequently seen. It is known that hemodialysis does not improve these conditions, which may appear or even develop under the treatment. In contrast, renal transplantation, when the graft is operational, may cure these osteopathies, which are in large part explained by a resistance induced by insufficient renal hydroxylation.

Several renal *tubular diseases* can result in rickets or severe osteomalacia. Treatment with large doses of vitamin D has long been used in rickets, with varying results. The therapeutic action of 1,25 $(OH)_2\text{-}D_3$ and 1α $OH\text{-}D_3$ is remarkable, particularly in patients with cystinosis.

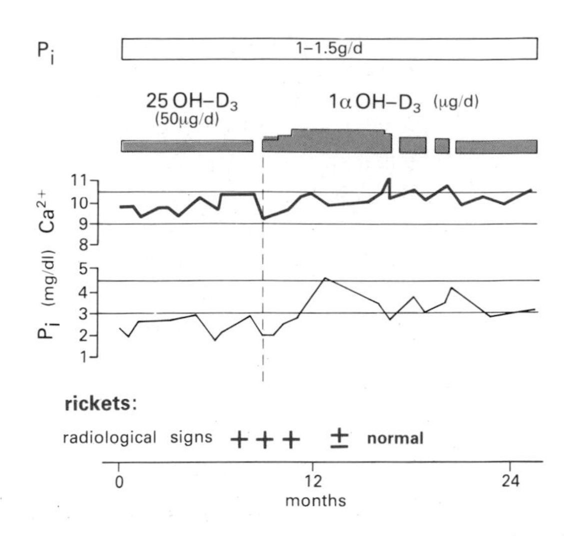

Figure XIV.21 Familial hypophosphatemia. Treatment consists of a supplement of phosphorus (P_i), of 1.0 to 1.5 g/d for four days. This treatment is associated initially with 25 $OH\text{-}D_3$, which corrects neither the hypophosphatemia nor the radiological signs of rickets. However, the replacement of 25 $OH\text{-}D_3$ with 1α $OH\text{-}D_3$ allows a return to a normal aspect of the skeleton and restitutes normal phosphatemia (P_i). Calcemia (Ca^{2+}) remains at the upper limit of normal.

Hereditary Vitamin D-Resistant Rickets

Hereditary hypophosphatemia is transmitted dominantly on the X chromosome. To some extent, it causes common rickets, with some differences, particularly histological. This disease is *resistant to usual doses of vitamin D.* The most efficient treatment associates an increased intake of phosphate, several times per day, with the administration of 0.5 to 2.0 μg per day of calcitriol or of 1α $OH\text{-}D_3$[55] (Fig. XIV.21).

There are also two types of *hereditary pseudodeficient hypocalcemic rickets. Type I* is called dependent rickets. It is transmitted in an autosomal recessive mode, and is manifested in the third month of life by

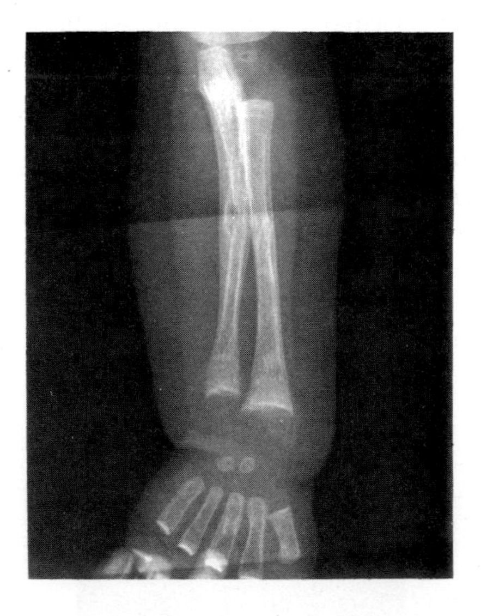

Figure XIV.22 Hereditary pseudodeficient hypocalcemic rickets (type II). Radiography of the forearm and hand shows cupular deformity of metaphyses, periostal reaction, and Looser's pseudofractures of the cubitus and radius. This vitamin D-resistant disease appears to be associated with hereditary abnormality of renal 1α-hydroxylation of 25 $OH\text{-}D_3$.

very severe osteomalacic rickets, with Looser striae, tetany, and life-threatening respiratory problems. Treatment with elevated doses of vitamin D and calcium has spectacular results. This disease is secondary to a hereditary deficiency of 25 $OH\text{-}D_3$ 1α-hydroxylase in the kidney (Fig. XIV.22). *Type II*, in contrast, is distinguished by elevated plasma calcitriol levels, and is refractory to physiological doses of vitamin D derivatives. It, also, is transmitted by an autosomal recessive mode, and is frequently accompanied by alopecia. A defect in the calcitriol receptor is thought to be the cause of this disorder[56].

Resistant Rickets Due to Tumors

Finally, there are cases of vitamin-resistant rickets due to *tumors*, characterized by extreme hypophosphatemia and severe bone lesions. The removal of the tumor may lead to complete recovery from rickets. Phosphaturetic substances have been found in these tumors, which are neither PTH nor calcitonin. One has recently been cloned[57].

4.4 Pathology and Vitamin D Hyperactivity

Several clinical syndromes, including hypercalcemia, diffuse calcinosis (corneal, renal, pulmonary, myocardial and vascular, digestive) and skeletal changes, seem to be either directly or indirectly related to excessive vitamin D activity. Intoxication by massive doses has been observed with calciferol, tachysterol, cholecalciferol, 25 OH-cholecalciferol, calcitriol, and 1α $OH\text{-}D_3$.

Vitamin D and 25 $OH\text{-}D_3$ have a *long half-life*, and their effects are prolonged. Excessively *high levels lead to hypercalcemia*, with polydipsia, excessive diuresis, absence of appetite, vomiting, and constipation. Fever, palor, somnolence, cephaly, abdominal pains, growth arrest and thinning are usually observed. Arterial hypertension is frequent. Radiological examination demonstrates skeletal demineralization or opaque metaphyseal bands, sometimes with metastatic calcifications, particularly in arterioles, and occasionally nephrocalcinosis. Calcemia, and sometimes phosphatemia, is elevated. The evolution of the disease is variable. If intoxication continues, renal insufficiency progresses, cachexia appears, and death may result from uremia, coronary problems or cerebral complication due to arterial hypertension. If intoxication is ceased, abnormal signs disappear or improve. Hypersensitivity to vitamin D persists for a period of variable duration (Fig. XIV.23).

Excessive doses of *calcitriol and* 1α $OH\text{-}D_3$

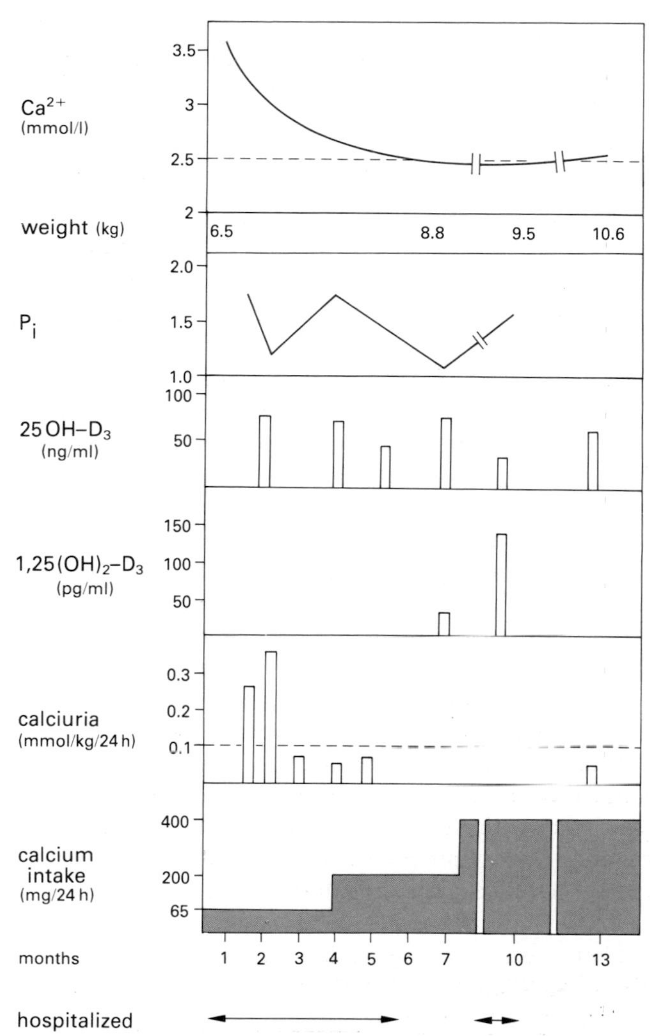

Figure XIV.23 Intoxication by massive doses of vitamin D_2 in a 10-month-old child. Calcemia, calciuria, 25 $OH\text{-}D_3$ and 1,25 $(OH)_2\text{-}D_3$ in serum are determined during low calcium diet and after stopping vitamin D intake.

54. Heubi, J. E., Tsang, R. C., Steichen, J. J., Chan, G. M., Chen, I. W. and De Luca, H. F. (1979) 1,25 dihydroxyvitamin D in childhood hepatic osteodystrophy. *J. Pediat.* 94: 977–982.
55. Chesney, R. W., Mazess, R. B., Rose, P. R., Hamstra, A. J., De Luca, H. F. and Breed, S. T. (1983) Long-term influence of calcitriol and supplemental phosphate in X-linked hypophosphatemic rickets. *Pediatrics* 71: 559–567.
56. Liberman, U. A., Eil, C. and Marx, S. J. (1983) Resistance to 1,25-dihydroxyvitamin D. *J. Clin. Invest.* 71: 192–200.
57. Suva, L. J., Winslow, G. A., Wettenhall, R. E. H., Hammonds, R. G., Moseley, J. M., Diefenbach-Jagger, H., Rodda, C. P., Kemp, B. E., Rodriguez, H., Chen, E. Y., Hudson, P. J., Martin, T. J. and Wood, W. I. (1987) A parathyroid hormone-related protein implicated in malignant hypercalcemia: cloning and expression. *Science* 237: 893–896.

result in *acute* symptoms. These compounds are used especially in hypoparathyroidism, hereditary hypophosphatemia, renal osteodystrophia, and pseudodeficient rickets of type I and type II. Diagnosis is made by the observation of hypercalcemia, and, since vitamin D derivatives have a short period of action, cessation of treatment is effective[58].

In addition, states of *exaggerated sensitivity* to near physiological doses of vitamin D, called *moderate hypercalcemia*, have been reported. The origin of this hypersensitivity is either idiopathic, and sometimes familial (benign hypercalcemia in children), or it may be induced by sarcoidosis[59] or thyroid insufficiency.

Finally, *idiopathic hypercalcemia* accompanied by elfin facies is sometimes observed in children shortly after birth; it associates dwarfism (with an increase in skeletal density, arterial hypertension, nephrocalcinosis, mental retardation), and hypercalcemia, which disappears at about two years of age. Frequently, stenosis of the aorta and/or of the pulmonary and renal arteries occurs. An endogenous disorder of vitamin D action and metabolism has been demonstrated. An exogenous factor acting during pregnancy may be involved, as suggested by the endemic and regional characteristics of the disease; this, however, is not proved, and, indeed, the administration of high doses of calcitriol during pregnancy results in normal fetal development[60].

5. Action of Other Hormones

5.1 Thyroid Hormones

Thyroid hormones act on the development of the skeleton, renewal of bone and the metabolism of calcium, and influence calcitriol serum concentrations[61]. They *increase urinary and plasma calcium*, and accelerate the replenishment of cellular calcium and magnesium. They also augment phosphatemia and reduce phosphaturia. In addition, T_3 and T_4 inhibit the intestinal absorption of calcium, stimulating the renewal of bone, but affect the resorption of bone more than they do its formation. Changes in bone cells induced by thyroxine are identical to those stimulated by PTH (fibro-osteoclasia). The action of thyroxine occurs even in the absence of the parathyroid gland.

A negative calcium balance accompanies *hyperthyroidism*. Urinary hydroxyproline is elevated. Various studies have demonstrated an increase in bone metabolism, favoring resorption over osteoblastic construction, occurring mostly in aged patients. A parathyroid adenoma is rarely associated with hyperthyroidism.

Thyroid insufficiency in the infant is characterized by a localized or generalized densification of the skeleton, an increased digestive absorption and a reduced urinary excretion of calcium. Severe hypercalcemia is rare.

5.2 Cortisol and Glucocorticosteroids

Spontaneous or therapeutic hypercorticism leads to osteoporosis, reduction in osteoblastic function, and the slowing of growth. It is accompanied by normal calcemia with increased plasma PTH, suggesting secondary hyperparathyroidism. Chronic adrenal insufficiency may lead to hypercalcemia. The relationship between glucocorticosteroids, metabolism of calcium, and calcified tissues is still poorly understood. Glucocorticosteroids *reduce intestinal absorption of calcium*, opposing the action of vitamin D[62]. *Osteoporosis* is associated with poor intestinal absorption of calcium, a reduction in the retention of calcium by the kidneys, and a reduction in bone formation, specifically in collagen synthesis, and an increase in its destruction. Remarkably, almost all patients with hypercalcemia, except when due to an excess of PTH, are improved by *the administration of glucocorticosteroids*, especially in cases which are associated with an excess of vitamin D. In contrast, experimental or pathological cortisone deficiency may lead to hypercalcemia; in humans, this has been reported in Addison's disease, in hereditary masculinizing hyperplasia of the adrenals, and after insufficiently treated bilateral adrenalectomy.

58. Chan, J. C. M., Reuben, Y., Alon, U. and Mamunes, P. (1983) Hypercalcemia in children with disorders of calcium and phosphate metabolism during long-term treatment with 1,25-dihydroxyvitamin D_3. *Pediatrics* 72: 225–233.
59. Chesney, R. W., Hamstra, A. J., De Luca, H. F., Horowitz, S., Gilbert, E. F., Hong, R. and Borcherding, B. S. (1981) Elevated serum 1,25-dihydroxyvitamin D concentration in the hypercalcemia of sarcoidosis: correction by glucocorticoid therapy. *J. Pediat.* 98: 919–922.
60. Marx, S. J., Swart, E. G., Hamstra, A. J. and De Luca, H. F. (1980) Normal intrauterine development of the fetus of a woman receiving extraordinary high doses of 1,25-dihydroxyvitamin D. *J. Clin. Endocrinol. Metab.* 51: 1138–1142.
61. Bouillon, R., Muls, E. and Demoor, P. (1980) Influence of thyroid function on the serum concentration of 1,25-dihydroxyvitamin D_3. *J. Clin. Endocrinol. Metab.* 51: 793–797.
62. Lindgren, J. V., Merchant, C. R. and De Luca, H. F. (1982) Effects of 1,25 $(OH)_2$ D_3 on osteopenia induced by prednisolone in adult rat. *Calcified Tissue Int.* 34: 253–257.

5.3 Estrogens, Prolactin, and Growth Hormone

The effect of gonadal steroids on bone tissue has been *studied extensively*, mainly because of the wide interest in the treatment of postmenopausal osteoporosis. *Estrogens* reduce the response of bone tissue to PTH. Short-term treatment with estrogens has several effects: a reduction in calcemia, phosphatemia and calciuria, an increased retention of calcium in bone, a reduction in urinary hydroxyproline, and, as shown by bone biopsies, a reduction of the resorptive surface of bone, with little modification of the zones of bone formation. Estrogens may increase the activity of 1α-hydroxylase and thus favor the synthesis of calcitriol, as shown in the chicken. During pregnancy in women, calcitriol levels are increased. The role of prolactin in the activation of 1α-hydroxylase in the kidney is clear. This effect of estrogens and prolactin on 1α-hydroxylation during periods of increased calcium need (pregnancy, lactation, formation of the egg shell) is logical. However, there are differences between birds and mammals. In the former, ovariectomy can inhibit renal activation of 25 OH-D_3, and the inhibitors of the final stage of follicle formation reduce interstitial CaBP. In rats, ovariectomy has no effect on 25 OH-D_3 or on calcitriol in plasma, nor does it effect CaBP in the intestinal mucosa[63].

Growth hormone has a potent effect on bone growth, via IGFs, and has a stimulatory effect on 1α-hydroxylation of 25 OH-vitamin D_3. IGF action on calcium and phosphate metabolism seems physiologically modest: elevation of calciuria, especially in children, and an increase in phosphatemia are observed. Pathologically, hypercalciuria and hyperphosphatemia are seen in acromegaly.

5.4 Other Hormones

Numerous molecules affect calcium equilibrium and the mineralization of calcified tissue.

Prostaglandins of the E series stimulate bone resorption and bone formation in tissue culture, however their physiological role is unclear. Pathologically, enhanced secretion of prostaglandins in certain types of malignant tumors is responsible for severe hypercalcemia. *Vasoactive intestinal polypeptide (VIP)* has a hypercalcemic effect. Other factors, including interleukin-2 and interferon, may also have paracrine activity. Some activated lymphocytes produce an *osteoclast activating factor* (OAF), consisting of at least two different molecules. Hypercalcemia observed in some leukemias and lymphosarcomas is due to the exaggerated secretion of OAF or similar molecules[64]. Studies conducted in humans and other animals with *osteopetrosis* have shown a deficiency of a *thymic factor*. Some successes in laboratory animals have led to the treatment of children with malignant osteopetrosis by transplantation of bone marrow.

6. Hormonal Regulation of Calcium Metabolism during Pregnancy, Lactation, and Ante- and Perinatal Development

6.1 Pregnancy [65]

Pregnancy is accompanied by a positive calcium balance, an increased absorption of calcium by the small intestine, and elevated serum PTH and calcitriol levels.

Maternal calcemia is *reduced* during pregnancy, a decrease which accompanies the reduction in albumin concentration in the serum. Ionized calcium levels are variable, sometimes elevated, and there is no explanation for the increase of plasma PTH. Parathyroid hormone does not cross the placental barrier.

Calcitonin is *increased* during pregnancy. Its concentration is high in the maternal plasma and in umbilical cord blood, but it does not cross the placental barrier. It may have a protective effect on the maternal skeleton.

The concentration of 25 OH-D_3 in maternal serum is similar to that observed in non-pregnant women. Vitamin D and 25 OH-D_3 *cross the placenta.* DBP is increased during pregnancy and may reach 1.2 mg/ml at term, a level four times higher than in non-pregnant women, and double that of cord blood. Thus, in contrast to its total concentration, the free form of 25 OH-D_3 is higher in umbilical cord blood than in the maternal circulation.

63. Pavlovitch, H., Clemens, T. L., Laouari, D., O'Riordan, J. L. H. and Balsan, S. (1980) Lack of effect of ovariectomy on the metabolism of vitamin D and intestinal calcium-binding-protein in female rats. *J. Endocrinol.* 86: 419–425.
64. Mundy, G. R. and Raisz, L. G. (1977) Big and little forms of osteoclast activating factor. *J. Clin. Invest.* 60: 122–128.
65. Gray, T. K., Lowe, W. and Lester, G. (1981) Vitamin D and pregnancy: the materno-fetal metabolism of vitamin D. *Endocrine Rev.* 2: 264–274.

Contrary to total calcitriol, free 1,25 $(OH)_2$-D_3 is lower in maternal serum than in the serum of the fetus. Very little calcitriol crosses the placental barrier, much less than 25OH-D_3.

6.2 Lactation[66]

During lactation, approximately 300 mg of *calcium* and 150 mg of *phosphorus* are *lost per day*. During the first six months of lactation, calcemia and magnesemia are elevated, and phosphoremia and alkaline phosphatase are normal. Calcitonin is unchanged and PTH is slightly reduced. On the other hand, the concentration of 1,25 $(OH)_2$-D_3 may attain 100 pg/ml in the serum, and is probably the major factor accounting for calcium homeostasis during lactation. The effectiveness of maternal milk in counteracting rickets is controversial, since human milk has a low concentration of vitamin D and 25 OH-D_3.

6.3 Fetus and Placenta[65]

Calcium accumulates in the fetus during the last months before birth. It contains 3 g at five months, 6 g at six months before birth, 9 g at seven months, 15 g at eight months, and 30 g at birth. Phosphorus increases from 2 g at five months to 18 g at birth. Fetal calcemia is greater than maternal calcemia. Phosphatemia is also high in the fetus. Urinary excretion of phosphate is weak, and the percentage of tubular reabsorption is high, approaching 100%. The fetus, therefore, is found to be hypercalcemic and hyperphosphatemic in relation to the mother. Nutritional or endocrine anomalies of mineral metabolism in the mother clearly have an influence on the availability of calcium and phosphorus to the fetus. The existence of *bidirectional transport* of calcium *across the placenta* has been shown. *High calcemia* in the human fetus suggests a process of active transport of calcium, from the mother *to the fetus*, at the placental level. CaBP, identified in the placenta, increases during pregnancy. The calcitriol receptor has been identified in the placenta, and it is believed that placental CaBP is vitamin D-dependent. In birds, CaBP, which transfers calcium from the shell to the embryo, is vitamin K-dependent.

In the normal state, a high fetal calcium concentration maintained by the placenta does not stimulate the parathyroids, which explains the hyperphosphatemia and low phosphaturia found in the fetus. In fact, at the end of pregnancy in humans, fetal hypercalcemia is involved more in the stimulation of calcitonin secretion than in the reduction of parathyroid activity. Calcitonin inhibits the ability of PTH to resorb fetal long bones.

With respect to vitamin D and its derivatives, the placenta synthesizes calcitriol, but in quantities insufficient to cover all fetal needs[68]. The fetal kidney and probably other tissues, particularly bone, also synthesize calcitriol. The fetus is therefore dependent on the mother for its supply of vitamin D and 25 OH-D_3, but can rely on its own metabolism for the synthesis of calcitriol.

In summary, *three factors* appear to be involved in prenatal calcium metabolism: 1. maternal mineral metabolism, with its nutritional components and its own endocrine regulation; 2. placental transfer of vitamin D_3 and its 25 OH derivative, and the synthesis of some calcitriol. In addition, the bidirectional exchange of calcium and phosphorus by an active mechanism maintains fetal calcemia between 10 and 20 mg/l, higher than maternal calcium levels; 3. early fetal production of parathyroid hormone, calcitonin and calcitriol, the ability to regulate their secretion, and the possession of specific receptors in the target organs, bone and kidney[69].

6.4 Neonatal Adaptation

In *umbilical cord blood* at parturition, calcemia is elevated (107 mg/l) and inorganic phosphorus is also increased (56 mg/l). Just *after birth*, there is a progressive decrease in calcemia (85–90 mg/l) and a progressive increase in the concentration of phosphorus (70–80 mg/l). All situations which make an appropriate diet difficult (premature, newborns of high-risk pregnancies, small newborns) will also reduce the intake of calcium. The skeleton of the newborn is able

66. Greer, F.R., Tsang, R.C., Searcy, J.E., Levine, R.S. and Steichen, J.J. (1982) Mineral homeostasis during lactation: relationship to serum 1,25-dihydroxyvitamin D, 25-hydroxyvitamin D, parathyroid hormone and calcitonin. *Amer. J. Clin. Nutr.* 38: 431–437.
67. Reeve, L.E., Chesney, R.W. and De Luca, H.F. (1982) Vitamin D of human milk: identification of biologically active forms. *Am. J. Clin. Nutr.* 36: 122–126.
68. Whitsett, J.A., Ho, M., Tsang, R.C., Norman, E.J. and Adams, K.G. (1981) Synthesis of 1,25 $(OH)_2$ D_3 by human placenta in vitro. *J. Clin. Endocrinol. Metab.* 53: 484–488.
69. Narbaitz, R., Stumpf, W.E., Sar, M., Huang, S. and De Luca, H.F. (1983) Autoradiographic localization of target cells for 1 alpha-25 dihydroxyvitamin D_3 in bones of fetal rats. *Calcified Tissue Int.* 35: 177–182.

to incorporate large quantities of calcium. At birth, parathyroid activity remains low for several days. In the newborn, either term or premature, serum *calcitonin* is markedly *elevated*, especially in the premature. This level is increased to even greater levels in the first two days of life, and begins to decrease on the seventh day. The level of *PTH*, undetectable at birth, increases at 48 h, and decreases thereafter. The level of 25 OH-D_3 is normal in premature infants and newborns at term, and correlates with calcium in the blood and umbilical cord. *Calcium* levels *decrease* in the first two days following birth, more markedly in the premature than in the infant at term, and return to normal at seven days.

Dietary factors, maternal milk as opposed to modified cow milk, and premature birth as opposed to birth at full term, affect intestinal absorption of calcium and phosphorus in the neonate. From birth, calcitriol plays a major role in intestinal absorption of calcium. The neonate born at term, and the premature infant, even at very young gestational ages, absorb vitamin D, produce calcitriol, develop intestinal receptors to these compounds, and synthesize intestinal CaBP. *Vitamin D requirements are three to four times greater in premature* infants than in infants born at term. Severe osteopenia, sometimes with numerous fractures, observed in premature infants nourished with maternal milk appears to be dependent on a variety of factors: immature kidney development, with deficient tubular reabsorption of phosphorus, and insufficient phosphorus intake, with hypercalciuria and very low phosphatemia[70]. This situation can be counteracted by the addition of phosphorus to milk, in the form of disodium or dipotassium phosphate.

70. Rowe, J., Rowe, D., Horak, E., Spackman, T., Saltzman, R., Robinson, S., Philipps, A. and Raye, J. (1984) Hypophosphatemia and hypercalciuria in small premature infants fed human milk. *J. Pediat.* 104: 112–116.

7. Summary

Calcium plays a *fundamental role in cellular activity*, especially as a second messenger for various hormonal signals and as an activator of several enzymes. It is also a major factor in the *mineralization of bone*. *Calcemia* (100 mg/l) is dependent on the equilibrium of calcium exchange occurring in the intestine, from dietary sources; in bones, where calcium is in the form of crystals of hydroxyapatite in the collagen matrix; and in kidneys, which are both regulatory and excretory organs. Several hormones are involved in these processes. The most important are PTH, calcitriol, and calcitonin. In *three main organs*, intestine, bone and kidney, these *three hormones* act in an integrated manner, with calcitriol being the primary regulator in the intestine. At the cellular level, PTH and calcitonin have membrane receptors that stimulate adenylate cyclase activity, whereas calcitriol enters target cells and binds to an intracellular receptor, the complex acting ultimately on gene expression. PTH and calcitriol increase the concentration of calcium in the soluble fraction of the cytoplasm, whereas calcitonin tends to reduce this concentration. However, a more complete understanding of the physiological importance of calcitonin may modify these concepts.

Parathyroid hormone is an 84 aa peptide produced by parathyroid cells; it is evolutionarily well conserved. The *concentration of* Ca^{2+} *controls* its secretion. The mechanism of this regulation involves a *calcium binding protein (CaBP)*, the synthesis of which is *dependent upon vitamin D*. There is no pituitary *stimulus* of PTH, nor does cAMP affect its biosynthesis. PTH acts primarily on bone and kidney. *It activates osteoclasts* indirectly, via osteoblasts, and from bone it liberates not only calcium but also magnesium, phosphate, and citrate. In the kidney, its *stimulatory effect on phosphodiuresis* is rapid. In order to examine parathyroid function, PTH is measured by radioimmunoassay, and cAMP may be measured, during PTH stimulation, along with calcium and phosphate, in blood and urine. Bone biopsies allow a quantitative histological evaluation.

Parathyroid insufficiency leads to hypocalcemia, hypocalciuria, and hyperphosphatemia. Hypocalcemia, in turn, leads to a state of neuromuscular excitability known as *tetany*. In addition to true hypoparathyroidism, Albright described *pseudohypoparathyroidism*, a genetic disease involving a defect in the activation, by PTH, of the G_s protein controlling adenylate cyclase activity. *Hyperparathyroidism* results in hypercalcemia accompanied by hypercalciuria, with elevated citraturia and hydroxyprolinuria, and reduced phosphatemia. The clinical syndrome involves polydipsia, polyuria, and metastatic calcifications. Primary hyperparathyroidism is due to an adenoma or to a genetic disorder. Secondary hyperparathyroidism is caused mostly by renal insufficiency, which leads to parathyroid reaction with fibrocystic osteitis; ectopic hyperparathyroidism may be due to a tumor.

Calcitonin is synthesized in cells of the APUD system, the parafollicular C cells of the thyroid gland in mammals. The sequence of the 32 aa peptide varies markedly between species. Calcitonin *secretion* is *controlled* by the concentration of *Ca^{2+}*, with an increase in calcium level causing an increase in calcitonin production. *No hypothalamic-hypophyseal hormone* is involved, but several GI hormones stimulate its secretion. Calcitonin inhibits the action of osteoclasts on bone resorption, and consequently *reduces calcemia* and phosphatemia. It *increases urinary* excretion of *calcium* and phosphate. Its *protective effect* on *bones* has been used successfully in the treatment of *Paget's disease.*

Calcitriol is a *metabolite of vitamin D_3*, or cholecalciferol. This vitamin appears to be a prohormone. Its synthesis in skin depends on ultraviolet irradiation, and it can be supplemented by dietary intake. Cholecalciferol is *hydroxylated*, at position 25, in the *liver*; it then undergoes 1-hydroxylation in the *kidney*, producing *calcitriol*, the *active metabolite* and true hormone for the regulation of calcium metabolism. Calcitriol *acts principally in the intestine*, inducing the *synthesis of CaBP* involved in increased calcium absorption. In bones, there is a permissive effect on the mobilization of calcium by PTH, and, in kidneys, PTH favors the reabsorption of calcium. *PTH stimulates* the synthesis of *calcitriol* by the activation of a 1α-hydroxylase. *Calcitriol acts with PTH in bone*, and stimulates *calcium absorption* in the *intestine*. The overall effect is an increase in calcemia. In addition to its action on the demineralization of mature bone, vitamin D augments mineralization of newly formed bone. Vitamin D affects mineralization indirectly, by increasing intestinal transfer of calcium and phosphorus, and probably also directly, by the effect of hydroxylated vitamin D metabolites.

In cases of vitamin D insufficiency, hypocalcemia, hypophosphaturia and hyperaminoaciduria are observed. Vitamin D deficiency in children leads to *rickets*, possibly with hypocalcemic accidents. In adults, *osteomalacia* is found predominantly in the diaphysis of long bones, and frequently in vertebrae. Vitamin D resistance which has clinical and biological signs similar to those of cholecalciferol deficiency may be due to chronic renal insufficiency or to chronic treatment with barbiturates, resulting in excessive inactivating metabolism. Hereditary vitamin D-resistant rickets has also been described: X-linked hypophosphatemia and two types of hypocalcemic diseases, of which type I is due to a deficiency of the renal 25 OH-D_3 1α-hydroxylase. Type II possibly involves a defect in the calcitriol receptor. *Intoxication by massive doses* of vitamin D or its derivatives leads to signs of *hypercalcemia*, with possible bone demineralization and renal alteration.

Several *other hormones* also have a potent *action on calcium* metabolism. Hyperthyroidism and hypothyroidism may be accompanied by hypercalcemia, and increased skeletal density can be observed in progressive myxedema. Glucocorticosteroid treatment can lead to osteoporosis. Estrogens reduce the response of bone to PTH, and their deficiency has been implicated in the development of postmenopausal osteoporosis. Growth hormone (GH) stimulates bone development, via IGFs, and an excess of GH leads to hypercalciuria. Prostaglandins E and OAF (osteoclast activating factor, of lymphatic origin) have PTH-like activity.

Fetal parathyroids produce PTH very early. Fetal calcemia is higher than maternal calcemia, and the fetus is relatively hyperphosphatemic. There is an active transport of calcium, from the mother to the fetus, through the placenta. Vitamin D and 25 OH-D_3 cross the placenta, whereas PTH, calcitonin and calcitriol do not. In the neonate, transitory hypocalcemia is common, and it may be accentuated in premature infants. This hypocalcemia is associated with hypercalcitoninemia, and may indicate great avidity of the skeleton for calcium during a period when the dietary intake is low.

General References

Aurbach, G. D., Marx, S. J. and Spiegel, A. (1981) Parathyroid hormone, calcitonin and the calciferol. In: *Textbook of Endocrinology* (Williams, R. H., ed), W. B. Saunders Co., Philadelphia.

Austin, A. L. and Heath, H. (1981) Calcitonin. Physiology and pathology. *N. Engl. J. Med.* 304: 269–279.

Cohn, D. V., Talmage, R. V. and Mathews, J. L. (1981) Hormonal control of calcium metabolism. *Excerpta Medica*, Amsterdam.

Cohn, D. V. and Elting, J. (1983) Biosynthesis, processing and secretion of parathormone and secretory protein-1. *Rec. Prog. Horm. Res.* 39: 181–209.

De Luca, H. F. and Anast, C. S. (1980) *Pediatric Diseases Related to Calcium.* Elsevier Science Publishers, New York.

Fournier, A., Garabedian, M., Sebert, J. L. and Meunier, P. J. (1984) *Vitamine D et maladies des os et du métabolisme minéral.* Masson, Paris.

Hath, D. and Marx, S. (1982) *Calcium Disorders.* Butterworths, London.

Haussler, M. R., Mangelsdorf, D. J., Komm, B. S., Terpening, C. M., Yamakoa, K., Allegretto, E. A., Baker, A. R., Shine, J., McDonnell, D. P., Hughes, M., Weigel, N. L., O'Malley, B. W. and Pike, J. W. (1988) Molecular biology of the vitamin D hormone. *Rec. Prog. Horm. Res.* 44: 263–305.

Kemper, B. (1984) Biosynthesis and secretion of parathyroid hormone. In: *Cell Biology of the Secretory Process* (Cantin, M., ed), Karger, Basel, pp. 443–480.

Norman, A. W. (1982) *Vitamin D, chemical, biochemical and clinical endocrinology of calcium metabolism.* Walter de Gruyter, Berlin.

Raisz, L. G. and Kream, B. E. (1983) Regulation of bone formation. *N. Engl. J. Med.* 309: 29–35 and 83–89.

Royer, P. (1981) Growth and development of bony tissue. In: *Scientific Foundations of Pediatrics* (Davis, J. A. and Dobbing, J., eds), 2nd ed., W. Heinemann, Med Books, London.

Urist, M. R. (1980) *Fundamental and Clinical Bone Physiology.* J. B. Lippincott Co., Philadelphia.

Chromosomal Localization of Genes of Hormones, Growth Factors, Receptors and Other Genes Related to the Endocrine System

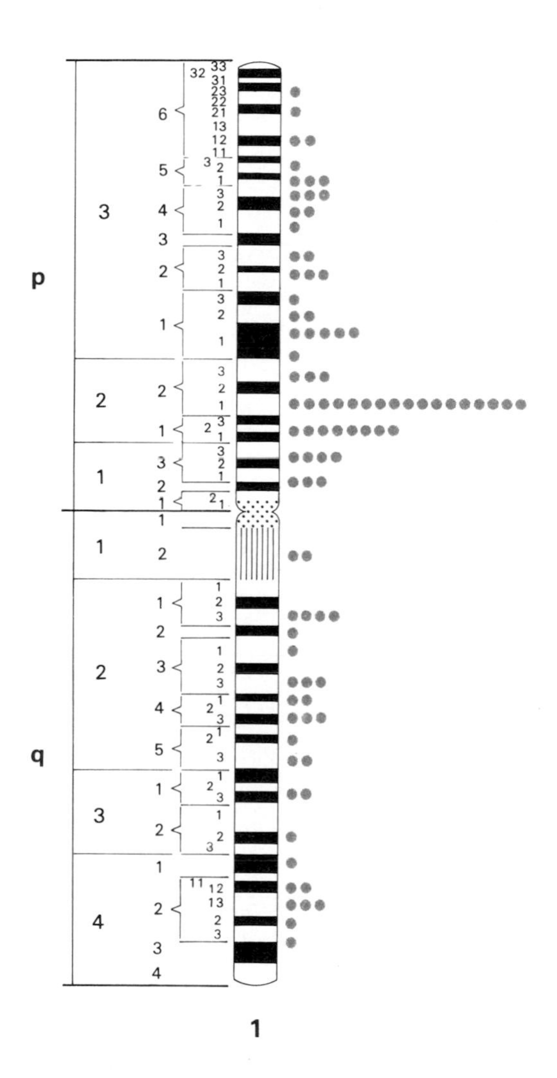

Human chromosome 1 *Banding and localization as determined by in situ hybridization to normal mitotic chromosome with a cDNA probe (NFG-β cDNA probe for chromosome 1 (p.22.1)) (p, short arm; q, long arm). The extent of recombination between two genes on the same chromosome can be used as a map distance to measure their relative location (map distance=*$\frac{\text{number of recombinants} \times 100}{\text{total number of progeny}}$*, with 1 unit, or centiMorgan, equal to 1% crossover).* (Human Gene Mapping 9. Ninth International Workshop on Human Gene Mapping *(1987) Paris,* Cytogenet. Cell. Genet. *46: 1–769*; H. G. M. 9.5 *(1988) New Haven,* Cytogenet. Cell Genet. *47: 1–258*).

Chromosomal Localization of the Genes of Hormones, Growth Factors,

Chromosome	Hormones and Growth Factors		Receptors		Endocrine Related Genes	
1	**ANF** **NGF-β** **β-TSH**	1p36.1 1p22.1 1p22			**Renin** **Transducin 2 (cone cells)** **B-*lym*** **N-*ras*** ***ski***	1q32–qter 1p32 1p11–p13 1q21.1–q24
2	**Glucagon** **IL-1** **Inhibin, α** **Inhibin, βB** **Proopiomelano-cortin** **TGF-α**	2q36–q37 2q13–q21 2q33–qter 2 cen–q13 2p23 2p13			**N-*myc*** **Ornithine decarboxylase 1**	2p23–p24 2p25
3	**Somatostatin**	3q28	**Retinol binding protein** **RA-R-β** **TR-β** **Transferin-R**	 3q24 3p22.3–p21.3 3q26.2–qter	***hap*** ***raf/mil*** **Transferrin** **Transducin 1 (rod cells)**	 3p25 3q21
4	**EGF** **FGF, basic** **IL-2**	4q25 4q21–qter 4q26–q27	**MR** **PDGF-R-A**	 4q11–q12	**α-fetoprotein**	4q11–q13
5	**A-FGF (HBGF-I)** **CSF-1** **Endothelial cell growth factor–A** **GM-CSF (CSF-2)**	5q31.3–q33.2 5q33.1 5q31.2–31.3 5q23–q31	**β-A-R** ***fms* (CSF-1-R)** **GR** **PDGF-R-B** **GH-R** **PRL-R**	5q31.2–31.3 5q33.2–q33.3 5q31 5q31.2–31.3 5p12–p13.1 5p13–p14	**HMG-CoA reductase**	5q13.3–q14
6	**Immune interferon** **α subunit of glycoproteins** **Prolactin** **Tumor necrosis factor** **VIP and PHM**	6q 6q12–q21 6p22.2–p21.3 6p21.3 6q16–q24	**ER** **γ-INF-R** **IGF2-R**	6q24–q27 6q 6q25–q27	**Cytochrome P450$_{21}$** **Insulin-dependent diabetes mellitus** **Major histo-compatibility complex** ***myb*** **Paget's disease** ***yes*-2**	6p21.3 6p21.3 6p21.3 6q23–q22 6p21.3
7	**Inhibin-β-A** **Erithropoietin** **Neuropeptide Y** **PDGF-A** **Substance P**	7p15–p13 7q22 7pter–q22 7p22–q21.1	**EGF-R** **T cell-R β subunit** **T cell-R γ subunit**	7p11.2 7q32–q35 7p15	***erb*-B-1** **Homeobox** **Ornithine decar-boxylase 2**	7p14–p12 7p21–p14 7q31–qter
8	**CRF** **Enkephalin** **GnRH**	8q13 8q23–q24 8p21–q11.2			**Fibronectin** ***mos*** ***myc*** **Plasminogen activ-ator (urokinase)** **Thyroglobulin**	 8q22 8q24 8p12 8q24
9	**Lipocortin 1** **Relaxin H1, H2** **β-interferon**	9q11–q22 9pter–q12 9p21			***abl*** **Orosomucoid**	9q34 9q34

Receptors, and of Other Genes Related to the Endocrine System.

Chromosome	Hormones and Growth Factors		Receptors		Endocrine Related Genes	
10			**IL-2-R**	10p14–p15	**Cytochrome** $P450_{17}$	
			α2-A-R	10q24–q26		
11	**Calcitonin**	11p15.4	**PR**	11q22–q23	***ets***	11q23–q24
	IGF-II	11p15.5			**H-*ras*-1**	11p15.5
	Insulin	11p15.5			***int*-2**	11q13
	PTH	11pter–p15.4			**Tyrosine hydroxylase**	11p15.5
	β-FSH	11p13				
12	**IGF-I**	12q22–q24.1			**Homeobox 3**	12q
	γ-interferon	12q24.1			**K-*ras***	12pter–p12
14			**T cell-R α, δ subunits**	14q11.2	**Chromogranin A**	
					fos	14q21–q31
15			**IGF-I-R**	15q25–q26	**c-*fes***	15q26.1
					Cytochrome $P450_{scc}$	
16			**GR (pseudogene)**		**PK-C β**	16p12–q11.1
17	**CSH**	17q22–q24	**EGF-R-2/*neu***	17q21–q22	***erb*-A-1**	17q11.2
	Gastrin	17cen–qter	**NGF-R**	17q21–q22	***erb*-A-2**	17q21.3
	GCSF/CSF-3	17q11.2–q12	**RAR**	17q21.1	***erb*-B-2**	17q21
	GH	17q22–q24	**TR-α**	17q11.2–q12	**Homeobox 2**	17q11–q22
	GH/PL gene cluster	17q21–q22			**PK-C α**	17q22–q24
	Pancreatic polypeptide	17q11.1–qter				
18	**Gastrin releasing peptide**	18q21			**Prealbumin**	18q11.2–q12.1
					***yes*-1**	18q21.3
19	**AMH**	19p13.3	**I-R**	19p13.2–p13.3	**PK-C γ**	19q13.2–q13.4
	β-LH	19q13.3	**LDL-R**	19p13.2–p13.1		
	β-CG	19q13.3				
	TGF-β	19q13.1–q13.3				
20	**Arginine vasopressin/ neurophysin-II**		**Interferon-R,α,β**	21q21–qter	**Chromogranin-B**	
	Dynorphin	20 pter–p12			**Multiple endocrine neoplasia, type 2 (Sipple's syndrome)**	20p12.2
	GRF				***src***	20q12–q13
	Oxytocin-neurophysin-I					
22	**PDGF-B**	22q12.3–q13.1	**TSH-R**	22q11–q13	***sis***	22q13.1
X			**AR**	Xq11–q12	**Gonadal dysgenesis**	Xp22–p21
					Steroid sulfatase	Xp22.3
					TBG	Xq21–q22
Y					**H-Y**	Yp11.2–qter
					Testis determining factor (TDF)	Ypter–p11.2

Index

Bold-face page numbers refer to a major description of the entry. The letter F after a page number indicates a figure; T after a page number indicates a table. Entries without page numbers define abbreviations.